W0037374

Formation of Active Ocean Margins

Editors

Noriyuki NASU, Ph.D., D.Sc.
Chief Editor
Professor, University of the Air, Chiba
Professor Emeritus and Former Director of the Ocean Research Institute, University of Tokyo

Kazuo KOBAYASHI, D.Sc.
Professor
Ocean Research Institute, University of Tokyo

Seiya UYEDA, D.Sc.
Professor
Earthquake Research Institute, University of Tokyo

Ikuo KUSHIRO, D.Sc.
Professor
Geological Institute, University of Tokyo

Hideo KAGAMI, D.Sc.
Associate Professor
Ocean Research Institute, University of Tokyo

Advances in Earth and Planetary Sciences

Formation of Active Ocean Margins

Edited by

N. Nasu, K. Kobayashi, S. Uyeda, I. Kushiro, and H. Kagami

Terra Scientific Publishing Company / Tokyo

D. Reidel Publishing Company

A MEMBER OF THE KLUWER ACADEMIC PUBLISHERS GROUP

Dordrecht / Boston / Lancaster / Tokyo

Library of Congress Cataloging-in-Publication Data

Formation of active ocean margins.

(Advances in earth and planetary sciences)
Papers from the OJI International Seminar on the Formation of Ocean Margins, held Nov. 21–23, 1983, at the Ocean Research Institute, University of Tokyo.
Bibliography: p.
1. Faults (Geology)--Congresses. I. Nasu, Noriyuki. II. OJI International Seminar on the Formation of Ocean Margins (1983: Ocean Research Institute, University of Tokyo) III. Series.
QE606.F67 1986 551.1'36 86-17050

ISBN-13: 978-94-010-8599-1 e-ISBN-13: 978-94-009-4720-7
DOI: 10.1007/978-94-009-4720-7

Published by Terra Scientific Publishing Company (TERRAPUB),
307 Shibuyadai-haim, 4-17 Sakuragaoka-cho, Shibuya-ku, Tokyo 150, Japan,
in co-publication with D. Reidel Publishing Company, Dordrecht, Holland

Sold and distributed in the U.S.A. and Canada
by Kluwer Academic Publishers,
101 Philip Drive, Assinippi Park, Norwell, MA 02061, U.S.A.
in Japan by Terra Scientific Publishing Company (TERRAPUB),
307 Shibuyadai-haim, 4-17 Sakuragaoka-cho, Shibuya-ku, Tokyo 150, Japan

In all other countries, sold and distributed
by Kluwer Academic Publishers Group,
P. O. Box 322, 3300 AH Dordrecht, Holland

Softcover reprint of the hardcover 1st edition 1986

Preface

The ocean floor spreading theory was proposed during 1961 and 62 by Robert Dietz and Harry Hess. This concept was a revolutionary one, and renewed the scientists thoughts on the dynamics of the ocean bottom. Then, for example, the coincidence of the Wadati-Benioff Zone with the subduction zone proposed by new concept was well understood.

Further development of the ocean floor spreading theory was the proposal of new concept "plate tectonics" proposed by Xavier LePichon and by a few others during 1967 and 68. This new idea could solve the various conflicts involved in the "ocean floor spreading theory".

Therefore, today, scientists understand that the plate tectonics theory was born by the ocean floor spreading theory, which is able to cover the weak points of the latter.

D/V Glomar Challenger started her Leg 1 on 20 July, 1968 from Orange, Texas to implement the Deep Sea Drilling Project. The timing almost coincided with the proposal period of the plate tectonics.

After carrying out a few legs of the drilling operations, the results obtained by D/V Glomar Challenger well proved the rightness of the newly proposed theories of the ocean floor spreading and the plate tectonics.

For us, the successful processes started by the ocean floor spreading theory, improved by the concept of plate tectonics and proved by the DSDP results have been a golden monument in the field of earth sciences probably for several centuries.

Because of the successful results, DSDP became the International Phase of Ocean Drilling, so-called IPOD, from Leg 45 which started on 30 November, 1975 from San Juan, Puerto Rico. At this time, Japan, U.K., France, West Germany and U.S.S.R. jointed into the U.S. operation for DSDP.

In 1977, I had the pleasure to serve as one of the co-chief scientists with Roland von Huene on Leg 57 which drilled the Japan Trench subduction area.

Finally, the sea going operation of D/V Glomar Challenger for DSDP terminated at Mobile, Alabama, on 8 November, 1983.

The deep sea drilling operation will be succeeded by the "Ocean Drilling Project" (ODP), using D/V JOIDES RESOLUTION.

The ship is larger, and the drilling is expected to be deeper. Ocean drilling is going into a new phase.

The OJI International Seminar on the Formation of Ocean Margins was held during 21–23 November, 1983 at the Ocean Research Institute, University of Tokyo in Japan to bridge the understanding between DSDP and ODP.

Because scientists' interests on the ocean margins have deepened during DSDP operations, and because much new knowledge in this field has been obtained, the organizing committee of the Seminar thought that it would be the time to call together all scientists and related persons who have deep connections with DSDP. As the result, scientists assembled from all over the world. The Seminar was a great success.

It was our pleasure to have the presence of Xavier LePichon, the major proposer of plate tectonics on a global scale with us at this Seminar.

Many presentation were concerned about active margins because the Seminar was held in Japan which locates on just such a typical active margin. However, other presentations, for example, concerning passive margins, even though few in number, are extremely important and we could understand after listening to these, why areas were called passive and not inactive margins. Even though tectonic activities are not so intense, they are still going on partially in passive margins areas.

Since the Seminar was so successful the organizing committee decided to put it in book form. This book is the outcome.

The title is "Formation of Active Ocean Margins" because many articles cover mainly the area of active margins. But, articles on passive margins and other topics should be counted as important also.

In view of the success of the OJI Seminar, the organizing committee took over as the editors of this book. We, the editors deeply wish to thank the contributors and reviewers of this book. Also, we acknowledge appreciation to the Ministry of Education, Science and Culture of Japan for the assistance in the form of the Grant in Aid for Publication of Scientific Result. Deep appreciation is extended to the publisher.

Furthermore, the editors wish to extend their appreciation to all participants of the OJI Seminar and to The Japan Society for the Promotion of Science, The Fujiwara Foundation of Science and to the Ocean Research Institute, University of Tokyo which assisted the Seminar and made it successful.

30 November, 1984

Noriyuki NASU
Editor in Chief

CONTENTS

CHAPTER 3: ARC MAGMATISM

CHAPTER 4: BACK-ARC TECTONICS

CHAPTER 5: ACCRETION TECTONICS (I)

CHAPTER 6: ACCRETION TECTONICS (II)

CHAPTER 7: ACCRETION TECTONICS (III)

CHAPTER 8: PELAGIC SEDIMENTS

CHAPTER 1: SUBDUCTION TECTONICS

Formation of Active Ocean Margins, edited by N. Nasu *et al.*, pp. 3–42.

PANGEA, GEOID AND THE EVOLUTION OF THE WESTERN MARGIN OF THE PACIFIC OCEAN

X. LE PICHON,[1] P. HUCHON,[1] and E. BARRIER[2]

[1]*Laboratoire de Géodynamique & CNRS LA 215, Université P. et M. Curie 4, place Jussieu 75230 Paris, France*
[2]*CFR Laboratoire mixte CEA/CNRS, Avenue de la Terrasse 91190 Gif sur Yvette, France*

Abstract. LE PICHON and HUCHON (1984) pointed out the probable relation between present geoid and Pangea. Within this context, we explore the way in which the Earth changed from an asymmetrical state of order one with a continental hemisphere and an oceanic one and an equatorial axis of symmetry to the present state where the continents are distributed more equally between both hemispheres. We use absolute motions with respect to the deep convection related geoid. These motions coincide with hot spots framework motions for the last 80 Ma. We show that continental drift activity has peaked 80 to 54 Ma ago and that this peak is not related to an increase in global oceanic accretion. The argument used is based in part on the evolution of the global sea-level. We analyze in more detail the dispersal of Eastern Pangea in terms of absolute motions, paying special attention to the evolution of marginal basins. Our analysis suggests that the motions of the continents may be in great part controlled by the deep convection responsible for the geoid configuration whereas the evolution of the shallow oceanic thermal boundary layer shows no obvious correlation with deep convection. We consequently propose that the independent motions of the continents may play a much more significant role than previously thought.

1. Introduction

LE PICHON and HUCHON (1983, 1984) have pointed out that the outline of Pangea, at least between 200 and 135 Ma ago, exactly lay along a great circle passing through the paleo-poles of rotation of the Earth.

This configuration has an axis of symmetry in the corresponding paleo-equatorial plane. Le Pichon and Huchon, after GOUGH (1977), also pointed out that the present geoid has a tennis-ball pattern with an axis of symmetry in the present equatorial plane. The positive belt is equatorial and the negative belt is polar (Fig. 1). HESS (1965) had previously suggested that the distribution of mid-ocean ridges and trenches could be due to a tennis-ball type convection. The remarkable low-order configuration of the geoid is tightly related to the axis of rotation of the Earth and appears to have persisted essentially unchanged at least since Upper Mesozoic (CHASE and SPROWL, 1983), as far as the low order (2 and 3) harmonics are concerned (HAGER, 1983).

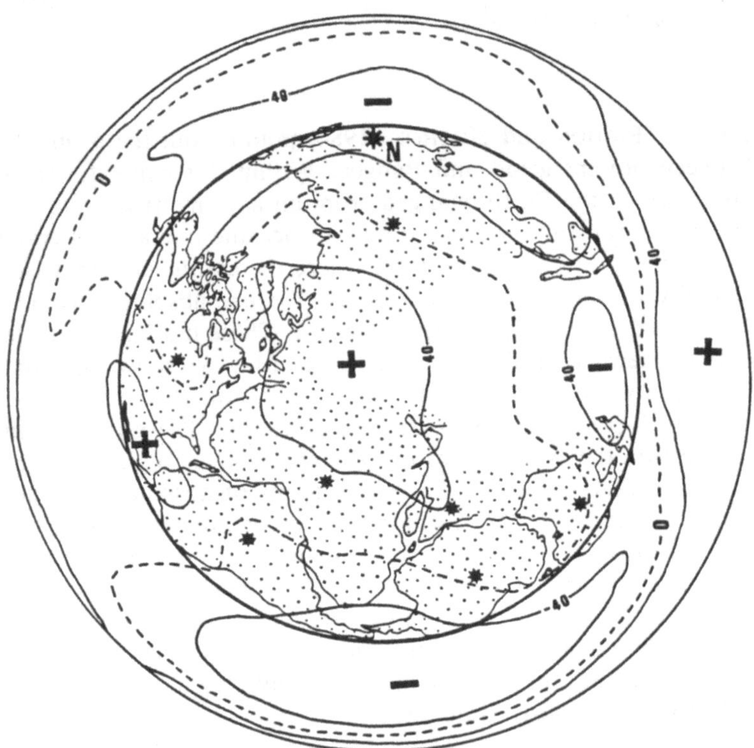

Fig. 1. Pangea configuration 200 Ma ago in the geoid frame of reference, superposed to the geoid anomalies (in meters) with respect to the hydrostatic geoid (after figures 2 and 6 of LE PICHON and HUCHON, 1984). Parameters of reconstruction are given in table 2. N is the paleomagnetic North pole (after IRVING, 1977). Stars show the centers of gravity of the continents (see table 2). Lambert equal area projection with pole of projection at 6N, 4E.

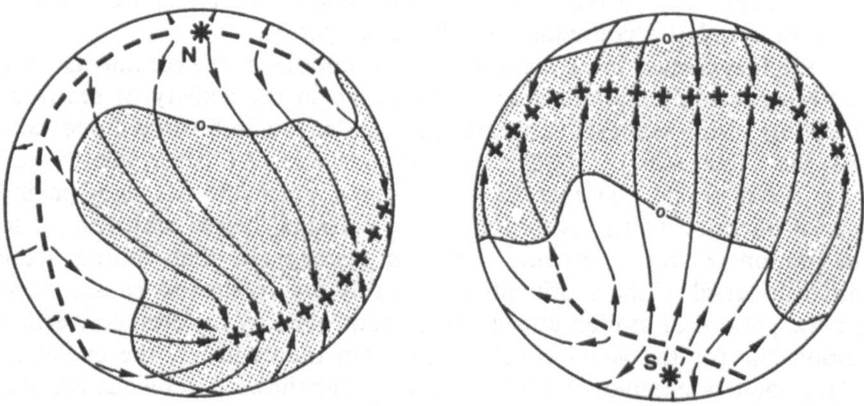

Fig. 2. Schematic view of the tennis-ball pattern of the geoid. Dotted: positive anomalies. Plus and minus show the axes of the positive and negative belts. Arrows show a hypothetical pattern of convection currents in the lower mantle which could correspond to the geoid. Orthographic projection.

This would fit the conclusions of BUSSE's (1983) theoretical work in which he proposes that lower mantle convection may be described by low-order spherical harmonics and that, among the harmonics of degree 2, the motion exhibiting an axis of symmetry in the equatorial plane will be selected. Thus, LE PICHON and HUCHON (1984) argued that the hemispheric Pangea configuration could be controlled by a steady-state lower mantle convection, responsible for the tennis-ball geoid and weakly coupled to the upper mantle one (Fig. 2). The negative polar belt of the geoid would then coincide with the peripheral subduction belt of Pangea. It would consequently be a descending cell in the upper mantle and thus possibly an ascending one in the lower mantle. If this hypothesis is correct, there is a unique way to relate the permanent convection related geoid to Pangea by superposing both the great circles and the equatorial planes (Fig. 1).

In this paper, we investigate the transition from the hemispheric super-continent configuration in Lower Mesozoic time to the complex Upper Mesozoic and Tertiary Pangea dispersal. LE PICHON and HUCHON (1984) argued that the main change occurred in Middle Cretaceous time during a major geologic catastrophe which appeared to have been accompanied by high rates of spreading (PITMAN, 1978), hot spot outbreaks (ANDERSON, 1982) and high sea level stands (VAIL *et al.*, (1977). Following ANDERSON (1982), they attributed the catastrophe to the excessive heating of the upper mantle due to the insulating continental cap. Here, we examine in more detail the consequences of this dispersal.

We first discuss the general pattern of motions of the continents since 200 Ma ago and compare it to the variation in the activity of sea-floor spreading, using sea-level variations as a guide. It is in Eastern Gondwana that the Cretaceous dispersal was most dramatic. We consequently consider the dispersal phase of Eastern Pangea, since 135 Ma ago, paying special attention to the evolution of the marginal basins with respect to the motions of the continents. Our analysis suggests that, although the initial dispersal of the continents is probably due to shallow causes, their general pattern of motion appears to be controlled by the deep convection responsible for the geoid configuration. On the contrary, the evolution of the shallow oceanic thermal boundary layer shows no obvious correlation with deep convection. We consequently propose that the independent motions of the continents may play a much more significant role that previously thought.

2. *Hot Spots and Geoid Frames of Reference*

We assume that the time constant of the deep convection responsible for the geoid tennis-ball pattern is large so that deep mantle convection can be considered steady-state since 200 Ma. The main argument for this assumption is the existence of a peripheral great circle polar subduction belt around Pangea between 200 and 135 Ma ago which it is tempting to relate to the polar belt of the geoid (Fig. 1). The peripheral subduction belt can be equated with an upper mantle descending cell, coupled to the underlying lower mantle cell (Figs. 1 and 2). To show the coincidence between the Pacific peripheral subduction belt and the geoid low, CHASE and SPROWL (1983) have used MORGAN's (1981) hot spots frame of reference. Unfortunately, MORGAN (1981) has shown that significant polar wandering (about 30° in 200 Ma) has occurred with respect to the hot spots frame of reference. But as the deep convection pattern is tied to the rotation axis of the Earth, any significant (larger than 10°) polar wandering with respect to the deep mantle must be accompanied by deep convection pattern adjustment. In order words, there should not be any detectable polar wandering with respect to the deep convection pattern.

To investigate the motions of the continents with respect to the deep convection pattern, we use what LE PICHON and HUCHON (1984) have called the "geoid frame of reference". The peripheral Pangea subduction belt is superposed to the present geoid polar low and the Earth's rotation axis is superposed to the corresponding paleomagnetic pole. It is further assumed that the absolute position of the deep convection pattern has not changed since 200 Ma, thus defining the "geoid frame of reference". If this hypothesis is correct, the "geoid frame of reference" should coincide with the "hot-spots frame of reference" of MORGAN (1981) for the period during which polar wandering with respect to the "hot-spots frame of reference" has been small, that is for the last 80 Ma. The 80 Ma reconstruction (Fig. 11 of LE PICHON and HUCHON (1984), Fig. 6 in this paper), which is the latest one for which the Pangea peripheral subduction belt is still fairly well defined, confirms that both frames of reference coincide at this time. The Lower Cretaceous (135 Ma, Figs. 9 and 10 of LE PICHON and HUCHON (1984), Fig. 5 in this paper) shows a minimum 15° difference between both frames of reference. This modest difference explains the reasonable fit found by CHASE and SPROWL (1983) between geoid and Pacific belt location using Morgan's hot-spots frame. Strictly speaking, if we do not have the hot spots frame of reference, the longitudes of the geoid frame of reference are not defined. However, we then choose for the geoid frame of reference the position which involves the minimum displacement with respect to the hot-spots frame of reference. The discrepancy between both frames of reference increases to 30° in Triassic, which is about the amplitude of polar wandering with respect to the hot spots frame (MORGAN, 1981). But the geoid frame of reference is certainly not defined to better than 10°. We conclude that the motions of continents with respect to the convection pattern in the lower mantle can be described reasonably well for the last 80 Ma, as they can be obtained with respect to the hot-spots frame of reference. But the accuracy degrades progressively as one tries to reconstruct their positions backward to 200 Ma without the help of the hot-spots frame. With this caveat, we now proceed to the discussion of the motions of continents.

3. *Rate of Motions of the Continents since 200 Ma Ago*

Ideally, we would like to know whether there has been variations in the rate of continental drift and variations in the activity of sea-floor spreading and whether these variations are related. We first discuss the variations in continental drift, taking as a measure of continental drift activity the average absolute velocity of the continents. We use a weighted average, the weight being proportional to the areas of the continents. The areas of the continents, submerged portions included, are after SCLATER *et al.* (1980, see Table 1), the total surface of the continents being 201.5×10^6 km^2 or 39.5% of the surface of the Earth (510.5×10^6 km^2). We compute the average velocity of the approximate center of gravity of a continent along a great circle joining two successive positions in time. The positions of the centers of gravity are given in Table 1. This procedure ignores the rotational component with respect to this center of gravity and consequently these velocities are generally minimal ones. Most of the time however, the pole of rotation is sufficiently far away that this simplification does not affect significantly the result. We use MORGAN's (1981) hot spots frame of reference since 80 Ma. Prior to that, we use both Morgan's

Table 1: Absolute velocities of continents ($mm \cdot yr^{-1}$) in the hot spots and geoid frames of reference

Period (Ma)	EUR	AFR	SAM	IND	AUS	ANT	NAM	Weighted average
Present	12.0	22.0	23.0	49.0	47.0	12.0	27.0	21.0 *
0–20	10.6	14.4	22.2	39.4	67.7	14.4	9.1	18.8
20–35	15.7	17.9	26.2	27.3	74.0	57.4	20.1	27.3
35–54	30.0	22.0	21.3	117.3	62.3	11.6	30.1	31.9
54–80	65.4	48.7	27.8	171.1	22.3	28.2	59.6	53.7
80–135	21.1	52.9	15.2	28.1	88.6	32.1	18.0	33.2
Id. (M)	5.8	20.8	27.0	32.2	48.4	28.6	31.6	22.1
135–200	29.8	13.8	20.0	15.0	5.5	15.5	38.5	23.2
Id. (M)	51.3	30.7	37.4	21.9	21.3	12.7	44.9	37.7
Weight	31.9	18.8	11.1	3.5	9.0	8.8	16.9	**
Present position of the center of gravity	50N 90E	10N 20E	10S 60W	20N 80E	20S 135E	85S 70E	50N 90W	

(M): in the hot spot frame of reference of Morgan (1981)
* after Minster et al., 1974
** (in percent) after Sclater et al., 1980.

Table 2: Parameters of plate motions in the geoid frame of reference for the 200 and 135 Ma reconstructions.

	125 Ma			200 Ma		
Continent	Lat (°N)	Long (°E)	α (°ccw)	Lat (°N)	Long (°E)	α (°ccw)
AFT	−49.6	124.5	45.5	−44.9	101.3	47.2
ANT	−67.7	356.2	59.7	−55.3	337.8	52.4
AUS	−82.2	199.6	62.2	−75.2	260.4	55.7
EUR	−42.5	72.3	29.8	−18.8	44.5	40.2
IND	−28.4	171.4	82.3	−35.6	168.2	75.9
NAM	29.9	58.7	25.7	31.1	25.4	45.9
SAM	4.0	353.0	19.1	−0.7	3.9	30.0

frame of reference and the geoid one as defined by LE PICHON and HUCHON (1984). The first one includes the polar wandering motion whereas the second one does not include it. Table 2 gives the parameters of plate motion used, based on MORGAN (1981) and on NORTON and SCLATER (1979) for the 200 Ma reconstruction.

The present average weighted instantaneous velocity of the continents is 21 $mm \cdot yr^{-1}$ (Table 1, after MINSTER *et al.*, 1974), taking separately the velocities of the centers of gravity of Eurasia, North America, South America, Africa, Antarctica, India and Australia. Table 1 gives the absolute velocities of the same continents during six intervals, 200–135, 135–80, 80–54, 54–35, 35–20, and 20–0 Ma. With respect to MORGAN (1981), we corrected the date of anomaly M12 from 125 to 135 Ma age after PALMER (1983). Figure 3b shows the variations of the weighted average velocity of the continents in the geoid frame of reference. As stated previously, the hot spots frame of reference includes a fairly large polar wandering motion between 200 and 135 Ma ago, which corresponds to a counter-clockwise rotation of the pole of rotation of the Earth of about $0.5° \cdot Ma^{-1}$ along the periphery of Pangea during Triassic and Jurassic (LE PICHON and HUCHON, 1984). This is the main cause of the relatively high 200–135 Ma velocity in the hot spots frame of reference. Note again that between 80 Ma and the present, the geoid frame of reference coincides with the hot spots one because the amplitude of polar wandering is quite limited (MORGAN, 1981).

The main conclusion, from Table 1 and Fig. 3b, especially if we disregard the polar wandering effect between 200 and 135 Ma ago, is the remarkable increase to a maximum of about 50 $mm \cdot yr^{-1}$ between 80 and 54 Ma ago, in Campanian, Maestrichtian and Paleocene followed by a rapid decrease back to the Jurassic level. This conclusion is unlikely to be changed by any further adjustment in the definition of the hot spots frame of reference, which is relatively well defined for the last 80 Ma, or in the definition of the magnetic anomaly time scale. Recently, SCHULT and GORDON (1984) have independently computed absolute velocities of continents using a different method of computation and a different reconstruction in the Indian ocean. They also obtained a maximum between 80 and 50 Ma although their maximum is smaller (35 $mm \cdot yr^{-1}$ instead of 50). This build-up in activity coincides with the Pangea dispersal phase discussed by LE PICHON and HUCHON (1984) but the peak in activity occurred near 65 Ma ago and not prior to 80 Ma as stated by them. Table 1 shows that this peak in activity is mostly controlled by the motions of Eurasia, Africa and North America as these continents represent about 70% of the total surface of the continents.

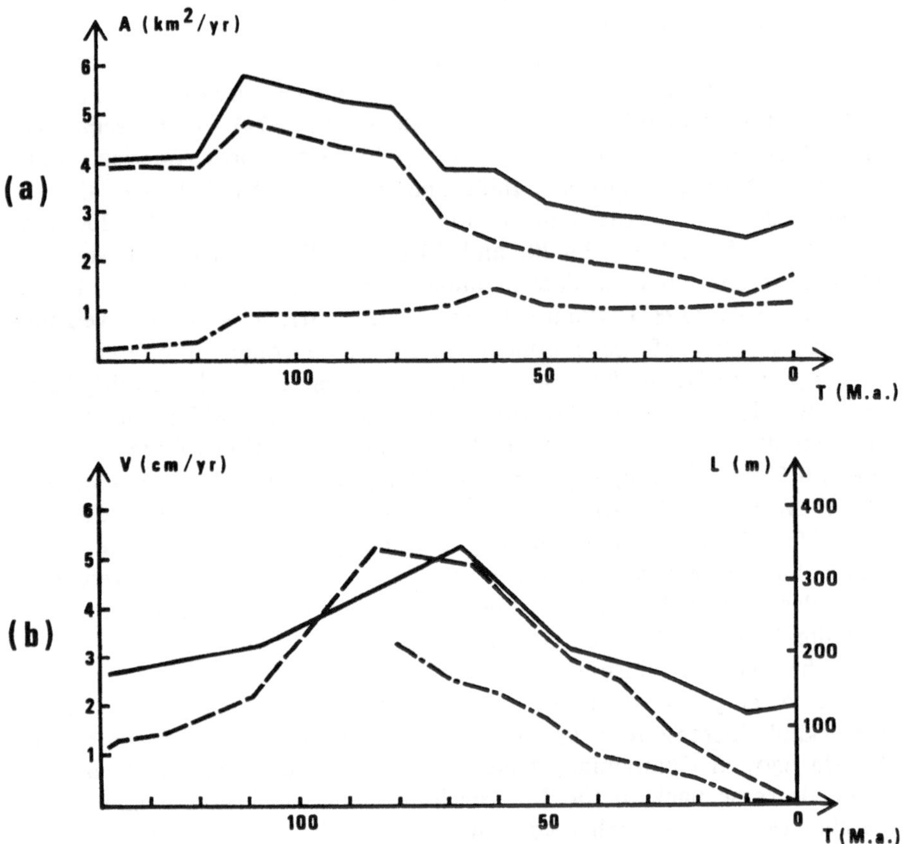

Fig. 3. (a) Accretion rates (from Kominz's (1984) data). Dashed line: ocean with triangular area-age distribution (Pacific, Tethys). Dash-dotted line: oceans with rectangular area-age distribution (Atlantic, Indian). Solid line: global accretion rate. (b) Solid line: weighted average velocity of continents over the geoid frame of reference (see table 1). Dashed line: smoothed curve of sea level changes based on VAIL *et al.* (1978) and calibrated by the estimates of PITMAN (1978) (taken from PARSONS, 1982). Dash-dotted line: sea level curve from KOMINZ (1984) using the time scale of LARSON *et al.* (1981). Note the concordance between the maximum velocity of continents and the maximum sea level, about 65–85 Ma ago. Note also the shift of about 30 Ma with respect to the peak in accretion rate.

4. *Sea-Floor Spreading Activity since 200 Ma Ago*

The present weighted average velocity of the plates at the surface of the Earth is 42 $mm \cdot yr^{-1}$, using a weight proportional to the areas of the plates (Forsyth and Uyeda, 1975) and taking the hot spots frame velocities of the 10 plates model of Minster *et al.* (1974). However, the velocities of the purely oceanic plates are large, 86 $mm \cdot yr^{-1}$ for the weighted average velocity of the three purely oceanic plates (Pacific, Cocos and Nazca). It follows that the present velocities of the plates attached to continents is much smaller, as noted by Minster *et al.* (1974). But Table 1 shows that this was not the case in Early Tertiary and Late Cretaceous, where the velocity of Eurasia or North America is comparable to the velocity of the present oceanic plates. Solomon *et al.* (1977) have determined the absolute plate motions in the Early Tertiary using the assumption that no net torque is exerted on the lithosphere and argued that the oceanic plates were not going faster in Early Tertiary so that a clear separation between oceanic and continental plates, based on velocity, could not be made at this time.

However, the velocities of the oceanic plates may not be a good guide to judge the activity of sea-floor spreading. Actually, at the present time, there is no systematic relationship between the velocity of a plate and its average heat flow. The average ages of the three main oceans, Pacific, Atlantic and Indian, are not significantly different, being respectively 68, 58 and 54 Ma (based on Sclater *et al.*, 1981) and, to a first approximation, the average heat flow is also about equivalent (Parsons, 1982). The average ages as well as the average heat flows are essentially independent of the velocities of the plates because the faster Pacific plate is correspondingly larger. Thus instead of considering the velocities of the plates, we use as an indicator of sea-floor spreading activity the global rate of accretion which is presently close to 3.45 $km^2 \cdot yr^{-1}$ (Sclater *et al.*, 1981; Parsons, 1982).

Hays and Pitman (1973), Pitman (1978) and more recently Kominz (1984) have argued that the global rate of accretion has decreased by a factor of about 2 since Upper Cretaceous time, during the Pangea dispersal phase. Figure 3a shows the variation of global rate of accretion after Kominz (1984), compared in Fig. 3b to the curve of continental drift activity. Note that the two peaks are offset by about 30 Ma. An important point shown in Fig. 3a is that the Cretaceous peak in accretion rate is due to the Pacific-Tethys accretion *whereas the Atlantic-Indian one is very constant since 110 Ma ago*. But the Pacific-Tethys curve is obtained by reconstructions of sea-floor which has been in great part subducted. For example, the rate of accretion in the Pacific 110 Ma ago was 4.8 $km^2 \cdot yr^{-1}$ according to KOMINZ (1984). But, actually, the amount of sea-floor of that age which has not been subducted, according to SCLATER *et al.* (1981) is 0.7 $km^2 \cdot yr^{-1}$. *Thus, the reconstruction of Kominz for this period is based for 15% on data and for 85% on extrapolation and may be affected by very large errors.*

An argument in favor of the existence of a large Pacific-Tethys peak in accretion rate is that it explains the Cretaceous peak in sea-level which is now qualitatively well established (VAIL *et al.*, 1977). Unfortunately the amplitude of the maximum Upper Cretaceous transgression is poorly known, although it is most probably more than 100 meters and less than 350 meters. The total accretion curve of KOMINZ (1984) shown in Fig. 3a results in a peak of 230 meters, with a minimum estimate of 45 meters and a maximum one of 365 meters. The question then is whether an important increase in sea-floor spreading activity in the Pacific and Atlantic is necessary to explain the sea-level curve. The variation in sea-level does not depend directly on the instantaneous global rate of accretion but rather on the distribution of ocean floor areas with ages. This distribution in turn depends on the variation of the global rate of accretion and on the distribution of subduction with respect to age. An important observation made by PARSONS (1982) is that subduction is close to being uniformly distributed with age. Parsons argues that the present distribution of ocean floor with ages can be explained by assuming a steady rate of accretion and a steady uniform distribution of subduction with age. This differential area versus age variation is of the form:

$$\mathrm{d}A/\mathrm{d}t = C_0(1 - t/t_\mathrm{m}) \tag{1}$$

where t is the age, A the area of the ocean, C_0 the accretion rate and t_m the maximum age of the ocean floor. $\mathrm{d}A/\mathrm{d}t$ decreases linearly with time and, for this reason, the variation of $\mathrm{d}A/\mathrm{d}t$ with time is called triangular. The conclusions of PARSONS (1982) are not compatible with those of KOMINZ (1984) who does not apply the proposed principle of uniform distribution of ocean floor subduction with age. In the following, we examine whether the Upper Cretaceous peak in sea-level may be explained without violating the uniform distribution of ocean floor subduction with age.

As shown by SCLATER *et al.* (1981), there is an important difference between the Pacific and the Atlantic-Indian oceans differential age-area distributions. The Pacific one is triangular and may be schematized by relation (1) whereas in the Atlantic-Indian oceans $\mathrm{d}A/\mathrm{d}t$ does not change with time and, for this reason, the variation of $\mathrm{d}A/\mathrm{d}t$ with time is called rectangular. It may be schematized by:

$$\begin{aligned} \mathrm{d}A/\mathrm{d}t &= C_o \qquad t > t_m \\ \mathrm{d}A/\mathrm{d}t &= 0 \qquad t < t_m. \end{aligned} \tag{2}$$

This is because there is no significant subduction in the Atlantic and minor subduction in the Indian oceans whereas subduction is distributed approximately uniformly with respect to age in the Pacific ocean. It is reasonable to assume that this was also true earlier in the Tethys ocean as well as in the Pacific ocean. Thus, the present differential age-area distribution for the world ocean is the sum of a Pacific triangular one and an Atlantic-Indian rectangular one. We propose that the sum of the two distributions has determined the variation of the differential are-age distribution with time.

One hundred and ten Ma ago, the Pacific hemisphere occupied close to half the Earth, which is 255.25×10^6 km^2 and the Tethys ocean occupied the remaining oceanic space which is 53.75×10^6 km^2 (using SCLATER *et al.*, 1980 surfaces throughout). Thus the 309×10^6 km^2 of oceanic space were presumably characterized by a triangular distribution. Since that time, 142.5×10^6 km^2 of oceanic space have been created in the Atlantic and Indian oceans which are known to be of the rectangular distribution type (we ignore the small amount created prior to 110 Ma). The average rate of creation there has been quite constant and close to 1.3 km^2·yr^{-1} (SCLATER *et al.*, 1981); this absence of variation is confirmed by the independent estimate of KOMINZ (1984). This creation has been compensated by the destruction of Tethys and of 92×10^6 km^2 of Pacific ocean, as the present surface of the Pacific ocean, marginal basins included, is 163×10^6 km^2.

We make the simplest assumption which is that the amount of heat which escaped through the ocean floor during Pangea time before dispersal, prior to 110 Ma ago, was not significantly different from the present one. With uniform distribution of subduction of ocean floor with respect to age, conservation of the surface of the oceans and steady-state, this assumption implies the same global accretion rate C_0 as now (3.45 km^2·yr^{-1}) and the same maximum age t_m (180 Ma) for the sea-floor. We then assume that, as dispersal time arrived, the new accretion occurring within Pangea 110 Ma ago vented the excess heat accumulated because of thermal insulation by continental lithosphere (ANDERSON, 1982). There is no dynamic reason to assume that the velocities of Pacific plates slowed instantaneously to maintain a constant global rate. It should take several tens of millions of years for the Earth to adjust to this new thermal situation. Consequently, the global accretion rate suddently increased 110 Ma ago from 3.45 to 4.75 $km^2 \cdot yr^{-1}$ by addition of 1.3 $km^2 \cdot yr^{-1}$ of new intra-Pangea accretion to the 3.45 $km^2 \cdot yr^{-1}$ of Pacific-Tethys accretion. And this increase was compensated by an advance of the continents over the Tethys-Pacific space. Since 110 Ma, Earth heat release adjusted itself progressively back to normal on a global level.

Knowing the variation of depth with age:

$$h = h_0 + a \ t^{1/2} \tag{3}$$

We choose $a = 300$ m (after Heestand and CROUGH, 1981) and $h_0 = 2500$ m, so that the average depth $\bar{h}$ in the ocean, assuming steady-state is:

$$\bar{h} = 8a/15 \ t_m^{1/2} + h_0 \tag{4}$$

and $\bar{h} = 4650$ m with $t_m = 180$ Ma, which is about the present average depth (PARSONS, 1982). Then, we can compute numerically the variation of world sea-level with age as sea-level is related to average ocean depth through the isostatic compensation relation (e.g., KOMINZ, 1984).

To summarize, our numerical computations assume that at all times ocean-floor subduction is uniformly distributed with respect to age in the Pacific and Tethys oceans. With steady-state, this results in a triangular distribution of age. We start 110 Ma ago, assuming that steady-state prevailed then and the Pacific and Tethys were triangular type oceans. Then, suddenly, 110 Ma ago, an additional 1.3 $km^2 \cdot yr^{-1}$ accretion started in the Atlantic and Indian oceans. Thus the global rate changed from 3.45 to 4.75 $km^2 \cdot yr^{-1}$. We choose a linear readjustment from 4.75 to 3.45 $km^2 \cdot yr^{-1}$ between 110 Ma ago and 30 Ma ago (Fig. 4b, solid line). The resulting area versus age curve in the Pacific is compared in Fig. 4c (dashed line) to the present one. It is reasonably similar but somewhat too high between 110 and 30 Ma ago. For the model proposed by KOMINZ (1984), we find a 230 meters maximum in sea-level, as she does (Fig. 4b, dotted line). But the excess in accretion rate between 110 and 80 Ma ago produces a large excess of sea-floor between 110 and 80 Ma compared to the observed one (Fig. 4c, dotted line). To eliminate this excess, one would have to assume that subduction is not uniformly distributed with age and instead preferentially destroys older ages, at the difference of what happens at present (PARSONS, 1982).

Note that the SCLATER *et al.* (1981) differential area versus age Pacific curve in Fig. 4c (+ signs) does not include the marginal basins and is thus smaller than the adopted value of 163×10^6 km^2 for the present surface of the Pacific ocean. A perfect fit consequently cannot be attempted but the break in the curve at 30 Ma imposes that the global accretion rate readjustment was accomplished by this time. The main point we wish to make however is that, even without any increase in total accretion rate within the Pacific-Tethys ocean, the dispersal of Pangea implies a sizeable sea-level rise which peaked in Uppermost Cretaceous. But its amplitude of only about 100 meters is two to three times smaller than many estimates although it falls into the range of uncertainties proposed for example by KOMINZ (1984).

HARRISON *et al.* (1981) have discussed causes of change of sea-level, other than igneous sea-floor depths. A significant one is the increase in total ocean area due to continental collision. We believe that HARRISON *et al.* (1981) have underestimated this effect. This is because they assumed that the total shortening is equal to 2.3×10^6 km^2, which they indicate to be the proposed area in Asia above 4 km. But the present rate of continental collision according to PARSONS (1981) is 0.25 km^2·yr^{-1}, with 0.152 km^2·yr^{-1} between Eurasia and India, 0.077 km^2·yr^{-1} between Eurasia and Arabia and 0.025 km^2·yr^{-1} between Eurasia and Africa. Of the 0.15 km^2·yr^{-1} between Eurasia and India, part may be absorbed by expulsion of blocks toward the Pacific ocean and results in no net shortening (TAPPONNIER *et al.*, 1982). Allowing for 0.05 km^2·yr^{-1} due to expulsion, there is still about 0.2 km^2·yr^{-1} which gives a loss of 9×10^6 km^2 since the first collision of India, with Asia, 45 Ma ago. Prior to that, collision was already occurring in western Tethys at a rate which is presumably quite steady since 70 Ma (BIJU DUVAL *et al.*, 1977). We take a rate of 0.1 km^2·yr^{-1}. This gives an additional 2.5×10^6 km^2. Note that ALLÈGRE *et al.* (1984) estimate that the present area of doubling of crust within Tibet and Himalaya is 4×10^6 km^2. As a sizeable amount of material may have been removed by subcrustal and subaerial erosion and as the area of continental collision extends beyond the Himalaya over the whole Pakistan-Iran-Turkey and over the Alps, such a figure is not impossible.

The effect of continental collision is to decrease the subduction rate and consequently to replace continental surface by old oceanic crust. It is then easy to compute the amount of regression produced by a continental collision at 0.1 km^2·yr^{-1} between 70 and 45 Ma and at 0.2 km^2·yr^{-1} between 45 Ma and Present. There are about 25 meters of regression between 70 and 45 Ma and 90 meters since 45 Ma. The total additional regression is about 115 meters since Upper Cretaceous to add to the 90 meters which were computed. We believe that continental collision may be the

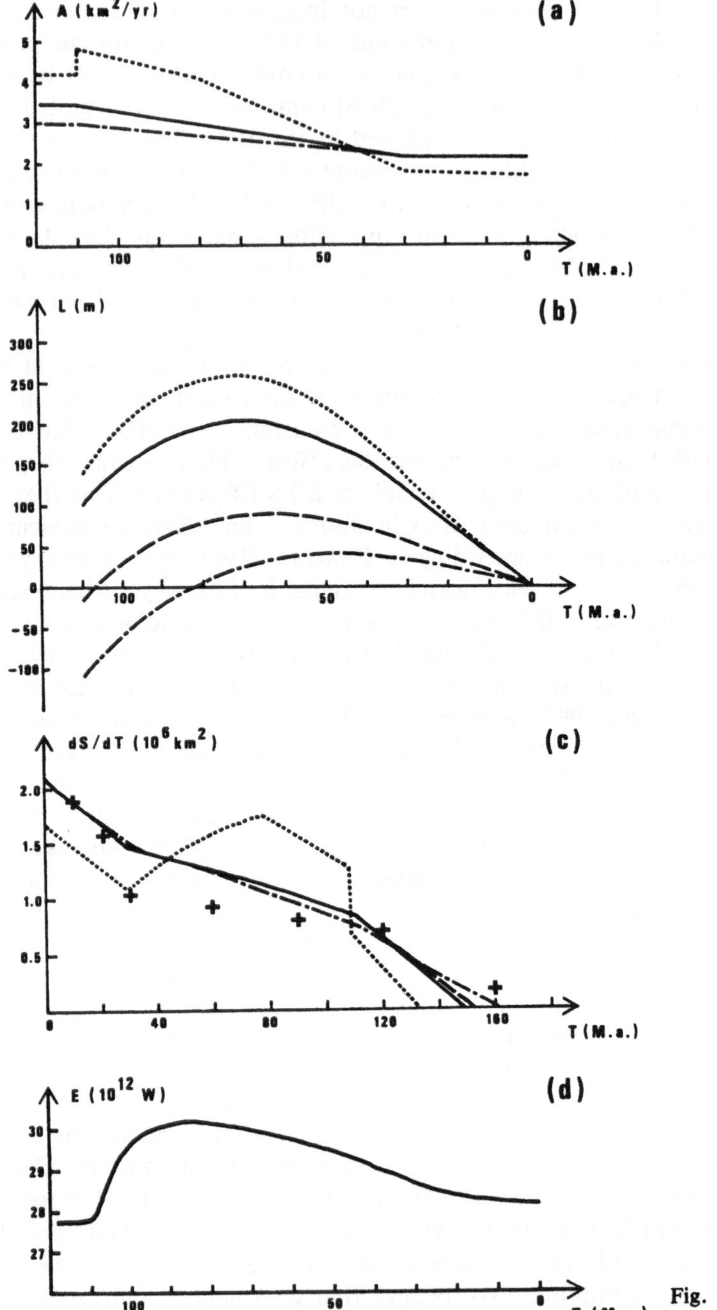

Fig. 4

cause of the rather large otherwise difficult to explain Oligocene regression (VAIL *et al.*, 1977).

Figure 4b (continuous line) shows the effect of introducing this continental collision rate within our numerical model. Figure 4b (continuous line) compares favorably with Fig. 3b.

We conclude that the total sea-level drop since 80 Ma, adding the continental collision effect to the Pangea disruption effect, is a recession of about 200 meters which may be a reasonable value. If it is, there is no need to advocate a large increase in Pacific-Tethys total accretion rate during Upper Cretaceous, in contrast with the steady accretion within Pangea. If however, 200 meters is considered too small a value, then an increase in activity of Pacific accretion should probably be assumed in Upper Cretaceous, increase which might then be associated to the hotspots outbreak at this time in the Pacific (ANDERSON, 1982; LE PICHON and HUCHON, 1984). However, increased accretion is not needed if uplift due to mid-plate volcanism is considered (SCHLANGER *et al.*, 1981). The present evidence, in our opinion, is not in favour of such an increase which should be considered more critically than it has been up to now in view of the very large uncertainties related to the reconstructions of the Pacific ocean-floor during Cretaceous (KOMINZ, 1984). That the 15% of remaining Cretaceous Pacific floor correspond to crust which has been produced with a high rate of spreading may be explained if subduction zones are randomly distributed with respect to surface as argued by PARSONS (1982). Our prefered curve for the variation in accretion rates since 110 Ma ago is shown in Fig. 4a (solid line).

Fig. 4. (a) accretion rates in the Pacific-Tethys oceans used in the numerical analysis. Dotted line: from KOMINZ (1984). Solid line: with initial accretion rate equal to the present one (3.45 $km^2 \cdot yr^{-1}$). Dash-dotted line: with a lower initial accretion rate (3 $km^2 \cdot yr^{-1}$). From 110 Ma ago, the accretion rate goes back linearly to the present value 30 Ma ago. (b) Resulting curves of sea level changes (normalized with respect to the present level). Dotted and dash-dotted lines refer to (a). Solid and dashed lines corresponds to the solid line accretion curve of (a) but solid line takes into account 0.10 $km^2 \cdot yr^{-1}$ of Africa and Arabia-Eurasia continental collision since 70 Ma ago and 0.10 $km^2 \cdot yr^{-1}$ of India-Eurasia continental collision since 45 Ma ago. (c) Resulting area versus age distributions in the Pacific ocean, to compare with the SCLATER *et al.* (1980) estimates (crosses). (d) Computed thermal energy output from the world ocean corresponding to the solid line case of (b) ad (c). The present level is 2% higher than the 110 Ma one because of the increase in ocean floor area due to continental collision.

5. *The Asymmetry in Thermal Venting of the Earth from Pangea to Present*

The previous discussion leads to our prefered hypothesis that the thermal venting of the Earth is steady-state since 200 Ma ago, except for the excess heat released during the break-up of Pangea. But this thermal venting is not equally distributed between the two hemispheres. The asymmetry is maximum during Pangea time and it is this asymmetry which forces the transition from Pangea to Present as proposed by ANDERSON (1982). If a tennis-ball type deep convection does indeed exist and can be considered steady-state over 200 Ma (LE PICHON and HUCHON, 1984), the continental and oceanic hemispheres during Pangea time are each centered above half of the corresponding upper mantle ascending cell, the common upper mantle descending cell being below the peri-Pangea subduction zone (Figs. 1 and 2). The amount of heat vented from each portion of the ascending cell is equal to the surface heat flux minus the crustal radiogenic production. In this section, we evaluate this mantle thermal venting separately for each hemisphere, using the adopted accretion rate curves of Fig. 4a.

The mantle heat flow below continents is taken as constant and equal to 0.7 HFU (SCLATER *et al.*, 1981). We ignore the radiogenic oceanic crust heat which is 0.2 HFU or less and take the variation of heat flow with age as:

$$\phi = 11\ t^{-1/2}\ \text{HFU} \qquad \text{where } t \text{ is in Ma.} \tag{5}$$

For the Pacific-Tethys triangular ocean, combining with (1), we have a mean heat flow:

$$\bar{\phi} = 29.3\ t_{\mathrm{m}}^{-1/2} \tag{6}$$

and for the Atlantic-Indian rectangular ocean, combining with (2), we have:

$$\bar{\phi} = 22\ t^{-1/2}\ \text{HFU}. \tag{7}$$

We then evaluate, for each hemisphere, as defined by the great circle in Figs. 1, 5, 6 and 7 and the corresponding reconstructions based on Table 2, the surface of continents in each hemisphere at the different periods since 110 Ma ago. We assume, as in the previous section that the rectangular ocean increases at a constant rate of 1.3 $km^2 \cdot yr^{-1}$ since 110 Ma

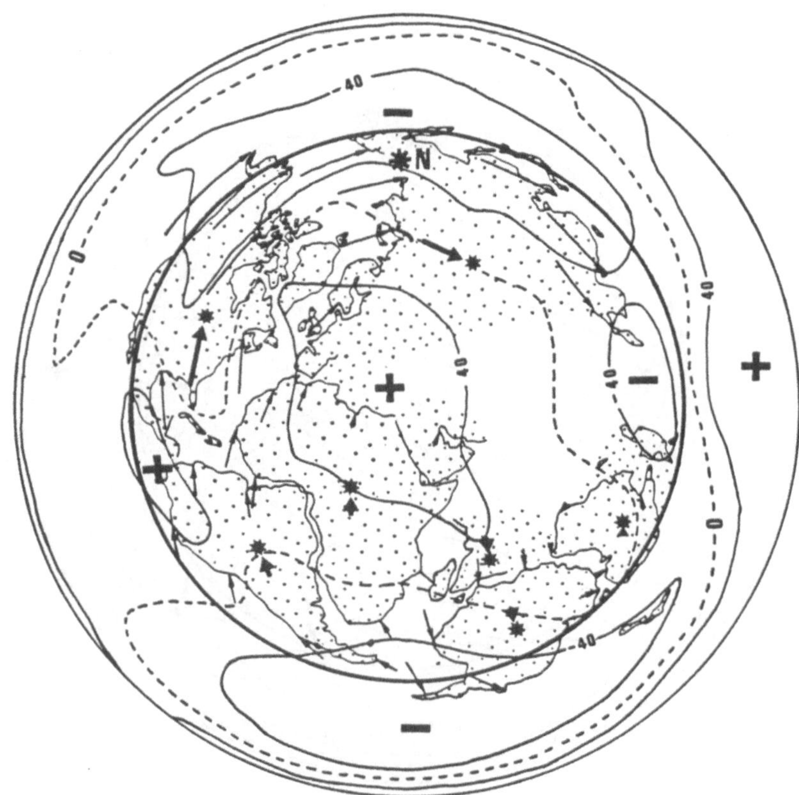

Fig. 5. Pangea configuration 135 Ma ago in the geoid frame of reference (same projection as figure 1). Arrows indicate the motions of continents from 200 to 135 Ma. Stars indicate the positions of the centers of gravity of the continents.

ago and we adopt $t_m = 180$ Ma during Pangea time as well as now.

Table 3 shows the corresponding estimates of mantle thermal venting in the Pangea and Pacific hemispheres. During Pangea time, Pangea thermal venting was only half the Pacific one, because continents covered 80% of the Pangea hemisphere, whereas at present venting of both hemispheres is approximately equal. This has been realized principally by shifting 40% of the surface of the continents outside of the Pangea hemisphere. The heat release in the Pangea hemisphere reached a maximum in Early Tertiary and is now essentially stationary. The heat release in the Pacific hemisphere decreased continually, but is now approximately steady. The total heat release follows rather closely the sea-level curve

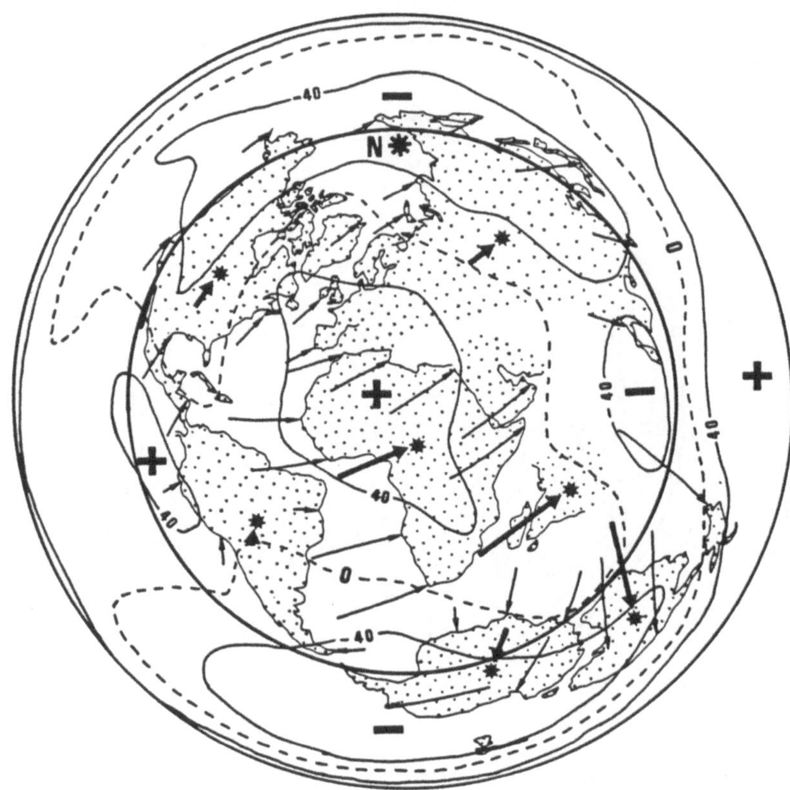

Fig. 6. Same as figure 5, but 80 Ma ago. Arrows indicate the motions of continents from 135 to 80 Ma.

increasing by about 10% in Late Cretaceous-Early Tertiary above the present level. Note that the values obtained for the present are essentially those of SCLATER *et al.* (1981), that is 8.04×10^{12} cal·s^{-1} (33.65×10^{12} W) compared to 8.3×10^{12} cal·s^{-1} (34.7×10^{12} W) (radiogenic heat excluded). Figure 4d shows the results of the more precise numerical computations corresponding to the case adopted previously (solid line in Figs. 4a and 4b) which includes continental collision. The results are similar to those of Table 3.

The asymmetry in heat venting 110 Ma ago is impossible to change with reasonable assumptions. For example, to eliminate it, one would need a t_m of 60 Ma and a C_0 of 1.8 km^2·yr^{-1} in Tethys and a t_m of 470 Ma and a C_0 of 1.1 km^2·yr^{-1} in the Pacific. It is much larger than the excess

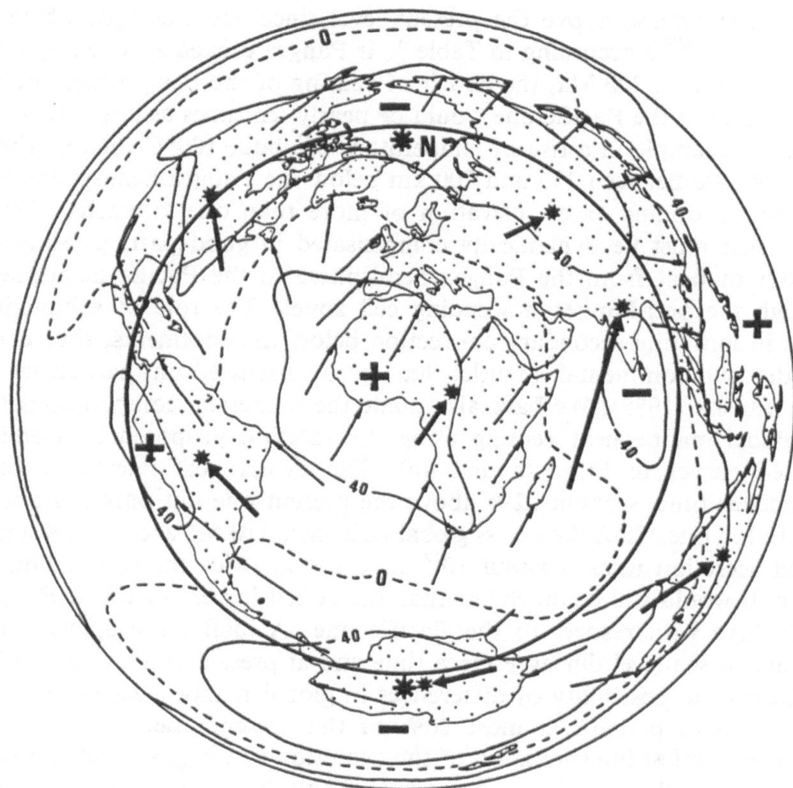

Fig. 7. Same as figure 5, but for the present situation. Arrows indicate the motions of continents form 80 Ma to present.

Table 3: Mantle thermal venting (in 10^{12}W) in the Pangea and Pacific hemispheres 110, 80, 54 Ma ago and at present.

Time (Ma)	110	80	54	0
Q_{Pangea}	10.80	15.74	17.87	15.07
$Q_{Pacific}$	23.35	21.43	18.62	18.58
Q_{Total}	34.15	37.17	36.49	33.65

global heat release above the present level since 110 Ma ago, which is about 6.5×10^{27} J according to Table 3. If Pangea existed as a hemispheric continent during 200 Ma, the deficit of venting of the Pangea hemisphere with respect to the Pacific one would be nearly ten times as large. It could result for example in a rise of temperature of about 100°C for the whole upper mantle between 250 and 700 km below the Pangea hemisphere and in a corresponding excess elevation of more than one kilometer. Thus, this deficit must be dynamically compensated in good part, possibly by transfer of heat from the Pangea hemisphere to the Pacific hemisphere through the peri-Pangea descending cell zones. The role of subduction zones in initiating secondary convection below the continents, thus cooling the sub-continental mantle, has been discussed in particular by CHRISTENSEN (1983). We have also made the numerical computations for the global mantle heat venting using the solution proposed by Kominz (dashed curves in Figs. 4a and 4b). The excess heat release in Late Cretaceous time is about 22% above the present one but only 8% above the 110 Ma one. Thus the excess global heat since 110 Ma ago using Kominz global accretion rates is about 10^{28} J, less than twice the one computed above. It would still be much less than the deficit in venting of the Pangea hemisphere with respect to the Pacific one. Actually, the approximate equality of sea-level during Pangea time and at present (VAIL *et al.*, 1977) eliminates the possibility of differences in global mantle heat venting exceeding 10 to perhaps at most 20% of the present one.

The simplest interpretation of the evolution of the global mantle heat venting since Pangea time, as given in Table 3, is that about 100 Ma were necessary to vent the excess heat accumulated below the Pangea hemisphere and that approximate equality between the two hemispheres has been realized since Early Tertiary, when 40% of the surface of the continents had been pushed out from the Pangea hemisphere. The excess heat accumulated below the Pangea hemisphere could have produced a rise in temperature of about 15° for the upper mantle between 700 and 250 km. This relatively small excess heat implies that transfer of heat from one hemisphere to the other through the descending peri-Pangea cell is quite efficient to correct the large asymmetry in mantle heat release during Pangea time.

6. *Pangea Dispersal and the Geoid*

We now wish to discuss two of the observations we made earlier. The first one is the large increase in the weighted mean velocity of the continents at the end of Cretaceous and the beginning of the Tertiary

(Table 1). The second one is the absence of any significant change in the total accretion rate within the new Pangea oceans (Atlantic and Indian oceans). These two statements appear to be exclusive but they are not. The cause of this paradox can be understood by looking at Figs. 5, 6 and 7 which show the "absolute" positions of the continents with respect to the present geoid. The opening of the North Atlantic ocean, between 80 and 54 Ma, results in large displacements of Eurasia and North America with a comparatively small amount of accretion (note that the increase in accretion rate in the Pacific-Tethys ocean, if it did occur, is 30 Ma older than the continental velocity maximum).

Either continents offer no special resistance to displacement, in spite of their thicker lithosphere, as has been argued for example by SOLOMON *et al.* (1977), or some driving mechanism must be able to displace the continents without affecting the total amount of heat released by the accretion process in the Atlantic and Indian oceans. We suggest that it is unreasonable to assume that the motions of large mostly continental plates like Eurasia may be entirely governed by the "ridge push" force. We consider consequently that the history of break-up of Pangea suggests that, after the break-up within a continent has occurred, the motions of the larger mostly continental plates are governed by deeper convection whereas the motions of the oceanic plates are principally controlled by "ridge push" and "slab pull" forces (FORSYTH and UYEDA, 1975).

Some indications that this might be so are given in Figs. 5 to 7. Although the shape of the geoid probably changes slowly through time, as is indicated by its present deviations from a tennis-ball pattern, the change in the low order harmonics (2 and 3) may not been very large since 80 Ma for reasons discussed by LE PICHON and HUCHON (1984). We see in Fig. 7 that Eurasia and North America, since 80 Ma ago, moved outward so that their centers of gravity now lie over the axis of the negative geoid belt. These absolute continent motions actually occurred in Early Tertiary and have since greatly slowed down. In the same way, Australia-Antarctica moved outward toward the present negative axis of the geoid between 110 and 80 Ma ago.

It is also intriguing that there is a correlation between the absolute velocity of India and the slope of the geoid (Fig. 8). However, this correlation may be coincidental as the slow-down is also related to continental collision. In addition, an obvious counter-example is the recent motion of Australia toward the axis of the Pacific maximum. But Australia is now attached to a large piece of oceanic lithosphere. Recently, SEIDLER *et al.* (1984) have independently shown that, in general, continent carrying plates tend to move down the geoid slope. Actually, the most complex

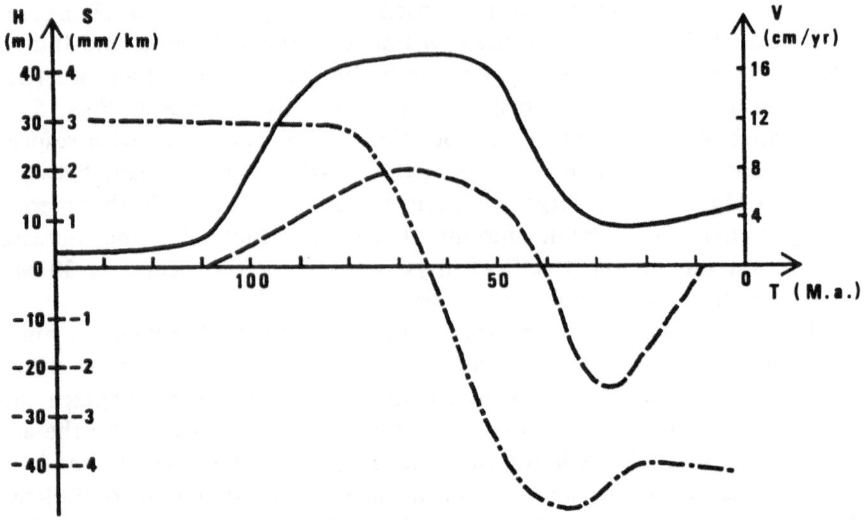

Fig. 8. Velocity of India (solid line) compared to the local slope of the geoid (dashed line) and the height of the geoid (dash-dotted line). Note that the increase in velocity 110 Ma ago correlates with an increase in the slope of the geoid. The slow-down 40 Ma ago is mainly related to the collision with Eurasia.

pattern of motions, during the dispersal of Pangea, occurred in the southeastern area, over former Eastern Gondwana. This is possibly due to the break-up in smaller fragments than elsewhere, probably because of greater heating through this part of Pangea (a possible cause of this greater heating is the fact that Tethys subduction zones were dipping to the north, toward Eurasia and consequently contributed to the cooling of Eurasia but not of Gondwana). We argue that smaller continental fragments get attached to larger oceanic plates which then may have more rapid and complex motions. In the next section, we examine in more detail the dispersal of Eastern Pangea and its evolution since 135 Ma ago, to investigate relations between absolute motions, geoid and marginal basins.

7. *Eastern Pangea Dispersal and West Pacific Marginal Basins*

Figures 9 to 14 compare six reconstructions of Eastern Pangea (at 135–80–54–35–20 Ma) within the geoid frame of reference which coincides with the hot spots one since 80 Ma. The positions of the continents are the same as those previously used and the parameters are mostly taken from MORGAN (1981). See Table 4 for detailed sources of the reconstruc-

Table 4: Sources of data for the reconstructions of Eastern Pangea for 135 Ma to present.

Parameters of plate motions: (see also tables 2 and 6)	Morgan, 1981 Patriat et al., 1982 Norton and Sclater, 1979
Indian ocean ridges:	Norton and Sclater, 1979 Patriat et al., 1982 Liu et al., 1983
Pacific ocean ridges:	Hilde et al., 1977
Collision of India with Eurasia:	Molnar and Tapponnier, 1975
Rotation of Indochina:	Tapponnier et al., 1982
Collision of North New Guinea:	Hamilton, 1979
Origin of the West Philippine basin:	Uyeda and McCabe, 1983
Ages of opening of marginal basins:	
Tasman Sea	Weissel and Hayes, 1977
Coral Sea	Weissel and Watts, 1979
West Philippines	Hilde and Lee, 1984
New Hebrides	Weissel et al., 1982
South Fiji	Malahoff et al., 1982
Caroline	Weissel, 1981
South China Sea	Taylor and Hayes, 1983
Parece Vela	Langseth and Mrozowski, 1980
Shikoku	Kobayashi and Nakada, 1978
Japan Sea	Kobayashi and Isezaki, 1976
Andaman Sea	Curray et al., 1979
Mariannas	Hussong et al., 1981
Okinawa	Weissel, 1981
Pattany trough	Helliguer and Sclater, 1983

tions and Table 6 for parameters of motion of the Pacific plate. The projection used is an oblique Mercator with the equator of projection along the polar great circle of the geoid. Distorsion is thus minimal along the periphery of Pangea. Figure 15b, with the same projection, compares the present geoid to the absolute motions of the continents (Fig. 15a) as depicted in Figs. 9 to 14. Note that in Figs. 9 to 14, the heavy arrows depict the absolute motions in the interval of time immediately preceding the date of the reconstructions.

A first observation is the contrast between extreme dispersal of Gondwana (over the future Indian Ocean) and absence of any fragmentation of Eurasia. By Early Tertiary (54 Ma ago), Eastern Gondwana had been broken into three main pieces, Australasia and Antarctica, each about 18×10^6 km^2 (somewhat smaller than South America) and Greater India which probably had about half that size. In contrast, North America and Africa are about twice and Eurasia four times that size.

A second observation concerns the motions of the continents, which are relatively well known whereas the motions of the purely oceanic plates in Tethys and Pacific oceans are highly conjectural, most of the floor of that age being now subducted (see the earlier discussion of KOMINZ (1984) reconstructions). Large mostly continental plates, as Eurasia and Antarctica, tend to move over the negative belt of the geoid which presumably corresponds to a descending cell at depth in the upper mantle (see above). Smaller continents attached to oceanic plates (as Australia) do not.

The first breaks, in Jurassic, in the Central North Atlantic, between Africa and North America, and in the Indian Ocean between Africa-South America and Madagascar-India-Australia-Antarctica (NORTON and SCLATER, 1979) had maintained Pangea within its great circle (Fig. 5). The net result of the motions was a closure of Tethys. In Early Cretaceous (Fig. 9), the open ocean between Australasia and Eurasia was less than 2000 km wide instead of about 4000 km in Early Jurassic. But, during Middle Cretaceous time, Pangea great circle broke away. Australia-Antarctica rifted apart from India (NORTON and SCLATER, 1979), rotating clockwise out of the Pangea great circle onto the negative belt of the geoid (Figs. 10 and 15). A direct consequence was that the connection between Tethys and Pacific oceans widened again, the distance increasing to a maximum of about 4000 km in Early Tertiary (Fig. 11), back to the same width as in Early Jurassic.

The motion of Eurasia out of Pangea great circle occurred later and was completed 54 to 35 Ma ago, the total motion over the Pacific being about 3000 km at the latitude of Japan. By this time, Eurasia was also installed over the axis of the negative belt of the geoid (Fig. 15).

In the mean time, India had rifted away from Madagascar, which belonged to the same plate as Africa, about 80 Ma ago (GOSLIN and PATRIAT, 1984; see Fig. 10). Thus, India started moving to the north whereas Australia-Antartica got stabilized in their position above the geoid negative belt (Fig. 15).

The next rifting occurred between Australasia and Antarctica. CANDE and MUTTER (1982) have proposed that this rifting started 70 Ma ago instead of 54 Ma ago, as had been previously accepted (see for example discussion in GOSLIN and PATRIAT, 1984) and as depicted in Fig. 11. In

Table 5: Summary of data on the absolute motions of the plates along the western margin of the Pacific ocean and on the openings of marginal basins. See table 4 for the references. The density of hatchuring schematically indicates the velocity. C shows the date of beginning of collision.

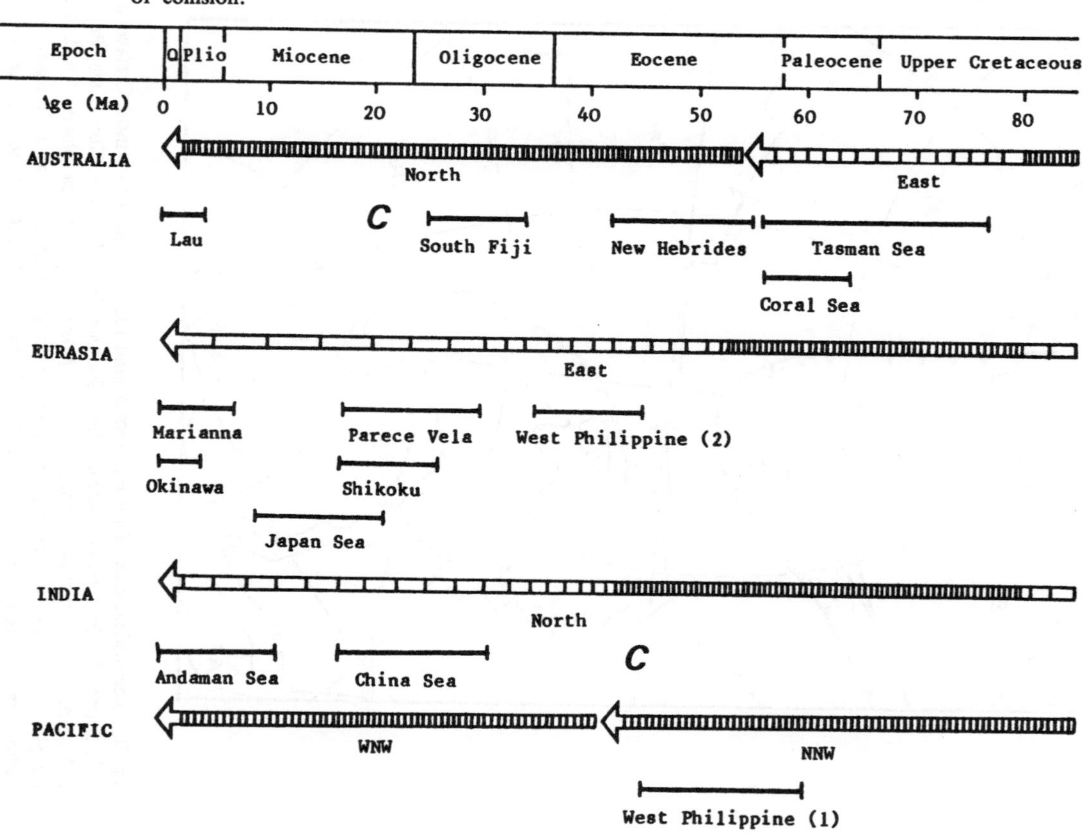

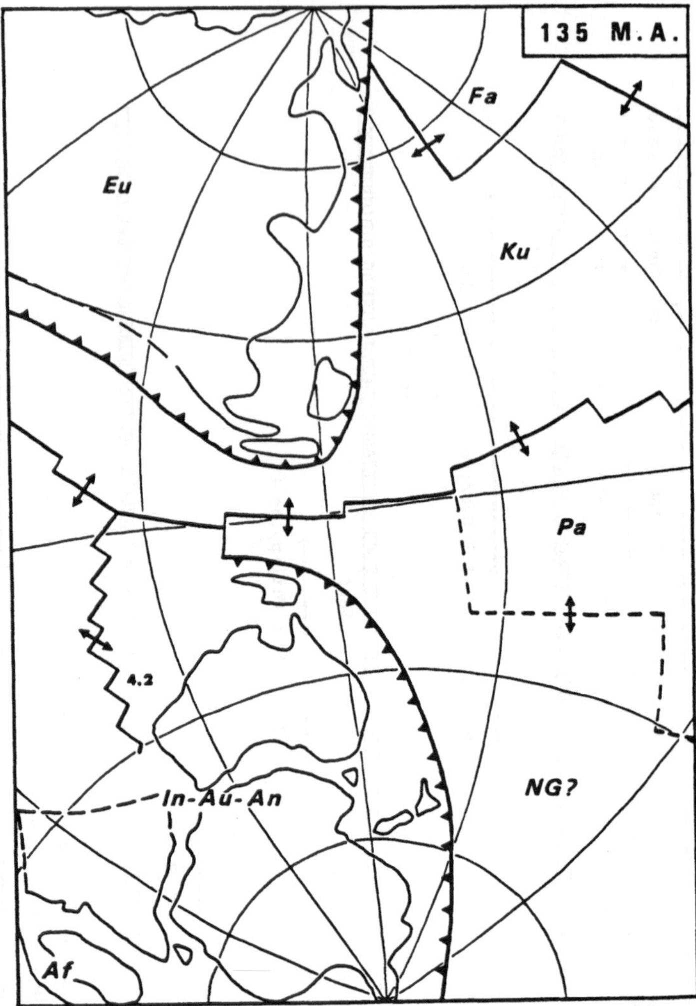

Fig. 9. Reconstructions of Eastern Gondwana 135 Ma ago (Lowermost Cretaceous) in the geoid frame of reference. Af: Africa, Eu: Eurasia, Fa: Farallon, Ku: Kula, Pa: Pacific, In-Au-An: Indo-Australo-Antarctic, NG: North New Guinea plates. Thick lines with triangles are subduction zones, based on geology. Thick lines with divergent arrows are medio-oceanic ridges (see table 4 for the references). Mercator oblique projection in which the equator of projection follows the axis of the negative polar belt of the geoid. Pole of projection at 6°N, 4°E. The central part of the map is thus undistorted whereas the left and right sides are.

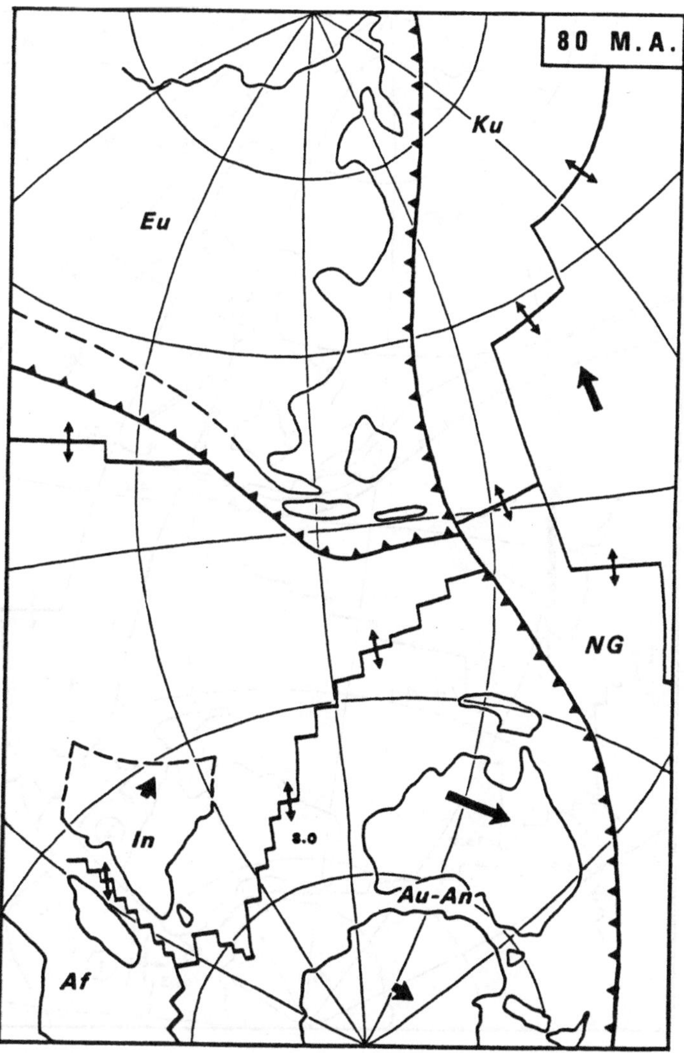

Fig. 10. Same as figure 9, 80 Ma ago (Late Cretaceous). Black arrows show the absolute motions with respect to the geoid frame of reference (from the previous to the current stage). The length is proportional to the velocity (not shown if less than 25 mm·yr^{-1}). Open arrows show relative motions. Indian and Australo-Antarctic plates are now separated.

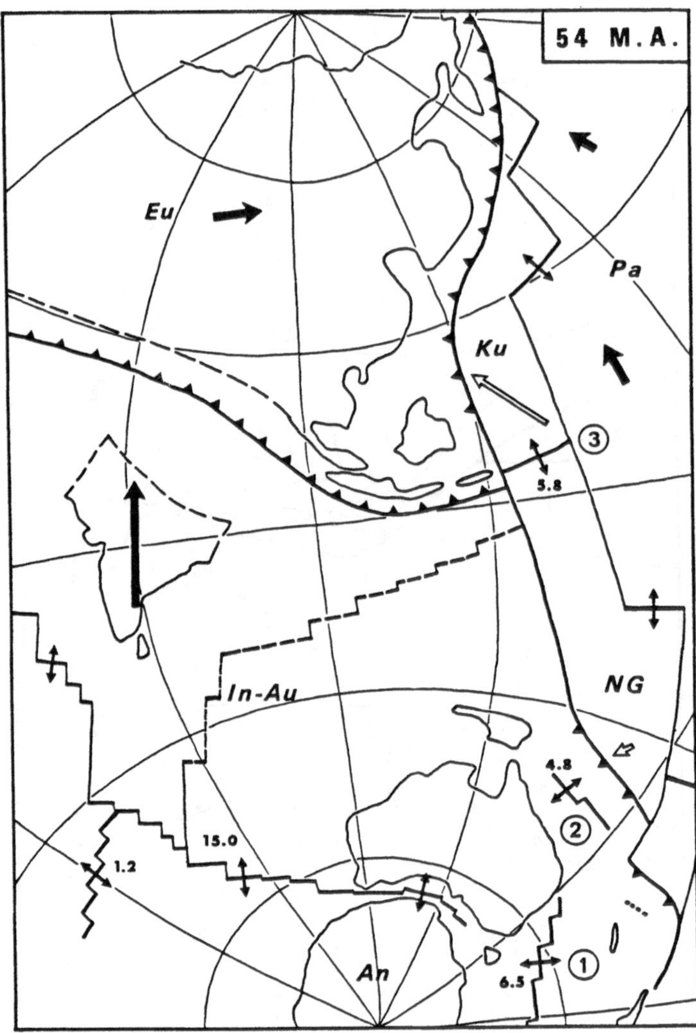

Fig. 11. Same as figure 10, 54 Ma ago (Early Eocene). As a result of the jump of the ridge from between Eurasia and Australia to between Australia and Antarctica, India and Australia now belong to the same plate whereas Antarctica is a separate one. Circled numbers: 1—Tasman Sea, 2—Coral Sea, 3—West Philippine Sea. Chronology of opening: see table 5. Small number indicates the total rate of opening (in cm·yr^{-1}).

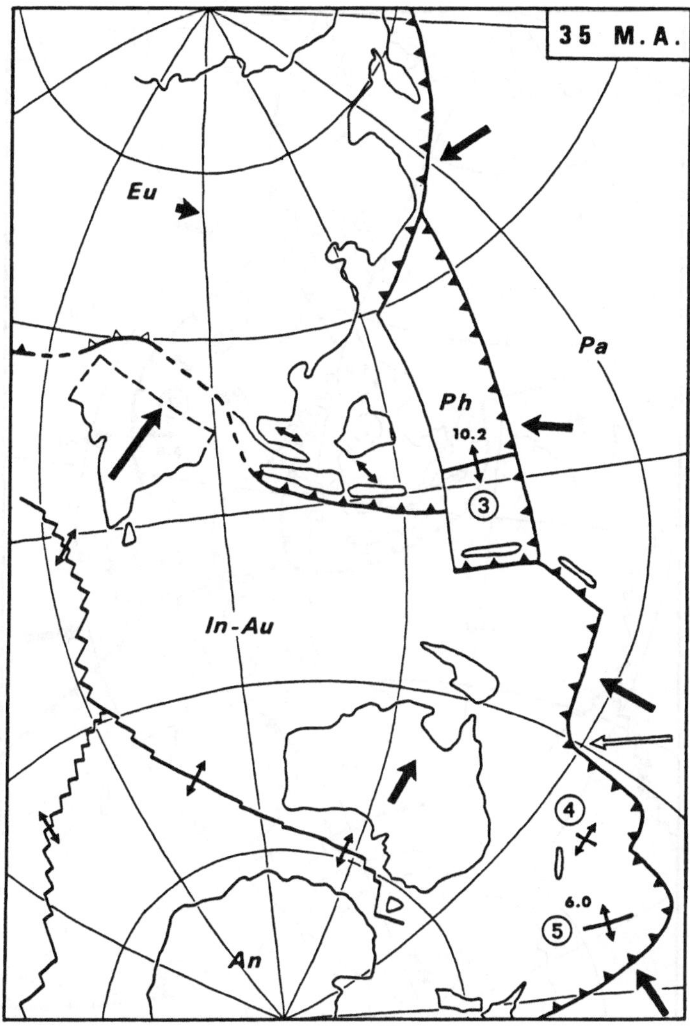

Fig. 12. Same as figure 11, 35 Ma ago (Lowermost Oligocene). India now collides with Eurasia. Circled numbers: 3—West Philippine Sea (trapped behind the new subduction zone developed in place of a transform because of the change in direction of motion of the Pacific plate with respect to Eurasia). 4—New Hebrides, 5—South Fiji.

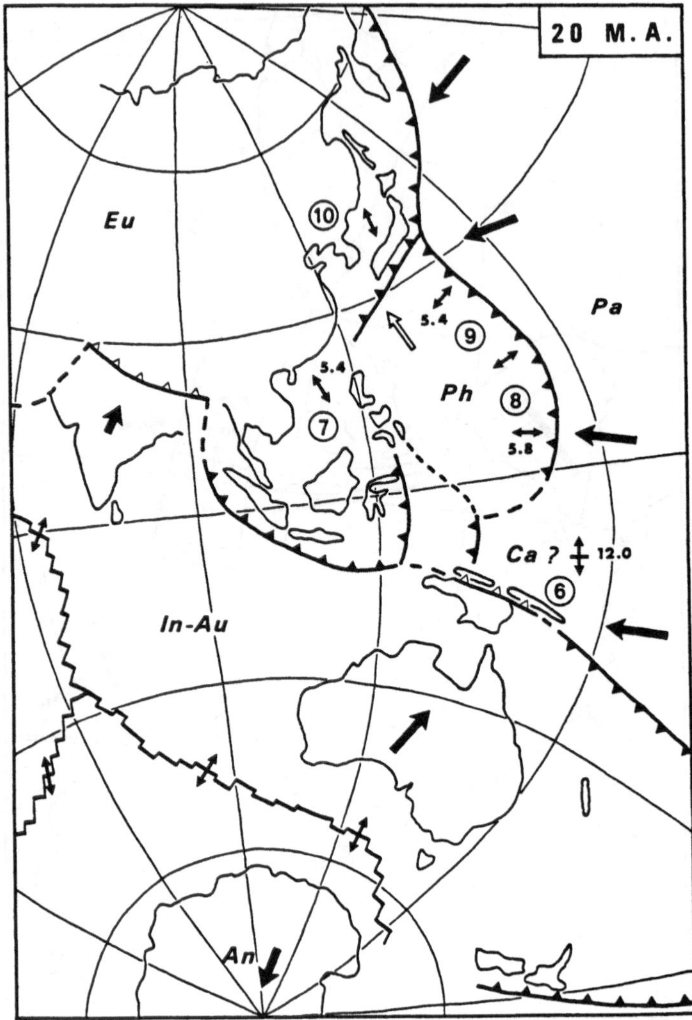

Fig. 13. Same as figure 12, 20 Ma ago (Early Miocene). South New Guinea collides with volcanic arcs which will form North New Guinea. Circled numbers: 6—Caroline (Ca) plate, 7—South China Sea. 8 and 9—Parece Vela and Shikoku basins. 10—Japan Sea.

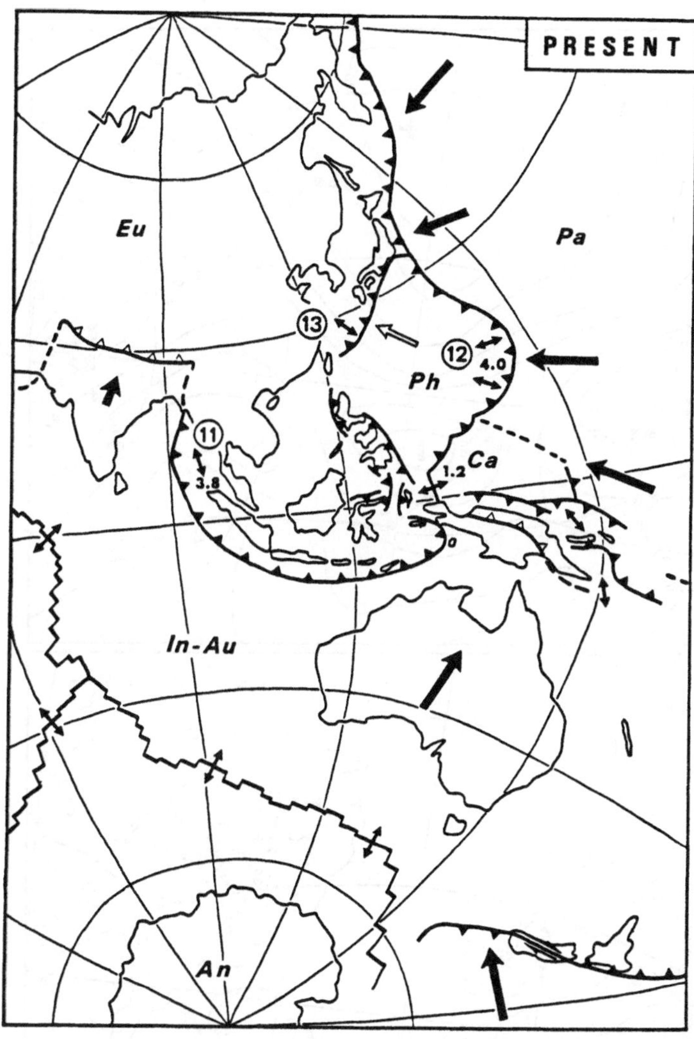

Fig. 14. Same as figure 13, Present situation. Circled numbers: 11—Andaman Sea, 12—Mariana trough, 13—Okinawa basin.

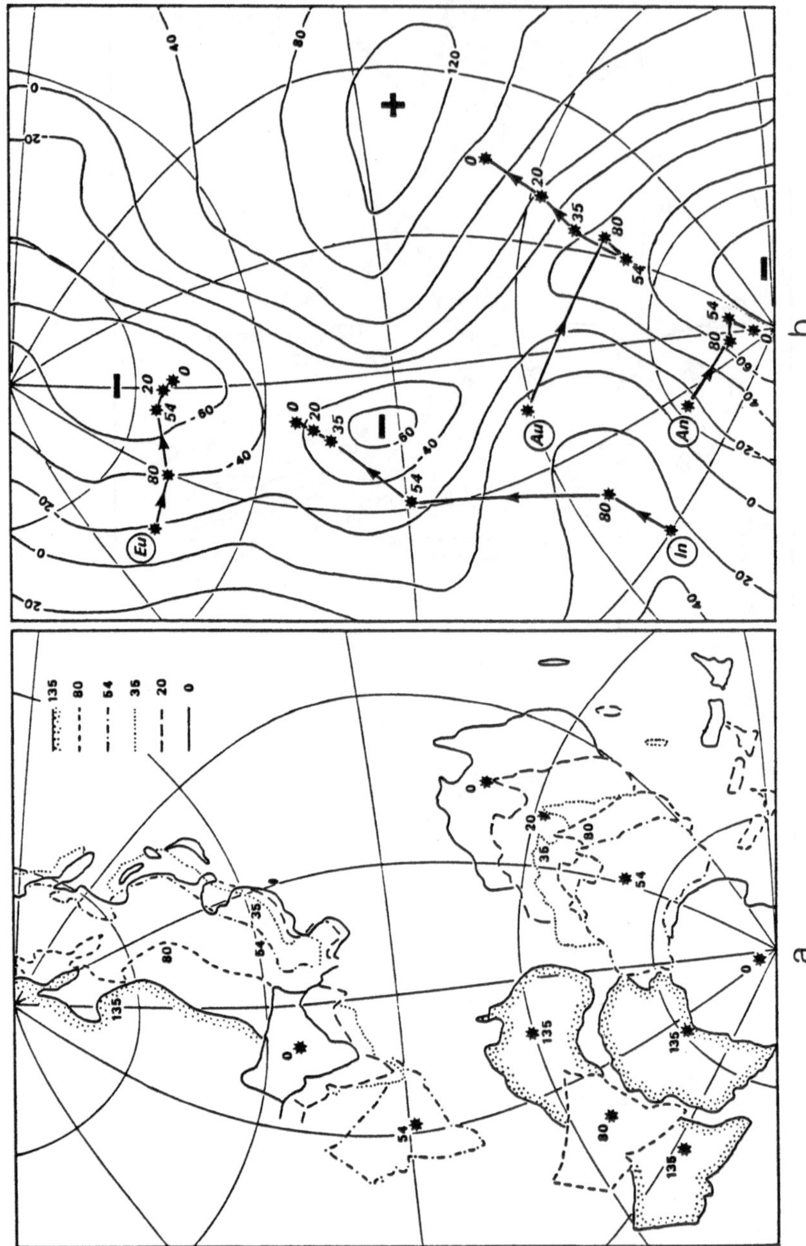

Fig. 15. (a) Successive positions of the continents, as depicted in figures 9 to 14. (b) Trajectories of the centers of gravity of the continents superposed to the geoid anomalies (in meters).

any case, rifting was slow prior to 43 Ma ago, date at which the accreting plate boundary between India and Australasia had completely died. From then on, Australasia belonged to the large India mostly oceanic plate moving to the north toward the Tethys northern subduction zone.

A third observation is that marginal basins were not present while the continents advanced over the Pacific (see Table 5 and Figs. 9 to 14). This is true of Australasia which ceased its motion toward the Pacific 80 Ma ago. The change in motion was followed by the opening of the Tasman Sea between 77 and 56 Ma (WEISSEL and HAYES, 1977) and the opening of the Coral Sea between 64 and 54 Ma (WEISSEL and WATTS, 1979, see [1] and [2] in Fig. 11). Thus, during this time, the edge of Australasia was still advancing over the Pacific ocean whereas the main continent had stopped. The same sequence of events occurred east of Eurasia after this continent slowed down its eastward motion 50 Ma ago. After entrapment of the West Philippine basin in Eocene time (UYEDA and MCCABE, 1984), the Philippine basin progressively evolved into an E-W intermittently spreading marginal basin with the Parace Vela basin opening between 30 and 17 Ma (LANGSETH and MROZOWSKI, 1980) and the Shikoku basin opening between 26 and 17 Ma (KOBAYASHI and NAKADA, 1978). Marginal basins have been characteristic of both the Australasia and Eurasia borders since (see Table 5). However, TAPPONNIER *et al.* (1982) have pointed out that the origin of some marginal basins is probably related to the expulsion of the Indo-chinese peninsula away from Eurasia, expulsion triggered by the collision of India. This is most probably the case of the oceanic accretion within the South China Sea between 32 and 17 Ma (TAYLOR and HAYES, 1983) and within the Andaman Sea since 11 Ma (CURRAY *et al.*, 1979), both accretions being preceded by a rather long phase of continental rifting.

A fourth and last observation is the remarkable change in the accreting boundary system produced by the closure of Tethys and the break-up of Gondwana. Prior to 54 Ma ago, a continuous east-west accreting boundary extended from the Tethys ocean to the Pacific ocean. After 43 Ma ago, the Tethys connection was broken. North-south accreting boundary systems then prevailed in the equatorial regions of both the Pacific and Indian oceans. This major change was pointed out by HILDE *et al.* (1977).

We have mostly adopted Hilde et al.'s solution for the Pacific accreting boundaries. However, we have adapted it to the absolute motions of the continents, as described above, and to the absolute motion of the Pacific plate as given by CLAGUE and JARRARD (1973) since 70 Ma, and by LANCELOT (1978) prior to 70 Ma (see Table 6). SUAREZ and MOLNAR

Table 6: Parameters of absolute motion of the Pacific plate.

	Age (Ma)	Lat (°N)	Lon (°E)	α (°ccw)	Reference
Phase 1	125–70	30	−97	0.69	Lancelot, 1978
Phase 2	70–43	11	−89	0.57	Suarez and Molnar, 1980
Phase 3	43–0	67	−43	0.50	Idem

(1980) have shown that the results of Clague and Jarrard are internally consistent for absolute motion data within the Pacific plate although they cannot be simply related by kinematics, through the Antarctic plate, to the surrounding continental plates prior to 40 Ma ago (see also JURDY, 1979 and the discussion in MORGAN 1981). Thus, it is only through these absolute motion vectors that the relative motions of the Pacific and surrounding continental plates prior to 40 Ma can be obtained. Prior to 70 Ma, the results of LANCELOT (1978) are quite uncertain although they do indicate the approximate direction of motion of the Pacific plate between 125 and 70 Ma ago.

The absolute motions vectors in Figs. 9 to 14 can be used to estimate the relative velocity vectors along the main Pacific subduction zones. It is clear that a subduction zone must have existed between Australasia and the Pacific ocean from 135 to 80 Ma because of the fast eastward progression of Australasia. In the same way, a subduction zone must have existed since 80 Ma between Eurasia and the Pacific ocean because of the eastward progression of Eurasia between 80 and 35 Ma and the westward component of motion of the Pacific plate since 43 Ma ago. SENO (1984) argued that no subduction could have existed there prior to 43 Ma ago, in part because he ignored the eastward motion of Eurasia.

However, the existence of a subduction zone between Australasia and the Pacific ocean from 80 to 54 Ma cannot be inferred on the basis of the relative motion between the Pacific plate and Australia (Fig. 11). Taking into account the Coral Sea opening and the slow Australia-Antarctica rifting, there is a slight opening and certainly no shortening along this boundary. To solve this problem, SENO (1984) has recently suggested a solution that we adopt which is the existence of a North New Guinea plate which might be the remnant of the southern plate in Fig. 9.

It is obvious, in any case, that these reconstructions are hypothetical and probably oversimplified with respect to the real geologic situations. But that north-south spreading along east-west accreting boundaries existed both in Tethys and West Pacific oceans in Cretaceous time seems beyond dispute. It is also well established that the accreting boundary system was completely reorganized between 48 and 43 Ma ago in the Indian ocean (GOSLIN and PATRIAT, 1984) and that the direction of the Pacific plate changed from NNW to WNW 43 Ma ago (MORGAN, 1972; CLAGUE and JARRARD, 1973). Thus the connection between closure of Tethys and reorganization of spreading in both oceans seems probable. As argued in particular by HILDE *et al.* (1977), this reorganization of spreading may be due to subduction of an accreting plate boundary system. Figures 11 and 12 suggest that it is the India-Australasia and Kula-Pacific systems.

This description of the dispersal of Eastern Gondwana in an "absolute" frame of reference may also be interpreted as suggesting the existence of two relatively independent types of motions at the surface of the Earth concerning either the mostly continental plates or the mostly oceanic ones. In Eastern Gondwana, the break-up of the continents in small pieces probably reflects an excess accumulation of mantle heat there, perhaps, as suggested above, because of the absence of southward directed subduction zones within the Tethys ocean. The motions of the continents after break-up appear to be dominated by a tendency for them to settle over the main polar negative belt of the geoid provided the plates to which they belong are mostly covered by continental crust (Fig. 15). We argue that, because of a greater lithospheric thickness, continents are influenced by deeper convection (below 200 to 250 km) and tend to move over the descending cell lying below the negative of the geoid. Plates mostly covered by oceanic crust, as for the India-Australia plate since 43 Ma, are not influenced by the deeper geoid convection.

8. *Summary and Conclusions*

We have discussed in more detail than LE PICHON and HUCHON (1984) the dispersal phase of Pangea which brought the Earth from a single hemispheric polar continent stage to the present dispersed continents stage. We have argued that the low-order configuration of the geoid has a time constant sufficiently large that the related deep mantle convection can be considered steady-state over 200 Ma. Our hypothesis is based on the coincidence between the peripheral Pangea subduction belt and the present geoid polar negative belt at 80 Ma, using the hot spots framework, and its near-coincidence at 135 Ma. We consequently discuss the dispersal of Pangea with respect to this steady-state deep mantle convection. As LE PICHON and HUCHON (1984), we suggest that the low-order geoid reflects convection within the lower mantle which is weakly coupled to upper mantle convection. If this is so, as there must have been a descending cell in the upper mantle along the peri-Pangea subduction belt, one would expect an underlying ascending cell in the lower mantle if mechanical coupling dominates. The alternative is whole mantle convection or thermal coupling.

We estimated continental drift activity by computing the weighted mean absolute velocity of the continents, the weight being proportional to the areas of the continents. There is an increase to a maximum of about 50 $mm \cdot yr^{-1}$ in Campanian, Maestrichtian and Paleocene, between 80 and 54 Ma ago. This is to be compared to the present 21 $mm \cdot yr^{-1}$ mean velocity. This peak is mostly due to the fast motion of Eurasia and North America out of the Pangea great circle. It is not accompanied by any change in accretion rate within the intra-Pangea oceans (Atlantic and Indian oceans) since 110 Ma ago. This fact suggests that the motions of the mostly continental plates are not governed by ridge-push or slab-pull.

We estimated the variation of the global accreting plate boundary activity based on its consequences on global sea-level variations. We proposed that there is no need to advocate a large increase in accretion rate within the Pacific-Tethys ocean which would be in contrast with the steady rate within the Atlantic-Indian one. It is possible to explain the Upper Cretaceous transgression with the formation of the Atlantic-Indian oceans and the Cenozoic regression with the progressive return to equilibrium of the global rate of accretion and the additional effect of continental collision. If this is so, global mantle thermal venting has been approximately steady-state since 200 Ma, the maximum increase being about 10% in Late Cretaceous-Early Tertiary. The total excess heat eliminated during Pangea dispersal is only about 6.5×10^{27} J, about ten times less than the heat which might have accumulated below Pangea because of the deficit of mantle thermal venting below continents. This fundamental asymmetry in mantle thermal venting between Pangea and the Pacific ocean hemisphere must have been dynamically compensated by exchange of heat through convection induced by sinking slabs (CHRISTENSEN, 1983). At present, the symmetry in mantle thermal venting between both hemispheres has been restored by transfer of 40% of the surface of the continents from the Pangea to the Pacific hemisphere.

We described in more detail the displacements of the continents with respect to the low-order geoid, especially within Eastern Pangea. We showed that the large mostly continental plates, except South America and Africa, moved out of Pangea onto the axis of the geoid negative polar belt after Pangea had been broken apart. These two exceptions might be explained by the fact that South America is situated on the positive bridge connecting the two positive caps and consequently could not move in this way, and that Africa is situated too far off from the geoid negative belt. On the other hand, once a continent is inserted within a much larger oceanic plate, its motions do not show any clear relation to geoid. This is the case of Australia after it rifted from Antarctica. We consequently argued that there are two independent types of motions at the surface of the Earth. The motions of the plates which are covered mostly by continental crust appear to be governed by the deep (below 250 km?) geoid-related convection, perhaps because of the thicker continental lithosphere. These continental plates tend to move over the negative geoid belt, which presumably corresponds to descending currents at depth in the upper mantle. The motions of the plates mostly covered by oceanic crust show no such relation because they are related to a different type shallower convection governed by ridge-push and slab-pull (FORSYTH and UYEDA, 1975). We suggest that the independent motions of the continents play a much more significant role than previously thought.

Finally, we point out that marginal basins did not and presumably could not develop on the Eurasian and Australasian Pacific boundaries as long as both continents advanced rapidly eastward over the ocean.

Acknowledgements

This paper was prepared for the Tokyo November 1983 OJI meeting at the invitation of Professor Noriyuki Nasu. We thank Seiya Uyeda for detailed comments. Research was supported by CNEXO and CNRS.

REFERENCES

ALLÈGRE, C., V. COURTILLOT, *et al.*, Structure and evolution of the Himalayan-Tibet orogenic belt, *Nature*, **307**, 5946, 17–22, 1984.

ANDERSON, D. L., Hot spots, polar wander, Mesozoic convection and the geoid, *Nature*, **297**, 391–393, 1982.

BIJU-DUVAL, B., J. DERCOURT, and X. LE PICHON, From the Tethys ocean to the Mediterranean seas: a plate tectonic model of the evolution of the western alpine system, in Intern. Symp. on Structural History of the Mediterranean basins, Split (Yougoslavia), 25-29 October 1976, B. Biju-Duval and L. Montadert Eds., Technip, Paris, 1977, 143–164, 1977.

BUSSE, F. H., Quadrupole convection in the lower mantle?, *Geophys. Res. Lett.*, **10**, 285–288, 1983.

CANDE, S. C. and J. MUTTER, A revised interpretation of the oldest sea floor spreading anomalies between Australia and Antarctica, *Earth Planet. Sci. Lett.*, **58**, 151–160, 1982.

CHASE, C. G. and D. R. SPROWL, The modern geoid and ancient plate boundaries, *Earth Planet. Sci. Lett.*, **62**, 314–320, 1983.

CHRISTENSEN, U., A numerical model of coupled subcontinental and oceanic convection, *Tectonophysics*, **95**, 1–23, 1983.

CLAGUE, D. A. and R. D. JARRARD, Tertiary Pacific plate motion deduced from the Hawaiian-Emperor chain, *Geol. Soc. Am. Bull.*, **84**, 1135–1154, 1973.

CURRAY, J. R., D. G. MOORE, L. A. LAWYER, F. J. EMMEL, R. W. RAITT, M. HENRY, and R. KIEKHEFER, Tectonics of the Andaman sea and Burma, Geological and geophysical investigations of continental slopes and rises, *Amer. Assoc. Petrol. Geol. Mem.*, **29**, 189–198, 1979.

FORSYTH, D. and S. UYEDA, On the relative importance of the driving forces of plate motions, *Geophys. J. R. astr. Soc.*, **43**, 163–200, 1975.

GOSLIN, J. and P. PATRIAT, Absolute plate motions and hypotheses on the origin of five aseismic ridges in the Indian ocean, *Tectonophysics*, **101**, 221–244, 1984.

GOUGH, D. I., The geoid and single-cell mantle convection, *Earth Planet. Sci. Lett.*, **34**, 360–364, 1977.

HAGER, B. H., Global isostatic anomalies for plate and boundary layer models of the lithosphere, *Earth Planet. Sci. Lett.*, **63**, 97–109, 1983.

HAMILTON, W., Tectonics in the Indonesian region, U. S. Geol. Survey Prof.Paper 1078, 345 pp., 1979.

HARRISON, C. G. A., G. W. BRASS, E. SALTZMAN, J. SLOAN II, J. SOUTHAM, and J. M. WHITMAN, Sea level variations, global sedimentation rates and the hypsographic curve, *Earth Planet. Sci. Lett.*, **54**, 1–16, 1981.

HAYS, J. D. and W. C. PITMAN III, Lithospheric plate motion, sea level changes and climatic and ecological consequences, *Nature*, **246**, 18–22, 1973.

HEESTAND, R. L. and S. T. CROUGH, The effect of hot spots on the oceanic age-depth relation, *J. Geophys. Res.*, **86**, 18–22, 1981.

HELLINGER, S. J. and J. G. SCLATER, Some comments on two-layer extensional models for the evolution of sedimentary basins, *J. Geophys. Res.*, **88**, 8251–8269, 1983.

HESS, H. H., Mid-oceanic ridges and tectonics of the sea-floor, in Submarine geology and geophysics, Colston paper 17, Butterworths, 317–333, 1965.

HILDE, T. W. C. and C. S. LEE, Origin and evolution of the West Philippine basin: a new interpretation, in Geodynamics of back-arc regions, R. L. Carlson and K. Kobayashi, Eds., *Tectonophysics*, **102**, 85–104, 1984.

HILDE, T. W. C., S. UYEDA, and L. KROENKE, Evolution of the western Pacific and its margin, *Tectonophysics*, **38**, 145–165, 1977.

HUSSONG, D., S. UYEDA, and SCIENTIFIC STAFF, Near the Philippines: Leg 60 ends in Guam, *Geotimes*, **22** (10), 19–22.

IRVING, E., Drift of the major continental blocks since the Devonian, *Nature*, **270**, 304–309, 1977.

JURDY, D. M., Relative plate motions and the formation of marginal basins, *J. Geophys. Res.*, **84**, 6796–6802, 1979.

KOBAYASHI, K. and N. ISEZAKI, Magnetic anomalies in Japan Sea and Shikoku basin and their possible tectonic implications, in *The Geophysics of the Pacific ocean Basins and Its Margins*, G. Sutton, M. H. Manghnami and R. Moberly Eds., Monogr. Am. Geophys. Union, 19, pp. 235–251, 1976.

KOBAYASHI, K. and M. NAKADA, Magnetic anomalies and tectonic evolution of the Shikoku inter-arc basin, *Adv. Earth Planet. Sci.*, **6**, 391–402, 1979.

KOMINZ, M. A., Oceanic ridge volumes and sea level change—an error analysis, in *Interregional Unconformities and Hydrocarbons Accumulations*, J. Schlee Ed., Amer. Assoc. Petrol. Geol. Memoir, 1984 (in press).

LANCELOT Y., Relations entre évolution sédimentaire et tectonique de la plaque Pacifique depuis le Crétacé inférieur, Mém. Soc. Geol. France, Nouv. Sér., 57, Mém. n°134-1-40, 1978.

LANGSETH, M. G. and C. L. MROZOWSKI, Geophysical surveys of Leg 59 sites, Deep Sea

Drilling Project, in L. W. Kroenke, R. W. Scott et al., Initial Reports of the D.S.D.P., 59, U. S. Government Printing Office, Whashington D. C., pp. 487–502, 1980.

LARSON, R. L., X. GOLOVCHENKO, and W. C. PITMAN III, Magnetic timescale, in plate Tectonic Map, Circum Pacific Map Project, Chairman J. A. Reinemund, 1981.

LE PICHON, X. and P. HUCHON, Pangée, géoide et convection, C. R. Acad. Sc. Paris, 296, série II, 1313–1320, 1983.

LE PICHON, X. and P. HUCHON, Pangea, geoid and convection, *Earth Planet. Sci. Lett.*, **67** (1), 123–136, 1984.

LIU, C. S., J. S. CURRAY, and J. M. MCDONALD, New constraints on the tectonic evolution of Eastern Indian ocean, *Earth Planet. Sci. Lett.*, **65**, 331–342, 1983.

MALAHOFF, A., R. FEDEN, and H. FLEMING, Magnetic anomalies and tectonic fabric of marginal basins north of New Zealand, *J. Geophys. Res.*, **87**, 4109–4125, 1982.

MINSTER, J. B., T. H. JORDAN, P. MOLNAR, and E. HAINES, Numerical modelling of instantaneous plate tectonics, *Geophys. J. R. astr. Soc.*, **36**, 541–576, 1974.

MOLNAR, P. and P. TAPPONNIER, Cenozoic tectonics of a continental collision, Science, **189**, 419–426, 1975.

MORGAN, W. J., Plate motions and deep mantle convection, in Studies in Earth and Space Science, *Geol. Soc. Am. Mem.*, **132**, 7–22, 1972.

MORGAN, W. J., Hotspots tracks and the opening of the Atlantic and Indian oceans, in The Sea, vol. 7, C. Emiliani Ed., J. Wiley and Sons, New York, 443–487, 1981.

NORTON, I. O. and J. G. SCLATER, A model for the evolution of the Indian ocean and the break-up of Gondwanaland, *J. Geophys. Res.*, **84**, 6803–6829, 1979.

PALMER, A. R., The decade of North American Geology—1983 Geologic time scale, *Geology*, **11**, 503–504, 1983.

PARSONS, B., The rates of plate consumption and creation, *J. Geophys. Res.*, **67**, 437–448, 1981.

PARSONS, B., Causes and consequences of the relation between area and age of the ocean floor, *J. Geophys. Res.*, **87**, 289–302, 1982.

PITMAN III, W. C., Relationship between eustacy and stratigraphic sequences of passive margins, *Geol. Soc. Amer. Bull.*, **89**, 1389–1403, 1978.

SCHLANGER, S. O., H. C. JENKYNS, and PREMOLI-SILVA I., Volcanism and vertical tectonics in the Pacific basin related to global Cretaceous transgressions, *Earth Planet. Sci. Lett.*, **52**, 435–449, 1981.

SCHULT, F. R. and R. G. GORDON, Root mean square velocities of the continents with respect to hot spots since the Early Jurassic, *J. Geophys. Res.*, **89**, 1789–1800, 1984.

SCLATER, J. G., C. JAUPART, and D. GALSON, The heat flow trough oceanic and continental crust and the heat loss of the Earth, Rev. Geophys. Space Phys., **18** (1), 269–311, 1984.

SCLATER, J. G., B. PARSONS, and C. JAUPART, Oceans and continents: similarities and differences in the mechanisms of heat loss, *J. Geophys. Res.*, **86**, 11535–11552, 1981.

SEIDLER, E., W. R. JACOBY, and M. LEMMENS, Plate motions, the driving mechanism and geopotential, Annales Geophysicae, **2**, 333–342, 1984.

SENO, T., Is there a North New Guinea plate?, 1984 (in press).

SOLOMON, S. C., N. H. SLEEP, and D. M. JURDY, Mechanical models for absolute plate motions in the Early Tertiary, *J. Geophys. Res.*, **82**, 203–212, 1977.

SUAREZ, G. and P. MOLNAR, Paleomagnetic data and pelagic sediment facies and the motion of the Pacific plate relative to the spin axis since the Late Cretaceous, *J. Geophys. Res.*, **85**, 5257–5280, 1980.

TAPPONNIER, P., G. PELTZER, A. Y. LE DAIN, and R. ARMIJO, Propagating extrusion tectonics in Asia: new insights from simple experiments with plasticine, *Geology*, **10**, 609–688, 1982.

Taylor, B. and D. E. Hayes, Origin and history of the South China basin, in the tectonic and geologic evolution of Southeast Asian seas and islands, Part 2, D. E. Hayes Ed., Monogr. Am. Geophys. Union, **27**, 23–56, 1983.

Uyeda, S. and R. McCabe, A possible mechanism of episodic spreading of the Philippine Sea, in Accretion tectonics in the circum-Pacific regions, M. Hashimoto and S. Uyeda Eds., Terra Scientific Publishing Company, Tokyo, 291–306, 1983.

Vail, P. R., R. M. Mitchum, and S. Thompson III, Seismic stratigraphy and global change of sea level—Part 4: global cycles of relative changes of sea level, in Seismic stratigraphy-applications to hydrocarbon exploration, C. E. Payton Ed., *Amer. Assoc. Petrol. Geol. Mem.*, **26**, 1977.

Weissel, J. K., Magnetic lineations in marginal basins of the west Pacific, Philos. Trans. R; Soc. London, Ser. A, **300**, 223–247, 1981.

Weissel, J. K. and R. N. Anderson, Is there a Caroline plate?, *Earth Planet. Sci. Lett.*, **41**, 143–158, 1978.

Weissel, J. K. and D. E. Hayes, Evolution of the Tasman Sea reappraised, *Earth Planet. Sci. Lett.*, **41**, 143–158, 1977.

Weissel, J. K. and A. B. Watts, Tectonic evolution of the Coral Sea basin, *J. Geophys. Res.*, **84**, 4572–4582, 1979.

Weissel, J. K., A. B. Watts, and A. Lapouille, A., Evidence for Late Paleocene to Late Eocene seafloor in the southern New Hebrides basin, *Tectonophysics*, **87**, 243–251, 1982.

Formation of Active Ocean Margins, edited by N Nasu *et al.*, pp. 43–58.

THICKNESS ANOMALIES OF THE LITHOSPHERE, DRIVING FORCE OF SUBDUCTION AND ACCRETION TECTONICS

Yoshibumi TOMODA, Hiromi FUJIMOTO, and Takeshi MATSUMOTO

Ocean Research Institute, University of Tokyo,
Minamidai, Nakano-ku, Tokyo, 164 Japan

Abstract. An approach to the physical mechanism of the lithospheric motion is presented. The approach consists of the following four steps.

1) Thickness anomaly distribution of the lithosphere in the Pacific was calculated by use of both seismic and gravimetric results.

2) Thickness anomaly of the lithosphere generally produces the driving force of the motion due to density difference of the structure. One of the typical examples is the force produced at the boundary of two lithosphere having different age.

3) Motion of highly viscous liquid is similar to the elastic deformation, when the displacement is replaced by the velocity of particle. The motion of the lithosphere is calculated when the viscosity and density structures are given.

4) Numerical simulation of the motion of lithosphere was conducted without any assumption but gravity acceleration for the driving force of the plate. The mechanism of lithospheric subduction and its interaction with, seamounts or rises were simulated by "marker and cell" method. It is expected that accretional process at the trench can be simulated by the method when reliable density structure is given.

1. Introduction

Based on the data from 1963 to 1980, new maps of gravity anomalies in the northwest Pacific were compiled (TOMODA and FUJIMOTO, 1982). Simplified map of free-air anomalies is shown in Fig. 1.

The characteristics of free-air anomalies in this region are summarized as follows; firstly there are large negative anomalies around the tren-

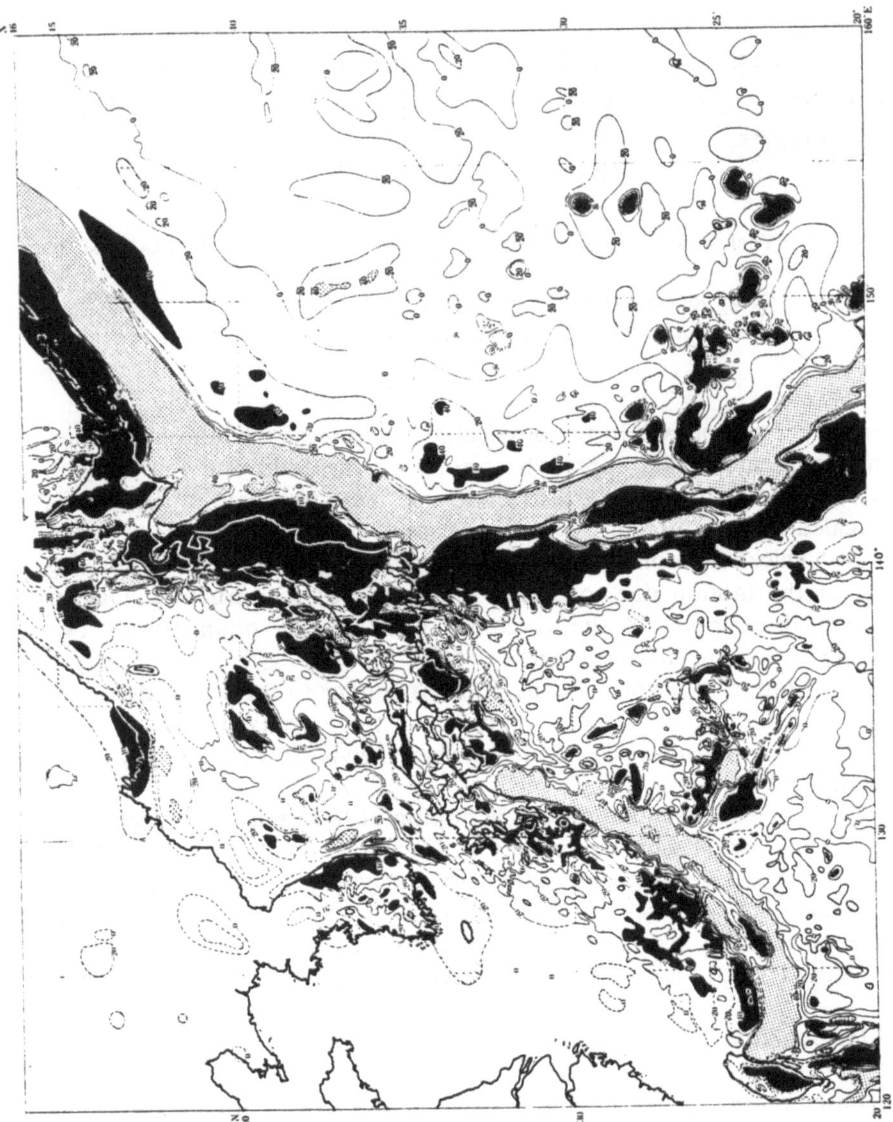

Fig. 1. Simplified map of free-air gravity anomaly in the western Pacific. ●: free-air anomaly is larger than +40 mgal. ○: free-air anomaly is smaller than −40 mgal.

ches, and these anomalies run parallel to the large positive anomalies of the island-arc.

Secondly, the axis of the minimum free-air anomalies lies at the landward side of the trench axis of bottom topography, especially in the Japan Trench.

Thirdly, in the vicinity of a seamount, there are negative free-air gravity anomalies in a wide area around the large positive anomalies caused by the body of the seamount.

Fourthly, there are wide positive free-air anomalies seaward of a trench, which are called "outer gravity high seaward of the trench". The outer gravity high which accompanies the Ryukyu Trench or the Nankai Trough is about one third in its horizontal scale compared with that accompanying the Japan Trench or Izu-Bonin Trench. The outer gravity high seaward of the trench is not observed in the vicinity of seamounts (e.g., seaward of the Japan Trench at 36°N) or seamount chain (e.g., junction point of the Ogasawara and Mariana Trenches), which is the fifth characteristics of free-air anomaly distribution.

What kind of geophysical processes exist to build up such characteristics of free-air anomalies? The first step of our research to estimate these geophysical processes is to investigate the variations in the structure of the lithosphere. The next step is to simulate what kind of lithospheric motion is expected from the density structure of the lithosphere and asthenosphere obtaind in the first step.

2. *Gravity Anomalies and Thickness of the Lithosphere*

If the velocity structure above the Moho is determined by the seismic method, the gravimetric effect due to the crustal structure can be estimated. Generally speaking, the estimated gravity anomalies do not agree with the actually observed gravity anomalies, if uniform density structure below the Moho is assumed. By use of the difference between the actual and the estimated gravity anomalies, which is called "residual gravity anomaly" (YOSHII, 1973), structure below the Moho can be estimated.

Figure 2 shows a structure section across the northern Pacific from Vancouver in the right-hand side to the Japan Trench and the Japan Sea in the left-hand side (TOMODA and FUJIMOTO, 1981). The Emperor Seamount Chain is shown in the middle part of the section. The lowest line shows the residual gravity anomaly, which indicates the depth of the lithosphere-asthenosphere boundary.

If the density of the lithosphere is larger than that of the asthenosphere, the larger residual gravity anomaly produces or leads to the thicker

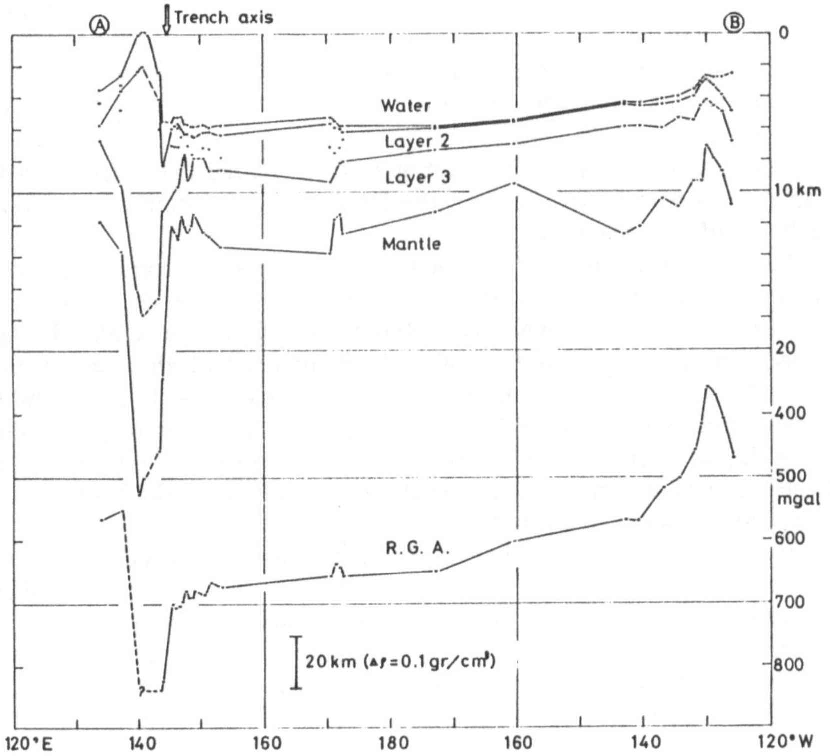

Fig. 2. Profiles of subterranean structure across the northern Pacific. The lowest line indicates the depth of the lithosphere-asthenosphere boundary estimated from the residual gravity anomaly.

lithosphere. Suppose that the density contrast between the lower part of the lithosphere and that of the asthenosphere is 0.1 gr/cm^3 (YOSHII, 1973). Then an increase of 100 mgals of residual gravity anomaly corresponds to an increase of lithospheric thickness by about 24 km.

In the Pacific area, crustal structure has been obtained at about 500 locations. Residual gravity anomaly has been calculated by use of the crustal structures, and the result is shown in Fig. 3 (TOMODA and FUJIMOTO, 1983). Sites of seismic refraction observation are shown by closed circles. High residual gravity anomaly is found near the subduction zones such as the Japan, Izu-Bonin, and Mariana trenches. This belt of thick lithosphere extends south to the Ontong-Java Plateau, and does not follow the Yap or Palau Trench. This belt coincides with strongly magnetized body lying at the depth of 50–100 km estimated from the distribution

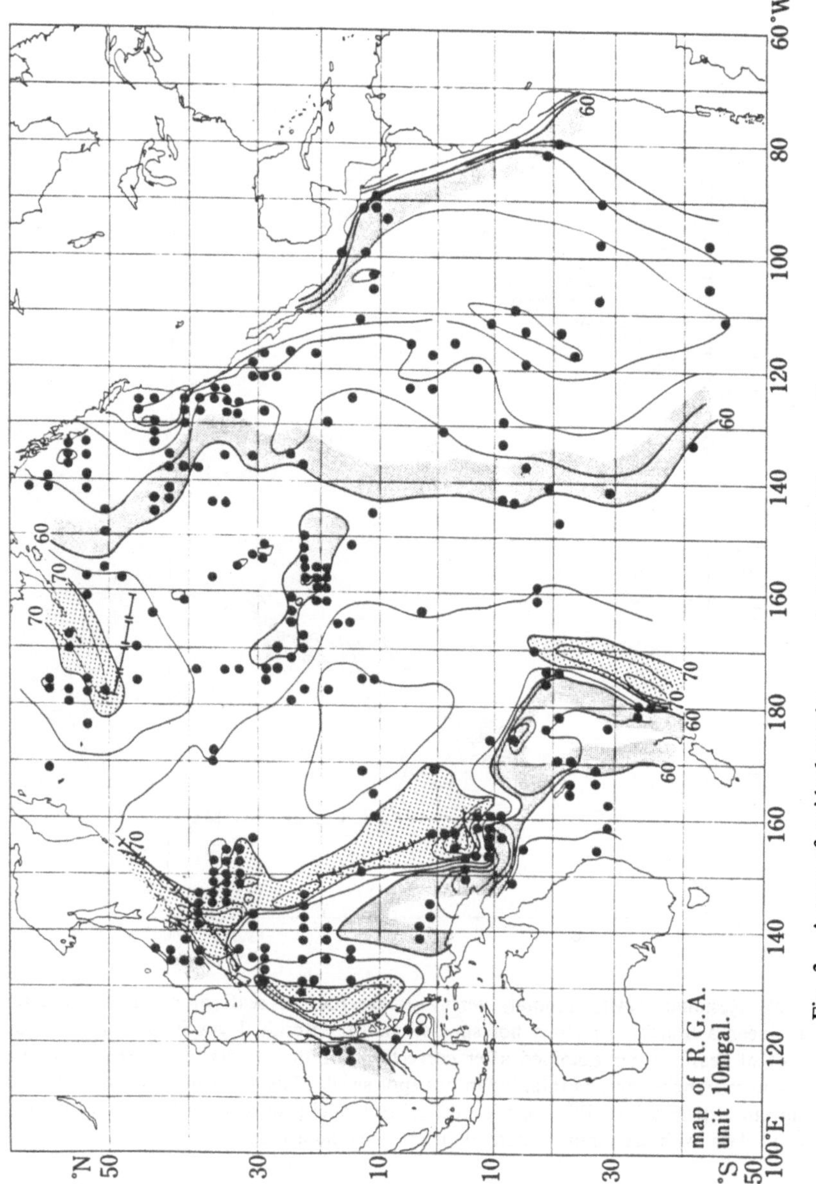

Fig. 3. A map of residual gravity anomaly in the Pacific area. Sites of seismic refraction observation are shown by closed circles. ⊣⊣⊣ : strongly magnetized body lying at the depth of 50–100 km estimated by NOMURA (1979).

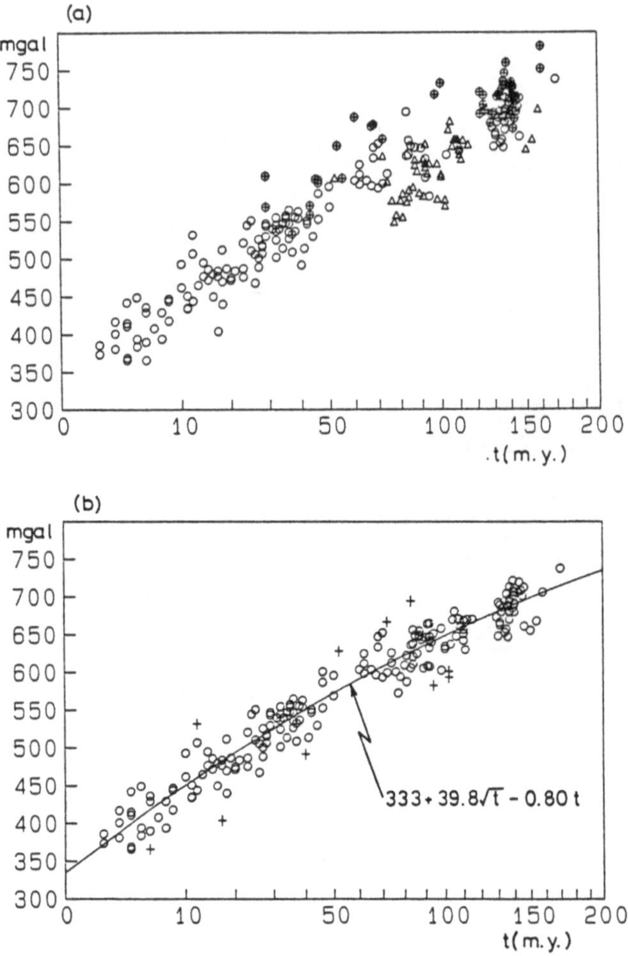

Fig. 4. (a) Residual gravity anomaly versus square root of seafloor age in the Pacific area. ⊕: near a trench, △: in a hot spot swell. (b) Residual gravity anomaly versus square root of seafloor age obtained after corrections of the effects of thickening of the lithosphere near trenches and its thinning in hot spot swells. The curve indicates lithospheric thickening rate expected by a lithospheric thickening model proposed by KONO and YOSHII (1975). +: data with deviations larger than twice standard deviation.

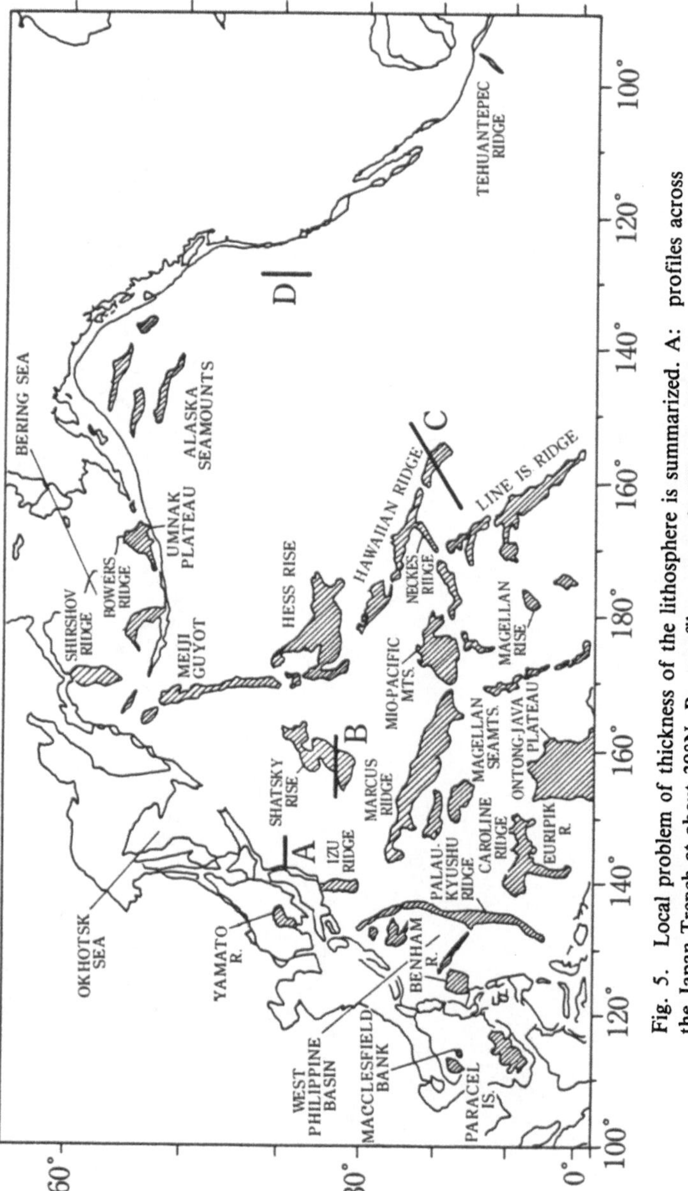

Fig. 5. Local problem of thickness of the lithosphere is summarized. A: profiles across the Japan Trench at about 39°N. B: profiles across the Shatsky Rise. C: profiles across the Hawaiian Ridge. D: profiles across the Mendocino Fracture Zone at 128°W.

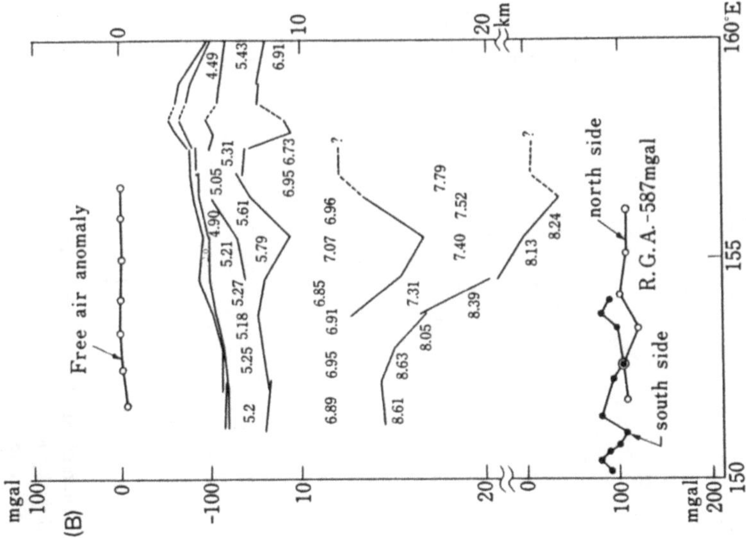
(B)
mgal
Free air anomaly
km
160°E
155
150
north side
south side
R. G. A. -587mgal
?

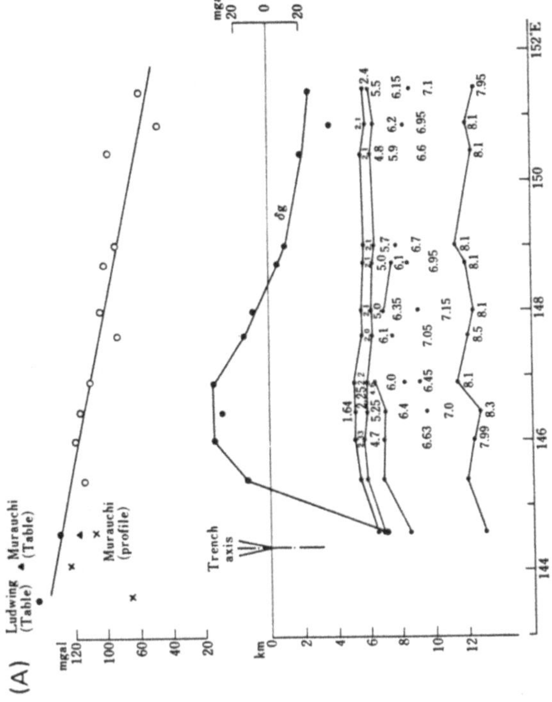
(A)
mgal
Ludwing (Table)
Murauchi (Table)
Murauchi (profile)
Trench axis
δg
km
152°E
150
148
146
144

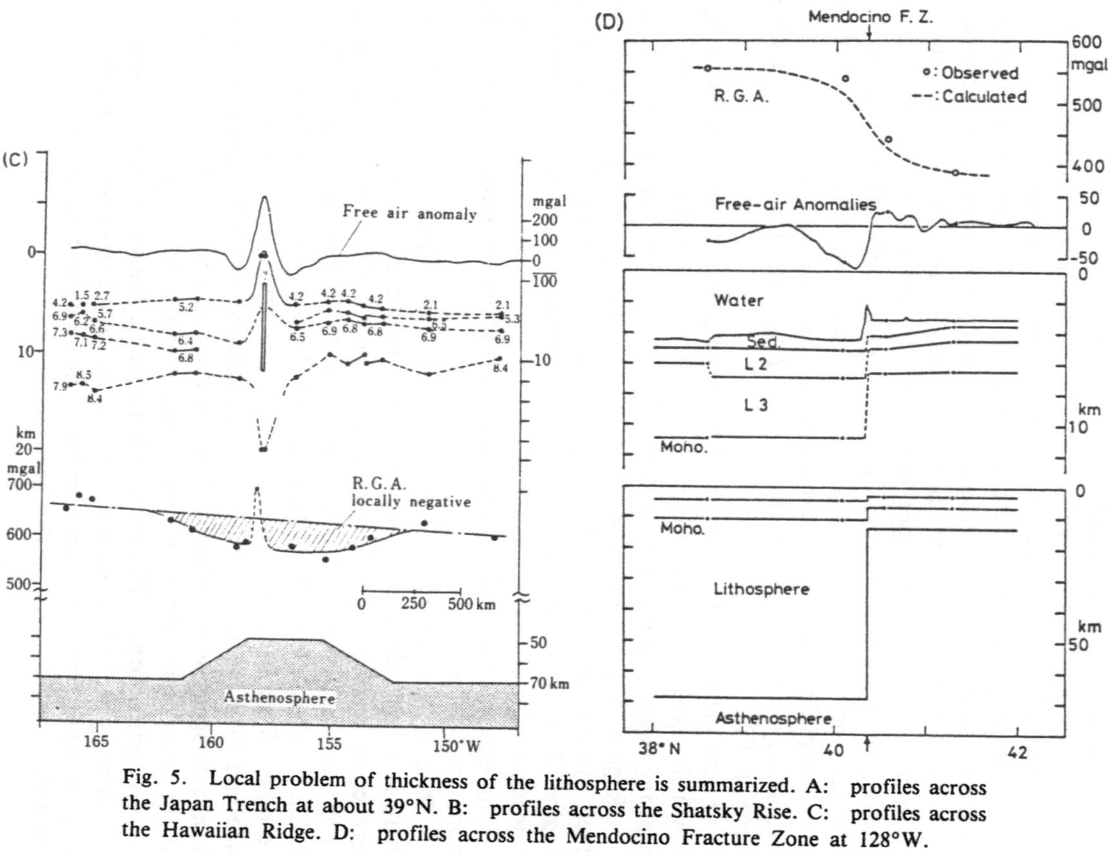

Fig. 5. Local problem of thickness of the lithosphere is summarized. A: profiles across the Japan Trench at about 39°N. B: profiles across the Shatsky Rise. C: profiles across the Hawaiian Ridge. D: profiles across the Mendocino Fracture Zone at 128°W.

of geomagnetic anomalies of intermediate wavelengths (NOMURA, 1979). Low residual gravity anomaly around the Hawaiian Ridge indicates thinning of the lithosphere.

Figure 4(a) shows the relationship between the residual gravity anomaly and seafloor age in the Pacific area (FUJIMOTO and TOMODA, 1985). The ordinate is proportional to the thickness of the lithosphere and the abscissa is proportional to the square root of seafloor age. It is clear that the thickness of the oceanic lithosphere is nearly proportional to the square root of age. Deviations from this relationship are recognized near trenches (⊕ in Fig. 4 (a)) or hot spot swells (Δ). In order to show the general tendency of the thickening rate of normal oceanic lithosphere, these deviations are excluded in Fig. 4(b). This thickening rate of the lithosphere is well explained by a thermal model of lithospheric thickening solved in unsteady state condition (KONO and YOSHII, 1975) as shown by the curve in Fig. 4(b).

Local problem of thickness of the lithosphere is summarized in Fig. 5, which shows the thickness anomaly of the lithosphere at several interesting regions such as trench, oceanic rise, ridge, and fracture zone.

Region A is an example of trench, where the lithosphere becomes thick toward the trench axis irrespective of the age of the ocean floor. Region B is an example of oceanic plateau, where the thickness of lithosphere is nearly constant irrespective of the bottom topography. Region C is an example where the thickness of lithosphere becomes thin by heating from the hot spot.

Region D is an example across the Mendocino Fracture Zone, in which there is a large difference in lithospheric thickness due to the large difference in seafloor age. If we estimate the thickness of the lithosphere in the northern side of the fracture zone as 10 km (KONO and YOSHII, 1975), thickness of the lithosphere calculated from residual gravity anomalies becomes 65 km in the southern side. Seafloor age is 3 m.y. in the northern side and 31 m.y. in the southern side. Because the lithospheric thickness decreases again at the Murray Fracture Zone at about 80 km to the south of the Mendocino Fracture Zone, variation in residual gravity anomalies correspond to larger difference in lithospheric thickness than is expected.

Among these four local examples, the result obtained at the fracture zone is impressive and important from the viewpoint of the origin of subduction or movement of the lithosphere.

3. *Numerical Experiment*

Plate or lithosphere is usually considered to be elastic, and several problems, such as bending of the plate at the subduction zone or bending of the crust due to the load of a seamount, have been discussed assuming that the plate is elastic.

However, behaviour of elastic body and highly viscous fluid is quite similar, if the displacement of elastic body is replaced by the velocity of viscous flow (Fig. 6). We suppose that what we see as the displacement of the lithosphere would be the displacement which is the product of velocity and elapsed time, and the phenomena which are explained by an elasticity model would successfully be explained by a viscosity model.

Hooke's Law

$$T_{ij} = \lambda \delta_{ij} \sum \frac{\partial \xi_k}{\partial x_k} + \mu \left(\frac{\partial \xi_i}{\partial x_j} + \frac{\partial \xi_j}{\partial x_i} \right)$$

Poiseuille's Law

$$T_{ij} = \lambda \delta_{ij} \sum \frac{\partial \xi_k}{\partial x_k} + \varkappa \left(\frac{\partial v_i}{\partial x_j} + \frac{\partial v_j}{\partial x_i} \right)$$

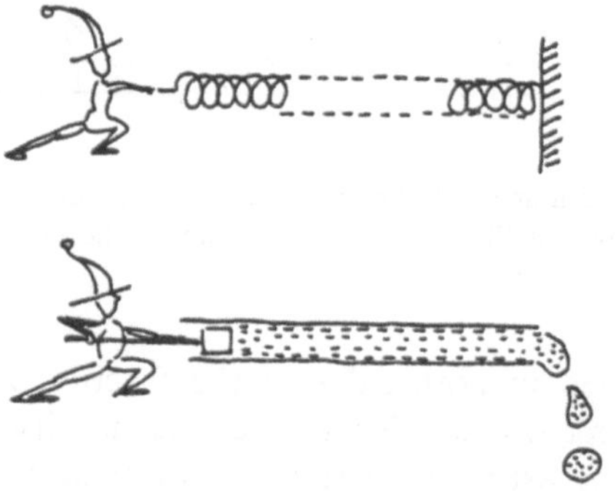

Fig. 6. Behaviour of elastic body and highly viscous fluid is quite similar, if the displacement of elastic body is replaced by the velocity of the viscous flow.

In order to demonstrate the physical processes of the motion of the lithosphere in the case of such a large difference in lithospheric thickness as observed across the Mendocino Fracture Zone, numerical experiments were performed on the assumption that both the lithosphere and the asthenosphere behave as viscous imcompressible fluid with different viscosities and densities.

Initial condition is shown in the upper part of Fig. 7. Although the depth of the simulated area is limited to be 140 km due to the memory size of an electronic computer used, effect of the boundary is small as long as the motion of the lithosphere is small. Viscosity is assumed to be 10^{23} poise in the lithosphere and 10^{21} poise in the asthenosphere. Density is assumed to be 3.3 gr/cm^3 in the lithosphere and 3.2 gr/cm^3 in the asthenosphere.

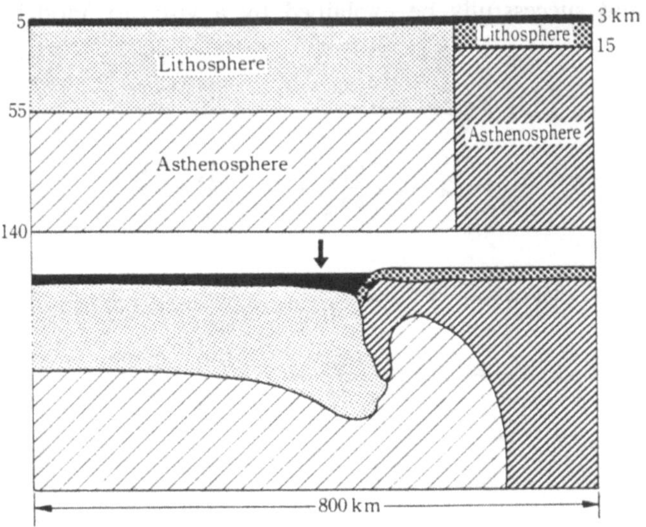

Fig. 7. (a) Initial state of numerical experiment given by a rough approximation of the subterranean structure shown in Fig. 5(D). (b) A result of the numerical simulation with the elapse time of 25 m.y.

Isostasy is assumed in both sides of the fracture zone. No external force is assumed except gravity. Therefore, the density contrast between the lithosphere and the asthenosphere is the only sorce of driving force. Effect of the lithospheric cooling is not taken into account because the simplest case is the basic.

Computation was carried out by use of a combined method of finite difference method and "marker and cell" method (HARLOW and WELCH, 1965). All the four boundaries are assumed free-slip and stress-free. Results of numerical computation is shown in the lower part of Fig. 7 (MATSUMOTO and TOMODA, 1983a). Elapsed time is 25 m.y.

The results show a downgoing motion of the thicker lithosphere and associated topographic depression. These results suggest that initiation of new subduction and formation of a new trench may be taking place here. Lithospheric thickening near the trench is also shown.

In the area of the thinner lithosphere, upward motion of light material of the asthenosphere takes place. This upward motion accounts for high terrestrial heat flow in an island arc and back-arc basin. This upward motion supports mass excess (e.g., topographic high) in this region, that is, positive free air gravity anomaly in this region is maintained by this upward motion. The area of the thinner lithosphere spreads due to the asthenospheric upward motion like a spreading of a back-arc basin, and the trench axis moves "seawards" consequently.

Figure 8 shows the situation after elapsed time of 74 m.y. In this model computation is carried out for a three-layer model. The uppermost layer represents the crust. We can see in the figure that the crust of the back-arc basin is seperated into two regions and a new seafloor is produced in the back-arc basin. Almost the same result is obtained even if the left-hand side boundary locates more leftward.

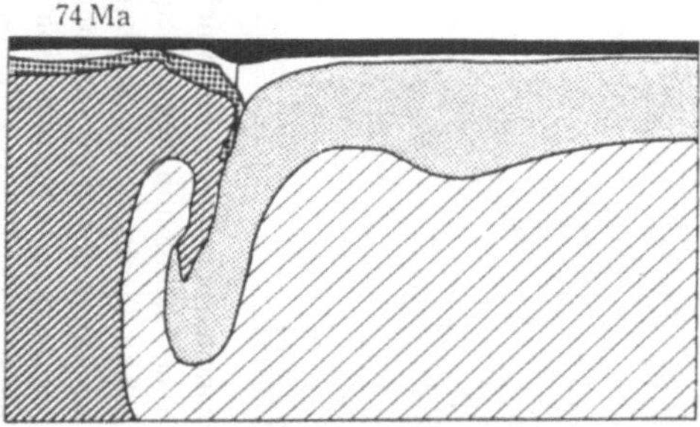

Fig. 8. A result of numerical experiment which possibly indicates the double-layered distribution of hypocenters of deep focus earthquakes at some subduction zone. Elapsed time is 74 m.y. Calculation area is 800 km (horizontally) × 200 km (vertically).

Downward motion of the thin lithosphere of the back-arc basin which is pulled by the downward motion of the thick oceanic lithosphere may account for the double-layered distribution of hypocenters of deep focus earthquakes at some subduction zones (e.g., HASEGAWA *et al.*, 1977).

Processes which would take place at a collision zone is shown in Fig. 9, in which a rectangular block represents a continental block which is going to arrive at the trench axis. Results of computation are shown for the elapsed time 16 m.y. through 45 m.y. (MATSUMOTO and TOMODA, 1983b.)

As illustrated in Fig. 7, trench axis moves towards the oceanic side,

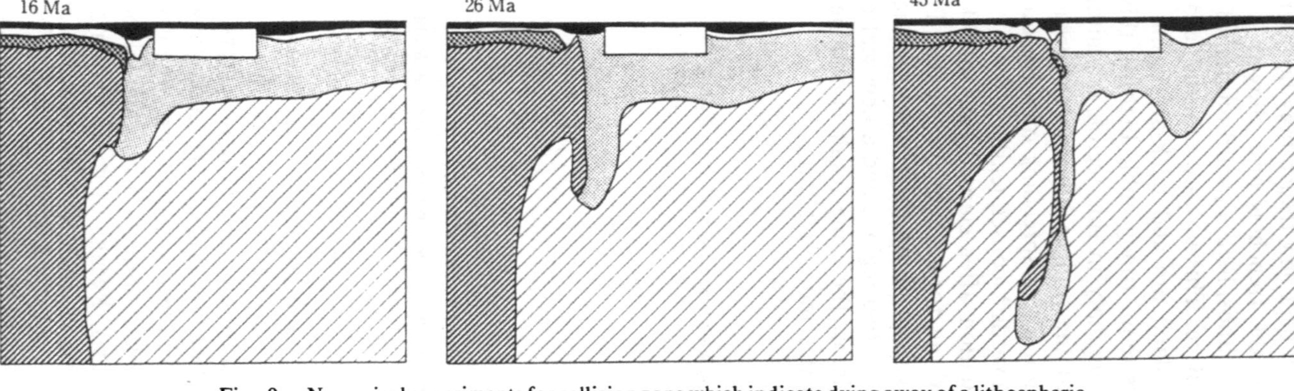

Fig. 9. Numerical experiments for collision zone which indicate dying away of a lithospheric subduction, a formation of new subduction, and consequently seaward jumping of the trench axis. Calculation area is 800 km (horizontally)$\times$300 km (vertically). Viscosity of seamount is assumed to be 10^2 times as large as that of the surrounding materials so as not to deform its shape during the calculation process. Initial condition is set as contacting of typical oceanic and island-arc blocks as shown in Fig. 7(a), in which seamount block is 150 km away from the contacting point.

that is to say, towards the subducting lithosphere according to the down going motion of lithosphere. When the trench axis arrives at the continental block, the movement of the trench axis stops, and a lithospheric subduction ceases at the elapsed time of 45 m.y. In such a situation, the subducting lithosphere has a vertical dip angle. This seems to correspond well to the distribution of hypocenters of the Izu-Bonin-Mariana region where the Wadachi-Benioff zone has vertical dips.

It is worthy of note that new convection takes place at the right-hand side of the continental block. It seems that the trench axis jumps to oceanward of the continental block.

4. *Summary*

As described in this paper, geophysical processes seem to be well explained by numerical simulations. At the present stage, computation technique and ability of the computer of our institute are not enough, and so our computation is not numerical simulation but numerical experiment. When we compare the results of computation with actually observed free-air anomalies and topography, discrepancy between the two is not so small. Calculated gravity anomalies qualitatively well agrees with real ones, but not quantitatively.

Our next stage is to refine our model by development of numerical computation technique. We believe that geophysical processes, which are our final target to resolve, will be clarified by numerical simulations with the aid of boundary condition given from gravimetric and topographic data as well as precise subterranean structure given from explosion seismology.

REFERENCES

FUJIMOTO, H. and Y. TOMODA, Lithospheric thickness anomaly near the trench and possible driving force of subduction, *Tectonophysics*, **112**, 103–110, 1985.

HARLOW, F. H. and J. E. WELCH, Numerical calculation of time-dependent viscous incompressible flow of fluid with free surface, *Phys. Fluids*, **8**, 2182–2189, 1965.

HASEGAWA, A., N. UMINO, and A. TAKAGI, A duble-planed structure of the deep seismic zone in the northeastern Japan arc, *Tectonophysics*, **47**, 43–58, 1977.

KONO, Y. and T. YOSHII, Numerical experiments on the thickening plate model, *J. Phys. Earth*, **23**, 63–75, 1975.

NOMURA, M., Marine geomagnetic anomalies with intermediate wavelengths in the Western Pacific Region, *Bull. Ocean Res. Inst., Univ. Tokyo*, **11**, 1–42, 1979.

MATSUMOTO, T. and Y. TOMODA, Numerical simulation of the initiation of subduction at the fracture zone, *J. Phys. Earth*, **31**, 183–194, 1983a.

MATSUMOTO, T. and Y. TOMODA, Numerical simulation of the mutual interaction between a trench and a seamount, *J. Phys. Earth*, **31**, 281–297, 1983b.

TOMODA, Y. and H. FUJIMOTO, Gravity anomalies in the western Pacific and geophysical interpretation of their origin, *J. Phys. Earth*, **29**, 387–419, 1981.

TOMODA, Y. and H. FUJIMOTO, Maps of gravity anomalies and bottom topography in the western Pacific and reference book for gravity and bathymetric data, *Bull. Ocean Res. Inst., Univ. Tokyo*, **8**, 158p, 1982.

TOMODA, Y. and H. FUJIMOTO, Roles of seamount, rise, and ridge in lithospheric subduction, in *Accretion Tectonics in the Circum-Pacific Regions*, M. Hashimoto and S. Uyeda (Editors), pp. 319–331, Terra Scientific Publishing Company, Tokyo, 1983.

YOSHII, T., Upper mantle structure beneath the north Pacific and the marginal seas, *J. Phys. Earth*, **21**, 313–328, 1973.

Formation of Active Ocean Margins, edited by N. Nasu *et al.*, pp. 59–73.

BENDING OF THE VISCOELASTIC LITHOSPHERE AND THE OUTER TOPOGRAPHIC RISE

Yoshiaki IDA

Ocean Research Institute, University of Tokyo, Nakano, Tokyo 164, Japan

Abstract. A theory to describe the deflection of a thin viscoelastic plate predicts that the height of the outer rise topography seaward of the deep-sea trench should change in the form of $\exp(-ax)\sin(bx)$ with a distance x normal to the trench. The constants a and b in this formula are evaluated for some observed bathymetric profiles. The distributions of horizontal force, vertical force, and bending moment and some other quantities in the oceanic lithosphere are calculated for these profiles. Although the calculation requires an independent estimate of Maxwell relaxation time, the result does not strongly depend on the ambiguity of the prescribed relaxation time. The distributions of the forces and bending moment are mainly determined by the ratio of b to a.

1. Introduction

The analysis of the outer rise topography seaward of the deep-sea trench is now a classical problem. WATTS and TALWANI (1974) and CALDWELL *et al.* (1976) showed that a simple elastic model could explain the observed distribution of outer rise deflection fairly well. The same topographic feature is also consistent with a purely viscous deformation of the plate that is treated as a Newtonian flow (DE BREMAECKER, 1977). Some other theories analysed the outer rise deflection, based on a more realistic rheology of the lithosphere. TURCOTTE *et al.* (1978), BODINE and WATTS (1979), and CAREY and DUBOIS (1981) considered elastic-plastic plate with plastic layer underlying the upper elastic layer. The lower layer is sometimes replaced by the medium governed by a power-law creep (GOETZE and EVANS, 1979; MCNUTT and MENARD, 1982).

Recently IDA (1984a) gave an analysis of outer rise deflection, assuming that the lithosphere behaves like a thin viscoelastic plate. This treatment takes both elastic and inelastic deformations into account in a simpler way, compared with those previous theories involving non-linear rheology. Although both theories are able to explain the observation equally well, the viscoelastic theory represents the deflection by a simple analytic formula, while the treatment of plasticity or power-law creep requires some numerical calculation to obtain the distribution of deflection.

MELOSH (1978) and MELOSH and RAEFSKY (1980) developed another theory of viscoelastic deformation for the subducting lithosphere, supposing that a viscous flow is induced in the lower lithophere by the upper plate motion. However, the physical picture represented by them differs from the ordinary treatment, including the present theory, in which a buoyancy force governs the lithospheric deflection. Therefore, it is difficult to compare their result with others.

There were mainly two reasons for which IDA (1984a) gave a new analysis to the classical problem of the outer rise deflection. One reason is a theoretical necessity to formulate the deformation of a thin viscoelastic plate without the assumption of Poisson's ratio equal to 1/2. Ordinary analyses of the viscoelastic deformation have been based on such an artificial assumption to avoid a mathematical difficulty. Another reason that requires a new analysis is more geophysical. Even if observed outer rise deflection is explained fairly well by a simple theory of purely elastic or purely viscous deformation, there remains a meaningful misfit between the observed and theoretical bathymetries. Namely the observed width over the topographic rise tends to be longer than is expected from elastic or viscous solution to which the data near the trench axis are fitted. If the theoretical curve is adjusted to the width of the rise, the observed height of the forebulge is lower, and the observed slope seaward of the trench axis is steeper, than is expected from these simple theories. IDA (1984a) explained such misfit by the effect of horizontal force, which is made significant due to the viscoelastic deformation.

In the present paper, we examine in more detail the distribution of forces that work on the oceanic lithosphere, using observed profiles of deflection. We pay a special attention to the nature of forces acting on the subducting slab at the trench, since this information is important to understand the interaction between the oceanic and cotinental plates, and also to specify the boundary condition to calculate the deformation of subducting slab. The present theory also evaluates the difference of the surface deflection from the displacement at the center of the plate. This difference is neglected in ordinary treatments including our previous paper (IDA, 1984a), but it may be meaningful when we call a relatively small deviation of topography in question.

2. *Bending of Viscoelastic Plate*

It is assumed in this paper that the topography of the outer rise is produced by a one-dimensional bending of a thin viscoelastic plate that is stationarily moving toward the trench at velocity U. Such deformation of plate is governed by the following set of simultaneous differential equations (IDA, 1984a).

$$M-\tau U(\mathrm{d}M/\mathrm{d}x)=-D\tau U(\mathrm{d}\beta/\mathrm{d}x)+(1-2\upsilon)D\zeta \tag{1}$$

$$-\tau U(\mathrm{d}\beta/\mathrm{d}x)=\zeta-c\tau U(\mathrm{d}\zeta/\mathrm{d}x) \tag{2}$$

$$(\mathrm{d}N/\mathrm{d}x)+\mathrm{T}\beta-g\Delta\rho w'=0 \tag{3}$$

$$(\mathrm{d}M/\mathrm{d}x)=-N \tag{4}$$

$$\beta=\mathrm{d}^2w/\mathrm{d}x^2 \tag{5}$$

$$w'=w+(h^2/8)\{\beta-3[(1-2\upsilon)/(1+\upsilon)]\zeta\}. \tag{6}$$

These six equations determine six variables M, β, ζ, N, w and w' as a function of a horizontal coordinate x, which is perpendicular to the trench axis and directed seaward. The meaning of the variables and constants appearing here is explained below with the physical basis of these equations.

Firstly M is the bending moment, β is the curvature, and ζ is the reduced pressure gradient (i.e., the vertical pressure gradient in the plate divided by twice rigidity). The constitutive relation for a thin viscoelastic plate, from which (1) and (2) are derived, is written in terms of these three variables. Ordinary treatment of viscoelastic plate bending is based on the assumption of Poisson's ratio equal to 1/2, but our constitutive relation is free from such an artificial assumption. (1) and (2) contain three constants, i.e., flexural rigidity D, Poisson's ratio υ and Maxwell relaxation time τ (i.e., viscosity divided by rigidity), which describe the viscoelastic material property. The parameter c in (2) is a dimensionless constant defined by

$$c=3(1-\upsilon)/(1+\upsilon). \tag{7}$$

It is also noted that the flexural rigidity is written as

$$D = h^3E/12(1-\upsilon^2) \tag{8}$$

where h and E are the thickness and Young's modulus of the plate, respectively.

In this paper, we distinguish the vertical displacement w' on the top surface of the plate, which also agrees with the displacement on the bottom, from the displacement w at the center, even if the difference between w and w' is usually neglected. The curvature β is defined by (5) in terms of w, while the buoyancy force is related to w', as is seen below. Using the formulation of deformation in the plate interior by IDA (1984b), we can show that these two displacements are related to each other by (6).

In (3) and (4), N is the vertical force acting across a cross section of the plate (positive upward for the force acting from the positive side of x). The conditions (3) and (4) represent the balance of the vertical force and moment, respectively. From the other condition of force balances, the horizontal force T (positive extensional) appearing in (3) turns out to be a constant independent of x. The last term of (3) represents the hydrostatic restoring force (i.e., the buoyancy effect) in which g is the gravitational acceleration and $\Delta\rho$ is the density difference between the asthenosphere and sea water.

The simultaneous differetial equations (1) to (6) have the following solution

$$w' = w_0' \exp(qx) \tag{9}$$

$$w = w_0 \exp(qx) \tag{10}$$

$$\beta = \beta_0 \exp(qx) \tag{11}$$

$$\zeta = \zeta_0 \exp(qx) \tag{12}$$

$$M = M_0 \exp(qx) \tag{13}$$

$$N = N_0 \exp(qx). \tag{14}$$

Here q is a complex wave number.

Substituting (9) to (14) into (1), (2), (4), (5) and (6), we have the following relations between the constant coefficients

$$w_0 = w_0'/\{1 + (h^2/8)q^2[1 + 3\tau Uq(1-2\upsilon)/(1+\upsilon)(1-c\tau Uq)]\} \tag{15}$$

$$\beta_0 = w_0 q^2 \tag{16}$$

$$\zeta_0 = -\beta_0 \tau U q/(1 - c\tau U q) \tag{17}$$

$$M_0 = \zeta_0 D[2(1-\upsilon) - c\tau U q]/(1 - \tau U q) \tag{18}$$

$$N_0 = -M_0 q. \tag{19}$$

Putting (15) to (19) into (3), we have now the following characteristic equation of the differential equations, as

$$D\tau U q^5 [2(1-\upsilon) - c\tau U q]/(1-\tau U q)(1 - c\tau U q) + T q^2 - g\Delta\rho\{1 + (h^2/8)q^2[1 + 3\tau U q(1-2\upsilon)/(1+\upsilon)(1 - c\tau U q)]\} = 0. \tag{20}$$

Non-vanishing solutions exist only when (20) holds.

3. *Analysis of Outer Rise Topography*

Equation (20) is satisfied by the following solution

$$q = -a + ib \tag{21}$$

where a and b are positive real constants, and i is the imaginary unit. Substituting (21) into (9) and taking the imaginary part of the resultant equation with a real w_0', we have the expression of deflection w' on the surface of the oceanic plate, as

$$w' = w_0' \ \exp(-ax) \ \sin(bx). \tag{22}$$

Here the trench is located at a point specified by $x = -x_t$.

The constants w_0', a and b in (22), as well as the trench position x_t, can be determined from the observations. Since the distribution of gravity seems to simply reflect the bathymetry (WATTS and TALWANI, 1974), we here use only the topographic data as independent information. The four constants were actually obtained by IDA (1984a) for some observed bathymetric profiles, as is given in Table 1. The data sources of the original profiles were WATTS and TALWANI (1974) and CAREY and DUBOIS (1981). The locations of trenches involving the adopted profiles are shown in Fig. 1.

Let us first specify the values of constants appearing in the theory. We have a sufficient knowledge on the elastic parameters E and υ from

Table 1. Topographic parameters of outer rise.

Trench		Profile	x_t (km)	w_0' (km)	a (/km)	b (/km)	b/a
Aleutian	3	AL1	33	6.96	0.0189	0.0041	0.217
	4	AL2	28	2.71	0.0235	0.0113	0.479
	5	AL3	28	4.52	0.0153	0.0074	0.487
Kuril	10	KU1	37	1.79	0.0209	0.0110	0.523
	12	KU2	35	6.30	0.0174	0.0053	0.307
	13	KU3	59	2.98	0.0162	0.0083	0.512
Bonin	17	BO1	31	7.67	0.0214	0.0069	0.321
	18	BO2	29	4.87	0.0251	0.0090	0.359
Philippine	27	PH1	54	2.78	0.0178	0.0090	0.508
	28	PH2	42	4.55	0.0140	0.0074	0.530
New Hebrides		NH1	59	0.69	0.0275	0.0134	0.489
Middle America		MA1	35	1.19	0.0262	0.0209	0.798
Peru		PE1	31	1.52	0.0206	0.0124	0.604

The constant w_0', a and b determine the surface deflection w' through (22), in which the trench is situated at $x=-x_t$. Data sources are WATTS and TALWANI (1974) for Aleutian, Kuril, Bonin and Philippine trenches, and CAREY and DUBOIS (1981) for New Hebrides, Middle America and Peru trenches. The figures immediately after the trench names are the profile No. given by the original authors.

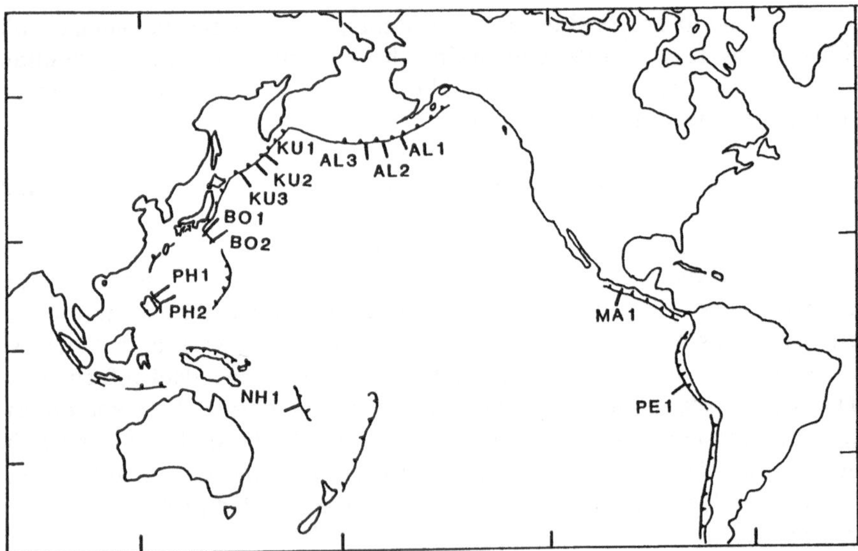

Fig. 1. Locations of trenches involving adopted outer rise profiles. The symbols for the data as well as the data sources are listed in Table 1.

the seismological observations. The velocity U of oceanic plate motion is also known at each trench, and we here use the estimates made by CHASE (1978). The adopted values of these constants are given in Table 2. On the other hand, we can not predict a priori the plate thickness h, Maxwell relaxation time τ and horizontal force T. If one of these three constants is further prescribed, however, the other two are determined from (20). In fact, the real and imaginary parts of (20) give two independent equations that constrain h, τ and T, since q is specified from observed a and b by (21). Here let us give a trial value of τ, and then obtain corresponding T and D from (20), and h from (8).

Even if viscoelastic relaxation plays an important role in various phenomena involving lithospheric deformation, we do not have a widely accepted estimate of relaxation time τ. In the previous paper, IDA (1984a) predicted that τ should be 1 m.y. or less so as not to make the lithospheric stress abnormally large. This constraint of τ is consistent with the value of 0.1 m.y. obtained by WALCOTT (1970), and 1 m.y. by SLEEP and SNELL (1976) and BEAUMONT (1978), but it is considerably less than the estimate of 10 m.y. by LAMBECK (1981). In the present paper, we consider two test values, $\tau = 1$ m.y. and 0.1 m.y. The other constants T and h are evaluated for these values of τ, as in Table 2.

Table 2. Plate thickness h and horizontal force T evaluated for assumed Maxwell relaxation time τ.

Profile	U (cm/y)	$\tau=0.1$ m.y.		$\tau=1$ m.y.	
		h (km)	T (10^{14}N/m)	h (km)	T (10^{14}N/m)
AL1	6	39	0.99	25	1.05
AL2	6	25	0.32	17	0.38
AL3	6	50	0.76	31	0.86
KU1	8	27	0.34	19	0.42
KU2	8	40	0.99	27	1.09
KU3	8	41	0.60	27	0.72
BO1	9	28	0.63	20	0.70
BO2	9	21	0.42	16	0.48
PH1	2	54	0.53	29	0.54
PH2	2	79	0.81	41	0.78
NH1	5	20	0.23	14	0.27
MA1	5	19	−0.01	12	0.05
PE1	6	29	0.23	19	0.31

$g=9.8$ m/s^2, $\Delta\rho=2400$ kg/m^3, $E=160$ GPa and $\upsilon=0.27$ are assumed. Velocity U of plate motion follows CHASE (1978).

Table 2 shows that plate thickness h systematically decreases and horizontal force T systematically increases with the increase of τ. It is noted, however, that these changes of h and T are rather quite small, compared with the change of τ over one order of magnitude. The difference from profile to profile seems more remarkable than the dependence on τ. A careful examination of Table 2 compared with Table 1 would reveal that the dependence on individual profiles mainly occurs through the ratio of b to a. For example, T is very small or even negative for MA1 that has the smallest value of b/a.

Once all these constants are fixed, we can obtain the coefficients w_0, β_0, ζ_0, M_0 and N_0 from (15) to (19), and thus the distribution of variables along the plate. It is noted here that, even if w_0' is assumed to be real, other coefficients are generally complex. For this reason, the curve of each variable has a different shape. Taking the imaginary parts of (9) to (14), we can now determine the variables at each point of x, corresponding to w' in (22). Some examples of the variable distribution are given in Figs. 2 to 5. Figures 2 and 3 are for profile BO2, corresponding to $\tau=0.1$ and 1 m.y., respectively, and Figs. 4 and 5 are for MA1. Each variable tends to vanish for very large x. The values of variables at the trench $x=-x_t$ are given for all the profiles in Table 3 ($\tau=0.1$) and Table 4 ($\tau=1$ m.y.).

Figures 2 to 5 first shows that the distributions of these variables generally depend only weekly on relaxation time τ. Apart from this general trend, it is remarkable that the magnitude of reduced pressure gradient ζ is substantially reduced for smaller τ. Since ζ is proportional to the stress associated with bending, this means a remarkable dependence of estimated bending stress on the assumed τ. We next pay an attention to the fact that the difference between BO2 and MA1 is significant particularly in the distribution of vertical force N, bending moment M and reduced pressure gradient ζ. These variables tend to have a minimum near the trench for MA1, while they monotonically decrease to the trench for BO2. Since BO2 and MA1 have relatively small and great ratio of b/a, respectively, we again find that the ratio b/a well characterizes the distribution of variables. It is noted here that, even if w_0' is not same between the two cases, w_0' simply determines the scale of these variables, and does not influence the shape of curves at all.

4. *Discussion*

The present theory predicts that the mechanical state of the lithosphere is well correlated with the parameter b/a. In this section, we study this

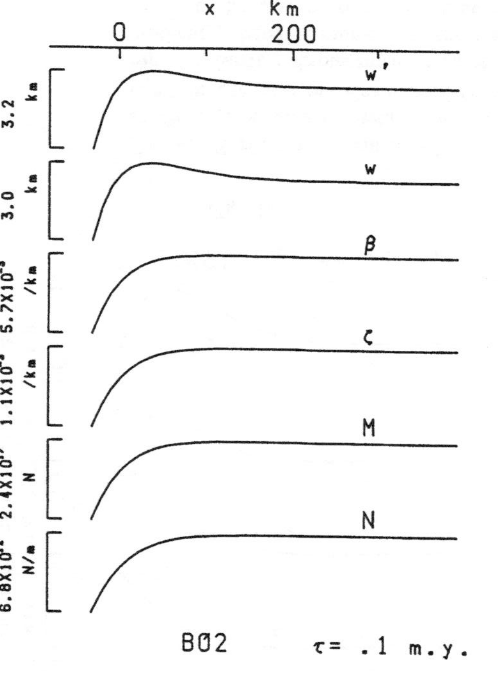

Fig. 2.

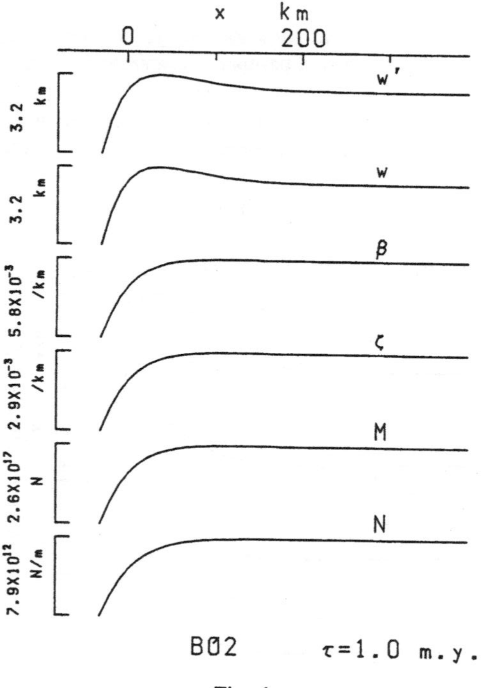

Fig. 3.

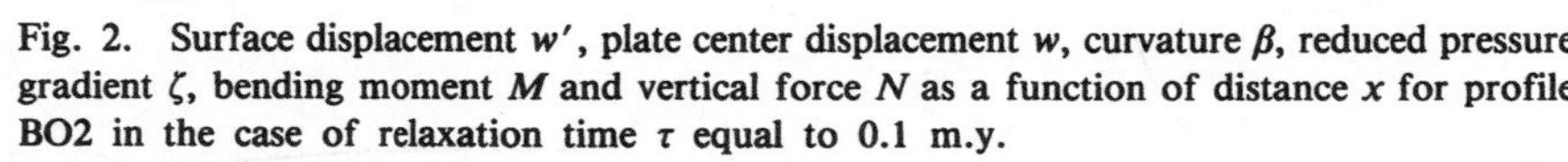

Fig. 2. Surface displacement w', plate center displacement w, curvature β, reduced pressure gradient ζ, bending moment M and vertical force N as a function of distance x for profile BO2 in the case of relaxation time τ equal to 0.1 m.y.

Fig. 3. Surface displacement w', plate center displacement w, curvature β, reduced pressure gradient ζ, bending moment M and vertical force N as a function of distance x for profile BO2 in the case of relaxation time τ equal to 1 m.y.

MA1 τ= .1 m.y.

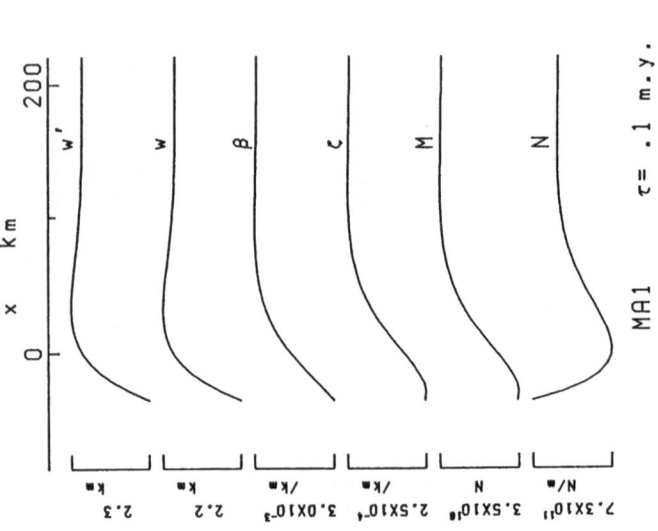

Fig. 4.

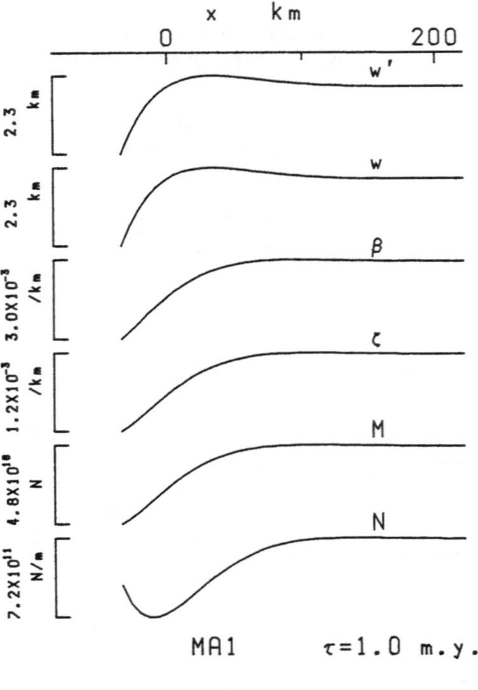

Fig. 5.

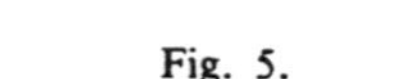

Fig. 4. Surface displacement w', plate center displacement w, curvature β, reduced pressure gradient ζ, bending moment M and vertical force N as a function of distance x for profile MA1 in the case of relaxation time τ equal to 0.1 m.y.

Fig. 5. Surface displacement w', plate center displacement w, curvature β, reduced pressure gradient ζ, bending moment M and vertical force N as a function of distance x for profile MA1 in the case of relaxation time τ equal to 1 m.y.

Table 3. Surface displacement w', plate center displacement w, curvature β, reducd pressure gradient ζ, bending moment M and vertical force N at the trench for the relaxation time τ of 0.1 m.y.

Profile	w' (km)	w (km)	β (10^{-3}/km)	ζ (10^{-3}/km)	M (10^{17}N)	N (10^{12}N/m)
AL1	−1.75	−1.37	−2.4	−0.29	−3.7	−8.7
AL2	−1.63	−1.42	−3.2	−0.43	−1.4	−3.4
AL3	−1.43	−0.97	−1.7	−0.17	−4.4	−7.1
KU1	−1.54	−1.38	−2.1	−0.30	−1.3	−2.6
KU2	−2.14	−1.72	−2.5	−0.36	−5.0	−10.3
KU3	−3.65	−3.19	−2.4	−0.29	−4.3	−6.3
BO1	−3.16	−2.75	−5.3	−0.95	−4.5	−11.4
BO2	−2.60	−2.35	−5.6	−1.11	−2.4	−6.8
PH1	−3.40	−2.51	−2.6	−0.10	−3.2	−5.1
PH2	−2.51	−1.25	−1.7	−0.06	−5.8	−7.9
NH1	−2.48	−2.34	−3.2	−0.35	−0.6	−1.3
MA1	−1.99	−1.87	−3.0	−0.24	−0.3	0.2
PE1	−1.08	−0.93	−1.6	−0.18	−1.0	−1.6

Table 4. Surface displacement w', plate center displacement w, curvature β, reduced pressure gradient ζ, bending moment M and vertical force N at trench for the relaxation time τ of 1 m.y.

Profile	w' (km)	w (km)	β (10^{-3}/km)	ζ (10^{-3}/km)	M (10^{17}N)	N (10^{12}N/m)
AL1	−1.75	−1.65	−2.5	−1.09	−4.1	−9.9
AL2	−1.63	−1.57	−3.3	−1.48	−1.6	−4.2
AL3	−1.43	−1.31	−1.8	−0.71	−4.7	−8.3
KU1	−1.54	−1.49	−2.1	−0.97	−1.5	−3.2
KU2	−2.14	−2.02	−2.6	−1.19	−5.4	−11.9
KU3	−3.65	−3.52	−2.5	−1.08	−5.0	−8.0
BO1	−3.16	−3.03	−5.5	−2.65	−4.9	−13.1
BO2	−2.60	−2.52	−5.8	−2.84	−2.6	−7.9
PH1	−3.40	−3.18	−2.8	−0.68	−3.5	−6.0
PH2	−2.51	−2.20	−1.9	−0.42	−6.4	−9.0
NH1	−2.48	−2.45	−3.2	−1.35	−0.8	−1.8
MA1	−1.99	−1.96	−2.9	−1.18	−0.5	−0.4
PE1	−1.08	−1.04	−1.6	−0.70	−1.1	−2.1

b/a dependence in more detail. Since the constants a and b are observable from the bathymetric data in the outer rise region, this study may provide useful tool to reveal the nature of forces acting in the lithosphere.

Figures 6 and 7 demonstrate the effect of b/a on some quantities

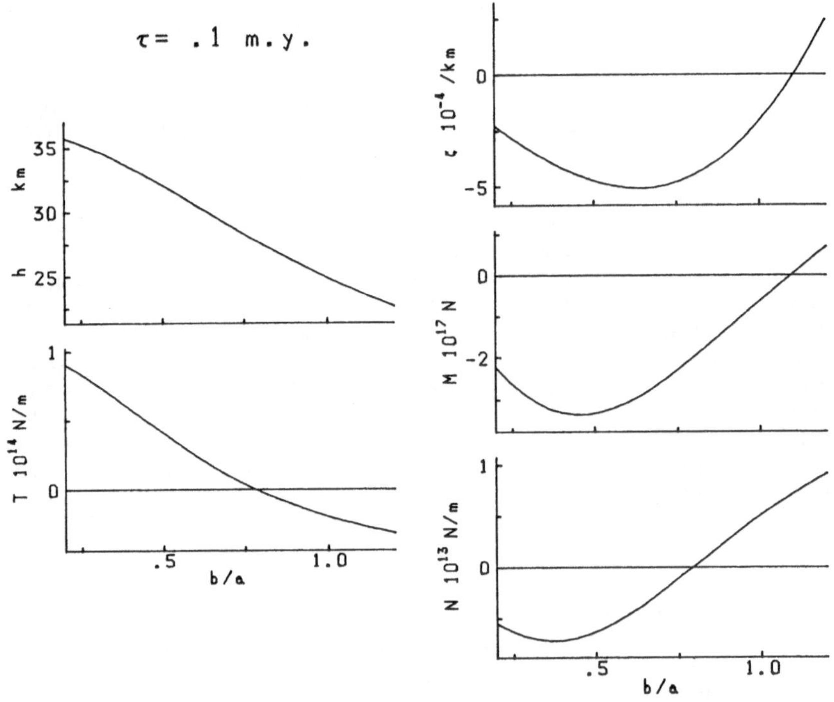

Fig. 6. Dependence on b/a of plate thickness h and horizontal force T (left) and reduced pressure gradient ζ, bending moment M and vertical force N at the trench (right) for relaxation time τ of 0.1 m.y. Other parameters are fixed as $a=0.02$/km, $w_0'=4$ km, $x_t=40$ km and $U=6$ cm/y.

that represent the rheological property and mechanical state of the lithosphere for the assumed relaxation time of 0.1 and 1 m.y., respectively. In these figures, the plate thickness h and horizontal force T, which are constant along the plate, are displayed on the left, and the values of reduced pressure gradient ζ, bending moment M and vertical force N at the trench are shown on the right. Here b/a varies following the change of b, and the other parameters are fixed at likely values mentioned in the figure captions.

A gross feature of the b/a dependence is almost same between the two cases of $\tau=0.1$ and 1 m.y. The constants h and T monotonically decrease as b/a increases, and the variables ζ, M and N at the trench have a minimum at a certain value of b/a. Even if the curves in Figs. 6 and 7 look qualitatively similar, the scale of each quantity more or

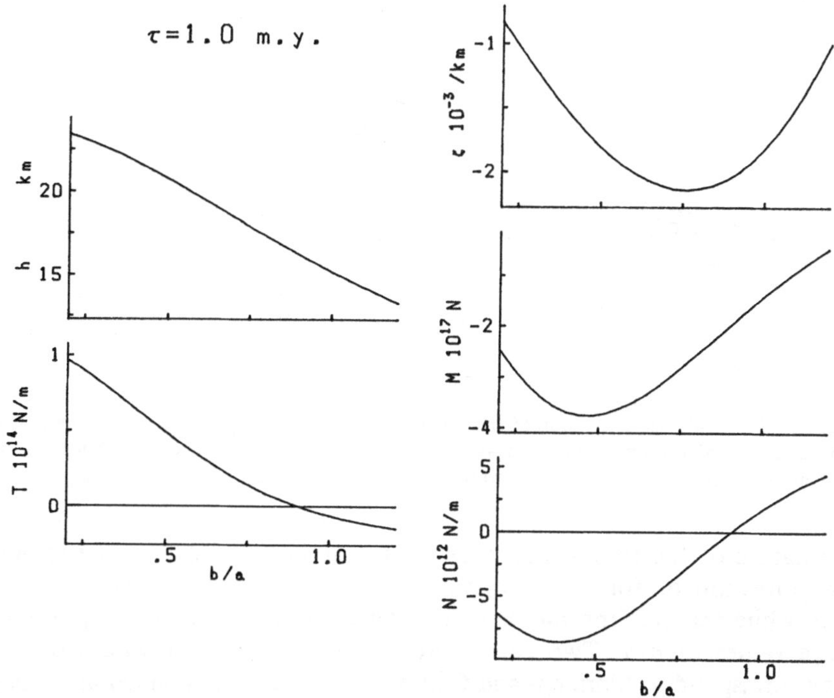

Fig. 7. Dependence on b/a of plate thickness h and horizontal force T (left) and reduced pressure gradient ζ, bending moment M and vertical force N at the trench (right) for relaxation time τ of 1 m.y. Other parameters are fixed as $a=0.02$/km, $w_0'=4$ km, $x_t=40$ km and $U=6$ cm/y.

less differs between the two cases. Particularly the magnitudes of ζ and N significantly depend on τ.

As is seen in Figs. 6 and 7, the observable b/a value thus contains an important information on the forces acting in the lithosphere. These forces may even change their sign within a realistic range of b/a. It is noted in these figures, that horizontal horce T and vertical force N cross the zero-level at almost same point of b/a. Namely both forces have opposite signs for smaller and larger values of b/a. Horizontal force is extensional for smaller values of b/a, but compressive for larger values. Vertical force at the trench is downward below and upward above the critical value of b/a. On the other hand, bending moment M is negative over almost entire range of b/a in these figures.

Summarizing these results, we may schematically display in Fig. 8

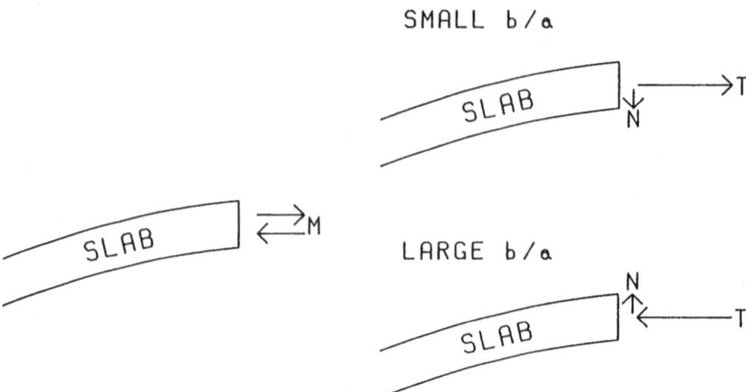

Fig. 8. Schematic demonstration of the forces working on the subduction slab at the trench. The direction of the horizontal and vertical forces, T and N, respectively, change below and above a critical value of b/a, while the bending moment M has the same sense.

the nature of the forces that work on the subducting slab at the trench. The direction of forces are opposite for smaller and greater values of b/a, while the bending moment has same sense except in exceptionally large values of b/a. Two cases with smaller and greater values of b/a may correspond to Mariana- and Chilean-type modes of subduction, respectively, proposed by UYEDA and KANAMORI (1979).

Since the lithospheric state to be predicted is rather quite sensitive to b/a, we may need a more careful examination of the assumptions employed in the present theory. Our analysis of the outer rise bathymetry, as well as other previous treatments, has considered a one-dimensional deflection of the plate which is assumed to continue infinitely along the trench axis. We have employed the treatment of a thin viscoelastic plate involving linear and uniform rheology. We have also neglected the effect that the lithosphere moves along a spherical surface of the earth. Furthermore the direction of plate motion is assumed to be normal to the trench axis. Neither of these assumptions exactly holds in nature, and the factors neglected here might cause a systematic shift of the b/a dpendence of various quantities in question. A future analysis based on a more realistic model is desired to draw a definite conclusion on the mechanical state of the lithosphere for each subduction zone.

Acknowledgements

I would like to thank Prof. S. Uyeda and Prof. M. Ohnaka for reading the manuscript and providing many helpful comments.

REFERENCES

BEAUMONT, C., The evolution of sedimentary basins on a viscoelastic lithosphere: theory and examples, *Geophys. J. R. Astr. Soc.*, **55**, 471–497, 1978.

BODINE, J. H. and A. B. WATTS, On lithospheric flexure seaward of the Bonin and Mariana trenches, *Earth Planet. Sci. Lett.*, **43**, 132–148, 1979.

CALDWELL, J. G., W. F. HAXBY, D. E. KARIG, and D. L. TURCOTTE, On the applicability of a universal elastic trench profile, *Earth Platet. Sci. Lett.*, **31**, 239–246, 1976.

CAREY, E. and J. DUBOIS, Behaviour of the oceanic lithosphere at subduction zones; Plastic yield strength from a finite-element method, *Tectonophysics*, **74**, 99–110, 1981.

CHASE, C. G., Extension behind arcs and motions relative to hot spots, *J. Geophys. Res.*, **83**, 5385–5387, 1978.

DE BREMAECKER, J. C., Is the oceanic lithosphere elastic or viscous?, *J. Geophys. Res.*, **82**, 2001–2004, 1977.

GOETZE, C. and B. EVANS, Stress and temperature in the bending lithosphere as constrained by experimental rock machanics, *Geophys. J. R. Astr. Soc.*, **59**, 463–478, 1979.

IDA, Y., Stress and relaxation time in the viscoelastic lithosphere inferred from the outer topographic rise, *J. Geophys. Res.*, **89**, 3211–3221, 1984a.

IDA, Y., Buckling of viscoelastic lithosphere and initiation of subduction, 1984b (in preparation).

LAMBECK, K., Lithospheric response to volcanic loading in the Southern Cook Islands, *Earth Planet. Sci. Lett.*, **55**, 482–496, 1981.

MCNUTT, M. K. and H. W. MENARD, Constraints on yield strength in the oceanic lithosphere derived from observations of flexure, *Geophys. J. R. Astr. Soc.*, **71**, 363–394, 1982.

MELOSH, H. J., Dynamic support of the outer rise, *Geophys. Res. Lett.*, **5**, 321–324, 1978.

MELOSH, H. J. and A. RAEFSKY, The dynamic origin of subduction zone topography, *Geophys. J. R. Astr. Soc.*, **60**, 333–354, 1980.

SLEEP, N. H. and N. S. SNELL, Thermal contraction and flexure of mid-continent and Atlantic marginal basins, *Geophys. J. R. Astr. Soc.*, **45**, 125–154, 1976.

TURCOTTE, D. L., D. C. MCADOO, and J. G. CALDWELL, An elastic-perfectly plastic analysis of the bending of the lithosphere at a trench, *Tectonophysics*, **47**, 193–205, 1978.

UYEDA, S. and H. KANAMORI, Back-arc opening and the mode of subduction, *J. Geophys. Res.*, **84**, 1049–1061, 1979.

WALCOTT, R. I., Flexural regidity, thickness, and viscosity of the lithosphere, *J. Geophys. Res.*, **75**, 3941–3954, 1970.

WATTS, A. B. and M. TALWANI, Gravity anomalies seaward of deep-sea trenches and their tectonic implications, *Geophys. J. R. Astr. Soc.*, **36**, 57–90, 1974.

Formation of Active Ocean Margins, edited by N. Nasu *et al.*, pp. 75–108.

NEOGENE TECTONIC EVOLUTION AND PLATE SUBDUCTION IN THE JAPANESE ISLAND ARCS

Nobuaki NIITSUMA and Fumio AKIBA

Institute of Geosciences, Shizuoka University, Shizuoka-shi 422, Japan
Japan Petrolium Exploration Co., Ltd., Tokyo, Japan

Abstract. Relationship between geologic events in the Japanese Island Arcs and plate subduction was examined on a new standard magnetomicro-biostratigraphic time scale which was established in the Neogene sediments at Site 584 of DSDP-IPOD, and applied a new stage division to the Neogene sediments in the Japanese Island Arcs. Geologic evolution of the Japanese Island Arcs has been directly controled by the processes of subduction of the Pacific and Philippine Sea plates. The back arc basins in the region, the Japan Sea and the Kurile Basin, opened in the middle Miocene during 15 Ma ~ 12 Ma. The bending of the zonal structure in central Japan was caused by the collision of the Izu and Honshu arcs.

Chronolostratigraphic Frame of the Japanese Neogene

A new diatom biostratigraphic zonation (Neogene North Pacific Diatom zone: NPD) and a new magnetobiostratigraphic time scale were established (NIITSUMA and AKIBA, in press) by the examination of drilled sediments at DSDP-IPOD Site 584 on the inner wall of the Japan Trench off Hachinohe in northeast Japan (KARIG, KAGAMI *et al.*, 1983). The new time scale is useful for the Neogene sediments in the Northern Pacific region where the calcareous microfossils are very rare and diatom fossils are dominant. Although the planktonic diatom flora is different from that of the Equatorial regions, this time scale applied to Equatorial region and classic European stage stratotypes based on the magnetostratigraphic sequence. This time scale is essensially similar to that of BARRON (1980), but differs from that of KOIZUMI (1977) in the late Miocene.

The geology of Japanese Neogene has been studied for one hundred years and knowleges on the tectonic evolution of Japanese Island Arcs have been accumulated. The distinct tectonic events and their sequence had already been determined until 1960s (TAKAI *et al*, 1963; MINATO *et al.*, 1965). In the earlier stages of the study, methods for global correlation had not been available, and correlation of tectonic events and their areal distributions had been difficult up to the late 1960s, when magnetomicrobiostratigraphic methods became applicable to global correlation of the Neogene strata.

Most commonly used stages for the description of the Neogene strata in Japan, were based on the stratigraphy in the Oga Peninsula of Akita Prefecture, Northeast Japan (HUZIOKA, 1959). Magnetobiostratigraphic examination of the Neogene marine sediments of the Oga Peninsula (KITAZATO, 1975) demonstrated drastic changes in the rate of sedimentation and a hiatus in the Late Miocene, Funakawa Stage, which had caused many confusions in the correlation of the Late Miocene.

Except the Hokkaido Island of northern Japan, important key areas, where the thick marine sediments are exposed, have been already examined magnetobiostratigraphically. Based on the correlation, the tectonic evolution of the Japanese Island Arcs was examined and its possible relation with the subduction of the Pacific Plate was discussed (NIITSUMA, 1978).

Using the new time scale, we can date the Neogene marine sediments in the northern part of Japan, including the Hokkaido Island, more precisely than before, and propose a new stage division in this article. The proposed stages are, from older to younger, Daijimakai, Nishikurosawakai, Onnagawakai, Shirasawakai, Ishidokai, Toyookakai, Kurotakikai, Sekikai, Akimotokai, and Shimosakai. The older three stages are based on the strata in the Oga Peninsula, next one stage is based on the pyroclastics in the western part of Sendai area, Miyagi Prefecture, and younger six stages are based on the marine strata and a hiatus in the Boso Peninsula, Chiba Prefecture (Fig. 1). The magnetomicrobiostratigraphic investigations have been made on the younger six stages directly (NIITSUMA, 1976; NAKAO, 1982 MS.), and biostratigraphic investigations on the older four stages in their type areas (AKIBA *et al.*, 1982; KITAZATO, 1975). The ending of the stage names "kai" is the Japanese term for "stage" of chronostratigraphic unit. We also propose an ending of the name "ki", e.g., Daijimaki, Akimotoki, and Shimosaki, for the corresponding geochronologic unit "age". The "ki" is the Japanese term for "age".

Relation between the new magneto-diatom biostratigraphic time scale, proposed stages division, planktonic Foraminiferal and nannofossil zones and classic European stages is summarized in Fig. 2. Figure 3 shows the distinct tectonic events in Japan on the time table.

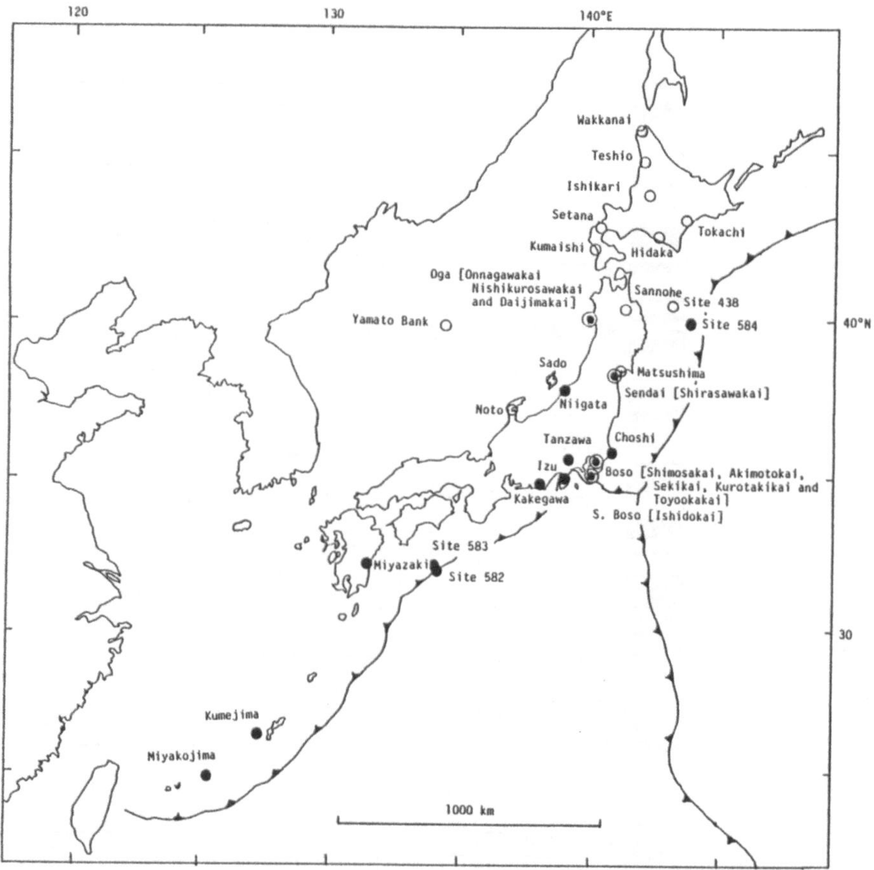

Fig. 1. Index map. ◉: locality of the type area of newly proposed stage, ○: area investigated microbiostratigraphically, and ●: magnetobiostratigraphically both by the authors and their collegues.

2. *Tectonic Evolution of Japanese Island Arcs*

The distinct tectonic events and their characters, and their relation with plate motions, will be described and discussed in the following sections for each Stage, from Present to older.

2.1 Shimosaki (0 ~ 0.5 Ma)

This age is corresponding Shimosakai and younger stages. The stage name "Shimosakai" is derived from Shimosa Group (SUZUKI and AOKI,

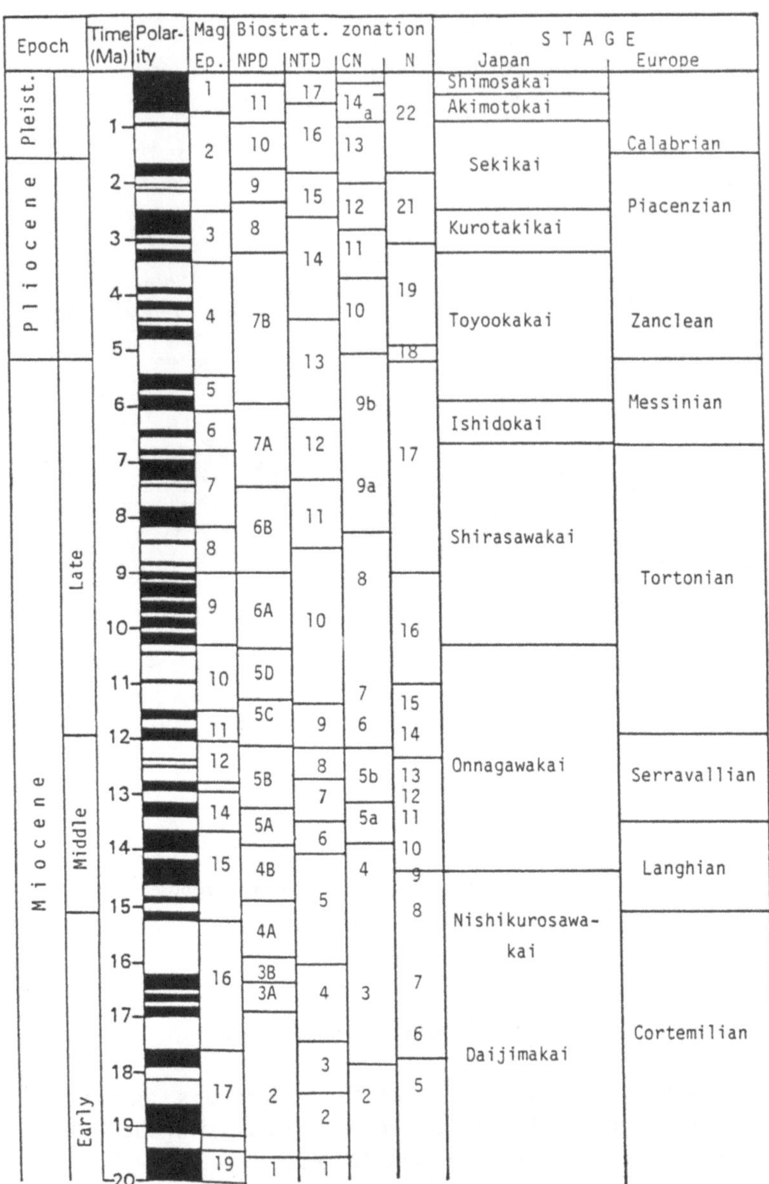

Fig. 2. Magnetobiostratigraphic time table (NIITSUMA and AKIBA, in press). NPD: Neogene North Pacific Diatom Zone, NTD: Neogene Tropical Diatom Zone, CN: Neogene Coccolith Zone, N: Neogene Planktonic Foraminiferal Zone.

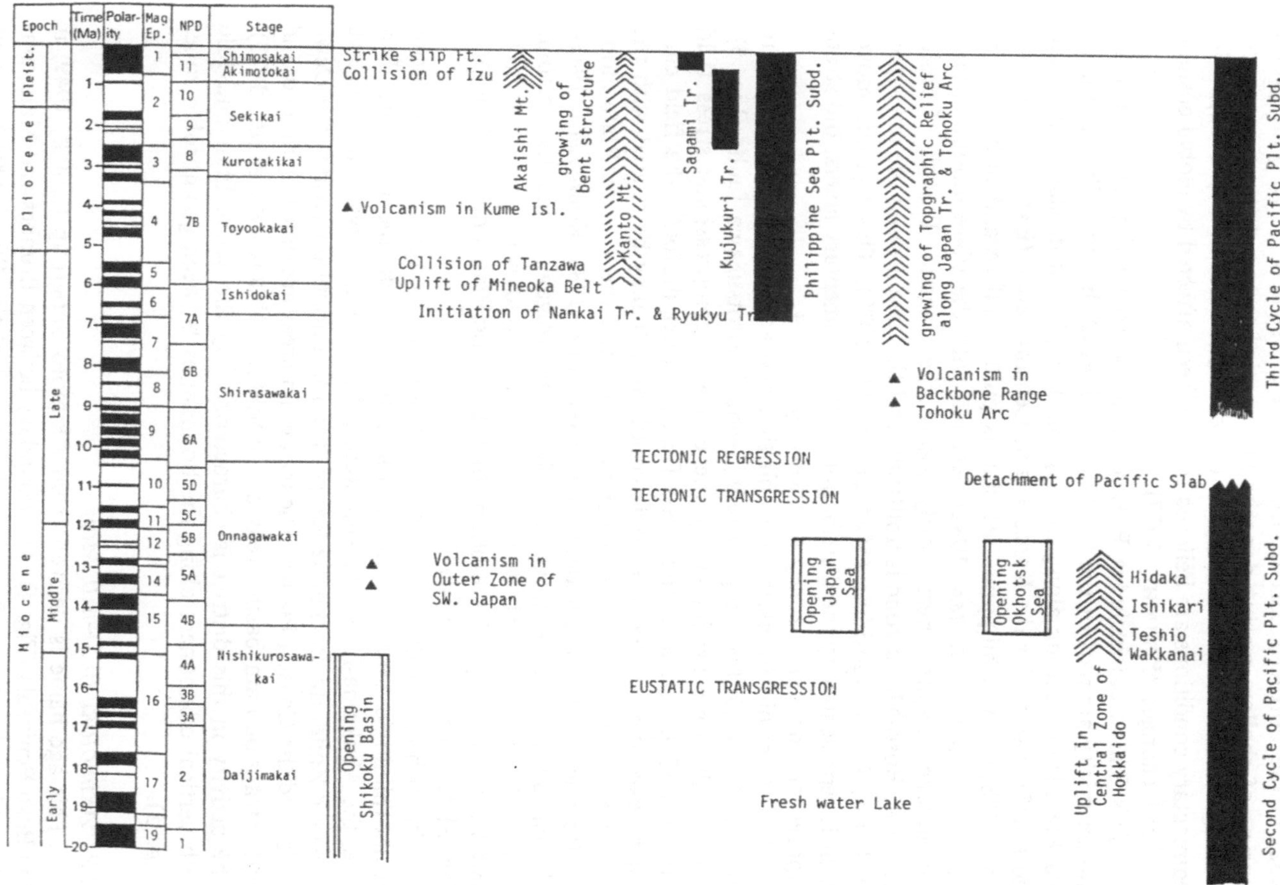

Epoch
Time (Ma)
Polar-ity
Mag Ep.
NPD
Stage
Pleist.
Pliocene
Miocene
Late
Middle
Early
Shimosakai
Akimotokai
Sekikai
Kurotakikai
Toyookakai
Ishidokai
Shirasawakai
Onnagawakai
Nishikurosawa-kai
Daijimakai
Strike slip Ft.
Collision of Izu
Akaishi Mt.
growing of bent structure
Kanto Mt.
Sagami Tr.
Kujukuri Tr.
Philippine Sea Plt. Subd.
growing of Topgraphic Relief along Japan Tr. & Tohoku Arc
Third Cycle of Pacific Plt. Subd.
Volcanism in Kume Isl.
Collision of Tanzawa
Uplift of Mineoka Belt
Initiation of Nankai Tr. & Ryukyu Tr.
Volcanism in Backbone Range Tohoku Arc
TECTONIC REGRESSION
TECTONIC TRANSGRESSION
Detachment of Pacific Slab
Volcanism in Outer Zone of SW. Japan
Opening Japan Sea
Opening Okhotsk Sea
Hidaka
Ishikari
Teshio
Wakkanai
Opening Shikoku Basin
EUSTATIC TRANSGRESSION
Uplift in Central Zone of Hokkaido
Fresh water Lake
Second Cycle of Pacific Plt. Subd.

1962) in the northern part of the Boso Peninsula (Fig. 4). This group consists of shallow marine deposits and marine terrace formations intercalating with volcanic ash layers (TOKUHASHI and ENDO, 1984). The sedimentary conditions are believed to have been affected by global eustatic sea level changes (KIKUCHI, 1977).

This Shimosaki includes Present, and the recent crustal movements and seismicity are part of the tectonic activities of this age. The Philippine Sea Plate is descending along the Sagami Trough and its slab is touching the Pacific slab under the Kanto Plain, Central Japan (EARTHQUAKE RES. INST., 1982). The Philippine Sea slab is pulled and recambently bent by the Pacific slab (NIITSUMA, 1982). Related with the Plate motion, Kanto Earthquakes happen periodically along the Sagami Trough. Choshi-Kujukuri-Boso-Miura zone is uplifting, and the area around the inner part of Tokyo Bay is subsiding (MATSUDA *et al.*, 1978). This tectonic movement is represented by the distribution of the sedimentary layers: old strata exposes on the uplift zone, as shown in Fig. 4.

The Akaishi Mountains are uplifting and active strike slip faults are cutting Chubu District, e.g., Atera, Neodai, Atotsugawa faults (Fig. 5). These faults were originated by the east-west compressional stress field (HUZITA, 1980). Because the stress field is rather uniform, the field cannot be explained by local tectonic conditions. It may be related with the collision between the North American Plate and the Eurasia Plate along the Itoigawa-Shizuoka Tectonic Line (NAKAMURA, 1983; KOBAYASHI, 1983). The calculated relative movement along the Itoigawa-Shizuoka Tectonic Line is 0.6 cm/year, which is small as compared with ca. 11 cm/year and 5 cm/year of relative motions of the Pacific Plate and the Philippine Sea Plate with the Eurasia Plate.

In Tohoku District, Northeast Japan, the tectonic activity forms geomorphologic zones parallel to the Japan Trench: the Deep Sea Terrace, the Abukuma-Kitakami Mountains, the Abukumagawa-Kitakamigawa Lowland Zone, the Backbone Range, the Intramountain Basins, the Dewa Hills, and the Coastal Planes. The mountain areas are uplifting and coastal and intra-mountain basin areas are subsiding (NAKAGAWA *et al.*, 1973). The activity in this district is characterized by the active fault and fold with vertical component along the boundaries between geomorphic zones (Fig. 5).

2.2 Akimotoki (0.5 ~ 1.0 Ma)

This age name is derived from Akimoto Subgroup (KOIKE, 1949) in the Boso Peninsula (Fig. 4). The boundary between Brunhes and Matuyama Magnetic Polarity Epochs is the middle part of this Stage. This Subgroup

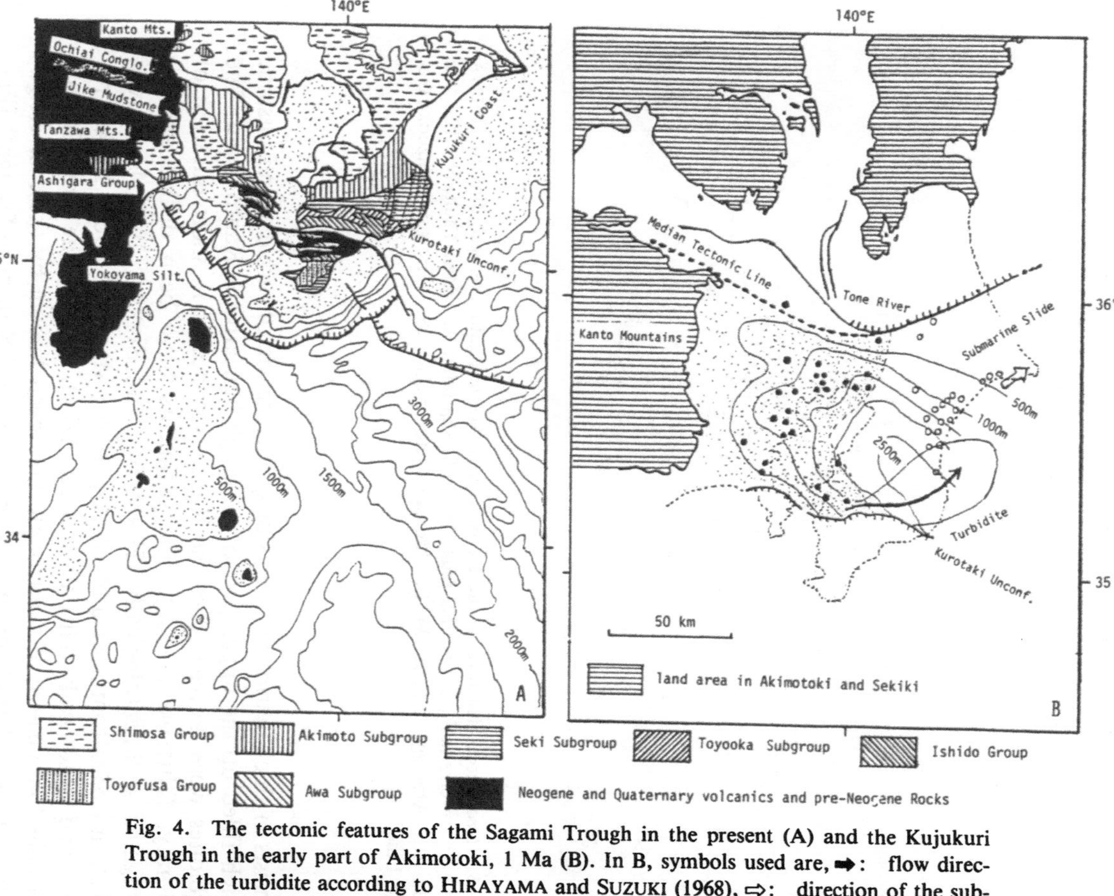

Fig. 4. The tectonic features of the Sagami Trough in the present (A) and the Kujukuri Trough in the early part of Akimotoki, 1 Ma (B). In B, symbols used are, ➡: flow direction of the turbidite according to HIRAYAMA and SUZUKI (1968), ⇨: direction of the submarine sliding according to NIITSUMA (1977); paleoenviroment ●: shelf and ○: slope according to KITAZATO (1977), ——: contour of thickness of Seki Subgroup at 500 m interval. Outline of geology of the Kanto District is shown in A.

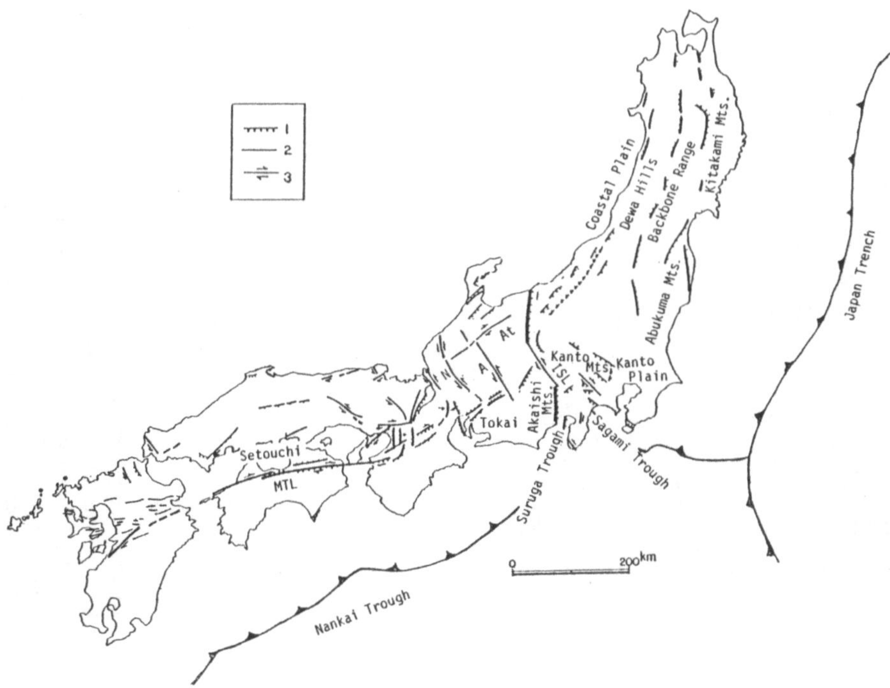

Fig. 5. Neotectonic features of the Honshu Arc. Active Faults are according to HUZITA (1980). 1: reverse fault, 2: normal fault, 3: strike slip fault, A: Atera Fault, N: Neodani Fault, At: Atotsugawa Fault, MTL: Median Tectonic Line, ISL: Itoigawa-Shizuoka tectonic Line.

consists of thick marine sediments of an alternation of turbidite and siltstone which filled up the upper part of a trough beneath the Kanto Plain. The turbidites were derived from the Kanto Mountains in short duration of 0.5 m.y. (NIITSUMA, 1982). The sediments have been drilled and investigated sedimentologically (HIRAYAMA and SUZUKI, 1968; NIITSUMA, 1977), micropaleontologically (AOKI, 1968; MATOBA, 1967; KITAZATO, 1977) and isotopically (NIITSUMA *et al.*, 1984). Because the morophology and paleobathymetry of the trough have been known by these investigations, a name of the trough was proposed as "Kujukuri Trough" (Fig. 4; NIITSUMA, 1982). The Kujukuri Trough had an original depth of 2000 ~ 3000 m.

A huge amount of gravels was released from the Tanzawa Mountains and filled up the Ashigara Trough (Ashigara Group) along the southern boundary of the Tanzawa Mountains. The released clastics from the Akaishi Mountains were deposited as a deltaic sediments (Ogasayama Group) upon the turbidite sequence of forearc basin (Kakegawa Group) of Sekikai in the Kakegawa area, south of the Akaishi Mountains. These gravels and clastics indicate that this stage is characterized tectonically by the collision of the Izu Peninsula and Honshu Arc. The collision also pushed up the Akaishi Mountains, and intensive erosion occurred. The clastics were deposited also as turbidite in the axis of the Nankai Trough off Shikoku (DSDP Site 582), 600 km west of the Akaishi Mountains (TAIRA and NIITSUMA, in press). The supply of turbidite to the Site 582 was suddenly decreased at the end of Akimotoki (NIITSUMA, in press). This fact suggests that the uplifting rate of the Akaishi Mountain area was decreased at the end of this age probably by the start of the strike slip faulting in Shimosaki, mentioned above.

In southwest Japan, an intra-arc basin chain was developed, and terrestrial sediments were deposited and interbedded with marine sediments (the Second Setouchi Supergroup; HUZITA, 1962) in the Setouchi, Osaka, lake Biwa, and Tokai districts.

2.3 Sekiki (1.0~2.5 Ma)

This age is corresponding Sekikai. The Stage name Sekikai is derived from Seki Subgroup (KOIKE, 1949) in the Boso Peninsula (Fig. 4). The boundary between Pliocene and Pleistocene is in the middle part of this stage. The Seki Subgroup consists of a thick sediments of an alternation of siltstone and turbidite, which were derived from the Kanto Mountains and deposited in the growing Kujukuri Trough which had 2000~3000 m depth in the central part (Fig. 4). Because the depth of the trough estimated by foraminiferal fauna have been kept constantly in spite of the accumulation of the sediments (MATOBA, 1967), the rate of the subsidence of the trough was comparable with or larger than the rate of sedimentation. The shape and dimension of the trough was similar to the present Sagami Trough and the deep axis was along the Tone River in the Kanto Plain, where the Median Tectonic Line exists. Because of the similarity of the shape and the location of axis, the Kujukuri Trough was a plate-boundary of the Philippine Sea Plate which is at present along the Sagami Trough (Fig. 4). Present seismic activity along the Tone River may represent the ancient slab descened through the boundary in that age.

The Neogene of the Izu Peninsula mostly consists of shallow marine and terrestrial pyroclastics and volcanic rocks deposited at the depth above continental shelf. In this age siltstone and turbidite of normal clastics were deposited at the depth below continental slope (Yokoyama Siltstone: KOYAMA, 1982). This siltstone indicates sinking Izu Peninsula into the trough. The trough continued to the Ashigara Trough whose axis was deeper than 1000 m. This chain of troughs was probably the northern boundary of the Philippine Sea Plate (Fig. 4).

2.4 Kurotakiki (2.5~3.2 Ma)

This age name is derived from Kurotaki Unconformity (UEDA, 1930) in the Boso Peninsula and the age represents the interval of gap by the unconformity (Fig. 4). The unconformity represents a most remarkable discontinuity in the Neogene sedimentary sequence of the Kanto District. A trough had laid to the south of the Boso area before this age. The trough had an east-west axis. After the age, the trough jumped toward north, and became the Kujukuri Trough.

Such a big discontinuity can be found not only in the Kanto District but also all over the Northeast Japan (NIITSUMA, 1978; in press). These drastic changes in the Northeast Japan in Kurotakiki can be related with the collision between the Izu and Kuril arcs with the Honshu Arc (NIITSUMA and AKIBA, in press). In the outer zone of the Southwest Japan, discontinuity in the sedimentary sequences is not detected, but the sedimentation rate increased at the end of this age (NIITSUMA, 1978).

2.5 Toyookaki (3.2~6.0 Ma)

The age corresponds Toyookakai. The name of the stage Toyookakai is derived from Toyooka Subgroup (KOIKE, 1949) in the Boso Peninsula (Fig. 4). The subgroup is characterized by submarine fan deposits below the continental slope with east-west strike. The turbidites composing the submarine fan were derived through a submarine cannyon from north, which is now under the Kanto Plain (TOKUHASHI, 1979). The submarine fan might have developed on the flat plane beneath the continental slope. In this case the flat plane was formed by the uplift of the Mineoka zone which consists of an ophiolitic sequence.

A huge amount of gravels was supplied into the Tanzawa and Fujikawa areas from a part of Honshu Arc, which is the present Kanto Mountains and Akaishi Mountains (Akebono Conglomerate and Ochiai Conglomerate: KANEKO *et al.*, 1983; TAMURA *et al.*, 1984). The gravels indicate a collision occurred between the Tanzawa Block and the Honshu Arc.

In the Kume Island of the Ryukyu Arc, basalt and andesite lava flows (Uegusukudake Formation) overlay the shallow marine sediments (Maja Formation) at ca. 4 Ma (NAKAGAWA *et al.*, 1976). This volcanic activity suggests that the slab from Ryukyu Trench became long enough for the initiation of the volcanic activity.

On the inner wall of the Japan Trench, as determined at DSDP Site 584 (NIITSUMA and AKIBA, in press), the deposition of diatomaceous muds terminated, and the inner wall steepened toward the Trench. The sediments were cut by many faults. The depth of the Japan Trench increased and the present tectonic activity started.

2.6 *Ishidoki (6.0~6.7 Ma)*

The name of the age is derived from Ishido Group (NAKAO, 1981 MS) in the southern part of the Boso Peninsula (Fig. 4). The Group consists of tuff bearing siltstone and turbidite. These sediments are folded and faulted strongly with wave lengths of about 1 km. The axial planes of the folds decline northward and the southern limbs of anticlines are overturned (Fig. 6). In spite of the strong folding, each layer can be traced over the fold and faults in the field. The sediments contain only Radiolaria in the lower part, Radiolaria and nannofossil in the middle part, and Radiolaria, nannofossil and Foraminifer in the upper part. The characteristic preservation pattern indicates gradual decrease of water depth

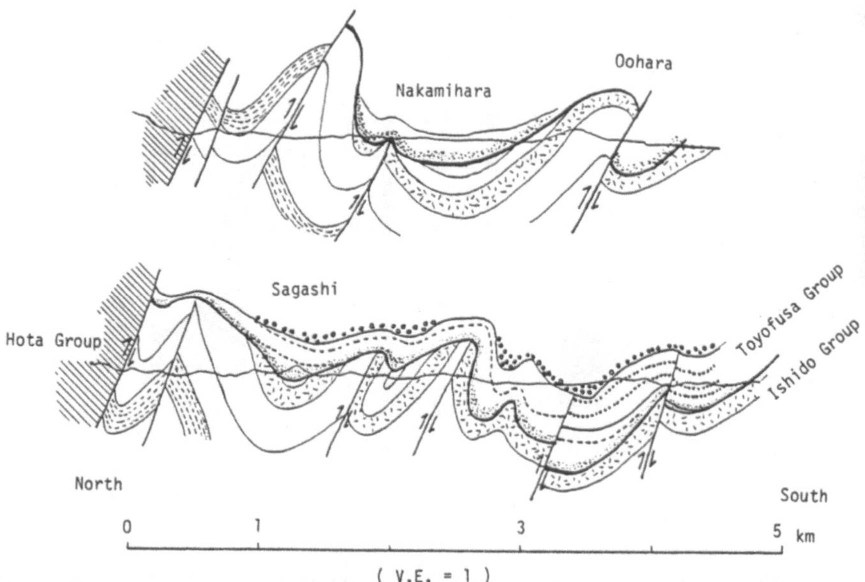

Fig. 6. Geologic cross section of Ishido Group (after NAKAO, 1982 MS).

beyond calcium carbonate compensation depth (CCD) (Fig. 7). In the present Shikoku Basin, the bottom of the axis of the Nankai Trough and the abyssal plain only are below CCD. These facts show that the sediments were deposited in a trough axis. Because the folding axes have an east-west strike and any rotations are not detected paleomagnetically, the trough axis should have had east-west direction. The direction and position indicate that the trough was the east extention of the Nankai Trough and the Ishido Group is situated at the northern end of an accretionary body which is in fault contact with so called ophiolitic and olistostrome basements on the north (Mineoka and Hota groups: NAKAJIMA *et al.*, 1981). This means that subduction of the Philippine Sea Plate was initiated along the Nankai Trough in this age.

Along the Nankai Trough and the Ryukyu Trench, deposition of thick

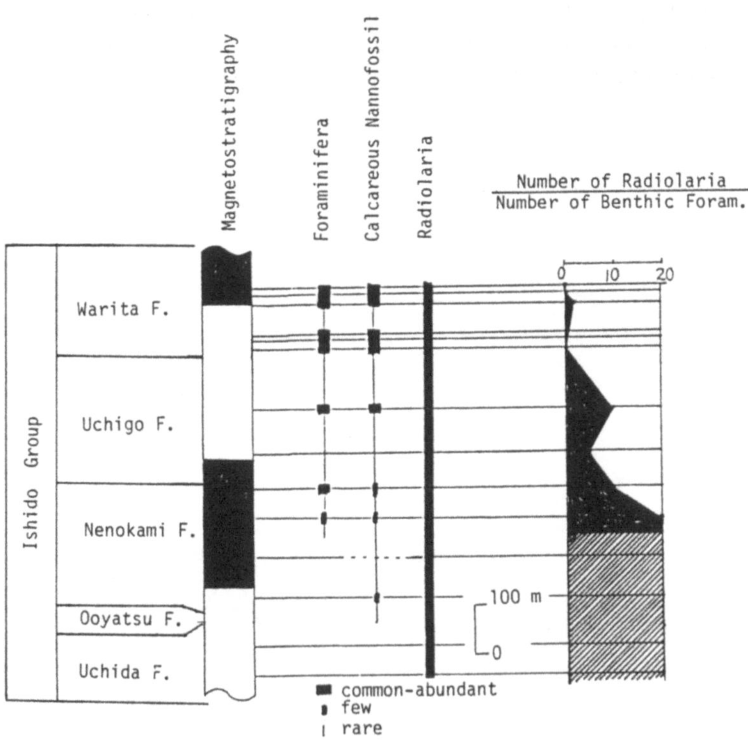

Fig. 7. Magnetostratigraphy and preservations of microfossils in Ishido Group (after NAKAO, 1982 MS).

marine sediments with turbidites (Sagara Group, Miyazaki Group, and Shimajiri Group) started in the forearc basins in this age (Fig. 8). The slab length along the Ryukyu Trench and Nankai Trough (YOSHII, 1979) needs a duration of 6–7 m.y. to descend in the constant plate motion of 1979 Minster-Jordan Model. These considerations also support our view

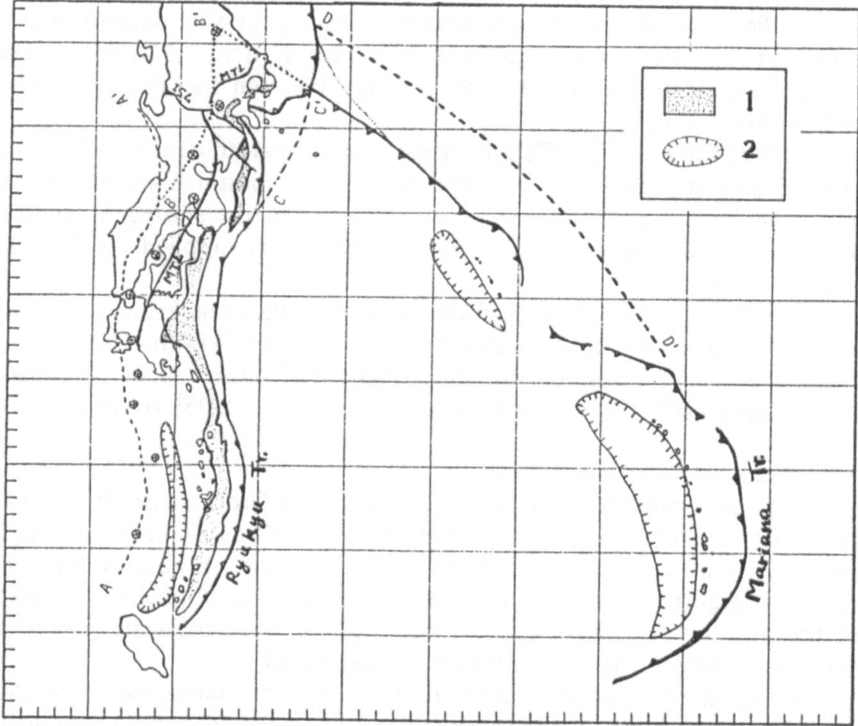

Fig. 8. Thick sedimentary sequences of Ishidokai, Toyookakai, Kurotakikai and Sekikai along the Ryukyu Trench and Nankai Trough. The map is in the Mercator projection with the axis passing through the pole of the relative motion of the Philippine Sea Plate to the Eurasia Plate (48.3°N, 162.5°E). One division of the horizontal axis corresponds with a distance of a plate motion for 1 m.y. and the interval of vartical lines corresponds with a plate motion for 5 m.y. 1: Ishidokai, Toyookakai, Kurotakikai and Sekikai, 2: rifted basin in the back-arc area, MTL: Median Tectonic Line, ISL: Itoigawa-Shizuoka Tectonic Line, ⊗: a leading edge of Wadati-Benioff zone (YOSHII, 1979 and EARTHQUAKE RES. INST., 1982) which is turned horizontally, A—A′: leading edge of a slab of the Philippine Sea Plate for 6 m.y. constant plate motion, B—B′: leading edge of a slab from the pre-retreated plate boundary (C—C′) for 6 m.y., D—D′: estimated plate boundary of the Philippine Sea Plate before the subduction along the Ryukyu Trench and Nankai Trough.

on the initiation of subduction at the Nankai Trough.

In the Ishido area of south Boso, marine sediments were deposited in this age, while no sediments were deposited in Toyookaki. In Toyookaki thick marine sediments were deposited north of ophilitic belts. The change of the site of deposition can be explained by the uplift of this ophiolitic belt. The uplifted belt dammed up the sediments accompanied with the accretionary process and formed a mid-slope terrace or deep sea terrace, where a submarine fan developed.

In the Tanzawa area, mudstones (Jike Mudstone) were unconformably deposited on the pyroclastics and volcanics (OHTA, 1982 MS). The mudstone indicate the existence of a trough between the Tanzawa block and Honshu Arc.

In the Hokkaido and Tohoku areas, the formation of the basic structure of their geomorphology—backbone range, coastal plain and intramountain basin—was initiated and marine sediments started to deposit in the basins and coastal area. Uplifting mountains supplied clastics into subsiding basins.

In Taiwan, there were remarkable changes in the sedimentary sequences and new sedimentary basins were created in the west Taiwan. The tectonic changes are later phase of Puli Movement and estimated to be caused by the compression from east (LIN and NAKAGAWA, 1975; HUANG, 1967).

2.7 Shirasawaki (6.7~10.2 Ma)

This stage name is derived from the Shirasawa Group in Sendai district, northeast Japan (AMANO, 1980). The Shirasawa Group consists of silicic pyroclastics including pumice flow deposits, which are part of the products of large scale silicic volcanism along the present Backbone Range of the Tohoku Arc (KITAMURA, 1959; Fig. 9). The volcanism produced calderas in which lake sediments were deposited.

In marine depositional basins of the Japanese Island Arcs, the rate of sedimentation was dropped and diatomite and glauconitic sediments were deposited or hiatus was formed (NIITSUMA, 1978). Supply of clastics from land decreased remarkably and the subsidence of the basins was small. Such condition was caused by the releasing from the stress field and low relief of land. These situations might be explained by the special condition of subducting plate which was just after the detachment along the Kuril, Japan and Izu-Mariana trenches, and the present Pacific slab initiated to descend.

On the inner wall of the Japan Trench, at DSDP Site 584, we found corresponded change in sedimentation represented by a very low sedimentation rate and hiatus, and abundant resting spores of neritic benthic diatom

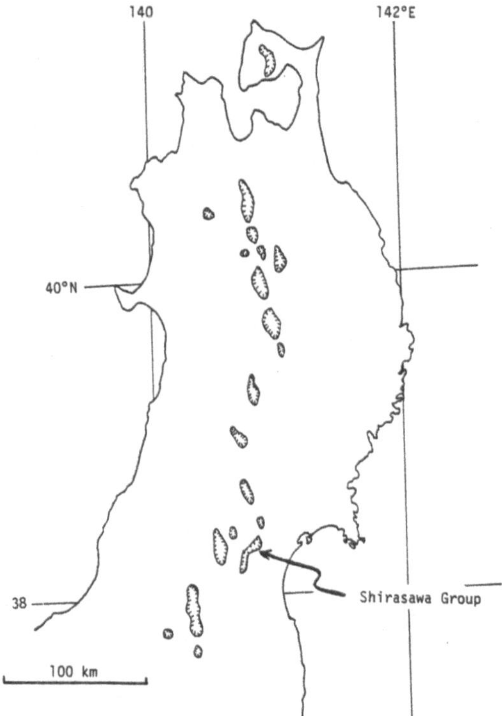

Fig. 9. Silicic volcanics accompaned with caldron collapse of Shirasawakai along the Backbone Range of the Tohoku Arc (after KITAMURA, 1983).

(NIITSUMA and AKIBA, in press). Submarine topographic relief was low and the sediment grains, such as resting spores, could reach the Trench.

This age is correlated with the time of Anomaly 5 of ocean magnetic lineations. In the East Pacific Rise, the present spreading axis was activated in different place from the old one during this age, at ca. 10 Ma (MAMMERICKX and KLITGORD, 1982). The detachment of the Pacific Slab in the western Pacific margins might affect not only the tectonics of near-by subduction zone, but also that of the spreading center of the Pacific Plate.

2.8 Onnagawaki (10.3 ~ 14.5 Ma)

This age name is derived from the Onnagawa Formation (HUZIOKA, 1959) in the Oga Peninsula (Fig. 10). The formation is characterized by

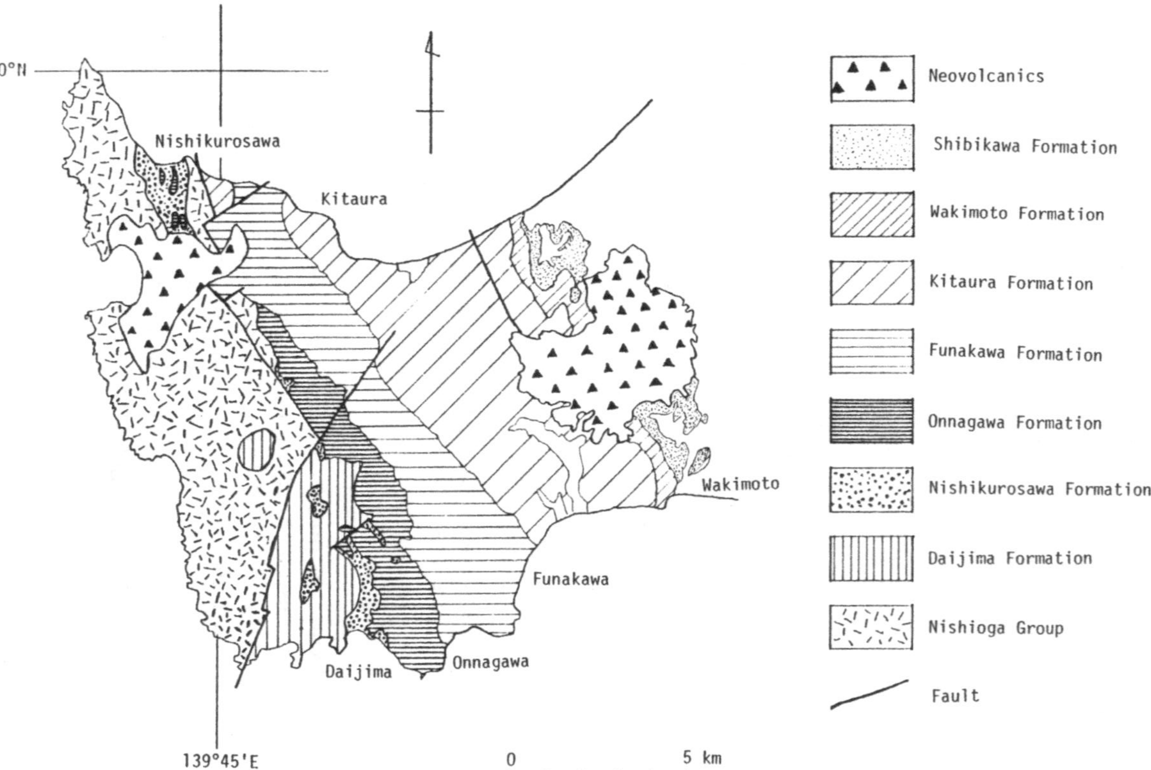

Fig. 10. Geologic sketch map of the Oga Peninsula, Northeast Japan (after KITAZATO, 1975; TAKAYASU and MATOBA, 1976).

the "hard" shale and diatomaceous mudstone. In this age pelagic sediments were most widely deposited in the Japanese Neogene. Most of the Japanese Island Arcs subsided below the sea level like the present Mariana Arc.

On the Japan Sea side of the Tohoku Arc, the sediments were intruded by dolerite and a rifted deep basin was formed (KONDA, 1974). The basin has filled with thick marine sediments which became the parent rock for oil and gas (ISHIWADA *et al.*, 1977). The "Kuroko" type ores were deposited in relation with silicic submarine volcanisms also in this age (SATO, 1974; KITAZATO, 1983). These geologic events, described above, indicate the opening of the Japan Sea during this age and were phenomena along the eastern margin of the opening marginal sea.

The oldest marine sediments deposited at 12 Ma (NPD 5C) in this age in the Japan Sea overlie diatomaceous and glauconitic clay. The diatoms in the clay were characterized by the life habitat of fresh water plankton. This sedimentary sequence was reported in the northern part of the Yamato Bank at the depth of 2300 m (BURCKLE and AKIBA, 1978) and also in the southwestern part of the Bank (Tsoy, personal comunication). The fresh water diatomites of the characterized flora have been reported around the Japan Sea area, southeastern and northeastern coasts of the Korean Peninsula, Sado Island, Noto Peninsula and southwestern Hokkaido Island (KOIZUMI, 1979). The life habitat of the fresh water plankton and the wide distributions of the diatom mean that there were large fresh water lakes deeper than a photic zone in the back arc area.

The transgressive marine sediments of this age covered the basement rocks with basal conglomerate in the Pohang area of southeastern Korean Peninsula (KIM, 1965). Because the sedimentary rocks in the Korean Peninsula are mostly terrestrial in origin except this marine deposits and Cambro-Ordovician formations, this transgression should be a distinct event in the long geologic history of Korea. This event may be considered to be one of the phenomena in the southwestern margin of the opening Japan Sea.

From these geologic records, we infer that the Pacific Plate was in the last stage of a subduction cycle (NIITSUMA, 1978), the "Mariana" type Island Arcs (UYEDA and KANAMORI, 1979) along Kuril-Japan-Izu-Bonin-Mariana trenches were caused by the largest length of descending slab in the subduction area, and marginal seas opened in a back arc area.

The mountain building in central Hokkaido is recorded in large amount of turbiditic conglomerate, e.g., Kawabata and Kotanbetsu formations (OKADA, 1978). This type of conglomerate is distributed along the central mountains of Hokkaido in north-south direction. The age of the conglomerates is younger in the south and older in the north; 15 Ma (NPD

4A) in Wakkanai area (north end of Hokkaido), 14.8 Ma (NPD 4A-4B) in Teshio area, 13.8 Ma (NPD 4B-5A) in Ishikari area, and 13 Ma (NPD 5A) in Hidaka area (south end of Hokkaido); as shown in Fig. 11. The mountain building is related with the compressional stress field in the junction area of the Kuril Arc and the central zone of Hokkaido (KIMURA *et al.*, 1983). The change of the position of deposition of the conglomerates indicates that the junction area shifted southward from 15 Ma to 13 Ma along the Abashiri Tectonic Line in accordance with a southward shift of the Kuril Arc against the central zone of Hokkaido and the opening of the Kuril Basin. If the Kuril Arc moved southward, the volcanic front also should have moved southward. We can find widely distributed Miocene volcanics in the northeastern part of Hokkaido (Fig. 11). The volcanics could be a evidence of the moved volcanic front of Kuril Arc.

These facts suggest that the Kuril Basin also opened in this Age. We can check the opening of the Kuril Basin in the geology of Tokachi area where the junction of the central mountains of Hokkaido and Kuril Arc and the boundary between the Hidaka Belt and Tokoro Belt of basement rocks existed. The eastern part of the area is the west end of Kushiro Coal Field, and, the coal measures of Paleogene age overlies unconformably the Cretaceous Nemuro Group and is overlain conformably by Neogene marine sediments. The gap by the uncomformity is determined in the Neogene marine sequence from 15 Ma to 12 Ma (Fig. 12). The interval exactly correlates with the proposed time of the opening of the Kuril Basin.

In the western part of the area, Neogene deep marine mudstone covers the Jurassic rocks. The mudstone had been deposited for the time interval of the unconformity in the eastern part and turbiditic conglomerates were interbedded in the upper part at the last stage of opening of the Kuril Basin when the same type of conglomerates was deposited on the west of the Central mountains, Hidaka area. After that, both parts of this area form one sedimentary basin and common sedimentary sequences were accumulated (Figs. 11 and 12). These geologic records indicate that the both parts of this area were under completely different circumstances before this age, juxtaposed with each other during this age and formed one sedimentary basin as in the present.

Recently a clockwise rotation of the Southwest Japan occurred in a short time interval from 15 Ma to 12 Ma has been recognized paleomagnetically by OTOFUJI and MATSUDA (1983), and HAYASHIDA and ITO (1983), who suggest that the rotation is a result of opening of the Japan Sea. The estimation agrees well with our geologic inferences, mentioned above.

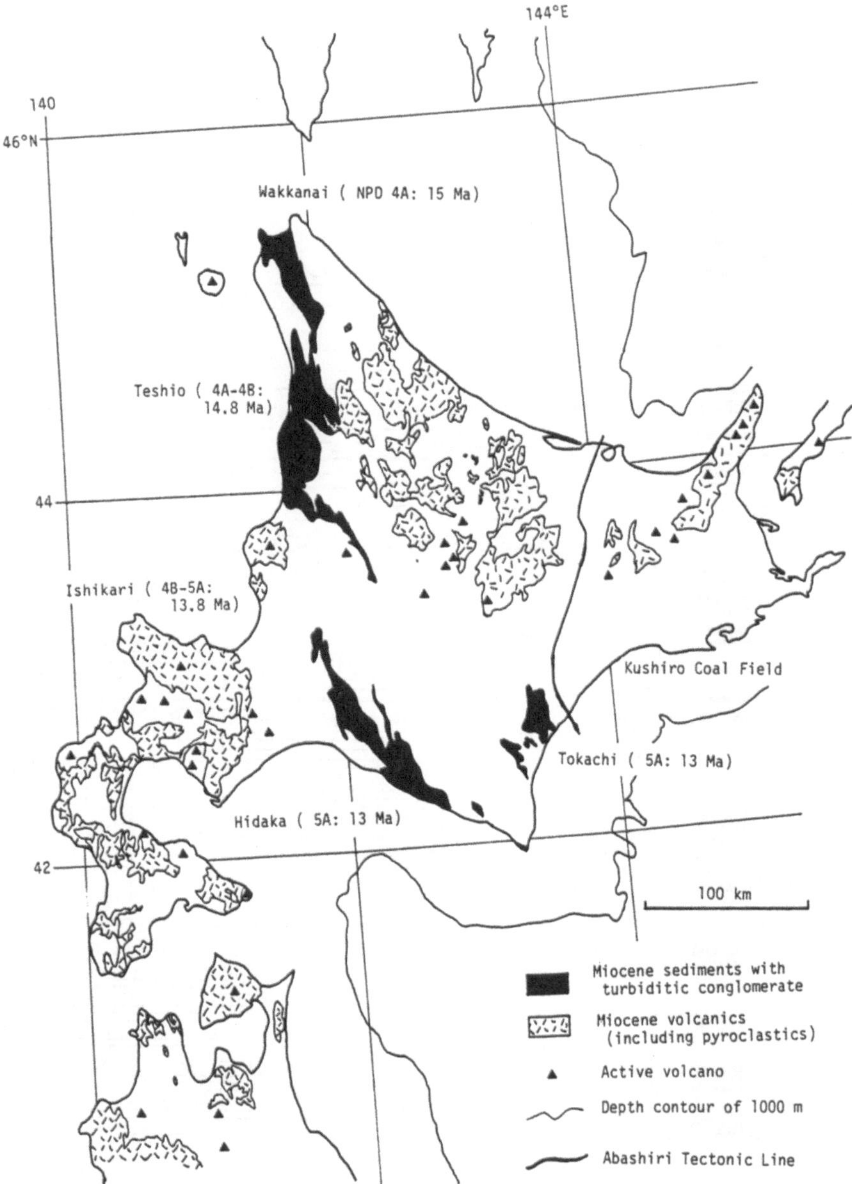

Fig. 11. Distribution and age of the sediments with turbiditic conglomerate of Nishikurosawakai and Onnagawakai. After 2,000,000 geologic map of Geological Survey of Japan.

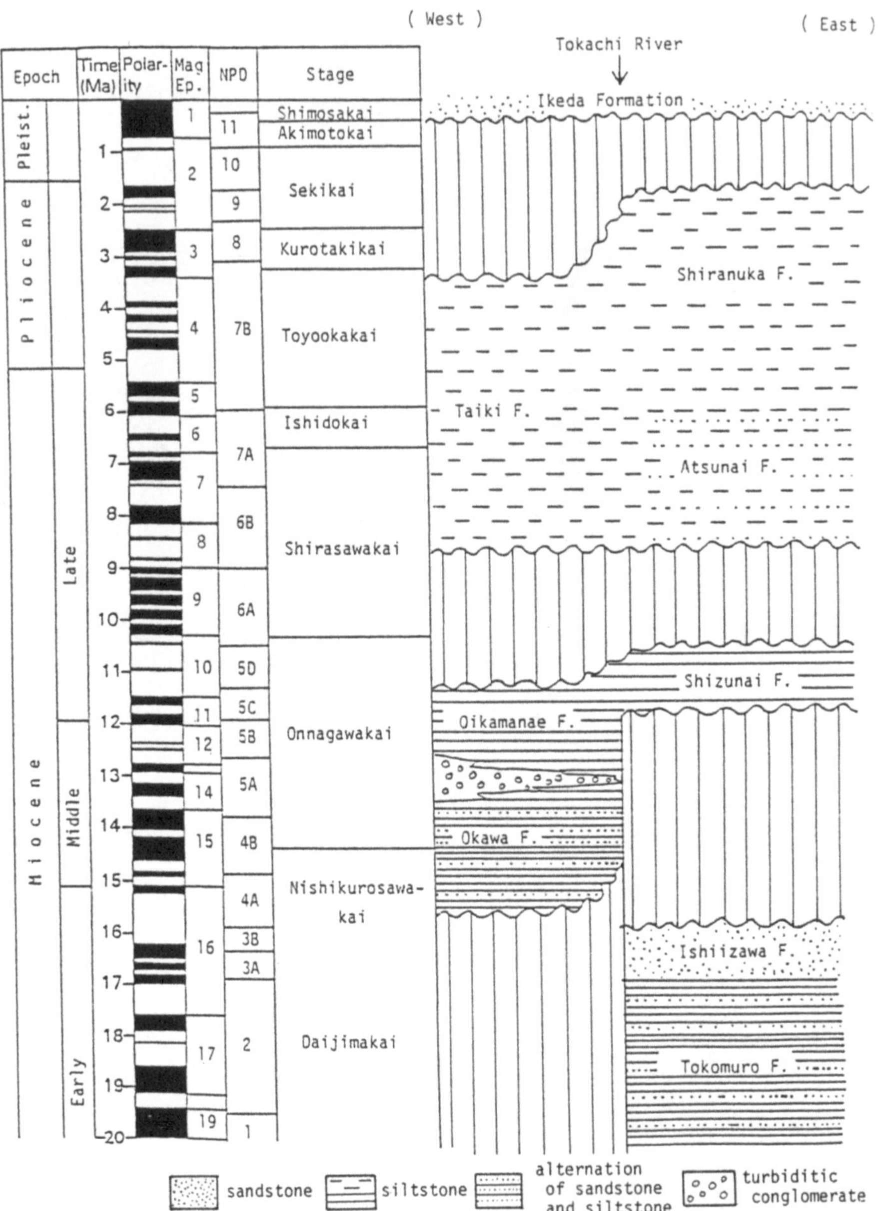

Fig. 12. Sedimentary sequences in the Tokachi Area, Hokkaido (see Fig. 11 for location). The eastern and western parts were seperated before Onnagawaki, and juxtaposed with each other in one sedimentary basin as in the present during Onnagawaki.

No more effects of southward movement of Kuril Arc may be excepted on the other areas, and only Kuril Trench retreated. However, in the Southwest Japan, there should be some tectonic effects of opening of the Japan Sea on the area to the south, because the Nankai Trough has been situated along active margin since Ishidoki, as mentioned before. In this age the Nankai Trough probably behaved as a trough along passive margin. The Philippine Sea Plate should have rotated together with the Southwest Japan, and subduct in the Philippine area.

In the Philippine Islands, volcanism occurred widely in Eocene and coincidenty with opening of the Philippine Basin. After this volcanism, there no deposition occurred there till coal measures and limestone were deposided widely in Late Oligocene to Early Miocene. Since Onnagawaki, clastic sediments have been deposited in several basins. Specially the thickness of the clastic sediments is upto 6500 m in the Panay Island. The lower part of the sedimentary sequence is Sewaragan Complex of metasandstone, indurated slate, basaltic lava and sill. The middle part of the sequence is siltstone of Onnagawakai and Shirasawakai, and the upper part is mudstone of Ishidokai (GONZALES, 1963; TAKAYANAGI *et al.*, 1977). The time of facies changes in the sediment and volcanic activity seems to be coincident with those in the Japanese Neogene sequences and these events should be affected by opening of the Japan Sea opening in the Onnagawaki and the initiation of the Philippine Sea Plate subduction along the Nankai Trough and Ryukyu Trench in Ishidoki.

The boundary between the Southwest Japan and the Philippine Sea Plate was a passive margin during the time of the Japan Sea opening, as mentioned above. However, some special igneous activities occurred in the outer zone of the Southwest Japan during the same period. The igneous activity was characterized by Valles type large scale silicic activity along the outer zone of the Southwest Japan and andesitic volcanism along the Median Tectonic Line. Because their radiometric ages are 14 Ma for the silicic igneous rocks and 13 ± 1 Ma for the andesitic volcanic rocks (SHIBATA, 1978; TATSUMI, 1983), the silicic igneous activity coincided with the rotation of the Southwest Japan and the andesitic volcanism happened just after the rotation. Both activities were distributed widely along the arc, but were limited to a short time interval and characterized by special chemical compositions (TAKAHASHI *et al.*, 1980; TATSUMI and ISHIZAKA, 1981). The special igneous activities were limited in the outer zone which was facing the Shikoku Basin, the eastern half of the Philippine Sea Plate. Because basaltic volcanism had continued in the Shikoku Basin (DSDP Sites 442, 443, 444; KLEIN *et al.*, 1978), until just before the rotation of the Southwest Japan since the opening of the Shikoku Basin, the special igneous activity in the outer zone might have been related with the hotness of the plate of the Shikoku Basin. Usually hot plate would not descend under another plate because of its low density. In this case, the Shikoku Basin did not descend, but the Southwest Japan thrusted over the Shikoku Basin by its rapid rotation. The crust of the Southwest Japan reacted with the hot plate of Shikoku Basin and might have produced special magmas.

2.9 Nishikurosawaki, Daijimaki, and older ages (14.5 ~ 25 Ma)

The names of ages, Nishikurosawaki and Daijimaki, are derived from the Nishikurosawa Formation and Daijima Formation (HUZIOKA, 1959) of Oga Peninsula, Akita Prefecture (Fig. 10). Nishikurosawa Formation is characterized by shallow marine sediments with the fossil fauna of warm water mass. Daijima Formation is characterized by terrestrial sediments with fossil flora. The transgression from the terrestrial Daijimakai to marine Nishikurosawakai can be determined widely in the Island Arcs (e.g., CHINZEI, 1978). Because the oxygen isotope stratigraphy shows decrease of polar icecaps in the boundary between Daijimakai and Nishikurosawakai (e.g., SAVIN, 1977), the transgression might be caused by the eustatic change as suggested by AMANO (1984).

In the consideration on the tectonics of these stages we have to be careful, because the original positions have been disturbed by the opening of the marginal seas, and age determination for these stages has not yet been precise enough to discuss the tectonic evolution as adequately as for the younger stages. However we can say that intra-arc rifting happened along the present Backbone range of the Northeast Japan in relation with back arc rifting in the present Japan Sea area which might have initiated large fresh water lakes. This event may be regarded as one of the phenomena in the "Mariana" stage of the previous cycle of plate subduction, in ca. 25 Ma. (NIITSUMA, 1978).

3. Geologic Record and Plate Motion

The Neogene and Quaternary tectonic events in the Japanese Island Arcs and their surrounding areas (Fig. 3) are summarized into followings: 1) collision of the North American Plate and Eurasia Plate along the Itoigawa-Shizuoka Tectonic Line in the Central Japan since 0.5 Ma in Shimosaki, 2) collision of the Izu Arc and Honshu Arc, caused by the subduction of the Philippine Sea Plate since 6 Ma in Toyookaki, 3) initiation of the descending of the Philippine Sea Plate along the Ryukyu Trench and Nankai Trough at 6.7 Ma in Ishidoki, 4) detachment of the slab of the Pacific Plate at 10 Ma in Shirasawaki, and 5) rotation of the Southwest Honshu Arc and Kuril Arc, caused by the opening of the Japan Sea and Kuril Basin at 14 Ma in Onnagawaki.

In the discussion we will use the 1979 Minster-Jordan Model for the plate motions, and the shape and length of the Wadati-Benioff zones which represent the descending slabs of the oceanic plates along the trenches. If the leading edge of the slab fits with the shape of the trenches after moving back the plates, we can estimate the initial age of the slab descending and movement of trenches without retreat. In the same way, we can turn the inclined Wadati-Benioff zone horizontally and plot the leading edge of the slab on a Mercator projection map with an axis passing through the poles of the relative plate motion. On this Mercator projection, latitudes are horizontal lines which are parallel to the plate motion and longitudes are vertical lines of which interval corresponds to the amount of the plate motion in a certain time interval.

Figure 8 illustrates the plate boundary and the leading edge of the descended slab of the Philippine Sea Plate on the Mercator projection with the axis of the relative motion to the Eurasia Plate. The positions of the leading edge (⊗) fit well the calculated leading edge after 6 m.y. plate motion, as shown by dotted line A-A′ along the Ryukyu Trench and the western part of the Nankai Trough. However, the positions do not fit along the Sagami Trough, Suruga Trough and the eastern part of the Nankai Trough. The bend of the zonal structure of the Southwest Japan, which is represented by the bend of the Median Tectonic Line (MTL), is parallel to the present northern boundary of the Philippine Sea Plate: the Sagami, Suruga and Nankai Troughs. The positions of the leading edge of the Wadati-Benioff zone are explained as those of the slab descended for 6 m.y. from the line connecting the western part of the Nankai Trough and the triple junction, as shown by broken line C-C′ in Fig. 8. This relation indicates a retreat of the Suruga, Sagami and Nankai Troughs toward northwest being accompanied with the bending of the zonal structure of the Southwest Japan since 6 Ma. The retreat should be related with the collision of the Izu Arc and Honshu Arc. If we adopt the plate motion of the Philippine Sea Plate since 6–7 Ma, the eastern boundary of the plate should be moved. Another broken line D-D′ shows the estimated eastern boundary of the Philippine Sea Plate in 6 Ma. D-D′ is an extension of the main part of the present Mariana Trench. It means that the position of the Mariana Trench has not been changed by the motion of the Philippine Plate, but the Mariana Trough opened (MATSUBARA and SENO, 1980). Such a back arc opening along the eastern boundary of the Philippine Sea Plate is also found in the Bonin Trench area also.

Figure 13 illustrates the position of the Tanzawa Block at 5 Ma in Toyookaki, the Izu Peninsula at 2 Ma in Sekiki, the present position of the Tanzawa Block and Izu Peninsula, Kujukuri and Ashigara troughs of Sekiki, and the trough chain of the northern boundary of the Philippine Sea Plate on the Mercator projection. Broken lines are estimated position of present northern boundary of the Philippine Sea Plate in 5 and 2 Ma. The bending shape of the present trough chain and zonal structure can be explained by the collided blocks since 6 Ma in Toyookaki.

Figure 14 shows the history of the plate boundaries in the northwestern Pacific area on the Mercator projection with axis as same as of relative motion of the Pacific Plate to the Eurasia Plate. A shows the present situation. B shows the plate boundaries after the opening of the Japan Sea and Kuril Basin, and before the descent of the Philippine Sea Plate. The eastern plate boundary of the Philippine Sea Plate was smoother than at the present one. In the area of the Philippine Islands, the descended slabs are restored horizontally. In the Philippine area, several descending slabs have been found from the Philippine Trench, Manila Trench, and Molucca Sea (CARDWELL *et al.*, 1980). The Eurasia Plate is descending to the east along the Manila Trench and the slab reaches 200 km of depth. The Philippine Sea Plate is descending to the west along the Philippine Trench and the slab reaches 100 km of depth. In the Molucca Sea, descending slab of saddle-shape is descending westward 600 km of depth and eastward 200 km. The top of the saddle like slab is located in the Molucca Sea and the slab does not connect with neither Eurasia nor Philippine Sea plates.

Because the Philippine Sea Plate has been descended along the Nankai Trough and Ryukyu Trench, it should have descended also in the Philippine area. However, the total length of the slabs from the Manila and Philippine Trenches is only a half or two thirds of the length calculated from the plate motion since Ishidoki, as shown in Fig. 14B. The calculated length and area of the short of the slab are shown by (−) in this figure. If we take account of the slab under the Molucca Sea (+), the length is too much for the plate motion.

Figure 14C illustrates a situation after the opening of the Shikoku Basin and before the opening the Japan Sea and Kuril Basin. The rotaion poles of the Southwest Honshu Arc and Kuril Arc can be calculated as 26° North latitude and 122° East longitude; and 50° North and 156° East, respectively by the best fit of the 1800 m depth contour in the Japan Sea and Kuril Basin. Dark areas in the Japan Sea and Kuril Basin show residual spaces of the fitting. The rotation of the Southwest Japan should induce the rotation of the Philippine Sea Plate, as shown in Fig. 14B.

Fig. 13. Positions of the Tanzawa Block at 5 Ma of Toyookaki and the Izu Peninsula at 2 Ma of Sekiki. The present position of the Tanzawa Block and Izu Peninsula, Kujukuri and Ashigara Troughs of Sekiki also are shown. The present trough chain of the northern boundary of the Philippine Sea Plate is shown by solid line with teeth. The estimated the present plate boundaries at 5 Ma and 2 Ma are shown by broken line.

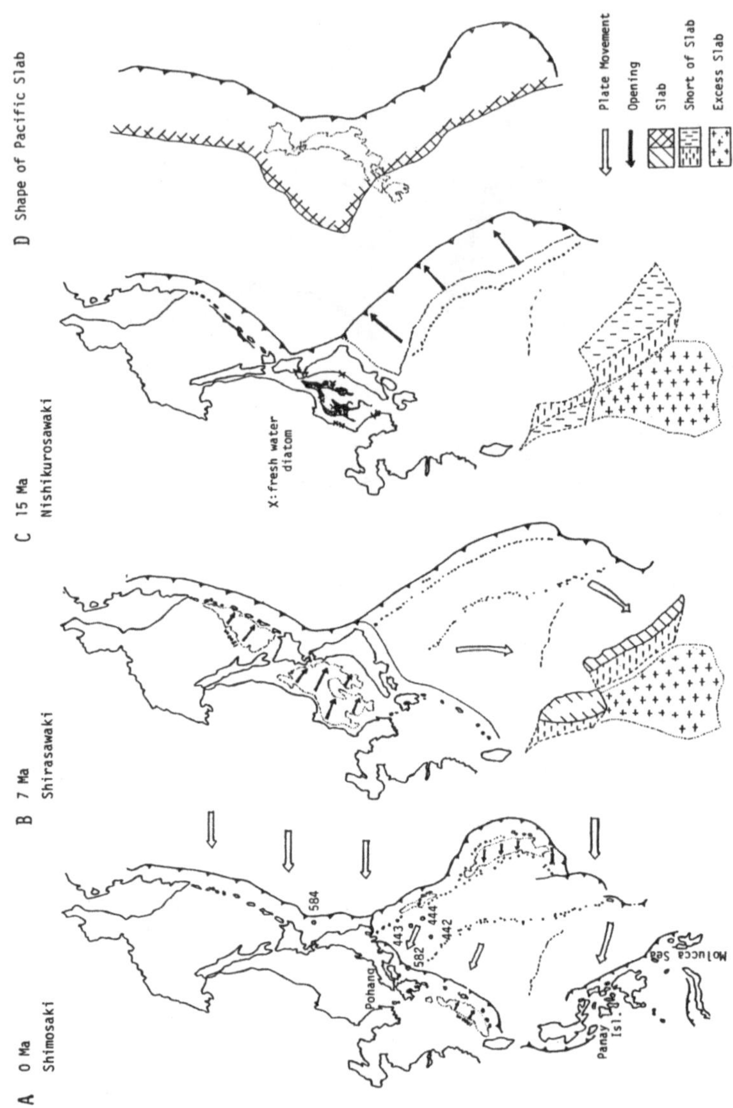
A 0 Ma
Shimosaki
B 7 Ma
Shirasawaki
C 15 Ma
Nishikurosawaki
D Shape of Pacific Slab
X: fresh water diatom
584
443
444
442
582
Pohang
Molucca Sea
Panay Isl.
Plate Movement
Opening
Slab
Short of Slab
Excess Slab

Fig. 14. Plate boundaries in successive ages in the northwestern Pacific area, by using the Mercator projection with axis through the pole of relative motion of the Pacific Plate to the Eurasia Plate (60.6°S, 101.1°E). A: Present. B: after the opening of the Japan Sea and Kuril Basin, and before the initiation of the subduction of the Philippine Sea Plate. C: after the opening of the Shikoku Baisn and before the opening the Japan Sea and Kuril Basin. D: horizontllly turned present slab along the Kuril, Japan, Izu, Bonin and Mariana Trenches. Dark area in C shows residual space of the 1800 m depth contor fitting for the opening of the Japan Sea (pole: 26°N, 122°) and Kuril Basin (pole: 50°N, 156°E). Number in (A) are DSDP Sites. × in (C): reported localities of the fresh water planktonic diatom flora of Nishikurosawakai and Daijimakai. The length of white arrow in (A) represents plate motion for 5 m.y.

The rotation should be supported by the subduction in the Philippine area. The slab lengths for the subduction are shown as (|) in Fig. 14C. The amount of the shorts of the slab is comparable with the amount of the excess length of slabs beneath the Molucca Sea (+).

The balances of the amount of the plate motion and the length of the observed Wadati-Benioff zone indicates that the plate motions which are estimated from geologic records are consistent with the present seismicities and plate motions.

Figure 14D shows the horizontally restored slab along the Kuril, Japan, Izu, Bonin and Mariana Trenches. The shape of the leading edge of the Pacific slab does not fit with the shape of the present Kuril-Japan-Izu-Bonin-Mariana Trenches when moved back the plate motion. However, the shape of the slab is similar to the pattern of the trench chain before the opening of Japan Sea and Kuril Basin (Fig. 14C). It suggests that the slab shape records the position of the trenches before the marginal sea opening.

4. Conclusion

Using the new time table based on the magnetomicrobiostratigraphy, we organized geologic evidence of the tectonic evolution of the Japanese Island Arcs that has been controled by plate subduction, collision, accretion, and, back-arc and intra-arc opening in a series. As a conclusion, we show the tectonic history of the region, as follows.

The Pacific Plate changed the direction of movement from north to west at about 40 Ma, and the first cycle of plate subduction started in the Japanese Island Arcs area. In the last stage of the first cycle, intra-arc and back-arc rifting occurred. Large fresh water lakes were formed in the present Japan Sea area, and spreading of marginal sea started in the Shikoku Basin. The slab of descended Pacific Plate was detached and the second cycle of subduction started at ca. 25 Ma.

In the late stage of the second cycle, the global eustatic change caused transgression, and the spreading of the Shikoku Basin were terminated in Nishikurosawaki at ca 15 Ma.

In the last stage of the cycle, the Japanese Island Arcs became "Mariana" type, and the Japan Sea and Kuril Basin opened in Onnagawaki, ca. 14 Ma. Opening of the Japan Sea induced subduction of the Philippine Sea Plate in the Philippine Islands area, and thick normal clastic sediments were accumulated. The opening also induced a clockwise rotation of the Southwest Japan and its outer zone thrust over the hot Shikoku Basin Plate. The thrusted crust reacted with the hot plate and Valles type silicic igneous activities occurred. In the Japan Sea area marine transgression yielded marine sediments over previous lake deposits and basement rocks. The eastern margin of the Japan Sea also rifted and dolerite intruded into the marine sediments.

The Kuril Trench and volcanic arc moved southward as a result of the opening of the Kuril Basin in Onnagawaki, 15–13 Ma. The position of the junction between the Kuril Arc and Central zone of Hokkaido Island also moved southward. The junction gave cause for mountain building by the compressional stress and the turbiditic conglomerates were deposited around the mountain area. The depositional area of conglomerates, also shifted south. The movement of the volcanic front left volcanics widely in the northeastern part of Hokkaido.

The Pacific slab was detached again and the third cycle of the Pacific Plate subduction was initiated in Shirasawaki, 10 Ma. The detachment produced a stress free condition and topography of low relief in the area. The rate of sedimentation was very low or non-depositional. Large scale silicic volcanisms occurred along the present Backbone Range of the Northeast Japan. The spreading center of the East Pacific Rise jumped and present spreading were started at that time.

The Philippine Sea Plate initiated to subduct along the Nankai Trough, Ryukyu Trench, Manila Trench, Philippine Trench and Molucca Sea, where accretionary processes started and forearc basins were developed in Ishidoki, 7 Ma. Andesitic volcanic activities of Ryukyu Arc were initiated when the descending slab had enough length. The collision between the Honshu and Izu Arcs started on the north of the Tanzawa Mountains. The Kanto Mountains were uplifted and supplied a large amount of clastics. The northern boundary of the Philippine Sea Plate extended from the Nankai Trough, through north of the Tanzawa Mountains, to the south of the Boso Peninsula in Toyookaki, 6–3 Ma.

The collision of the Tanzawa Mountains proceeded and the zonal structure of basement rocks which can be traced throughout the Southwest Japan, was bent northward. The bend induced subsidence of the Kujukuri Trough in the Kanto Plain area and uplift of the Kanto Mountains in Kurotakiki and Sekiki, 3-1 Ma. The Kanto Mountains supplied huge amount of clastics for turbidites into the Kujukuri Trough. The Plate boundary of the Philippine Sea Plate at that time extended from the Kujukuri Trough through Ashigara Trough on the south of the Tanzawa Mountains to the Nankai Trough. Izu Peninsula located on the northern boundary of the Philippine Sea Plate.

Along the Japan Trench, topographic reliefs started to develop in Kurotakiki, 3 Ma; the Trench itself was deepened, coastal area and intramountain basin subsided, and the Backbone Range and coastal ranges were uplifted. The growth of reliefs has continued to the Present.

The Izu Peninsula collided with the Tanzawa Mountains and the Akaishi Mountains were uplifted. The Akaishi Mountains began to supply a large amount of clastics to the Nankai Trough in Akimotoki, 1–0.5 Ma. The Kujukuri Trough was filled up by the turbidites from the Kanto Mountains. The bending of the geologic zonal structure in central Japan developed further northwestward.

The boundary between the Eurasia Plate and the North American Plate jumped to Itoigawa-Shizuoka Tectonic Line from Hokkaido, and large scale strike slip faults developed in the central Japan. The faulting reduced the rate of uplift in the Akaishi Mountains in Shimosaki, 0.5–0 Ma. Then the northern boundary of the philippine Sea Plate jumped south to a line connecting the Sagami Trough, Ashigara Trough, Suruga Trough and Nankai Trough, as in the Present.

Acknowledgment

We wish to express our gratitude to H. Nakagawa of Tohoku University and S. Uyeda and R. McCabe of Earthquake Research Institute, University of Tokyo, for valuable discussions and reviewing the manuscript. We also express our thanks to H. Kagami of Ocean Research Institue, University of Tokyo, N. Kitamura, Y. Takayanagi, M. Oda, and T. Maruyama of Tohoku University, S. Maiya of Japan Petrolium Exploration, H. Kitazato of Shizuoka University, A. Taira of Tokyo University, K. Amano and M. Takahashi of Ibaraki University, G. Kimura of Kagawa University, K. Nakamura of Earthquake Research Institute, University of Tokyo, T. Seno of Building Research Institute, and Y. Saito of National Science Museum, for their valuable comments and discussions.

REFERENCES

AKIBA, F., Y. YANAGISAWA, and T. ISHII, Neogene diatom biostratigraphy of the Matsushima area and its environs, Miyagi Prefecture, Northeast Japan, *Bull. Geol. Surv. Japan*, **33**, 215–239, 1982.

AMANO, K., Geology of the Ou Backbone Range in Miyagi and Yamagata Prefectures Northeast Honshu, Japan, *Inst. Geol. Paleont. Tohoku Univ., Contr.*, **81**, 1–56, 1980.

AMANO, K., Late Cenozoic paleogeography and eustatic changes in Northeast Japan, *Kaiyokagaku*, **16**, 32–38, 1984.

BARRON, J., Lower Miocene to Quaternary diatom biostratigraphy of Leg 57, off northeastern Japan, Deep Sea Drilling Project, *Init. Rep. DSDP.* **56/57**, 641–685, 1980.

AOKI, N., Benthic foraminiferal zonation of the Kazusa Group, Boso Peninsula, *Paleont. Soc. Japan, Trans. Proc. N. S.*, **70**, 238–266, 1968.

BURCKLE, L. H. and F. AKIBA, Implications of late Neogene fresh-water sediment in the Sea of Japan, *Geology*, **6**, 123–127, 1978.

CARDWELL, R. K., B. L. ISACKS, and D. E. KARIG, The spatial distribution of earthquakes, focal mechanism solutions, and subducted lithosphere in the Philippine and Northeastern Indonesia Island, in *The Tectonic and Geologic Evolution of Southeast Asian Seas and Islands*, Geophys. Monograph, vol. 23, pp. 1–35, 1980.

CHINZEI, K., Molluscan faunas in the Japanese Islands: An ecologic and zoogeographic synthesis, *Veliger*, **21**, 155–170, 1978.

EARTHQUAKE RESEARCH INSTITUTE, UNIVERSITY OF TOKYO, Seismicity Maps of the Kanto District 1971–80, *Rep. Coord. Committee Earthquake Predict.*, **27**, 63–73, 1982.

GONZALES, B. A., Foraminiferal analyses on measured section along the Tarao (Jarao) and Tanian rivers, southwestern Iloilo, *Philippine Bur. Min., Rep. Invest.*, **456**, 1–35, 1963.

HAYASHIDA, A. and Y. ITO, Paleoposition of Southwest Japan at 16 Ma: implication from paleomagnetism of the Miocene Ichishi Group, *Earth Planet. Sci. Lett.*, **68**, 335–342, 1983.

HIRAYAMA, J. and Y. SUZUKI, Analysis of layers—An example in flysch-type alternations, *Chikyu Kagaku*, **22**, 43–62, 1968.

HUANG, T., Late Terriary planktonic Foraminifera from southern Taiwan, *Tohoku Univ., Sci. Rep. 2nd Ser. (Geol.)*, **38**, 165–192, 1967.

HUZIOKA, K., Explanatory Text of Geological Map of Japan, scale 1: 50,000, edited by Toga and Funakawa, 61pp., Geol. Surv. Japan., 1959.

HUZITA, K., Tectonic development of the Median Zone (Setouchi) of Southwest Japan, since the Miocene, *J. Geosci., Osaka City Univ.*, **6**, 103–144, 1962.

HUZITA, K., Role of Median tectonic line in the Quaternary tectonics of the Japanese Islands, *Mem. Geol. Soc. Japan.*, **8**, 129–153, 1980.

ISHIWADA, Y., Y. IKEBE, K. OGAWA, and T. ONITSUKA, A consideration on the scheme of sedimentary basins of Northeast Japan, *Geological papers dedicated to Prof. K. Huzioka*, 1–7, 1977.

KANEKO, T., H. ISHIGURO, J. TAMURA, and N. NIITSUMA, Lexicon of stratigraphic names of Cenozoic Erathem in the Southern Fossa-Magna Region, Central Japan, *Geosci. Rep. Shizuoka Univ.*, **9**, 1–228, 1983.

KARIG, D. E., H. KAGAMI, and DSDP LEG 87 SCIENTIFIC PARTY, Varied responses to subduction in Nankai Trough and Japan Trench forearcs, *Nature*, **304**, 148–151, 1983.

KIKUCHI, T., Pleistocene sea level changes and tectonic movements in the Boso Peninsula, Central Japan, *Geogr. Rep. Tokyo Metro. Univ.*, **12**, 77–103, 1977.

KIM, B. Y., The stratigraphic and paleontologic studies of the Tertiary (Miocene) of the Pohang area, Korea, *Soul Univ. J., Sci. Tech. Ser.*, **15**, 32–121, 1965.

KIMURA, G., S. MIYASHITA, and S. MIYASAKA, Collision Tectonics in Hokkaido and Sakhalin, in *Accretion Tectonics in the Circum Pacific Regions*, edited by Hashimoto, M. and Uyeda, S., pp. 123–134, Terrapub, Tokyo, 1983.

KITAMURA, N., Tertiary Orogenesis in Northeast Honshu, Japan, *Inst. Geol. Paleont. Tohoku Univ., Contr.*, **49**, 1–98, 1959.

KITAMURA, N., Tectonics and volcanic activities in the Backbone Range, Northeast Japan, *Mem. Nat. Sci. Mus.*, **16**, 3–12, 1983.

KITAZATO, H., Geology and geochronology of the younger Cenozoic of Oga Peninsula, *Inst. Geol. Paleont. Tohoku Univ., Contr.*, **75**, 17–49, 1975.

KITAZATO, H., Vertical and lateral distributions of benthic foraminiferal fauna and the fluctuation of warm and cold waters in the Middle Pleistocene of the Boso Peninsula, Central Japan, *Tohoku Univ., Sci. Rep., 2nd Ser. (Geol.)*, **47**, 7–41, 1977.

KITAZATO, H., Submarine topography of the Northeast Honshu Arc during the early Middle Miocene Nishikurosawa stage, based on the benthic foraminifera, *Minning Geol. (Japan) Special Issue*, **11**, 263–270, 1983.

KLEIN, G., K. KOBAYASHI, H. CHAMLEY, D. M. CURTIS, H. J. B. DICK, D. J. ECHOLES, D. M. FOUNTAIN, H. KINOSHITA, N. G. MARSH, A. MIZUNO, G. V. NISTERENKO, H. OKADA, J. R. SLOAN, D. M. WAPLES, and S. M. WHITE, Off-ridge volcanism and sea-floor spreading in the Shikoku Basin, *Nature*, **273**, 746–748, 1978.

KOBAYASHI, Y., Condition on initiation of Plate subduction, *Chikyu*, **3**, 510–518, 1983.

KOIKE, K., Geology of the middle part of the Boso Peninsula (II), *Physiogr. Sci. Res. Inst., Tokyo Univ.*, **3**, 1–6, 1949.

KOIZUMI, I., Neogene diatom correlations in the North Pacific region, *Proc. First Internat. Congress Pacific Neogene Stratigraphy.*, pp. 235–253, 1977.

KOIZUMI, I., The geologic history of the Sea of Japan—based upon sediments and microfossils, *Circular "Nihonkai"*, **10**, 69–90, 1979.

KONDA, T., Bimodal volcanism in the Northeast Japan Arc, *J. Geol. Soc. Japan.*, **80**, 81–89, 1974.

KOYAMA, M., Stratigraphy of the upper Cenozoic strata in the northeastern part of Izu Peninsula, Central Japan, *Geosci. Rep. Shizuoka Univ.*, **7**, 61–85, 1982.

LIN, C. C. and H. NAKAGAWA, Problems on Cenozoic of Taiwan and Ryukyu Islands, in *Geologic Problems on Philippine Sea Area*, pp. 27–42, Geol. Soc. Japan., 1975.

MAMMERICKX, J. and K. D. KLITGORD, Northern East Pacific Rise: Evolution from 25 m.y.B.P. to Present, *J. Geophys. Res.*, **87**, B8, 6751–6759, 1982.

MATSUDA, T., Y. OTA, M. ANDO, and N. YONEKURA, Fault mechanism and recurrence time of major earthquakes in southern Kanto district, Japan, as deduced from coastal terrace data, *Geol. Soc. Am. Bull.*, **89**, 1610–1618, 1978.

MATOBA, Y., Younger Cenozoic foraminiferal assemblages from the Choshi District, Chiba Prefecture, *Tohoku Univ., Sci. Rep., 2nd Ser. (Geol).*, **38**, 221–263, 1967.

MATSUBARA, Y. and T. SENO, Paleogeographic reconstruction of Philippine Sea at 5 m.y.B.P., *Earth Planet. Sci. Lett.*, **51**, 406–414, 1980.

MINATO, M., M. GORAI, and M. HUNAHASHI, (eds.), The geologic development of the Japanese Islands, *Tsukiji Shokan*, 442pp., Tokyo, 1965.

MINSTER, J. B. and T. H. JORDAN, Rotation vectors for the Philippine and Rivera Plates, *EOS*, **60**, 958, 1979.

NAKAGAWA, H. and TOHOKU REGION QUATERNARY RESEARCH GROUP, Quaternary crustal movement in Tohoku Region (Northeast Japan), in *The Crust and Upper Mantle of the Japanese Area, pt II (Geol. Geochem.)*, pp. 114–120, Japan. Nation. Commit. Upper Mantle Proj., Geol. Surv. Japan, 1973.

NAKAGAWA, H., N. NIITSUMA, M. MURAKAMI, and M. WATANABE, Brief Report of magnetostratigraphy in the Miyakojima and Kumejima, Okinawa Prefecture, *Geol. Stud. Ryukyu Isl.*, **1**, 55–63, 1976.

NAKAJIMA, T., H. MAKIMOTO, J. HIRAYAMA, and S. TOKUHASHI, Geology of the Kamogawa district. Quadrangle Series, scale 1: 50,000, *Geol. Surv. Japan*, 107pp, 1981.

NAKAMURA, K., Possible nascent trench along the eastern Japan Sea as the convergent boundary between Eurasian and North American Plates, *Bull. Earthquake Inst., Tokyo Univ.*, **58**, 711–722, 1983.

NAKAO, S., Geology of the Awa-Furukawa district in the southern part of the Boso Peninsula, Gradu. Thesis, Inst. Geosci. Shizuoka Univ., 80pp, 1982MS.

NIITSUMA, N., Magnetic stratigraphy of the Boso Peninsula, *J. Geol. Soc. Japan.*, **82**, 163–181, 1976.

NIITSUMA, N., Remanent magnetization of slumped marine sedimentary rocks, *Rock Mag.*

Paleogeophys., **4**, 44–52, 1977.

NIITSUMA, N., Magnetic stratigraphy of the Japanese Neogene and the development of the island arcs of Japan, *J. Phys. Earth*, **26**, Suppl. 367–378, 1978.

NIITSUMA, N., Touchstone of the Plate Tectonics—Southern Fossa Magna, Central Japan, *Chikyu*, **4**, 326–333, 1982.

NIITSUMA, N., Paleomagnetic results, Nankai Trough and Japan Trench, DSDP IPOD Leg 87., *DSDP Initi. Rep.*, **87** (in press).

NIITSUMA, N., N. FUJII, and H. KITAZATO, Estimation on the paleoeclogy of the Foraminifera by means of Carbon and Oxygen isotope of their shells, *Geosci. Rep. Shizuoka Univ.*, **10**, 113–122, 1984.

NIITSUMA, N. and F. AKIBA, Magnetostratigraphy and diatom biostratigraphy of Site 584, DSDP Leg 87 and Implications on the tectonic evolution of Japanese Island Arcs, *DSDP Initi. Rep.*, **87** (in press).

OHTA, H., Geology of the northeastern part of the Tanzawa Mountainland in the Reach Area of the Nakatsu River and the Hayato River, Gradu. Thesis, Inst. Geosci. Shizuoka Univ., 84pp., 1982MS.

OKADA, H., Sedimentary patterns in apparent back-arc basin: a case study of the Neogene sequence in north western Hokkaido, Japan, *J. Phys. Earth*, **26**, Suppl., 477–490, 1978.

OTOFUJI, Y. and T. MATSUDA, Paleomagnetic evidence for the clockwise rotation of Southwest Japan, *Earth. Planet. Sci. Lett.*, **61**, 349–359, 1983.

SATO, T., Distribution and geological setting of the Kuroko deposits, *Mining Geol. (Japan)*, Spec. Issue, **6**, 1–9, 1974.

SAVINN, S. M., The History of the Earth's Surface Temperature during the Last 100 Million Years, *Ann. Rev. Earth Planet. Sci.*, **5**, 319–355, 1977.

SHIBATA, K., Contemporanity of Tertiary granites in the outer zone of Southwest Japan, *J. Geol. Surv. Japan*, **29**, 551–554, 1978.

SUZUKI, T. and N. AOKI, On the stratigraphy and the foraminiferal fossils of the Jizodo and the Yabu formations, northwest of Mobara, Boso Peninsula, *J. Geol. Soc. Japan*, **68**, 497–506, 1962.

TAIRA, A. and N. NIITSUMA, Turbidite sedimentation in Nankai Trough as interpreted from magnetic fabric, grain size and detrital mode analyses, DSDP IPOD LEG 87, *DSDP Initi. Rep.*, **87** (in press).

TAKAHASHI, M., S. ARAMAKI, and S. ISHIHARA, Magnetite-series/Ilmenite-series vs. I-type/S-type granitoids, *Mining Geol.*, Spec. Issue, **8**, 13–28, 1980.

TAKAI, F., T. MATSUMOTO, and R. TORIYAMA, (eds.), *Geology of Japan*, 279pp, Univ. Tokyo Press., Tokyo, 1963.

TAKAYANAGI, Y., T. TAKAYAMA, and M. ODA, Note on the Late Cenozoic planktonic Foraminifera and Calcareous Nannofossils from Panay, Philippines, *Geol. Palaeont. Southeast Asia*, **18**, 77–86, 1977.

TAKAYASU, T. and Y. MATOBA, Oga Peninsula, *Guidebook for Exc. 1, 1st Intern. Congr. Pacific Neogene Stratigr., Tokyo 1976*, 78pp., 1976.

TAMURA, J., T. KANEKO, and N. NIITSUMA, Geology of the southern part of the Koma Mountains, western part of Yamanashi Prefecture, Central Japan, *Geosci. Rep. Shizuoka Univ.*, **10**, 23–54, 1984.

TATSUMI, Y., High magnesian andesites in the Setouchi Volcanic Belt, Southwest Japan and their possible relation of the evolutionary history of the Shikoku Inter-Arc Basin, in *Geodynamics of the Western Pacific-Indonesian Region*, AGU-GSA Geodynamic Ser. 11, pp. 231–341, 1983.

TATSUMI, Y. and K. ISHIZAKA, Existence of andesitic primary magma: an example from Southwest Japan, *Earth. Planet. Sci. Lett.*, **53**, 124–130, 1981.

TOKUHASHI, S., Three dimensional analysis of a large sandy Flysch body, Mio-Pliocene Kiyosumi Formation, Boso Peninsula, Japan, *Mem. Fac. Sci. Kyoto Univ., Ser. Geol. Mineral.*, **46**, 1–60, 1979.

TOKUHASHI, S. and H. ENDO, Geology of the Anesaki District, Quadrangle Serieses scale 1: 50,000, *Geol. Surv. Japan*, 136pp, 1984.

UEDA, H., Geology of Northern Part of Boso Peninsula, *J. Geol. Soc. Japan*, **37**, 250–253, 1930.

UYEDA, S. and H. KANAMORI, Back-arc opening and the mode of subduction, *J. Geophys. Res.*, **84**, 1049–1061, 1979.

YOSHII, T., Compilation of geophysical data around the Japanese Island (I), *Bull. Earthq. Res. Inst.*, **54**, 75–117, 1979.

Formation of Active Ocean Margins, edited by N. Nasu *et al.*, pp. 109–129.

TWO TYPES OF ACTIVE MARGINS: CONVERGENT-COMPRESSIONAL MARGINS AND CONVERGENT-EXTENSIONAL MARGINS

Jean AUBOUIN, Jacques BOURGOIS, and Jacques AZEMA

Département de Géotectonique, Université Pierre et Marie, Tour 26-0—1er Etage, 4, Place Jussieu, 75230 PARIS Cédex 05, France

Abstract. Based on IPOD results it is shown that accretionary process is not necessarily associated with subduction. In Barbados trench (Leg 78A), the top of the sedimentary sequence—Upper Miocene to Present—is accreted, whereas the main part of the lower plate sedimentary cover—Cretaceous to Miocene—goes down into the subduction without accretion. In Middle America trench (Leg 67, 84), there is no accretion: the margin is built on the Central America basement cut off by normal faulting, as in passive margins. Thus, the tectonics of active margins can be either compressional or extensional, that allows to define convergent compressional margins and convergent extensional margins.

At first approach, C.C. margins seems related to island arcs and C.E. margins to cordilleran margins with their large shelves analogous to passive margins shelves; beyond Middle America trench—South of Tehuantepec—Peru-Chile trench is probably another example of C.E. margins.

However some islands arcs could have C.E. margins; for instance Japan could show the two types of active margins, C.C. margin along Nankaï trough, C.E. margin along Japan trench.

1. *Introduction*

Two major types of continental margins are distinguished in the classical view of plate tectonics, Passive and Active. Passive margins develop as a result of rifting and spreading along an ocean ridge and are marked by extensional and occasional transcurrent structure within a single plate. Active margins develop as a result of the subduction of an oceanic plate beneath a continental or island arc and are often marked by compressional and transcurrent structure along the boundary between two plates.

Compressional stress is emphasized along active margins and the evidence cited for compression are earthquake focal mechanisms and the global plate motion scheme. The concept of the *accretionary prism* (Fig. 1) is related to the view of compressional stress and considerable friction between plates. A succession of wedges accumulate at the front of the continental margin in proportion to the amount of the seafloor subducted and the corresponding amount of sediment scraped off. A gradual uplift and rotation of the imbricate stack results from continued operation of this process and tectonic slices of oceanic material are superposed in a reverse stratigraphic order, with the older at the top and the younger at the bottom of the prism (SEELY, 1979).

As the concept of an accretionary prism evolved, marine seismic data were cited to support the general model in the Middle America, Mariana and Japan trenches (KARIG, 1971; SEELY *et al.*, 1974; MOORE, WATKINS *et al.*, 1979; MOORE *et al.*, 1981), where landward dipping reflectors were proposed as evidences of imbricated oceanic slabs.

The combined study of seismic reflection records and Deep Sea Drilling Project (DSDP) cores from active margins, however, showed that the reflections are not necessarily from fault planes (MOORE *et al.*, 1981) and the interpretation of other active margin geophysical features was also questioned. The DSDP results show that the interpretations of geophysical data without supplimental geological data are very uncertain. Landward

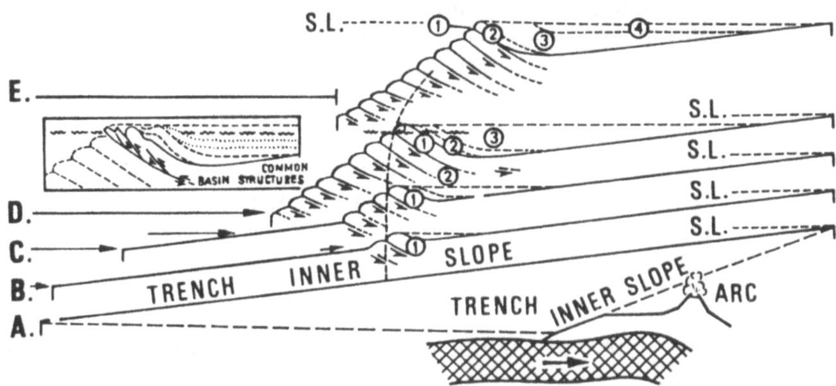

Fig. 1. The Accretionary model (after SEELY, 1979). The two main results of the accretionary process are: the general uplift in the fore-arc position; the reverse order of the tectonic wedges.

dipping reflectoes can not only be bedding planes, but also older tectonic features, or artifacts of the geophysical technique; sedimentation is very complex, and tectonic histories contain both compressional and extensional episodes. Without geological information beneath the slope deposits, which are generally of at least Miocene age, the interpretation of geophysical data only in accord with the classical accretionary model is likely to be uncertain.

2. *Summary of IPOD Results*

Active margins were drilled in the West-Pacific (Japan trench, Legs 57, 88: Mariana trench, Leg 60), in the East Pacific (Middle America trench, Legs 66, 67, 84) and the Caribbean (Barbados ridge, Leg 78A). Due to the length of the drillstring aboard Glomar Challenger (around 7000 m) only Middle America trench was drilled down to the basement of the trench floor and the continental slope (Legs 67, 84); Barbados ridge was drilled on the lower part of the inner wall (Leg 78A); whereas Japan Trench was drilled midslope and upslope on the inner wall.

Only on the Barbados margin (DSDP Leg 78A) drilling gave direct evidence of accretion (BIJU-DUVAL, MOORE *et al.*, 1981) (Fig. 2). The top (Upper Miocene to Quaternary) of the Upper Cretaceous to Quaternary oceanic sediment sequence is repeated and although the contact with the main subduction zone was not drilled, disturbances in the stratigraphic order were recovered and high pore pressures (300 psi) were measured.

Drilling on the Mexican margin (DSDP Leg 66) of the Middle America Trench (Acapulco Trench) provided indirect arguments for accretion (MOORE, WATKINS *et al.*, 1979). Coarse grained sediment was drilled on the margin, ranging from Upper Miocene to Quaternary. Quaternary sand was first recovered in the trench (Site 485) and then older (Upper Miocene) sand was drilled upslope (Site 492), and Pliocene sand was recovered at a site in between (midslope, Site 491). This disposition was interpreted as resulting from, at least, three tectonic slices in reverse stratigraphic order (MOORE, WATKINS *et al.*, 1979), as in the accretionary model. Clearly, this is only one interpretation, which implies that coarse grained sediments are always confined in trench deposits. However sand could also be trapped and ponded in basins on the margin, as shown in many places around the Pacific e.g. Peru Chile trench (THORNBURG and KULM, 1981; SCHWELLER *et al.*, 1981; SHEPHERD and MOBERLY, 1981).

Various drilling results can only be explained with models other than classical accretionary model.

Instead of uplift during subduction, subsidence was observed on the

Japan (LANGSETH *et al.*, 1981; VON HUENE *et al.*, 1982; KARIG, KAGAMI *et al.*, 1983) and Mariana Trench margins (HUSSONG and UYEDA, 1981).

Instead of very young sediment at the toe of the continental slope, old material was recovered in the Middle America Trench (Upper Certaceous and Eocene) during Leg 67 (AUBOUIN, VON HUENE *et al.*, 1979; VON HUENE, AUBOUIN *et al.*, 1980), and was subsequently recovered all along the cross section of the slope during Leg 84 (AUBOUIN, VON HUENE *et al.*, 1982). The Leg 84 results are essential (Figs. 3, 4). In four holes the same pre-early Eocene (and probably pre-Campanian) ophiolitic basement was recovered under true pre-early Eocene unconformity (two holes) and possible pre-late Cretaceous (Upper Campanian) unconformity (two holes). So, the ophiolitic basement of the Guatemalan slope belongs to the substratum of the Central America because similar ophiolites crop out in Costa Rica. There, pre-Campanian ophiolites of the Santa Elena peninsula are overthrust (AZÉMA and TOURNON, 1980; BOURGOIS *et al.*, 1982) on the Nicoya complex of Upper Jurassic-Cretaceous age (DENGO, 1973; KUIJPERS, 1980). And, as a consequence, no net accretion has occurred in the Middle America Trench off Guatemala, at least since the very beginning of the Eocene (around 60 MY B.P.), if not since the pre-Campanian (around 70 MY B.P.).

The ophiolitic complex was drilled at the toe of the slope to approximately 10–20 m above the subduction zone where high pore pressures (350 psi) caused collapse of the hole and abandonment of drilling.

In the Miocene to Quaternary sediment of the trench landward slope, sand was recovered, demonstrating that sandy facies are not restricted to the trench. The Miocene sequence at Site 568 in the midslope area is thicker and coarser than the presently ponded turbidites filling the trench axis.

Thus the combined results of Leg 67 and 84 demonstrate that: 1) landward dipping reflections in seismic records are ancient discontinuities and not features formed during the present tectonic regime (Figs. 3, 8); 2) there has been no net accretion of the sediment entering the Middle America Trench off Guatemala since at least early Eocene time and probably Late Cretaceous; and 3) the assumption that sandy sediment is a facies deposited only in the trench and not on the slope is violated in a gross way and sandy facies cannot be used as a paleonvironmental indicator. So, only Barbados (Leg 78A) and Middle-America drillings (Legs 67, 84) gave direct results on the structure of the trenches and their slopes; as, at first approach, these results appear contradictory, they have to be discussed.

3. *Guatemala Trench as a Model of Convergent-Extensional Margin*

3a. The Middle America Trench off Guatemala is as well studied as any in the world (Fig. 3, 4). It is the only trench drilled to igneous oceanic crust (Sites 499 and 500 of Leg 67) and the only landward trench slope drilled to basement in 4 places (Sites 566, 567, 569 and 570 of Leg 84); the Middle America Trench is the first, and so far, the only trench to have been precisely surveyed by SEABEAM (AUBOUIN *et al.*, 1982) and Deep Tow (MOORE *et al.*, 1982). These data allow us to propose the Guatemala Trench as a model strongly supported by facts.

3a1. The Cocos plate of pre-early Miocene age, has a horst and graben structure inherited from sea floor spreading at East Pacific rise, as shown by Cyamex diving (RANGIN et FRANCHETEAU, 1981). The horst and graben enter the trench obliquely, dividing it into a succession of diamond shaped basins separated by bottle-necks, corresponding, respectively, to graben and horst of the ocean plate (AUBOUIN *et al.*, 1982). Beyond the trench, the Cocos plate sinks into the subduction zone beneath the continental margin.

All these features appear on the seismic reflection records (LADD *et al.*, 1982), the SEABEAM (AUBOUIN *et al.*, 1982) and the Deep Tow (MOORE *et al.*, 1982) surveys. All features are confirmed by drilling (Figs. 3, 4). The oceanic crust of the Cocos plate that was drilled at the foot of the continental slope (Leg 67, Site 500), is 150 m higher than in the trench itself (Site 499), and show clear extensional features, specifically normal faults in the cores of the otherwise undisturbed Miocene chalk.

Thus, as the Cocos plate flexes down into the subduction zone, the tensile stress from plate bending causes an extensional horst and graben structure which breaks along the original plate grain inherited from the spreading ridge, rather than breaking parallel to the axes of flexure.

3a2. The landward slope is built on an ophiolitic basement which was drilled in five holes (Figs. 3, 4). In two of them, in the midslope area (Sites 569, 570), a pre-early Eocene unconformity was recovered; in two others, at the base of the slope (Sites 494, 567), an Upper Campanian-Maestrichtian limestone was drilled above the ophiolites, but the nature of the contact is questionable. Nevertheless, the slope is pre-Eocene in age, if not pre-Upper Campanian.

During Leg 84 off Guatemala (four) sedimentary sequences were recovered, which are separated by unconformities: Upper Campanian-Maestrichtian limestones; Eocene siliceous limestones and sandstones; (Upper Oligocene ?) Lower Miocene to Present mudstones and muds divided into two subsequences: (Upper Oligocene)—Middle Miocene and Upper

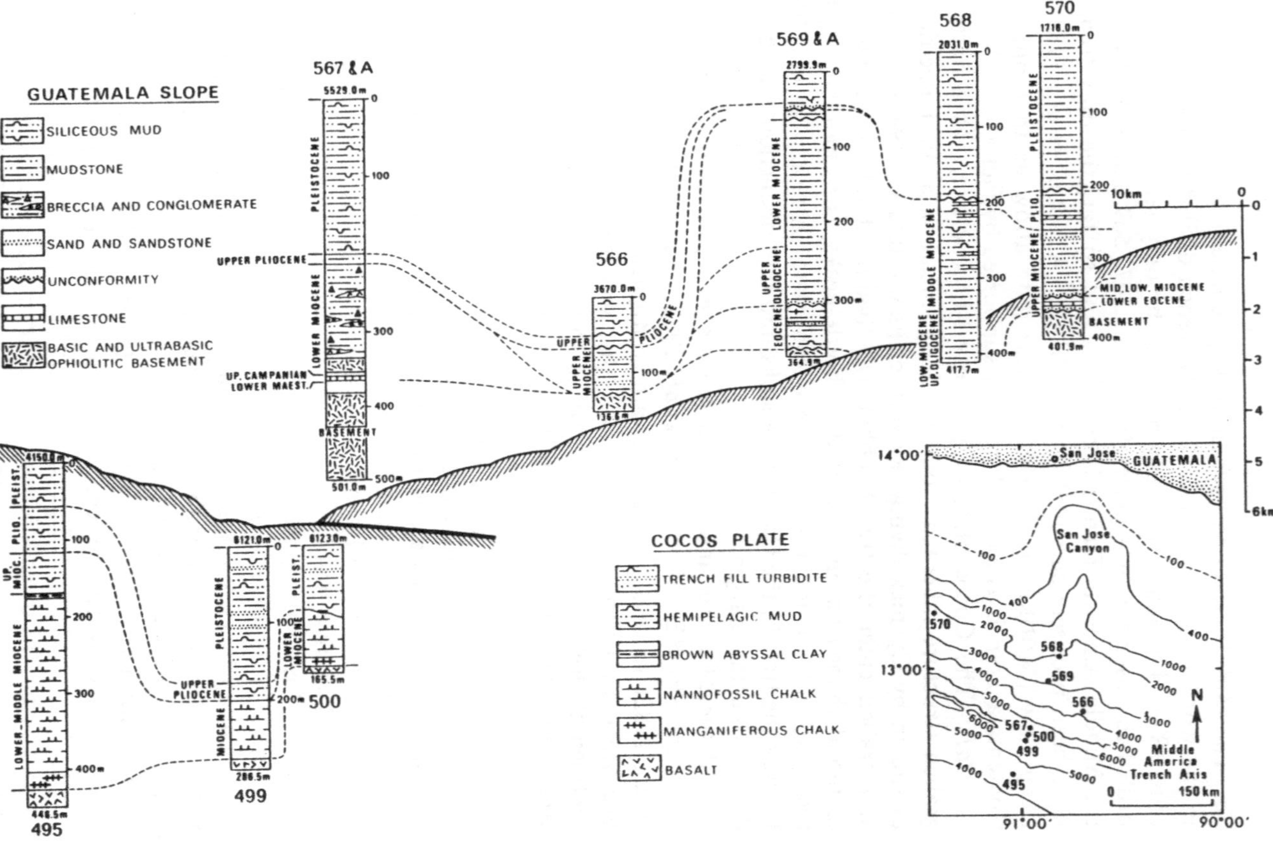
GUATEMALA SLOPE
SILICEOUS MUD
MUDSTONE
BRECCIA AND CONGLOMERATE
SAND AND SANDSTONE
UNCONFORMITY
LIMESTONE
BASIC AND ULTRABASIC OPHIOLITIC BASEMENT
567 & A
566
569 & A
568
570
PLEISTOCENE
UPPER PLIOCENE
LOWER MIOCENE
UP. CAMPANIAN LOWER MAEST.
BASEMENT
UPPER MIOCENE
PLIOCENE
EOCENE
UPPER OLIGOCENE
MIDDLE MIOCENE
LOW. MIOCENE UP. OLIGOCENE
PLIO.
MID. LOW. MIOCENE
LOWER EOCENE
10km
6km
495
499
500
PLEIST.
UP. MIOC.
LOWER-MIDDLE MIOCENE
MIOCENE
COCOS PLATE
TRENCH FILL TURBIDITE
HEMIPELAGIC MUD
BROWN ABYSSAL CLAY
NANNOFOSSIL CHALK
MANGANIFEROUS CHALK
BASALT
San Jose
GUATEMALA
San Jose Canyon
N
Middle America Trench Axis
150 km
14°00'
13°00'
91°00'
90°00'

Fig. 3. Simplified stratigraphic columns summarizing the results of Legs 67 and 84 (after AUBOUIN, VON HUENE *et al.*, 1983). Note that the oceanic plate, the trench floor, the continental margin were drilled to their basement. The main fact is the presence of the pre-Cretaceous basement everywhere on the slope, namely at the very toe (Site 494, 567) that disagree with the accretionary model (Fig. 1).

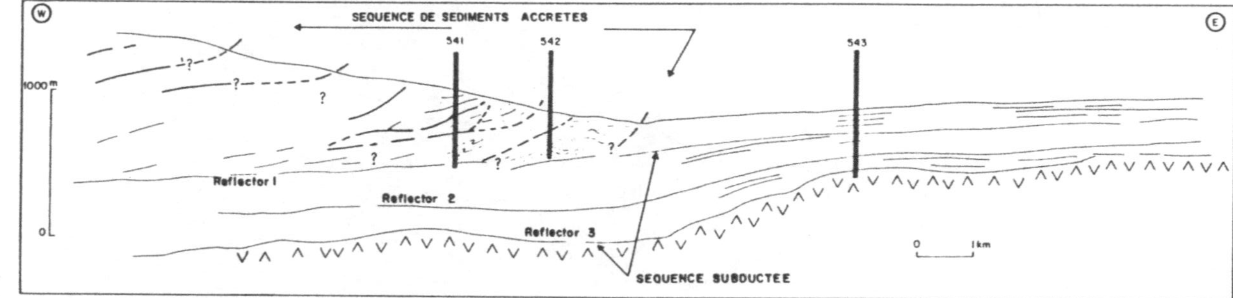

Fig. 2. Cross section of Barbados ridge along the Leg 78A transect (after BIJU-DUVAL, CASEY, MOORE *et al.*, 1981). Note that only the top of the sedimentary cover (Upper Miocene to Present) is accreted. The main part of the cover (Upper Cretaceous to Middle Miocene) pass into the subduction zone with the oceanic plate.

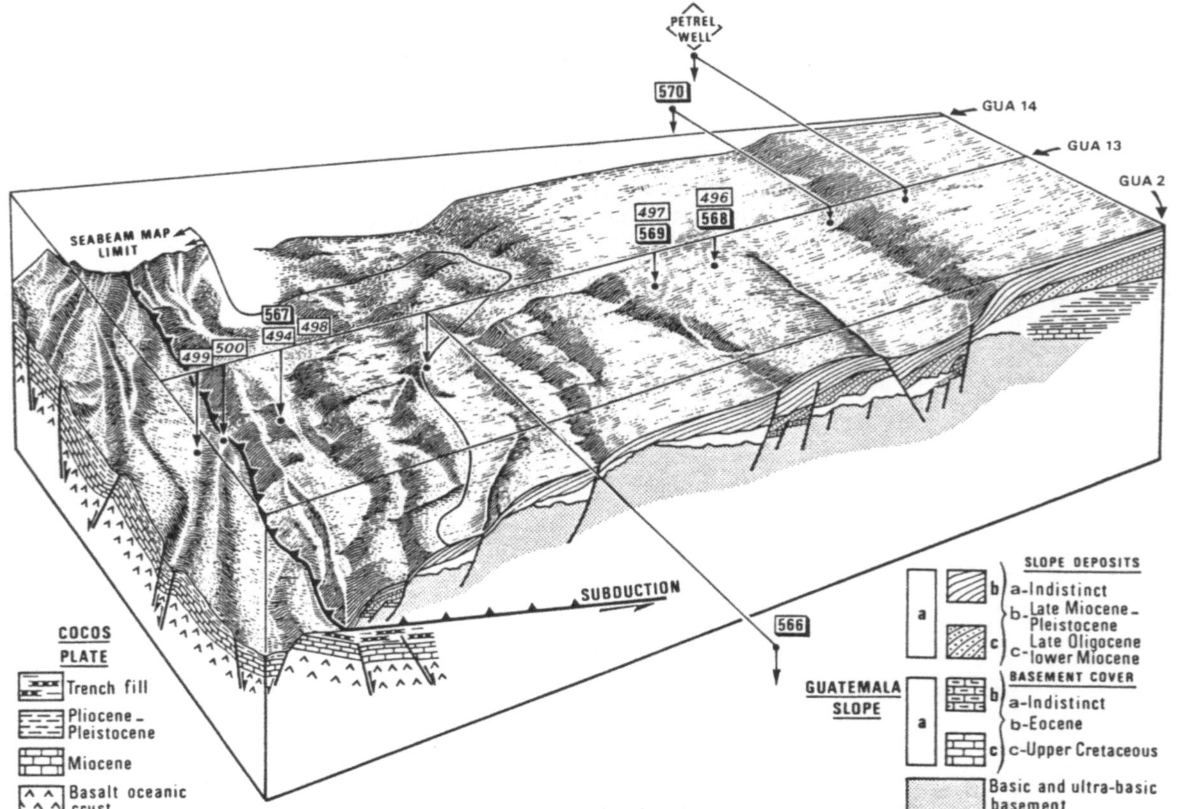

Fig. 4. Block diagram showing the convergent zone between Cocos Plate and Caribbean (Central America) Plate in Middle America Trench off Guatemala (after AUBOUIN, VON HUENE *et al.*, 1982). This block diagram includes all the data of seismic, Seabeam, Deep Tow surveys and Legs 67, 84 drilling results.

Miocene to Present. Upper Cretaceous and Eocene represent the sedimentary cover of the basement; (Upper Oligocene ?) Lower Miocene to Present are the slope deposits. As a consequence, the history of Guatemala margin can be divided into (at least) four episodes:

a pre-Upper Cretaceous formation of the ophiolitic basement;

a strong Paleocene uplift documented by the direct unconformity of early Eocene upon the ophiolitic basement in the midslope (Sites 569, 570);

a strong (Upper Oligocene)—Lower Miocene uplift at the shelf-slope break accompanied by a first large downfaulting of the slope; during this episode the slope took its present situation;

a second downfaulting episode from Pliocene to Present, especially in downslope area; the details of the present slope structure are related to this episode.

All the pre (Upper Oligocene) Lower Miocene structure in the basement and its Upper Cretaceous-Eocene sedimentary cover are unrelated with the present Middle America Arc and Trench which was born in (Upper Oligocene) Lower Miocene time. This can be related with the fact that Middle America Trench cut off all the structures of the Mexican Sierra Madre (AUBOUIN, 1974); and that the first substantial layers of volcanic ashes, indicating development of the present Arc-Trench system, appears only in the (Upper Oligocene)—Lower Miocene, at the base of the slope deposits.

The geological history of Middle America margin is polyphased; only the features related to (Upper Oligocene) Lower Miocene to Present history are related to the present framework of subduction along Middle America trench. This conclusion has to be taken in account when interpreting seismic records; namely, a part of the features must be related to older tectonic episodes without any relationship with the present Arc-Trench system.

It is therefore necessary to look for on land datas as keys for interpreting the Guatemala margin pre-Miocene structure and history.

3b. Costa Rica as a key to the structure of Middle America Trench off Guatemala

Middle America trench is divided in two parts by the Tehuantepec ridge (Fig. 5): on the northern side, Acapulco trench runs directly along the Mexican coast; on the southern side, a large shelf characterizes the Guatemala margin.

3b1. Guatemala shelf is underlain by a thick sedimentary sequence of Upper Cretaceous to Present age, interpreted as sedimented in a fore arc basin. This sequence, drilled in the Exxon Petrel well off Guatemala

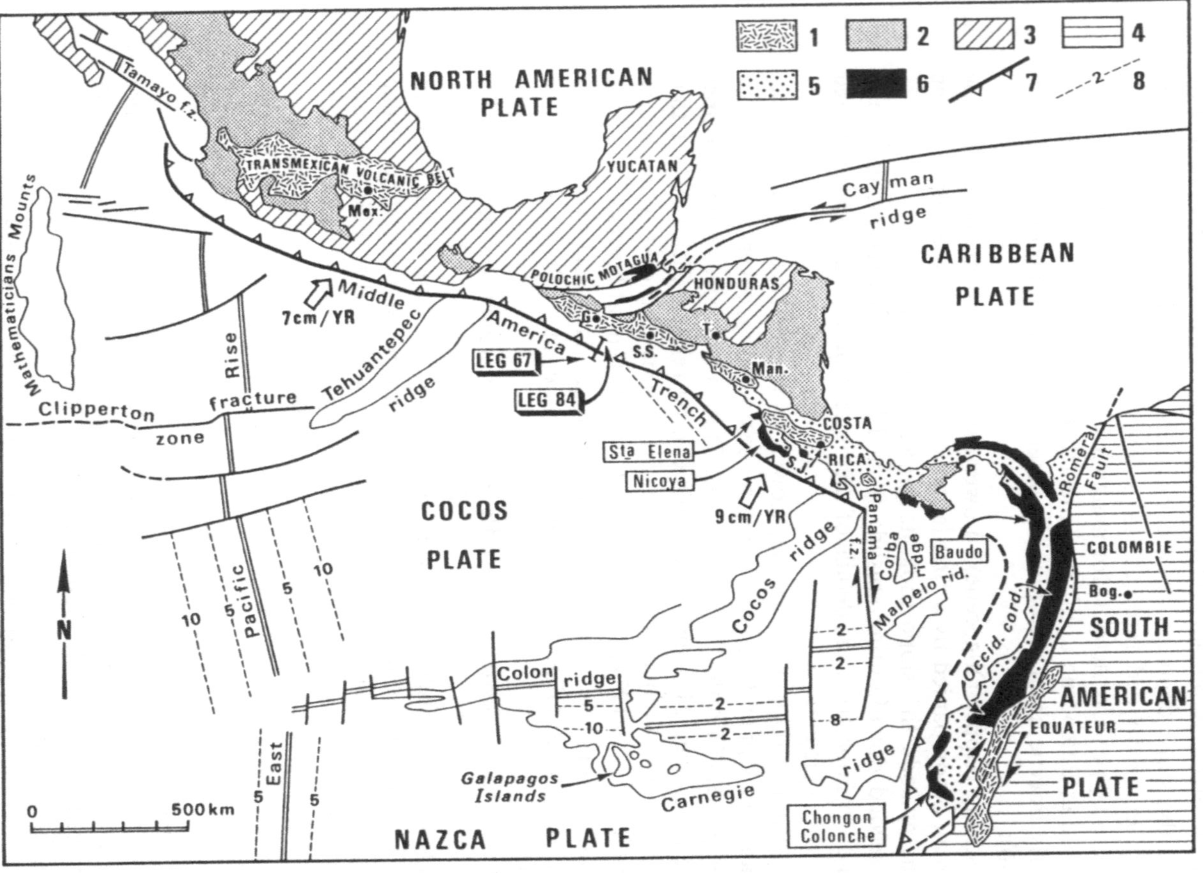

Fig. 5. Tectonic setting of the Santa Elena and Nicoya Peninsulas (Costa Rica) and of Leg 67 and 84 off Guatemala. Present-day plate motions from MINSTER and JORDAN (1978). 1, Pliocene and Pleistocene volcanism; 2, Oligocene and Miocene; 3, North American plate; 4, South American plate; 5, Cenozoic formations of ophiolitic Andes and southern Central America; 6, Mesozoic and Cenozoic ophiolitic complexes; 7, subduction zones; 8, magnetic anomalies.

(KARIG *et al.*, 1978), showed the same unconformities as recorded on the Guatemala slope (Upper Campanian, Early Eocene, Upper Oligocene to Lower Miocene, Upper Miocene to Lower Pliocene unconformities) but with some peculiarities: the thickness of the sediments; their clastic facies, either they are deep sea deposits as Upper Cretaceous to Eocene sequences, or shelf facies as Miocene to Present series; the presence of a thick paleocene sequence (thousands of meters) which is missing on the slope where Early Eocene rests directly on the ophiolitic basement.

Guatemala shelf and Guatemala slope have suffered the same tectonics episodes: pre-Upper Campanian, pre-Early Eocene that we know unrelated to the present Arc-Trench system; pre-(Upper Oligocene) Lower Miocene and pre-Upper Miocene—Lower Pliocene, that are related to the present Arc-Trench system. The specific history of the shelf basin is marked by a general subsidence on the shelf and general uplifting at the shelf break.

Guatemala shelf runs along Central America up to Costa Rica where Santa Elena and Nicoya peninsulas merge above sea level. The alineated magnetic anomalies running along the shelf from Tehuantepec to Nicoya peninsula has been considered as an axial outcrop of the shelf basement and so, a key for interpretation of Guatemala margin.

3b2. The pre-Upper Cretaceous basement of Costa Rica crops out in the Santa Elena and Nicoya Peninsulas. From south to north the basement includes (DENGO, 1973; AZEMA and TOURNON, 1980; KUIJPERS, 1980; BOURGOIS *et al.*, 1982): the Esperanza, the Matapalo and the Santa Elena units. The Esperanza unit is Albian to Santonian in age and consists mainly of pillow basalt and massive basalt flows. The Matapalo unit includes Callovian to Cenomanian radiolarites and exhibits a massive basalt flow, and dolerite basement. The Santa Elena unit contains ultramafic and mafic rocks in which harzburgite is the major component.

The most important tectonic features of the Nicoya Complex are (BOURGOIS *et al.*, 1984) the large Santa Elena and Matapalo nappes (Fig. 6). Nappe emplacement was from north to south during upper Santonian time.

The sedimentary cover of the Nicoya Complex comprises: 1) the upper Campanian-Maestrichtian Sabana Grande formation that consists of shallow water and non-marine sediments to the north, and of deep water sediment to the south, 2) the Paleocene sediment indicates deposition in a deep water environment, and comprises the Rivas (p.p.), Las Palmas and Samara formations, 3) the post Paleocene (?) sediment consists of the Barra Honda and Montezuma formations of Middle to Upper Eocene age. Two unconformities are the important geological features of the Late

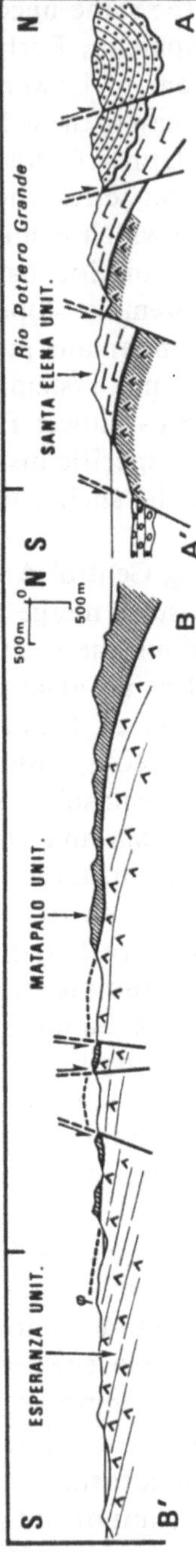

Fig. 6. Tectonic cross section of Santa Elena and Nicoya peninsulas (after BOURGOIS *et al.*, 1984). AA′, located in 1, Fig. 7A; BB′, located in 3, Fig. 7A.

Upper Cretaceous Lower Eocene history of Costa Rica. The lower is at the base of the Sabana Grande formation and marks a major tectonic episode for the Mesozoic formations; the upper unconformity at the base of the Barra Honda and Montezuma formations is not as prominent as the lower one.

Basement and sedimentary cover are disrupted by normal faulting occurring in two stages (Fig. 7).

3b3. It is easy to correlate the sedimentary cover of Nicoya and Santa Elena peninsulas with the sedimentary sequence of Guatemala shelf and to compare the tectonic history with Guatemala shelf and slope where the same episodes occur. The Nicoya and Santa Elena pre-Upper Campanian ophiolitic basement can therefore be related to the Guatemala margin pre-Upper Campanian ophiolitic basement.

The basement of Guatemala margin, shelf and slope, is the basement of Central America. Since it is present at the very toe of the slope (site 494, 567), *there was no accretion along Guatemala trench since Late Cretaceous.*

Three features must be used for interpreting the seismic records: layering as usual, faulting as evident, and old (pre Campanian) thrusting that we know unrelated with the present Arc-Trench system. Figure 8B is a first attempt to such an interpretation with respect to the geophysical analysis of the profile given on Fig. 8A. *Most of the landward dipping reflectors are old tectonic thrust planes; they are not, at all, accretionary contacts in recent sediments.*

3c. *The present status of Guatemala Arc and Trench system: the model of a convergent extensional margin*

The present Guatemala trench initiated in Late Oligocene—Lower Miocene time by a first normal faulting; a second episode of normal faulting occurred from Upper Miocene to Present time, that appears clearly on the Seabeam survey (AUBOUIN *et al.*, 1982) and the seismic profiles (LADD *et al.*, 1982).

From the beginning there was no accretionary process working along Guatemala trench. On the contrary, Cocos plate fell obliquely into subduction, reactivating its own normal fault network; the continental slope accompanied the downgoing of Cocos plate by a normal downfaulting (AUBOUIN, VON HUENE *et al.*, 1982; AUBOUIN *et al.*, 1982). Thus, along this part off Guatemala, the trench has the morphology and the structure of a graben.

So we propose (AUBOUIN, BOURGOIS *et al.*, 1984) *the Guatemala trench as a model of Convergent Extensional Active Margin (CE Active Margin) by opposition to the classical Convergent Compressional Active*

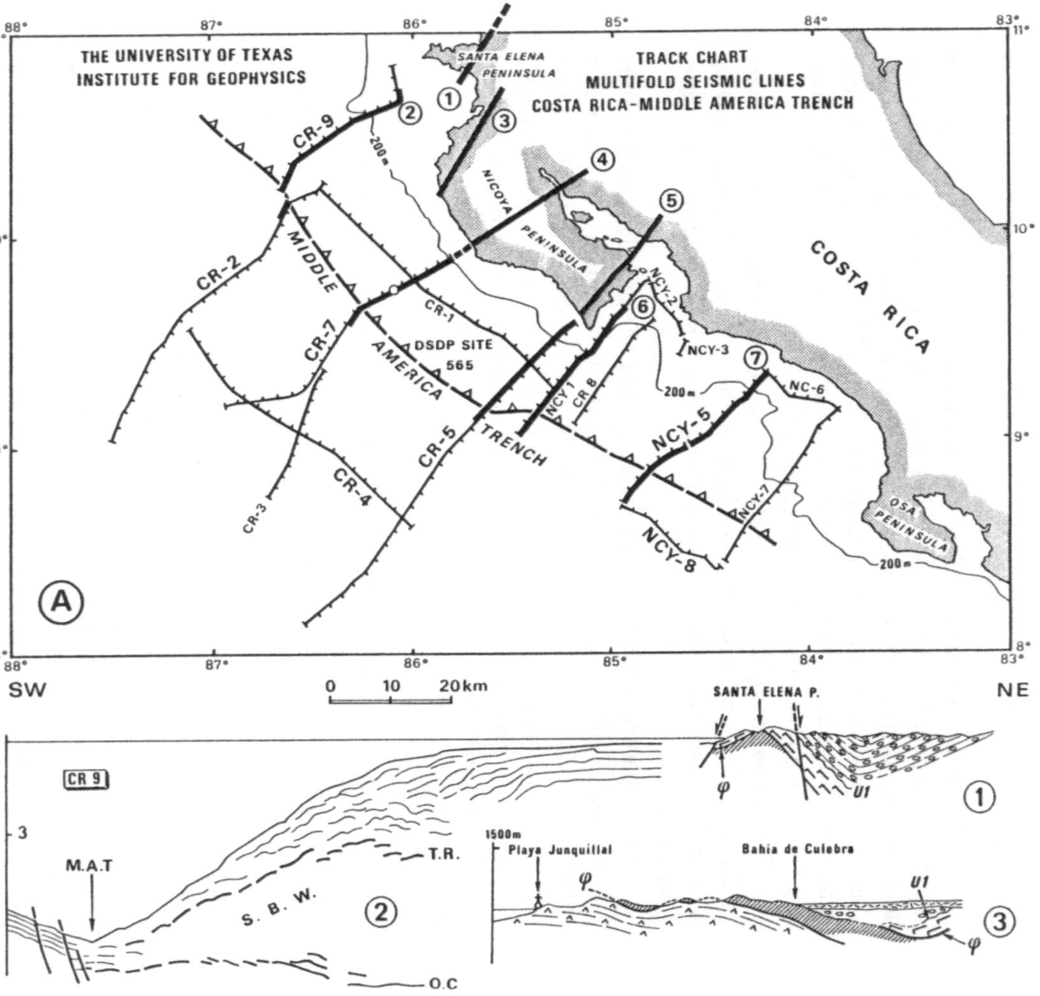

THE UNIVERSITY OF TEXAS
INSTITUTE FOR GEOPHYSICS
TRACK CHART
MULTIFOLD SEISMIC LINES
COSTA RICA-MIDDLE AMERICA TRENCH
SANTA ELENA PENINSULA
NICOYA PENINSULA
COSTA RICA
OSA PENINSULA
MIDDLE AMERICA TRENCH
DSDP SITE 565
CR-1
CR-2
CR-3
CR-4
CR-5
CR-7
CR 8
CR-9
NCY 1
NCY-2
NCY-3
NCY-5
NCY-7
NCY-8
NC-6
200 m
A
SW
NE
0 10 20km
SANTA ELENA P.
U1
CR 9
3
M.A.T
T.R.
S. B. W.
O.C
1500m
Playa Junquillal
Bahia de Culebra

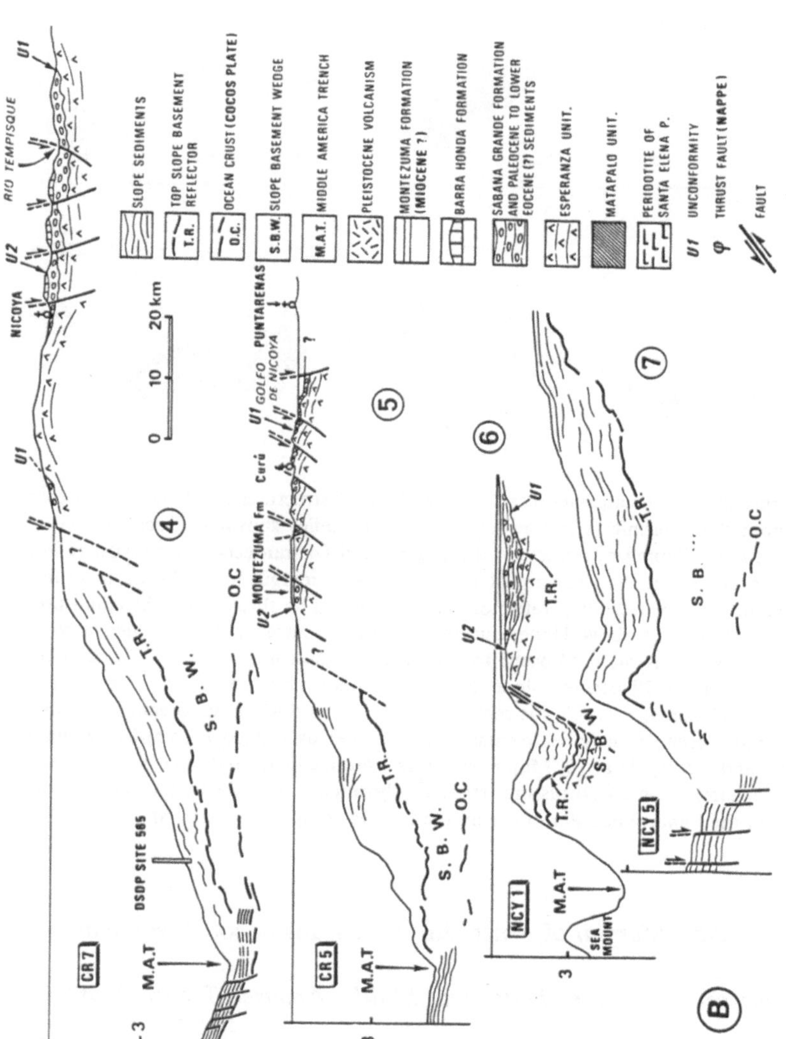

Fig. 7. Generalized profiles from Middle America trench to Santa Elena peninsula (after BOURGOIS *et al.*, 1984) (offshore data from Buffler, in press, and Crowe and Buffler, in press). A. Location of the profiles. B. Synthetized cross sections, with on-land and offshore data.

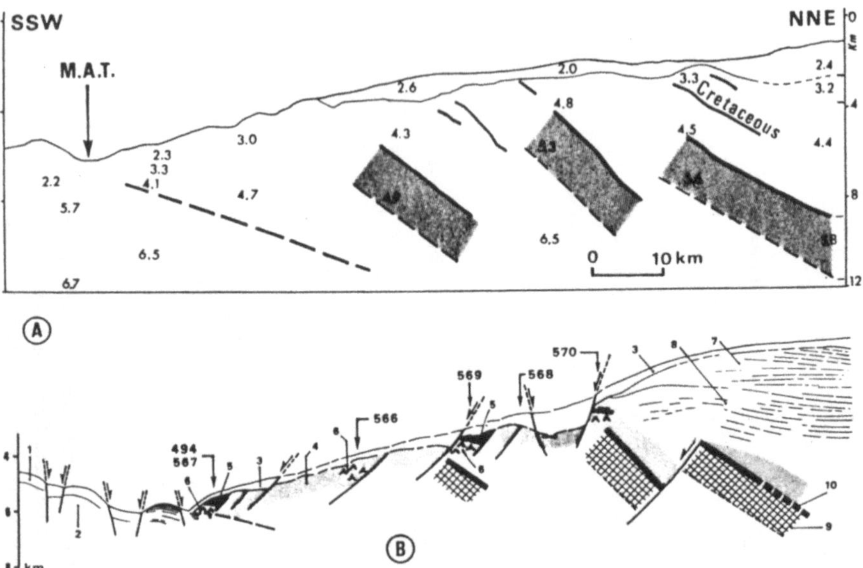

Fig. 8. Interpretation of Guatemala margin, in the light of structures cropping out in Costa Rica. A. Summary of seismic reflection and refraction studies showing velocity structure and major landward dipping reflectors on the Leg 67 and 84 transects (after LADD *et al.*, 1982). B. Interpretation of the structure of the Guatemalan margin (after AUBOUIN *et al.*, 1982). The major landward dipping reflectors are interpreted as thrust faults bringing pre-Campanian ophiolitic slabs above Upper Jurassic—Upper Cretaceous oceanic Nicoya complex as on land in Costa Rica. 1, Miocene to Pleistocene sedimentary cover of Cocos plate; 2, Basalts; 3, Pliocene to Pleistocene slope deposits; 4, Acoustic basement; 5, Upper Senonian to Miocene sediments; 6, Ophiolitic rocks; 7, Oligocene and Miocene forearc sediments; 8, Paleocene and Eocene forearc sediments; 9, Interpreted as a Nicoya complex remnant from magnetic and gravity data; 10, Major reflectors from magnetic and gravity datas interpreted as a pre-Campanian overthrusting: from on-shore data it could be the same as the overthrusting of the Santa Elena Peridotite on the Nicoya complex (see Fig. 6).

Margin (CC Active Margin) of which Barbados transect could be the model.

4. *Comparison between Barbados and Middle America Trench Transects*

Following Legs 67, 78A, 84 data, it appears that subduction gives two superficial responses:

a. as in Barbados ridge, the top of the sedimentary cover of the Lower plate (Upper Miocene to Quaternary) is detached and repeated by tectonic wedges that build an accretionary prism. The main part of the sedimentary cover (Upper Cretaceous to Middle Miocene) is bound up with the oceanic crust and pass into the subduction. Nothing is known of a possible underplating. Detachement plane plays the main role in the accretionary process.

Barbados, where such a tectonic repetition was drilled can be proposed as the model of Convergent Compressional Active Margin (CC Active Margin). Java, Nankaï and Aleutian trenches could be of the same type.

b. as in Middle America trench, the total sedimentary cover is bound up with the oceanic crust and pass into subduction. Together with the diving of the lower plate, the continental margin breaks up and collapses by normal faulting. Nothing is known of a possible underplating. Normal faulting play the main role in the non accretionary process.

Middle America trench where such a structure was proved can be proposed as the model of Convergent Extensional Active Margin (CE Active Margin).

At first approach *Cordilleran Margins*, with their large shelves, could belong to this type, as the Peru-Chile margin where normal faulting was recorded by seismic profiling (THORNBURG and KULM, 1981; SCHWELLER *et al.*, 1981; SHEPHERD and MOBERLY, 1981) (Fig. 9). Although the processes are different, *such margins have a structure alike to passive margins* with their tilted blocks and their large shelf basins.

Convergent Extensional Active Margins are more general than Cordilleran margins, since Mariana (HUSSONG and UYEDA, 1981) and Japan trenches (VON HUENE *et al.*, 1982) margins could belong to this type.

c. Beyond the differences, there are similarities between the two types: the subduction of the oceanic crust with its sedimentary cover correspond, in the two cases, to a crustal shortening with related seismicity and volcanism;

the decoupling between the lower and upper plate is related to high pore pressures (300 to 350 psi) that caused, in the two cases, failure of the drilling just above the main subduction contact (VON HUENE and LEE, 1984).

The main difference can be in the sediment thickness on the lower plate. With a great thickness of sediments (over 2 Km) accretionary process is working. Along the Barbados ridge the accretionary style changes from South to North in relation with sediment sources on the South American continent (Amazon and Orinoco deep sea fans).

Tectonics in the sedimentary cover of the lower plate is probably dependant on thickness, stratigraphic position and frequency of undercompacted

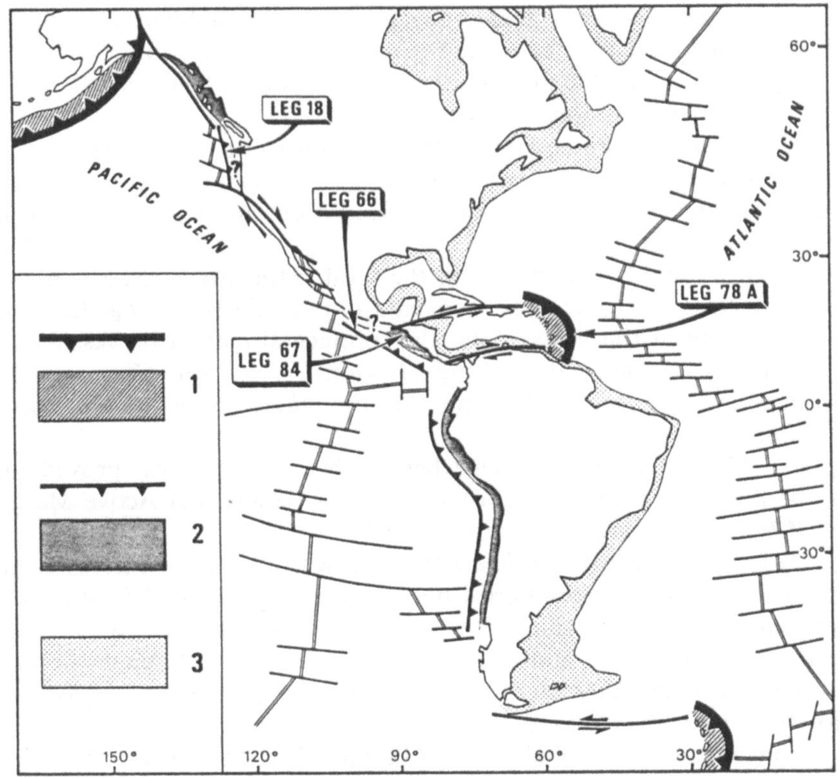

Fig. 9. Convergent extensional margins in the Eastern Pacific (after AUBOUIN *et al.*, 1983). 1. Convergent Compressive Active Margins (i.e.: C. C. Active Margins); 2. Convergent Extensional Active Margins (i.e.: C. E. Active Margins); 3. Passive Margins.

layers that are the possible detachement planes. Tectonic in the basement is certainly controlled by the nature, the age of the crust of the two plates, the rate of convergence and the dipping of subduction plane.

Finally, the present morphology of a trench system reflects the history of the subduction which could have changed during geological time as observed along Guatemala margin.

REFERENCES

AUBOUIN, J., Mer Caraïbe et mer Méditerranée: réflexion sur une comparaison structurale, *7ème Conf. Caraïbe*, Pointe à Pitre, BRGM éd. Orléans, pp. 41–49, 1974.

AUBOUIN, J., J. BOURGOIS, and J. AZÉMA, A new type of active margin: the convergent-extensional margin, as examplified by Middle America trench off Guatemala, *Earth*

Planet. Sci. Letter, **67**, 211–218, 1984.

AUBOUIN, J., J. F. STEPHAN, J. ROUMP, and V. RENARD, The Middle America Trench as an example of a subduction zone, *Tectonophysics*, **86**, 113–132, 1982.

AUBOUIN, J., R. VON HUENE, J. AZÉMA, W. T. COULBOURN, D. S. COWAN, J. A. CURIALE, C. A. DENGO, R. W. FAAS, W. HARRISON, R. HESSE, J. W. LADD, N. MUZILEV, T. SHIKI, P. R. THOMPSON, and J. WESTBERG, Premiers résultats des forages profonds dans le Pacifique au niveau de la fosse du Guatemala (fosse d'Amérique Centrale)(Leg 67 du "Deep Sea Drilling Project": mai-juin 1979), *Comptes Rendus de l'Académie des Sciences*, Paris, série D, t. 289, 1215–1220, 1979.

AUBOUIN, J., R. VON HUENE, M. BALTUCK, R. ARNOTT, J. BOURGOIS, M. FILEWICZ, R. HELM, K. KVENVOLDEN, B. LIENERT, T. MCDONALD, K. MCDOUGALL, Y. OGAWA, E. TAYLOR, and B. WINSBOROUGH, Leg 84 of the Deep Sea Drilling Project; Subduction without accretion: Middle America Trench off Guatemala, *Nature*, **297**, 548–560, 1982.

AZÉMA, J. and J. TOURNON, La péninsule de Santa Elena, Costa Rica: un massif ultrabasique charrié en marge pacifique de l'Amérique Centrale, *Comptes-Rendus de l'Académie des Sciences*, Paris, série D, t. **290**, 9–12, 1980.

BIJU-DUVAL, B., J. C. MOORE, G. BLACKINTON, J. A. BERGEN, G. E. CLAYPOOL, D. S. COWAN, R. T. GUERRA, C. H. J. HEMLEBEN, M. S. MARLOW, J. H. NATLAND, C. J. PUDSEY, G. W. RENZ, M. TARDY, M. E. WILLIS, D. WILSON, and A. A. WRIGHT, Premiers résultats des forages IPOD implantés lors de la croisière 78A du Glomar Challenger au NE de la ride de la Barbade (arc des Petites Antilles): tectonique frontale d'un prisme d'accrétion, *Comptes Rendus de l'Académie des Sciences*, Paris, série II, t. **293**, 621–628, 1981.

BOURGOIS, J., J. AZÉMA, J. TOURNON, H. BELLON, B. CALLE, E. PARRA, J. F. TOUSSAINT, G. GLAÇON, H. FEINBERG, P. DE WEVER, and I. ORIGLIA, Ages et Structures des complexes basiques et ultrabasiques de la façade pacifique entre 3°N et 12°N (Colombie, Panama et Costa Rica), *Bulletin de la Société Géologique de France*, **3**, 545–554, 1982.

BOURGOIS, J., J. AZÉMA, P. BAUMGARTNER, J. TOURNON, A. DESMET, and J. AUBOUIN, The geologic history of the Caribbean—Cocos plate boundary with special reference to the Nicoya ophiolite complex (Costa Rica) and DSDP results (Legs 67 and 84 off Guatemala): a synthesis, *Tectonophysics*, **108**, 1–32, 1984.

DENGO, G., Estructura geologica, historia tectonica y morfologia de America central, Instituto Centroamericano de Investigacion y Tecnologia Industrial (ICAITI), Mexico, Buenos Aires (Centro Regional de Ayuda Tecnica), 2ème éd., 52 pp., 1973.

HUSSONG, D. M. and S. UYEDA, Tectonics in the Mariana arc: results of recent studies, including D.S.D.P. Leg 60, in R. Blanchet et L.-Montadert, Geologie des marges continentales, *26ème Int. Geol. Cong.*, Paris, Oceanologica acta, n°sp., pp. 203–212, 1981.

KARIG, D. E., Structural history of the Mariana Island arc system, *Geological Society of America Bulletin*, **82**, 323–344, 1971.

KARIG, D. E., R. K. CAROWELL, G. F. MOORE, and D. G. MOORE, Late Cenozoic subduction and continental margin truncation along the Northern Middle America trench, *Geological Society of America Bulletin*, **89**, 265–276, 1978.

KARIG, D. E., H. KAGAMI, F. AKIBA, C. J. BRAY, J. P. CADET, J. CHARVET, W. COULBOURN, K. FUJIOKA, H. KINOSHITA, M. LAGOE, T. H. LANG, J. K. LEGGETT, G. A. LOMBARI, N. LUNDBERG, T. MACHIHARA, R. MATSUMOTO, P. MUKHOPADHYAY, N. NIITSUMA, A. J. SMITH, C. L. STEIN, and A. TAIRA, Varied responses to subduction in Nankaï Trough and Japan Trench forearcs, *Nature*,

148–151, 1983.

KUIJPERS, E., The geologic history of the Nicoya ophiolite complex, Costa Rica, and its geotectonic significance, *Tectonophysics*, **68**, 233–255.

LADD, J. W., A. K. IBRAHIM, K. J. MCMILLEN, G. V. LATHAM, and R. VON HUENE, Interprétation of seismic reflection data of the Middle America Trench offshore Guatemala, in J. Aubouin, R. von Huene *et al.*, Init. Repts. DSDP Leg 67, pp. 675–689, U.S. Govt. Printing Office, Washington, D.C., 1982.

LANGSETH, M. G., R. VON HUENE, N. NASU, and H. OKADA, Subsidence of the Japan Trench forearc region of Northern Honshu, in R. Blanchet et L. Montadert, Geologie des marges continentales, *26ème Int. Geol. Congr.*, Paris, Océanologica acta, n°sp., pp. 173–179, 1981.

MOORE, J. C., J. C. WATKINS, T. H. SHIPLEY, S. B. BACHMAN, F. W. BETHTEL, A. BUTT, B. M. DIDYK, J. K. LEGGETT, N. LUNDBERG, K. J. MCMILLEN, N. NIITSUMA, L. E. SHEPARD, J. F. STEPHAN, and H. STRADNER, Progressive accretion in the Middle America Trench, Southern Mexico, *Nature*, **281**, 638–642, 1979.

MOORE, J. C., J. C. WATKINS, and T. H. SHIPLEY, Summary of accretionary processes, Deep Sea Drilling Project, Leg 66: offscraping underplating and deformation of the slope apron, in J. S. Watkins, J. C. Moore *et al.*, Init. Rept. DSDP, 66, pp. 825–836, U.S. Govt. Printing Office, Washington, D.C., 1981.

MOORE, G. F., P. LONSDALE, and R. VON HUENE, Nearbottom observations of the Middle America Trench off Guatemala, in J. Aubouin, R. von Huene *et al.*, Init. Rept. DSDP Leg 67, pp. 707–718, U.S. Govt. Printing Office, Washington, D.C., 1982.

RANGIN, C. and J. FRANCHETEAU, Fine scale morphological and structural analysis of the East Pacific Rise, 21°N (Rita project), in X. Le Pichon, J. Debyser, F. Vine éd., Géologie des Océans, *26ème Int. Geol. Congr.*, Paris, Oceanologica acta, n°sp, pp. 15–24, 1981.

SCHWELLER, W. J., L. D. KULM, and R. A. PRICE, Tectonics, structure and sedimentary framework of the Peru-Chile Trench, in L. D. Kulm, J. Dymond, E. J. Dasch, D. M. Hussong ed., Nazca plate: crustal formation and Andean convergence, *Geological Society of America, Memoir*, **154**, 323–349, 1981.

SEELY, D. R., The evolution of structural highs bordering major forearc basins, in Geological and geophysical investigation of continental margins, édited by J. S. Watkins *et al.*, *Am. Assoc. Petrol. Geol. Mem.*, **29**, Tulsa, Oklahoma, 245–260, 1979.

SEELY, D. R., P. R. VAIL, and G. G. WALTON, Trench slope model, in *Burk and Drake*, ed., the Geology of continental margins, pp. 248–260, Springer Verlag, New York, 1974.

SHEPHERD, G. L. and R. MOBERLY, Coastal structure of the continental margin, northwest Peru and southwest Ecuador, in L. D. Kulm, J. Dymond, E. J. Dasch, D. M. Hussong ed., Nazca Plate: Crustal formation and Andean convergence, *Geological Society of America, Memoir*, **154**, 351–391, 1981.

THORNBURG, T. and L. D. KULM, Sedimentary basins of the Peru continental margin: structure, stratigraphy and cenozoic tectonics from 6°S to 16°S latitude, in L. D. Kulm, J. Dymond, E. J. Dasch, D. M. Hussong ed., Nazca Plate: Crustal formation and Andean convergence, *Geological Society of America, Memoir*, **154**, 393–422, 1981.

VON HUENE, R., J. AUBOUIN, J. AZÉMA, G. BLACKINTON, J. A. CARTER, W. T. COULBOURN, D. S. COWAN, J. A. CURIALE, C. A. DENGO, R. W. FAAS, W. HARRISON, R. HESSE, D. M. HUSSONG, J. N. LADD, N. MUZYLOV, T. SHIKI, P. R. THOMPSON, and J. WESTBERG, Leg 67: the Deep Sea Drilling Project Mid-America Trench transect off Guatemala, *Geological Society of America Bulletin*, **91**, 421–432,

1980.

von Huene, R., M. Langseth, N. Nasu, and O. Okada, Summary of Cenozoic tectonic history along the IPOD Japan Trench transect, *Geological Society of America Bulletin*, **93**, 829–846, 1982.

von Huene, R. and H. Lee, The possible significance of pore fluid pressures in subduction zones, in Watkins, Drake, ed., *American Association Petroleum Geologists Memoir*, **34**, 781–791, 1984.

Formation of Active Ocean Margins, edited by N. Nasu *et al.*, pp. 131–151.

THE AGE AND MODE OF KINEMATICS OF THE FORMATION OF FOREARC BASINS AND TRENCH, JAPAN TRENCH REGION

Shozaburo NAGUMO

Earthquake Research Institute, University of Tokyo, Tokyo, Japan

Abstract. Volcanoes, forearc basins, a trench, and deep seismic planes are the typical tectonic features of the forearc region of the arc-trench system. This paper presents a view that the age and the mode of kinematics of the formation of the structure can be different between the forearc basin and the trench. IPOD drillings and their site-surveys revealed that the Japan Trench was formed mainly during the Quaternary while its forearc basin was formed during the Paleogene. The kinematics of the formation of the forearc basin is probably a folding motion associated with the uplifting motion of the island arc, while that of the trench inner slope is probably a overthrusting motion associated with the deep seismic plane. The kinematics of the formation of the forearc basin can be described by the Biot's viscous-fluid-thin-layer folding model. The kinematics of the formation of the trench slope break and the inner slope can be described by an overthrusting motion of the island arc mass along the deep seismic plane over the subducting oceanic plate.

1. *Introduction*

An alignment of arc-volcanoes, forearc basins, a trench and deep seismic planes is a typical tectonic feature of the arc-trench system (DICKINSON and SEELY, 1979). One may regard this tectonic feature as a product of continuous steady-state subduction since Cretaceous to Quaternary, as being anticipated by the consumption of an oceanic plate (SEELY, 1979). Most arc-volcanoes which we see today are, however, Quaternary volcanoes, and the shape of islands and the heights of the mountain ranges are also products of Quaternary deformations (GEOLOGICAL SURVEY OF JAPAN, 1977, chap. 23). IPOD drillings and its site-surveys revealed that the present Japan Trench was formed during the periods from Pliocene to Quaternary, while the erosion of Oyashio-Paleo-land and the formation of forearc basin occurred during Paleogene time (ISHIWADA *et al.*, 1977; VON HUENE *et al.*, 1980; NAGUMO, 1980). These findings have lead people to think that the occurrence of subduction in Japan Trench region is episodic rather than steady (VON HUENE *et al.*, 1980), or, more drastically, an event of Quaternary (NAGUMO, 1980).

The recognition of the absence of subduction motion in a long history of tectonics and also the recognition of the associated thick sedimentation environment lead the writer to think that the Japan Trench may have behaved as a passive margin during the period of the absence of subduction. It also leads him to think that transition from an active margin to a passive margin or vice-versa could have occurred in the same region. In other words, the present subduction region is not necessarily a subduction region throughout the long geologic history since Cretaceous. Such interpretation like this is different from the previous ones (SEELY, 1979; KARIG and SHARMAN, 1975) for the tectonics in the forearc region.

Thus, the purpose of this paper is to present a view that the age and mode of kinematics of the formation of the structure can be different between forearc basin and trench. In other words, even though the line-up of forearc basin and trench exhibits a feature of the present-day arc-trench system, their formations are not contemporaneous. First, in Section 2, I present evidences which lead to such a view, and secondly in Sections 3 and 4, I present kinematical models which account for such tectonics. As a supplement, a tectonic history of the Japan Trench forearc region is briefly outlined in Section 5. In Section 6, some discussions are presented for the differences of interpretations about the subduction history in the Japan Trench region between VON HUENE *et al.* (1982) and the present paper.

The data used in this paper are not new. They are all cited from already published papers. Some of the important interpretations used in this paper are also not new. They are already published or personally communicated. What the writer did is only to bundle them in his responsibility into a scheme of an understanding.

2. *Japan Trench*

In this section, I will show several evidences which show that (1) the present-day bathymetric feature of the Japan Trench was formed mainly during Quaternary, and (2) the forearc basin under the deep-sea terrace was formed during Paleogene.

2.1 Trench formation

The bathymetric features of the oceanic trench are characterized by trench slope break, inner (landward) slope, mid-slope terrace, trench axis including trench floor, and outer (seaward) slope (DICKINSON and SEELY, 1979). The evidences which show that these bathymetric features were formed during Quaternary will be seen at the trench slope break and the inner slope.

The first evidence is seen at the inner slope (Fig. 1). As stated clearly by ISHIWADA *et al.* (1977), the thick uniform Neogene (Pliocene and Miocene) strata cover the upper half of the inner slope region, and are abruptly truncated or thin out at the mid-slope terrace. The feature similar to this is seen on the multi-channel seismic sections JNOC 1 and 2 (VON HUENE *et al.*, 1980). The mid-slope terrace is accompanied with thrust faults. The Neogene strata which continue to cover the lower half of the inner slope region are strongly disturbed by numerous faults, and the sediments are highly deformed (CARSON and VON HUENE, 1982). Since the formation of the mid-slope terrace and the deformation of the Neogene sediments are direct manifestation of subduction motion, it is evident that the subduction motion is a matter of Quaternary.

The second evidence is seen at the trench slope break (Fig. 1). ISHIWADA *et al.* (1977) remarked Quaternary uplifting motion there, stating that "in the neighbourhood of the trench slope break, Neogene strata are uplifted as a whole, without much thickness variation, and very recent sedimentation develope westward (landward) forming a small basin, the top of which forms a deep-sea terrace". From this, it is evident that the uplifting motion at the trench slope break is a matter of Quaternary. This uplifting motion seems to be directly related to the seismic activity along the deep seismic plane.

The third evidence is seen in the correspondence between the uplift of the trench slope break and the deep seismic planes (Fig. 2). The surfaceward extension of the upper plane of the dual seismic planes is straight rather than curved. This feature was delineated by the aftershocks of 1978 Miyagi-oki earthquake (FACULTY OF SCIENCE, TOHOKU UNIVERSITY, 1979). The main shock of this earthquake occurred along the upper plane of the deep seismic planes, and most aftershocks occurred within the crust above the upper plane (Fig. 2A). The lower boundary of the aftershock distribution delineated the surfaceward extension of the upper plane of the deep seismic planes. The location where the surfaceward projection of this plane intersects the seafloor is the place of the trench slope break (Fig. 2B). This earthquake activity shows that the formation of the trench

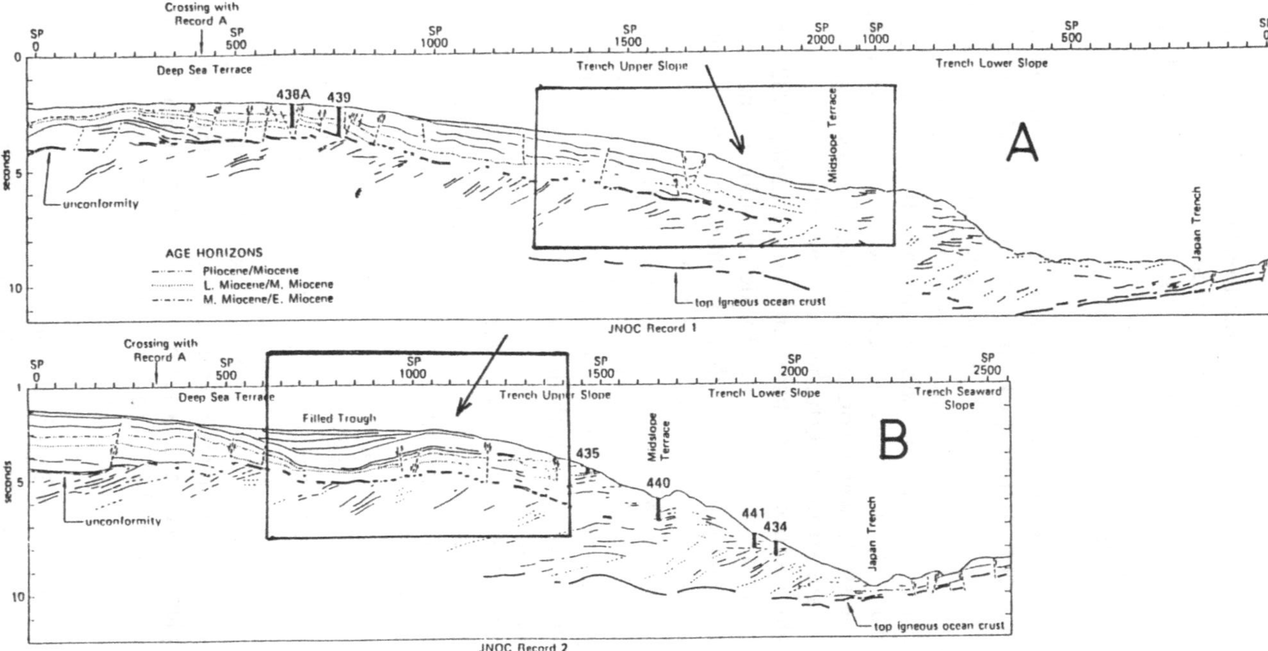

Fig. 1. Evidences which show that the Japan Trench was formed during Quaternary. Solid line boxes indicate the places. (A) JNOC Record 1 showing the Neogene (Pliocene) strata are trancated at the mid-slope terrace. (B) JNOC Record 2 showing the Neogene (Pliocene) strata are uplifted at the trench slope break. The tracings of multi-channel seismic record sections were cited from VON HUENE *et al.* (1982).

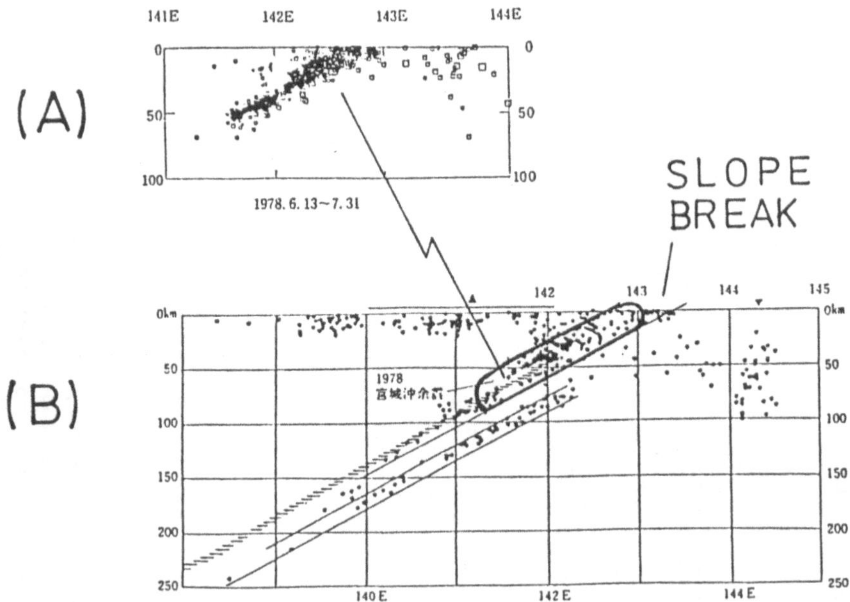

Fig. 2. (A) Aftershock distribution of the 1978 Miyagi-oki earthquake which delineated the surfaceward extension of the upper plane of the deep seismic planes (after FACULTY OF SCIENCE, UNIVERSITY OF TOHOKU, 1979). (B) Dual seismic planes in the northeast Japan (39°–40° N) defined by hypocenter distribution (after HASEGAWA *et al.*, 1978; solid lines are added by the writer). The aftershocks of the 1978 Miyagi-oki earthquake distribution above the upper plane, and its surfaceward projection intersects the seafloor at the trench slope break.

slope break is a presently active phenomenon.

In Fig. 3 is shown the projection of the dual seismic planes on the profile of velocity structure in the latitude between 39°–40° N (NAGUMO, 1980). The vertical exaggeration is about 1 : 1.2, almost natural scale. The intersection of the upper plane and the seafloor is at the trench slope break region, about 143°40′E. The studies of focal mechanisms have shown that the movement of rock strata along the upper plane is an overthrust (UMINO and HASEGAWA, 1975). Therefore, it will be very natural to think that the formation of the trench slope break is still a presently active phenomenon and the uplifting motion there has been caused by the overthrusting motion of the landward mass over the subducting oceanic plate. Here in the Japan Trench, the location of the slope break does not correspond to the oldest front of the accreted subduction complex, which is shown in the model by SEELY (1979).

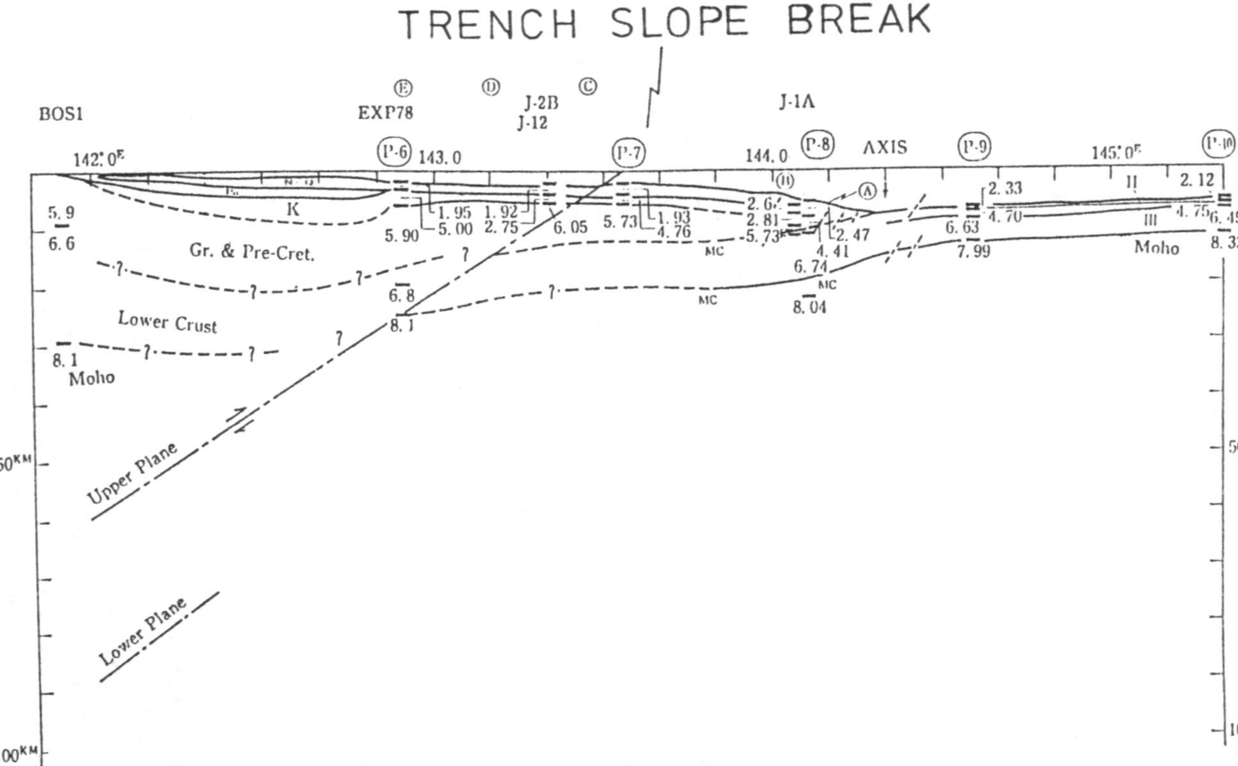

Fig. 3. Projection of the upper and lower planes of the deep seismic planes (HASEGAWA *et al.*, 1978) on the profile of the velocity structure (vertical exaggeration is about 1 : 1.2, after NAGUMO, 1980) showing that the surfaceward extension of the upper plane intersects the seafloor at the trench slope break, and the overthrust motion along the upper seismic plane.

Another feature to which the writer call attention in this Fig. 3 is the curvature of the subducting plate. The curvature has a sharp angular bend rather than a smooth gentle curve commonly referred to in many papers. The writer inferred this angular shape from the appearance seen on the multi-channel seismic sections of JNOC 1 and 2, and Shell lines. The top surface of the subducting oceanic plate right under the trench slope toe shows sudden depression. The average level of the subduction interface under the inner slope is nearly horizontal, being acompanied by many thrust faults. The presentation by the natural scale does show the above mentioned features.

2.2 Forearc basin formation

The evidences which show that the forearc basin was formed mainly during Paleogene period are seen on the multi-channel seismic sections. A large scale folding of thick Cretaceous strata is seen in Fig. 4, which is a sketch of the lines JNOC-1 and its landward extension one (VON HUENE *et al.*, 1982). Beneath the deep-sea terrace, thick uniform strata overlie the Paleogene basin on the landward side, and Cretaceous/Late-Oligocene unconformity on the seaward side. The angular unconformity is the famous relic of the Oyashio-Paleo-land discovered by IPOD drillings site 494 (VON HUENE *et al.*, 1980). This seismic section evidently shows that the region of the forearc basin was in an environment of thick sedimentation during Cretaceous period, and the basin form appeared during the Paleogene period.

The uplift and erosion of the Oyashio-Paleo-land, and subsidence and sedimentation in the forearc basin would be a contemporaneous motion, and the kinematics of this motion would be, basically, a large scale folding. By now there is a different model which shows that the forearc basin was formed by a progressive upward and outward growth of subduction complex, and a structural high (SEELY, 1979). Here in the Japan Trench, however, this is not the case. As regards the anomalous thickness of the Cretaceous strata under the present inner (landward) slope, VON HUENE *et al.* (1982) interpreted them as a product of overlapping of strata caused by thrust motion. Such overlapping motion is very likely to have occurred. However, that would not be a progressive growth of subduction complex.

During the Neogene period, contrary to the Paleogene period, a thick sedimentation developed over the whole present forearc region; both the Paleogene sedimentary basin and the angular unconformity were covered by Neogene sedimentation (Fig. 4). This thick Neogene strata extends down to the mid-slope terrace without much internal deformations (Fig. 1). Such

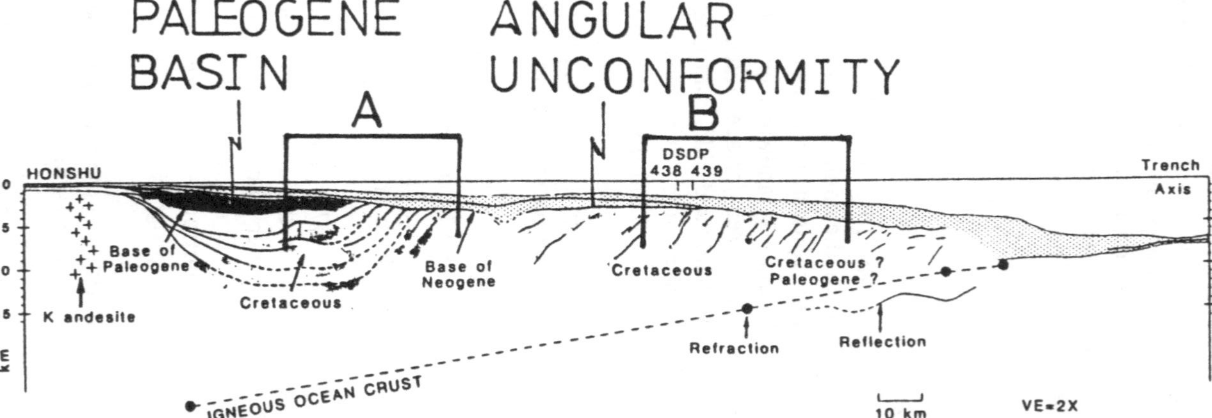

Fig. 4. Evidences which show that the forearc basin was formed during the Paleogene period. Solid line frame A indicates folding motion of the thick Cretaceous strata. Solid line frame B indicates probable overlapping of thrust sheets. A sketch of the multi-channel seismic sections is after VON HUENE *et al.*, 1982.

distribution of thick Neogene strata indicates that this region has been in an environment of stable sedimentation during the Neogene period.

2.3 Repitition of uplift and subsidence

From the features stated above, it should be noted that the region near the present trench slope break has been subjected to repitition of uplift and subsidence (Fig. 5). During the Cretaceous period, the motion was subsidence, and produced thick sedimentation. During the Paleogene period, it was uplift, and produced the Oyashio-Paleo-land. During the Neogene it was again subsidence, and produced thick sedimentation. During the Quaternary, it was again uplift, and produced the present bathymetric feature of trench slope break. Such repitition of uplift and subsidence will not be produced by a steady continuous subduction motion. Instead, it does show that the mode of kinematics should have changed with geologic ages.

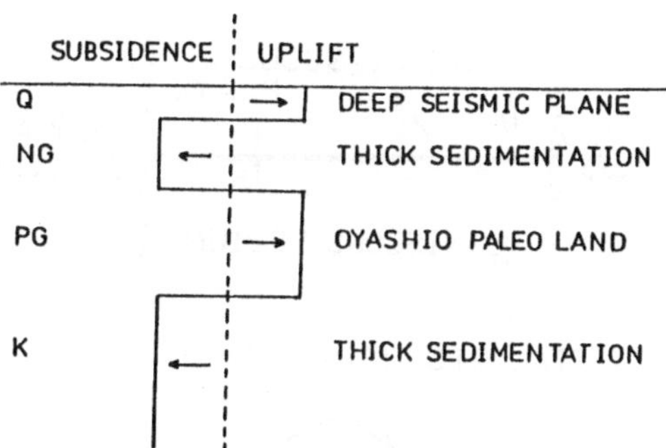

Fig. 5. Schematic diagram of the repitition of subsidence and uplift in the trench slope break region of the Japan Trench. Subsidence in Cretaceous (K) and Neogene (NG) periods, uplift during Peleogene (PG) and Quaternary (Q) periods.

3. Mode of Kinematics—Forearc Basin—

In the following two sections, I will present kinematical models which will explain the formation of the structure of the forearc basin and the trench. I think that the mode of kinematics could be different to different ages. Since the formation of the forearc basin is not related to the present subduction motion, I inclined to think that the kinematics of the forearc basin formation could be independent of subduction, and could be caused by a folding motion. In this section, the writer will show how Biot's folding theory (BIOT, 1961; BIOT *et al.*, 1961; BIOT, 1965) of viscous-thin-plate model can explain the development of forearc basin from the Cretaceous time to the Paleogene time.

The model is as such that a viscous-thin-plate (viscosity η, thickness h) is embedded in a viscous medium (viscosity, η_1), and is subjected to a compressional stress P (Fig. 6A). When an initial disturbance is given at a time $t=t_0$ (Fig. 6B), the deformation of the thin plate developes with time, t, as seen in Figs. 7C and 7D. Such a motion is called as buckling, that is, a deflection perpendicular to the direction of the applied stress. In the early stage of deformation, the uplift occurs in the central part and a subsidence results adjacent to the uplift (Fig. 6C). As the time goes on, the deformation developes, and another uplift appears outwards adjacent to the subsidence region, forming gradually an oscillatory folding structure (Fig. 6D). The uplift at the central region will correspond to the uplift in the central arc during the Cretaceous, and subsidence will

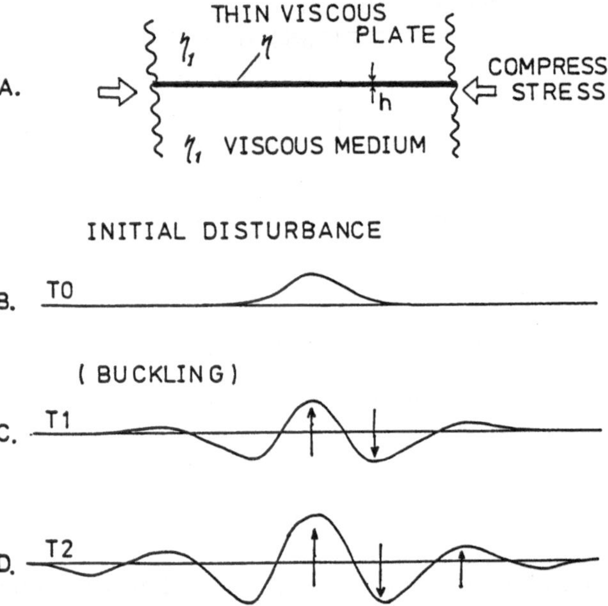

Fig. 6. Boit's folding model. (A) A thin-viscous-plate (viscosity η, thickness h) is embedded in a viscous medium (viscosity η_1) and is subjected to compressional stress P. (B) Given initial deflection of the plate. (C) Deflection of the plate after a certain time T1; the central part uplifts and subsidence appears adjacent to it. (D) Deflection of the plate at a certain later stage T2; another uplifting motion appears outwards adjacent to the subsidence region.

correspond to the forearc basin, and the outward uplift will correspond to the Oyashio-Paleo-land. Such combination of uplift, subsidence and another uplift outwards could possibly explain the actual tectonic features.

In order to fit the model to the Paleogene basin formation, we need a sedimentation environment. A modification of the thin-plate model so as to fit the tectonic history is schematically shown in Fig. 7. Only one side of the symmetrical solution is used. The viscous-thin-plate may correspond to a thin competent layer within the crust. The viscous medium may correspond to thick crust and sedimentary rocks. A sea-water layer is added so as to satisfy the sedimentation and erosion environments. Figure 7B may correspond to the Early Cretaceous, which is the time of the initial disturbance. Figure 7C may correspond to the Late Cretaceous and shows a thick sedimentation. Figure 7D may correspond to the Late Paleogene. The deformations from Figs. 7C to 7D form a Paleogene basin, and uplift and erosion of the Oyashio-Paleo-land.

A fitting shown in Fig. 7 is only qualitative one. The curves of the thin plate are merely an example selected from Biot's calculations (Biot *et al.*, 1961). In order to fit the model to the actual structure, we need further quantitative examinations. We have to search proper values for various parameters such as viscosity ratio (η/η_1), ratio of the thickness of the plate, h, and the width of the initial disturbance, a, and the amount of compressional stress P. In addition to this, we have to take into account the effect of gravity for the folding motion, the effect of sedimentation loading in the subsidence region, and the effect of erosion in the uplifted region. Probably, we will need more rigorous treatment by the theory of the mechanics of incremental deformation (Biot, 1965). In order to fit the time scale, we may require changes of compressional stress with time. The effect of sedimentation loading may accelerate the development of the basin. The effect of the gravity may reduce the amount of the uplift at the central part. An asymmetric solution may be required for explaining the ratio of subsidence in the forearc basin and the uplift in the central arc. Even though these quantitative examinations are left open for further studies, it is quite striking to see that (1) such a simple model appears to be able to explain a long sequence of tectonic history from the Cretaceous to the Late Paleogene, and (2) the initial disturbance which is required in the model seems to correspond to the large granite intrusion in the arc. In other words, the granite intrusion during the Cretaceous period and horizontal compressional stress might have generated the forearc tectonics from the Cretaceous to the Late Paleogene.

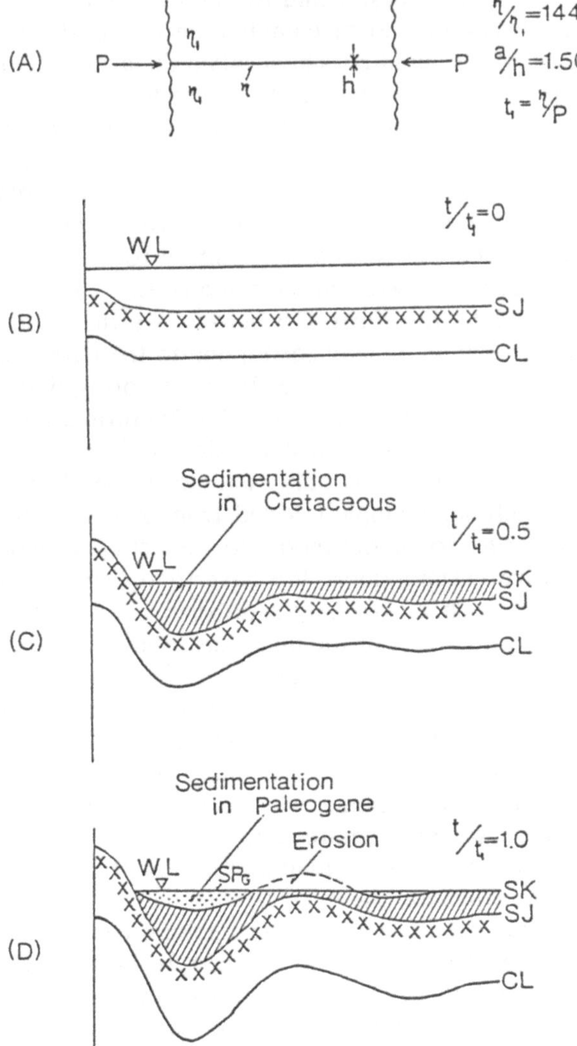

Fig. 7. Schematic illustration of the forearc basin development by Biot's folding model. (A) Model and parameters. a: dimension of the bell shape initial deflection. (B) Given initial deformation. The thin-viscous-plate is equated to a competent layer (CL) within the crust. Water layer (WL: water level) is added above the crust so as to produce sedimentation environments (SJ: surface at Late Jurrassic). (C) Deformation at a certain intermediate stage, showing thick sedimentation during Cretaceous (SK: surface at Late Cretaceous). (D) Deformation at a later stage showing the development of the Paleogene sedimentary basin and the erosion of the uplifted region (SPG: surface at Late Paleogene).

4. Mode of Kinematics—Trench—

A conceptual illustration of a process for forming a trench slope break is shown in Fig. 8. Along a potential slip plane (Fig. 8A) the landward rock mass thrusts up the seaward rock mass (Fig. 8B), and overthrusted landward rock mass forms a trench slope break and landward slope (Fig. 8C). In order to apply the above model to the real structure in the Japan Trench region, a small modification was made as shown in Fig. 9. The slip plane of the overthrusting motion is assumed at the interface of island-arc mass and oceanic plate. An overthrusting motion of the island-arc mass over the oceanic plate generates the bathymetric features of slope break, inner (landward) slope, and trench axis. The subsidence of the trench axis and the downward bending of the oceanic plate in the outer (seaward) slope region are also produced by the loading effect of the overthrusted mass. The slip plane of the thrust motion may extend downwards to the dual deep seismic planes. The formation of the mid-slope terrace is omitted for the sake of simplicity. It could be modeled by additional thrust faulting motions between the major thrust plane and the trench axis.

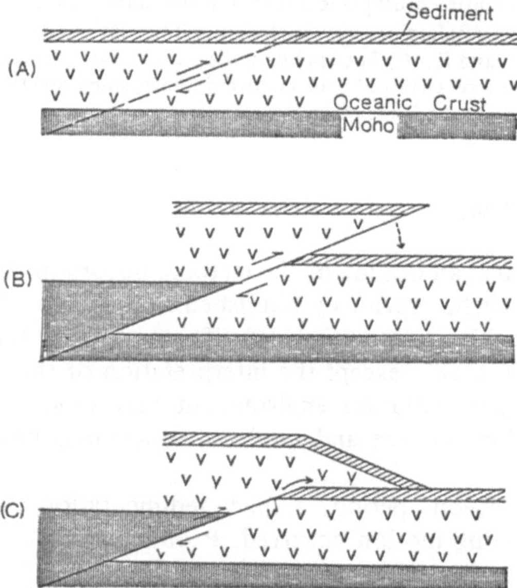

Fig. 8. A schematic illustration of that the trench slope break is formed by overthrust. (A) Initial state. (B) Explanatory illustration of relative displacement. (C) Overthrusted mass forms trench slope break.

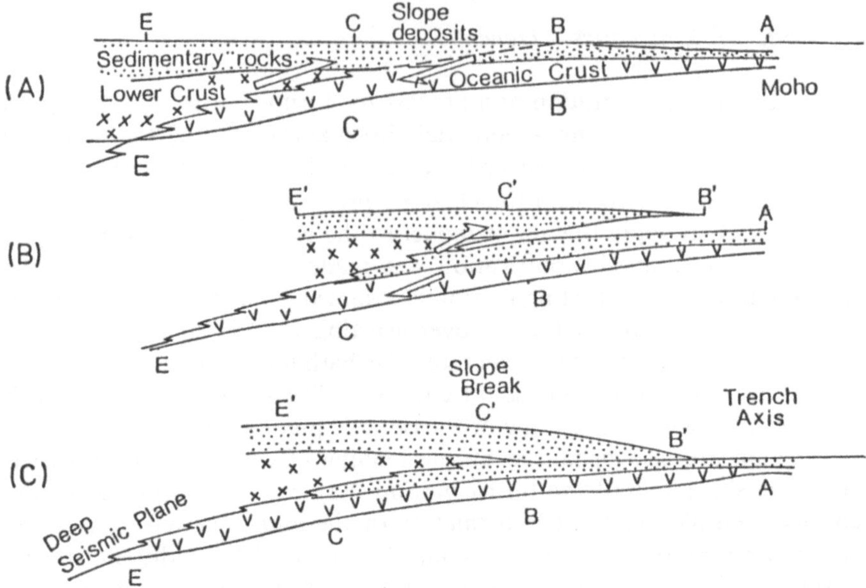

Fig. 9. A schematic illustration of the formation of the Japan Trench by overthrust model. (A) Initial stage. A potential thrust plane is along the boundary of the lower crust of the Japan arc and the ocean crust. (B) Explanatory illustration of the relative displacement. Points B, C, and E move to B′, C′, and E′. (C) The landward mass overthrusts the ocean plate and forms bathymetric features of trench slope break, landward (inner) slope, and trench axis. The thrust plane extends towards deep seismic plane.

5. *Tectonic History*

As a summary, an outline of the tectonic history in the forearc region of the Japan Trench, previously published in Japanese (NAGUMO, 1980), will be cited in this section (Fig. 10). The history is similar to that of VON HUENE *et al.* (1982) except the interpretation of the relation between arc volcanism, sedimentation environment and subduction. Both these workers think that the age and mode of kinematics have changed with geologic ages.

During Cretaceous period, a thick sedimentation developed. In the central arc, uplifting motion occurred. Probably, this corresponds to the granite intrusion. The mode of the kinematics in the forearc region was basically down warping. The sedimentation environment was a passive margin type.

During the Paleogene period, a basin was formed in the landward

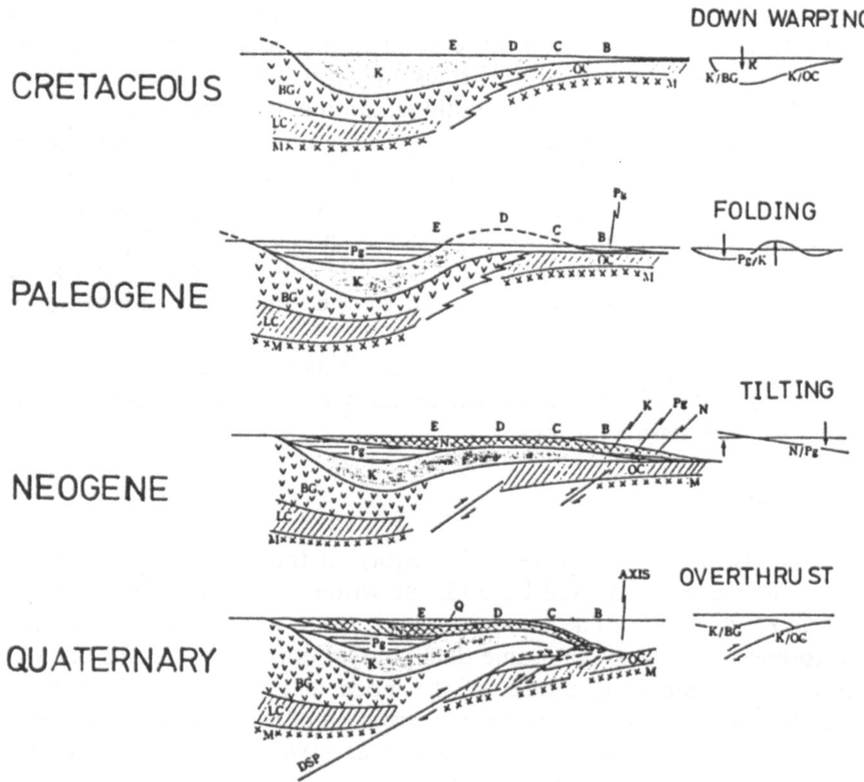

Fig. 10. Schematic illustration of the tectonic history and its basic movement in the forearc region of the Japan Trench (after NAGUMO, 1980). B: position of mid-slope terrace. C: trench slope break. D and E: deep-sea terrace. DSP: deep seismic plane. Q, N, Pg, K: sedimentary strata of Quaternary, Neogene, Paleogene and Cretaceous, respectively. BG: pre-tertiary basement and granitic rocks. LC: lower crust. OC: ocean crust. M: Moho discontinuity.

side of the forearc region, and the Oyashio-Paleo-land was formed in the seaward side. The mode of kinematics was basically a folding motion. This is probably caused by a buckling motion due to a horizontal compressional stress. However, this model will not exclude a possibility of another uplifting agency at the Oyashio-Paleo-land. As VON HUENE *et al.* (1982) have described, the dacite volcanism at the Oyashio-Paleo-land might caused uplifting motion there. The anomalous thickness of Cretaceous strata under the Oyashio-Paleo-land might have been caused by subduction motion as interpreted by VON HUENE *et al.* (1982). In this case, the occurrence of subduction would be at the later stage of the folding motion. The sedimentation environment had been a passive margin type since the Cretaceous, and at the later stage, it might turn to an active margin type.

During the Neogene period, thick uniform sedimentation developed in the broad forearc region. The sedimentation environment was a passive margin type. The mode of kinematics appears to be tilting with subsidence. However, intensive arc volcanism, called "Green-tuff volcanism", occurred during the Miocene period in the arc. If we correlate the Green-tuff volcanism to subduction, the kinematics of subduction will be quite different to that of Quaternary. Because the Green-tuff volcanism is accompanied with tilting and subsidence and resulted in a thick sedimentation in the forearc region.

During the Quaternary period, we meet subduction and active margin. The basic mode of kinematics will be an overthrust which is associated with the deep seismic planes.

6. *Discussion*

Reconstruction of the tectonic history in the Japan Trench forearc region have been attempted by both the writer (NAGUMO, 1980) and VON HUENE *et al.* (1982). The latter is more precise and more realistic than the former. Because of the same data source, both these models are very similar. Both allow the change of the mode of kinematics with geologic ages. The only differences are the interpretation of the role of arc volcanism in the mode of kinematics. VON HUENE *et al.* (1982) equated arc volcanism directly to the presence of subduction.

For identifying subduction, the writer, however, has put more weight on the role of sedimentation environment and bathymetric feature of deep-sea trench rather than on the volcanic activity, and considered that a passive margin type sedimentation environment is indicative of non-subduction. As for arc volcanism during passive margin type period, I presumed that some other possible petrogenesis processes should exist besides subduction. Thus, I was lead to think that transition from passive margin to active margin or vice-versa could have taken place in the same region during the long geological history. In the following, I will show where and how these two interpretations differ.

Cretaceous subduction

Based upon the Cretaceous andesite volcanism in the central arc (Pacific coast of northeast Japan), VON HUENE *et al.* (1982) inferred subduction during the Cretaceous period, and explained the anomalous thickness of Cretaceous sedimentary strata in the forearc region as products of overlapping of strata caused by subduction. In contrast, instead of subduction, I presumed down warping folding motion and explained the passive margin type sedimentation environment. The points from where these interpretations differ are the evaluation of the andesite volcanism and the cause of the anomalous thickness of sedimentary strata.

As regards the Cretaceous andesite volcanism, I presumed some other petrogenesis processes besides subduction. The reasons are as follows. Firstly, the Cretaceous andesite volcanic rocks in the northeast Japan are associated with granite, and the amount of these andesite volcanic rocks is by far less than that of granite, whose igneous activity was the predominant one in the northeast Japan during Cretaceous period (GEOLOGICAL SURVEY OF JAPAN, 1977, Part 4). Secondly, the period of andesite volcanic activity in the Pacific coast of northeast Japan which is used as an evidence of subduction during the Cretaceous period (VON HUENE *et al.*, 1982) is Early Cretaceous, while the granite intrusion activity prevailed throughout the Cretaceous period, namely from the Early Cretaceous to the Late Cretaceous.

From these features, the granite igneous activity appear to be more directly related to the tectonic motion of the central arc during the Cretaceous, and the andesite volcanism may play less important role in the tectonic motion. In other words, the tectonic motion during the Cretaceous appear to be controlled mainly by the large scale granite intrusion, and not by short-lived andesite volcanic activity.

In addition to such mechanical consideration, another petrogenetic consideration may be required to the andesite magma generation, which was associated with a large amount of granite. The petrological process which had taken place during the Cretaceous in the northeast Japan should be such that generates a large amount of granite and a small amount of andesite. From a view point of petrogenesis, there seems to be several possible processes for generating andesite magma. A text book (KUSHIRO and ARAMAKI, 1973) tells us processes such as (1) fractional recrystallization of basalt magma, (2) partial melting of upper mantle rocks under the existence of H_2O, and (3) subduction process, that is, partial melting of oceanic crust subducted under the central arc. Therefore, it would not be so awkward to presume some other tectonic processes besides subduction for the Cretaceous arc volcanism in the northeast Japan.

As regards the thickness of the Cretaceous sedimentary strata, VON HUENE *et al.* (1982) explained them by the overlapping of strata caused by thrust faults. In the region under the upper slope of the landward slope of the Japna Trench, I agree with them. However, in the region under the deep-sea terrace, I think that the landward dipping strata would represent folding strata as being shown in Fig. 4. Probably, the region of thrusts faults would be limited in the seaward part of the large scale folding structure. The thrust faults and overlapping strata would be interpreted either by subduction or by the compressional stress which produced folding strata. Especially, at the later stage of buckling motion, when the degree of deformation exceeds a certain limit, it is very likely that many thrust faults would develop within the folding strata. If this is the case, the age of the thrust faults, thereby thickening of strata, would be the Paleogene or the Late Cretaceous, not throughout the Cretaceous period.

Late Oligocene subduction

Based upon the dacite volcanic rocks found at the Oyashio-Paleoland, VON HUENE *et al.* (1982) equated subduction to this volcanic activity. The folding motion based on the Biot's model would not reject the possibility of such volcanic activity. Since the folding motion developes very fast, almost explosively, at its later stage as remarked by BIOT (1961), it is quite probable that the folding motion terminated in a catastrophe with volcanism and subduction. However, the duration of the volcanism and subduction would be probably very short because of its explosive nature.

Green-tuff volcanism—Miocene subduction—

The relation between the Miocene volcanism (Green-tuff volcanism) and subduction is the most difficult one to understand. I inferred that the basic mode of the kinematics during the Neogene would be tilting motion that is accompanied with subsidence (NAGUMO, 1980). This interpretation is based upon mainly such sedimentation characteristics that thick Neogene strata extend almost uniformly down to the place of the present mid-slope terrace. However, this does not explain the Green-tuff volcanism in the central arc. VON HUENE *et al.* (1982) equated the volcanism to subduction. However, there are several differences in the mode of kinematics of the arc-trench system between the Green-tuff volcanism and the present-day subduction. The major differences are (1) sedimentation environment in the forearc region, and (2) the vertical tectonics in the central arc. As stated above, during the Neogene, the broad forearc region from the present deep-sea terrace to the present landward slope region had been in a steady massive sedimentation environment, that is a passive margin type, being characterized by a slope basin with high sedimentation rate. Such an environment is quite different from that of Quaternary subduction. The vertical tectonics in the central arc during the Green-tuff were subsidence, while those during Quaternary were large uplift. The Green-tuff volcanism occurred in an environment of shallow water with successive sedimentation, while the Quaternary arc volcanism is associated with large uplifting motions which formed the present-day mountain ranges. Therefore, when we think of such differences of structural motions, it is not easy to accept a view that a subduction process similar to the Quaternary one occurred during the Green-tuff period.

Pliocene subduction

In this paper, Quaternary subduction is emphasized. However, the initiation of subduction can be traced back to Pliocene. The gentle tilting motion which resulted in a slope basin may be an initial phase of subduction motion. However, as stated in Section 2, the appearance of the bathymetric features such as deep-sea trench, steep landward slope, and mid-slope terrace are mature aspects of subduction, and will be a matter of Quaternary.

Passive margin—active margin transition

As far as the sedimentation environments is concerned, we have seen that transition from the passive margin type to the active margin type occurred in the forearc region of the Japan Trench. It was passive margin type in the Cretaceous and Neogene periods. On the other hand, as far as volcanic and igneous activities are concerned, the Japan arc was an active margin type in these periods; large granite intrusions during the Cretaceous period, and Green-tuff volcanism during the Miocene period. Therefore, the ocean margin in the Japan arc region possessed two faces; a passive margin type for sedimentation environment and an active margin type for volcanic activity. Therefore, the tectonic features in these ages should be quite different from those seen in the Quaternary. In other words the bathymetric feature of deep-sea trench such as seen in the Quaternary would not exist during the Green-tuff volcanism and Cretaceous volcanism. This may imply that the bathymetric feature of the deep-sea trench, which is the most direct manifestation of subduction motion, is not necessarily required for the generation of arc volcanism.

From these considerations, I inclined to think that transition from passive margin to active margin or vice-versa can occur in the same ocean margin. Any ocean margin will not be constrained to a single framework of tectonic motion throughout its long geologic history. Such a view as this was strongly influenced by the concept of "the Green-tuff tectonic movement", and "the Island-arc tectonic movement" in the development of Japanese islands (FUJITA, 1973). The Green-tuff tectonic movement was defined as a new Miocene orogeny which is independent from the preceding

Late Cretaceous/Paleogene orogeny, and the Island-arc tectonic movement was defined as a new Pliocene-Quaternary orogeny which is independent from the preceding Green-tuff orogeny.

The phenomena similar to those that the forearc basin was formed during the Paleogene period, and the present-day trench was formed during the Quaternary or the Pliocene-Quaternary period appear to be also seen in some other ocean margins, for examples, in the eastern Aleutian, Kodiak island region (VON HUENE, 1979), and off Guatemala, Middle America Trench (SEELY, 1979), and etc. The writer awaits the development of future studies.

7. *Summary*

A view was presented that, in the Japan Trench region, the forearc basin was formed during the Paleogene and its mode of kinematics was a folding motion, and the present-day deep-sea trench was formed during the Quaternary and its accompanied subduction is a matter of Quaternary. This view is in accord with a more general view that the mode of tectonic movements can change with geologic ages.

Instead of steady-state subduction motion, episodic occurrence of subduction motion was presumed by VON HUENE *et al.* (1982). This paper presumed more drastically that the subduction motion in Japan Trench is a matter of the Quaternary.

It is inferred that the mode of kinematics during the Cretaceous and Paleogene periods was a folding motion, and this motion was associated with the granite igneous activity in the central arc. Biot's folding theory appears to be promising to explain such folding motion.

Acknowledgements

The writer is greatly indebted to Dr. Yasufumi Ishiwada for his works on the marine petroleum geology. He is also very grateful to Dr. Roland von Huene for his kind advices and valuable comments and discussions.

REFERENCES

BIOT, M. A., Theory of folding of stratified visco-elastic media and its implications in tectonics and orogenesis, *Geol. Soc. Am. Bull.*, **72**, 1595–1620, 1961.

BIOT, M. A., H. ODÉ, and W. L. ROEVER, Experimental verification of the theory of folding of stratified visco-elastic media, *Geol. Soc. Am. Bull.*, **72**, 1621–1632, 1961.

BIOT, M. A., Mechanics of incremental deformations, John Wiley & Sons Inc., New York, 1965.

CARSON, B. and R. VON HUENE, Small-scale deformation structures and physical properties related to convergence in Japan Trench slope sediments, *Tectonics*, **1**, 277–302,

1982.

DICKINSON, W. R. and D. R. SEELY, Structure and stratigraphy of forearc regions, *Am. Assoc. Petrol. Geol.*, **63**, 2–31, 1979.

FACULTY OF SCIENCE, TOHOKU UNIVERSITY, Earthquake off Miyagi Prefecture, June 12, 1978, *Report of Coordinating Committee for Earthquake Prediction*, **21**, 55–59, 1979.

FUJITA, Y., *Development of Japanese Island Arc—Green-Tuff Orogeny—*, Tsukiji-shoten, Tokyo, 1973.

GEOLOGICAL SURVEY OF JAPAN, *Geology of Japan*, 3rd edition, chap. 23, 1977.

HASEGAWA, A., N. UMINO, and A. TAKAGI, Double-planed deep seismic zone and upper mantle structure in the northeastern Japan Arc, *Geophys. J. R. Astr. Soc.*, **54**, 281–296, 1978.

ISHIWADA, Y., Y. IKEBE, K. OGAWA, and T. ONIZUKA, A consideration on the scheme of sedimentary basins of northeast Japan, in *Professor Kazuo Huzioka Memorial Volume*, 1–7, 1977.

KUSHIRO, I. and S. ARAMAKI, editors, Petrogenesis of andesite magma, in *Material Science of the Earth II*, Chikyu Kagaku (Earth Science), Vol. 3, chapt. 4, Iwanami-shoten, Tokyo, 1973.

NAGUMO, S., Seismic activity and geological structure across the Japan Trench and forearc, in *Zisin*, edited by R. Sugiyama *et al.*, pp. 25–40, Tokai University Press, Tokyo, 1980.

SEELY, D. R., The evolution of the structural highs bording major forearc basins, *Am. Assoc. Petrol. Geol. Mem.*, **29**, 245–260, 1979.

VON HUENE, R., Structure of the outer convergent margin off Kodiak islands, *Am. Assoc. Petrol. Geol. Mem.*, **29**, 261–272, 1979.

VON HUENE, R., M. LANGSETH, N. NASU, and H. OKADA, Japan Trench transect, *Initial Reports of the Deep Sea Drilling Project*, Vols. 56 and 57, pp. 473–488, 1980.

VON HUENE, R., M. LANGSETH, N. NASU, and H. OKADA, A summary of Cenozoic tectonic history along the IPOD Japan Trench transect, *Geol. Soc. Am. Bull.*, **93**, 829–846, 1982.

CHAPTER 2: FOREARC TECTONICS

Formation of Active Ocean Margins, edited by N. Nasu *et al.*, pp. 155–177.

KINEMATICS AND MECHANICS OF DEFORMATION ACROSS SOME ACCRETING FOREARCS

D. E. KARIG

Department of Geological Sciences, Cornell University, Ithaca, NY 14853, U.S.A.

Abstract. Accretion of sediments is one common response in the forearc regions of convergent plate margins. Accretion is associated with a range of structural styles, and with marked changes in physical and mechanical properties. As accreted material becomes progressively incorporated into the interior of the forearc, both the nature of deformation and the constitutive mechanical relations appear to change.

Sediments entering the toe of the forearc (accretionary prism) are deformed as folds or thrust sheets that range in size and style from coherent thrust sheets to melange. Progressive deformation arcward across the forearc has not been resolved by seismic profiling or drilling, but can be constrained by interpolation between cumulative deformation states in the toe region and in emergent accretionary complexes in several arc systems. Further constraints are supplied by the structure of slope strata as expressed by seismic profiling. Both sources indicate continuous steepening of accreted strata to near vertical dips, both by arcward rotation and by tight to isoclinal folding.

The instantaneous distribution of deformation across the lower forearc can now be quantitatively estimated in several better constrained arcs. Interpretation of horizontal shortening from structural sections agrees well with mass flow calculations in showing that instantaneous shortening rate is greatest at or near the deformation front (toe of prism), and decreases rapidly arcward to very low values. In the seismically coherent toe of the Nankai prism, within 15 km of the deformation front about half of the deformation is accommodated along seismically defined thrusts and the rest by distributed deformation within the thrust sheets.

Thrusting and associated stratal rotation define a gross simple shear, whereas the distributed deformation is more a flattening, resulting from pure shear. The dominance of simple shear near the prism toe is probably a result of proximity to the interplate boundary, which is a controlling surface of simple shear. The gross flattening that dominates deformation away from the shear boundary is accommodated in accretionary complexes as isoclinal folding and quite likely by bedding-parallel extension of near-vertical strata.

1. Introduction

Accretion of sediments at the toe of the forearc is one of several mechanical responses that can occur at shallow levels along convergent plate margins. Not only is this a relatively common response but it also leads, by its very nature, to the preservation of the record of convergence in accretionary complexes. In contrast to the structural geometry produced by tectonic erosion or non-accretion, that resulting from sediment accretion is often resolved on seismic profiles, which undoubtedly has been one reason for the concentration of research efforts in this type convergent margin. Because of these factors, a relatively larger amount of information is available concerning deformation at accreting forearcs. Even so, the results of the studies in these forearcs cannot be simply or unambiguously correlated with observations concerning the deformation in emergent accretionary complexes, which preserve the cumulative history of deformation across accretionary prisms.

For example, it has long been suspected that sediments are stripped off the descending plate in thrust or fold units, and that most subduction related deformation is concentrated near the prism toe. However, there must be a sufficient fraction of deformation distributed across the prism to effectively destroy the early structural geometry imparted near the toe and to create the very steeply dipping and highly disrupted fabric commonly observed in accretionary complexes. A major objective is to understand the progressive deformational paths of accreted sediments and to couple these with the changing physical and mechanical properties that are presumed to accompany deformation.

With the application of modern investigative techniques to the deep water environment of trenches, the progressive changes in the character of sediments as they become incorporated into the accretionary prism can be more quantitatively and realistically described. The internal geometry of accreted units, the quantitative distribution of cumulative and instantaneous deformation, and the relationship of these parameters with physical properties of sediments are now topics capable of being addressed. The purpose of this paper is to discuss the geometry and distribution of deformation in some better-resolved prism toes, to review the nature of deformation in slope sediments overlying these prisms and to speculate concerning the changing styles of deformation across accreting forearcs. This presentation draws heavily on data arising from the Deep Sea Drilling Project in the Nankai Trough, but also stresses data from the Sunda and Makran arcs.

2. *Structural Styles of Accretion*

Sediments are accreted to the toe of the accretionary prism in a range of structural styles, some members of which are now quite well-delineated. Accretion in relatively large and coherent thrust sheets is one of the most common modes and certainly the one most clearly outlined by reflection seismology and drilling. Two examples in which accreted thrust sheets are quite large and structurally coherent have been described in the Sunda arc (KARIG *et al.*, 1980) and Nankai Trough (AOKI *et al.*, 1983; KARIG, KAGAMI *et al.*, 1983).

In a survey of the Sunda arc, incoming sediment was interpreted as being accreted in thrust sheets, about 1 km thick, 8–10 km long, and bounded by 20–25° dipping faults (KARIG *et al.*, 1980; MOORE and CURRAY, 1980). The only folding that could be seismically imaged was related to the geometry of thrust surfaces. Brief note was also made of diffuse deformation occurring immediately in front of the frontal thrust, which tectonically thickened and uplifted trench fill strata.

Better resolved reflection profiles and deep sea drilling in the Nankai Trough permit a more detailed and quantitative interpretation of this style of accretion. Deformation in the Nankai prism occurs both on discrete thrusts and as diffuse sub-horizontal shortening within thrust sheets. Diffuse deformation actually begins in a protothrust zone extending 5 km seaward of the frontal thrust. The resultant thickening and bedding parallel shortening occurs in part along 60°N dipping seismic discontinuities that may be large versions of the kink-band like structures observed in DSDP Hole 583 (LUNDBERG and KARIG, in press).

The 500 to 800 m thick sediment fill of the Nankai Trough is stripped off initially in imbricate thrust sheets along frontal thrusts that develop 3 to 4 km apart (AOKI *et al.*, 1983; NASU *et al.*, 1983; KARIG, in press). The frontal thrust rises on a 30° dipping ramp. Structural reconstruction indicates that the original ramp dips of the older, more arcward imbricates were also 20° to 30°. All these imbricate thrusts rise from a basal decollement that remains in the upper hemipelagic section of the Shikoku Basin strata for more than 25 km behind the deformation front (Fig. 1; KARIG, in press).

The response of the frontal thrust sheet, despite its relatively unlithified condition, is qualitatively very similar to that of highly lithified foreland thrust sheets (e.g. DAHLSTROM, 1977). Bedding lengths are not preserved, so that restoration of structural sections is not as simple as in classic foreland thrust belts. Nevertheless, both flanks of a hanging wall anticline can be identified (Fig. 1; KARIG, in press), although quantitatively modified by diffuse deformation. No folding other than this response to thrust ramp geometry has been recognized.

Displacement along the frontal thrust, as determined by offset of seismic and paleontologic horizons, differs from that inferred from morphology alone because there are rapid and large lateral changes in erosion and deposition modifying the leading edge of the thrust sheet (KARIG, in press). The disparity between morphologic relief across thrust noses and displacement interpreted from ramp corner offsets on the seismic reflection profile becomes very large for some of the older thrusts (Fig. 1) and has also been attributed to local erosion and redeposition (KARIG, in press). The magnitude of fault scarp relief is, however, though to be a qualitative measure of the rate of continued displacement along the older faults.

The lack of scarp relief indicates that some thrusts quickly become inactive, and that units of two to three original thrust sheets, 5 to 7 km long, become the basic kinematic element behind the frontal thrust (KARIG, in press). These amalgamated sheets are wedge shaped in section, reflecting not only the original arcward thickening of the trench fill segment, but also the internal deformation, both diffuse and on deactivated thrusts. Displacement on the unit-bounding thrusts continue to increase along the study profile at least 15 km arcward from the deformation front. Construction of structural geometries further inbound is prevented by loss of seismic coherence. Over this 15 km interval, displacements cause arcward rotation of strata above ramps. Fifteen km from the deformation front (S.P. 1600, Fig. 1), thrust displacement has become sufficient to overlap the thrust surfaces and to cause rotation of the overlying thrust as well (Fig. 1). Diffuse deformation also continues to increase arcward across this section as shown by progressive degradation of seismic coherence and by overall thickening of the stratal units (KARIG, in press).

Other styles of deformation accompanying accretion of sediments at trenches require more speculative interpretation because relevant seismic profiles display less structural definition. As the thickness of the accreted section diminishes, the individual thrust sheets appear to become shorter and are apparently more internally disrupted. In the central Lesser Antilles arc, structural sections constructed from deep sea drilling and related

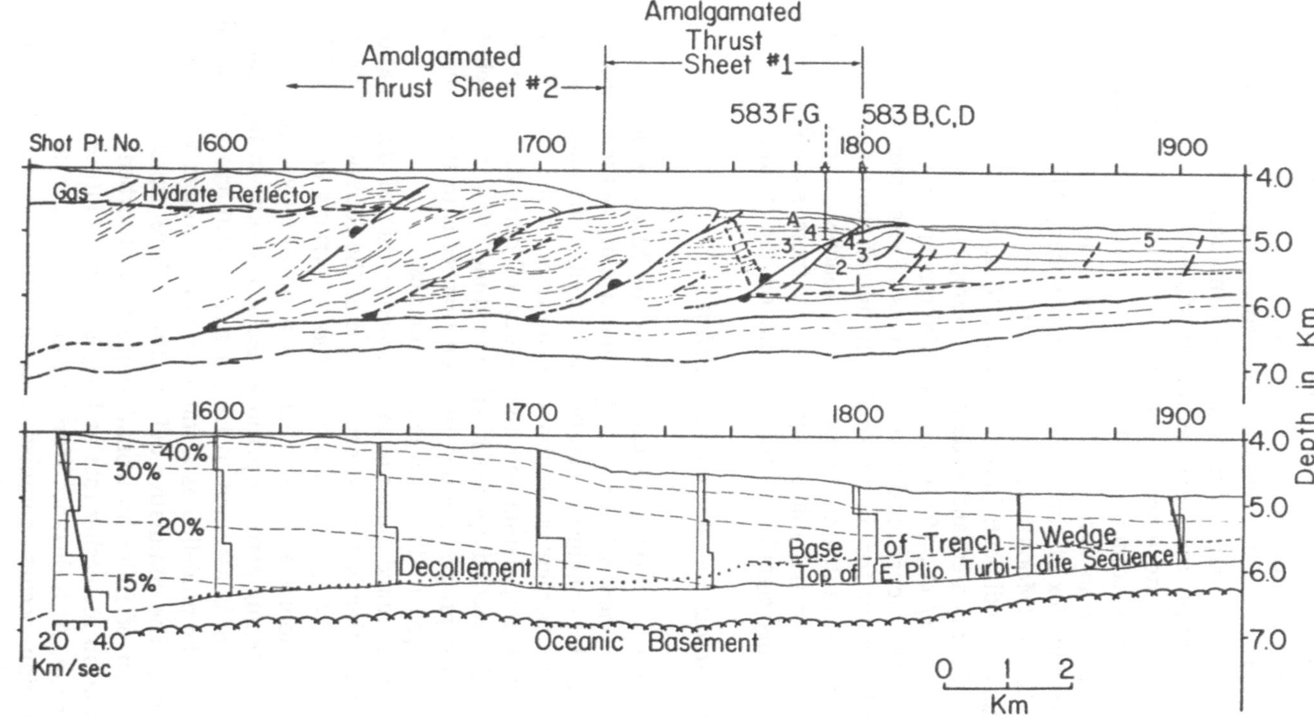

Fig. 1. A. Seismic depth section across the toe of the Nankai accretionary prism, from seismic profile 55-3-1 (NASU *et al.*, 1982) as interpreted by KARIG (in press). The displaced half circles represent estimates of offset along major thrust faults. B. Porosity distribution across the prism toe, from DSDP results and seismic velocities, using the empirical relationships discussed in BRAY and KARIG (1985). Interval velocities along the profile (from NASU *et al.*, 1982) are shown together with two bounding linear velocity-depth gradients, derived from the interval velocities, DSDP results and refraction velocities (YOSHII *et al.*, 1973).

seismic data (MOORE and BIJU-DUVAL, 1984) show that accretion of 200 m of pelagic sediments takes the form of thrust sheets 1 to 2 km long. The internal structure of these sheets was not resolved on seismic profiles, in part because of severe internal mesoscale deformation, as implied by high and variable dips in the cored sediments (COWAN *et al.*, 1984).

A logical, but still undocumented extension of this trend is that even thinner accreted sediment columns would lead to formation basalt-bearing melange near the toe of an accretionary prism. Such an origin was proposed for the Oyo complex of Nias, which represents early Miocene accretion in the Sunda arc (MOORE and KARIG, 1980).

In some trenches, sediments initially respond to accretion by folding. Perhaps the best documented and widespread examples are the northern Cascade arc (CARSON *et al.*, 1974) and the Makran arc (WHITE, 1977), but this response also occurs locally in other arcs. These folds are open and upright, and account for only a small amount of shortening before being broken by thrust faults, either arcward or trenchward in vergence. It is unlikely that the initial folding displayed by these examples could be identified following significant displacement on the thrusts.

3. *Deformation Across the Accretionary Prism*

Deformation must continue arcward of the 10–20 km wide zone at the toe of the prism where seismic profiling and drilling outline a thrust-dominated fabric. This conclusion can be reached either by analyzing the structural fabric of emergent accretionary complexes at the rear of the trench slope or by extrapolating the distribution of deformation in the toe of the prism. The second approach will be outlined first, as a prelude to a discussion of the progressive deformation that leads to the observed cumulative deformation in the emergent complexes.

It is generally accepted that the instantaneous rate of deformation resulting from subduction is greatest near the toe of the accretionary prism (e.g. SEELY *et al.*, 1974, KARIG, 1974; MOORE, WATKINS *et al.*, 1979), but it is not often appreciated, especially in a quantitative sense, how the cumulative shortening across a prism varies as a result. In this study, the accretionary prism is assumed to have a steady state cross sectional shape and thus can be modelled as a standing wave, through which sediment masses move. The cumulative deformation is here expressed as the arcward summation of finite horizontal shortening absorbed within the prism, arcward at an increasing rate as long as any instantaneous deformation is occurring. At a distance from the deformation front where no further deformation occurs, cumulative deformation increases linearly arcward.

A major conceptual problem is that, although a sediment mass may be deformed at the greatest rate near the prism toe, it moves through this zone relatively rapidly and spends a proportionately much longer time in the more slowly deforming regions of the prism. Thus, a seemingly disproportionate amount of deformation accumulates in the zone of very low instantaneous deformation. These points can be demonstrated by an analysis of deformation in the section of the Nankai Trough studied during DSDP leg 87 (KARIG, in press).

This analysis began with estimates of cumulative shortening, which can be more directly extracted from the data than can instantaneous deformation, because the relationships between deformation and time can be avoided. Cumulative shortening was estimated in two ways. In the first, the displacements on individual thrust faults and shortening due to diffuse deformation were summed arcward from the deformation front (Fig. 2). This summation extended to a point 15 km inboard (S.P. 1600) where seismic coherence became insufficient for quantitative analysis. At S.P. 1600, an estimated 8 km of total shortening has been absorbed.

A second approach to the distribution of cumulative deformation utilizes the observed cross-sectional geometry of the accretionary prism and the rate of sediment influx through the deformation front (Fig. 3). Before methods using this approach can be applied however, several assumptions and corrections must be made.

The first assumption, that of mass balance, requires that no sediment mass crosses the upper or lower surfaces of the prism over the area of interest. This assumption is satisfied along the base of the prism if it can be shown that the surface of decollement, along which material is transferred into the prism, does not change stratigraphic level. On the reference seismic profile (Fig. 1 and NASU *et al.*, 1982, p. 22) the decollement is observed to descend to a position just above the lower Pliocene turbidites at the frontal thrust and remains above these turbidites to at least S.P. 1350 (not shown). This point is 27.5 km behind the deformation front and more than encompasses the range of original lengths of sediment now deformed in the first 15 km of prism toe, thus satisfying the assumption. The situation on the surface of the prism is less simple because slope sedimentation adds, and erosion over thrust toes removes mass. Much of the eroded sediment is redeposited immediately downslope and the addition of hemipelagic sediment to the slope tends to compensate for the fraction recycled to the trench. The local lateral transfer of mass on the slope is assumed to be smaller than the effects of smoothing the data.

The steady state assumption requires that the prism geometry remain constant over the duration of interest. Because the accretion represented in the first 15 km of the prism occurred in only slightly more than 1 my, during which time the rate of sediment influx to the trench was relatively constant (KARIG and ANGEVINE, in press), this assumption is considered to be satisfied.

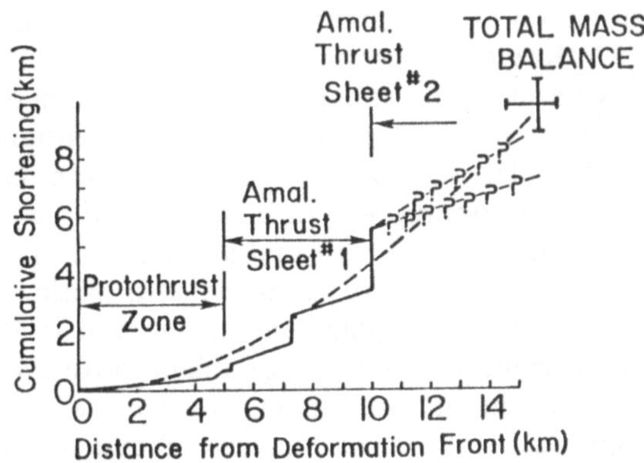

Fig. 2. Cumulative horizontal shortening as a function of position (x) across the toe of the Nankai accretionary prism. The solid line represents the sum of estimated shortening within individual thrust slices and displacement on the thrust faults. The dashed line represents the shortening derived from mass flow considerations as well as being a reasonable smooth fit to the raw data (from KARIG, in press).

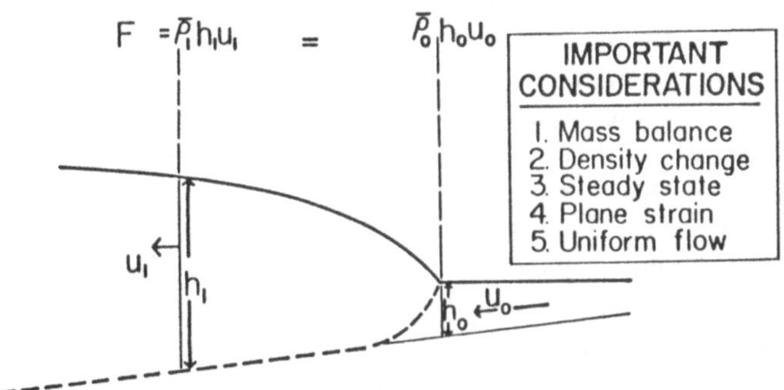

Fig. 3. A simple continuum model of mass flow through the toe of an accretionary prism with a steady-state geometry. This model is used and discussed in the text to estimate the distribution of cumulative and instantaneous deformation across the toe of an accretionary forearc. The velocity (u) is that relative to the prism geometry, $\bar{\varrho}$ is the average dry bulk density, and h is the height of the deforming sediment column.

A condition of plane strain reduces the problem to two dimensions and is closely approximated in the Nankai prism because the system is quite linear and the structural pattern constant for several tens of km on either side of the reference section.

Ths horizontal velocity of sediment units is assumed to be constant with depth at any distance from the deformation front to simplify the calculation. This pure flattening condition is clearly not closely approximated near the prism toe, where at least half the shortening occurs on discrete thrust faults, but the effect of these shear discontinuities is minimized by the smoothing of the geometry and deformation data (Fig. 2).

Bulk sediment volumes are not preserved as material passes through the prism because sediment porosities are sharply reduced during deformation. This problem is solved by use of in-situ porosity to reduce the bulk volume to the volume of the sediment fraction, or dense rock equivalent (DRE) volume. For most of the prism, porosity values must be derived indirectly, in this case from seismic refraction and reflection data coupled with an empirical V_p *vs* density relationship (BRAY and KARIG, 1985). Seismic velocities were converted to in-situ bulk densities using the curve of HAMILTON (1978) for shale-rich clastics corrected for the effect of the high sand/shale ratio in the Nankai Trough (BRAY and KARIG, 1985). Porosities were derived from bulk densities assuming an average grain density of 2.7 gr/cm^3 and plotted on the prism section (Fig. 1).

The thickness of the sediment section at the deformation front (S.P. 1980), down to the Pliocene turbidites is almost 1,000 m, with a DRE thickness of 0.65 km. The thickness has probably been somewhat larger in the past because the trench wedge has been narrowing with time (KARIG and ANGEVINE, in press). On the other hand, Karig and Angevine also present evidence for an increasing subduction rate over this period, which would tend to counterbalance the shrinking trench wedge. Moreover, the average thickness of the input section is greater than that at the deformation front if the sediments are stripped off the downgoing plate are wedge shaped sheets of significant width. For these reasons an input thickness of 0.8 km (DRE) is used, except over the protothrust zone, where the values are faired smoothly to the observed influx (0.65 km, DRE). This input is felt to be conservatively large, which would minimize the calculated shortening values.

With these boundary conditions, and the assumption of conservation of DRE volumes, the arcward thickening of the accretionary prism can be converted to shortening ratios for elemental volumes of accreted sediment. Horizontal shortening thus calculated reflects both dewatering and tectonic thickening, but because both processes contribute to total shortening, the geometry of dewatering need not be determined. Integration of the elemental shortening across the prism then provides a cumulative shortening curve (Fig. 2).

The cumulative shortening over the 15 km of prism toe using this approach is almost 9 km (Fig. 2). The very good agreement with the estimate derived from the structural geometry strengthens the quite different assumptions used in each approach. If, for instance, displacement on the thrusts was more nearly as implied by the morphologic relief than by the seismically defined structural geometry, the observed thickening of the prism would require a very rapidly decreasing input thickness over the past 1–2 my.

4. *Distribution of Instantaneous Deformation*

The second method used to calculate the distribution of cumulative deformation is essentially an application of the mass continuity equation, which can also be used to derive instantaneous deformation of sediment masses within the accretionary prism. The instantaneous deformation of a sediment element in this simplified model is equivalent to $\partial u_x/\partial x$, the rate of change of velocity in the direction perpendicular to the arc.

In a reference frame fixed to the cross-sectional geometry of the accretionary prism (assumed to be constant over the duration considered), accreted sediment enters the prism at the deformation front with the subduction velocity (u_0). This is the orthogonal component of plate convergence plus the outgrowth rate of the prism shape from the interior of the upper plate. As the sediment mass element moves far into the prism (the distance from the deformation front (x) becomes large) its velocity (u_x) does not go to zero but decreases to the outgrowth rate (u_a).

Because the recent subduction rate is not always a well-known quality, as is the case in the Nankai Trough (KARIG and ANGEVINE, in press), it is more useful to present the calculations as the ratio of velocities, u_x/u_0. This ratio, which is a measure of the fraction of the total instantaneous shortening remaining arcward of a point x from the deformation front, can be calculated from the deformation data without assuming a subduction rate. The slope to this curve is the relative instantaneous deformation.

The subduction rate does enter the calculation of the distribution of instantaneous deformation, however, because the fraction of deformation remaining arcward of a point x must go to zero when x becomes large. The ratio u_x/u_0 goes to u_a/u_0 as x becomes large, but if the ratio $u_x - u_a/u_0 - u_a$ (which is a conversion to a reference frame fixed to the interior of the upper plate) is plotted instead, the fractional instantaneous deformation is made to vary between 1 and 0 (Fig. 3).

Despite the neccessity of assuming a subduction velocity, the resultant relationship is not overly sensitive to the velocity chosen because it is effective as the ratio u_a/u_0 which is quite small.

The first continuum mechanics approach uses the steady state one-dimensional form of the continuity equation $\partial(\varrho u_x h)/\partial x = 0$, which is simply integrated to $(u_x/u_0) = (\bar{\varrho}_0 h_0/\bar{\varrho}_x h_x)$, where $\bar{\varrho}$ is the average dry bulk density. Because the ratio of densities times heights ($\bar{\varrho}h$) is numerically identical to the ratios of DRE thickness at these points, values of u_x/u_0 can be obtained immediately from the previous calculations of cumulative deformation (Fig. 4).

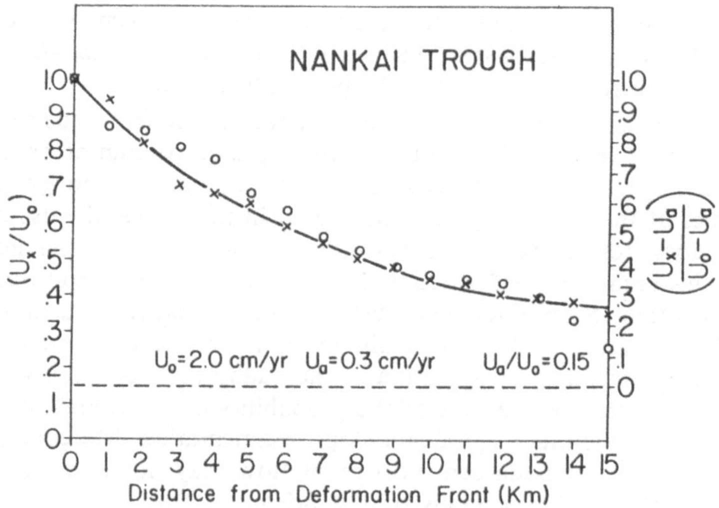

Fig. 4. Distribution of relative instantaneous shortening across the toe of the Nankai accretionary prism. The slope to this curve is the instantaneous shortening at distance x from the deformation front. Values on the left (u_x/u_0) represent normalized ratios of particle velocity within the prism with respect to the subduction velocity. Values on the right take into account the outgrowth velocity of the prism; instantaneous relative shortening approaches 0 as the particle velocity (u_x) approaches the outgrowth velocity (u_a).

A second method by which u_x/u_0 can be determined uses the definition of the shortening ratio (ϕ), and of particle velocity (u_x). The shortening ratio is a function analogous to strain, but is referred to the present, or deformed state of the material, and is defined as $\phi = l_0 - l_d/l_d$ where

l_0 is original length and l_d is the present length. In continuum notation, the shortening ratio is a particle displacement, $\phi = \partial x_p / \partial x$. Similarly, u is the velocity of a particle; $u = \partial x_p / \partial t$. In the steady state any particle at this position will have this time-independent velocity. Then through partial differentiation, $(\partial\phi/\partial t) = (-\partial u_x/\partial x)$, which can be rewritten as $(\partial u/\partial t) = (-u\partial\phi/\partial x)$. Recasting, $(\partial\phi/\partial t) = -(1/u)(\partial u/\partial x\ /) = (-\partial \ln u/\partial x)$. Upon integration, $\phi = -\ln u + c$, but at $x = 0$, $\phi = 0$ and $u = u_0$. Thus With values of ϕ taken from a smoothed curve fit to the observed cumulative shortening across the Nankai prism (Fig. 2), this equation gives values of u_x/u_0 very close to those of the first approach (Fig. 4), thus strengthening confidence in the continuous mechanics assumptions.

Before a curve describing the cumulative percentage of instantaneous shortening across the toe of the Nankai prism can be constructed, the outgrowth rate (u_a) and the subduction rate (u_0), must be obtained. For the western Nankai Trough estimated values are $u_0 = 2.0$ cm/yr and $u_a = .3$ cm/yr (KARIG and ANGEVINE, in press). The value of $u_a/u_0 = 0.15$ is not well constrained but is small and apparently typical.

The resulting plot of instantaneous deformation across the toe of the Nankai prism (Fig. 4) shows that shortening is maximum at or near the deformation front and decreases very rapidly inboard. Although smoothed values of the shortening ratio were used in this curve, the use of local values does not change this conclusion (KARIG, in press). At least 1/3 of the instantaneous deformation is occurring in the proto-thrust zone and 3/4 within 15 km of the deformation front. The other 1/4 is presumably spread in an arcward decreasing distribution across the rest of the lower inner slope. Arcward extrapolation of the quantitative continuity assumptions is not warranted because of the probability of additional mass influx through subcretion and significant slope sedimentation. Moreover, steady state conditions and uniform horizontal flow may not even be approximated in the thicker sections of the prism.

The same calculation of the distribution of deformation across the Sunda arc in a well studied area off Nias (KARIG *et al.*, 1980; MOORE and CURRAY, 1980; KIECKHEFER *et al.*, 1980; BRAY and KARIG, 1985) resulted in an extremely similar curve for the instantaneous deformation, despite a much higher rate of subduction (Fig. 5). The degree of coincidence may be fortuitous but the similar cross-sections of toes of accreting forearcs suggest generally similar distributions of deformation.

5. *Deformation Style in Emergent Tertiary Accretionary Prisms*

The deformational fabric impressed on the accretionary prism behind the most rapidly deforming toe area can be observed in the sediments of younger, less deeply eroded examples of accretionary complexes. Those examples with which I am most familiar are in the Sunda arc (MOORE and KARIG, 1980); the Shimanto complex of Shikoku (TAIRA *et al.*, 1982), and the Iranian Makran (MCCALL and KIDD, 1982; D. E. KARIG, unpublished field notes). All of these show remarkable similarities in structural style, which appears also to be shared in large degree with complexes in Barbados (e.g. TORRINI *et al.*, in press), Timor (KENYON, 1974), and the eastern Aleutians (BYRNE, 1982). Nias and the Makran lie within the actively deforming zone of the forearc, trenchward of the forearc basin, whereas the Shimanto complex emerges behind the forearc basin. Although more difficult to relate to depositional sites, the units in the Shimanto complex should have similar structural trajectories as do the others because there is essentially no deformation in the forearc basin.

On Nias and in the Makran (especially in the Iranian section where I visited), slope sediments occur in synclinoria (Fig. 6). These rest with tectonized, but perhaps originally depositional contacts on melange that is interpreted as accreted trench or oceanic basin strata. These synclinoria are comprised of large angular folds, with very tight or sheared out hinges,

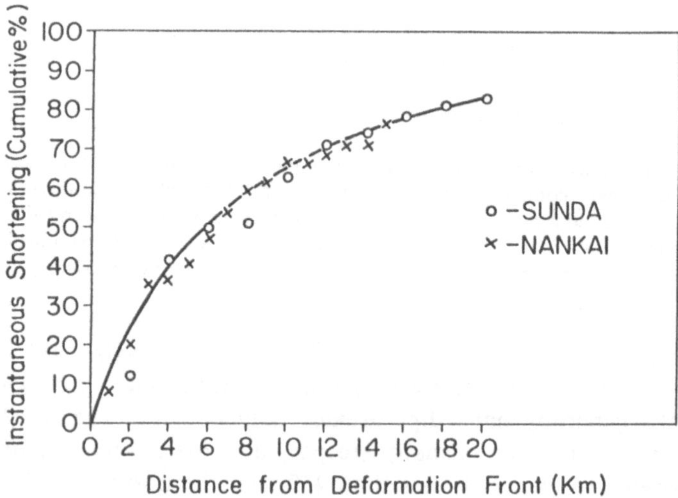

Fig. 5. Comparative curves of the percentage of instantaneous deformation across the toes of the Nankai and Sunda arcs (from KARIG, in press and BRAY and KARIG, 1985). Both data sets are normalized respective to the subduction rates and zeroed to the ratio of subduction to outgrowth velocities (u_0/u_x). Note the extreme similarity in these two curves despite a large difference in subduction rate.

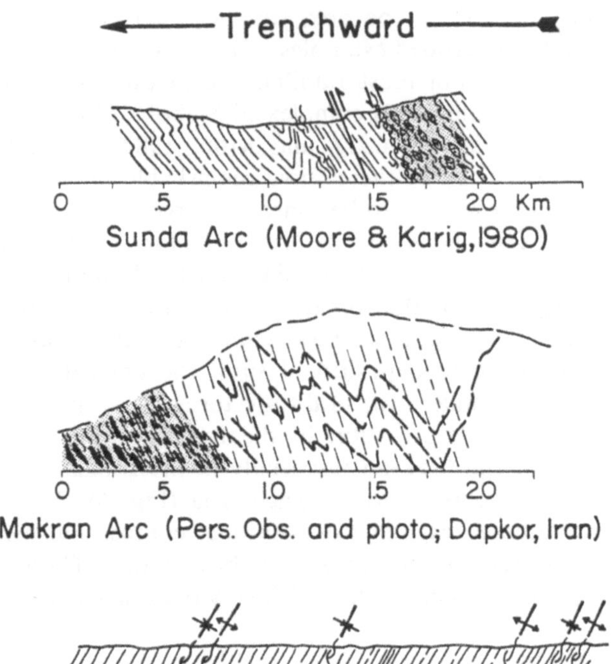

Fig. 6. Structural sketch sections through slope basin strata and subjacent accreted rocks in three accretionary complexes showing similarity in structural fabric. In these examples fold limbs, spaced axial plane cleavage and foliation in the subjacent accreted material are tending toward parallelism. This behavior is interpreted as a result of a sub-horizontal flattening, dominating deformation behind the toe of the accretionary prism.

particularly at deeper structural levels (McCALL and KIDD, 1982; MOORE and KARIG, 1980). Most folds are asymmetric, with longer arc-facing limbs. Both limbs generally have dips greater than 60°, but the shorter, trenchward facing limbs are usually steeper, and often are overturned. Interlimb angles are generally less than 45° near the bottom of these basins and increase upward. Fold wavelengths are several hundred meters and amplitudes are about the same. Upward, stratal dips decrease and the synclinoria become modified to broad open synclinal basins (e.g. FARHOUDI and KARIG, 1977; McCALL and KIDD, 1982) very similar to slope basins defined offshore by seismic profiles (HARMS *et al.*, in press).

On the Muroto Peninsula, the more coherent units of the Shimanto complex dip very steeply and young predominantly northwestward, but there are significant sections, to 500 m thick, of southerly younging beds (Fig. 6 and TAIRA *et al.*, 1980). These least deformed strata within the Shimanto complex most likely represent both accreted and slope sediments. Poles to bedding and a few preserved fold hinges suggest that the panels of strata with opposing dips represent shallow plunging, tight to isoclinal folds in which shearing has destroyed hinges and modified limb geometries.

The clastic sediments at the deeper structural levels of the Makran synclinoria show a very well developed, near-vertical spaced cleavage, which, from reports (HUNTING SURVEY CORP., 1960; MCCALL and KIDD, 1982) and personal observations is coplanar with axial surfaces of the angular folds (Fig. 6). This cleavage also transects complex structures in the intensely deformed pelitic strata that occupy a transitional position between the syclinorium and the clearly subjacent melange "basement". This superposition and the apparent lack of pronounced cleavage fanning within the large folds lead to the conclusion that cleavage formed late in the progressive deformation of the complex.

A spaced cleavage has also locally developed in the shale-rich sediments within coherent units of the Shimanto complex (UMEMURA and UEMATSU, 1980; pers. obs. 1983). This cleavage postdates an early, soft sediment deformation and appears sub-parallel to the axial surfaces of the late stage, isoclinal folds.

In both the Shimanto and Makran complex, cleavage in the slope sediments is also sub-parallel to the foliation within the more highly disrupted melange or broken formations. Although this cleavage was not recognized in the slope sediments of Nias, similar deformational geometry is suggested by approximate parallelism of the melange foliation with axial planes of the folds in slope strata (MOORE and KARIG, 1980). Thus, there appears to be a strong tendency toward a condition of vertical orientation and parallelism of several planar fabric elements that initially had very different attitudes. This pattern cannot be a simple extrapolation of the rotation and thrusting that dominates deformation at the toe of the prism. Instead the response seems to require that large amounts of horizontal shortening occur as a gross flattening, with relatively little regional rotation in a stress field having a relatively constant orientation of principal axes. Structural analysis in the Ghost Rocks complex of the eastern Aleutians has also led BYRNE (1982) to stress the importance of horizontal shortening without rotation as a response to deformation in trench and slope sediments.

6. *Deformation of Slope Strata*

The deformation path or paths by which sediments travel from the toe to the top of the inner trench slope cannot be traced directly by progressive changes in structure within accretionary prisms, because of the arcward loss of seismic coherence and the lack of deep drill holes. Some constraints should be provided, however, by the behavior of slope sediments as interpreted from seismic profiles. Because, in general, only the uppermost km or so of slope strata is seismically imaged with adequate resolution, the profiles record partial cumulative deformation paths across the trench slope. Conceptually these partial paths might be overlapped to produce a cumulative deformation path for slope sediments, but the data are seldom so obliging. Seismic coherence, even at very shallow depth, is often very poor, perhaps as a result of near-surface processes such as sediment remobilization (e.g. KARIG, 1983) and hydrofracturing (ARTHUR *et al.*, 1980). Moreover, slope sediments are sometimes discontinuous, or at least highly variable in thickness. Nevertheless, there are observations that do suggest changing deformational styles across the slope.

Near the toe of the Sunda accretionary prism, slope sediments accumulate in thrust-backed wedges, within which strata steepen with depth to at least 15° (KARIG *et al.*, 1980). There is little evidence of a trench-facing syclinal limb. Slope basins on the shallower slope, immediately west of Nias show significantly different characteristics. Progressive arcward rotation of strata is still obvious, but internal folding and/or faulting is recognized (STEVENS, 1983). Processed seismic profiles (MOORE and CURRAY, 1980) indicate that these arcward dipping strata are resolved arcward by a synclinal fold rather than ending against thrust faults.

A quite similar progression of internal geometry is very clearly expressed in the slope basins of the Makran arc. The lowermost slope basins are primarily rotating sediment wedges (WHITE, 1982). However, some evidence for the development of a synclinal axis in the deeper sections of these basins is provided by the arcward flattening of the more steeply dipping strata beneath the thrust zone forming the inner edge of the basins, where acoustic coherence on the available profiles is lost. Nevertheless, there is no room for a significant trench-facing limb if the bounding thrust had an initial ramp dip of 20–30° and has since rotated 10° or more with the basin strata.

Arcward rotation continues to characterize the shallow strata in the series of slope basins to at least 70 km behind the deformation front (WHITE and LOUDEN, 1983). Although these authors cite the nearly constant spacing of morphologically defined inter-basin ridges as evidence for only minor shortening over this distance, their seismic profiles show basement ridges within the basins that do suggest shortening, as should result from rotation and thrusting of the suprajacent slope strata. Moreover, strata in the deeper sections of the acoustically defined basins are deformed into more symmetric synclines, suggesting an additional mechanism of shortening within these basins.

Seismic profiles and geologic mapping across the shallow water and emergent section of the active Makran slope show further changes in the structural style of slope deformation, although Plio-Pleistocene trenchward prograding clastic wedges (HARMS *et al.*, 1982) mask and diffuse the geometry of the slope basins. The gross near-surface geometry of these shallow water to non-marine basins is one of very open, more nearly symmetric synclines without systematic landward rotation. Between these synclines are sharp antiformal zones and reverse faults (e.g. FALCON, 1974), some of which are structually active at least 100 km north of the coastline (KARIG, pers. obs., 1978).

The geometries of structures in the slope strata across the inner slope of the Nankai Trough are not well expressed in seismic profiles, but in general they parallel the progression observed in the other arcs. A small slope basin 17 km behind the deformation front near DSDP site 583 (NASU *et al.*, 1982) and several near the base of the slope on profiles to the southwest (AOKI *et al.*, 1983) are rotated sediment wedges, without noticeable folding. Further upslope, the thickening carapace of slope sediments on the several published (AOKI *et al.*, 1983; NASU *et al.*, 1982) and other unpublished profiles show that open symmetrical folds are more common than slope basins with rotated fill.

This review of deformation at shallow levels across the three lower trench slopes demonstrates the marked change in style, from one dominated by rotation and thrusting near the toe to one of symmetric folding at the rear. However, data from the toe of the Nankai prism and from the slope basins, show that, despite the dominance of the two respective styles at the end points, both are active across the slope, but differ in proportional intensity.

Such a progressive deformation scheme, operating across the lower slope at an arcward decreasing rate, produces slope basins with a characteristic structural geometry. Stratal dips shallow upward as folds change from tight angular forms, cut by many shears, to symmetric very open syclines. Modification of the simply rotated slope strata near the toe of the prism to sheared angular folds in accretionary complexes is interpreted as a result of subsequent impression of a flattening-dominated deformation operating to the rear of the structurally most active toe of the accretionary prism.

Deformation of the subjacent accreted trench sediments is dominated by thrusting and rotation near the prism toe, but there is a significant component of flattening, as expressed in the diffuse deformation within the thrust sheets. If the gross kinematic framework within accreted sediments at shallow levels of the prism is similar to that defined by the slope sediments, we could also expect the influence of flattening in accreted sediments to be increasing arcward. This should lead to vertical attitudes of bedding and foliation and possibly also to isoclinal folding.

The geometry and structures related to rotation and to flattening are not well known at this time. Rotational deformation results from the bedding plane step thrusts, as well as from other type thrusts, and produces the arcward fanning of slope basin sediments. Structural features that might be associated with flattening include the angular folds with uniform limb thicknesses and steeply dipping axial plane cleavage. It is possible that some bedding parallel extensional structures could develop under a flattening deformation in steeply dipping beds well behind the deformation front. It remains to be seen whether a significant degree of flattening can be due to anisotropic volume loss during reduction in porosity. A flattening deformation may also be responsible for the passive rotation and approach toward parallelism of the fabric elements in the deeper levels of slope basins and subjacent accreted material in emergent accretionary complexes.

One other possible effect of flattening may be the intimately intersheared relations at the contact between accreted and slope sediments. This observation was previously attributed to shear along shallow dipping surfaces near the toe of the prism, but might alternatively occur later, when foliations are much steeper. Some negative evidence supporting the latter origin is the lack of significant deformation of this lithologic boundary where observed within 15 km of the deformation front, the zone of highest instantaneous deformation rate.

A major concern is whether the distribution of deformation determined across the Nankai and Sunda forearcs is compatible with observed deformation of slope strata at the rear of the actively deforming prism. From earlier computations I concluded that 3/4 of the instantaneous shortening, and half the total shortening absorbed by an accreted element, is presently accommodated within 15 km of the deformation front.

If valid, this distribution of deformation severely limits the shortening that should occur in slope sediments, most of which are deposited more than 15 km behind the deformation front. The maximum shortening ratio of slope sediments that were deposited 15 km behind the front can be obtained by calculating a differential value of ϕ between $x=15$ and very large values of x (where $u_x=u_a$). From the relation $\phi_x=\ln(u_0/u_x)$ and values of velocity presented earlier (Fig. 1), the horizontal shortening ratio (ϕ) for such slope sediments would be 0.43. Even approximate estimates of shortening ratios in strata of the deeper levels of slope basins are significantly greater than 0.5, which brings into question either the calculated distribution of deformation or the assumptions implicit in making this extrapolation.

At this point, I prefer to question the assumptions. One of these is the steady-state outgrowth rate, which produces a relationship between age of accretion and distance from the deformation front that becomes nearly linear with increasing distance (KARIG *et al.*, 1980). This in turn implies continuous accretion with constant input volumes. Discontinuous accretion over the time represented by the development of the prism would lead to greater deformation than calculated.

Another assumption is that deformation varies very smoothly across the upper levels of the accretionary prism, rather than being concentrated in certain areas. In short, the calculated distribution of deformation near the toe of the prism cannot presently be quantitatively extrapolated to explain the cumulative deformation near the rear of the prism, but the attempt is valuable in pointing out which parameters need to be better constrained.

7. *Conclusion and Overview*

Sediments, dominantly trench fill clastics, are initially stripped off the subducting oceanic plate in thrust sheets having geometries very similar to those in foreland thrust belts. This thrusting and resultant rotation of sediments thus defines a type of simple shear. Also operative near the toe of the prism is a diffusely distributed deformational style that causes shortening without significant bulk rotation of bedding. This style might be viewed as a flattening or coaxial deformation resulting from pure shear.

The changing style of deformation in slope sediments arcward across the prism leads to the interpretation that bulk non-rotational flattening becomes the dominant form of deformation more than 10 to 20 km behind the deformation front. Two causes for this general change in response can be suggested. One reason is the increase in distance of accreted elements from the interplate boundary, which is the controlling surface of simple shear (Fig. 7). A second reason could be the mechanical anisotropy in sediments imparted by the original layering. As layering is rotated to angles

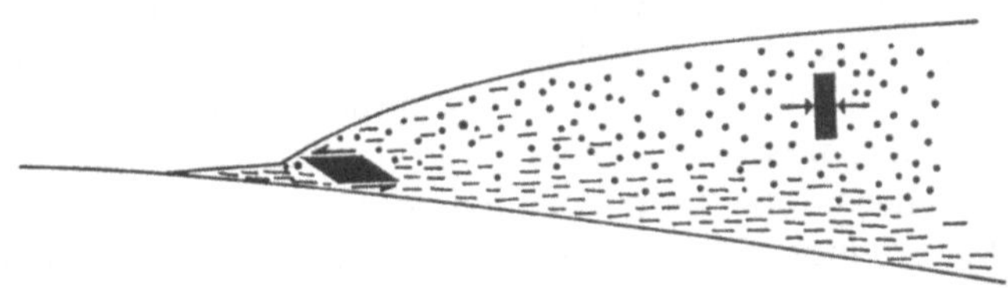

Fig. 7. Schematic diagram of deformation mechanisms within accretionary prisms, inspired by Fig. 14 of COWAN (1982). A rotational deformation, resulting from simple shear, dominates the areas near the zone of interplate slip, but extends upward as indicated by the dashed pattern. A flattening, or irrotational deformation occurs throughout the prism, but dominates away from the plate interface, as shown by the dotted pattern.

much more than 30° from the horizontal, and fault surfaces are strongly folded, the resistance to continued movement along the faults increases until it exceeds the rock strength in other failure modes. By either mechanism, the result is that rock masses within an accretionary prism are rotated over a large angle through a stress field within which the orientation of the maximum compressive stress changes relatively little.

The distribution of deformation is very similar across the toe of the most carefully analyzed sections of the Nankai and Sunda prisms. Qualitatively similar distributions have been obtained for several other prisms accreting clastic sediments (e.g. MOORE and BIJU-DUVAL, 1984; Karig, unpublished data) although lack of acoustic velocity data and/or control of boundary conditions precludes adequate analyses. It is interesting that the very different convergence rates between the Nankai and Sunda examples do not seem to affect the distribution of deformation (Fig. 5). Nevertheless, one might logically expect that the distribution is a function of lithology of accreted material. Non-accreting or tectonically eroding arcs may have very different distributions of deformation but the limited data (KARIG and RANKEN, 1983; VON HUENE *et al.*, in press) suggests that instantaneous deformation in these cases also is very strongly concentrated toward the base of the inner slope. No evidence has yet been presented for "locked trenches" (HELWIG and HALL, 1974) along convergent margins where normal ocean crust is being subducted.

Before the instantaneous deformation pattern deduced near the toe of an accreting prism can be adequately integrated with the deformational history of emergent accretionary complexes, more information concerning the age of accretion across the prism and better understanding of the deformation within slope basins must be obtained. Even so, we are finally approaching the stage where studies of gross morphology and structure of prisms, of material rheologies, and of more detailed structures can be integrated and where data from each different approach supply powerful insights and constraints for the others.

Acknowledgements

This work developed from the Deep Sea Drilling Project and several other studies supported by grants from the National Science Foundation. I am particularly indebted to D. L. Turcotte for assistance concerning continuum mechanics and also to J. Casey Moore and Cynthia Bray among others who have offered constructive criticism.

REFERENCES

AOKI, Y., T. TAMANO, and S. KATO, Detailed structure of the Nankai Trough from migrated seismic sections, *Am. Assoc. Petroleum Geol. Mem.*, **34**, 309–324, 1983.

ARTHUR, M. A., B. CARSON, and R. VON HUENE, Initial tectonic deformation of hemipelagic sediment at the leading edge of the Japan convergent margin, in *Scientific Party*, Init. Repts. DSDP, Vol. 56, 57, pt. 1, pp. 569–613, U. S. Gov't. Printing Office, Washington, 1980.

BRAY, C. J. and D. E. KARIG, Porosity in accretionary prisms and some implications, *J. Geophys. Res.*, **90**, 768–778, 1985.

BYRNE, T., Structural evolution of coherent terranes in the Ghost Rocks Formation, Kodiak Island, Alaska, in *Geol. Soc. London Sp. Publ.*, 10, pp. 229–244, 1982.

CARSON, B., J. YUAN, P. B. MYERS, Jr., and W. D. BARNARD, Initial deep-sea sediment deformation at the base of the Washington continental slope: a response to subduction, *Geology*, **2**, 561–564, 1974.

COWAN, D. S., Deformation of partly dewatered and consolidated Franciscan sediments near Piedras Blancas Point, California, in *Geol. Soc. London, Sp. Publ.*, 10, pp. 439–458, 1982.

COWAN, D. S., J. C. MOORE, S. M. ROESKE, and N. LUNDBERG, Structural features at the deformation front of the Barbados Ridge Complex, DSDP Leg 78A, in Init. Repts. DSDP, Vol. 78A, B. Biju-Duval and J. C. Moore *et al.*, pp. 535–548, U. S. Gov't Printing Office, Washington, 1984.

DAHLSTROM, C. D. A., Structural geology in the eastern margin of the Canadian Rocky Mountains, Wyoming Geol. Assn. Guidebook, 29th Ann. Field Conf., pp. 407–439, 1977.

FALCON, N. L., An outline of the geology of the Iranian Makran, *Geograph. J.*, **140**, 284–291, 1974.

FARHOUDI, G. and D. E. KARIG, Makran of Iran and Pakistan as an active arc system, *Geology*, **5**, 661–668, 1977.

HARMS, J. C., H. N. CAPPEL, and D. C. FRANCIS, Geology and petroleum potential of the Makran coast, Pakistan; Proc. 1982 Offshore S.E. Asia Conference, Singapore, Feb. 9–12, 1982 (in press).

HELWIG, J. and G. A. HALL, Steady-state Trenches?, *Geology*, **2**, 309–316, 1974.

HUNTING SURVEY CORPORATION, Reconnaissance survey of part of west Pakistan, Toronto, Maracle Press, 1960.

KARIG, D. E., Evolution of arc systems in the Western Pacific, *Am. Rev. Earth Planet. Sci.*, **2**, 51–75, 1974.

KARIG, D. E., Deformation in the forearc, Implications for mountain belts in *Mountain Building Processes*, edited by K. J. Hsu, pp. 59–72, Academic Press, New York, 1983.

KARIG, D. E., The framework of deformation in the Nankai Trough, in Init. Repts. DSDP, Vol. 87, H. Kagami and D. E. Karig *et al.*, U. S. Gov't Printing Office, Washington, D. C. (in press).

KARIG, D. E., G. F. MOORE, J. R. CURRAY, and M. B. LAWRENCE, Morphology and shallow structure of the lower trench slope off Nias Island, Indonesia, in *Tectonic and Geologic Evolution of Southeast Asian seas and Islands*, Am. Geophys. Un. Geophys. Mono., 23, edited by D. E. Hayes, pp. 179–208, 1980.

KARIG, D. E. and B. RANKEN, Marine geology of the forearc region, southern Mariana Island Arc, AGU Geophys. Mono., 27, pp. 266–280, 1983.

KARIG, D. E., H. KAGAMI, and DSDP LEG 87 SCIENTIFIC PARTY, Varied responses to subduction in Nankai Trough and Japan Trench forearcs, *Nature*, **304**, 148–151, 1983.

KARIG, D. E. and C. L. ANGEVINE, Geologic constraints on subduction rates in the Nankai Trough, in Init. Repts. DSDP, Vol. 87, H. Kagami and D. E. Karig *et al.*, U. S. Gov't Printing Office, Washington, D. C. (in press).

KENYON, C. S., Stratigraphy and sedimentology of the late Miocene to Quaternary Deposits of Timor, unpublished Ph.D. Thesis, Imperial College, London, 291 pp., 1974.

KIECKHEFER, R. M., G. G. SHOR, Jr., and J. R. CURRAY, Seismic refraction studies of the Sunda Trench and forearc basin, *J. Geophys. Res.*, **85**, 863–889, 1980.

LUNDBERG, N. and D. E. KARIG, Structural features from the Nankai Trough lower slope, DSDP sites 582 and 583, in Init. Repts. DSDP, Vol. 87, H. Kagami and D. E. Karig *et al.*, U. S. Gov't Printing Office, Washington (in press).

MCCALL, G. J. H. and R. G. W. KIDD, The Makran, southeastern Iran, The anatomy of a convergent plate margin active from Cretaceous to Present, in *Trench-forearc Geology, Sedimentation and Tectonics on Modern and Ancient Active Plate Margins*, *Geol. Soc. Lond. Sp. Publ.*, 10, edited by J. K. Leggett, pp. 387–397, 1982.

MOORE, G. F. and J. R. CURRAY, Structure of the Sunda Trench lower slope off Sumatra from multichannel seismic reflection data, *Marine Geophys. Res.*, **4**, 319–340, 1980.

MOORE, G. F. and D. E. KARIG, Structural geology of Nias Island, Indonesia, Implications for subduction zone tectonics, *Am. J. Sci.*, **280**, 193–223, 1980.

MOORE, J. C., J. S. WATKINS, and others, Progressive accretion in the Middle America Trench, southern Mexico, *Nature*, **281**, 638–642, 1979.

MOORE, J. C. and B. BIJU-DUVAL, Tectonic synthesis, Leg 78A DSDP, Structural evolution of offscraped and underthrust sediment, northern Barbados Ridge, in Init. Repts. DSDP, Vol. 78A, B. Biju-Duval and J. C. Moore *et al.*, pp. 601–621, U. S. Gov't Printing Office, Washington, D. C., 1984.

NASU, N. *et al.*, Multi-channel seismic reflection data across Nankai Trough, IPOD-Japan Basic Data Series, n. 4, 34 pp., Ocean Research Inst., Tokyo, 1982.

SEELY, D. R., P. R. VAIL, and G. G. WALTON, Trench slope model, in *The Geology of Continental Margins*, edited by C. A. Burk and C. L. Drake, pp. 249–260, Springer-Verlag, New York, 1974.

STEVENS, S. H., Sedimentation and deformation of Sunda Trench slope basins, Indonesia, unpub. M. Sc. Thesis, U. C. San Diego, 1983.

TAIRA, A., M. TASHIRO, M. OKAMURA, and J. KATTO, The geology of the Shimanto Belt in Kochi Prefecture, Shikoku, Japan, in *Geology and Paleontology of the Shimanto Belt*, edited by A. Taira and H. Tashiro, pp. 319–389, 1980.

TAIRA, A., H. OKADA, J. H. WHITAKER, MCD.,and A. J. SMITH, The Shimanto belt of Japan, Cretaceous-lower Miocene active margin sedimentation, in *Trench-forearc Geology, Sedimentation and Tectonics on Modern and Ancient Active Plate Margins*, *Geol. Soc. Lond. Sp. Publ.*, 10, edited by J. K. Leggett, pp. 5–26, 1982.

TORRINI, R., R. C. SPEED, G. S. MATTIOLI, and J. B. SAUNDERS, Tectonic relationships between forearc basin strata and the accretionary complex at Bath, Barbados, *Geol. Soc. Am. Bull.* (in press).

UMEMURA, H. and Y. UEMATSU, Tectonic deformation of clastic dikes in Gyodo-Misaki, Kochi Prefecture, with particular reference to their relations with slaty cleavage, Research reports of the Kochi University, *Natural Sci.*, **29**, 19–34, 1980 (in Japanese).

VON HUENE, R., M. LANGSETH, N. NASU, and H. OKADA, A summary of Cenozoic tectonic history along the IPOD Japan Trench Transect (in press).

WHITE, R. S., Recent fold development in the Gulf of Oman, *Earth Planet. Sci. Lett.*, **36**, 85–91, 1977.

WHITE, R. S., Deformation of the Makran accretionary sediment prism in the Gulf of Oman (N.W. Indian Ocean), in *Trench and Forearc Sedimentation and Tectonics in Modern and Ancient Subduction Zones, J. Geol. Soc. London, Sp. Publ.*, **10**, edited by J. S. Leggett, pp. 357–372, 1982.

WHITE, R. S. and K. E. LOUDEN, The Makran continental margin, Structure of a thickly sedimented convergent plate boundary, *Am. Assoc. Petroleum Geol. Mem.*, **34**, 499–520, 1983.

YOSHII, T., W. J. LUDWIG, N. DEN, S. MURAUCHI, M. EWING, A. HOTTA, P. BUHL, T. ASANUMA, and N. SAKAJIRI, Structure of southwest Japan margin off Shikoku, *J. Geophys. Res.* **78**, 2517–2525, 1973.

Formation of Active Ocean Margins, edited by N. Nasu *et al.*, pp. 179–191.

SOME IMPLICATIONS REGARDING TECTONIC MECHANISMS FROM THE STRUCTURAL DIVERSITY ALONG SEVERAL MODERN CONVERGENT MARGINS

Roland VON HUENE

U.S. Geological Survey, Middlefield Road, Menlo Park, California 94025, U.S.A.

Abstract. An analytical approach commonly used in the geosciences is to compare the physical and kinematic development of structure with the dynamics of causative features. Along convergent margins, geological and geophysical data are becoming sufficiently quantitative to begin using such an approach. In a deformational environment dominated by plate convergence, not only convergence but also differences of sediment condition and the relief of the subducting seafloor are effective controls of structural style.

1. Introduction

A remarkable diversity of tectonic structure along active convergent margins has been recognized during the past decade from studies based on multichannel seismic reflection records and drill cores. In the past 3 years, previous difficulties of seismic resolution and penetration in deep ocean trenches have been partly overcome and subsurface structure 5–6 km deep has been imaged by carefully processing multichannel seismic reflection records. Also, the results of the International Program of Ocean Drilling (IPOD) on active margins show a varied tectonic history even when rates of plate convergence have been constant during the period in question. Single trenches are now known to contain a spectrum from materially conservative to erosional tectonic processes. Thus recent studies of the fronts of convergent margins indicate variation of tectonic style in space and time. This structural *variation* in space and time is paradoxial in an environment dominated by *constant* plate convergence. However, it presents an opportunity to investigate causative tectonic processes by correlating the time/space parameters of structural development with those of modern processes. The tectonic processes along modern convergent margins involve features of different dimension. For instance, gross plate convergence involves subduction of large oceanic and continental plates over long time periods whereas the seamount colliding with a margin produces a relatively small and quickly developed structure. Deposits of pelagic and hemipelagic sediment uniformily cover broad areas whereas the toe of a large slump may be locally restricted in the trench axis. Thus if the dimensions of a structure are known, these dimensions can be compared with those of a feature entering the trench or on the converging oceanic plate. This approach may seen selfevident yet the lack of morphological and structural data with sufficient resolution have hampered its application.

A correlation between structure and process requires quantitative data; the gross structural features and broad temporal control are available, in some well-studied convergent margin areas. Thus the importance of major processes can be approximated with certain modern data sets. I recently constructed a spectrum of tectonic processes along convergent margins (VON HUENE, 1984b), as shown in Table 1. The relative vigor of any part of the spectrum is largely controlled by 3 conditions along the convergent boundary: 1) the rate and direction of convergence, 2) the topographic relief of the ocean floor entering the trench, and 3) the abundance and composition of sediment input. Of these three conditions the rate of plate convergence is obviously a dominent and driving one. Relative plate motion is commonly constant or varies gradually along a single trench. The second condition, topography of the ocean floor entering a trench, can be very effective provided the topography is not smoothed over by sediment. Commonly the seafloor is broken into 30 km to 50 km long horst and graben striking parallel or at a highly oblique angle to the trench axis (cf. AUBOUIN *et al.*, 1981). In addition, seamounts, oceanic plateaus, and masses so large that they cannot be subducted must frequently have been subducted in the past (BEN AVRAHAM and NUR, 1982). The third condition related to sediment abundance is perhaps more important than has been recognized. Sediment sources frequently enter the trench at points and the sediment is distributed from tens to hundreds of kilometers axially by currents. However, axial sediment transport is generally not sufficient to mask a sediment flooded and starved trench segment (i.e. Antilles, Java, Peru-Chile trenches).

Several trenches have been crossed with multichannel instruments and have been drilled during the IPOD so that sufficient data are available to make some first approximations regarding the significance of the three

Table 1. Tectonic processes along the front of convergent margins ordered from accretion to erosion on the horizontal axis and from strong to weak coupling on the vertical axis (from VON HUENE, 1984b).

Accretion	Sediment subduction	Erosion	
Accretion by *tectonic addition* of sediment and crustal rock to the front of a margin; associated with compressional structure	Sediment *trapped in pockets* of igneous ocean floor and thus guarded from offscraping	Erosion from *abrasion* by drag of positive features against the underside of the upper plate	Strong coupling
Accretion by *underplating*: subducted sediment detached from the descending plate and added from below to the upper plate	Sediment subduction by the passage of *deep-ocean sediment* beyond the front of the margin, commonly beneath a décollement	Erosion from *fragmentation* of the underside of the upper plate by hydrofracture and transport of slurry with subducting matter down a subduction zone	
Accretion by *deposition* of sediment and products of mass wasting that become tectonically incorporated into an accretionary complex	Sediment subduction including the upper *terrigenous debris* ponded in a trench axis	Erosion by *failure* of margin front through mass wasting, with gravity transport to trench floor and incorporation into subducting sediment	Weak coupling

conditions that control many tectonic processes at the fronts of active convergent margins. Since the Oji Conference, much of the material I presented has been published (VON HUENE, 1984a, b). Therefore, I will review only briefly the published examples of structural variation along single margins where the rate and direction of convergence is essentially the same everywhere. In this way, two of the three conditions controlling deformation are isolated and the resulting structure can be illustrated.

2. *Examples of Structural Diversity*

One of the first margins where structural diversity was outlined by both multichannel seismic records and IPOD drilling was the Middle America Trench (Fig. 1). Off southern Mexico the trench was drilled during Leg 66 and the staff interpreted the geologic and geophysical data as showing a 10-m.y. old accretionary body against a structurally truncated continental edge (SHIPLEY *et al.*, 1980; WATKINS, MORE *et al.*, 1981). They estimated that about 30% of the sediment entering the trench is subducted and that an addition 30% of the sediment is added to the underside of the accretionary body by underplating. In contrast, the Middle

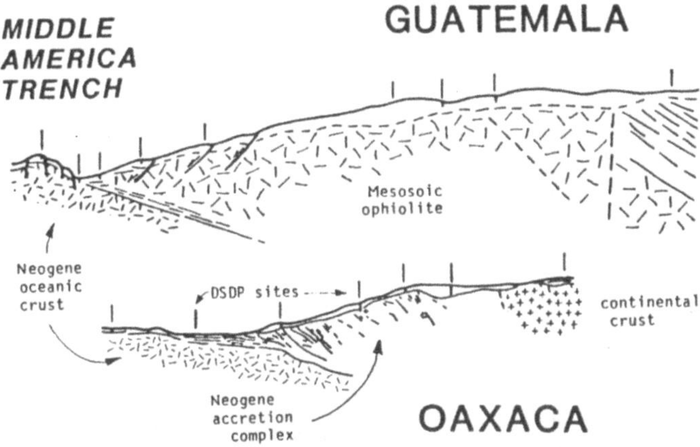

Fig. 1. Diagrammatic sections across the Middle America Trench showing the structural differences between the IPOD transects off southern Mexico and Guatemala. The Mexican transect shows a 10 m.y. old accretionary complex backed against the truncated Mexican continental crust of Mesozoic and older rocks. Thus the present tectonic regime was preceded by another one in which tectonic erosion dominated. The Guatemalan transect shows no accretionary complex and instead, slope deposits of Eocene age are underlain by a Mesozoic ophiolitic complex. In the area of the transect the front of the margin appears to be collapsing.

America transect farther south off Guatemala was marked by the recovery of Quaternary to Eocene rather than Quaternary to upper Miocene slope deposits and the staffs of Legs 67 and 84 interpreted this result as indicating the absence of accretion and net subduction of all sediment entering the trench since early Eocene time (AUBOUIN, VON HUENE *et al.*, 1982). The two IPOD transects across the Middle America Trench, show both modern subduction-accretion and non-accretion and the principle difference in conditions appears to be a local high rate of sedimentation near a large canyon in the trench axis off Mexico, and sparce sediment off Guatemala. The topography of the converging ocean crust is similar in both areas of the trench as shown by SEABEAM surveys (AUBOUIN *et al.*, 1981). Thus in the Middle America Trench, a sediment flooding of the trench axis appears to promote accretion whereas sediment starvation, and incomplete burial of topography appears to promote non-accretion. This correlation between accretion and sediment flooded trenches was extended on a Pacific wide basis by HILDE (1983).

A similar condition of sediment flooding and starvation are revealed along the Antilles margin (BIJU-DUVAL *et al.*, 1982, WESTBROOK *et al.*, 1982). In the south, a large sediment source from South America floods the margin and a wide accretionary complex has developed, whereas in the north, far from the principal sediment source, the sediment starved segment is associated with a narrower accretionary complex (Fig. 2). The two complexes contrast in structural style as well; much of the southern complex is repeatedly imbricated whereas in the northern area a well developed decollement is the principal structure. The change of structure from areas of thick to thin sediment is sharp. It correlates with sediment abundance but is not entirely without influence from the effects of topography on the converging oceanic crust. The subduction of an aseismic ridge corresponds to a change of deformational style. Thus along both the Antilles margin, sediment abundance and the topography of the oceanic plate seem to work in an interactive manner to influence structural style.

Considerable multichannel seismic data along the Aleutian Trench has been recently reprocessed (MCCARTHY and SCHOLL, in press; VON HUENE *et al.*, 1983; VON HUENE, 1984a, b). Here the convergence rate and direction change gradually from about 7 mm/yr at the eastern end to strike slip motion along the western end but plate motion changes little over 1000 km long segments. Along the eastern 600 km of the trench, a relatively complete set of geophysical data record structural differences despite an essentially constant rate of plate convergence (Fig. 3). Well developed decollement and thrust packets are recorded along one line (record A, Fig. 3) in a network off southern Kodiak Island. These thrust packets

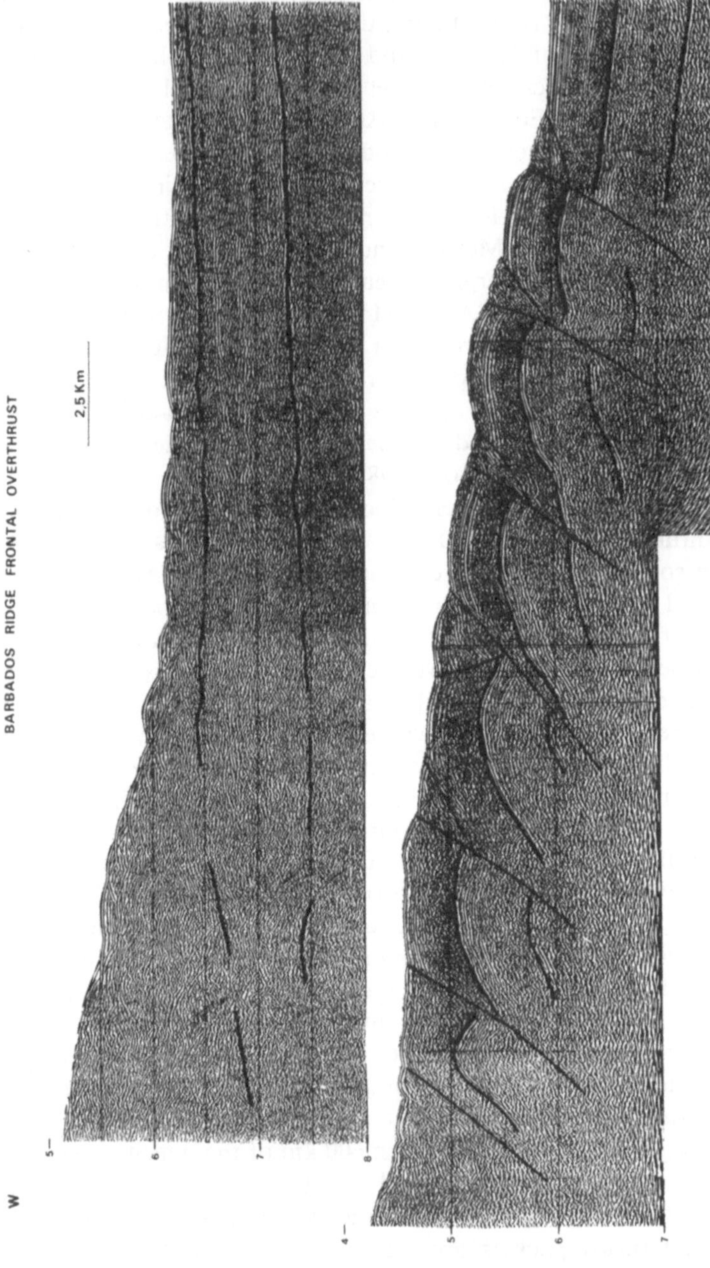

Fig. 2. Two contrasting structural styles in records from the northern and southern Lesser Antilles. Record A, from the northern area shows a decollement with little deformed sediment below and accreted sediment above. Record B from the southern area shows imbrication (from BIJU-DUVAL *et al.*, 1982).

become progressively more deformed landward over a distance of about 40 km. About 220 km northeast, another record illustrates a mildly faulted but thickened sedimentary sequence at the front of the margin (record B, Fig. 3). The sediment appears to be subducting beneath the mid-slope area, but it cannot be clearly imaged because it plunges beyond the 5 km depth of seismic resolution. Still another structural style is seen approximately 290 km farther northeast (record C, Fig. 3). Here, a broad landward-dipping sequence of reflections alternates with a seaward-dipping sequence and therefore both subduction and obduction are indicated. Finally, at the northeast end of the eastern Aleutian Trench, the time stratigraphic information projected into a seismic record from an industry drill hole indicates a major decollement not obvious in the seismic data alone (record D, Fig. 3). The Eocene contact can be followed downslope along reflectors that crop out near the trench axis (PLAFKER *et al.*, 1982). Below the outcropping reflectors, other reflectors can be traced to DSDP hole 178 where they are Pleistocene. Thus, an oceanic section capped by a layer of Pleistocene sediment is being subducted beneath an Eocene and younger continental-margin section along a master decollement essentially concordant with bedding. Perhaps more significant is the indication that the Eocene section is now truncated at the front of the margin and a significant mass of rock is missing. The removal of the Eocene and younger rock is curious considering that only a short distance away much sediment has been accreted.

The four seismic records along a 600 km long segment of the eastern Aleutian Trench show structure indicating imbrication, decollement, sediment subduction, tectonic erosion, and obduction (lower inset, Fig. 3), and in a few hundred kilometers, much of the spectrum of convergent margin structural diversity has been produced without change in plate motion. Along the eastern Aleutian Trench, conditions of sedimentation and topographic relief of the lower plate must be the major causes of structural diversity.

The remarkable structural diversity along seismic records spaced 100 km to 200 km apart stimulates curiosity regarding the diversity of structure at a spacing of tens of kilometers. The record shown previously (record A, Fig. 3), which exhibits decollement structure, is flanked by other records across the margin (Fig. 4). The other records show a closer spacing of folds and thrust faults above the basal decollement. Interestingly, some of the initial deformation has produced seaward dipping reverse faults that are apparently overwhelmed by the landward dipping thrust faults as deformation progresses. In addition, the thickness of subducted sediment below the decollement and even the width of the accretionary zone

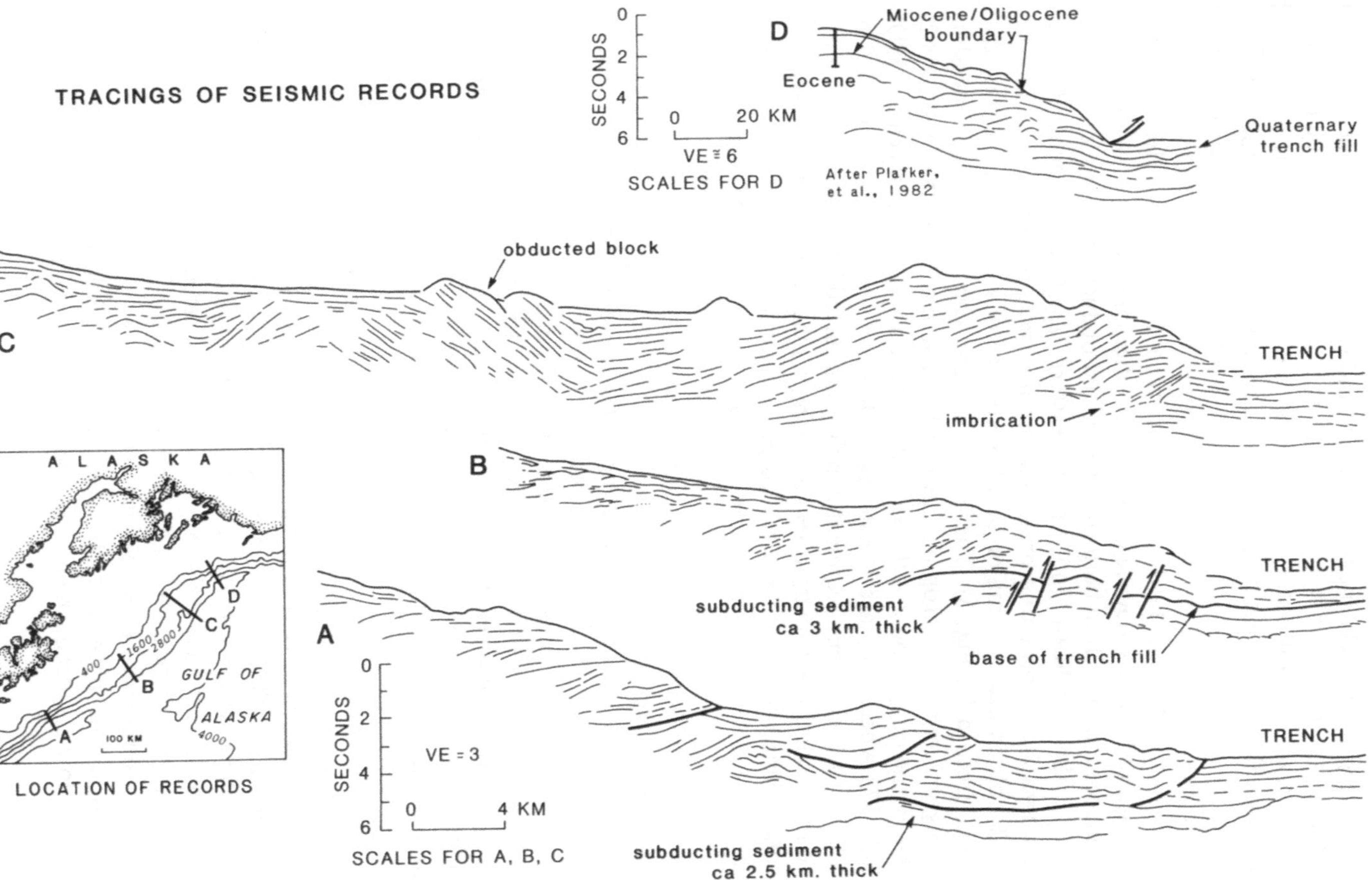
TRACINGS OF SEISMIC RECORDS
D
SECONDS
0
2
4
6
0
20 KM
VE ≅ 6
SCALES FOR D
Miocene/Oligocene boundary
Eocene
Quaternary trench fill
After Plafker, et al., 1982
C
obducted block
TRENCH
imbrication
B
TRENCH
subducting sediment ca 3 km. thick
base of trench fill
A
SECONDS
0
2
4
6
VE = 3
0
4 KM
SCALES FOR A, B, C
TRENCH
subducting sediment ca 2.5 km. thick
A L A S K A
400
1600
2800
GULF OF ALASKA
4000
100 KM
A
B
C
D
LOCATION OF RECORDS

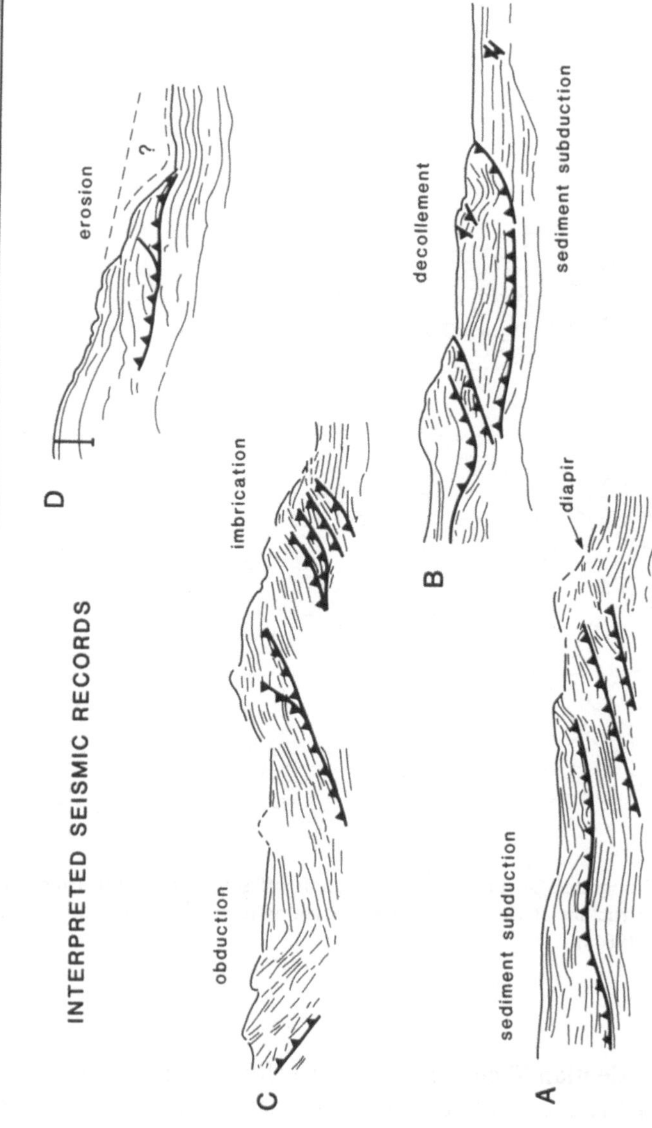

Fig. 3. Tracings of 24-channel seismic records across the landward slope of the eastern Aleutian trench showing the diversity of structure at intervals of hundreds of kilometers. Lower inset shows interpreted records and indicates the tectonic processes involved (after VON HUENE, 1984b).

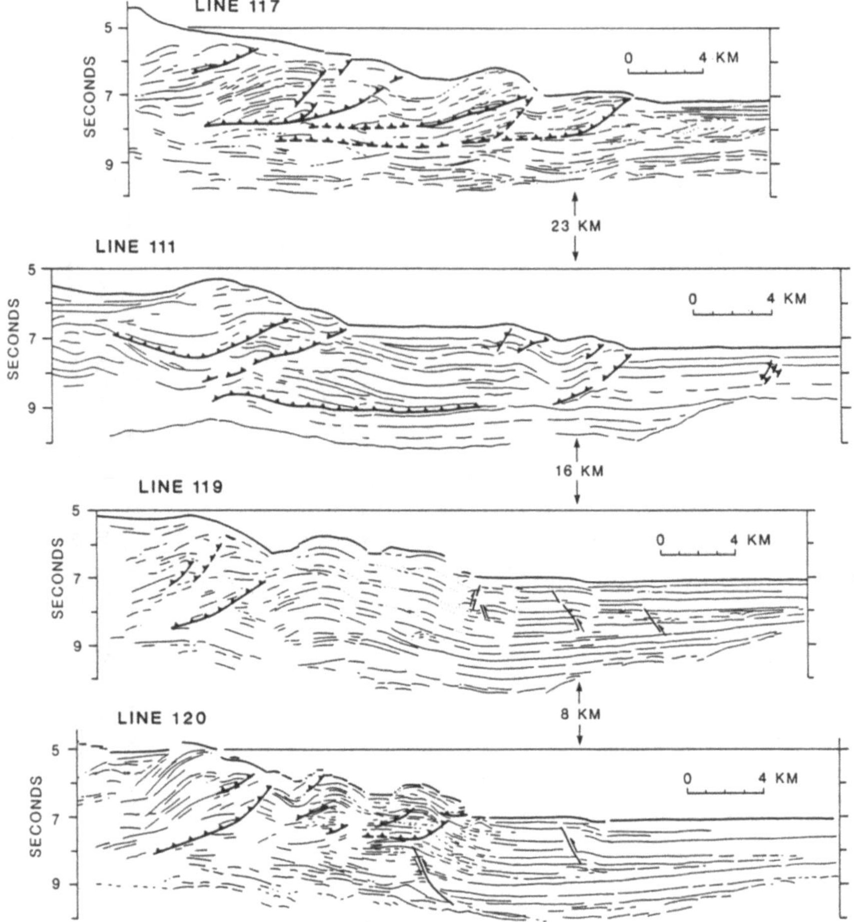

Fig. 4. Tracings of migrated 24-channel seismic records off southern Kodiak Island showing the diversity of structure at the front of the margin at intervals of tens of kilometers (after VON HUENE, 1984a).

vary.

The eastern Aleutian Trench is filled with thick sediment (Figs. 3 and 4). Consequently the relief of the igneous oceanic basement is so deeply buried that the horst and graben structure which develops as the oceanic plate is flexed into the trench is not seen at the seafloor. The seismic records have not yet been converted to depth sections and therefore the topography of the buried igneous oceanic crust is not well known. Rough estimates indicate a basement ridge beneath the record displaying the wide thrust packets and the long decollement and a basement trough beneath the records showing closer spacing of thrust faults. However, the relation between basement relief and structure in the scraped off sediment is unclear and the correlation between basement topography and structure requires more deep seismic data. In this area of the Aleutian Trench, the great thickness of sediment in the trench (up to 5 km) and lack of surface horst and graben relief on the oceanic plate suggest that structural diversity is closely related to sediment diversity. Locally, glacial channels empty into the trench. The trench sediments show differing seismic character away from the glacial channels. Therefore fault spacing and character may be influenced by thickness and texture of the subducting sediment.

3. *Summary*

The relative significance of causative conditions on structural diversity can be examined using well migrated seismic data and the results of IPOD drilling. I defined the principal causative conditions as, 1) the rate and direction of convergence, 2) the topographic relief of the ocean floor entering the trench, and 3) the abundance and composition of sediment in the trench axis. The rate and direction of convergence can be isolated as a variable by studying the structural diversity along a single trench where convergence is constant. Structural diversity is known from several trenches and the spectrum of tectonic processes from accretion to non-accretion are found along the Middle America and Aleutian trenches. In the Middle America Trench off Mexico, where the supply of sandy sediment is locally abundant, accretion predominates although perhaps an equal amount of sediment is subducted (WATKINS, MOORE *et al.*, 1981), whereas off Guatemala, where the muddy sediment is not sufficiently thick to smooth out topography, non-accretion predominates (AUBOUIN, VON HUENE *et al.*, 1982). The Antilles (BIGU DUVAL *et al.*, 1982, MARLOW *et al.*, 1974) and the Peru-Chile margins (VON HUENE and KULM, 1984 in press) show a similar rough correlation between sediment abundance that smooths or subdues topography, and the vigor of accretion. The structurally diverse eastern Aleutian Trench, however, has little horst and graben topography entering the trench because sediment is more than a kilometer thick yet it displays a broad structural diversity at scales of hundreds of kilometers. This diversity, even at a scale of tens of kilometers, involves the spacing of thrust faults and the thickness of the subducting sediment. Therefore diversity in sediment character appears to most strongly affect diversity in structure. With more complete seismic records, the geophysical and geological data along convergent margins may soon allow a "one-to-one" correlation. None-the-less, some approximations can be made from existing data to indicate the relative significance of controls on deformational style.

There is not much doubt that the direction and rate of plate convergence is a dominent tectonic control along the fronts of modern subduction zones. Perhaps equally effective in controlling the spacing of thrust faults in zones of imbrication, the depth of the decollement, and thus the amount of sediment accreted or subducted, is the character of the sedimentary section along a trench axis. Sediment variability probably imparts a variability in the friction within the subduction zone. Trench sediment 1 km thick is about 50% interstitial water and has a low permeability (cf. AUBOUIN, VON HUENE *et al.*, 1982). Interstitial water content and permeability are important controls of pore water pressure, and elevated pressure greatly reduces friction along faults as well as the strength of sediment. Thus the texture and thickness of sediment in a trench can influence the tectonic coupling between the upper and lower places of a subduction zone. Similarly, topographic relief of the igneous oceanic basement is also likely to influence structural style, particularly where the feature is not smoothed over by sediment. Certainly a positive feature entering a trench and colliding with the landward slope is bound to leave some kind of a structural scar (see KOBAYASKI, this volume).

This preliminary analyses of seismic data, some of which is presently being processed, suggests the potential of a comparative approach for tectonic studies based on closely spaced networks of well processed multichannel records. Such surveys in areas where the conditions of plate motion, seafloor topography, and sediment condition are varied would provide further insight regarding tectonic mechanisms along convergent margins.

REFERENCES

J. AUBOUIN, J. F. STEPHAN, V. RENARD, and P. LONSDALE, A Seabeam survey of the Leg 67 area, Middle American Trench off Guatemala, *Nature*, **294**, 146, 1982.

J. AUBOUIN, R. VON HUENE *et al.*, Initial reports of the Deep Sea Drilling Project, vol. 67, U.S. Government Printing Office, Washington, D.C., 1982.

Z. BEN AVRAHAM and A. NUR, Continental accretion and orogeny; from oceanic plateaus to allochthonous terranes, *Science*, **213**, n. 4503, 47–54, 1982.

B. BIJU-DUVAL, P. LEQUELLEC, A. MASCLE, V. RENARD, and P. VALERY, Multibeam bathyretric survey and high resolution seismic investigations on the Barbados Ridge complex (Eastern Caribbean): A key to knowledge and interpretation of an accrretionary wedge, *Tectonophysics*, **86**, 275–304, 1982.

T. HILDE, Sediment subduction versus accretion around the Pacific, *Tectonophysics*, **99**,

381–397, 1983.

M. S. MARLOW, L. E. GARRISON, R. G. MARTIN, J. B. A. TRUMBULL, and A. K. COOPER, Tectonic transition zone in the northeastern Caribbean, *J. Res. U.S. Geol. Survey*, **2**, 289–302, 1974.

J. MCCARTHY and D. W. SCHOLL, Mechanisms of subduction accretion along the Central Aleutian Trench, *Bull. Geol. Soc. Amer.*, **96**, 691–701, 1985.

G. PLAFKER, T. R. BRUNS, G. R. WINKLER, and R. G. TYSDALE, Cross section of the eastern Aleutian Arc from Mount Spurr to the Aleutian Trench near Middleton Island, Alaska, Geol. Soc. Amer. Map Chart Ser. MC-28-P, 1982.

T. H. SHIPLEY, K. J. MCMILLEN, J. S. WATKINS, J. C. MOORE, H. SANDOVAL-OCHOA, and J. L. WORZEL, Continental margin and lower slope structures of the Middle America Trench near Acapulco, Mexico, *Marine Geol.*, **35**, 65–82, 1980.

R. VON HUENE, J. MILLER, M. FISHER, and G. SMITH, An Eastern Aleutian Trench seismic record, in *Seismic Expression of Structural Styles—A Picture and Work Atlas*, ed. A. W. Bally, Amer. Assoc. Pet. Geol. Stud. Geol., 15, 1983.

R. VON HUENE, Structural diversity along modern convergent margins and the role of overpressured pore fluids in subduction zones, *Bull. Soc. Geol. France*, **XXVI**, n. 2, 11–23, 1984a.

R. VON HUENE, Tectonic processes along the front of modern convergent margins—Research of the Past Decade, *Ann. Rev. Earth Planet. Sci.*, **12**, 359–381, 1984b.

R. VON HUENE, L. D. KULM, and J. MILLER, Structure of the frontal part of the Andean convergent margin, *J. Geophy. Res.* (in press).

J. S. WATKINS, J. C. MOORE *et al.*, Accretion, underplating, subduction, and tectonic evolution, Middle America Trench, Southern Mexico: results of DSDP Leg 66, *Oceanol. Acta*, **4**, 213–234, 1982.

G. K. WESTBROOK, M. J. SMITH, J. H. PEACOCK, and M. H. POULTER, Extensive underthrusting of undeformed sediment beneath the accretionary conplex of the Lesser Antilles subduction zone, *Nature*, **300**, 625–628, 1982.

Formation of Active Ocean Margins, edited by N. Nasu *et al.*, pp. 193–219.

INTERNAL STRUCTURES OF THE ACCRETIONARY WEDGE IN THE NANKAI TROUGH OFF SHIKOKU, SOUTHWESTERN JAPAN

Hideo KAGAMI

Ocean Research Institute, University of Tokyo, Minami-dai, Nakano-ku, Tokyo 164, Japan

Abstract. The Nankai accretionary wedge is actively accreting with sediment supply from the Akaishi Mountains, and shows a intricate tectonic wedge consisting of more than four zones; protothrust zone with a velocity range between 1.7–2.1 km/s, imbricate thrust zone with a velocity range between 2.5–3.2 km/s, multiple decollement zone with a velocity range between 3.5–3.9 km/s, and earthquake thrust zone with a velocity range between 4.0–4.9 km/s. There still exists a deeper portion of the wedge under the Tosa Terrace, but it can not be clarified because of thick overlying sediments in the Tosa forearc basin.

The complex model is considered for stability conditions of the Nankai wedge. The steady condition of sliding over the basal decollement is only observed in the protothrust and imbricate thrust zones, where high basal fluid pressure ratio is observed. The primary dewatering from the wedge sediments takes place in these zones by kinetic processes such as gravitational loading or tectonic stress. The expelled water can be released through imbricate spray thrusts connecting from the basal decollement, which are shown by their strong reflectivity on seismic image.

While, new decollements are forming higher in the wedge in the seaward multiple decollement zone, where the basal and internal fluid pressure ratio become equal. Comparing with a referenced basin sequence in Japan, thermal gradient, porosity and diagenetic condition can be estimated for the wedge. It is found that a relative increase in fluid pressure ratio in the multiple decollement zone is caused by the secondary dewatering process as a result of interlayer water discharge by smectite/illite alteration. A large-scale horizontal movement by multiple decollements can be maintained through such a dewatering, associated with a stability condition postulated by DAHLEN (1984).

Further landwards, earthquake stresses are being accumulated within the wedge and their occasional releases as low-angle thrust faults have shown a nature of seismic deformation in the earthquake thrust zone, where a increase in basal friction and internal fluid pressure ratio is expected. Particularly important findings are that upon accretion of trench-fill sediments with a velocity of 1.7 km/s, they quickly become consolidated to have a nature of seismic deformation and that the earthquake thrust zone changes its position relative to the trough axis from place to place. Such a rapid increase of velocity in accreted sediments causes advancement of the earthquake thrust zone toward the trough axis resulting in wedge shortening. While, the landward retreat of the earthquake thrust zone may, in the long run, cause a sediment accretion.

1. Introduction

It was in the late 1970s that the internal structure of accretionary wedges was beginning to be investigated at many convergent margins by using multichannel seismic reflection profiling. By the beginning of 1980 it became possible to obtain high-quality records for the Nankai Trough (OKUDA *et al.*, 1979; TAMANO *et al.*, 1981; NASU *et al.*, 1982; AOKI *et al.*, 1982; KAGAMI *et al.*, 1982; TAMANO *et al.*, 1983). As a result of the analysis of these relatively accurate records, the Nankai Trough has become one of the regions of the world where the internal structures of accretionary wedges are best understood.

In the summer of 1983, marine geological and geophysical investigations of the Nankai Trough off Shikoku were carried out by the voyages of the Hakuho-maru of the Ocean Research Institute, University of Tokyo (KH 83-2) and of No. 25 Kaiko-maru, the latter of which was made available through Grant in Aids on the special project research entitled "The ocean characteristics and their changes" for the period of 1981–1983.

On the Hakuho-maru, a nine litre air-gun and a twelve channel streamer were used with air pressure of 80 kg/cm^2. Also a nine litre air-gun with an air pressure of 45 kg/cm^2 and a six channel streamer were used on the No. 25 Kaiko-maru. The records obtained were processed at the Computer Center of the University of Tokyo by using HITAC M-280H. The resultant seismic sections will be discussed in the next section.

2. *Interpretation of the Seismic Sections*

The bathymetry and locations of the seismic lines are shown in Fig. 1. There are three lines which run parallel to the axis of the trough. The line furthest seaward, L 15, runs over oceanic crust of the Shikoku Basin. This profile shows a basement ridge at shot point (SP) 500 and a graben at SP 650 (Fig. 2). Within the graben, "the lower turbidite and earlier sequences" identified at Site 297 (KARIG, INGLE *et al.*, 1975) are clearly recognized between 7.0 and 7.5 s in two-way traveltime. This fact means that the development of the graben goes back to the past when the Shikoku Basin opened. The graben continues to the northwest and is again observed at SP 1400 on line L 12 where it becomes deeper as a consequence of reactivation of the old structure at the subduction zone. The basement ridge at SP 1200 on line L 12 represents a part of the central rift zone of the Shikoku Basin (ISEZAKI and YASUKAWA, 1985), and it seems to effectively trap trench-fill turbidite.

Line L 3 runs parallel to the topographic contours along the edge of the accretionary toe. The profile between SP 0 and 650 shows a complex reflection pattern composed of oblique and lenticular reflections, while that between SP 650 and 1300 shows much smoother reflections. This indicates that the imbricate thrust zone east of the area deformed without vertical displacement, while the western imbricate thrust zone deformed with vertical displacement. A graben is observed on the oceanic crust between SP 700 and 1150. The extension of this graben is tracable to Line L 1 at SP 400 (Fig. 3). From its distribution, it is thought to trend northwest, which is almost parallel to local lineation of the magnetic anomalies (ISEZAKI and YASUKAWA, 1985).

Line L 2 crosses the middle part of the accretionary wedge. There is a topographic high between SP 1200 and 1400, which is called the Minami-muroto knoll (SATO and SAKURAI, 1975). Two types of reflections are identified at the knoll. One shows almost all the upper layers on the profile to be dipping to the northwest with an angle of about 14°. A large-scale dip like this (Fig. 2, SP 1000 on Line L 2) might be caused by imbricate thrusts, because their direction of dip agrees with the general trend of subduction. The dipping reflectors terminate near the valley at SP 850 and the valley may be a tectonic line representing either a basement depression or a transform fault. The other type shows nearly horizontal reflectors at the base of the sediment column at about 5.0 s in two-way traveltime near SP 1400. These represent multiple decollements. They are also recognized on the cross lines such as at 55-4 (Fig. 6). Local development of the multiple decollements may cause thickening of the accretionary

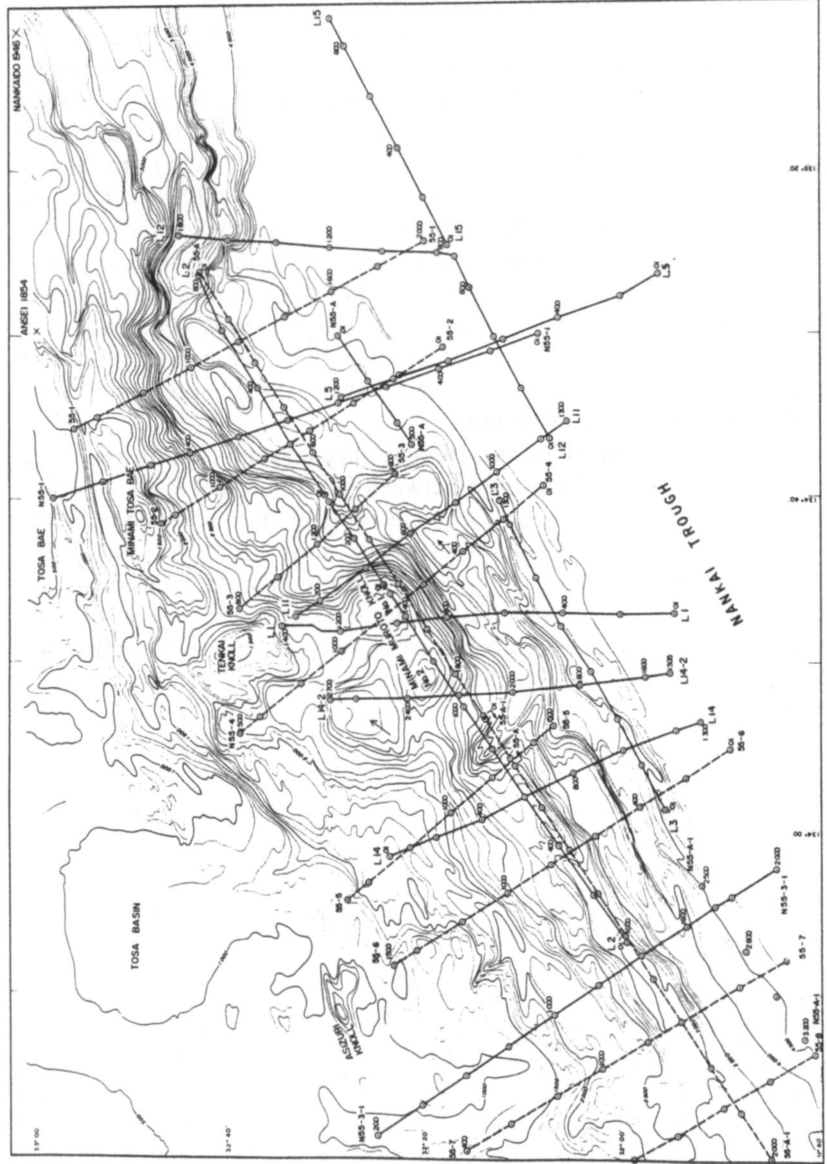

Fig. 1. Submarine topography and existing multichannel lines along the Nankai Trough off Shikoku. L-lines represent the present survey.

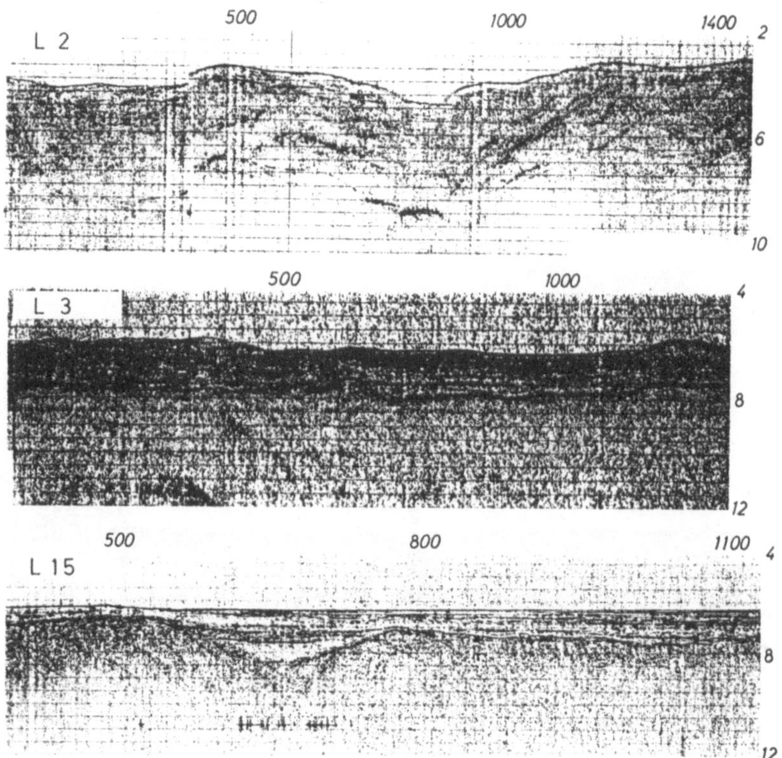

Fig. 2. Processed sections running parallel to the trough axis. Profiles L 2 and L 3 twelve channel data obtained onboard the Hakuho-maru and L 15 is six channel data on No. 25 Kaiko-maru. The upper numerals are shot point number. The right-side numerals are two-way traveltime in second.

wedge, thus contributing to formation of the dipping reflectors in the upper layers.

Next, we shall discuss perpendicular line crossing the Nankai Trough (Fig. 3). Line L 14-2 runs from the bottom of the trough through No. 2 Minami-muroto knoll at SP 2400 to an accretionary basin behind the knoll at SP 2600. At SP 1600, the trench-fill turbidite is slightly deformed to from a box-fold. This part is called the protothrust zone. At about 1650, the bottom shoals suddenly, where the first thrust is developing, known as the imbricate thrust boundary. From there inward, the sequence is called the imbricate thrust zone. Thus, a train of imbricate thrust blocks continues until SP 1850, where the bottom rises rapidly. Deduced from the topography only, this change of the slope is known as the major thrust

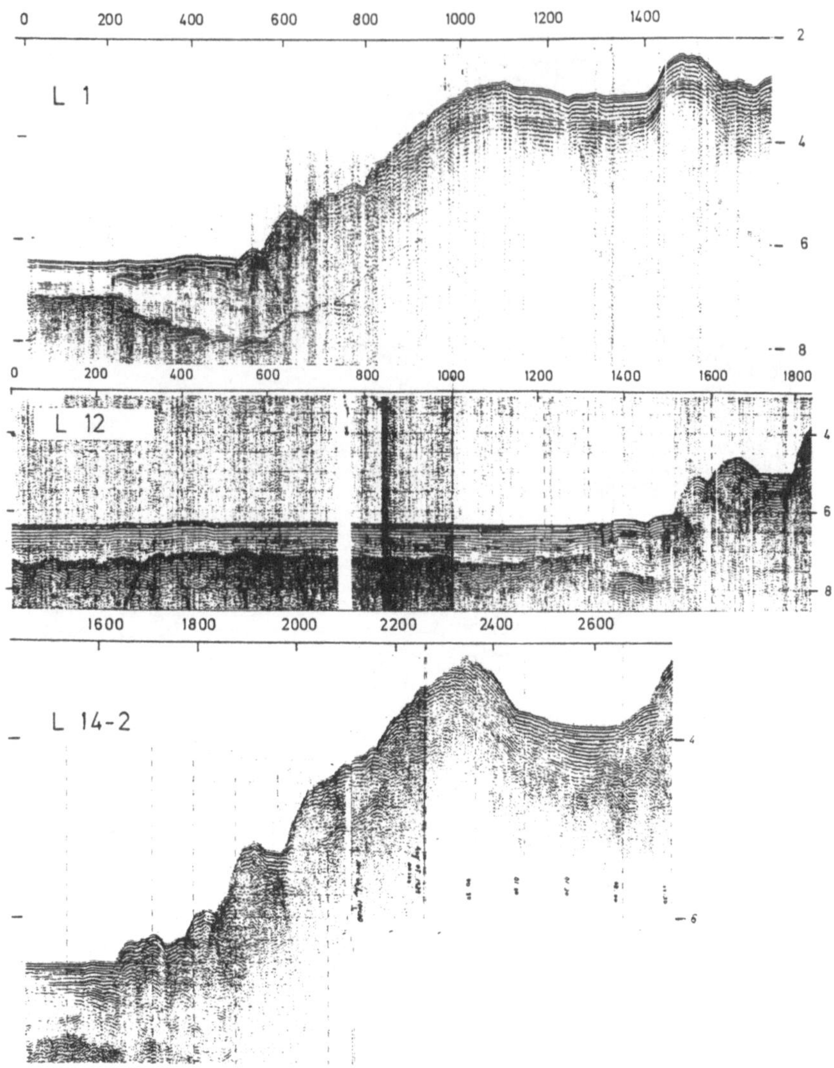

Fig. 3. Onboard monitor records crossing perpendicular to the trough axis.

boundary, although it is not yet ascertained by real thrusts. At SP 1950, a valley is observed. This is the same one already observed at SP 850 on line L 2. The accretionary basin at SP 2600 is formed by the northwestward tilting of the Minami-muroto knoll. Eight hundred meter sediment is observed in the basin. The southern floor of the basin is uplifting, indicating the knoll is still actively tilting to northwest.

The line L 1 extends from the Nankai Trough floor through No. 1 Minami-muroto knoll at SP 1000 to the foot of Tenkai knoll at SP 1400. The deformation front begins at around SP 200. A very gentle fold is observed in the protothrust zone between 200 and 500. On the oceanic crust, a graben is clearly recognized between SP 300 and 600. A total 1500 m thickness of sediments was trapped in the graben of which 900 m is composed of the trench-fill turbidite. Thick sedimentary layers are observed at a plateau between No. 1 Minami-muroto knoll and Tenkai knoll (SP 1200–1400). The strata seem horizontal, but actually dip northwest-west with cross-reference to line L 2.

The line L 12 is composed of two parts; One between SP 0 and 750, runs on the ocean crust and is parallel to the trough axis, which corresponds to the southwestern extension of line L 15. The other one runs northward from Sp 750 to 1800. At SP 1270, a deformation front is detected and a box-fold is observed from SP 1310 to 1360 where the imbricate thrust zone begins to develop. On the oceanic crust, a deep graben is observed between SP 1350 and 1550. In the graben a thick set of sediments of more than 1800 m have been trapped.

The line L 11 indicates the CDP section of three folds (Fig. 4). At around SP 1120, the imbricate thrust zone starts to develop, however, the existence of horizontal reflectors at about SP 1000–1050 make this section quite different from the previous ones. This is called a duplex thrust, a horizontal displacement is taking place by multiple low-angle thrusts or decollements (BOYER and ELLIOTT, 1982).

3. *Four Zones of the Nankai Accretionary Wedge*

The accretionary wedge of the Nankai Trough has been classified into four zones according to internal structure and seismic velocity. They are the protothrust, imbricate thrust, multiple decollement and earthquake thrust zones from the trough axis toward the land (KAGAMI *et al.*, 1983; KAGAMI, 1985; KAGAMI *et al.*, 1985).

The protothrust zone is characterized by developing box-folds and high angle (=60°) thrusts in the trench-fill turbidite sequence, and incipient growth of the primary decollement in the lower Pleistocene-upper Pliocene hemipelagic clay sequence. Thickening of the hemipelagic clay sequence overlying the primary decollement, with a velocity increase from 2.1 to 2.7 km/s toward the imbricate thrust boundary, is demonstrated by KAGAMI *et al.* (1982) and AOKI *et al.* (1985) (Fig. 5). The box-folds

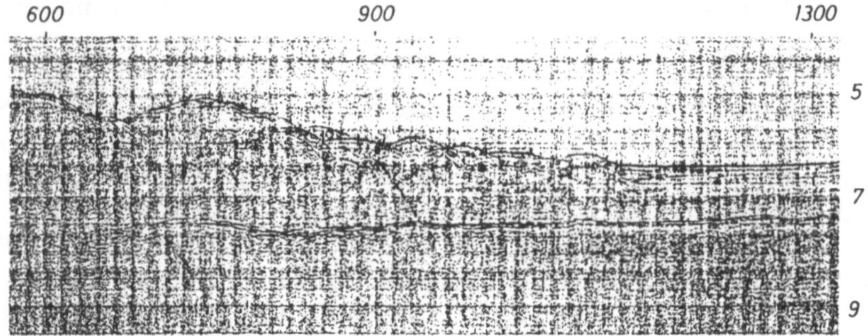

Fig. 4. The CDP section of profile L 11.

and high-angle thrusts were formed by the compressive push from below due to thickening of the hemipelagic clay sequence. The scaly clay type foliations were probably built up through the transportation of hemipelagic sediments laterally by the decollement, although no direct evidence was recovered from the Leg 87 drilling. From the information observed on the seismic records, however, we believe that this type of thickening, underplating by decollement, is occurring in this zone.

The imbricate thrust zone can be divided into two subzones according to the different modes of the thrust. The western subzone is characterized by a typical development of the imbricate thrust. The thrusts dip 40°, and converge into the basal decollement at the lower extension. Each imbricate block has a width of 5 km and a thickness of about 2 km (Fig. 6, Profile 55-7). The horizontal stratifications within the imbricate block near the imbricate thrust boundary show almost no vertical displacement. On the landward side, they gradually incline up to 40° as the blocks rotate landward by thrust-faulting, and finally the vertical displacement was estimated to be 1000 m (KAGAMI, 1984).

The eastern subzone is characterized by the existence of a duplex thrust, or development of imbricate thrusts without vertical displacement even in the landward blocks. In case of the typical development of a duplex thrust, the basal decollement does not grow (Fig. 6, Profile 55-4). Probably the duplex thrusts are so effective for compensation of plate movement between Eurasia and Philippine Sea plates, that no basal decollements are needed. However, the basal decollement is observed in case of the imbricate thrust without vertical displacement, which is more dominant in the eastern area (Fig. 6, Profile 55-1). We interpret this profile as that of duplex thrust originally with the mode of thrust having changed to

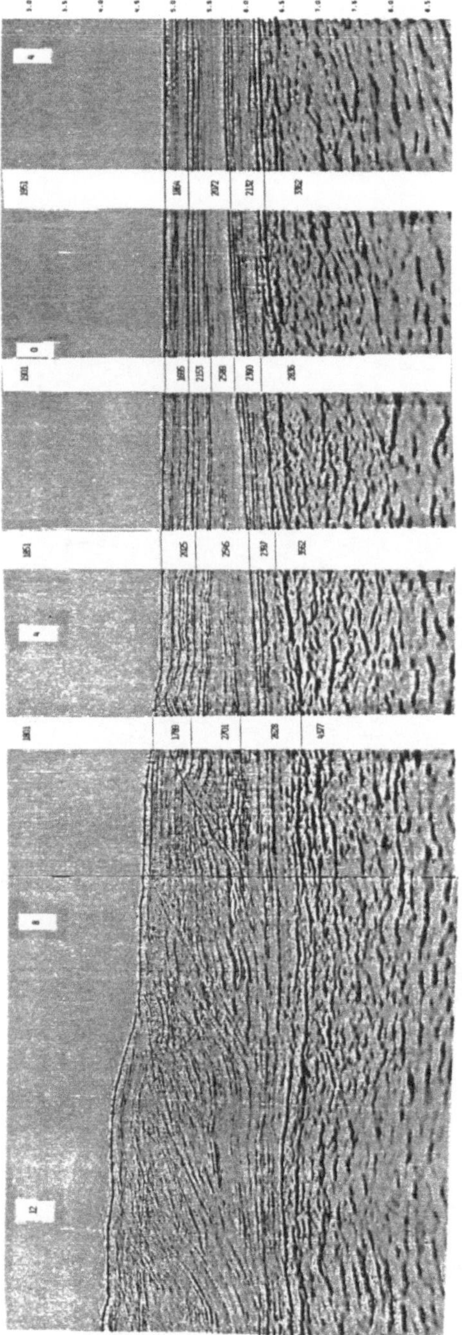

Fig. 5. Migrated depth section of N 55-3-1. Lateral change of velocity (km/s) is shown in the sediments of the protothrust zone.

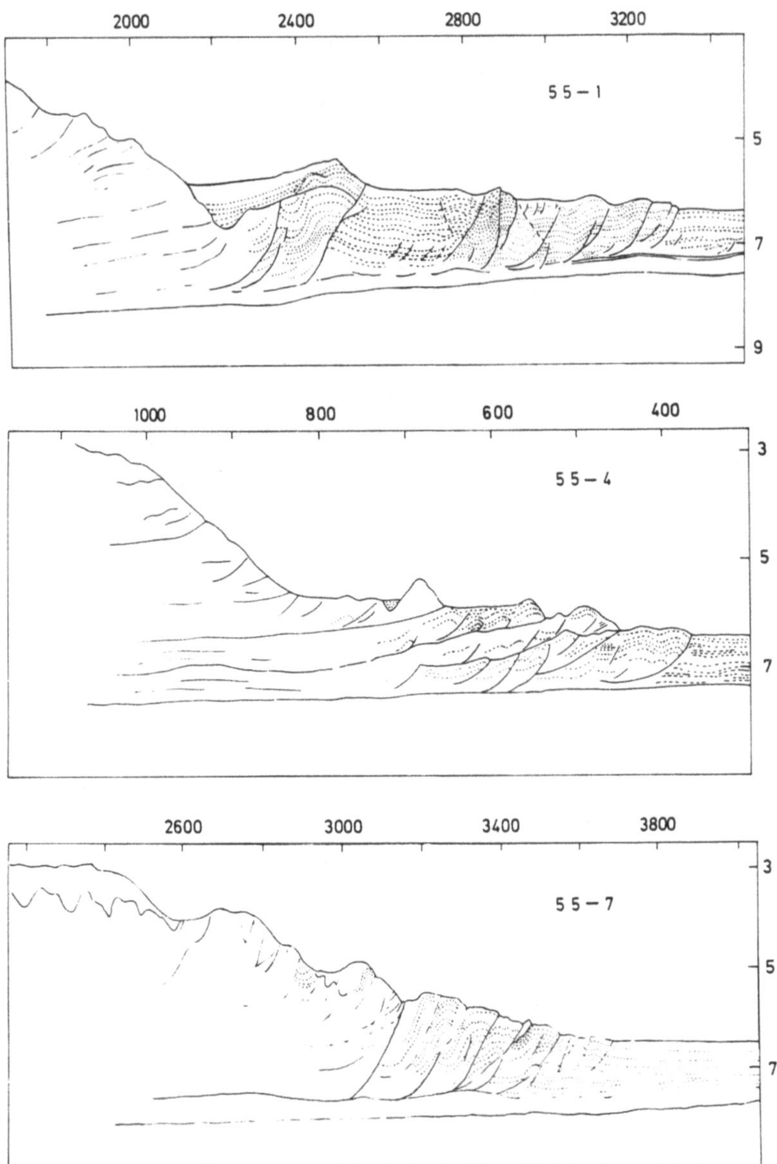

Fig. 6. Sketch of migrated time sections of 55-1, 55-4 and 55-7.

the present imbricate thrust only recently. The rapid sediment supply 190 cm/10^3yr along the Nankai Trough continued for 0.7-0.4 Ma B.P. It has decreased to 20 cm/10^3yr since then (KAGAMI, KARIG *et al.*, 1983). Besides rapid accumulation, these trench-fill turbidites were preferentially trapped in the eastern subzone by basement ridges, observed on profiles L 1 and L 12. The thick accumulation of trench-fill turbidites in the eastern subzone might have caused rapid accretion by effective lateral transportation through duplex thrusts. From our drilling results of the change in the rate of sedimentation, we think the main phase of sedimentation and also of accretion has already ended. This may explain change of the mode of thrusts in the eastern subzone. The magnitude of horizontal displacement is not estimated from our profiles, because no tracable marker horizons were identified. It is estimated, however, that the horizontal displacement per duplex thrust could be about 1000 m.

The multiple decollement zone is marked by a sudden increase in the slope angle. Therefore, the boundary between the imbricate thrust and multiple decollement zone is sometimes called the major thrust boundary, even though no major thrusts are recognized there. The steep slope gradient may be caused by thickening of the accreted sediments as observed in Profile L 2. According to DAHLEN (1984), the large surface slope angle of the accretionary wedge is initiated by a decrease in basal fluid pressure. This matter will be discussed later. The primary decollement disappears in this zone, after which multiple decollements begin to develop. The areas where multiple decollements develop actively, are arranged in en echelon pattern perpendicular to the subduction direction along the outer margin of this zone. The further landward structure of this zone is rather obscure, but steep thrusts in the back-side and folding in the upper horizon are the main structures observed in the profiles.

The earthquake thrust zone occurs at the outer margin of the Tosa Terrace, where old slope sediments outcrop on the sea floor leaving avoided recent slope sedimentation. This is known as one of the active fault zones in Japan (RESEARCH GR., 1980). Within this zone, major historical earthquakes, Hoei in 1707, Ansei in 1854, and Nankaido in 1946 have occurred every 120 years on an average. The uplift rate of the landward of the earthquake thrust faults is estimated to be 60–170 cm per 120 years (ANDO, 1975), although this upheaval had been partly cancelled during the subsequent intra-seismic period. The residual uplift has been accumulated in a long time to form the outer ridge of the Tosa terrace. This is the reason for the slope being steepest in this zone among the four zones in the accretionary wedge.

Time of initiation of uplift is not clear, but if it goes back with initiation of accretion it might have been sometime during lower middle Miocene, when the T formation was deposited on the slope, with igneous activity occurring at the Tosabae bank and the outer margin of Tosa terrace (OKUDA, 1984).

4. *Velocity Structure of the Accretionary Wedge*

The velocity structure of the wedge is drawn by using stacking velocity (NASU *et al.*, 1982) of Profile N55-3-1 (Fig. 7). The stacking velocity profile is compared with that of the refraction data (YOSHII *et al.*, 1970). Yoshii and others identified trench sediments to be of 2.0–2.3 km/s, accreted sediments to be of 2.0–3.5 km/s (B layer), and the inner accreted sediment to be of 4.6 km/s (A layer). They thought that the A layer may be accreted and deformed during an older cycle of subduction, because of its higher velocity. This layer corresponds to zone 4 of the present paper. The B layer can be divided into two; the outer B layer is represented by 2.8 km/s, and the inner B layer is 3.5 km/s, corresponding to zones 2 and 3 respectively of the present paper. Consequently, the refraction result coincides well with our reflection data.

Zonation of fault structures discussed in the previous section is well supported by the velocity zones. Velocity patterns in the protothrust zone has been discussed (KAGAMI *et al.*, 1982; AOKI *et al.*, 1985). A two-storied horizon is recognized; a trench-fill turbidite with a 1.7–2.1 km/s velocity and a underlying hemipelagic clay with a 2.1–2.7 km/s velocity. These accreted sequences overlie the lower turbidites and the vitric claystone sequences of the Shikoku Basin formation with a 2.1–2.6 km/s velocity. Within this first zone, the primary decollement develops in the hemipelagic clay sequence with the velocity increase from 2.1 to 2.7 km/s. The thickness of the sequence doubled in the distance of about 8 km, and the sediments change from unconsolidated (porosity, $\phi > 60$) to semiconsolidated ($\phi \fallingdotseq 40$). These changes in thickness and compaction might have been formed by decollements through development of scaly foliation in the clayey sediments. The underlying Shikoku Basin formation shows lower velocity which undergoes dewatering and plays an important role for developing underplating in the next imbricate thrust zone.

The second zone falls in the imbricate thrust zone. A two-storied velocity structure consists of the upper horizon with a 1.9–2.4 km/s velocity and the lower horizon with a 2.5–3.2 km/s velocity. The upper horizon partially contains a slope sediment, which is indicated from the drilling result at Site 583A and by a velocity of 1.9 km/s. However, it apparently consists of semiconsolidated trench-fill turbidites, since most of the veloci-

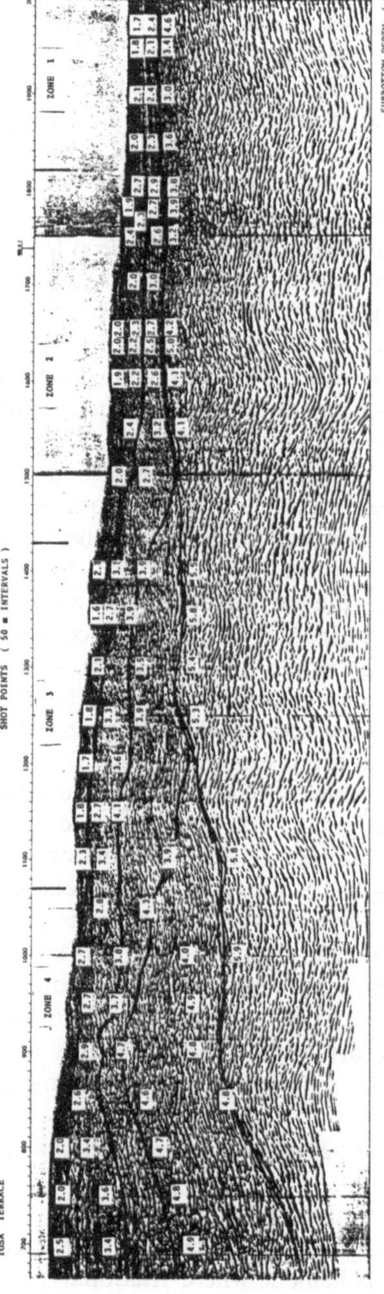

Fig. 7. Velocity structure and tectonic zones observed along Profile N 55-3-1. Zone 1; protothrust zone, Zone 2; imbricate thrust zone, Zone 3; multiple decollement zone, Zone 4; earthquake thrust zone.

ty range are from 2.2 to 2.4 km/s. The lower horizon contains hemipelagic sediment and lower turbidite sequences. Here, the decollements undercut the lower horizon and come close to the oceanic basement, thus creating underplating (MOORE *et al.*, 1982; WESTBROOK and SMITH, 1983). The interstitial water within the horizon will be removed through spray faults branching from the decollement and therefore the velocity increases to 3.2 km/s ($\phi \fallingdotseq 30$) at the landward end of the zone. Most of the primary decollement will diminish at that point.

The third zone correlates with the multiple decollement zone. Three horizons are recognized; the upper one is 500 m thick and its velocity ranges from 1.6 to 2.1 km/s, which is considered to represent a slope sediment; the middle horizon is about 2 km thick with a velocity range between 2.7 and 3.4 km/s indicating semiconsolidated trench-fill turbidites; and the lower horizon is 2–3 km thick with a velocity range between 3.5 and 3.9 km/s which is probably composed of consolidated lower turbidite and vitric clay sequences. YOSHII *et al.* (1970) also postulated that the horizon was composed of sedimentary material from its velocity. At the seaward part of the lower horizon, multiple decollements are observed in the velocity range of 3.5 km/s ($\phi \fallingdotseq 20$).

The fourth zone falls in the earthquake thrust zone, and consists of three horizons. The upper horizon, 1–2 km thick, consists of the K and T formations with a velocity range of 2.6–2.9 km/s, which have been directly traced from the continental shelf (OKUDA, 1984). It is noteworthy to say that no recent slope sediments were observed in this portion of the slope. This is because the historical earthquake thrusts have been occurring periodically (THATCHER, 1984). The middle horizon is 1–2 km thick, with a velocity range of 3.7–3.8 km/s which can be correlated to consolidated trench-fill turbidites. The lower horizon is 4–5 km thick with a velocity range of 4.0–4.9 km/s. This is probably composed of consolidated accretionary sediments, because there is a gradual increase of thickness without any breaks from 4–5 to 5–7 km and of velocity from 3.5–3.9 to 4.0–4.9 km/s between zone 3 and zone 4. The porosity of this horizon reaches 5 to 15 per cent which is almost a boundary between plastic and elastic deformation. Therefore, this is the reason why historical earthquake thrusts have only been occurring in this zone.

5. *Secondary Pore Water Pressure for Development of Multiple Decollements*

The wedge-like cross-sectional shape of submarine accretionary complexes overlying a basal decollement where frictional sliding is occurring can be modelled in the light of Coulomb failure (DAHLEN, 1984). The magnitude of this critical taper was a function of the internal and basal coefficients of friction μ and μ_b, and the internal and basal Hubbert-Rubey fluid pressure ratios λ and λ_b. The wide variety of tectonic styles observed along convergent margins, including subduction erosion, active accretion, and even extension and normal faulting might be controlled by relatively small spatial variations in either μ_b or λ_b (DAHLEN, 1984). Increasing the basal friction increases the taper of the wedge, whereas increasing the strength of the internal wedge decreases its taper.

Surface and basal angles of the Nankai accretionary wedge are taken along Profile N55-3-1 and listed in Tables 1 and 2. Basal depth of the wedge was determined from refraction data (DEN *et al.*, 1968; YOSHII *et al.*, 1973) and multichannel seismic data (NASU *et al.*, 1982). The surface slope angle increases from 1° in the protothrust zone, 3.5 to 4.5° in the imbricate thrust zone, 6.5° in the multiple decollement zone to 1 to 5° in the earthquake thrust zone. The basal dip angle gradually changes from 2–3° in the protothrust zone and seaward portion of the imbricate thrust zone, to 5° in the landward portion of the imbricate thrust zone and the multiple decollement zone, and to 6 to 8° in the earthquake thrust zone. The observed geometry is plotted on the basal fluid pressure diagram drawn for the active fold-and-thrust belt of western Taiwan (Fig. 8).

Since physical property parameters of the observed geometry of the Taiwan fold-and-thrust belt are nearly similar to those of the imbricate thrust zone of the Nankai accretionary wedge, they are determined using those found in Taiwan. In the first model, we assume that the basal friction and internal fluid pressure ratio is constant throughout the wedge. Thus, values of basal fluid pressure ratio, λ_b, are obtained from Fig. 8 and listed in Table 1. The values of λ_b change gradually from 0.90 in the protothrust zone and reach a minimum ($\lambda_b = 0.70$) in the multiple decollement zone. It is also low ($\lambda_b = 0.71$) in the seaward portion of the earthquake thrust zone. This means that the basal and interior fluid pressures become nearly equal and this will be the preferred attitude of any new internal deformations. In the same way, values of internal coefficient of friction, μ, are obtained. They are 1.1 in the imbricate thrust zone and reaches a minimum ($\mu = 0.9$) in the multiple decollement zone. It is also low ($\mu = 1.0$) in the seaward portion of the earthquake thrust zone. This result, again, shows that the basal and interior strengths become nearly equal in this zone causing internal deformation.

In the second model, we assume basal friction and basal fluid pressure ratio will change laterally. In the Nankai wedge, landward decreases in basal fluid pressure ratio, λ_b, most probably take place, associated with

TABLE 1. PARAMETERS OF PHYSICAL PROPERTIES IN THE NANKAI ACCRETIONARY WEDGE

MODEL 1: CONSTANT BASAL FRICTION AND INTERNAL FLUID PRESSURE RATIO

	PROTOTHRUST ZONE	IMBRICATE THRUST ZONE		MULTIPLE DECOLLEMENT ZONE	EARTHQUAKE THRUST ZONE	
SHOT POINT	1820	1650	1500	1100	700	
SURFACE SLOPE ANGLE α	1°	4.5°	3.5°	6.5°	5°	1°
BASAL DIP ANGLE β	2 -- 3°		5°		6°	7 -- 8°
INTERNAL FRICTION μ	1.5	1.1	1.1	0.9	1.0	1.1
BASAL FRICTION μ_b	0.85	0.85	0.85	0.85	0.85	0.85
INTERNAL FLUID PRESSURE λ	0.67	0.67	0.67	0.67	0.67	0.67
BASAL FLUID PRESSURE λ_b	0.90	0.77	0.75	0.70	0.71	0.77
STABILITY CONDITION	$\mu_b(1-\lambda_b) < \mu(1-\lambda)$					

TABLE 2. PARAMETERS OF PHYSICAL PROPERTIES IN THE NANKAI ACCRETIONARY WEDGE

MODEL 2: CHANGING BASAL FRICTION AND FLUID PRESSURE RATIO

	PROTOTHRUST ZONE	IMBRICATE THRUST ZONE	MULTIPLE DECOLLEMENT ZONE	EARTHQUAKE THRUST ZONE
SHOT POINT	1820	1650 1500	1100	700
SURFACE SLOPE ANGLE α	1°	4.5^{o} | 3.5^{o}	6.5^{o}	5° | 1°
BASAL DIP ANGLE β	$2 -- 3^{o}$		5°	6° | $7 -- 8^{o}$
INTERNAL FRICTION μ	1.0	1.1	1.2	≤ 1.4
BASAL FRICTION μ_b	0.7	0.85	1.2	1.4
INTERNAL FLUID PRESSURE λ	0.7	0.67	0.7	0.68
BASAL FLUID PRESSURE λ_b	0.9	0.80	0.7	0.68
STABILITY CONDITION	$\mu_b (1 - \lambda_b) < \mu (1 - \lambda)$		$\mu_b(1-\lambda_b)=\mu(1-\lambda)$	$\mu_b (1 - \lambda_b) > \mu (1 - \lambda)$

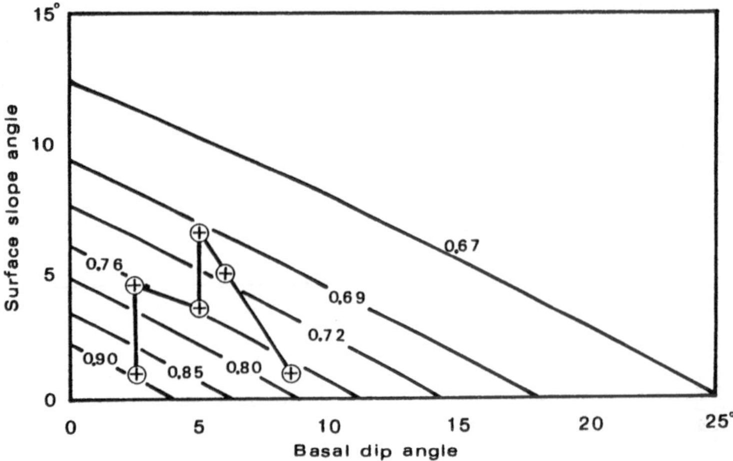

Fig. 8. Basal fluid pressure values of the Nankai wedge are estimated from Taiwan conditions having $\mu=\mu_b=0.85$ and $\lambda=0.67$ on surface slope versus basal dip angles (DAHLEN, 1984). Values are shown in Tables 1 and 2.

landward increase in the basal friction coefficient, μ_b, because they occur in the combination $\mu_b(1-\lambda_b)$. This equation is derived from the following. The basal traction τ_b resisting frictional sliding on the base of the wedge is written; $\tau_b=\mu_b(1-\lambda_b)\ \rho$ gH, where ρ is the wedge density and H is the thickness of the wedge (DAVIS *et al.*, 1983). For a portion of the wedge with a throughgoing basal decollement, it must necessarily have $(1-\lambda_b)\mu_b<(1-\lambda)\mu$.

Initially, parameters are taken from the Taiwan wedge, because its angles ($\alpha=4°$, $\beta=4.2°$) are quite similar values to those of the imbricate thrust zone of the Nankai wedge. Therefore, the following parameters are chosen, $\mu_b=0.85$, $\mu=1.1$, $\lambda_b=0.80$ and $\lambda=0.67$ for the imbricate thrust zone (Table 2). Then, values of μ_b are determined following a model with available values such as 0.7 in the protothrust zone, 1.2 in the multiple decollement zone and 1.4 in the earthquake thrust zone. Similarly values of λ_b are determined as 0.9 in the protothrust zone, 0.7 in the multiple decollement zone and 0.68 in the earthquake thrust zone. Variations in λ and μ are obtained in the following way. Values of λ and μ in the multiple decollement zone are determined using $\mu_b(1-\lambda_b)=\mu(1-\lambda)$, because stability of the foregoing taper is no longer maintained and the angles of the wedge in the multiple decollement zone are beyond the relation between α and β empirically shown as $\alpha=(5.7\pm0.2)-(0.66\pm0.14)\beta$.

Values of λ and μ in the earthquake thrust zone probably follow $\mu_b(1-\lambda_b)>\mu(1-\lambda)$ and are listed in Table 2.

Thus, it is clear from the second model that the protothrust and imbricate thrust zones are represented by basal decollement, and the multiple decollement and earthquake thrust zones are by internal deformation. Further, it is interesting to note that a relative increase of internal fluid pressure ratio occurs in the multiple decollement zone as shown in Table 2, although there are involved many assumptions. Considering the same effect observed in the first model, it is important to think about the cause of internal deformation, especially the mechanism of dewatering in the multiple decollement zone.

To understand the condition of the accretionary wedge which is under tectonic stress resulting from plate convergence, the velocity versus depth curve for a basinal sediment of the Niigata oil field is compared as a reference, where velocity is expressed in the following equation (CHUJO, 1962); V (m/s) $=1650+0.6\ Z$, (Z is depth in meter). On Fig. 9, ranges of velocity of the Nankai accretionary wedge are plotted using the lowest horizon's values; 2.1–2.7 km/s for the protothrust zone, 2.5–3.2 km/s for the imbricate thrust zone, 3.5–3.9 km/s for the multiple decollement zone, and 4.0–4.9 km/s for the earthquake thrust zone. The depth-velocity curve thus constructed can be used for comparison with porosity and diagenetic change in the reference basin sequence. Such data from the Niigata oil field are plotted in Fig. 9 (AOYAGI and ASAKAWA, 1984).

A paleogeothermal gradient during the Neogene in the Niigata Basin is summarized as ranging between 2.1 and 3.6°C/100 m. Using an average geothermal gradient of 2.7°C/100m, the corresponding transformation depth as defined by the first appearance is clarified from the fact that the transformation from smectite to mixed-later smectite/illite is located at a depth of 2800 m with a temperature of 104°C, and mixed-layer to illite at a depth of 3800 m with a temperature of 137°C (AOYAGI and ASAKAWA, 1984). These data are also plotted in Fig. 9. The geothermal gradient at sites 582 in the protothrust zone and 583 in the imbricate thrust zone of the Nankai wedge was measured to be 4.5°C/100 m (KINOSHITA and YAMANO, 1985). Therefore, transformation levels tend to become shallower in the Nankai wedge than in the Niigata Basin, although we do not know the exact value in the multiple decollement zone.

An interesting result has been obtained from change of clay minerals through a plot of the per cent illite and smectite versus depth in the Gulf Coast area where a systematic increase in illite from 20 to 65 per cent was apparent in the depth interval from about 2500 m to 3700 m (BURST, 1969). Therefore, significant diagenesis from smectite to illite appears to

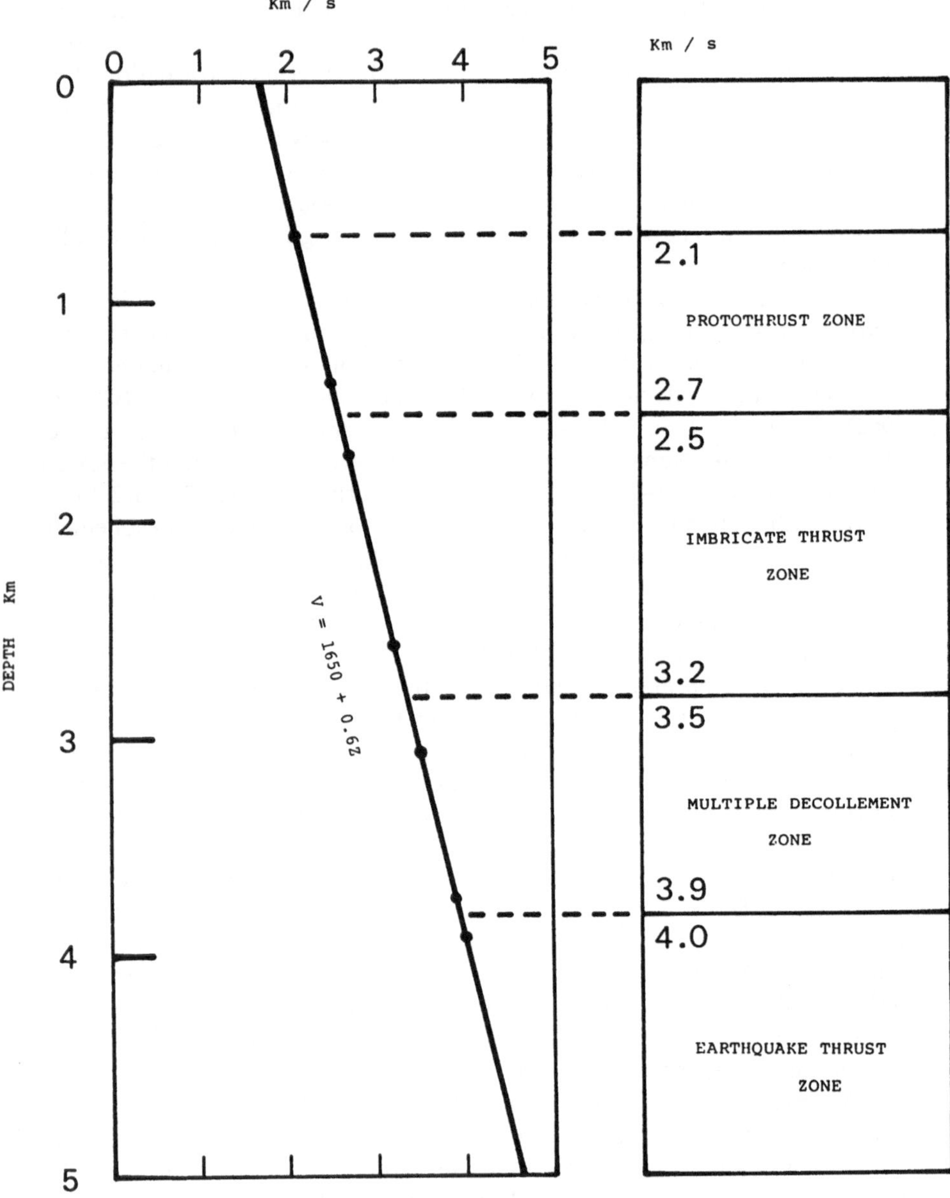
VELOCITY IN BASIN SEDIMENTS
Km / s
0 1 2 3 4 5
0
1
2
3
4
5
DEPTH Km
V = 1650 + 0.6Z
VELOCITY STRUCTURE IN
THE ACCRETIONARY WEDGE
Km / s
2.1
PROTOTHRUST ZONE
2.7
2.5
IMBRICATE THRUST
ZONE
3.2
3.5
MULTIPLE DECOLLEMENT
ZONE
3.9
4.0
EARTHQUAKE THRUST
ZONE

Fig. 9.

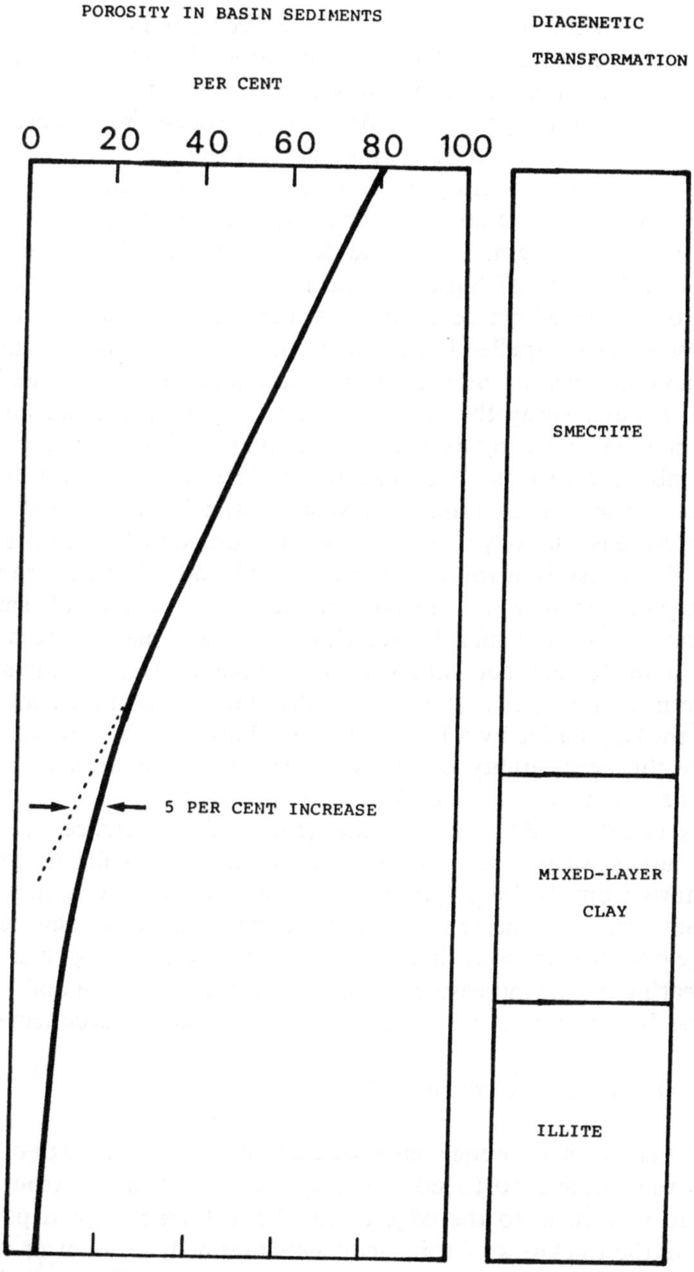

POROSITY IN BASIN SEDIMENTS

PER CENT

0 20 40 60 80 100

5 PER CENT INCREASE

DIAGENETIC

TRANSFORMATION

SMECTITE

MIXED-LAYER

CLAY

ILLITE

Fig. 9. Comparison between velocity structure in the Nankai wedge and velocity, porosity and diagenetic transformation depth in the reference basinal sediment in Japan.

be confined to this 1200 m interval which corresponds to a temperature interval between 93 and 110°C. This diagenesis involves transfer of large amounts of bound-water from smectite to the surrounding areas. A particularly important aspect of the dewatering process from subsurface clay sediments is summarized following BURST (1969).

In stage one dewatering, pore water and a small amount of excessive interlayer water are removed by the action of overburden pressure and stress of tectonic origin. This initial dehydration, which will be completed in the first 1000s m of burial for basin strata, reduces the water content to approximately 30 per cent. As far as the absolute volume is concerned, the most water is expelled in this stage. In case of the accretionary wedge, this is accomplished in the protothrust and imbricate thrust zones by water-release through spray thrust faults connecting to the basal decollement which may be shown by its strong reflectivity on the seismic image along some imbricate thrusts. It is also responsible for high heat flow along the axis of the Nankai Trough (KINOSHITA and YAMANO, 1985). In stage two, pressure is relatively ineffective as a dehydration agent, rather a physio-chemical process is involved. If geothermal and chemical gradients are efficient to mobilize the interstitial water, diagenesis of smectite to dehydrated illite and mixed-layer clays may take place. The amount of water in movement constitutes 10–15 per cent of the compacted bulk volume in this stage. In stage three, the final water increment is forced out of the clay lattice by a kinetic process. This is a slow process however, because the permeability of the water-egress system deteriorates rapidly after the water content falls below 10 per cent.

As noted by BRAY and KARIG (1985), the reference basin section shows nearly 80 per cent porosity at the surface, while th protothrust zone starts from the lower porosity level (Fig. 9), indicating that processes associated with tectonic accretion increase the efficiency of dewatering early in stage one. It is obvious that the second fluid-release stage has an important bearing on the increase of porosity and fluid pressure and thus affecting the tectonics that we described in the multiple decollement zone.

6. *Discussion and Conclusion*

The accretionary wedge observed along the Nankai Trough off Shikoku has a wedge-shape 50 to 60 km long starting from the trough axis at a depth of 4800 m to the edge of the Tosa Terrace at a depth of 1100 m where the thickness of total sediments attained is 10 km or more. The wedge can be divided into four tectonic zones according to their velocity ranges and mode of thrusting namely; the protothrust zone, imbricate thrust zone, multiple decollement zone and earthquake thrust zone, from the trough axis landward (KAGAMI, 1985).

The thrust system observed in these zones decreases a slip angle of thrusts from 60° in the protothrust zone, 30–40° in the imbricate thrust zone, to less than several degrees in the multiple decollement zone in response to increase in the surface slope angle of the wedge. These progressive changes in the slip angles of thrusts can be expected when considering the Coulomb wedge theory of DAHLEN (1984).

From the velocity distribution, the wedge can be divided into two deformation zones; the seismic or elastic deformation zone and the non-seismic deformation zone. The protothrust, imbricate thrust and multiple decollement zones are included in the non-seismic deformation zone where the velocity ranges from 1.7 to 3.9 km/s (Fig. 10). The earthquake thrust zone is the only active zone of seismic or elastic deformation, the velocity of which ranges from 4.0 to 4.9 km/s. Earthquake stresses can only be accumulated in this zone which have been released every 120 years. It becomes clear for the earthquake thrust zone in controlling tectonic styles including active accretion and wedge shortening. Tectonically induced velocity increase causes advancement of the earthquake thrust zone toward the trough axis resulting in wedge shortening.

Considering spacial variation in the physical properties of either the interior wedge (μ and λ) or the basal decollement (μ_b and λ_b), the accretionary wedge can be divided into three different conditions (Table 2). When a wedge maintain higher basal fluid pressure ratio and lower basal friction such that $\mu_b(1-\lambda_b)<\mu(1-\lambda)$, the steady state accretionary wedge with a basal decollement can be formed as we see in the protothrust and imbricate thrust zones. When a surface slope angle of the wedge increases landward, and the fluid pressure ratio becomes equal such that $\mu_b(1-\lambda_b)=\mu(1-\lambda)$, it can not maintain the basal decollement and new ones must be formed higher within the existing wedge. This case is observed in the multiple decollement zone. Further, when a wedge experiences a drop in basal fluid pressure ratio and a increase in basal friction such that $\mu_b(1-\lambda_b)>\mu(1-\lambda)$, the wedge undergoes internal deformation. This case is represented by the earthquake thrust zone.

In discussing the stability conditions of the wedge, we noticed a relative increase in fluid pressure ratio in the multiple decollement zone (Tables 1 and 2), and this can best be explained by the secondary dewatering process. The velocity of the accretionary wedge was compared with the reference basin sequence in Japan. It is obvious in Fig. 9 that the secondary dewatering zone corresponding to the alteration zone of mixed-layer smectite/illite affects tectonics in the multiple decollement zone. From our

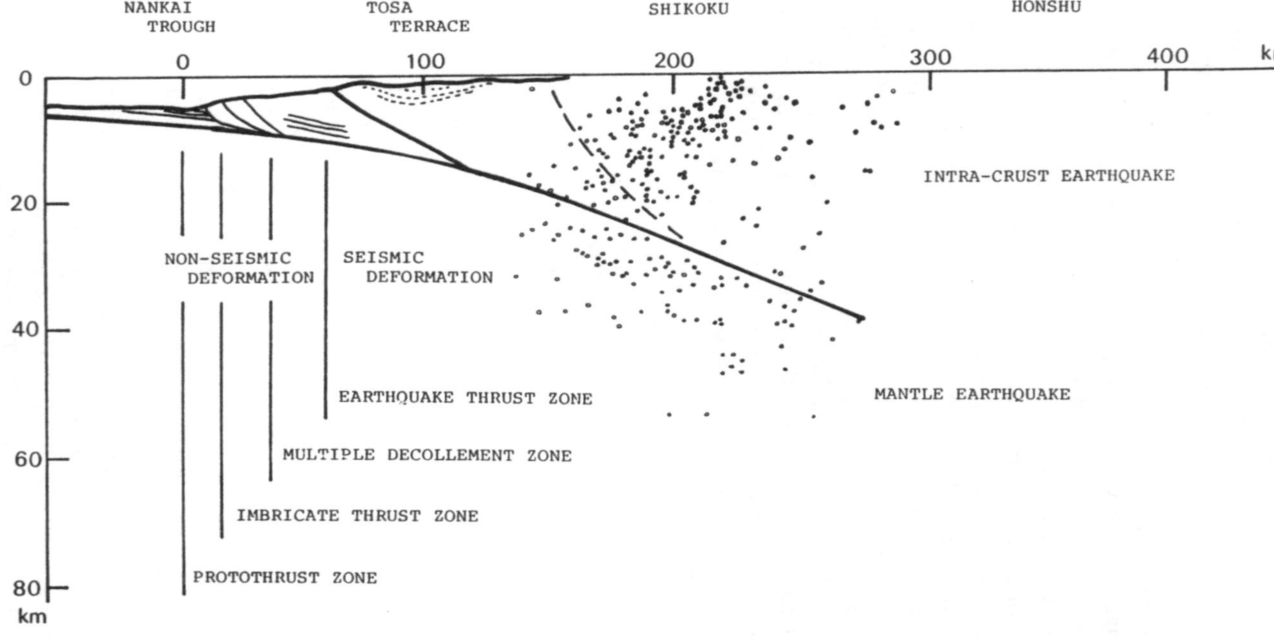

Fig. 10. Tectonic zones of the Nankai accretionary wedge and distribution of earthquakes detected from the land.

observations in the Nankai Trough, it is concluded that the multiple decollement zone is principally related to the stability condition in physical properties, on condition that the secondary dewatering process also play a part in its development since the Shikoku Basin sequence contains about 30 per cent smectite among clay minerals in the upper part and 50 to 90 per cent in th lower part (CHAMLEY, 1980).

It is interesting to note that the shape of the dewatering curve with depth has a similar form to that for the porosity curve with depth. Therefore, the amount of secondary dewatering can be estimated from the porosity curve as shown in Fig. 9. The primary porosity curve is almost straight following a regression line. The observed porosity at about 3000 m depth deviates about 5 per cent from the regression line, which probably indicates an approximate amount of the secondary dewatering.

According to DAVIS *et al.* (1983), the decrease in surface slope beyond 55 km landward from the trench in the Peruvian continental margin reflects a reduced basal resistance, where the base of the wedge is extended below the brittle-plastic transition. An abrupt drop in surface slope beyond 55 km interior from the trough axis at the earthquake thrust zone of the Nankai wedge, however, is interpreted as uplifting of the outer margin of Tosa Terrace caused by the frontal thrust of inter-plate earthquakes (Fig. 10).

Acknowledgements

Submarine topographic data used here were provided by the Hydrographical Bureau of Japan, and the seismic profiles in Figure 6 provided by Dr. T. Toba of JAPEX are gratefully acknowledged. Constructive criticism of the manuscript was provided by K. Aoyagi, S. Uyeda, H. Tokuyama and J. D. Buckley.

REFERENCES

M. ANDO, Source mechanics and tectonic significance of historical earthquakes along the Nankai trough, Japan, *Tectonophys*, **27**, 119–140, 1975.

Y. AOKI, T. TAMANO, and S. KATO, Detailed structure of the Nankai Trough from migrated seismic sections, *A. A. P. G. Mem.*, **34**, 309–322, 1982.

Y. AOKI, H. KINOSHITA, and H. KAGAMI, An evidence of low velocity layer beneath the accretionary prism of Nankai trough, inferred from synthetic sonic log, Kagami, H., D. E. Karig *et al.*, Init. Rept, DSDP, 87, U.S. Govt. Printing Office, Washington, 1985 (in press).

K. AOYAGI and T. ASAKAWA, Paleotemperature analysis by authigenic minerals and its application to petroleum exploration, *A. A. P. G. Bull.*, **68**, 903–913, 1984.

S. E. BOYER and D. E. ELLIOTT, Thrust systems, *A. A. P. G. Bull.*, **66**, 1196–1230, 1982.

C. J. BRAY and D. E. KARIG, Porosity of sediments in accretionary prisms and some implications for dewatering processes, *J. Geophys. Res.*, **90**, 768–778, 1985.

J. F. BURST, Diagenesis of Gulf Coast clayey sediments and its possible relation to petroleum migration, *A. A. P. G. Bull.*, **53**, 73–93, 1969.

H. CHAMLEY, Clay sedimentation and paleoenvironment in the Shikoku Basin since the middle Miocene, deVries Klein, G., Kobayashi, K. *et al.*, Int. Rept, DSDP, 58: Washington, U.S. Govt. Printing Office, pp. 669–682, 1980.

J. CHUJO, An interpretation of sonic profiling records, *Taiheiyo*, **3**, 34–43, 1962.

F. A. DAHLEN, Noncohesive critical Coulomb wedge: An exact solution, *J. Geophys. Res.*, **89**, B12, 10125–10133, 1984.

D. DAVIS, J. SUPPE, and F. A. DAHLEN, Mechanics of fold-and-thrust belts and accretionary wedges, *J. Geophys. Res.*, **88**, 1153–1172, 1983.

N. DEN, S. MURAUCHI, H. HOTTA, T. ASANUMA, and K. HAGIWARA, A seismic refraction exploration of Tosa deep-sea terrace off Shikoku, *J. Physics Earth*, **16**, 7–10, 1968.

N. ISEZAKI and K. YASUKAWA, Three component magnetic anomaly in the Nankai trough, Kajiura, K., Ocean Characteristics and Their Changes, pp. 310–317, Koseisha-Koseikaku, Tokyo, 1985.

H. KAGAMI, Accreted sedimentary wedges at trenches observed on multichannel reflection seismic records, *J. Geography*, **93**, 455–463, 1984.

H. KAGAMI, The accretionary prism of the Nankai Trough off Shikoku southwestern Japan, Kagami, H., D. E. Karig *et al.*, Int. Repts, DSDP, 87, U.S. Govt. Printing Office, Washington, 1985 (in press).

H. KAGAMI, H. TOKUYAMA, Y. S. KONG, C. IGARASHI, and N. NASU, Multichannel seismic reflection survey in the Nankai Trough off Ashizuri, southwest Japan, *Kaiyo-Kagaku*, **14**, 351–357, 1982.

H. KAGAMI, S. SHIONO, and A. TAIRA, Plate subduction and formation of the accretionary prism along the Nankai Trough, *Kagaku*, **53**, 429–438, 1983.

H. KAGAMI, D. E. KARIG, F. AKIBA, C. J. BRAY, J. P. CADET, J. CHARVET, W. COULBOURN, K. FUJIOKA, H. KINOSHITA, M. LAGOE, T. H. LANG, J. K. LEGGETT, G. A. LOMBARI, N. LUNDBERG, T. MACHIHARA, R. MATSUMOTO, P. MUKHOPADHYAA, N. NIITSUMA, A. J. SMITH, C. L. STEIN, and A. TAIRA, In the orient Leg 87 drills off Honshu and SW Japan, *Geotimes*, **28**, 15–18, 1983.

H. KAGAMI, H. TOKUYAMA, and E. NISHIYAMA, Structure of the accretionary wedge of the Nankai Trough off Shikoku, southwestern Japan, Kajiura, K., Ocean Characteristics and Their Changes, pp. 287–296, Koseisha-Koseikaku, Tokyo, 1985.

D. E. KARIG, J. C. INGLE, Jr. *et al.*, Site 297, Init. Repts, 31, pp. 275–316, U.S. Govt. Printing Office, Washington, 1975.

H. KINOSHITA and M. YAMANO, On the heat flow anomaly in the Nankai Trough area, Kagami, H., Karig, D. E. *et al.*, Init. Repts, DSDP, 87, U.S. Govt. Printing Office, Washington, 1985 (in press).

J. C. MOORE, J. S. WATKINS, T. H. SHIPLEY, K. J. MCMILLEN, S. B. BACHMAN, and N. LUNDBERG, Geology and tectonic evolution of a juvenile accretionary terrane along a truncated convergent margin, *Geol. Soc. Am. Bull.*, **93**, 847–861, 1982.

N. NASU, Y. TOMODA, K. KOBAYASHI, H. KAGAMI, H. KINOSHITA, S. NAGUMO *et al.*, Multichannel seismic reflection data across Nankai Trough, IPOD-Japan Basic Data Series, 4, 34 pp.

Y. OKUDA, Tectonic evolution of the continental margin off southwest Japan during the late Cenozoic, *Rept. Tech. Res. Center, JNOC*, **19**, 33–93, 1984.

Y. OKUDA, M. KUMAGAI, and T. TAMAKI, Tectonic development of the continental slope and its peripheral area off southwest Japan in relation to sedimentary sequences in sedimentary basins, *J. Japan Assoc. Petroleum Tech.*, **44**, 47–55, 1979.

RESEARCH GROUP FOR ACTIVE FAULTS, *Active Faults in Japan*, 363 pp., Univ. Tokyo press, Tokyo, 1980.

T. SATO and M. SAKURAI, Submarine geological structure at the continental slope off Shikoku, southwestern Japan, *Proceed. 82nd Meeting Geol. Soc. Japan on Geological Problems in the Philippine Sea*, pp. 5–10, 1975.

T. TAMANO, T. TOBA, and Y. AOKI, Exploration of trench slopes in Japan and Nankai trenches by reflection seismic methods, *Geophys. Exploration*, **34**, 204–221, 1981.

T. TAMANO, T. TOBA, and Y. AOKI, Development of fore-arc continental margins and their potential for hydrocarbon accumulation, *Petroleum Geology*, **PD3**, 1–11, 1983.

W. THATCHER, The earthquake deformation cycle at the Nankai trough, southwest Japan, *J. Geophys. Res.*, **89**, B5, 3087–3101, 1984.

G. K. WESTBROOK and M. J. SMITH, Long decollements and mud volcanoes: evidence from the Barbados Ridge complex for the role of high pore fluid pressure in the development of an accretionary complex, *Geology*, **11**, 279–283, 1983.

T. YOSHII, N. DEN, H. HOTTA *et al.*, Crustal structure of Tosa deep-sea terrace and Nankai Trough, Hoshino, M. and H. Aoki, *Island Arc and Ocean*, pp. 93–103, Tokai University Press, Tokyo, 1970.

T. YOSHII, W. J. LUDWIG, N. DEN, S. MURAUCHI, M. EWING, H. HOTTA, P. BUHL, T. ASANUMA, N. SAKAJIRI, Structure of southwest Japan margin off Shikoku, *J. Geophys. Res.*, **78**, 2517–2525, 1973.

Formation of Active Ocean Margins, edited by N. Nasu *et al.*, pp. 221–255.

SEDIMENT ACCRETION AND SUBDUCTION IN THE MIDDLE AMERICA TRENCH

Thomas H. SHIPLEY[1] and Gregory F. MOORE[2]

[1]*Institute for Geophysics, University of Texas at Austin, Austin, Texas 78712, U.S.A.*

[2]*Department of Geosciences, University of Tulsa, Tulsa, Oklahoma 74104, U.S.A.*

Abstract. DSDP drilling has confirmed that both sediment offscraping and subduction occur in the Middle America Trench. To determine why sediment is accreted only along some parts of this arc, we collected a dense network of high resolution seismic reflection and Sea Beam bathymetric data which provide a detailed description of the sedimentation patterns, morphologies and structures within the lower part of the trench slope region.

Within the turbidite filled trench off Mexico, 20 to 200 m high ridges and back tilted benches are common. The ridges are the bathymetric expression of deeper small thrust faults which are usually rooted near the base of the turbidite fill. The fold and thrust fault trends are more often parallel to the oceanic plate fabric. Deformation may be localized here because turbidite dewatering pathways intersect the lower porosity hemipelagic sediments on the footwall of the step faults.

In the Guatemalan area, the trench contains only minor muddy turbidite fill and has an outer slope of horsts and grabens. No structures are observed in the trench or lower slope which resemble those off Mexico and the main decollement is not identified in the seismic data, but is probably rooted in or at the top of the oceanic section consisting of pelagic and hemipelagic sediments.

In the Costa Rican area, turbidite fill is absent, and a thick blanket of pelagic and hemipelagic sediment on the oceanic plate is mainly subducted beneath the deformation front, However, in places the upper third of the section appears to be stripped off and accreted to the lower slope. Between 3 and 5 km landward of the inner slope the seismic identity of the subducting section abruptly ends. The overlying slope is smooth and regular with well-defined to slightly contorted reflections from within the slope section.

While offscraped deposits are a significant component of some accretionary wedges (perhaps the broken formation in ancient sequences), accretion only seems to occur when trench fill or oceanic plate sections of high porosity exist in the trench and appropriate discontinuities such as onlapping or normal faulting inhibit fluid migration and overpressures result. Thus, the evolution of a particular convergent margin and the composition of the accretionary wedge will be very sensitive to the sedimentary history and plate fabric.

The differences in the accretionary process along the Middle America Trench may result largely from variations in plate fabric and trench sedimentation along the arc, particularly since when trench fill is present it is always accreted to the lower slope. DSDP drilling shows that turbidites have been supplied sporadically to the trench off Mexico since early Pliocene but perhaps never in significant quantities off Guatemala or Costa Rica. Further, the Mexico drilling area, because of the presence of trench turbidites, is somewhat anomalous. Fifty kilometers from the drilling transect in a more typical area of minor turbidite fill, the morphology of the lower slope is similar to that off Guatemala.

Within the MAT we find little evidence for slumping, either on a small-scale or large scale, or for sediment erosion. Therefore, we prefer in interpret the Guatemala area as a nearly steady-state sediment subduction regime. The Costa Rican area is also interpreted as a mainly steady-state sediment subduction margin but probably with minor offscraping and underplating of the oceanic sections.

1. Introduction

There have been many investigations of modern continental margin subduction zones describing the morphology and speculating on the structure and tectonics. While we have been searching for simpler unifying processes to help understand ancient sequences exposed onshore, increasingly detailed studies of modern trenches have produced the unsatisfactory notion of unique and complex tectonic histories for each margin. DSDP drilling has shown convincingly the absence of sediment accretion in the Marianas (HUSSONG *et al.*, 1982) and Guatemala (VON HUENE *et al.*, 1980; AUBOUIN *et al.*, 1982a, b) but abundant evidence for accretion off Oregon-Washington (KULM *et al.*, 1973, 1974), Mexico (WATKINS *et al.*, 1982a, b; MOORE *et al.*, 1982b, c), Barbados (MOORE *et al.*, 1982a) and Shikoku (MOORE and KARIG, 1976; KARIG *et al.*, 1983). Less direct data on the processes of accretion and subduction have come from geophysical studies of convergent margins but the interpretations of regional geophysical data has been far from perfect in light of subsequent drilling. We believe that part of the problem of interpretation results from using inappropriately scaled geophysical data sets.

We conducted a "high resolution" geophysical survey of several portions of the Middle America Trench, including the area of sediment accretion off Mexico, the area of sediment erosion and/or subduction off Guatemala, and off Costa Rica (Fig. 1). These regions already have extensive multichannel seismic coverage and moderately dense bathymetry, magnetics, gravity and refraction coverage, mostly collected prior to the drilling program. By placing our surveys around these existing drilling transects we have an opportunity to develop and test interpretation methods in this environment and to add the third dimension to the drilled transects. During our study, off Guerrero and Oaxaca, Mexico, Guatemala and the Nicoya Peninsula, Costa Rica in April, 1982, we collected narrow multibeam echo sounder swath data (Sea Beam), water-gun seismic reflection data and other conventional underway geophysical data. These surveys provide the basis to compare and contrast the Mexico and Guatemala morphology and structure in light of the drilling results, and to further compare these areas with Costa Rica which has yet another geological setting.

The drilling transects concept has produced excellent site specific evolutionary models. However, every margin studied has produced a significantly different history and unique model. We believe a unifying simplified model will only come from integration of data along significant portions of trenches, that is extrapolating the drilling results to three dimensions. In this study we will concentrate only on sediment accretion processes along the base of the trench slope and illustrate the variations 10 to 100 km along strike away from the well-constrained drilling transects. We believe this analysis shows the temporary and oscillatory nature of sediment accretion along the Middle America Trench and reveals some of the pitfalls of models based on singular cross sections.

2. *Terminology*

The purpose of this paper is neither to introduce new terminology nor to review the existing sediment and tectonic terms used in describing the geology. We will selectively use the following terms which are derived mainly from MOORE (1975), SCHOLL *et al.*, (1980) and MOORE *et al.*, (1982b) and applied to processes active on the lower slope of the trench.

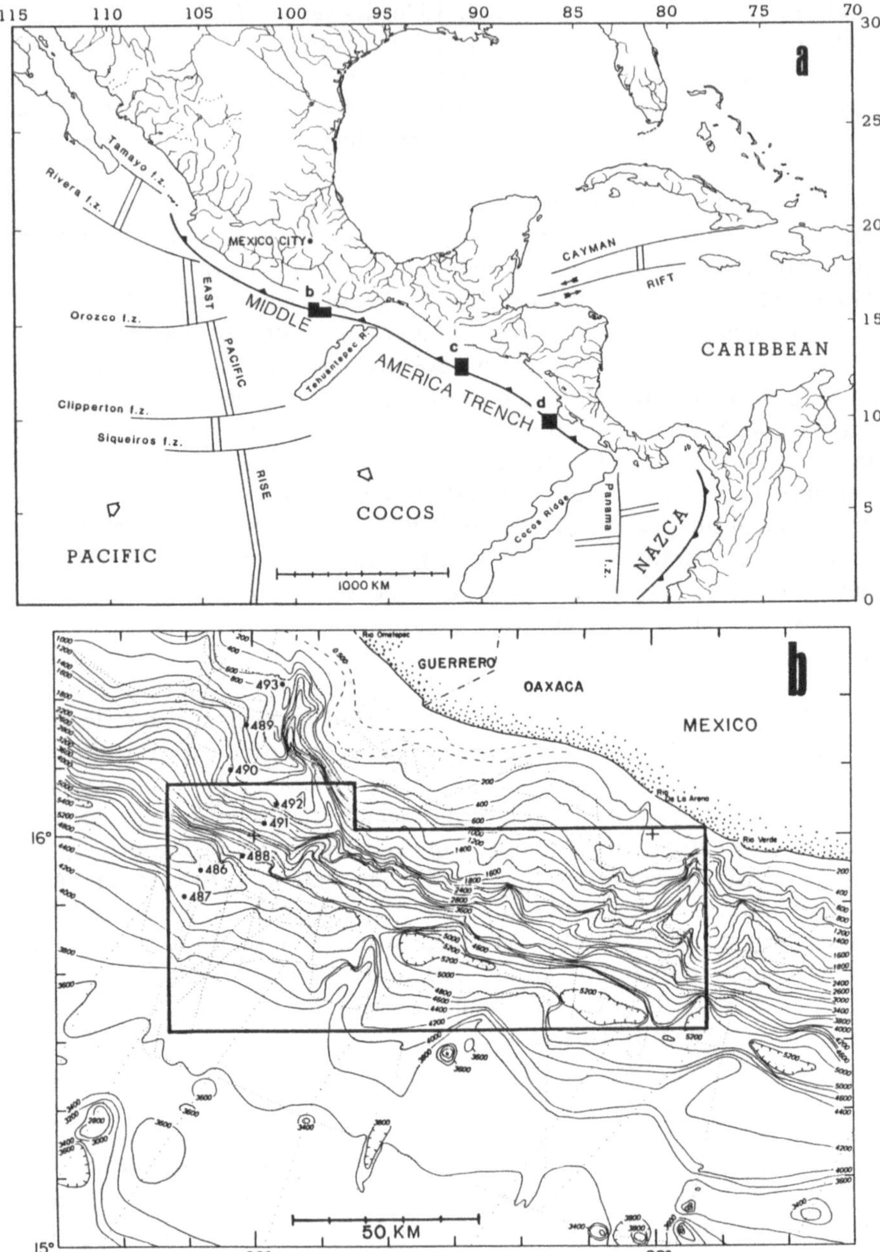

Fig. 1. (a) Location of the Middle America Trench and its relationship to Central America, the East Pacific Rise, Cocos and Tehuantepec Ridges. (b)–(d) Conventional bathymetric contour charts of the areas studied in detail by earlier University of Texas programs.

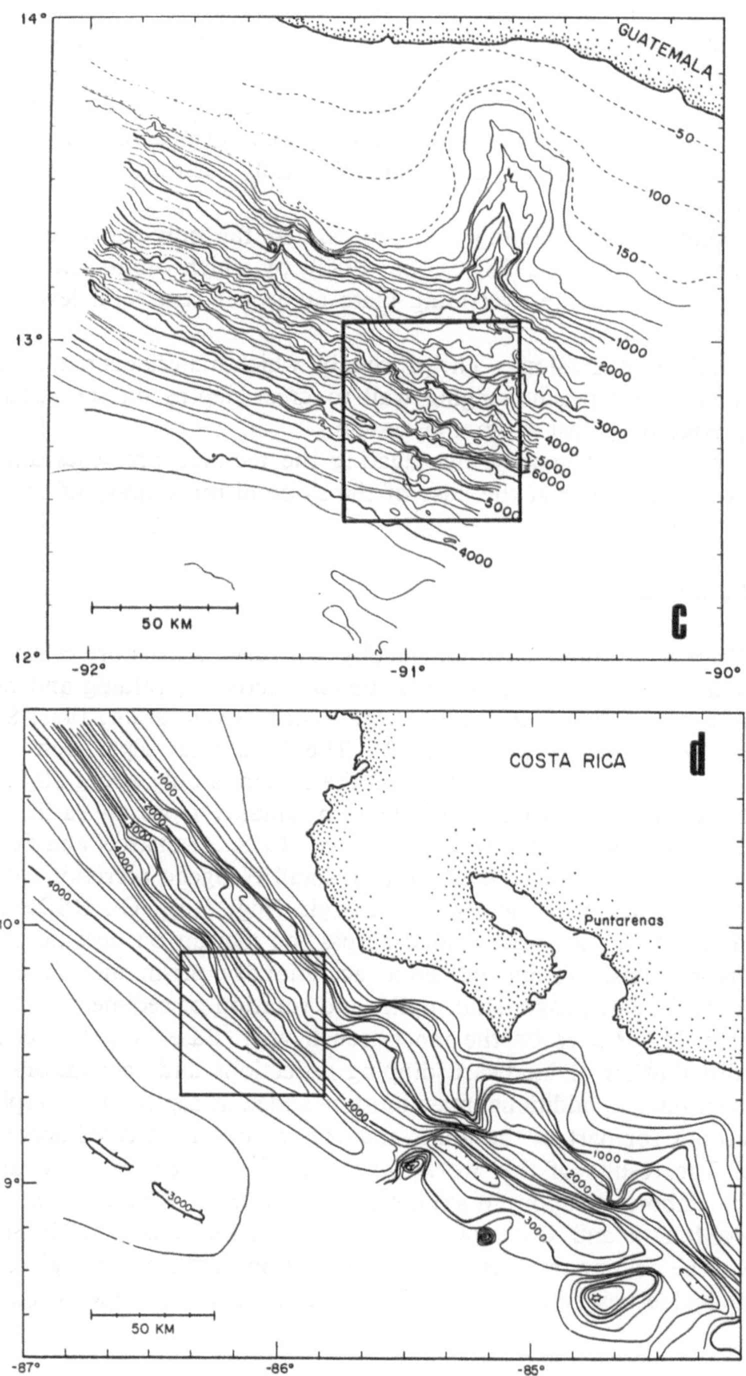
14°
GUATEMALA
50
100
150
13°
1000
2000
3000
4000
5000
6000
5000
4000
50 KM
C
12°
-92°
-91°
-90°
COSTA RICA
d
1000
2000
3000
4000
4000
10°
Puntarenas
3000
1000
2000
3000
9°
3000
50 KM
-87°
-86°
-85°

Sediment Offscraping: Process by which oceanic plate and/or other sediments accumulating in the trench axis are accreted to the overriding plate at or near the base of the slope.

Sediment Subduction: A steady-state process by which trench and oceanic plate sediments continue relatively undeformed 10's of kilometers beneath the deformational front. Probably associated with excess pore-pressures.

Sediment Underplating: A process, so far only based on models (COWAN and SILLING, 1978; CLOOS, 1982), by which subducted sediments are transferred to the overriding plate at deeper structural levels in the accretionary prism.

Subduction Erosion: The process by which material is removed from the accretionary prism hanging wall to great depths on the subducting plate, effectively lost from the system.

We will be primarily investigating the tectonic processes and their results as they occur at the base of the slope in the vicinity of the deformation front.

3. *Background*

Prior to drilling, extensive geophysical surveys offshore central Mexico, were interpreted to indicate sediment accretion, folding and faulting near the base of the inner slope of the trench (KARIG *et al.*, 1978; SHIPLEY *et al.*, 1980, 1982; SHIPLEY, 1982). The horizontal separation between crystalline and continental crust and the trench is only about 30 km. The narrow continental margin may have originated from an earlier period of subduction or translation (KARIG *et al.*, 1978; ANDERSON and SCHMIDT, 1983) but all of these studies reveal a small Neogene accretionary wedge along this portion of the Mexican margin [Fig. 1(b)]. On DSDP Leg 66 a transect of 8 sites were drilled across the margin on oceanic crust, in the trench axis, continental slope accretionary prism and outer shelf. Evidence for age progression, uplift and rotation of sediments was found at three DSDP sites on the inner trench wall. Based on the occurrence and distribution of landward dipping reflections and calculations of the sediment budget, additional mass must be added at depth (by underplating?) to account for part of the uplift and volume of the inferred accretionary zone. The results of this work have been extensively reported and will not be elaborated upon (WATKINS *et al.*, 1982a, b; MOORE *et al.*, 1982b, c; MCMILLEN and BACHMAN, 1982; MCMILLEN *et al.*, 1982; SHIPLEY, 1982). The drilling transect and the description of the geology given above is applicable near the region offshore of the Rio Ometepec (Fig. 4). As we will show, just to the southeast the trench fill is thinner and most of the obvious structures associated with the sediment offscraping are not found.

The Guatemalan portion of the Middle America Trench was drilled on DSDP Legs 67 and 84. In many respects Guatemala represents a typical continental convergent margin with a mature forearc basin, outer high and wide accretionary prism [Fig. 1(c)]. However, in detail, the drilling transect demonstrated that there is little, if any, recent sediment offscraping. The upper level of the accretionary prism is composed of rocks of ophiolitic affinities overlain by a thin veneer of slope sediments. The basement rocks were emplaced prior to the Eocene, and while their exact distribution is not well documented, they apparently extend seaward to within a few kilometers to the trench axis, where Miocene oceanic crust is being subducted (LADD *et al.*, 1978, 1982; IBRAHIM *et al.*, 1979; AUBOUIN *et al.*, 1982a, b; VON HUENE *et al.*, 1980). Thus this margin which once was interpreted as a type example of the imbricate thrust model (SEELY *et al.*, 1974), actually had no Neogene sediment offscraping, and is now undergoing sediment subduction and possibly erosion. The surprising difference between the geophysical and drilling results may simply highlight our inability to interpret seismic reflection records in undrilled regions. On the other hand, the seismic line interpreted by SEELY *et al.* (1974) was northwest of the drilled section and it may be that structural styles vary over short distances. We will evaluate this problem later.

The Costa Rican margin offshore of the Nicoya Peninsula, where Cretaceous oceanic crust and sediments are exposed, is included in this investigation because it represents yet another significant variation in tectonic style along the Middle American Margin, [Fig. 1(b)]. The sedimentary sequence on the oceanic plate and continental slope are well-defined and regular, and the gross geometry of the margin is delineated by reflections from the top of the oceanic crust which are trace from the trench, landward to within a few kilometers of the shoreline (CROWE and BUFFLER, 1983; SHIPLEY *et al.*, 1982). A portion of the slope section was drilled, and consists of Holocene to late Miocene mud and mudstone, but the presumed accretionary prism was not reached (AUBOUIN *et al.*, 1982b). There is no direct evidence for Neogene accretion in the seismic data discussed by CROWE and BUFFLER (1983) and SHIPLEY *et al.*, (1982) but there is a suggestion of deformation of the overlying slope sequence. Our higher resolution seismic study will show that some sediment accretion may be occurring here.

4. *Observations*

Sea Beam swath mapping provides about 16 soundings perpendicular to the ship's track every 10 seconds during the surveys. At trench axis depths, this results in sonification of a rectangle approximately 250 m by 40 m. Four to seven sequential measurements were averaged to produce an approximately 250 m × 250 m area on the seafloor characterized by a single depth required for the contour algorithm and then computer plotted. Crosstrack tie lines show that the reproducibility of Sea Beam at the 10-m contour level is good to excellent with only minor contour shape distortion. After replotting with the final satellite navigation the individual swaths were manually aligned for best fit of adjacent swaths since unique bathymetric features often provided better reference points. For more information on the Sea Beam system see RENARD and ALLENOU (1979) and references therein. At the same time bathymetric data were collected, high resolution seismic reflection data were obtained to supplement the existing low detail, deep penetrating multichannel seismic grid. We generally used two 1.3 liter (80 cubic inch) water-guns as a source and recorded the data digitally (FRENCH and HENSON, 1978). Occasionally two 9.0 liter (550 cubic inch) air guns were used for deeper penetration. The digital recording allows redisplay, mixing, filtering and migration to help resolve some of the morphologic structures observed in the Sea Beam data.

Morphologic highs, lows, escarpments, lineaments, benches, and basins may have their origin in either depositional, erosional or tectonic processes. In the region of the trench axis the continuous recording of Sea Beam swath bathymetry and seismic data provides a powerful and accurate 3-dimensional image of the subsurface structure.

Three sets of charts are shown for each of the three surveys. One is a 200-m contour of the area earlier surveyed by the Uiversity of Texas (Fig. 1). The second is the Sea Beam bathymetry at a 20-m contour interval (Figs. 2, 6, 10) and the third is a 200-m contour extract of the Sea Beam data showing structural and sedimentological features and the location of various profiles discussed in this paper (Figs. 4, 7, 11).

4.1 *Mexico*

The Mexico bathymetric data displayed at the 20-m contour interval (Fig. 2) reveals complexities of the lower slope, not evident in earlier studies [Fig. 1(b)]. As noted in a Sea Beam survey reported by RENARD *et al.* (1980) and ROUMP *et al.* (1982) the orientation of the oceanic crust fabric is subparallel to the trench deformational front by about 5° to 30°. The distinct crustal lineations have relief of 100 m or more with steps almost invariably down to the trench; there are no horsts. A definite cross-trend is illustrated by a gentle persistent northward increase in depth. A linear series of seamounts 3 to 5 km wide and 500 m to 800 m high bisects the trench near 98°40′W and separates oceanic crust with slightly different fabric trends, N58°W to the west and N50°W to the east, similar to trends in the magnetic anomaly lineations (SHIPLEY *et al.*, 1980). The seamounts might be the trace of a pseudo fault (HEY *et al.*, 1980).

The trench axis is composed of a series of small, irregular basins formed by the intersection of the deformation front with the 100 m scarps of the oceanic plate step faults. If the basins were stripped of sediment, they would show a series of depressions each separated by a scarp progressively deeper toward the northwest. The dispersal of sediments along the trench is controlled by the underlying basement morphology. Many small canyons begin below the shelf break, and thus the trench sediments are mainly muddy turbidites supplied to the trench at a few places along its axis (MCMILLEN and HAINES, 1982; MCMILLEN *et al.*, 1982). One exception is the Rio Ometepec Canyon (location indicated in Fig. 4) which extends all the way from the shoreline to the trench. This portion of the trench contains poorly sorted to well-sorted sands derived from the shelf, some of which are deposited in a submarine fan at the base of slope (LONSDALE and MOORE, in prep; MCMILLEN *et al.*, 1982). The sediment supply rate is high enough so that the step fault structures are filled in the western third of the survey area. The seamount chain forms a sill that blocks sediment transport along the trench axis to the east, dividing the trench into two distinct provinces, one a partially sediment-filled trench and the other a more sediment starved trench.

The deformation front defined by the abrupt decrease in gradient of the continental slope, is easily located. This front is more complicated than suggested by RENARD *et al.* (1980) or ROUMP *et al.* (1982) from their limited survey. In detail it is not linear but, almost serrated. Interestingly, a 3 km landward reentrant in the front is coincident with the Ometepec submarine fan. LONSDALE and MOORE (in prep.) have observed channels cut through recent faults on the fan in submersible diving in this area. On the lower inner slope of the trench, linear benches are common; in the western part of the area they are usually less than 10 km in length and a few kilometers wide, and in the eastern area they are fewer and less well defined. There are several larger basins in the eastern area higher up on the slope in 3000 m to 4000 m water depths. The relationship between the canyons and the slope basins and benches is clearly defined by the bathymetric data. Excluding the Rio Ometepec, three other canyons

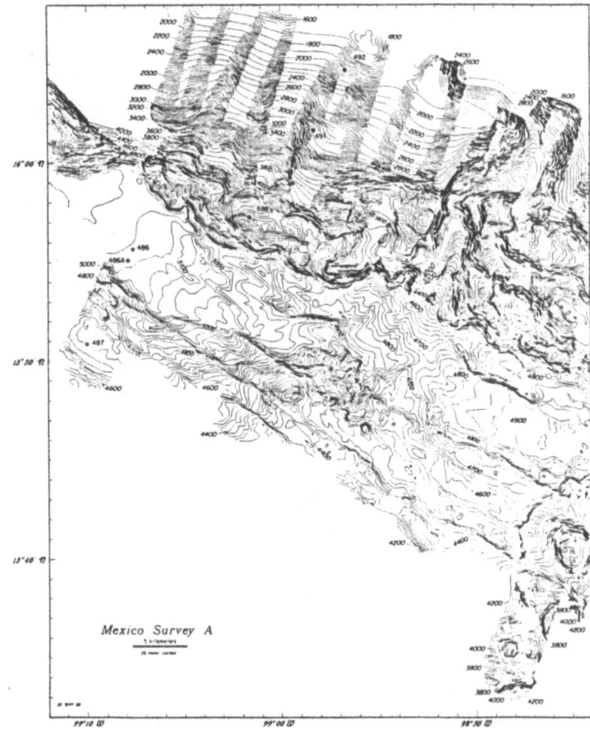

are incised in the upper and mid-slope region but not in the lower slope. One does end in a large mid-slope basin, and another is identifiable at the 20 m contour to within 5 km of the trench (Fig. 4). Generally the lower slope benches in the western region are not fed by canyons and are not filled with turbidites.

The subsurface structure is obtained by integrating the seismic reflection data with the contour charts. For the purposes of this paper we will concentrate on two sub-sets of the data. One is in the vicinity of the Rio Ometepec Canyon and the DSDP drilling transect and the other is a slope basin and trench area in the more sediment starved eastern area. In the western region two migrated seismic lines illustrate the structure and the relationship between the structure and morphology (Fig. 3). The locations of the profiles are shown in Fig. 4, which summarizes the tectonic and sedimentologic features of the area. These two profiles are separated by only 7 km, yet the correlation of structures across this distance would

Fig. 2. Sea Bean bathymetric survey off Mexico at a 20-m contour interval. The 100-m contour is interpolated between swath gaps. Filled dots are DSDP drill holes.

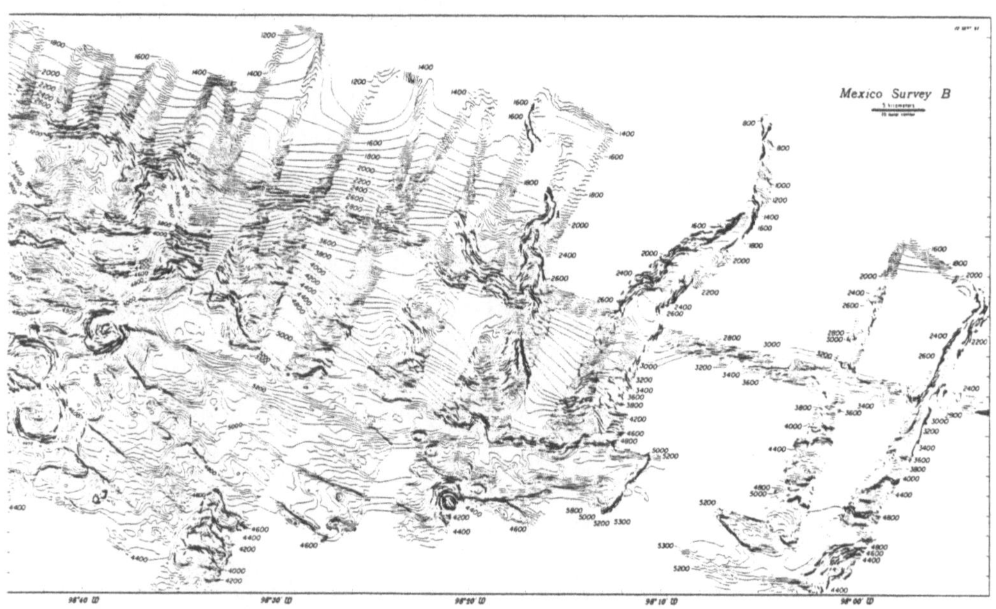

not be unequivacol, without the Sea Beam coverage.

Seaward of the trench the step-faulted basement is identified by the large reflectivity contrast between the pelagic and hemipelagic sequences and the volcanic basement. The low reflectivity of the oceanic section and spatial resolution are such that we can not detect the likely recent movement on these normal faults. The trend of the steps in the basement is fairly constant but the magnitude varies (Fig. 3). Within the trench the reflectivity of the trench fill is higher and internal structure is better defined. The onlapping character of the sediments is not obvious in these sections as it is in others to the west (SHIPLEY, 1982). The reflections define gentle anticlinal structures [e.g., Fig. 3(b), 0642 z] usually not more than a km in cross section and usually with much less than 200 m amplitude. Most of the fold axes in the trench are only 4 to 5 km long, the longest identified is only 8 km in length (some are inferred to be slightly longer). At this stage in our analysis we are unable to define the subsurface ade-

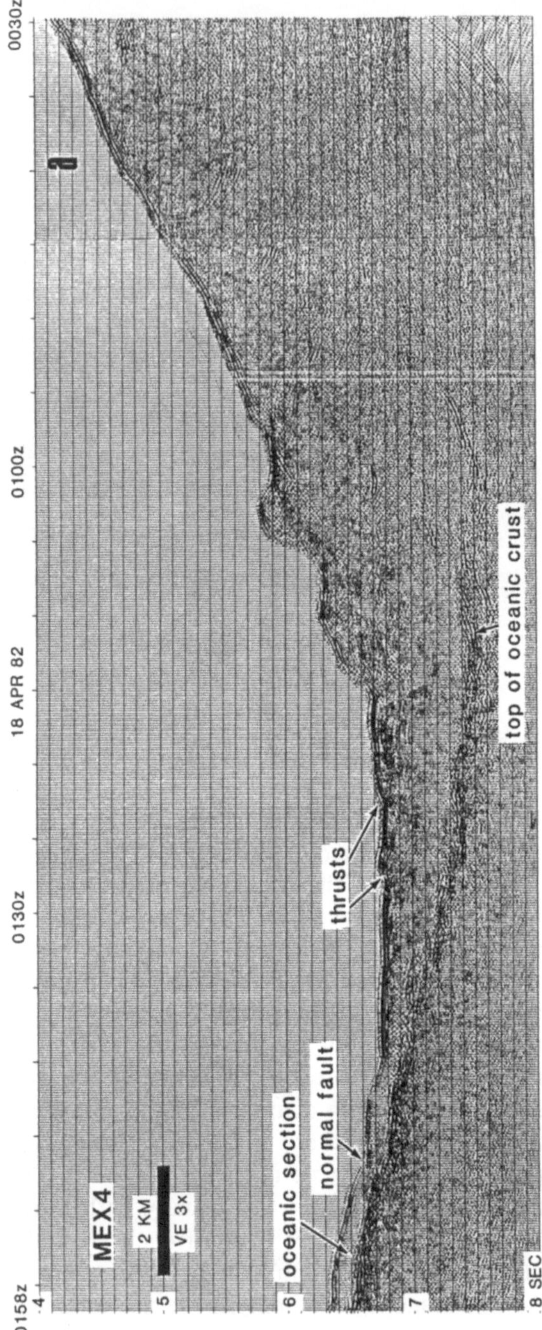
0030z
0100z
18 APR 82
0130z
0158z
MEX4
2 KM
VE 3x
oceanic section
normal fault
thrusts
top of oceanic crust
4
5
6
7
8 SEC

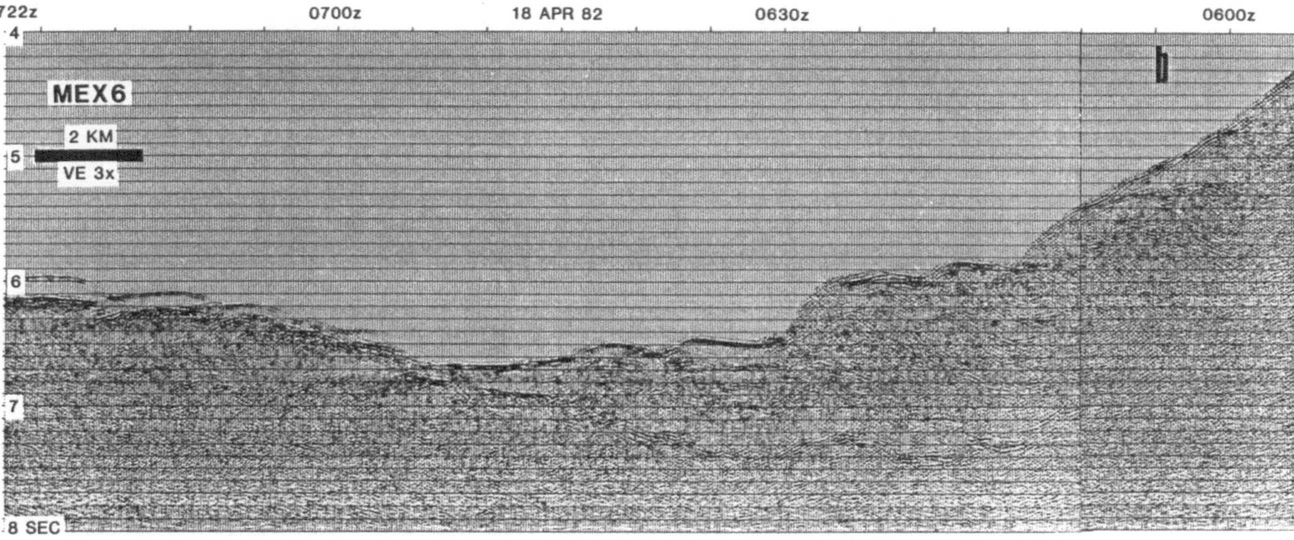

Fig. 3. Water-gun seismic profiles MEX 4 and MEX 6. Profiles have been filtered, migrated and amplitude scaled. Location shown in Fig. 4. Note the structure of 0103 z in MEX 4. With reference to Fig. 4 it clearly originates as a side-swipe from the anticline to the northeast.

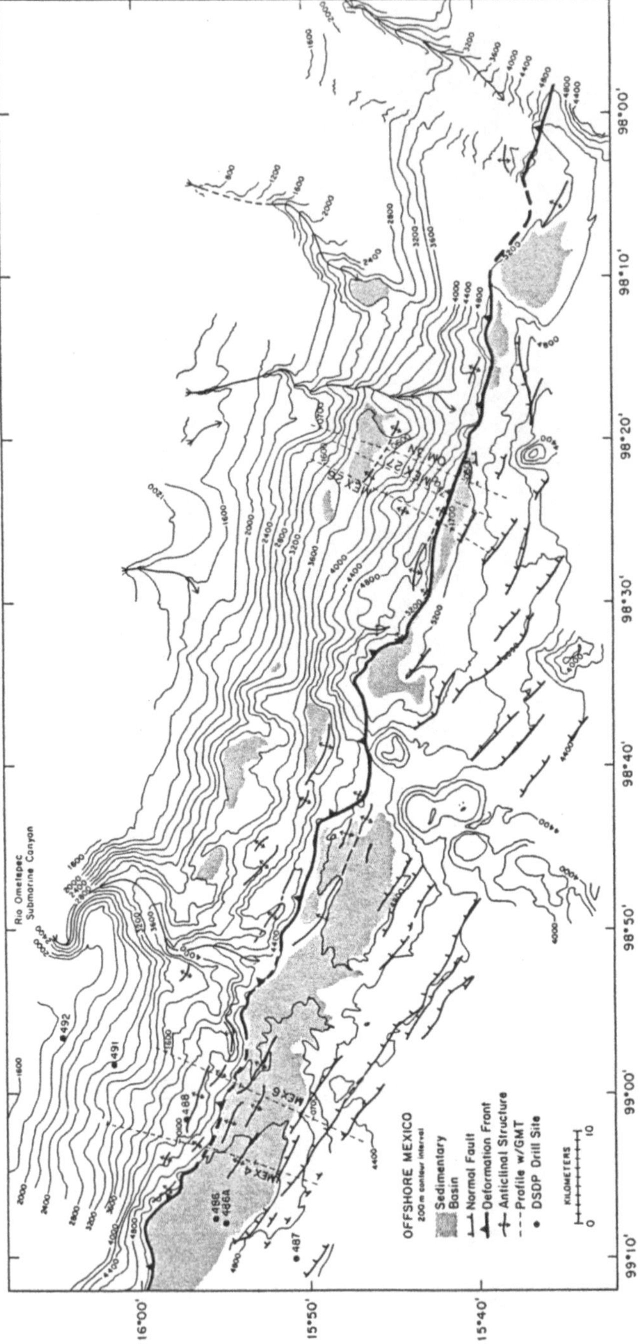

Fig. 4. Chart illustrating the structural features observed in the seismic and Sea Beam data offshore Mexico.

quately to constrain the geometry of the thrust surface at depth but it is shallow and within the turbidites.

The benches, several hundred meters up the slope from the trench, are anticlinal features on their seaward edge at very shallow subbottom depths. At greater subbottom depth the structure seems to be monoclinal, gently landward dipping. The internal geometry of these benches is poorly imaged, perhaps indicative of more disruption of the primay bedding than within the trench. The benches have similar dimensions to the structures in the trench and are probably just slightly older accreted packets of trench material. At DSDP Site 488, located between MEX 4 and MEX 6, the upper section is composed of muddy slope deposits (apparently the folded anticlinal section) overlying the sandy trench material (the monoclinal section).

Three more seismic profiles in the eastern sediment starved segment of the survey area off Mexico are shown in Fig. 5. The separation between each of these lines is about 3 km. This area includes one of the rare sediment filled slope basins which is offshore of a concentration of several minor rivers with limited drainage basins (ENKEBOLL, 1982). This region has only a thin trench fill, about 20% of that in the western profiles MEX 4 and MEX 6 and within the axis there is no folding or disruption of bedding at a scale we can observe. On the lower slope there are only minor inflections in the bathymetry with ambiguous internal structure. Interestingly, in OM3N, the low resolution multichannel profile [Fig. 5(c)], several distinct landward dipping events, extend down dip 2 to 3 km. Careful examination of adjacent lines MEX 26 and 27 suggest that the dipping reflections correlate from OM3N to lines MEX 26 and 27. If this sequence is offscraped material, it is nearly exposed at the seafloor, because the slope cover is very thin.

The slope basin, about 8 km × 3 km has a clear inflection point on its seaward side which becomes a well defined ridge towards the southeast. MEX 27 [Fig. 5(b) at 1923 z] best shows the internal landward dipping structure of this inflection point. Overlying this dipping sequence is a basin filling sequence with seaward dipping and basin axis thickening geometry. This geometry suggests rather straight forward infilling with little subsequent tectonic movement.

As we will demonstrate, the differences in structures on the lower slope and trench between the sediment filled and sediment starved trenches area as fundamental as differences between Guatemala and Costa Rica.

4.2 Guatemala

The Sea Beam bathymetry offshore Guatemala, shown in Fig. 6, il-

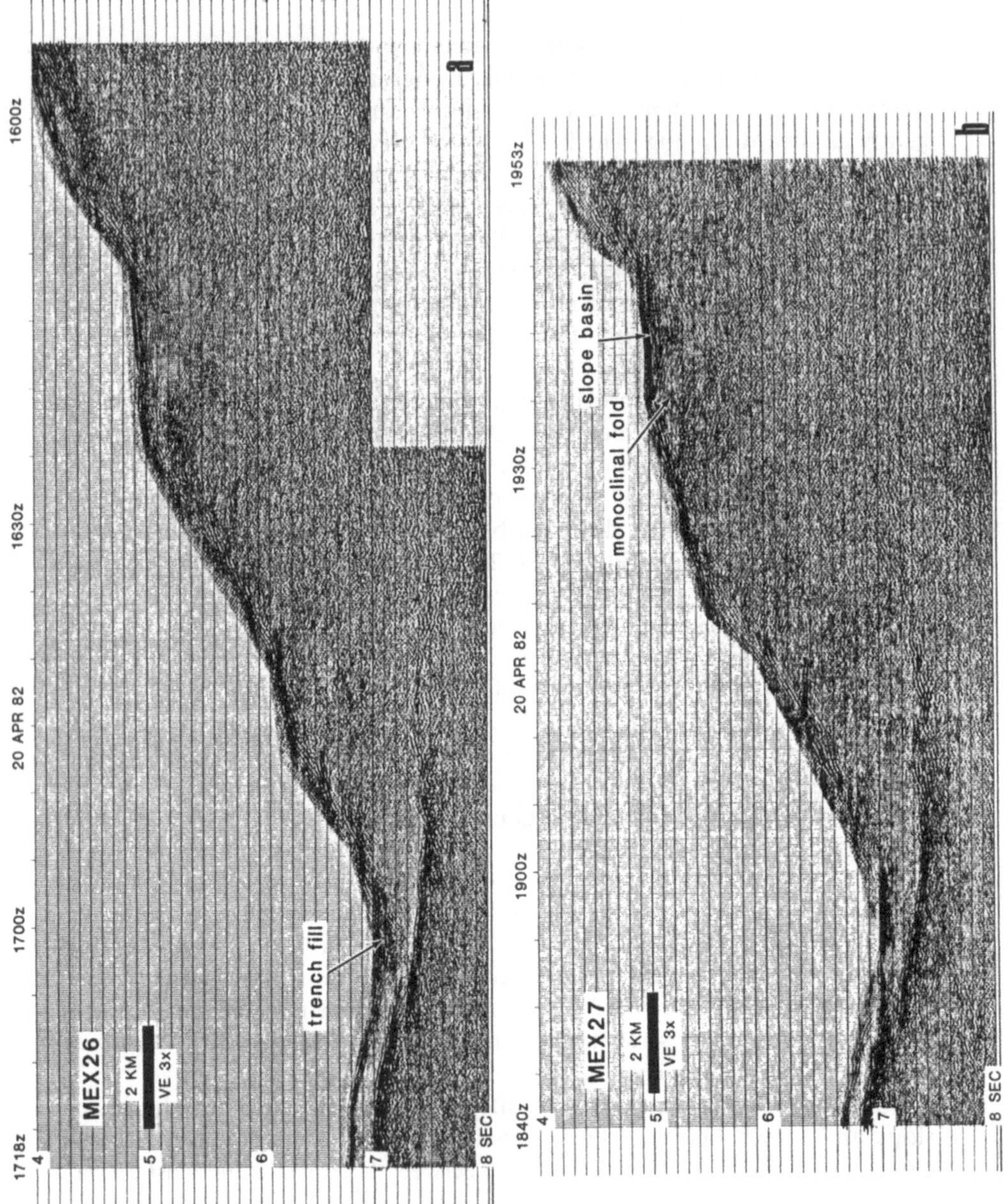
a
1600z
1630z
20 APR 82
1700z
1718z
MEX26
2 KM
VE 3x
trench fill
4
5
6
7
8 SEC
b
1953z
slope basin
monoclinal fold
1930z
20 APR 82
1900z
1840z
MEX27
2 KM
VE 3x
4
5
6
7
8 SEC

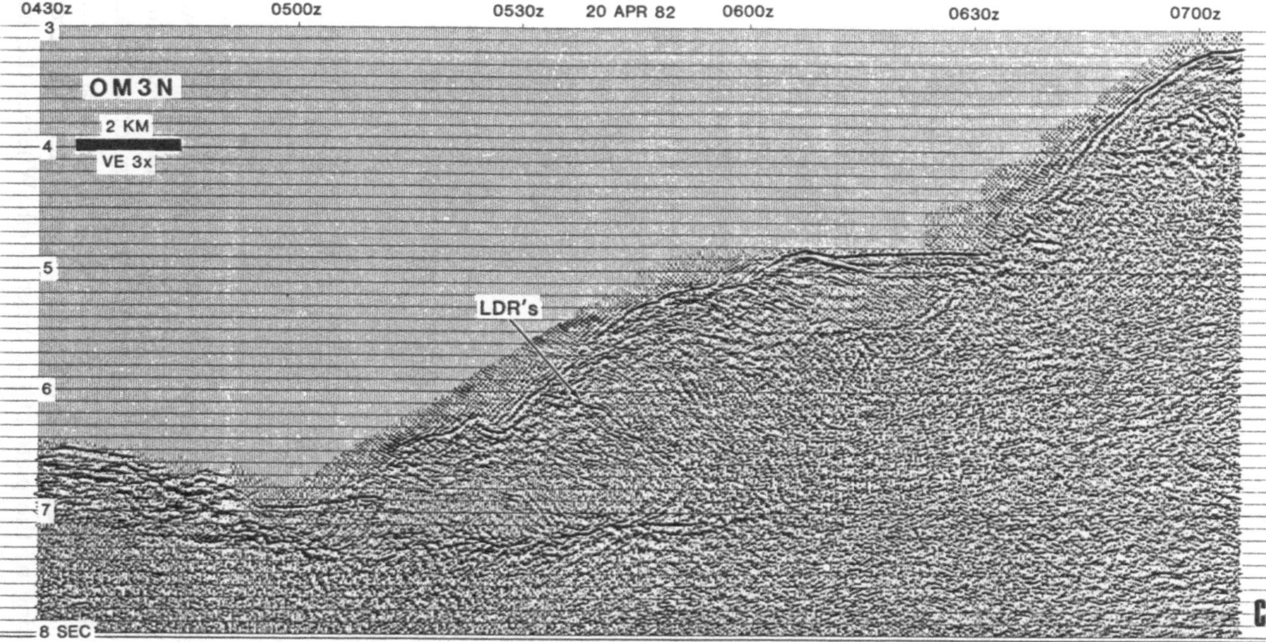

Fig. 5. Seismic profiles MEX 26, MEX 27 and OM3N. MEX 26 and MEX 27 are high resolution water gun sections while OM3N is a low pressure air gun, multichannel seismic line. All have been filtered, migrated and amplitude scaled. Locations are shown in Fig. 4. LDR is landward dipping reflection.

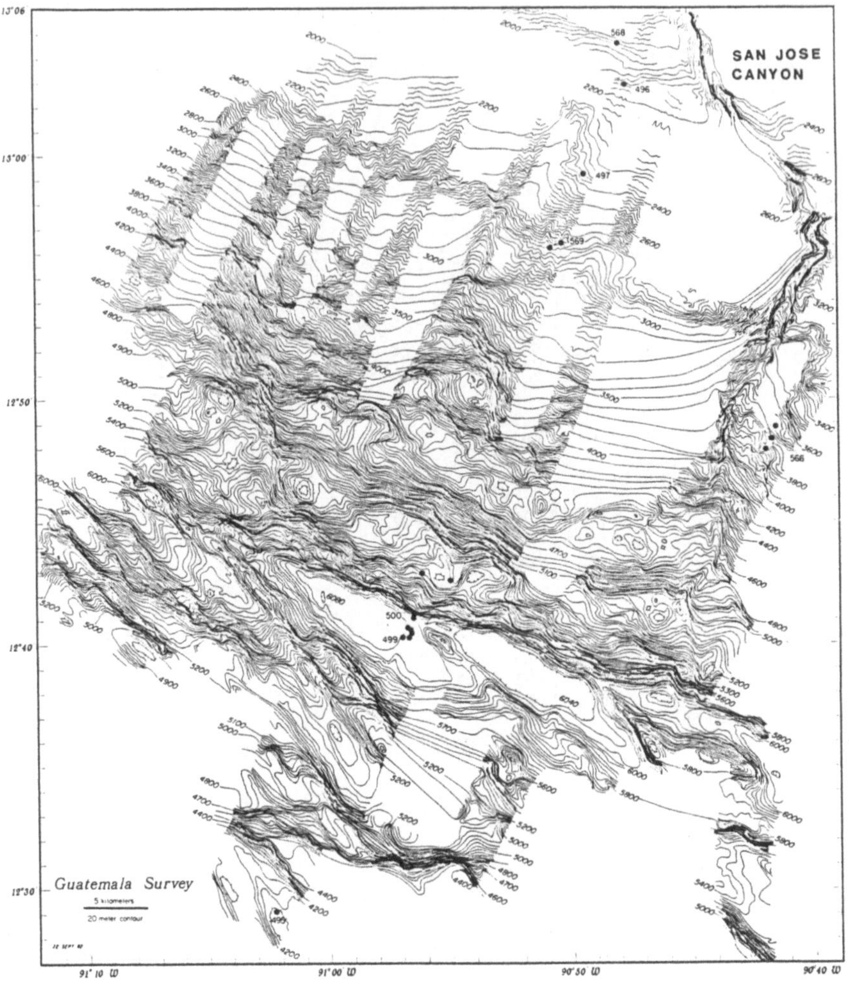

Fig. 6. Sea Beam bathymetric survey off Guatemala, conventions as in Fig. 2.

lustrates the complexity of both the oceanic plate fabric and the lower continental slope region. As discussed by AUBOUIN *et al.* (1982b) and RENARD *et al.* (1980) the oceanic fabric has about 150 m of relief and a 2–3 km wavelength trending N35°W. Our data coverage does not extend seaward far enough to observe if the horst-graben structure is accentuated due to bending of the plate at the outer swell. Whereas the step faults acted as sills off Mexico, here the horst's intersections with the continental slope are sills forming small trapezoidal shaped depressions. Locally some larger depressions are partially filled with up to 250 m of turbidite sediments but generally the sediment accumulation rates are low and the basins remain unfilled and unconnected along the trench. The deformation front is well delineated where it abuts the flat-floored basins but it is less obvious elsewhere, as west of 91°05′W, where the relief of the inner and outer walls are similar and meet in a "V" shaped depression—the trench axis. The deformation front is somewhat more regular than off Mexico. We do not find fold or thrust structures in the sediment filled or unfilled trench like those in the western Mexico area. In fact the trench is strikingly similar to the morphology of the eastern Mexico area.

The inner wall morphology is somewhat less regular, or at least less trench parallel, than off Mexico. It contains several broad 2–3 km wide and 10 to 20 km long benches below 4500 m depth. These are *arcuate* shaped regions, both seaward and landward convex, characterized by more widely space contours and small closed highs less than 2 km in diameter and 20 to 60 m high. Large canyons are not as common as off Mexico. The San Jose submarine canyon is the only major canyon system to create a large valley within the shelf-forearc basin [Figs. 1(c) and 7]. Even so, it carries only minor muddy turbidites to the trench, and the incised portion of the canyon does not quite reach the trench (ENKEBOLL, 1982; MCMILLEN *et al.*, 1982). One other small valley evident on the Sea Beam chart may have supplied the very thin laminated sections within the slope bench in 3600 m of water shown in Fig. 7.

Three migrated seismic profiles are shown in Figs. 8 and 9. These sections, each separated by 3 km, illustrate the relationship between the morphology and sub-seafloor structures. GUA 10 is near the transect of DSDP drill holes (Leg 67 and 84) and University of Texas multichannel sections UTGUA 13 and UTGUA 18 (LADD *et al.*, 1982). Our profiles show many of the same features as the multichannel profiles described by VON HUENE *et al.* (1982). Like the morphology,the seismic reflection data show the absence of deformational structures within the trench and inner wall of the trench. The slope section is well defined as a zone of low amplitude, generally slope sub-parallel, reflections (VON HUENE *et al.*, 1982; VOLPE *et al.*, 1985). The arcuate shaped benches seem to contain the thickest slope sections, although they do not appear to be basin-filling sequences, but rather slope parallel reflections. Piston cores from these basins were all muds with few fine sand layers (MCMILLEN *et al.*, 1982).

We have been unable to seismically identify the top of the oceanic

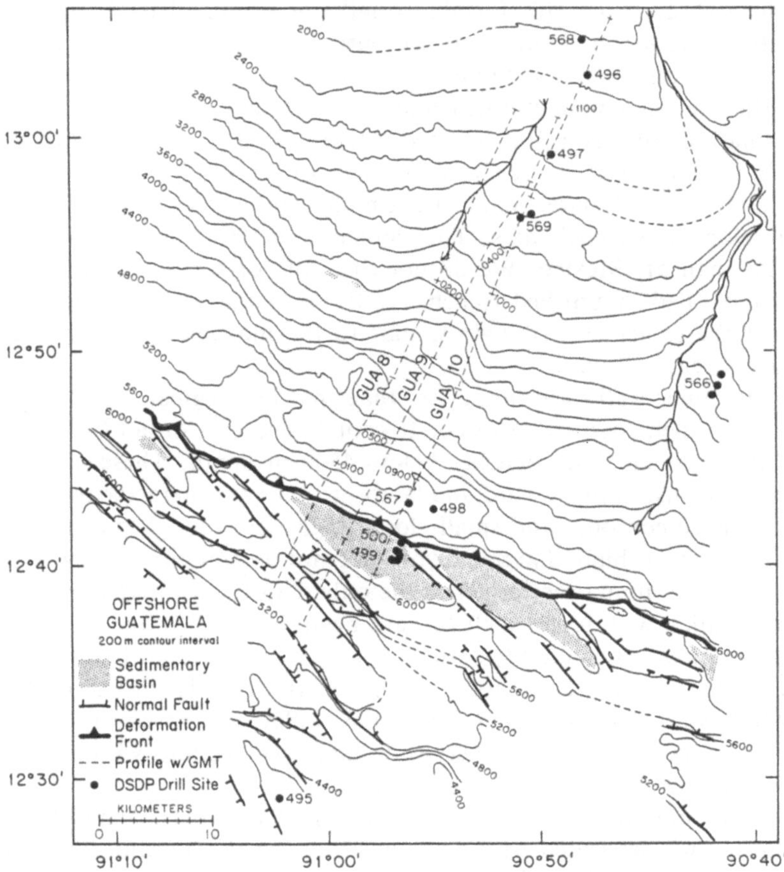

Fig. 7. Major structural features observed in the Sea Beam and seismic reflection data.

crust landward of deformation front. Within the trench, the boundary between the turbidite fill and oceanic plate sediment is easily defined but the reflections from the oceanic crust are diffuse and the boundary obscure. The parallel high amplitude reflections which characterize the turbidite sequence abruptly terminate less than 1 km beneath the deformation front, in contrast with the western Mexico area. This suggests that the turbidite fill is fairly recent, or more likely the deformation is very intense at the base of the slope.

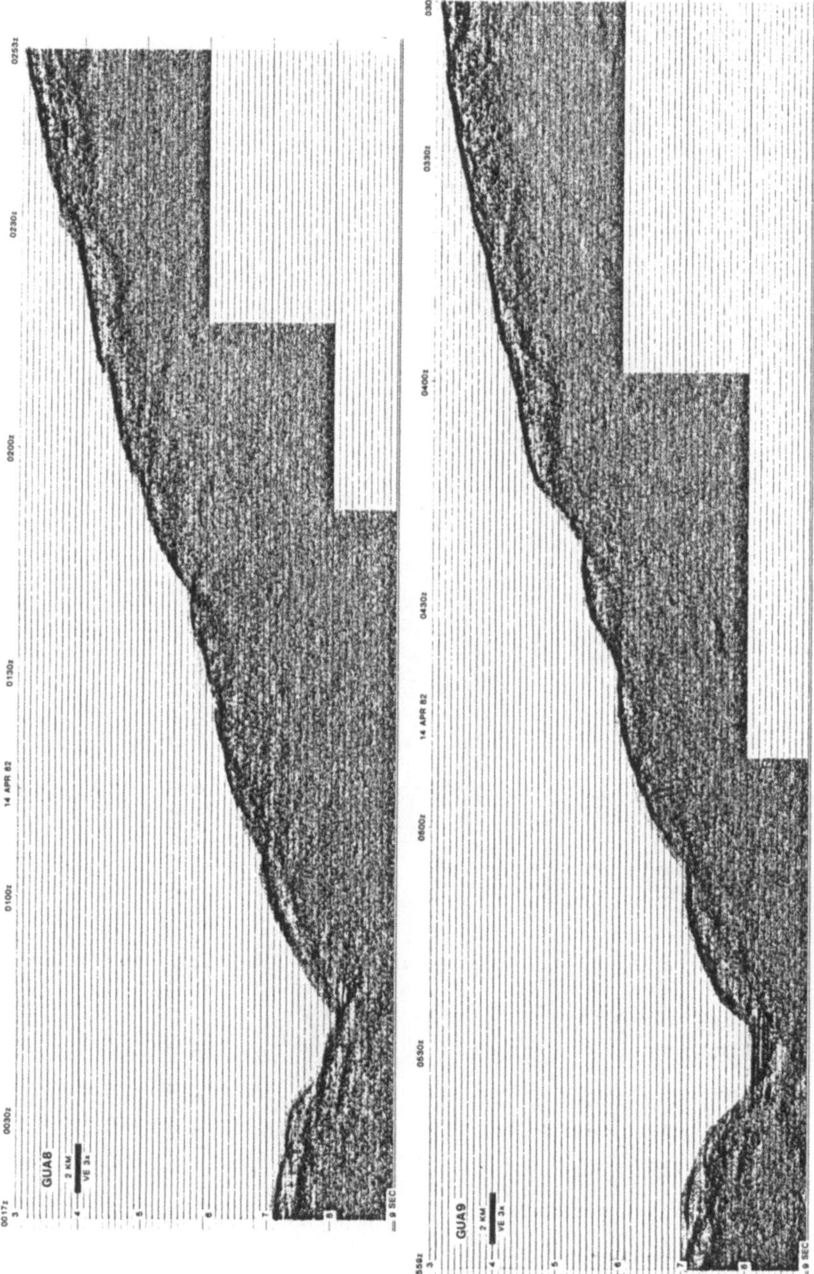

Fig. 8. GUA 8 and GUA 9 filtered, migrated and amplitude scaled seismic reflection profiles. Location shown in Fig. 7.

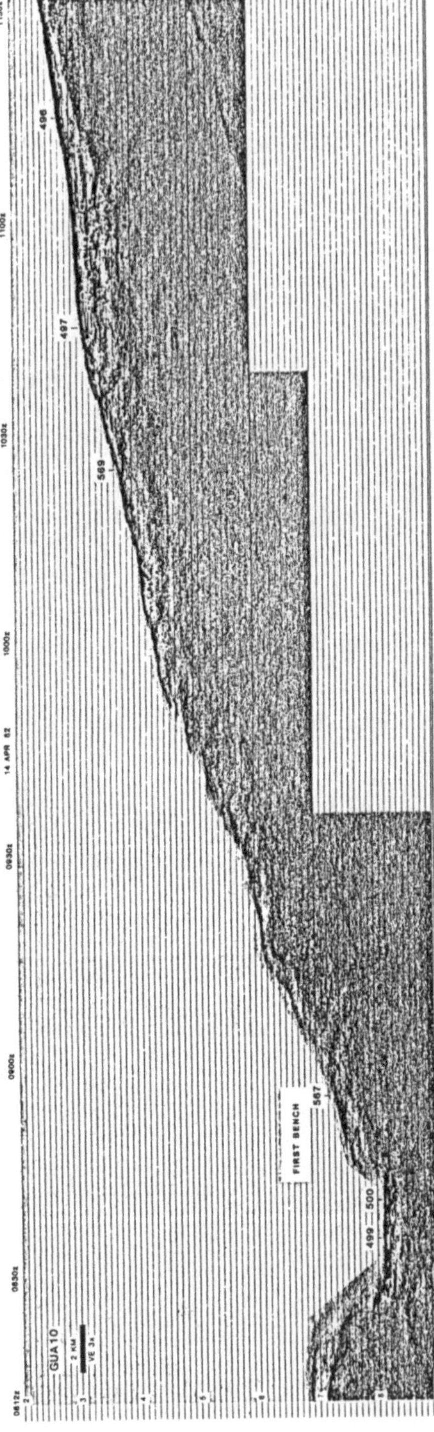

Fig. 9. GUA 10 filtered, migrated and amplitude scaled seismic profile. This section is close to the DSDP Leg 67 and 84 drilling transect.

4.3 *Costa Rica*

The Costa Rican portion of the trench is smooth and regular (Fig. 10). The plate fabric is characterized by a series of scarps, down to the trench, having relief typically of 100 m, and is within a few degrees of being parallel to the deformation front. Within 3 or 4 km of the axis, the landward slope is very steep, then has a gentle slope for the next 10 km before becoming somewhat steeper again. Few closed contours are evident and no submarine canyons are found on the Nicoyian continental margin slope.

The seismic profiles illustrate the reflective oceanic section (Fig. 12) that extends far beneath the continental slope on all the sections. In UTCR

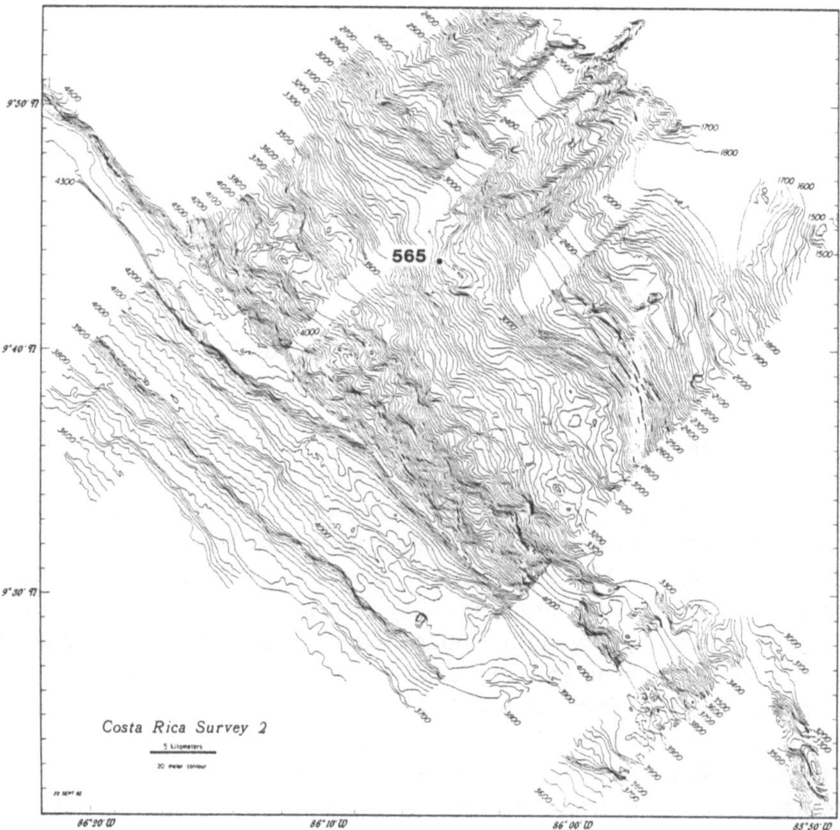

Fig. 10. Sea Beam bathymetric survey off the Nicoya Penninsula of Costa Rica.

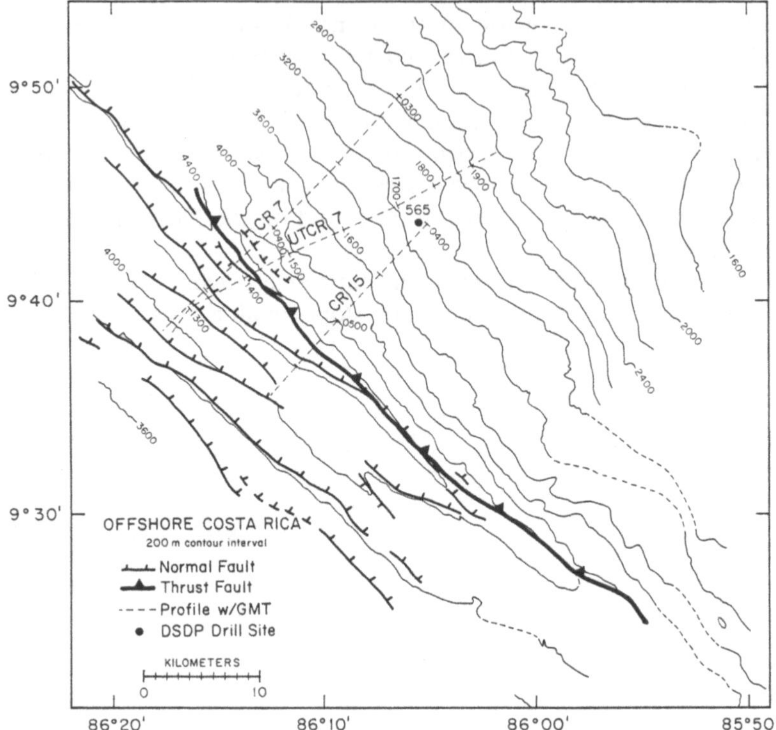

Fig. 11. Structural features and profile locations offshore Costa Rica.

7, a multichannel section which used explosive Maxipulse charges as the sound source, the crust is traced to within a few kilometers of the shoreline (CROWE and BUFFLER, 1983; SHIPLEY *et al.*, 1982). There is only one small pocket of trench fill less than 50 meters thick and less than 600 m wide on CR 7 at 0420 z. There are also no structures within the trench except those produced by high angle faulting associated with the plate fabric.

The zone defined by an irregular diffraction producing surface at the base of the slope apron and above the oceanic plate is the so-called accretionary prism. Velocities increase from the range of 2 km/sec in the slope section to 4 km/sec and higher in the accretionary prism (CROWE and BUFFLER, 1983). Even after migration to suppress the diffraction tails, little structure is observed. In the upper portion of the prism there is a faint landward dipping "grain".

The slope apron overlying the accretionary prism pinches out on the lower slope. It thickens significantly on the slope above 4000 m, becoming up to 700 to 800 m thick in the mid-slope. Even though the slope section progressively thickens upslope, the basic morphology of the inner wall, the steep upper slope, gentler middle slope and steep lower slope is the subdued expression of the top of the prism.

5. *Discussion*

A wide variety of bathymetric features were encountered in the Sea Beam surveys, and with the exception of oceanic plate fabric, these features are parallel or subparallel to the trench and seldom more than 10 km in length. Earlier characterizations of lower trench slopes produced less accurate definition of the morphology because track lines were typically spaced greater than 10 km, thus the same feature might not appear on two adjacent tracks or be incorrectly correlated across the tracks. Further, conventional wide-beam echo-sounding, with about a 30° beam width, sonifies a huge area of seafloor and the unmigrated returns are superimposed on each other, obscuring the bathymetry.

Seismic reflection profiles collected along these tracks have the same problems with focusing, compounded by the subsurface complexities. Narrow beam swath mapping now provides a tool for correctly interpreting the seismic data. As often as not, the seismic lines image structures out of the plane of the section and if interpreted as simple two-dimensional sections will be inaccurate [for example Fig. 3(a) at 0104 z]. Since the standard marine survey techniques are appropriate only for gross structural studies of subduction zones, models based on signular cross sections should be expected to have many inaccuracies of small scale structure and processes. The broader surveys are a necessity for a description of the regional framework, but predictions based on these surveys, at the scale of drill holes, have not been particularly good. These discrepancies encountered between seismic reflection interpretations and actual drilling samples are due to the problem of scale.

An example of this scale problem is illustrated offshore Mexico in the sediment-filled trench. Within the trench, several very small channels were observed [as in Figs. 3(a) and 3(b)] along several University of Texas multichannel seismic lines (SHIPLEY *et al.*, 1980). These were interpreted as turbidite channels by sedimentologists (MCMILLEN *et al.*, 1982), and as the bases of thrust faults by others (SHIPLEY *et al.*, 1980, 1982). From the series of lines then available, the level fo the trench turbidite fill on adjacent (10+ km) crossings of the trench were interpreted as evidence for either a fairly continuous gradient for the trench floor or as a series

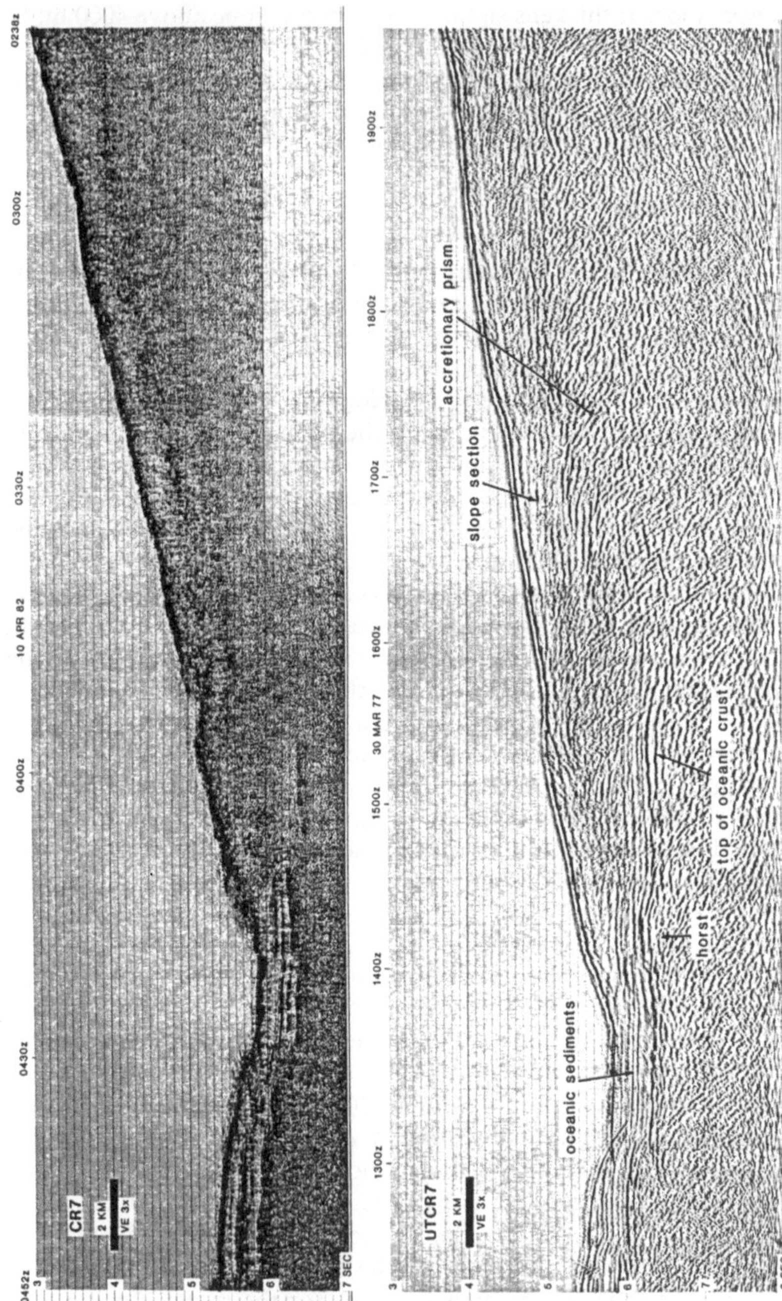
CR7
2 KM
VE 3x
0238z
0300z
0330z
10 APR 82
0400z
0430z
0452z
7 SEC
UTCR7
2 KM
VE 3x
1900z
1800z
1700z
1600z
30 MAR 77
1500z
1400z
1300z
8 SEC
accretionary prism
slope section
top of oceanic crust
horst
oceanic sediments

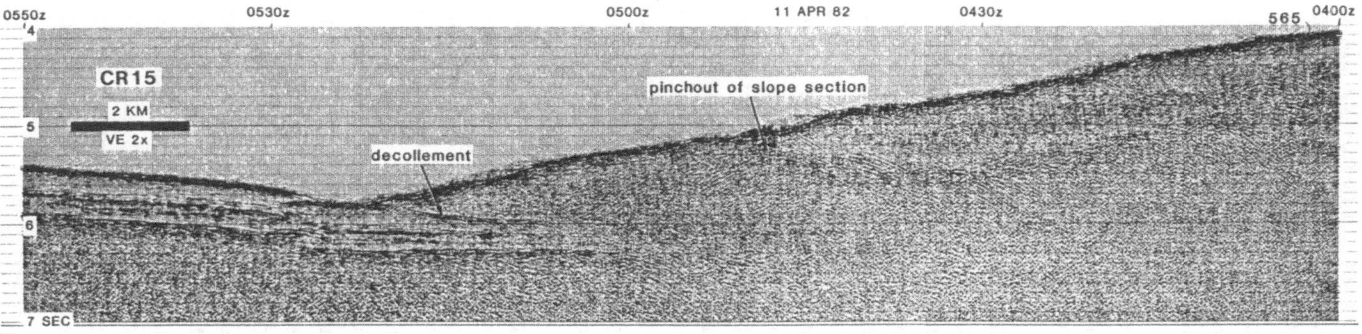

Fig. 12. Filtered, migrated, amplitude normalized seismic profiles off Costa Rica. CR-7 used a water-gun source, CR-15 used two 9.0 liter cubic inch air guns with deconvolution and UTCR 7 is a multichannel profile which used maxipulse explosives as the source.

of ponds each spilling sediments down into the next flat-floored plain. With our higher resolution surveys we now know that with the exception of a 3 km segment near 98°58′W, the axial channels with more than 20–30 m or relief follow the base of fold or thrust trends; they may enhance the relief of the fold if they are erosional but their location, if not their occurrence at all, is controlled by the structure. Also the newer surveys show that the trench is fairly continuous with only a few obvious places where trench axis flow is blocked. These Sea Beam data provide a somewhat different understanding of the relationship between the trench depositional and structural regime. We will now discuss some of the other small structures and their spatial settings.

The plate fabric has little influence on the morphology of the inner trench slope. The only locations where we have seen obvious effects are with the trench sedimentation pattern and perhaps in the position of the deformation front. For instance, the linear chain of volcanoes being subducted beneath Mexico (Fig. 2) seems to offset the location of the trench axis. However, this may be secondary, the reason for the dislocation may be the slightly greater depth of the oceanic plate to the southeast of the volcanoes. Other reentrants in the front, toward the northwest where the trench sediments are thicker, may be less significant because the definition of the front becomes more subjective as small thrusts develop seaward of the base of slope. Even off Guatemala, there is little evidence for fabric inheritance, yet the plate relief is quite high. We have been able to trace small horst blocks beneath the inner wall in several areas. One is shown in Figs. 12(a) and 12(b), where a horst block is accomodated beneath the slope off Costa Rica. Note that there is no evidence for downdropping of the lower slope material into the adjacent graben. In fact, the decollement appears to nearly mimic the relief. Admittedly, in this example, there is only about 200 m of relief. We do not observe morphologic or seismic evidence in our surveys to suggst that the fabric has any fundamental effect on the subduction/accretion processes.

The significance of slope basins and benches in the Middle America Trench seems to be different than that discussed by MOORE and KARIG (1976) where sediment filled basins are used as indicators of deformation and rotation. Here they are mostly devoid of sediment fill and mostly found in the lower slope region of Mexico and Guatemala. Off Mexico the benches contain only minor sediment fill. Earlier work, on the attitude and distribution of landward dipping reflections (LDR's) suggest that most rotation of the thrust packets is accomplished within 10–15 km of the trench (SHIPLEY, 1982). If the LDR's are a meaningful indicator of rotational history, then these lower slope benches, which might become sediment filled with age and appropriate sediment supply, may not record the same history as slope basins of Nias or others with thick sediment accumulation. The dip of sedimentary units within these basins have a significant component of depositional dip which has to be accounted for in any structural analysis (Fig. 5). Most of the benches off Mexico are the result of drag folds on thrust faults or are relief created by fault blocks.

The Guatemala lower slope is also interesting because of a number of arcuate-shaped (both seaward and landward facing) zones, not necessarily benches or flat spots, but less steep slopes with some closed contours. These are also not sediment filled, but simply inflections in the slope. The slope apron of slope sub-parallel reflections blankets the slopes, but is thicker on the less steep slopes and significantly thinner on the very steep slopes. The origin of these slope irregularities is unknown but they do appear similar to features on the slope off Costa Rica where the relief is somewhat less and features are better imaged. Off Costa Rica, the slope goemetry is in response to the morphology of the underlying surface (the top of the "accretionary zone") and that surface does not appear to be actively disrupting the overlying slope section in any organized manner (Fig. 12). We believe that off Guatemala and Costa Rica the slope irregularities are not related to recent tectonics as they are off Mexico.

What do we know about sediment subducted beneath the deformation front? It is difficult to seismically trace sediment beneath the front off Mexico, because the oceanic section is not highly reflective, and we usually do not trace the bedded turbidites more than a few kilometers. In a few exceptional cases the pelagic sequence is traced perhaps 5 km, and apparently much of the turbidite fill is scraped off into thrust packets. The fate of the oceanic sediments remains obscure. Off Guatemala turbidites are never traced more than 1 km beneath the front where they abruptly terminate (Figs. 8, 9). The fate of oceanic sediments is also not defined. Off Costa Rica, we had an unusual opportunity to investigate the fate of the subducted sediments because of the higly reflective nature of the pelagic and hemipelagic sediments. While we are still trying to enhance these reflections by reprocessing (SHIPLEY and MOORE, 1986), some typical profiles are shown in Fig. 12. In CR 7 a horst block and overlying sediment section extend undisturbed about 1 km beneath the front, while in the graben at about 0400 z only the lowermost 2/3 or 3/4 of the section is still seismically identifiable. Does this place constraints on the dip of the decollement? Is the upper part of the section offscraped? In UTCR 7 the same horst block is observed a little further beneath the front, but the overlying oceanic section is significantly reduced in apparent thickness and then extends to about 1530 z at a fairly constant thickness.

This suggests that the decollement is nearly horizontal on this line and also removing only a small portion of the oceanic section. Alternatively, the variations in thickness may be due to dewatering and a velocity increase, and the decollement is above the section. But in CR 15, we observe an interesting phenomena, which if extrapolated to the other sections, would indicate modest offscraping as the reason for the thinning and not just a velocity effect. There the upper half of the section appears thinned much more than the underlying portion. There also appears to be termination of some reflections against the sloping reflector [near 0515 z, Fig. 12(c)]. We interpret these observations as follows: the sloping reflector is the decollement which is separating and accreting the upper portion of the oceanic section from the downgoing plate.

We now include these last observations into a discussion of the underlying processes of accretion. First the accretionary processes. It appears that trench sediments are conducive to offscraping. Some circumstantial evidence, from the orientation of the folds and thrusts within the trench, parallel to the plate fabric and not the deformation front, suggests that the sedimentation pattern is affecting the eventual offscraping. In Fig. 13 we illustrate what might be happening. The compaction of the turbidite fill of interbedded sands and muds will, upon loading of the wedge near the base of the deformation front, produce fluid migration into the sandy layers. At the seaward onlapping position, the fluids are sealed against the lower porosity hemipelagic and pelagic sequence and there may be

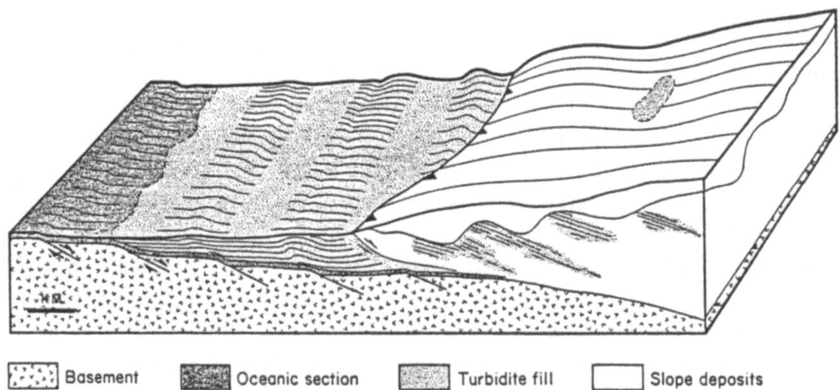

Fig. 13. Diagram illustrating localization of thrusts related to fluid migration against the lower permeability oceanic section. The same failure might occur in faulted oceanic sections, even if turbidites are not present. The orientation of the deformation fron is not necessarily parallel to or controlled by the accretion process.

fluid migration along the onlap surface. Folds and faults are more likely to be localized at these discontinuities than elsewhere where the high pressures would be bedding prallel. The localities of the structure parallel to the ocean fabric may also be enhanced simply because of the discontinuites in sediment thickness that develops related to the throw of the normal faults and the onlapping fill. In either case the plate fabric and trench sediments and their geometry are all important in offscraping. The importance of high fluid pressures has recently been discussed by VON HUENE and LEE (1983) and WESTBROOK and SMITH (1983).

In the case of Guatemala we have less information, but where turbidites are present they onlap the underlying oceanic section, although the sediments never occupy more than one graben. Even so, we would expect a similar processes to occur as off Mexico but the scale will be smaller. While there has not been long term offscraping off Guatemala, based on drilling at a site about 4 km from the trench, it may be occurring now. At the scale of our surveys, we observe no structures in the trench, but MOORE *et al.* (1982) in a Deep Tow survey did find low-relief, trench parallel ridges (perhaps folds?) just seaward of the base of the deformation front within the trench pond. Unfortunately, their seismic system had no penetration to examine the subsurface structure. We argue that offscraping depends greatly on the development of elevated pore pressures which will occur only where there are significant discontinuities in the sedimentary section, such as onlapping in a starved trench or faulting of the oceanic section to act as the barrier to fluid migration. Then if appropriate porosity contrasts exist, the trench fill and/or parts of the oceanic section could be stripped off, as in Costa Rica.

Finally, the question of subduction erosion is an interesting concept. We may only speculate because our data is not fully processed yet. The idea of sediment erosion is a difficult one to test. We do know it is not occurring off parts of Mexico, and apparently not off Costa Rica. But what is happening in the eastern portion of the Mexico survey area and perhaps fairly recently in the Guatemala area? While we know there has been no net accretion in the Neogene off Guatemala, conceptually, we prefer to think of a steady-state subduction process versus one of net erosion. We have difficulty with the mechanism of erosion at the toe-of-slope by sediments which are far less rigid than the highly deformed and dewatered section above. The oceanic plate normal faults do not erode the slope as in the "buss-saw" model (HILDE and SHARMAN, 1978; SCHWELLER *et al.*, 1981) at least when the horsts are of about 200 m or less. We have found no evidence for slumps, scars, slides or debris flows within the surveyed areas. If they exist, they are very small, <500

m features, right at the deformation front. As a mechanism for passive erosion (AUBOUIN *et al.*, 1984), volumetrically they would be a small component of the total volume that is postulated to have to be removed since the early Miocene off Guatemala. If the large arcuate shaped features on the slope off Guatemala were related to larger scale rotational blocks, we would expect to see bulges in the deformation front, which we do not. These benches may be older features of the slope now covered by slope material.

Thus, we believe that sediment offscraping and subduction are intimately related to plate fabric and the vertical succession of sedimentary sequences, either of the oceanic section or trench fill, which will produce the conditions needed for accretion, otherwise sediment subduction will ensue. We find no compelling evidence for sediment erosion.

Acknowledgements

We thank the governments of Costa Rica, Guatemala and Mexico and their ship board representatives, D. Siu, H. Leandro, L. Arias and H. Sandoval, for permission to work on their continental margins. P. Crampton provided excellent support for the seismic acquisition equipment. J. Charters was responsible for much of the Sea Beam manipulation and plotting software development. P. Henkart supervised initial seismic prcessing at Scripps. Students, A. Volpe, J. Roump, D. Reed and C. deMoustier were much help at sea and Roump and Volpe were involved in various post-cruise processing and interpretation. This work was funded by NSF OCE 80-24402 and NSF OCE 83-14401. University of Texas contribution No. 598.

REFERENCES

ANDERSON, T. H. and V. A. SCHMIDT, The evolution of Middle America and the Gulf of Mexico—Caribbean Sea region during Mesozoic time, *Geol. Soc. Am. Bull.*, **94**, 941–966, 1983.

AOKI, Y., T. TAMANO, and S. KATO, Detailed structure of the Nankai Trough from migrated seismic sections, *Am. Assoc. Petro. Geol. Mem.*, **34**, 309–322, 1983.

AUBOUIN, J., R. VON HUENE *et al.*, *Init. Rept. Deep Sea Drill. Proj.*, Vol. 67, 1982a.

AUBOUIN, J., R. VON HUENE *et al.*, Subduction without accretion: Middle America Trench off Guatemala, *Nature*, **297**, 458–460, 1982b.

AUBOUIN, J., J. BOURGOIS, and J. AZEMA, A new type of active margin: The convergent-extensional margin, as exemplified by the Middle America Trench off Guatemala, *Earth Planet. Sci. Lett.*, **67**, 211–218, 1984.

CLOOS, M., Flow Melanges: Numerical and geologic constraints on the origin in the Franciscan subduction complex, California, *Geol. Soc. Am. Bull.*, **93**, 330–345, 1982.

COWAN, D. S. and R. M. SILLING, A dynamic scaled model of accretion at trenches and its implications for the tectonic evolution of subduction complexes, *J. Geophys. Res.*, **83**, 5389–5396, 1978.

CROWE, J. C. and R. T. BUFFLER, Regional seismic reflection profiles across the Middle

America Trench and convergent margin of Costa Rica, in A.W. Bally, ed., studies in Geology Series #15, Seismic expression of structural styles, *Am. Assoc. Petro. Geol.*, 1983.

ENKEBOLL, R. H., Petrology and provenance of sands and gravels from the Middle America Trench and trench slope, southwestern Mexico and Guatemala, in *Init. Rept. Deep Sea Drill. Proj.*, Vol. 66, pp. 521–530, 1982.

FRENCH, W. S. and C. G. HENSON, Signature measurements on the water gun marine seismic source, *Proc. Offshore Tech. Conf.*, **1**, 631–638, 1978.

HEY, R. N., F. K. DUENNEBIER, and W. J. MORGAN, Propagating rifts on midocean ridges, *J. Geophys. Res.*, **85**, 3647–3558, 1980.

HILDE, T. W. C. and G. F. SHARMAN, Fault Structure of the descending plate and its influence on the subduction process, *Trans. Am. Geophys. Un.*, **59**, 1182, 1978.

HUSSONG, D., S. UYEDA *et al.*, *Init. Rept. Deep Sea Drill. Proj.*, Vol. 60, 1982.

IBARAHIM, A. K., G. V. LATHAM, and J. W. LADD, Seismic refraction and reflection measurements in the Middle America Trench offshore Guatemala, *J. Geophys. Res.*, **84**, 5643–5649, 1979.

KARIG, D. E., R. K. CARDWELL *et al.*, Late Cenozoic subduction and continental margin truncation along the northern Middle America Trench, *Geol. Soc. Am. Bull.*, **89**, 256–276, 1978.

KARIG, D. E., H. KAGAMI *et al.*, Varied responses to subduction in Nankai Trough and Japan Trench forearc, *Nature*, **304**, 148–151, 1983.

KARIG, D. E., G. F. MOORE, J. R. CURRAY, and M. B. LAWRENCE, Morphology of the lower trench slope off Nias island, Sunda arc, in *The Tectonic and Geologic Evolution of Southeast Asian Seas and Islands*, Geophys. Monogr. Ser., Vol. 23, pp. 179–208, 1980.

KULM, L. D., R. VON HUENE, R. *et al.*, *Init. Rept. Deep Sea Drill. Proj.*, Vol. 18, 1973.

KULM, L. D. and G. A. FOWLER, Oregon continental margin structure and stratigraphy: A test of the imbricate thrust model, in *The Geology of Continental Margins*, pp. 261–283, 1974.

LADD, J. W., A. K. IBRAHIM *et al.*, Tectonics of the Middle America Trench offshore Guatemala, *International Symp. of the Guatemala February 4 Earthquake and Reconstruction Process*, **1**, 1978.

LADD, J. W. and A. K. IBRAHIM *et al.*, Interpretation of seismic reflection data of the Middle America Trench offshore Guatemala, in *Init. Rept. Deep Sea Drill. Proj.*, Vol. 67, pp. 675–689, 1982.

MCMILLEN, K. J. and S. B. BACHMAN, Bathymetric and tectonic evolution of the southerm Mexico active margin, Deep Sea Drilling Project Leg 66, in *Init. Rept. Deep Sea Drill. Proj.*, Vol. 66, pp. 815–822, 1982.

MCMILLEN, K. J. and T. R. HAINES, Late Quaternary sediments of the southern Mexico margin, in *Init. Rept. Deep Sea Drill. Proj.*, Vol. 66, pp. 437–444, 1982.

MCMILLEN, K. J., R. H. ENKEBOLL *et al.*, Sedimentation in different tectonic environments of the Middle Americaon Trench, Southern Mexico and Guatemala, in *Geol. Soc. London Sp. Publ.*, No. 10, pp. 107–120, 1982.

MOORE, G. F. and D. E. KARIG, Development of sedimentary basins on the lower trench slope, *Geology*, **4**, 693–697, 1976.

MOORE, G. F., P. LONSDALE, and R. VON HUENE, Near-bottom observations of the Middle America Trench, in *Init. Rept. Deep Sea Drill. Proj.*, Vol. 67, pp. 707–718, 1982.

MOORE, J. C., Selective subduction, *Geology*, **3**, 530–532, 1975.

MOORE, J. C. and BIJU-DUVAL *et al.*, Offscraping and underthrusting of sediment at the deformation front of the Barbados Ridge, Deep Sea Drilling Project Let 78A, *Geol.*

Soc. Am. Bull., **93**, 1065–1077, 1982a.

MOORE, J. C. and D. E. KARIG, Sedimentology, structural geology, and tectonics of the Shikoku subduction zone, southwest Japan, *Geol. Soc. Am. Bull.*, **87**, 1259–1268, 1976.

MOORE, J. C., J. S. WATKINS, and T. H. SHIPLEY, Summary of accretionary processes, Deep Sea Drilling Project Leg 66, in Offscraping, Underplating, and Deformation of the Slope Apron, in *Init. Rept. Deep Sea Drill. Proj.*, Vol. 66, pp. 825–836, 1982b.

MOORE, J. C., J. S. WATKINS *et al.*, Geology and tectonic evolution of a juvenile accretionary terrane along a truncated convergent margin: Synthesis of results from Leg 66 of the Deep Sea Drilling Project, Southern Mexico, *Geol. Soc. Am. Bull.*, **93**, 847–861, 1982c.

RENARD, V. and J.-P. ALLENOU, Sea Beam, multi-beam echo-sounding in *Jean Charcot*: description, evaluation and first results, *Intern. Hydrog. Rev. Monaco*, 56, 35–67, 1979.

RENARD, V. *et al.*, First results of a Seabeam survey of the Mid-America trench, *Academie des Sciences Comptes Rendus, 291, Ser. D.*, 137–143, 1980.

ROUMP, J., V. RENARD *et al.*, Preliminary version of the bathymentry of the area around DSDP Site 488, Leg 66, in the Middle America Trench, in *Init. Rept. Deep Sea Drill. Proj.*, Vol. 66, Appendix II, 1982.

SCHOLL, D. W. *et al.*, Sedimentary masses and concepts about tectonic processes at underthrust ocean margins, *Geology*, **8**, 564–568, 1980.

SCHWELLER, W. J., L. D. KULM, and R. A. PRINCE, Tectonics, structure, and sedimentary framework of the Peru-Chile Trench, *Geol. Soc. Am. Mem.*, **154**, 323–349, 1981.

SEELY, D. R., The significance of landward vergence and oblique structural trends on trench inner slopes, in *Maurice Ewing Series*, Vol. 1, pp. 187–198, 1977.

SEELY, D. R., P. R. VAIL, and G. G. WALTON, Trench slope model, in *The Geology of Continental Margins*, pp. 249–260, 1974.

SHIPLEY, T. H., Seismic facies and structural framework of the southern Mexico continental margin, in *Init. Rept. Deep Sea Drill. Proj.*, Vol. 66, pp. 775–790, 1982.

SHIPLEY, T. H., K. J. MCMILLEN *et al.*, Continental margin and lower slope structures of the Middle America Trench near Acapulco, Mexico, *Mar. Geol.*, **35**, 65–82, 1980.

SHIPLEY, T. H. and G. F. MOORE, Sediment accretion, subduction and dewatering at the base of the trench slope off Costa Rica: A seismic reflection view of the decollement, *J. Geophys. Res.*, 1986 (in press).

SHIPLEY, T. H. *et al.*, Tectonic processes along the Middle America Trench inner slope, in *Geol. Soc. London Sp. Publ.*, No. 10, pp. 95–106, 1982.

UNDERWOOD, M. B. and D. E. KARIG, Contrasting patterns of trench and trench-slope sedimentation, *Geology*, **8**, 432–436, 1980.

VOLPE, A. M., T. H. SHIPLEY, and G. F. MOORE, A high resolution geophysical survey of DSDP Leg 84 Site 570, in *Init. Rept. Deep Sea Drill. Proj.*, Vol. 84, 851–860, 1985.

VON HUENE, R., J. AUBOUIN *et al.*, Leg 67: The Deep Sea Drilling Project Mid-America Trench transect off Guatemala, *Geol. Soc. Am. Bull.*, **91**, 414–420, 1980.

VON HUENE, R., J. LADD, and I. NORTON, Geophysical observations of slope deposits, Middle America Trench off Guatemala, in *Init. Rept. Deep Sea Drill. Proj.*, Vol. 67, 719–732, 1982.

VON HUENE, R. and H. LEE, The possible significance of pore fluid pressures in subduction zones, *Am. Assoc. Petro. Geol. Mem.*, **34**, 781–791, 1983.

WATKINS, J. S., K. J. MCMILLEN *et al.*, Tectonic synthesis, Leg 66 DSDP transect and vicinity, in *Init. Rept. Deep Sea Drill. Proj.*, Vol. 66, 837–849, 1982a.

WATKINS, J. S., J. C. MOORE *et al.*, *Deep Sea Drill. Proj.*, Vol. 66, 1982b.

WESTBROOK, G. K. and M. J. SMITH, Long decollements and mud volcanoes: Evidence from the Barbados Ridge Complex for the role of high prefluid pressure in the development of an accretionary complex, *Geology*, **11**, 279–283, 1983.

Formation of Active Ocean Margins, edited by N. Nasu *et al.*, pp. 257–272.

THE ORIGIN OF SOME COMMON TYPES OF MELANGE IN THE WESTERN CORDILLERA OF NORTH AMERICA

Darrel S. COWAN

Department of Geological Sciences, University of Washington, Seattle, Washington 98195, U.S.A.

Abstract. Three distinct types of late Mesozoic and Cenozoic melange are widespread in the western Cordillera. One type consists of sequences of interbedded sandstone and mudstone which record incipient to advanced disruption and fragmentation of strata. Another type consists of similarly disrupted, thin layers of black mudstone, green tuff, ribbon chert, and minor sandstone. The third type comprises fragments and blocks of diverse rock types in a scaly mudstone matrix. The melanges probably resulted from either stratal disruption along faults; gravitationally driven downslope extension of partly consolidated sediments; deposition of muddy debris-flow materials (olistostromes); or emplacement of mud diapirs. Recent high-quality seismic reflection profiles across modern accretionary wedges define several possible settings in which these types of melanges could develop, including: (1) thrust faults near the base of the inner trench-slope; (2) zones of underplating where ocean-floor and trench materials are added to the underside of the wedge; (3) diapirs generated beneath the trench slope; and (4) slope aprons and slope basins affected by gravity-driven slumping and spreading. Although Cordilleran melanges are commonly interpreted to have formed on or within accretionary wedges, the actual convergent-margin setting in which most examples originated is poorly constrained by available structural, stratigraphic, and sedimentologic data.

1. Introduction

Soon after HSU (1968, 1969) described some of the chaotic mudstones in the Franciscan complex in California as "melange", the Franciscan was interpreted as a subduction complex formed as a plate of oceanic lithosphere descended beneath the western margin of North America (e.g., ERNST, 1970). Melanges are generally characterized by either stratal disruption or variably shaped blocks enveloped in a fine-grained, generally mud-rich matrix. They have come to be viewed as practically synonymous with subduction, even though the processes through which they originated were, and still are, controversial. Moreover, in the late 1960's and early 1970's so little was known about accretionary wedges (or prisms) that geologists were unable to specify where melanges actually formed within them.

In the past decade, we have learned much about not only the nature of accretionary wedges, but also the fabric of melanges. High-quality multichannel seismic reflection profiles, and drilling at active convergent plate boundaries sponsored by the Deep Sea Drilling Project (DSDP) and International Phase of Ocean Drilling (IPOD), have provided important data on the overall geometry and internal structure of many wedges. We can now identify specific environments where sediments are likely to be deformed. Several Mesozoic and Cenozoic melanges, especially in western North America and Japan, have been described in detail, and their fabrics have been ascribed to specific sedimentary and tectonic processes. In view of the advances of the past decade, it is timely for this paper to review the question of how melanges along the Pacific margin of North America are related to subduction and to ask whether the mesoscopic structural styles of melanges are diagnostic of particular environments within an idealized accretionary wedge.

2. *Melanges in the Western Cordillera of North America*

The term melange is now generally used in a descriptive sense for mappable rock bodies that have a "block-in-matrix" fabric (e.g., SILVER and BEUTNER, 1980). In some cases, this fabric clearly resulted from the progressive disruption of well-layered sedimentary sequences. At present, the term is being applied to several types of rock units that meet the general definition above but which differ from one another in several respects. Three types of melange, which can be differentiated on the basis of mesoscopic fabric and overall lithology, are particularly abundant in the mountain belts along the Pacific margin of North America (Fig. 1). Each type probably formed in a covergent-margin setting. The following brief summaries are distilled from a more extensive discussion by COWAN (1985).

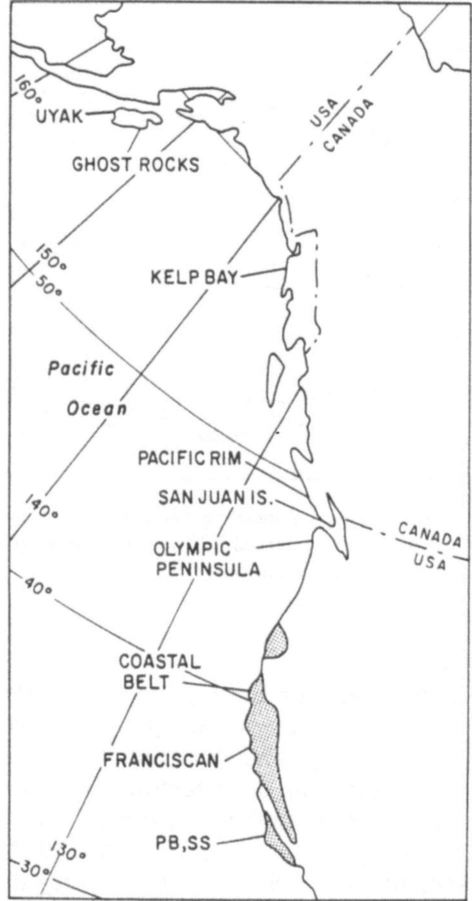

Fig. 1. Map of part of the Pacific margin of North America showing the location of melanges discussed in this paper. Outcrop area of the Franciscan is shaded. PB, SS indicates location of Franciscan melange at Piedras Blancas Point and San Simeon.

2.1 Progressively disrupted sequences of interbedded sandstone and mudstone

These melanges, which I call "type I", originated as sequences of inter-layered sandstone and mudstone. Most outcrops and even some hand-specimens show evidence for the progressive disruption of sand layers by layer-parallel extension (Fig. 2). Tight to isoclinal folds that could have

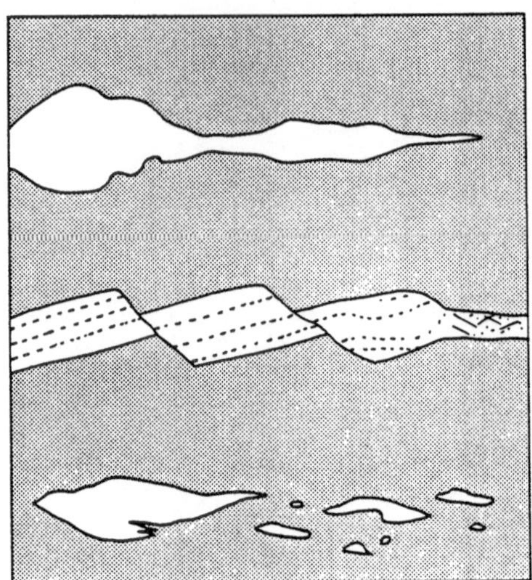

Fig. 2. Schematic diagram of type I melange depicting the progressive disruption and fragmentation of sandstone layers interbedded with mudstone (shaded). Layer-parallel extension is the dominant style of deformation.

led to extension by transposition are rare or absent. Extension in the less ductile sand was accommodated by a variety of structures, including pinch-and-swell, complete boudinage, and extensional shear fractures. Complete disruption led to the isolation of sandstone fragments, which became enveloped in and dispersed through a mudstone matrix.

Detailed studies of type I melanges (COWAN, 1982; BACHMAN, 1982; UNDERWOOD, 1984; BYRNE, 1985) have led to very different alternative hypotheses for the origin of the mesoscopic fabric. COWAN (1982), who studied part of the late Mesozoic Franciscan complex at Piedras Blancas (Fig. 1), concluded that partly dewatered, "soft" sediments on a slope were deformed by gravitationally driven vertical collapse and sub-horizontal, layer-parallel spreading. This stratal disruption may represent an arrested stage in the progressive development of a muddy debris flow. According to COWAN's (1982) hypothesis, type I melanges may be precursors to olistostromes consisting of sandstone clasts in a mudstone matrix.

BYRNE (1985) studied Upper Cretaceous (?)-lower Tertiary melanges in the Ghost Rocks formation, and BACHMAN (1982) and UNDERWOOD (1984) studied the Coastal Belt of the Franciscan (Fig. 1). All interpreted the stratal disruption in type I melange to have formed in faults or fault zones in an accretionary complex at an active convergent margin. BYRNE (1985) presents evidence that the sands were consolidated and deformed in a deeply buried, "tectonic" setting rather than at the shallow levels envisioned by COWAN (1982) for deformation at Piedras Blancas.

2.2 *Progressively disrupted sequences of interbedded mudstone, tuff, chert, and sandstone*

These melanges, like type I, have a mesoscopic fabric recording the progressive desruption of originally stratified sediments, but in this case, deformation affected a lithologically distinctive suite comprising black mudstone, green tuff, radiolarian ribbon chert, and minor sandstone (Fig. 3). Beds of chert and sandstone and some layers of tuff display pinch-and-swell structure, and boudinage; other tuff layers are irregularly swirled and folded. The mud and tuff were probably "soft", partly consolidated sediments at the time of deformation.

Few of these melanges, which I call "type II", have been studied in detail. MOORE and WHEELER (1978) interpreted the stratal disruption in the melange of the Uyak Complex (Fig. 1) as having occurred along the "master shear zone" or thrust at the base of an accretionary wedge.

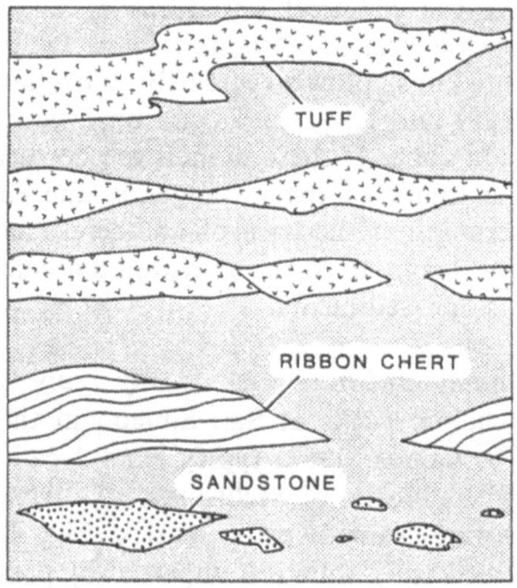

Fig. 3. Schematic diagram of type II melange. Some layers of green basaltic tuff are irregularly deformed; others display pinch-and-swell structure and boudinage. Interbedded radiolarian ribbon chert and minor sandstone are also progressively disrupted by layer-parallel extension. Interbedded mudstone is shaded.

This hypothesis was partly based on CONNELLY's (1978) interpretation that most of the sediments in the melange were deposited on the abyssal ocean floor, carried into a trench, and subsequently scraped off of the descending plate. However, recent studies on correlative rocks elsewhere along the Pacific margin suggest that alternative hypotheses need to be considered. HEIN and KARL (1983) point out that the sediments in type II melange of the Kelp Bay group (Fig. 1), which include ribbon chert interbedded which graywacke turbidites and pillow lavas, are atypical of sediments that have been drilled in open-ocean environments. Type II melange also occurs in the Pacific Rim Complex and in the San Juan Islands (Fig. 1). Geologic evidence from both of these localities (BRANDON *et al.*, 1983; BRANDON, 1984) indicates that the sediments were originally deposited on a lithologically diverse basement terrane rather than on oceanic crust.

A modern analogue of the depositional environment in which the sediments in type II melange accumulated has not yet been identified. The general characteristics of this environment probably included: high organic productivity to account for siliceous ooze, which became chert (HEIN and KARL, 1983); proximity to a volcanic arc constructed on a continental margin, to explain the provenance of interbedded graywacke-type sands; and sporadic eruptions of submarine lavas to yield pillowed flows and tuffs that have the chemical signatures of mid-ocean ridge basalt (DECKER, 1980; BRANDON *et al.*, 1983). I favor the hypothesis that the sediments accumulated in shallow structurally controlled basins near a continental margin (HEIN and KARL, 1983) or on a continental slope landward of a subduction zone. If these models are correct, it seems unlikely that the mesoscopic fabric of the melanges is due entirely to imbricate thrusting and offscraping at the base of an accretionary wedge. Instead, the stratal disruption may record the gravitational collapse and spreading of partly lithified slope sediments.

2.3 Chaotic polymict mudstone with locally scaly fabric

Some of the chaotic rocks at San Simeon in the Franciscan (HSU, 1968; COWAN, 1978; CLOOS, 1982) typify what I call type III melanges. Fragments (also called blocks or inclusions) of diverse shapes and sizes are enveloped in a fine-grained, peltic matrix (Fig. 4) which locally has a crude to well-defined scaly foliation imparted by penetrative, slickensided slip surfaces. Blocks of all sizes range from equant to lenticular; some have irregular margins and appear to have been intruded by mobile mudstone matrix. Blocks include a variety of rock types. Most common are graywacke sandstones; less common are pillowed or massive basaltic

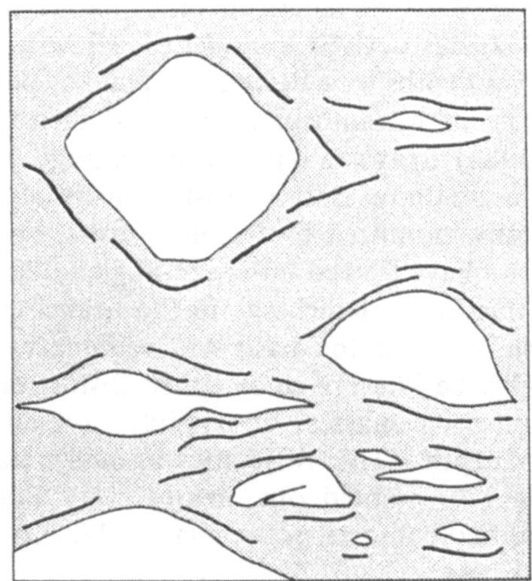

Fig. 4. An idealized example of type III melange, consisting of variably shaped blocks of diverse rock types surrounded by a scaly, dark gray pelitic matrix (shaded).

greenstones and radiolarian chert; and least common are limestones and metamorphic rocks, including blueschist and amphibolite. Type III melanges may each be characterized by a different population of block types, and some definitely do not contain blueschist.

Unlike melanges of types I and II, which preserve strata in various arrested stages of disruption and fragmentation, a type III melange contains no direct *in situ* evidence of how or where the vast majority of its inclusions were derived. Apparently, fragments from one or more source terranes were mixed together and dispersed through a mobile pelitic matrix. Subsequent to fragmentation, mixing, and dispersal, the entire chaotic mass moved away from these source terranes.

There are only a few known geological processes that can explain the fabric and history of this type of melange. PAGE (1978) notes that in many ways, the Franciscan melanges resemble well-documented submarine debris-flow deposits, or olistostromes, in Taiwan and Italy, in which fragments from diverse sources have been dispersed through and transported in a mobile matrix. Diapirism is another possible process. Some Tertiary melanges on the Olympic Peninsula presently occur in diapirs that have migrated upward through shelf strata overlying a Tertiary accretionary prism (RAU and GROCOCK, 1974). WILLIAMS *et al.* (1984) describe what appear to be type III melanges in mud diapirs in Irian Jaya, Indonesia. CLOOS (1982) has devised a model of "flow melange" in which shales, carried to great depths by a descending plate of oceanic lithosphere, can sample fragments of metamorphic rock and then flow to shallower depths. COWAN (1985) draws a kinematic analogy between type III melange and scaly serpentinite. Both consist of materials with a low shear strength that are easily mobilized by tectonic forces. Some melange, like serpentinite, may even have flowed into high-angle fault zones.

The origin of this kind of melange in the Franciscan has been very difficult to establish, and on the basis of available evidence, most examples could probably be interpreted as either olistostromes, diapiric and otherwise mobile, intrusive masses, or possibly flow-melanges that have migrated to shallow depths. The difficulties in interpretation arise partly because, as PAGE (1978) pointed out, nearly every Franciscan melange has tectonically modified contacts, so its relationships to neighboring rock units are typically obscure.

2.4 Summary

Detailed field studies have shown that several types of melanges occur in the western Cordillera. The structural styles and mesoscopic fabrics of three common types probably resulted from a variety of processes that operated on a local scale, including: (1) stratal disruption along thrust faults; (2) gravitationally driven, downslope extension of partly consolidated strata; and (3) generation and emplacement of mobile muddy debris flows and diapir-like masses. Some studies have attributed these processes to specific settings within accretionary wedges at active convergent margins. These phenomena are depicted in Fig. 5 and discussed in the next section. However, the actual environment of deformation or emplacement of many Cordilleran melanges is speculative and poorly constrained by field evidence alone, partly because of poor exposures, faulted or unexposed contacts, and superimposed Cenozoic deformation. It is possible that some could have originated in other settings, besides an accretionary wedge, along the broadly convergent western margin of North America.

3. Accretionary Wedges in Modern Forearc Settings

In light of available information, where within accretionary wedges are melanges likely to develop? Many active convergent plate-margins are marked by simple, or "classic" accretionary wedges, which consist of materials scraped off of or otherwise transferred from oceanic lithosphere descending beneath the fore-arc region. On the basis of a number of seismic-

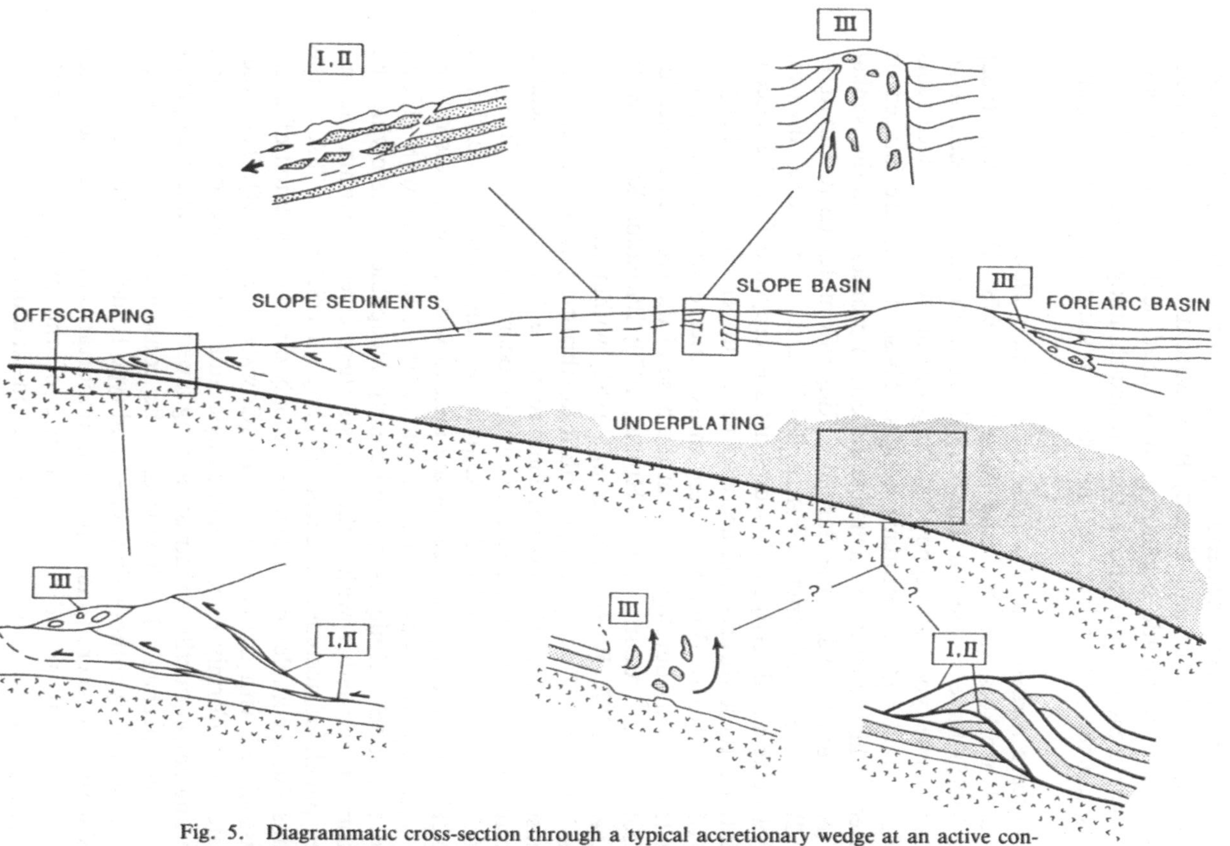

Fig. 5. Diagrammatic cross-section through a typical accretionary wedge at an active convergent margin. The three types of Cordilleran melanges discussed in this paper could in theory form in a number of specific environments on or within the wedge, as indicated in the figure. The olistostromes of type III deposited in the forearc basin are analogous to the Lichi melange in Taiwan discussed by PAGE and SUPPE (1981). Thrust duplex model in the zone of underplating modified from BRANDON (1984).

reflection profiles, the results of deep-sea drilling, and some theoretical analyses, an "idealized" accretionary wedge (for example, KARIG and SHARMAN, 1975; MOORE and KARIG, 1976; MOORE *et al.*, 1981; KARIG, 1983) is known to include a prism of offscraped sediments at the base of the inner trench slope, and a trenchward-thinning blanket of slope sediments punctuated by thicker accumulations of sediments in slope basins.

3.1 Imbricate thrusts near the base of the inner slope

Several workers (MOORE and WHEELER, 1978; BACHMAN, 1982; BYRNE, 1985) have suggested that the disruption and fragmentation of strata in some type I and type II melanges resulted from displacements along thrust faults in accretionary wedges. These thrusts are imaged or interpreted on high-quality multichannel reflection profiles from the Nankai trench (NASU *et al.*, 1982), the Barbados Ridge Complex (BIJU-DUVAL *et al.*, 1982), the Gulf of Oman (WHITE, 1982) and other active forearcs. Panels of pelagic and terrigenous sediments are scraped off of the descending plate on these thrusts and accreted to the upper plate, or "hanging wall", of the forearc (Fig. 5). These thrusts merge with a sole thrust, or decollement, that develops either at the base of the sedimentary section or, in some cases (Aleutians, VON HUENE, 1979; Barbados Ridge, BIJU-DUVAL *et al.*, 1982, WESTBROOK *et al.*, 1982) within it.

As KARIG (1983) emphasizes, imbricate thrusting is a fundamental mechanism of accretion. However, the nature of the deformation associated with accretionary thrusts and decollements in modern forearcs is still largely conjectural. It is not certain, for example, whether lower-slope thrusts are typically discrete slip surfaces or are instead wide zones of distributed simple shear as hypothesized by BYRNE (1985) and BACHMAN (1982). Seismic reflection techniques are inherently incapable of resolving the mesoscopic fabric either within the thrusts themselves or within the imbricated panels of accreted sediment. Drilling on DSDP Leg 78A on the Barbados Ridge complex has provided the only direct observation of structures associated with accretionary faults (COWAN *et al.*, 1984). The holes penetrated a uniformly ductile hemipelagic and pelagic section, whereas in type I melange, ductility contrasts among layers of sand and mud greatly influenced mechanical behavior and the consequent structural style. Nevertheless, the hypothesis that type I and type II melanges formed along thrusts within an accretionary prism is important, and it can be tested by: (1) comparing the distinctive fabric of the melanges to fabrics characterizing major thrust faults in other on-land orogenic belts, and (2) trying to establish the large-scale, or "seismic-profile" scale structural setting of melanges so that geological cross-sections can be compared directly with seismic profiles.

3.2 Diapirism

Diapirism, involving upward intrusion of mobile mud-rich materials, has been documented in several modern forearcs (Fig. 5), including the Barbados Ridge complex (BIJU-DUVAL *et al.*, 1982), and in the Makran, both offshore (WHITE, 1982) and onshore (J. Leggett, personal communication, 1983). Equant (in map view) to elongate diapirs are locally common on the continental shelf from northern California to British Columbia landward of the locus of active accretion marking the descent of the Juan de Fuca plate beneath the margin of North America (TIFFIN *et al.*, 1972; RAU and GROCOCK, 1974; FIELD *et al.*, 1980). One of these diapirs is exposed in sea-cliffs on the Olympic Peninsula, where it consists of type III melange. By analogy, other ancient type III melangs may have been emplaced in diapirs either in the zone of active offscraping at the base of the slope, or farther upslope beneath the apron of slope sediments. The diapiric sediments in active forearcs probably represent either mobilized, overpressured materials rapidly underthrust beneath the toe of the accretionary wedge, or possibly underconsolidated slope sediments in the cores of anticlines. Seismic profiles show that diapirs off the coast of Washington extend to depths of at least 2 km beneath the continental shelf (RAU and GROCOCK, 1974).

3.3 Gravitational phenomena

KARIG (1983) suggests that small gravitational features such as slumps and muddy debris flows are most common near the base of the lower slope. Examples have been described from the Sunda trench (KARIG *et al.*, 1980), Aleutian trench (PIPER *et al.*, 1973), and the Mid-America trench of Mexico (SHIPLEY, 1981), and elsewhere. The deposits probably consist of previously offscraped sediments (and perhaps oceanic crust) and lower-slope sediments. The surface morphology and acoustic characteristics of these features suggest that they are internally disrupted and possibly chaotic, but there is no other direct information concerning their mesoscopic fabric. Several type III melanges in the Cordillera have been interpreted as debris-flow deposits or olistostromes (for example, COWAN, 1978; PAGE, 1978), and by analogy with gravitationally generated deposits on modern accretionary prisms, some of these ancient examples may have accumulated near the base of the inner slope. KARIG (1983) notes that debris-flow deposits derived from the lower slope and deposited in the trench are likely to be incorporated into the accretionary wedge as imbricate thrusting continues. PAGE (1978) invokes just such a sequence of events to explain why the Franciscan melanges that have many of the earmarks of olistostromes apparently have tectonic, rather than depositional, contacts with adjacent stratified units.

There are still major questions regarding the importance and scale of gravity-driven processes on the upper trench-slope. KARIG *et al.* (1980) found little evidence for either large-scale gravitational deformation of slope sediments or deposition of debris-flow deposits in slope basins on the upper slope of the Sunda trench. In contrast, MOBERLY *et al.* (1982) interpreted possible gravity slides in forearc and slope basins near the Peru–Chile trench. Many kilometer-scale slumps, slides, and debris flows affecting slope sediments on the continental-margin forearc off northwestern California have been identified on seismic reflection profiles (FIELD *et al.*, 1980). Some zones containing abundant slumps exceed 200 km^2 in area. Although the small-scale internal structure of these modestly displaced sequences is not known, it is clear that bedding has been variably disturbed. COWAN (1982; 1985) attributed the fabric of type I melanges to gravitational spreading of partly consolidated sediments, and I suggest here that some of these melanges formed in a setting analogous to the modern forearc off northwestern California (Fig. 5). A similar scenario can be envisioned for type II melanges (Fig. 5), although slope and slope-basin sediments in this case would have comprised heterogeneous sections of mudstone, siliceous ooze, tuff, and sand rather than just interbedded mud and sand.

3.4 Underplating (subcretion)

At least one other large-scale process operating within accretionary wedges must be considered as potentially capable of generating all the types of melange discussed in this paper (Fig. 5). Mass-balance calculations for the Sunda (KARIG *et al.*, 1980) and Mid America (WATKINS *et al.*, 1981) accretionary wedges indicate that not all ocean-floor (pelagic, hemipelagic, trench) sediments are scraped off the descending plate in the zone of active imbricate thrusting at the base of the slope. Some materials are carried landward and added (accreted) to the underside of the accretionary wedge. The actual processes responsible for the "underplating" (MOORE *et al.*, 1981) or "subcretion" (KARIG and KAY, 1981) of sediments and possibly oceanic crust are conjectural. For example, experimental and theoretical modelling by COWAN and SILLING (1978) and CLOOS (1982) is compatible with the hypothesis that underplated materials may deform by distributed ductile flow and become transformed into upwardly mobile, "intrusive" type II melange. On the other hand, BRANDON (1984) proposes that, deep within some active accretionary wedges, stacked thrust slices are arranged in duplexes. In this case, deformation concentrated along imbricate thrust faults could conceivably result in the styles of deformation that typify type I and type II melanges.

3.5 Summary

Melanges are likely to develop in a variety of environments within accretionary wedges at active convergent margins. Geophysical data and deep-sea drilling have defined the "seismic-scale" architecture (on the order of a few kilometers) of accretionary wedges, but they have not yet provided much information about small-scale styles of deformation in the forearc. In theory, the mesoscopic fabrics of some melanges are representative of the small-scale structures that develop within accretionary wedges. Before we can confidently attribute a particular fabric to a specific environment on or within a wedge, however, we need far more information than is currently available regarding the original depositional setting and subsequent deformational history of Cordilleran melanges.

Although some Cordilleran melanges undoubtedly formed within accretionary wedges, a recent development is food for thought regarding alternative tectonic settings. PAGE and SUPPE (1981) studied the Pliocene Lichi melange in Taiwan. Field relations prove that the melange is largely an olistostrome or debris-flow deposit. It consists of fragments, blocks, and slabs of diverse rock types enveloped in a matrix of scaly mudstone. PAGE (1978) had earlier noted that the Lichi strongly resembles Franciscan melanges in many respects, and I would classify the Lichi as type III on the basis of its mesoscopic fabric alone. Since the late Cenozoic plate-tectonic setting of Taiwan is relatively well constrained, PAGE and SUPPE (1981) could also prove that the Lichi olistostrome was deposited in a forearc basin, which has since been partially telescoped along arcward-verging thrusts. The Lichi, then, is an example of a melange that manifestly was not deposited or deformed in any of the accretionary-wedge settings so commonly hypothesized for some similar units in western North America. The Lichi example illustrates how a correct interpretation of the origin of a melange utilizes not only its fabric but also independent paleogeographic constraints.

REFERENCES

BACHMAN, S. B., The Coastal Belt of the Franciscan: youngest phase of northern California subduction, in *Trench-Forearc Geology*, edited by J. K. Leggett, Geol. Soc. London Spec. Pub. 10, pp. 401–417, 1982.

BIJU-DUVAL, B., P. LEQUELLEC, A. MASCLE, V. RENARD, and P. VALERY, Multi-beam bathymetric survey and high resolution seismic investigations on the Barbados Ridge complex (eastern Caribbean): a key to the knowledge and interpretation of an accretionary wedge, *Tectonophys.*, **86**, 275–304, 1982.

BRANDON, M. T., A study of deformational processes affecting unlithified sediments at active margins: the results of a field study and a structural model, Ph.D. dissertation, University of Washington, 159 pp., Seattle, 1984.

BRANDON, M. T., D. S. COWAN, J. E. MULLER, and J. A. VANCE, Pre-Tertiary geology of the San Juan Islands, Washington and southeast Vancouver Island British Columbia, *Field Trip Guidebook*, Geological Assn. of Canada Victoria Section, 65 pp., 1983.

BYRNE, T., Structural geology of melange terranes in the Ghost Rocks Formation, Kodiak Islands, Alaska, in *Geol. Soc. Am. Spec. Paper*, edited by L. A. Raymond, 1985.

CLOOS, M., Flow melanges: Numerical modeling and geologic constraints on their origin in the Franciscan subduction complex, California, *Geol. Soc. Am. Bull.*, **93**, 330–345, 1982.

CONNELLY, W., Uyak Complex, Kodiak Islands Alaska: A Cretaceous subduction complex, *Geol. Soc. Am. Bull.*, **89**, 755–769, 1978.

COWAN, D. S., Origin of blueschist-bearing chaotic rocks in the Franciscan Complex, San Simeon, California, *Geol. Soc. Am. Bull.*, **89**, 1415–1423, 1978.

COWAN, D. S., Deformation of partly dewatered and consolidated Franciscan sediments near Piedras Blancas Point, California, in *Trench-Forearc Geology*, edited by J. K. Leggett, Geol. Soc. London Spec. Pub. 10, pp. 439–457, 1982.

COWAN, D. S., Structural styles in Mesozoic and Cenozoic melanges in the western Cordillera of North America, *Geol. Soc. Am. Bull.*, **96**, 451–462, 1985.

COWAN, D. S., J. C. MOORE, S. M. ROESKE, N. LUNDBERG, and S. LUCAS, Structural features at the deformation front of the Barbados Ridge complex, Deep Sea Drilling Project Leg 78A, in *Initial Repts. Deep Sea Drilling Proj.*, **78A**, 535–548, 1984.

COWAN, D. S. and R. M. SILLING, A dynamic, scaled model of accretion at trenches and its implications for the tectonic evolution of subduction complexes, *J. Geophys. Res.*, **83**, 5389–5396, 1978.

DECKER, J. E., Jr., Geology of Cretaceous subduction complex, western Chichagof Island, southeastern Alaska, Ph.D. dissertation, Stanford University, Stanford, 135 pp., Calif., 1980.

ERNST, W. G., Tectonic contact between the Franciscan melange and the Great Valley sequence—crustal expression of a late Mesozoic Benioff zone, *J. Geophys. Res.*, **75**, 886–901, 1970.

FIELD, M. E., S. H. CLARKE, Jr., and M. E. WHITE, Geology and geologic hazards of offshore Eel River Basin, northern California continental margin, *U. S. Geological Survey Open-File Report 80-1080*, 80 pp., 1980.

HEIN, J. R. and S. M. KARL, Comparisons between open-ocean and continental margin chert sequences, in *Siliceous Deposits in the Pacific Region*, edited by A. Iijima, J. R. Hein, and R. Siever, pp. 25–43, 1983.

HSU, K. J., Principles of melanges and their bearing on the Franciscan-Knoxville paradox, *Geol. Soc. Am. Bull.*, **79**, 1063–1074, 1968.

HSU, K. J., Preliminary report and geologic guide to Franciscan melanges of the Morro Bay—San Simeon area, California, *Spec. Pub. 35*, California Div. Mines and Geology, 46 pp., 1969.

KARIG, D. E., Deformation in the forearc: implications for mountain belts, in *Mountain Building Processes*, edited by K. J. Hsu, pp. 59–71, Academic Press, New York.

KARIG, D. E., G. F. MOORE, J. R. CURRAY, and M. B. LAWRENCE, Morphology and shallow structure of the lower trench slope off Nias Island, Sunda Arc, in *The Tectonic and Geologic Evolution of Southeast Asian Seas and Islands*, Geophy. Monograph. 23, edited by D. E. Hayes, pp. 179–208, Am. Geophys. Union, Washington, D.C., 1980.

KARIG, D. E. and R. W. KAY, Fate of sediments on the descending plate at convergent margins, *Phil. Trans. Royal Soc. London, Series A*, **301**, 233–251, 1981.

KARIG, D. E. and G. F. SHARMAN III, Accretion and subduction in trenches, *Geol. Soc. Am. Bull.*, **86**, 377–389, 1975.

MOBERLY, R., G. L. SHEPHERD, and W. T. COULBOURN, Forearc and other basins, continental margin of northern and southern Peru and adjacent Ecuador and Chile, in *Trench-Forearc Geology*, edited by J. K. Leggett, Geol. Soc. London Spec. Pub. 10, pp. 171–189, 1982.

MOORE, G. F. and D. E. KARIG, Development of sedimentary basins on the lower trench slope, *Geology*, **4**, 693–697, 1976.

MOORE, J. C., J. S. WATKINS, and T. H. SHIPLEY, Summary of accretionary processes, Deep Sea Drilling Project Leg 66: offscraping, underplating, and deformation of the slope apron, in *Initial Repts. Deep Sea Drilling Proj.*, **66**, 825–836, 1981.

MOORE, J. C. and R. L. WHEELER, Structural fabric of a melange, Kodiak Islands, Alaska, *Am. J. Sci.*, **278**, 739–765, 1978.

NASU, N. and others, Multichannel seismic reflection data across Nankai Trough, *IPOD-Japan Basic Data Series*, No. 4, 34pp., Ocean Res. Inst., Univ. of Tokyo, 1982.

PAGE, B. M., Franciscan melanges compared with olistostromes of Taiwan and Italy, *Tectonophys.*, **47**, 223–246, 1978.

PAGE, B. M. and J. SUPPE, The Pliocene Lichi melange of Taiwan: Its plate tectonic and olistostromal origin, *Am. J. Sci.*, **281**, 193–227, 1981.

PIPER, D. J. W., R. VON HUENE, and J. R. DUNCAN, Late Quaternary sedimentation in the active eastern Aleutian trench, *Geology*, **1**, 19–22, 1973.

RAU, W. W. and G. R. GROCOCK, Piercement structure outcrops along the Washington Coast, *Information Circular 51*, Washington Div. of Geology and Earth Resources, 7 pp., 1974.

SHIPLEY, T., Seismic facies and structural framework of the southern Mexico continental margin, in *Intial Repts. Deep Sea Drilling Proj.*, **66**, 775–790, 1981.

SILVER, E. A. and E. C. BEUTNER, Penrose Conference Report: Melanges, *Geology*, **8**, 32–34, 1980.

TIFFIN, D. L., B. E. B. CAMERON, and J. W. MURRAY, Tectonics and depositional history of the continental margin off Vancouver Island, British Columbia, *Canadian J. Earth Sci.*, **9**, 280–296, 1972.

UNDERWOOD, M. B., A sedimentologic perspective on stratal disruption within sandstone-rich melange terranes, *J. Geology*, **92**, 369–385, 1984.

VON HUENE, R., Structure of the outer convergent margin off Kodiak Island, Alaska, from multichannel seismic records, in *Geologic and Geophysical Investigations of Continental Margins*, Mem. 29, edited by J. S. Watkins, L. Montadert, and P. Dickerson, pp. 261–272, American Assn. of Petroleum Geologists, Tulsa, Okla., 1979.

WATKINS, J. S., K. J. MCMILLEN, S. B. BACHMAN, T. H. SHIPLEY, J. C. MOORE, and C. ANGEVINE, Tectonic synthesis, Leg 66 transect and vicinity, in *Initial Repts. Deep Sea Drilling Proj.*, **66**, 837–349, 1981.

WESTBROOK, G. K., M. J. SMITH, J. H. PEACOCK, and M. J. POULTER, Extensive underthrusting of undeformed sediment beneath the accretionary complex of the Lesser Antilles subduction zone, *Nature*, **300**, 625–628, 1982.

WHITE, R. S., Deformation of the Makran accretionary sediment prism in the Gulf of

Oman (north-west Indian Ocean), in *Trench-Forearc Geology*, edited by J. K. Leggett, Geol. Soc. London Spec. Pub. 10, 357–372, 1982.

WILLIAMS, P. R., C. J. PIGRAM, and D. B. DOW, Melange production and the importance of shale diapirism in accretionary terrains, *Nature*, **309**, 145–146, 1984.

Formation of Active Ocean Margins, edited by N. Nasu *et al.*, pp. 273–290.

FORE-ARC TECTONICS IN THE NORTHERN MARIANA ARC

Donald M. HUSSONG and Patricia FRYER

Hawaii Institute of Geophysics, University of Hawaii, 2525 Correa Road, Honolulu, Hawaii 96822, U.S.A.

Abstract. SeaMARC II long-range side-scan sonar and swathmap bathymetry data on the northern Mariana fore-arc between 19° and 20°N reveal a complex region of large seamounts, depressions, and fields of small sediment mounds. The large seamounts are the surface expression of diapirism of serpentinized plutonic rocks that uplift the pre-existing seafloor and eventually erupt in massive cold and viscous flows. Large areas of the fore-arc also appear to have collapsed, perhaps in response to changes in the same tectonic forces that help mobilize the diapirs as well as to possible tectonic erosion of the underside of the overriding fore-arc by irregularities in the subducting oceanic plate. Fields of smaller mounds on the fore-arc are likely associated with sediment diapirism initiated by the tectonic disruption of deeper formations during the collapse and uplift episodes on the fore-arc. The mounds may also be caused by fluids percolating up through the fore-arc after being released by deserpentinization before the subducting oceanic crust reaches the 500°C isotherm.

1. Introduction

Most recent models of the structure and tectonic development of the inner trench walls and fore-arcs of both continental regions and oceanic island arcs invoke uplift and accretion of oceanic plate sediments and crust (e.g., SEELEY, 1979) or subsidence, brought about by tectonic erosion of the overriding plate (e.g., VON HUENE *et al.*, 1980), to explain observed features. In most cases it is apparent that fore-arcs can be the product of episodes of tectonic erosion and accretion brought about by variable

Hawaii Institute of Geophysics Contribution No. 1536.

modes of convergence and subduction (UYEDA, 1982). Furthermore, the tectonic style of a given trench system may vary between erosion and accretion along its length as well as through time.

The Mariana Islands arc system (Fig. 1) is a good example of a purely oceanic island arc arising where the Pacific oceanic plate is being overridden by the Philippine oceanic plate. UYEDA (1982) uses the Mariana convergent margin as the model for one type of subduction, the "Mariana-type" margin characterized by a pervasive tensional stress regime, seafloor spreading in the back-arc region, subsidence of the inner trench wall, and tectonic erosion of the outer fore-arc wedge. The lack of an appreciable accretionary prism throughout most of the central (17–20°N) Mariana fore-arc has been noted, based on geophysical surveys (LATRAILLE and HUSSONG, 1980; HUSSONG and FRYER, 1981; MROZOWSKI *et al.*, 1981), dredging (BLOOMER and HAWKINS, 1980), and deep drilling (HUSSONG and UYEDA, 1981), suggesting that tectonic erosion may be dominant in this portion of the arc. On the other hand, in the southern Mariana arc (south

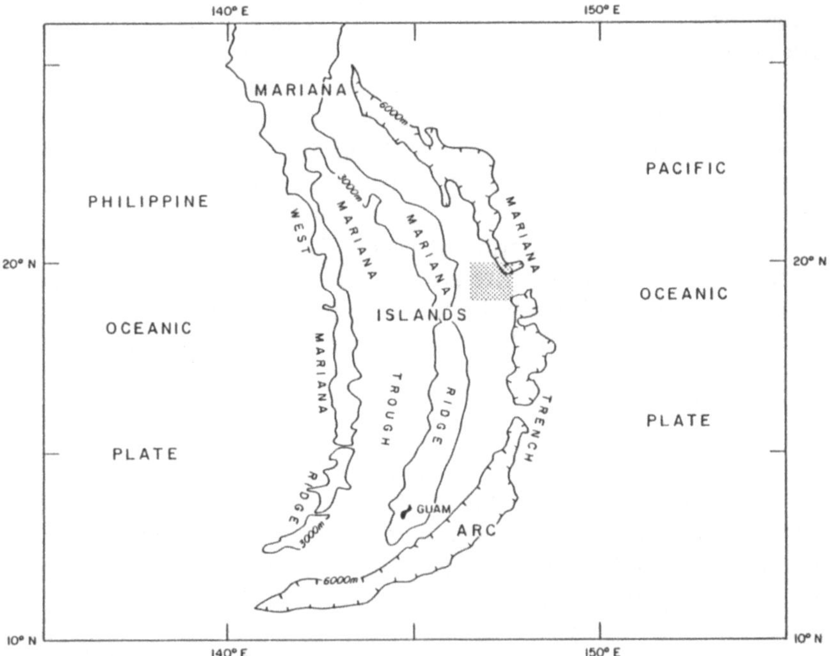

Fig. 1. Bathymetric map of the Mariana Island Arc region, contoured at 0.5 km (from GEBCO Map 5.06). The hatchured area denotes the SeaMARC II survey region.

of 14°S) KARIG and RANKIN (1983) note evidence for uplift and construction of the fore-arc that could be attributed to accretion of oceanic plate materials onto the overriding fore-arc. Further complicating the once simple interpretation of the structure of the fore-arc is the occurrence of large (up to 2 km relief), often very conical, seamounts near the trench slope break. The morphology of these seamounts is reminiscent of volcanic seamounts that occur elsewhere in the region. HUSSONG and FRYER (1981) pointed out, however, that while these seamounts resemble volcanoes, they occur too close to the trench axis. Generation of arc magmas is not possible along or just above the underlying subduction shear zone. Subsequent sampling of these outer fore-arc seamounts by the authors and by a Scripps Institution of Oceanography cruise (HAWKINS *et al.*, 1979) yielded a variety of serpentinized and tectonized gabbros and ultra-mafic rocks, which discounted the possibility that these seamounts are volcanic. BLOOMER (1983) interprets the fore-arc seamounts as the products of diapirism brought about by serpentinization of arc-derived ultramafic rocks along deep faults caused by tectonic erosion associated with subduction in the landward trench wall.

As part of a study of the entire northern Mariana arc and back-arc we completed a SeaMARC II side-scan sonar and bathymetric mapping survey of the fore-arc from 19° to 20°N, an area which includes two particularly spectacular fore-arc seamounts. The interpretation of these data is still in a preliminary stage but has progressed enough for us to present a description of some major fore-arc features and to speculate on some of the processes that might be forming these features. More detailed interpretation will follow in later publications after complete reduction of the data.

2. *Data Acquisition*

The SeaMARC II (Sea Mapping And Remote Characterization) is a shallow-towed, long-range, 12-kHz, side-scan sonar system that provides an acoustic image of the seafloor that is 10 km wide along the ship's track. The resultant side-scan image is geometrically corrected so that the reflectors are properly positioned relative to the ship, enabling a photo mosaic of a survey area to be assembled from adjacent tracks. In addition, the SeaMARC II system includes dual transducer arrays between which the angle of incidence of returning signals can be measured so that the range and bearing of reflectors can be calculated. These location data are then used to construct a grid of the depth and position of seafloor reflectors that is then converted to a bathymetric swath map along the ship's track. The bathymetric data are presently calculated out to a time range which is less than the second multiple of the earliest bottom reflection, effectively providing bathymetry in a swath which is approximately 3.4 times water depth.

Monitor records of the geometrically corrected side-scan data are photographed and made into a mosaic at sea. All the side-scan and phase (angle-of-incidence) data are also recorded on digital tape (at the somewhat overwhelming rate of one 1600 BPI tape every 30 minutes) for further processing on land. The post-cruise processing presently includes navigation corrections based on the side-scan images, followed by gain, static, and contrast corrections of the side-scan images. Navigation corrections are the most tedious and troublesome portion of this data reduction procedure. When the side-scan images are completed the calculations required to produce bathymetric data are made. More detailed descriptions of the SeaMARC II data acquisition and processing are available in BLACKINTON *et al.* (1983) and HUSSONG and FRYER (1983).

The SeaMARC II processing procedures are still under development and are constantly improving, but at this time the data we consider to be final have the following characteristics:

1. the wavelength of the side-scan signals is of the order of 10 cm, so backscattered reflections are most sensitive to relief on that scale;

2. with a beam width of 2° and a pixel size of 5 m perpendicular to the ship's track, the system has a resolution that permits recognition of the shapes of objects like small seafloor mounds or flows that have dimensions of about 100–200 m or more, although linear features like faults and channels are readily identifiable when they are only a few meters across;

3. although dependent on water depth and bottom characteristics, the bathymetry data are accurate to approximately 100 m while shapes of smaller features are repeatable to levels of the order of a few tens of meters.

3. *Regional Setting*

A sketch of the major features related to the tectonic setting of this northern Mariana fore-arc survey is presented as Fig. 2. The sketch was prepared from our surveys, including Hawaii Institute of Geophysics cruises KK 810626 and KK 830116, U.S. Navy bathymetric data, and the GEBCO map of the area.

The converging Pacific oceanic plate in this area is covered by large seamounts and guyots, in particular the Dutton Ridge (described in detail

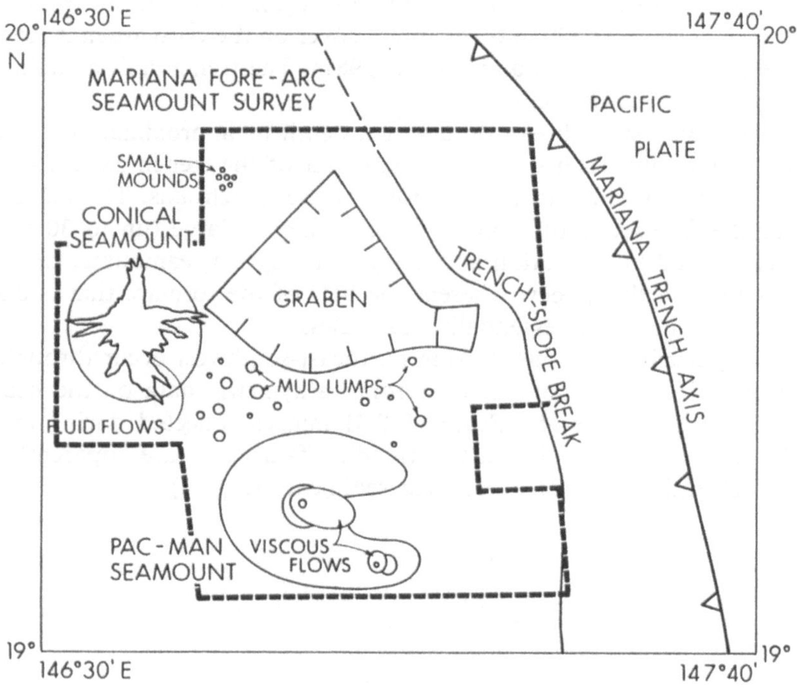

Fig. 2. Sketch of major morphologic features in the SeaMARC II survey area. The limits of the survey are shown by the dashed line.

in SMOOT, 1983), which is composed of an east-west trending series of large guyots rising to within 1200 m of sea level. As it is likely that the present western end of the Dutton Ridge, which is now entering the subduction zone, was preceded by even more westerly guyots that have already been subducted, it can be assumed that any influence on fore-arc structure incurred during subduction of large seamounts could be observed in the present fore-arc (e.g., FRYER and SMOOT, in press). The somewhat oblique plate convergence (from MINSTER and JORDAN, 1978) would cause the collision point with an east-west ridge to slowly sweep north along the trench, suggesting that features in the SeaMARC II survey area might have been influenced by the subduction of large seamounts from the Dutton Ridge (FRYER and SMOOT, in press). Smaller seamounts near 20°50′N and 18°30′N are in the trench axis and are being subducted; the more northerly of these features was also surveyed with the SeaMARC II system (FRYER and HUSSONG, 1985, this volume). Although the top of Layer 2 is difficult to observe in this region, our seismic reflection records indicate over 500 m of sediment cover on the crust when it reaches the trench axis (HUSSONG and FRYER, 1981). The trench axis contains no appreciable turbidite ponds.

The trench slope break occurs at a depth of approximately 4.2 km, which is typically reached about 25 km west of the trench axis. The axis of arc volcanism is about 200 km west of the trench axis. The outer 100 km of the fore-arc is disrupted by the numerous large (up to 30 km in diameter and 2 km elevation above regional depths) seamounts, that will be discussed in this paper, whereas the half of the fore-arc that is closer to the volcanic arc is essentially featureless.

Bathymetric lineations striking northeast-southwest occur throughout the region. Although not obvious from bathymetric data on the heavily sedimented inner fore-arc, SeaMARC II images collected during transit to the survey area reveal similarly trending faults that are apparently actively deforming the recent surface sediments (Fig. 3).

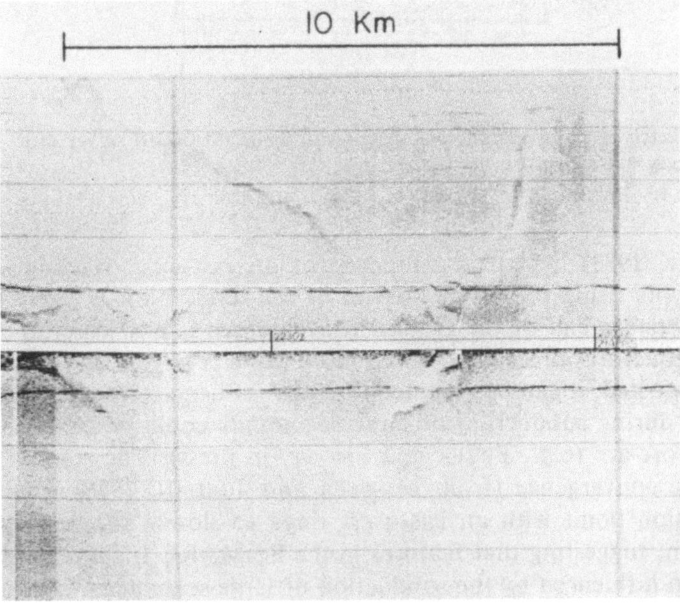

Fig. 3. SeaMARC II side-scan sonar image showing large faults on the Mariana fore-arc near 18°57′N and 146°28′E. There is no appreciable bathymetric offset across these faults. Because the faults occur in thick sediments and extend through to the surface, we suggest that they are recently active.

4. *Fore-Arc Morphology*

The corrected SeaMARC II side-scan mosaic is shown as Fig. 4. This survey, which covers over 5800 square kilometers of ocean floor, took 75 hours to complete at a speed of about 7 knots. The images are printed so that the higher amplitude reflections are darker gray to black. The data are discarded for the first 200 meters from each side of the ship's track and the resultant strip along the track is represented by a mottled gray stripe. Navigation is controlled by satellite, LORAN C, and the overlap of the side-scan data to provide an absolute position accuracy of better

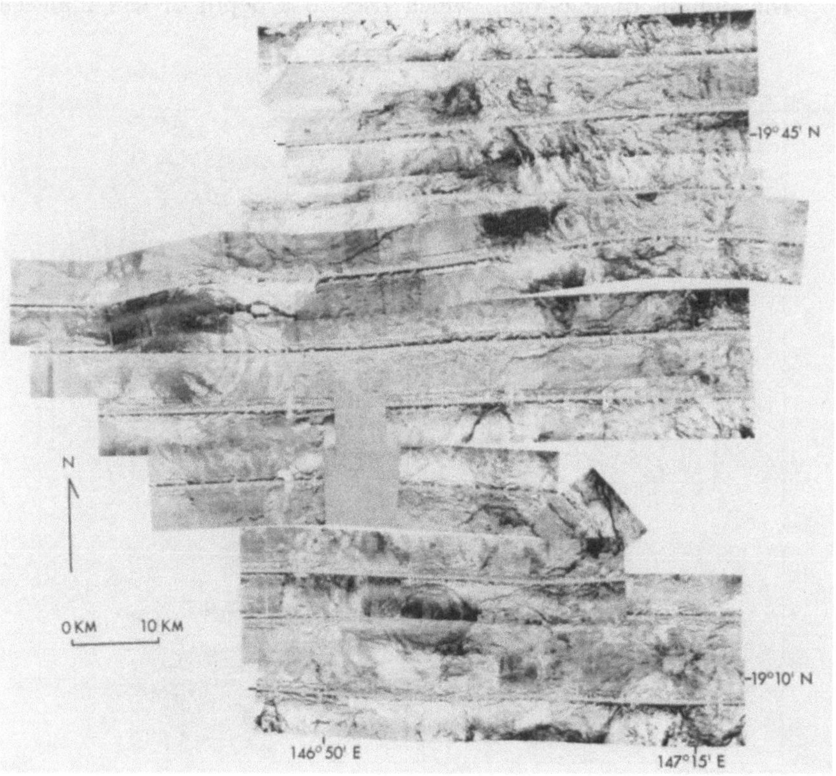

Fig. 4. Mosaic of ten side-scan sonar tracks across the survey area. Each swath is ten kilometers wide and has been corrected for ship's navigation. The ship's track shows as a grey swath in the center of each swath. Data are printed with high-amplitude echoes in dark. Major features can be identified using Fig. 2.

than 150 m.

The accompanying SeaMARC II bathymetry is shown as Fig. 5. The contours have been smoothed by hand from computer-generated plots that are also at a 100 m interval. The depths are calculated assuming a sound speed in water of 1500 m per second.

4.1 Pac-Man Seamount

The most striking feature in the survey region is a seamount rising 1500 m above the seafloor and that has a base over 30 km in diameter (Fig. 6). The seamount has an open-jawed shape, apparently caused by the collapse of its eastern flank, which has prompted us to call it Pac-Man Seamount.

The summit of Pac-Man, which rises to a depth of less than 2700

Fig. 5. Bathymetry derived from SeaMARC II data in the survey area. Depths are contoured at 100 meters, assuming sound speed in water of 1500 meters per second. Depths are considered accurate to better than 100 meters except in regions with dashed contours. Boxed areas are enlarged as Figs. 6 and 8.

m, is a broad dome scarred by numerous small (relief of probably only a few meters) ridges and fractures. The upper portion of the eastern slope of Pac-Man is covered by a highly reflective, textured apron that we interpret to be a series of large flows emanating from an area located at a

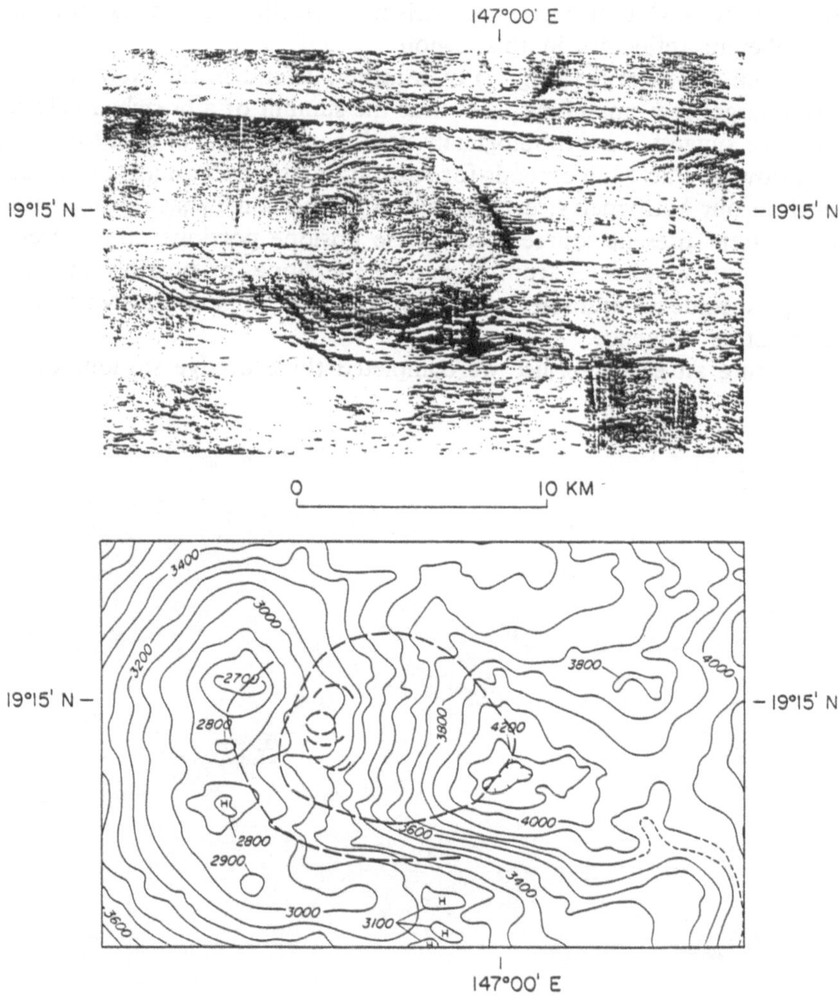

Fig. 6. Enlarged region around Pac-Man Seamount showing side-scan image as well as the bathymetry derived from the sonar data. Note the offset of the flow feature from the bathymetric summit of the seamount.

depth of 3050 m, 4 km east-south-east of the summit. The SeaMARC II bathymetry shows that the flow is over 100 m thick. The thickness of the flow, the fabric on the sonar image, and the character of dredge samples collected from the flow (see below), indicate it to be a thick, probably highly viscous effusion of blocky material. The flow covers an oval region of the steeply sloping flank of the seamount and extends down over 8 km to a 4200 m depression. Seismic reflection records do not show any coherent reflectors in this region.

Aerial photographs (Fig. 7) of Matthew Island in the New Hebrides show a large flow emanating from near the summit of a breached volcano, which we suggest is morphologically analogous to that of Pac-Man. The thick flow on Pac-Man extends for substantial distances across slope and, based on our bathymetry, even seems to extend somewhat up-slope from the apparent present source area. These upslope portions may represent earlier eruptive episodes from higher on the slope, or possibly they may indicate subsidence and steepening of the eastern flank of Pac-Man during the eruptive sequence.

During an earlier cruise we completed three dredge stations on the

Fig. 7. Matthew Island, New Hebrides. This island, although it is of different origin than Pac-Man (Matthew is a dacitic volcano), has similar surface morphology as Pac-Man and is useful for visualizing the summit, viscous flank flow, and collapsed side of the fore-arc seamount. Photo from MAILLET and MONZIER, 1982.

eastern flanks of Pac-Man, at least one of which sampled the flow. This dredge yielded blocky, highly altered and serpentinized ultramafic and mafic plutonic rocks. These samples and the morphology of the flow suggest the flow is a cold and viscous eruption of deep-seated island arc rocks that were serpentinized, probably remobilized by tectonic activity and serpentinization and brought up by subsequent diapirism (FRYER *et al.*, in press).

A smaller flow sequence is observed at a depth of 3400 to 3500 m on the lower southeastern flank of Pac-Man (note Fig. 2). This flow comes from what appears to be a small conical vent, extends for 3 to 5 km, and covers a tongue in the local bathymetric trends that rises 200–300 m above regional depths. This flow has not been sampled, but its morphology suggests that it may be a smaller scale analogy to the upper slope flow. The more muted appearance of the smaller flow may indicate that it is older and has been covered with a thin layer of sediments that partially masks the small-scale relief to which SeaMARC II is most responsive.

4.2 Conical Seamount

A second large seamount with a notably conical shape, a summit depth of less than 3100 m, and a base about 20 km in diameter, is situated a little over 40 km northwest of Pac-Man. Flows emanating from the summit of this seamount (Fig. 8) are apparently more fluid than those from Pac-Man and have sinuous and braided courses that extend farther downslope. This conical seamount is also not as fractured in appearance as Pac-Man, and it has no associated depressions, collapse features, or flank vents. The side-scan image suggests that the cone may be partially composed of concentric low-relief ridges, best observed on the southeastern flank. These ridges may relate to vertical motion of the seamount caused by uplift related to diapiric intrusion.

We do not know what kind of material comprises the flows on the flanks of the seamount. A dredge station near the summit of the conical seamount yielded only semi-consolidated vitric siltstones. Either the flows are composed of this material or they are composed of remobilized unconsolidated sediments. It is possible that uplift related to diapiric intrusion possibly coupled with emanations of fluids from the diapiric body could activate mass flow of the sediments on the flanks of the seamount. As these sediments sluff off the upper slopes they would expose the more consolidated material beneath.

The morphology and limited samples from the conical seamount suggest that it is a similar feature to Pac-Man but is in an earlier stage of development. The deeper igneous material that produces the viscous flow

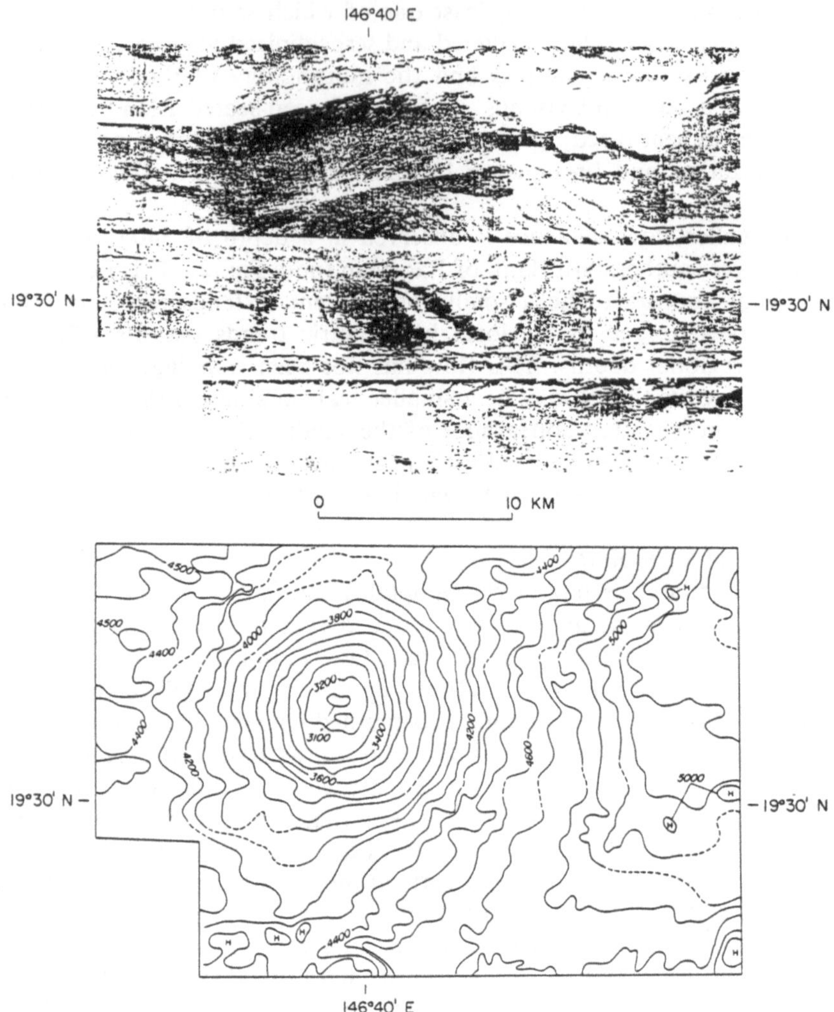

Fig. 8. Side-scan image and SeaMARC II bathymetry of the conical fore-arc seamount in the survey area.

on Pac-Man has not reached the surface at the conical seamount, but the latter is being actively uplifted by deep-seated diapirism.

4.3 Collapse features

In the center of the SeaMARC II survey area a large depression can be interpreted as resulting from the collapse of over 1000 sq km of the outer fore-arc from a regional average depth of about 4100 m to as deep as 5900 m. The steepest side of this collapse strikes parallel to the trench axis. No predominant fracture patterns appear around the depression that could be used to relate it to lateral motion, such as would produce a pull-apart basin. It seems that the collapse has occurred in response to removal of material from below, perhaps by tectonic erosion of the lower surface of the overriding plate by the subducting Pacific plate. Certainly the large oceanic seamounts that are converging on the trench along this portion of the arc must accentuate any disruptions of the fore-arc caused by irregularities in subduction geometry (FRYER and SMOOT, in press; FRYER and HUSSONG, 1985, this volume).

4.4 Mounds

Many small sediment mounds are observed throughout the side-scan image in Fig. 4. Some of the largest of these mounds are in the saddle between Pac-Man and the conical seamount, are only a few kilometers in lateral extent (Fig. 9), and have relief on the order of 150 to 200 m. More typical are even smaller mounds that occur in small fields throughout the image. A particularly good example of these smaller mounds is shown in the upper left-hand corner of the image (expanded illustration in Fig. 10), where a field about 3 km in extent contains approximately 30 mounds, each about 100 m in diameter. With such low relief, their elevation can not be reliably determined by SeaMARC II.

An attempt to dredge a field of small mounds located slightly north of the region in Fig. 10 yielded only thinly laminated clayey siltstones with no evidence of alteration.

With no direct evidence, we can only speculate on two possible causes of the fields of mounds. The most likely cause of at least some of the mounds is overpressuring of underlying sediment horizons caused by local tectonic activity. The collapse of large areas of the fore-arc which causes depressions such as on the eastern flank of Pac-Man, and the very large depression that occupies the entire center of the survey region, would cause tilting, flow, and instability of deeper sediment layers that might readily produce mud lumps and mud volcanoes. In fact, the most obvious groupings of the larger mounds are adjacent to the southwest and northeast bounds of the large fore-arc depression.

A second possibility is that some of the smaller fields of mounds may be caused by outflow of fluids that have percolated through the overriding plate after having been squeezed out of the oceanic sediments and dehydrated from crustal rocks that are being subducted. The location of

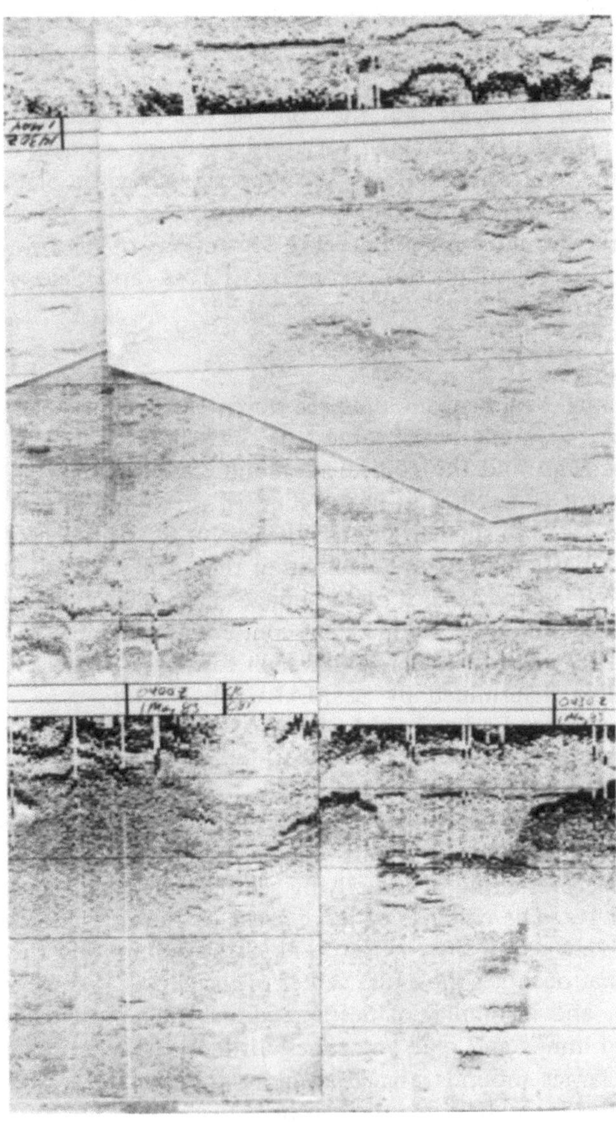

Fig. 9. SeaMARC II image of large mud lumps, Mariana fore-arc.

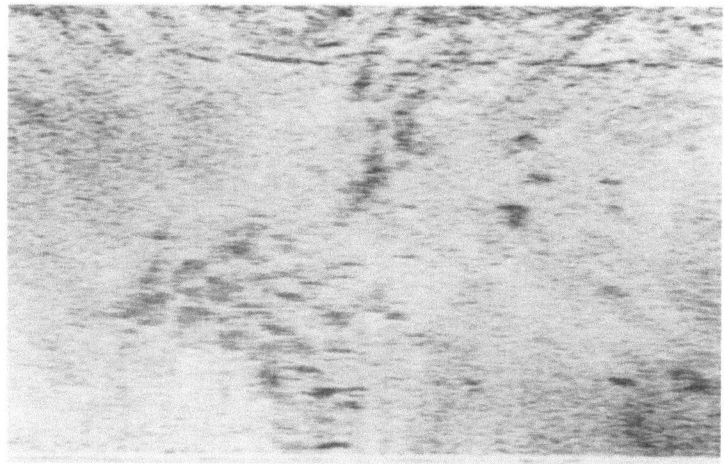

Fig. 10. SeaMARC II image of a field of small mounds on the outer Mariana fore-arc.

these small mounds is about 40 to 60 km from the trench axis, so the top of the subducting slab would be expected to be around 10 to 15 km beneath the seafloor. It is unlikely that fluids derived from the subducted Pacific oceanic crust could migrate directly to the overlying seafloor, but these fluids may find indirect paths to the seafloor. Most interstitial water would be expected to be squeezed out of the sediments before reaching an overburden depth of about 6 km (Fig. 11). Another major source of water would be dehydration of serpentinized minerals, which will take place before temperatures reach 500°C. Generalized isothermal models for subducting slabs (such as from TOKSOZ *et al.*, 1973) configure the 500°C isotherm so that it is near vertical, and thus intersects most of the uppermost subducting slab, in the mid-fore-arc region (Fig. 11). The fluids resulting from dehydration of the subducting slab may contribute to serpentinization of overlying deep seated island arc rocks (FRYER *et al.*, in press) and eventually rise by diapiric emplacement of this serpentinite. Alternatively they may rise by other mechanisms as free fluids.

5. Discussion and Conclusions

The SeaMARC II side-scan images show features consistent with a diapiric origin for the fore-arc seamounts in the northern Mariana Arc. Furthermore, the two largest seamounts observed in this survey seem to

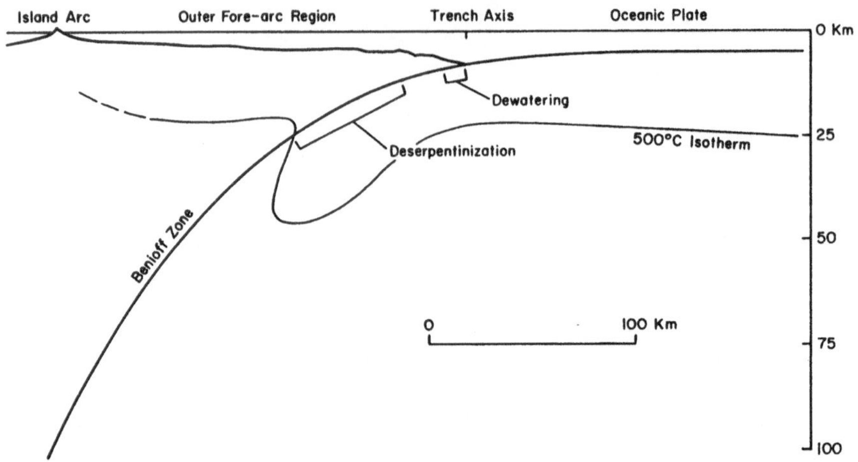

Fig. 11. Generalized and simplified cross-section of the subduction zone. Although not modeled specifically for the Mariana arc, this geometry and 500°C isotherm (adapted from TOKSOZ *et al.*, 1973) is presented to demonstrate the possible relationship between fluids driven out of the subducting slab at horizontal distances from the trench axis similar to the location of the observed fields of small fore-arc mounds (such as those in Fig. 10).

represent different stages of diapiric uplift. We interpret Pac-Man as a more developed feature where highly altered plutonic rocks from deep in the fore-arc (geochemical affinities identified by J. Hawkins, personal communication) are erupting on the seafloor. These cold eruptions of crystalline rocks produce thick viscous flows. In contrast, the conical seamount in our survey area is interpreted to be a less developed feature. Flows emanating from near the summit may be composed of remobilized sediments. These flows are more fluid so have greater lateral extent and appear braided and sinuous compared with the Pac-Man flows. Eventually the diapir beneath the conical seamount may reach the surface and produce flows of mafic and ultramafic material much like those observed on Pac-Man.

It should also be noted that some of the larger fore-arc seamounts seem to be located along the strike of large bathymetric offsets in the northern Mariana fore-arc. We suspect that these offsets are an expression of the large fractures that cut across the arc and are colinear with arc volcanoes and back-arc fracture zones (FRYER and HUSSONG, 1982). Furthermore, there are less extensive fracture traces within and colinear with linear groups of fore-arc seamounts. The distribution of these seamounts is probably controlled by these local fracture patterns. The SeaMARC

II images (Fig. 3) further suggest that some of these fractures are still active.

The smaller sediment mounds that are common on the fore-arc are probably largely caused by density inversions in the sediments triggered by tectonic events such as that which produced the large collapse feature dominating the middle of our survey area. It is tempting, however, to speculate that some of these mounds, particularly the fields of very small mounds seen only in the side-scan data, mark the outflow of fluids originating from subducted hydrated material.

REFERENCES

J. G. BLACKINTON, D. M. HUSSONG, and J. G. KOSALOS, First results from a combination side-scan sonar and seafloor mapping system (SeaMARC II), Offshore Technology Conference, OTC 448, 307–311 (+10 Figs.), 1983.

S. H. BLOOMER, Distribution and origin of igneous rocks from the landward slopes of the Mariana Trench: Implication for its structure and evolution, *J. Geophys. Res.*, **88**, B9, 7411–7428, 1983.

S. BLOOMER and J. W. HAWKINS, Arc-deived plutonic and volcanic rocks from the Mariana Trench wall (abstract), *Eos. Trans. AGU*, **61**, 1143, 1980.

P. FRYER and D. M. HUSSONG, Arc volcanism in the Mariana Trough (abstract), *Eos. Trans. AGU*, **63**, 1135, 1982.

P. FRYER and D. M. HUSSONG, SeaMARC II studies of subducting seamounts, this volume.

P. FRYER and N. C. SMOOT, Morphology of ocean plate seamounts in the Mariana and Izu/Bonin subduction zone, *Marine Geology* (in press).

P. FRYER, E. L. AMBOS, and D. M. HUSSONG, Origin and emplacement of serpentinite diapirs in the Mariana fore-arc, *Nature* (in press).

GEBCO MAP 5.06, Canadian Hydrographic Service, 1979.

J. HAWKINS, S. BLOOMER, C. EVANS, and J. MELCHIOR, Mariana arc-trench system: Petrology of the inner trench wall (abstract), *Eos. Trans. AGU*, **60**, 968, 1979.

D. M. HUSSONG and P. FRYER, Structure and Tectonics of the Mariana arc and forearc: Drillsite selection surveys, *Initial Rep. Deep Sea Drill. Proj.*, **60**, 33–44, 1981.

D. M. HUSSONG and S. UYEDA, Tectonic processes and the history of the Mariana arc–A synthesis of the results of DSDP leg 60, *Initial Rep. Deep Sea Drill. Proj.*, **60**, 909–929, 1981.

D. M. HUSSONG and P. FRYER, Back-arc seamounts and the SeaMARC II seafloor mapping system, *Eos. Trans. AGU*, **64**, 45, 627–632, 1983.

D. E. KARIG and B. RANKIN, Marine geology of the forearc region, southern Mariana Island arc, in *The Tectonic and Geologic Evolution of Southeast Asian Seas and Islands: Part 2*, Geophys. Monogr. Ser., vol. 27, edited by D.E. Hayes, pp. 266–280, AGU, Washington, D.C., 1983.

S. L. LATRAILLE and D. M. HUSSONG, Crustal structure across the Mariana Island arc, in *The Tectonic and Geologic Evolution of Southeast Asian Seas and Islands*, Geophys. Monogr. Ser., vol. 23, edited by D. E. Hayes, pp. 209–222, AGU, Washington, D. C., 1980.

P. MAILLET and M. MONZIER, Volcanisme et petrologie des Iles Matthew et Hunter: donnes preliminaires, *Contribution a l'etude geodynamics du sud-ouest Pacifique, Equipe geologie-geophysique, Noumea*, Travaux et documents de L'O.S.T.O.M., N. 147, pp.

187–216, Editions de l'office de la recherche scientifique et technique outre-mer, Paris, 1982.

J. B. MINSTER and T. H. JORDAN, Present-day plate motions, *J. Geophys. Res.*, **83**, B11, 5331–5354, 1978.

C. L. MROZOWSKI, D. E. HAYES, and B. TAYLOR, Multichannel seismic reflection surveys of leg 60 sites, Deep Sea Drilling Project, *Initial Rep. Deep Sea Drill. Proj.*, **60**, 57–70, 1981.

D. R. SEELY, The evolution of structural highs bordering major forearc basins, in *Geological and Geophysical Investigations of Continental Margins*, Mem. 29, pp. 245–260, edited by J. S. Watkins, L. Montadert, and P. W. Dickerson, American Association of Petroleum Geologists, Tulsa, Okla., 1979.

N. C. SMOOT, Geological Notes: Guyots of the Dutton Ridge at the Bonin/Mariana Trench juncture as shown by multi-beam surveys, *J. Geol.*, **91**, 211–220, 1983.

M. N. TOKSOZ, N. H. SLEEP, and A. T. SMITH, Evolution of the downgoing lithosphere and the mechanism of deep focus earthquakes, *Geophys. J. R. Astron. Soc.*, **35**, 285–310, 1973.

S. UYEDA, Subduction zones: An introduction to comparative subductology, *Tectonophysics*, **81**, 188–159, 1982.

R. VON HUENE, M. LANGSETH, N. NASU, and H. OKADA, Summary, Japan Trench transect, *Initial Rep. Deep Sea Drill. Proj.*, **56/57**, 473–488, 1980.

Formation of Active Ocean Margins, edited by N. Nasu *et al.*, pp. 291–306.

SEAMARC II STUDIES OF SUBDUCTING SEAMOUNTS

Patricia FRYER and Donald M. HUSSONG

Hawaii Institute of Geophysics, University of Hawaii, 2525 Correa Road, Honolulu, Hawaii 96822, U.S.A.

Abstract. A number of seamounts are located in or very near the axes of the Mariana and Izu/Bonin trenches. Effects of subduction related tectonic processes on the oceanic plate are expressed in the morphology of some of these seamounts. Fractures, presumably caused by bending of the oceanic lithosphere prior to subduction beneath the arc, are observed on the seafloor near the trench axes. These fractures cut across the smaller seamounts on the subducting plate, and displacement along fractures within these seamounts corresponds to that on fractures on the surrounding ocean floor. The larger seamounts and seamount chains abutting the trenches are apparently less influenced by plate fracturing prior to subduction. Seamounts do not appear to be accreted onto the outer fore-arc wedge. Rather, as they are subducted they cause vertical tectonic movement on the Mariana fore-arc. The outer half (within 100 km of the trench axis) of the Mariana fore-arc shows these features particularly well. Restriction of major vertical movements to the outer half of the fore-arc may be related to rheological properties of the fore-arc wedge. By comparison, the Izu-Bonin fore-arc shows much less tectonic disturbance. The most obvious difference between the Mariana and Izu-Bonin regions is the lack of numerous seamounts on the Pacific plate adjacent to the Izu-Bonin arcs.

High resolution swath map bathymetry and side-scan images (SeaMARC II) of one of the seamounts, and portions of the inner trench wall, show details of convergence effects.

1. *Introduction*

Seamounts in the western Pacific (Fig. 1), particularly the Magellan

Hawaii Institute of Geophysics Contribution No. 1537.

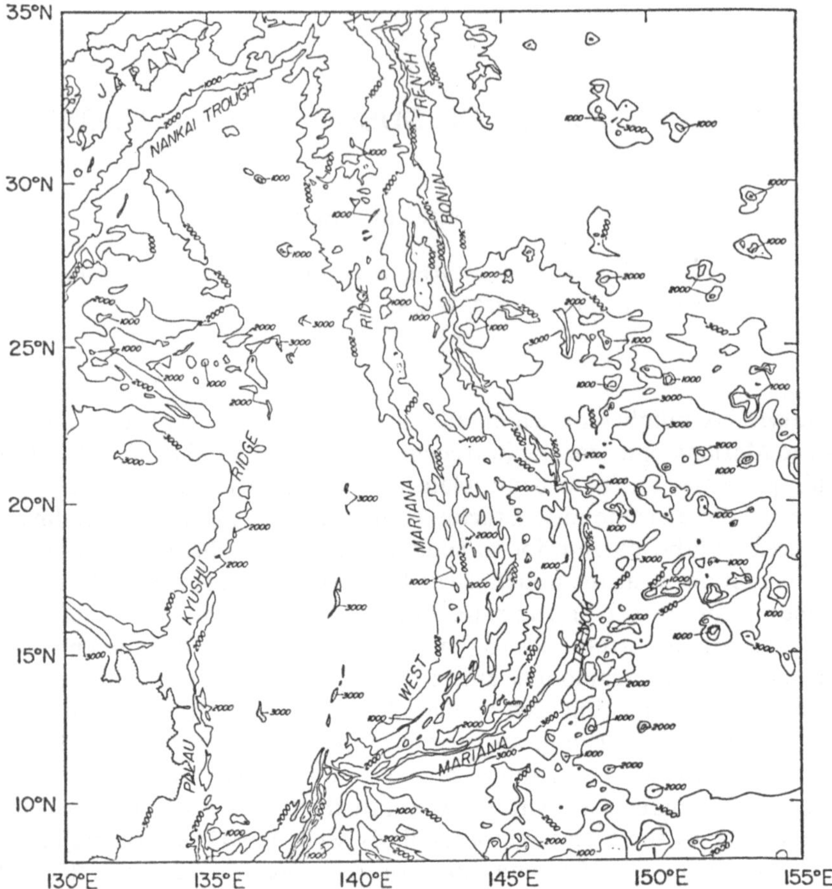

Fig. 1. Bathymetry map of the Western Pacific showing locations of the mariana and Izu-Bonin arcs. Note the positions of the major seamount groupings on the oceanic plate. The Magellan Seamount group extends from 13°N to about 23°N.

Seamounts, clustered near the Mariana/Bonin trenches from 15°–26°N, are converging with the Mariana and Izu-Bonin arcs at a rate of about 9 cm per year (MINSTER and JORDAN, 1978). Since plate motion in the western Pacific has been much the same for the past 40 my, we assume that some of the seamounts of the group have already been subducted. The manner in which these seamounts are subducted is little known. Furthermore, the effects of subduction of these seamounts on the fore-arc regions of the Mariana and Izu-Bonin area has not been studied in detail.

Our research for the past two years, under the auspices of the Office of Naval Research and the National Science Foundation, has concentrated on the morphology and composition of the Mariana fore-arc region and of the oceanic plate immediately east of the Mariana Trench. Development of the SeaMARC II seafloor mapping system (BLACKINTON *et al.*, 1983) has facilitated this research by providing a tool capable of producing an image of the sea floor that looks much like an aerial photograph. In addition, the data collected by SeaMARC II can be used to produce a bathymetry map over the area surveyed.

One of the most recent cruises using the SeaMARC II system surveyed the Mariana fore-arc and trench near 19°N to 21°N (Fig. 2). Images from this cruise are presented in this paper. The objectives of surveying this are were: (1) to study the effects that convergence of the oceanic plate with the Mariana arc would have on the seamounts resting on the plate; and (2) to study the effects that subduction of plate seamounts on the downgoing plate might have on the fore-arc region of the Mariana arc.

2. *SeaMARC II Data Collection*

The SeaMARC II system is a long-range side-scan sonar system that produces geometrically correct acoustic images of the seafloor. In addition, when the range and direction to a reflector are accurately determined these data can be converted to position and depth, so that bathymetry can be calculated for the ensonified swath of seafloor.

The system consists of a transducer array that is encased in a streamlined, neutrally buoyant housing. The system operates at 10 kHz on the starboard side and 11 kHz on the port side. The housing is towed at a shallow depth (100 m below sea level) in order to provide the stability demanded by the narrowness (2°) of the beam pattern. Yaw or pitch of the transducer array must be less than 1.5°. In water depths greater than 1 km the system produces side-scan images 10 km wide. The width of the swath decreases in steps in water shallower than 1 km. The minimum swath width is 1 km and the minimum depth that can be surveyed is 50 m beneath the array. At present the width of the bathymetric swath is limited to 3.4 times water depth. Beyond this distance, the second echo from the bottom begins to be detected. The bathymetry information is very difficult to distinguish amid the noise from the combined bottom multiples.

The images produced by SeaMARC II are particularly sensitive to reflectivity changes caused by backscatter from seafloor relief on the order

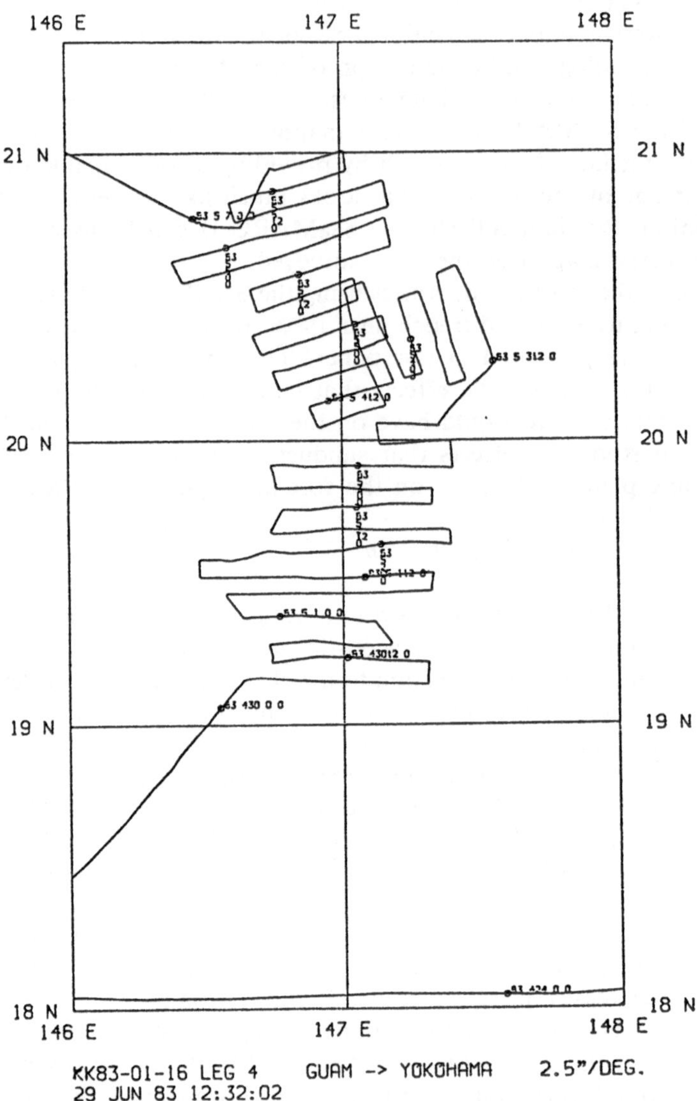

Fig. 2. Track chart of cruise KK83011604 surveys of the Mariana fore-arc and the Pacific plate from 19° to 21°N.

of 10 cm. That is because the wavelength of a 12 kHz signal is approximately 12 cm in water. Pattern recognition from SeaMARC II images, however, is a function of the beam width of the signal (2°), range, ping repetition rate, and the ship's speed. Typically, linear features such as channels or faults are recognizable if they have dimensions of only a few to several tens of meters. Three-dimensional features, such as small mounds or sediment ponds must have dimensions on the order of 100 m or more before they can be recognized.

3. *SeaMARC II Survey of a Subducting Seamount*

A small seamount (about 25 km diameter) is situated in the axis of the Mariana Trench at about 20°45′N. During a survey of this seamount we made six traverses of the region. The entire seamount and a portion of the oceanic plate on which it rests, the inner trench wall and part of the outer fore-arc adjacent to the trench are covered by this SeaMARC II survey (Fig. 3). The side-scan image of the region, produced from the SeaMARC II data, shows a number of striking features (Fig. 4a). The broad, dark strip running north to south through the middle of the image is the face of the inner trench wall (see sketch in Fig. 4b). The eastern (right) edge of this dark strip roughly coincides with the trench axis. To the east of the trench axis is the Pacific plate. The small seamount lies within the trench axis in the center of the image. Upon first looking at the side-scan image, the body of the seamount is difficult to distinguish from the surrounding sea floor. The body of the seamount is generally darker and is marked by numerous fractures which appear as lineations oriented roughly N-S. The sediment covered oceanic plate surrounding the base of the seamount is somewhat lighter in appearance than the broken flanks of the seamount, although lineations also occur on the seafloor around the seamount.

At the base of the inner trench wall in the southernmost portion of the side-scan image are a number of small mounds (see Fig. 4b and southernmost profile in Fig. 5). They are less than 100 m high and have diameters of a few hundred meters. Their origin may be from overpressuring of sediments by local tectonic movement. Fluids in the sediments being subducted with the oceanic plate will be driven off by reordering, compaction, and dehydration by a depth of about 6 km (Burst, 1976; Pittman, 1979). It is possible that some of these fluids may find their way back to the trench axis along the decoupling zone between the plates. Percolation of these fluids through sediments near the trench axis may cause formation of the small mounds seen on the side-scan image. The fore-arc

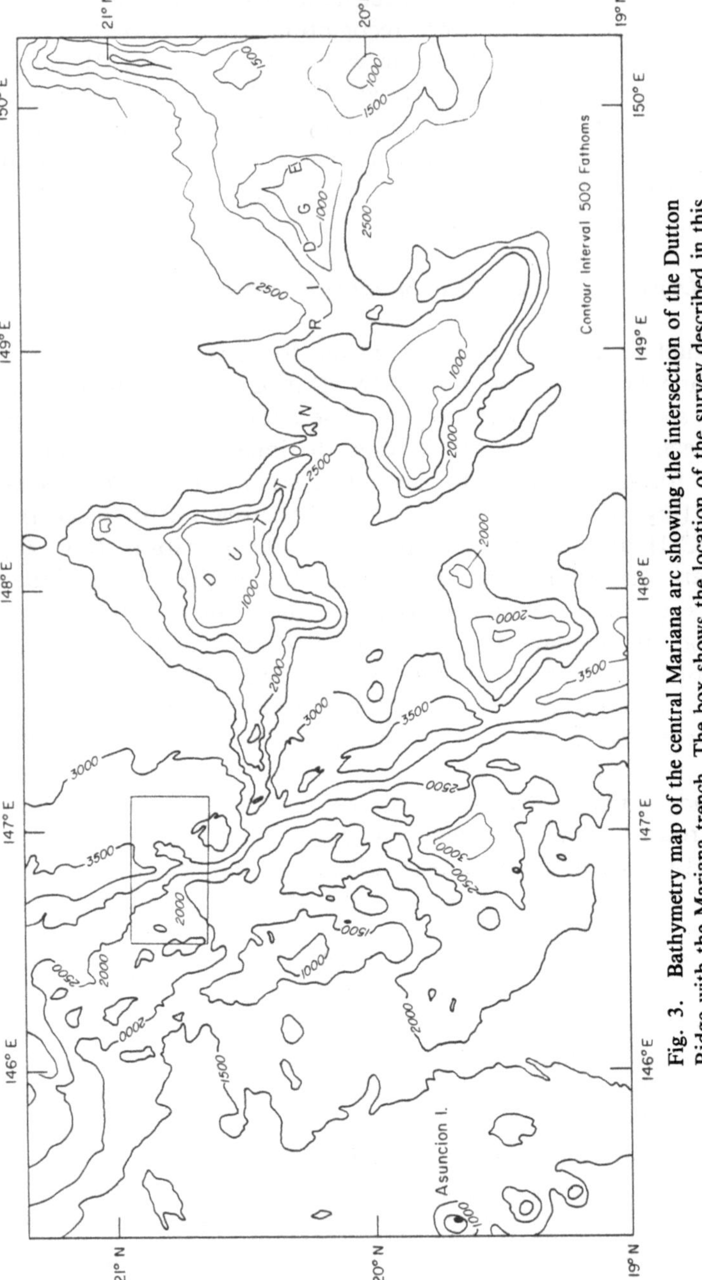

Fig. 3. Bathymetry map of the central Mariana arc showing the intersection of the Dutton Ridge with the Mariana trench. The box shows the location of the survey described in this paper. Coordinates are shown within the Figure (Figure from SMOOT, 1983).

portion of the side-scan image, to the left of the trench axis shows a few regions of similar small mounds. They are possibly formed by either out-flow of fluids from beneath the fore-arc wedge, or more probably, by overpressuring of sediments by local tectonic movement.

A portion of a small seamount can be seen at the lower edge of the left side of the side-scan image. This feature appears to be nearly circular in outline. The maximum diameter of the feature on the image is 8 km. It appears to have a secondary mound on its flank which is truncated by the edge of the side-scan swath. Similar features occur elsewhere on the Mariana fore-arc, although most are considerably larger. Their origin is suggested to be by diapiric emplacement of serpentinized, arc-derived ultramafic and mafic plutonic rocks (BLOOMER, 1983; HUSSONG and FRYER, 1984, this volume).

The most striking feature of the side-scan image occurs to the east of the trench axis; a series of dark fractures run NW-SE oblique to the strike of the inner trench wall. Most likely, these fractures occur in response to the bending of the oceanic plate prior to subduction. Magnetic lineations occur on the Pacific plate near the Magellan Seamounts. They are oriented 35° to 45° true, and have been identified by HUSSONG and FRYER (1981). The trend, amplitude, and wavelength of these anomalies are similar to the Japanese lineations described by HILDE *et al.* (1976). Likely these lineations were produced by Mesozoic seafloor spreading from the Kula-Pacific Ridge. The strike of the fractures seen on the oceanic plate in this side-scan image suggest that the fractures are not evidence for reactivation of fracturing along the Mesozoic spreading direction, rather they may be related to local subduction tectonics. The fractures on the plate are colinear with fracture traces on the southern flank of the seamount. However, the major fracture traces through the seamount appear to have been offset by a right-lateral displacement.

The nature of the fractures on the plate and the seamount show up very well on unprocessed side-scan images of the surveyed area (Fig. 5). The unprocessed side-scan images represent side-scan data recorded in slant range rather than in distance. They therefore are a very distorted view of the ocean floor. Despite the distortion in raw side-scan images, they provide a powerful tool for presenting a sense of three-dimensional perspective of key features. The profile of the seafloor shown on the raw side-scan images is identical to that produced using standard seismic reflection profiling equipment (e.g., Fig. 6). The image on the raw side-scan shows the character of the seafloor looking out to the side, away from the ship's track, for 5 km. The distorted view produced is reminiscent of oblique aerial photographs of rough terrain. Prominent fracture traces can be seen

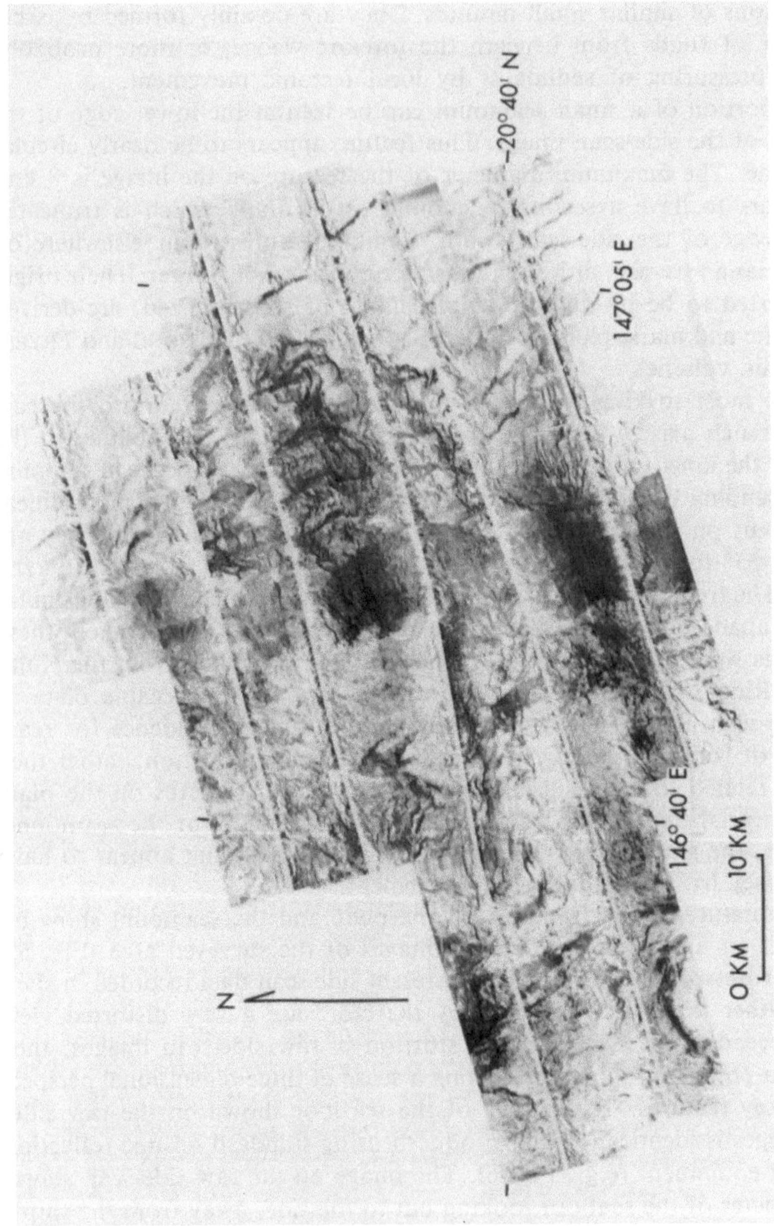

Fig. 4a. SeaMARC II side-scan mosaic of a small seamount situated in the Mariana trench axis. Swath width is 10 km. Grey lines mark position of ship's track.

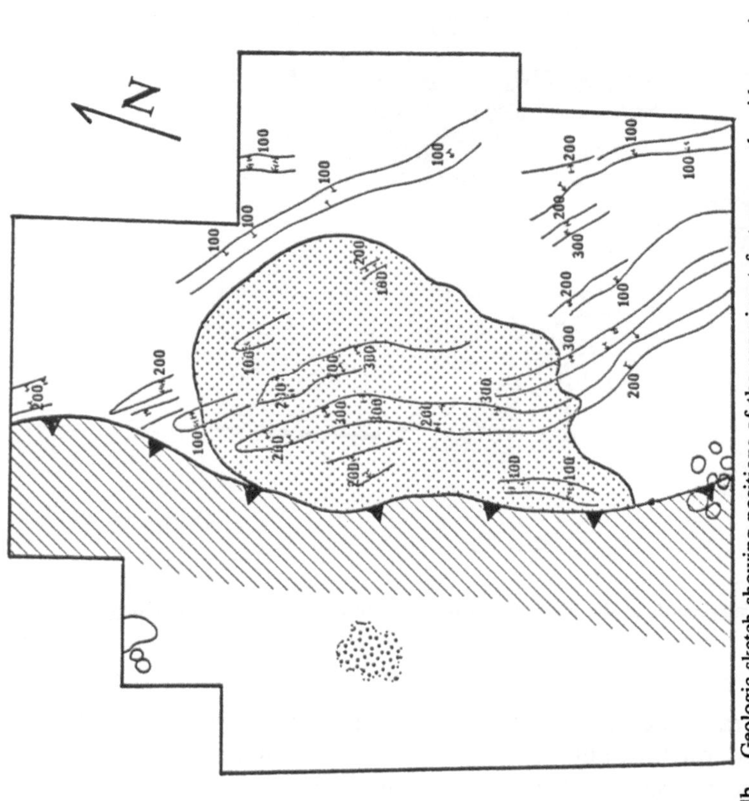

Fig. 4b. Geologic sketch showing positions of the prominent features on the side-scan image shown in Fig. 4a. The cross hatched area corresponds to the dark strip on the center of the side-scan image and corresponds to the inner-trench wall. The trench axis is shown by a thick line with solid triangles. The body of the subducting seamount is shown by the "x" pattern. The field of small open circles shows the location of a field of small mounds. The larger circles in the southern and northern portions of the figure represent larger mounds. Fracture bounded ridges on the plate and through the body of the seamount are shown by solid lines, with throw on the faults shown by arrows and number of meters of displacement. The Pacific plate is the region to the right of the trench axis and the Mariana fore-arc is located to the left of the trench axis.

Fig. 5. Raw side-scan records of the traverses across the seamount situated in the trench axis.

on each of the raw side-scan images of the seamount survey area. The amount of throw on the faults is roughly the same on both the ocean floor and on the seamount edifice itself. This factor is consistent with the suggestion that the seamount is fracturing in response to the same stresses which cause the upper surface of the oceanic plate to fracture as it bends prior to being subducted. Of course, the absolute amount of throw on the faults can be determined only beneath the ship on the raw side-scan record, and not at all on the processed side-scan images. The bathymetry data from the survey is needed to allow measurements of relief along fracture faces to be made.

The bathymetry data is produced from the side-scan data by determining pairs of values of depth and horizontal distance from the ship's

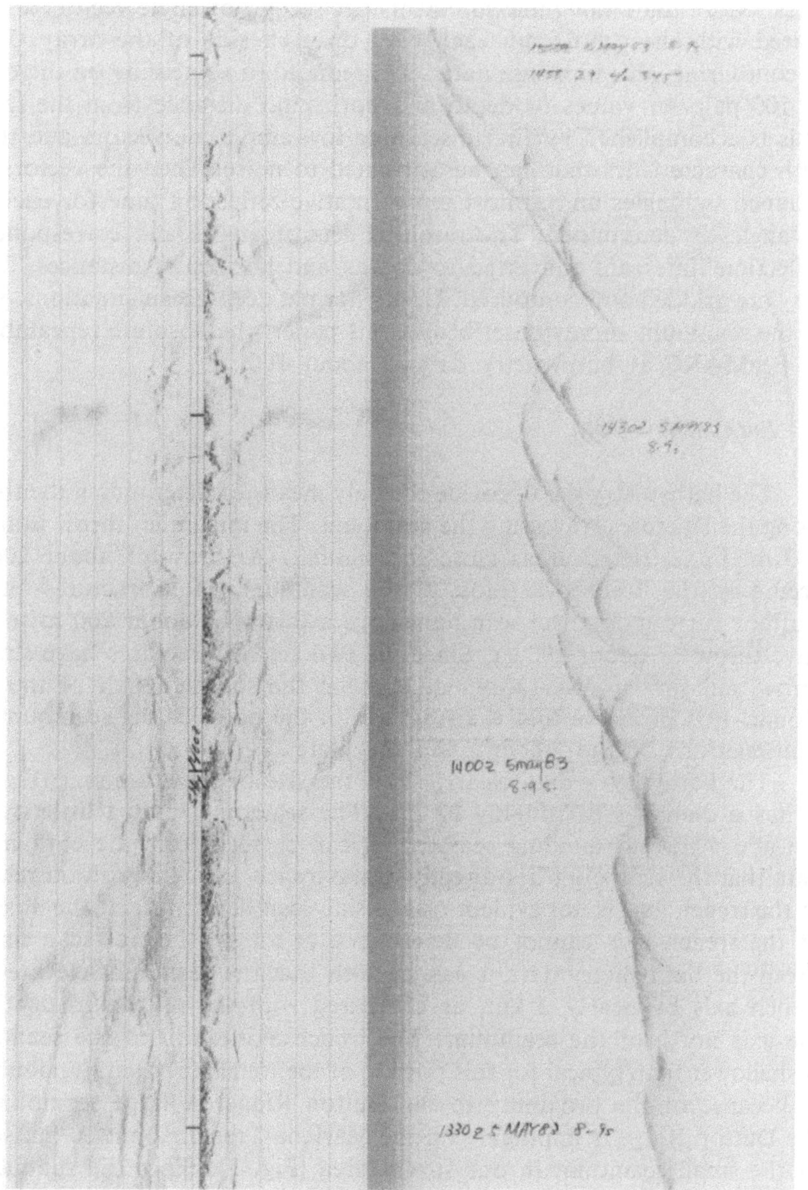

Fig. 6. Seismic reflection profile (3.5 kHz) and processed side-scan data of a portion of the southern traverse across the seamount survey area. The lineations on the processed side-scan image correspond to the uplifted fault blocks apparent on the seismic reflection profile.

track. More than 4,000 measurements per second of angle vectors are acquired with the echo from each ping on each side of the array. Prior to contouring, this immense data set is reduced to something on the order of 100 pairs of values of depth and horizontal distance from the track. This is accomplished by first discarding low amplitude vectors and those with characteristics that can be attributed to noise. Then the vectors are grouped by angles and a most representative reflection time for each set of angles is determined. The resulting sets of angles and corresponding reflection times are converted to depths and horizontal distances. Then they are gridded and contoured. Under normal deep-ocean situations, such as the seamount survey described in this paper, the absolute repeatability of SeaMARC II bathymetry data is about 100 m.

4. *Bathymetry Data*

The bathymetry data provide the only means of determining the throw along the fractures traversing the seamount. The maximum throw is about 300 m, in restricted areas close the summit. A throw of about 200 m occurs on the faults over most of the seamount. Most fractures on the seafloor surrounding the seamount have a throw of about 200 m, a few have throw of about 100 m. Since the two sets of fractures have similar throw and similar orientation, at least on the southern half of the seamount, it is probable that the fractures in the body of the seamount are continuations of the fractures on the plate.

The bathymetry data clearly show the size of the seamount (Fig. 7). It has a diameter of roughly 22 km. The seamount rises a little over 2 km above the surrounding seafloor. It can be seen from the bathymetry data that the seamount lies directly in the trench axis. Its position relative to the trench axis is not evident on the side-scan data. In fact, the position of the trench axis cannot be determined at all on the side-scan image. From the bathymetry data it can be seen that the seamount elevates the trench axis by nearly 2 km, as compared with the maximum depth of the axis north of the seamount. The trench axis south of the seamount is shallower than typical for this portion of the trench (6700 m), undoubtedly because of the proximity to the Dutton Ridge. A large seamount of the Dutton Ridge is impinging on the Mariana Trench about 20 km south of the small seamount in our survey area (Fig. 3). Since the subducting seamount has elevated the trench axis by nearly the same amount as the total relief on the seamount, we suggest that roughly half of the seamount has already been subducted.

The effect of this small seamount on the overriding fore-arc wedge

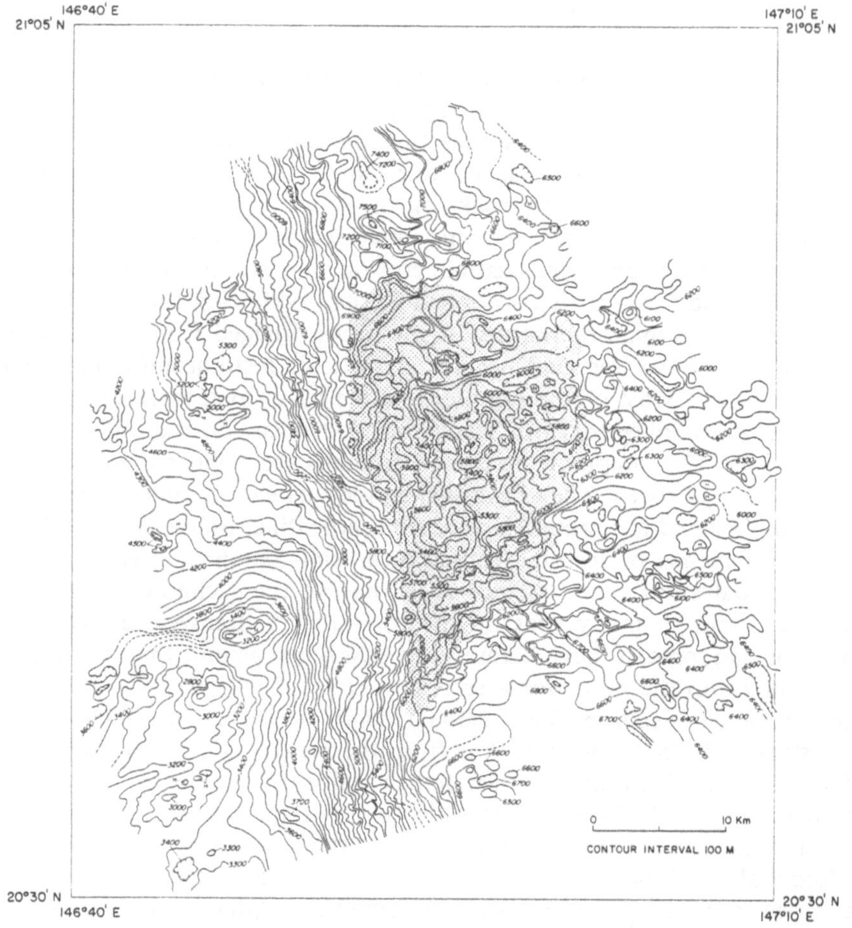

Fig. 7. Bathymetry map, based on the side-scan data, of the survey area. Contoured at a 100 m interval, annotated every 200 m.

is difficult to quantify. However, some aspects of the interaction between the seamount and the fore-arc can be noted in the survey area. The raw side-scan records (Fig. 5) show a change in the nature of the inner-trench wall adjacent to the subducting seamount. The inner-trench wall in the southernmost traverse, where the seamount is absent, shows a smooth face lacking in any obvious slumps, terrace, or fault features. By contrast, the inner-trench wall on the second traverse from the south shows a series of terraces which we believe are related to differential uplift of outer portion of the fore-arc as the seamount is subducted. This traverse was run very close to the summit of the subducting seamount.

There is no evidence from dredging, DSDP drilling data, seismic profiling surveys, or any other source that indicates that the outer part of the fore-arc consists of sediment accreted to the arc from the oceanic plate. The occurrence of arc rocks on the inner trench wall, and the apparently ubiquitous vertical tectonic movement on the movement on the outer fore-arc, suggest that tectonic erosion, not accretion, predominates in the Mariana fore-arc (HUSSONG and FRYER, 1981; HUSSONG and UYEDA, 1981). If we examine the nature of the rocks dredged from the Mariana fore-arc closely we see evidence for small scale tectonic movement over broad regions. Most of the samples that we have dredged from the fore-arc have been collected from seamounts that occur on the outer part of the fore-arc. The samples from these seamounts are highly tectonized mafic and ultramafic rocks. They have slickenside surfaces in many cases. They are highly altered to uralite, chlorite, and serpentine. Many of them shown internal evidence of tectonic disturbance, kink banding, mylonitization, and internal shear planes. Since rocks having these characteristics were dredged from so many sites on the fore-arc (19 dredge sites recovered rocks having some of these features) it is possible to assume that the material that makes up the outer fore-arc is in a state of mobilization on a scale not previously realized. The regional depth of the inner-trench wall at a distance of about 10 to 15 km from the trench axis is roughly 4 km. The minimum depth of the inner trench wall as shown in the bathymetry map on Fig. 7, is 2800 m. This depth is attained at the summit of a rise that is situated directly south-west of the subducting seamount. The additional relief of about 1200 m, over the expected regional depth, is ascribed to uplift of the fore-arc in response to subduction of the subducting seamount.

5. *Discussion*

The seamount described in this paper is one of a number of seamounts situated within or very close to the axis of the Mariana Trench. This seamount is particularly small (22 km in diameter). Others, larger in size, have been noted by Fryer and Smoot (in press). These larger seamounts also appear to be broken up depending on their size, position relative to the trench axis, and the degree of fracturing of the oceanic plate on which they rest. Apparently the larger the seamount the less it fractures prior to subduction. Despite break-up of small seamounts (less than about 30 km in diameter), they apparently have an observable effect on the overriding fore-arc wedge. Larger seamounts are probably subducted at least partially intact and must therefore influence the overriding fore-arc wedge far more. We speculate that there would be a limit to the extent to which these subducting masses could influence the overriding fore-arc. Presumable, as the seamounts are subducted they would eventually enter a regime within the mantle that would behave more plastically than the overlying material. If the fore-arc of the Mariana system is about 40 my old (Hussong and Uyeda, 1982) the elastic layer of Watts *et al.* (1980) should be roughly 30 km thick. Below that depth the mantle should behave in a more plastic manner, and accommodate the excess mass of the subducting seamount without large-scale vertical movement of the overlying material. The geometry of the surface of the descending slab, as defined by the top of the Benioff zone, would place the top of a large (100 km diameter, 4 km high) seamount at a depth of about 30 km at a distance of about 100 km from the trench axis. Hussong and Fryer (1981; 1984, this volume) and Hussong and Uyeda (1981) have noted evidence for pervasive tectonic movement, particularly on the outer 100 km of the fore-arc wedge. We suggest that the outer 100 km of the fore-arc is probably responding to the subduction of large (greater than 30 km) seamounts on the oceanic plate by vertical tectonic movement. By comparison, few seamounts occur on the ocean plate north of the Ogasawara Plateau. The outer portion of the Izu-Bonin fore-arc wedge shows far less evidence of the kind of vertical tectonics that are present on the Mariana fore-arc (Fryer and Hussong, 1983; Taylor and Smoot, 1984).

6. *Conclusions*

The seamount survey presented in this paper serves to illustrate the nature of the response of small seamounts on the oceanic plate to subduction processes. As a small seamount approaches a subduction zone, the oceanic plate on which it rests bends and fractures. The seamount apparently fractures on the same scale as the oceanic plate on which it rests. Very likely larger seamounts fracture less than small ones and are therefore subducted at least partially intact. The subduction of small seamounts probably has far less effect on the overriding fore-arc wedge that the subduction of larger seamounts. The latter is likely responsible, in large part, for the vertical tectonic movement seen on the fore-arc wedge of the Mariana and southern Izu-Bonin arcs.

Acknowledgements

For assistance in acquisition of this set of data we are particularly indebted to David Cuddy, Bill Robinson, and the captain, crew and scientific staff of the R/V KANA KEOKI. Karen Mansfield and Diane Hills reduced the side-scan and bathymetric data for the survey. This work was supported by the Office of Naval Research and the National Science Foundation. Hawaii Institute of Geophysics Contribution No. 1537.

REFERENCES

J. G. BLACKINTON, D. M. HUSSONG, and J. KOSALOS, First results from a combination of side-scan sonar and seafloor mapping system (SeaMARC II), Offshore Technology Conference, OTC 4478, 307–311, 1983.

S. H. BLOOMER, Distribution and origin of igneous rocks from the landward slopes of the Mariana trench: Implications for its structure and evolution, *J. Geophys. Res.*, **88** (B9), 7411–7428, 1983.

J. F. BURST, Argillaceous sediment dewatering, in *Annual Rev. of Earth and Planet. Sci.*, edited by Fred A. Donath, **4**, pp. 293–318, 1976.

P. FRYER and D. M. HUSSONG, Arc volcanism in the Mariana Trough (abstract), *Eos, Trans. Am. Geophys. Un.*, **63**, 1135, 1983.

P. FRYER and D. M. HUSSONG, Tectonic features of the northern Mariana arc, (abstract), *Eos, Trans. Am. Geophys. Un.*, **64**, 239, 1983.

P. FRYER and N. C. SMOOT, Processes of seamount subduction in the Mariana and Izu-Bonin Trenches, *Marine Geology*, **64**, 77–90, 1985.

T. W. C. HILDE, N. ISEZAKI, and J. M. WAGEMAN, Mesozoic seafloor spreading in the North Pacific, in *The Geophysics of the Pacific Ocean Basin and Its Margin*, American Geophysical Union Monograph, **19**, edited by G. H. Sutton, M. H. Manghnani, R. Moberly, and E. U. McAfee, pp. 205–226, Washington, D. C., AGU, 1976.

D. M. HUSSONG and P. FRYER, Structure and tectonics of the Mariana arc and fore-arc: Drill site selection surveys, in Hussong, D. M., S. Uyeda *et al.*, *Init. Repts. DSDP*, **60**, Washington, U.S. Govt. Printing Office, 1981, 1983, this volume.

D. M. HUSSONG and S. UYEDA, Tectonic processes and the history of the Mariana arc: A synthesis of the results of Deep Sea Drilling Project Leg 60, in Hussong, D. M., S. Uyeda *et al.*, *Init. Repts. DSDP*, **60**, Washington, U.S. Govt. Printing Office, 1981.

J. B. MINSTER and T. H. JORDAN, Present day plate motions, *J. Geophys. Res.*, **83**, 5331–5354, 1978.

E. D. PITTMAN, Recent advances in sandstone diagenesis, in *Annual Rev. of Earth and Planet. Sci.*, edited by Fred A. Donath, **7**, pp. 39–62, 1979.

N. C. SMOOT, Guyots of the Dutton ridge at the Bonin/Mariana trench juncture as shown by multi-beam surveys, *J. Geol.*, **91**, 211–220, 1983.

B. TAYLOR and N. C. SMOOT, Morphology of Bonin fore-arc submarine canyons, *Geology*, **12**(12), 724–727, 1984.

A. B. WATTS, J. H. BODINE, and M. S. STECKLER, Observations of flexure and the state of stress in the oceanic lithosphere, *J. Geophys. Res.*, **85**, B11, 6369–6376, 1980.

Formation of Active Ocean Margins, edited by N. Nasu *et al.*, pp. 307–342.

DREDGED SAMPLES FROM THE OGASAWARA FORE-ARC SEAMOUNT OR "OGASAWARA PALEOLAND"—"FORE-ARC OPHIOLITE"

Teruaki ISHII

Ocean Research Institute, University of Tokyo, Minamidai, Nakano-ku, Tokyo, 164, Japan

Abstract. Six hundred rocks including boninite, serpentinite, gabbro, basalt, and many volcaniclastic rocks, and four hundred rocks including harzburgite, dunite, gabbro, dolerite, basalt and volcaniclastic rocks have been dredged during KH80-3 and KH82-4 cruises, respectively, from the Ogasawara fore-arc seamounts which are located about 20–60 km west of the junction of the Ogasawara Trench and the Mariana Trench.

The dredged samples from the crest (about 1100 m deep) contain many subrounded to rounded gravels of igneous rocks including relatively fresh basaltic, doleritic and gabbroic rocks. The rounded gravels might be products of, in situ, inshore wave erosion, that is, the present Ogasawara fore-arc seamount was exposed above the sea surface. This land is called "Ogasawara paleoland".

Observation of more than three hundred polished thin sections from the paleoland have revealed the followings; (1) harzburgite is most predominant and dunite is subordinate in the ultramafic rocks, (2) lherzolite is not observed, (3) metagabbro was derived from (two-) pyroxene gabbro, and (4) some pyroxene gabbro contain inverted pigeonite.

The above observations suggest that these rocks may have been composed of an ophiolite body and that this ophiolite was formed by fore-arc volcanism, and should be called "fore-arc ophiolite".

1. Introduction

Many mafic and ultramafic igneous complexes are distributed throughout Japanese islands and are regarded as "ophiolites", for example, Yakuno in Maizuru belt (ISHIWATARI, 1978), Horokanai in Kamuikotan belt (ISHIZUKA, 1981), Setogawa (OHASHI and SHIRAKI, 1981), Miyamori in Kitakami mountains (OZAWA, 1983) and Mineoka (TONOUCHI and KOBAYASHI, 1983; OGAWA, 1983). Ophiolite is generally considered to be oceanic lithosphere originated at a midoceanic ridge and tectonically emplaced along plate boundaries (COLEMAN, 1977). On the other hand, MIYASHIRO (1973, 1975) discussed chemical diversity of ophiolitic volcanics and suspected oceanic ridge origin of some ophiolites.

Many dredged igneous rocks with ophiolitic assemblages have been reported from fore-arc region of Mariana Trench (HONZA and KAGAMI, 1977; IGCP WORKING GROUP, 1977; HAWKINS *et al.*, 1979; BLOOMER, 1983; BLOOMER and HAWKINS, 1983). On the other hand, there had been no report on the ophiolitic rocks along the Ogasawara Trench.

A new type of ophiolite located in the fore-arc region at Ogasawara area will be reported. The origin of this ophiolite will be discussed on the basis of geological and geophysical observations.

2. *Sample Locations*

A large topographic high is observed in the Ogasawara fore-arc region (Fig. 1). It is located about 20 km west of the junction of the Ogasawara Trench and the Mariana Trench (Fig. 2a) and about 110 km southeast by east from the Ogasawara Islands. This topographic high is 30 km×60 km in diameter of the base and 1800 m in height from the base. It will be called Ogasawara Paleoland (OPL) in this paper (the reason will be mentioned later). The crest is 1050 m deep. A very small topographic high is also observed about 20 km west of the paleoland. The dimension of the small high is 5 km×7 km in diameter at the base and about 600 m in height from the base. This high will be called a submarine volcano (SMV) in this paper.

About one thousand rock samples have been dredged from the above-mentioned fore-arc topographic highs (Fig. 2) during the Hakuhō Maru Cruise & KH80-3 and KH82-4 of the Ocean Research Institute, University of Tokyo. Three sites (Station No. 32, 33 and 34) and two sites (Station No. 3 and 4) were selected for the investigations of dredge hauls in the KH80-3 and KH82-4 cruises, respectively. Station No. 33 is from the small seamount (SMV), whereas others are from the large one (OPL).

Precise position, depth of each station (Table 1) and other information are given in the operation logs of dredge hauls in the Preliminary Report of the Hakuhō Maru Cruise & KH80-3 (KOBAYASHI, 1981) and KH82-4 (KOBAYASHI, 1983), for stations 32+33+34 and 3+4, respectively.

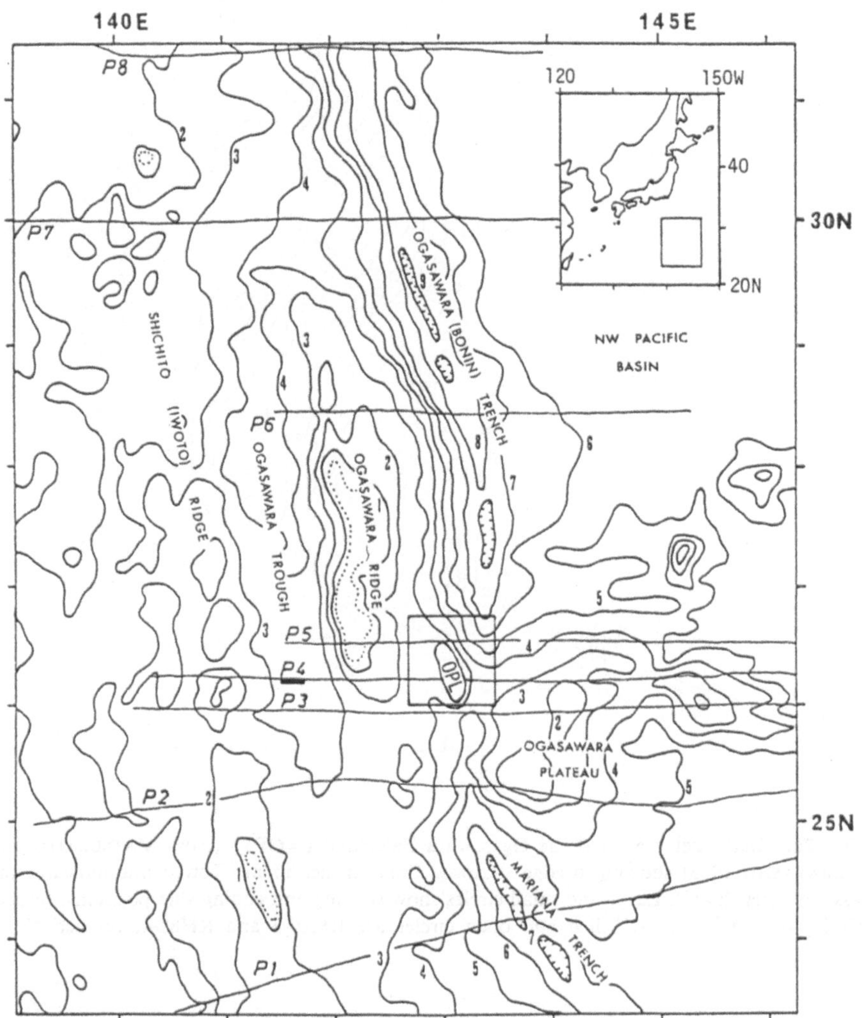

Fig. 1. Bathymetry around the Ogasawara (=Bonin) Ridge (from MATSUMOTO and TOMODA, 1982). Contourlines in Km. Inside of square and line P4 are in Figs. 2a and 10, respectively. Ogasawara Paleoland (=OPL) is also shown.

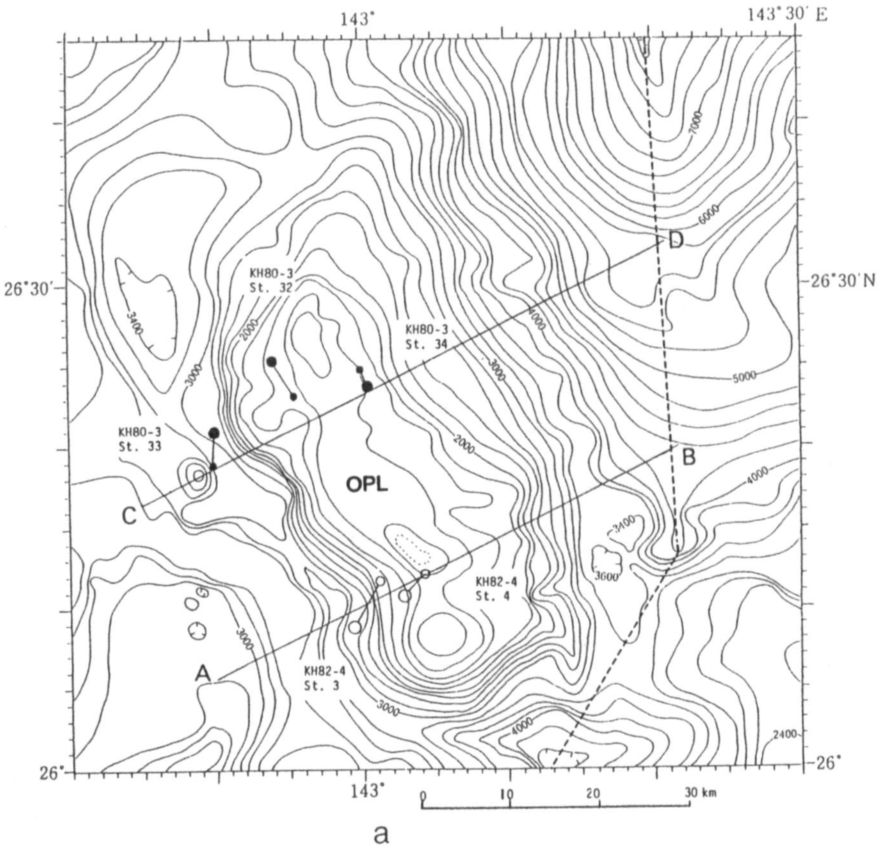

Fig. 2a. Bathymetry around the Ogasawara Paleoland (=OPL) (from MATSUMOTO and KOBAYASHI, 1983) showing dredge stations. Contourlines in m. Dotted line indicates the assumed trench axis. Large and small circles show starting and ending ship positions, respectively, of dredge hauls. Filled and open circles are KH80-3 and KH82-4, respectively.

Nalwalk chain-bag dredges with small pipe-dredge (8 liter capacity) were operated to collect rock samples. Because most of the dredged rock-samples are more or less covered with soft sediment and/or Mn-coating, these rocks were at first separated from soft sediments, and cut into two or more pieces for observation and description of visual features inside each sample. Washed samples were classified into several groups according to lithological characteristics. After numbering the samples (in the

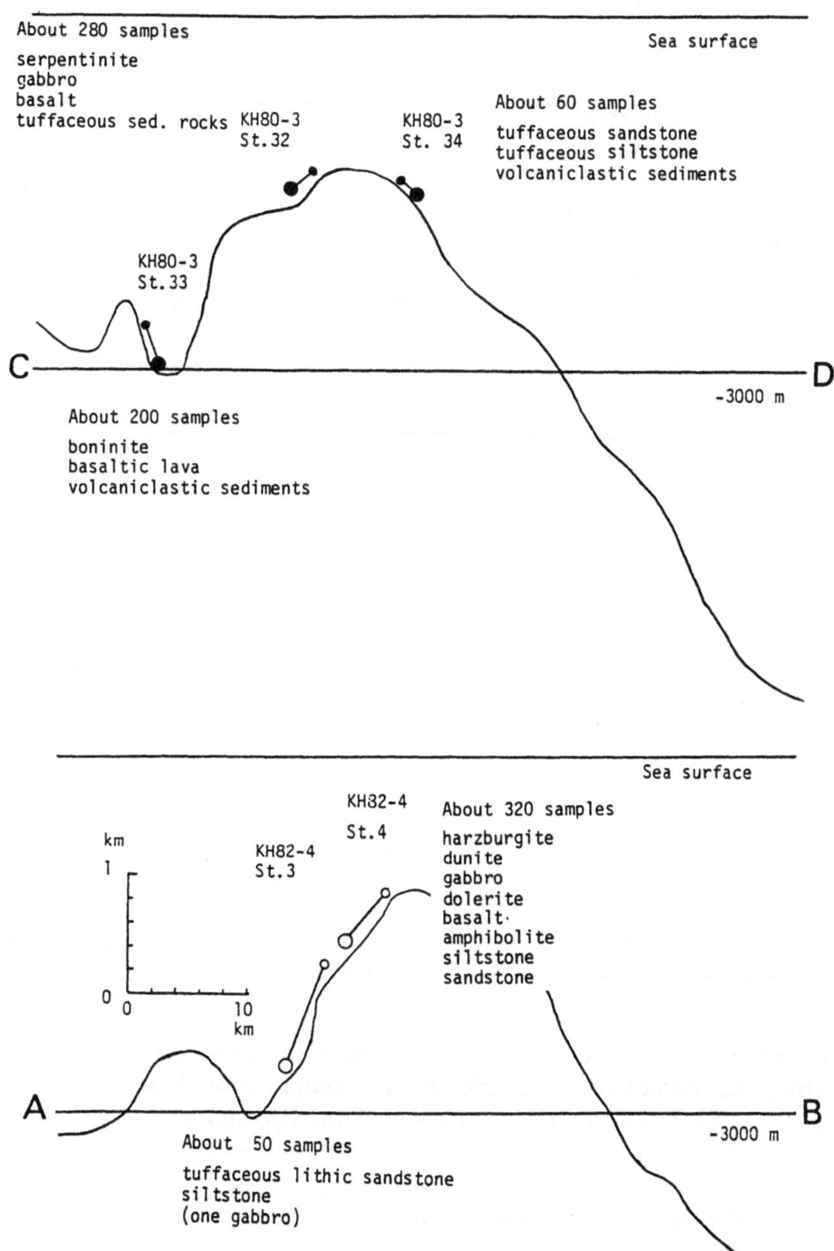

Fig. 2b. Topographic cross sections along AB and CD in Fig. 2a, on which dredged samples and stations are also shown. Symbols are the same as in Fig. 2a.

Table 1a. Lithologic distribution of dredged samples of KH80-3.

Station	Lithology	Number of samples(%)		Weight of samples(%)	
	harzburgite	13	(4.7)	6480	(13.0)
	gabbro	6	(2.2)	65	(0.1)
KH80-3-32	basic lava	14	(5.0)	4845	(9.7)
26°24.9'N	tuff-breccia	16	(5.8)	2900	(5.8)
142°54.0'E	tuffacious sandstone	114	(41.0)	28870	(57.8)
1510m deep	scoria	44	(15.8)	3245	(6.5)
	pumice	71	(25.5)	3525	(7.1)
	Total	278	(100.0)	49930 g	(100.0)
	basaltic lava	155	(78.3)	37845	(82.1)
KH80-3-33	boninite	1	(0.5)	850	(1.8)
26°20.9'N	tuff-breccia	13	(6.6)	2880	(6.2)
142°49.8'E	tuff-lapillituff	28	(14.1)	4575	(9.9)
3120m deep	scoria	1	(0.5)	5	(0.01)
	Total	198	(100.0)	46155	(100.0)
	altered basalt lava	2	(3.5)	445	(0.9)
	tuffaceous siltstone	6	(10.5)	15550	(31.6)
KH80-3-34	tuffaceous sandstone	37	(64.9)	30630	(62.3)
26°24.3'N	tuff	10	(17.5)	2340	(4.8)
143°00.1'E	scoria	1	(1.8)	155	(0.3)
1500m deep	pumice	1	(1.8)	30	(0.1)
	Total	57	(100.0)	49150	(100.0)

order of size), roundness (POWERS, 1953), weight, and thickness of Mn-coating, the lithology and remarks of each sample were observed on board and described in the Preliminary Report of the Hakuhō Maru Cruise (ISHII *et al.*, 1981; ISHII *et al.*, 1983).

3. *Description of Rocks and Chemical Analyses*

Lithologic descriptions of each dredged sample were given in the cruise report of the R/V Hakuhō Maru (ISHII *et al.*, 1981; ISHII *et al.*, 1983).

Table 1b. Lithologic distribution of dredged samples of KH82-4.

Station	Lithology	Number of samples(%)		Weight of samples(%)		Density [g/cm^3]
	gabbro	1	(2.0)	150 g	(0.4)	-
KH82-4-3	tuffaceous sandstone	36	(73.5)	29932	(83.9)	-
26°08.8'N	siltstone	9	(18.4)	987	(2.8)	-
142°59.2'E	claystone	1	(2.0)	4550	(12.5)	-
2730m	scoria	2	(4.1)	151	(0.4)	-
	Total	49	(100.0)	35670 g	(100.0)	-
	harzburgite	138	(42.8)	45018 g	(37.2)	2.43
	dunite	35	(10.9)	9738	(8.0)	2.31
KH82-4-4	gabbro	35	(10.9)	26965	(22.2)	2.85
26°10.8'N	dolerite	16	(5.0)	9698	(8.0)	2.76
143°02.9'E	basalt	10	(3.1)	3800	(3.1)	2.85
1630m	amphibolite	5	(1.6)	1647	(1.4)	2.94
	orthopyroxenite	1	(0.3)	191	(0.2)	3.00
	siltstone	25	(7.8)	6091	(5.0)	-
	sandstone	19	(5.9)	10440	(8.6)	-
	monomict breccia	20	(6.3)	4986	(4.1)	-
	polymict breccia	8	(2.5)	1846	(1.5)	-
	scoria	7	(2.9)	551	(0.5)	-
	pumice	2	(0.6)	297	(0.2)	-
	Total	321	(100.0)	121268 g	(100.0)	-

The compiled lithologic distributions of those samples are shown in Table 1. Representative rocks of each station are also shown in the topographic cross sections (Fig. 2b) along A-B and C-D in Fig. 2a. Scoria and pumice are excluded in Fig. 2b, because they were dredged from all stations. Wet chemical analyses of these rocks and a Mn-crust were made by H. Haramura and are shown in Tables 2a through 2c. Rare earth elements (REE) of bulk rocks were analysed by Y. Minai with neutronactivation method (MINAI, 1982) and are also shown in Fig. 3.

Chemical compositions of rock-forming minerals were determined with a JEOL electron probe X-ray microanalyzer (EPXMA) Model JCXA-733 (Ocean Research Institute, University of Tokyo). The analytic method is presented by OTSUKI (1983). Analysed minerals are pyroxene, olivine,

Table 2a. The major-element analyses of ultramafic rocks from the Ogasawara Paleoland (Analyst: H. Haramura).

	KH80-3-32		KH82-4-4			
	Harzburgite	Harzburgite	Harzburgite	Dunite	Harzburgite	Dunite
Sampl. No.	KH80-3 32-001	KH80-3 32-002	KH82-4 04-001	KH82-4 04-023	KH82-4 04-219	KH82-4 04-105
Anal. No.	1	2	3	4	5	6
SIO2	37.72	38.06	41.14	37.52	40.53	36.56
TIO2	0.03	0.04	0.02	0.05	0.07	0.05
AL2O3	0.50	0.29	0.10	0.93	0.10	0.10
FE2O3	7.31	8.57	6.60	8.19	6.13	8.79
FEO	0.81	0.73	1.57	1.53	3.02	1.75
MNO	0.07	0.09	0.08	0.08	0.08	0.08
MGO	33.05	35.32	37.14	36.74	40.01	36.88
CAO	0.07	0.05	0.17	0.10	0.72	0.02
NA2O	0.13	0.07	0.15	0.16	0.15	0.28
K2O	0.02	0.02	0.02	0.02	0.02	0.02
P2O5	0.06	0.07	0.05	0.04	0.05	0.06
NIO	0.21	0.23	0.28	0.22	0.31	0.24
CR2O3	0.27	0.28	0.18	0.60	0.31	1.31
H2O+	11.83	12.57	11.23	12.77	8.03	11.95
H2O-	7.95	3.82	1.37	1.00	0.70	1.80
Total	100.03	100.21	100.10	99.95	100.23	99.89

Table 2b. The major-element analyses of basic rocks from the Ogasawara Paleoland area (Analyst: H. Haramura).

	KH80-3-32				KH80-3-33				KH82-4-04	
	Basalt	Basalt	Basalt	T-Andesite	Boninite	Basalt	Basalt	B-Andesite	Gabbro	Gabbro
Sample No.	KH80-3 32-010	KH80-3 32-012	KH80-3 32-013	KH80-3 32-202	KH80-3 33-004	KH80-3 33-001	KH80-3 33-005	KH80-3 33-060	KH82-4 04-051	KH82-4 04-302
Anal. No.	1	2	3	4	5	6	7	8	9	10
SIO2	48.39	48.45	49.39	58.46	52.55	47.51	49.14	52.87	47.29	43.16
TIO2	1.24	1.52	1.39	1.05	0.25	0.78	0.77	0.78	0.24	0.25
AL2O3	13.22	12.22	12.89	15.98	12.34	15.45	15.97	15.42	19.24	23.81
FE2O3	5.76	5.57	6.66	3.17	3.79	6.39	6.30	4.19	3.11	2.75
FEO	7.98	8.75	8.07	4.66	4.05	3.95	3.66	5.53	2.85	2.31
MNO	0.26	0.26	0.22	0.24	0.16	0.16	0.12	0.13	0.11	0.06
MGO	7.17	7.40	7.23	1.81	11.12	7.78	7.23	6.81	8.61	7.61
CAO	9.01	9.29	10.41	4.04	8.90	8.43	8.80	6.48	15.16	11.34
NA2O	2.86	2.69	2.25	6.03	1.51	3.26	2.88	3.51	1.30	2.30
K2O	0.52	0.56	0.05	3.75	0.37	1.15	0.42	0.36	0.07	0.23
P2O5	0.09	0.09	0.13	0.55	0.03	0.08	0.06	0.06	0.03	0.03
H2O+	2.51	2.68	1.04	0.24	3.93	3.39	2.22	2.71	1.62	4.99
H2O-	0.70	0.90	0.03	0.0	0.85	2.24	2.91	1.32	0.22	1.05
Total	99.71	100.38	99.76	99.98	99.85	100.57	100.48	100.17	99.85	99.89
Ni(ppm)	82	68	78	17	328	44	48	53	13	204
Cr(ppm)	66	78	73	3	744	35	63	39	513	23
CIPW NORM										
Q	1.11	1.31	5.68	0.0	9.26	0.0	4.30	6.29	0.0	0.0
OR	3.18	3.42	0.30	22.22	2.30	7.16	2.60	2.21	0.42	1.45
AB	25.08	23.51	19.29	51.15	13.44	29.05	25.56	30.89	11.22	18.10
AN	22.49	20.26	25.25	5.48	27.14	25.41	30.84	26.27	47.40	57.50
NE	0.0	0.0	0.0	0.0	0.0	0.0	0.0	0.0	0.0	1.43
WO	9.70	11.16	10.94	4.60	7.97	7.55	6.07	2.82	12.16	0.93
EN	6.18	6.94	7.20	2.23	6.18	6.28	5.14	1.95	9.67	0.76
FS	2.89	3.56	2.97	2.30	0.94	0.32	0.13	0.65	1.11	0.07
EN	12.32	12.10	11.05	1.34	22.95	7.27	13.74	15.69	9.93	0.0
FS	5.75	6.19	4.56	1.38	3.47	0.37	0.36	5.23	1.14	0.0
FO	0.0	0.0	0.0	0.67	0.0	4.81	0.0	0.0	1.60	13.62
FA	0.0	0.0	0.0	0.76	0.0	0.27	0.0	0.0	0.20	1.32
MT	8.65	8.34	9.78	4.61	5.78	9.76	9.58	6.32	4.60	4.25
IL	2.44	2.98	2.68	2.00	0.50	1.56	1.53	1.54	0.47	0.51
AP	0.22	0.22	0.31	1.28	0.07	0.20	0.15	0.14	0.07	0.07

Abbreviations : T-Andesite = trachyandesite, B-Andesite = basaltic andesite.

Table 2c. The chemical analyses of Mn-crust from the Ogasawara Paleoland (Analyst: H. Haramura).

Sample No.	KH80-3 32-001M
Anal. No.	1
SiO_2	14.44
TiO_2	1.21
Al_2O_3	5.00
Fe_2O_3	21.74
FeO	< 0.03
Mn_2O_3	0.87
MnO_2	17.93
MgO	1.53
CaO	2.60
Na_2O	2.01
K_2O	0.51
$H_2O(-)$	20.99
$H_2O(+)$	10.22
P_2O_5	0.68
NiO	0.23
Cr_2O_3	0.002_2
Co_2O_3	0.23
CuO	0.03_5
ZnO	0.07_0
PbO	0.15
Total	100.47

chromian-spinel, plagioclase, titanomagnetite, hornblende, glass as well as metamorphic minerals; selected results are presented in Tables 3 through 5 and are shown in Figs. 4 and 5.

4. Rocks from the Large Seamount or Ogasawara Paleoland (OPL)

4.1 KH82-4-4

The station KH82-4-4 is located on the southwest crest of the paleoland

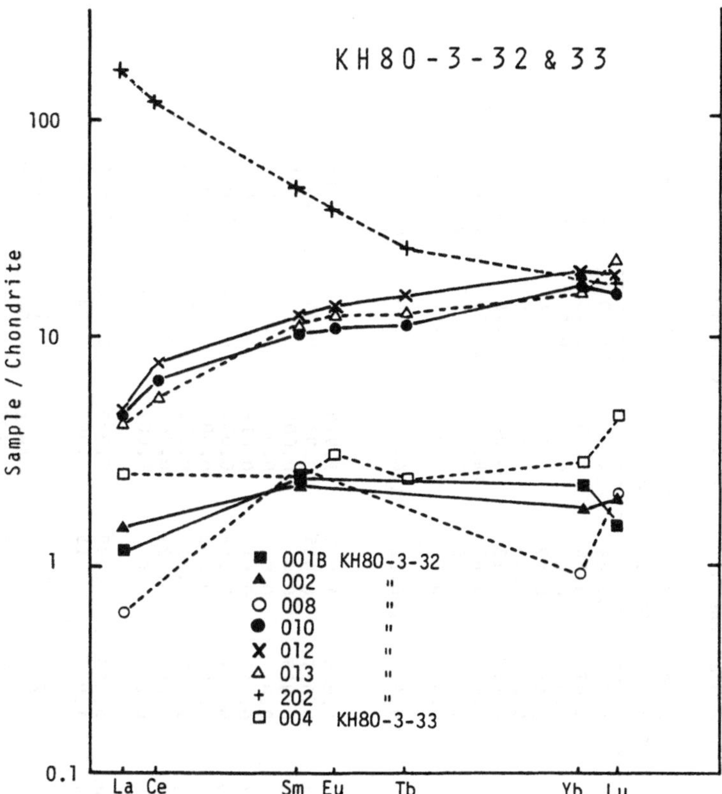

Fig. 3. Chondrite normalized rare earth element abundances for representative dredged rocks of KH80-3-32 and KH80-3-33 (Y. Minai, unpublished data). Symbols: solid square-001B, solid triangle-002, open circle-008, solid circle-010, cross-012, open triangle-013 and plus-202 of KH80-3-32, and open square-004 of KH80-3-33.

at a water depth of 1630 to 1200 meters. Rock types and their frequencies in the dredged samples are summarized in Fig. 6. They are plenty gravels and cobbles of serpentinized ultramafic rocks (harzburgite, dunite and orthopyroxenite), mafic rocks (pyroxene gabbro, allivalite, dolerite and basalt), a few metamorphic rocks (amphibolite), sedimentary rocks (tuffaceous siltstone+sandstone and tuffaceous monomictbreccia+polymictbreccia) and pyroclast (scoria and pumice). Serpentine minerals in ultramafic rocks are mainly lizardite and chrysotile, and a trace amount of antigorite (UEHARA and NAKA, 1983). Harzburgite is the most predomi-

Table 3a. Selected analyses of pyroxenes in ultramafic rocks from the Ogasawara Paleoland.

	Harzburgite						
	KH82-4-4-001			KH82-4-4-004		KH82-4-4-101	
Sample No.	Enstatite	Enstatite	Enstatite	Enstatite	Diopside **	Enstatite	Diopside**
Anal. No.	OPX4-0 01-066	OPX4-0 01-084	OPX4-0 01-085	OPX4-0 04-012	AUG4-0 04-011	OPX4-1 01-083	AUG4-1 01-084
SIO2	56.91	57.06	56.71	57.48	54.53	56.30	54.50
AL2O3	1.45	1.50	1.47	0.88	0.90	1.13	0.92
TIO2	0.02	0.0	0.0	0.0	0.0	0.0	0.03
CR2O3	0.63	0.69	0.64	0.39	0.62	0.44	0.53
FEO *	4.97	4.92	5.21	5.53	1.76	5.59	1.74
MNO	0.14	0.08	0.15	0.12	0.04	0.16	0.02
MGO	34.52	34.16	34.01	35.67	18.11	35.25	18.03
CAO	1.25	1.32	1.30	0.75	24.45	0.61	24.75
NA2O	0.02	0.0	0.0	0.0	0.03	0.02	0.16
TOTAL	99.91	99.73	99.49	100.82	100.44	99.50	100.68
O = 6.000							
SI	1.960	1.966	1.963	1.963	1.970	1.951	1.966
AL	0.040	0.034	0.037	0.035	0.030	0.046	0.034
AL	0.018	0.027	0.023	0.0	0.008	0.0	0.005
TI	0.001	0.0	0.0	0.0	0.0	0.0	0.001
CR	0.017	0.019	0.018	0.011	0.018	0.012	0.015
FE	0.143	0.142	0.151	0.158	0.053	0.162	0.052
MN	0.004	0.002	0.004	0.003	0.001	0.005	0.001
MG	1.772	1.755	1.755	1.816	0.975	1.821	0.969
CA	0.046	0.049	0.048	0.027	0.946	0.023	0.957
NA	0.001	0.0	0.0	0.0	0.002	0.001	0.011
Z	2.000	2.000	2.000	1.999	2.000	1.997	2.000
WXY	2.003	1.994	1.998	2.015	2.003	2.023	2.012
CA	2.4	2.5	2.5	1.4	47.9	1.1	48.3
MG	90.3	90.2	89.8	90.7	49.4	90.8	49.0
FE	7.3	7.3	7.7	7.9	2.7	8.1	2.7
MG/MG+FE	0.925	0.925	0.921	0.920	0.948	0.918	0.949

* Total Fe as FeO
** Clot of diopside

Table 3b. Selected analyses of pyroxenes in gabbro from the Ogasawara Paleoland.

	Aug-opx gabbro		Augite-pigeonite gabbro				
	KH80-3-32-305B		KH82-4-4-309				
Sample No.	Augite	Bronzite	Augite	Pigeonite **	Pigeonite **		
			Bulk	Bulk	Bulk	Host	Lamella
Anal. No.	AUG32-305B-4	OPX32-305B-7	AUG4-309-031	PIG4-309-029	PIG4-309-057	OPX4-309-033	AUG4-309-034
SIO2	50.93	52.58	53.01	54.21	55.18	53.81	52.47
AL2O3	2.63	1.34	1.99	1.07	1.14	1.14	2.07
TIO2	0.80	0.49	0.71	0.47	0.39	0.45	0.89
CR2O3	0.17	0.02	0.05	0.03	0.02	0.03	0.06
FEO*	7.69	16.37	5.64	11.80	11.65	12.70	5.72
MNO	0.22	0.38	0.18	0.31	0.30	0.29	0.18
MGO	14.61	25.53	16.68	28.25	27.63	29.02	16.27
CAO	21.46	1.62	21.53	2.56	3.72	1.13	22.27
NA2O	0.35	0.03	0.21	0.01	0.04	0.0	0.17
TOTAL	98.86	98.36	100.00	98.71	100.07	98.57	100.10
O = 6.000							
SI	1.914	1.946	1.943	1.959	1.968	1.949	1.929
AL	0.086	0.054	0.057	0.041	0.032	0.049	0.071
AL	0.031	0.004	0.029	0.004	0.016	0.0	0.018
TI	0.023	0.014	0.020	0.013	0.010	0.012	0.025
CR	0.005	0.001	0.001	0.001	0.001	0.001	0.002
FE	0.242	0.507	0.173	0.357	0.348	0.385	0.176
MN	0.007	0.012	0.006	0.009	0.009	0.009	0.006
MG	0.818	1.408	0.911	1.522	1.469	1.566	0.891
CA	0.864	0.064	0.846	0.099	0.142	0.044	0.877
NA	0.026	0.002	0.015	0.001	0.003	0.0	0.012
Z	2.000	2.000	2.000	2.000	2.000	1.997	2.000
WXY	2.015	2.012	2.001	2.006	1.998	2.017	2.007
CA	44.9	3.2	43.8	5.0	7.3	2.2	45.1
MG	42.5	71.2	47.2	77.0	75.0	78.5	45.8
FE	12.6	25.6	9.0	18.0	17.7	19.3	9.0
MG/MG+FE	0.772	0.735	0.841	0.810	0.809	0.803	0.835

* Total Fe as FeO
** Inverted pigeonite

Table 4. Selected analyses of olivine in ultramafic rocks from the Ogasawara Paleoland.

	Harzburgite			Dunite	
Sample No.	KH82-4-4-001			KH82-4-4-105	
Anal. No.	53	59	82	14	15
SiO_2	40.98	40.98	40.77	40.53	40.43
FeO*	7.52	7.39	7.58	7.82	8.01
MnO	0.12	0.11	0.08	0.12	0.11
MgO	50.56	50.15	51.41	50.59	50.13
CaO	0.04	0.03	0.01	0.24	0.23
NiO	0.39	0.39	0.47	0.39	0.37
Total	99.61	99.05	100.33	99.69	99.28
O = 4.000					
Si	0.999	1.003	0.988	0.990	0.992
Fe	0.153	0.151	0.154	0.160	0.164
Mn	0.002	0.002	0.002	0.002	0.002
Mg	1.837	1.830	1.857	1.842	1.834
Ca	0.001	0.001	0.000	0.006	0.006
Ni	0.008	0.008	0.009	0.008	0.007
Total	3.000	2.994	3.010	3.009	3.006
Mg/(Mg+Fe)	0.923	0.923	0.924	0.920	0.918

* Total Fe as FeO

nant among the dredged ultramafic rocks, and dunite is subordinate.

The assemblage of the dredged igneous rocks is the same as that of so called ophiolites. Some harzburgites show marked foliation and lineation, which are common in the ultramafic tectonite (COLEMAN, 1977), but most of the harzburgite do not show porphyroclastic textures.

Primary minerals of serpentinized harzburgite are olivine, enstatite, chromian spinel and diopside. Diopside occurs sparsely at boundaries between enstatite crystals as a clot. It may be exsolution product from enstatite under subsolidus condition. Primary minerals of serpentinized dunite are olivine and chromian spinel. Major chemistries of relatively fresh samples of the Cr-spinel bearing harzburgite and dunite are shown in Table 2a. Selected chemical compositions of pyroxenes, olivine and chromian spinel are given in Tables 3a, 4 and 5. Both harzburgite and dunite are high in MgO content (Table 2a); their spinel is high in Cr content (Fig. 5a) and olivine and pyroxenes are high in MgO contents (Fig. 4). Spinel in dunite is more chrome-rich ($Cr/Cr+Al>0.61$) than that in harzburgite ($Cr/Cr+Al>0.55$), but the olivines have similar $Mg/Mg+Fe$ ratios of 0.92. The analyzed spinels from the paleoland are so Cr-rich that they are plotted

Table 5. Selected analyses of spinel in ultramafic rocks from the Ogasawara Paleoland.

	Harzburgite						Dunite	
Sample No.	KH80-3-32-001			KH82-4-4-001			KH82-4-4-105	
Anal. No.	2	61	134	63	77	78	10	12
Al2O3	21.15	19.08	19.38	16.93	17.01	16.48	20.02	19.81
TiO2	–	0.00	–	0.02	–	0.02	0.12	0.13
FeO*	18.46	18.15	17.49	21.24	19.48	19.36	18.35	18.28
MnO	0.22	0.23	0.17	0.36	0.27	0.28	0.18	0.22
MgO	11.63	11.97	12.90	10.80	11.28	11.26	12.90	12.84
NiO	0.07	0.08	0.07	0.12	0.04	0.09	0.08	0.08
Cr2O3	47.37	50.00	48.79	48.97	51.76	50.74	45.65	46.25
V 2O3	0.08	0.09	0.11	0.08	0.06	0.05	0.18	0.17
Total	98.97	99.60	97.90	98.52	99.89	98.29	97.49	97.78
O=32,000								
Al	6.296	5.692	5.774	5.217	5.134	5.068	6.050	5.977
Ti	—	0.000	—	0.004	—	0.004	0.023	0.024
Fe^{2+}	3 .596	3.462	2. 866	3. 280	3. 686	3. 612	3. 076	3. 102
Fe^{3+}	0 .302	0.377	0. 795	0. 381	0. 479	0. 603	0. 797	0. 756
Mn	0.047	0.049	0.037	0.079	0.058	0.062	0.038	0.047
Mg	4.377	4.518	4.880	4.211	4.301	4.377	4.931	4.901
Ni	0.014	0.015	0.014	0.025	0.009	0.020	0.017	0.017
Cr	9.460	10.006	9.752	10.121	10.482	10.463	9.253	9.360
V	0.016	0.018	0.022	0.018	0.013	0.011	0.037	0.036
Cr/(Cr+Al)	0.600	0.637	0.628	0. 660	0. 671	0. 682	0.568	0. 610
$Mg/(Mg+Fe^{2+})$	0.549	0.566	0.629	0. 563	0. 539	0. 548	0.616	0. 612

* Total Fe as FeO

outside the abyssal spinel peridotite field and inside the Alpine-type dunite-harzburgite field in $Mg/Mg+Fe^{2+}$ vs. $Cr/Cr+Al$ diagram (Fig. 5a) and Cr-Al-Fe diagram (Fig. 5b).

Augite-bronzite gabbro (two-pyroxene gabbro) including olivine is the most predominant in the gabbroic rocks and augite-gabbro, which is free from orthopyroxene or pigeonite, is subordinate. Some two-pyroxene gabbros contain inverted pigeonite of both Stillwater and Kintoki-san type (Ishii and Takeda, 1974) as a Ca-poor pyroxene phase. The minimum temperature 1180°C of the magma of this pigeonite bearing gabbro is estimated for the most magnesian pigeonite (Pig4-309-057 in Table 3b) by pigeonite geothermometer (Ishii, 1975). The estimated temperature is higher than that of island arc magma (i.e., up to 1125–1155°C (Ishii,

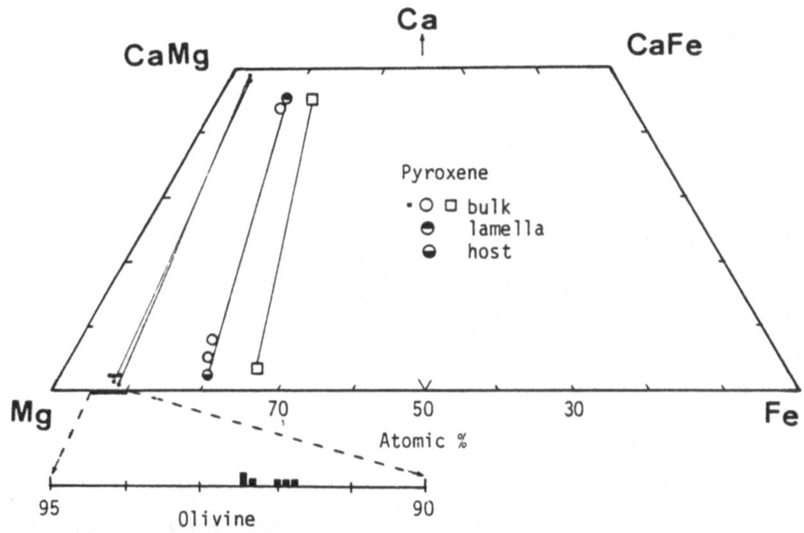

Fig. 4. Ca-Mg-Fe and Mg-Fe plots of analysed pyroxenes and olivines, respectively, for dredged rocks of KH80-3-32 and KH82-4-4. Symbols: small solid square-pyroxene in harzburgite, open circle and open square-bulk composition of pyroxene in gabbro, half solid circle-host (bronzite) and lamella (augite) of inverted pigeonite of gabbro, large solid square-olivine in harzburgite and dunite, and solid line-tie line.

1981)). Ca-poor pyroxene (orthopyroxene or pigeonite) bearing gabbro is distinct from the gabbro dredged from oceanic fracture zones, where Ca-poor pyroxene bearing gabbro is scarce.

Olivine normative rocks are the most predominant in the basic rocks including gabbro, dolerite and basalt, and quartz normative ones are subordinate (Table 2b, also see NAKA and UEHARA, 1973). Nepheline normative basic rocks are rare, which may be fragments of oceanic seamount (BLOOMER, 1983).

Metamorphic and/or secondary minerals observed in the basic rocks, are chlorite, actinolite, green hornblende, white mica, talc, prehnite, zeolites (analcime and natrolite) and epidote (clinozoisite) (NAKA and UEHARA, 1973).

4.2 KH80-3-32

The station KH80-3-32 is located on the northwest crest of the paleoland at a water depth of 1510 to 1450 meters. The dredged samples are gravels and cobbles of volcaniclastic sedimentary rocks (tuffaceous sil-

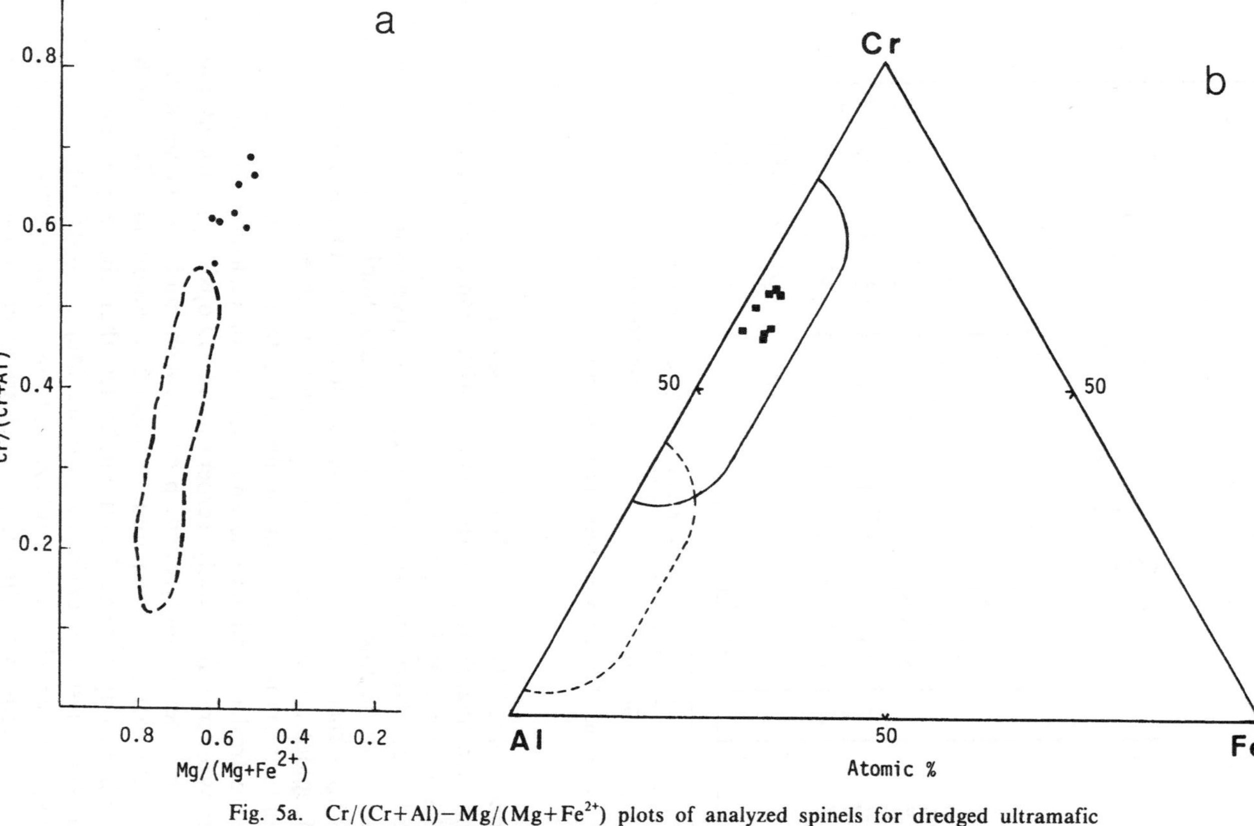

Fig. 5a. $Cr/(Cr+Al)-Mg/(Mg+Fe^{2+})$ plots of analyzed spinels for dredged ultramafic rocks of KH80-3-32 and KH82-4-4. Solid line indicates the field for abyssal spinel peridotites (Dick and Bullen, 1984).

Fig. 5b. Cr-Al-Fe^{3+} plots of the analysed spinels. Solid and dotted lines indicate the fields for spinels from Alpine-type dunite-harzuburgite and from Alpine-type lherzolites and lherzolite nodules in basaltic rocks, respectively (Arai and Fujii, 1979).

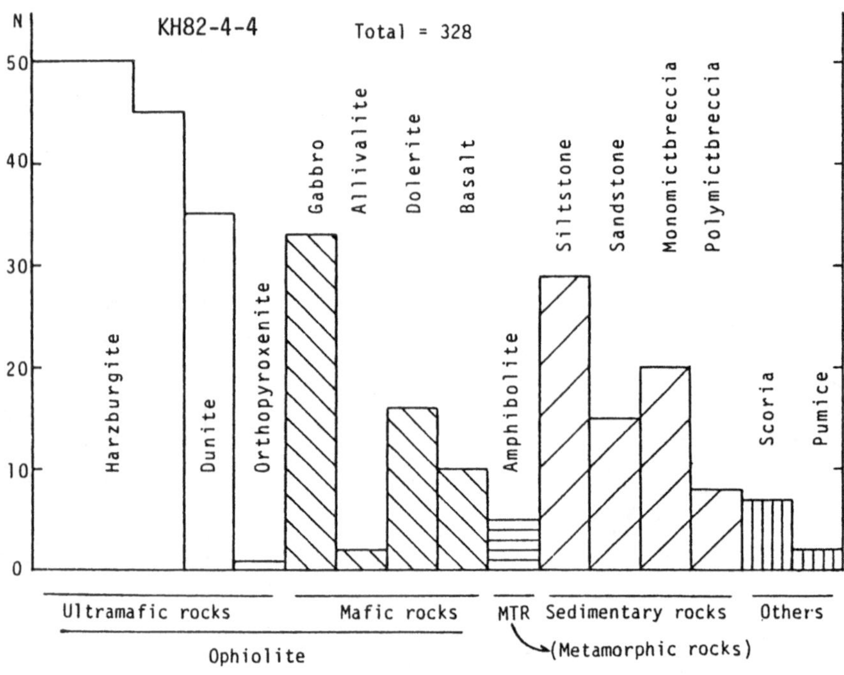

Fig. 6. Frequency of rock species dredged at KH82-4-4.

ty sandstone and tuff breccia), relatively fresh basic rocks (gabbro, basaltic rocks and trachyandesite), serpentinites with Mn-crust, and fresh pyroclastics (pumice and scoria).

Euhedral green-hornblende and/or Ti-augite crystals are common in the volcaniclastic sediments as well as scoria and pumice. Some of the sediments are serpentine sandstone. One cobble of metamorphic rock consisting of epidote, green-hornblende and albite is observed.

Serpentinites consist of serpentine minerals after olivine and fastific orthopyroxene, fresh chromian spinel and small clot of fresh diopside. Olivine and orthopyroxene are totally serpentinized. Diopside occurs at the boundaries between orthopyroxene pseudomorphs and may be secondly exsolved from orthopyroxene under subsolidus conditions. The mesh texture with orthopyroxene bastite indicates that the serpentinite is originated from Cr-spinel bearing harzburgite. Major and REE chemistries of the harzburgite are shown in Table 2a and Fig. 3, respectively. Selected chemical compositions of chromian spinel are given in Table 5 and are plotted in Fig. 5. The serpentinized harzburgite is high in MgO

(MgO/MgO+FeO=0.88) and low in REE contents, and spinels are high in Cr content (Cr/Cr+Al>0.60). Bulk compositions of the strongly serpentinized harzburgite and spinel compositions from KH80-3-32 are similar to those from KH82-4-4.

Three relatively fresh basaltic cobbles were selected for analyses of the major elements and REE, which results are shown in Table 2b and Fig. 3, respectively. These rocks consist of phenocrysts and microphenocrysts of plagioclase and augite with an intersertal groundmass of plagioclase, augite, titanomagnetite and devitrified dark brown glass. The plagioclase-augite phyric basalts are very similar in major elements and REE chemistries, but not in potassium and water contents; variations may be due to difference of weathering conditions. It is recognized that low Al_2O_3 and Cr content, low concentration of light REE (i.e., La and Ce) in comparison with those of abyssal tholeiites, and that the chemistries are similar to those of primitive island-arc tholeiite (ISHII, 1981).

A few gabbroic cobbles consist of plagioclase, magnesian diopside, bronzite and opaque minerals. Chemistries of these pyroxenes are shown in Table 3b and Fig. 4.

One gravel (about 1500 g weight) of very fresh basic lava with a thick scoriaceous crust was analyzed (Table 2b and Fig. 3). Its major element chemistry is similar to that of trachyandesite from Iwō-zima (KUNO, 1962). The REE distribution of the trachyandesite shows those of typical alkali-rocks (Fig. 3). It is expected that the provenance of the trachyandesite is very closely located to the paleoland.

All dredged samples from the station KH80-3-32 are more or less covered with Mn-crust (up to 7 mm thick). One Mn-crust (about 5 mm thick) coating harzburgite (KH80-3-32-001), was selected for the wet chemical analyses of major and minor elements (Table 2c). The Mn-crust is classified into hydrogenous origin according to the Bonatti's method (BONATTI *et al.*, 1972).

4.3 KH80-3-34 and KH82-4-3

The station KH80-3-34 and KH82-4-3 are located on the northeast crest and southwest flank of the paleoland, respectively. The dredged samples from the two sites are gravels and cobbles of volcaniclastic sedimentary rocks (tuffaceous sandstone, tuffaceous siltstone and tuff) and a few igneous rocks (altered basalt, gabbro, scoria and pumice). The lithologic characteristics of the sedimentary and igneous rocks are similar to those of samples from the stations KH82-4-4 and KH80-3-32.

5. *Rocks from the Small Seamount or Submarine Volcano (SMV)*

5.1 KH80-3-33

The station KH80-3-33 is located on the northeast crest of the small topographic high at a water depth of about 3150 meters. The dredged samples are gravels and cobbles of moderately altered angular to subangular basalts (pillow lavas), pyroclastic rocks (tuff, lapilli-tuff and tuff-breccia) and one boninite.

Basalts have phenocrysts and microphenocrysts of plagioclase, augite and bronzite in altered intersertal groundmass of plagioclase, augite, titanomagnetite and devitrified dark brown glass. Three plagioclase-augite-bronzite basalts analysed (Table 2b) are similar in major elements except for potassium and water contents.

Chilled marginal glass (up to 5 mm thick) is sometimes observed on basalt and in the matrix of tuff-breccia. The glass is very fresh and contains fresh phenocrystic minerals and shows calc-alkali andesite composition.

One gravel is compact glassy volcanic rock, free from plagioclase. It has phenocrysts and microphenocrysts of olivine, bronzite and Cr-spinel in groundmass of olivine, feathery augite, opaque minerals and fresh brown glass. The major elements and REE pattern of the rock (Table 2b and Fig. 3) show the following chemical features, different from basalts analysed, i.e., high concentration of MgO (11.1 wt %), SiO_2 (52.6 wt %), Ni (328 ppm) and Cr (744 ppm), and low concentration of REE and flat REE pattern. Petrographical and chemical characteristics indicate that the rock is high magnesian andesite close to boninite, however it is relatively poor in silica in comparison with boninites reported by SHIRAKI and KURODA (1977).

From the petrological characteristics of the dredged samples, it is considered that the small topographic high was probably a submarine volcano consisting of plagioclase rich two-pyroxene phyric basalts and some boninites, i.e., island arc type volcanic rock affinity. This high is therefore called as a submarine volcano in this paper.

6. *Summary of the Dredged Rocks and Geological Model*

Five dredge stations are divided into three groups according to lithologic characteristics.

Group I: Stations KH80-3-32 and KH82-4-4 are crest area of the Ogasawara Paleoland. The dredged samples are composed mainly of members of ophiolite assemblage, i.e. harzburgite, dunite, gabbro, dolerite, basalt and metamorphic rocks, with subordinate amounts of volcaniclastic sedimentary rocks.

Group II: Stations KH80-3-34 and KH82-4-3 are flank area of the paleoland. The dredged samples are volcaniclastic sedimentary rocks (tuffaceous lithic sandstone and tuffaceous siltstone), similar to those from group I. Two altered basalts and one gabbro, probably members of the ophiolite, were dredged.

Group III: Station KH80-3-33 is flank area of a small submarine volcano (SMV). Dredged samples are boninite, plagioclase rich two-pyroxene phyric basalt and volcaniclastic sediments, that is, they are island arc type volcanic rock affinity.

Based on the lithologic summary of the dredged samples, geological cross sections along AB and CD in Fig. 2a have been assumed as shown in Fig. 7. The crest area of the paleoland consists of the group I rocks with ophiolitic assemblage, and is covered by thin sedimentary rocks. On the other hand the flank areas of the paleoland are covered by the group II rocks, i.e. thick volcaniclastic sedimentary rocks containing a few mafic rocks derived from the crest areas. The submarine volcano consists of the group III rocks with the island-arc type volcanic rock affinity.

7. *Origin of Rounded Gravels*

The dredged samples from the crest areas of the Ogasawara fore-arc seamount (stations KH80-3-32 and KH82-4-4) contain many subrounded to rounded gravels of igneous rocks including serpentinites, basaltic rocks, doleritic rocks and gabbroic rocks (Table 1 and Fig. 6). These rocks are members of the ophiolitic assemblage. Dacite, rhyolite and granite, representative rock-types in the island-arc and continental crusts, have not been found from the seamount.

The roundness of each sample was described with the Powers' method (POWERS, 1953) in the cruise reports (ISHII *et al.*, 1981; ISHII *et al.*, 1983). One of the typical rounded gravel of pyroxene gabbro (KH82-4-4-051) is shown in Fig. 8. It is very difficult to regard these gravels as ice raft products, or detrital products from remote provenances or bottom current erosion products under the sea surface.

These characteristics suggest that the rounded gravels were products of, in situ, inshore wave erosion and that the present Ogasawara fore-arc seamount was exposed above the sea surface. Topographic feature of the seamount with a relatively flat surface (Fig. 2) supports the above suggestions. This land shall be called "Ogasawara Paleoland" in this paper.

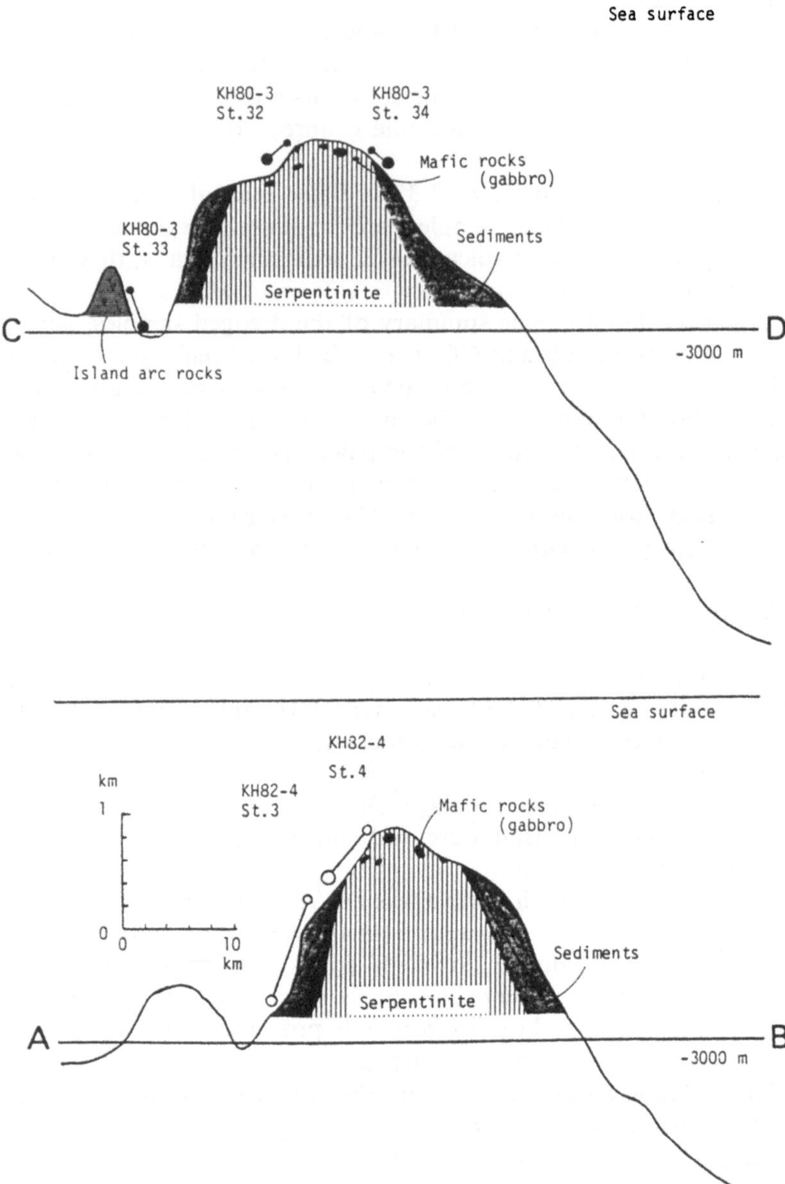

Fig. 7. Hypothetical geologic cross sections of the Ogasawara Paleoland along AB and CD in Fig. 2a. Symbols: Vertical lines-serpentinite, fine dots-sediments, solid mafic rocks and horizontal lines-island arc rocks. (See, Fig. 2a).

Fig. 8. Photograph of one of typical rounded gravels of pyroxene gabbro (KH82-4-4-051).

8. *Age Determinations*

The age assignments of the dredged sedimentary rocks were investigated using calcareous nannofossils by H. Okada, with the same paleontological method as that mentioned by OKADA (1980). The ^{10}Be method for dating of Mn-crust was made by T. Inoue with the same method as that described by TANAKA and INOUE (1979).

8.1 Calcareous nannofossils dating

Pliocene to Pleistocene calcareous nannofossils were recovered (H. Okada, personal communication) in the tuffaceous siltstone and silty sandstone from the two sites of the crest area of the paleoland.

Distribution of calcareous nannofossils and geologic age assignments of sedimentary rocks are shown in Tables 6a and 6b, respectively. Middle early Pliocene (CN 11a) to earliest Pleistocene (CN 13a) and latest Miocene (CN 10a)(?) through Pliocene to later Pleistocene (CN 14b)(?) calcareous nannofossils were recovered from the stations KH80-3-34 and KH82-4-4, respectively. Hence it will be reasonable to conclude that the paleoland has been situated at the sites of sedimentation from early Pliocene to the present.

Table 6a. Distribution of calcareous nannofossils from the Ogasawara Paleoland. (by H. Okada).

Station	KH80-3-34										KH82-4-4															
Sample Number	002	003	005	006	007	009	011	016	017	020	555	556	559	560	561	566	567	571	582	586	592	593	594	595	596	601
Overall abundance *2	A	A	A	R	C	C	R	R	C	A	A	A	R	F	A	C	R	C	C	A	R	R	R	R	R	F
Preservation *3	X	M	M	G	G	M		G	M	M	M	P		P	G	P		P	M	G	P					G
Etching *4	0	2	2	1	1	2		1	2	2	2	3		3	1	3	3	3	2	1	3					1
Overgrowth *4	0	0	0	0	0	0		0	0	0	0	0		0	0	0	0	0	1	0	2					0
Nannozone CN	13a	12d	12a	X	12ab	11		X	12a-c		14a	14a?	14?	-12	14a	14a?	10-12	14a	11	14a	10-12	11?	12ab		10-12	14a
Calicidiscus leptoporus	C *1	A	C		A	C	O	O	C	C	A	C		O	A	A	O	A	C	A		O	O	O		A
C. macintyrei	R	C	C		C	C			F	C				O					F							
Ceratolithus acutus																							O			
C. cristatus		R	F							F																
C. primus																			R?			O				
C. tricorniculatus			F ?																							
C. rugosus			F						C																	
Ceratolithus spp.																	O								O	
Coccolithus pelagicus	F	F	C			F			F	C	C	F		O	R	C	O	F	F	F			O			
Conusphaera mexicana															r											
Coronocyclus nitscense			F																							
Crenalithus doronicoides	A	A	A	O	A	A		O	A	A				O					A	C			O			C
C. Productellus	F	F																								
Cyclicargolithus floridanus																r		r								
Discoaster asymmelricus			R																							
D. brouweri.V. triradiatus			R			F			F																	
D. brouweri		R	F		F	R		O	C	C	r								F				O			
D. intercalaris			C							F																
D. pentaradiatus			C						F	A								f	F							
D. quinqueramus					f																					
Discoaster sp.												f		O			O		C			O	O	O		O
D. surculus			C							C												O	O			
D. tamalis			R							R																

D. variabilis			F						F	F									F	f	O		O	
Emiliania annula	C	C	C		C	C				C		C?			C				C	F	C			
E. huxleyi																								
E. ovata	A	A			A	F		O	C		F ?	C?			C	F?		C	F	C				F
Florisphera profunda V. profunda	A	A	C		A	C			A			A			C			F	C	C				
F. profunda V. elongata	F																			F				
Geophyrocpasa caribbeanica																								
G. oceanica											A	C			A			C		A				C
G. oceanica var.A.								O			F	C	O		F	A				C				C
Gephyrocpusa spp. (small)	C										C	C			C	A		C		C				A
Helicosphaera cartei	F	F			F										C					C				
H. hyalina	F														F									
H. neogranulata																								
H. sellii																				R				
H. wallichii	R																			R				
Neosphaera coccolithomorpha	F										R													
Reticulofenestra pseudoumbilica						F																		
Rhabdosphaera clabigera	C	F														F			C	f	O			
Sphenolithus abies																				F				
S. moriformis																			C					
Syracosphaera pulchra(endotheca)	R																	f	F				O	
Syracosphaea spp. (endotheca)	F																			F				
Thoracosphara sp.		F																						
Umbilicosphaera sibogae															R	F				F				
															R									

*1 A= abundant (more than 10 % of assemblage); C= common (more than 1 %, less than 10%); F= few (more than 0.1 % less than 1 %); R= rare (less than 0.1 %); O= existence. Reworked specimens; c= common; f= few; r= rare.

*2 A= abundant; C= common; F= few; R= rare.

*3 X= excelent; G= good; M= moderate; P=poor.

*4 O= none; 1= slight; 2= moderate; 3= strong.

Table 6b. Geologic age assignments of sedimentary rocks from the Ogasawara Paleoland (by H. Okada).

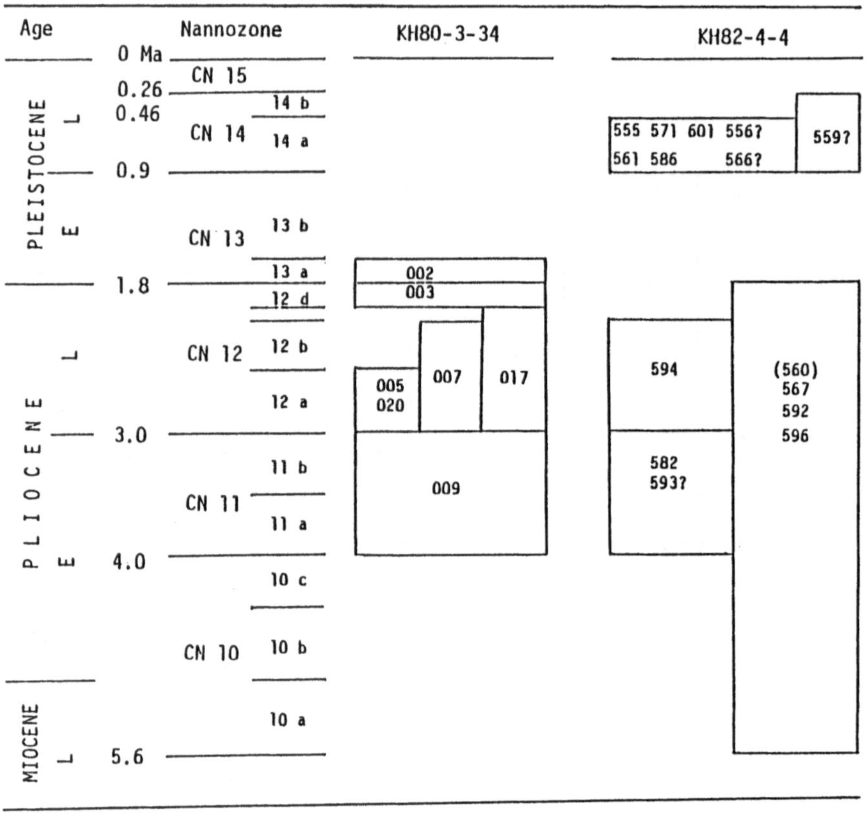

Age		Ma	Nannozone		KH80-3-34	KH82-4-4
		0 Ma	CN 15			
PLEISTOCENE	L	0.26	CN 14	14 b		
		0.46		14 a		555 571 601 556? 561 586 566?; 559?
	E	0.9	CN 13	13 b		
				13 a	002	
PLIOCENE	L	1.8	CN 12	12 d	003	(560) 567 592 596
				12 b	007; 017	594
				12 a	005 020	
	E	3.0	CN 11	11 b	009	582 593?
				11 a		
		4.0	CN 10	10 c		
				10 b		
MIOCENE	L			10 a		
		5.6				

8.2 ^{10}Be dating of Mn-crust

The selected sample for ^{10}Be dating is Mn-crust (KH80-3-32-001M, see Table 2c) of serpentinized harzburgite (KH80-3-32-001, see Table 2a) from the crest area of the paleoland.

The 4 mm thick Mn-crust was separated into three layers; outer, middle and inner layers, of which average depth from the surface is 0.25, 1.25 and 3.0 mm, respectively. The investigated value of $^{10}Be/^{9}Be \times 10^{-7}$ versus the depth of each sample is shown in Fig. 9. The calibrated inclination of 0.78 mm/10^6 yr corresponds to growth rate of the Mn-crust. Assuming that growth rate of the Mn-crust is constant, it is concluded that the Mn-crust began to precipitate at about 5×10^6 yr before present. It is noteworthy that the ^{10}Be dating age is consistent with the paleontological age of early Pliocene mentioned previously.

9. *Geophysical and Seismological Studies*

The geological model shown in the cross section (Fig. 7) is supported by the geophysical and seismological data obtained by MATSUMOTO and TOMODA (1982, 1983) and TAMAKI *et al.* (1981), respectively.

9.1 *Gravity and magnetic anomalies*

Five topographic peaks with the same height (about 1000 m deep) occur along the P4 line (26°15′N latitude) in Fig. 1. They are two Guyots

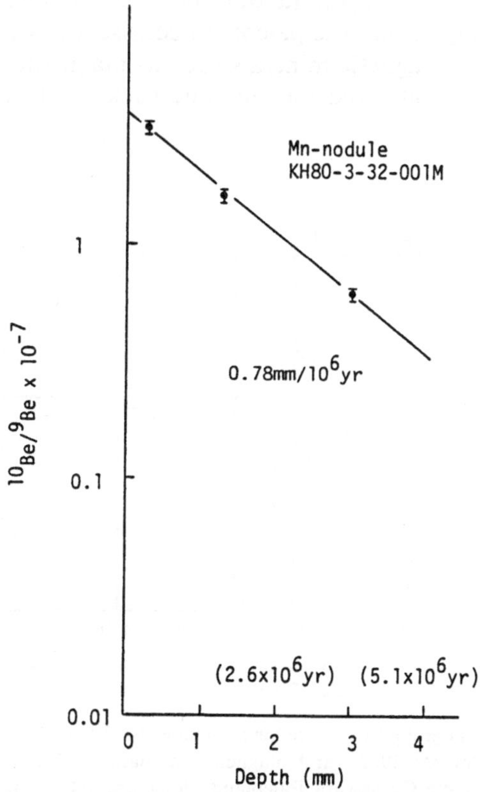

Fig. 9. Atomic ratio of $^{10}Be/^{9}Be \times 10^{-7}$ versus distance of each sample from the surface of Mn-crust (KH80-3-32-001M, T. Inoue unpublished data).

in the Ogasawara Plateau area (OPT), the Ogasawara Paleoland (OPL), old volcanic ridge (OR; Ogasawara ridge) and active volcanic ridge (SIR; Shichito Iwoto ridge). Free-air gravity anomaly (G′ in Fig. 10) of the paleoland is about 50 milli-gal. The value is relatively low in comparison with those of another four peaks ranging from 130 to 200 milli-gals.

Estimated average densities of serpentinized harzburgites and dunites are 2.43 g/cm^3 and 2.31, respectively (Table lb). Densities of dredged gabbro, dolerite and basalt are 2.85, 2.76, 2.85, respectively.

If the paleoland was composed mainly of serpentinized ultramafic rocks with density of about 2.4, and other four peaks were composed mainly of basic rocks with density of about 2.8, the above gravity data are understandable.

Magnetic anomaly (MA in Fig. 10, MATSUMOTO and TOMODA, 1983) of the paleoland is about 100 γ. The value is relatively low in comparison with those from the two Guyots (about 700 γ) in the Ogasawara plateau area. These data suggest that the paleoland consists of weakly or randomly magnetized rocks. Magnetic minerals such as magnetite and maghemite are scarce in the serpentinized ultramaphic rocks that magnetization of

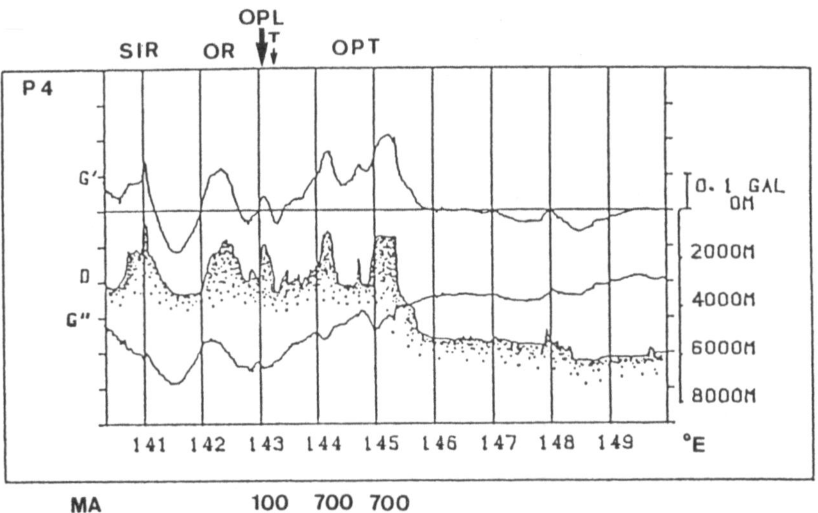

Fig. 10. Profiles of topography (D), free air anomalies (G′) and Bouguer anomalies (G″) (MATSUMOTO and TOMODA, 1982), and magnetic anomalies (MA, γ) (MATSUMOTO and TOMODA, 1983), across the Ogasawara Paleoland along line P4 in Fig. 1. Abbreviations: SIR-Shichito Iwoto Ridge, OR-Ogasawara Ridge, OPL-Ogasawara Paleoland, T-trench, OPT-Ogasawara Plateau.

the rock may be very weak. If the paleoland consists mainly of these rocks, the measured magnetic anomaly is explicable. The geophysical data of both the gravity and magnetic anomalies are interpreted very well on the assumption that the paleoland mainly consists of serpentinized ultramafic rocks.

9.2 Seismic reflection profiling

Geophysical and geological investigations of the Ogasawara (Bonin) and northern Mariana were carried out by Geological Survey of Japan (HONZA *et al.*, 1981). Figure 11 (TAMAKI *et al.*, 1981) shows single channel seismic reflection profilings along the 26°00′N latitude, crossing southern end of the paleoland.

It is noteworthy that the flank area of the paleoland as well as the Ogasawara ridge and the Ogasawara Plateau are covered by sediments. On the other hand the crest area of the paleoland is not covered by sediments. This data also support the geological model of the paleoland (Fig. 9) assumed from dredged samples.

10. Discussion and Conclusions

The geological and geophysical characteristics of the Ogasawara seamounts are summarized in the followings.

1. The dredged samples from the crest of larger high with flat summit (Fig. 2) are mainly subrounded to rounded gravels with ophiolitic assemblages (harzburgite, dunite, gabbro, dolerite and basalt). The gravels are formed by wave erosion. Because the high was once exposed above the sea surface, it shall be called the Ogasawara Paleoland (OPL).

2. The dredged samples from the smaller high with conical shape are angular to subangular, island-arc type rocks (basaltic pillow lava, pyroclastic rocks and boninite). The high shall be called the submarine volcano (SMV).

3. Among the dredged ultramafic rocks from OPL, harzburgite is the most predominant with subordinate dunite. The harzburgites and dunite are distinct from peridotite of midoceanic ridges, according to the spinel-chemistry. Spinel in the former is more Cr-rich than that in the latter, but it is similar to the spinel of Alpine-type dunite-harzburgite.

4. Two-pyroxene (augite + orthopyroxene or pigeonite) gabbro is the most predominant in the gabbroic rocks from OPL, and clinopyroxene (augite) gabbro is subordinate. Some gabbros show very high magmatic temperature (more than 1180°C) estimated by inverted pigeonite. The gabbros are distinct from those of midoceanic ridges, where Ca-poor pyrox-

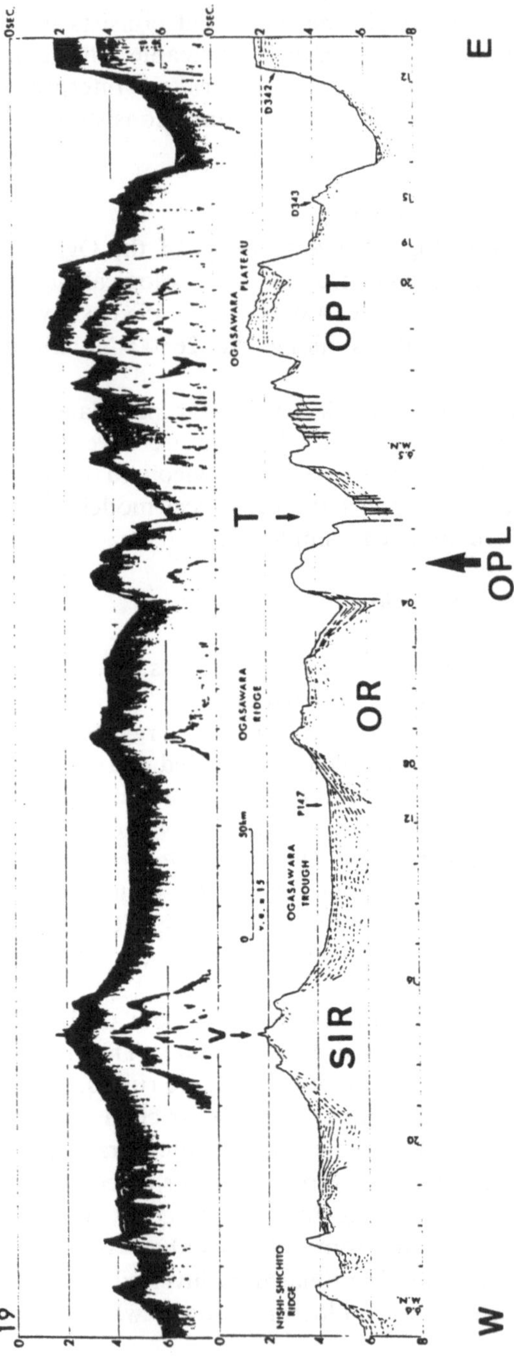

Fig. 11. Profile of single channel seismic reflection across southern end of the Ogasawara Paleoland along the 26°00′N latitude (TAMAKI *et al.*, 1981). Abbreviations are the same as those in Fig. 10.

ene bearing gabbro is rare.

5. Among the dredged basic rocks from OPL, island arc basalts and a very fresh trachyandesite are confirmed according to their chemical analyses. Some metabasites are nepheline normative. The alkali basalts may be interpreted to be fragments of oceanic seamount.

6. The dredged samples from flank areas of OPL consist largely of euhedral crystal-bearing volcaniclastic sedimentary rocks. They are dredged from all sites.

7. The Mn-crust from OPL is classified into hydrogenous origin.

8. Middle early Pliocene to Pleistocene calcareous nannofossils were recovered in the tuffaceous sedimentary rocks from OPL.

9. Estimated growth rate by ^{10}Be method of the 4 mm thick Mn-crust from OPL is small (0.78 mm/10^6 yr) and the Mn-crust began to precipitate 5×10^6 yr before present. The result is consistent with the above paleontological dating.

10. Free-air gravity and magnetic anomalies of OPL are relatively small. The both geophysical data are interpreted very well by the model, that OPL consists mainly of serpentinized ultramafic rocks with low density and weak magnetization.

11. Single channel seismic reflection profiling shows that the flank and crest areas of OPL are covered by thick and thin sediments, respectively.

Generally speaking, upper mantle consists of lherzolite with mineral assemblage of olivine (=Ol)+orthopyroxene (=Opx)+Ca-rich clinopyroxene (=Cpx)±garnet±spinel, and can be approximated by forsterite (=Fo)−diopside (=Di)−silica (=SiO_2) system (KUSHIRO, 1972). Increase of degree of partial melting under proper conditions of pressure and water content, mantle peridotite changes from lherzolite (mineral assemblage of stage A: Fo+En+Di) through Di-poor lherzolite (stage B: Fo+En+poor-Di), harzburgite (stage C: Fo+En) and En-poor harzburgite (stage D: Fo+poor-En) and to dunite (stage E: Fo). According to the above order of refractory residues of primary peridotite, induced basaltic magma varies its own composition from Di-saturated magma (stage A and B) through moderately Di-depleted magma (stage C) to strongly Di-depleted magma (stage D and E).

Harzburgite-rich ultramafic rocks dredged from OPL may be refractory residues of fore-arc magmatism from primary peridotite of wedge mantle and correspond to the stage C and/or stage D of the above partial melting sequences. On the other hand, lherzolite-rich ultramafic rocks from mid-oceanic ridges may correspond to the stages A or B. The ratio of harzburgite to lherzolite in the upper mantle may be larger in the island arcs than in the oceanic regions (KUSHIRO, 1983). It is suggested that relatively high degree of partial melting may have taken place in the upper mantle beneath the island arcs (KUSHIRO, 1983).

Opx-rich gabbros as well as primitive island-arc basalts (relatively Ca-poor in bulk chemistry, see Table 2b) dredged from OPL may be derived from the stages C and/or D magma, of which refractory residue may be harzburgite (stage C and/or D).

Non-effusive rocks (for example, doleritic, gabbroic and ultramafic rocks) associated with boninite have not been observed in Ogasawara Islands. It is expected that boninite (being free from Cpx-phenocryst and plagioclase) dredged from SMV may have originated from stage C and/or D magma, that is, refractory residue of the boninite magma is also harzburgite (stage C and/or D). Recent experimental studies on island arc basalts (TATSUMI *et al.*, 1983) and andesites (TATSUMI, 1981) in Japanese Islands indicate that the residues after generation of primary tholeiite and high-magnesian andesite magmas are harzburgite.

Assemblages of dredged igneous rocks from both OPL and SMV, excluding alkalic rocks, can be interpreted very well on the assumption that harzburgite is cogenetic with gabbros and basalts and boninite. They should be called "fore-arc ophiolite". The above hypothesis does not make big petrological contradictions to present data (summary No. 1 through No. 5). The ultramafic portions of some ophiolites which consist largely of harzburgite may represent the residual upper mantle materials beneath the island arcs (KUSHIRO, 1983).

The Ogasawara Plateau (OPT) on the Pacific plate is now colliding with the Philippine Sea plate in the Ogasawara Island region (KOBAYASHI, 1983), where OPL and the Ogasawara Ridge (OR) are closely located from OPT within 100 km (Figs. 1, 10, and 11). OPL consists of ophiolitic rocks and OPT may consist of oceanic seamount-type alkali rocks. OR may consist of boninites (and their derivatives) and/or island-arc tholeiites, which are distributed in Chichijima and Hahajima, respectively, of the northern part of the OR. In the process of continual subduction of the Pacific plate beneath the Philippine Sea plate, OPT will eventually collide with OPL being accompanied with OR. Ophiolitic igneous complexes with alkali rocks and/or island arc rocks, which are sometimes observed in the Japanese islands, may be collision products. The present observations and discussions support that some ophiolites were formed in island arcs (MIYASHIRO, 1973, 1975).

Existence of fore-arc ophiolite or the new-type ophiolite in OPL will give strong constraints for a discussion of the origin of on-land ophiolite, that is, the following two geological processes should be considered, first:

emplacement of mantle peridotite from upper wedge mantle into the surface of fore-arc crust, and second: settling of the fore-arc ophiolite to the level of the present island arc crust.

The time and space of fore-arc magmatism were very limited and the activity occurred at the earliest stage of subduction beginning at about 42 MaBP (BLOOMER, 1983; KOBAYASHI, 1983). The fore-arc ophiolite may have originated from characteristic igneous activities (including volcanism, plutonism, mantle diapir and serpentinite diapir) caused by subduction of the old Pacific plate beneath the very young and still hot Philippine Sea plate with thin lithosphere (YOSHII, 1975).

The present studies are concluded as follows; the Ogasawara Paleoland is composed largely of cogenetic igneous rocks of ultramafics (Cpx-poor), gabbros (Ca-poor pyroxene rich) and island arc volcanic rocks and rarely of alkalic basalts of seamount fragments from the Pacific plate. The former rocks are called "fore-arc ophiolites" and are regarded as the first igneous products during the early stages of subduction.

Acknowledgements

The author thanks to Dr. H. Okada for his paleontological analyses; to Mr. H. Haramura and Dr. T. Urabe for their wet chemical analyses of rocks; to Dr. T. Inoue for his ^{10}Be dating on Mn-crust and to Dr. Y. Minai for his REE analyses of rocks. The author wishes to thank Professors I. Kushiro, S. Uyeda, K. Nakamura, S. Takayanagi, T. Vallier and B. Tyler for reviews of the manuscript.

Thanks are due to Professors Noriyuki Nasu, K. Kobayashi, H. Kagami, N. Yonekura, and Y. Ida and Mr. Y. Saito for their encouragement during the research; to the crew of the R/V Hakuhō-Maru for their help in collecting the samples during KH80-3 and KH82-4 cruises; to Prof. K. Konishi, Dr. A. Omura, Dr. J. Naka, Mr. K. Futakuchi and Mr. H. Ohara for their on board studies on samples and to Drs. K. Ozawa, K. Shiraki, M. Otsuki, H. Tokuyama, T. Matsumoto and R. J. Arculus for their discussion and suggestion in optical observations, electron probe micro analyses and data reductions.

Thanks are also due to Mrs. Akemi Hatanaka for her help in preparation of polished thin sections, to Miss Yoshiko Moriwaki and Mr. H. Kayane in typewriting, to Miss Sayuri Washida for her help on shore observation of rocks, and to Mr. E. Tsuchida and Mr. M. Watanabe for their photographic works. The study was partly supported by a grant-in-aid for scientific research, Ministry of Education, Japan, No. 57540472.

REFERENCES

S. ARAI and T. FUJII, Petrology of ultramafic rocks from Site 395, in Init. Rep. Deep Sea Drill. Proj. 45, edited by W. G. Melson and P. D. Rabinowitz, pp. 587–594, U.S. Govt. Printing Office, Washington, D.C., 1979.

S. H. BLOOMER, Distribution and origin of igneous rocks from the landward slopes of the Mariana trench: implication for its structure and evolution, *J. Geophys. Res.*, **88**, 7411–7428, 1983.

S. H. BLOOMER and J. W. HAWKINS, Gabbroic and ultramafic rocks from the Mariana Trench: An island arc ophiolite, in *The Tectonic and Geologic Evolution of Southeast Asian Seas and Islands*, Part II, Geophys. Monogr. Ser., vol. 27, edited by D. E. Hayes, pp. 294–317, AGU, Washington, D.C., 1983.

E. BONATTI, T. KRAMER, and H. S. RYDELL, Classification and genesis of submarine iron-manganese deposits, in *Ferro-Manganese Deposits on the Ocean Floor*, edited by D. H. Horn, pp. 149–166, Lamont-Doherty of Columbia University, Palisades, N.Y., 1972.

R. G. COLEMAN, *Ophiolites*, 229 pp., Springer-Verlag, New York, 1977.

H. J. B. DICK and T. BALLEN, Chromian spinel as a petrogenetic indicator in abyssal and alpine-type peridotites and spatially associated lavas, *Contrib. Mineral. Petrol.*, **86**, 54–76, 1984.

J. W. HAWKINS, S. BLOOMER, C. EVANS, and J. MELCHIOR, Mariana arc-trench system: Petrology of the inner trench wall (abstract), *Eos, Trans. AGU*, **60**, 968, 1979.

E. HONZA and H. KAGAMI, A possible accretion accompanied by ophiolite in the Mariana trench, *J. Geography, Tokyo Geograph. Soc.*, **86**, 80–91, 1979.

E. HONZA, E. INOUE, and T. ISHIHARA, Cruise Report No. 14, Geological investigation of the Ogasawara (Bonin) and northern Mariana arcs April-August 1979 (GH79-2, 3 and 4 Cruises), edited by E. Honza, E. Inoue and T. Ishihara, 170 pp., Geol. Survey Japan, 1981.

IGCP WORKING GROUP, 'Ophiolites', Initial report of the geological study of oceanic crust of the Philippine Sea floor, Ofioliti 2, 137–168, 1979.

T. ISHII, The relations between temperature and composition of pigeonite in some lavas and their application to geothermometry, *Mineral. J.*, **8**, 48–57, 1975.

T. ISHII, Pyroxene geothermometry of basalts and an andesite from the Palau-Kyushu and West Mariana ridges, Deep Sea Drilling Project Leg 59, in Init. Rep. Deep Sea Drill. Proj. 59, edited by L. Kroenke and R. Scott, pp. 693–707, U.S. Govt. Printing Office, Washington, D.C., 1981.

T. ISHII and H. HARAMURA, Data on major-element compositions of DSDP leg 59 tholleiitic basalts, calcalkalic andesite, and glass obtained by wet chemical analysis, XRF, and EPMA methods, in Init. Rep. Deep Sea Drill. Proj. 59, edited by L. Kroenke and R. Scott, pp. 707–718, U.S. Govt. Printing Office, Washington, D.C., 1981.

T. ISHII, S. MCCALLUM, and S. GHOSE, Petrological and thermal histories of a lunar breccia 73217 as inferred from pyroxene crystallization sequences, exsolution phenomena, and pyroxene geothermometry, *Proc. Lunar Planet. Sci. Conf. 13th, J. Geophys. Res.*, **88**, Suppl. A631–A644, 1983.

T. ISHII and H. TAKEDA, Inversion, decomposition and exsolution phenomena of terrestrial and extraterrestrial pigeonites, *Mem. Geol. Soc. Japan*, **11**, 19–36, 1974.

T. ISHII, K. KONISHI, J. NAKA, K. FUTAKUCHI, and H. OHARA, Description of samples from Ogasawara fore-arc seamount or "Ogasawara Paleoland", in Preminary Report of the Hakuhō Maru Cruise KH82-4, edited by K. Kobayashi, pp. 173–186, Ocean Research Inst, Univ. of Tokyo, 1983.

T. ISHII, K. KONISHI, and A. OMURA, Description of samples, in Preliminary Report of the Hakuhō Maru Cruise KH80-3, edited by K. Kobayashi, pp. 105–165, Ocean Research Inst., Univ. of Tokyo, 1981.

A. ISHIWATARI, A preliminary report on the Yakuno ophiolite in the Maizuru zone, Inner

Southwest Japan, *Earth Science (Chikyu Kagaku)*, **32**, 301–310, 1978 (in Japanese with English abstract).

H. ISHIZUKA, Geology of the Horokanai Ophiolite in the Kamuikotan Tectonic Belt, Hokkaido, *Japan. J. Geol. Soc. Japan*, **86**, 119–134, 1980 (in Japanese with English abstract).

K. KOBAYASHI, Preliminary Report of the Hakuhō Maru Cruise KH80-3, edited by K. Kobayashi, 201 pp., Ocean Research Inst. Univ. of Tokyo, 1981.

K. KOBAYASHI, Fore-arc volcanism and cycles of subduction, in *Arc Volcanism: Physics and Tectonics*, edited by D. Shimozuru and I. Yokoyama, pp. 153–164, Terra Scientific Publishing Company, Tokyo, 1983.

K. KOBAYASHI, Preliminary Report of the Hakuhō Maru Cruise KH82-4, edited by K. Kobayashi, 267 pp., Ocean Research Inst. Univ. of Tokyo, 1983.

H. KUNO, Iwō-zima, in Catalogue of the Active Volcanoes of the World including Solfatara Fields; Part 11, Japan, Taiwan and Marianas, edited by H. Kuno, pp. 259–265, Int. Assoc. Volcanology, 1962.

I. KUSHIRO, Determination of liquidus relations in synthetic silicate systems with electron probe analysis: the system forsterite-diopside-silica at 1 atmosphere, *Amer. Mineral.*, **57**, 1260–1271, 1972.

I. KUSHIRO, Generation of magmas and upper mantle materials in Japanese islands, in abstract of OJI international seminar on the formation of ocean margins, 44 pp., Ocean Research Inst., Univ. of Tokyo, 1983.

T. MATSUMOTO, Gravimetry and bathymetry, in Preliminary Report of the Hakuhō Maru Cruise KH82-4, edited by K. Kobayashi, 7 pp., Ocean Research Inst., Univ. of Tokyo, 1983.

T. MATSUMOTO and K. KOBAYASHI, Bathymetric map of Ogasawara fore-arc region and adjacent area, in Preliminary Report of the Hakuhō Maru Cruise KH82-4, edited by K. Kobayashi, foldout map, Ocean Research Inst., Univ. of Tokyo, 1983.

T. MATSUMOTO and Y. TOMODA, Gravity anomalies and tectonics in Bonin region, Proceeding of the General Meeting of the IAG, Tokyo, pp. 243–250, 1982.

T. MATSUMOTO and Y. TOMODA, Magnetic anomalies on and around the Bonin Rise, *Zisin, J. Seismol. Soc. Japan* (Second Series), **36**, 77–82, 1983 (in Japanese with English abstract).

Y. MINAI, Geochemical studies of the ocean floor rocks and sediments. Doctoral Dissertation, 279 pp., Univ. of Tokyo, Tokyo, 1982 (in Japanese).

A. MIYASHIRO, The troodes ophiolitec complex was probably formed in an island arc, *Earth Planet. Sci. Lett.*, **19**, 218–224, 1973.

A. MIYASHIRO, Classification, characteristics, and origin of ophiolites, *J. Geol.*, **83**, 249–281, 1975.

J. NAKA and S. UEHARA, Igneour rocks dredged from sites KH82-4-3 and -4, in Preliminary Report of the Hakuhō Maru Cruise KH82-4, edited by K. Kobayashi, pp. 187–194, Ocean Research Inst., Univ. of Tokyo, 1983.

Y. NAKAMURA and I. KUSHIRO, Compositional relations of coexisting orthopyroxene, pigeonite and augite in a tholeiitic andesite from Hakone volcano, *Contr. Mineral. Petrol.*, **26**, 265–275, 1970.

Y. OGAWA, Mineoka ophiolite belt in the Izu forearc area-Neogene accretion of oceanic and island arc assemblages on the Northeastern Corner of the Philippine sea plate, in *Accretion Tectonics in the Circum-Pacific Regions*, edited by M. Hashimoto and S. Uyeda, pp. 245–260, Terra Scientific Publishing Company, Tokyo, 1983.

F. OHASHI and K. SHIRAKI, High-magnesia and high- silico volcanic rock in the Setogawa ophiolite, *J. Japan Assoc. Mineral. Petrol. Econ. Geol.*, **76**, 69–79, 1981 (in Japanese with English abstract).

H. OKADA, Colcareous nannofossils from Deep Sea Drilling Project sites 442 through 446, Philippine sea, in Init. Rep. Deep Sea Drill. Proj. 58, edited by G. V. Klein and K. Kobayashi, pp. 549–563, U.S. Govt. Printing Office, Washington, D.C., 1980.

M. OTSUKI, Application of a JCXA-733 to the natural minerals (I), *Nihon-denshi News*, **23**, 50–53, 1983 (in Japanese).

K. OZAWA, Relationships between tectonite and cumulate in ophiolites: the Miyamori ultramafic complex, Kitakami Mountains, northeast Japan, *Lithos*, **16**, 1–16, 1983.

M. C. POWERS, A new roundness cole for sedimentary particles, *J. Sediment. Petrol.*, **23**, 117–119, 1953.

K. SHIRAKI and N. KURODA, The boninite revisited, *J. Geography*, **86**, 174–190, 1977 (in Japanese with English abstract).

K. SHIRAKI, F. OHASHI, and N. KURODA, Chrome-spinels in a basanitic lava from Nanzaki volcano, Izu peninsula, *J. Japan Assoc. Mineral. Petrol. Econ. Geol.*, **74**, 114–121, 1979 (in Japanese with English abstract).

K. TAMAKI, M. TANAHASHI, Y. OKUDA, and E. HONZA, Seismic reflection profiling in the Ogasawara (Bonin) arc and the northern Mariana arc, in Cruise Report 14, edited by E. Honza, E. Inoue and T. Ishihara, pp. 83–91, Geological Survey of Japan, 1981.

S. TANAKA and T. INOUE, ^{10}Be dating of north pacific sediment cores up to 25 million years B, *Earth Planet. Sci. Lett.*, **45**, 181–187, 1979.

Y. TATSUMI, Melting experiments on a high magnesian andesite, *Earth Planet. Sci. Lett.*, **54**, 356–365, 1981.

Y. TATSUMI, M. SAKUYAMA, H. FUKUYAMA, and I. KUSHIRO, Generation of arc basalt magmas and thermal structure of the mantle wedge in subduction zones, *J. Geophys. Res.*, 5815–5825, 1983.

S. TONOUCHI and K. KOBAYASHI, Paleomagnetic and geotectonic investigation of ophiolite suites and surrounding rocks in south-central Honshu, Japan, in *Accretion Tectonics in the Circum-Pacific Regions*, edited by M. Hashimoto and S. Uyeda, pp. 261–288, Terra Scientific Publishing Company, Tokyo, 1983.

S. UEHARA and J. NAKA, Serpentines dredged from KH82-4, in Preliminary Report of the Hakuhō Maru Cruise KH82-4, edited by K. Kobayashi, pp. 195–204, Ocean Research Inst. Univ. of Tokyo, 1983.

T. YOSHII, Regionality of group velocities of Rayleigh waves in the Pacific and thickening of the plate, *Earth Planet. Sci. Lett.*, **25**, 305–312, 1975.

Formation of Active Ocean Margins, edited by N. Nasu *et al.*, pp. 343–364.

ACCRETION IN THE NANKAI TROUGH

Eiichi HONZA and Fumitoshi MURAKAMI

Geological Survey of Japan, Higashi, Yatabe, Tsukuba, Ibaraki 305, Japan

Abstract. We presented here subduction and accretion in the east part of the Nankai, the Suruga and the Sagami Troughs based on multi-channel seismic data. Imbricated thrust and overthrust faults turning decollements in the lower extension are commonly observed in the Nankai and the Suruga Troughs. Underplating by decollements in the lower horizon of the accretionary wedge tends to increase in thickness landward. Total increase of thickness in accretion on descending subducted basement is supplied by increase of underplating associated with decollements. Offscraping is approximately in the same thickness throughout accretion. Thrust faults are discontinuous and show wavy patterns with a few to several kilometers long. Internal velocity tends to increase to lower part within a thrusted unit. There is a high velocity even in the initial unit near the Trough.

1. Introduction

Concept of accretionary wedge for convergent margin has been improved by the interpretation of multi-channel seismic data and by the drilling results of the Deep Sea Drilling Project (DSDP) in the Pacific Rim.

The first, imbricated thrust model for accretionary wedge was proposed on the basis of seismic data in the Sunda and the Middle America Trenches (BECK and LEHNER, 1974; SEELY *et al.*, 1974; KARIG and SHARMAN, 1975).

Recent DSDP drilling series in the convergent margin has been resulted to establish much clear model for accretionary wedge in the Japan Trench (VON HUENE, LANGSETH, NASA, OKADA *et al.*, 1982), in the Mariana Trench (HUSSONG, UYEDA *et al.*, 1982), in the Middle America Trench (VON HUENE, AUBOUIN *et al.*, 1982; WATKINS, MOORE *et al.*, 1982) and in the Nankai Trough (KARIG *et al.*, 1983). From these drillings, some different models were also documented on the process not only for accretionary wedge, but also that for consumption in the subduction zone. The consumption type is posturated in the Mariana Trench. Different process during arc development is suggested in the Japan Trench, having been subduction erosion in earlier stage and turned to form accretionary wedge in the latest stage (HONZA, 1982; VON HUENE *et al.*, 1982). There may be different process between the north and the south Middle America trench. A progressive accretionary wedge is suggested in the northern part (MOORE *et al.*, 1982), while large scale sediments subduction is suggested in the southern part (VON HUENE, AUBOUIN *et al.*, 1982). Studies on tectonic process and on formation mechanism in the subduction zone are required to clarify these variations both in time and in places.

Here we present possible mechanism to form accretionary wedge in the east part of the Nankai Trough based on multi-channel seismic data which are resulted under the conducts of Geological Survey of Japan and Hydrographic Office of Japan.

2. *Geological Setting*

The Nankai Trough is a part of the northern consumption margin of the Philippin Sea Plate which extends from the Sagami Trough to the Ryukyu Trench from east to southwest respectively and is subducting under the Seinan (SW) Japan Arc. The Nankai Trough is shallower than the neighbouring the Japan and the Bonin trenches and increases gradually its depth to west, approximately 3,000 m in the northeast margin and 5,000 m in the southwest margin. Several troughs and basins are observed in the continental slope. Outer margin of them are bordered by ridges or highs. Outer side slope of the ridges is extended to the relatively steep inner trench slope of the Nankai Trough (Fig. 1).

Subduction angle of oceanic basement in the Nankai Trough is rather gentle than that in the neighbouring trenches. It is from 18 to 25 degrees in average of the total seismic plain in the surveyed area (SHIONO, 1977; FICTH and SHOLTZ, 1971) and is from 4 to 8 degrees in the shallower part within the profiles presented here. This feature is preferable for study on mechanism of accretion at trench, for deformation process by accretion can be well traced on account of gentle slope of the subducted oceanic basement.

The surveyed area is an eastern part of the Nankai Trough, extending to the Suruga Trough in the northeast margin. The Suruga Trough is narrower and deeper with steep walls on both sides. Two outer highs, tilted blocks to west, in the west side of the Suruga Trough are arranged

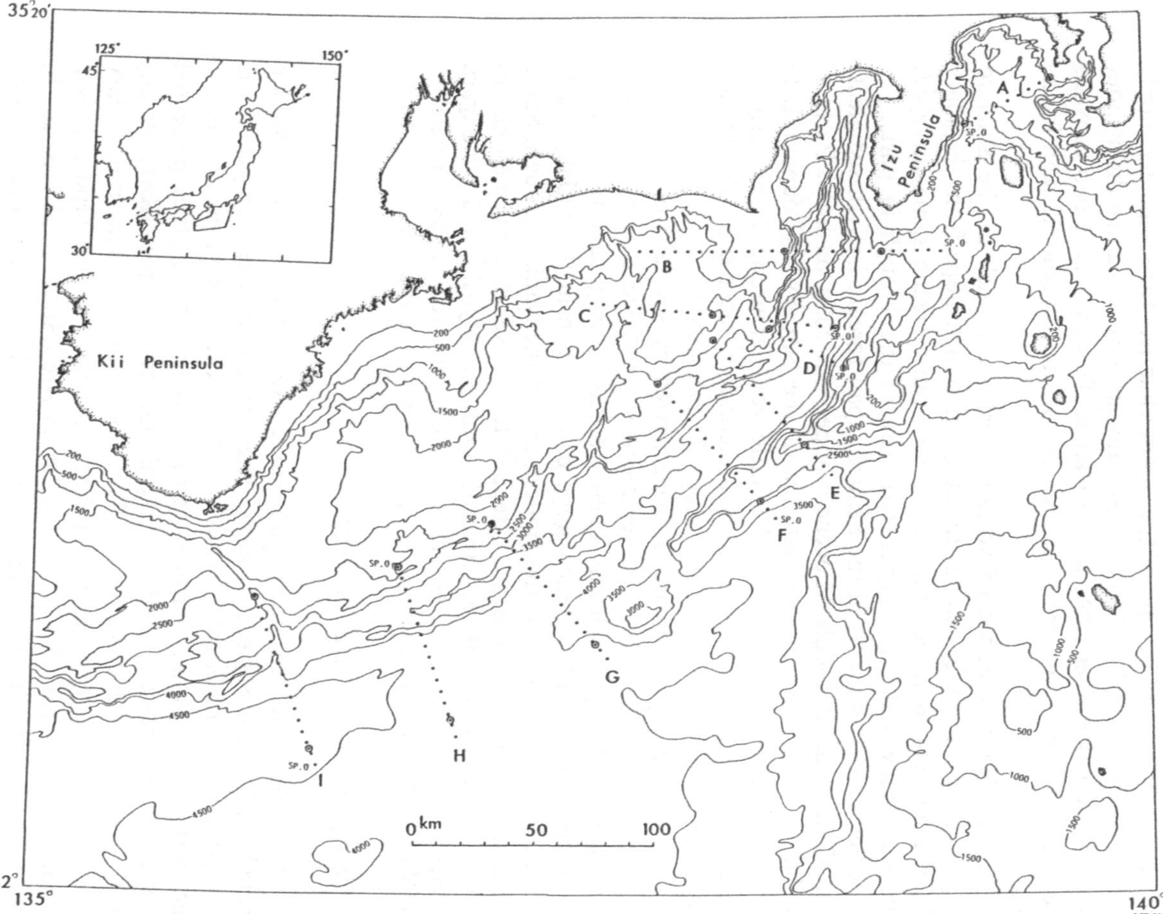

Fig. 1. Bathymetry (HYDROGRAPHIC OFFICE OF JAPAN, 1982) and tracks of multi-channel seismic profiling survey (dotted lines) in the east part of the Nankai Trough. Profiles presented here are dotted in double circle.

on the extension from the uplifted high which consists of trench slope break in the Nankai Trough. Several canyons from the outer shelf to the mid-slope trough may be one of the major channels for sediments supply to the mid-slope troughs. Two of these canyons have a channel throughout the slope over the outer ridge to the Nankai Trough. These two canyons also border the east and the west margins of the Kumano Trough in the mid-slope.

Zenisu Ridge consisted of volcanic rocks, a branch of the Shichito Ridge in the Bonin Arc, extends to the outer trench slope from the south of Izu Peninsula to more than half of the outer trench slope surveyed. This ridge may make a complex structure in the Trough in parts.

One of the basements in the Seinan Japan forearc is flysh sediments from Cretaceous to middle Miocene cropped out coast onshore. Parts of the basements which are correlated with the Setogawa Group in Late Cretaceous to Paleogene age and a little younger group from early to middle Miocene are documented to be accretionary wedge in a paleo-forearc which might develop in the earlier stage (TAIRA, 1981).

The Seinan Japan forearc is characterized by several troughs where are a few thousands meters thick sediments since Miocene, unconformably overlain on the Setogawa Group. These troughs are originated by the uplift of the outer ridge on the trench slope break, which might initially occur during Miocene obliquely to the modern axis of the Nankai Trough. Some of the sediments in troughs are extended to the Neogene formations onshore. They are a part of the Miyazaki Group in Kyushu and the Sagara Group in Shizuoka. The uplifted outer ridges are rearranged parallel to the Nankai Trough, at least, since Pliocene (OKUDA *et al.*, 1978, 1979).

3. *Interpretation of Seismic Profiles*

A multi-channel seismic profiling survey was carried out in the Nankai Trough and in the Sagami Trough. They consist of one transect in the northern part of the Sagami Trough (profile A) and eight profiles in the east part of the Nankai Trough (profiles B–I). The profiles presented here are also reported by Hydrographic Office of Japan (KATO *et al.*, 1983). The Sagami and the Suruga Troughs are in the northern boundary of the Bonin Arc. Profiles in the both troughs are also reported as a northern consumption margin of the Bonin Arc (HONZA and TAMAKI, in press). Here a detailed examination of the profiles is presented to study mechanism of accretion in the Nankai Trough.

3.1 Seismic Profile in the Sagami and the Suruga Troughs (profiles A and B)

It is demonstrated that the subduction in the Sagami Trough is so oblique that sometime occurs eduction in parts (SIMAZAKI *et al.*, 1981). Basements which constitutes Izu Peninsula dips toward east and are overlain by thick horizontally layered sediments on it in the Sagami Trough (profile A in Figs. 2 and 5). Thrust faults may occur at the foot of banks west off Miura Peninsula. Subduction in the Sagami Trough is not clear, but banks might be uplifted by thrust faults (KIMURA, 1975; SHIMAZAKI *et al.*, 1981) making a deformed zone in the inner (east) side of the thrust faults (HONZA and TAMAKI, in press).

Subduction in the Suruga Trough is well demonstrated in this profile (profile B in Figs. 2 and 5). The seaside basement is subducting under the sedimentary sequence in the west side of the Trough. The seaward basement is the same constituents as is in the Izu Peninsula and is consisted of volcanic rocks which consists of volcanic chain of the Bonin Arc. Thrust faults are a dominant feature to form accretionary wedge. They are observed even immediately on the inner side of the subducted basement below the trough. Some uplifted layers by thrust faults are eroded on the surface to form a smooth surface with miner impression by structure in the foot of the west side slope. A thrust fault develops through the upper part of subducted basement. This may suggest an initial supply of the basement to form parts of the imbricated thrust faults.

3.2 Seismic Profiles in the Northern Margin of the Nankai Trough (profiles C–F)

Rather high angle thrust faults are a dominant feature to form accretionary wedge in the profiles C–E (Figs. 3, 5 and 6). Thrust faults occur from the foot of the inner trench slope. Some of them develope throughout both of the sediments and the subducted basement and some of thrust faults occur only in the subducted basement in the trough as are shown in the profile E. Highs on the lower part of the inner trench slope in the profiles C and D are an extension of the Omaezaki Spur which is terminated at the Nankai Trough to the south. Thrust faults tend to develop on both sides of the high in the profile C, while upper layers in the high are not deformed. Thrust faults are observed in a little seaward side of the trough in the profile F. The upper layers inner side of the thrust faults are uplifted and are also eroded on surface. There are low angle thrust faults on the subducted oceanic basement which has rather smooth and gentle dipped plain in the profile E.

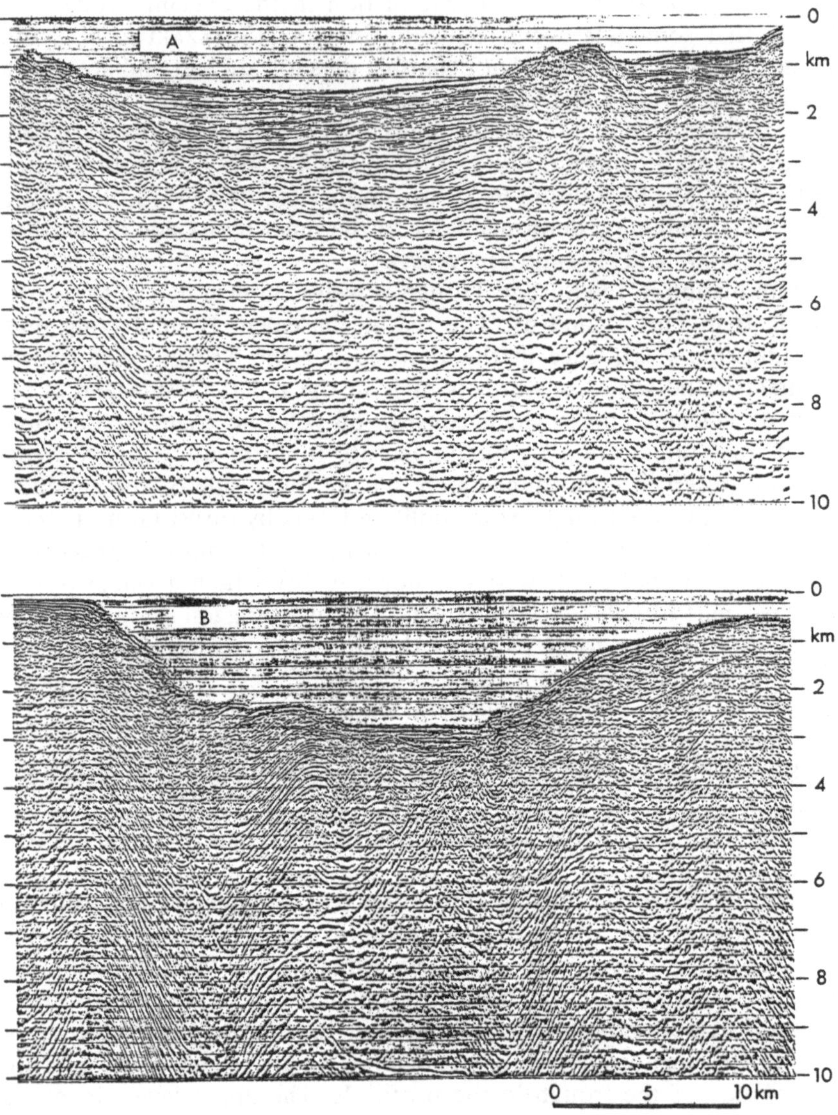

Fig. 2. Migrated depth sections A and B.

3.3 Seismic Profiles in the Nankai Trough South of Kii Peninsula (profiles G-I)

Low angle thrust faults are a dominant feature in the accretionary wedge in the profiles G and I (Figs. 4 and 7). Rather high angle thrust faults are a dominant in the profile H. Accretionary wedge is formed by low angle thrust faults at first in the profile G and is initially formed by rather high angle thrust faults in the profile I. Low angle thrust faults occur in the inner trench slope several kilometers inner side from the high angle thrust faults. These low angle thrust faults are discontinuous as to show a small scale wavy pattern with a few to several kilometers units. This feature is well shown in the profile I. Thrust faults in the accretionary wedge turn to lower angle on near surface showing overthrusted feature on the sea side imbricated unit in some of them. This feature is the same one which is suggested as a possible mechanism of accretionary wedge by SEELY and others (1974). There is an exception in the initial thrust faults near the trough where is no overthrusted feature on near surface. The low angle thrust faults turn to decollement on their extension to the lower horizon of accretionary wedge. It is difficult to ascertain the reflectors are of boundary of bedding or of faulted plain. However, some of them are distinguishable from the trace of reflectors. It is obvious that there are decollements especially in the lower horizon of accretionary wedge. The thickness of underplated sheet by decollement are in ratios from one third to one fifth of accretionary wedge within a few tens of kilometers from the trough. They are generated even in the foot of the inner trench slope immediately inner side of the trough where high angle thrust faults are dominant in the upper horizon.

Some of overthrust faults in the lower part of the initial imbricated units offset to the neighboured thrust fault of landward side, and overthrust faults in the landward side are discontinuous with wavy pattern offset. Terrigenous sediments which have horizontally layered pattern are distributed in the depressions of thrusted blocks approximately 15 km inner part from the trough. Turbidites overlay unconformably on the hemipelagic sediments in the Nankai Trough. Some thrust faults or reverse faults are seen on the outer trench slope in the Profile G and H.

In multi-channel seismic profiling, it is possible to have stacked (CDP) velocities and associated calculated internal velocities. It is commonly not accurate ones than that by refraction method, for horizontal distance in measurement is limited in this method. However, we can refer them to show additional signals for bottom constituents.

The Shikoku Basin consists of three remarkable layers of the upper sediments with velocity of 1.6–1.7 km/s, the lower sediments with veloci-

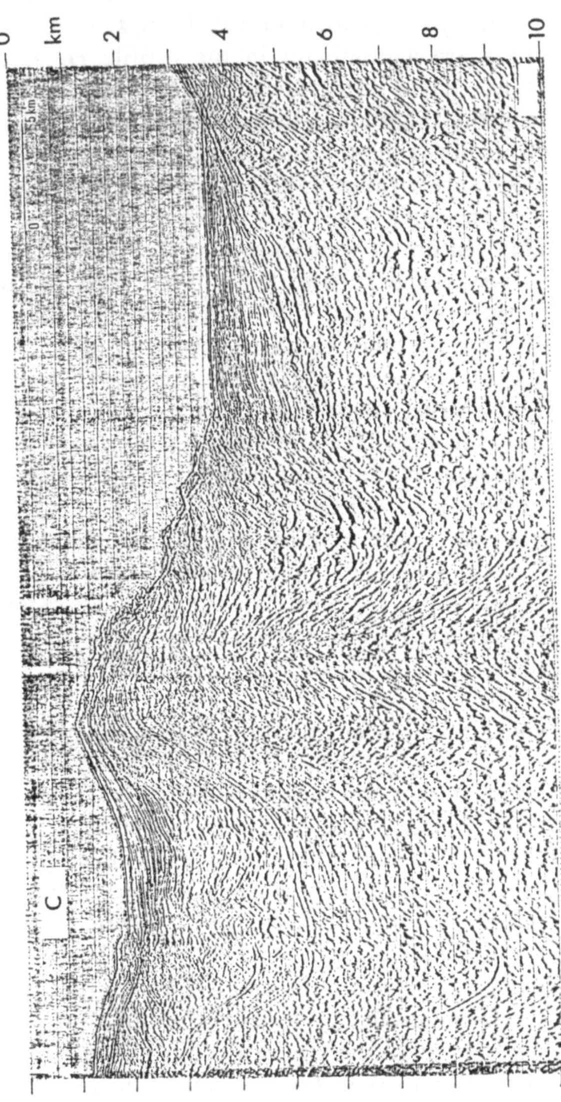
0
km
2
4
6
8
10
C
0 5 km

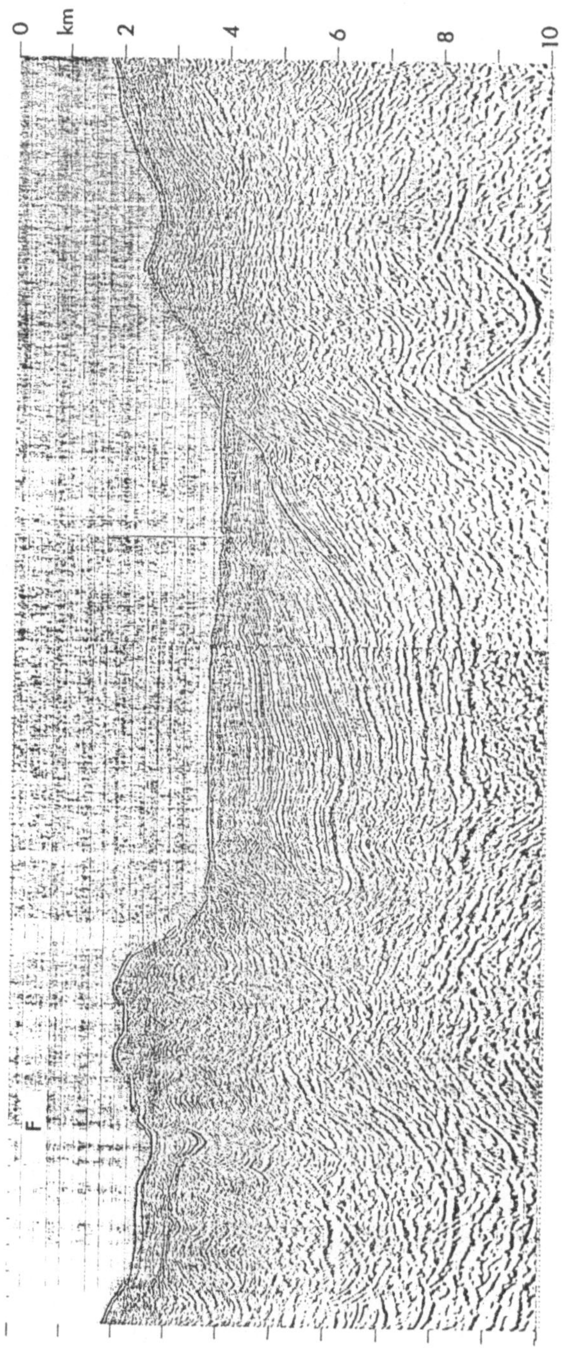

Fig. 3. Migrated depth sections C and F.

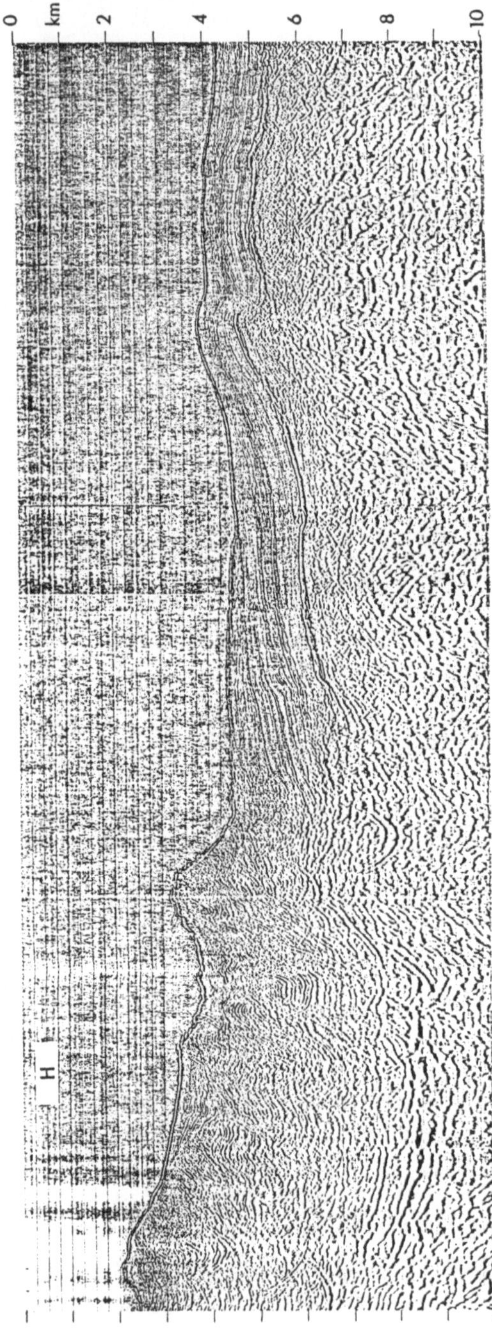
0
km
2
4
6
8
10
H

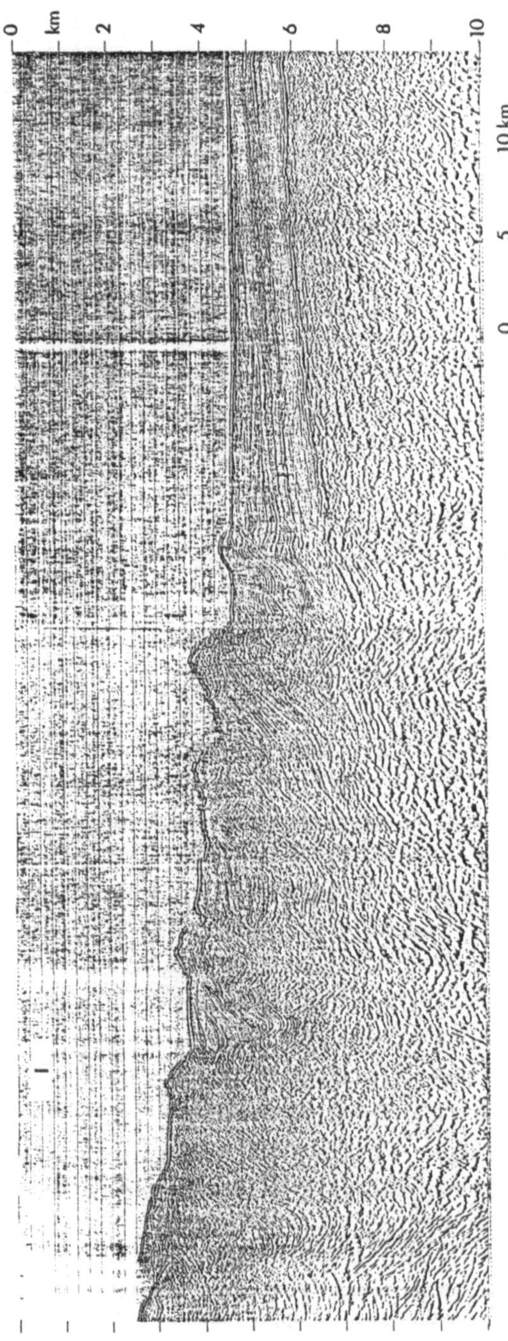

Fig. 4. Migrated depth sections H and I.

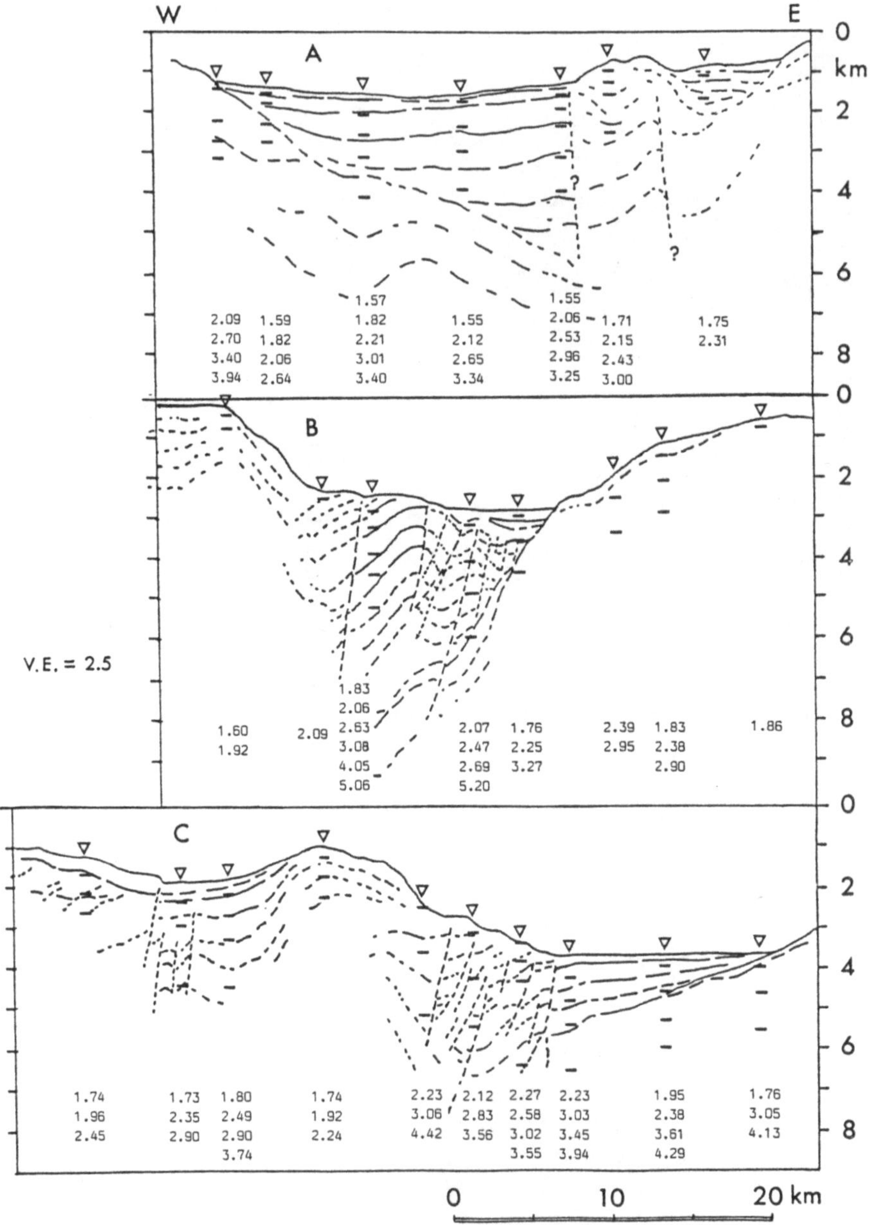

Fig. 5. Interpreted profiles of migrated depth sections A, B and C. Triangles are sites for internal velocities which are shown in the lower or upper parts.

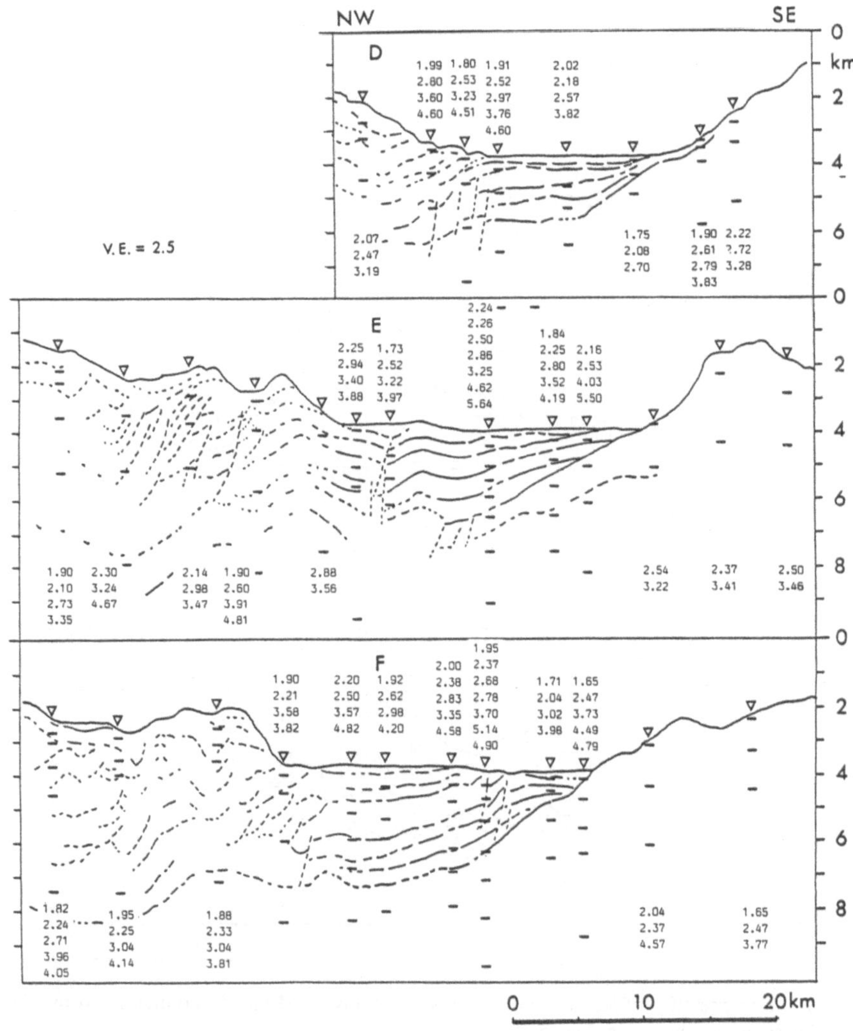

Fig. 6. Interpreted profiles of migrated depth sections D, E and F. Triangles and numbers are the same as shown in Fig. 5.

ty of 2.0–2.4 km/s and acoustic basement with velocity of 3.2–3.8 km/s or higher. Turbidites in the Nankai Trough has velocities of 1.7–2.0 km/s in the upper part and 2.1–2.6 km/s in the lower part. Pelagic sediments overlain by turbidites in the trough are in a little higher velocity

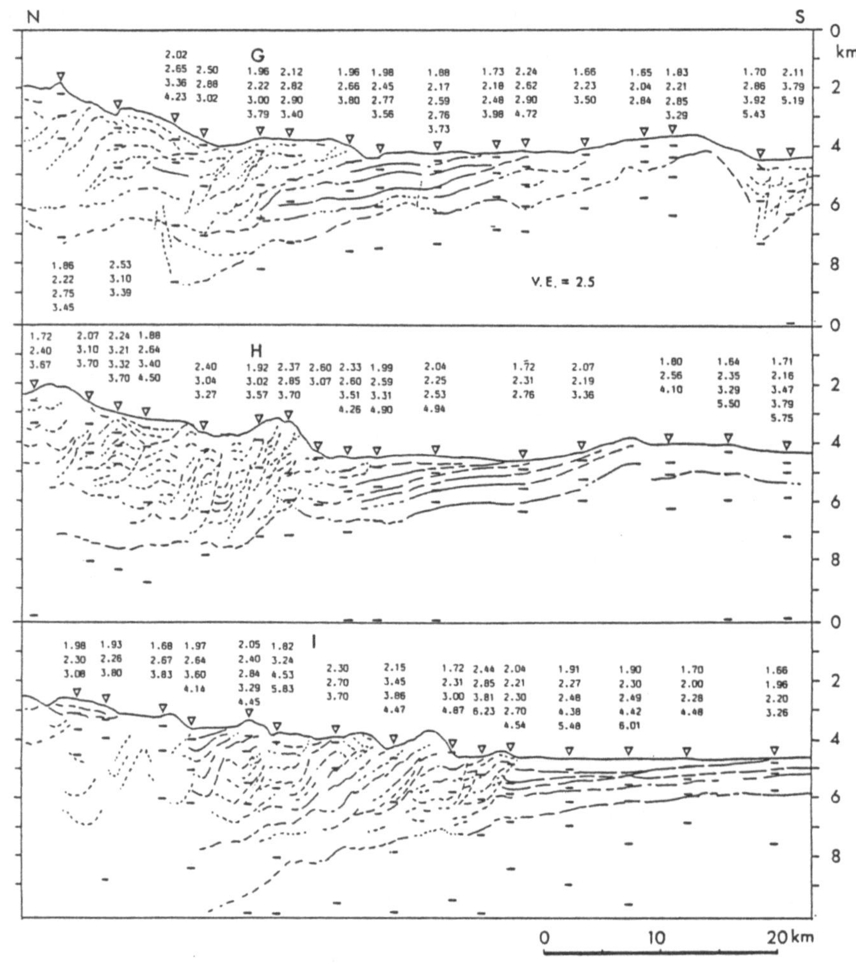

Fig. 7. Interpreted profiles of migrated depth sections G, H and I. Triangles and numbers are the same as shown in Fig. 5.

than that in the basin and are in the velocity with 2.3–2.9 km/s. Velocity of the layers within thrust faults tends to increase toward deeper part. For example, velocities are 1.96, 2.82, 3.00 and 3.02 km/s from the upper to the lower part in descending order in a layer of the profile G and are 2.15, 3.70 and 4.53 km/s in the same order as the former one in a layer of the profile I. From these velocities, it is noted that the velocity of the layer increase toward deeper part up to 50 to 100 percent of the surficial velocity. There may be more or less contribution of overburden pressure for velocity increase, however neighbouring layers are not in the same velocity, and are variable. This suggests that the horizontal compression is dominant force in the accretionary wedge. Landward extension of the turbidites on the foot of the inner trench slope where some thrust faults are developed in them are also in a higher velocity than that in the trough. The velocities throughout the underplated sediments with the decollements are in the same level as is seen in the foot of the inner trench slope. This fact may suggest that the initial compression by thrust fault occurs in the foot of the inner trench slope and once the sediments are displaced by thrust faults, there may be not much increase of compressional force below the inner trench slope till the next stage deformation may occur in the deeper part of the landward side.

4. Formation Mechanism of Accretionary Wedge in the Nankai Trough

A few models for accretion in the Nankai Trough are distinguished in the profiles presented here. The one is a dominant formation of rather higher angle imbricated thrust faults. In this case, some of thrust faults may develop through the subducted basement as are shown in the profiles B and C. This process may contribute to the underplating of oceanic basement in the accretionary wedge. The next is dominant in rather lower angle imbricated thrust faults turning to overthrust faults near surface and to decollements in the deeper part. The third is a mixture of both high and low angle thrust faults or overthrust faults. Imbricated thrust faults are blocked in units of high and low angle thrust faults. These features may be controlled by both dip angle of the subducted basement and its morphology. Layered sediments which overlay on the hemipelagic sediments in the Shikoku Basin consist of turbidites (KINOSHITA *et al.*, 1978). Thick turbidites may offer a preferable condition for offscraping and underplating in accretion (COWAN and SILLING, 1978; MOORE *et al.*, 1982).

Four structural provinces are distinguished in the accretionary wedge of the eastern part of the Nankai Trough (KAGAMI, in this volume; KAGAMI *et al.*, 1983). Almost the same structural features are documented on both sides of the Trough, except a little difference on the formation of decollements. In the west part of the Nankai Trough, initial decollement generated near trough can trace for 100 km along the trough and extends 25 km landward from the trough (KAGAMI *et al.*, 1983). In the east part of the Nankai Trough, there is no evidence of extinction of initial decollement, but show thickening of layers by decollements.

Accretionary wedge in the east part of Nankai Trough is variable in profiles remarkably changing their angle of landward dipping reflectors (LDRs). It is inferred from the interpretation of these profiles that angles of distinguishable thrust or overthrust faults are controlled by both subduction angle and morphology of subducted basement. If subduction angle is high, high angle thrust faults are dominant feature in the accretionary wedge (profile B). If there is a high, in spite of small or large one, thrust faults on the high is complicated to form many high angle thrust faults on both sides of the high (profile C, E, F and H). On a smooth basement, relatively uniform imbricated thrust faults with relatively long segments are formed.

Imbricated thrust faults are discontinuous with wavy patterns. In many cases of landward imbricated units approximately 10 km from the foot of inner trench slope, layers near surface are overthrusted and have depressions where some amount of terrigenuous sediments are ponded. This overthrusted feature is concordant with an offscraping model initially documented by SEELY and others (1974). Apparently, offscraped units are in an uniform thickness in total throughout the accretionary wedge, except a few initial units near the trough. Total landward thickening of accretionary wedge by the down dipping of subsided slub is supplied by landward thickening of underplated sheets (Fig. 8). Another possible mechanism for landward thickening of the accretionary wedge is differential viscous flow supplied from the lower accreted layer which is based on the experimental data (COWAN and SILLING, 1978). However, there are also increase of decollements in the underplated sheets in the Nankai Trough.

5. *Discussion and Concluding Remarks*

Forearc is an outer part of an arc from volcanic chain (SEELY, 1979; HONZA, 1981). Several forearc geomorphologic models were proposed in late 1970s to early 1980s (DICKINSON and SEELY, 1979; SEELY, 1979; SHOLL *et al.*, 1980; HONZA, 1981). They are based on the tectonics at trench, ridge and basin. One of the distinguished category in these models is accretionary wedge at trench. Forearc without accretionary wedge at trench is categorized as consumption forearc or that of subduction erosion (KULM and SCHWELLER, 1977). This type of forearc is noted in the Mariana, and a part of the Middle America and the Chili Trenches (HUSSONG, UYEDA *et al.*, 1982; VON HUENE, AUBOUIN *et al.*, 1982; KULM and SCHWELLER, 1977). The Bonin Trench may be in the same category, nevertheless there

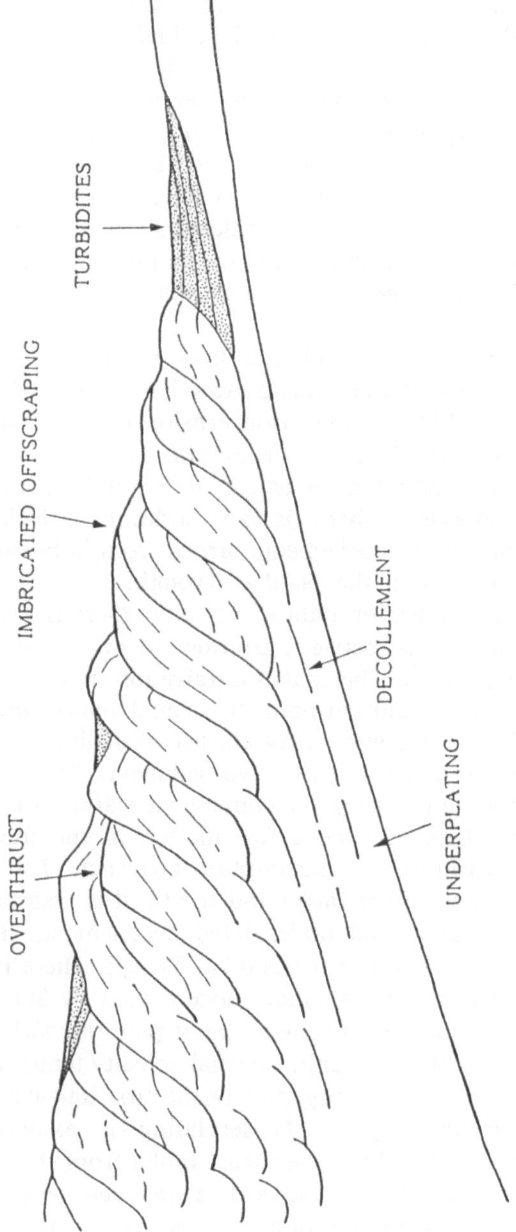

Fig. 8. A model of accretion in the Nankai Trough. Imbricated offscraping by thrust and overthrust faults are dominant in the upper horizon and underplating by decollements are dominant in the lower horizon.

are a little possible accretion on the foot of the inner trench slope (HONZA and TAMAKI, in press). Many trenches in the Pacific Rim are associated with accretionary wedge. Some of them make a high on the trench slope break prevailing in the whole of the inner trench slope. Some types of forearc is distinguished on the basis of fundamental elements which constitute arc and of geomorphology of accretionary wedge (HONZA, 1981).

Subduction complex in the trench is formed by horizontal compressional stress at trench. This is well suggested in the imbricated thrust faults in them. However, there is sometimes underplated sheet on the subducting oceanic basement associated with imbricated thrust faults overlain (COWAN and SILLING, 1978; WATKINS *et al.*, 1981; CLOOS, 1982; MOORE *et al.*, 1982).

Landward thickening of underplating may be different from the underplating process noted in the Middle America Trench (WATKINS *et al.*, 1981; MOORE *et al.*, 1982). Subduction erosion process noted in the south of the Middle America Trench (VON HUENE, AUBOUIN *et al.*, 1982) may apparently show the same feature as the underplating process in the Nankai Trough, nevertheless there is an additional thickening of underplating by generation of decollements turned from imbricated thrust faults of the upper horizon in the Nankai Trough.

In almost of seismic reflection data in forearcs, there is a less reflective area in the landward side between continental mass and trench or former and accretionary wedge. This makes a transition zone between them and may suggest some deformation of both continental and oceanic material by subduction and that in accretion may consist of sediments deposited before onset of accretion or earliest accreted sediments (WATKINS *et al.*, 1981). From these surveyed results in the convergent margins of both sides of the Pacific Ocean, there may be a few models in the forearc. The first is a subduction erosion or a consumption at trench. The second is association with imbricated thrust faults and the third is association with both imbricated thrust faults and underplated decollements (Fig. 9).

We presented here accretion in the Nankai Trough where imbricated thrust faults are concordant with a model documented by SEELY (1974). Thrust faults are discontinuous and show wavy patterns with a few to several kilometers long. Within an unit, internal velocity tends to increase to lower part. There is a high velocity even in the first unit neighbouring the thrust fault near the trough. Underplating is associated with decollements which are turned from the thrust faults from the upper part and tends to increase in thickness landward. Total increase of thickness in accretionary wedge on descending subducted basement under arc is supplied by increase of underplating. Offscraping is approximately in the same

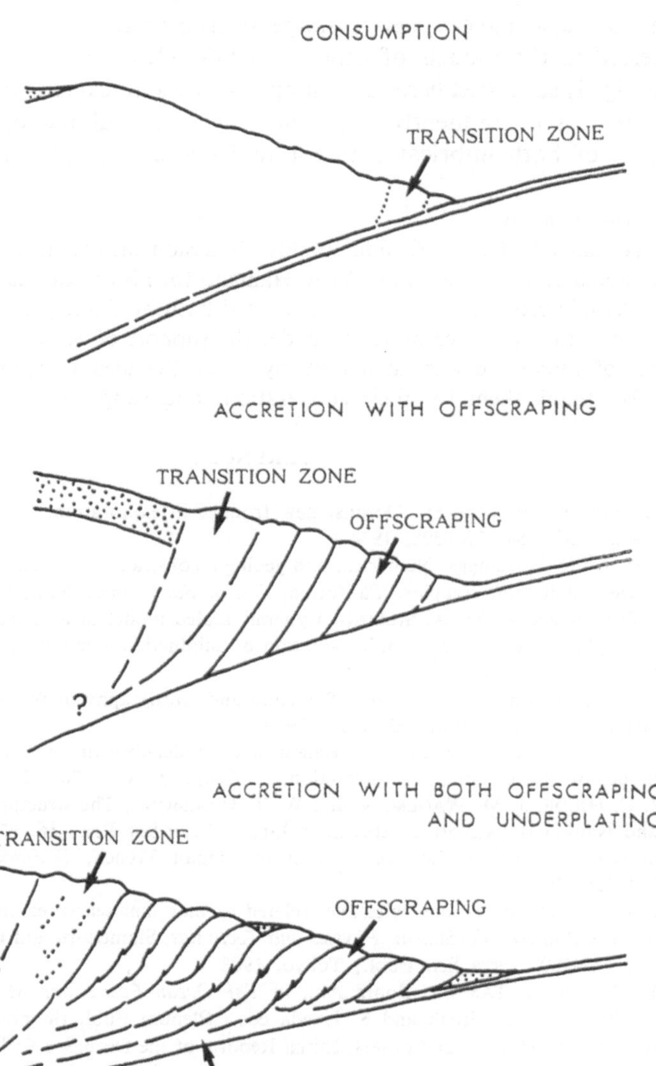

Fig. 9. Forearc models on basis of formation mechanism in the convergent margin. Consumption or subduction erosion is associated with subduction of oceanic and sometime continental materials. Some of accretion are only associated with offscraping and some of that with both offscraping and underplating.

thickness throughout the lower and upper slope of accretion. Accretion itself may also have a wedge shape profile in the Nankai Trough as is suggested in the models of other trenches. However, the deeper part of the wedge is suggested here to be supplied by the thickening of decollements. This process consequently imply decreasing of subduction angle by the increase of both imbricated thrust faults and underplated decollements.

Acknowledgements

We thank to Dr. Y. Okuda for his discussion on the history of the Seinan Japan Forearc. We thank to Dr. Yozo Hamano for his critical reading and suggestions. Seismic profiling survey was conducted both by Geological Survey of Japan and Hydrographic Office of Japan under the support of Science and Technology Agency of Japan and was carried out by Japan Petroleum Exploration Corporation. We thank them for their cooperations and supports.

REFERENCES

R. H. BECK and P. LEHNER, Oceans, new frontier in exploration, *Amer. Assoc. Petrol. Geol., Bull.*, **58**, 376–395, 1974.

M. CLOOS, Flow melanges; Numerical and geologic constraints on their origin in the Franciscan subduction complex, California, *Geolo. Soc. Amer. Bull.*, **93**, 330–345, 1982.

D. S. COWAN and R. M. A. SILLING, Dynamic scaled model of accretion at trenches and its implications for the tectonic evolution of subduction complexes, *J. Geophys. Res.*, **83**, 5389–5396, 1978.

W. R. DICKINSON and D. R. SEELY, Structure and stratigraphy of forearc regions, *Amer. Assoc. Pet. Geol. Bull.*, **63**, 2–31, 1979.

T. J. FITCH and C. H. SCHOLZ, Mechanism of under-thrusting in southwest Japan: A model of convergent plate interaction, *J. Geophys. Res.*, **76**, 7260–7292, 1971.

T. W. C. HILDE, J. M. WAGEMAN, and W. T. HAMMOND, The structure of Tosa Terrace and Nankai trough off southeastern Japan, *Deep-Sea Res.*, **16**, 67–75, 1969.

E. HONZA, Subduction and accretion in the Japan Trench, *Oceanol. Acta*, sp. no., 251–258, 1981.

E. HONZA, Evolution of arc volcanism related to marginal sea spreading and subduction at trench, in *Arc Volcanism: Physics and Tectonics*, Shimozuru and I. Yokohama eds., pp. 177–189, Terra Sci. Publ., Tokyo, 1983.

E. HONZA and K. TAMAKI, Bonin Arc, in *The Ocean Basins and Margins*, v.7. A. E. M. Nairn, F. G. Stehli and S. Uyeda eds., Plenum Publ. (in press).

D. HUSSONG, S. UYEDA, and others, Initial Reports of the Deep Sea Drilling Project, v.60, Washington, D.C., U.S. Government Printing Office, 1982.

HYDROGRAPHIC OFFICE OF JAPAN, Central Nippon, Bathymetric chart no. 6313, 1:1,000,000, 1982.

H. KAGAMI, The structure of the accretionary wedge in the Nankai Trough, this volume.

H. KAGAMI, K. SHIONO, and A. TAIRA, Subduction and accretion of plate in the Nankai Trough, *Kagaku*, **53**, 429–438, 1983 (in Japanese).

D. E. KARIG, H. KAGAMI, and DSDP Leg 87 SCIENTIFIC PARTY, Varied responses to subduction in Nankai Trough and Japan Trench forearcs, *Nature*, **304**, 148–151, 1983.

D. E. KARIG and G. F. SHARMAN THE 3RD, Subduction and accretion in trenches, *Geol.*

Soc. Am. Bull., **86**, 377–389, 1975.

S. KATO, T. SATO, and M. SAKURAI, Multi-Channel seismic reflection survey in the Nankai, Suruga and Sagami Troughs, *Rept., Hydrog. Res.*, n. 18, 1–23, Hydrographic office of Japan, 1983 (in Japanese with English abstract).

M. KIMURA, Geological structure in the south Kanto, in *Marine Geology*, Nasu ed., pp. 155–181, Univ. of Tokyo Press, 1975.

H. KINOSHITA, Y. OKUDA, R. GRAPES, and E. INOUE, Sediments and rocks from the continental slope and deep-sea bottom off the outer zone of southwest Japan, in *Investigations of the Continental Margin of Southwest Japan*, Cruise Rept., no. 9, E. Inoue ed., pp. 22–29, 1978.

L. D. KULM and W. J. SCHWELLER, A preliminary analysis of the subduction processes along the Andian continental margin, 6 to 45 S, in *Island Arcs, Deep Sea Trenches and Back-Arc Basins*, M. Talwani and W. C. Pitman, the 3rd eds., Am. Geophys. Union, Maurice Ewing Ser. 1, pp. 285–301, 1977.

J. C. MOORE, J. S. WATKINS, T. H. SHIPLEY, K. J. MCMILLEN, S. B. BECHMAN, and N. LUNDBERG, Geology and tectonic evolution of a juvenile accretionary terrain along a truncated convergent margin: Synthesis of results from Log 66 of the Deep Sea Drilling Project, Southern Mexico, *Geol. Soc. Am. Bull.*, **93**, 847–861, 1982.

Y. OKUDA, M. KUMAGAI, and K. TAMAKI, Tectonic development of the continental slope and its peripheral area off Southwest Japan in relation to sedimentary sequences in sedimentary basins, *J. Jap. Assoc. Petrol. Tech.*, **44**, 279–290, 1979.

Y. OKUDA, K. TAMAKI, and M. JOSHIMA, Seismic reflection survey results, in *Investigation of the Continental Margin of the Southwest Japan*, Cruise Rept. n. 9, E. Inoue ed., pp. 10–16, Geol. Surv. Japan, 1978.

D. R. SEELY, the Evolution of structural highs bordering major forearc basins, in *Geological and Geophysical Investigations of Continental Margins*, edited by J. S. Watkins, L. Montadert and P. W. Dickerson, *Am. Assoc. Pet. Geol. Mem.*, 29, pp. 245–260, 1979.

D. R. SEELY, P. R. VAIL, and G. G. WALTON, Trench slope model, in *The Geology of Continental Margins*, Burk, C. A., and Drake, C. L., eds., pp. 249–260, Springer Verlag, New York, 1974.

D. W. SCHOLL, R. VON HUENE, T. L. VALLIER, and D. G. HOWELL, Sedimentary masses and concepts about tectonic processes at underthrust ocean margins, *Geology*, **8**, 564–568, 1980.

K. SHIMAZAKI, K. NAKAMURA, and N. YONEKURA, The Suruga and the Sagami Troughs: Movement deduced from geodesy and morphology, and their relation to plate movement, *The Earth Monthly (Tokyo)*, **31**, 455–463, 1981 (in Japanese).

K. SHIONO, Focal mechanism of major earthquakes in southwest Japan and their tectonic significance, *J. Phys. Earth*, **25**, 1–26, 1977.

T. TAIRA, Formation mechanism of the Shimanto Group, *Kagaku*, **51**, 516–523, 1981 (in Japanese).

R. VON HUENE, J. AUBOUIN, and SHIPBOARD SCIENTIFIC MEMBERS, Leg 67, DSDP, Mid-America Transect off Guatemala, *Geol. Soc. Amer. Bull.*, **91**, 421–432, 1982.

R. VON HUENE, M. LANGSETH, N. NASU, and H. OKADA, A summary of Cenozoic tectonic history along the IPOD Japan Trench transect, *Geol. Soc. Amer. Bull.*, **93**, 829–846, 1982.

R. VON HUENE, N. NASU, and SHIPBOARD SCIENTIFIC MEMBERS, Initial reports of the Deep Sea Drilling Project, Leg 56, 57, pt. 1., 629 pp., Washington, D.C., U.S. Government Print Office, 1980.

J. S. WATKINS, J. C. MOORE, and SHIPBOARD SCIENTIFIC MEMBERS, Initial reports of the Deep Sea Drilling project, v.66, Washington, D.C., U.S. Government Print Office, 1982.

J. S. WATKINS, J. C. MOORE, T. H. SHIPLEY, S. B. BACKMAN, F. W. BEGHTEL, A. BUTT, B. M. DIDYK, J. K. LEGGET, N. LUMDERG, K. J. MCMILLEN, N. NIITSUMA, L. E. SHEPHERD, J. F. STEPHAN, and H. STRADNER, Accretion, underplating, subduction and tectonic evolution, Middle America Trench, South Mexico: result from DSDP Leg 66, *Oceanol. Acta*, sp. no., 213–224, 1981.

CHAPTER 3: ARC MAGMATISM

Formation of Active Ocean Margins, edited by N. Nasu *et al.*, pp. 367–397.

ARC MAGMATISM—AN UNRESOLVED PROBLEM OF SOURCES, MATERIAL FLUXES, TECTONIC EVOLUTION AND THERMOCHEMICAL REGIONS OF SUBDUCTION ZONES

Richard J. ARCULUS

Department of Geological Sciences, University of Michigan, 1006 C.C. Little Bldg., Ann Arbor, MI 48109, U.S.A.

Abstract. The nature and volumetric significance of the different potential sources of arc magmas is a subject of dispute. The resolution of this problem is important for understanding the extent of lithosphere recycling and the contribution of arc magmatism versus sediment subduction to continental growth and destruction respectively. It seems that the mantle wedge overlying the subducted slab is the major source component, but heterogeneous and systematic shifts in isotopic characteristics and key trace element ratios suggest that at least one other component is involved. Considerable data has accumulated that indicates the crust of the arc itself is significant in the petrogenesis of arc lavas, whereas the contribution from the slab appears to be trivial volumetrically. Estimated production rates of arc magmatism are of the order of 1–3 km^3yr^{-1}, and equivalent to some estimates of sediment subduction rates. If the subducted sediment is not recycled in arc magmas, then a net growth of the mantle reservoir of incompatible lithophile elements including radiogenic K, U and Th is implicated. Long-term (1Gyr) recycling of subducted sediment via hotspot magmatism is a possibility.

Characteristics like the high Ba/Nb of arc lavas are not easily reconciled with purely crust and mantle wedge involvement, and a degree of fluid recycling through a wedge-slab convective system might be considered, with retention of Nb, Ta and Ti in a refractory phase in the upper layers of the subducted slab. Processes leading to the potential juxtaposition of heterogeneous mantle source components are the migration of arcs from neighboring continental lithosphere and the formation of back-arc basins.

Evidence that sub-oceanic and sub-continental lithosphere have different chemical and mechanical behavior suggests that chemical and isotopic variation in the spectrum of arc types from intraoceanic to continental margin may reflect the variable involvement of different lithospheric components.

1. Introduction

The magmatism and tectonism associated with active ocean margins in the plate collision zones of island arc and Andean-type volcanic belts, have been among the major contributors towards continental growth and structural modification during the past 1 Gyr of terrestrial evolution. Apart from the intense practical concern with which the association of dangerously explosive volcanism, destructive earthquakes and the development of some major ore deposit types at active ocean margins is viewed, there is also critical interest in the role of these regions in the more general dynamic and chemical cycles of the Earth.

Despite considerable research and the application of sophisticated geochemical and geophysical analytical techniques, much remains uncertain about the nature of the processes occurring along subduction zones and in the genesis of arc magmas. For example, the nature and significance of different potential magma sources is one of the subjects of dispute, and this is not a trivial issue because the extent of lithosphere (oceanic crust plus entrained sediments, hydrospheric and biospheric components) recycling is really the point of debate.

The nature and volumetric proportions of material fluxes in subduction zones is also a controversial topic. Some argue that there is a volumetric deficiency of exposed sediments in forearc terrains, and that erosion of both these sediments and possibly cratonized crust must have taken place (e.g. FYFE, 1979) to be partially recycled in arc magmas. Others suggest that any removal of sediments from forearc regions is restricted to an underplating of this material at relatively shallow levels (<30 km), and that minimal sediment is recycled into the upper mantle (KARIG and KAY, 1981). The mass balance of volatile emissions from volcanoes of H_2O, CO_2, F, and Cl for example, with respect to the possible subduction of these components in the oceanic lithosphere is also by no means resolved (ITO *et al.*, 1983).

The extent to which the tectonic evolution of arc systems is related to the type of magma erupted from the constituent volcanoes, and the spatial and temporal evolution of the various magma types is problematic (e.g. ARCULUS and JOHNSON, 1978). Geochemical differences between continental margin arcs (e.g. the Andean chain), those arcs floored by continental crust but separated by a back-arc basin from the nearest continental landmass, and those arcs apparently surrounded by oceanic lithosphere with no continental substrate (e.g. the Marianas-Izu Bonin system) suggest that the nature of the crust of the arc (COULON and THORPE, 1981; MATSUHISA and KURASAWA, 1983) and the subcrustal lithosphere as much as the subducted slab are important (see also LEEMAN, 1983), but our knowledge of the geometry of lithosphere distribution and evolution in ocean margins is fairly rudimentary.

Finally in this introductory catalogue of uncertainties is the puzzle presented by the thermal and dynamic regimes of subduction zones. Many geoscientists have pointed out the anomalous association of heat release manifested by volcanism in arcs with the insertion of a heat sink in the form of a cold, subducted lithosphere beneath the arc. Furthermore, where there is an extensive distribution of volcanoes normal to the axis of a trench, such as in the Japanese islands, the most voluminous and active volcanism is typically located over the apparently colder and shallower parts of the arc-trench system (KUSHIRO, 1983).

2. *Arc Magma Sources*

The chemical and isotopic characteristics of arc magmas have been examined in great detail with the aim of identifying the nature of the magma sources involved, evolutionary processes and controls on eruptive style. Recognition of the zonation of magma type and general relationship to the underlying Wadati-Benioff zone in arc-trench systems is of longstanding, and may be exemplified by the studies of SUGIMURA (1960) and KUNO (1959) in Japan, and VAN BEMMELEN (1949) in Indonesia. With the introduction of the concepts of plate tectonics in the 1960's, there was early enthusiasm for attributing the nature of arc magmas to wholesale melting of the upper layers of the subducted slab (e.g. GREEN and RINGWOOD, 1968). However, most petrologists would now accept that the upper mantle wedge overlying the subducted lithosphere is volumetrically the major source of arc magmas with contributions from the subducted slab being obscure or trivial in the same terms, but of great significance in both the triggering of magma genesis, and vital in the sense of the introduction of subtle trace element and isotopic characteristics (see GILL, 1981, for a review). It is worthwhile examining the reasons for this consensus viewpoint, and pointing out the difficulties and uncertainties that still impede our full understanding of the nature of the sources of magmas in arcs.

a) *Major element chemistry* The primary observation that may be used to discriminate between a peridotitic upper mantle or metabasaltic (amphibolitic to eclogitic) source is the near-identity of basalts in both arc and non-collisional zone environments. A comprehensive review of the chemistry of basalts in island arcs was presented by PERFIT *et al.* (1980), and attention drawn to the remarkable similarity between mid-ocean ridge basalt (MORB) types, and the vast majority of basalts in arcs which are tholeiitic in character. On a historical note, it should be emphasized that this similarity was implicitly stated by KUNO (1968) in his suggestion that the variety of primary magmas found in arcs is equivalent to the variety found in other environments.

One of the difficulties in the process of inversion of erupted magma chemistry to the unravelling of the nature of the mantle source compositions is the apparently fractionated character of most arc basalts. In this regard also, arc basalts and MORB share a common major characteristic which is a median Mg/(Mg + Fe) ratio in the vicinity of 0.5 (see however, KUSHIRO, 1983) which suggests that both magmatic types have mostly experienced extensive fractionation and approach 1 bar cotectic equilibrium. Consequently, few are representative of primary magmas that could have been in equilibrium with a peridotitic source (PERFIT *et al.*, 1980).

Two possibilities to account for the similarity of arc and ocean ridge tholeiite chemistry have been generally proposed. The first of these which found early support with the development of the plate tectonic paradigm was melting of the subducted slab, and in particular the basaltic layer 2.

However, only complete melting of portions of layer 2 of the subducted lithosphere without any intervening fractionation from source to surface could account for the similarity in composition of tholieiitic magmas in arc and spreading ridge environments. It is theoretically feasible of course, that the minimally hydrous *liquidus* of a subducted basaltic layer could be at a lower temperature than the dry *solidus* of a peridotitic upper mantle overlying the slab, and thus complete melting of the subducted layer 2 without melting the overlying wedge might be invoked. Note that even though the wet solidus of basaltic compositions at pressures of about 40 kbar is in the range 900° C compared with the dry solidus of peridotite at the same pressure of about 1600° C, it is necessary for *complete* fusion of the basaltic layer 2 for the duplication of erupted tholeiite compositions in arc volcanoes, if no intervening fractionation and mantle interaction takes place. Dynamic isolation of the slab from the wedge during complete fusion of layer 2 seems improbable.

It can be argued that it is inherently more likely that only partial fusion of the subducted slab takes place, and that this melt fraction interacts with the overlying mantle wedge prior to a second stage of melting of a compositionally transformed wedge at shallower levels in the mantle (e.g. NICHOLLS and RINGWOOD, 1973).

This type of model is a hybrid between the end-member possibility of melting of the subducted slab with no interaction between melt and overlying wedge, and the hypothesis that arc magma sources are confined to a MORB-like upper mantle wedge with no contribution whatsoever from the slab. It might be expected that the upper mantle wedge would become modified to some extent if invaded and altered by interaction with a *partial melt* of the subducted layer 2 (plus, perhaps some component from any portions of layer 1 that survive subduction). Given the similarity in terms of major element geochemistry between low-K tholeiitic basalts in arc and oceanic ridge environments, it would seem that the upper mantle wedge is not extensively modified by such partial melt infiltration processes if the hypothesis is accepted that arc magmas are derived by partial fusion of the wedge.

It is also worth noting that the spectrum of magma compositions in arcs includes some highly alkaline and potassic basalt types that do not seem likely to be partial or even total fusion products of a metamorphosed MORB source in the subducted slab. An alternative explanation for the local development of both sodic and potassic lavas in arc environments is a modification of the hybrid source model described above wherein the metasomatism of the mantle wedge by a silica-alkali-rich hydrous fluid (e.g. GILL, 1981) rather than silicate partial melt from the subducted slab is implicated. Granted that a spectrum of alkalinity is present in arc magmas and that such a model might be locally applicable to some melt types, it has not been demonstrated that a *persistent* shift towards higher silica modes (cf. RINGWOOD, 1975) or more alkaline characteristics are present in arc lavas compared with those generated at mid-ocean ridges or oceanic hot-spots. Thus pervasive and significant modification of the indigenous alkali abundances in the wedge prior to subduction does not seem to be present. With respect to the silica mode of arc lavas however, the non-shift with respect to MORB is not compelling evidence against an alteration of the olivine-orthopyroxene ratio in the mantle wedge as a result of silica metasomatism. Given any type of eutectic-like melting process in the wedge, then a change in the relative proportions of olivine and orthopyroxene will not result in a major change in the melt composition generated by fusion of the wedge until all the olivine is transformed to orthopyroxene or unless loss of residual olivine occurs during melting of the wedge (see YODER, 1976). Neither of these processes seems likely given the modal predominance of olivine in peridotitic sources.

Finally with respect to major element characteristics, it is emphasized that there is abundant experimental evidence together with chemical constraints that deny the possibility that the vast bulk of evolved rocks in arcs such as andesites, dacites and rhyolites, could be unmodified melts derived directly from either an upper mantle peridotite or subducted lithosphere source (GILL, 1981). Although there is evidence that some inappropriately labelled "high-Mg andesites" (including boninites) are primary melts of hydrous peridotite sources (NICHOLLS and RINGWOOD, 1973; KUSHIRO and SATO, 1978; TATSUMI, 1981), these are relatively rare intermediate rock types in arcs, and the majority of intermediate and silicic volcanics appear to be derived by complex combinations of fractional crystallization, partial assimilation and mixing processes within the crust.

b) *Trace Element Geochemistry* There are however, some critical differences between basalts found in arcs, spreading ridge and oceanic island hot-spot environments that suggest at least one additional source component, or process, must be involved in the genesis of magmas at actively colliding ocean margins. The most striking of these, and the one that has resisted all convincing explanations, is the persistently low abundance of the Ti group elements (e.g. Ti, Nb, and Ta) in arc basalts (e.g. CHAYES and VELDE, 1965; SUN, 1980; WOOD, 1979). This difference is not so obvious when arc tholeiitic basalts of relatively high Mg/(Mg + Fe) are compared (PERFIT *et al.*, 1980; GILL, 1981) which suggests, in the case of Ti at least, that fractionation events en route to the surface may be partly responsible for the divergence in relative abundances in more fractionated arc and ridge magmas. Such explanations of course implicitly invoke subtle differences in inherited volatile complements or other redox changes to account for the earlier precipitation of phases such as titanomagnetite in the case of arc basalts than in MORB. These subtle differences may be inherited from source conditions in the mantle or the result of some kind of contamination event along the lines envisaged by OSBORN (1959).

Nevertheless, even strongly alkaline, mafic, silica-undersaturated and feldspathoid-bearing magmas in arcs share the property of low Ti-group abundances, and in the case of these magma types, this feature cannot be dismissed as resulting from either large degrees of partial melting of a refractory source, or extensive Fe-Ti oxide fractionation (ARCULUS, 1976; FODEN and VARNE, 1980). Thus it is possible that the low-Ti group abundance in arc basalts relative to MORB and hot-spots is ubiquitous in all arc magma types although muted in the case of Ti in tholeiites.

Other indications of secondary source involvement are the anomalous relative enrichments of incompatible lithophile trace elements such as Ba and Sr, and more variably Rb and Pb in arc basalts (e.g. KAY, 1980; SUN, 1980; PERFIT *et al.*, 1980), resulting in high Ba (or Sr)/La (or Nd) ratios. In fact, one of the most effective discriminants between island arc and MORB magmas is the high Ba/Nb ratios of the former (PEARCE and NORRY, 1979) (see Fig. 1).

We are faced therefore with the curious situation of relative over abundances of some trace elements and under abundances of others with respect to basaltic magma types from non-collision environments. These patterns are not easy to explain for reasons that follow.

It might seem relatively straightforward to account for additions of elements such as the mobile alkaline earths (Ba and Sr) to the mantle wedge by metasomatic transport in a fluid generated by devolatilization processes in the slab. However, such additions cannot simultaneously explain relative depletions of trace elements such as Nb and Ta. Even if these Ti-group elements are retained in a refractory phase in the subducted lithosphere such as rutile, ilmenite or sphene (e.g. GREEN, 1980), no distortion of ratios such as Nd/Nb should be expected to result in the upper

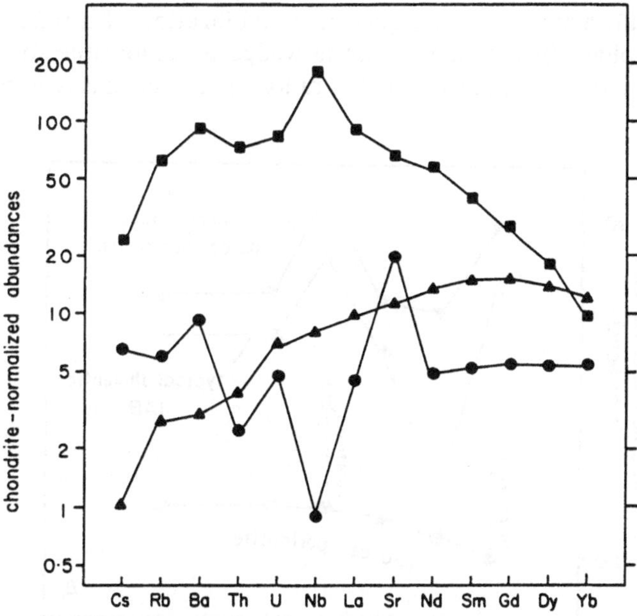

Fig. 1. Chondrite-normalized abundances of selected trace elements in representative island arc tholeiite (filled dots), N-type MORB (triangles) and hot-spot, alkali basalt (squares). Data are from SUN (1980) and ARCULUS and JOHNSON (1981).

mantle wedge as Nd, together with the other rare earth elements (REE), does not appear to be either mobilized or selectively enriched in arc magmas with respect to MORB and hot-spot lava types (see also MORRIS and HART, 1983). Thus Nd/Nb ratios comparable with MORB would be expected to result from melting an upper mantle source with no secondary additions of *either* Nd *or* Nb (Fig. 2).

Some petrogenetic models that have been constructed to account for the Sr and Nd isotopic character of arc basalts compared with MORB do implicitly invoke both Sr and REE transport from slab to the wedge (e.g. MCCULLOCH and PERFIT, 1981). However, the remarkably unfractionated and even strongly depleted REE abundaces in arc tholeiitic basalts from localities such as New Britain (DEPAOLO and JOHNSON, 1979) or the Izu Peninsula (FUJIMAKI, 1975; NOHDA and WASSERBURG, 1981) do not reveal any enrichments in the REE in the mantle source regions with respect to MORB, and in some samples are even less enriched in REE than MORB. Furthermore, no systematic shift in Nd or Sr isotopic character with respect to relative variations of bulk Sr/Nd ratios are observed in arcs like New Britain, contrary to the result anticipated if REE and Sr in metasomatic fluids infiltrate the wedge (see below).

It has alternatively been argued that stabilization of a refractory Nb-Ta-bearing phase in the upper mantle wedge is responsible for the high Nd/Nb and more dramatically Ba/Nb ratios in arc magmas (e.g. SAUNDERS

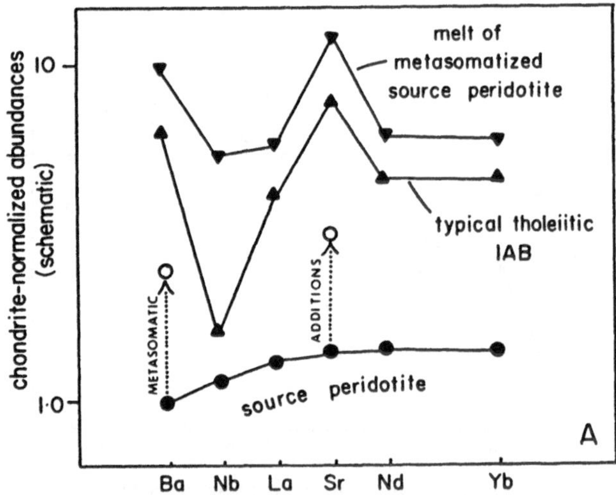

Fig. 2a Schematic effects of melting of a source peridotite with metasomatic addition of Ba and Sr but no Nd compared with the patterns typical of 1AB (see Fig. 1 also).

et al., 1980). Such models suppose that Ba (and Sr)-bearing fluids are derived from the subducted slab, and during subsequent melting of the fluxed peridotite in the mantle wedge under conditions of comparatively high water and oxygen fugacity (fH_2O and fO_2 respectively), the alkaline and rare earths are partitioned more extensively than the Ti-group elements into the melt. The deficiencies of this type of model are in part theoretical as no experimental evidence has yet been generated to justify the supposition that saturation of a peridotite with a Nb-Ta-bearing phase, if achieved under high fH_2O and fO_2 conditions, will result in the appropriate trace element characteristics in melts subsequently generated from such assemblages.

In the case of increases in fH_2O, it must be supposed that stabilization of a Ti-bearing phase such as kaersutitic amphibole or phlogopite

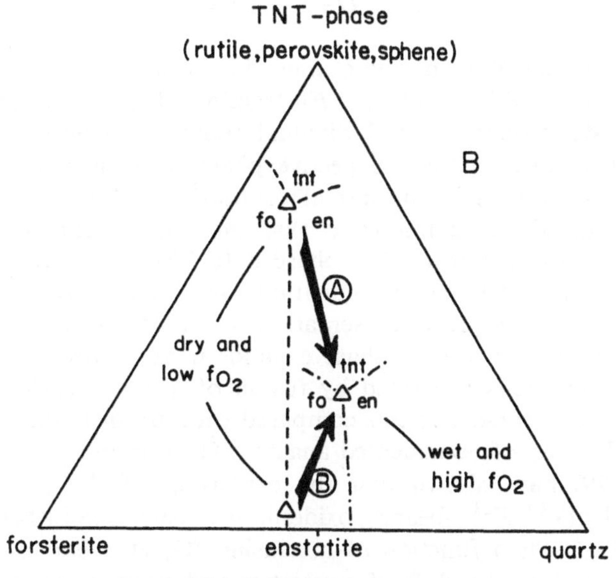

Fig. 2b Schematic phase diagram illustrating the hypothetical expansion of a TNT-bearing phase under conditions of elevated fH_2O or fO_2 in the upper mantle wedge. Note that if the TNT phase is previously present in the wedge, then expansion of the primary liquidus field results in a diminution of TNT components in the melt (A). Conversely, nucleation of the phase can result in an increase of the TNT components of a given partial melt (B) during anatexis if the phase was previously absent.

is critical. Similarly, an increase in fO_2 should result eventually in a sufficient increase in the Fe^{3+}/Fe^{2+} ratio of the mantle wedge for saturation to occur with a separate Fe^{3+}-compatible oxide such as ilmenite or magnetite. Providing no bulk compositional changes apart from H_2O or O_2 additions take place, then it is theoretically possible that the bulk distribution coefficient (D) for a given Ti-group element (= concentration of Ti-Nb-Ta (TNT) in solid assemblage/ concentration of TNT in melt) would increase, and thus melts generated from a TNT/phase-bearing peridotite would be less enriched in these elements than melts produced in the absence of a TNT phase.

For example, the appropriate equation describing relative enrichments or depletions of given elements during batch equilibrium melting (SHAW, 1970) is

$$\frac{C^{\mathrm{L}}}{C^{\mathrm{O}}} = \frac{1}{D+F(1-P)} \tag{1}$$

where C^{L} = concentration of the element in the melt, C^{O} = original concentration in the solid assemblage, F = fraction of melt, and P = the sum of all the modal proportions of individual phases contributing to the melt composition divided by their respective phase/melt element distribution coefficients. Variations in the modal proportions of phases other than TNT are assumed to be minor so that the most significant effect of saturating the wedge with a TNT phase is to increase considerably both D and P in this batch melting equation, and thereby reduce $C^{\mathrm{L}}/C^{\mathrm{O}}$.

An alternative graphical presentation in Fig. 2 can be used to argue that if the stability of a TNT-phase expands under hydrous or oxidizing conditions, then melts generated by fusion of a TNT-bearing peridotite will be depleted in these elements compared with dry or reducing anatectic conditions. By way of documented analogy, Osborn and co-workers (e.g. OSBORN, 1979) have shown that the expansion of the primary phase volumes of the Fe^{2+}-Fe^{3+}-bearing oxides in the system Mg_2SiO_4-$CaAl_2Si_2O_8$-SiO_2-FeO-Fe_2O_3 as a function of increasing fO_2 at 1 bar total pressure are accompanied by the shift of peritectics and eutectics to progressively less Fe-rich melt compositions.

Despite the instructive character of these diagrams, it is also possible to argue that they are misleading in that melting of a peridotite in the *absence* of a TNT-bearing phase is considered as one end-member situation, and that conversely melting in the *presence* of any such phase under broadly comparable conditions of pressure, temperature and percent fusion must inevitably result in a relative shift in the position of the eutectic/peritectic/cotectic melting point *towards* the TNT-phase composition (Fig. 2). Without some kind of increase in Ti content of the bulk peridotite and hence C^O in (1), it is difficult to envisage saturation of the wedge with a phase like perovskite or sphene. Furthermore, silica activities in the upper mantle peridotitic wedge, as buffered by olivine and orthopyroxene, seem too high for the stable existence of perovskite (CARMICHAEL *et al.*, 1974, p. 52, 109).

Experimental evidence in support of the supposition that melts generated from assemblages in which minor phases such as phlogopite or apatite are present are enriched in the components of the minor phase, can be found in the experiments reported by EDGAR *et al.* (1976) and Watson and colleagues (e.g. WATSON, 1980). As an example, consider the fact that fusion of phlogopite-bearing peridotite results in melts that are richer in K than those produced from phlogopite-free assemblages (EDGAR *et al.*, 1976). Further, saturation of a basalt with a phase like apatite results in considerably higher P concentrations in the melt than in the case of a basalt that is unsaturated with this phase (WATSON, 1980). In both types of experimental study of course, a change in bulk composition of the source (addition of K as well as H_2O, and addition of P) is required for saturation with a minor phase to be achieved.

Of the possible TNT-bearing phases that have been identified such as rutile, ilmenite, sphene, phlogopite or perovskite (e.g. GREEN, 1980; MORRIS and HART, 1983), there are few comparable direct experimental sets of data available on the nature of the anatectic products of peridotitic asemblages in which they are included. The study by EDGAR *et al.* (1976) convincingly demonstrated the Ti- and K-enrichment of melts in equilibrium with a phlogopite-bearing peridotite compared with phlogopite-free compositions. Similarly, GUST (1982) has shown that melts of rutile-bearing eclogite under both wet and dry conditions are enriched in Ti compared with melts generated from similar bulk compositions in the absence of rutile.

A final but important observation is appropriate for our assessment of the significance of models that appeal to the survival of a TNT-bearing phase in the upper mantle during anatexis being responsible for the relative depletion of these elements in equilibrium melts. One would suppose that low degrees of melting of a TNT-phase-bearing peridotite source such as might occur during the generation of alkalic magmas remote from subduction zones, would also result in relative depletions of these elements in the derivative melts. Cessation of subduction or transport of the modified peridotite away from a subduction zone could presumably allow the involvement of the TNT-bearing assemblage in melting events in non-collision environments. However, it is noteworthy that non-arc, hot-spot type alkalic magmas are characterized by relative enrichments of the TNT elements in a manner complementary to arc lavas (e.g. WOOD, 1979; SUN, 1980; Fig. 1).

Any petrogenetic model that accounts for the relative depletions of the TNT group in arc basalts should sensibly address complementary overabundances in hot-spot magmas. A possible explanation for the relative enrichments in hot-spots basalts is that the fusion events take place at depths in the mantle where a separate TNT-bearing phase is not stable, and all of this group are preferentially partitioned into the melt.

If a MORB-type mantle source is adopted for arc basalts and the relative depletions of the TNT-group elements in derivative melts are regarded as the result of the refractory behavior of a minor phase produced *isocompositionally* in the peridotite excepting for a hydration/oxidation event, it is hard to envisage how the TNT-group elements could be enriched dramatically with respect to MORB in subsequent melting events where the TNT-bearing phase is no longer stable or present.

The enrichment of the TNT-group elements in hot-spot magmas together with isotopic evidence that suggests different or additional mantle source components are involved compared with MORB would seem to require an alternative model than merely the isochemical production and obliteration of a TNT-bearing phase in the source peridotite. A possible coherent solution to this problem has been proposed by ARCULUS and POWELL (1984) and incorporates aspects of both types of subduction zone metasomatic/refractory phase models described above. The general approach is outlined below, subsequent to discussion of some other petrogenetic and more general constraints on the processes involved.

c) *Isotopic constraints* The characteristics of isotopic ratios have fueled the debate over the nature of island arc sources. The combined evidence from Sr, Nd, Pb, Hf, Th-U, O, He and Be isotopes can be summarized first of all, in a non-controversial manner, as follows. No island arc magmas have yet been found with isotopic values identical to those of MORB, suggesting the involvement of either a separate mantle source, or a secondary component in addition to that present in the sources of MORB (e.g. DEPAOLO and WASSERBURG, 1977; HAWKESWORTH *et al.*, 1977; 1979a; DEPAOLO and JOHNSON, 1979; NOHDA and WASSERBURG, 1981). Furthermore, a far greater range of isotopic values is present in island arc magmas than in MORB, comparable with those observed in isolated oceanic island or hot-spot volcanoes (e.g. HAWKESWORTH *et al.*, 1979b; WHITE and HOFMANN, 1982) implying heterogeneity of sources or diversity in the proportion of end-member components involved, or both (Fig. 3).

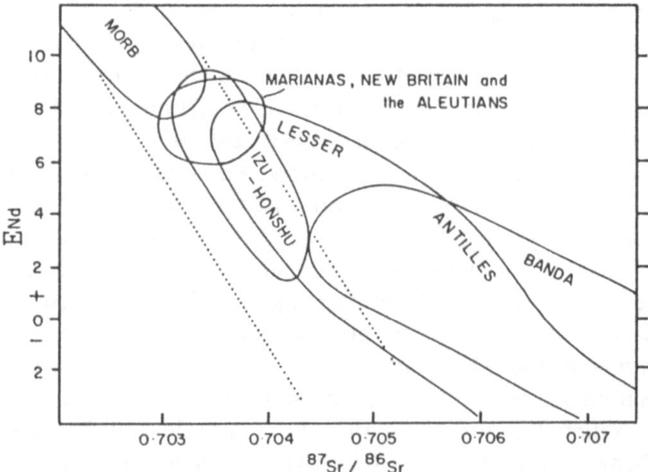

Fig. 3. Variation of $^{143}Nd/^{144}Nd$ (as Nd) versus $^{87}Sr/^{86}Sr$ in selected island arc volcanics. Dotted lines are the mantle array. Data from sources referenced in text.

The detailed interpretation of these isotopic patterns is by no means straightforward. For example, the Sr and Nd isotopic data for arcs such as the Aleutians (McCulloch and Perfit, 1981) and New Britain (DePaolo and Johnson, 1979) are located within the "mantle array", and comparable with some major hot-spot data from the islands of Hawaii and Iceland. Detailed study of the variation of Sr/La with respect to $^{87}Sr/^{86}Sr$ ratios does not reveal the type of correlation expected from the addition of a secondary and enriched radiogenic component to a primary mantle source (Arculus and Johnson, 1981; cf. McCulloch and Perfit, 1981). It has been argued (e.g. Sun, 1980) that Pb isotopic data for the Aleutians, as well as the gross correlation observed regionally of pelagic sediment Pb isotopic values with those of neighbouring arcs (Karig and Kay, 1981), are compatible with the addition of $<1\%$ subducted sediment to a MORB-type source. Comparison of Pb-Sr-Nd isotopic ratios for lavas of the arc, back-arc basin and subducted Atlantic lithosphere in the vicinity of the South Sandwich islands has been presented by Cohen and O'Nions (1982). The involvement of a relatively radiogenic component in both arc and back-arc basin lavas is suggested by these authors, and a relatively larger proportion of subducted sedimentary component seems to be involved in the genesis of the arc than the back-arc magmas.

In contrast to these studies, examples are known where Pb isotopic data show no displacement form those values typical of MORB and some hot-spot localities, and no indication of a component of subducted Pb in sediments. For example, the intra-oceanic arcs of the Marianas (MEIJER, 1976) and Tonga-Kermadec (OVERSBY and EWART, 1972), and the more problematic tectonic location of the New Britain arc (M. T. Thirlwall and R. W. Johnson, pers. comm., 1983) are environments where non-radiogenic and in some cases homogeneous Pb isotopic values are present. Note that in the case of New Britain and the majority of the Marianas data, values of Nd-Sr isotopic ratios are essentially identical to those of the Aleutians (Fig. 4) so that displacement of the Nd-Sr isotopic values from those typical of MORB cannot logically be attributed to sediment in the case of the Aleutians without the same cause being invoked for New Britain and the Marianas.

The Banda arc is one oft-quoted example of the unequivocal involvement of subducted sediment in the upper mantle source regions of the magmas (MARGARITZ *et al.*, 1978; WHITFORD *et al.*, 1977). The extensive variation of Sr-Nd-O isotopes unquestionably points to a mixture of source components, but the involvement of subducted, Australian-derived continental sediments is a more contentious issue. For example, the most radiogenic samples come from Ambon island in the northeast of the arc

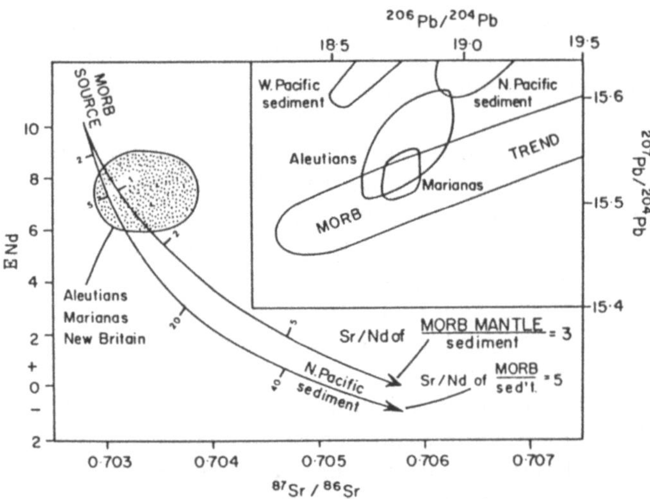

Fig. 4. Schematic effect of contamination (tick marks in wt.% contaminant) added to mantle source of MORB. Dotted region is locality of all Aleutian and New Britain data, and majority of Marianas. In contrast with Pb isotopic data for Aleutian lavas, which show possible 2-component mixing (SUN, 1980), no such effect is apparent in the Marianas data (MEIJER, 1976) or New Britain data (M. T. Thirlwall and R. W. Johnson, pers. comm., 1983).

where the appearance of cordierite phenocrysts in some of these samples is an important feature. The upper pressure stability limit of cordierite is less than 10 kbar and so the presence of this phase in arc lavas cannot be the result of survival of cordierite in subducted sediment to depths of the order of 30 kbar or more, followed by reascent in melts passing through the upper mantle wedge. Cordierite phenocrysts are also present in rhyolites from Wetar in the southwest of the arc remote from Ambon (HUTCHISON, 1982) and their presence is a strong indication that contamination events within the crust of the arc itself are primarily responsible for the isotopic arrays interpreted as consequent to mixing of separate components.

It has become increasingly clear that the evidence for a strictly oceanic-type crust in the Banda arc is equivocal, and that continental detritus is prominent on the floor of the Banda Sea to the north of the arc (J. B. Gill, pers. comm., 1984). Consequently, arguments that the *only* source of continental—derived sediments that seem to appear in Banda arc volcanics must be the subducted Indian Plate (MARGARITZ *et al.*, 1978) may not be correct. Other ambiguities in the interpretation of the location of the sediment contaminant have been highlighted in a more extensive study of $^{87}Sr/^{86}Sr$ vs. ^{18}O variations of Banda volcanics by MCCULLOCH *et al.* (1983). Contrary to the scatter of similar data previously interpreted as defining a convex-downward curvature on a plot of ^{18}O vs. $^{87}Sr/^{86}Sr$ and implying source mixing within the mantle wedge (MARGARITZ *et al.*, 1978), McCulloch and coworkers have shown an opposing trend which together with Nd and Pb isotopic data for the same sample set is strongly indicative of intra-crustal mixing events within the arc itself, and not at depths within the upper mantle.

Isotopic disequilibrium (activity ratios) in the Th-U decay series has been detected in a number of arc-type volcanoes such as Arenal and Poas in Central America (ALLÈGRE and CONDOMINES, 1976; 1982). Recent additions (of the order of 10^5 years) of U to the source regions of these magmas seems to have occurred, and one possible addition is material in the subducted lithosphere. However, emphasizing the point with respect to the overall isotopic heterogeneity of arc magmas, is the contrast in Th-U isotopic disequilibrium observed in lavas of the Aleutians and Cascades compared with Central America and the Marianas (Fig. 5), and the overlap of ($^{230}U/^{232}Th$) behavior of these arcs with spreading center and hot-spot activity (NEWMAN *et al.*, 1982). There does not appear to be a simple correlation between the type of arc crust and the nature of Th-U isotopic disequilibrium, but possibly the amount of sediment subducted beneath the Marianas and Central America relative to the Aleu-

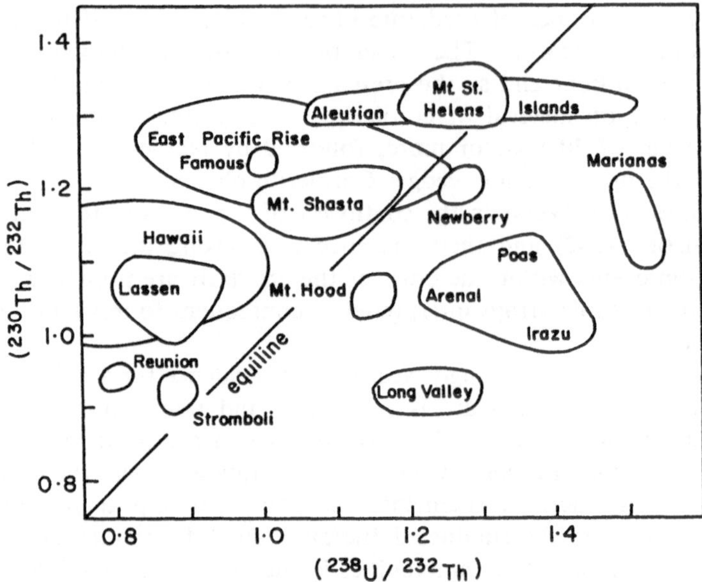

Fig. 5. Th-U activity ratios presented for island-arc, and representative spreading-center and hot-spot activity. Data from sources referenced in text.

tians and Cascades is significant.

The detection of ^{10}Be in island arc lavas and the apparent absence of this nuclide in volcanics from isolated hot-spots like Hawaii (BROWN *et al.*, 1982) has been regarded as the definitive proof of the involvement of recently (i.e. ≤ 10 myr) subducted pelagic sediment. Nevertheless, it might be as well to recall that the comparatively sporadic eruption styles of arc compared with hot-spot volcanoes and the possibility of incorporation of crustal sediment horizons developed prior to construction of the volcanic edifice, or during eruptive quiescence, might be an alternative explanation for the presence of trace quantities of ^{10}Be. An anomaly in this respect is that no ^{10}Be has been detected in the volcano of Usu in southern Hokkaido (BROWN *et al.*, 1982) despite the development of trace element systematics (e.g. high Ba/La ratios) (H. Fujimaki, pers. comm., 1981) that might otherwise be regarded as diagnostic of sediment involvement (e.g. KAY, 1980). In addition, it is doubtful whether an appeal to the location of Usu with respect to the volcanic front in Hokkaido as an explanation for the lack of ^{10}Be is sustainable, since the trace of the front at this sharp break in trench orientation from the Kurile to Honshu arcs is problematic.

The major problem therefore is that unambiguous evidence for the involvement of a subducted component from the lithosphere is hard to obtain given that assimilation processes during the rise of the magmas through the arc crust can give rise to similar trace element and isotopic patterns (ARCULUS and JOHNSON, 1981). The significance of combined assimilation and fractional crystallization (AFC) (DEPAOLO, 1981) in the crust has been emphasized in the genesis of MORB, continental flood basalt and active ocean margin volcanism (O'HARA, 1980; TAYLOR, 1980; CARLSON *et al.*, 1981; THOMPSON, 1982; COULON and THORPE, 1981; ARCULUS *et al.*, 1983). It is necessary to remove the affects of interactions of the arc magmas via AFC processes with the arc crust before identification of a subducted component is possible.

Our lack of understanding of the material transfer processes from the subducted slab and the subsequent melting events in the upper mantle wedge is another stumbling block that is impeding progress in identifying the sources of arc magmas. For example, although in detail the correlations of Sr, Nd and Pb isotopic ratios with Sr/La, Pb/La and Ba/La in any individual arc segment seem incompatible with subducted sediment involvement (ARCULUS and JOHNSON, 1981), there does seem to be the expected correlations if a broader view is taken of several arcs such as those forming the complex collision zones of the Papua New Guinea-Solomon Islands region (M. R. Perfit and M. T. McCulloch pers. comm., 1983).

Summarizing all these data, it is obvious that there is considerable complexity in the detailed geochemistry of active margin volcanics that has defied a simple resolution of the nature of the sources involved and subsequent fractionation events. Despite these difficulties, the general consensus seems to be that the subducted slab must be involved to some extent, even if the amount of material contributed from this source either as a complex fluid phase or partial melt amounts to less than 1% by mass of the major source comprising the upper mantle wedge. This limited flux may be the necessary trigger for the initiation of melting of the overlying mantle and also account for the subtle and awkwardly unresovable trace element and isotopic characteristics of arc magmas (GILL, 1981). An important corollary of this type of hypothesis is the consequences for the geochemistry of the upper mantle wedge in a temporal sense. It seems most improbable that totally efficient extraction of a contribution from the subducted lithosphere can take place, and that some persistent alteration of the upper mantle must be imposed by the subduction event. It might be predicted that subsequent melting in a non-arc environment of portions of the upper mantle that had previously suffered the leaking passge of subducted lithosphere should result in volcanics with similar trace element patterns and isotopic values to those observed in arcs. Possibly the gradation between MORB and arc-type values observed in back-arc basin rocks from the South Sandwich Islands and the Antarctic Peninsula, are a reflection of this memory effect in the mantle wedge (COHEN and O'NIONS, 1982; SAUNDERS *et al.*, 1980).

To some extent, this prediction is confirmed on a broader scale by the overlap of Sr-Nd-Hf-Pb isotopic values of hot-spot volcanics like the Azores (HAWKESWORTH, 1979b), Samoa and the Society Islands (WHITE and HOFMANN, 1982; PATCHETT, 1983) with some of the arc isotopic data, and ancient subduction events have been proposed to account for this overlap of characteristics. The location of the source of this subduction component, whether it be in the fossil mantle wedge or retained in a deeply recycled oceanic lithosphere, is a matter of active speculation (e.g. HOFMANN and WHITE, 1982; RINGWOOD, 1982).

3. *Material Fluxes*

The whole question of the growth of continents, versus a steady-state volume or even a present-day diminution of continental crust revolves around the material fluxes at active oceanic margins. Isotopic arguments have been shown to be ambiguous in this regard (ARMSTRONG, 1981), and other, less precise observations must be used. An additional hazard is that extrapolation of present-day tectonic styles and sediment accumulation rates back into geologic time may be an error. However, it is worthwhile summarizing the major points of view on this matter given the difficulty experienced by geochemists in resolving the role of the subducted flux in arc magma sources.

Recent important reviews of the general and specific problems in this area have been presented by ARMSTRONG (1981), FYFE (1970), ZARTMAN and DOE (1981) and ITO *et al.* (1983). The suggestion that the total continental crustal volume has been essentially constant through most of geological time is based on the constancy of continental freeboard (WISE, 1974) and the equivalence of Archean to present-day crustal thicknesses (DAVIES, 1979). There is no doubt however, that significant peripheral welts of island arcs have been added to cratonic nuclei at least during the Proterozoic and Phanerozoic. Granted this major source of crustal addition, then a comparable amount of crust must be subducted (or otherwise removed) and added to the mantle in order to preserve a constant crustal volume.

ARMSTRONG (1981), for example, estimates a subducted sediment flux at present of 1 to 3 $km^3 yr^{-1}$ is required, which is equivalent to about 20% of the total flux into the oceans, where this total flux is of the order of 10 to 15 $km^3 yr^{-1}$ or 1.4×10^{16} gm (MILLIMAN and MEADE, 1983). ITO *et al.* (1983) have reviewed the production rates of magmas at active oceanic margins (both island and continental arc) and arrive at a figure in the range of $5\text{–}10 \times 10^{15}$ gm yr^{-1}, which is equivalent to 1.8 to 3.6 $km^3 yr^{-1}$ assuming an average rock density of 2.8 gm cc^{-1}. Note the critical feature that the flux of subducted sediment proposed by ARMSTRONG (1981) is of the same order of magnitude as the production rate of arc magmas.

Considerable argument exists over the fate of sediment transported to subduction zones (ARMSTRONG, 1981; KARIG and KAY, 1981). Generalist observations center on the paucity of oceanic pelagic sediments in crustal sequences, and the relative lack of sediment and apparent crustal erosion at some oceanic margins (e.g. SCHOLL *et al.*, 1977). In contrast, KARIG and KAY (1981) and KARIG *et al.* (1983) have shown that essentially all of the sediment transported to the trench is either accreted onto or subcreted beneath the forearc sedimentary prism in the Sunda arc and Nankai trough. Note that subcretion beneath the prism does not necessarily constitute recycling into the mantle. The preservation in exhumed subduction complexes of metasedimentary sequences with relict high-pressure mineralogy must be taken as evidence that some portion of the subducted sediment reaches depths of the order of 30 to 60 km (ERNST, 1977; HOLLAND, 1979). Thus despite disagreement over the volumes of sediment involved, and the depths to which they are transported and remain for periods longer than 10 myr, there seems little doubt that some fraction must be recycled into the mantle. If the view is adopted that volumetrically minimal sediment contribution is involved in arc magma genesis, then in order to preserve a constant crustal volume, 1 to 3 $km^3 yr^{-1}$ of sediment must be added to the mantle. Anything smaller must result in net continental crustal growth with the current arc compositions and production rates being the most important factors controlling that growth.

It seems at odds with our knowledge of the high-pressure fusion resistance of sedimentary protoliths (e.g. STERN and WYLLIE, 1973) that the bulk of subducted sediment is not being returned to the surface in arc magmas within about 10 myr of disappearance. However, isotopic evidence is in support of long-term (i.e. >1 Gyr) recycling through the mantle (HOFMANN and WHITE, 1982; COHEN and O'NIONS, 1982) so that perhaps further efforts should be expended in order to understand the stability of sedimentary rocks at pressures greater than about 10 kbar.

The question of volatile fluxes through arc-trench systems would appear to be a fundamental problem, and some curious facts have emerged which suggest serious imbalances in emission compared with subduction rates. Although there may be room for debate over the total volatile concentrations and the variability of occurrence of such species as H_2O and CO_2, there seems little doubt that magmas in island arcs are more volatile-rich than those produced at oceanic ridges (GARCIA *et al.*, 1979). The generally explosive nature of arc volcanism, and the first appearance of amphibole in host compositions raging from basaltic (rarely) to andesitic (more commonly—see GILL, 1981) are also evidence of higher volatile content than that typical of basalt-related activity at spreading centers. The most logical source of these volatiles of course is the subducted oceanic lithosphere, which would seem entirely capable of supplying the necessary flux required. As an example, in addition to the volatile budget of H_2O, CO_2, Cl and F quenched in the lavas during extrusion at mid-ocean ridges, further hydration and chemical alteration takes place during transport to subduction zones. If these additional volatiles are not returned to the surface in oceanic margin volcanism, then a net loss from the hydrosphere to the mantle must result.

ITO *et al.* (1983) have estimated that about 10 times as much H_2O is disappearing in the altered and possibly sediment-laden subducted lithosphere than is erupted in arc magmas. The important figures are the estimated H_2O production in arcs of about 1×10^{14} gm yr^{-1} compared with a subducted mass of about 1×10^{15} gm yr^{-1}, comprising 1.1×10^{14} gm yr^{-1} as indigenous ridge-quenched complement, and 8.8×10^{14} gm yr^{-1} through alteration. In contrast to the potential discrepancy between inlet and outlet fluxes of H_2O, a balance seems to exist for Cl, which may suggest highly active halogen-bearing fluids are circulating in the mantle wedge above the subducted slab. If the H_2O dissolved in arc magmas is recycled and essentially dominated by a sea water component, then total isotopic reequilibration with normal mantle overlying the subducted plate must have occurred, because the δD and $\delta^{18}O$ values for primary arc magmas are not appreciably different from those of other mantle-derived melts (see GILL, 1981 for a review). Furthermore, transport of such large volumes of H_2O into the Earth's mantle of hydrospheric origin must result in the secular change of the values of δD and $\delta^{18}O$ in the mantle. The implication is that effective convective stirring of the mantle wedge cycles comparatively pristine peridotite into the zones overlying active dehydration bands in the subducted slab.

Other authors have reviewed the H_2O flux in subduction zones, and have suggested that apart from the portion transferred directly back to the surface via arc magmas, extensive hydration of the mantle wedge overlying the slab must take place (FYFE and MCBIRNEY, 1975), and have related surface forearc and rear-arc topography to the effects of this hydration. However, there is little direct evidence that reveals the preservation of relict hydrated lithosphere. For example, the δD and $\delta^{18}O$ values of phlogopite and amphibole in upper mantle-derived nodules in alkali basalt and kimberlite, show no effect of recycled ocean water with δD and $\delta^{18}O = 0$ (BOETTCHER and O'NEIL, 1980).

Another potential isotopic tracer of the volatile flux in subduction zones is the relative abundance of the helium isotopes ($^3He/^4He$), as pointed out by CRAIG *et al.* (1978). In view of the abundance of Th and U in sediments, and also U in the altered layer 2 of the subducted slab, and the average age of this material by the time it is located about 100 km or more beneath arc volcanoes, then it is puzzling that despite production of ^4He by decay of Th and U, the $^3He/^4He$ ratios of volatile emissions in regions of arc volcanism overlap those of spreading centers. In other words, there does not appear to be a major flux of ^4He from the subducted slab to the sources of volatiles of arc magmas. It would appear that ^4He either escapes from the slab prior to dehydration, or remains trapped within the slab in the phases that host U and Th. It is worth noting that $^3He/^4He$ data for one of the hot-spots (Samoa) that shares the enriched Sr-depleted Nd isotopic character of some arc volcanics shows a far greater proportion of primordial ^3He (up to 18 times the atmospheric ratio of $^3He/^4He$) than is found in either MORB or arc environments (RISON and CRAIG, 1982). If hot spots like Samoa represent recycled subducted lithosphere, then it is curious that radiogenic ^4He is in such short supply in Samoa.

Thus in terms of mass balance and isotopic character, understanding the volatile flux of arc-trench systems presents some difficulties. There appears to be an excess of supply over emission, but little record isotopically of a hydrospheric or subducted lithosphere component. The critical question must be what has stopped the hydrosphere from diminishing drastically if the present negative flux has persisted for any length of time without some other counter-flow mechanism (ITO *et al.*, 1983).

4. *Tectonic and Thermochemical Regimes*

Limited direct observation and the circularity of many thermal-petrological arguments mean that crucial aspects of the tectonic environment and thermal regimes of arc magma genesis are poorly constrained. For example, it has become apparent that there is a major division of active margin tectonic type between arcs with and without associated back-arc basins (UYEDA, 1982). In addition to the complexities that are introduced to the problem of magma ascent and fractionaton by variable stress states in the crust of the arc (MARSH, 1982) as a consequence of these contrasts in tectonic style, it should be noted that the position of an arc with respect to the nearest neighboring continental lithosphere may also be significant. There is increasing recognition of the contrasting chemical and thermomechanical behavior of subcontinental and suboceanic lithosphere (e.g. DAVIES, 1979; JORDAN, 1981).

The nature of marginal or back-arc basin development in the western Pacific documented by KARIG (1974) has been suggested by ARCULUS (1981) to be a significant factor in the creation of source heterogeneities in the lithosphere beneath arcs. The prediction is that arcs that have comparatively recently been rifted from continental masses such as Japan from east Asia, should have some proportion of trailing continental lithosphere, whereas those intraoceanic arcs like the Marianas and Izu-Bonin must be less likely to have any complication of subcontinental and suboceanic lithosphere source mixing. The considerably more extensive isotopic variability both along and across strike in Japan (NOHDA and WASSERBURG, 1981) compared with the Marianas- Izu Bonin arcs is a possible reflection of this lithosphere contrast. Attribution of such variability to the involvement of different proportions of subducted pelagic sediments appears to be less probable as an explanation than either crustal contamination, or subcrustal lithosphere variability. However, the complex tectonic evolution of the western Pacific arcs where a back-arc basin, underlain by MORB-like mantle is juxtaposed between the eastwardly migrating continental lithospheric sliver and the main Asian lithosphere renders analysis of across-arc isotopic variability in locations like the Northeast Japan arc difficult.

A number of studies of the thermal profiles of subduction zones have been published with consideration of such variables as frictional heating, rates of descent, internal radioactive heating and cooling via endothermic dehydration reactions (e.g. OXBURGH and TURCOTTE 1976; TOKSÖZ *et al.*, 1971; HASEBE *et al.*, 1970; ANDERSON *et al.*, 1978). A recent comprehensive examination of the potential stabilities of hydrous phases in the subducted slab by DELANEY and HELGESON (1978) has indicated preservation of these phases to greater depths at least within the interior of the slab than previously proposed (e.g. WYLLIE, 1973).

Without the constraint of a depth at which the leading upper surface of the slab crosses a dehydration reaction boundary or solidus, it is of course much more difficult to provide independent estimates of the thermal regime of the subduction zone. While accepting this difficulty, it is a critical feature of all of these thermal models that a depression of the isotherms by the presence of subducted lithosphere is developed (Fig. 6). Important considerations for magma genesis are the extent and stability of this thermal perturbation, the seemingly inevitable convective system in the overlying asthenospheric portion of the mantle wedge, and the extent of volatile recycling through the subducted slab-mantle wedge system.

Heat transfer from the overlying mantle wedge to the interior of a subducted slab is likely to be very slow in comparison with vertical subduction rates of the order of 5 cm yr^{-1}, given the low thermal conductivity of the slab. Therefore the persistence of the isothermal depression around a slab seems consistent with the mechanisms proposed for the subduction process of thermal contrast at shallow depths in the mantle (<400 km) corresponding to density contrasts as a result of the elevation of the phase boundaries normally associated with phase transitions in the mantle (see however, RINGWOOD, 1982).

There is little doubt that some devolatilization of the slab must take place at a range of depths depending on the mineralogy and thermal profile of the slab (DELANEY and HELGESON, 1978). The fate of fluids lost from the slab however, is not so clear, and depends in part on the fugacities of the volatile species which are functions of the relative proportions of volatiles, the reactions of the volatiles with phases present in the overlying wedge and the *P-T* paths traversed by the fluids. As an instructive exercise, calculation of the fH_2O as a function of *P* and *T* by DELANEY and HELGESON (1978), has shown that in the pressure range of the order 20 to 30 kbar or more, decreasing *P* and increasing *T* results in a larger decrease in fH_2O than decreasing *T* under isobaric conditions, or both decreasing *P* and *T*. The significance of the thermal cusps (Fig. 6) in the overlying mantle wedge may therefore be dramatic in favoring enhanced volatile loss upwards from the slab where there is an increase in temperature coincident with pressure decrease (see also MARSH, 1979a; 1979b). Buffering of the f_i's by stabilization of amphibole-phlogopite-carbonate in the mantle wedge must also be considered (EGGLER, 1978). A more direct measure of the maximum chemical potential gradients for the path of H_2O is also given in Fig. 6, showing that variation $RT\ln fH_2O$ is greatest *parallel* to the slab and back along the slab-wedge interface. Consequently, the rate of slab-wedge transport downwards into the mantle, and the infiltration of hydrous fluids upwards through the wedge (which can be assumed to be relatively anhydrous with respect to the upper layers of the slab) may be the most important factors governing volatile circulation in subduction zones.

The possibility arises of fluids being released from greater depths pass-

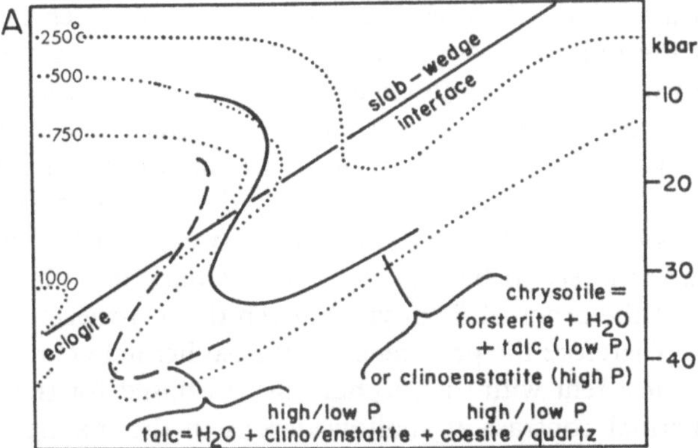

Fig. 6a. Thermal profile of a subduction zone with isotherms after HASEBE *et al.* (1970) and representative dehydration fronts from DELANEY and HELGESON (1978). Note that the temperatures in the wedge and depth of penetration of a given isotherm downwards along the slab may be higher and deeper respectively if the petrologic model of TATSUMI *et al.* (1983) is correct.

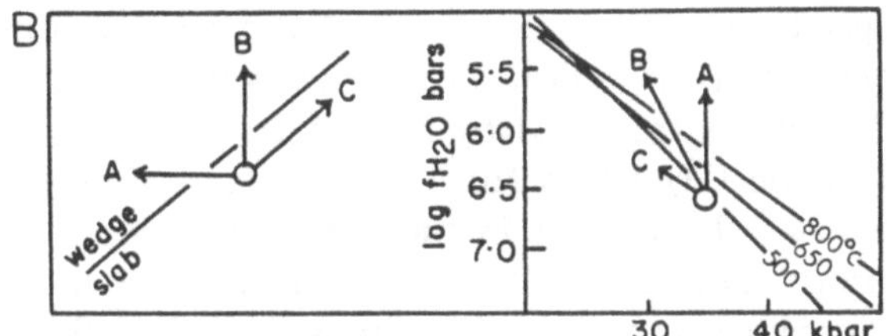

Fig. 6b. Variation of fH_2O as a function of P-T showing the variable effect of P-T on fH_2O for different migration paths in the subduction zone labelled A, B and C in both parts of the figure.

ing along the wedge-slab interface interacting with both materials enroute to shallower levels prior to stabilization again in hydrous minerals. A type of circular system can be envisaged with leakage upwards into the wedge keeping pace with the steady supply from the slab. Speculative geochemical consequences could be the progressive leaching and enrichment in the fluid of the alkaline earths with respect to the rare earths, with the majority of the alkaline earths being derived from the wedge, as indicated by Sr-Nd systematics (see previous sections). The refractory and stable mineralogy of a basaltic composition slab experiencing melt + fluid interaction in the pressure range around 30 kbar is clinopyroxene-garnet-rutile (GUST, 1982).

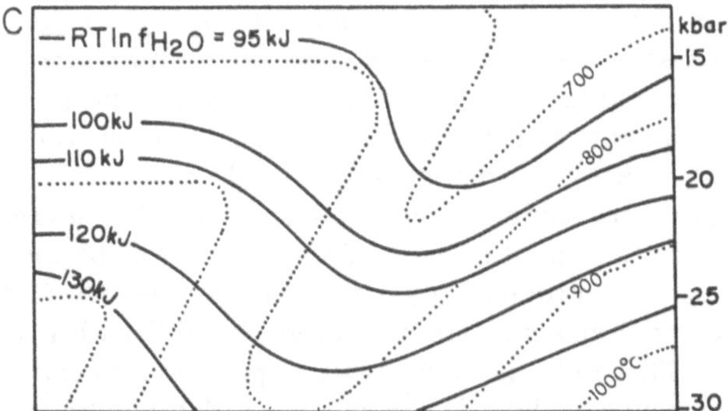

Fig. 6c. Variation of RTlnfH_2O in a subduction zone given the thermal profile of a type illustrated in Fig. 6a. Thermodynamic data for construction of 6b and 6c are from DELANEY and HELGESON (1978) and T. J. B. Holland and R. Powell (pers. comm., 1983).

In order to account for the depletion of the Ti group elements in the mantle source regions of arc lavas, it is possible that if these elements are also selectively leached from the wedge from relatively unfavorable crystallographic sites in phases like clinopyroxene or garnet, that precipitation occurs in the eclogitic facies of the subducted slab where the saturation limit is exceeded and suitable nucleation sites in rutile exist. Although the TNT group elements are generally regarded as immobile in geochemical processes, their relative enrichment and apparent coupling with incompatible, mobile alkaline earths and alkalies in upper mantle materials belonging to the Marid suite (DAWSON and SMITH, 1977) suggests a degree of mobility under some upper mantle conditions.

A final consideration is the potential of the slab-wedge interface as a volatile release path for fluids derived from greater depths in the mantle wedge. For example, FODEN and VARNE (1980) have suggested such a mechanism in order to account for the enrichment of apparently mantle-derived, incompatible lithophile trace elements in the alkaline arc rocks of Indonesia. It should be noted however, that the characteristic arc-type depletion of the Ti group elements in these Indonesian volcanics is suggestive of their removal and stabilization in a refractory phase, most plausibly within the slab and perhaps within a rutile-bearing eclogite. It should be recalled that the characteristic trace element signature of enhanced Ba or Sr/La ratios in island arc volcanics is also present in other rock types such as some continental flood basalts, lower continental crustal eclogites and granulites, and even spinel peridotites of upper mantle derivation (ARCULUS and JOHNSON, 1981). In general however, these other rock types are not depleted in the Ti group elements with respect to the REE in the same manner as arc volcanics. The relative mobilities of the alkaline earths and the Ti group elements under conditions appropriate to the lower crust (about 10 kbar) and subduction zone dehydration fronts (say 30 kbar) are obviously areas that merit further study.

Subsequent recycling of the subducted slab by internal radioactive heating (due to the retention of Th, U and K) over periods of the order of 1 Gyr has been proposed by HOFMANN and WHITE (1982). The lack of dramatic enrichments in arc basalts of U and Th with respect to other incompatible lithophile trace elements is possibly indicative of retention of these radioactive nuclides in the subducted slab (ARCULUS and JOHNSON, 1981). Possible candidates for high pressure stability (i.e., greater than 20 kbar) are members of the epidote group such as zoisite (GUST, 1982).

The relative enrichment of the Ti group elements in oceanic island, hot-spot volcanics has been emphasized by SUN (1981) and WOOD (1979), and is consistent with the behavior of a refractory Ti-bearing phase in the slab in the regions of arc magma generation, and subsequent destablization of this phase at greater depths during hot-spot magma genesis. Nevertheless an addition of Ti group elements to the subducted slab seems necessary in view of the unfractionated nature of REE/Nb (or Ta) ratios of the MORB protoliths, and may be indicative of some kind of fluid recycling in the mantle wedge and exchange with the slab.

In conclusion, it must be apparent from the foregoing that much uncertainty still remains over the nature of the chemical and dynamic evolution of active ocean margins and the subjacent mantle. The balance of the sources involved, the material flux and the processes of fluid convection in the mantle wedge-subducted slab systems are clearly important problems. Interrelated research must explore in more detail the temperature and dynamic regimes of subduction zones, the processes of magma transport through the wedge and the convection of the upper mantle within the wedge.

Acknowledgements

Firstly my gratitude is offered to the Organizers of the Oji International Seminar on Active Ocean Margins for making my participation in the meeting in Japan possible. The temporary refuge and critical discussion provided at the Universities of Leeds and Aston by Dr.s Roger Powell and Andy Chambers during the gestation of the outlines presented in this paper is much appreciated. The Birmingham Settlement gave logistic support during manuscript preparation. Mike O'Hara and an anonymous reviewer pointed out many inconsistencies in the ideas expressed in an early version of this paper, and are thanked for their critical care.

REFERENCES

C. J. ALLÈGRE and M. CONDOMINES, Fine chronology of volcanic processes, *Earth Planet. Sci. Lett.*, **28**, 395–406, 1976.

C. J. ALLÈGRE and M. CONDOMINES, Basalt genesis and mantle structure studied through Th-isotopic geochemistry, *Nature*, **299**, 21–24, 1982.

R. N. ANDERSON, S. E. DELONG, and W. M. SCHWARZ, Thermal model for subduction with dehydration in the downgoing slab, *J. Geol.*, **86**, 731–739, 1978.

R. J. ARCULUS, Geology and geochemistry of the alkali basalt-andesite association of Grenada, Lesser Antilles island arc, *Geol. Soc. Amer. Bull.*, **87**, 612–624, 1976.

R. J. ARCULUS and R. W. JOHNSON, Criticism of generalized models for the magmatic evolution of arc-trench systems, *Earth Planet. Sci. Lett.*, **39**, 118–126, 1978.

R. J. ARCULUS and R. W. JOHNSON, Island arc magma sources: a geochemical assessment of the roles of slab-derived components and crustal contamination, *Geochem. J.*, **15**, 109–133, 1981.

R. J. ARCULUS and R. POWELL, A cyclic fusion-freezing-fluid fluxing-fusion model for the sources of island arc magmas, in Proceedings of the Conference on Open Magmatic Systems, M. A. Dungan, T. L. Grove and W. Hildreth, eds., pp. 3–5, I.S.E.M., Dallas, 1984.

R. J. ARCULUS, R. W. JOHNSON, B. W. CHAPPELL, C. O. MCKEE, and H. SAKAI, Ophiolite-contaminated andesites, trachybasalts and cognate inclusions of Mount Lamington, Papua New Guinea: anhydrite-amphibole-bearing lavas and the 1951 cumulodome, *J. Volc. Geotherm. Res.*, **18**, 215–247, 1983.

R. L. ARMSTRONG, Radiogenic isotopes: the case for crustal recycling on a near steady-state-no-continental-growth Earth, *Phil. Trans. R. Soc. Lond.*, **A-301**, 443–472, 1981.

L. BROWN, J. KLEIN, R. MIDDLETON, I. SACKS, and F. TERA, ^{10}Be in island arc volcanoes and implications for subduction, *Nature*, **299**, 718–720, 1982.

A. L. BOETTCHER and J. R. O'NEIL, Stable isotope, chemical and petrographic studies of high pressure amphiboles and micas: evidence for metasomatism in the mantle source regions of alkali basalts and kimberlites, *Am. J. Sci.*, **280-A**, 574–621, 1980.

R. W. CARLSON, G. W. LUGMAIR, and J. D. MACDOUGALL, Columbia River volcanism: the question of mantle hetrogeneity or crustal contamination, *Geochim. Cosmochim. Acta*, **45**, 2483–2499, 1981.

I. S. E. CARMICHAEL, F. J. TURNER, and J. VERHOOGEN, *Igneous Petrology*, 739 pp., McGraw-Hill, New York.

F. CHAYES and D. VELDE, Distinguishing basaltic lavas of circum-oceanic and oceanic-island type by means of discriminant functions, *Am. J. Sci.*, **263**, 206–222, 1965.

R. S. COHEN and R. K. O'NIONS, Identification of recycled continental material in the mantle from Sr, Nd and Pb isotope investigations, *Earth Planet. Sci. Lett.*, **61**, 73–84, 1982.

C. COULON and R. S. THORPE, Role of continental crust in petrogenesis of orogenic volcanic associations, *Tectonophysics*, **77**, 79–93, 1981.

H. CRAIG, J. E. LUPTON, and Y. HORIBE, A mantle helium component in circum-Pacific volcanic gases: Hakone, the Marianas and Mt. Lassen, in *Terrestrial Rare Gases*, eds. E. C. Alexander Jr. and M. Ozima, pp. 3–16, Center for Academic Publications, Japan Sci. Press, Tokyo, 1978.

G. F. DAVIES, Thickness and thermal history of continental crust and root zones, *Earth Planet. Sci. Lett.*, **44**, 231–238, 1979.

J. B. DAWSON and J. V. SMITH, The MARID (mica-amphibole-rutile-ilmenite-diopside) suite of xenoliths in kimberlite, *Geochim. Cosmochim. Acta*, **41**, 309–323, 1977.

J. M. DELANEY and H. C. HELGESON, Calculation of the thermodynamic consequences of dehydration in subducting oceanic crust to 100 kb and 800° C, *Am. J. Sci.*, **278**, 636–686, 1978.

D. J. DEPAOLO, Trace element and isotopic effects of combined wall rock assimilation and fractional crystallization, *Earth Planet. Sci. Lett.*, **57**, 47–62, 1981.

D. J. DEPAOLO and G. J. WASSERBURG, The sources of island arcs as indicated by Nd and Sr isotopic studies, *Geophys. Res. Lett.*, **4**, 465–468, 1977.

D. J. DEPAOLO and R. W. JOHNSON, Magma genesis in the New Britain island arc: constraints from Nd and Sr isotopes and trace element patterns, *Contrib. Mineral. Petrol.*, **70**, 367–379, 1979.

A. D. EDGAR, D. H. GREEN, and W. O. HIBBERSON, Experimental petrology of a highly potassic magma, *J. Petrology*, **17**, 339–356, 1976.

D. H. EGGLER, The effect of CO_2 upon partial melting of peridotite in the system Na_2O-CaO-Al_2O_3-MgO-SiO_2-CO_2 to 35 kb with an analysis of melting in a peridotite H_2O-CO_2 system, *Am. J. Sci.*, **278**, 305–343, 1978.

W. G. ERNST, Mineral parageneses and plate tectonic settings of relatively high-pressure metamorphic belts, *Fortschr. Mineral.*, **54**, 192–222, 1977.

J. D. FODEN and R. VARNE, The petrology and tectonic setting of Quarternary to Recent volcanic centres of Lombok and Sumbawa, Sunda Arc, *Chem. Geol.*, **30**, 201–226, 1980.

H. FUJIMAKI, Rare earth elements in volcanic rocks from Hakone volcano and northern Izu Peninsula, *J. Fac. Sci. Univ. Tokyo*, **19**, 81–93, 1975.

W. S. FYFE, The geochemical cycle of Uranium, *Phil. Trans. R. Soc. Lond.*, **A-291**, 433–445, 1979.

W. S. FYFE and A. R. MCBIRNEY, Subduction and the structure of andesitic volcanic belts, *Am. J. Sci.*, **275-A**, 285–297, 1975.

M. O. GARCIA, N. W. K. LIU, and D. W. MUENOW, Volatiles in submarine volcanic rocks from the Mariana island arc and trough, *Geochim. Cosmochim. Acta*, **43**, 305, 1979.

J. B. GILL, *Andesites: Orogenic Andesites and Plate Tectonics*, 399 pp., Springer-Verlag, 1981.

T. H. GREEN, Island arc and continent-building magmatism-a review of petrogenetic models based on experimental petrology and geochemistry, *Tectonophysics*, **63**, 367–385, 1980.

T. H. GREEN and A. E. RINGWOOD, Genesis of the calcalkaline rock suite, *Contrib. Mineral. Petrol.*, **18**, 105–162, 1968.

D. A. GUST, Experimental, petrologic and geochemical studies on the origins of andesite, Ph.D. thesis, Australian National Univ., 1982.

K. HASEBE, N. FUJII, and S. UEDA, Thermal processes under island arcs, *Tectonophysics*, **10**, 335–355, 1970.

C. J. HAWKESWORTH, R. K. O'NIONS, R. J. PANKHURST, P. J. HAMILTON, and N. M. EVENSON, A geochemical study of island-arc and back-arc tholeiites from the Scotia Sea, *Earth Planet. Sci. Lett.*, **36**, 253–262, 1977.

C. J. HAWKESWORTH, R. K. O'NIONS, and R. J. ARCULUS, Nd and Sr isotope geochemistry of island arc volcanics, Grenada, Lesser Antilles, *Earth Planet. Sci. Lett.*, **45**, 237–248, 1979a.

C. J. HAWKESWORTH, M. J. NORRY, and J. C. RODDICK, $^{143}Nd/^{144}Nd$ and $^{87}Sr/^{86}Sr$ from the Azores and their significance in LIL-enriched mantle, *Nature*, **280**, 28–31, 1979b.

A. HOFMANN and W. M. WHITE, Mantle plumes from ancient oceanic crust, *Earth Planet. Sci. Lett.*, **57**, 421–436, 1982.

T. J. B. HOLLAND, High water activities in the generation of high-pressure kyanite eclogites in the Tauern Window, Austria, *J. Geol.*, **87**, 1–27, 1979.

C. S. HUTCHISON, Indonesia, in *Andesites: Orogenic Andesites and Related Rocks*, ed. R. S. Thorpe, pp. 207–224, Wiley, London, 1982.

E. ITO, D. M. HARRIS, and A. T. ANDERSON, Jr., Alteration of oceanic crust and geologic cycling of chlorine and water, *Geochim. Cosmochim. Acta*, **47**, 1613–1624, 1983.

T. H. JORDAN, Continents as a chemical boundary layer, *Phil. Trans. R. Soc. Lond.*, **A-301**, 359–373, 1981.

D. E. KARIG, Evolution of arc systems in the western Pacific, *Ann. Rev. Earth Planet. Sci.*, **2**, 51–75, 1974.

D. E. KARIG and R. W. KAY, Fate of sediments on the descending plate of convergent margins, *Phil. Trans. R. Soc. Lond.*, **A-301**, 233–251, 1981.

D. E. KARIG, H. KAGAMI, and DSDP LEG 87 SCIENTIFIC PARTY, Varied responses to subduction in Nankai trough and Japan trench forearcs, *Nature*, **304**, 148–151, 1983.

R. W. KAY, Volcanic arc magmas: implications for a melting-mixing model for element recycling in the crust-upper mantle system, *J. Geol.*, **88**, 497–522, 1980.

R. W. KAY, S-S SUN, and C-N LEE-HU, Pb and Sr isotopes in volcanic rocks from the Aleutian Islands and Pribilof Islands, Alaska, *Geochim. Cosmochim. Acta*, **42**, 263–273, 1978.

H. KUNO, Origin of Cenozoic petrographic provinces of Japan and surrounding areas, *Bull. Volc.*, **20**, 37–76, 1959.

H. KUNO, Differentiation of basaltic magma, in *Basalt—The Poldervaart Treatise on the Origin of Basaltic Rocks*, eds. H. H. Hess and A. Poldervaart, pp. 623–688, Interscience, New York, 1968.

I. KUSHIRO, On the lateral variations in chemical compositions and volume of Quaternary volcanic rocks across Japanese arcs, *J. Volc. Geotherm. Res.*, **18**, 435–447, 1983.

I. KUSHIRO and H. SATO, Origin of some calcalkaline andesites in the Japanese islands, *Bull. Volc.*, **41-4**, 576–585, 1978.

W. P. LEEMAN, The influence of crustal structure on compositions of subduction-related magmas, *J. Volc. Geotherm. Res.*, **18**, 561–588, 1983.

M. T. MCCULLOCH and M. R. PERFIT, $^{143}Nd/^{144}Nd$, $^{87}Sr/^{86}Sr$ and trace element constraints on the petrogenesis of Aleutian island arc magmas, *Earth Planet. Sci. Lett.*, **56**, 167–179, 1981.

M. T. MCCULLOCH, W. COMPSTON, M. ABBOTT, and A. CHIVAS, Neodymium, strontium, lead and oxygen isotopic and trace element constraints on magma genesis in the Banda island-arc Wetar, *Geol. Soc. Australia Abstracts*, **9**, 152–153, 1983.

M. MARGARITZ, D. J. WHITFORD, and D. E. JAMES, Oxygen isotopes and the origin of high $^{87}Sr/^{86}Sr$ andesites, *Earth Planet. Sci. Lett.*, **40**, 220–230, 1978.

B. D. MARSH, Island arc volcanism, *Am. Scientist*, **67**, 161–172, 1979a.

B. D. MARSH, Island arc development: some observations, experiments and speculations, *J. Geol.*, **87**, 687–713, 1979b.

B. D. MARSH, On the mechanics of igneous diapirism, stoping and zone melting, *Am. J. Sci.*, **86**, 808–855, 1982.

Y. MATSUHISA and H. KURASAWA, Oxygen and strontium isotopic characteristics of calcalkaline volcanic rocks from the central and western Japan arcs: evaluation of contribution of crustal components to the magmas, *J. Volc. Geotherm. Res.*, **18**, 483–510, 1983.

A. MEIJER, Pb and Sr isotopic data bearing on the origin of volcanic rocks from the Mariana island arc system, *Geol. Soc. Amer. Bull.*, **87**, 1358–1369, 1976.

J. D. MILLIMAN and R. H. MEADE, World-wide delivery of river sediment to the oceans, *J. Geol.*, **91**, 1–21, 1983.

J. D. MORRIS and S. R. HART, Isotopic and incompatible element constraints on the genesis of island arc volcanics from Cold Bay and Amak Isand, Aleutians, and implications for mantle structure, *Geochim. Cosmochim. Acta*, **47**, 2015–2030, 1983.

S. R. NEWMAN, C. FINKEL, and J. D. MACDOUGALL, Comparison of $^{230}Th/^{238}U$ disequilibrium systematics in young continental and oceanic volcanic rocks, Fifth Int. Conf. on Geochron., Cosmochron. and Isotope Geology, Japan, pp. 265–266, 1982.

I. A. NICHOLLS and A. E. RINGWOOD, Effect of water on olivine stability in tholeiites and production of silica-saturated magmas in the island arc environment, *J. Geol.*, **81**, 285–300, 1973.

S. NOHDA and G. J. WASSERBURG, Nd and Sr isotopic study of volcanic rocks from Japan, *Earth Planet. Sci. Lett.*, **52**, 264–276, 1981.

M. J. O'HARA, Nonlinear nature of the unavoidable long-lived isotopic, trace and major element contamination of a developing mama chamber, *Phil. Trans. R. Soc. Lond.*, **a-297**, 215–227, 1980.

E. F. OSBORN, The complementariness of orogenic andesite and alpine peridotite, *Geochim. Cosmochim. Acta*, **33**, 307–324, 1964.

E. F. OSBORN, The reaction principle in *The Evolution of the Igneous Rocks, Fiftieth Anniversary Perspectives*, H. S. Yoder ed., pp. 133–169, Princeton Univ. Press, 1979.

V. M. OVERSBY and A. EWART, Lead isotopic compositions of Tonga-Kermadec volcanics and their petrogenetic significance, *Contrib. Mineral. Petrol.*, **37**, 181–210, 1972.

E. R. OXBURGH and D. L. TURCOTTE, The physico-chemical behavior of the descending lithosphere, *Tectonophysics*, **32**, 107–128, 1976.

P. J. PATCHETT, Importance of the Lu-Hf isotopic system in studies of planetary chronology and chemical evolution, *Geochim. Cosmochim. Acta*, **47**, 81–91, 1983.

J. A. PEARCE and M. J. NORRY, Petrogenetic implications of Ti, Zr, Y and Nb variation in volcanic rocks, *Contrib. Mineral. Petrol.*, **69**, 33–47, 1979.

M. R. PERFIT, D. A. GUST, A. E. BENCE, R. J. ARCULUS, and S. R. TAYLOR, Chemical characteristics of island-arc basalts: implications for mantle sources, *Chem. Geol.*, **30**, 227–256, 1980.

A. E. RINGWOOD, *Composition and Petrology of the Earth's Mantle*, 618 pp., McGraw Hill, New York, 1975.

A. E. RINGWOOD, Phase transformations and differentiation in subducted lithosphere: implications for mantle dynamics, *J. Geol.*, 1982.

W. RISON and H. CRAIG, Helium 3: coming of age in Samoa, *EOS*, **63**, 1144, 1982.

A. D. SAUNDERS, J. TARNEY, and S. D. WEAVER, Transverse geochemical variations across the Antarctic Peninsula: implications for the genesis of calc-alkaline magmas, *Earth Planet. Sci. Lett.*, **46**, 344–360, 1980.

D. W. SCHOLL, M. S. MARLOW, and A. K. COOPER, Sediment Subduction and Offscraping at Pacific Margins, *Amer. Geophys. Union*, M. Ewing series, 1, pp. 199–210, 1977.

D. M. SHAW, Trace element fractionation during anatexis, *Geochim. Cosmochim. Acta*, **34**, 237–243.

C. R. STERN and P. J. WYLLIE, Melting relations of basalt-andesite-rhyolite-H_2O and a pelagic red clay at 30 kilobars, *Contrib. Mineral. Petrol.*, **42**, 313–323, 1973.

A. SUGIMURA, Zonal arrangement of some geophysical and petrological features in Japan and its environs, *J. Fac. Sci. Univ. of Tokyo*, **12**, 133–153, 1960.

S-S. SUN, Lead isotopic study of young volcanic rocks from mid-ocean ridges, oceanic islands and island arcs, *Phil. Trans. R. Soc. Lond.*, **A-297**, 409–445, 1980.

Y. TATSUMI, Melting experiments on a high-magnesia andesite, *Earth Planet. Sci. Lett.*, **54**, 357–365, 1981.

Y. TATSUMI, M. SAKUYAMA, H. FUKUYAMA, and I. KUSHIRO, Generation of arc basalt magmas and thermal structure of the mantle wedge in subduction zones, *J. Geophys. Res.*, **88**, 5815–5825, 1983.

H. P. TAYLOR, Jr., The effects of assimilation of country rocks by magmas on $^{18}O/^{16}O$ and $^{87}Sr/^{86}Sr$ systematics in igneous rocks, *Earth Planet. Sci. Lett.*, **47**, 243–254, 1980.

R. N. THOMPSON, Magmatism of the British Tertiary volcanic province, *Scott. J. Geol.*, **18**, 49–107, 1982.

M. N. TOKSOZ, J. W. MINEAR, and B. R. JULIAN, Temperature field and geophysical effects of a downgoing slab, *J. Geophys. Res.*, **76**, 1113–1138, 1971.

S. UYEDA, Subduction zones: an introduction to comparative subductology, *Tectonophysics*, **81**, 133–159, 1982.

R. W. VAN BEMMELEN, *The Geology of Indonesia*, The Government Printing Office, The Hague, 1949.

E. B. WATSON, Apatite and phosphorus in mantle source regions: an experimental study of apatite/melt equilibria at pressures to 25 kbar, *Earth Planet. Sci. Lett.*, **51**, 322–335, 1980.

W. M. WHITE and A. HOFMANN, Sr and Nd isotope geochemistry of oceanic basalts and mantle evolution, *Nature*, **296**, 821–825, 1982.

D. J. WHITFORD, W. COMPSTON, and I. A. NICHOLLS, Geochemistry of the late Cenozoic lavas from eastern Indonesia: role of subducted sediments in petrogenesis, *Geology*, **5**, 571–575, 1977.

D. U. WISE, Continental margins, freeboard and the volumes of continents and oceans through time, in *The Geology of Continental Margins*, eds. C. A. Burke and C. L. Drake, pp. 45–58, Springer-Verlag, 1974.

D. A. WOOD, A variably veined sub-oceanic upper mantle: genetic significance for mid-ocean ridge basalts from geochemical evidence, *Geology*, **7**, 499–503, 1979.

P. J. WYLLIE, Experimental petrology and global tectonics—a preview, *Tectonophysics*, **17**, 189–209, 1973.

H. S. YODER, *Generation of Basaltic Magma*, 265 pp., National Academy of Sciences, Washington, D.C., 1976.

Formation of Active Ocean Margins, edited by N. Nasu *et al.*, pp. 399–410.

GEOCHEMICAL CHARACTERISTICS OF BACK-ARC BASIN BASALT

Hiroaki SATO[1] and Tatsuhide TOHARA[2]

[1]*Department of Earth Sciences, Faculty of Science, Kanazawa University, Kanazawa, Ishikawa 920, Japan*
[2]*Department of Geology and Mineralogy, Faculty of Science, Hokkaido University, Sapporo, Hokkaido 060, Japan*

Abstract. Geochemical data on back-arc basin basalt (BABB) are compiled and compared with normal mid-oceanic ridge basalt (N-type MORB), oceanic island basalt (OIB), and island arc volcanic rocks. The primary magma composition of BABB was estimated, and it is characterized by rather low MgO (10–13 wt.%) and FeO* (7–9 wt.%), and high Al_2O_3 (15–17.5 wt.%) contents compared with that of OIB, which may mainly be due to shallower depth (ca. 10 kb) of segregation of BABB than OIB from mantle peridotite. BABB can be classified into three types according to the incompatible element abundances and the isotopic composition of Sr. BABB from the West Philippine Sea has a low concentration of incompatible elements and $^{87}Sr/^{86}Sr$ ratio within the compositional range of N-type MORB, suggesting that the source of BABB from the West Philippine Sea is not affected by subducted material. BABB from the Lau Basin has a high $^{87}Sr/^{86}Sr$ ratio and incompatible element abundances comparable to those of the associated Tonga Arc volcanic rocks. BABB from the Scotia Sea, Mariana Trough, Shikoku and Parece Vela Basins has intermediate geochemical characteristics between N-type MORB and the associated island arc tholeiite.

1. *Introduction*

Recent DSDP and dredge samples have afforded a number of chemical and petrographical data on back-arc basin basalt (hereafter abbreviated as BABB). Back-arc basins commonly have oceanic crustal structure, and BABB may have been formed by processes similar to mid-oceanic ridge volcanism. On the other hand, the formation of back-arc basins is always accompanied by subduction, and it is possible that the subducted material is to some degree incorporated in the genesis of BABB. In this paper, we first discuss the composition of primary magmas of BABB and some other subalkalic rocks, and then discuss the variability of BABB from six back-arc basins, i.e., the Scotia Sea, Lau Basin, Mariana Trough, Shikoku and Parece Vela Basins, and the West Philippine Sea. Enough geochemical data are available on the basalts from these back-arc basins for the classification and for the comparison with basalts from other tectonic environments.

2. *Primary Magma Composition of BABB and Other Subalkalic Rocks*

Primary magma compositions have been identified either by phase equilibra relationships (e.g., O'HARA, 1965; KUSHIRO, 1973) or by element partition relationships (e.g., SATO, 1977). Here we adopt the latter approach, and present a method to estimate the composition of primary magmas of some subalkalic rock suites. Primary magma composition can be estimated in the following way (a similar approach has been given by TOHARA, 1978).

1. Determine the Fo content of olivine in the source peridotite by the NiO-Fo relationship of phenocryst olivine.

2. Obtain the olivine-control line in the MgO-FeO* (total iron as FeO) diagram.

3. Find the intersect of the olivine-control line and the constant FeO*/MgO line which is derived from the following equation;

$$K=[(100-\mathrm{Fo})/\mathrm{Fo}]_{\mathrm{olivine}}[(\mathrm{total\ Fe})*0.90/\mathrm{Mg}]_{\mathrm{magma}}=0.30$$

where K denotes the partition coefficient, and is almost independent of the temperature and composition of the melt (ROEDER and EMSLIE, 1970). The slight pressure dependence of the partition coefficient (TAKAHASHI and KUSHIRO, 1983) was neglected in the following discussions. The Fe(3)/(total Fe) ratio of the primitive magma was assumed to be 0.1, which is the value of fresh glass of MORB (SATO *et al.*, 1978) and Hawaiian tholeiite (SWANSON and FABBI, 1973).

Figure 1 illustrates the NiO versus Fo relationships of olivine phenocrysts in some primitive rocks. The NiO content of olivine commonly decreases as the Fo content decreases, and delineates a smooth curve for each rock. The correlation of NiO and Fo of olivine is well explained by the fractional crystallization of olivine (+sulfides), though magma mix-

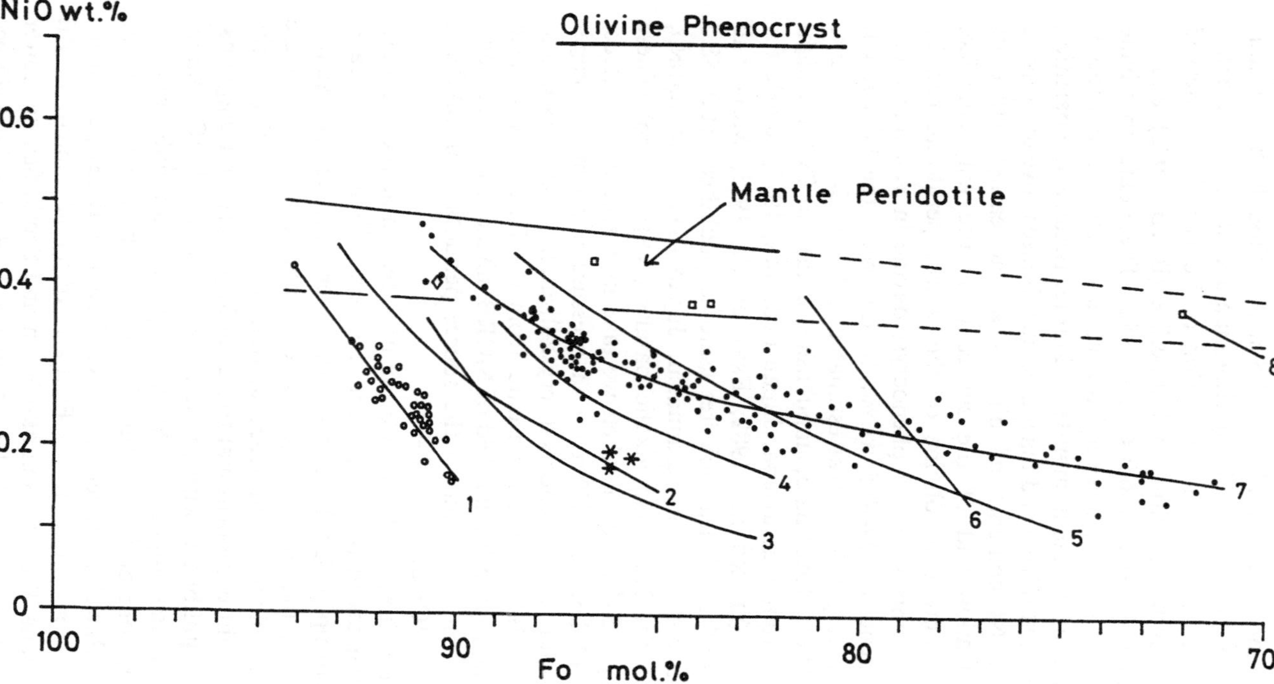

Fig. 1. NiO-Fo relations of olivine phenocrysts in representative volcanic rocks. Line 1 and open circle: boninite (WALKER and CAMERON, 1983; Sato, unpublished data), line 2: cumulates in high-magnesian andesite (TOHARA, 1977), line 3 and diamond: picritic and phyric MORB (MUIR and TILLEY, 1964; LE ROUXE *et al.*, 1980), line 4: ol-phyric MORB (ibid), line 5: high-alumina basalt (SATO and BANNO, 1983), line 6: high-magnesian andesite (ibid), line 7 and dot: Kilauea tholeiite (Sato, unpublished data), line 8 and open square: Hawaiian alkalic rocks (WHITE, 1966). The compositional range of mantle peridotite is also shown.

ing may shift the trends toward slightly higher NiO content. The highest NiO content of olivine is commonly 0.4–0.5 wt.%, though the Fo content varies from 94 for boninite to 72 for Hawaiian alkalic basalt. A rather constant NiO content versus a wide variation of Fo has been observed in the olivine in mantle xenoliths in kimberlite (BOYD, 1973) and in Hawaiian alkali basalt (WHITE, 1966; GOTO, 1982). These data on olivine suggest the presence of heterogeneity in Fo content in contrast to a rather uniform NiO content of olivine in mantle peridotite. Because the partition coefficient of Ni between olivine and magmatic melt is insensitive to pressure (TAKAHASHI, 1980), Ni contents of olivines at the earliest stage of phenocryst crystallization and of the source peridotite may coincide. Therefore, the Fo content of olivine of the source peridotite can be estimated from the composition of phenocryst olivine in volcanic rocks that have NiO contents of 0.4–0.5 wt.%. In this way, the Fo content of olivine in the source peridotite was estimated to be ca. 94 for boninite (WALKER and CAMERON, 1983; Sato, unpublished data), ca. 93 for a high-magnesian andesite from Matsuyama, Japan (TOHARA, 1977), 90–92 for oceanic-ridge basalts (LE ROUXE, 1982; FREY *et al.*, 1973; MUIR and TILLEY, 1964), 90–91 for Kilauea tholeiites (Sato, unpublished data), 88.5 for a high-alumina basalt from Takamatsu, Japan (SATO and BANNO, 1983), and 87–72.5 for Hawaiian alkalic basalts (WHITE, 1966) (These values are lower by 0.5–1.0 Fo % when we take into account the effect of pressures of 8–17 kb, the difference in pressure between the stages of magma segregation and phenocryst crystallization: TAKAHASHI and KUSHIRO, 1983). For BABB, available data from the Parece Vela Basin basalt plot within the compositional field of MORB olivine, and it is assumed that the Fo content of olivine in the source peridotite for BABB is similar to those of MORB, i.e., 92–90.

Using these Fo values of olivine, we can now determine the composition of primary magmas in MgO variation diagrams. Figure 2 illustrates FeO*-MgO relationship of BABB and several other volcanic rock suites. Most of the basic magmas with a MgO content more than 10 wt.% are saturated with only olivine (and trace amount of spinels). The rocks delineate nearly straight compositional trends in MgO variation diagrams, because olivine in equilibrium with the magma has a high MgO content and is rather uniform in composition. Such compositional trends are commonly called olivine-control lines. Figure 2 shows olivine control lines (or bands) of FeO* for Hawaiian tholeiite, MORB, boninite, BABB, and high-magnesian andesite from Shikoku. As FRYER *et al.* (1982) noted, BABB glass from the Mariana Trough, Shikoku Basin and Scotia Sea has lower FeO* and MgO trend than MORB glass, though BABB glass from Parece

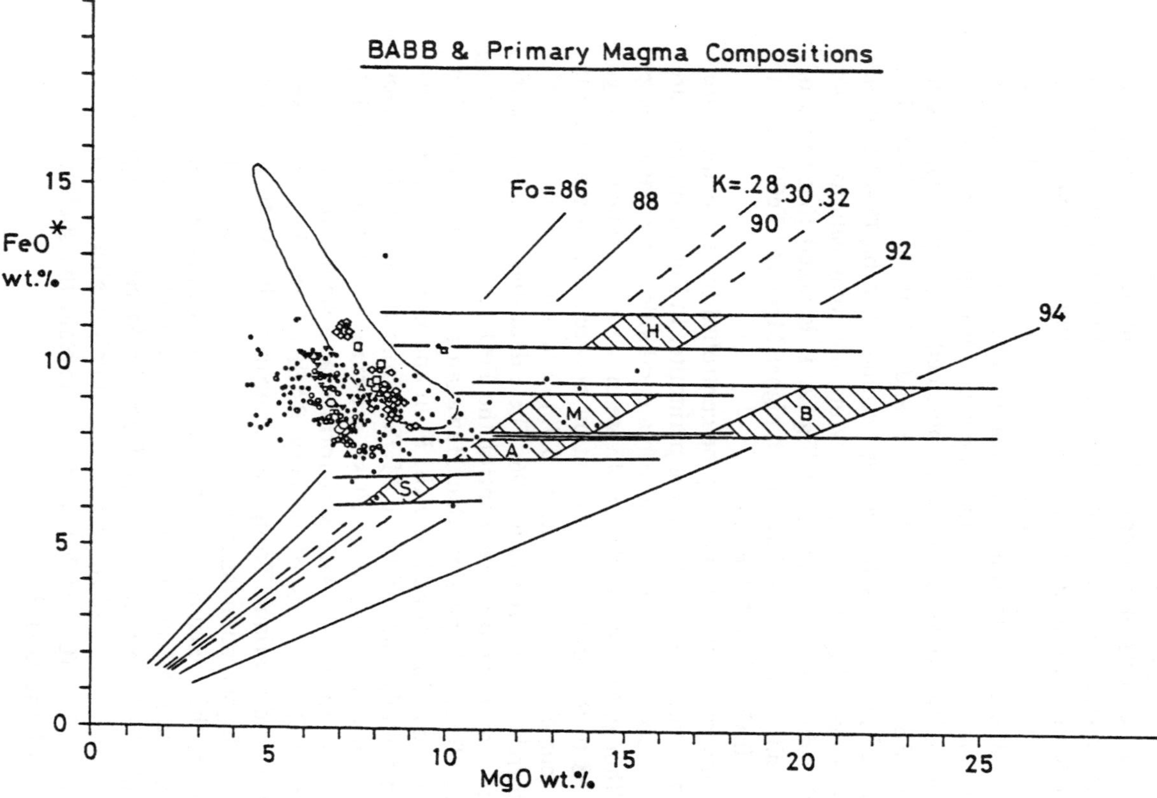

Fig. 2. FeO*-MgO relation of back-arc basin basalt and some subalkalic primary magmas. Dot: bulk rock composition of BABB, open Symbols: glass composition of BABB (circle: Mariana Trough, hexagon: Scotia Sea, triangle: Lau Basin, diamond: West Philippine Sea, square: Parece Vela Basin, reversed triangle: Shikoku Basin). Shaded areas are the estimated compositional range of primary magmas, H: Hawaiian tholeiite, B: boninite, M: MORB, A: BABB (Scotia Sea, Mariana Trough and Shikoku Basin), S: high-magnesian andesite from Shikoku. Constant FeO*/MgO lines, see text.

Vela Basin, Lau Basin and West Philippine Sea plots within the compositional field of MORB as illustrated in Fig. 2. The most magnesian glass of the Mariana Trough (MgO = 8 wt.% and FeO* = 7.5 wt.%) coexists with only olivine, and its olivine-control line should start from this composition. The intersect of the olivine control line and the constant FeO*/MgO line as determined by the above partition equation gives the possible primary magma composition for each rock suite. The advantage of using the MgO-FeO* partition relationship rather than Ni partitioning is that the partition coefficient is much more uniform and precisely determined for the former relationship. The obtained primary magma composition of BABB is 10–13 wt.% of MgO, and 7–9 wt.% of FeO*. The Al_2O_3 content of primary BABB was also estimated to be 15–17.5 wt.% from an Al_2O_3-MgO diagram. The primary magma compositions of other volcanic rock suites are also shown in Fig. 2. It is noted that those of Hawaiian tholeiite composition have much higher MgO and FeO* content than those of BABB and MORB.

Primary magma compositions may be controlled by the temperature, pressure, and composition of the source peridotite. In Fig. 3, the pressure dependence of FeO* in melt coexisting with a peridotitic mineral assemblage is shown. Although the Fo content of olivine in these experimental charge varies from 88 to 95, an overall correlation between pressure and FeO* is observed (correlation coefficient is 0.73). For the data of STOLPER (1980), the correlation coefficient is 0.96. These data suggest that pressure is one of the major factors that control the FeO* contents of primary magmas. Therefore, it is possible that Hawaiian tholeiite was formed at higher pressure than BABB and MORB. FUJII and BOUGAULT (1983) have shown experimentally that primitive MORB from the FAMOUS region (FeO* = 8.0 wt.%) was formed at 10 kb pressure. From Figs. 2 and 3, it is deduced that primary BABB was formed at a similar pressure. The slightly lower FeO* contents of primary BABBs from the Scotia Sea, Shikoku Basin and Mariana Trough as compared with MORB may be ascribed to the higher abundance of H_2O in these BABBs. FRYER *et al.* (1982) noted that the presence of water expands the stability field of olivine, thus decreases the FeO* content of the melt in equilibrium with olivine. The H_2O content of fresh BABB glass from the Mariana Trough is ca. 1 wt.% (GARCIA *et al.*, 1979), which is distinctly higher than the contents in MORB glass (0.3 wt.%), and may explain the slightly smaller FeO* content of the BABB.

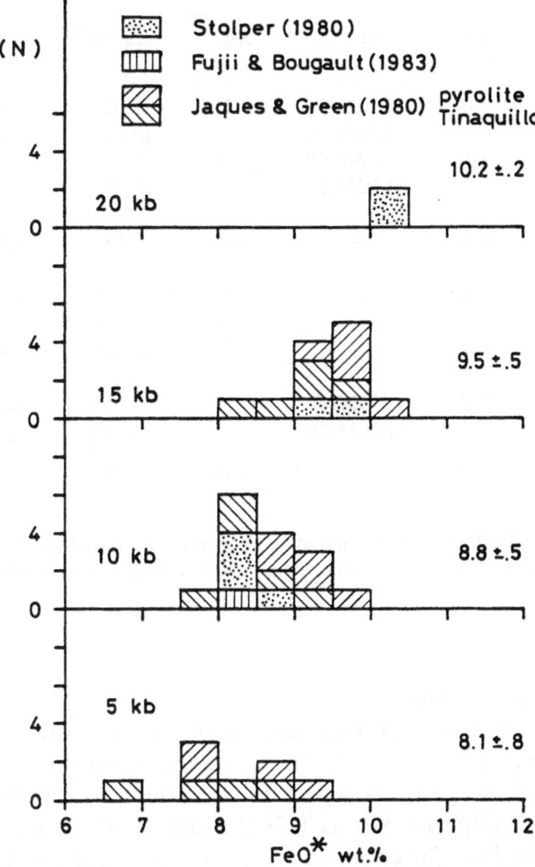

Fig. 3. Pressure dependence of the FeO* content of melt in equilibrium with a peridotite mineral assemblage.

3. *Variability of BABB*

BABB is commonly thought to have an intermediate composition between N-type MORB and IAT (e.g., TAYLOR and KARNER, 1983). However, BABB shows geochemical variations, and back-arc basins can be classified into at least three types according to the chemistry of basalt. Table 1 summarizes some of the diagnostic features of basalts from six back-arc basins (Fig. 4).

Table 1. Diagnostic features of back-arc basin basalt.

	FeO** wt, %	$^{87}Sr/^{86}Sr$	K_2O (glass)	$(La/Yb)_N$	Tectonics
Scotia Sea	7–8	.7030(2) [.7039(1)]	0.27	1.62 [0.51]	spreading shallow basin
Lau Basin	8–9	.7040(9)] [.7039(3)]	0.28	1.38 [0.77]	spreading underlain by slab
Mariana Trough	7.5–8.5	—	0.19	0.83*	spreading
Shikoku Basin	7.5–8.5	—	0.19	0.83*	spreading
Parece Vela Basin	8–9	—	0.20	0.81*	spreading
W. Philippine Sea	8–9	.7026(1) [.7035(1)]	0.11	0.51* [0.73*]	entrapped or spreading ?
N-type MORB	8–9	.7026	0.10	0.48	

*$(La/Tb)_N$.

**Total iron as FeO of primary magmas as estimated from Fig. 2. Data in bracket are those of associated island-arc volcanic rocks. Data in parentheses denote standard error.

3.1 West Philippine Sea

Among BABB from six back-arc basins, only those from the West Philippine Sea have low concentrations of incompatible elements within the compositional range of N-type MORB. The $^{87}Sr/^{86}Sr$ ratio of BABB from the West Philippine Sea is 0.7026, and is lower than those from other back-arc basins. Other geochemical features, such as the FeO* content of primary magmas, the K_2O content of fresh glass, and the REE pattern of BABB from West Philippine Sea all show similarities with N-type MORB. However, the origin of the West Philippine Sea has not been firmly established. It may represent a trapped ocean (UYEDA and KANAMORI, 1979), or it may be the result of back-arc spreading (SENO and MARUYAMA, 1983). The magnetic anomaly pattern of the West Philippine Sea has been identified as Eocene, which contradicts the entrapped ocean origin (TAYLOR and KARNER, 1983). The geochemical features of the BABB from the West Philippine Sea suggest that the composition of their source region was not affected by subduction.

3.2 Scotia Sea, Mariana Trough, Shikoku Basin and Parece Vela Basin

BABB from these back-arc basins commonly has intermediate

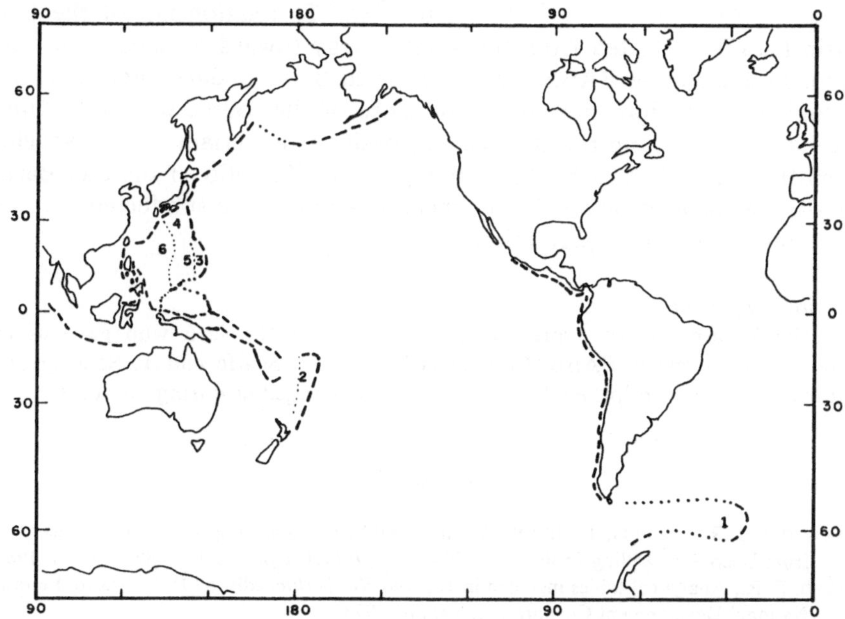

Fig. 4. Location map of back-arc basins discussed in this work. 1. Scotia Sea, 2. Lau Basin, 3. Mariana Trough, 4. Shikoku Basin, 5. Parece Vela Basin, 6. West Philippine Sea.

geochemical characteristics between MORB and the associated IAT. The FeO* content of the primary BABB is often lower, and $^{87}Sr/^{86}Sr$, K_2O, and the $(La/Yb)_N$ ratio are commonly higher than N-type MORB. It is noted that the $(La/Yb)_N$ ratio of the BABB is higher than those of the associated IAT in the Scotia Sea, while the reverse relationship is observed in the Mariana Trough. The $^{143}Nd/^{144}Nd$ versus $^{87}Sr/^{86}Sr$ relation of BABB from the Scotia Sea lies on the mantle array, while that of basalt from the associated South Sandwich Islands shifts from the mantle array toward a higher $^{87}Sr/^{86}Sr$ ratio (HAWKESWORTH *et al.*, 1979).

3. *Lau Basin*

The FeO* content of the primary BABB from the Lau Basin is 8–9%, and is within the range of MORB. On the other hand, the K_2O content and $(La/Yb)_N$ ratio are distinctly higher than those of N-type MORB. The $^{87}Sr/^{86}Sr$ ratio of the Lau Basin basalt is high (0.7032–0.7056) and similar to that of the associated Tonga arc volcanics

(GILL, 1976). The $^{143}Nd/^{144}Nd$ versus $^{87}Sr/^{86}Sr$ relationship of the Lau Basin basalts is shifted from the mantle array toward a higher $^{87}Sr/^{86}Sr$ ratio. Because these BABB are not affected by secondary alteration, the isotopic compositions may represent those of the source materials. Only the Lau Basin among the six back-arc basins is underlain by a Wadachi-Benioff zone. It is possible that the high $^{87}Sr/^{86}Sr$ ratio of the Lau Basin basalts was brought about by the incorporation of the subducted oceanic crust into the source peridotite.

Acknowledgements

Thanks are due to Professors K. Konishi and N. Nasu who gave us an opportunity to review BABB. Comments by Drs. C. Scarfe and T. Seno greatly improved the manuscript. Prof. M. Yamasaki encouraged us during the study.

REFERENCES†

*ARMSTRONG, R. L. and G. T. NIXON, Chemical and Sr-isotopic composition of igneous rocks from Deep Sea Drilling Project Legs 59 and 60, *Initial Rep. D.S.D.P*, **59**, 719–727, 1981.

BOYD, F. R., Analytical tables included in *Lesotho Kimberlite*, edit. by P. H. Nixon, Lesotho National Development Corporation, 350 pp., 1973.

BRYAN, W. B. and J. G. MOORE, Compositional variations of young basalts in the Mid-Atlantic Ridge rift valley near 36°49′N, *Bull. Geol. Soc. Am.*, **88**, 556–570, 1977.

*DICK, H. J. B., N. G. MARSH, and T. D. BULLEN, Deep Sea Drilling Project Leg 58 abyssal basalts from the Shikoku Basin; Their petrology and major-element geochemistry, *Initial Rep. D.S.D.P.*, **58**, 843–872, 1980.

*FRYER, P., J. M. SINTON, and J. A. PHILPOTTS, Basaltic glasses from the Mariana Trough, *Initial Rep. D.S.D.P.*, **60**, 601–609, 1982

FRYER, P. and J. M. SINTON, Generation and characteristics of basaltic glasses from the Mariana Back-Arc Basin, *J. Geophy. Res.* (in press).

FUJII, T. and H. BOUGAULT, Melting relations of a magnesian abyssal tholeiite and the origin of MORBs, *Earth Plan. Sci. Lett.*, **62**, 283–295, 1983.

GARCIA, M. O., N. W. K. LIU, and D. W. MUENOW, Volatiles in submarine volcanic rocks from the Mariana Island arc and trough, *Geochim. Cosmochim. Acta*, **43**, 305–312, 1979.

*GILL, J. B., Composition and age of Lau Basin and Ridge volcanic rocks: Implications for evolution of an interarc basin and remnant arc, *Geol. Soc. Am. Bull.*, **87**, 1384–1395, 1976.

GOTO, A., The spinel lherzolite xenoliths from Salt Lake Crater, Hawaii, Unpublished Thesis, Department of Earth Sciences, Kanazawa Univ., 1982.

*HART, S. R., W. E. GLASSLEY and D. E. KARIG, Basalts and sea floor spreading behind the Mariana Island Arc, *Earth Plan. Sci. Lett.*, **15**, 12–18, 1972.

*HAWKESWORTH, C. J., R. K. O'NIONS, R. J. PUNKHURST, P. J. HAMILTON, and N. M. EVENSEN, A Geochemical study of island-arc and back-arc tholeiites from the Scotia Sea, *Earth Plan. Sci. Lett.*, **36**, 253–262, 1979.

†Asterisks indicate the works that served as data sources for Fig. 2 and Table 1.

*HAWKINS, J. W., Petrology and geochemistry of basaltic rocks of the Lau Basin, *Earth Plan. Sci. Lett.*, **28**, 283–297, 1976.

JAQUES, A. L. and D. H. GREEN, Anhydrous melting of peridotite at 0–15 kb pressure and the genesis of tholeiitic basalts, *Contrib. Mineral. Petrol.*, **73**, 287–310, 1980.

KUSHIRO, I., Origin of some magmas in oceanic and circum-oceanic regions, *Tectonophysics*, **17**, 211–222, 1973.

LE ROEX, A. P., A. J. ERLANK, and H. D. NEEDHAM, Geochemical and mineralogical evidence for the occurrence of at least three distinct magma types in the "FAMOUS" Region, *Contrib. Mineral. Petrol.*, **77**, 24–37, 1981.

*MARSH, N. G., A. D. SAUNDERS, J. TARNEY, and H. J. B. DICK, Geochemistry of basalts from the Shikoku and Daito basins, Deep Sea Drilling Project Leg 58, *Initial Rep. D.S.D.P.*, **58**, 805–842, 1980.

*MATTEY, D. P., N. G. MARSH, and J. TARNEY, The geochemistry, mineralogy, and petrology of basalts from the West Philippine and Parece Vela Basins and from the Palau-Kyushu and West Mariana Ridges, Deep Sea Drilling Project Leg 59, *Initial Rep. D.S.D.P.*, **59**, 753–800, 1981.

*MELSON, W. G., T. L. VALLIER, T. L. WRIGHT, G. BYERLY, and J. NELEN, Chemical diversity of abyssal volcanic glass erupted along the Pacific, Atlantic, and Indian Ocean sea-floor spreading centers, *Geophys. Monograph*, **19**, 351–367, 1976.

MUIR, I. D. and C. E. TILLEY, Basalts from the northern part of the rift zone of the Mid-Atlantic Ridge, *J. Petrol.*, **5**, 409–434, 1964.

O'HARA, M. J., Primary magmas and the origin of basalts, *Scottish J. Geol.*, **1**, 19–40, 1965.

ROEDER, P. L. and R. F. EMSLIE, Olivine-liquid equilibrium, *Contrib. Mineral. Petrol.*, **29**, 275–289, 1970.

SATO, H., Nickel content of basaltic magmas: identification of primary magmas and a measure of the degree of olivine fractionation, *Lithos*, **10**, 113–120, 1977.

SATO, H., K. AOKI, K. OKAMOTO, and F. FUJITA, Petrology and chemistry of basaltic rocks from Hole 396B, IPOD/DSDP Leg 46, *Initial Rep. D.S.D.P.*, **46**, 115–141, 1978.

SATO, H. and S. BANNO, NiO-Fo relation of magnesian olivine phenocryst in high-magnesian andesite and associated basalt-andesite-sanukite from northeast Shikoku, Japan. *Bull. Volc. Soc. Japan*, Ser. 2, **28**, 141–156, 1983.

*SAUNDERS, A. D. and J. TARNEY, The geochemistry of basalts from back-arc spreading center in the East Scotia Sea, *Geochim. Cosmochim. Acta*, **43**, 555–572, 1979.

SENO, T. and S. MARUYAMA, Paleogeographic reconstruction and origin of the Philippine Sea, *Tectonophysics*, **102**, 53–84, 1984.

STOLPER, E., Phase diagram for mid-ocean ridge basalts: preliminary results and implications for petrogenesis, *Contrib. Mineral. Petrol.*, **74**, 13–27, 1980.

SWANSON, D. A. and B. P. FABBI, Loss of volatiles during fountaining and flowage of basaltic lava of Kilauea Volcano, Hawaii, *J. Res. U.S. Geol. Survey*, **1**, 649–658, 1973.

TAKAHASHI, E., Olivine/liquid nickel partitioning at high pressures, *EOS*, **61**, 397, 1980.

TAKAHASHI, E. and I. KUSHIRO, Melting of a dry peridotite at high pressures and basalt magma genesis, *Am. Mineral.*, **68**, 859–879, 1983.

TAYLOR, B. and G. D. KARNER, On the evolution of marginal basins, *Review Geophy. Space Phys.*, **21**, 1727–1741, 1983.

TOHARA, T., Petrological study of Tertiary volcanic rocks and ultramafic inclusions from Matsuyama area, northwestern Shikoku, Unpublished Master's Thesis, Faculty of Earth Sciences, Kanazawa University, 155 pp., 1978.

UYEDA, S. and H. KANAMORI, Back-arc opening and the mode of subduction, *J. Geophys. Res.*, **84**, 1049–1061, 1979.

WALKER, D. A. and W. E. CAMERON, Boninite primary magmas: evidence from the Cape Vogel Peninsula, PNG, *Contrib. Mineral. Petrol.*, **83**, 150–158, 1983.

WHITE, R. W., Ultramafic inclusions in basaltic rocks from Hawaii, *Contrib. Mineral. Petrol.*, **12**, 245–314, 1966.

*WOOD, D. A., J. L. JORON, N. G. MARSH, J. TARNEY, and M. TREUIL, Major- and trace-element variations in basalts from the north Philippine Sea drilled during Deep Sea Drilling Project Leg 58; A comparative study of back-arc basin basalts with lava series from Japan and Mid-ocean Ridges, *Initial Rep. D.S.D.P.*, **58**, 873–894, 1980.

*WOOD, D. A., D. P. MATTEY, J. L. JORON, N. G. MARSH, J. TARNEY, and M. TREUIL, A geochemical study of 17 selected samples from basement cores recovered at Sites 447, 448, 449, and 451, Deep Sea Drilling Project Leg 59, *Initial Rep. D.S.D.P.*, **59**, 743–752, 1981.

*WOOD, D. A., N. G. MARSH, J. TARNEY, J. L. JORON, P. FRYER, and M. TREUIL, Geochemistry of igneous rocks recovered from a transect across the Mariana Trough, Arc, Fore-arc, and Trench, Sites 453 through 461, Deep Sea Drilling Project Leg 46, *Initial Rep. D.S.D.P.*, **60**, 611–645, 1982.

*ZAKARIADZE, G. S., L. V. DMITRIEV, A. V. SOBOLEV, and N. M. SUSCHEVSKAYA, Petrology of basalts of Holes 447A, 449, and 450, south Philippien Sea Transect, Deep Sea Drilling Project Leg 59, *Initial Rep. D.S.D.P.*, **59**, 669–680, 1981.

Formation of Active Ocean Margins, edited by N. Nasu *et al.*, pp. 411–422.

A MIOCENE FOREARC MAGMATISM AT SHIONOMISAKI, SOUTHWEST JAPAN

Yasuyuki MIYAKE[1] and Kunihiko HISATOMI[2]

[1]*Department of Geology and Mineralogy, Faculty of Science, Kyoto University, Kyoto, 606, Japan*

[2]*Institute of Earth Sciences, Faculty of Education, Wakayama University, Wakayama, 640, Japan*

Abstract. Kumano Group of Miocene age was deposited in a forearc basin of Southwest Japan arc-trench system. The strata were composed of the sediments from the northern hinterland which were trapped by an upheaval zone. The Shionomisaki igneous complex was formed just in this upheaval zone with intimate relation with the upheaval movement. The upheaval zone of Miocene age corresponds to a structural high in the modern continental slope off Southwest Japan. Along with other concealed igneous masses, we designate the igneous zone along the structural highs "Nankai Structural High Magmatic Zone". The petrologic features of the Shionomisaki igneous complex are similar to those of abyssal igneous rocks, and are distinct from the other igneous rocks in the Outer Zone of Southwest Japan. The petrochemical resemblances between the Shionomisaki basic rocks and the off-ridge intrusives in Shikoku Basin are recognized.

1. Introduction

The Shionomisaki area is situated within the forearc region of Southwest Japan. The main constituents of the Southwest Japan arc-trench system are shown in Fig. 1. In the Chugoku district, Quaternary volcanoes are arranged, such as Mt. Takura (T in Fig. 1) , Mt. Ooginosen (O), Mt. Daisen (Ds), Mt. Sambe (S), Mt. Aono (A) and volcanoes in the Abu district (Ab). These volcanoes are characterized by alkali basalt and high-K andesite-dacite, and the volcanism started in the Miocene age (NAGAO, 1976). Nankai Trough is not so deep (4,500 m) as other trenches in the world,

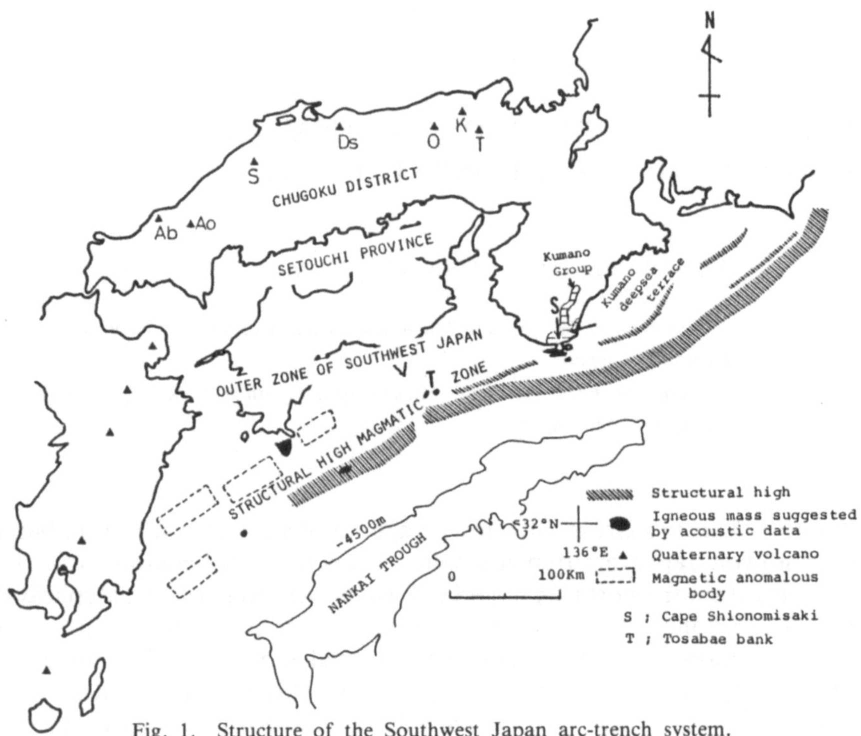

Fig. 1. Structure of the Southwest Japan arc-trench system.

but it was deeper than observed topographically, having been buried by the sediments of the Miocene to Recent (GEOLOGICAL SURVEY OF JAPAN, 1977). The Shionomisaki area is situated within the trench-side area of the modern volcanic arc of the SW Japan.

As discussed by DICKINSON and SEELEY (1979), a structural high (or a trench slope break) is an essential factor of a continental slope of convergent margin, trapping terrigenous sediments to form a terrace of a forearc basin. Several structural highs are well-developed in the continental slope off Southwest Japan (HILDE *et al.*, 1969; OKUDA *et al.*, 1979) and are shown in Fig. 1 after GEOLOGICAL SURVEY OF JAPAN (1977). It is noticeable that the Shionomisaki area is just situated on the onshore extension of one of these structural highs (Fig. 1). Within forearc basins of arc-trench systems in the world, not a few igneous bodies are found (VON HUENE *et al.*, 1978; BURK, 1965; TYSDAL *et al.*, 1977; EWART and BRYAN, 1972 etc.). However, rather small number of them were clarified to have been formed in situ. And every model concerning the geotectonics of an arc-trench system, for example that of DICKINSON and SEELEY (1979), has never illustrated that any volcanisms take place in a forearc basin, although DICKINSON and SEELEY (1979) themselves described the "anomalous" igneous bodies in several forearc basins.

The main purpose of this paper is to demonstrate that the Shionomisaki igneous complex, one of the examples of forearc igneous rocks, was formed autochthonously within a Miocene forearc basin of Southwest Japan arc-trench system, just on a structural high in intimate relation with the upheaval movement. Then, we will compare the petrologic features of this complex with other Miocene igneous rocks in and off Southwest Japan.

2. *Geologic Setting of the Kumano Group and the Shionomisaki Igneous Complex*

2.1 *Morphology of the Kumano basin*

The Lower to Middle Miocene Kumano Group, which overlies the Cretaceous to Lower Miocene Shimanto Supergroup, is a thick marine sequence of muddy sediments. The total thickness of the group attains more than 4,000 m. The group is lithologically divided into three formations, namely, the Shimosato, the Shikiya and the Mitsuno Formations in ascending order (Fig. 2; HISATOMI, 1981).

The Shimosato Formation, 600 to 1,800 m thick, consists mostly of mudstone and muddy alternation of sandstone and mudstone. The formation decreases in thickness and grain-size toward the north, and pinches out at the northern end of its outcrop. The Shikiya Formation, 800 to 1,200 m thick, consists of bedded or massive mudstone, and partly of muddy alternation and angular clast-bearing mudstone in the southern part of the basin. The thickness of the formation gradually decreases toward the north. The Mitsuno Formation, 1,500 m thick, consists of alternation of sandstone and mudstone, sandstone and angular clast-bearing mudstone (HISATOMI, 1981).

The geologic age of the Kumano Group is safely assigned to Blow's N8 and N9, based on the benthonic and planktonic foraminiferal fossils, which are of the upper half of Blow's N8, reported at the horizon 200 m above the base of the Shikiya Formation (NISHIMURA and MIYAKE, 1973; IKEBE *et al.*, 1975). And this horizon lies about 150 m below the base of the effusive rocks of the Shionomisaki igneous complex. Therefore, it is safe to conclude that the activity of the complex started at the latest Early Miocene.

The morphology of the paleobasin of the Kumano Group was

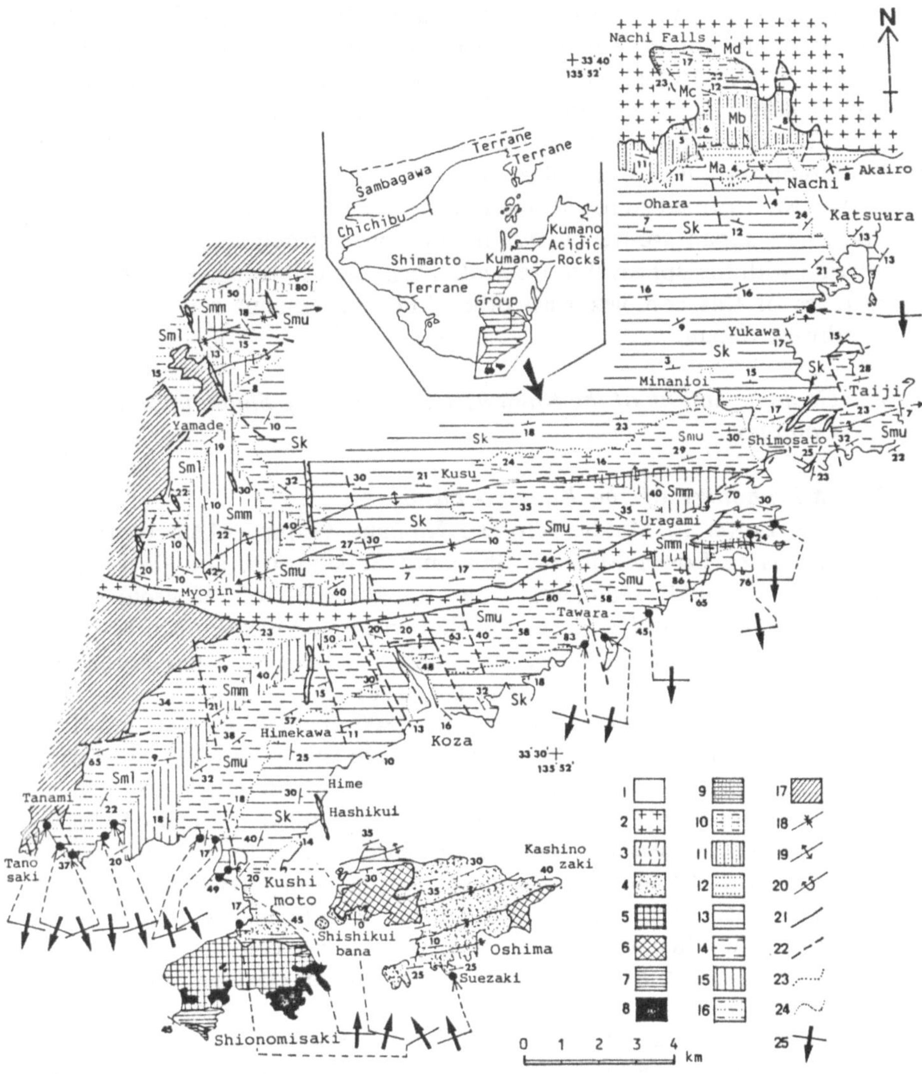

Fig. 2. Geologic map. 1: alluvium, 2: acidic pyroclastic rock and granite porphyry (Kumano Acidic Rocks), 3: quartz porphyry (Kumano Acidic rocks), 4–8: Shionomisaki Igneous Complex, 4 rhyolitic pyroclastic rocks, 5: quartz porphyry and granophyre, 6: hypersthene rhyolite, 7: basaltic lava, 8: basic plutonic rocks, 9–12: Mitsuno Formation, 9: Member Md, 10: Member Mb, 12: Member Ma, 13: Shikiya Formation, 14–16: Shimosato Formation, 14: Member Smu, 15: Member Smm, 16: Member Sml, 17: Muro Group, 18: syncline, 19: anticline, 20: overturned anticline, 21: fault, 22: inferred fault, 23: formation boundary, 24: member boundary, 25: paleoslope deduced from slump structure. Compiled after HISATOMI and MIYAKE (1981), MIYAKE (1981) and HISATOMI (1984).

reconstructed by HISATOMI (1981) as shown in Fig. 3A, based on the lithology and thickness of the Kumano Group mentioned above, as well as the following sedimentological evidences:

(1) Most of paleocurrent directions deduced from current sole marks, ripple marks and other unidirectional sedimentary structures, indicate the constant supply from the north or the northwest.

(2) Slump structures developed in the southernmost part of the study, that is, the Shionomisaki area, show the northward-inclining paleoslope (Fig. 2). Contrarily, those in the rest of the area indicate the southward-inclining paleoslope.

(3) The characteristic features which are the diagnostic features of the slope or the base-of-slope environments, such as the constant paleocurrent direction, the development of slump structures and debris flow deposits, are developed in the southernmost part of the basin.

The Kumano basin was situated on an E-W trending and southward-inclining slope in the late Early Miocene age. The clastic sediments, which were supplied from the shelf area situated to the north, flowed down on the slope as low density turbidity currents (HISATOMI, 1984). The clastics were dammed up by an E-W trending upheaval zone stretching through the Shionomisaki area, to form a thick sedimentary pile, which decreases in thickness and grain-size of deposits northword. The trend of the upheaval zone is supposed to be ENE-WSW, judging from the paleoslopes deduced

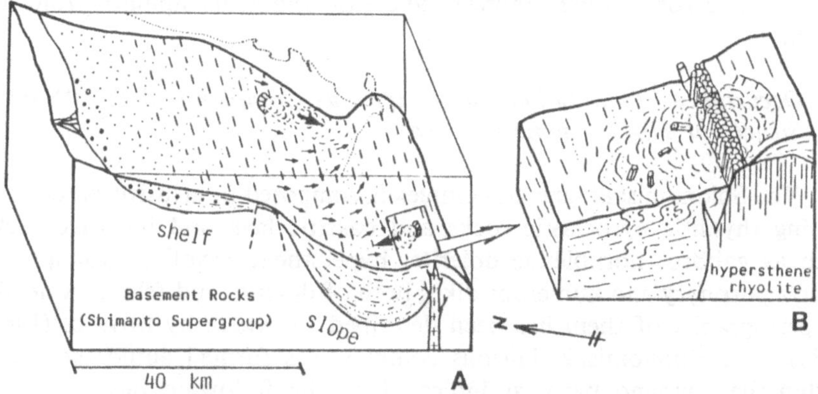

Fig. 3. A: Schematic reconstruction of the Kumano paleobasin. Solid arrows and broken arrows represent turbidity and bottom current, respectively. Heavy arrow shows the direction of dislocation of slump sheet. Dotted line shows the present topography, After HISATOMI (1984). B: Enlarged figure around the upheaval zone. The dimensions of the rhyolite blocks and the rhyolite mass are exaggerated. The further explanations are given in the text.

from the slump structures in the Shionomisaki area (Fig. 3; HISATOMI, 1984). The tectonic setting of the Kumano paleobasin and the upheaval zone shown in Fig. 3 are the equivalents of the basins of shelf and upper continental slope and the structural high in the modern arc-trench systems.

2.2 Magmatism on the upheaval zone

HISATOMI and MIYAKE (1981) discussed the upheaval zone stretching through the Shionomisaki area, and the geological history of the area close to the upheaval zone as follows (Fig. 3B):

(1) At first, a sequence of mudstone, sandstone and conglomerate which yields the planktonic foraminiferal assemblage of upper half of Blow's N8 was formed.

(2) Hypersthene rhyolite mass intruded into the sediments of the Kumano Group. A part of this hypersthene rhyolite was exposed on the sea floor, and blocks of them (max. 4 m in length), cracked along the columnar joints, were included in the surrounding sediments, preserving their original shapes.

(3) The sediments which included the rhyolite blocks were, in turn, dislocated from south to north to form a slump sheet 40–50 m thick.

(4) Another hypersthene rhyolite mass intruded into the slump sheet. Subsequently, the activity of the Shionomisaki igneous complex continuously took place. However, these are not illustrated in Fig. 3B.

These occurrences clearly show that the igneous activity was closely related with and keeping pace with the uplifting and the formation of the upheaval zone, which trapped the clastic sediments supplied from the north.

3. Occurrence of the Shionomisaki Igneous Complex and the "Nankai Structural High Magmatic Zone"

The Shionomisaki igneous complex is composed of effusive rocks comprising rhyolitic pyroclastic rocks and basaltic lava and intrusive rocks such as gabbro, hornblende dolerite, hypersthene rhyolite, granophyre, quartz porphyry and numerous dikes both of dolerite and felsite. A detailed petrography of them has been described elsewhere by MIYAKE (1981, 1983). The Shionomisaki igneous complex was formed autochthonously within the Kumano basin as indicated by the following facts.

1) The rhyolitic pyroclastic rocks, mainly distributed in Oshima Island, consist of alternations of tuff breccia (5–30 m) and crystal tuff (<2 m), and these pyroclastic rocks are of subaqueous origin. Moreover, the interfingering relation between the fossiliferous mudstone of the Kumano Group

and the rhyolitic pyroclastic rocks occur in the western part of Oshima Island (see Fig. 9 and Plate I-6; MIYAKE, 1981).

2) The hypersthene rhyolite contacts with the mudstone giving the thermal metamorphism to the latter, along the western coast of the Oshima Island. It also intruded into the alternation of mudstone and sandstone at the northwestern coast of the Shionomisaki area, as a plug (see Fig. 5 and Plate I-5, 6 in HISATOMI and MIYAKE, 1981).

3) In the northern part of the Shionomisaki area, the alternation of mudstone and sandstone, the rhyolitic pyroclastic rocks and the basaltic lava are bedded conformably in ascending order.

4) The alternation of mudstone and sandstone in the northwestern part of the Shionomisaki area contains many angular gravels of the hypersthene rhyolite which had fallen down from the precursory mass (see Fig. 3B and Plate I-3 in HISATOMI and MIYAKE, 1981).

Moreover, as discussed above, the hypersthene rhyolite is included in the slump sheet as a gravel and it, in turn, intruded into the slump sheet, these facts suggesting that the magmatism occurred simultaneously with the upheaval movement, which caused the slump structure.

Many igneous bodies are distributed not only in the Shionomisaki area, but also along the modern structural highs of the SW Japan continental slope (Fig. 1). On the west of the Shionomisaki igneous complex, a submarine igneous mass is suggested by the acoustic data (GEOLOGICAL SURVEY OF JAPAN, 1977). Furthermore, OSHIMA *et al.* (1980) and GEOLOGICAL SURVEY OF JAPAN (1978) reported the positive magnetic anomalies along the structural highs trending from the Shionomisaki area WSW to the Tosabae bank and ENE into the Kumano deep-sea terrace. Also on the outer ridge, the igneous masses are suggested by acoustic data (GEOLOGICAL SURVEY OF JAPAN, 1977), and at one of these sites, HONZA *et al.* (1984) reported a dredged fragment of basalt. WATANABE and HATTORI (1980) reported the positive magnetic anomalies as shown in Fig. 1, and attributed them to the subterranean basic masses. It is to be noticed that these anomalies are arranged parallel to the trend of the structural highs. All of these oceanographic evidences suggest that igneous masses exist along these structural highs. And it is supposed that the structural highs off Southwest Japan were once the magmatic zone as is the case of the Shionomisaki area. We designate this Miocene igneous zone along the modern structural highs off Southwest Japan, as "Nankai Structural High Magmatic Zone". Formerly, such terms as "structural high" or "trench slope break" have been used only to mean the topographic high within continental slopes or the upheaval of the basement rocks beneath there. However, the nature and the structure of these basement rocks have not been clarified well. JAKEŠ and MIYAKE (1984) proposed that the magma which was formed within forearc region underplates the soft sediments to form upwelling of the surface. We believe that so-called structural highs off SW Japan was formed in intimate relation with the magmatism.

4. *Petrologic Characters of the Shionomisaki Igneous Complex, in Comparison with other Miocene Igneous Rocks*

The Shionomisaki igneous complex has similar petrologic characters to the abyssal igneous associations, as demonstrated by MIYAKE (1983, 1985). The major-element chemistry shows low K_2O (generally lower than 0.5 wt.%) in basic rocks comparable to that of abyssal tholeiite, positive correlation between TiO_2 and FeO/MgO ratio (Fig. 4), higher Ti and Cr contents than island arc tholeiite (after PEARCE, 1975), and low K_2O in acidic members.

Nevertheless, the chondrite-normalized REE (rare earth element) patterns of the Shionomisaki basic rocks are not common with the so-called "Normal-type MORB" of, for instance, SUN *et al.* (1979). The REE pat-

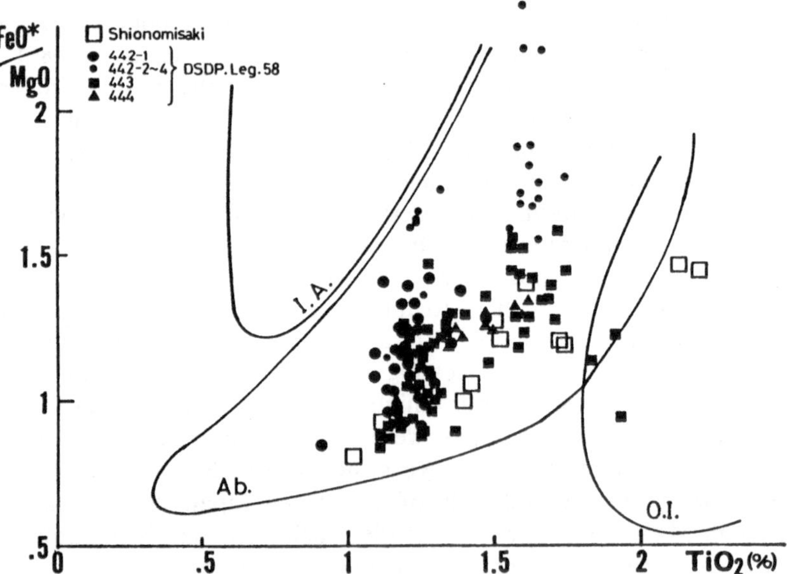

Fig. 4. TiO_2—FeO* (total Fe as FeO)/MgO figure. The fields of island arc tholeiite (I.A.), abyssal tholeiite (Ab) and oceanic island tholeiite (O.I) are after GLASSLEY (1974).

terns of Shionomisaki rocks are shown in Fig. 5. The data are reported in MIYAKE (1985). Each of REEs increases as crystallization fractionation proceeds. The pattern shows that light REEs (LREEs) are not depleted and have rather higher chondrite-normalized ratios than heavy REEs (HREE), contrary to that N-type MORB shows LREE depleted pattern. T-type MORB shows LREE non-depleted pattern (SUN *et al.*, 1979), whereas, the Shionomisaki rocks generally show anomalously low La, which is not the case even in the T-type MORB. Moreover, that the Shionomisaki basic rocks contain considerable amount of amphibole (MIYAKE, 1981) is a distinct feature from abyssal tholeiites.

In the same age as the magmatism in the Shionomisaki area, magmatism took place within Outer Zone of Southwest Japan, in Setouchi Province and in Shikoku Basin off Southwest Japan. The magmatism in both Outer Zone and Setouchi Province is mainly of acidic one and is contemporaneous within short interval in age (about 14 my.; NAKADA and TAKAHASHI, 1979). One of them, Kumano Acidic Rocks were formed after the elevation of the Kumano Basin (ARAMAKI and HADA, 1965). The magmatism in Shikoku Basin in middle Miocene is confirmed by the Deep Sea Drilling Project Leg. 58. At three sites (442, 443, 444), basaltic intrusive rocks were obtained, and their ages are nearly the same and are 16–13 my. These magmatisms are considered to be widespread off-ridge magmatism (KOBAYASHI and NAKADA, 1978; SLOAN *et al.*, 1978). Petrologic

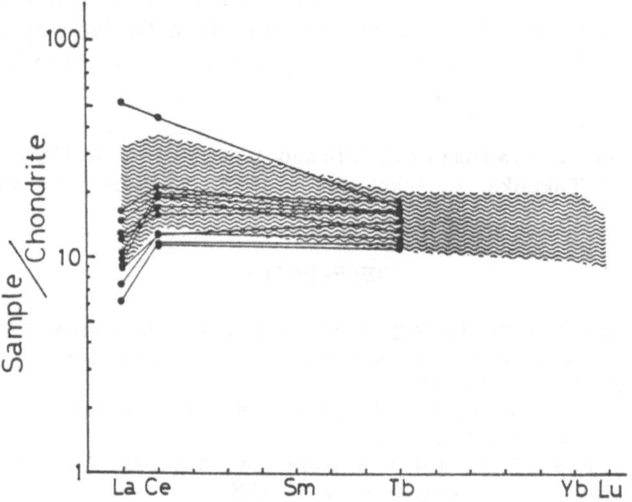

Fig. 5. Chondrite normalized REE pattern. Hatched zone shows the range of the Shionomisaki basic rocks, and the solid circles show the Shikoku Basin rocks.

characters among these igneous rocks of middle Miocene in and off Outer Zone of Southwest Japan are compared, below.

As mentioned by MIYAKE (1981), the acidic rocks of Shionomisaki igneous complex have distinct character from the acidic rocks of Outer Zone of Southwest Japan or Setouchi Province. NAKADA and TAKAHASHI (1979) compiled the igneous rocks from Outer Zone and Setouchi Province and regarded them of common petrographic province, showing such regional variation as that K_2O increases towards south. On the contrary, the acidic rocks of the Shionomisaki igneous complex have very low K_2O content (less than 1.7 wt.%) and low K_2O/Na_2O ratio (less than 0.6), whereas, the acidic rocks of Outer Zone have K_2O as high as 2-7 wt.% and K_2O/Na_2O ratio 0.9–2.1.

Petrochemistry of the basaltic rocks from the Shikoku Basin is reported by MARSH *et al.* (1980) and WOOD *et al.* (1980). Most of these basalts are tholeiite but some from Site 444 are alkali basalt. The tholeiites are plotted on TiO_2 versus FeO/MgO figure (Fig. 4). After the discrimination of GLASSLEY (1974), these rocks are abyssal tholeiite in composition as is the case of the Shionomisaki basic rocks. The REE patterns show very close resemblance to those of Shionomisaki rocks. As shown in Fig. 5, REE patterns of Shikoku Basin rocks other than alkali basalt have flat or slightly decreasing pattern toward HREE. These features are common with the Shionomisaki rocks. Thus, the magmatism of Shionomisaki has affinities with that of the off-ridge magmatism in Shikoku Basin of the same age. The common features of REE among the Shionomisaki rocks, one of the Structural High Magmatism, and the off-ridge magmatism will be further evaluated in a separate paper (Miyake, in prep.).

Acknowledgements

The authors acknowledge Prof. S. Banno, Emeritus Prof. K. Nakazawa, Drs. I. Nakayama, T. Tokuoka, P. Jakeš and Prof. K. Kobayashi for their fruitful discussions.

REFERENCES

ARAMAKI, S. and S. HADA, Geology of the central and southern parts of the Acidic Igneous Complex (Kumano Acidic Rocks) in southeastern Kii Peninsula (J), *Bull. Geol. Soc. Japan*, **71**, 494–512, 1965.

BURK, C. A., Geology of the Alaska Peninsula—Island arc and continental margin, *Mem. Geol. Soc. Amer.*, **99**, 250, 1965.

DICKINSON, W. R. and D. R. SEELEY, Structure and stratigraphy of forearc regions, *Bull. Amer. Assoc. Petrol. Geologists*, **63**, 2–31, 1979.

EWART, A. and W. B. BRYAN, Petrography and geochemistry of the igneous rocks from Eua, Tongan Islands, *Bull. Geol. Soc. Amer.*, **83**, 3281–3298, 1972.

GEOLOGICAL SURVEY OF JAPAN, 1:1,000,000 Geological map off Outer Zone of Southwest Japan, 1977.

GEOLOGICAL SURVEY OF JAPAN, Fundamental study on the mining resources under the shelf, Report on the method of airborne magnetic survey (1), 1978 (in Japanese).

GLASSLEY, W., Geochemistry and tectonics of the Crescent volcanic rocks, Olympic Peninsula, Washington. *Bull. Geol. Soc. Am.*, **85**, 785–794, 1974.

HILDE, T. W. C., J. M. WAGEMAN, and W. T. HAMMOND, The structure of Tosa terrace and Nankai trough off Southwest Japan, *Deep-Sea Res.*, **16**, 67–75, 1969.

HISATOMI, K., Geology and sedimentology of the Kumano Group in the southeastern part of the Kumano basin, Kii Peninsula, *J. Geol. Soc. Japan*, **87**, 157–174, 1981 (in Japanese).

HISATOMI, K., Sedimentary environment and basin analyses of the Miocene Kumano Group in the Kii Peninsula, Southwest Japan, *Mem. Fac. Sci. Kyoto Univ., Ser. Geol. Mineral.*, **50**, 1984.

HISATOMI, K. and Y. MIYAKE, Upheaval movement and igneous activity in the Sionomisaki area, Kii Peninsula, Southwest Japan, *J. Geol. Soc. Japan*, **87**, 629–639, 1981 (in Japanese).

HONZA, E., M. ARITA, Y. KINOSHITA, Y. OKUDA, and K. TAMAKI, Geological survey in the sea around Japan—The survey by the Hakurei-maru in 1983, *Chishitsu News*, **355**, 26–48, 1984 (in Japanese).

IKEBE, N., M. CHIJI, and Y. MOROZUMI, Lepidocyclina horizon in the Miocene Kumano Group in reference to planktonic foraminiferal biostratigraphy, *Bull. Osaka Mus. Nat. Hist.*, **29**, 81–89, 1975 (in Japanese).

JAKEŠ, P. and Y. MIYAKE, Magma in forearcs: implication for ophiolite generation, *Tectonophysics*, **106**, 349–358, 1984.

KOBAYASHI, K. and M. NAKADA, Magnetic anomalies and tectonic evolution of the Shikoku Inter-arc Basin, *J. Phys. Earth*, **26**, 391–402, 1978.

MARSH, N. G., A. D. SAUNDERS, J. TARNEY, and H. J. B. DICK, Geochemistry of basalts from the Shikoku and Daito Basins, Deep Sea Drilling Project Leg 58, in *Initial Reports of the Deep Sea Drilling Project*, vol. 58, edited by L. N. Stout, pp. 805–842, US. Government Printing Office, Washington, 1980.

MIYAKE, Y., Geology and petrology of the Shionomisaki igneous complex, Wakayama Prefecture, Japan, *J. Geol. Soc. Japan*, **87**, 383–403, 1981 (in Japanese).

MIYAKE, Y., Abyssal affinities of the Shionomisaki igneous complex, formed within the Miocene forearc basin Southwest Japan, Doctor thesis, Kyoto Univ., Kyoto, 1983.

MIYAKE, Y., MORB-like tholeiites formed within the Miocene forearc basin Southwest Japan, *Lithos*, **18**, 23–34, 1985.

NAGAO, T., Late cenozoic volcanism in the Chugoku and the Shikoku provinces, Western Japan, *Earth Sci.*, **30**, 110–121, 1976 (in Japanese).

NAKADA, S. and M. TAKAHASHI, Regional variation in chemistry of the Miocene intermediate to felsic magmas in the Outer Zone and the Setouchi Province of Southwest Japan, *J. Geol. Soc. Japan*, **85**, 571–582, 1979.

NISHIMURA, A. and Y. MIYAKE, Occurrence of Lepidocyclina and Miogypsina in the Kumano Group, *Commun. Paper, Joint Research for the Shimanto Geosyncline*, **2**, 37–38, 1973 (in Japanese).

OKUDA, Y. M. KUMAGAI, and K. TAMAKI, Tectonic development of the continental slope and its peripheral area off Southwest Japan in relation to sedimentary sequences in sedimentary basins, *J. Japan Assoc. Petroleum Technol.*, **44**, 279–290, 1979 (in Japanese).

OSHIMA, S., T. TOZAKI, and K. ONODERA, Geomagnetic anomalies at sea around southwest

Japan, *Rep. Hydrographic Res.*, **15**, 33–54, 1980 (in Japanese).

PEARCE, J. A., Basalt geochemistry used to investigate past tectonic environments on Cyprus, *Tectonophysics*, **25**, 41–67, 1975.

SLOAN, J. R., D. M. WAPLES, S. M. WHITE, D. M. FOUNTAIN, H. KINOSHITA, N. G. MARSH, A. MIZUNO, G. V. NISTERENKO, H. OKADA, G. D. KLEIN, K. KOBAYASHI, H. CHAMLEY, D. M. CURTIS, H. J. B. DICK, and D. J. ECHOLS, Off-ridge volcanism and sea floor spreading in the Shikoku Basin, *Nature*, **273**, 746–748, 1978.

SUN, S. S., R. W. NESBITT, and A. Y. SHARASKIN, Geochemical characteristics of mid-ocean ridge basalts, *Earth Planet. Sci. Lett.*, **44**, 119–138, 1979.

TYSDAL, R. G., J. E. CASE, J. E. WINKLER, and S. H. B. CLARK, Sheeted dikes, gabbro and pillow basalt in flysch of coastal southern Alaska, *Geology*, **5**, 377–383, 1977.

VON HUENE, R., N. NASU, and OTHER DSDP STAFF LEG. 57, Japan trench transected, *Geotimes*, **23**, 16–21, 1978.

WATANABE, S. and H. HATTORI, Magmatic properties and densities on rocks near Miyazaki city for evaluation of regional magnetic anomalies in south-eastern Kyushu, *Bull. Geol. Surv. Japan*, **31**, 105–136, 1980 (in Japanese).

WOOD, D. A., J. L. JORON, N. G. MARSH, J. TARNEY, and M. TREUIL, Major- and trace-element variations in basalts from the north Philippine Sea drilled during Deep Sea Drilling Project Leg 58: A comparative study of back-arc-basin basalts with lava series from Japan and Mid-ocean Ridges, in *Initial Reports of the Deep Sea Drilling Project*, vol. 58, edited by L. N. Stout, pp. 873–894, US. Government Printing Office, Washington, 1980.

Formation of Active Ocean Margins, edited by N. Nasu *et al.*, pp. 423–442.

VOLCANOGENIC SEDIMENTS OF THE JAPAN TRENCH AREA AND TERTIARY EXPLOSIVE VOLCANISM OF THE TOHOKU ARC

Kantaro FUJIOKA

Ocean Research Institute, University of Tokyo, Nakano, Tokyo 164, Japan

Abstract. During the cruises of DSDP Legs 56 and 57 many cores were successfully recovered at seven sites around the Japan Trench region off the Tohoku Arc. Of these sites two cores penetrated sediments as old as the Cretaceous. The site 438 is located on the upper slope basin of inner slope of the Japan Trench forearc region. The site 436 is on the outer slope of the Japan Trench. The two sites are located on the different plates, the Eurasian and the Pacific Plates, respectively.

The site 438 experienced mainly vertical movements during the Tertiary, whereas the site 436 experienced horizontal movements being closely connected with the motion of the Pacific Plate through the Cretaceous. Sedimentation rates at these two sites are also different. The site 438 represents two distinct maxima of the accumulation of the terrigenous components between 16–14 Ma.b.p. and 5–2 Ma.b.p., respectively. However, at site 436, terrigenous materials increase gradually from about 14 Ma.b.p. to the present. Ash frequency patterns and sedimentation rates are identical at both the sites. These results reflect the different geologic history and plate motion relative to the Tohoku Arc during the Tertiary time.

The discrepancy observed between landward and seaward records of explosive volcanism may be explained as follows; landward pattern with two maxima is common to the history of the Tohoku Arc. Ealier maximum corresponds to the acidic volcanic activity in the Daijima and Nishikurosawa stages which may be related to the formation of the Kuroko deposits. The latter maximum corresponds to the Funakawa and Kitaura stages when extensive acidic volcanic activity including the intrusion of granites (the so-called Tertiary granites) occurred in relation to the Dewa disturbance. On the other hand, the apparent increase in the frequency of the volcanic ash layer observed in the seaward reference site is governed by the Pacific Plate approaching gradually towards the volcanic source area, the Japanese Islands.

1. *Introduction*

Since the first report of volcanic ash layers from the floor of the Atlantic Ocean (BRAMLETTE and BRADLEY, 1941), ash layers have been recorded by many marine geologists and used as critical time markers in ocean-floor strata (BOWLES *et al.*, 1973; NINKOVICH and SCHACKLETON, 1975). Since the start of the Deep Sea Drilling Project (DSDP), evidence for global explosive volcanism throughout Tertiary times has been gathered by many workers (KENNETT, 1981). In the 1970's, Kennett's group, who were compiling data on volcanic ash layers and ashy sediments obtained by the DSDP demonstrated that explosive volcanism increased world wide during the Quaternary and discussed the close relation between volcanic activity, geomagnetic reversals, climatic change and faunal extinctions. Their work influenced many volcanologists and marine geologists. NINKOVICH and DONN (1976) found in contrast, that the older sediments recovered by DSDP contained far fewer volcanic ash layers. Additionally, they observed that some commonly piston cores of the same age strata were very variable in their content of volcanic ash layers. They proposed a moving plate model in which DSDP sites were approaching volcanic source areas, so that patterns obtained by KENNETT and THUNELL (1975) show an apparent increase toward Quaternary, as sites shifted closer to volcanic sources.

This author and his co-workers have been interested in this problem and started to investigate the volcanic ash layers at two different sites in the Japan Trench region off the Tohoku Arc; landward reference site 438 and 439, and seaward reference site 436. These sites were chosen because both occur on the lee side of the present volcanic front of the Tohoku Arc and both penetrated Cretaceous sediments. The former is on the Eurasian plate which is the same plate as the volcanic source region, but the latter is on the Pacific plate, so that they have experienced different geologic histories through Cretaceous time.

Additionally, when KENNETT and THUNELL (1975) made histograms of volcanic ash layers, DSDP was not coring continuously. They evaluated the frequency patterns using the artificial factors which were observed in the number of volcanic ash layers above and below the vacant intervals.

Cores obtained from the Japan Trench are continuous, which obviates the needs to evaluate the number of the volcanic ash layers using artificial factors.

In this paper, frequency diagrams of volcanic ash layers for the two sites are compared to patterns of volcanicity evident in the onshore Tertiary geology of Japan.

2. *Location and Topography of the Drilling Sites*

Figure 1 shows the location of the DSDP Legs 56 and 57 drilling sites on a bathymetric chart of the Japan Trench forearc region. Sites 438 and 439 together represent a continuous record from the deep sea terrace at about water depth 1600 m. Note that in the following discussion "Site 438" refers to the continuous coring of both sites 438 and 439. The deep sea terrace is inclined gently seaward, and is underlain by thick

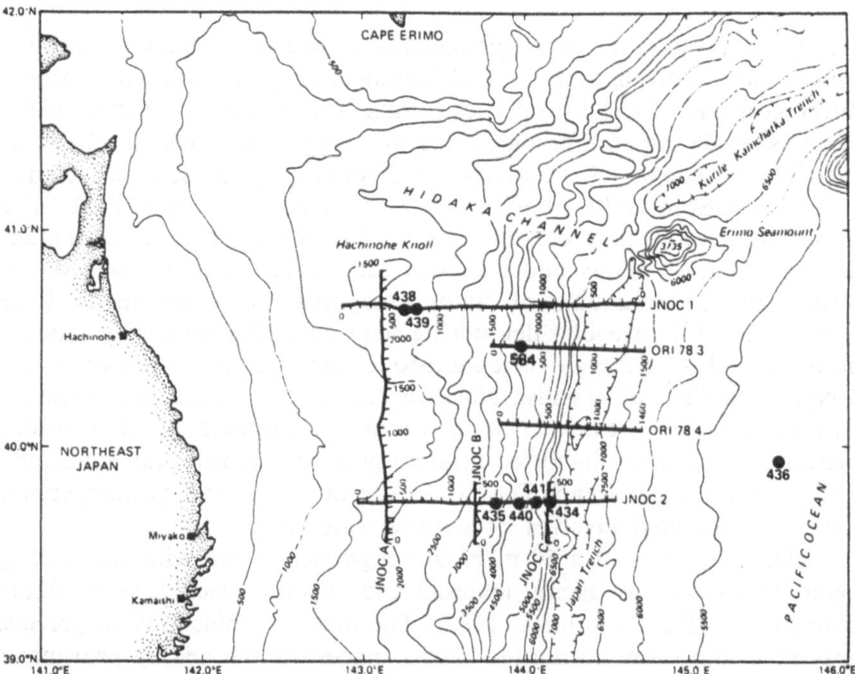

Fig. 1. Bathymetry and Location of DSDP sites off Tohoku Arc. Bathymetry around the Japan Trench is drawn from the bathymetric chart of the Hydrographi Office Maritime Agency. The contour is 500 m. Lines of JNOC and ORI are the ship tracks of the multichannel seismic profiles. Sites other than site 436 are on these seismic lines.

sediments younger than Cretaceous (Nasu *et al.*, 1980). Site 436 is located on the steep slope between the outer swell and the trench axis. The water depth of this site corresponds to that of the lower trench inner slope; about 5200 m. Eastward beyond this site are a group of horst and graben structures (Ludwig *et al.*, 1966). These two sites reach as deep as Cretaceous and are reference sites both landward and seaward of the Japan Trench. Other cores obtained in this region all penetrated no sediment older than upper Miocene. Site 440 forms a sedimentary pond in which thick Quaternary turbidites occur site 434 on the lower slope of the trench inner slope, repeat the same diatom zones four times showing slump deposits and are cut by dewatering veins, micro-fractures and faults. Site 441 sediments have a lot of offsets of the fractures therefore recovery of sediments was poor. By the above mentioned reasons, discussion on the explosive volcanic activity will be focused only at sites 438 and 436.

3. *Lithologies of the Reference Sites*

Figure 2 shows the lithologic columns of two sites. Sediments through Cretaceous are divided into several lithologic units according to lithologic differences and physical properties (Shipboard Scientific Party, 1980a, b). At site 438, seven lithologic units were identified (Von Huene *et al.*, 1980). Lithologic unit 1 comprises mainly olive-gray coarse sand including erratic pebbles in various horizons. Unit 2 consists mainly of dark olive-gray diatomaceous mud or mudstone, including various kinds of diatoms and volcanic ash layers and classified into three subunits based on the differences of the amount of diatoms, tephras and so on. In the lower portion of unit 2, irregular dewatering veins and calcareous micro-nodules occur. Unit 3 is dard gray vitric sandstone composed of lithic wacke. Unit 4 consists mainly of greenish black turbidite beds several centimeters to tens of centimeters thick. The sandstone is composed of lithic wacke. Beneath this unit are ill-sorted massive gray to greenish gray sandstones which correspond to unit 5. They are calcareous sandstones yielding various kind of articulated bivalves of shallow-water type.

Unit 6 consists of a 50 m basal conglomerate bed with clasts of igneous rocks such as dacite, andesite and rhyolite reaching a maximum diameter of 65 cm (Fujioka, 1980). The matrix is blue-gray to greenish gray mud and show no marine fossils suggesting the nearby provenance (Nasu *et al.*, 1980; Fujioka, 1980). ^{39}Ar–^{40}Ar dating of the volcanic boulders gives ages of 22–24 Ma. (Yanagisawa *et al.*, 1980; Moore and Fujioka, 1980). Under these sequences, unit 7 comprises dark gray hard siltstones and corresponds to the hyperbolic seismic patterns of the acoustic

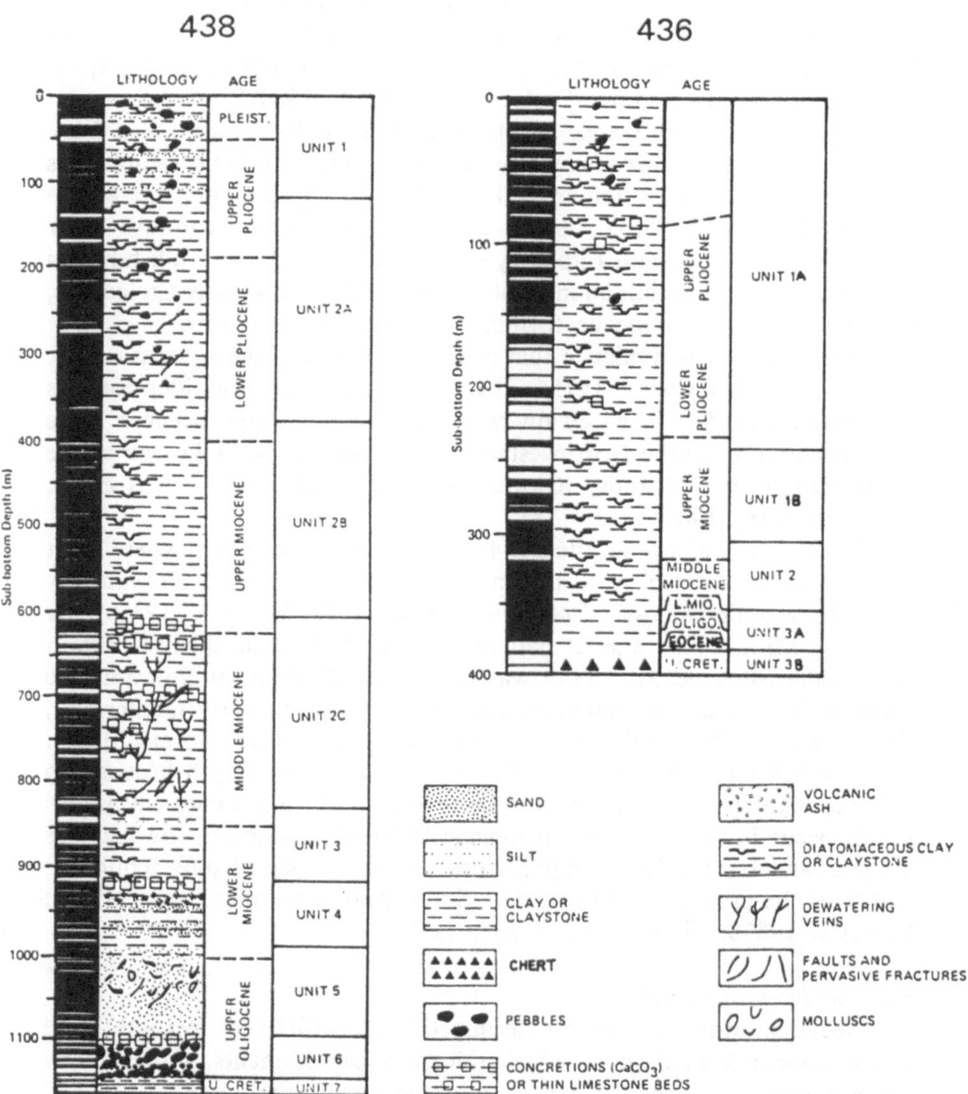

Fig. 2. Lithologic columns of the reference sites. Left side shows the lithologic changes of site 438 and the right side those of site 436. Unit 2 of the site 438 is divided into three subunits A, B and C based on the minor lithologic change. Unit 1 of the site 436 is divided into two subunits A and B depending on the amount of the predominating siliceous fossils. Black and white bands show the recovery rate of the cores: Black; recovered.

basement in the multi-channel seismic profiles of the Japan Trench (NASU *et al.*, 1980; VON HUENE *et al.*, 1980). The unconformity between units 6 and 7 represents a long hiatus. Based on the foraminifer biostratigraphy, the age of unit 7 is upper Cretaceous (KELLER, 1980).

At the seaward reference site 436 lithostratigraphic units 1 and 2 are dark olive-gray diatomaceous mud or mudstone rich in diatoms, radiolarians and volcanic ash layers similar to the landward reference site (SHIPBOARD SCIENTIFIC PARTY, 1980a). In unit 3A, there are few microfossils other than Ichtioliths, fish teeth. Lithology changes downward into reddish-brown pelagic claystone. Beneath a long hiatus between Eocene to early Cretaceous, unit 3B is a chert sequence forming the acoustic basement.

This sequence is similar to that of the other sites of the western Pacific (LANCELOT and LARSON, 1975). The age of the chert was identified as Cenomanian to Albian based on radiolarians. The hiatus between these units is common in the Northwestern Pacific region and is known as the Great Pacific unconformity (OKADA and SAKAI, 1979; LANGSETH *et al.*, 1980; KOIZUMI, 1981).

During the Paleogene, the site 438 area was a land region where short term volcanism took place at the end of Oligocene. The site subsequently began to subside rapidly (VON HUENE *et al.*, 1982; FUJIOKA, 1983b). By the early Middle Miocene, it had reached a continental shelf depth. In the Middle Miocene, the depth was more than 2500 m; at this time the finest sediments accumulated. In the late Pliocene to early Quaternary, the site was uplifted to the present water depth of 1600 m (KELLER, 1980).

Site 436 was beneath the carbonate compensation depth because no calcareous microfossils except for the reworked shallow facies calcareous fossils occur in the cores. From Eocene to Middle Miocene, site 436 was located far from the land judging from the occurrence of only Ichtioliths and from the presence of brown clay (SHIPBOARD SCIENTIFIC PARTY, 1980a; FUJIOKA, 1983b; OKADA and SAKAI, 1979).

3.1 Sedimentation rates

Sedimentation rates were determined by two different methods. One is the age-depth method in which thickness of sediments is divided by the age deduced from the biostratigraphic zonation; i.e.

$$S = T/A \text{ (m/Ma)} \tag{1}$$

(S: Sedimentation rate; T: Thickness of sediments; A: Absolute age). This rate takes no account of sediment compaction after deposition.

The other involves the use of water contents and bulk density of the sediments. In this method, the following equations will be assumed (ARTHUR *et al.*, 1980).

$$(1-Cw)*\rho_s=\rho_b-Cw*\rho_w \tag{2}$$

(Cw: Water contents of the sediments; ρ_s: Grain density of dry sediments (Mg/M^3); ρ_b: Bulk density of sediments (Mg/m^3); ρ_w: Density of the interstitial water (1.01 Mg/m^3)). The sedimentation rate, R is in the following equation.

$$R=S*\rho_b-Cw*\rho_w \tag{3}$$

In Eq. (3), if one multiplies the contents of each component in the defined intervals using the smear slide data of the sediments, one can obtain the yield of each component of the sediments (Mg/m^2/Ma).

Figures 3A and 3B show the sedimentation rates calculated by these methods. The rate line of site 438 shows two distinct steep slopes (Fig. 3A). In between these two periods, there is gentler slope. This result fits the rates which NIITSUMA (1978) obtained in onshore Tertiary systems using micropaleontologic and paleomagnetic methods. However, at site 436, the inclination of the line is steeper toward Quaternary time. At both sites there are small hiatuses. Sediments usually contain several consituents such as terrigeneous, biogenic, volcanogenic and authigenic components. Figure 3B shows the sedimentation rates obtained by the second method. In both methods, at site 438 bulk sedimentation rates show maxima at 16–14 Ma and 5–2 Ma, respectively.

During these times, sediments were rich in terrigeneous materials and biogenic silica. Between the two maxima (14–15 Ma), there are hiatuses, however, and the sedimentation rate of this interval is less than half of those of the two maxima. At site 436, sedimentation rates are almost zero before 14 Ma and tend to increase gradually, until 5 Ma ago. However, absolute rates of sedimentation are less than half those in the forearc region, at site 438, where terrigeneous input was higher. From the superposition of hemipellagic clays relatively rich in terrigeneous detritus on pelagic clays of open ocean character it appears that site 436 was located east of its present position before 14 Ma and has subsequently moved on the Pacific Plate gradually toward its present position. If this is the case, the pattern of the sedimentation rates will be easily explained (FUJIOKA, 1981; FUJIOKA, 1983b, c).

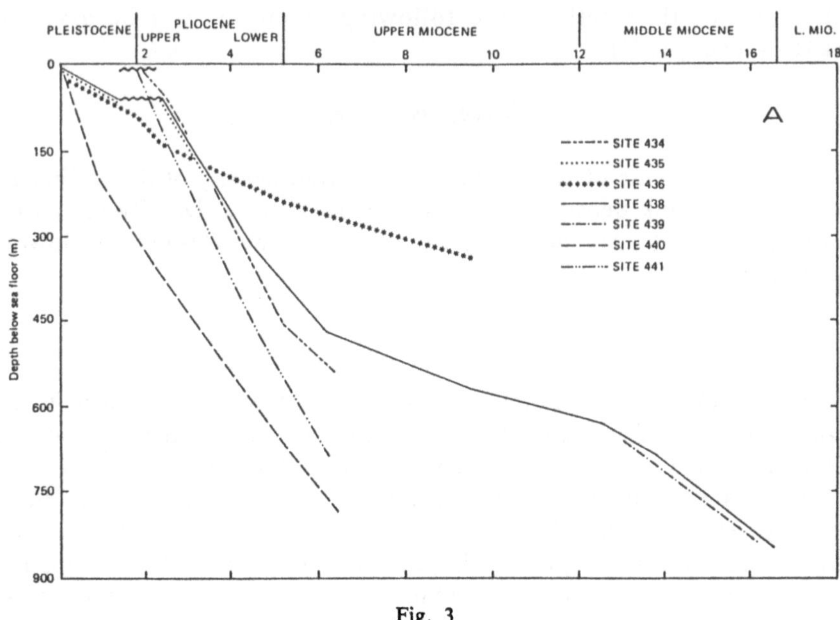

Fig. 3

4. *Mode of Occurrence and Chemistry of Volcanic Glass Shards*

Volcanic glass shards are included as a constituent material in the sediments both of landward and seaward sites. They are divided visually and microscopically into three major categories shown in Fig. 4. They are of AAW type defined by MATSUDA and NAKAMURA (1970).

(A) *Dispersed component*

Usually, volcanic glass shards dispersed in the sediment are difficult to identify with the naked eye. However, smear slide microscopy shows that the component of volcanic glass shards sometimes exceeds ten percent. This kind of volcanic glass shard is of "dispersed type" and its origin of this type is quite complex. One cause is that the supply rate of the volcanic glass is less than that of the other constituent materials. When benthonic organisms disturbed sediments on the sea floor, this type of volcanic glass shards also be formed.

(B) *Pocket or Pod*

In cut cores lenticular ash bodies are commonly encountered, and may represent originally distinct volcanic ash layers. After deposition, ben-

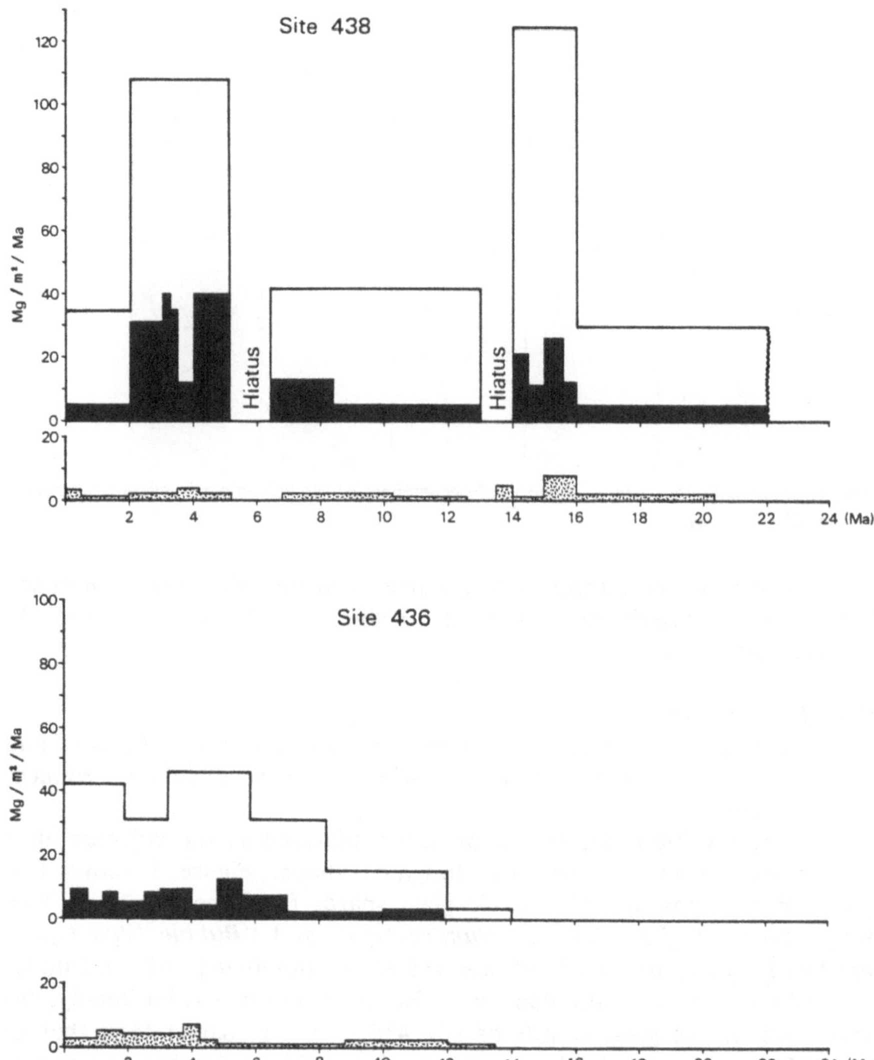

Fig. 3. (A) Sedimentation rate curves of the DSDP sites. Data were calculated based on the thickness of the sediments and the biostratigraphic ages. (B) Sedimentation rate calculated for each component of sediments. Open; Terrigenous component, Solid; Siliceous component, Dotted; Volcanogenic component.

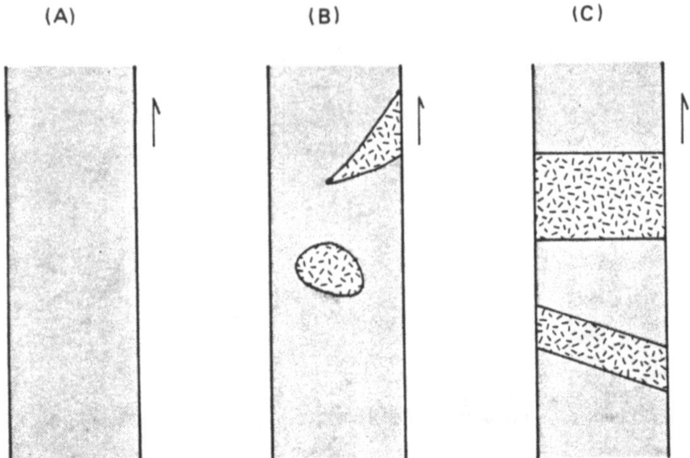

Fig. 4. Schematic three types of mode of occurrences of the volcanic ash layers encountered in the DSDP cores.

thonic organisms or strong bottom currents disturb the layer somewhat. If this kind of activity continues, then sediments will change to type (A) volcanic glass shard.

(C) *Volcanic ash layer*

This type shows a relatively uniform thickness across the core and is a very good bedding marker in cores where there are no definite sedimentary structures.

Under the binocular or petrographic microscope, the volcanic glass shards are translucent, and of elongated shape. Figure 5 shows two photomicrographs of the volcanic glass shards from the two sites. Two major types of glass shards, "*Pumice type*" and "*Bubble Wall type*", are identified. Glass shards of site 438 are predominately of the Pumice type whereas those of site 436 are of Bubble wall type. Additionally, the grain size of the glass shards of site 438 is much coarser than that of site 436. Brown volcanic glass is rare or absent. Those few brown volcanic shards observed are round.

The color of the volcanic glass shards reveal their chemistry to be of dacitic and/or rhyolitic composition. Refractive indices range from 1.489 to 1.511 and they are constant in any one volcanic ash layer. They are similar to each other in both sites 438 and 436 (FUJIOKA *et al.*, 1980; FUJIOKA, 1983b). Volcanic ash layers include as phenocrysts plagioclase, quartz, alkali-feldsper, pyroxene, hornblende, and biotite. They additionally

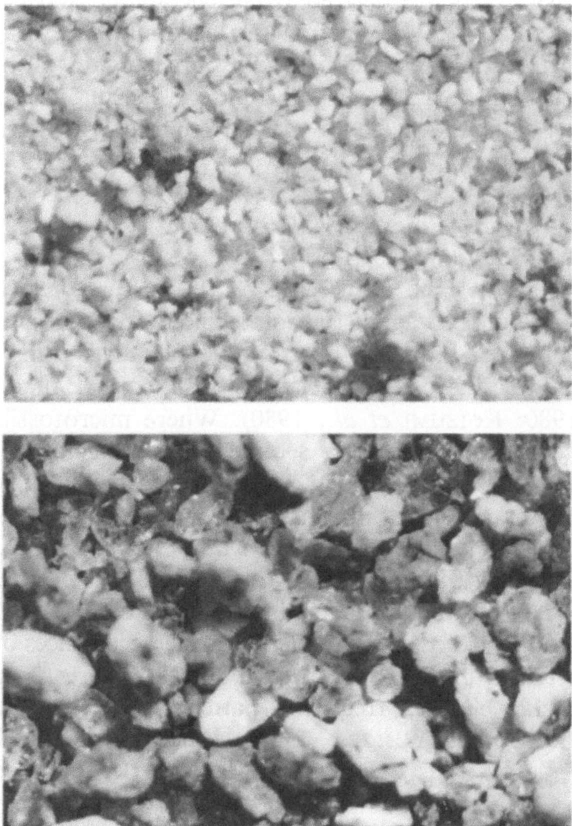

Fig. 5. Binocular photographs of the volcanic ash layers. Upper: JAT 70 at site 438. Lower: JAT 214 at site 436. Note the differences of the grain sizes between both samples. Scale is about 2.7 mm from left to right.

contain clay minerals, lithic fragments and microfossils as accessory constituents. However, the volume of accessory materials is mostly less than 10% of the sediments of types (B) and (C).

The volcanic glass shards were analysed using EPMA. Samples were washed by distilled water, at first, dried and separated by heavy liquids. They were impregnated by epoxy resin. Analyses were carried out by JXA-5 and JCXA-733 under 1×10^{-8} mA, 15 KV, 10 μm beam conditions. The chemical compositions of the glass shards mostly fall into the rhyolitic category (SiO_2 more than 70%), (FUJIOKA *et al.*, 1980). Results show that the volcanic ash layers are derived from calc-alkalic and tholeiitic series magma (MAYASHIRO, 1974). Taking into account the average grain size of the glass (fine sand to fine silt) the source region of these volcanic glass shards may be the Tohoku Arc (FUJIOKA, 1983c). The distances from site 438 and 436 to the present volcanic front of the Tohoku Arc are about 200 km and 300 km, respectively.

5. *Frequency of the Volcanic Ash Layers*

The frequency of the occurrence of the volcanic ash layers of site 438 and 436 were plotted against absolute age determined by microfossils. In the sediments there are many microfossils such as diatoms, radiolarians, foraminifers and nannofossils, facilitating precise age (BARRON, 1980; BARRON *et al.*, 1980; KOIZUMI *et al.*, 1980). Where microfossils are sparse, the age was calculated using the sedimentation rate, except the hiatus. Fugure 6 shows the age-frequency diagram so obtained. The pattern for site 438 shows two distinct maxima. The oldest peak is around 16–15 Ma, and the younger peak occurs 5–2 Ma, respectively (CADET and FUJIOKA, 1980). During the Quaternary, a further small peak exists. At site 436, there are no volcanic ash layers in sediments older than 11 Ma. From about 11 Ma to the present, the frequency of the occurrences of the volcanic ash layers increases gradually (FUJIOKA, 1980). Figures 3 and 6 show a close relation ship between frequency pattern of volcanic ash layers and the sedimentation rates (FUJIOKA, 1981). A normal relation between type (A) volcanic glass versus age is suggested by ARTHUR *et al.* (1980).

6. *Discussion*

The distances which volcanic ashes can be dispersed depend on various factors; principally, chemistry of magma. Maximum distances are approximately 1000 km (MACHIDA, 1981; MACHIDA and ARAI, 1976). How, then, does one interpret the frequency patterns observed at the two different study sites in the Japan forearc? Light can be shed on the opposing arguments on patterns of global volcanicity by Kennett's group and Ninkovich's group if the frequency patterns observed at these two study sites are compared carefully. This is because (1): Coring rates were quite high and continuous, (2): One site is on the Pacific Plate and the other is on the Eurasian Plates, (3): Both sites are located lee side from the volcanic front of the Tohoku Arc.

As shown in Fig. 6, the patterns observed at sites 438 and 436 are different, even though both sites are less than 200 km apart. Had, therefore,

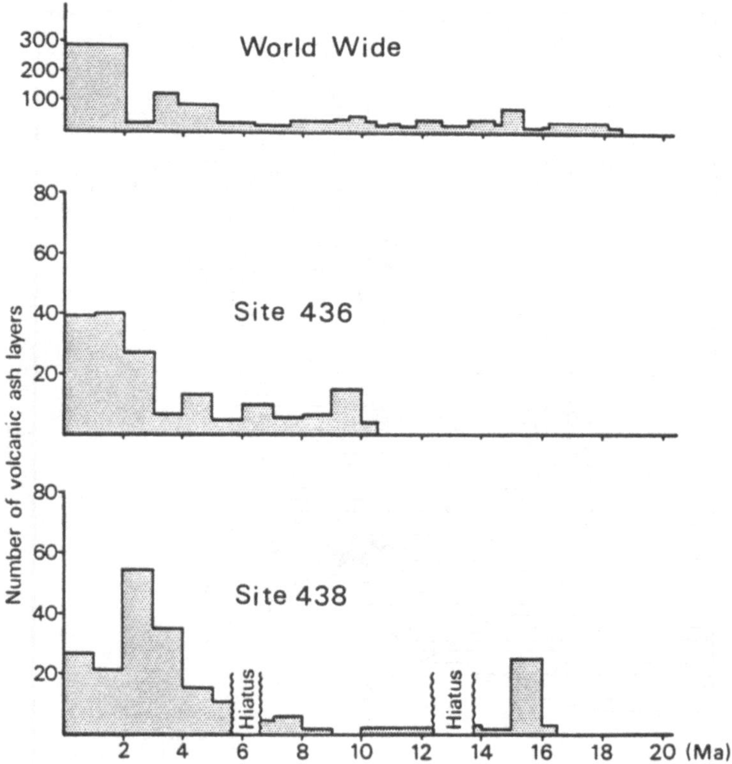

Fig. 6. Frequency-age diagram of the volcanic ash layers. Both type (B) and (C) are counted and shown in this figure. The upper is refered from KENNETT and THUNELL (1975).

the distance between the two sites been constant through geologic time, the patterns should have been more or less similar to each other? At site 436, after about 11 Ma.b.p., volcanic ash layers increase in frequency in parallel with the sedimentation rates at this site. It seems clear that as site 436 approached the Japanese Islands, the frequency pattern of the volcanic ash layers increased. Additionally, the grain size becomes coarser upward and the lithologic column changes upward also. The frequency pattern of site 436 is probably typical of patterns of sites approaching volcanic source regions (FUJIOKA, 1983c).

In Fig. 6, the volcanic maxima were during Pliocene (5–2 Ma) and Middle Miocene (16–15 Ma). Therefore, the phase proposed by Kennett's group should be examined carefully again with respect to the Tohoku Arc. Figure 7 shows the "volcanic ash zone" scheme proposed by NINKOVICH

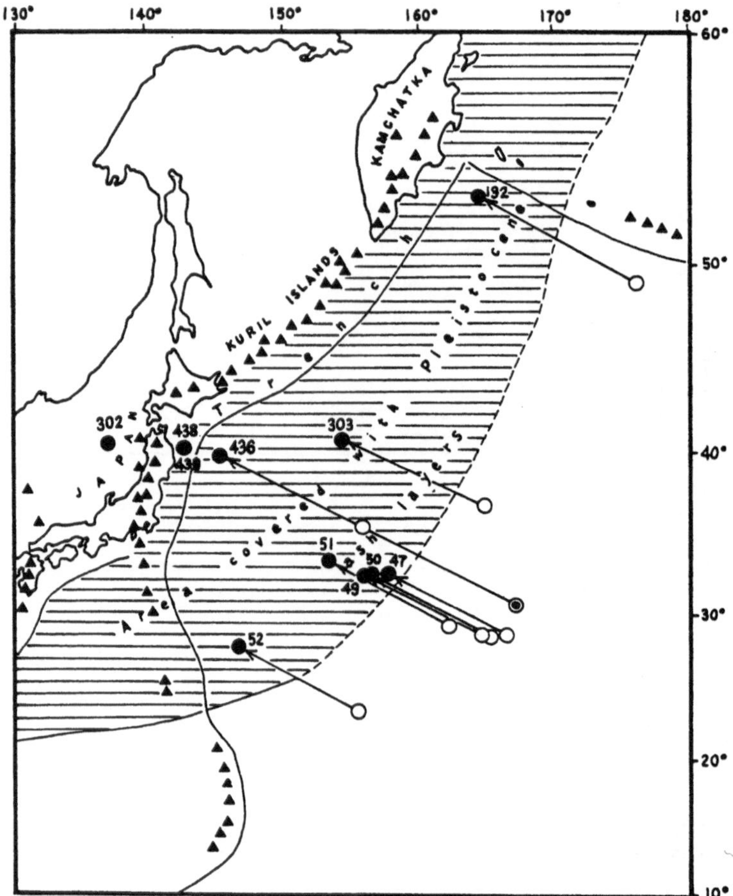

Fig. 7. Volcanic ash zone around the Japanese Islands. This figure was modified from NINKOVICH and DONN (1976). Arrows show the direction of the movement of the Pacific Plate. Open circle shows the position of the sites 10 Ma.b.p., solid circles the present position and double circle 20 Ma.b.p.

and DONN (1976) and this figure includes both sites. If we assume that the Pacific Plate moves 10 cm/a, site 436 will have entered into the volcanic ash zone about 11 Ma, as the sedimentary record indicates.

At site 438, the volcanic maxima were 16–15 Ma and 5–2 Ma.b.p. How does this correspond with the volcanic activity in the onshore Tohoku Arc through the Neogene Era? The Neogene System of the Tohoku Arc includes a significant volcanogenic component in the sediments. The history of volcanic activity is summarized by OZAWA (1963), SUGIMURA *et al.*

(1963), ISHIHARA (1974), HORIKOSHI (1975), HUZIOKA *et al.* (1977), FUJIOKA (1983a). Using the standard stages of the Neogene System of the Tohoku Arc (KITAZATO, 1975; TAKAYASU and MATOBA, 1976; FUJIOKA *et al.*, 1981), acidic volcanism predominated during the Daijima and Nishikurosawa stages and in Funakawa and Kitaura stages (HUZIOKA, 1963; KONDA, 1974).

According to the micropaleontological work (TSUCHI, 1979) and the absolute age determination (KONDA and UEDA, 1980; KANEOKA, 1983), these two stages correspond to 16–14 Ma and 5–2 Ma.b.p. These results are in good agreement with the two maxima of the volcanic ash layers observed at site 438.

Therefore, sediments of site 438 clearly record the results of explosive volcanic activity onshore in the Tohoku Arc. Because marine sediments preserve microfossils more or less continuously through the sediment section, the history of the explosive volcanism of an arc can perhaps better be studied using the marine sediments instead of the onshore record which has been subject to post-depostional weathering, erosion, metamorphism and orogenic movements (FUJIOKA, 1983b, c).

At the two sites, sedimentation rates and the frequency of the occurrence of the volcanic ash layers are concordant with each other. Additionally, the sedimentation rates of site 438 are similar to those of onshore Tohoku Arc described by NIITSUMA (1978). These two evidences will be discussed in relation to the geologic history of the onshore Neogene system of the Tohoku Arc (Fig. 8).

About 16–14 Ma, site 438 was supplied by terrigeneous and also siliceous sediment, and the water depth of site was deeper than that at present. On the other hand, in the onshore region (Daijima and Nishikurosawa stages) explosive volcanic activity was strong in relation to the Kuroko formation (HORIKOSHI, 1975; SATO, 1978). The paleogeography of the Tohoku Arc at that time was similar to the present day Izu-Ogasawara trench-arc systems (FUJIOKA, 1983a; KITAZATO, 1983).

About 5–2 Ma, terrigeneous, siliceous, and volcanogenic sediments were transported into the site 438 region which began to be uplifted gradually at the end of this period (SHIPBOARD SCIENTIFIC PARTY, 1980; VON HUENE and ARTHUR, 1982). In the onshore Tohoku Arc, this period corresponds to the Funakawa-Kitaura stages (KITAZATO, 1975; TAKAYASU and MATOBA, 1976). The region was covered by extensive acidic volcanic materials and then began to be uplifted from the Sekiryo (Back-bone range) to the west (HUZIOKA, 1968; KITAMURA, 1959; IKEBE, 1962; HUZIOKA, 1963; MATSUDA *et al.*, 1967; AMANO, 1981). As a consequence of this uplift, terrigeneous materials were transported to site 438. But, the phase of uplifting of the Tohoku Arc and site 438 were slightly different.

Episodic volcanic activity in Tohoku Arc may relate to the subduction of the Pacific Plate under the Tohoku Arc (UYEDA and KANAMORI, 1979). The mode of subduction of the Pacific Plate in the period of the older peak of the acidic volcanism was "Mariana Type". A tensional stress field predominated and the land area was mostly submerged in this period considering the paleogeography, bathymetry and stress field (NAKAMURA, 1969; NAKAMURA and UYEDA, 1980; TAKEUCHI, 1981; FUJIOKA, 1983a; SUGI *et al.*, 1983).

However, the mode of subduction from 5–2 Ma, was "Chilian Type" and resulted in uplift of the Tohoku Arc under an west east compressional field (SUGI *et al.*, 1983). Regional acidic volcanic activity may be caused by the subduction in close relation to the tectonics of the arc-trench systems.

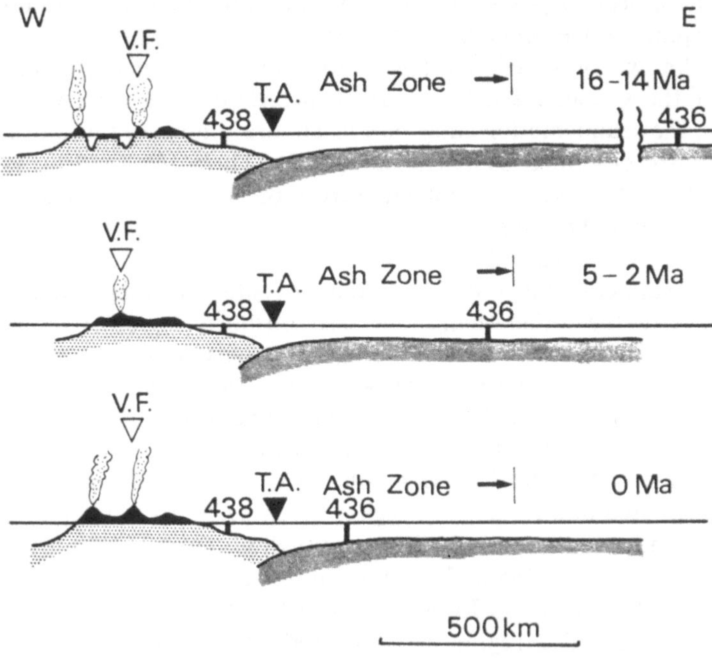

Fig. 8. Topographic cross sections of the Tohoku Arc at three different ages. Topography was estimated by many references (FUJIOKA, 1983a). V.F.; Volcanic front, T.A.; Trench axis.

7. *Summary and Conclusion*

During the cruise of IPOD Legs 56 and 57, two long continuous cores recovering sediments as old as Cretaceous were obtained at two different sites of the Japan Trench region. Lithology of sediments, sedimentation rates, and frequency of the occurrences of the volcanic ash layers yield the following important results.

(1) Acidic volcanic activity of the Tohoku Arc did not occur continuously through Neogene. Activity was strongest during 16–15 Ma (Daijima and Nishikurosawa stages) and 5–2 Ma (Funakawa and Kitaura stages).

(2) The frequency pattern seen at site 436 suggests that site 436 was moving towards the Tohoku Arc on the Pacific Plate and that the frequency of Quaternary ash is controlled by the approach of the site towards land.

(3) Therefore, the model for the global increase of the explosive volcanism during Quaternary proposed by KENNETT and THUNELL (1975) does not appear to be valid in the Tohoku Arc.

(4) Sedimentation rates of each site are consistent with the frequency of the volcanic ash layers indicating the subduction of the Pacific Plate underneath the island arc.

(5) Two distinct maxima observed at site 438 may be correlated with the acidic volcanic activities forming the Kuroko deposits and acidic volcanic activity relating to the uplift of the Tohoku Arc. The former was during "Mariana Type" subduction mode and the latter during "Chilian Type", respectively.

Acknowledgements

The present author would like to express his sinsere thanks to the following people during the preparation of this manuscript. Prof. N. Nasu, former Director of the ORI, University of Tokyo who gave him continuous encouragements during the process of the present work. Drs. K. Kobayashi and H. Kagami of the ORI Univ. of Tokyo gave him invaluable discussions on the volcanogenic sediments. Drs. S. Uyeda, K. Nakamura and S. Aramaki, ERI Univ. of Tokyo gave him invaluable suggestions about the volcanogenic sediments of the Japan Trench and critical discussion on this problem. Drs. Y. Nakamura, Coll. Arts & Sci., Univ. of Tokyo, E. Horikoshi, Toyama Univ. and K. Chinzei and late Skuyama, Fac. Sci. Univ. of Tokyo, I. Moriya, Kanazawa Univ. and A. Ozawa, Geol. Surv. of Japan discussed about the volcanology of the Japanese Quaternary and Tertiary volcanoes. Drs. J. K. Legget, Imp. Coll. Univ. London and G. De V. Klein, Illinoi.

REFERENCES

AMANO, K., Geology of the Ou backbone ranges in Miyagi and Yamagata prefectures, northeast Honshu, Japan, *Sci. Rep. Tohoku Univ., Inst. Geol. Pal., Contr.*, **81**, 1–56, 1981.

ARTHUR, M. A., R. VON HUENE, and C. G. Jr. ADELSECK, Sedimentary evolution of the Japan forearc region off northern Honshu Legs 56 and 57, Deep Sea Drilling Project, Initial Rept. of the DSDP, 56–57, 1980.

BARRON, J. A., Lower Miocene to Quaternary diatom biostratigraphy of Leg 57, off northeast Japan, Deep Sea Drilling Project, Initial Rept. of the DSDP, 56–57, pp. 641–685, 1980.

BARRON, J. A., H. E. Jr. HARPER, G. KELLER, R. A. REYNOLDS, T. SAKAI, B. L. SHAFFER, and P. R. THOMPSON, Biostratigraphic summary of the Japan Trench transect, Legs 56 and 57, Deep Sea Drilling Project, Initial Rept. of the DSDP, 56–57, pp. 505–520, 1980.

BOWLES, F. A., R. N. JACK, and I. S. E. CARMICHAEL, Investigation of deep-sea volcanic ash layers from Equatorial Pacific cores, *Geol. Soc. Am. Bull.*, **84**, 2371–2388, 1973.

BRAMLETTE, M. N. and W. H. BRADLEY, Geology and biology of north Atlantic deep-sea cores between Newfoundland and Ireland: Part I. Lithology and geological interpretation, U.S. Geol. Surv., Prof. Pap., 196-A, pp. 1–55, 1941.

CADET, J. P. and K. FUJIOKA, Neogene volcanic ashes and explosive volcanism: Japan Trench transect, Leg 57, Deep Sea Drilling Project, Initial Rept. of the DSDP, 56–57, pp. 1027–1041, 1980.

FUJIOKA, K., Conglomerates of volcanic rocks of Deep Sea Drilling Project Site 439, Initial Rept. of the DSDP, 56–57, pp. 1075–1082, 1980.

FUJIOKA, K., Explosive volcanicity of the Tohoku Arc during Late Cenozoic Era, 1981 IAVCEI Symposium Japan, pp. 98–99, 1981.

FUJIOKA, K., Where were the "Kuroko deposits" formed looking for the present day analogy, *Mining Geol., Spec. Issue*, **11**, 55–68, 1983a.

FUJIOKA, K., History of the explosive volcanism of the Tohoku Arc from the core sediment samples of the Japan Trench, *J. Volcanol. Soc. Jap.*, **28**, 41–58, 1983b.

FUJIOKA, K., Geology of volcanogenic sediments of the Japan Trench area and explosive volcanism of the Tohoku Arc, Ms. Doctral Thesis of Univ. of Tokyo, 1983c.

FUJIOKA, K., T. FURUTA, and F. ARAI, Petrography and geochemistry of volcanic glass: Leg 57 Deep Sea Drilling Project, Initial Rept. of the DSDP, 56–57, pp. 1049–1066, 1980.

FUJIOKA, N., T. OHGUCHI, S. MAIYA, M. USUDA, and K. BABA, On the sediments of Daijima-Nishikurosawa age in the northeast Inner Japan area, *J. Jap. Assoc. Petrol. Tech.*, **46**, 159–174, 1981.

HORIKOSHI, E., Genesis of Kuroko-stage deposits from the tectonic point of view, *J. Volcanol. Soc. Jap.*, **20**, 341–353, 1975.

HUZIOKA, K., Geology of the Green Tuff region in Japan, *Mining Geol.*, **13**, 358–375, 1963.

HUZIOKA, K., The Dewa disturbance in Akita oil-field, *J. Jap. Assoc. Petrol. Tech.*, **33**, 283–297, 1968.

HUZIOKA, K., A. OZAWA, T. TAKAYASU, and Y. IKEBE, Geology of the Akita district, Quadrangle Series Scale 1 : 50000, Geol. Surv. Japan, 75 pp., 1977.

IKEBE, Y., Tectonic developments of oil-bearing Tertiary and migrations of oil, in Akita oil fields, Japan, *Rep. Res. Inst. Undergr. Resource Min. Coll. Akita Univ.*, **26**, 1–59, 1962.

ISHIHARA, S., Magmatism of the Green Tuff tectonic belt, northeast Japan, *Mining Geol.* (Spec. Issue), **6**, 235–249, 1974.

KANEOKA, I., On the radiometric ages of volcanic rocks from the northeastern part of the Honshu Island, *Japan. Min. Geol.* (Spec. Issue), **11**, 69–78, 1983.

KELLER, G., Benthic foraminifers and paleobathymetry of the Japan Trench area, Leg 57, Deep Sea Drilling Project, Initial Rept. of the DSDP, 56–57, pp. 835–865, 1980.

KENNETT, J. P., Marine tephrochronology, *The Sea*, Ed. C. Emiliani, 7, pp. 1313–1436, 1981.

KENNETT, J. P. and R. C. THUNELL, Global increase in Quaternary explosive volcanism, *Science*, **187**, 497–503, 1975.

KITAMURA, N., Tertiary orogenesis in northeast Honshu, Japan, *Sci. Rep., Tohoku Univ., Inst. Geol. Pal., Contr.*, **49**, 1–98, 1959.

KITAZATO, H., Geology and geochronology of the younger Cenozoic of Oga Peninsula, *Tohoku Univ., Inst. Geol. Pal., Contr.*, **75**, 17–49, 1975.

KITAZATO, H., Submarine topography of the northeast Honshu arc during the early Middle Miocene Nishikurosawa stage, based on the benthic foraminifera, *Mining Geol.* (Spec. Issue), **11**, 263–270, 1983.

KOIZUMI, I., Hiatus of the north Pacific ocean floor based on the diatom biostratigraphy, *Mar. Sci.*, **13**, 95–100, 1983 (in Japanese).

KOIZUMI, I., J. A. BARRON, and H. E. HARPER, Diatom correlation of Legs 56 and 57 with onshore sequences in Japan, Initial Rept. of the DSDP, 56–57, pp. 687–693, 1980.

KONDA, T., Bimodal volcanism in the northeast Japan arc, *J. Geol. Soc. Japan*, **80**, 81–89, 1974.

KONDA, T. and Y. UEDA, K-Ar age of the Tertiary volcanic rocks in the Tohoku area, *Japan. J. Jap. Assoc. Min. Petrol. Econ. Geol.* (Spec. Issue), **2**, 343–346, 1980.

LANCELOT, Y. and R. L. LARSON, Sedimentary and tectonic evolution of northwestern Pacific, Initial Rept. of the DSDP, 32, pp. 925–939, 1975.

LANGSETH, M. G., R. VON HUENE, N. NASU, and H. OKADA, Subsidence of the Japan Trench forearc region of Northern Honshu, *Oceanol. Acta.* (Spec. Issue), 173–179, 1981.

LUDWIG, W. J., J. I. EWING, M. EWING, S. MURAUCHI, N. DEN, S. ASANO, H. HOTTA, M. HAYAKAWA, T. ASANUMA, K. ICHIKAWA, and I. NOGUCHI, Sediment and structure of the Japan Trench, *J. Geophys. Res.*, **71**, 2121–2137, 1966.

MACHIDA, H., Tephrochronology and Quaternary studies in Japan, in *Tephra Studies*, Eds. S. Self and R. S. J. Sparkes, pp. 161–191, D. Reidel Pub. Co., Holland, 1981.

MACHIDA, H. and F. ARAI, The very widespread tephra, The Aira-Tn ash, *Kagaku*, **46**, 339–347, 1976 (in Japanese).

MATSUDA, T. and K. NAKAMURA, Sedimentary features of water-laid volcanic deposites and their classification based on mode of emplacement, *Mining Geol.*, **20**, 29–42, 1970.

MATSUDA, T., K. NAKAMURA, and A. SUGIMURA, Late Cenozoic orogeny in Japan, *Tectonophysics*, **4**, 349–366, 1967.

MIYASHIRO, A., Volcanic rock series in island arcs and active continental margins, *Am. J. Sci.*, **274**, 321–355, 1974.

MOORE, G. W. and K. FUJIOKA, Age and origin of dacite boulder conglomerate anomalously near the Japan Trench, Initial Rept. of the DSDP, 56–57, pp. 1083–1088, 1980.

NAKAMURA, K., Island arc tectonics, a hypothesis, Papers for symposium "Problems concerning Green Tuff" at the 1969 annual meeting of Geol. Soc. Jap., pp. 31–38, 1969.

NAKAMURA, K. and S. UYEDA, Stress gradient in arc-backarc regions and plate subduction, *J. Geophys. Res.*, **85**, 6419–6428, 1980.

NASU, N., R. VON HUENE, Y. ISHIWADA, M. LANGSETH, T. BRUNES, and E. HONZA, Interpretation of multichannel seismic reflection data, Legs 56 and 57, Japan Trench transect, Deep Sea Drilling Project, Initial Rept. of the DSDP, 56–57, pp. 489–503, 1980.

NIITSUMA, N., Magnetic stratigraphy of the Japanese Neogene and the development of the island arcs of Japan, *J. Phys. Earth* (Suppl.), **26**, S367–S378, 1978.

NINKOVICH, D. and W. L. DONN, Explosive Cenozoic volcanism and climatic interpretations, *Science*, **194**, 899–906, 1976.

NINKOVICH, D. and N. J. SHACKLETON, Distribution, stratigraphic position and age of ash layer "L" in the Panama Basin region, *Earth Planet. Sci. Lett.*, **27**, 20–34, 1975.

OKADA, H. and T. SAKAI, Deep sea drilling of the Japan Trench-I, *Mar. Sci.*, **11**, 756–762, 1979.

OZAWA, A., Neogene orogenesis, igneous activity and mineralization in the central part of northeast Honshu, (1) On the Neogene igneous activity, *J. Jap. Assoc. Min. Petrol. Econ. Geol.*, **50**, 167–184, 1963.

SATO, T., Kuroko deposits—genesis and evolution—, *Kagaku*, **48**, 185–204, 1978 (in Japanese).

SHIPBOARD SCIENTIFIC PARTY, Sites 438 and 439, Japan deep sea terrace, Leg 57, Initial Rept. of the DSDP, 56–57, pp. 23–191, 1980a.

SHIPBOARD SCIENTIFIC PARTY, Site 436, Japan Trench outer rise, Leg 56, Initial Rept. of the DSDP, 56–57, pp. 399–446, 1980b.

SUGI, N., K. CHINZEI, and S. UYEDA, Vertical crustal movements of the northeast Japan since Middle Miocene, A.G.U.-G.S.A. Geodynamic Series, 1983.

SUGIMURA, A., T. MATSUDA, K. CHINZEI, and K. NAKAMURA, Quantitative distribution of late Cenozoic volcanic materials in Japan, *Bull. Volcanol.*, **26**, 125–140, 1963.

TAKAYASU, T. and Y. MATOBA, Oga Peninsula, Guide book for Excursion 1, 1-CPNS, Tokyo, 78 pp., 1976.

TAKEUCHI, A., Temporal changes of regional stress field and tectonics of sedimentary basin, *J. Geol. Soc. Jap.*, **87**, 737–751, 1981.

TSUCHI, R., Biostratigraphy and ages of the Japanese Neogene Systems, IGCP-114, National Working Group of Japan, 1979.

UYEDA, S. and H. KANAMORI, Back-arc opening and the mode of subduction, *J. Geophys. Res.*, **84**, 1049–1061, 1979.

VON HUENE, R., M. LANGSETH, N. NASU, and H. OKADA, Summary, Japan Trench transect, Initial Rept. of the DSDP, 56–57, pp. 473–488, 1980.

VON HUENE, R., M. LANGSETH, N. NASU, and H. OKADA, A summary of Cenozoic tectonic history along the IPOD Japan Trench trensect, *Geol. Soc. Am. Bull.*, **93**, 829–846, 1982.

VON HUENE, R. and M. A. ARTHUR, Sedimentation across the Japan Trench off northern Honshu Island, in *Trench-Forearc Geology*, Geol. Soc. London, Spec. Publ., Vol. 10, Ed. J. K. Leggett, pp. 27–48, 1982.

YANAGISAWA, M., Y. TAKIGAMI, M. OZIMA, and I. KANEOKA, $^{40}Ar/^{39}Ar$ ages of boulders drilled at site 439, Leg 57, Deep Sea Drilling Project, Initial Rept. of the DSDP, 56–57, pp. 1281–1284, 1980.

Formation of Active Ocean Margins, edited by N. Nasu *et al.*, pp. 443–465.

ON THE EPISODIC VICISSITUDE OF TECTONIC STRESS FIELD OF THE CENOZOIC NORTHEAST HONSHU ARC, JAPAN

Akira TAKEUCHI

Department of Earth Sciences, Toyama University, Gofuku 3190, Toyama 930, Japan

Abstract. Twenty-two dike swarms in the southern part of Northeast Japan, ranging in age from the end of Eocene to Pliocene, were examined to reconstruct the paleostress field for a mechanical study on the geologic history of the volcanic arc. The changes in orientation of stress field of the area include at least four re-arrangements of the horizontal maximum principal stress $\sigma_{H\ max}$. Magnitude ratio of horizontal stress to vertical one had inverted abruptly at the Miocene-Pliocene boundary.

The re-orientations of tectonic stress field on NE-Honshu Arc especially in the inner province have been almost synchronized with the oscillatory swing motion of the Pacific plate in reference to the Hawaiian hot spot: At the climax phases of the 'Pacific swing' as it dose at present, lithosphere of the arc might have behaved as a thin plate in a uniform stress field with transarc $\sigma_{H\ max}$, while at the anticlimax phases the arc might have behaved as a very thick massif, resulting in backarc extension and subsidence.

The vicissitude of stress state on the arc is so contemporary with the phases of and particular events of arc magmatism as well as with the global magmatic episodes. Thus, it is suggested that culmination in activity of the asthenospheric movement caused lithospheric thinning effective to the changes in the mechanical behavior of the arc.

1. *Introduction*

The present crustal stresses have been measured by various techniques such as direct measurements of rock stresses, focal mechanism solution of earthquakes, geodetic measurements of crustal strain, and so on. To reconstruct an ancient stress field, distribution of faults and igneous dikes offer very useful information, if we can determine the age of their deformation.

As a mechanical approach to understand the geologic development of the southern part of Northeast Honshu Arc, the "dike method" was adopted for the purpose of reconstruction of the past tectonic stress field. Some of the basic results from such field-works have been already published (TAKEUCHI, 1980, 1981; NAKAMURA and UYEDA, 1981).

Recently, I and my colleagues conducted the radiometric age determination of dike rocks in the study area by K-Ar method (TSUNAKAWA *et al.*, 1983). Based on these field and age data, the late Cenozoic history of tectonic stress field in Northeast Honshu Arc has been constructed, revealing a synchronous relation among the temporal variation in tectonic stress field, volcanic activity, and structural evolution in the inner arc basin. Here, the geology is to be compared with the episodic or oscillatory motion ('swing') of Pacific Plate by JACKSON *et al.* (1975).

2. *Method*

The dike method is basically based on the following logics (TAKEUCHI, 1980, 1984):

It is theoretically admitted that (1) a dike is considered as the direct result of magmafracturing very similar to the artificial hydrofrac techniques, and that (2) such a kind of opened fracture would be formed within the principal plane of the least compressive stress σ_3 and parallel to the horizontal maximum compression $\sigma_{H\,max}$. For simplicity, vertical principal component σ_V of crustal stresses is assumed to be due only to gravitational loading.

Suppose that a certain directivity, i.e. a preferred orientation, was recognized in the spatial pattern of dike distribution. This itself indicates that the crustal stress field at the time of intrusion was in a deviatoric state where the preferred orientation of dikes reflects the principal stress directions. Therefore, dikes commonly show the tectonophysically significant character that would reflect the condition of somewhat ambient tectonic stress field even if their spatial distribution was restricted to local areas under a control of volcanogenic stresses.

The dike data are processed by three steps as follows (TAKEUCHI, 1980):

Field data At first we have to perform field works to collect the dike data for orientation of dike/host rocks interfaces, with observing lithology, configuration and thickness as well as internal structures such as flow lineations, contemporal microfaults and cooling joints, vesicles, chilled margin, and so on.

Directivity analysis Rose-diagram at 10 degrees interval is used for the analysis frequency distribution of the wall-strikes of each individual dike swarm. The peak of predominant frequency in the azimuthal distribution is taken immediately as the time averaged $\sigma_{H\,max}$-direction at the duration of intrusion, although the degree of concentration (*F*-value ranging form 0 to 1: TAKEUCHI, 1981) is to vary under the control of physical properties such as pressure and viscosity of magma fluid, strength and ductility of the wall rock, toughness and stress intensity factor of induced fractures and so on. We also have to take into account the following influencing factors effective to diverging and polymodal peaks in the azimuthal distribution: Irregularities on dike propagation direction caused by pre- or co-existing structures in the host rock, spatial and temporal instability of stress state, and other artificial factors concerning field surveys.

Assessment and representation The last step is to asses the spatial uniformity and temporal durability in the distribution pattern of the inferred stress orientations. Using an age-azimuth chart, it was examined whether the inferred $\sigma_{H\,max}$ orientations from individual dike-swarms have anything in common with each other. Through the procedure of assortment inferred stress systems were divided into some groups in terms of their age and geography, and were schematically represented by the stress trajectory maps. Here, a concept of "stress province" (LIU, 1979; ZOBACK and ZOBACK, 1979; TAKEUCHI, 1980; KOBAYASHI, 1982) is used, the definition of which is unit area of geographical division of stress field, and within which there distribute a group of evidences that reveals a uniform stress orientation of same age.

Noteworthy, it is impossible to distinguish the magnitude ratio of σ_H to σ_V as long as the $\sigma_{H\,max}$-directivity is only used. In other words, we cannot speak about the difference among three types of stress field, that is, extensional (T-type), neutral (B-type) and compressional (P-type) stress field in terms only of $\sigma_{H\,max}$-direction. Comparing with stress systems inferred by the other methods using the distribution pattern of faults and folds, for example, are useful to identify the types of stress systems.

Despite such a defect of the dike method, however, it should be noticed that such a $\sigma_{H\,max}$-directivity of dike system would commonly appear in a wide range of depth, even if the crustal stress field were in the state where the ratio σ_V/σ_H changes with depth as well as in the other stable state of either $\sigma_{H\,max}<\sigma_V$ or $\sigma_V<\sigma_{H\,max}$. This is one of the major validities of the dike method as well as its easiness of dating the inferred stress system.

3. *Geologic Background of the Study Area*

The study area covers the southern part of NE Honshu extending from latitude 36.5°N to 38.5°N (Fig. 1). The basement of later Cenozoic sedimentary strata consists of the Echigo, Asahi and Abukuma mountainlands composing the Joetsu and Abukuma metamorphic belts and the Uetsu-Ashio zone.

Acidic igneous activities intensely occurred during the period from the Cretaceous to the Paleogene in the basement. K-Ar dates measured

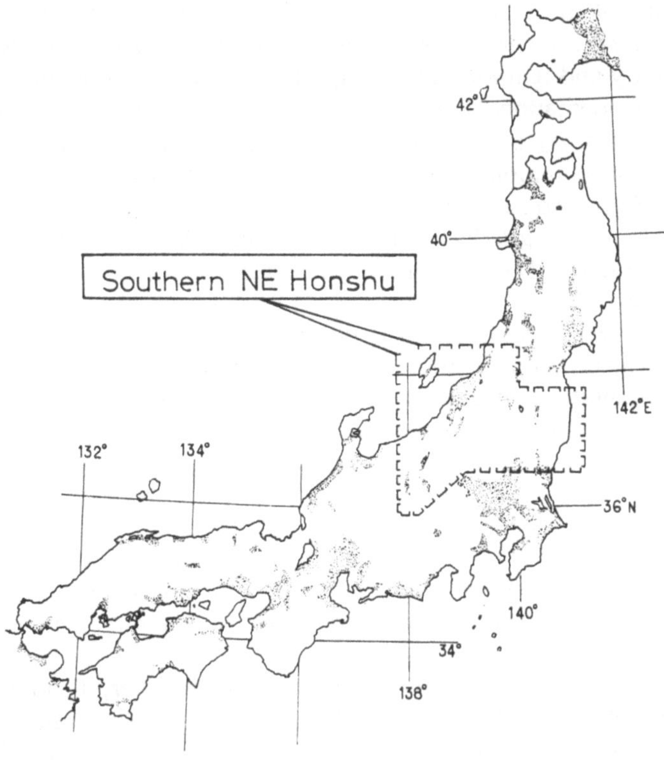

Fig. 1. Index map. Enclosed area indicates the southern part of Northeast Honshu, central Japan. Dotted parts represent aluvial plains and basins.

by the present and previous works from igneous rocks distributed in the whole area of NE Honshu are compiled as shown in Fig. 2. In this diagram, number of age determinations for intrusive and effusive rocks are shown separately. On the basis of the data for effusives, a volcanostratigraphic division is used as a reference chronology in this study (left column in Fig. 2). The chronostratigraphy includes several hiatus or intermissions of both sedimentation and volcanic activity.

The igneous activities since the latest Eocene were divided into six stages in terms of lithology and geography. Spatial distribution of intrusive and effusive activities mainly of acidic rocks during Stage I was restricted and biased within the westernmost side of Honshu Island.

As the result of volcanism and subsequent marine transgression from the end of Paleogene (Stage II in Fig. 2) to the middle Mocene (Stage III), thick piles of volcaniclastics and normal clastic sediments were widely distributed in the inner belt of Honshu, that is, so-called the "Green Tuff" region.

Intrusive activity centred at the stage IV (typically around 12–13 Ma) when the inner zone was subjected under such a extensional stress field. Emplacement of so-called the "Miocene Granitoids" occurred mainly around 7–9 Ma, and other intrusives around 12–13 Ma (see Fig. 2).

The oil- and gas-field along the Japan Sea coast has developed in the intra-arc trough with normal faulting in the basement endured during the period from the middle Mid-Miocene to Late-Miocene (KATAHIRA, 1969).

In the Pacific side of the NE Honshu the volcanism was rather weak during the Miocene stages and later periods. During the Pliocene and later periods, igneous activity including dike formation decreased, and the volcanic rocks of this period show a sparse distribution in the whole NE Honshu Arc.

4. Dike Data

There are number of igneous dikes composed of various rock-types in the study area. Their geographical distribution is not uniform but tends to cluster as isolated swarms. The duration of individual igneous activity can be estimated on the basis of such evidences as; (a) the stratigraphic horizon of the wall rocks, (b) the horizon of unconformity underlain by the dike swarm, and (c) the age of the effusive rocks with petrography similar to the dike rocks. (a) gives the oldest possible age, while (b) and (c) give the youngest and contemporary, respectively.

The selected swarms each of which has almost obvious geologic

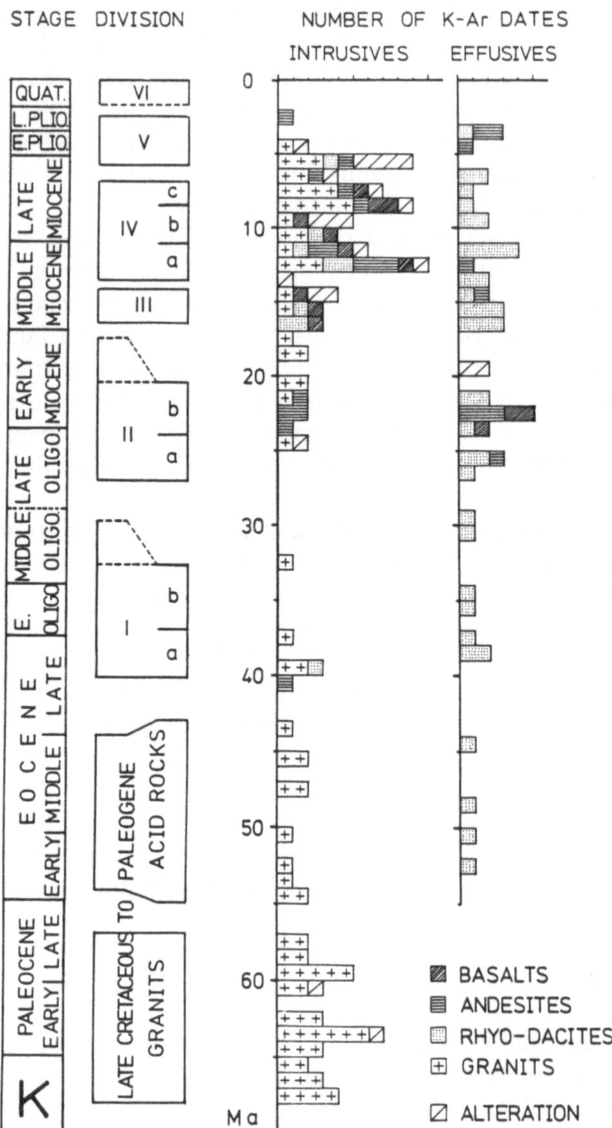

Fig. 2. Histograms for radiometric (K-Ar age) datings. Histograms showing the frequency distribution of age determination by K-Ar method for Tertiary intrusive (middle) and effusive (right) rocks, compiled from TAKEUCHI (1980), SHIBATA and NOZAWA (1982), TSUCHI (1983), KANEOKA (1983), TSUNAKAWA *et al.* (1983). Left column is the time-stratigraphic division of volcanic stages after TAKEUCHI (1981) with a slight modification.

evidences for the age of intrusion, and from which the data on the azimuthal distribution for the stress-field analysis have been sampled, are listed in Table 1. K-Ar age determinations have been done with the twelve sites of dike swarms above (TSUNAKAWA *et al.*, 1983). Detailed description about the dike-data used in the present paper are given in TAKEUCHI (1980), although a few corrections to his geological estimation on the age of magmatism with the new data are necessary.

The rose-diagrams for the azimuthal distribution of dike-walls at each location of these swarms are shown in Fig. 3 (compiled from Figs. 8a, b, TAKEUCHI, 1980).

5. *Analytical Result*

There are many varieties among the patterns of rose-diagram. It is clearly seen in each diagram that the azimuths of dikes tend to concentrate to a single trend, although the dike swarms of No. 1, 6, 7, 8, 11, 13, 14, and 15 show polymodal pattern, and their maximum peaks are rather less sharp (Fig. 3). Table 1 is a compilation of the results from the orientation analysis of dike swarms with K-Ar dates.

The stress systems obtained from the maximum-peak direction seem non-uniform, and were divided into several groups in terms of their ages and geography:

(A) Geographic division: The outer and inner provinces

Two types of directional pattern are recognized in the study area as shown in Fig. 1. Roughly saying, one type is characterized by the $\sigma_{H\,max}$-direction having a N-S trend and another by that of E-W trend. It can be seen from the figure that the latter type of direction pattern had been distributed in the both inner and outer belts of NE Honshu Arc. The former type, on the other hand, might have appeared only in the Green Tuff region. Therefore, it is necessary to recognize the region as an different stress province divided from the outer belt of the arc.

The boundary between the inner and outer provinces is defined by a NNW-SSE line passing through the central line of Abukuma Highland and could be considered as the volcanic front of the Green Tuff region. Each dike swarm in the outer province reveals the preferred orientation of a WNW-ESE trend.

(B) Chronological division

To examine the temporal change in stress orientation, the inferred $\sigma_{H\,max}$-direction are plotted along the horizontal line for the age of dike swarms in the age-athimuth chart (Fig. 4). Since we have no radiometric age determination available for the dikes during Stage I (late Eocene to

Table 1. Dike-wall data. A list of the results form the orientation analysis of dike data and K-Ar age determination of dike rocks (TAKEUCHI, 1980, 1981; TUNAKAWA *et al.*, 1983).

LOCALITY	LITHOLOGY	TREND E of N	AGE (K-Ar date)	HOST ROCKS
1. Takanuki	andesite	-50°	22.5 - 23.0 Ma	Pre-Tertiary gneiss
2. Ryozen a)	tholeiitic basalt	55°	earliest Miocene	Pre-Tertiary granite
b)	basalt & andesite	20°	21.3 - 22.1	granite & volcanics
3. Tadami a)	rhyolite	-45°	23 ± 1	Miocene volcanics
b)	rhyolite	45°	E./M. Miocene	ditto
c)	basalt & andesite	5°	12, 15-16	ditto
4. Tohachiyama	propylite	20°	Early Miocene	Pre-Tertiary granite
5. Atsumi	porphyrite	30°	Early Miocene	Pre-Neogene volcanics
6.a) Yahiko	dolerite	90°	14.2 - 15.0	Miocene shale
b) Maze	basalt	0°	7.1 - 8.2	ditto
7. Koriyama	andesite	-60°	12.3 - 13.7	Miocene sandstone
8. Ogi	basalt	-5°	12.0 - 12.3	Miocene shale
9. Kakudayama	andesite	80°	11.5 - 13.6	Miocene volcanics
10. Tanigawadake	rhyolite & andesite	25°	11.9 - 12.1	Tertiary granite
11. Minami Aizu	rhyolite	10°	10.1 - 10.5	Miocene volcanics
12. Shimokura	andesite	45°	7.7 - 8.2	Miocene volcanics
13. Tochiku	porphyrite	-20°	7.4 -~8.8	Miocene sandstone
14. Motojuku	andesite	30°	Late Miocene	Miocene volcanics
15. Shigarami	andesite	-60°	5.1 - 5.6	Neogene mudstone
16. Yoneyama	andesite	-75°	2.5 - 2.8	Pliocene volcanics
a. Aikawa	andesite	85°	Oligocene, 13-14	Tertiary volcanics
b. Asahi	alkali rhyo-dacite	-85°	Oligocene	Tertiary volcanics
c. Tsugawa	rhyolite	-40°	Early Miocene	granite, Mesozoic shale
d. Kokuzo	basalt	0°	Early Miocene	Miocene volcanics
e. Otanigawa	rhyolite	0°	Middle Miocene	Miocene volcanics
f. Kirizumi	andesite	80°	6.5 - 8.0	Miocene volcanics

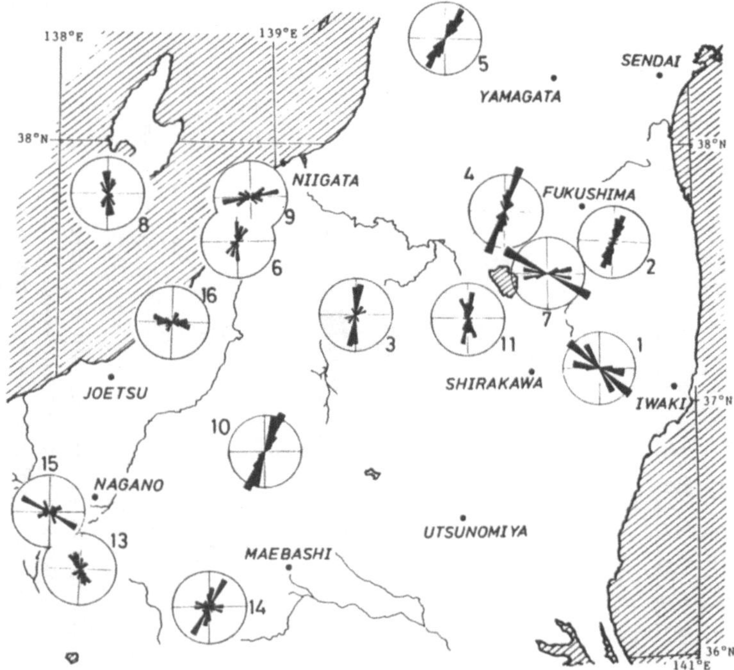

Fig. 3. Rose-diagram map of dike swarms. Azimuthal pattern of dike-walls are shown at the location of dike swarm, after TAKEUCHI (1980, Figs. 8a, b). Radius of the circle with each rose-diagram indicates the frequency of 30 percent. Numerals are correspond to those in Table 1.

early Oligocene) at the present, at least four reorientation can be recognized since Neogene. The earliest reorientation occurred at about 23 Ma, and most remarkable tansposition is around 15 Ma.

The Cenozoic history of stress field can be divided into five stages as follows:

(1) latest Eocene to middle Oligocene (37?–30? Ma), E-W $\sigma_{H\,max}$.

(2) late Oligocene (27–23), NW-SE $\sigma_{H\,max}$.

(3) Early-Miocene (23–18), N-S $\sigma_{H\,max}$.

(4) late Early-Mio- to earlier Mid-Miocene (16–12), unstable $\sigma_{H\,max}$ orientations in either E-W, NW-SE, or N-S trends.

(5) later Mid Mio- to Late Miocene (12-8), N-S $\sigma_{H\,max}$. Followed by unstable state among NNW-SSE through NE-SW to ENE-WSW trends of $\sigma_{H\,max}$ (8-6).

(6) latest Late Miocene and later (6-), WNW-ESE to E-W $\sigma_{H\,max}$.

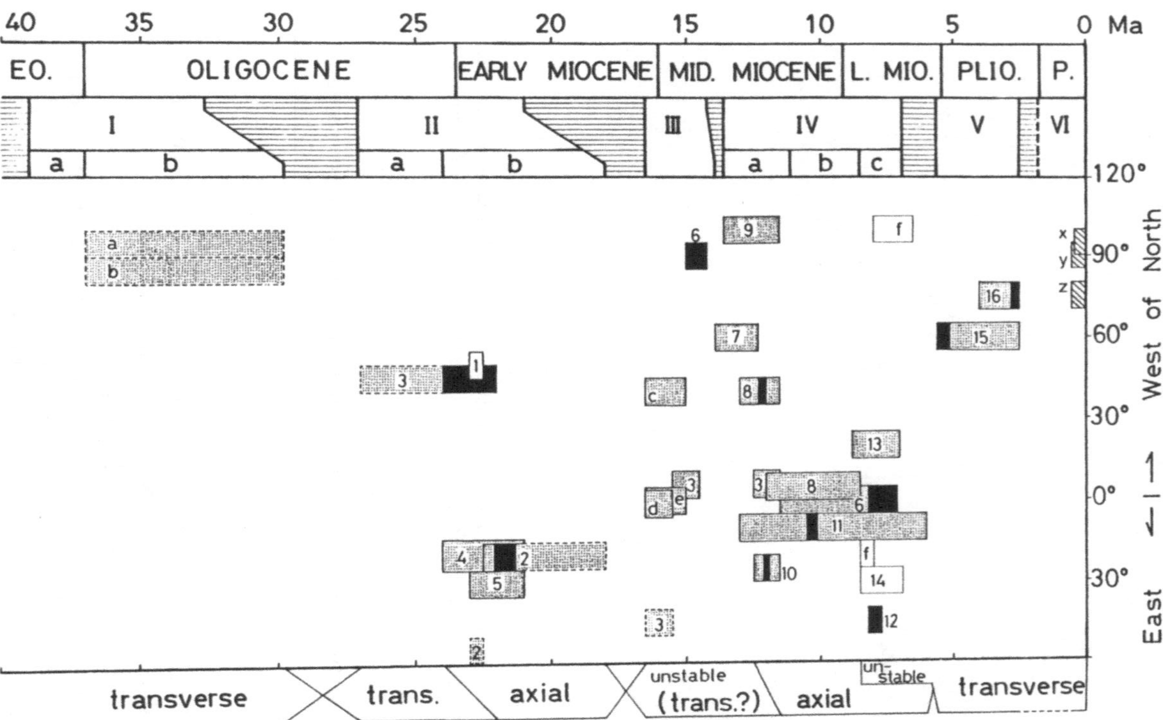

Fig. 4. Age-azimuth chart of inferred $\sigma_{H\ max}$-directions in the inner belt of southern Northeast Honshu Arc. The azimuths of $\sigma_{H\ max}$ are plotted along the horizontal axis for the age of the dike system. Dotted box: possible full-range of volcanic activity by geological estimation (TAKEUCHI, 1980, 1981, with a slight modification). Filled: K-Ar ages (TSUNAKAWA *et al.*, 1983). Error range is shown by the width of each box. Open box: dike swarms located on the outer belt of NE Honshu Arc. Bottom: a simplified representation of the phases and reorientations of the inner stress province.

Combined with the dike data, distribution of major faults is also taken into account to characterize any stress systems as one of extensional (T), neutral (B) and compressional (P) fields. Selected data sets on strike, slip sense and age of major faults in the area are listed in Table 2, although determination of nature, especially age, of fault activity is generally difficult to be confirmed. They are not always sufficient but useful when such evidences as an unconformity overlying them and differences in thickness and/or facies of sedimentary sequences on both side of fault zone are recognizable.

Identification of type of stress system and examination on history of stress field were already presented in TAKEUCHI (1980, 1981) in detail. The results are re-examined and corrected here according to the modern time-stratigraphy mentioned already. Stress trajectory map in the late Cenozoic NE Honshu Arc (TAKEUCHI, 1980, 1981; TSUNAKAWA and TAKEUCHI, 1983) are simply illustrated in Fig. 5. The selected $\sigma_{H\,max}$ directions with K-Ar data are shown in the figure.

(1) Map-1 (27–23 Ma)

During the period from Late Oligocene to Early Miocene, sedimentation of volcaniclastics due to fissure eruptions in and along the graben-like depressions trending E-W or NW-SE direction had occurred in several areas in the inner zone. The evidences for this tectonic activity have been reported in the following areas (Faults A, B and C in Table 2): The Asahi Mountainland district where the Sumikawa, Budo and Kitaoguni Formations distribute (TAKAHAMA, 1976; MIZUGUCHI, 1978), the Osado district of Sadogashima Island (SAKAI and OBA, 1970), and the mid-stream district of River Aburumagawa in the Echigo Mountainland (IJIMA, 1974). The graben-like depression on the Pre-Tertiary basement in the last district had been covered by the greenish tuff and dacite dating the next stage.

Such an extensional tectonism could be attributed to a T-type stress field with the largest σ_V and the $\sigma_{H\,max}$ of NW-SE trend. The trend of horizontal compression is highly oblique to or almost perpendicular to the elongation axis of NE Honshu.

(2) Map-2 (22–18 Ma)

A reorientation of stress field occurred at about 22 Ma and the $\sigma_{H\,max}$-axes oriented in an N-S trend. The stress trajectories runs parallel to the axial trend of NE Honshu. In this stage an extensional tectonism with fault block movement was carried out at the central zone of inner belt (Faults D, E, F, G1, H and I). One of the typical example of large normal faulting recognized along the western margin of Abukuma Highlands. This line had been a normal dip-slip fault with vertical displacement of some 1500 m during the Early- to Middle Miocene (YASHIMA

	Locality	Trend	Sense	Age(Stage)	Remarks(name or source)
A.	Asahi Mts.	NW/WNW	**Normal**	lower Aikawa	graben-like basin
B.	Sado Is.	E/ ENE	N.+ D.	lower Aikawa	..........outcrop in gallery
C.	Echigo Mts.	NW-SE	Normal	lower Aikawa	graben overlain by Greentuff
D.	Niigata Pl.	N-S	Normal	pre-Nanatani	..............oil/gas boring
E.	Asahi Mts.	N-S	Normal	middle Aikawa	graben-like depression
F.	Sado Is.	NNE	N.?	upper Aikawa?	.............seismic profile
G1.	Echigo Mts.	NNE	Normal	pre-Nanatani	Nagatani Fault
H.	N.Fossa SB.	N-S	N.?	Uchimura	trough & volcanic ridge
I.	Abukuma Mts.	NNE	Normal	Daij.-- Fun.	Yanagawa-Shirasawa T.Line
J.	Abukuma Mts.	NNW	Normal	Ryozen --	Ogai,Nabeyari & Miyauchi F.
K.	Sado Is.	NW-SE	Normal	Nanatani	Oda-Toji Tec. Zone (graben)
L.	Tadami	NW-SE	N.?	Nanatani	Tadami-Ina T. Zone (graben)
M.	Niigata Pl.	NNE	Normal	Nan./Tedad.	Shibata-Koide Tectonic Line
N1.	N.Fossa SB.	NNE	Normal	Ter./Shiiya	differential subsidence
O1.	Nagano FB.	NNE	Normal	Ter.--Shiiya	.............seismic Profile
P.	Yahiko Mt.	NNE	Reverse	post-Shiiya	growth fault......oil boring
Q.	Niitsu Hill	NNE	Reverse	Shi./Nishiy.	.................oil boring
R.	Kushigata Mts.	NNE	Reverse	post-Shiiya	...................outcrop
S.	Aizu Basin	NNE	Reverse	post-Shiiya	intermontane basins
N2.	N.Fossa SB.	NNE	Reverse	post-Shiiya	Saigawa Fault
O2.	Nagano FB.	NNE	Reverse	post-Shiiya	.............seismic profile
T1.	Abukuma	NNW	Reverse	--13, --4 Ma	Futaba Reverse Fault
T2.	Abukuma	NNW	Sinistral	post-Taga	active Futaba Fault
G2.	Echigo Mts.	NNE	Reverse	Quaternary	Nagatani Fault
U.	Nasuno	NNE	Reverse	Quaternary	Sekiya Fault

and OIDE, 1966).

(3) Map-3 (16–12 Ma)

$\sigma_{H\,max}$-trajectory during the rather short period of early Mid Miocene around 14 Ma shows East-Westery strike across the arc. Since few significant evidence for faulting in this stage is known in the Sea of Japan side (Faults J, K), the state of crustal stresses might be not so deviatoric that characteristics of the stress field are not clear.

(4) Map-4 (12–8 Ma)

The orientation of the stress field during the period from late Mid Miocene to early Late Miocene is almost the same as in Map-2: Faults L, M, N1 and O1.

(5) Map-5 (6–0 Ma)

P-type field and regionally uniform state are inferred with regard to the reverse faulting (e.g. TAKEUCHI, 1981: MIZOUE *et al.*, 1982) in the whole area (Faults P, Q, R, S, N2, O2, T1, T2, G2 and U). The pattern of stress trajectories resembles well to that of the Present σ_1-trajectories (e.g. ANDO *et al.*, 1979; WESNOUSKY *et al.*, 1983) inferred from geodetic measurements, focal mechanism solutions and Quaternary volcanic fissures.

6. *Discussion*

6.1 *Relations of stress field to tectonism and magmatism since 30 Ma*

For the beginning of discussion, I intended to refer to what could be expected to have occurred and/or what had happened indeed in response to the change in regional stress field.

First, the most significant change in the late Cenozoic stress field that seriously affected the tectonic regime of inner belt of the arc occurred around 6 Ma after the Late Miocene unstable state around 7.5 Ma. Before the change, there have existed an extensional stress field all over the study

Table 2. Fault Data. A list of selected data sets on the characteristics of major faults in the southern part of Northeast Honshu. (for further explanation, see text).

Abbreviations: D (dextral), S (sinistral), N (normal); Nan (Nanatani), Ter (Teradomari), Nishiy (Nishiyama); Mts (mountainlands), Pl (Plain), SB (sedimentary basin), FB (fault-bounded basin).

References: A: TAKAHAMA and YOSHIMURA (1969), TAKAHAMA (1976), B: SAKAI and OBA (1970), C: IJIMA (1974), D, M: KATAHIRA (1970, 1974a, b), E: MIZUGUCHI(1978), F: NIIGATA PREF. (1978), G1, G2: TAKEUCHI (1979), H, N1, O1, P, Q, N2, O2: TAKEUCHI (1978), I, J: YASHIMA (1962), YASHIMA and OIDE (1966), K: SHIMAZU *et al.* (1973), L: SHIMADA and HIRABAYASHI (1972), F, M, P, Q, R: IKEBE *et al.* (1972), P, Q, G2: CHIHARA (1980), R: UEMURA and TAKAHASHI (1974), S: SUZUKI*et al.* (1977), T1, T2: TSUNEISHI (1966, 1974), SATO *et al.* (1982), T2, U: RESEARCH GROUP FOR ACTIVE FAULTS (1980).

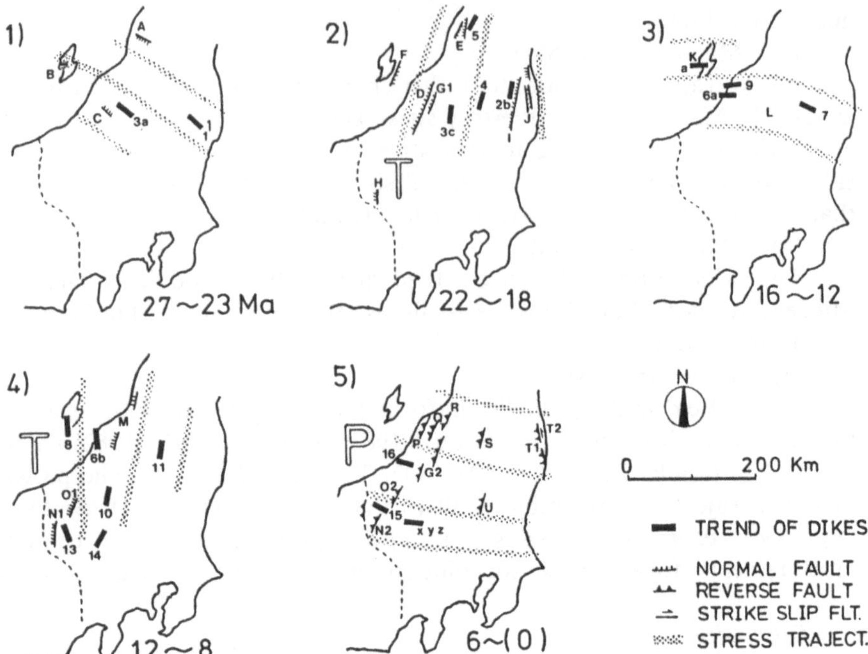

Fig. 5. Stress trajectory maps of southern Northeast Honshu in the late Cenozoic period, after TSUNAKAWA and TAKEUCHI (1983) with a slight addition. $\sigma_{H\,max}$-trajectory is drawn by smoothing the inferred stress orientations from the selected dike-swarms with K-Ar dates. Selected major faults with age estimation are also shown for indicating types of stress fields. T: Extentional stress field, where $\sigma_V > \sigma_{H\,max} \gg \sigma_{H\,min}$, and normal or gravity faulting is preferable. P: Compressional, $\sigma_{H\,max} \gg \sigma_{H\,min} > \sigma_V$, reverse or thrust faulting. Labeled alphabets and numerals for dikes and faults are the same as those in Tables 1 and 2.

area regardless of the $\sigma_{H\ max}$ orientation, and the intra-arc sedimentary basin had developed through the Miocene. At the time of reorientation, arrangement of principal stress axes of the regional stress field completely renewed, with the reversal change in magnitude relationship between σ_V and $\sigma_{H\,max}$, into the low gradient and high stress condition where $\sigma_{H\,max} = \sigma_1$ and $\sigma_V = \sigma_3$. Significantly, it is not until the stress field change that a distinct compressional stress field dominated the entire region of NE Honshu Arc as well as the inner part. Therefore, the latest change of stress field at the end of Late Miocene would be a particular case and is different from the previous reorientations where the low stress condition of $\sigma_v = \sigma_1$ had been held on. This is the major event that characterizes the history of inner stress province. Such a drastic change of tectonic stress field had not appeared in the outer zone at all.

The tectonic regime after 6 Ma is characterized by a couple of phases of compressional crustal movement with reverse faulting and tilting in the basement, resulting into the intermontane basins and intense folds in the oil fields (TAKEUCHI, 1981).

From this point of view, it is likely that the former unstable state of stress field around 15 Ma with $\sigma_{H\,max}$-direction not only in a NW-SE but also both an E-W and a N-S trends was just a short disturbance to the extensional field with north-southery $\sigma_{H\,max}$ during the somewhat lengthy period through the Miocene.

In turn, volcanic activities also seem to be affected by or correlated with the vicissitude of stress field:

(1) Volcanic activity at the Stage II from the late Late Oligocene to the early Early Miocene was more violent than that at present (SUGIMURA and UYEDA, 1973). The output of volcanic effusives per unit time reached nearly ten times more than that of the Quaternary period. It is very interesting that the higher effusive activities around 24 Ma, 15 Ma and at the Recent tend to occur when and where the $\sigma_{H\,max}$-trajectory, and also the feeder dikes, oriented almost perpendicular to the axis of the arc or the volcanic front: It is just before the sudden decrease of effusive activity that the $\sigma_{H\,max}$ direction had finished to turn almost 90 degrees from a NW-SE trend to a NNE-SSW one at about 22 Ma.

(2) By the way, volcanic activity during the unstable stress field around 15 Ma mentioned before is characterized by the uncommon trap activity called as "regional basalt" that output through the entire region. Stratiform "Kuroko" deposits were formed along the inner side of backbone ranges of the arc around 15 Ma (E. HORIKOSHI, personal com.).

(3) As to intrusive activity, Miocene granitoids emplaced twice at 25–20 Ma and 13–5 Ma when the $\sigma_{H\,max}$-trajectory oriented parallel to the arc. It was during the latter period (see Fig. 2) that remarkable vein-mineralization and alteration associated with a variety of intrusives had occurred through the inner belt of the arc.

Consequently, it is suggested that; (1) changes in the stress field in NE Honshu during the late Cenozoic were not only short-ranged and so simple reorientations among horizontal stresses but also drastic in magnitude between horizontal and vertical ones, and that (2) each phase in the variation of stress field was closely associated with such different regimes of magmatism.

6.2 Swing motion of Pacific Plate and reorientation of stress field on Northeast Honshu Arc

Change in angle of convergence between overriding arc massif and undergoing oceanic plate is a possible important control factor of tectonic stress field of island arc type (DEWEY, 1980).

JACKSON *et al.* (1975) suggested that the Pacific plate revealed an oscillatory fluctuation from the average trend of plate motion, 70 degrees west of north. Jackson's curve indicates that the largest oceanic plate has moved northwesterly with swinging, if the Hawaiian hot spot is taken as a reference point fixed to the deep mantle (SENO, 1983). Present motion of such a swing motion reveals as alignment of magma conduits to the active Hawaiian volcanoes, Mauna Loa and Loihi (KLEIN, 1982).

Relative plate motion concerning Pacific Plate and NE Honshu Arc can be estimated as about 300 degrees east of north using seismic slip vectors of shallow earthquakes along the deep seismic zone (KANAMORI, 1977; SENO, 1977; ANDO *et al.*, 1979). This is the only reason why the convergence at the Cenozoic Japan Trench as well as there at the Present is often envisioned as somewhat high angle, and the large stress magnitude ($\sigma_{H\,max} \gg \sigma_V$) and the state of laterally low gradient in the Present stress field on the Arc are attributed to this (EGUCHI, 1983; TAKAGI *et al.*, 1984). It is still a matter of discussion whether the seismic coupling along the interface between the plate and the arc is strong (EGUCHI, 1983; ISHII *et al.*, 1983) or weak (TADA, 1984; NAKAMURA, this volume). Anyway, relative motion of the convergence between East-Asian Block and Pacific Plate would have changed with time in accordance with change in absolute motion of this block as well as with the Pacific swing.

In the NE Honshu Arc, stress field orientation changes periodically in the way that both principal stress axes tend to direct parallel and perpendicular to the trench/arc axis. Although the origin of the cyclic swing motion is still unknown at present, the timing of the reorientations in the arc stress field and the periodicity of the apparent swing of the converging plate are phenomenally correlated each other completely:

(1) At the climax phase of deflection in a clockwise episode of Pacific Plate motion, the $\sigma_{H\,max}$ trajectories on NE Honshu Arc directed in a trend almost perpendicular to the arc axis.

(2) The extensional T-type stress province with $\sigma_{H\,max}$ of an E-W trend appeared only at the period of counterclockwise episodes.

The timing of volcanic activities in NE Honshu Arc also seems phenomenally synchronized with the swinging motion of Pacific Plate: two periods of basalt activity occurred at two climaxed of clockwise episodes (C1 and C2 in Fig. 6). Explosive rhyo-dacitic activities shown by ash layers in pelagic and hemipelagic sediments of the Northwest Pacific (CADET and FUJIOKA, 1980) occurred just at the time of inflection points or shoulders of the curve. To the contrast, counterclockwise (anticlimax) episodes coincide with relative quiescence or intermission of effusive activity and with remarkable intrusive one (see Fig. 2). Hawaiian volcanism itself also seems almost synchronized with the swing motion of Pacific Plate (VOGT, 1979).

6.3 Possible origin of the arc stress field

On the analogy of the Himalayan tectonics (VILOTTE *et al.*, 1982), the geohistorical study on the arc stress field mentioned already provides an idea for the Present conditions both of rheological behavior of the arc and of the mechanical coupling between the arc and oceanic plate.

When the arc were put in a state of 'plane strain with laterally fixed boundary', stress gradient across the arc would become large and an extentional stress province develope behind compressional one. This model corresponds to the idea of "paired stress filed" (TAKEUCHI, 1980; TAKEUCHI *et al.*, 1979).

In another case of 'plane strain with free boundary', or of 'plane stress conditions', stress gradient across the arc should be small and horizontally uniform compression with uplifting dominates the almost entire arc.

According to the terminology of VILOTTE *et al.* (1982), the former state can be referred to "very thick massif model", because vertical strain tend to be diluted away by the thickness. On the other hand, the latter can be called as "thin plate model", because it is very easy to bend in response to external force and the induced vertical stress is likely released into zero. The latter case might correspond to a state of lithospheric decoupling even if the coupling at the shallower part of subduction zone was seismically strong enough. Such a thin plate condition might be attributed to a large geothermal gradient caused by remarkable magmatism associated with a global synchronism.

JACKSON *et al.* (1975), REA and SCHEIDEGGER (1979), and VOGT (1979) demonstrated some evidences for globally coeval magmatic episodes, affecting both intraplate magmatism (hot spot activity) and island arc volcanism along subduction zones as well as magmatism along divergent plate boundary. MASUDA (1984) pointed out the coeval relationship between the Jackson's curve and the global eustacy. Because such a global scale through the different tectonic circumstaces in the lithosphere should be attributed to a deeper origin, the global magmatic episodes could be related to asthenospheric ones something like an episodic change in the pattern of deep mantle convection or dilation of the Earth.

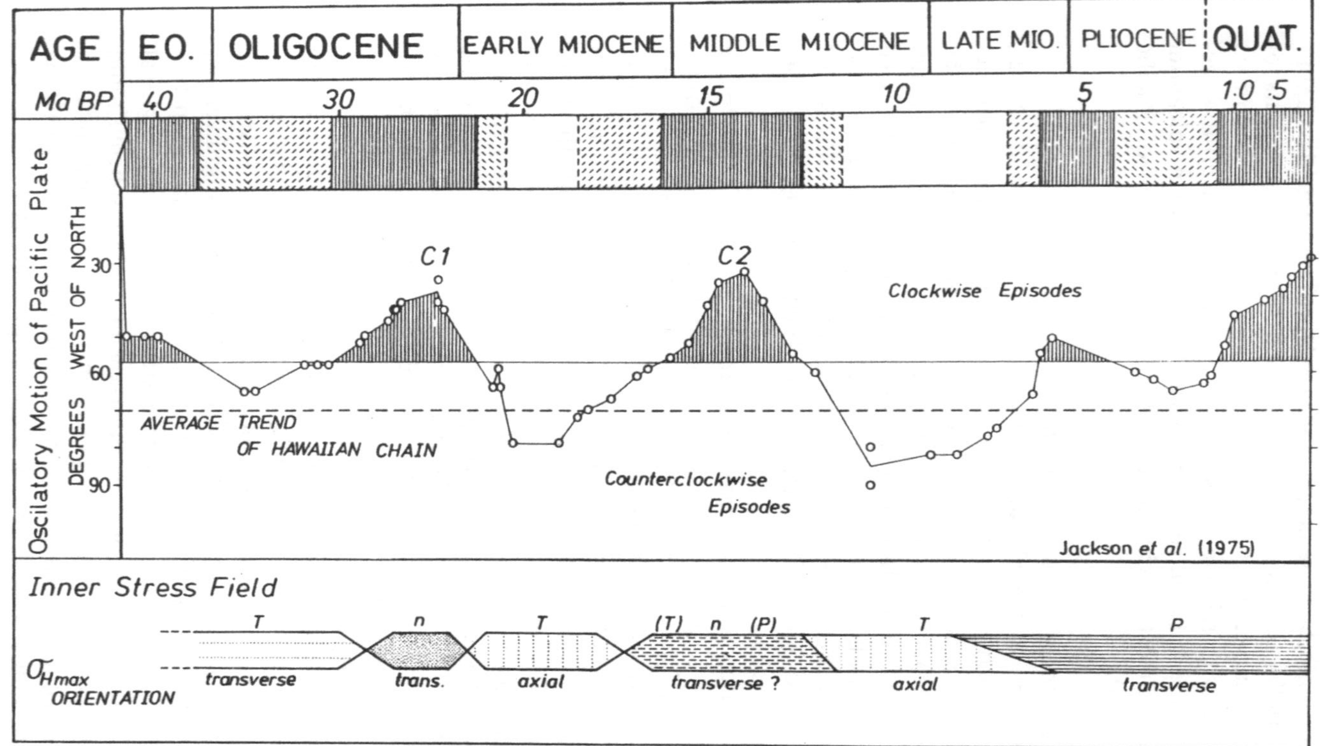

Fig. 6. Jackson's curve and arc stress reorientations Apparent swing motion of Pacific Plate (JACKSON *et al.*, 1975) and regional stress orientation at the Northeast Honshu convergent margin (TAKEUCHI, 1980, 1981, 1984; TSUNAKAWA and TAKEUCHI, 1983) are illustrated in order to show their synchronous relationship. Dashed line represents the average trend of the Hawaiian volcanic chain. Pacific plate moves along the direction with fluctuation in reference to Hawaii Hot Spot. Vertically shaded part of the graph indicate the climax phases of "clockwise episodes". Lower part of the figure shows the phases and reversals in orientation of tectonic stress field on the inner zone of NE Honshu Arc (same as Fig. 4).

By the way, the important question arises here, whether the Present state of NE Honshu Arc is in a plane stress condition or not. Significantly, strain rate in horizontal shear strain field of NE Honshu seems lower than in that of western Central Japan and western Central Hokkaido (NAKANE, 1973). Taking into the fact that the Quaternary effusive activity is one of culminations in the periodicity of the Cenozoic volcanism in the arc, it is quite reasonable to consider that the Present lithosphere of the arc is heated and uplifted by the underlying mantle and behaves as a thin plate in a plane stress condition.

From the viewpoint of uniformitalianism, we could expect that; (1) inherently, a volcanic arc without free lateral termination would be dominated by a stress field with higher stress gradient due to subduction or collision, (2) a stress field with $\sigma_{H\ max}$ trajectories across the arc axis would have appeared at the time of "climax episode" of Jackson's curve, namely of culmination in the globally synchronized volcanism.

7. *Summary*

In this paper stress field history and tectonic evolution of Northeast Honshu Arc was discussed. The stress history of the study area for the last 40 million years includes four reorientations and five phases. The major change in magnitude and orientation of regional stress field occurred around 8–6 Ma and affected the tectonic regime of the inner zone of the arc seriously. During the period from about 27 to 7 Ma, the inner zone had been subjected to T-type filed with a remarkable intra-arc trough, i.e. the Uetsu sedimentary basin, had developed by normal faulting in the basement. At the time when the arrangement of regional tectonic stresses was renewed, this zone evolved abruptly into a distinct field of intense compressional tectonics as the results of reverse faulting and tilting of the basement blocks.

The temporal coincidence of reversals in orientation of tectonic stress field on NE Honshu Arc together with the oscillatory motion of the Pacific plate and global synchronism of magmatism were as well demonstrated as lately suggested by MASUDA (1984) from the point of view that the change of oceanic plate motion might be recorded in the history of sedimentary basins developed in the arc-trench system.

However, it seems better that the drastic change of tectonic stresses in the inner stress province is more foundamental and is attributed rather to another origin something like the evolution of downgoing slab such as the Kanamori-Uyeda cycle than to the episodic change in angle between trench axis and relative plate motion.

It is significant that the episodic vicissitude of regional tectonic stress field and resulting tectonics in the arc system reflects both the apparent swing motion of the Pacific plate and the global magmatic synchronism. This problem will provide a key for mechanical understanding of Pacific-type ocean margin. And, the important question to be solved is by which the absolute swinging has been carried out, the Pacific plate or the Hawaiian hot spot.

REFERENCES

E. M. ANDERSON, *The Dynamics of Faulting an Dyke Formation with Application to Britain* (2nd ed.), pp. 22–58, Oliver and Boyed, 1951.

M. ANDO, T. MATSUDA, and K. ABE, Stress field of the upper crust of the Japanese Islands, Programme and abstracts, Seismol. Soc. Japan, No. 1, 66, 1979 (in Japanese).

R. E. BISCHKE, A model of convergent plate margins based on the recent tectonics of Shikoku, Japan, *J. Geophys. Res.*, **79**, 4845–4857, 1974.

J. CADET and K. FUJIOKA, Neogene volcanic ashes and explosive volcanism: Japan trench transect, leg 57, Deep Sea Drilling Project, Init. Rept. D.S.D.P., LVI-LVII, U.S. Government Printing Office, pp. 1027–1041, 1980.

K. CHIHARA, Seismological informations on the 1964 Niigata Earthquake and its geological interpretation, in *Earthquake: Dialogues between Seismologists and Geologists*, Eds. R. Sugiyama, M. Hayakawa and M. Hoshino, pp. 121–148, Tokai Univ. Press, 1980 (in Japanese with English abstract).

J. F. DEWEY, Episodicity, sequence, and style at convergent plate boundaries, *Geol. Assoc. Canada, Spec. Paper*, **20**, 553–574, 1980.

T. EGUCHI, Tectonics along the eastern margin of the Sea of Japan, *Earth Monthly*, **6**, 55–60, 1983 (in Japanese).

K. FUJIOKA, History of explosive volcanism of the Tohoku arc from the core sediment sample of the Japan trench, *J. Volcanol.*, **28**, 41–58, 1983.

S. IJIMA, Nature of the Middle Miocene unconformity of the mid-stream region of the R. Aburuma-gawa, Shinano-gawa River Group, Rept. Geol. Surv. Japan, Ser. 250-1, pp. 145–154, 1974 (in Japanese with English abstract).

Y. IKEBE, K. MASATANI, and T. KATAHIRA, Some considerations on the "Green Tuff" in Niigata Sedimentary Basin, in *Izu Peninsula*, Eds. M. Hoshino and H. Aoki, pp. 41–47, Tokai Univ. Press, 1972 (in Japanese with English abstract).

H. ISHII, T. SATO, K. TACHIBANA, K. HASHIMOTO, E. MRAKAMI, M. MISHINA, S. MIURA, K. SATO, and A. TAKAGI, Crustal strain, crustal stress and microearthquake activity in the northeastern Japan arc, *Tectonophysics*, **97**, 217–230, 1983.

E. D. JACKSON, H. R. SHAW, and K. E. BARGAR, Caluculated geochronology and stress field orientation along the Hawaiian chain, *Earth Planet. Sci. Lett.*, **26**, 145–155, 1975.

I. KANEOKA, On the radiometric ages of volcanic rocks from the northeastern part of the Honshu island, Japan, Mining Geology Spec. Issue, No. 11, pp. 69–78, 1983 (in Japanese with English abstract).

T. KATAHIRA, Basement structure and geologic development in the Kitakambara plain, Niigata Prefecture, *J. Japan. Assoc. Petrol. Technol.*, **39**, 167–178, 1969 (in Japanese with English abstract).

T. KATAHIRA, Geological development of the Nagaoka plain and its surrounding areas, Niigata Prefecture, Japan, *J. Japan Assoc. Petrol. Technol.*, **35**, 59–66, 1970 (in Japanese with English abstract).

T. KATAHIRA, Stratigraphy of the Neogene Tertiary in the central and northern parts of Niigata Prefecture: Petroleum geology of Neogene Tertiary in the Chuetsu and the Kaetsu regions, Niigata, Japan, *J. Japan Assoc. Petrol. Technol.*, **39**, 167–178, 1974a (in Japanese with English abstract).

T. KATAHIRA, Hydrocarbon deposits found in the Green Tuff in the Niigata sedimentary basin: Petroleum geology of the Neogene Tertiary in the Chuetsu and the Kaetsu regions, Niigata, Japan—2, *J. Japan. Assoc. Petrol. Technol.*, **39**, 337–356, 1974b (in Japanese with English abstract).

J. P. KENNET, *Marine Geology*, 183 pp., Prentice-Hall, 1982.

G. KIMURA and K. TAMAKI, Tectonic framework of the Kuril Arc since its initiation, this volume.

F. W. KLEIN, Earthquake at Loihi Submarine Volcano and the Hawaiian hot spot, *J. Geophys. Res.*, **87**, 7719–7726, 1982.

F. MASUDA, Sedimentry basins in arc-trench system as a high-sensitive recorder of oceanic plate motion, *Mining Geology*, **34**, 1–20, 1984 (in Japanese with English abstract).

M. MIZOUE, T. YOKOTA, and I. NAKAMURA, High angle reverse faulting in the interaxial zone of active folds in the inner belt of Northeast Japan, *Bull. Earthq. Res. Inst.*, **57**, 359–377, 1982.

K. MIZUGUCHI, The study of Hitokasumi Conglomerate Bed and the movement of basement from the view point of structural geology, Master thesis of Fac. Sci. Niigata Univ., 1978ms (in Japanese).

K. NAKAMURA, Volcanoes as possile indicators of tectonic stress orientation—principles and proposal, *J. Volcanol. Geotherm. Res.*, **2**, 1–16, 1976.

K. NAKAMURA, The active tectonic belt along the eastern margin of Sea of Japan, this volume.

K. NAKAMURA and S. UYEDA, Stress gradient in arc—back arc region and plate subduction, *J. Geophys. Res.*, **85**, 6419–6428, 1981.

K. NAKANE, Horizontal tectonic strain in Japan (I), (II), *J. Geod. Soc. Japan*, **19**, 190–199, 200–208, 1973 (in Japanese).

NIIGATA PREFECTURE, Explanation of the Geologic Map (1:200,000) of Niigata Prefecture, Japan, 493 pp., Naigai, 1978.

D. K. REA and K. F. SCHEIDEGGER, Eastern Pacific spreading rate fluctuation and its relation to Pacific area volcanic episodes, *J. Volcanol. Geotherm. Res.*, **5**, 135–148, 1979.

RESEARCH GROUP FOR ACTIVE FAULTS, Active faults in Japan: Sheet Maps and inventories, Univ. Tokyo Press, 363 pp., 1980.

Y. SAKAI and M. OBA, Geology and ore deposits of the Sado Mine, *Mining Geol.*, **20**, 149–165, 1970 (in Japanese with English abstract).

H. SATO, K. OTSUKI, and K. AMANO, Neogene tectonic stress field of Northeast Honshu Arc, *Structural Geol.* (*J. Tect. Res. Group Japan*), No. 27, 55–79, 1982 (in Japanese).

T. SENO, The instantaneous rotation vector of the Philippine Sea plate relative to the Eurasian plate, *Tectonophys*, **42**, 209–226, 1977.

T. SENO, Age of subducting oceanic plate and its tectonism: an examination, *Marine Sci.*, **15**, 761–766, 1983 (in Japanese).

K. SHIBATA and T. NOZAWA, Radiometric age map 1. granitic rocks, 2. volcanic rocks, scale 1:4,000,000 in *Geological Atlas of Japan*, pp. 66–73, Geol. Survey Japan, 1982.

I. SHIMADA and T. HIRABAYASHI, Geologic structure of the Kuroko (Black Ore) metalliferous province in the western Aizu district, Fukushima Prefecture, *Mining Geol.*, **22**, 329–346, 1972 (in Japanese with English abstract).

M. SHIMAZU, Y. KANAI, T. TOYAMA, K. ICHIHASHI, J. MINAGAWA, and N. TAKAHAMA, Geological development and Igneous activity in the Sado Island, *Memories Geol. Soc. Japan*, No. 9, 147–157, 1973 (in Japanese with English abstract).

A. SUGIMURA and S. UYEDA, *Island Arcs, Japan and Its Environs*, 247 pp., Elsevier, 1973.

K. SUZUKI, T. YOSHIDA, and K. MANABE, The geologic development of Inland basins in the southern part of the Tohoku district, Japan, *Memories Geol. Soc. Japan*, No. 14, 45–64, 1977 (in Japanese with English abstract).

T. TADA, Horizontal crustal movement in the northeastern Japan and its characteristics, Programme and abstracts, *Seismol. Soc. Japan*, No. 1, **270**, 1984 (in Japanese).

N. TAKAHAMA, The Neogene system in the western slope of the Asahi Massif in the northern part of Niigata Prefecture, Japan, *Memories Geol. Soc. Japan*, No. 13, 211–228, 1976 (in Japanese with English abstract).

N. TAKAHAMA and N. YOSHIMURA, green Tuff in the northern part of Niigata in Japan: Preliminary report, 70th Geol. Soc. Japan, pp. 105–115, 1969 (in Japanese).

A. TAKEUCHI, The Pliocene stress field and tectonism in the Shinetsu region, central Japan, *J. Osaka City Univ.*, **21**, 37–52, 1978.

A. TAKEUCHI, On the fault history of Nagatani Fault (tentative name), western margin of Mikawa Village, Niigata Prefecture, Abstracts Geol. Soc. Japan, **378**, 1979 (in Japanese).

A. TAKEUCHI, Tertiary stress field and tectinic development of the southern part of the Northeast Honshu arc, Japan, *J. Osaka City Univ.*, **23**, 1–64, 1980.

A. TAKEUCHI, Temporal changes of regional stress field and tectonics of sedimentary basin, *J. Geol. soc. Japan*, **87**, 737–751, 1981 (in Japanese with English abstract).

A. TAKEUCHI, The Cenozoic stress field history of Japan—the "dike method" for paleostress field analysis—, ch. 16 in *Tectonic Belts of Asia*, K. Huzita ed., pp. 313–326, 1984 (in Japanese with English abstract).

A. TAKEUCHI, K. NAKAMURA, Y. KOBAYASHI, and K. HORI, Cenozoic stress field of Central Honshu, Japan, as derived from dike swarms—an introduction for paleostressology, *Earth Monthly*, **1**, 447–452, 1979 (in Japanese).

R. TSUCHI, Neogene bio- and chronostratigraphy in Japan, *J. Japan. Assoc. Petrol. Technol.*, **48**, 35–48, 1983 (in Japanese with English abstract).

H. TSUNAKAWA and A. TAKEUCHI, Paleostress field and igneous activity on the Japanese islands, *Kagaku*, **53**, 624–631, 1983 (in Japanese).

H. TSUNAKAWA, A. TAKEUCHI, and K. AMANO, K-Ar ages of dikes in Northeast Japan, *Geochem. J.*, **17**269–275, 1983.

Y. TSUNEISHI, Geologic structure of the Hirono Area in the Abukuma Mountains, *Bull. Earthq. Res. Inst.*, **44**, 749–764, 1966.

Y. TSUNEISHI, Block structure in the eastern margin of Abukuma Mountains, in *Geotectonic Studies on the Tertiary Crustal Disturbance in Northeast Japan*, pp. 37–41, 1974 (in Japanese).

T. UEMURA and A. TAKAHASHI, Kinematic picture of basement rocks and folding of overlying layer: an example of Kushigata Mountain Range, Niigata Prefecture, Japan, Rept. Geol. Soc. Japan, Ser. 250-2, pp. 1–19, 1974 (in Japanese with English abstract).

J. P. VILOTTE, M. DAIGNIERES, and R. MADARIAGA, Numerical model of intraplate deformation: Simople mechanical models of continental collision, *J. Geophys. Res.*, **87**, 10709–10728, 1982.

P. R. VOGT, Global magmatisc episodes: New evidence and implications foe the steady-state mid-oceanic ridge, *Geology*, **7**, 93–98, 1979.

S. G. WESNOUSKY, H. SHOLZ, K. SHIMAZAKI, and T. MATSUDA, Earthquake frequency distribution and the mechanics of faulting, *J. Geophys. Res.*, **88**, 9331–9340, 1983.

R. YASHIMA, Volcanic rocks in the Ryozen formation—2, Sci. Rept. Fukushima Univ., No. 11, pp. 31-68, 1962 (in Japanese with English abstract).

R. YASHIMA and K. OIDE, Volcanism and rock alteration in the early Miocene field along the Pacific coast of Northeast Japan—for the correlation of rocks between Green Tuff and Non-Green Tuff Regions, *Monograph Assoc. Geol. Collab. Japan*, **12**, 103–111, 1966 (in Japanese with English abstract).

CHAPTER 4: BACK-ARC TECTONICS

Formation of Active Ocean Margins, edited by N. Nasu *et al.*, pp. 469–481.

AGE OF SUBDUCTING LITHOSPHERE AND BACK-ARC BASIN FORMATION IN THE WESTERN PACIFIC SINCE THE MIDDLE TERTIARY

Tetsuzo SENO

International Institute of Seismology and Earthquake Engineering, Building Research Institute, Ministry of Construction, Tsukuba, Ibaraki Pref. 305, Japan

Abstract. Since the Pacific plate changed its motion with respect to hotspots at 43 Ma, back-arc basin formation in the western Pacific has occurred mainly in two episodes: late Oligocene to early–middle Miocene and late Miocene to the present. On the basis of the North New Guinea plate hypothesis (SENO, 1984), which maintains that there was another plate south of the Pacific plate during the early Tertiary in the western Pacific region, the age distribution of the Pacific ocean floor is reconstructed for the past 48 Ma. During the Paleogene, a young ocean floor, which was created by the ridge system between the Pacific and North New Guinea plates, existed west of the present older western Pacific ocean floor and was entering the western Pacific subduction zones. The older Pacific ocean floor approached the western Pacific subduction zones around 25 Ma in this reconstruction. This suggests that the change in the subducting lithosphere's age from young to old and the resulting increase of negative buoyancy of the slab possibly caused the retreat of the trenches in the western Pacific and the back-arc spreading during the late Oligocene to early–middle Miocene in the western Pacific region.

1. *Introduction*

Age of subducting lithosphere affects various tectonic features in subduction zones. Seismicity in subduction zones is one of such features. The length of the Wadati-Benioff zone is proportional to the age of the subducting lithosphere (VLAAR and WORTEL, 1976; MOLNAR *et al.*, 1979; SHIONO and SUGI, 1985). At intermediate depths, FUJITA and KANAMORI

(1981) showed that occurrence of double seismic zones depends on the age of the subducting lithosphere and convergence rate. Seismic coupling in the thrust zone between two converging plates is also affected by the age of the subducting lithosphere. RUFF and KANAMORI (1980) showed that the maximum magnitude of interplate thrust type earthquakes in each subduction zone is roughly a function of the age of the subducting lithosphere. Recently, PETERSON and SENO (1984) showed that the seismic moment release rate per unit length of subduction zone depends on the age of the subducting lithosphere. Among all subduction zones which the same oceanic plate is entering, the seismic moment release rate becomes smaller as the subducting lithosphere becomes older. The reason why the older subducting lithosphere causes weaker seismic coupling at the plate interface is explained by the fact that the older, thus colder, slab is denser, and has a greater tendency to sink into the asthenosphere, resulting in reduction of the normal stress across the thrust zone.

The stress regime in the back-arc area may also be affected by the age of the subducting lithosphere. MOLNAR and ATWATER (1978) noted that the back-arc basins of recent geological time formed around the Pacific behind arcs where older ocean floors are now subducting. ENGLAND and WORTEL (1980) refined their idea by calculating slab pull forces dependent on age of the lithosphere. UYEDA and KANAMORI (1979) pointed out a relation between interplate seismicity and the back-arc stress regime; in the Mariana type subduction zones, earthquakes do not occur and the back-arc basin is opening, but in the Chilean type, earthquakes occur and no back-arc basin is opening. Because the seismicity is related to the age of the subducting lithosphere, the above relation is consistent with the idea that the back-arc stress regime is related to the age of the subducting lithosphere.

However, it might be difficult to relate these two factors simply. Note that there are several back-arc basins where relatively young ocean floors are subducting, such as the New Hebrides, Ryukyu and Scotia arcs. Note also that the seismic moment release rate and age of the subducting lithosphere are only related among subduction zones which belong to one plate (PETERSON and SENO, 1984). This indicates that the seismicity is not a simple function of age of the subducting lithosphere, but other factors, which characterize the dynamics of each oceanic plate and possibly those of upper plates, also control the seismicity. Similarly we know that the back-arc stress regime would not be a simple function of the age of the subducting lithosphere (e.g., TAYLOR and KARNER, 1983; CARLSON and MELIA, 1984). However, among subduction zones which belong to one oceanic plate, the age of the subducting lithosphere could be one possible factor which controls the back-arc stress regime.

In this paper, I try to refine MOLNAR and ATWATER's (1978) idea by reconstructing the age distribution of the Pacific ocean floor since the middle Tertiary and seek a relationship between the age of the subducting lithosphere and episodes of back-arc basin formation. This is needed because the age of the subducting lithosphere when those back-arc basins formed is obviously different from the present age.

2. *Episodes of Marginal Basin Formation in the Western Pacific*

Figure 1 shows the age distribution of marginal basin crust in the western Pacific region (SENO, 1984). Marginal basins in the Indonesian–Philippine region are not presented in this figure to avoid complexity. Tectonic evolution in the latter region is likely to be much more complex than the frontal area in the circum-Pacific region treated in this study because it has been affected by the interaction with both the Pacific and Indo–Australian plates. A few marginal basins which have Mesozoic age, such as the Tasman Sea and the Bering Sea, are also not shown. Except for these Mesozoic marginal basins, marginal basins in the western Pacific have Tertiary and Quaternary ages. Cenozoic marginal basins in the Western Pacific are categorized into three groups based on their ages of formation: late Paleocene–mid Eocene, Oligocene–mid Miocene, and late Miocene–the present. The oldest marginal basins, such as the West Philippine, Coral Sea, and southern New Hebrides Basins, formed before the change of the Pacific plate motion with respect to hot-spots at 43 Ma (CLAGUE and JARRARD, 1973). These basins have magnetic anomaly lineations roughly in an east-west direction, and SENO (1984) and SENO and MARUYAMA (1984) proposed that they formed by back-arc spreading due to the southward subduction of the hypothesized North New Guinea plate beneath the northern margin of Australia. The younger marginal basins have magnetic anomaly lineations roughly perpendicular to the directions of plate convergence (JURDY, 1979) and most of them formed by back-arc spreading (KARIG, 1971; TAYLOR and KARNER, 1983).

Between the two episodes of marginal basin formation since the Pacific plate motion change, the earlier one is more significant. During the Oligocene to mid Miocene, many of the back-arc basins in the western Pacific opened. An especially important point to be noted is the fact that the Shikoku Basin, the Japan Sea and the south Kurile Basin formed almost simultaneously during the late Oligocene to mid-Miocene as SENO and MARUYAMA (1984) pointed out. Although the precise ages of formation of the Japan Sea and south Kurile Basin are not known, depths of the

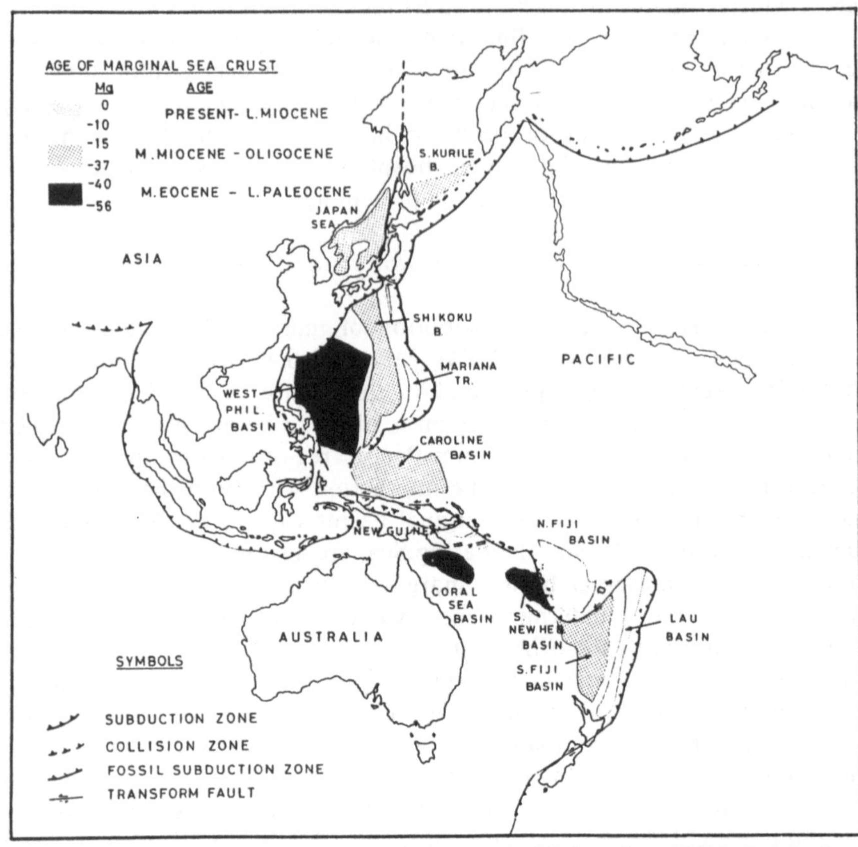

Fig. 1. Age of the marginal basin crust in the western Pacific from SENO (1984). Only basins of Cenozoic age are shown. Basins in the Philippine-Indonesian region are not shown to avoid complexity.

basement, heat flow values and stratigraphy of sediments indicate that they are likely to have formed during the late Oligocene to mid-Miocene (SCLATER, 1972; KOBAYASHI, 1983; KIMURA and TAMAKI, 1985a, b). Recent extensive paleomagnetic data for the Tertiary rocks collected from Honshu, Japan show a significant rotation of both southwestern and northeastern Honshu during the early–mid Miocene (HAYASHIDA and ITO, 1984; OTOFUJI *et al.*, 1985; TOSHA, 1983), which was likely to be caused by the Japan Sea opening. The landward margin of the south Kurile, Japan Sea and Shikoku Basins are continuous and the south Kurile and Shikoku

Basins are fan-shaped symmetrically against the Japanese islands as shown in Fig. 1. Based on this, SENO and MARUYAMA (1984) suggested that a simultaneous retreat of the Kurile–Japan–Bonin Trenches may have resulted in the opening of these back-arc basins.

In the next section, based on the reconstruction of the age of the Pacific ocean floor since the middle Tertiary, I will show that a change in the age of the subducting lithosphere occurred during this significant episode of back-arc basin formation.

3. *Age of the Pacific Ocean Floor since 48 Ma*

The spatial distribution of ages of the Pacific ocean floor at four stages since 48 Ma is reconstructed in the following manner. First, the present age distribution of the Pacific ocean is rotated back to 48, 38, 25, and 18 Ma using the known Pacific plate motion with respect to hot-spots (ENGEBRETSON, 1982). The ages of the present Pacific ocean are taken from the magnetic anomaly lineations and the geomagnetic reversal time scale shown in the Plate Tectonic Map of the Circum–Pacific Region (HALBOUTY *et al.*, 1981). In Fig. 2, isochrons at 150, 135, 120, and 72 Ma, which correspond to anomalies M29, M17, M4, and 32, are reproduced in a simplified form.

Reconstruction of tectonic features of the western Pacific at 48 Ma (Fig. 3) is from SENO (1984), who introduced a North New Guinea plate south of the Pacific plate in the western Pacific during the early Tertiary on the basis of geological and geophysical evidence in the southwest Pacific. The North New Guinea plate was separated by a ridge-transform fault system from the Pacific plate. SENO (1984) located the ridge system at 48 Ma more or less arbitrarily. In this paper, assuming that the ridge system was subducted beneath the Australian margin at 43 Ma and that the half spreading rate was 7 cm/yr, I located the ridge system at 48 Ma as shown in Fig. 3. The annihilation of the ridge system at 43 Ma is a probable cause of the change in the Pacific plate motion at this time (SENO and PETERSON, 1984; YAMANO and UYEDA, 1985; SENO, 1984). Although this reconstructed position of the ridge system is still variable, the following results are not severely affected by the uncertainty of the position. In Fig. 3, the crust of the Pacific ocean of 25 Ma age at 48 Ma is indicated by the isochron on the basis of the above two assumptions. This isochron, along with other Pacific isochrons in Fig. 2, is transferred to those at 38, 25, and 18 Ma stages using the Pacific plate motion with respect to hot-spots (Figs. 4, 5, and 6). Reconstructions of marginal basins and subduction zones in the western Pacific in these figures are from SENO and PETERSON (1984).

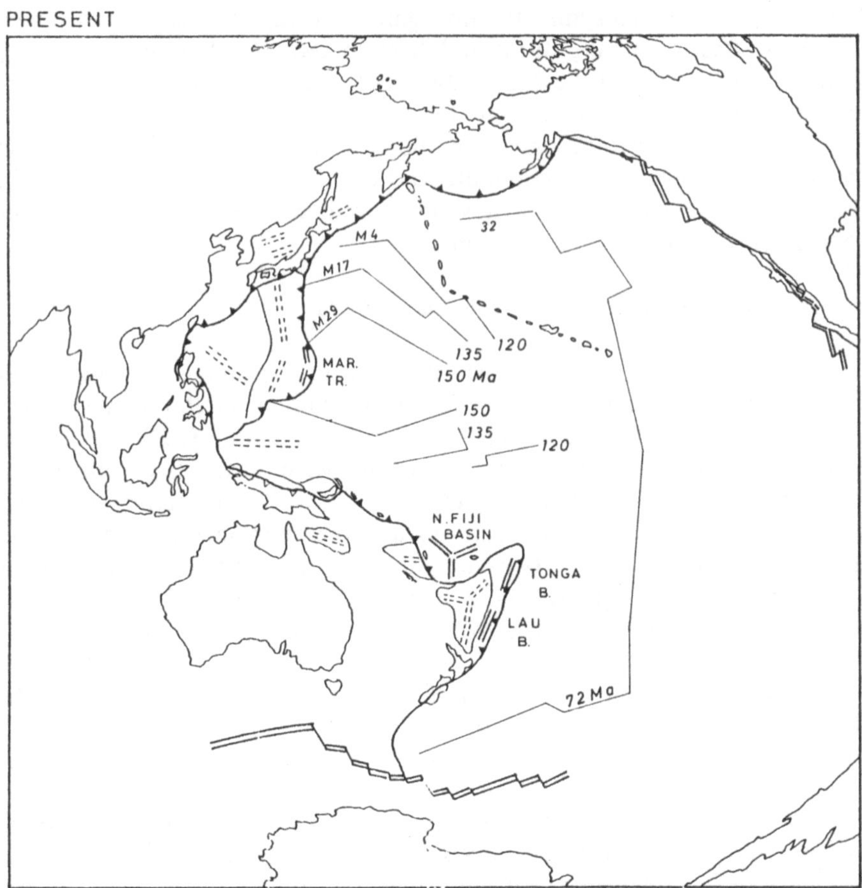

Fig. 2. Age distribution of the Pacific ocean floor at present. Isochrons at 150, 135, 120, and 72 Ma are from the magnetic lineations of the Plate-Tectonic Map of the Circum-Pacific Region (HALBOUTY *et al.*, 1981). Beneath the Mariana arc, the oldest ocean floor is subducting.

4. *Correlation*

From the reconstructions in Figs. 3–6, we can see that around 25 Ma, a young ocean floor of 48 Ma age, which was created by the mid-oceanic ridge system between the Pacific and North New Guinea plates, is entering and older Pacific ocean floor is approaching the western Pacific subduction zones. Although the boundary between the old and young Pacific ocean floors is difficult to depict and has a large uncertainty, the time that this

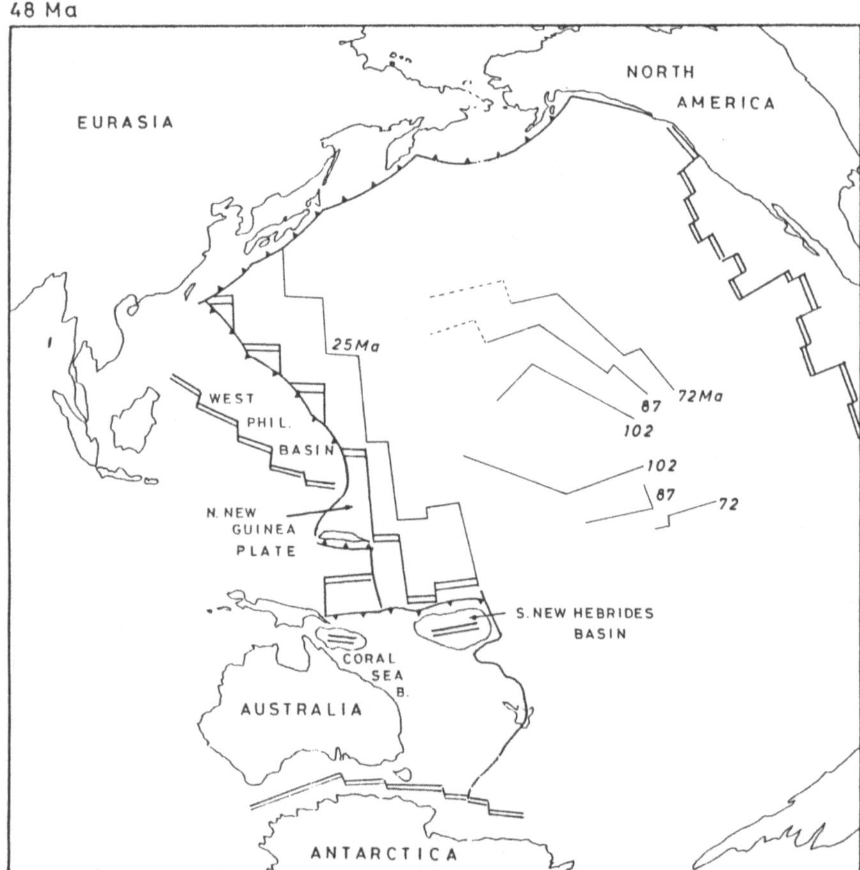

Fig. 3. Reconstruction of the western Pacific at 48 Ma based on the North New Guinea plate hypothesis (SENO, 1984). Isochron of 25 Ma, which shows the young crust formed by the mid-oceanic ridge system between the North New Guinea and Pacific plates during the early Tertiary. Other Pacific isochrons are those in Fig. 2 rotated back to 48 Ma using the Pacific plate motion with respect to hot-spots by ENGEBRETSON (1982).

boundary reached the western Pacific subduction zones would be expected to be around the late Oligocene to middle Miocene. This time corresponds to the major episode of the formation of back-arc basins in the western Pacific. Because the older Pacific ocean continued to subduct in the western Pacific subduction zones, it is not easy to understand why the opening of marginal basins in the western Pacific ceased during the mid to

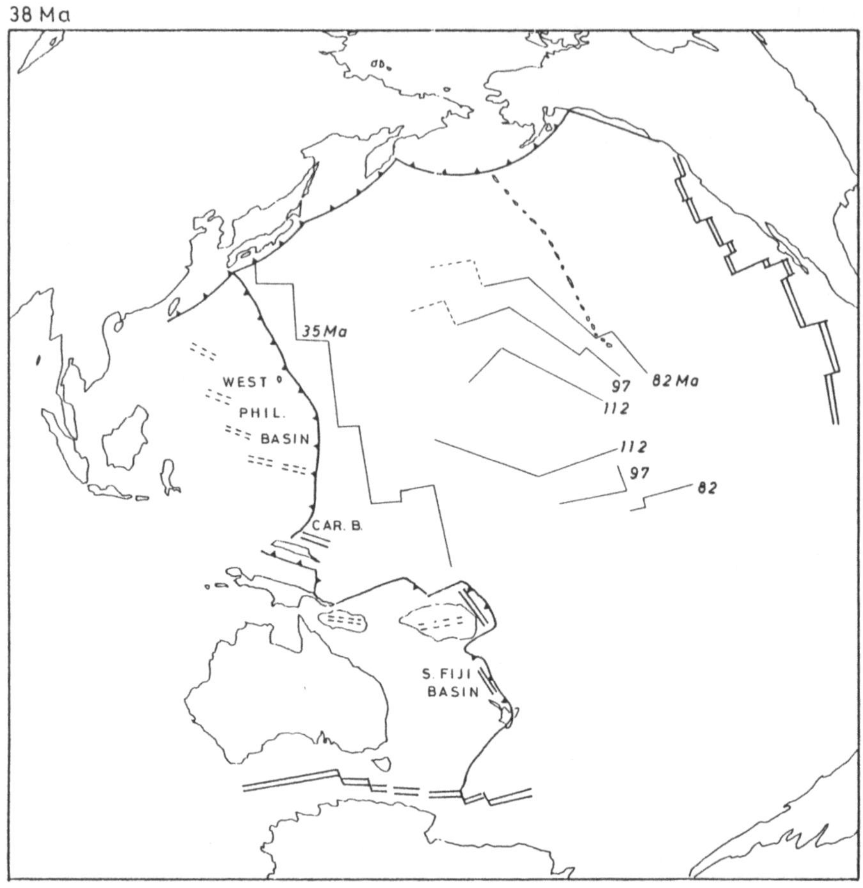

Fig. 4. Age distribution of the Pacific ocean floor at 38 Ma obtained by rotating the isochrons of Fig. 3 using the Pacific plate motion with respect to hot-spots (ENGEBRETSON, 1982). Reconstruction of other tectonic features are from Seno and Peterson (1984). The following figures are similarly reconstructed.

late Miocene. However, it is noted that the oldest Pacific ocean floor approached the Mariana arc around the late Miocene (Fig. 6) when the Mariana Trough started to open.

The reader might wonder why such an abrupt change in age can occur in the middle of a plate. In the above case, this is simply the result of the supposed early Tertiary spreading in the southwestern Pacific. Although the cause of this spreading is not known, the latest Cretaceous–early

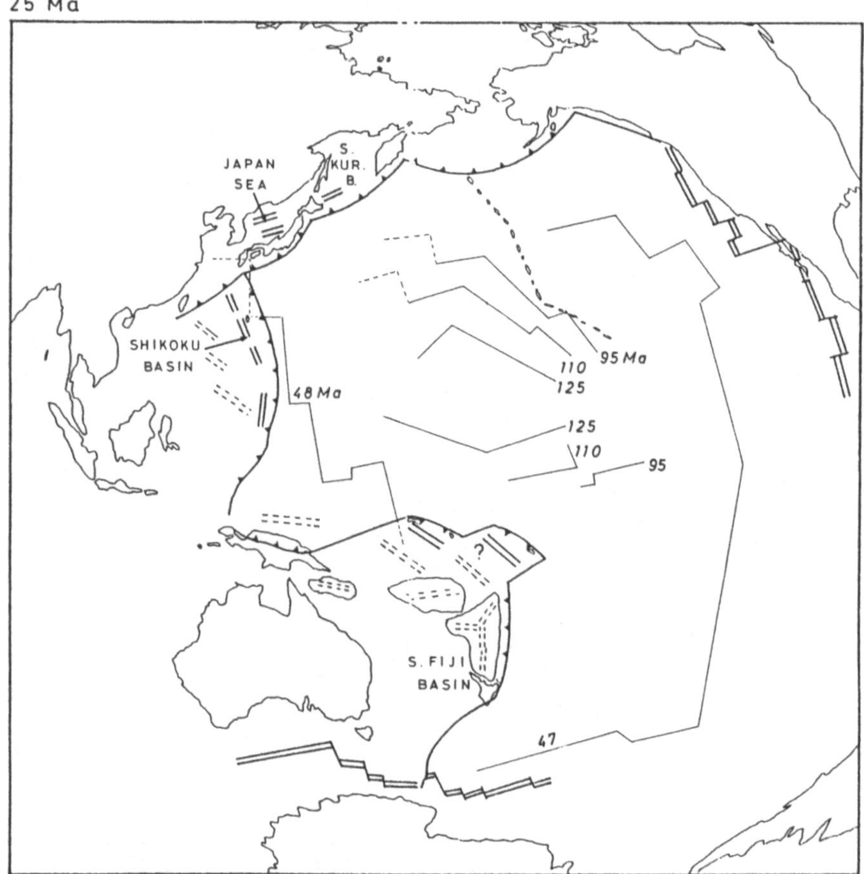

Fig. 5. Age distribution of the Pacific ocean floor at 25 Ma. The older Pacific ocean floor is approaching the western Pacific subduction zones and many back-arc basins started to open around this time.

Tertiary subduction of the Kula-Pacific ridge beneath the northeastern margin of Eurasia (TAIRA, 1981; KIMURA and TAMAKI, 1985a) may be responsible for this creation of a new spreading center in the south.

5. *Discussion and Conclusions*

In this study, I focused on the role of the age of the subducting lithosphere on back-arc basin formation. However, as TAYLOR and

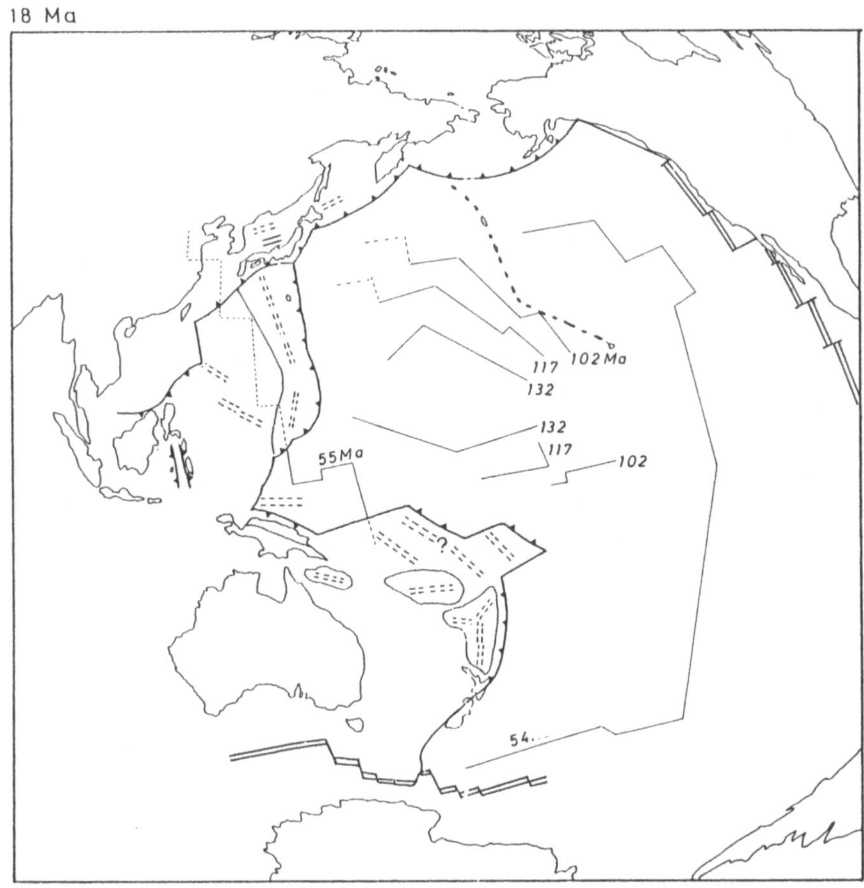

Fig. 6. Age distribution of the Pacific ocean floor at 18 Ma.

KARNER (1983) discussed, no dominant single factor seems to control back-arc spreading. Also, CARLSON and MELIA (1984) showed that both the subduction hinge retreating and advancing occur in the presently active back-arc basins. Recently, SCHOLZ (1984) refined plate-kinematic considerations by UYEDA and KANAMORI (1979) and DEWEY (1980), proposing that two opposed migrating arcs require lateral migration with respect to one another and cause the mantle drag force against slab to reduce the normal stress across the thrust interface, resulting in back-arc spreading. Although this situation seems applicable to some extent to presently active back-arc basins, it does not seem applicable to older episodes of back-arc spreading such as the opening of the Japan Sea, south Kurile, and south Fiji Basins because there was no opposed arc in these areas. Thus the plate-kinematic situation proposed by Scholz is not the dominant factor which controls back-arc spreading.

Because in some cases back-arc spreading occurs when young lithosphere is subducting and in other cases it does not occur even when old lithosphere is subducting, it seems meaningless at first to discuss the correlation between the age of subducting lithosphere and episodes of back-arc spreading. However, it would be expected that if other factors are more or less uniform, the effect of age of the subducting lithosphere may be more evident. In the Philippine Sea subduction zones, the Izu-Bonin back-arc is not spreading as extensively as the Mariana back-arc at present. Because the plate kinematic situation for these arcs is similar, i.e., the retreat of the upper Philippine Sea plate, this might be explained by differences in the age of the subducting lithosphere between these two arcs (Fig. 2), although the difference in age is not large. The lack of active back-arc spreading behind the Nankai Trough and the Philippine Trench and minor active spreading behind the Ryukyu Trench may be due to the difference in the age of the subducting lithosphere beneath these arcs, as is the difference in seismic coupling along these arcs (PETERSON and SENO, 1984). Thus we believe that it is not meaningless to point out the correlation with the age distribution of ocean floor unless some other dominant factor is proposed.

For the Japan Sea and south Kurile Basin, KIMURA and TAMAKI (1985b) proposed that the collision of the Indian subcontinent with Asia and the succeeding retreating movement of the Khanka-Blea massif in eastern China caused the opening of these basins. To consider this as a dominant cause of the opening, it would be necessary to discuss quantitatively the motion of the Khanka-Blea Block in the Tertiary. At present, the early–middle Miocene clockwise and counterclockwise rotations of southwest and northeast Japan revealed by the paleomagnetic data favor the trench retreat for the mechanism of spreading of the Japan Sea than the retreat of the Asian continent they proposed.

Change in age of the Pacific ocean floor during the Oligocene to early Miocene proposed in this paper depends entirely on the North New Guinea plate hypothesis (SENO, 1984). It would be desirable for the age distribution of the already subducted ocean floor to be revealed by other methods. One possible way is to know the age of the oceanic material which has been accreted to the western Pacific subduction zones as melanges. Much younger age would be expected for the ophiolitic rocks in the melanges accreted during the Paleogene in the Nankai Trough and the Izu–Bonin–

Mariana Trenches than for the case that the North New Guinea plate did not exist. Ophiolitic rocks of Paleocene to Eocene ages have been found in the southern Shimanto Belt, which is a Paleogene–Miocene accretionary prism in southwestern and central Japan (SAKAI and KANMERA, 1981; TAIRA *et al.*, 1983; OGAWA, 1983). This may be a fragment of the North New Guinea plate, because such young oceanic material is not expected from the Pacific plate. More detailed study of the middle Tertiary melanges in the western Pacific margins is necessary before further discussion is possible.

The reconstructed age of the Pacific ocean floor shows a change in age of the subducting plate from young to old around the early Miocene. The coincidence of this period with the episode of significant back-arc spreading during the late Oligocene–mid Miocene in the western Pacific suggests that the abrupt change in the age of the subducting lithosphere resulted in the increase of negative buoyancy of the slab and caused the trench retreat and back-arc spreading. Further elucidation of the age distribution of the Pacific ocean floor by other methods is necessary to conclude that the age of the subducting lithosphere is an important factor affecting the back-arc stress regime.

Acknowledgements

I express my thanks to Patric Coleman, Kazuo Kobayashi and Bob Geller for critisism and Kensaku Tamaki, Gaku Kimura, Yujiro Ogawa and Akira Hayashida for discussion.

REFERENCES

R. L. CARLSON and P. J. MELIA, Subduction hinge migration, *Tectonophysics*, **102**, 399–411, 1984.

D. A. CLAGUE and R. D. JARRARD, Tertiary Pacific plate motion deduced from the Hawaiian-Emperor chain, *Geol. Soc. Am. Bull.*, **84**, 1135–1154, 1973.

J. F. DEWEY, Episodicity, sequence and style at convergent plate boundaries, in *The Continental Crust and Its Mineral Resources*, Rep. 20, ed. by D. Strangeway, pp. 553–574, Geol. Assoc. Canada, Tronto, Ont., 1980.

D. ENGEBRETSON, Relative motions between oceanic and continental plates in the Pacific basin, Ph.D. Thesis, Stanford University, 221 pp., 1982.

P. ENGLAND and R. WORTEL, Some consequences of the subduction of young slabs, *Earth Planet. Sci. Lett.*, **47**, 403–415, 1980.

K. FUJITA, and H. KANAMORI, Double seismic zones and stresses of intermediate depth earthquakes, *Geophys. J. R. Astron. Soc.*, **66**, 131–156, 1981.

M. T. HALBOUTY *et al.*, Plate-Tectonic Map of the Circum-Pacific Region, Circum-Pacific Council for Energy and Mineral Res., AAPG, 1981.

A. HAYASHIDA and Y. ITO, Paleoposition of southwest Japan at 16 Ma: implication from paleomagnetism of the Miocene Ichishi group, *Earth Planet. Sci. Lett.*, **68**, 335–342, 1984.

D. M. JURDY, Relative plate motions and the formation of marginal basins, *J. Geophys. Res.*,

84, 6796–6802, 1979.

D. E. KARIG, Origin and developement of marginal basins in the western Pacific, *J. Geophys. Res.*, **76**, 2542–2561, 1971.

G. KIMURA and K. TAMAKI, Tectonic framework of the Kurile arc since its initiation, this volume, 1985a.

G. KIMURA and K. TAMAKI, Collision, rotation, and back-arc spreading: The case of the Okhotsk and Japan Seas, *Tectonics*, 1985b (in press).

K. KOBAYASHI, Opening of the Japan Sea and the displacement of the Japanese islands, *Miner. Geol., Spec. Pub.*, **11**, 23–36 (in Japanese).

P. MOLNAR and T. ATWATER, Interarc spreading and cordileran tectonics as alternates related to the age of subducted oceanic lithosphere, *Earth Planet. Sci. Lett.*, **41**, 330–340, 1978.

P. MOLNAR, D. FREEDMAN, and J. S. F. SMITH, Length of intermediate and deep seismic zones and temperatures in downgoing slabs of lithosphere, *Geophys. J., R. Astr. Soc.*, **56**, 41–54, 1979.

Y. OGAWA, Mineoka ophiolite belt in the Izu forearc area—Neogene accretion of oceanic and island arc assemblages on the northeastern corner of the Philippine Sea plate, in *Accretion Tectonics in the Circum-Pacific Regions*, ed. by M. Hashimoto and S. Uyeda, pp. 245–260, Terra Sci. Publ. Co., Tokyo, 1983.

Y. OTOFUJI, A. HAYASHIDA, and M. TORII, When did the Japan Sea open?: Paleomagnetic evidence from southwest Japan, this volume, 1985.

E. T. PETERSON and T. SENO, Factors affecting seismic moment release in subduction zones, *J. Geophys. Res.*, **89**, 10233–10248, 1984.

L. RUFF and H. KANAMORI, Seismicity and the subduction process, *Phys. Earth Planet. Inter.*, **23**, 240–252, 1980.

C. H. SCHOLZ, Slab drag, extensional arcs, and seismic coupling, preprint, 1984.

J. G. SCLATER, Heat flow and elevation of the marginal basins of the western Pacific, *J. Geophys. Res.*, **77**, 5705–5719, 1972.

T. SENO, Was there a North New Guinea plate?, *Geologic Evolution, Resources and Geologic Hazard*, Rep. Geol. Surv. Japan, No. **263**, 29–42, 1984.

T. SENO and S. MARUYAMA, Paleogeographic reconstruction and origin of the Philippine Sea, *Tectonophysics.*, **102**, 53–84, 1984.

T. SENO and E. T. PETERSON, Fluctuated motions of the Pacific plate and collision episodes in the southwestern Pacific, *EOS*, **65**, 1100, 1984.

S. SHIONO and N. SUGI, Life of an oceanic plate: Cooling time and assimilation time, *Tectonophysics*, **112**, 35–50, 1985.

A. TAIRA, The Process of formation of the Shimanto Belt, *Kagaku*, **51**, 516–523, 1981 (in Japanese).

A. TAIRA, H. OKADA, J. H. McD. WHITAKER, and A. J. SMITH, The Shimanto Belt of Japan: Cretaceous-lower Miocene active-margin sedimentation, *Geol. Soc. Lond. Spec. Publ.*, **10**, 5–26, 1981.

B. TAYLOR and G. KARNER, On the evolution of marginal basins, *Rev. Geophys. Space Phys.*, **21**, 1727–1741, 1983.

TOSHA, Ph.D. Thesis, University of Tokyo, 1983.

S. UYEDA and H. KANAMORI, Back-arc opening and the mode of subduction, *J. Geophys. Res.*, **84**, 1049–1061, 1979.

N. J. VLAAR and M. J. R. WORTEL, Lithospheric aging, instability and subduction, *Tectonophysics*, **32**, 331–351, 1976.

M. YAMANO and S. UYEDA, Possible effects of collisions on plate motions, *Tectonophysics*, **119**, 223–244, 1985.

Formation of Active Ocean Margins, edited by N. Nasu *et al.*, pp. 483–496.

SOFUGAN TECTONIC LINE, A NEW TECTONIC BOUNDARY SEPARATING NORTHERN AND SOUTHERN PARTS OF THE OGASAWARA (BONIN) ARC, NORTHWEST PACIFIC

Makoto YUASA

Marine Geology Department, Geological Survey of Japan, Yatabe, Tsukuba, Ibaraki 305, Japan

Abstract. Longitudinal variations along the Ogasawara Arc are discussed. Included are differences in topographic, geological and structural features and in distribution of hypocenters and back-arc depressions between the northern and southern parts of the arc. The arc is divided into two parts by a tectonic gap, the Sofugan Tectonic Line. The Tectonic Line represents a boundary within a plate in which different parts are subducting at different modes, or a lateral fault with respect to different movements of the Shikoku and Parece Vela Basins.

1. *Introduction*

A broad geomorphological term, the Ogasawara Arc, is used here to indicate a set of N-S trending ridges lying to the west of the Ogasawara Trench. From west to east, the arc consists of the Izu, Shichito and Ogasawara (Bonin) Ridges (Fig. 1). The arc is underlain mainly by volcanic rocks. The length (1500 km) and width (400 km) of the arc are comparable to those of the Honshu Arc. Most of the ridges lie below sea level except for a number of small islands such as the Izu-shichito (Seven Izu Islands) and Ogasawara Islands including Kazan-Retto (Volcano Islands).

Previous studies (e.g., KUNO, 1966) indicate transversal systematic changes in the chemical compositions of volcanic rocks in relation to distances from the trench axis and to the depth of the Benioff Zone. Along-arc variations of chemical compositions as recognized in the Lesser An-

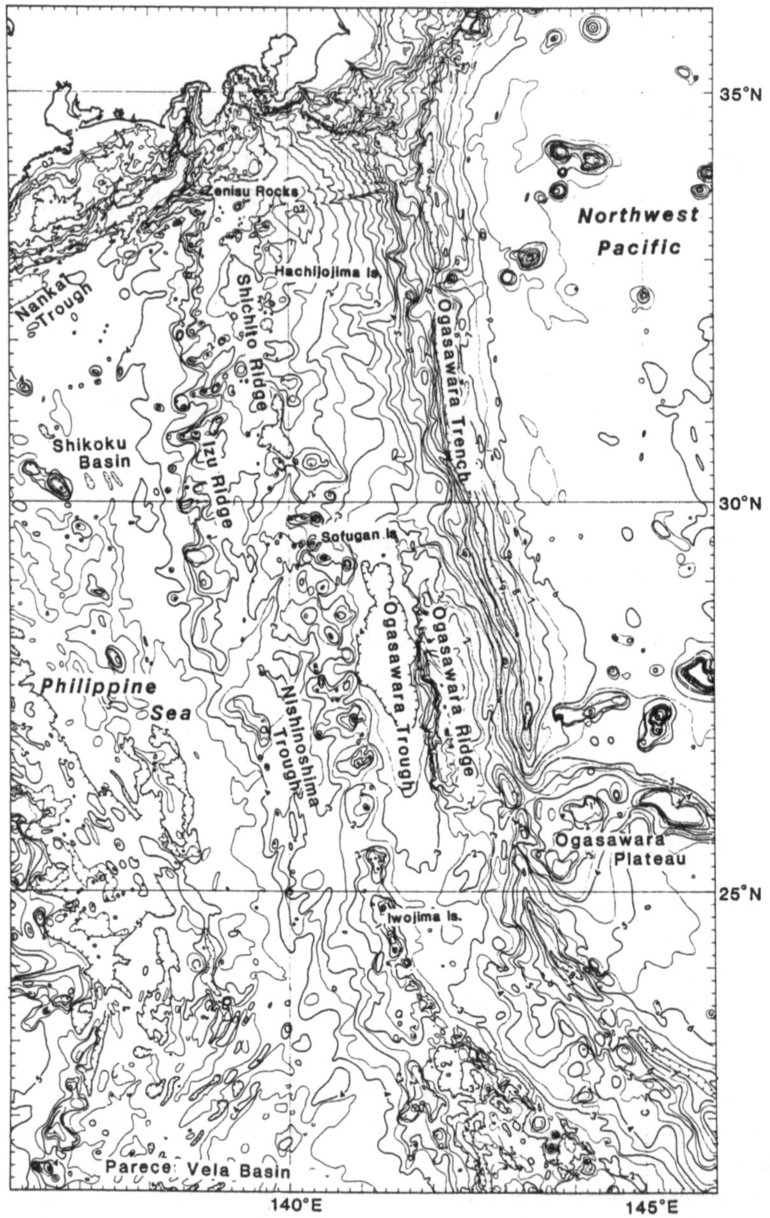

Fig. 1. Bathymetric map of the Ogasawara Arc region (digitized contour line data of Jpan Oceanographic Data Center of Maritime Safety Agency of Japan). Contour interval 500 m.

tiles Arc (BROWN *et al.*, 1977) are also recognized in this arc (YUASA and TAMAKI, 1982).

In this paper geological and geophysical contrasts between northern and southern parts of the Ogasawara Arc and a hitherto unrecognized tectonic gap are described.

2. Geological Setting

The geology of the Ogasawara Arc is shown in Fig. 2 (INOUE and HONZA, 1983).

Geology of the Izu Ridge, the western part of the arc, is less well known than the other ridges in this arc system. Small reefs, the Zenisu Rocks at the northern edge of the Izu Ridge consist of Miocene to Pliocene altered volcanic rocks (ISSHIKI, 1980). Miocene to Pliocene sediments cover most of the ridge (INOUE and HONZA, 1983). A pelitic schist fragment was dredged from an isolated seamount on the ridge (point D369 in Fig. 2). The fist-sized schist is angular in shape with thin manganese oxide coating. Mineral assemblage of the schist is white mica + chlorite + albite + quartz. Origin of the schist has not been explained. No recent volcanic activity is known from this ridge. Fairly young volcanic activity might have been widespread, however, based on an isotopic age data (2.2 ± 1.1 Ma, Table 1) of a block of dredged basalt (point D368 in Fig. 2).

The Shichito Ridge, central axis of the Ogasawara Arc, is situated on a recent volcanic front with many Quaternary volcanoes; from Oshima in the north to Minami-Iwojima Islands in the south. The bimodal chemistry of rocks of this ridge is well-known, especially in the northern islands (e.g., TSUYA, 1973; KUNO, 1962; ONO, 1962). In the southernmost two islands, Iwojima and Minami-Iwojima, alkaline volcanic rocks are known (ISSHIKI, 1976; FUKUYAMA, 1983). YUASA and TAMAKI (1982) pointed out that the volcanic rocks of the Schichito Ridge are divided into two groups on the basis of their alkalinity, i.e., low-alkali tholeiite in the northern part (Oshima to Sofugan Islands) and high-alkali tholeiite (~alkaline rock) in the southern part (Nishinoshima to Minami-Iwojima Islands) (Fig. 3). Although it is widespread in the southern half, no volcanic activity has been recognized at the northern end of the southern half of the ridge (north of the Nishinoshima Island). Northern half of this ridge is topographically displaced westward with respect to the southern half.

The Ogasawara Ridge, eastern part of the arc, consists mainly of Paleogene volcanic and sedimentary rocks. Oligocene high-magnesian andesite, boninite, occurring in Chichijima Island and *Nummulites boninensis* of Eocene from Hahajima Island are well known. Isotopic ages of boninite

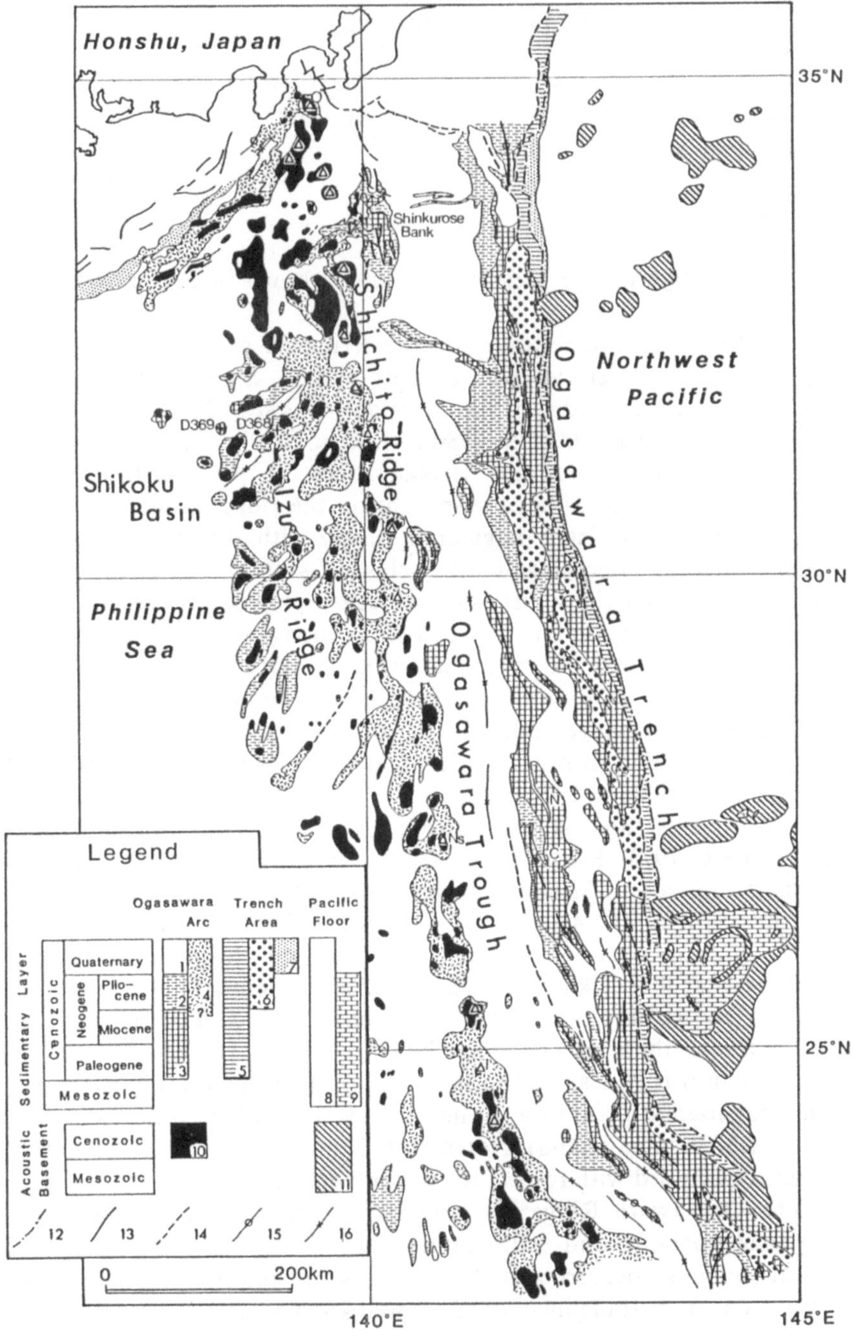
Honshu, Japan
35°N
Shinkurose Bank
Shichito Ridge
Northwest Pacific
Ogasawara Trench
D369
D368
Shikoku Basin
Izu Ridge
30°N
Philippine Sea
Ogasawara Trough
Legend
Ogasawara Arc
Trench Area
Pacific Floor
Sedimentary Layer
Cenozoic
Quaternary
Neogene
Plio-cene
Miocene
Paleogene
Mesozoic
Acoustic Basement
Cenozoic
Mesozoic
25°N
0
200km
140°E
145°E

Fig. 2.

from Chichijima and Nakodojima Islands are 40 Ma (TSUNAKAWA, 1983) and 27 Ma (HONZA *et al.*, 1981), respectively. The Ogasawara Ridge is associated with very high positive free-air gravity anomalies amounting to +400 mgal (ISHIHARA, 1983).

HONZA and TAMAKI (in press) found several small basement highs just east of the Shichito Ridge. They named this series of basement highs the "Shinkurose Ridge". The age of the ridge is estimated to be Miocene to Paleogene by HONZA *et al.* (1982). Gravity free-air anomalies of the Shinkurose Bank, northern end of this ridge are higher than +200 mgal (ISHIHARA, 1983).

Structural framework of the arc is shown in Fig. 4. En echelon structures shown by basement highs are recognized in the northwestern part of the arc, especially within Izu Ridge. Judging from topographic evidence, Quaternary volcanic front appears to be slightly offset at the south of Sofugan Island. To the east of the recent volcanic front lie older volcanic and sedimentary rocks. The high gravity anomaly of the Shinkurose Ridge seems to correspond to that of the Ogasawara Ridge. The Shinkurose Ridge

Table 1. K-Ar age of basalt from Izu Ridge (at D368).

K-Ar age of whole rock

K%	^{40}Ar rad (10^{-6} ccSTP/g)	Atmospheric ^{40}Ar (%)	Age
0.58	0.005	7.4	2.2±1.1 Ma
0.58	0.005	8.4	

constant: $\lambda_\beta = 4.962 \times 10^{-10}/y$, $\lambda_\epsilon = 0.581 \times 10^{-10}/y$

$^{40}K/K = 0.01167$ atom%

(analysed by Teledyne Isotopes, USA)

Fig. 2. Marine geologic map of the Ogasawara Arc region (compiled by INOUE and HONZA, 1983). 1: Unconsolidated sediments, 2: Semi-consolidated sediments and sedimentary rocks, 3: Sedimentary rocks (including Paleogene volcanic rocks of Ogasawara Ridge), 4: volcanic sediments, 5: Unstratified sediments including accretional sediments, 6: Thin sediment cover on the trench slope area, 7: Sediments in trench bottom (composed mainly of turbidites), 8: Pelagic sediments composed mainly of brown clay, 9: Sediments composed of calcareous ooze and limestone on seamount, 10: Volcanic rocks (Quaternary volcanic island in open triangle), 11: Oceanic basalt and acoustic basement stratified, 12: Trench axis, 13: Fault, 14: Concealed or inferred fault, 15: Anticlinal axis, 16: Synclinal axis, C: Chichijima Is., H: Hahajima Is., I: Iwojima Is., M: Minami-Iwojima Is., N: Nakodojima Is., Ns: Nishinoshima Is., O: Oshima Is., S: Sofugan Is., Z: Zenisu Rocks. D368: 31°30.3′N, 139°01.4′E, 1140 m depth, D369: 31°30.5′N, 138°20.4′E, 2310 m depth. Northernmost part of the arc simplified and very minor occurrences of strata omitted.

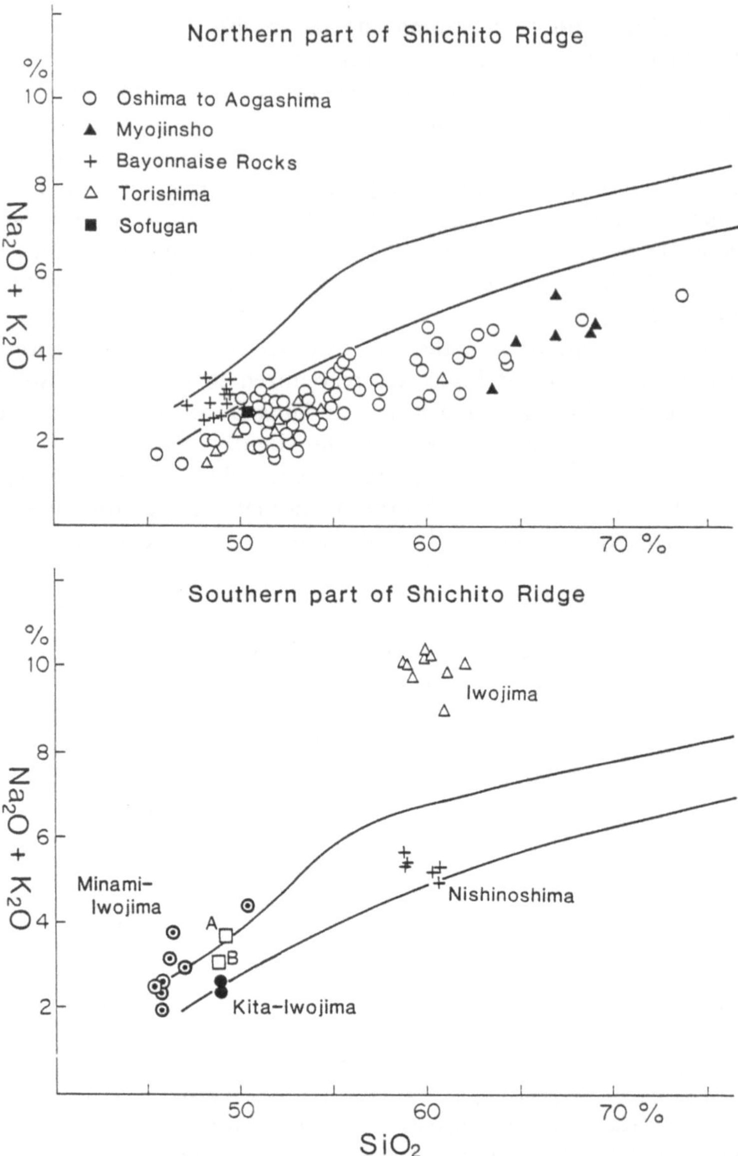

Fig. 3. Alkali-silica diagram of volcanic rocks from northern (upper) and southern (lower) parts of the Shichito Ridge (water free basis; YUASA and TAMAKI, 1982). Additional data are as follows. A: D351 (seamount, 27°59.7′N, 140°07.0′E, 3400 m depth), B: D344-7 (seamount, 26°15.1′N, 140°45.2′E, 1876 m depth), and six plots of Minami-Iwojima Volcano by FUKUYAMA (1983). An analysis from Bayonnaise Rocks is overprinted by that of Sofugan rock (solid square) and other crosses are from dredged rocks from the seafloor near the Bayonnaise Rocks. Localities of the Islands shown in Fig. 4.

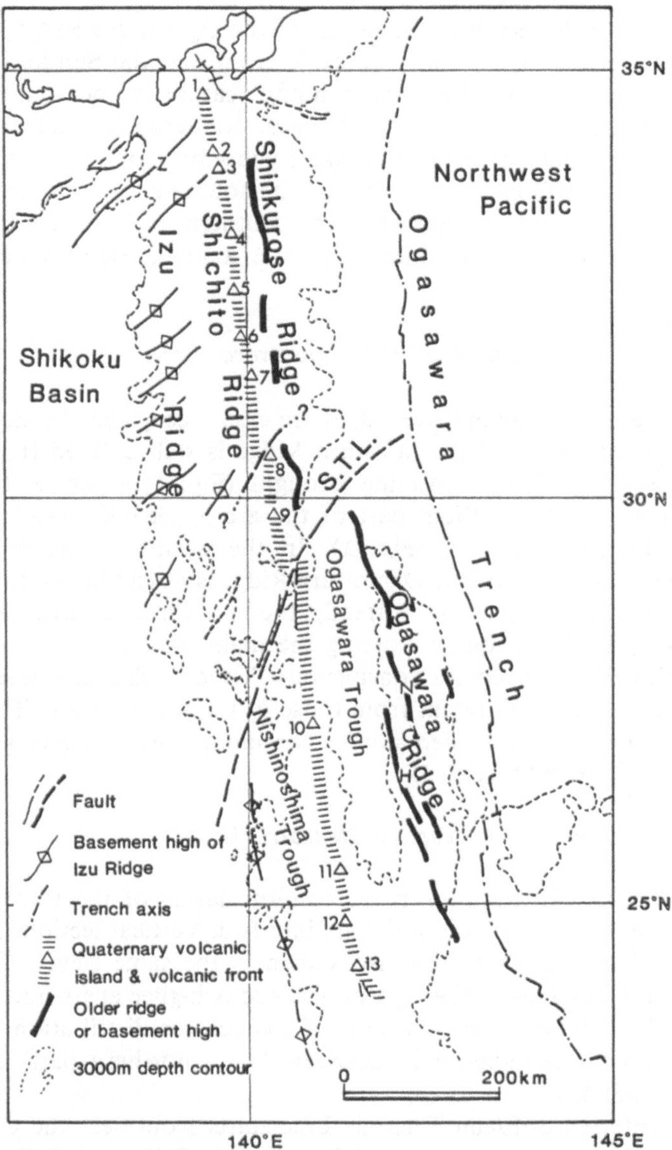

Fig. 4. Structural framework of the Ogasawara Arc region. 1: Oshima Is., 2: Miyakejima Is., 3: Mikurajima Is., 4: Hachijojima Is., 5: Aogashima Is., 6: Myojinsho Reef (Bayonnaise Rocks are at 8 km WSW of the Reef), 7: Smith-jima Is., 8: Torishima Is., 9: Sofugan Is., 10: Nishinoshima Is., 11: Kita-Iwojima Is., 12: Iwojima Is., 13: Minami-Iwojima Is., C: Chichijima Is., H: Hahajima Is., N: Nakodojima Is., Z: Zenisu Rocks, 1–13: on the Shichito Ridge, Z: on the Izu Ridge, N, C, H: on the Ogasawara Ridge. Note that two faults shown by thick broken lines are inferred and that they seem to cause in left lateral displacements of the volcanic front. One of them is named Sofugan Tectonic Line (S.T.L. on the map).

is correlated to the northern extension of the Ogasawara Ridge, although separation between Quaternary volcanic front and the Shinkurose Ridge is less significant than the corresponding feature in the southern part.

As will be mentioned later, submarine topography, back-arc depressions and distributions of earthquake hypocenters of the northern part of the arc are different from those of the southern part. A tectonic gap is inferred separating the northern and southern parts of the arc. I propose a name Sofugan Tectonic Line after a small island situated close to the gap.

3. *Topographic Features of the Ogasawara Arc*

The parallel trending three ridges are easily recognized in the southern part of the arc where the Ogasawara Ridge is well defined (Figs. 1 and 4). Ogasawara and Nishinoshima Troughs (Fig. 1) separate the ridges. Water depth of the northern part of the arc is generally shallower than that of the southern part (Fig. 1). In the northern part of the arc, topographic extension of the Ogasawara Ridge is obscured and three ridges are no longer recognized. Instead, the en echelon arrangement of topographic highs obliquely crossing the main structural trend from Izu to Shichito Ridges becomes conspicuous. Topographic characteristics of the northern arc are different from those of the southern arc. The boundary between the two is located along a continuous depression zone trending at about N30° to 50° E (Fig. 4).

4. *Distribution of Hypocenters Beneath the Arc*

Hypocenter distribution in the eastern margin of the Philippine Sea (Fig. 5) supports separation of the arc into two. Vertical sections for northern part of Mariana Arc are also shown in the same figure. Frequency of micro-earthquakes in the Ogasawara Arc is higher at its northern part than at the southern part. Pattern of hypocenter distribution is clearly different among sections, e.g., between A-A′ (northern part) and C-C′ (southern part).

The inferred Sofugan Tectonic Line starts from near the Ogasawara Trench in the east between the vertical sections B-B′ and C-C′ and continues to the west to reach Izu Ridge (I) crossing the section C-C′. Deep hypocenters shown in the section C-C′ are taken from an area north of the Sofugan Tectonic Line. Boundary of different earthquake frequencies corresponds to the Sofugan Tectonic Line. Hypocenters shown in vertical profiles are continuous for different depths in the northern arc, whereas

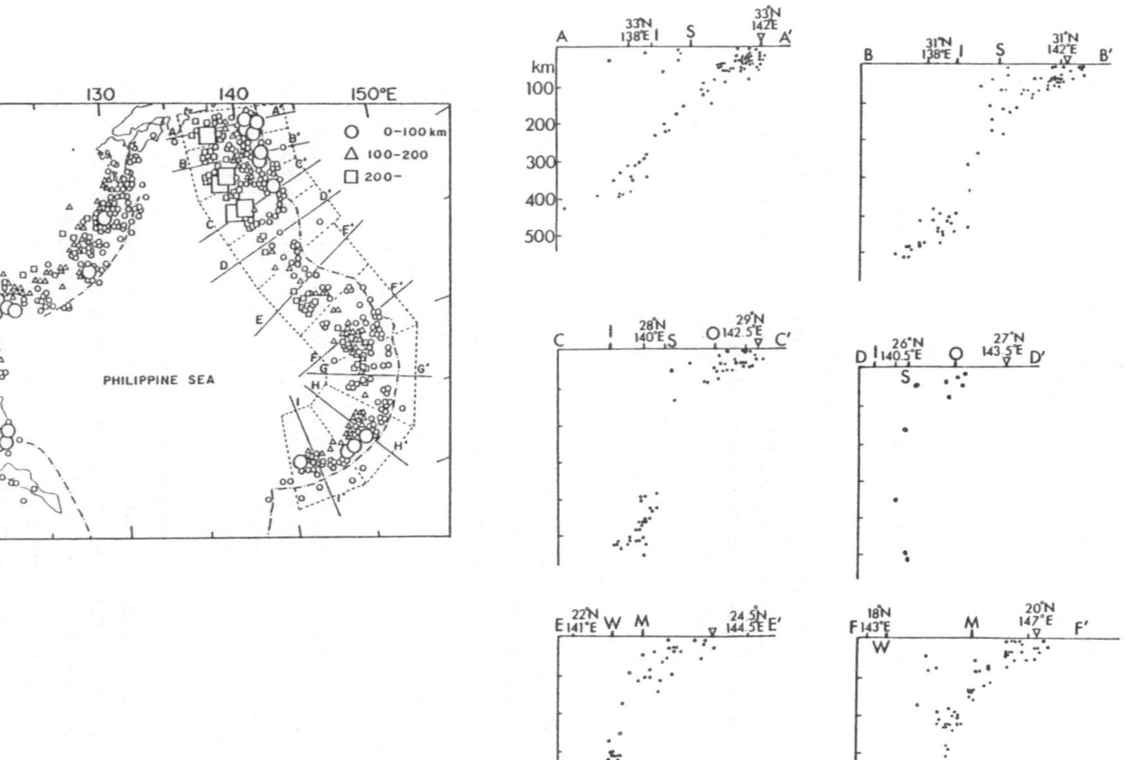

Fig. 5. Distribution of hypocenters in the eastern margin of the Philippine Sea (plan on the left and projections on vertical profiles on the right) (IZUTANI *et al.*, 1975). Large symbols show ten earthquakes and small symbols show one earthquake. Positions of the vertical sections onto which the hypocenters of earthquakes enclosed by dotted lines are projected are shown in solid lines. Vertical sections A-A′ and B-B′. C-C′ and D-D′, and E-E′ and F-F′ correspond to northern Ogasawara Arc, southern Ogasawara Arc, and northern Mariana Arc, respectively. Symbols on the right figures are as follows; I: Izu Ridge, S: Shichito Ridge, O: Ogasawara Ridge, W: West Mariana Ridge, M: Mariana Ridge, reversed triangle: trench axis

mid-depth earthquakes are absent in the southern arc. Mid-depth earthquakes occur in the northern Mariana Arc south of the Ogasawara Arc. This difference is attributed to difference in mode of subduction of the Pacific plate in the northern and southern parts of the Ogasawara Arc.

5. Distribution of Back Arc Depression

TAMAKI *et al.* (1981) found a series of topographically conspicuous depressions to the west of the Shichito Ridge, where the back-arc spreading is possibly taking place (Fig. 6). They named the depressions from north to south, Hachijo, Smith, Torishima and Nishinoshima Depressions. The back-arc depressions are not recognized on the topographic profiles between Nishinoshima and Minami-Iwojima Islands. They concluded that the southernmost part of the Ogasawara Arc has a nature similar to the Chilian-type arc with no back-arc spreading as the Ogasawara Plateau colliding with the Ogasawara Arc, although the Ogasawara Arc as a whole has a similality to Mariana-type arc with back-arc spreading. They did not recognize existence of a transitional zone between the Ogasawara and Mariana arc systems.

When the Ogasawara Arc is divided into two parts by the Sofugan Tectonic Line, the Nishinoshima Depression is located just south of the Tectonic Line. Structure of the Nishinoshima Depression is, however, rather complicated compared to the other three (Fig. 6). Existence of the depression on the west of Shichito Ridge was pointed out by KARIG and MOORE (1975). In the figure shown by them, Nishinoshima Depression is regarded as a depression oblique to the volcanic front rather than the back-arc depression (Fig. 7). As the position of the depression is near to that of Sofugan Tectonic Line, the Nishinoshima Depression is interpreted to be a depression accompanied with Sofugan Tectonic Line. The back-arc depressions other than the Nishinoshima Depression occur only in the northern part of Ogasawara Arc.

The absence of mid-depth earthquakes in the southern arc is interpreted to be due to discontinuous subducting slab. Deep-earthquakes are due to the remnant plate prior to the recent subducting plate which is shown by shallower earthquake activity. The dip represented by the shallower activity is relatively gentle in a similar manner to the Chilian-type subduction. This is the reason why there is no back-arc depression in the southern part of the arc.

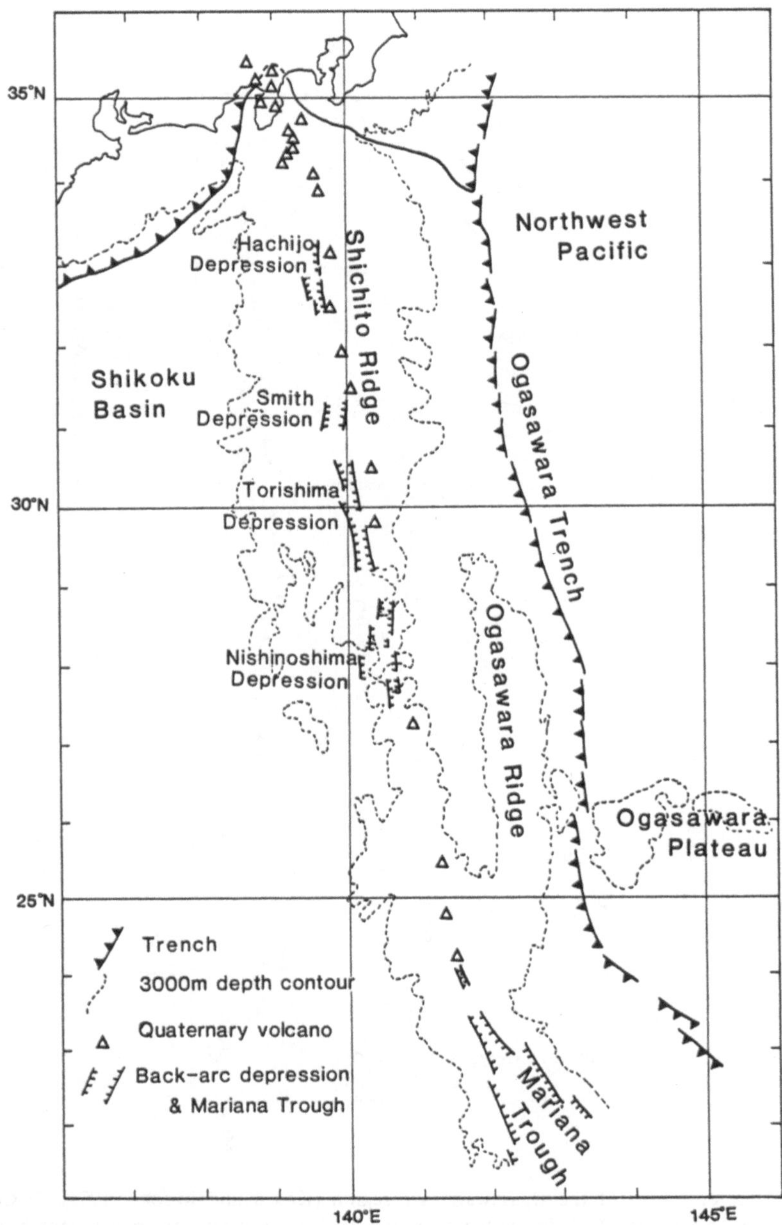

Fig. 6. Distribution of back-arc depressions (TAMAKI *et al.*, 1981).

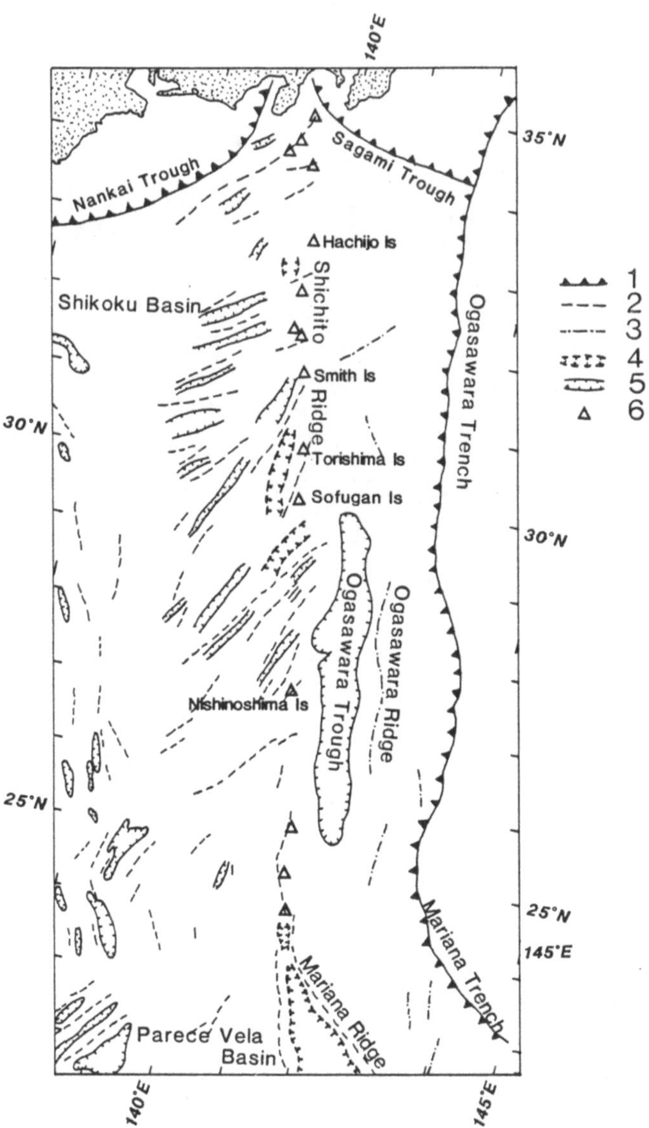

Fig. 7. Tectonic map of the Ogasawara Arc region (KARIG and MOORE, 1975). 1: Consuming plate boundary, 2: Ridge or tectonic high 3: Ridge segment on trench slope break, 4: Suspected young extensional basin, 5: Old trough, 6: Recent volcanic center.

6. *The Sofugan Tectonic Line*

The Sofugan Tectonic Line represents a boundary within a plate in which different parts are subducting at different modes. The Sofugan Tectonic Line is situated across the Ogasawara Arc from trench slope to the Izu Ridge. With respect to the eastern extension of the Tectonic Line, no data are available to trace it to the western Pacific Ocean floor over the trench except for a displacement of the trench axis. Geomorphologic map around the Philippine Sea by IWABUCHI (1982) shows trends of minor ridges from Shikoku to Parece Vela Basins. The trends change from NNW-SSE in Shikoku Basin to NE-SW in Parece Vela Basin near at 24°N. This change of the trend seems to correspond to the western extension of Sofugan Tectonic Line.

The origin of the different structural trends between the Shikoku and Parece Vela Basins is possibly interpreted by difference in opening directions of both basins. The Parece Vela Basin was rotated clockwise with respect to the Shikoku Basin. The movement such as mentioned above, can produce a lateral fault movement between them. The Sofugan Tectonic Line is identified to be the eastern extension of the fault. The Sofugan Tectonic Line itself is not found on air-gun seismic profiles. The exact location and nature of the Tectonic Line remain to be studied.

Acknowledgements

I would like to express my gratitude to the critical reading of the manuscript and constructive discussions by Drs. K. Fujioka (Ocean Res. Inst., Univ. of Tokyo), Y. Kawachi (Univ. of Otago) and T. Watanabe (Shimane Univ.).

REFERENCES

G. M. BROWN, J. G. HOLLAND, H. SIGURDSSON, J. F. TOMBLIN, and R. J. ARCULUS, Geochemistry of the Lesser Antilles volcanic island arc, *Geochim. Cosmochim. Acta*, **41**, 785–801, 1977.

H. FUKUYAMA, Geology of Minami-Iwo Volcano, Izu-Bonin Islands, Japan, *J. Geogr.*, **92**, 55–67, 1983 (in Japanese with English abstract).

E. HONZA, E. INOUE, and T. ISHIHARA (ed.), *Geological Investigation of the Ogasawara (Bonin) and Northern Mariana Arcs, GH79-2, -3, and -4 Cruises, April-August 1979*, Geol. Surv. Japan, 170 pp., 1981.

E. HONZA, K. TAMAKI, M. YUASA, M. TANAHASHI, and A. NISHIMURA, *Geological Map of the Northern Ogasawara Arc, Scale 1:1,000,000*, Geol. Surv. Japan, 1982.

E. HONZA and K. TAMAKI, Bonin Arc, in *Ocean Basin and Margins*, A. E. M. Nairn *et al.*, eds., Plenum (in press).

H. HOTTA, A crustal section across the Izu-Ogasawara Arc and Trench, *J. Phys. Earth*, **18**, 125–141, 1970.

E. INOUE and E. HONZA, *Marine Geological Map around Japanese Islands, Scale*

1:3,000,000, Geol. Surv. Japan, 1983.

T. ISHIHARA, Gravity field around Japan—sea gravimetry by the Geological Survey of Japan, *Marine Geodesy*, **7**, 227–256, 1983.

N. ISSHIKI, Geology and petrography of Iwo-jima (Sulphur Island), Volcano Islands, *Rept. Nat. Res. Cent. Disaster Prevention*, No. 23, 5–16, 1976 (in Japanese with English abstract).

N. ISSHIKI, *Geology of the Mikurajima, Inambajima and Zenisu Districts, Quadrangle Series, Scale 1:50,000*, Geol. Surv. Japan, 35 pp., 1980 (in Japanese with English abstract).

Y. IWABUCHI, Distribution of seamounts around Japanese Islands and seamount chains, *Earth Monthly*, **14**, 70–75, 1982 (in Japanese).

Y. IZUTANI, J. KOYAMA, S. HORIUCHI, and T. HIRASAWA, Tectonical implications of the Izu-Bonin-Mariana Arc from composite mechanical solutions, *Sci. Rep. Tohoku Univ., Ser. 5, Geophys.*, **23**, 1–12, 1975.

D. E. KARIG and G. F. MOORE, Tectonic complexities in the Bonin island arc system, *Tectonophysics*, **27**, 97–118, 1975.

H. KUNO, *Catalogue of the Active Volcanoes of the World Including Solfatara Fields, Part XI, Japan, Taiwan and Marianas*, Intern. Assoc. Volcanol., Roma, 332 pp., 1962.

H. KUNO, Lateral variation of basalt magma type across continental margins and island arcs, *Bull. Volcanol.*, **29**, 165–222, 1966.

S. MURAUCHI, N. DEN, S. ASANO, H. HOTTA, T. YOSHII, T. ASANUMA, K. HAGIWARA, K. ICHIKAWA, T. SATO, W. J. LUDWIG, J. I. EWING, N. T. EDGAR, and R. E. HOUTZ, Crustal structure of the Philippine Sea, *J. Geophys. Res.*, **73**, 3143–3171, 1969.

K. ONO (ed.), *Chemical Composition of Volcanic Rocks in Japan*, Geol. Surv. Japan, 441 pp., 1962 (in Japanese with English abstract).

K. TAMAKI, E. INOUE, M. YUASA, M. TANAHASHI, and E. HONZA, On the possibility of active back-arc spreading of Ogasawara Arc, *Earth Monthly*, **3**, 421–431, 1981 (in Japanese).

H. TSUNAKAWA, K-Ar dating on volcanic rocks in the Bonin Islands and its tectonic implication, *Tectonophysics*, **95**, 221–232, 1983.

H. TSUYA, On the volcanism of the Huzi volcanic zone, with special reference to the geology and petrology of Idu and the Southern Islands, *Bull. Earthqu. Res. Inst., Tokyo Imp. Univ.*, **15**, 215–357, 1937.

M. YUASA and K. TAMAKI, Basalt from Minami-Iwojima Island, Volcano Islands, *Bull. Geol. Surv. Japan*, **33**, 531–540, 1982 (in Japanese with English abstract).

Formation of Active Ocean Margins, edited by N. Nasu *et al.*, pp. 497–515.

REGIONAL GEOLOGY OF THE BERINGIAN CONTINENTAL MARGIN

Michael S. MARLOW and Alan K. COOPER

Pacific-Arctic Branch of Marine Geology, U.S. Geological Survey, Menlo Park, California 94025, U.S.A.

Abstract. The continental margin of North America west of Alaska extends about 1500 km northwestward from the Alaska Peninsula to eastern Siberia; this continental edge is informally known as the Beringian margin. Now isolated from the Pacific Ocean by the Aleutian Island arc, the Beringian margin formerly fronted the northward moving Pacific lithospheric plates during Mesozoic time. Convergence of these oceanic plates beneath eastern Siberia resulted in formation of the northeast-trending Okhotsk-Chukotsk volcanic belt of Late Cretaceous and earliest Tertiary age. Late Mesozoic and early Tertiary plate movement along the southern part of the northwest-trending Beringian margin was mainly by oblique subduction and strike-slip motion. The southern margin is now underlain by the westward extension of the Peninsular terrane of southern Alaska, and exposed accretionary deposits are absent here along the continental slope. However, multichannel seismic reflection data show a thick (9–10 km) sediment wedge buried beneath the continental margin, which may reflect an earlier period of Mesozoic convergence along the margin.

Multichannel seismic reflection data suggest that the central part of the Beringian margin may be underlain by a fossil trench buried beneath the base of the continental slope. This former trench is filled with as much as 7.5 km of trench deposits, and is overlain by 4.0 km of flat-lying beds that underlie the abyssal floor of the Bering Sea basin. Dredge, seismic reflection, and magnetic data also show that arc-type volcanic masses underlie the northern Beringian margin. Calc-alkalic rocks dredged from three sites on one arc structure give K-Ar ages of 50.2, 51.7, 52.8, 53.7, 54.1, and 54.3 m.y., or early Eocene. These arc-type rocks may have resulted from final subduction of Kula (?) plate along the northern margin in earliest Tertiary time. However, at least one arcuate mass extends from the slope into deepwater and is superimposed above presumed oceanic crust. The juxtaposition of an early Tertiary island arc over oceanic crust and against the continental margin is enigmatic.

An alternate origin for the igneous rocks exposed along the margin is that they formed in Eocene time near the tip of the Alaska Peninsula along the ancestral Aleutian arc. A fragment of the arc was then detached from the end of the Eocene Aleutain arc and rafted to the northwest by transform motion along the margin. Emplacement in their present position would have occurred 45 to 50 m.y. ago just before the formation of the western part of the Aleutian arc, which then isolated the Bering Sea from north Pacific plate motion. A third explanation of the volcanic ridges is that they formed in place by volcanism along a leaky transform boundary.

Seismic reflection data show that the outer Beringian shelf is underlain by a major unconformity separating bedrock from an overlying section of flat-lying strata as much as 2 km thick. The association of this unconformity, which is horizontal and appears to be a wave-cut terrace, with overlying marine conglomerate beds of Oligocene age suggests that the margin had begun to subside by that time. Also, during early Tertiary time, the more southerly Aleutian Island arc formed and isolated or trapped Pacific oceanic crust in the abyssal Bering Sea. Formation of the Aleutian arc effectively isolated the Beringian margin from the direct consequences of Pacific-North American plate interaction. The presence of thick sedimentary masses beneath the outer Bering Sea shelf and adjacent abyssal Bering Sea demonstrates that the Beringian margin has been a passive, collapsing margin since early Tertiary time.

1. Introduction

The North American continental margin west of Alaska includes the large shallow shelf extending 600 km from the coast of Alaska to the shelf edge; the continental slope dropping from 200 meters to about 2800 meters of water depth; and a narrow continental rise extending from the base of the slope out into the abyssal Aleutian Basin (Fig. 1). The outer shelf edge and continental slope are incised by several large submarine canyons, including Pribilof, Zhemchug, and Navarinsky Canyons. These canyons debouch much of their sediment load onto broad submarine fans at the base of the continental slope. The base of the continental slope is in places underlain by thick sedimentary masses, which may have significant hydrocarbon potential (COOPER *et al.*, 1979).

Because of national interest in the resource potential of the Bering

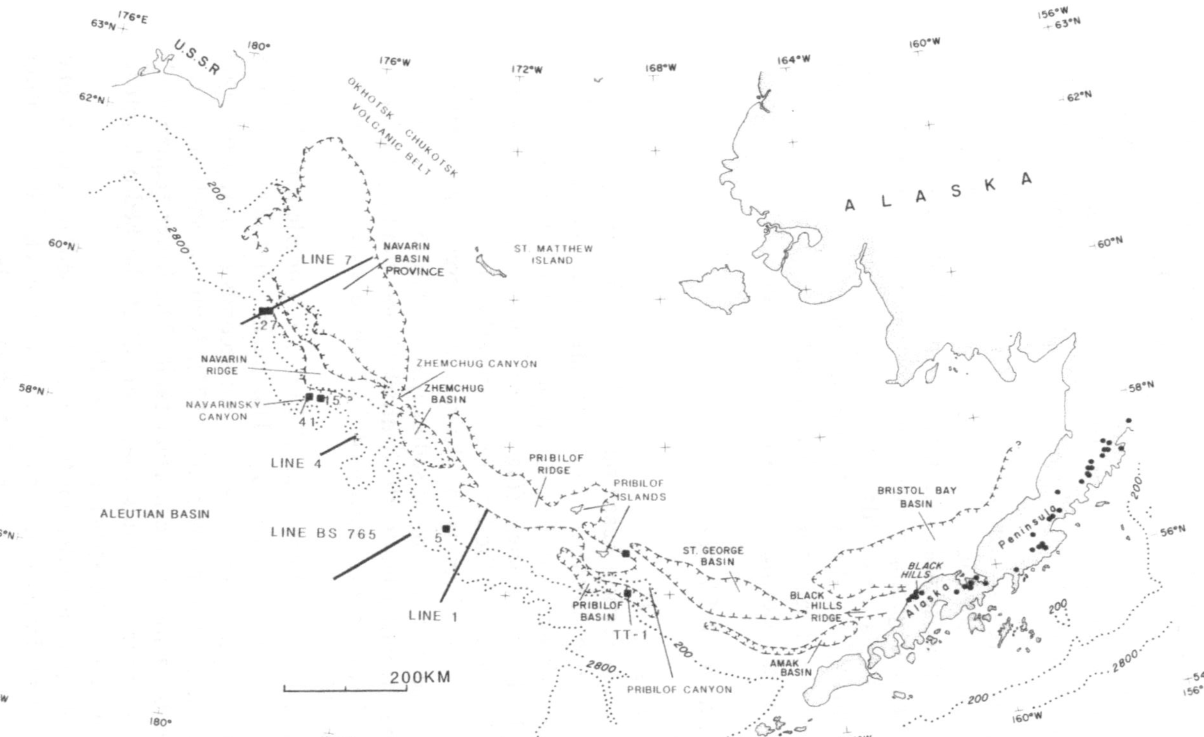

Fig. 1. Index map of the Beringian continental margin extending from the Alaska Peninsula to eastern Siberia. Major sediment-filled basins of the outer shelf are shown by inward-facing hachured lines, and major bedrock ridges beneath the shelf are outlined by outward-facing hachured lines (from MARLOW *et al.*, 1976 and MARLOW and COOPER, 1980a). Location of seismic reflection lines are marked Line 1, etc., and offshore dredge localities with station numbers are shown by black squares. Black circles shown on the Alaska Peninsula are fossil localities in the Naknek Formation (see text). Bathymetric contours are in meters. Albers equal-area projection.

Sea, the U.S. Geological Survey has conducted many oceanographic surveys across the continental margin since 1965. These surveys collected geophysical (seismic reflection and refraction, gravity, magnetic, and bathymetric) and geological sample data from the continental margin. Detailed discussions of the results of these surveys can be found in SCHOLL *et al.* (1975); MARLOW *et al.* (1976); MARLOW and COOPER (1980a); MARLOW *et al.* (1983a).

In this paper we present generalized cross sections of the southern and northern sections of the margin deduced from seismic reflection and refraction data as well as sample information from dredging operations. We discuss the regional geologic differences between the southern and northern margin and relate the differences to plate tectonic models of the Bering Sea. Finally, we discuss the conversion of the margin from an active plate margin during the Mesozoic and earliest Tertiary to an inactive, passive margin during most of the Cenozoic.

2. *Southern Continental Margin*

2.1 *Cross section—Line 1*

Figure 2 is a cross section based on a seismic reflection profile near the Pribilof Islands. The southern continental margin is characterized by a series of bedrock ridges extending from the Alaska Peninsula to the northwest beneath the outer shelf (Fig. 1; see Fig. 8 of MARLOW and COOPER, 1980a). Southeast of the Pribilof Islands the bedrock ridges were rifted during the Cenozoic, forming the elongate St. George basin, a graben that in places is filled with 10 km of sedimentary strata (MARLOW *et al.*, 1977). West of the Pribilof Islands, the bedrock ridges stand high beneath the outershelf and are mildly deformed by limited extensional rifting (MARLOW *et al.*, 1976).

The shelf along the northeast end of Line 1 is underlain by Dalnoi basin, a shallow half-graben (Fig. 2). The basin fill thickens to the southwest, reaching a maximum of about two seconds (two-way time) or two kilometers of strata (MARLOW *et al.*, 1976). Near the shelf edge, the basin is bounded by a high-angle normal fault. The lower sequence all along the basin is broken by faults that also offset the bedrock into a series of block or horst-like structures (Fig. 2).

The outer shelf edge is underlain by a mantle of sediment up to one kilometer thick, which drapes conformably over a bedrock ridge exposing acoustic basement beneath the outer shelf. As discussed below, the bedrock is a western extension of the Mesozoic Peninsular terrane exposed in southern Alaska and along the Alaska Peninsula. The bedrock has been

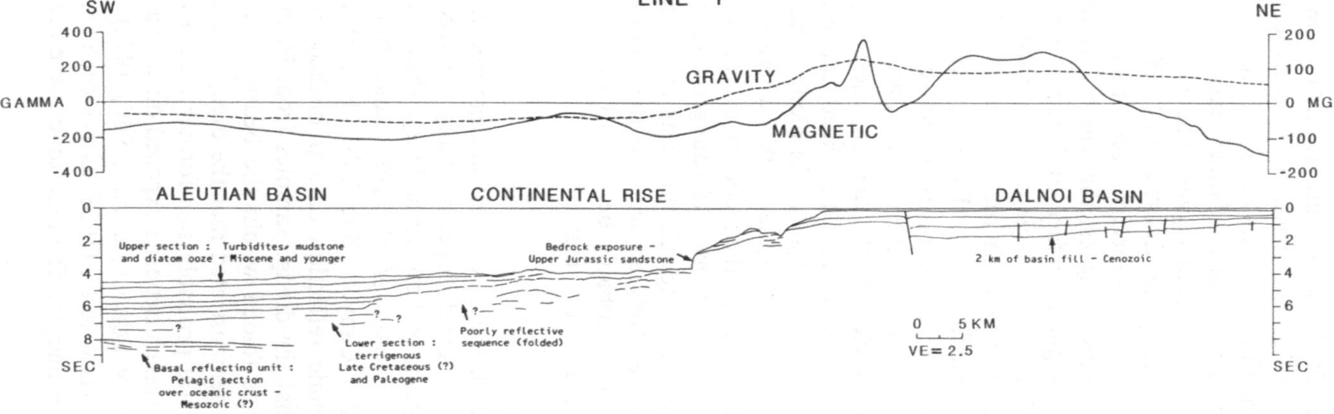

Fig. 2. Interpretative drawing of seismic reflection Line 1 with gravity and magnetic profiles across the southern Beringian margin. See Fig. 1 for location. Travel time is two-way time in seconds. The bold reflector beneath the Aleutian Basin is from a strong reflector thought to immediately overlie oceanic basement.

sampled from exposures in the canyons and steep walls of the continental slope (HOPKINS *et al.*, 1969; MARLOW and COOPER, 1980a).

The southwestern end of Line 1 crosses the eastern edge of the Aleutian Basin in water more than 3000 m deep, where 5 km or more of undeformed sedimentary strata overlie distinct basal reflectors (Fig. 2). Refraction-velocity and magnetic data suggest that the basement is oceanic crust of Mesozoic age (Kula (?) plate; COOPER *et al.*, 1976a, b). Immediately overlying the basement is highly reflective section a few hundred meters thick. Other profiles (not shown) suggest that the highly reflective section blankets and in-fills the irregular surface of igneous oceanic crust, suggesting that the infilling layer is a pelagic deposit, possibly similar to the cherty units that blanket and smooth the surface of Mesozoic igneous crust in the present day Pacific Ocean (EWING *et al.*, 1968). Overlying the pelagic section are moderately to poorly reflective strata that thin and dip gently to the southwest away from the margin. Because of this geometry and because the deposits are close to the margin, these deposits are interpreted to be mainly Upper Cretaceous (?) and Paleogene terrigenous debris. The upper section beneath the Aleutian basin along Line 1 may be correlative with the section at DSDP Hole #190, where Creager, SCHOLL *et al.* (1973) drilled through more than 600 m of turbidites, mudstone, and diatomaceous ooze of late Miocene and younger age. Strata equivalent in age and lithology presumably underlie the basin along the southwestern end of Line 1. Strata immediately adjacent to the continental slope are poorly reflective and arched, probably because of folding along a deformational front at the base of the margin.

2.2 *Bedrock dredge data*

Bedrock ridges beneath the outer shelf are exposed near the Pribilof Islands and along the continental slope; dredging has recovered sedimentary rocks of Kimmeridgian, or Late Jurassic age (Sites DR1 and 5, Fig. 1; VALLIER *et al.*, 1980; JONES *et al.*, 1981) and mudstones of Late Cretaceous age in Pribilof Canyon (Site TT-1, Fig. 1; HOPKINS *et al.*, 1969). Multichannel seismic reflection data published by MARLOW and COOPER (1980b) show that the dredge samples from Pribilof Canyon are from a stratified sequence folded within the bedrock beneath the outer margin. MCLEAN (1979) demonstrated that the dredged mudstones are not intensely deformed and are probably not former trench deposits similar to deformed Cretaceous trench deposits exposed on Kodiak Island, a supposition once proposed by SCHOLL *et al.* (1966, 1975), MOORE (1972, 1973a, b, 1974), and MARLOW *et al.* (1976) as evidence for Mesozoic subduction along the southern margin. The mudstones were probably deposited in an outer shelf basin, which was later folded into the bedrock fabric of the outer margin.

In the Pribilof Islands, bedrock is exposed as a serpentinite body along the southern shore of St. George Island (BARTH, 1956). The serpentinite mass is intruded by a quartz diorite dike, which has been dated by K-Ar methods as 50–57 m.y. old, implying that the serpentinite body is Eocene or older (HOPKINS and SILBERMAN, 1978). SCHOLL *et al.* (1975) suspect from seismic reflection data that the intruded serpentinite mass is Mesozoic in age and is part of the bedrock underlying the outer margin. Along nearby Line 1 (Fig. 2) there is a 400 gamma magnetic anomaly across the outer edge of the shelf. The high magnetic gradient may reflect a buried ultramafic body within the bedrock beneath the shelf edge, similar to the serpentinite mass exposed on St. George Island.

The bedrock ridges beneath the southern shelf can be traced on seismic reflection records from near the onshore Jurassic exposures of the Naknek formation near the tip of the Alaska Peninsula west to the Jurassic dredge sites near the Pribilof Islands and adjacent continental slope (Sites DR1 and 5, Fig. 1). The onshore Jurassic rocks of the Naknek Formation exposed on the Alaska Peninsula are part of the Peninsular terrane, which extends along the Peninsula to the east into southern Alaska. The offshore extension of these Jurassic rocks west to the continental slope suggests that much of the outer southern Bering Sea shelf is part of the allochthonous Peninsular terrane exposed on the Alaska Peninsula and in southern Alaska (MARLOW *et al.*, 1979; MARLOW and COOPER, 1983).

3. *Northern Continental Margin*

In this section we present two cross sections and associated dredge data from the northern margin (Figs. 3, 4, and 6) that show distinct features characteristic of the northern margin and that differ from those of the southern margin. The dredge, seismic reflection, and magnetic data suggest that the northern margin, unlike the southern margin, is not underlain by the Peninsular terrane of southern Alaska.

3.1. *Line 4*

The first section, Line 4, crosses the base of the margin near Zhemchung Canyon (Figs. 1, 3, and 4). At the southwestern or seaward end of the line a strong, smooth reflector (reflector R) occurs at a sub-bottom depth of about 3 seconds (two-way time). This strong basal reflecting sequence, like the deep reflecting sequence observed on the southwest end of Line 1 (Fig. 2), is probably a pelagic layer overlying oceanic crust of

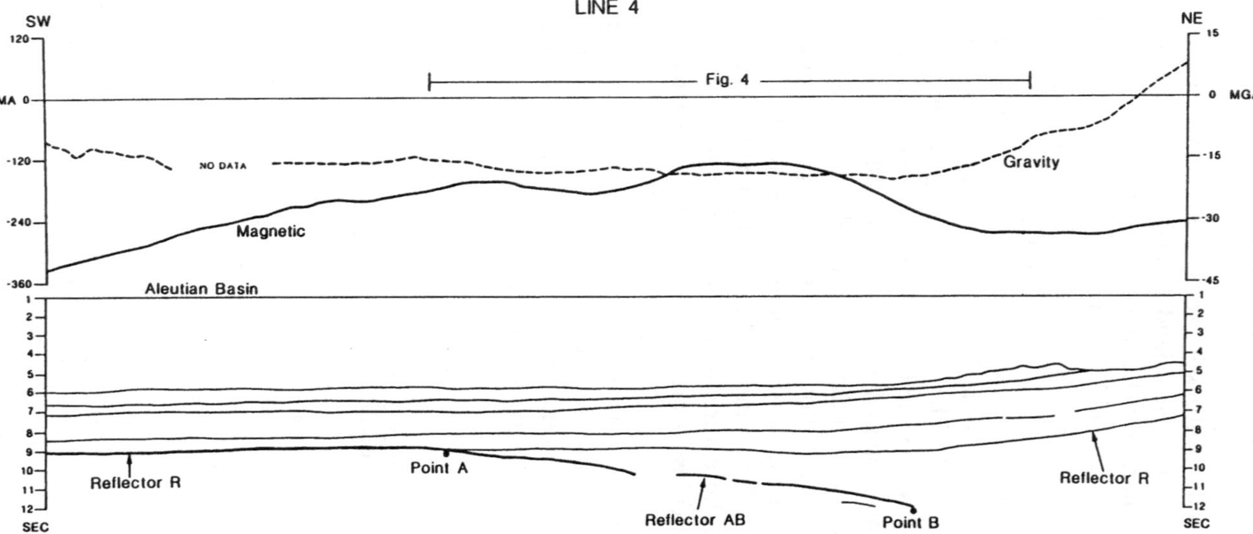

Fig. 3. Interpretative drawing of seismic reflection Line 4 with gravity and magnetic profiles across the northern beringian margin. See Fig. 1 for location. Travel time is two-way time in seconds. Note that the bold reflector labeled R can be traced up towards the margin, unlike the strong reflector recorded on Line BS-765 (Fig. 5). A deep reflector extends from point A to point B, dipping down towards the margin beneath reflector R.

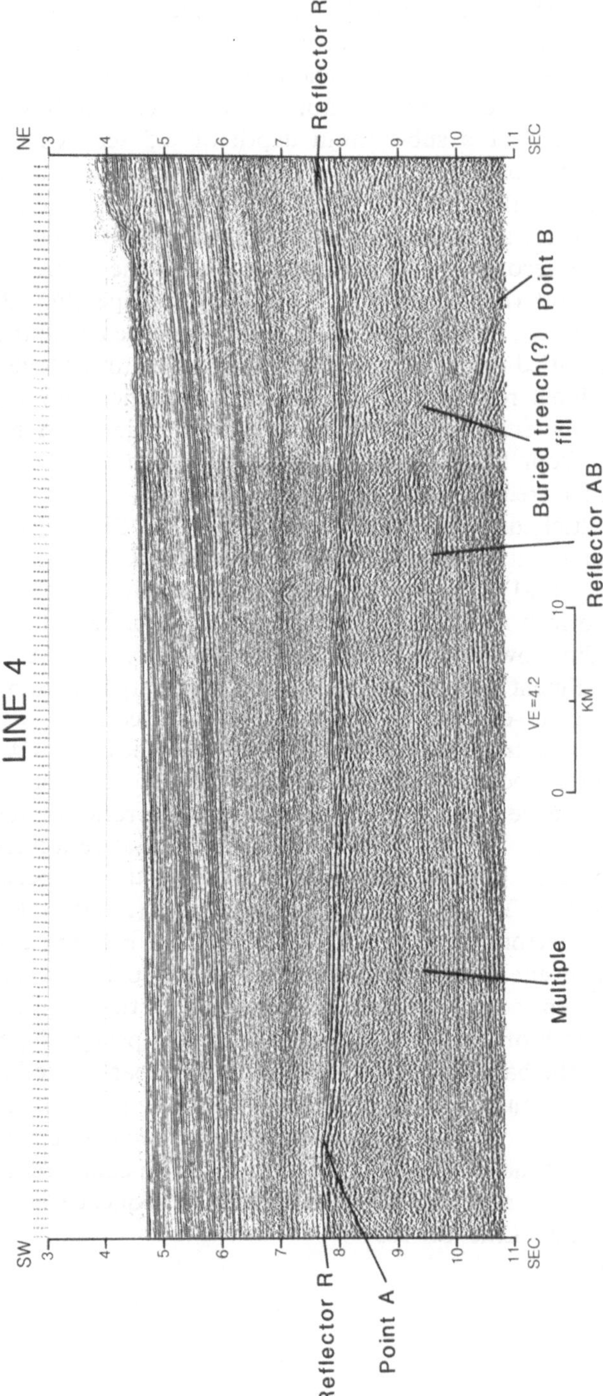

Fig. 4. Detail of seismic reflection profile for part of Fig. 3, showing reflector R and the deep reflector beneath R dipping down towards the margin between points A and B.

Mesozoic age (COOPER *et al.*, 1976a). Near the middle of the profile the reflector R rises slightly to a sub-bottom depth of slightly less than 3 seconds (Point A, Fig. 3), where a second deeper reflector begins, dipping towards the margin. The lower reflector can be traced landward to the bottom of the record at a sub-bottom depth of 6.5 seconds (Point B, Fig. 3), some 50 km from where the two reflectors merge seaward at Point A. Between the two reflectors an acoustically transparent section thickens landward and reaches a maximum recorded thickness of 3 seconds (4–5 km). The upper 3 seconds (4 km) of section across the entire profile is highly reflective and rises towards the continental slope (Fig. 3).

The deeply buried, acoustically transparent sequence above the reflector AB may be a buried relict trench deposit downbowed toward the margin. The upper prominent reflector R, thought to be pelagic deposit capping oceanic crust at the seaward or western end of the line, can be traced sub-horizontally from Point A (Fig. 3) towards the margin and above the so-called relict trench fill.

Another multichannel seismic reflection line, Line BS-765 crosses the southern Beringian margin some 150 km southeast of Line 4 (Figs. 1 and 5; COOPER *et al.*, 1981). This crossing shows a prominent oceanic basement reflector (horizon A′B′) and overlying pelagic layer beneath the Aleutian Basin that dips toward the margin. These reflectors can be followed beneath thick sediment wedges at the base of the margin to sub-bottom depths of 9–10 km (5–6 s two-way time). The thick sediment wedge and down-bowed oceanic crust have been cited by COOPER *et al.* (1981) and BEN-AVRAHAM and COOPER (1981) as evidence of late Mesozoic and early Tertiary convergence along the central margin. The reflector R of Line 4 (Figs. 3 and 4) may be equivalent to the deep, basal reflectors above horizon A′B′ of Line BS-765, although we have no tie line between the two margin crossings. These two reflective sequences (reflector R of Line 4 and the basal reflectors above horizon A′B′ of Line BS-765) may represent some special event such as pelagic deposition throughout an ancestral Aleutian Basin, even though the crustal framework of the basin may have been far to the south of its present position. If such a pelagic depositional event did blanket the basin, then the acoustically transparent section below reflector R recorded on Line 4 (Figs. 3 and 4) is older than the thickened and buried sequence overlying the equivalent basal reflectors above horizon A′B′ observed on Line BS-765 (Fig. 5). Thus, the so-called relict trench deposit along Line 4 may pre-date the thickened sequence at the base of the margin along Line BS-765.

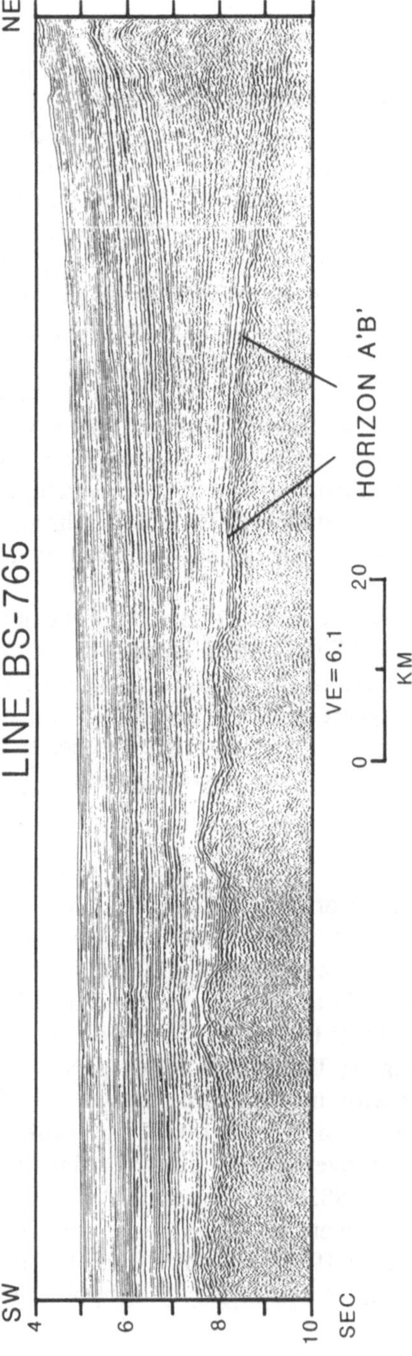

Fig. 5. Multichannel seismic reflection Line BS-765 shot by exxon Corporation and adapted from COOPER *et al.* (1981). This line crosses the southern continental margin near Line 1 (see Fig. 1). Travel time is two-way time in seconds. Note the deeply buried, strong reflector beneath the Aleutian Basin that dips landward beneath the margin.

3.2 *Line 7*

Line 7 crosses the northern continental margin west of St. Matthew Island and some 300 km north of Zhemchug Canyon (Fig. 1). An interpretive drawing of Line 7 (Fig. 6) shows that the south-western end of the profile crosses the eastern edge of the Aleutian basin in water more than 3000 m (4.0 s) deep, where 4–5 km of undeformed sedimentary strata overlie a distinct acoustic basement. Again, like Lines 1 and 4, magnetic and refraction-velocity data suggest that the basement is oceanic crust of Mesozoic age (Kula (?) plate; COOPER *et al.*, 1976a, b).

Below the upper part of the continental slope the rock sequence is characterized acoustically by scattered and discontinuous reflectors (Fig. 6). A strong basal reflector, part of the sub-shelf bedrock, can be traced from the northeast seaward to the middle of the continental slope where the basement crops out. Dredged conglomerates and a flat-top subshelf basement suggest that the bedrock surface of the outer shelf is a wave-base unconformity cut across broadly folded rock sequences in early Tertiary time. These dredge and seismic reflection data, discussed more fully below, attest that the subshelf basement has subsided at least 1500–1700 m since about Oligocene time.

Bedrock beneath the shelf can be followed to the northeast below the thick sedimentary section filling Navarin basin (Fig. 6). Near the north-eastern end of the profile strata in the basin are nearly 12 km (7 s) thick. Dredge data (see below) suggest that the upper 3–4 km of the beds in the basin are younger than early Eocene in age. Deeper strata are poorly reflective and diverge in dip from the overlying strata (MARLOW *et al.*, 1976; MARLOW, 1979).

3.3 *Dredge data*

Rocks dredged from the northern margin include a diverse suite of Mesozoic sedimentary and Cenozoic sedimentary and igneous samples. Muddly limestones of Late Jurassic and Late Cretaceous age (Site 22; Fig. 1) and sandy siltstone of Late Cretaceous age (Site 27, Fig. 1; JONES *et al.*, 1981) were recovered from the bedrock slope of the northern margin in 1978 (Sites 22 and 27, Fig. 1; The Mesozoic samples are probably from slope outcrop below the major unconformity shown on Line 7 (Fig. 6; JONES *et al.*, 1981). However, a second sampling cruise in 1982 along the same deepwater slope recovered only early Tertiary sandstones and mudstones (MARLOW *et al.*, 1983b).

A suite of igneous rocks were recovered from an arcuate ridge just north of Zhemchug Canyon in 1978 and 1982 (Sites 15, 40, and 41, Fig. 1). K-Ar dates on these samples are 50.2 m.y. (rhyolite or dacite), 51.7

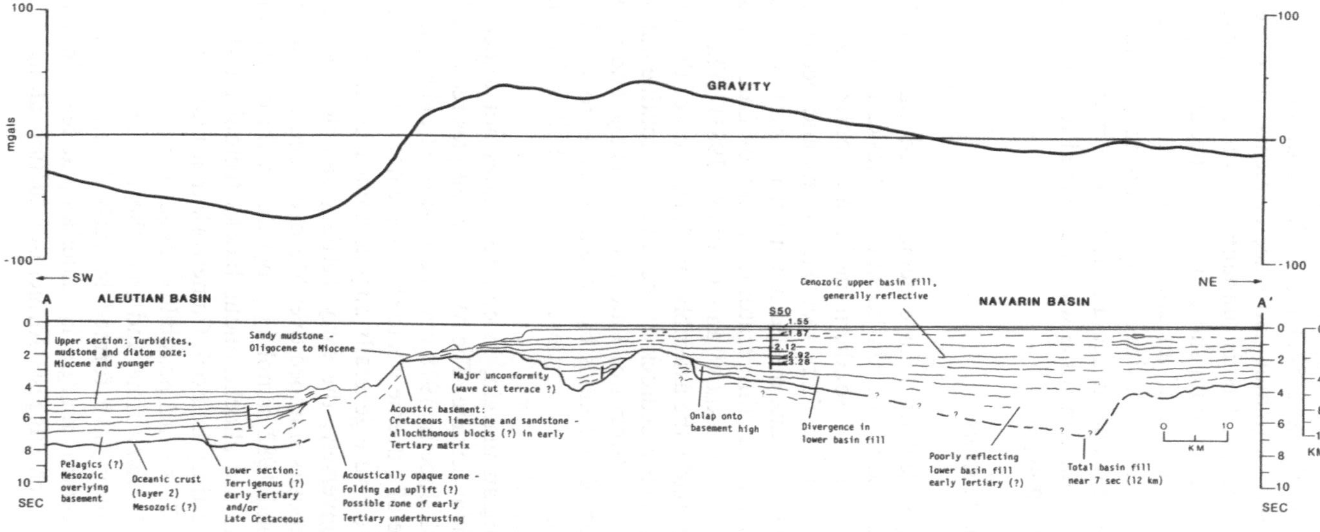

Fig. 6. Interpretative drawing of seismic reflection Line 7 and gravity profile across the northern Beringian margin. See Fig. 1 for location. Travel time is two-way time in seconds. this line crosses the margin near two dredge stations, 22 and 27, whose locations are shown in Fig. 1. Note the strong reflector beneath the Aleutian Basin can be traced as a flat reflector to near the base of the margin into an acoustically opaque zone. Note also the thick sedimentary fill in Navarin basin that extends to a subshelf depth of about 12 km. Adapted from MARLOW *et al.* (1982).

m.y. (rhyolite or dacite), 52.8 m.y. (rhyolite or dacite), 53.7 m.y. (rhyolite or dacite), 54.1 m.y. (basalt), and 54.3 m.y. (dacite or latite), or early Eocene. Geochemical analyses (major and trace elements) of these rocks show calc-alkaline or island arc trends (unpublished data). Bathymetric, magnetic, and seismic reflection data show that the igneous samples were collected from the northern end of an arcuate ridge that extends from the slope into deep water, where the ridge overlies presumed oceanic crust. The ridge is thus earliest Eocene in age and is built in part on an older oceanic section; the ridge may, in some way, be related to early Tertiary plate motions along the margin.

4. *Discussion and Conclusions*

Dredging and geophysical data have demonstrated that the Peninsular terrane extends from southern Alaska and the Alaska Peninsula offshore to the west beneath the southern beringian continental margin (MARLOW and COOPER, 1980a). Plate reconstructions and geologic data suggest that the Peninsular terrane moved north across the Pacific during the Mesozoic and had docked against cratonic Alaska by Late Cretaceous time (CSEJTEY *et al.*, 1982; MARLOW and COOPER, 1983). The time of docking presumably also marks the assembly and formation of the southern Beringian margin. The Jurassic bedrock of the southern continental margin contains locally infolded mudstones of Late Cretaceous age and is at least locally intruded in the Pribilof Islands by Eocene igneous rocks. The offshore Peninsular terrane extends northwest from the Pribilof Islands to Zhemchug Canyon (Fig. 1).

The northern Beringian margin, in contrast, does not appear to be underlain by the Peninsular terrane. The Mesozoic samples recovered from the northern continental slope in 1978 may be ice-rafted debris, although no striations or evidence of extreme weathering were noted in the samples, and the samples recovered were generally angular indicating they had been broken off outcrop; conversely, the early Tertiary samples recovered in 1982 may be upslope deposits slumped over Mesozoic bedrock, which was not reached in the second 1982 sampling program. A third possibility is that the Mesozoic rocks are allochthonous blocks tectonically-folded or thrust within an early Tertiary bedrock framework that underlies the northern continental slope. This third possibility suggests that the continental slope formed in early Tertiary time and that the adjacent shelf basins such as Navarin basin may have formed in early Tertiary time.

The northern and central margin is underlain, at least in part, by relict trench deposits, which suggests former subduction along this part of the margin (Figs. 1, 3, and 4; COOPER *et al.*, 1981). Also, the northern margin is capped in part by volcanic, island arc masses of Eocene age, which may also reflect early Tertiary plate motions along the northern margin. Alternatively, the Eocene igneous rocks along the northern margin may be related to volcanic outpourings along a leaky transform margin that would be analogous to the southern California borderland (HIGGINS, 1976). The northern margin may also in part be composed of allochthonous Mesozoic sedimentary and igneous masses embedded in an early Tertiary matrix. The allochthonous fabric of the northern margin suggests that its formation may have continued into early Tertiary time, incorporating Mesozoic rocks into the continental margin by a combination of obduction and transform motion along the margin. These allochthonous bodies could include relict trench deposits and fragments of remnant island arcs. We should note here that the bathymetric trend of the northern margin is directed more east-west than the northwest-trending southern margin, which may reflect a fundamental difference in the structural style of the two segments.

In Fig. 7 we have constructed in schematic fashion a possible model for the Mesozoic and Cenozoic formation of the Beringian margin adapted from MARLOW *et al.* (1982). Before 60 m.y. ago, the margin was obliquely underthrust by the Kula (?) plate. Between 50 and 60 m.y., oroclinal bending of western Alaska and resultant clockwise rotation of the margin caused a change in plate motion to transform or strike-slip movement between the North American and Kula (?) plates. Transform motion may have carried a fragment of the Eocene Aleutian volcanic arc to the northwest more than halfway up the margin, emplacing the fragment along the continental slope. Formation of the western part of the Aleutian arc between 40 and 50 m.y. ago and shifting of plate motion south to near the present Aleutian Trench isolated a piece of Kula (?) plate, thus forming the Bering Sea. Isolation of the Bering Sea resulted in sediment loading of the adjacent remnant Kula (?) plate (now the Aleutian basin) and deactivation of plate motions along the Beringian margin. Continued sediment loading of the Aleutian basin since eary Tertiary time has caused the collapse and rifting of the outer Beringian margin, forming the large and thickly buried sedimentary basins of the outer Bering Sea shelf.

Acknowledgements

We thank P. Carlson and J. Eaby for their thoughtful reviews. T. L. Vallier and L. B. G. Pickthorn provided the petrographic analyses and age dates for the many of the dredge samples. W. Patton, III suggested some of the plate motions depicted in Fig. 7, but we assume all responsibility for the figure.

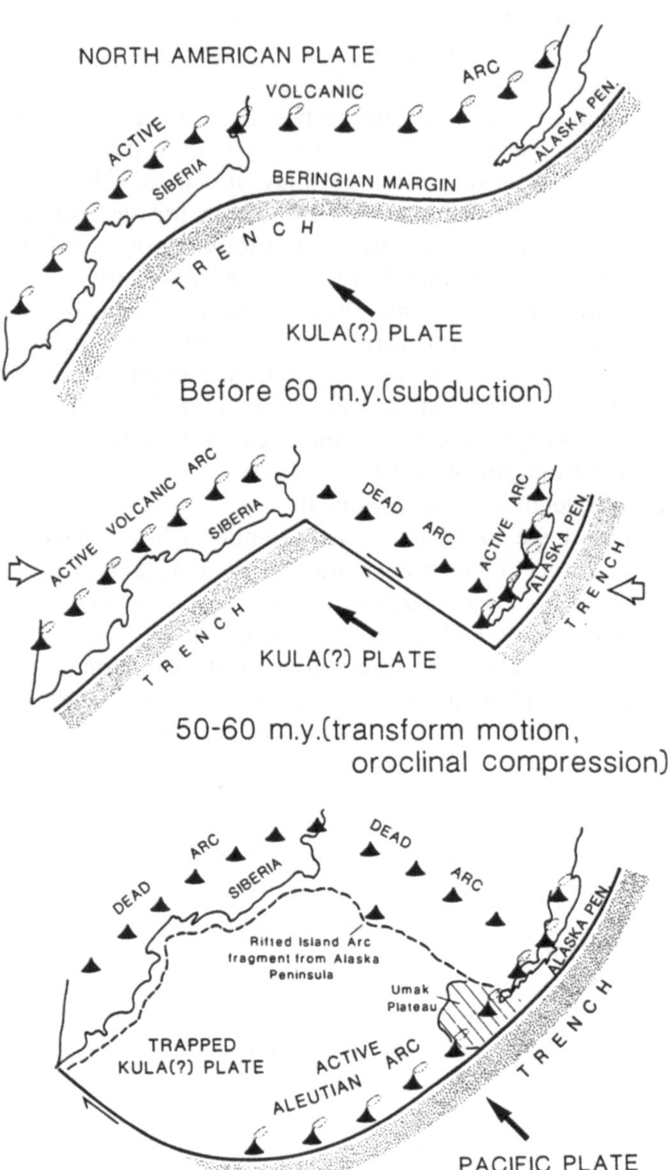
NORTH AMERICAN PLATE
ACTIVE
VOLCANIC
ARC
SIBERIA
BERINGIAN MARGIN
ALASKA PEN.
TRENCH
KULA(?) PLATE
Before 60 m.y.(subduction)
ACTIVE VOLCANIC ARC
SIBERIA
DEAD
ARC
ACTIVE ARC
ALASKA PEN.
TRENCH
KULA(?) PLATE
TRENCH
50-60 m.y.(transform motion,
oroclinal compression)
DEAD
ARC
SIBERIA
DEAD
ARC
ALASKA PEN.
Rifted Island Arc
fragment from Alaska
Peninsula
Umak
Plateau
TRAPPED
KULA(?) PLATE
ACTIVE
ARC
ALEUTIAN
TRENCH
PACIFIC PLATE

Fig. 7.

REFERENCES

T. W. F. BARTH, Geology and petrology of the Pribilof Islands, Alaska, *U.S. Geol. Survey Bull.*, **1028-F**, 101–160, 1956.

Z. BEN-AVARAHAM and A. K. COOPER, Early evolution of the Bering Sea by collision of oceanic rises and North Pacific subduction zones, *Geol. Soc. Amer. Bull.*, **92**, 485–495, 1981.

A. K. COOPER, M. S. MARLOW, and D. W. SCHOLL, Mesozoic magnetic lineations in the Bering Sea marginal basin, *J. Geophysical Res.*, **81**, 1916–1934, 1976a.

A. K. COOPER, D. W. SCHOLL, and M. S. MARLOW, Plate tectonic model for the evolution of the eastern Bering Sea basin, *Geol. Soc. Amer. Bull.*, **87**, 1119–1126, 1976b.

A. K. COOPER, D. W. SCHOLL, M. S. MARLOW, J. R. CHILDS, G. D. REDDEN, K. A. KVENVOLDEN, and A. J. STEVENSON, Hydrocarbon potential of Aleutian Basin, Bering Sea, *AAPG Bull.*, **63**, 2070–2087, 1979.

A. K. COOPER, M. S. MARLOW, and Z. BEN-AVRAHAM, Multichannel seismic evidence bearing on the origin of Bowers Ridge, Bering Sea, *Geol. Soc. Amer. Bull.*, **92**, 474–484, 1981.

J. S. CREAGER, D. W. SCHOLL *et al.*, Init. Rep. Deep Sea Drill. Proj., v. 19, 913 pp., U.S. Govt. Printing Office, Washington, 1973.

B. CSEJTEY, Jr. D. P. COX, R. C. EVARTS, G. D. STRICKER, and H. L. FOSTER, The Cenozoic Denali fault system and the Cretaceous accretionary development of southern Alaska, *J. Geophys. Res.*, **87**, 3741–3754, 1982.

J. EWING, M. EWING, T. AITKEN, and W. LUDWING, North Pacific sediment layers measured by seismic profilings, in *The Crust and Upper Mantle of the Pacific Area*, L. Knopf, C. L. Drake, and P. J. Hart, eds., pp. 147–173, Geophys. Monogr., 12, Am. Geophys. Union, 1968.

R. E. HIGGINS, Major-element chemistry of the Cenozoic volcanic rocks in the Los Angeles basin and vicinity, in *Aspects of the Geologic History of the California continental borderland*, D. G. Howell, ed., AAPG Misc. Publ., 24, pp. 216–227, 1976.

D. M. HOPKINS, D. W. SCHOLL, W. O. ADDICOTT, R. L. PIERCE, P. B. SMITH, J. A. WOLFE, D. GERSHANOVICH, B. KOTENEV, K. E. LOHMAN, J. H. LIPPS, and J. OBRADOVICH, Cretaceous, Tertiary, and early Pleistocene rocks from the continental margin in the Bering Sea, *Geol. Soc. Amer. Bull.*, **80**, 1471–1480, 1969.

Fig. 7. Schematic outline of the evolution of the Beringian margin since about 60 m.y. ago. Before 60 m.y. ago, the Bering Sea did not exist and the North Pacific rim extended from Siberia into a proto Alaska. Subduction occurred in southern Alaska, along the early Beringian margin, and in eastern Siberia, forming the Okhotsk-Chukotsk volcanic belt shown as an active volcanic arc (MARLOW *et al.*, 1983a). Between 50 and 60 m.y. ago transform motion began along the Beringian margin, perhaps as a result of oroclinal bending of Alaska, and active volcanism ceased along the margin. Active volcanism continued in eastern Siberia and near the present Alaska Peninsula. Between 40 and 50 m.y. ago formation of the central and western Aleutian arc formed the Bering Sea by entrapment of a remnant of oceanic crust, perhaps in response to accretion of Umnak Plateau to the margin (BEN-AVRAHAM and COOPER, 1981). Just before the formation of the Aleutian arc, a fragment of the ancestral eastern Aleutian was rifted from near the tip of present Alaska Peninsula and was rafted by transform motion to the northerrn Beringian margin (dredge sites 15, 40, and 41, Fig. 1 and see text; adapted from MARLOW *et al.*, 1982).

D. M. JONES, M. J. KINGSTON, M. S. MARLOW, A. K. COOPER, J. A. BARRON, F. H. WINGATE, and R. E. ARNAL, Age, mineralogy, physical properties, and geochemistry of dredge samples from the Bering Sea continental margin, U.S. Geol. Survey Open File Report 81-1297, 68 pp., 1981.

M. S. MARLOW, Hydrocarbon prospects in Navarin basin province, northwest Bering Sea shelf, *Oil and Gas Journal*, October, 190–196, 1979.

M. S. MARLOW and A. K. COOPER, Mesozoic and Cenozoic structural trends under southern Bering Sea shelf, *AAPG Bull.*, **64**, 2139–2155, 1980a.

M. S. MARLOW and A. K. COOPER, Multichannel seismic-reflection profiles collected in 1976 in the southern Bering Sea shelf, U.S. Geol. Survey Open-File Rept. 80-389, 2 pp., 1980b.

M. S. MARLOW, and A. K. COOPER, Wandering terranes in southern Alaska: The Aleutia microplate and implications for the Bering Sea, *J. Geophys. Res.*, **88**, 3439–3446, 1983.

M. S. MARLOW, D. W. SCHOLL, A. K. COOPER, and E. C. BUFFINGTON, Structure and evolution of Bering Sea shelf south of St. Lawrence Island, *AAPG Bull.*, **60**, 161–183, 1976.

M. S. MARLOW, D. W. SCHOLL, and A. K. COOPER, St. George basin, Bering Sea shelf: a collapsed Mesozoic margin, in *Island arcs, deep sea trenches, and back-arc basins*, M. Talwani and W. C. Pitman, III, eds., Am. Geophys. union Maurice Ewing Ser., v. 1, pp. 211–220, 1977.

M. S. MARLOW, D. W. SCHOLL, A. K. COOPER, and D. L. JONES, Mesozoic rocks from the Bering Sea: The Alaska-Siberia connection, Geol. Soc. Am. Abstr. Programs, v. 11, pp. 90, 1979.

M. S. MARLOW, A. K. COOPER, D. W. SCHOLL, and H. MCLEAN, Ancient plate boundaries in the Bering Sea region, in *Trench-forearc geology: sedimentation and tectonics on modern and ancient active plate margins*, J. K. Leggett, ed., Geol. Society of London Special Publication, No. 10, pp. 201–211, 1982.

M. S. MARLOW, A. K. COOPER, and J. R. CHILDS, Tectonic evolution of Gulf of Anadyr and formation of Anadyr and Navarin basins, *AAPG Bull.*, **67**, 646–665, 1983a.

M. S. MARLOW, T. L. VALLIER, A. K. COOPER, J. A. BARRON, and F. H. WINGATE, a description of dredge samples collected in 1982 from the Bering Sea continental margin west of Navarin basin, U.S. Geological Survey Open File Report 83-325, 4 pp., 1983b.

H. MCLEAN, Pribilof segment of the Bering Sea continental margin: a reinterpretation of Upper Cretaceous dredge samples, *Geology*, **7**, 307–310, 1979.

J. C. MOORE, Uplifted trench sediments-southwestern Alaska-Bering shelf edge, *Science*, **175**, 1103–1105, 1972.

J. C. MOORE, Cretaceous continental margin sedimentation, southwestern Alaska, *Geol. Soc. Amer. Bull.*, **84**, 596–614, 1973a.

J. C. MOORE, Complex deformation of Cretaceous trench deposits, southwestern Alaska, *Geol. Soc. Amer. Bull.*, **84**, 2005–2020, 1973b.

J. C. MOORE, The ancient continental margin of Alaska, in *The geology of continental margins*, C. A. Burk and C. L. Drake, eds., pp. 811–816, 1974.

D. W. SCHOLL, E. C. BUFFINGTON, and D. M. HOPKINS, Exposure of basement rock on the continental slope of the Bering Sea, *Science*, **153**, 992–994, 1966.

D. W. SCHOLL, E. C. BUFFINGTON, and M. S. MARLOW, Plate tectonics and the structural evolution of the Aleutian-Bering Sea region, in R. B. Forbes, ed., Contributions to the geology of Bering Sea basin and adjacent regions: GSA Special Paper, 151, pp. 1–32, 1975.

T. L. Vallier, M. B. Underwood, D. L. Jones, and J. V. Gardner, Petrography and geological significance of Upper Jurassic rocks dredged near Pribilof Islands, southern Bering Sea continental shelf, *AAPG Bull.*, **64**, 945–950, 1980.

Formation of Active Ocean Margins, edited by N. Nasu *et al.*, pp. 517–549.

SEDIMENTATION PATTERNS IN RELATION TO RIFTING, ARC VOLCANISM AND TECTONIC UPLIFT IN BACK-ARC BASIN OF THE WESTERN PACIFIC OCEAN

George deVries KLEIN

Department of Geology, University of Illinois at Urbana-Champaign, 245 Natural History Building, 1301 W. Green Street, Urbana, IL 61801, U.S.A.

Abstract. Nine sediment systems and types are recognized in cores recovered by the Deep Sea Drilling Project from back-arc basins in the western Pacific. These include submarine fans, debris flows, silty basinal turbidites, biogenic pelagic carbonates, resedimented carbonates, biogenic pelagic solica, pyroclastic, hemipelagic clays, and pelagic clays. Most of these systems are emplaced independent of basinal rifting processes, with deposition of pelagic clay being so dependent through post-rifting subsidence. Hemipelagic clay and silty turbidite systems are controlled by sediment yield off continental sources and climatic change, whereas biogenic pelagic systems depend on ocean circulation and latitudinally-defined productivity zones for accumulation. Pyroclastic systems are the product of regional and local volcanic periodicity coupled with seasonally-controlled atmospheric circulation. Resedimented carbonate systems occur in response to slope instability caused by chemical solution.

Submarine fan and debris flow systems are derived from andesitic volcanic arcs and obducted land sources. A direct correlation is established between short-term periodicity of turbidite deposition on fans and rates of known tectonic uplift, volcanicity and sediment yield in associated sources. Deposition of both fan and debris flow systems depends on large rates of tectonic uplift in sources followed by establishment of mature drainage systems which controls the volume of sediment yield and fan accumulation in back-arc basins. Absence of such fan systems in back-arc basins implies that source terrains were characterized by minimal rates of tectonic uplift.

Because back-arc basins are characterized by temporal and spatial sediment variability formed by an equally variable number of depositional processes, an association of specific sediment systems is suspect at best within this particular tectonic domain. In comparison to rifted passive continental margins, back-arc basins contain sediments deposited by slope-dependent, gravity processes, pelagic processes and volcanogenic-controlled processes of sedimentation. Passive margins are dominated more by fluvial, coastal, shelf and slope processes of sedimentation. These differences in processes of sedimentation are dependent on water depth, which in turn is controlled by differing subsidence histories. Thus, passive margins tend to preserve a larger volume of non-marine, shoreline and shelf sediments, whereas back-arc basins tend to preserve and receive a larger volume of deeper-water sediments. Within each setting, a large degree of variability in sediment distribution is documented.

1. *Introduction*

The association of sedimentary and tectonic events has been of interest to geologists since BERTRAND (1897) suggested that an association existed between sedimentary facies and tectonic elements. Bertrand's work was based on examination of the rock record in the Swiss and French Alps and it laid the foundation for tectonic stratigraphic and sedimentological associations such as proposed by KAY (1948), KRUMBEIN and SLOSS (1963) and PETTIJOHN (1975), amongst others. More recently, with the advent of plate tectonics, new associations of sedimentary facies to tectonic domain have been suggested (DOTT and SHAVER, 1974; DICKINSON, 1974; READING, 1978) as differences between active and passive continental margins emerged. These modern summaries, although significant and useful, again focused on the association of sedimentary features and tectonics in time and space. Most of the examples they are their predecessors reviewed occurred in complex tectonic terrains characterized by at least two or more tectonic events, including superimposed are resurgent tectonic events.

Several important questions still exist regarding attempts to associate sedimentary facies (and the record of their processes of deposition and diagenesis) to tectonic events. First and foremost, how do tectonic processes influence sedimentary processes? Second, why do tectonic processes influence sedimentary processes? Third, can one establish a causative correlation between sedimentary processes and events and tectonic processes and events in tectonic domains that are being formed today, or have been subjected only to a single cycle of tectonism and deformation? Are sedimentary processes and associated depositional systems dependent in their temporal and spatial distribution on tectonic processes, or do they act independent of them and if so, why?

It is the purpose of this paper to address these problems in a single tectonic domain that is still active during its first cycle of development, or has been subjected to a single cycle of tectonism only that is now complete. The tectonic domain selected is the back-arc basin province of the western Pacific Ocean. Sediment data is obtained from examination and compilation of data and observation from cores recovered by the Deep Sea Drilling Project (DSDP).

1.1 Back-arc basin history

The development of back-arc basins on over-riding plates of active continental margin convergence zones has been studied by several workers and include geophysical field campaigns (KARIG, 1970, 1971, 1975; HAYES, 1980, 1983), theoretical studies such as HSUI and TOKSOZS (1980), and programs of drilling by DSDP. Essentially what has emerged from this work is that back-arc basins form by a rifting process, indicated mostly from a striped magnetic anomaly pattern with spreading centers in the basin center being characterized by youngest ages and older magnetic ages occurring along the basin edges (WATTS and WEISSEL, 1975; WATTS *et al.*, 1977; WEISSEL, 1981). Both the acoustic and real basement of the basins consist of oceanic crust almost identical to mid-ocean ridge tholeitic basalt (HAWKINS, 1977). The basins are characterized by a history of large heat flow which appears to correlate well with times of rifting or for several million years after rifting ceased (KARIG, 1971). Older basins, therefore, are characterized by normal heat flow whereas the youngest basins show large rates of heat flow. Rifting processes are expressed both symmetrically as well as asymmetrically in the magnetic anomaly pattern. The duration of rifting tends to range from 5 to 16 million years (WEISSEL, 1981) and in a few cases, multiple rifting events are known.

1.2 Deep sea drilling project (DSDP) sites

The back-arc basins of the western Pacific have been penetrated by DSDP with more sites relative to area than any other first-cycle tectonic domain. Altogether, 34 sites were drilled by DSDP in these basins (Fig. 1). These sites have been described by FISCHER, HEEZEN *et al.* (1971) for Leg 6, BURNS, ANDREWS *et al.* (1973) for Leg 21, ANDREWS, PACKHAM *et al.* (1975) for Leg 30, KARIG, INGLE *et al.* (1975) for Leg 31, KLEIN, KOBAYASHI *et al.* (1980) for Leg 58, KROENKE, SCOTT *et al.* (1981) for Leg 59, and HUSSONG, UYEDA *et al.* (1981) for Leg 60. These references

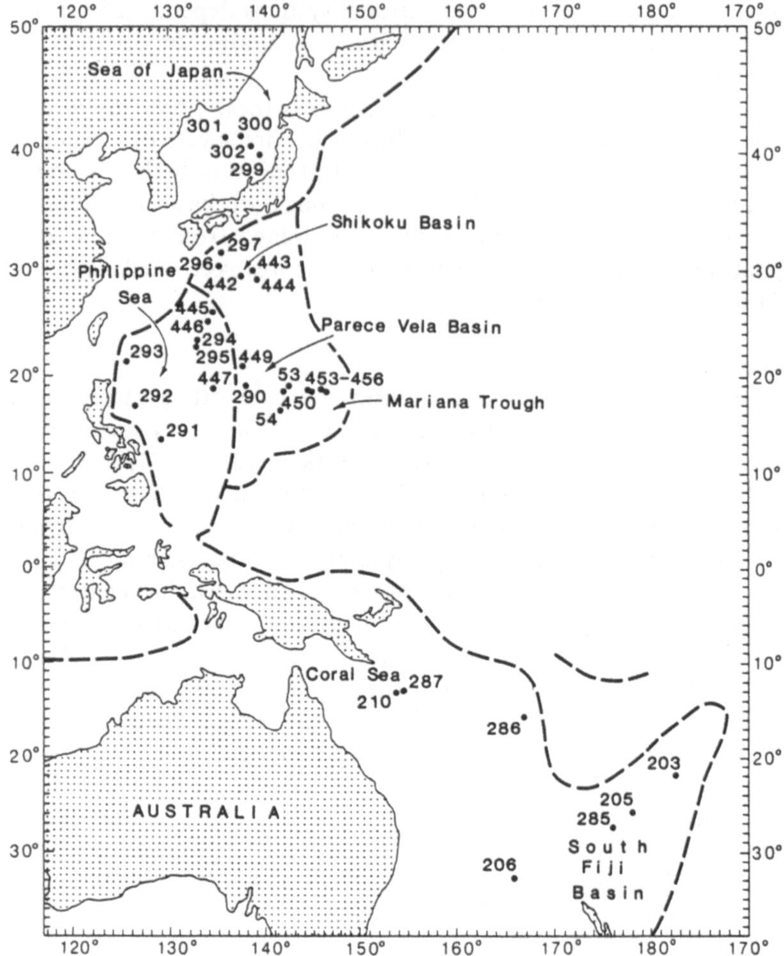

Fig. 1. Location of Deep Sea Drilling Project Sites in back-arc basins of the Westrn Pacific.

summarize and synthesize regional geology with drilling results and the reader is referred to them for such details.

The depositional history of sedimentation in these basins is less well-known and was developed on a leg-by-leg basis in an uneven fashion. Thus unevenness can be attributed to variable recovery of sediment and to interests of the shipboard party. Thus BOUMA (1975) described turbidite deposition in the Toyama submarine fan, KLEIN (1975a, b) described depositional facies from the southwest Pacific, and WHITE *et al.* (1980) reported on the lithofacies and sedimentary petrology of Leg 58 sediments from the Northwest Philippine Sea. RODOLFO (1981) provided correlation diagrams of sediment types and tectonic history for the Parece Vela Basin. No overall synthesis of sediment data from these 34 sites exists. This paper aims to provide the synthesis in terms of depositional systems.

Sediment systems were defined in DSDP cores according to their common processes of formation determined from lithology, sediment texture, association of sedimentary structures, vertical sequences of structures and lithologies, biogenic structures and mineral composition. These features were identified from core examination by the author on board ship as well as on shore at the DSDP West Coast Repository, La Jolla, California. Shore-based and shipboard descriptions in the individual leg volumes cited earlier supplemented many of the observations made herein. The approach is an extension to all DSDP back-arc basin cores of the facies summaries of KLEIN (1975a, b) and WHITE *et al.* (1980). Detailed descriptions of these depositional systems are provided in KLEIN (1985) to which the reader is referred.

Stratigraphic intervals in which depositional systems occur were tabulated and their age was recalibrated using shore-based and shipboard paleontological zonations. Recalibration of these age determination for the depositional systems into the well-known numerical time-scales of VAN COUVERING and BERGREN (1977), HARDENBOL and BERGREN (1978) and LABREQUE *et al.* (1977) and are provided also in KLEIN and LEE (1984, their Table 1). Nannoplankton were used to establish the age of the base and top of depositional systems in cores using the BUKRY (1975, 1981) time scale. Supplemental foraminiferal zonations were used following the revised foraminifera time-scale of THIEDE, VALLIER *et al.* (1981). Admittedly, not all depositional systems could be dated accurately where nannoplankton and foraminifera were absent.

Nine depositional systems are recognized in the back-arc basins of the western Pacific. Their age and depth intervals by site are summarized in KLEIN and LEE, (1984, their Table 1) to which the reader is referred.

2. *Relationship of Depositional Systems to Geodynamic Events*

2.1 *Sedimentation and rifting processes*

The primary tectonic process forming back-arc basins is rifting and it occurs in a mode and scale similar to sea-floor spreading processes in mid-ocean ridges. The primary evidence for this rifting process is mapping of magnetic anomalies which show a symmetrical or asymmetrically-

Table 1. Times of rifting, pacific back-arc basins.

Basin	DSDP/IPOD Sites	Age of Rifting	Reference
Sea of Japan	299, 300, 301, 302	Miocene (21–9 MA)	Kobayashi and Isezaki, 1976
Shikoku Basin	297, 442, 443, 444	Mid Oligocene to Early Miocene (25–14 MA)	Kobayashi and Nakada, 1978; Shih, 1980
Daito Basin	445, 446	Late Paleocene to Late Eocene (57–40 MA)	Klein and Kobayashi, 1980, 1981
Parece-Vela Basin	53, 54, 449, 450	Mid Oligocene to Early Miocene (30–17 MA)	Langseth and Mrozowski, 1981 Mrozowski and Hayes, 1979
Mariana Trough	453, 454, 455, 456	Late Miocene to Present (7–0 MA)	Hussong and Uyeda, 1981
West Philippine Sea	290, 291, 292, 293 294, 295, 447, 448	Early to Late Eocene (52–37 MA)	Louden, 1977; Watts *et al.*, 1977
Coral Sea Basin	210, 287	Paleocene (62–56 MA)	Weissel and Watts, 1979
New Caledonia Basin	206	No data	No Reference
East Caroline Basin	63	Oligocene (31.5–28.5 MA)	Weissel, 1981
New Hebrides Basin	286	Earliest to Latest Eocene (55–42 MA)	Lapouille, 1978; Weissel *et al.*, 1982
South Fiji Basin	205, 285	Oligocene (33–26 MA)	Weissel and Watts, 1975; Davey, 1982
Lau Basin	203	Plio/Pleistocene to Present (5–0 MA)	Lawver *et al.*, 1976

disposed striped pattern, and represent progressively older magnetic anomaly ages away from the spreading center to the basin flanks (WATTS *et al.*, 1977; WEISSEL, 1981). Table 1 summarizes the magnetic anomaly ages of each of the back-arc basins drilled by DSDP.

Correlation diagrams were constructed to compare the timing of depositional systems, rifting processes and andesitic volcanism (Figs. 2 to 8) in each of these back-arc basins so as to determine a causal connection between each of these processes. These diagrams show the following relations in each basin between sedimentation and rifting processes.

Sea of Japan: In the Sea of Japan (Fig. 2), rifting occurred during Miocene time (21 to 9 MA) and ceased (KOBAYASHI and ISEZAKI, 1976) before any record of sedimentation was documented from DSDP drill sites, although none of these sites reached basement. Post-rifting sedimentation is variable and included submarine fan, biogenic pelagic silica, hemipelagic clay and silty turbidite deposition.

Shikoku Basin: In the Shikoku Basin (Fig. 3), rifting began in Mid-Oligocene time (25 MA) whereas sedimentation began in Earliest Miocene time (22.5 MA) on the western and northern parts of the basin. Here sedimentation consists dominantly of hemipelagic clay, although some submarine fan deposition occurred at Site 297 in the northern Shikoku Basin before rifting ceased 14 MA. Sedimentation continues into the present, dominated by hemipelagic clay. This clay, derived from the Chinese

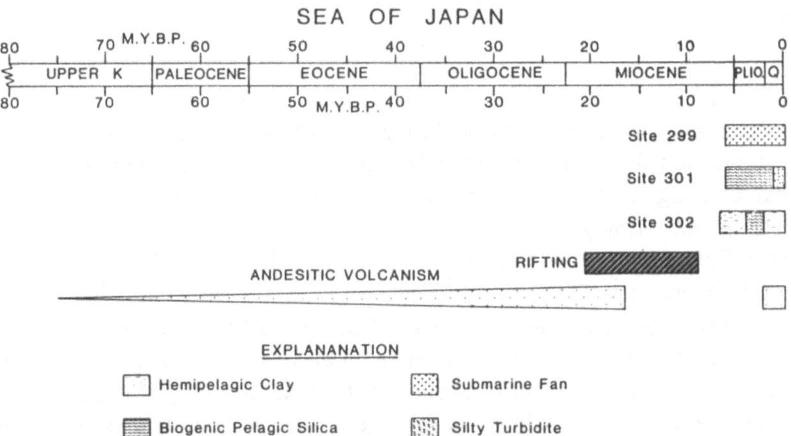

Fig. 2. Correlation diagram showing distribution of depositional systems, times of rifting (KOBAYASHI and ISEZAKI, 1976) and duration of andesitic volcanism (SUGIMURA *et al.*, 1963; LEBAS, 1982) in Sea of Japan and Japanese Islands, respectively (B-sites penetrating basement).

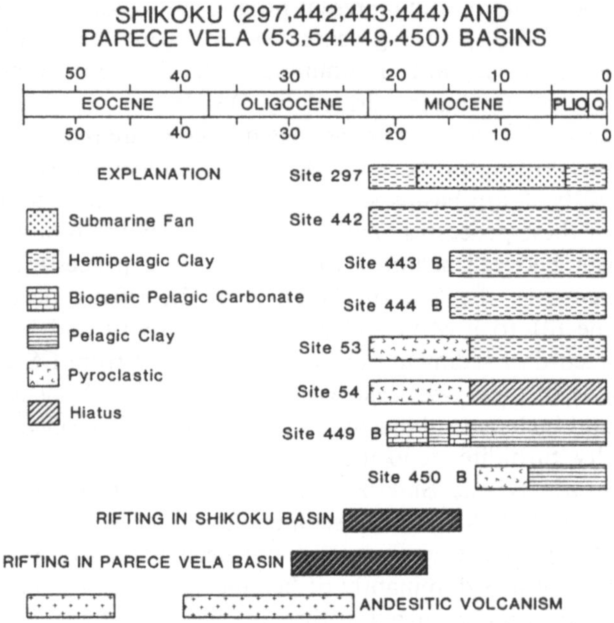

Fig. 3. Correlation diagram showing distribution of depositional systems in DSDP sites, times of rifting in Shikoku Basin (KOBAYASHI and NAKADA, 1978; SHIN, 1980) and Parece Vela Basin (LANGSETH and MROZOWSKI, 1981) and duration of volcanism in adjoining arc terrains (SHIRAKI *et al.*, 1978; SUGIMURA and UYEDA, 1963). (B-sites penetrating basement).

mainland by turbid layer flow and wind, shows a compositional change with increased illite content from Miocene to Present, reflecting a general global cooling or a northward drift of the basin (CHAMLEY, 1980a, b; WHITE *et al.*, 1980).

Parece Vela Basin: Rifting in the Parece Vela Basin (Fig. 3) started in Mid-Oligocene time (30 MA) and ceased in Early Miocene time (17 MA). The oldest sediments are Earliest Miocene (22.5 MA) and consist of pyroclastic depositional systems at Sites 53 and 54. Slightly younger biogenic carbonates, interbedded and capped with pelagic clay, was recovered at Site 449. Pyroclastic sedimentation and biogenic sedimentation occurred simultaneously with rifting. The pyroclastic deposition represents part of a regional pattern of volcanism known from the Pacific (KENNETT *et al.*, 1977) which was independent of basinal rifting events. Biogenic sedimentation is a function of foraminiferal and nannoplankton productivity and represents a time of deposition when the Parece Vela Basin was located in a more biogenically-productive equatorial latitude (cf. KINOSHITA, 1980) rather than being a function of rifting processes. Pelagic clay deposition at Sites 449 and 450 followed basinal rifting but required basinal subsidence below the CCD; this subsidence is dependent on heat loss in the basin as the heat flow regime shifted from a greater rate during rifting to a normal rate after rifting ceased. This basin subsidence is the only tectonic process which controls the sedimentation style in this basin.

West Philippine Basin: Rifting in the West Philippine Basin (Fig. 4) occurred from Early (52 MA) to Late Eocene (37 MA). Sedimentation was observed to occur simultaneously with the latest stages of rifting at only three sites; the other sites recovered sediment deposited after rifting ceased. Debris flow, biogenic pelagic silica and biogenic pelagic carbonate depositional systems were deposited during the latest stages of rifting. Each of these depositional systems owe their origin to processes other than rifting, however, with debris flow emplacement favored by large sediment yield from island arc sources (see discussion below) and biogenic sedimen-

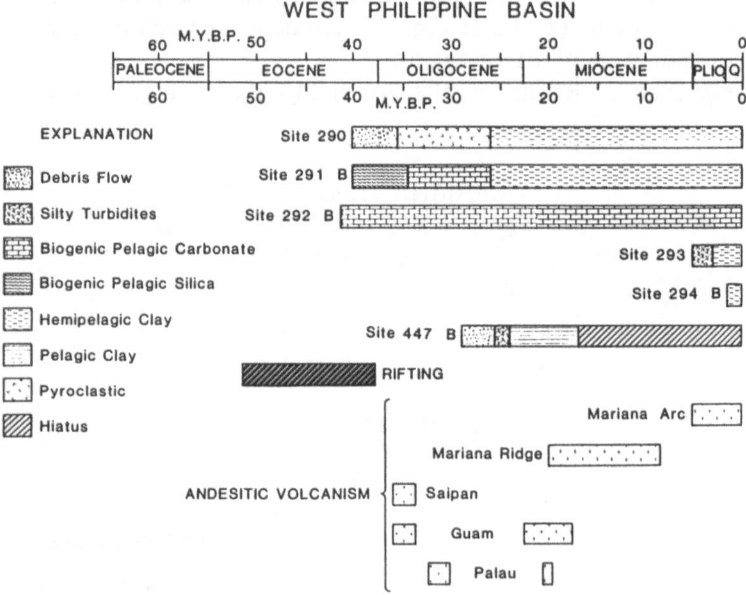

Fig. 4. Correlation diagram showing distribution of depositional systems in DSDP Sites and times of rifting in West Philippine Sea (LOUDEN, 1977; WATTS *et al.*, 1977), and times of andesitic volcanism in adjoining arc sources (SCOTT and KROENKE, 1982; HUSSONG and UYEDA, 1981; MEIJER *et al.*, 1983). (B-sites penetrating basement).

tation being controlled by latitudinally-defined productivity zones and ocean current circulation. In this basin, no particular depositional system appears to be caused by rifting processes as such, although being deposited simultaneously during late rifting events.

Mariana Trough: Rifting in the Mariana Trough started in Late Miocene time (7 MA) and continues to the present (Fig. 5). The oldest sediments are Pliocene (5 MA), and although the time-span of deposition is short, hemipelagic clay, debris flow, submarine fan and biogenic pelagic carbonate depositional systems are represented. Although their deposition is simultaneous with rifting, the primary control on sedimentation is other processes. Here, hemipelagic clays are deposited by turbid layer transport from distant sources, whereas the biogenic carbonate deposition is dependent on organic productivity and ocean current circulation. Debris flow and submarine fan systems owe their origin to sediment yield off tectonically-uplifted andesitic arc sources (as discussed below).

New Hebrides Basin: Rifting in this basin (Fig. 6) began in Earliest Eocene time (55 MA) and ended in Late Eocene time (42 MA). During the latest stage of rifting, both submarine fan and debris flow sedimentation occurred from sediment yield off New Caledonia (KROENKE, 1982). The other depositional systems (Table 1; Fig. 6) were deposited following rifting by processes independent of tectonics.

South Fiji Basin: Rifting in the South Fiji Basin (Fig. 7) was limited

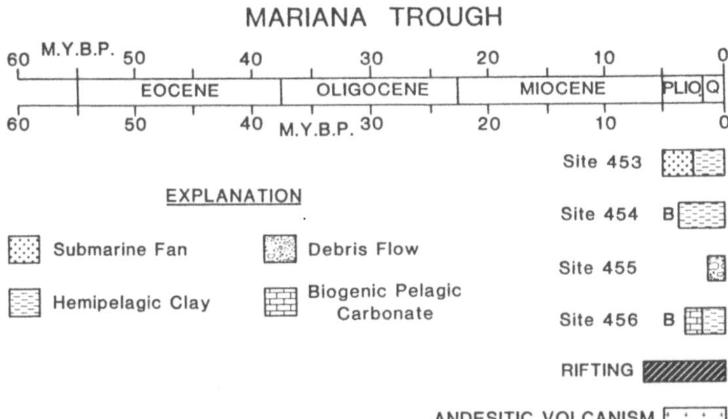

Fig. 5. Correlation diagram showing distribution of depositional systems in DSDP Sites and rifting (HUSSONG and UYEDA, 1981) in Mariana Trough, and duration of andesitic volcanism in adjoining arc sources (SCOTT and KROENKE, 1982; HUSSONG and UYEDA, 1981; MEIJER *et al.*, 1983).

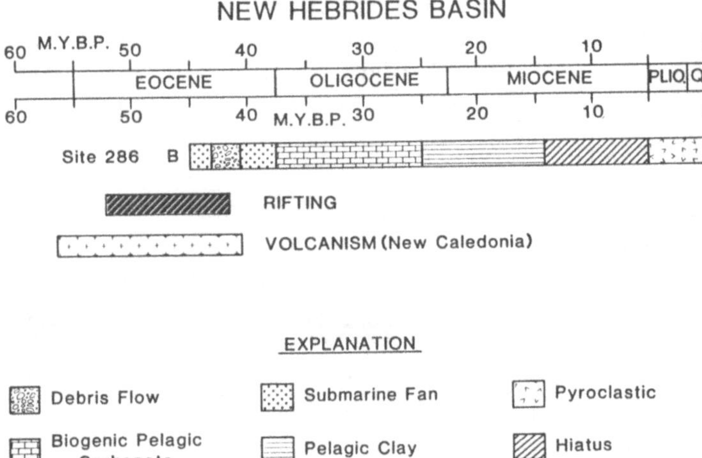

Fig. .6. Correlation diagram showing distribution of depositional systems at Site 286 and times of rifting in New Hebrides Basin (LAPOUILLE, 1978; WEISSEL *et al.*, 1982) and duration of andesitic volcanism in New Caledonia (LILLIE and BROTHERS, 1970) which acted as source for clastic wedges. (B-sites penetrating basement).

to the Oligocene (32-26 MA) and was simultaneous with times of pyroclastic and biogenic pelagic carbonate deposition only. Other depositional systems present include pelagic clay and submarine fans. The primary controls on their deposition, again, are independent of rifting processes with the exception of pelagic clay.

Lau Basin: Rifting in the Lau Basin (Fig. 7) started in Earliest Pliocene (5 MA) and continuous to the present. Concurrent sedimentation consists of pyroclastic and biogenic depositional systems only (Site 203).

Other Basins: Figure 8 compares times of rifting in the Daito Basin, East Caroline Basin and Coral Sea Basin with timing of depositional systems. Data from the New Caledonia Basin (Site 206), from which no rifting age determination exists (Table 3), are shown also. These basins are so grouped because no age determinations exists for potential volcanic arc sources adjoining the basins. In all these basins, variable sedimentation events may or may not occur simultaneously with rifting (Fig. 8) because the controlling processes of sedimentation are independent of rifting processes.

2.2 *Summary of sedimentation-rifting relations*

Even cursary examination of Figs. 2 through 8 shows that in the back-

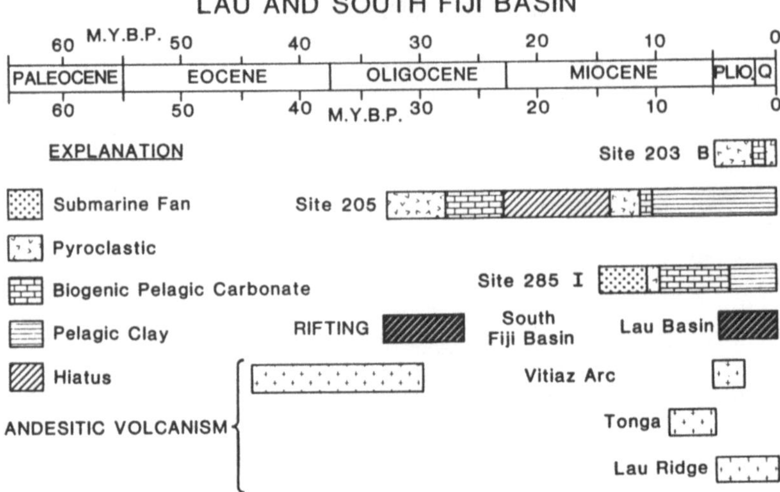

Fig. 7. Correlation diagram showing distribution of depositional systems and times of rifting in Lau (LAWVER *et al.*, 1975) and South Fiji (WEISSEL and WATTS, 1975; DAVIES, 1982) basins, and andesitic volcanism in adjoining arc sources (EWART *et al.*, 1977; GILL *et al.*, 1983). (B-sites penetrating basement).

arc basins of the western Pacific, rifting events are occurring simultaneously with a large variety of sedimentation events but the two processes operate independently. In this case, a coincidence represented by an association of sedimentation and rifting events does not produce a correlation of sedimentation and rifting processes. The role of rifting in these basins is to provide a sink where sediment may accumulate. Once the rifted basin forms, sedimentation styles vary depending on ocean current circulation, latitudinal arrangement of biogenic productivity zones (controlling biogenic silica and carbonate deposition), position of continental land masses, regional volcanism and sediment yield from tectonically-uplifted arc sources.

Rifting history controls the elevation of the depositional surface where sedimentation occurs, a factor crucial for preservation of biogenic carbonate depositional systems and pelagic clays. If the basinal surface occurs at an elevation above the CCD, preservation of biogenic carbonate systems in appropriate latitudes is favored. However, once rifting ceases in the basin, the heat flow rate dissipates causing basinal subsidence including subsidence below the CCD and thus favoring deposition of pelagic red clay with low sediment accumulation rates. In continental land sources adjoining such a subsided basin, however, pelagic clay deposition would

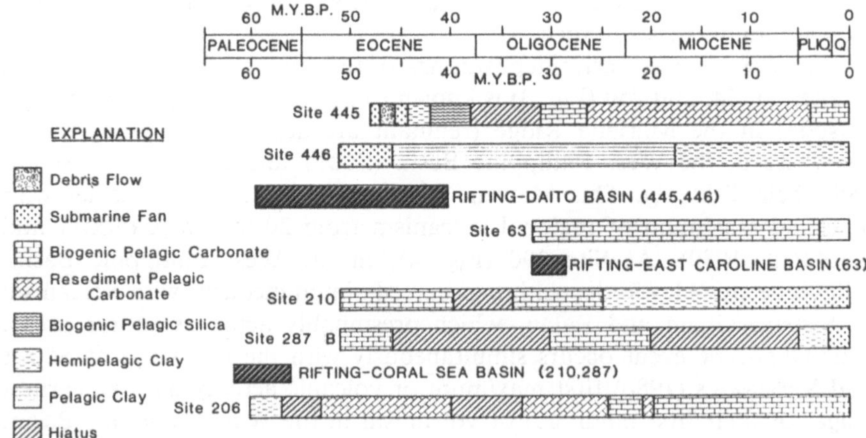

Fig. 8. Correlation diagram showing distribution of depositional systems and times of rifting in Daito Basin (KLEIN and KOBAYASHI, 1980, 1981), the East Caroline Basin (WEISSEL, 1981), The Coral Sea Basin (WEISSEL and WATTS, 1979) and the New Caledonide Basin (Site 206). (B- sites penetrating basement).

be masked by hemipelagic clay (as known from the Shikoku Basin).

In summary, rifting processes in the back-arc basins control sedimentation processes only by providing a sediment sink, and controlling the elevation of the depositional surface in response to heat flow history. This second process determines the preservation of biogenic pelagic carbonate depositional systems and the deposition of pelagic red clays.

2.3 Volcanism and pyroclastic sedimentation

Although Figures 2 through 8 show that andesitic volcanism precedes back-arc rifting pyroclastic deposition depends on regional volcanism. Figures 3, 4, 6, and 7 show timing of pyroclastic depositional systems in the Parece Vela, West Philippine, New Hebrides, Lau and South Fiji basins. Comparison of the timing of pyroclastic deposition in these basins with andesitic volcanism (Figs. 3, 4, 5, 7) in adjoining island arcs demonstrated that andesitic volcanism occurs simultaneously with pyroclastic deposition only in the West Philippine Basin during the Early and Middle Oligocene (Fig. 4), and in the South Fiji Basin (Site 205) during the Middle Oligocene (Fig. 7). RODOLFO (1981) suggested a similar correlation between Site 450 pyroclastic deposition in the Parece Vela Basin and volcanism on the Mariana Ridge. In the remaining basins where pyroclastic systems occur, no such relationship exists (Figs. 3, 6) because these pyroclastics were derived, probably, from sources other than island arcs adjoining these basins.

Timing of pyroclastic depositional systems in the Parece Vela Basin at Sites 53, 54, and 450 (Fig. 3) is known to coincide with regional volcanic episodes in the Mariana Ridge (remnant arc adjoining this basin) and elsewhere in the West Philippine Basin (SCOTT and KROENKE, 1980, p. 289, their Table 1). This Miocene pyroclastic deposition coincides with a second maximum of regional volcanism from 20 to 9 MA (SCOTT and KROENKE, 1980). At Site 290 (Fig. 4), in the West Philippine Basin, Oligocene pyroclastic deposition occurred simultaneously with volcanism in Saipan, Guam and Palau, which presumably acted as source areas. This Oligocene event occurs simultaneously with the late stage of SCOTT and KROENKE's (1980) first maximum of volcanic activity and the earliest stages of their first minimum of volcanism in the West Philippine Basin.

Neogene volcanism in the Circum-Pacific has occurred in a periodic fashion (KENNETT *et al.*, 1977) with major episodes known from the Quaternary and Middle Miocene and subordinate episodes known from the Late Miocene and Early Pliocene. This periodicity caused deposition of Quaternary and Pliocene pyroclastics in the Lau Basin (Site 203), Late Miocene pyroclastics in the South Fiji Basin (Sites 205, 285) and Quaternary and Pliocene pyroclastics in the New Hebrides Basin (Site 286). The potential sources for these pyroclastics are known areas of concurrent volcanism (KENNETT *et al.*, 1977) such as Fiji (for Sites 203, 205, 285), the New Hebrides (for Site 286) and possibly New Zealand.

Although ash beds occur throughout the back-arc basin cores as thin layers, the preservation of thick pyroclastic beds is much more rare, being confined to 5 basins (Figs. 3, 4, 6, and 7). With active volcanism occurring both in adjoining arcs and regionally (KENNETT *et al.*, 1977; SCOTT and KROENKE, 1980), some explanation is required to account for this distribution. In the Lesser Antilles Arc, a chance separation exists between volcaniclastic, gravity-deposited sediment and wind-transported ash (SIGURDSSON *et al.*, 1980). On the west side of that arc, the Granada Basin received volcaniclastic turbidites and debris flows only, whereas on the eastern side of that arc, larger volumes of ash and pyroclastics occur (and gravity sediments are absent) because prevailing wind systems flow from west to east; this dominant easterly-flowing wind system causes ash to occur only east of the Lesser Antilles Arc (SIGURDSSON *et al.*, 1980). Thick pyroclastic accumulations occur close to the arc and most likely, the recovery of thick pyroclastics in DSDP cores in pacific back-arc basins is favored by proximity of site location to arc source.

Meterological maps of the western Pacific (THE TIMES ALTAS OF THE

WORLD, 1977) indicate that during January, the back-arc basin provinces north of the equator and dominated by cold-air outbreaks from Siberia and the northeast trade winds, whereas south of the equator, westerly-flowing trade winds (southeast trades) occur (Fig. 9). During July, the southeast monsoon systems provide a northward-flowing system of winds north of the equator, whereas the southeast trade winds continue to develop westerly-flowing wind systems south of the equator (Fig. 10). In subtropical to subarctic latitudes upper atmosphere (about 10 km high) westerly winds (jet stream) transports a great amount of volcanic ashes from continental margins and island arcs to the back-arc basins such as Sea of Japan Basin and Shikoku Basin.

These Pacific wind patterns provide some clues to potential dispersal for some pyroclastic and ash beds in the back-arc basins. Thus, the Japanese Island Arc comprises a major source of ash during January (Fig. 9) into the Shikoku Basin, Parece Vela Basin and the northern part of the West

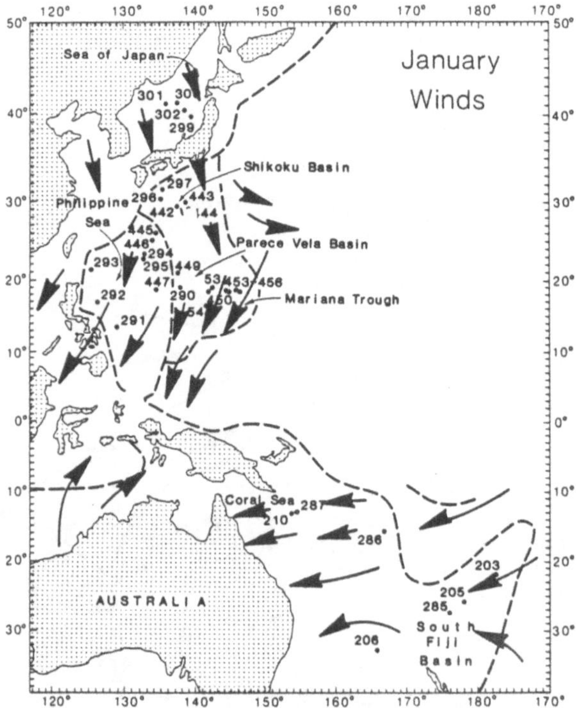

Fig. 9. Map of DSDP sites in western Pacific back-arc basins showing present-day wind directions during January. (Wind directions after TIMES ALTAS OF THE WORLD, 1977).

Philippine Basin. Ticker accumulations are developed locally by a combination of gravity flow and perhaps wind dispersal in the past off the Kyushu-Palau Ridge (a remnant arc); this dispersal may account for thick pyroclastics at Site 290 as would similar modes of dispersal off the Mariana Ridge into the Parece Vela Basin. Thicker pyroclastic accumulations in the New Hebrides Basin (Site 286), the South Fiji Basin and the Lau Basin (Site 203) are derived from a similar combination of gravity flow and airborne systems involving westerly-flowing air masses over the New Hebrides (for Site 286) and the Lau Ridge. During July (Fig. 10), the wind systems north of the equator move from a southerly direction so ash layers in the Sea of Japa would be dispersed there from the Japanese Islands, whereas Guam, Saipan and the southern Mariana Arc would act as sources for ash beds in the Shikoku and Parece Vela basins and the

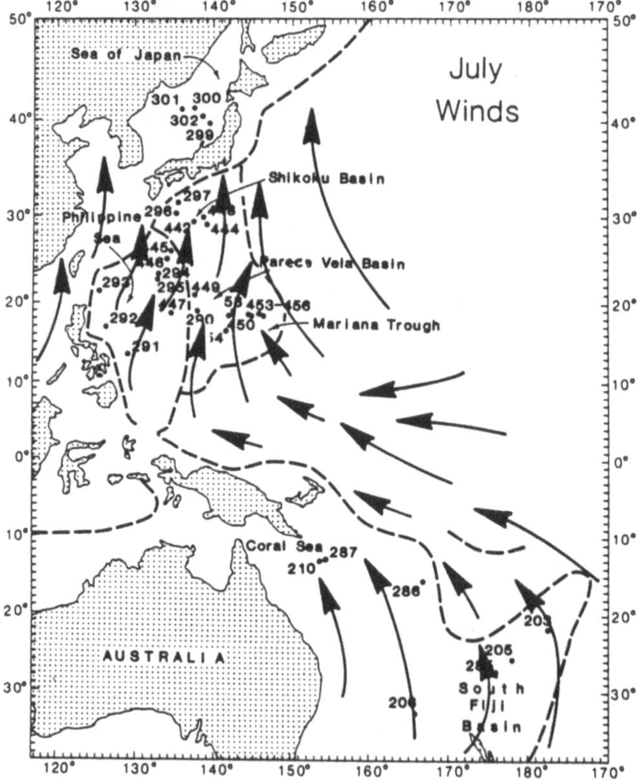

Fig. 10. Map of DSDP sites in western Pacific-back-arc basins showing present-day wind directions during July. (Wind directions after TIMES ALTAS OF THE WORLD, 1977).

Mariana Trough. South of the equator, the wind systems during July are directed west-northwest and as a consequence, ash derived from New Zealand and the Tonga Arc would be dispersed into the South Fiji and Lau Basins. The New Hebrides Basin would receive only small volumes of ash during that season. These wind dispersal directions may be the cause of developing significant ash accumulations during the present in some of these back-arc basin sites. During Cenozoic time, it would be expected on paleogeographic and paleoclimatological grounds (PARRISH, 1981; MARSAGLIA and KLEIN, 1983) that wind dispersal patterns were similar while the basins formed and shifted in response to larger-order regional tectonic trends (cf. KINOSHITA, 1980).

In summary, deposition and dispersal of ash beds and other pyroclastic sediments is controlled by three processes in the back-arc basins of the western Pacific. Regional and tectonically-controlled periodicity of volcanism (KENNETT *et al.*, 1977) account for the availability of volcanic ash and pyroclastics, proximity to island arc and seamount sources account for most of the thicker units, whereas seasonally-controlled atmospheric circulation patterns control ash dispersal into the basins. Combination of these wind dispersal patterns with the larger-order periodicity of volcanicity reported by KENNETT *et al.* (1977) appears to explain the limited distribution of pyroclastic depositional systems in the back-arc basins.

2.4 *Tectonic uplift of island arc sources and basin sedimentation*

Both submarine fan and debris flow depositional systems are derived from island arc sources and associated shallow-water zones as indicated by the occurrence of andesitic volcanic rock fragments and resedimented shallow-water fossils (KLEIN, 1975a, b; KARIG and MOORE, 1975; WHITE *et al.*, 1980). Deposition of these submarine fan and debris flow systems is restricted time-wise to either the earliest stage or the last stage of basin rifting (Figs. 2 to 8). Deposition of this system is independent of rifting processes (Figs. 2 through 8) but is known to occur simultaneously with the last stages of andesitic volcanism and after such andesitic volcanism ceased (Table 2) in known island arc sources (Figs. 2 to 8).

Three basins, the Coral Sea and Shikoku basins and the Sea of Japan contain submarine fan systems which adjoin sediment sources whose geological history is well known. Analysis of these fan systems should indicate what processes determine the timing and spatial distribution of submarine fan and debris flow depositional systems in back-arc basins.

Sites 210 and 287 (BURNS, ANDREWS *et al.*, 1973; ANDREWS, PACKHAM *et al.*, 1975; KLEIN, 1975a, b, 1984) penetrated the medial-distal and distal

Table 2. Age of andesitic ARC volcanism, western pacific.

Arc Source (Probable Basin Sink- in parenthesis)	Age of Volcanism (absolute ages-in parenthesis)	Reference
Japanese Islands (Sea of Japan)	Quaternary Early Miocene (major) Early Cretaceous to Oligocene (progressive increase)	LaBas, 1982 Sugimura & Uyeda, 1973
Bonin Islands (Shikoku Basin) (Parece Vela Basin)	Middle Eocene Oligocene to Early Miocene	Shiraki *et al.*, 1978 Sugimura & Uyeda, 1973
Kyushu-Palau Ridge (Daito Ridge and Basin Province)	Middle Eocene (47.5±4)	Ozima *et al.*, 1977
West Mariana Ridge	Miocene (20–9)	Scott & Kroenke, 1980
Mariana Arc	Pliocene (5–0) to Holocene (2.3–0.4)	Scott & Kroenke, 1980; Hussong & Uyeda, 1981 Meijer *et al.*, 1983
Saipan	Early Oligocene (35.7±0.5)	Mijer *et al.*, 1983
Guam	Early Oligocene (35.6±0.9)	Meijer *et al.*, 1983
Palau	Late Oligocene (32.2)	Meijer *et al.*, 1983
(Mariana Trough & West Philippine Basin)	Early Miocene (20.1±0.5)	
Malekula Island (New Hebrides Basin-?)	Miocene and Post-Miocene	Colley & Ward, 1974
Vitiaz Arc/Tonga/ Kermadec Is.	Pliocene to Holocene	Ewart *et al.*, 1977
(Lau Basin; South Fiji Basin)	Eocene to Oligocene (42–30) Mid Miocene to Pliocene (9–3.5)	Gill *et al.*, 1983
New Caledonia (New Hebrides Basin)	Early to Late Eocene (57–40)	Lillie & Brothers, 1970

portion of a diachronous submarine fan system in the Coral Sea Basin. Internally, these fans contained graded turbidite cycles which start with a sharp basal scour, overlain by graded fine sand which grades into clayey silts and silty clays and are capped by clays. Interbedded with these terrigeneous graded turbidite cycles are nannoplankton foraminiferal ooze, some of which are graded and resedimented, but clearly of basinal derivation. The repetitive nature of these turbidite graded cycles at Site 210 is such that their depositional periodicity over the last 10 million years averages to one turbidite event every 4,000 years (BURNS, ANDREWS *et al.*, 1973, p. 374).

These terrigenous turbidites were derived from the ancestral Owen-Stanley Range of Papua (BURNS, ANDREWS *et al.*, 1973; ANDREWS, PACKHAM *et al.*, 1975; KLEIN, 1975a, b, 1984). The Owen-Stanley Range owes its origin to obduction of the Pacific Plate onto the Indian-Australian Plate from Eocene to Pliocene time; maximum uplift was suggested for Miocene and Early Pliocene time (DAVIES and SMITH, 1971).

Table 3 summarizes periodicity data of turbidite graded cycles observed in cores recovered from Sites 210 and 287. The approach taken was to count the number of such cycles per core and subdivide, time-stratigraphically, the number of cycles for each site using the numerical Neogene time-scale of VAN COUVERING and BERGREN (1977). A correction was made to account for coring history at each site because some coring operations were continuous, whereas others involved alternate coring and washing, or a combination of both procedures (see BURNS, ANDREWS *et al.*, 1973, p. 373; ANDREWS, PACKHAM *et al.*, 1975, p. 139; and KARIG, INGLE *et al.*, 1975, p. 355 for coring summary details; an explanation is provided also in Table 3). This correction was justified because identical turbidites occur above and below the washed intervals. Thus a percentage correction was added to the number of cycles to account for wahed intervals (Table 3).

Table 3 shows that at Site 210, the shortest periodicity of deposition of graded turbidite cycles occurred during the Lower Pliocene, following the Miocene uplift, and coinciding, time-wise, with Early Pliocene uplift of the Owen-Stanley Range by DAVIES and SMITH (1971). The time-delay in turbidite deposition during Miocene time is attributed to the medial-distal location of this site, as well as a delay in establishing a mature enough river system to generate a sufficiently large sediment yield into the basin. The short-term turbidite periodicity during Early Piliocene time correlates with tectonic uplift in the Owen-Stanley Range. Progressive increase in periodicity of turbidite deposition occurred during the Late Pliocene and Pleistocene, indicating either a cessation of uplift or decreasing rates of tectonic uplift and associated decreasing sediment yields (cf. YOSHIKAWA, 1974) in the Owen-Stanley Range. At Site 287, a more distal fan site, turbidite cycles were deposited no earlier than Late Pliocene and continued to the present but show an decrease in turbidite periodicity from

Table 3. Periodicities of graded turbidite cycles, coral sea basin and Sea of Japan.

Age	Site 210			Site 287			Site 299		
	No. of cycles	Corrected no. of cycles*	Periodicity of turbidite events (yr)	No. of cycles	Corrected no. of cycles*	Periodicity of turbidite events (yr)	No. of cycles	Corrected no. of cycles*	Periodicity of turbidite events (yr)
Late Pleistocene	58	58	17,240	38	66	15,150	161	—	6,211
Early Pliocene	50	60	13,300	8	16	50,000	52	—	15,380
Late Pliocene	63	126	11,100	7	14	100,000	?	?	?
Early Pliocene	185	370	4,860	—	—	—	69	86	102,230
Late Miocene	122	244	28,680	—	—	—			

* Corrections according to drilling procedure. At Site 210, coring was continuous to depth of 90 m, followed by alternating coring and washing. Thus, early Pleistocene was corrected 20%, and the rest was corrected by an additional 100%. Site 287 was cored by alternating coring and washing; thus correction was by an additional 100%. Site 299 was cored continuously to depth of 247.0 m, followed by alternating coring and washing to depth of 513 m, followed by continuous coring, requiring a correction of 25% for early Pliocene-Miocene. See site-report coring summaries for details.

Late Pliocene through Late Pleistocene time. This change may well be caused also by larger rates of denudation and sediment yield associated with increased relief-triggered by sea level fall during world-wide Pleistocene glaciation. It is noteworthy that this more distal site records such a trend, but Site 210 (medial distal fan) does not.

In the Sea of Japan, the only submarine fan system to be drilled by DSDP is the Toyama Submarine Fan at Site 299 (KARIG, INGLE *et al.*, 1975; BOUMA, 1975; NASU, 1981) where outer levee, channel and overbank sediments were recovered. There, individual turbidites are also organized into cycles of progressively fine-grained sediment (BOUMA, 1975, p. 494, his Table 1) and consist of a distinct basal contact, overlain by sandy-silty clay, grading up into less silty clay and capped by clay. These sediments were derived from the Hida Range of Honshu Island, Japan, and encompass a time-stratigraphic interval from perhaps as early as late Miocene into the Present.

The geology of Japan is extremely complex and no attempt to summarize it is made herein. The history of tectonic uplift, volcanicity and sediment yield in the Hida Range is known since Late Miocene time, however, and it is germaine to this discussion. During Miocene and Pliocene time, tectonic uplift rates for the Hida Range of Honshu were negligible (SUGI *et al.*, 1983; MATSUDA *et al.*, 1967). During Quaternary time, however, the Hida Range experienced tectonic uplift rates ranging from 1 to 1.5 km (or 555 to 833 m per million years). This tectonic uplift is concurrent with a marked increase in andesitic volcanism (SUGIMURA *et al.*, 1963). Surveys of sediment yield in present rivers in Honshu shows that such data obtained from the upstream-most reservoirs are largest in the Hida Range in comparison to small sediment yields in adjoining areas of low relief and smaller rates of tectonic uplift (YOSHIKAWA, 1974; OHMORI, 1978; TANAKA, 1982). The rate of sediment accumulation determined in post-Miocene sedimentary basins of Honshu Island also parallels the rate of tectonic uplift of the Hida Range in particular and the rest of Honshu (NIITSUMA, 1978).

Table 3 summarizes the turbidite periodicity from the Toyama submarine fan at Site 299. The data was corrected in the same way as for the Coral Sea Basin. Turbidite periodicity shows a progressive shortening from Pliocene time to Late Pleistocene time, and parallels the tectonic uplift history of the Hida Range sediment source. Large rates of turbidite periodicity in the Late Miocene and Pliocene are correlated to relatively small rates of tectonic uplift, whereas short-term turbidite periodicity during the Quaternary are correlated to the larger rates of tectonic uplift in the Hida Range. Thus the deposition of the Toyama submarine fan correlates time-wise with a time of known large rates of tectonic uplift, andesitic volcanism and large rates of sediment yield in the Hida Range.

Submarine fan deposition of Late Early to Middle Miocene age was observed also in the northern part of the Shikoku Basin at Site 297; this fan sedimentation was derived from the southern part of Honshu island (KARIG, 1975; KARIG, INGLE *et al.*, 1975). This fan deposition occurred coevally with a period of maximum volcanism in Japan (SUGIMURA *et al.*, 1963) and a major period of tectonic uplift on southern Honshu Island (MATSUDA *et al.*, 1975; SUGI *et al.*, 1983). Fan deposition ceased when sediment yield off the Japanese Island Arc was diverted into the Nankei Trough during Early Pliocene time (KARIG, 1975; KARIG, INGLE *et al.*, 1975). The simultaneous deposition of this submarine fan into the Shikoku Basin during a time of known large tectonic uplift rate and increased volcanism parallels the development of the Toyama submarine fan.

In summary, analysis of turbidite periodicity and timing of fan deposition in the Coral Sea Basin, the Sea of Japan and the Shikoku Basin demonstrates that deposition of submarine fans there is governed by increased rates of tectonic uplift which may (Honshu Island) or may not (Owen-Stanley Range) be coupled with volcanism. Both areas indicate that increased rates of tectonic uplift provide sufficiently large volumes of sediment yield to permit deposition of submarine fan systems (and associated debris flows) in back-arc basins. A delay in fan deposition with respect to initial times of tectonic uplift is caused by the time necessary to develop a large drainage system by which means sediment yield is discharged into the back-arc basins.

Both debris flow and submarine fan systems have been observed in DSDP sites in the West Philippine Basin, Mariana Trough, New Hebrides Basin, South Fiji Basin and the Daito Basin (Figs. 4 through 8). The sources of all these fans, except in the Daito Basin, are known and the timing of andesitic volcanism in these sources is shown also in Figs. 4 through 8. A correlation can be established between fan deposition and andesitic volcanism in source terrains in these cases except in the South Fiji Basin where a time-lag is indicated. Because no data exists on past or current rates of tectonic uplift in these andesitic source terrains, no direct correlation can be established between tectonic uplift rate and fan deposition in these examples. Nevertheless, coeval fan deposition and andesitic volcanism suggests the interpretation of tectonic uplift process as requisite in these sources before fan deposition can occur in these cases. This deposition would be similar to fan deposition in the Coral Sea and Shikoku basins and the Sea of Japan. The absence of submarine fan depositional systems in other back-arc basins implies that the adjacent island arcs which could have acted as sources failed to do so because they did not undergo a history of significant or large rates of tectonic uplift. For instance, Pagen Island in the Mariana Arc reached a present-day elevation of 800 meters over the past 2 million years (HUSSONG, UYEDA *et al.*, 1981), requiring a minimal uplift rate of 400 meters per million year, a rate which is half of the present-day rate of tectonic uplift determined for Rota Island in the Mariana Arc (YONEKURA, 1982). Site 456 adjoins Pagen Island, yet no submarine fan system occurs there although debris flow occurs at Site 455, which is located closer to Pagen Island (Fig. 5). It appears that a tectonic uplift rate of approximately 400 meters per million years is the minimum rate necessary to generate the requisite sediment yield to develop submarine fans.

In summary, submarine fan and debris flow deposition in back-arc basins requires large rates of uplift, with or without associated andesitic volcanism, in their source areas, as demonstrated in the Coral Sea and Shikoku basins, and the Sea of Japan. A time-delay between fan deposition and the onset of tectonic uplift in known sources is controlled by the time required to develop a sufficiently mature drainage basin which will generate sufficient sediment yield into the back-arc basins. The absence of submarine fans in back-arc basins implies that potential sources, mostly inland arcs adjoining th basins, were not subjected to large rates of tectonic uplift.

3. *Brief Comparison of Back-Arc Basin Sediments to Passive Margin Sediments*

The large degree of variability of deep-water sediment systems in back-arc basins summarized above contrasts sharply with the style of deposition reported from the Atlantic continental margin by several workers (MANSPEIZER *et al.*, 1978; EVANS, 1978; SCHLEE, 1981, amongst others) were a more ordered pattern of deposition is reported in response to progressive rifting (see also SHERIDAN, 1981). During earliest stages of rifting, non-marine fluvial, eolian and lacustrine sediments accumulated (as indicated during deposition of the Triassic-Jurassic basins of eastern North America). As rifting continued, the proto-Atlantic formed and coastal carbonate and sabkha sediments as well as reefs formed; some are associated with evaporites (EVANS, 1978). With additional widening of the ocean, sediment yield off continental sources developed a thick clastic wedge which tongued laterally into reefs (SHERIDAN, 1981). These clastic wedges consist primarily of deltaic sediments, but also contain both narrow and wide continental shelf sedimentary systems (SCHLEE, 1981). They tongue laterally into continental slope and rise muds and silts, which grade laterally into debris flow, turbidite submarine fan systems, and contourites. Large abyssal plains filled with turbidites are also widespread in the Atlantic.

The present Pacific back-arc basins share similarities to the present Atlantic in terms of deep-water sediment systems only. They lack non-marine sequences and shallow water coastal plain and coastal sabkha and continental shelf sequences reported from the Atlantic margin.

These differences may be attributed to tectonic history in the following way. The opening of the Atlantic involved rifting of a large supercontinent and as rifting occurred, marine waters transgressed over a continental landmass. As the seafloor subsided in response to heat loss, the margins were deepened and as sediment yield off continents decreased, the rate of sediment accumulation diminished to less than the subsidence rate. The depositional surface thus subsided to deep-water elevations, favoring deep marine sedimentation. Back-arc basins formed by a similar rifting history, but the back-arc basins in the western Pacific involved a rifiting of crust whose original depth of deposition occurred below sea level, followed by rapid subsidence. Thus the original elevation of the locus of rifting in passive and active margins may account for many of the differences in sedimentological patterns observed between back-arc basins and passive continental margins. It is, however, noteworthy that areas of back-arc basin development are known where more continental-like settings occur above or at sea level. These include the back-arc basins of the Great Basin of the USA (EATON, 1982), the Permian alluvial fan, and coal beds of the Sydney Basin, Australia (HOBDAY, 1982) and the Pliocene back-arc basin fills of Hokkaido, Japan, where coastal deltas occur (KLEIN *et al.*, 1979). In the former two cases, back-arc opening occurred but basinal subsidence did not reach deep-water marine elevations. In the Hokkaido example, regional uplift favored deltaic deposition. Their presence, however, suggests that there may exist back-arc basin sedimentary sequences which share depositional systems in common with both the present western Pacific and modern Atlantic margins.

4. *Conclusions*

The sediment fills of the back-arc basins of the western Pacific Ocean are characterized by a large variety of depositional systems whose distribution is controlled by an equally variable number of sediment processes. Processes controlling deposition in these basins include latitudinally-controlled biogenic productivity and oceanic circulation (for biogenic carbonate and silica pelagic depositional systems), continental sediment yield and climate (for hemipelagic clay and silty turbidite systems), regional volcanism and wind patterns (for pyroclastic systems and ash beds), basin subsidence (for pelagic clay systems), and either large rates of uplift in source terrains or a coupling of andesitic volcanism and large rates of uplift of andesitic volcanic island arc sources (for submarine fan and debris flow systems). Figure 11 shows a schematic summary of these depositional systems in a series of back-arc basins flanking a subduction zone extending from pole to pole, and it shows also the effect of continental land masses, latitude, uplifting arcs and sources and basinal subsidence below the CCD to indicate the variety of combinations possible within a single or several back-arc basins.

Because of this variety in sedimentation processes and deposition, can

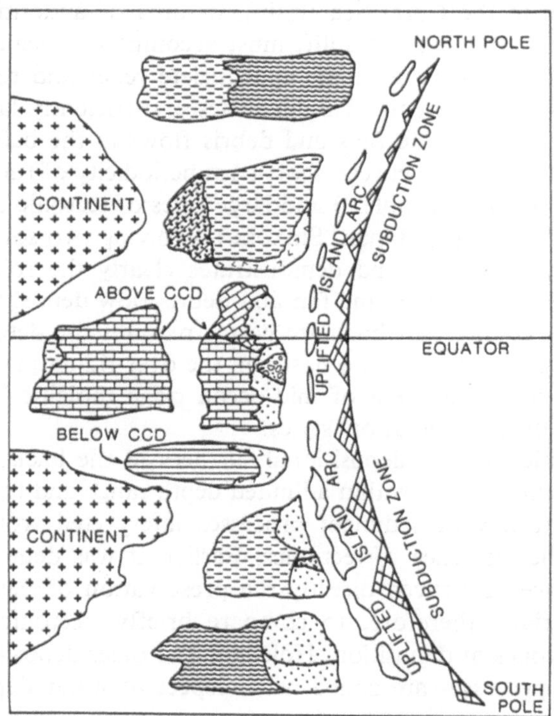

Fig. 11. Synthetic summary of distribution of depositional systems in back-arc basins along a longitudinal subduction zone extending from pole to pole. Latitudinal control of biogenic pelagic sediments is shown as is occurrence of fans and debris flows near island arcs characterized by great uplift rates (G). Occurrence of continentally-derived hemipelagic clays, randomly-distributed pyroclastics, silty turbidites and tectonically-controlled pelagic clays also shown. Symbolic coding as per Fig. 2 through 8.

back-arc basins be characterized by sediment associations for identification of similar successions in the rock record? It is the conclusion of this paper that a direct link between sediment type and tectonic setting in present-day back-arc basins is non-existent because the sedimentation processes operate independently of rifting processes during basin development. Tectonics controls sedimentation in these basins only in two ways. First, it determines the elevation of the depositional surface, and the direct response of this elevation is the preservation of pelagic biogenic carbonate systems if the basin floor occurs above the CCD, and the occurrence of pelagic clay systems if the basin floor occurs below the CCD. Second, if tectonic uplift rates in source areas are large, a sediment yield is generated which accumulates in the back-arc basins as debris flow and submarine fan systems. Such uplift occurs in areas of plate convergence by obduction (as shown in the Coral Sea Basin), or area of andesitic volcanism. In andesitic arcs, the tectonic uplift must accompany volcanism followed by a time-lag required to generate appropriate relief and relatively well-established drainage systems. This provides a sufficiently large sediment yield to develop submarine fans and debris flows in the back-arc basins. A close correlation exists between short-term periodicity of turbidite deposition on submarine fans (Table 3) and large rates of tectonic uplift in three basins (Coral Sea, Sea of Japan, Shikoku) involving two sources (Eastern Papua, Japanese Islands) and demonstrates clearly the role of tectonic uplift as a varible on submarine fan and debris flow deposition. Recognition of ancient counterpart back-arc basins requires the determination of th geographic position of such a basin on the over-riding plate of a paleo-subduction zone, a basement of mid-ocean ridge tholeitic basalt as well as variability of sedimentation styles.

The elevation of the depositional surfaces in the back-arc basins of the western Pacific occurs within a limited depth range and for that reason, perhaps, the variety of sediment processes interpreted from the DSDP cores masks the influence of tectonics. Earlier, it was demonstrated that basin subsidence influenced directly the preservation of pelagic clays. It seems appropriate, therefore, to compare briefly, tectonic setting with known depth zones of deposition. Holocene and older depositional systems and sediment processes are zoned with respect to water depth (Fig. 12). These depth ranges occur in a variety of overlapping tectonic settings (Fig. 12). Non-marine, coastal and shelf systems tend to show a greater degree of preservation in rifting basins, cratons, and passive margins. Slope systems are confined to relatively equal proportions of both passive and active continental margins, whereas continental rise systems appear to be more prevalent along active margins. Deep-water sediment systems are confined

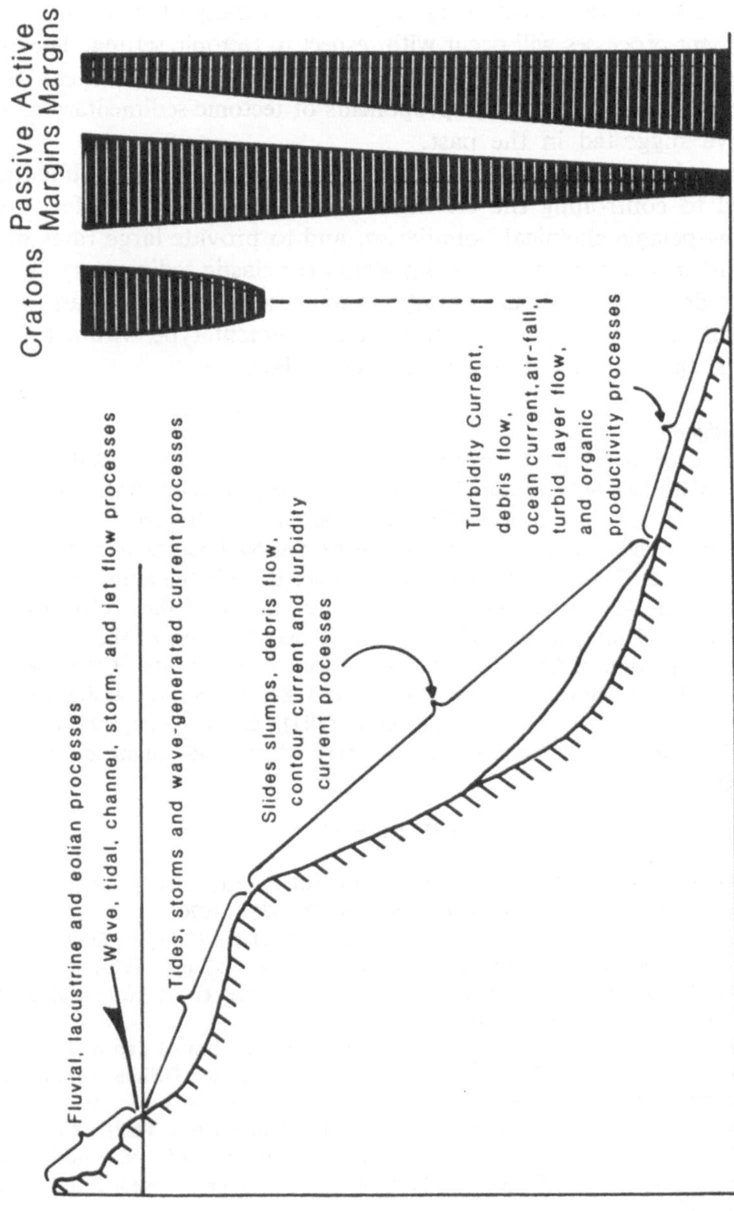

Fig. 12. Comparison of depositional processes and depth zonation with tectonic setting. Width of bands for tectonic setting show relative degree of preservation of depositional systems in each tectonic setting shown.

to trenches, back-arc basins, fore-arc basins, mid-ocean ridges, and deep, subsiding ocean basins. Because the distribution of tectonic setting overlaps so many zones of water depth (Fig. 12), a large variety of sediment systems and sediment processes will occur with respect to tectonic setting. Therefore, the association of sediment types with a specific tectonic element or domain shows greater variety than proponents of tectonic-sedimentary associations have suggested in the past.

In conclusion, the role of tectonics on back-arc basin sedimentation is limited to controlling the elevation of the depositional surface above and below pelagic chemical boundaries, and to provide large rates of tectonic uplift in source areas to yield material for clastic sedimentary systems. The large degree of sediment variability in these basins makes an association of all sediment types or a particular sediment type within back-arc basins, in the traditional sense, suspect at best.

Acknowledgements

Support for this research came from the National Science Foundation through its Grant OCE-81-09447 and the University of Illinois Research Board; the author is most grateful for this support. The Deep Sea Drilling Project, which provided access to cores and samples, is funded also by the National Science Foundation through its Grant G-482. This paper was completed while the author held an appointment as an Associate, Center for Advanced Study of the University of Illinois at Urbana-Champaign, and as a Senior Research Fellow of the Japan Society for the Promotion of Science at the Ocean Research Institute of the University of Tokyo during the Spring and early Summer, 1983. The author wishes to extend his thanks to Drs. Noriyuki Nasu (Director, ORI), Kazuo Kobayashi and Hideo Kagami for amenities extended during this stay which made manuscript completion easier.

REFERENCES

F. AHNERT, Functional relationships between denudation, relief, and uplift in large mid-latitude drainage basins, *Am. Jour. Sci.*, **268**, 243–263, 1970.

J. E. ANDREWS, G. H. PACKHAM, *et al.*, Initial Reports of the Deep Sea Drilling Project, v. 30, Washington, U.S. Government Printing Office, 753 pp., 1975.

M. BERTRAND, Structures des alpes Francais et recurrence de certain facies sedimentaries, C. R. VI Int. Geol. Cong., pp. 163–177, 1897.

A. H. BOUMA, Deep-sea fan deposits from Toyama Trough, Sea of Japan, in Karig, D. E., Ingel, J. C., Jr., 1975, Initial Reports of the Deep Sea Drilling Project, v. 31, Washington, D. C. U.S. Government Printing Office, pp. 471–488, 1975.

D. BUKRY, Coccolith and silicoflagellate stratigraphy, northwestern Pacific Ocean, Deep Sea Drilling Project, Leg 32, in Larson, R. L., Moberly, R. *et al.*, 1975, Initial Reports of the Deep Sea Drilling Project, v. 32, Washington, U.S. Government Printing Office, pp. 845–872, 1975.

D. BUKRY, Cenozoic coccoliths from the Deep Sea Drilling Project, in Warme, J. E.,

Douglas, R. G. and Winterer, E. L., 1981, The Deep Sea Drilling Project: A decade of progress, Soc. Econ. Paleontologists and Mineralogists Spec. Pub., 32, pp. 335–354, 1981.

R. E. BURNS, J. E. ANDREWS *et al.*, Initial Reports of the Deep Sea Drilling Project, v. 21, Washington, U.S. Government Printing Office, 931 pp., 1973.

H. CHAMLEY, Clay sedimentation and paleoenvironment in the Shikoku Basin since the Middle Miocene, in Klein, G. deV., Kobayashi, K. *et al.*, 1980, Initial Reports of the Deep Sea Drilling Project, v. 58, Washington, U.S. Government Printing Office, pp. 669–682, 1980a.

H. CHAMLEY, Clay sedimentation and paleoenvironment in the area of the Daito Ridge since the early Eocene, in Kein, G. deV., Kobayashi *et al.*, 1980, Initial Report of the Deep Sea Drilling Project, v. 58, Washington, U.S. Government Printing Office, pp. 683–649, 1980b.

H. COLLEY and A. J. WARDEN, Petrology of the New Hebrides, *Geol. Soc. Am. Bull.*, **85**, 1635–1646, 1974.

F. J. DAVEY, The structure of the South Fiji Basin, *Tectonophysics*, **87**, 185–241, 1982.

H. L. DAVIES and I. E. SMITH, Geology of eastern Papua, *Geol. Soc. Am. Bull.*, **82**, 3299–3312, 1971.

W. R. DICKINSON (Editor), Tectonics and sedimentation, *Soc. Econ. Paleontologists and Mineralogists Spec. Pub.*, **22**, 204, 1974.

R. H. DOTT, Jr. and R. H. SHAVER (Editors), Modern and ancient geosynclinal sedimentation, Soc. Econ. Paleontologists and Mineralogists Spec. Pub., 19, 380, 1974.

G. P. EATON, The basin-and-range province, origin and tectonic significance, *Ann. Rev. Earth and Planet. Sci.*, **10**, 409–440, 1982.

A. A. ECKDALE, Trace fossils in Deep Sea Drilling Project Leg 58 cores, in Klein, G. deV., Kobayashi, K. *et al.*, 1980, Initial Reports of the Deep Sea Drilling Project, v. 58, Washington, U.S. Government Printing Office, pp. 610–606, 1980.

R. EVANS, Origin and significance of evaporites in basins around Atlantic margin, *Am. Assoc. Petroleum Geol. Bull.*, **62**, 223–234, 1978.

A. EWART, R. N. BROTHERS, and A. MATTERN, An outline of the geology and geochemistry and the possible petrogenetic evolution of the volcanic rocks of the Tonga-Kermadec New Zealand Island Arc, *J. Volc. and Geothermal Res.*, **2**, 205–250, 1977.

A. G. FISCHER, B. C. HEEZEN *et al.*, Initial Reports of the Deep Sea Drilling Project, v. 6, Washington, U.S. Government Printing Office, 1329 pp., 1971.

J. B. GILL, A. L. STORK, and P. M. WHELAN, Volcanism accompanying back-arc basin development in the southwest Pacific, *Tectonophysics*, 1983 (in press).

J. HARDENBOL and W. A. BERGREN, A new Paleogene numerical time scale, *Am. Assoc. Petroleum Geologists Stud. Geol.*, **6**, 213–234, 1978.

J. W. HAWKINS, Origin and evolution of marginal basins and recognition of their remnants in orogenic belts (Abs), *Geol. Soc. Am.* Abs. with Programs, **9**, 1006–1007, 1977.

D. E. HAYES (Editor), The tectonic and geologic evolution of south-east Asian seas and islands, v. 1, Am. Geophys. Union Mon., 23, 326 pp., 1980.

D. E. HAYES (Editor), The tectonic and geologic evolution of southeast Asian seas and islands, Am. Geophys. Union Mon., 27, 396 pp., 1983.

A. T. HSUI and N. M. TOKSOZS, Back-arc spreading: trench migration, continental pull or induced convection?, *Tectonophysics*, **74**, 89–98, 1980.

D. M. HUSSONG and S. UYEDA, Tectonic processes and the history of the Mariana Arc: A synthesis of the results of Deep Sea Drilling Project Leg 60, in Hussong, D. M.,

Uyeda, S. *et al.*, Initial Reports of the Deep Sea Drilling Project, v. 60, Washington, U.S. Government Printing Office, pp. 909–929, 1981.

D. M. HUSSONG, S. UYEDA *et al.*, Initial Reports of the Deep Sea Drilling Project, v. 60, Washington, U.S. Government Printing Office, 929 pp., 1981.

D. E. KARIG, Ridges and basins of the Tonga-Kermadec island arc system, *J. Geophys. Res.*, **75**, 239–254, 1970.

D. E. KARIG, Origin and development of marginal basins in the western Pacific, *J. Geophys. Res.*, **76**, 2542–2561, 1971.

D. E. KARIG, Basin genesis in the Philippine Sea, in Karig, D. E., Ingle, J. C., Jr. *et al.*, Initial Reports of the Deep Sea Drilling Project, v. 31, Washington, U.S. Government Printing Office, pp. 857–880, 1975.

D. E. KARIG and G. D. MOORE, Tectonically-controlled sedimentation in marginal basins, *Earth and Planet. Sci. Lett.*, **26**, 233–238, 1975.

D. E. KARIG, J. C. INGLE, Jr. *et al.*, Initial Reports of the Deep Sea Drilling Project, v. 31, Washington, U.S. Government Printing Office, 927 pp., 1975.

G. M. KAY, North American geosynclines, *Geol. Soc. Am. Mem.*, **48**, 143, 1948.

J. P. KENNETT, A. R. MCBIRNEY, and R. C. THUNELL, Episodes of Cenozoic volcanism in the Circum-Pacific region, *J. Volc. and Geothermal Res.*, **2**, 145–163, 1977.

H. KINOSHITA, Paleomagnetism of sediment cores from Deep Sea Drilling Leg 58, Philippine Sea, in Klein, G. deV., Kobayashi, K. *et al.*, 1980, Initial Reports of the Deep Sea Drilling Project, v. 58, Washington, U.S. Government Printing Office, pp. 765–768, 1980.

G. deV. KLEIN, Sedimentary tectonics in southwest Pacific marginal basins based on Leg 30 Deep Sea Drilling Project Cores from the South Fiji Hebrides and Coral Sea basins, *Geol. Soc. Am. Bull.*, **86**, 1012–1018, 1975a.

G. deV. KLEIN, Depositional facies of Leg 30 Deep Sea Drilling Project sediment cores, in Andrews, J. E., Packham, G. H. *et al.*, 1975, Initial Reports of the Deep Sea Drilling Project, v. 30, Washington, U.S. Government Printing Office, pp. 423–442, 1975b.

G. deV. KLEIN, Relative rates of tectonic uplift as determined from episodic turbidite deposition in marine basins, *Geology*, **12**, 38–40, 1984.

G. deV. KLEIN, The control of depositional depth, tectonic uplift and volcanism on sedimentation processes in the back-arc basins of the western Pacific: *J. Geology*, **93**, 1–25, 1985.

G. deV. KLEIN and K. KOBAYASHI, Geological summary of the North Philippine Sea based on Deep Sea Drilling Project Leg 58 results, in Klein, G. deV., Kobayashi, K. *et al.*, 1980, Initial Reports of the Deep Sea Drilling Project, v. 58, Washington, U.S. Government Printing Office, pp. 951–962, 1980.

G. deV. KLEIN and K. KOBAYASHI, Geological summary of the Shikoku Basin and northwestern Philippine Sea, Leg 58 DSDP/IPOD drilling results, Oceanologica Acta, v. 4, No. SP., pp. 181–192, 1981.

G. deV. KLEIN, K. KOBAYASHI *et al.*, Initial Reports of the Deep Sea Drilling Project, v. 58, Washington, U.S. Government Printing Office, 1017 pp., 1980.

G. deV. KLEIN and Y. I. LEE, Depositional Systems and sandstone diagenesis in back-arc basins, western Pacific Ocean: Tectonophysics, **102**, 119–152, 1984.

G. deV. KLEIN, H. OKADA and K. MITSUI, Slope sediments in small basins associated with a Neogene active margin, western Hokkaido Island, Japan, in Doyle, L. E., and Pilkey, O. H., editors, 1979, Geology of continental slopes, *Soc. Econ. Paleon. and Min. Spec. Pub.*, **27**, 359–374, 1979.

K. KOBAYASHI and N. ISEZAKI, Magnetic anomalies in the Sea of Japan and the Shikoku Basin, possible plate tectonic implication, in Sutton, G. H., Manghanani, M. H., and

Moberly, R., editors, 1976, The geophysics of the Pacific Ocean Basin and its margins, Am. Geophys. Union. Mon., 19, pp. 235–251, 1976.

K. KOBAYASHI and M. NAKADA, Magnetic anomalies and tectonic evolution of the Shikoku interarc basin, *J. Phys. Earth.*, **26**, S391–S402, 1978.

L. W. KROENKE, The Cenozoic tectonic development of the southwest Pacific (Abstract), *EOS*, **63**, 1119–1120, 1982.

L. W. KROENKE, R. W. SCOTT *et al.*, Initial Reports of the Deep Sea Drilling Project, v. 59, Washington, U.S. Government Printing Office, 820 pp., 1981.

W. C. KRUMBEIN and L. L. SLOSS, Stratigraphy and sedimentation, 2nd ed, San Francisco, W. H. Freeman and Co., 660 pp., 1963.

J. L. LABREQUE, D. V. KENT, and S. C. CANDE, Revised magnetic polarity time scale for the Late Cretaceous and Cenozoic time, *Geology*, **5**, 330–335, 1977.

M. G. LANGSETH and C. L. MROZOWSKI, Geophysical surveys for Leg 59 sites, Deep Sea Drilling Project, in Kroenke, L. W., Scott, R. W. *et al.*, 1980, Initial Reports of the Deep Sea Drilling Project, v. 59, Washington, U.S. Government Printing Office, pp. 487–502, 1980.

A. LAPOUILLE, Southern New Hebrides Basin and western South Fiji Basin as a single marginal basin, *Australian Soc. Explor. Geophys. Bull.*, **9**, 130–133, 1978.

D. E. LAWSON, Mobilization, movement and deposition of active subaerial sediment flows, Matanuska Glacier, Alaska, *J. Geology*, **90**, 279–300, 1982.

L. A. LAWVER, J. W. HAWKINS, and J. G. SCLATER, Magnetic anomalies and crustal dilation in the Lau Basin, *Earth and Planet. Sci. Lett.*, **33**, 27–35, 1976.

M. J. LEBAS, Quaternary to Recent volcanism in Japan, *Geol. Assoc. Proc.*, **93**, 179–194.

A. R. LILLIE and R. N. BROTHERS, The geology of new Caledonia, *New Zealand J. Geology and Geophys.*, **13**, 145–183, 1970.

K. E. LOUDEN, Paleomagnetism of DSDP sediments, phase shifting of magnetic anomalies and rotations of the West Philippine Basin, *J. Geophys. Res.*, **82**, 2989–3002, 1977.

W. MANSPEIZER, J. H. PUFFER, and H. L. COUSMINER, Separation of Morocc and eastern North America, a Triassic-Lassic stratigraphic record, *Geol. Soc. Am. Bull.*, **89**, 901–920, 1978.

K. M. MARSAGLIA and G. deV. KLEIN, The Paleogeography of Paleozoic and Mesozoic Storm Depositional systems, *J. Geology*, **91**, 117–142, 1983.

T. MATSUDA, K. NAKAMURA, and A. SUGIMURA, Late Cenozoic orogeny in Japan, *Tectonophysics*, **4**, 349–366, 1967.

MEIJER, AREND, REAGAN, MARK, ELLIS, HOWARD, SHAFIQULLAH, MUHAMMAD, SUTTER, JOHN, DAMON, PAUL, and KLING, STANLEY, Chronology of volcanic events in the eastern Philippine Sea, in Hayes, D. E., editor, 1983, The tectonic and geological evolution of the southeast Asian seas and islands, Part 2, Am. Geophys. Union Mon., 27, pp. 349–359, 1983.

C. L. MROZOWSKI and D. E. HAYES, The evolution of the Parece Vela Basin, eastern Philippine Sea, *Earth and Planet. Sci. Lett.*, **46**, 49–67, 1979.

M. NASH, The sediments of Toyama deep sea fan, Unpub. MS thesis, Univ. of Tokyo, 106 pp., 1981.

N. NIITSUMA, Magnetic stratigraphy of the Japanese Neogene and the development of the island arcs of Japan, *J. Phys. Earth*, **26**, S367–S378, 1978.

H. OHMORI, Relief structure of the Japanese mountains and their stages in geomorphic development, *Dept. of Geography, Univ. of Tokyo Bull.*, **10** 31–84, 1978.

M. OZIMA, I. KANEOKA and H. UJIIE, ^{40}Ar-^{39}Ar age of rocks and the development mode of the Philippine Sea, *Nature*, **267**, 816–818, 1977.

J. T. PARRISH, Global atmospheric circulation in the Mesozoic and Cenozoic (Abs), *Am. Assoc. Petroleum Geol. Bull.*, **65**, 969, 1981.

J. T. PARRISH, Upwelling and petroleum source beds with reference to the Paleozoic, *Am Assoc. Petroleum Geol. Bull.*, **66**, 750–774, 1982.

F. J. PETTIJOHN, *Sedimentary Rocks*, 3rd ed., 628 pp., Harper and Row, New York, 1975.

H. C. READING (Editor), *Sedimentary Environments and Facies*, 557 pp., Elsevier, New York, 1978.

K. S. RODOLFO, Sedimentological summary: clues to arc volcanism, arc sundering, and back-arc spreading in the sedimentary sequences of Deep Sea Drilling Project Leg 59, in Kroenke, L. W., Scott, R. W. *et al.*, 1981, Initial Reports of the Deep Sea Drilling Project, v. 59, pp. 621–623, 1981.

J. S. SCHLEE, Seismic stratigraphy of Baltimore Canyon Trough, *Am. Assoc. Petroleum Geol. Bull.*, **65**, 26–53, 1981.

R. W. SCOTT and L. W. KROENKE, Evolution of back-arc spreading and arc volcanism in the Philippine Sea: Interpretation of Leg 59 DSDP results, in Hayes, D. E., 1982, The tectonic and geological evolution of southeast Asian seas and islands, Am. Geophy. Union Mon., 24, pp. 283–291, 1980.

R. E. SHERIDAN, Recent research on passive continental margins, in Warme, J. E., Douglas, R. G., and Winterer, E. L., 1981, The Deep Sea Drilling Project: A decade of progress, *Soc. Econ. Paleontologists and Mineralogists Spec. Pub.*, **32**, 39–56, 1981.

T. C. SHIH, Magnetic lineations in the Shikoku Basin, in Klein, G. deV., Kobayashi, K. *et al.*, 1980, Initial Reports of the Deep Sea Drilling Project, v. 58, Washington, U.S. Government Printing Office, pp. 783–788, 1980.

K. SHIRAKI, N. KURODA, S. MARUYAMA, and H. URANO, Evolution of the Tertiary volcanic rocks in the Izu-Mariana Arc, *Bull. Volc.*, **41**, 548–562, 1978.

H. SIGURDSSON, R. S. J. SPARKS, S. N. CAREY, and T. C. HUANG, Volcanogenic sedimentation in the Lesser Antilles Arc, *J. Geology*, **88**, 523–540, 1980.

N. SUGI, K. CHINZEI, and S. UYEDA, Vertical crustal movements of northeast Japan since Middle Miocene, in Hilde, T. W. C. and S. Uyeda, editors, Geodynamics of the Western Pacific Indonesian Region, Am. Geophys. Union Geodynamics Mon., pp. 317–330, 1983.

A. SUGIMURA and S. UYEDA, Island Arcs: Japan and Its Environs, 247 pp., Elsevier, Amsterdam, 1973.

A. SUGIMURA, T. MATSUDA, K. CCHINZEI and K. NAKAMURA, Quantitative distribution of Late Cenozoic volcanic materials in Japan, *Bull. Volc.*, **26**, 124–140, 1963.

M. TANAKA, A map of regional denudation rate in Japanese Mountains, *Japan. Geomorph. Union Trans.*, **3-2**, 159–167, 1982.

THE TIMES ALTAS OF THE WORLD, *London*: *Times Books*, 224 pp., 1977.

J. THIEDE, T. L. VALLIER *et al.*, Initial Reports of the Deep Sea Drilling Project, v. 62, Washington, U.S. Government Printing Office, 1120 pp., 1981.

J. A. VAN COUVERING, and W. A. BERGREN, The biostratigraphical basis of the Neogene time-scale, in Kauffman, E., and Hazel, J. E., editors, 1977, *New Concepts and Methods of Biostratigraphy*, pp. 283–362, Hutchinson and Ross, Inc., Strudsburg, PA, Dowden, 1977.

A. B. WATTS and J. K. WEISSEL, Tectonic history of the Shikoku marginal basin, *Earth and Planet Sci. Lett.*, **25**, 239–250, 1975.

A. B. WATTS, J. K. WEISSEL, and R. L. LARSON, Sea-floor spreading in marginal basins of the western Pacific, *Tectonophysics*, **37**, 167–181, 1977.

J. K. WEISSEL, Magnetic lineations in marginal basins of the Pacific, *Royal Soc. London*

Phil. Tras., Ser. A., **300**, 223–347, 1981.

J. K. WEISSEL and A. B. WATTS, Tectonic complexities in the South Fiji marginal basin, *Earth and Planet. Sci. Lett.*, **28**, 121–126, 1975.

J. K. WEISSEL and A. B. WATTS, Tectonic evolution of the Coral Sea Basin, *J. Geophys. Res.*, **84**, 4572–4582, 1979.

J. K. WEISSEL, A. B. WATTS, and A. LAPOUILLE, Evidence for Late Paleocene to Late Eocene seafloor in the southern Hebrides Basin, *Tectonophysics*, **87**, 243–25, 1982.

S. M. WHITE, H. CHAMLEY, D. M. CURTIS, G. deV. KLEIN, and A. MIZUNO, Sediment synthesis, Deep Sea Drilling Project Leg 58, Philippine Sea, in Klein, G. deV., Kobayashi, K. *et al.*, 1980, Initial Reports of the Deep Sea Drilling Project, v. 58, Washington, U.S. Government Printing Office, pp. 963–1014, 1980.

N. YONEKURA, Changing islands in the Pacific, *Kagaku*, **52**, 575–583, 1982 (in Japanese).

T. YOSHIKAWA, Denudation and tectonic movement in contemporary Japan, *Univ. of Tokyo Dept. of Geography Bull.*, **6**, 1–14, 1974.

Formation of Active Ocean Margins, edited by N. Nasu *et al.*, pp. 551–566.

WHEN WAS THE JAPAN SEA OPENED?: PALEOMAGNETIC EVIDENCE FROM SOUTHWEST JAPAN

Yo-ichiro OTOFUJI[1], Akira HAYASHIDA[2], and Masayuki TORII[3]

[1]*Department of Earth Sciences, Kobe University, Kobe 657, Japan*
[2]*Department of Earth Sciences, Doshisha University, Kyoto 602, Japan*
[3]*Department of Geology and Mineralogy, Kyoto University, Kyoto 606, Japan*

Abstract. Formation of the Japan Sea is viewed from paleomagnetic direction data reported from the southwestern part of Japan. Reliable paleomagnetic data-set is compiled to investigate exact timing of clockwise rotation of Southwest Japan. The following guidelines are adopted for selection of the data: (1) Samples were from the area about 800 km long between Amakusa and Toyama, where the clockwise declination shift in the Cretaceous and Paleogene rocks was already known. (2) Stability of remanent magnetization was examined by alternating field and thermal demagnetization treatments. (3) Paleomagnetic direction was obtained after tilting correction of strata. (4) Age of the rocks was determined through radiometric dating, correlation with a paleomagnetic polarity time scale, or biostratigraphic data of marine microfossils. Plot of the declinations, corrected for the apparent polar wonder effect, in respect to the age shows that Southwest Japan was abruptly rotated through about 50° at about 15 Ma. The best fit curve is obtained to describe the rotational motion; climax of the rotation is at 14.9 Ma and its duration is 0.6 m.y. Since the rotational motion of Southwest Japan is attributed to the back-arc spreading between Japanese Islands and the Asian Continent, it is concluded that the Japan Sea was opened at about 15 Ma.

1. *Introduction*

Marginal basins behind the trench-arc systems from Aleutian to New

Zealand characterize fundamental east-west geomorphological asymmetry of the Pacific Ocean margins (MOORE, 1973; UYEDA and KANAMORI, 1979). For the past decade, many geophysicists have elaborated to delineate maps of magnetic anomaly profiles, which document history of these marginal basins. Magnetic anomaly stripes suggesting sea floor spreading were identified in many marginal basins, and reliable ages were assigned for the oceanic crust of the marginal basins (SCLATER *et al.*, 1981).

The Japan Sea is one of the marginal basins in the northwestern Pacific. Its origin is, however, still quite nebulous. The Japan Sea is believed to have originated from sea floor spreading which occurred in a relatively recent geologic time, based on geological and geophysical evidence obtained from its basin (HILDE and WAGEMAN, 1973; ISEZAKI, 1975; LUDWIG *et al.*, 1975; WATANABE *et al.*, 1977). Magnetic anomalies in the Japan Sea, however, are much less pronounced than those observed in other ordinary oceans, so that the accurate age for the opening of the Japan Sea was not determined using the magnetic lineation (ISEZAKI and UYEDA, 1973). Deep-sea drilling in the Japan Sea would be an alternative to ascertain age of the basement. Unfortunately, the past trials in the Deep Sea Drilling Project failed in penetrating the entire sediments down to the oceanic basement, because detection of hydrocarbons in a layer in the holes prevented deeper drilling (KARIG, INGLE *et al.*, 1975).

Origin of the Japan Sea can be indirectly investigated through the work on the Japanese Islands, instead of the direct study in the sea, because the evidence for opening of the Japan Sea may be preserved on land surrounding it. Paleomagnetic studies of the Cretaceous and Paleogene rocks suggested that Southwest Japan has rotated clockwise through about 30° in the late Cenozoic time (SASAJIMA *et al.*, 1968; KAWAI *et al.*, 1971). Recent investigations further led to more precise determination of timing of the rotational motion (OTOFUJI and MATSUDA, 1983, 1984; HAYASHIDA and ITO, 1984; TORII, 1983). An aim of this paper is to evaluate the timing of the rotation through compilation of the paleomagnetic data from Southwest Japan, and consequently to estimate the timing of formation of the Japan Sea.

Southwest Japan is defined, in this paper, as an area where the declinations of the Cretaceous to Paleogene period show the clockwise deflection (YASKAWA, 1975; SASAJIMA, 1981); the area extends from Amakusa (31°N, 130°E) to Toyama (36°N, 137°E). The selected data is distributed widely in Southwest Japan from San'in (35° N, 132° E) to Shidara (35° N, 138° E), as shown in Fig. 1.

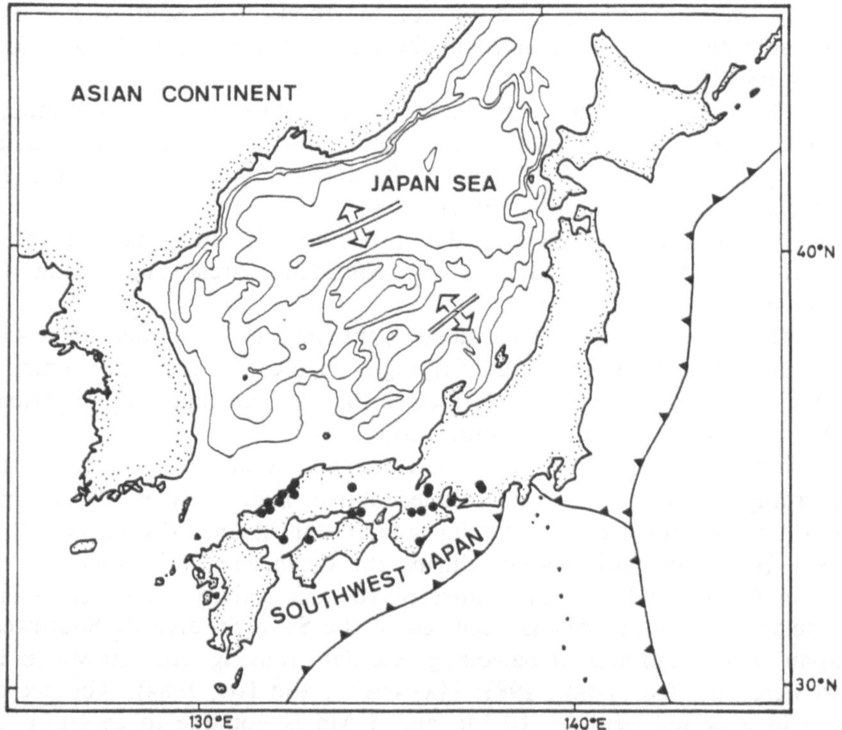

Fig. 1. Map showing the island arc—back arc basin system around Japan. Dots in Southwest Japan show locations where the reliable paleomagnetic data was obtained. Base map shows 1000-m interval contours and possible position of remnant spreading centers in the Japan Sea (after CIRCUM-PACIFIC MAP PROJECT, 1981). The present trench system around the Japanese Islands is also shown.

2. *Data Selection*

The rotation process of Southwest Japan is revealed through the declination change of remanent magnetization as a function of ages. Accuracy of the result depends on reliability of both age assignments and paleomagnetic directions. The following criteria were adopted to select reliable paleomagnetic data:

(1) Age: Radiometric ages are assigned for igneous rocks. Ages of sedimentary rocks are estimated either by magnetostratigraphic correlation to a standard polarity time scale, or using marine microfossils such as planktonic foraminifera (BLOW, 1969). Paleomagnetic data of the last

40 m.y. were taken into account, for the rotation of Southwest Japan was estimated to have occurred in the late Cenozoic time (OTOFUJI and MATSUDA, 1983).

(2) Demagnetization: Stability of remanent magnetization should have been examined by both alternating field and thermal demagnetization experiments. The data obtained without thermal demagnetization test was excluded from the compilation.

(3) Dispersion of magnetic directions: The site mean paleomagnetic directions which have a radius of 95% confidence circle smaller than 30° were selected.

(4) Tilt correction: It should be confirmed that the rock units suffered from little tectonic tilting. If tilting is appreciable, the bedding planes of the strata should be clearly defined so that the paleomagnetic directions can be corrected for the tectonic tilting.

Table 1 is a summary of the paleomagnetic data which were selected according to the guidelines mentioned above. Data between 20 Ma and 40 Ma come only from the San'in area (OTOFUJI and MATSUDA, 1983, 1984). Huge magmatic activity during the Oligocene was limited in this area in Southwest Japan. Subsequently, volcanic activity and development of small sedimentary basins occurred in the Setouchi area of Southwest Japan, where numbers of paleomagnetic data ranging from 20 Ma to 10 Ma were reported (TORII, 1983; HAYASHIDA and ITO, 1984). Absence of data in a period between 10 Ma and 5 Ma is not due to cessation of volcanic activity, but just to lack of reliable paleomagnetic data.

The paleomagnetic directions listed in Table 1 are values at a representative point of Southwest Japan (35°N, 134°E). Deflection of the declination is caused by both rotation of Southwest Japan and apparent polar wander. We calculated the expected field directions at the representative point using the apparent polar wonder curve of Eurasia (IRVING, 1977), and obtained rotation angles of Southwest Japan by subtracting the expected declination from the paleomagnetically observed values. The correction of the polar wander effect was made only for the data older than 30 Ma, because the paleomagnetic pole in younger periods is coincident with the present geographic pole (HARRISON and LINDH, 1982).

3. Rotation of Southwest Japan

Figure 2 shows plots of selected paleomagnetic directions of Southwest Japan, divided into four time intervals: 0–10 Ma, 10–20 Ma, 20–30 Ma and 30–40 Ma. While no systematic change is observed in inclination, it is shown from this figure that the easterly deflected declination turned to the north direction during an interval of 10–20 Ma.

Rotation angles of Southwest Japan, determined from the declination values, are plotted with respect to ages in Fig. 3. The plot of rotation angle vs. age for the last 40 m.y., combined with the Cretaceous to Paleogene paleomagnetic data, clearly delineates the following features:

(1) No rotation of Southwest Japan occurred until 20 Ma.

(2) A rotation occurred abruptly at about 15 Ma.

(3) Duration of the rotational motion was very short.

(4) Cessation of the motion was not gradual but sudden.

Relation of rotation angle and age can be expressed by some appropriate curves. We searched the analytically best fit function which could express the timing of the rotational motion objectively. The function should fulfill the following constraints:

(1) Prior to the rotational motion, the rotation angle, $R(t)$, is a constant value, R_0. After the end of the motion, $R(t)$ is equal to 0.

(2) The function does not retrogress as the time progresses.

(3) The function has a single value with respect to the time.

(4) The function has two parameters indicating the climax time (t_0) and the duration (T) of the rotation.

(5) If the duration is zero, the function should be a step function, that is, $R(t)=0$ for $t<t_0$, and $R(t)=R_0$ for $t>t_0$.

We chose the following function:

$$R(t)=R_0\left\{1-\frac{1}{\exp\left((t-t_0)/T\right)+1}\right\}$$

The parameter R_0 corresponds to the net angle through which Southwest Japan was subjected to rotation. The physical basis of this function is briefly described in the appendix.

The function has four principal characters (Fig. 4):

(1) The function provides a symmetric curve with respect to a point $t=t_0$ and $R(t)=R_0/2$.

(2) When $T=0$, the function represents a step function.

(3) $\mathrm{d}R(t)/\mathrm{d}t$ has a maximum value at $t=t_0$, corresponding to the climax of the rotational motion.

(4) The four fold of T corresponds to the length in time between the two points where the tangent line at $t=t_0$ crosses the two lines, $R(t)=0$ and $R(t)=R_0$.

Thus the rotational motion is described by the three parameters, R_0, t_0, and T.

The best fit function for the compiled set of the data was obtained

Table 1. List of paleomagnetic data (35°N, 134°E) from Southwest Japan.

Locality	SLAT	SLON	Age	D	I	N	α_{95}	R	ΔR	F	ΔF	ref
(40–30 Ma)												
Arifuku	35.0	132.2	35	86.3	46.3	4	38.1	76.6	43.5	12.6	28.5	1
Hirefuriyama	34.7	131.9	33	49.1	46.6	4	21.2	39.4	32.5	12.3	21.7	1
Yoboshiyama	34.6	131.9	33	46.4	34.1	2	8.2	36.7	11.8	24.8	9.4	1
Okami	34.8	131.9	33	64.8	60.5	3	16.5	55.1	35.8	−1.6	17.1	1
Kawamoto	35.0	132.5	33	61.6	52.7	7	6.4	51.9	12.4	6.2	7.8	1
Hamada	34.9	132.1	32	78.5	38.0	5	16.6	68.8	22.3	20.9	17.2	1
(30–20 Ma)												
Kawauchi	35.0	132.5	28	52.2	33.5	6	15.3	52.2	18.4	21.0	15.3	1
Hata	35.1	132.5	21	69.9	49.5	6	14.5	64.3	23.2	5.0	15.0	1

(20–10 Ma)												
Ichishi	34.7	136.4	16.4	44.6	47.4	11	11.9	44.6	17.7	7.1	11.9	2
N. Shidara	35.1	137.6	16	47.0	47.4	4	14.0	47.0	20.9	7.1	14.0	3
Morozaki	34.7	136.8	16.0	38.2	56.1	6	12.7	38.2	23.2	−1.6	12.7	4
Muro	34.6	136.5	15.3	60.7	64.9	21	3.7	60.7	8.8	−10.4	3.7	3
S. Shidara	35.1	137.6	15.3	5.1	56.7	10	8.9	5.1	16.4	−2.2	8.9	3
Kumano	33.8	136.0	15.2	57.5	48.8	3	20.6	57.5	32.3	5.7	20.6	5
Tonosho	34.5	134.3	15	42.5	60.5	4	11.1	42.5	23.0	−6.0	11.1	3
Tsuyama	35.1	134.1	15	23.4	37.3	4	11.4	23.4	14.4	17.2	11.4	3
Mt. Nijo	34.5	135.7	14.7	−8.8	47.6	5	17.5	−8.8	26.5	6.9	17.5	3
Takanawa	33.9	132.7	14.2	1.7	42.5	4	12.1	1.7	16.5	12.0	12.1	3
E. Yamaguchi	33.9	132.3	12.6	0.3	60.0	21	5.6	0.3	11.3	−5.5	5.6	3
Shodo-shima	34.5	134.3	12.4	7.3	49.1	17	6.1	7.3	9.3	5.4	6.1	3
(10–0 Ma)												
Sayama	34.9	136.2	2.9	−4.8	47.8	19	7.6	−4.8	11.4	6.7	7.6	6
Gamo	35.0	136.2	2.1	−3.5	52.2	26	6.1	−3.5	10.0	2.3	6.1	6

SLAT/SLON: representative latitude (°N)/longitude (°E) of locality. Age: radiometric or biostratigraphic age in Ma. D/I: mean declination/mean inclination in degrees. N: number of site or formation. α_{95}: radius of the 95% confidence circle. R, ΔR: rotation of declination and confidence angle. F, ΔF: flattenning of inclination and confidence angle. ref: reference number of data source, 1; OTOFUJI and MATSUDA (1983, 1984), 2; HAYASHIDA and ITO (1984), 3; TORII (1983), 4; HAYASHIDA (1984), 5; TAGAMI (1982), 6; HAYASHIDA and YOKOYAMA (1983).

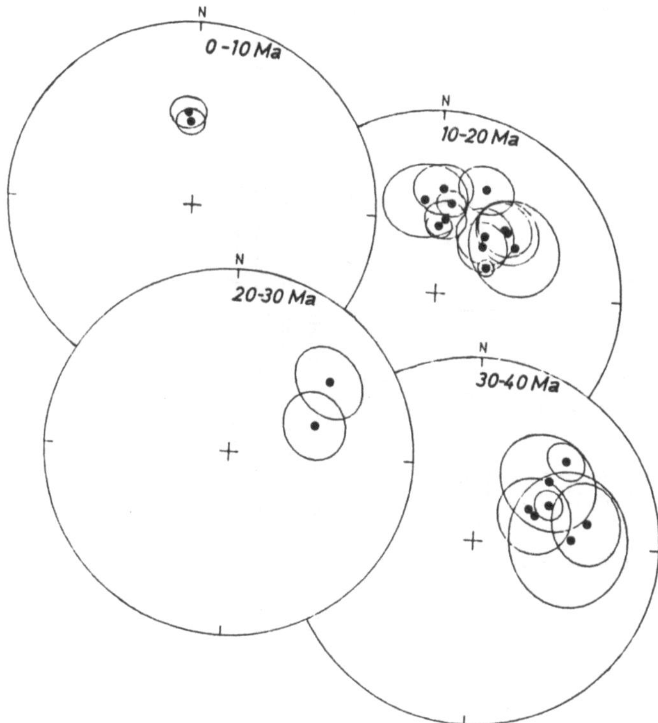

Fig. 2. Summary of reliable paleomagnetic data from Southwest Japan. Formation mean directions and 95% confidence circles are shown on equal-area projection for four time intervals: 0–10 Ma, 10–20 Ma, 20–30 Ma and 30–40 Ma. Projections are on the lower hemisphere.

by the method of a non-linear least square fitting, where the sum of squares of residuals of the rotation values weighted by 95% confidence angles was minimized. Error of the age estimation was not taken into account, because its resolution was assumed more accurate than that of the rotation angles. The other reason is that reliability of the age estimation was of various quality. The best fit curve obtained is drawn in Fig. 3. Values of the parameters were determined as follows: $R_0=47.1$ (°), $t_0=14.9$ (Ma), and $T=0.14$ (m.y.). Thus, we conclude that Southwest Japan rotated through 47° at 14.9 Ma for a transient duration of 0.6 m.y. (4T).

Our estimation of the timing of rotation is plausible for explanation of Miocene events in Southwest Japan. Recent geological studies elucidated that some of the geological features in Southwest Japan were changed in

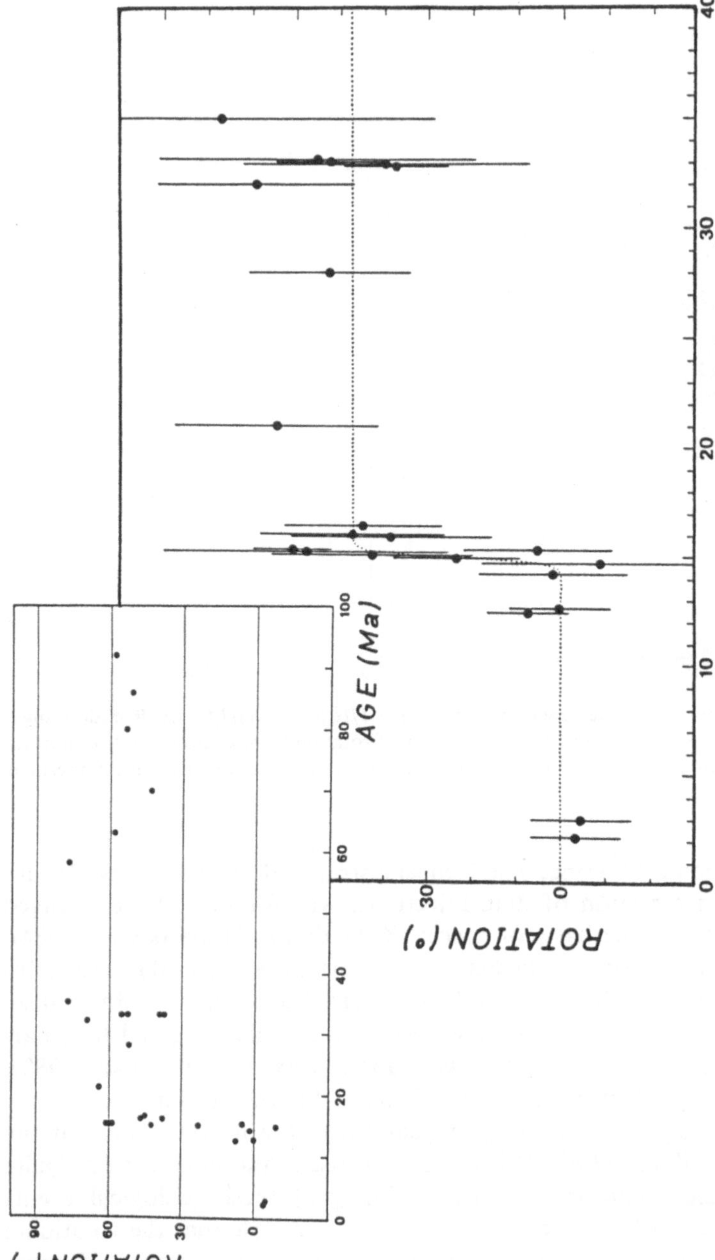

Fig. 3. Plot of rotation angles as a function of age. The error bar shows uncertainty in rotation at 95% confidence level: $\mathrm{Sin}^{-1}(\sin \alpha_{95}/\cos I)$, where I and α_{95} are inclination and radius of 95% confidence circle of the formation mean. Dotted line shows the best fit curve (see the text).

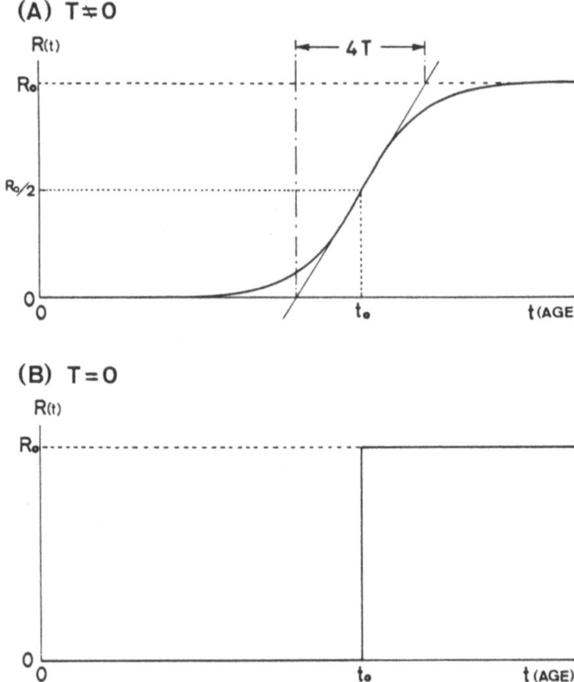

Fig. 4. Characteristics of the function: $R(t)=R_0\{1-1/(\exp((t-t_0)/T)+1)\}$. Rotation angle, $R(t)$, reaches $R_0/2$ at t_0, when the angular velocity of rotation is maximized. Four fold of the parameter T represents the duration of rotational motion. (A): T is finite. (B): T is infinitesimal.

the early to middle Miocene. The Neogene stress field in Southwest Japan, estimated from direction of dike injections, was found to have changed at about 15 Ma on the basis of precise K-Ar dating (TSUNAKAWA, 1984). Small sedimentary basins of the Setouchi area (ITOIGAWA, 1981) were formed before the middle Miocene and disappeared at about 15 Ma. Subsequently, at about 14 Ma, short-lived volcanic activity occurred in a narrow zone which extends east and west about 1000 km (TATSUMI, 1983). The forearc magmatism in the Outer Zone of Southwest Japan (SHIBATA and NOZAWA, 1967) and off-ridge volcanism in the northern part of the Shikoku Basin (KLEIN *et al.*, 1978) also occurred in the same period. Quite a good agreement appears to exist in timing of these geological events and the rotation of Southwest Japan. It is conceivable that the rotational motion is a main cause of the contemporaneous occurrence of these geological phenomena. Both of the rotation and other geological events may be explained by a comprehensive tectonic situation of Southwest Japan at about 15 Ma.

4. *When Was the Japan Sea Opened?*

The rotation of Southwest Japan has been explained as a consequence of the opening of the Japan Sea (e.g., UYEDA and MIYASHIRO, 1974). Several other causes, however, can be postulated to explain the rotational motion of land blocks.

According to the plate tectonics, plate boundaries are classified into the three tectonic features: (a) subduction zone, (b) transform fault, and (c) spreading center (LE PICHON *et al.*, 1973). Possibility of the past existence of each plate boundary between Southwest Japan and the Eurasia will be considered here.

(a) Subduction zone: Collision of the Kyushu Palau ridge, an aseismic ridge in the Philippine Sea (NUR and BEN-ABRAHAM, 1982), is a possible trigger of the rotation of Southwest Japan; the ridge pushed Southwest Japan, causing its clockwise rotation around the pole at the eastern end. The northern part of Southwest Japan was consumed along the subduction zone between Southwest Japan and the Eurasia plate in this hypothetical process [Fig. 5(a)]. The geological evidence, however, suggests that neither a subduction zone nor an orogenic belt existed in the Cenozoic time between the Korean Peninsula and Southwest Japan (LEE, 1974; KIM, 1974; UM and REEDMAN, 1975; TOMITA *et al.*, 1975). Therefore, possibility of the rotation accompanied by the plate convergence is eliminated.

(b) Transform fault: The clockwise deflection of paleomagnetic declinations could be attributed to rotations of several small fragments of Southwest Japan, which might occur along a transfrom fault bordering the Eurasia plate [Fig. 5(b)]. Such rotations of small rounded blocks between the Pacific and North American plates were proposed to explain the tectonics of western North America (BECK, 1976), and called the ball-bearing motion. It is evident, however, that Southwest Japan does not consist of round-shaped blocks but of zonally arranged terranes extending east and west. The zonal arrangement is supposed to have formed far before the Neogene, and, therefore, possibility of a ball-bearing motion of Southwest Japan at about 15 Ma is eliminated.

(c) Spreading center: When a back-arc basin is opened with different rate of spreading along the spreading center, the island arc may be rotated [Fig. 5(c)]. The Japan Sea is believed to have been created by back-arc spreading, based on the following evidence: (1) Analysis of magnetic

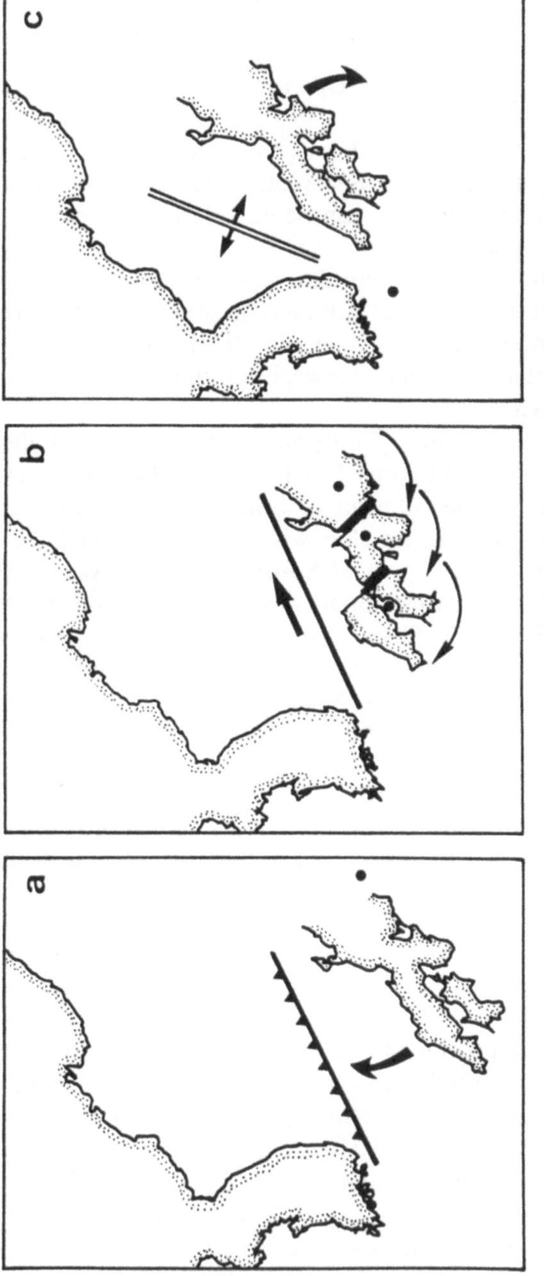

Fig. 5. Possible models for clockwise rotation of Southwest Japan. (a) Subduction toward the Asian continent. (b) Ball-bearing motion along a transform fault. (c) Back-arc spreading in the Japan Sea.

anomalies suggested that the Japan Sea has remnant spreading centers, which trend almost parallel to the Japanese Islands (ISEZAKI, 1975). (2) Distribution of heat flow and crustal structure (LUDWIG *et al.*, 1975; WATANABE *et al.*, 1977) indicated that the spreading occurred at a relatively young period in the Cenozoic. (3) Southwest Japan and the Korean Peninsula are fitted after counter-clockwise rotation of Southwest Japan by the angle of about 55°, which is comparable to the paleomagnetic estimation of the rotaional angle (47°). These results imply that the fan-shape spreading in the southwestern Japan Sea caused the clockwise rotation of Southwest Japan.

We can estimate the timing of the opening of the Japan Sea, based on the paleomagnetic results from Southwest Japan. Assuming that the spreading activity rifted the Japanese Islands from the Asian Continent and the Korean Peninsula, the opening of the Japan Sea must be synchronized with the rotation of Southwest Japan. The opening of the Japan Sea is, therefore, dated at 14.9 Ma, and its duration is estimated as short as 0.6 m.y.

Stratigraphical and paleontological studies of the Neogene sequences around the Japan Sea support the middle Miocene opening of the Japan Sea. INGLE (1975) showed that marine transgression and subsidence of the Japan Sea coast occurred at about 15 Ma around the early/middle Miocene boundary. Evidence for the middle Miocene transgression is well preserved in the Oga Peninsula and the Sado Island on the northeastern coast of the Japan Sea, and in the Pohan area of the Korean Peninsula. Similar stratigraphic records are widely observable along the Japan Sea coast of Southwest Japan (TSUCHI, 1981). Thickness of the sediments accumulated on the oceanic basement of the Japan Sea (KARIG *et al.*, 1975) is compatible to cover the period since the middle Miocene.

5. Conclusions and Their Restriction

Our compilation of the reliable paleomagnetic data from Southwest Japan allowed to reach the following conclusions:

(1) Southwest Japan was rotated clockwise at 14.9 Ma.

(2) Rotation was caused by back-arc spreading of the Japan Sea.

(3) The Japan Sea was opened at 14.9 Ma and duration of the opening was as short as less than 1 m.y.

(4) In a strict sense, the timing of the opening determined here is that of the southwestern part of the Japan Sea.

Acknowledgements

We are much indebted to Professor Sadao Sasajima who influenced our ideas and studies throughout of this work. We thank an anonymous reviewer for constructive criticism of this manuscript.

APPENDIX

A simplified model is considered in which Southwest Japan stayed at a stable position for a long time and began to rotate toward another stable position by an angle R_0 (°). If the rotational velocity is symmetric with respect to the climax of rotation at $t=t_0$, the simplest differential equation for the rotational motion is represented as:

$$\frac{\mathrm{d}R}{\mathrm{d}t}=R(R_0-R)/T$$

and the boundary condition is:

$$\frac{\mathrm{d}^2R}{\mathrm{d}t^2}=0 \quad \text{at} \quad t=t_0$$

where T is a parameter indicating the angular velocity of rotation. Solution of the equation is:

$$R(t)=R_0\left\{1-\frac{1}{\exp\left((t-t_0)/T\right)+1}\right\}.$$

REFERENCES

BECK, M. E. Jr., Discordant paleomagnetic pole positions as evidence of regional share in the Western Cordillera of North America, *Am. J. Sci.*, **276**, 694–712, 1976.

BLOW, W. H., Late Middle Eocene to Recent planktonic foramini feral biostratigraphy, Proc. 1st Internat. Conf. Planktonic Microfossils, Geneva, pp. 199–422, 1969.

CIRCUM-PACIFIC MAP PROJECT, Plate Tectonic Map of the Circum-Pacific Region, Northwest Quadrant, Am. Assoc. Petrol. Geologists, 1981.

HARRISON, C. G. A. and T. LINDH, A polar wandering curve for North America during the Mesozoic and Cenozoic, *J. Geophys. Res.*, **87**, 1903–1920, 1982.

HAYASHIDA, A., A paleomagnetic study of Miocene strata in the Chita Peninsula, central Japan, in preparation.

HAYASHIDA, A. and Y. ITO, Paleoposition of Southwest Japan at 16 Ma: Implication from paleomagnetism of Miocene Ichishi Group, *Earth Planet. Sci. Lett.*, **68**, 335–342, 1984.

HAYASHIDA, A. and T. YOKOYAMA, Paleomagnetic chronology of the Plio-Pleistocene Kobiwako Group to the southeast of Lake Biwa, central Japan, *J. Geol. Soc. Japan*,

89, 209–221, 1983.

HILDE, T. W. C. and J. M. WAGEMAN, Structure and origin of the Japan Sea, in *The Western Pacific—Island Arcs, Marginal Seas, Geochemistry—*, edited by P. J. Coleman, pp. 415–434, University of Western Australia Press, 1973.

INGLE, J. C., Jr., Summary of Late Paleogene-Neogene insular stratigraphy, paleobathymetry, and correlations, Philippine Sea and Sea of Japan region, in Initial Reports of the Deep Sea Drilling Project, Vol. 31, edited by J. C. Ingle, Jr. and D. E. Karig, pp. 837–855, U.S. Government Printing Office, Washington, D.C., 1975.

IRVING, E., Drift of the major continental blocks since the Devonian, *Nature*, **270**, 304–309, 1977.

ISEZAKI, N., Possible spreading centers in the Japan Sea, *Mar. Geophys. Res.*, **2**, 265–277, 1975.

ISEZAKI, N. and S. UYEDA, Geomagnetic anomaly pattern of the Japan Sea, *Mar. Geophys. Res.*, **2**, 51–59, 1973.

ITOIGAWA, J., Mizunami area, in *Neogene of Japan—Its Biostratigraphy and Chronology—*, edited by R. Tsuchi, pp. 62–64, IGCP-114 National Working Group of Japan, 1981.

KARIG, D. E., J. C. INGLE, Jr., *et al.*, Initial Reports of the Deep Sea Drilling Project, Vol. 31, 927 pp., U.S. Government Printing Office, Washington, D.C., 1975.

KAWAI, N., T. NAKAJIMA, and K. HIROOKA, The evolution of the Island Arc of Japan and the formation of granites in the Circum-Pacific Belt, *J. Geomag. Geoelectr.*, **23**, 267–293, 1971.

KIM, O. J., Geology and tectonics of South Korea, United Nations ESCAP, *CCOP Tech. Bull.*, **11**, 18–37, 1974.

KLEIN, G. V., K. KOBAYASHI, H. CHAMLEY, K. M. CURTIS, H. J. B. DICK, D. J. ECHOLS, D. M. FOUNTAIN, H. KINOSHITA, N. G. MARSH, A. MIZUNO, G. V. NISTERENKO, H. OKADA, J. R. SLOAN, D. M. WAPLES, and S. M. WHITE, Off-ridge volcanism and seafloor spreading in the Shikoku Basin, *Nature*, **273**, 746–748, 1978.

LEE, S. M., The tectonic setting of Korea with relation to plate tectonics, United Nations ESCAP, *CCOP Tech. Bull.*, **8**, 39–53, 1974.

LE PICHON, X., J. FRANCHETEAU, and J. BONNIN, *Plate Tectonics*, Developments in Geotectonics, Vol. 6, Elsevier, New York, 1973.

LUDWIG, W. J., S. MURAUCHI, and R. E. HOUTZ, Sediments and structure of the Japan Sea, *Geol. Soc. Am. Bull.*, **86**, 651–664, 1975.

MOORE, G. W., Westward tidal lag as the driving force of plate tectonics, *Geology*, **1**, 99–100, 1973.

NUR, A. and Z. BEN-AVRAHAM, Oceanic plateaus, the fragmentation of continent, and mountain building, *J. Geophys. Res.*, **87**, 3644–3661, 1982.

OTOFUJI, Y. and T. MATSUDA, Paleomagnetic evidence for the clockwise rotation of Southwest Japan, *Earth Planet. Sci. Lett.*, **62**, 349–359, 1983.

OTOFUJI, Y. and T. MATSUDA, Timing of rotational motion of Southwest Japan inferred from paleomagnetism, *Earth Planet. Sci. Lett.*, **70**, 373–382, 1984.

SASAJIMA, S., Pre-Neogene paleomagnetism of Japanese Islands (and vicinities), in *Paleoreconstruction of the Continents*, Geodyn. Ser., Vol. 2, edited by M. W. McElhinny and D. A. Valencio, pp. 115–128, A.G.U., Washington, D.C. and G.S.A., Boulder, Colorado, 1981.

SASAJIMA, S., J. NISHIDA, and M. SHIMADA, Paleomagnetic evidence of a drift of the Japanese Main Island during the Paleogene period, *Earth Planet. Sci. Lett.*, **5**, 135–141, 1968.

SCLATER, J. G., B. PARSONS, and C. JAUPART, Oceans and continents: Similarities and differences in the mechanism of heat loss, *J. Geophys. Res.*, **86**, 11535–11552, 1981.

SHIBATA, K. and T. NOZAWA, K-Ar ages of granitic rocks from the outer zone of Southwest Japan, *Geochem. J.*, **1**, 131–137, 1967.

TAGAMI, T., Paleomagnetism and fission-track ages of the Kumano acidic rocks, *NOM (News of Osaka Micropaleontologists)*, **9**, 23–32, 1982.

TATSUMI, Y., High-magnesian andesites in the Setouchi volcanic belt, Southwest Japan, and their possible relation on the evolutionary history of the Shikoku inter-arc basin, in *Geodynamics of the Western Pacific Indonesian Region*, Geody. Ser., Vol. 11, edited by T. W. C. Hilde and S. Uyeda,pp. 331–341, A.G.U., Washington, D.C. and G.S.A., Boulder, Colorado, 1983.

TOMITA, S., A. YAMASHITA, K. ISHIBASHI, T. MIKI, R. TAKAHASHI, T. SHUTO, H. URATA, I. HASHIMOTO, E. HONZA, and C. IGARASHI, Submarine geology west of the Tsushima Islands, *Sci. Rept. Dept. Geol., Kyushu Univ.*, **12**, 77–90, 1975.

TORII, M., Paleomagnetism of Miocene rocks in the Setouchi Province: Evidence for rapid clockwise rotation of Southwest Japan at middle Miocene time, Ph. D. Thesis of Kyoto University, 1983.

TSUCHI, R. and IGCP-114 NATIONAL WORKING GROUP OF JAPAN, Bio- and Chronostratigraphic correlation of Neogene sequences in the Japanese Islands, in *Neogene of Japan—Its Biostratigraphy and Chronology—*, edited by R. Tsuchi, pp. 91–104, IGCP-114 National Working Group of Japan, Shizuoka, 1981.

TSUNAKAWA, H., Neogene stress field of the Japan arc and its relation to the igneous activity, submitted to *Tectonophysics*, 1984.

UM, S. H. and A. J. REEDMAN, *The Geology of Korea*, 139 pp., Gological and Mineral Institute of Korea, Seoul, 1975.

UYEDA, S. and H. KANAMORI, Back-arc opening and the mode of subduction, *J. Geophys. Res.*, **84**, 1049–1061, 1979.

UYEDA, S. and A. MIYASHIRO, Plate tectonics and the Japanese Islands: A synthesis, *Geol. Soc. Am. Bull.*, **85**, 1159–1170, 1974.

YASKAWA, K., Palaeolatitude and relative position of Southwest Japan and Korea in the Cretaceous, *Geophys. J. R. astr. Soc.*, **43**, 835–846, 1975.

WATANABE, T., M. G. LANGSETH, and R. N. ANDERSON, Heatflow in back-arc basins of the Western Pacific, in *Island Arcs, Deep Sea Trenches and Back-Arc Basins*, edited by M. Talwani and W. C. Pitman III, pp. 137–161, A.G.U., Washington, D.C., 1977.

Formation of Active Ocean Margins, edited by N. Nasu *et al.*, pp. 567–591.

FORMATION OF THE OKINAWA TROUGH

Masaaki KIMURA

Department of Marine Sciences, College of Science, University of the Ryukyus, Nishihara, Okinawa 903-01, Japan

Abstract. Origin and history of the Okinawa Trough and its surrounding areas were discussed based upon data of seismic reflection, seismic refraction, dredge hauls and drilling compiled from various sources. As a result, it was concluded that major phase of rifting of the Okinawa Trough started in Late Pliocene to Early Pleistocene time. The lithospheric plate in the northern half of the Okinawa Trough has not yet spread but its axial part has been rifted en echelon. On the other hand, the axial zone of the southern most part might have been opened to a width of about 40 km since the Early Pleistocene with an estimated half-spreading rate of about 1~2 cm/yr. The Okinawa Trough is defined as an elongated topographic low which is a narrow graben with a width of about 100 km. The volcanic front (Tokara Volcanic Ridge), west of the Ryukyu Ridge, is regarded as a rifted eastern margin of the Okinawa Trough, while the Tunghai Slope, west of the Okinawa Trough is included in the western, rifted margin of the trough. Judging from the mode of sedimentation of the Upper Miocene to Lower Pleistocene Shimajiri Group, the Okinawa Trough was formed after the deposition of the Shimajiri Group.

1. Introduction

Based upon geologic evidence of Ryukyu Islands fringing the eastern margin of the Okinawa Trough, KONISHI and SUDO (1973) suggested that the Okinawa Trough (Fig. 1) has been spreading since Middle Miocene time. Using the single channel airgun reflection profiles, HERMAN *et al.* (1978) proposed that there is an unconformity between the upper and lower sediments (their Unit A and B) in the southern Okinawa Trough, and that main spreading occurred before deposition of Unit B after the formation of the unconformity, sometime after Late Miocene- Early Pliocene

Fig. 1. Survey lines including single and multi-channel reflection survey lines from various sources. Detailed surveys were carried out by the Geological Survey of Japan and the Hydrographic Department, MSA of Japan (1) and Hydrographic Department, MSA of Japan (2). The Inserted index map around the Okinawa Trough area is adopted from MOGI (1972).

time, and that a very slow spreading started again in Quaternary time. The two-ship refraction survey by LEE *et al.* (1980) in the southern Okinawa Trough showed that the upper mantle is relatively shallow compared to the surrounding continental masses. They also found relatively young basalt beneath the central axial part. They postulated that crustal separation occurred beneath the southern Okinawa Trough from Pliocene to Recent time.

In this paper the author summarizes available seismic reflection, refraction, dredge and oil-test drilling data covering the entire Okinawa Trough. Some geologic evidence for the Quaternary rifting of the Okinawa Trough is presented.

2. *Data*

The main source of seismic single channel reflection data (Figs. 1 and 2) is the Geological Survey of Japan. The data were compiled by HONZA (1976) and KIMURA *et al.* (1979, 1980). Data collected by the Ocean Research Institute, University of Tokyo (KAGAMI, 1975), NOAA and Lamont-Doherty Geological Observatory of Columbia University were also compiled. Single channel airgun data around Kyushu and Ryukyu Islands by the Hydrographic Department, Maritime Safety Agency of Japan were also included in the present analysis. Multi-channel seismic reflection profiles obtained by the Japan National Oil Corporation (JNOC) were also used. Seismic refraction data were mainly taken from "A Geophysical Atlas of the East and South Asian Seas" (HAYES *et al.*, 1978) and from a paper by LEE *et al.* (1980). Samples dredged by the Geological Survey of Japan and the Ocean Research Institute, University of Tokyo were studied. Offshore drilling data for oil research carried out by JNOC and other petroleum companies were partly referred to (Fig. 3).

Submarine topography in and around the Okinawa Trough is contoured using the Bathymetric Chart of Ocean (G1304-06, G1404-06) of the Hydrographic Department, MSA of Japan, checked by the author's own data. Unpublished data of the ocean bottom relief in the Philippine Basin by the Department is used here by their courtesy.

3. *Results*

3.1 *Submarine topography*

A submarine topographical map (Fig. 4) is shown, in which the contour interval is 100 m for areas shallower than 2000 m and 1000 m for areas deeper than 2000 m in water depth except a part of the Okinawa

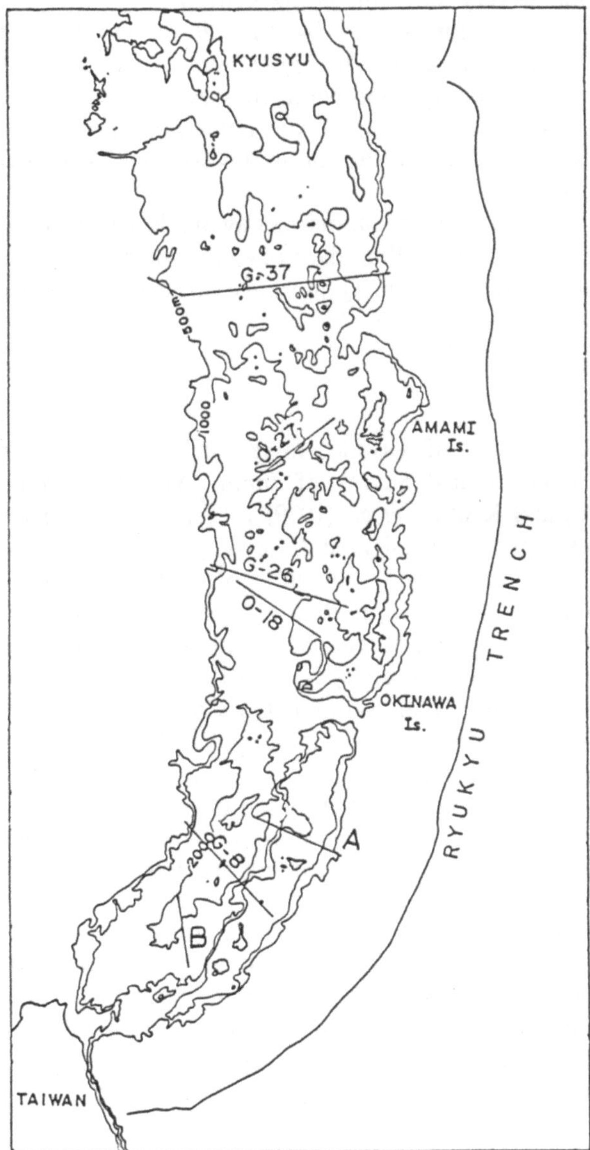

Fig. 2. Location map of seismic reflection profiling records shown in this paper. As for the survey numbers, "G-37, -26, and -8" represents those by the Geological Survey of Japan, "0-27-18" the Ocean Research Institute, University of Tokyo and "A and B" JNOC.

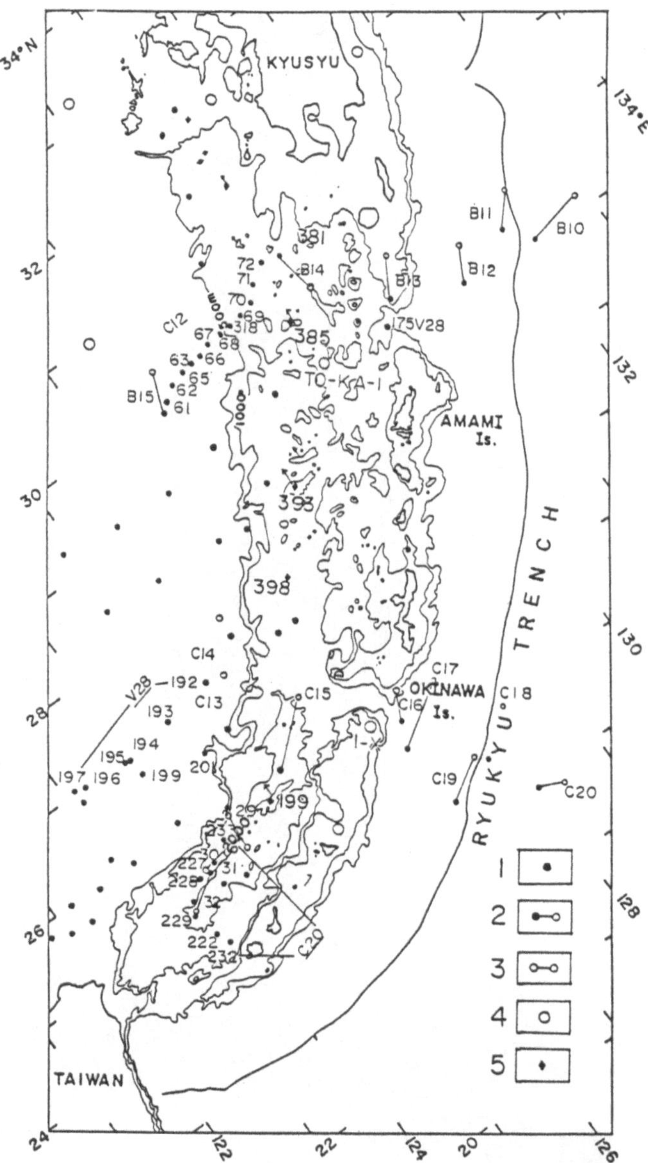

Fig. 3. Location map of radio sonobuoy, drilling and dredging, carried out by various organization. 1: Radio-sonobuoy station, 2: Two-ship refraction station. Reversed (filled circle correspond to left side of section which shows true thickness), 3: Unreversed, 4: Submarine drilling station, 5: Dredging station obtaining basement rock.

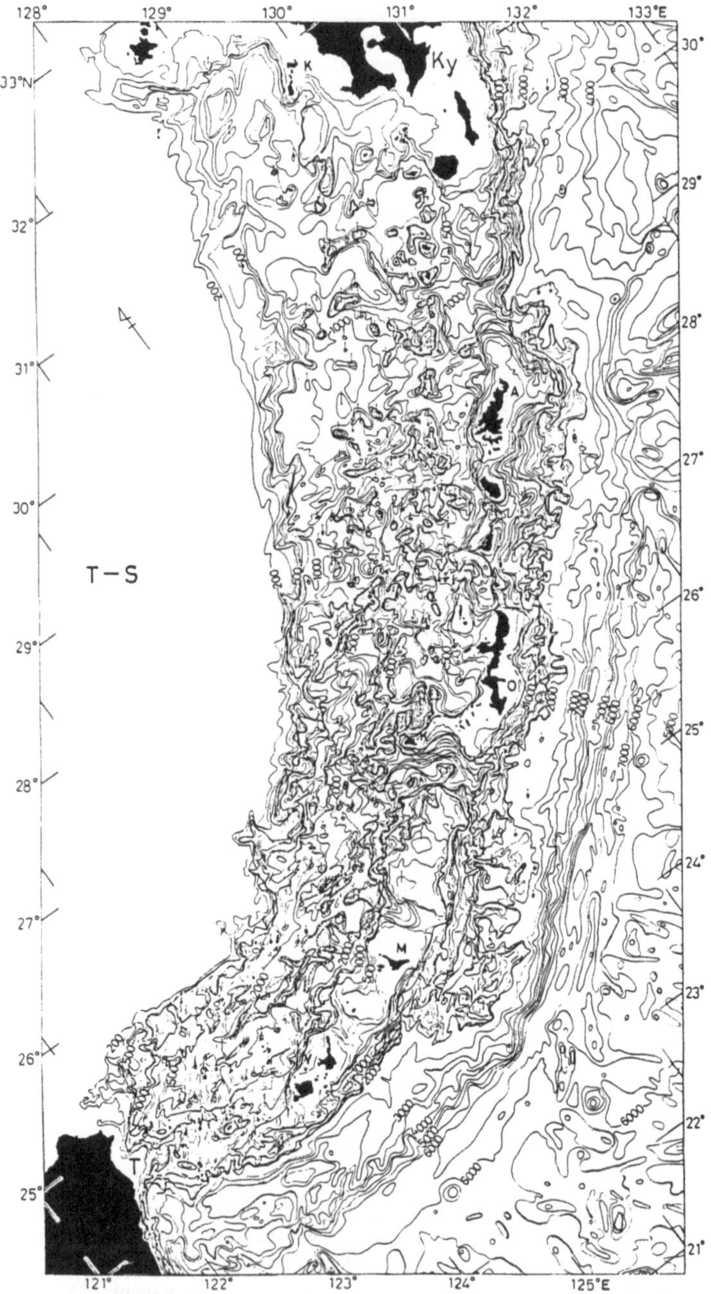
128°
129°
130°
131°
132°
133°E
33°N
32°
31°
30°
29°
28°
27°
26°
25°
30°
29°
28°
27°
26°
25°
24°
23°
22°
21°
121°
122°
123°
124°
125°E
Ky
K
A
O
M
T
T-S

Fig. 4

Trough. The Okinawa Trough is regarded as a back arc depression. Water depths of the trough become deeper from north to south; from about 500–700 m to 2300 m. The width of the trough is about 100 km. The entire length of the trough exceeds 1000 km.

In general, the Ryukyu Arc can be divided into three parts along its strike. The northern part is located north of the Tokara structural strait (30°N), and the middle part is between 30°N and Kerama Gap (Miyako Depression) (26°N). The southern part is south of 26°N. The Okinawa Trough is also divided into three parts at almost the same latitudes. Water depths range from 1500 to 1900 m in the northern half of the trough, and from 1500 to 2300 m in the south of 26°N. In the northern and southern parts, topography of the bottom of the trough is smooth, whereas it is rough in the middle of the trough. The northernmost part of the trough is called Danjo Basin and in the southern part there exist two large basins named Miyako and Yaeyama Basins (KIMURA, 1983). In the middle to southern Okinawa Trough, narrow depressions trending WSW-ENE develop en echelon (Fig. 5). Such a feature is not clearly seen in the northern part.

In the northern half of the Ryukyu chain, two parallel arcs can be recognized; the inner, volcanic arc (Tokara Ridge) and the outer, inactive arc (Ryukyu Ridge). In contrast, only one arc topography exists in the southern half of the Ryukyu chain. In the north these three topographic features; Okinawa Trough, Tokara Ridge and Ryukyu Ridge are separated by steep escarpments having relative heights of several hundred meters.

3.2 *Structural framework*

A tectonic map (Fig. 6) was drawn to show deep and active faults deduced from reflection seismic, dredging and drilling data. Tectonic provinces are defined from west to east as I (Tunghai Shelf), II (Tunghai Slope), III (Okinawa Trough), IV (Tokara Ridge) and V (Ryukyu Ridge) (Fig. 7). The Okinawa Trough is bounded by deep faults. The eastern boundary fault is here named the "Tokara Ridge Fault". This is essentially covered by younger sediments (Layer B_1 and A), but they are offset

Fig. 4. Submarine topography in and around the Okinawa Trough. The contour interval is 100 m for areas shallower than 2000 m, and is 1000 m for areas deeper than 2000 m. Contours are drawn from the bathymetric charts of ocean (G 1304-06, G 1404-06) of Hydrographic Department, MSA of Japan. The ocean bottom relief in the Philippine Basin is retaken from unpublished data of the Japanese Hydrographic Department, MSA of Japan. D: Danjo Basin, Ky: Kyushu, K: Koshiki-jima, A: Amami-Oshima, O: Okinawa-jima, M: Miyako-jima, Y: Yaeyama Islands, T: Taiwan, T-S: Tunghai Shelf.

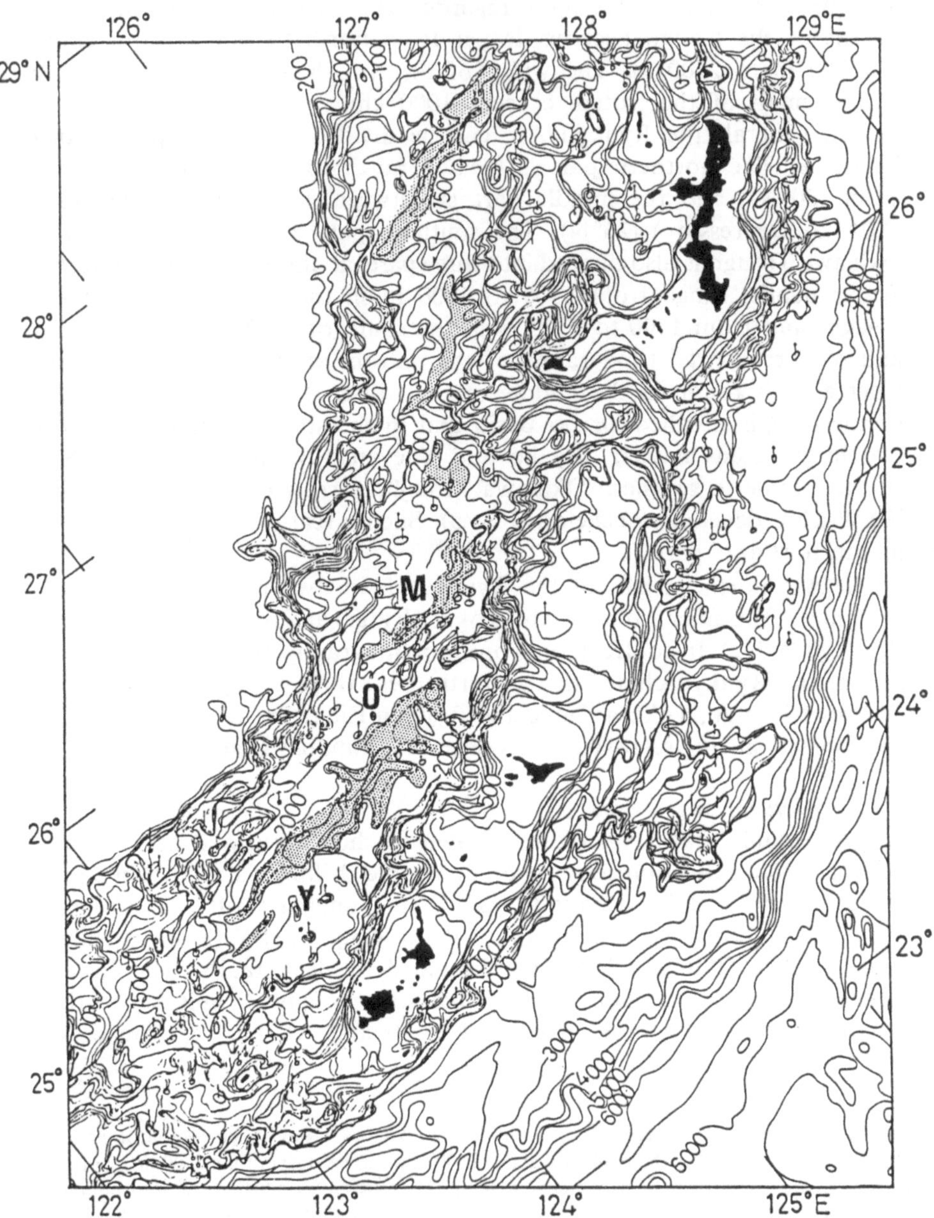

Fig. 5. Submarine topography showing central depressions (dotted areas) in en echelon. M: Miyako Basin, Y: Yaeyama Basin, O: Onodera Seamount.

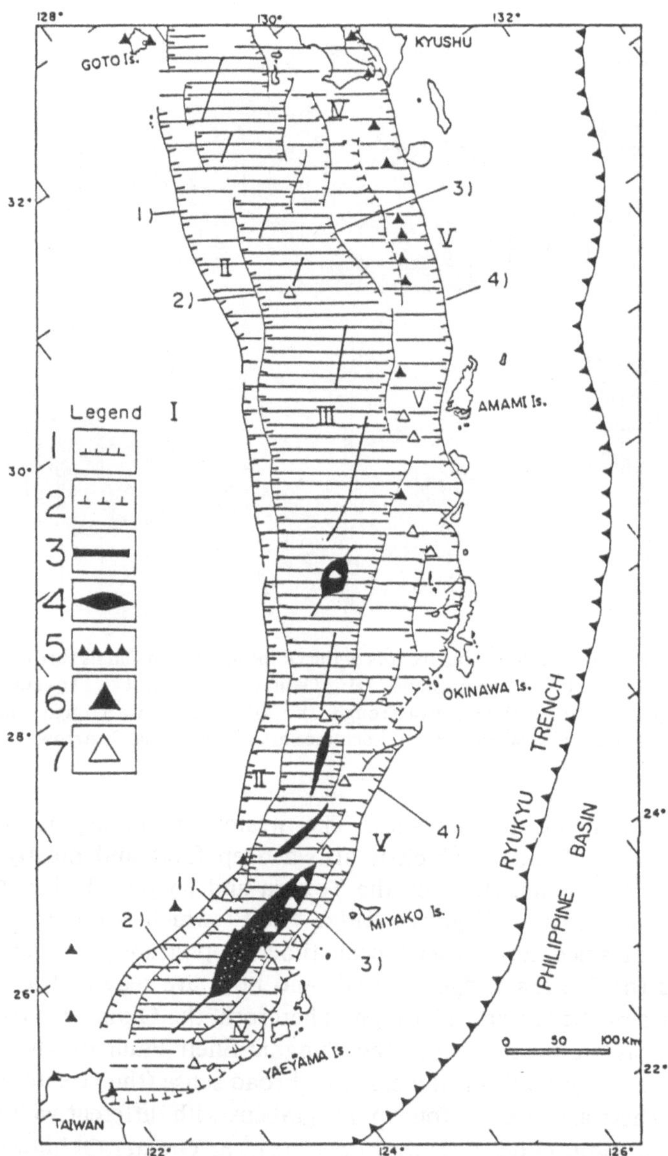

Fig. 6. Tectonic fromework of the Okinawa Trough. 1: Quaternary fault, 2: Estimated Quaternary fault, 3: Central Graben, 4: Area occupied by Quaternary volcanics, 5: Trench, 6: Quaternary volcano (ensured; mainly on the air), 7: Quaternary volcano (newly recognized by geologic and geophysical data; submarine), I: Tunghai Shelf, II: Tunghai Slope, III: Okinawa Trough, IV: Tokara Ridge, V: Ryukyu Ridge, 1): Tunghai Shelf Fault, 2): Tunghai Slope Fault, 3): Tokara Ridge Fault, 4): Ryukyu Ridge Fault.

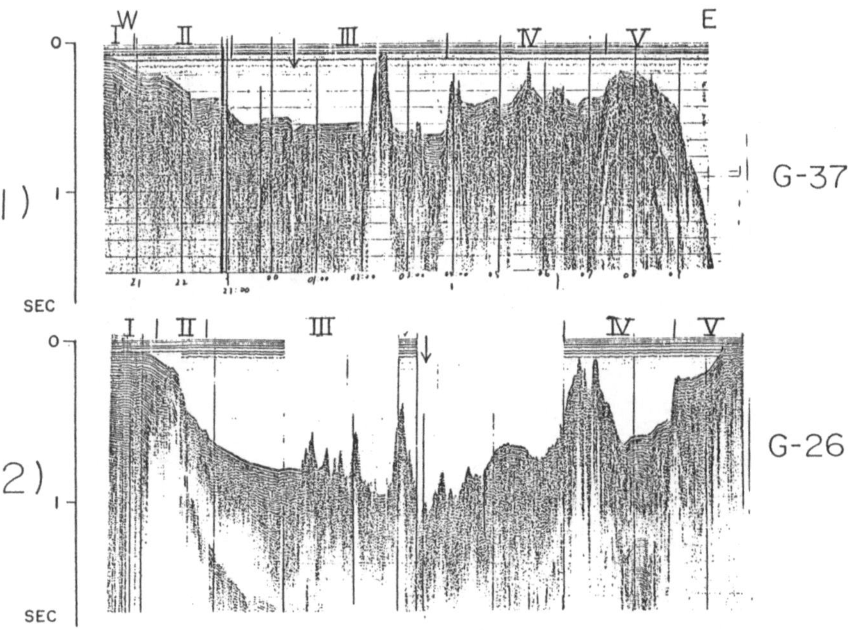

Fig. 7. Seismic reflection profiling records of the Okinawa Trough and its surrounding areas. I: Tunghai Shelf, II: Tunghai Slope, III: Okinawa Trough, IV: Tokara Ridge, V: Ryukyu Ridge. 1): Northern Okinawa Trough, 2): Middle Okinawa Trough. There are less sediments in the Central Graben (arrow). Locations are shown in Fig. 2. Arrows represent the Central Grabens.

by minor faults at many places. The western boundary is called the "Tunghai Slope Fault " which is also a deep fault and mostly covered by younger sediments. Between the Tokara and Ryukyu Ridges there exists a fault named the "Ryukyu Ridge Fault" which is regarded to have been formed since Early Pleistocene time. Topography of the Tunghai Slope and the Tokara Ridge is rough and both are regarded as marginal rifted zones of the Okinawa Trough. Therefore, the Okinawa Trough may be defined as the area between the Tunghai Shelf Fault on the west and the Ryukyu Ridge Fault on the east in a broad sense (the Greater Okinawa Trough). There are three or four major grabens with different widths behind the Ryukyu Ridge (Figs. 6, 7) as follows. (1) The Greater Okinawa Trough, the widest depression and bounded by the Tunghai Shelf Fault and the Ryukyu Ridge Fault. (2) Okinawa Trough; bounded by the Tunghai Slope Fault and the Tokara Ridge Fault. (3) Central Graben; the central axial low along the entire length of the Okinawa Trough (Figs. 6, 7 and 8).

Essentially in the margins of grabens (1) and (2) sediments of Layer

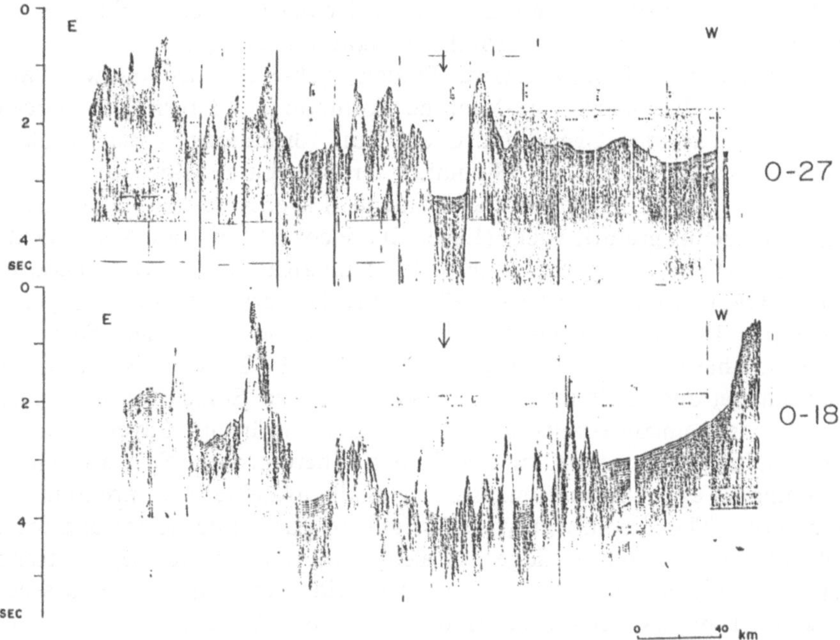

Fig. 8. Seismic reflection profiles showing Central Grabens in the middle part of the Okinawa Trough. Locations are shown in Fig. 2. Arrows show Central Graben. Vertical scale represents two way travel times.

C are faulted and covered by Layers B_2 and A, while in the Central Graben (3) the uppermost sedimentary layer (Layer A) is faulted. This means that the Central Graben is the most active now. Two grabens, the Greater Okinawa Trough and the Okinawa Trough are mostly parallel to the Ryukyu Ridge, whereas the Central Graben is not parallel to the ridge but en echelon. The Central Graben has topographic features typical of median graben, with rift valley on the seafloor and in many cases volcanic plug beneath the rift. The Central Graben probably originates from uprising magma. The existence of the Central Graben of the Okinawa Trough had been suggested by KIMURA *et al.* (1975) in the middle Okinawa Trough and the Central Grabens were found in the southern part by HERMAN *et al.* (1978). Similar topographic features are thereafter found along the entire length of the trough. Figures 7 and 8 show these profiles. Seismicity study (EGUCHI, 1982) shows that the Central Grabens are active and are extensional in the NS direction. The Tokara Ridge, with water depths ranging from about 500 to 1000 meters, is characterized by a volcanic ridge which has been active since Neogene time (so-called Green Tuff activity) in the northern and middle Okinawa Trough. However, such a ridge is not topographically clear in the southern Okinawa Trough.

Stratigraphy of the Okinawa Trough (Table 1) is as follows; Layer E: Acoustic basement (pre-Miocene formations), Layer D: Miocene sediments (Yaeyama Group and Green Tuff formations in the middle to northern parts), Layer C: Shimajiri Group (Late Miocene to Early Pleistocene), Layer B: Pleistocene sediments, Layer A: Holocene sediments. Late Cretaceous granitic rocks (Layer E) are covered by Late Miocene lave flow (6 Ma) and sediments, both of the Shimajiri Group, with unconformity (TOKA-1) (NASH, 1979a, 1979b). Pre-Tertiary sandstone (Layer E) covered by the Shimajiri Group (Layer C) is found at Onodera Seamount in the southern Okinawa Trough (St. 199, Fig. 3). Layer D is determined by drilling at one site in the southern Ryukyu Ridge. Sonic profiling records show a large unconformity under the Shimajiri Group (Layer C) in the surrounding areas of the Okinawa Trough. Therefore, the Shimajiri Group including its upper formations is easily distinguishable from pre-Shimajiri formations. The unconformity is of Middle to Late Miocene age and found not only in the Tokara Ridge but also in the Ryukyu Ridge, arc-trench gap and Tunghai Shelf. Layer B can be divided into B_1 and B_2 in ascending order by the difference of dips in trough basins.

It is difficult to distinguish Layer A in multi-channel air-gunning profiling records because only a slight difference exists between Layer A and B. Layer A overlies Layer B conformably in the trough, and it unconformably overlies older layers in the surrounding land areas.

3.3 Crustal sections of the Okinawa Trough based upon the seismic refraction data

The crustal structure is shown in Fig. 9. Six crustal units 1 to 6 are defined. Correlation between seismic reflection and refraction profiles are impossible in a strict sense but the following correlation may be reasonable. Unit 1 (2.1 km/sec) is correlatable to Layer A+B. Unit 2 (2.2–3.2 km/sec) is Layer C. However, in the trough basin in the southern Okinawa Trough, Layer B may contain both Units 1 and 2 represented in Fig. 9, B. Unit 3 (3.3–4.4 km/sec) is consolidated sediment older than Upper Miocene time. Unit 4 (4.5–6.1 km/sec) is composed of crystalline and granitic rocks in the continental regions and Layer 2 in the oceanic basin, Unit 5 (6.4–7.3 km/sec) is the gabbroic layer and Unit 6 (8.2 km/sec) is the upper mantle. In Fig. 9, 8.2 km/sec shows the mantle and 6.5 km/sec at St. 31–32 is regarded as the gabbroic layer (LEE *et al.*, 1980). 7.2 and 6.4 km/sec at St. C15 and St. C20–230 in Fig. 9, B should be included in the gabbroic layer. St. 233 and 222 showing large velocities of 6.4 and 8.7

Table 1. Stratigraphic correlation table for the Okinawa Trough and its surrounding areas. Ages are from HARLAND *et al.* (1982) and KASAHARA and SUGIMURA (1978). Stratigraphic position of the Kuchinotsu Group is referred from OKAGUCHI and OTSUKA (1980).

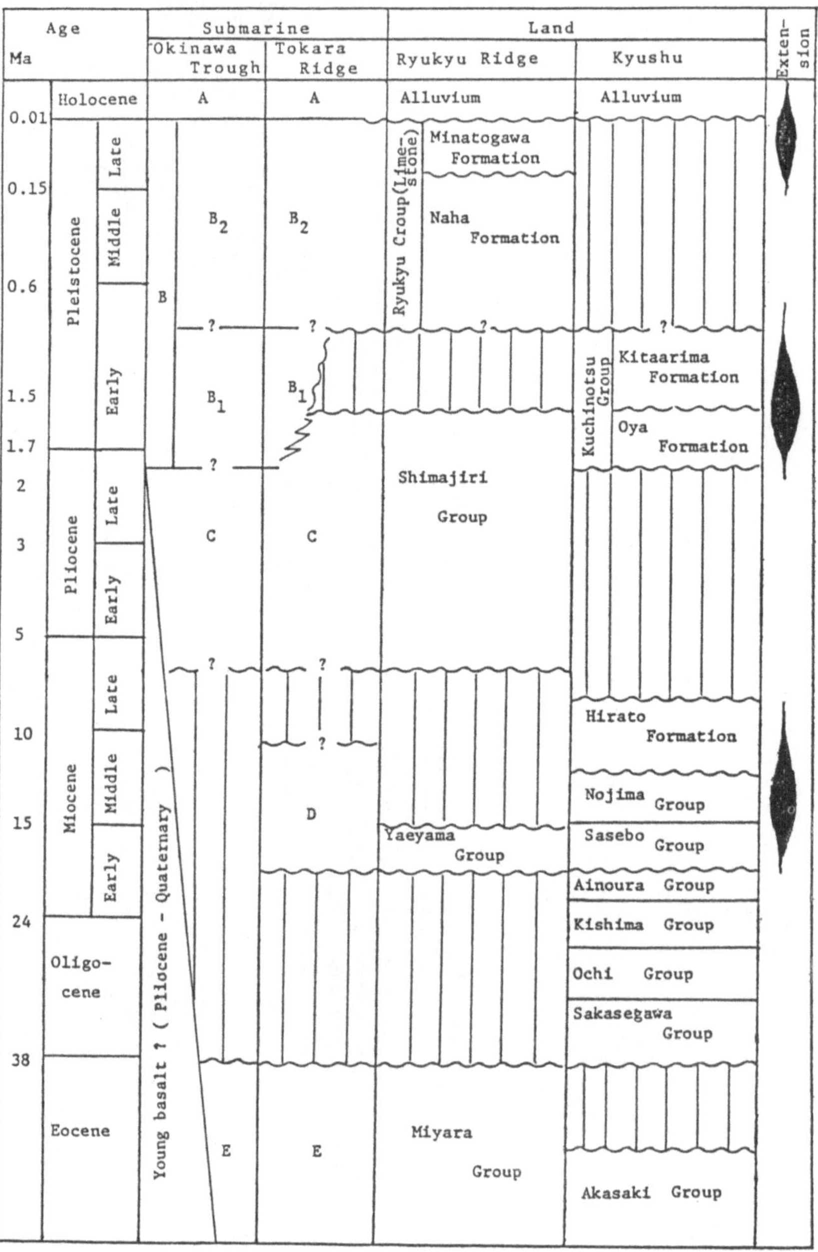

Age Ma	Age		Submarine: Okinawa Trough	Submarine: Tokara Ridge	Land: Ryukyu Ridge	Land: Kyushu	Extension
0–0.01	Holocene		A	A	Alluvium	Alluvium	
0.01–0.15	Pleistocene	Late	B (B_2)	B_2	Ryukyu Group (Lime-stone): Minatogawa Formation	—	
0.15–0.6	Pleistocene	Middle	B (B_2)	B_2	Ryukyu Group (Lime-stone): Naha Formation	—	
			?	?	?	?	
0.6–1.7	Pleistocene	Early	B (B_1)	B_1	—	Kuchinotsu Group: Kitaarima Formation; Oya Formation	
			?				
1.7–5	Pliocene	Late, Early	C	C	Shimajiri Group	—	
			?	?			
5–10	Miocene	Late	—	— ?	—	Hirato Formation	
10–15	Miocene	Middle	D	D	—	Nojima Group	
15–24	Miocene	Early	—	—	Yaeyama Group	Sasebo Group; Ainoura Group	
24–38	Oligo-cene		—	—	—	Kishima Group; Ochi Group; Sakasegawa Group	
38–	Eocene		E	E	Miyara Group	Akasaki Group	

Young basalt ? (Pliocene - Quaternary)

Sub-division ages: 0.01, 0.15, 0.6, 1.5, 1.7, 2, 3, 5, 10, 15, 24, 38

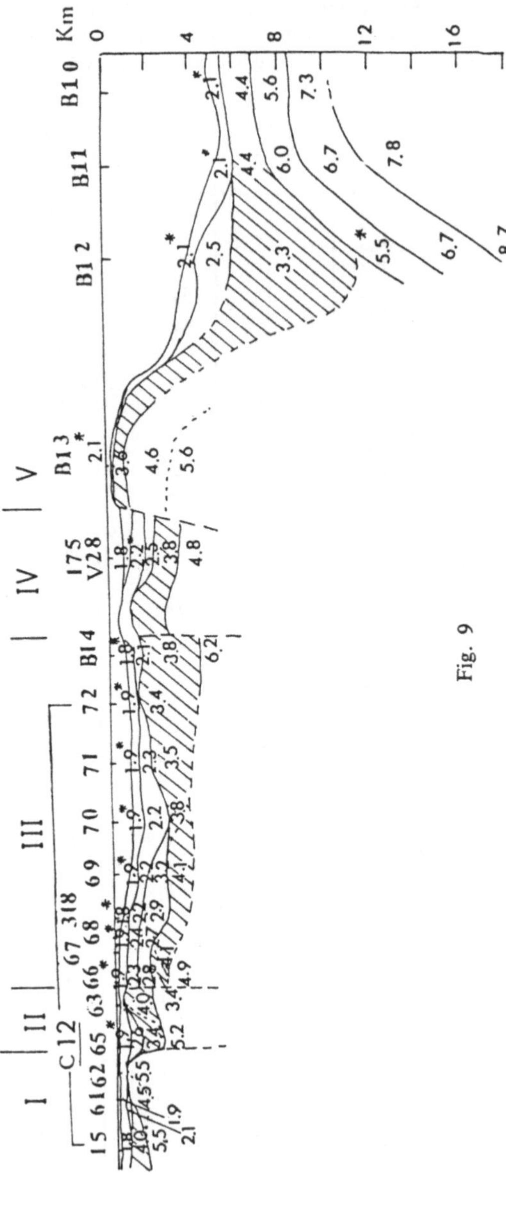

Fig. 9

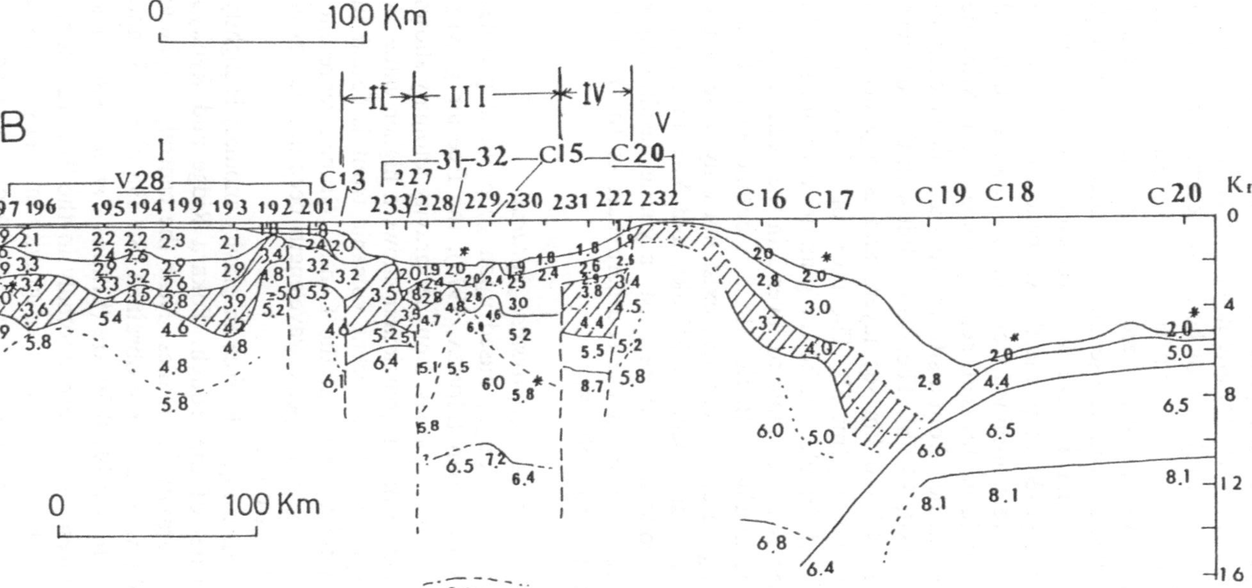

Fig. 9. Crustal cross sections compiled from radio-sonobuoy and two-ship refracion data. Numerals in the crustal sections represent the crustal sonic velocity (km/sec). Station numbers are shown in Fig. 3. I: Tunghai Shelf, II: Tunghai Slope, III: Okinawa Trough, IV: Tokara Ridge, V: Ryukyu Ridge, A: Northern and middle Okinawa Trough, B: Southern Okinawa Trough.

(?) km/sec which are regarded as the gabbroic layer, too. Whether the 4.6–6.0 km/sec layer in Fig. 9, B the granitic layer or a relatively shallow basaltic layer (Layer 2 in the ocean) cannot be easily determined. LEE *et al.* (1980) mentioned that 5.5 and 5.8 km/sec layer at Sts. 31–32 are correlatable to the basaltic layer, because there exists a magnetic anomaly and oceanic basalt was collected from the seamount located in the trough basin. LEE *et al.* (1980) stated that the layer with velocity of 4.8 km/sec in the Okinawa Trough is composed of primarily Miocene rocks. Miocene sedimentary rocks, however were obtained from the Okinawa Trough and this velocity is anyway too large for sedimentary rocks. Late Miocene pyroclastics and lavas were drilled from a site showing a sonic velocity of 4.6 km/sec at TOKA-1 (NASH, 1979b). Therefore, Unit 4 in the southern Okinawa Trough includes 4.7–6.0 km/sec layers and is regarded as a relatively shallow basaltic layer (Unit 5'). The layer of 4.7–6.0 km/sec beneath the Okinawa Trough should not be the granitic layer but should be the basaltic layer which is correlatable to layer 2 in the oceanic basin. If the layer of 5.5 km/sec in Fig. 9, B, St. 31–32 is composed of basalt (LEE *et al.*, 1980), a young basaltic layer may exist not only beneath the narrow, Central Graben but also beneath Miyako and Yaeyama basins based upon Fig. 9, B. In conclusion, there exists a young basaltic layer (Unit 5') beneath the trough basin in the southern part of the trough, having a width amounting to about 60 km.

3.4 Magnetic anomaly and other geophysical features

In the southern Okinawa Trough, based upon geophysical profiles by the Geological Survey of Japan (HONZA, 1976), the free air gravity anomaly is lower than surrounding areas. Bouguer gravity anomaly shows fairly high values. The central part in particular shows a higher value. A magnetic anomaly corresponding to a recent volcanic front is found in some profiles. The magnetic anomaly at the Central Graben represents a positive anomaly, and its peak to peak wavelength becomes narrower at the southern end of the trough.

In the middle part of the trough, the free air gravity anomaly is slightly low compared to both sides of the trough. Tokara Ridge and Ryukyu Ridge show high free air gravity anomalies. Near the central, magnetic anomaly, rhyolite and pillow basalt were obrained (St. 398 in Fig. 3) and the K-Ar age of the rhyolite is 0.79±0.39 Ma (SHIBATA *et al.*, 1984). The Recent volcanic front represents a sharp anomaly in the Tokara Ridge. In the northern Okinawa Trough, the free air gravity anomaly shows a slightly low value compared to the surrounding areas. A positive magnetic anomaly is often recognised in the Central Graben. The Bouguer gravity anomaly amounts to 100 mgl.

4. *Discussion*

4.1 *Origin of the Okinawa Trough*

In the southern part of the Okinawa Trough, it is postulated that Unit 5' (4.6–6.0 km/sec layer in Fig. 9, B) represents basalt including pyroclastics and a sedimentary mixture emplaced in conjunction with the extensional tectonics. This is concluded on the evidence of magnetic lineations and of relatively fresh pillow basalts dredged from a piercement structure in one of the central rifrs. Gravity data also supports the postulation. Its crustal structure is similar to that of the Mariana Trough which is a classic example of an interarc basin (KARIG, 1971). The 4.5–5.4 km/sec layer composed of volcanics 1 km thick on average is covered by surface sediments in the Mariana Trough (BIBEE *et al.*, 1980). On the other hand, the Okinawa Trough profile (Fig. 9) shows that a 2 km thick low speed layer with the 1.8–3.0 km/sec covers directly with high speed layer in the axial zone of the trough basin (III in Fig. 9, B).

Figure 9, B shows that the mantle is shallower beneath the trough basin that at its margins and that it seems to lack a granitic layer in the entire basin about 40 km wide. III in Fig. 9, B does not show real width of the basin. It strongly supports that spreading of an 40 km wide basin has occurred in the subsurface of the southern Okinawa Trough Basins (Fig. 9, B). The crustal section suggests that the southern part of the Okinawa Trough has spread since deposition of Unit 3, if Units 3 and 4 (Granitic layer) are lacking beneath the trough basin. Probably Unit 3 and Unit 4 would have been separated at the same time. Unit 3 is correlatable to older strata than Late Miocene time. Therefore, the crustal section suggests that the spreading started after deposition of Middle to Late Miocene sediments. Correlation between the seismic reflection record and the seismic refraction crustal section (Fig. 9, B) shows that the Pleistocene layer (Layer B_1) makes contact with the upper basalt.

When did the spreading start? Multi-channel seismic reflection profiles (Fig. 10) show that a thick Layer C (Shimajiri Group) decreases its thickness from both sides toward the Okinawa Trough, and that the general strike of Layer C is parallel to the Tokara Ridge and Ryukyu Ridge, and the bedding plane of Layer C dips toward the east. It shows that source materials of Layer C came from the west. The boundary fault, Ryukyu Ridge Fault, offsets Layer C. Layer C should be thick in the okinawa Trough if the sea-floor spread before and during the depostition of the group. Therefore, it is regarded that the spreading began after deposition

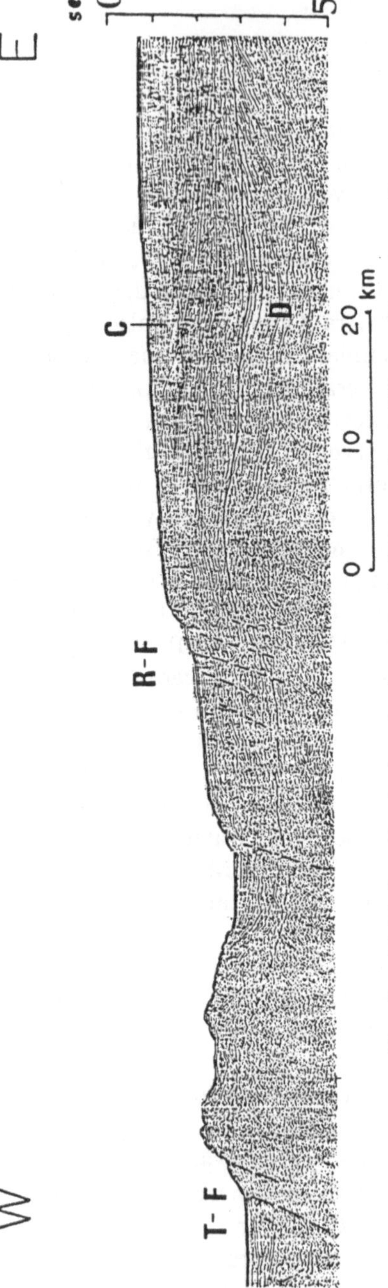

Fig. 10. Multi-channel seismic reflection record of the southern Okinawa Trough. "sec" shows two-way travel time. Line A (in Fig. 2) showing that the Shimajiri Group (Layer C) is offset by the Ryukyu Ridge Fault (R-F) and the Tokara Ridge Fault (T-F).

of the Shimajiri Group. Parts of major faults around the Okinawa Trough are exposed in the land areas (Tokara Ridge Fault in Amakusa Islands and Ryukyu Ridge Fault in Okinawa-jima). It has been clarified that these faults have been formed since Early Pleistocene. Layer C is thinner in the central part of the trough than at its margin. However, Layer B is thicker in the trough basins than at the margins in this area. It shows that the sedimentation mode has been controlled by submarine topography since Early Pleistocene. Such a relation favors a proposition that opening of the Okinawa Trough occurred at some time after deposition of Layer C and before or during deposition of Layer B. The real multi-channel reflection profile shows that sediments in the Yaeyama basin in the southern Okinawa Trough overly the Layer C at the margin of the basin. This shows that spreading started during deposition of Layer B. There exists a huge block of reefal limestone in the Ryukyu Group (Ryukyu Limestone) of Pleistocene age (post-Shimajiri Group) in the Ryukyu Islands whereas no reefal limestone is found in the Shimajiri Group. The author postulates that the Ryukyu Ridge was separated from the continent in Early Pleistocene time due to formation of the Okinawa Trough and then vast detrital materials did not reach the Ryukyu Ridge. The coral reef started to develop to form Pleistocene Ryukyu Limestone.

The boundary between Layers B and C shows unconformable relation in the Tunghai Shelf, Tunghai Slope, Tokara Ridge and Ryukyu Ridge. The unconformity between Layer C and B is correlatable to that between the Shimajiri Group and the Ryukyu Group on land (Table 1). If so, the unconformity was formed during Early Pleistocene time. The lower part of the sediments in the trough basin in Fig. 11 pinches out toward the center and is separated from the axial part of the trough. This suggests that the spreading occurred during the deposition of the lower part of Layer B (B_1) in the basin. And that, the upper part of Layer B (B_2) in the basin was continuously deposited. It is offset by Central Graben faults. This implies that the spreading speed was faster during deposition of the lower part of sediments and ceased or was very slow during sedimentation of the upper part of the sediments. The main body of the Shinzato Formation, upper part of the Shimajiri Group, shows N. 21 (planktonic foraminiferal zonation of BLOW (1969)) but the uppermost of the Shinzato Formation includes N. 22 (UJIIÉ and OKI,1974, NATORI, 1976). The Shinzato Formation is unconformably covered by the Chinen Formation which is the basal part of the Ryukyu Limestone. The uppermost part of Layer C is correlated to the Kuchinotsu Group in Kyushu. Correlation of Pleistocene chronology on land around the Okinawa Trough suggests that the unconformity between Layer B_1 and C was formed during 2 to

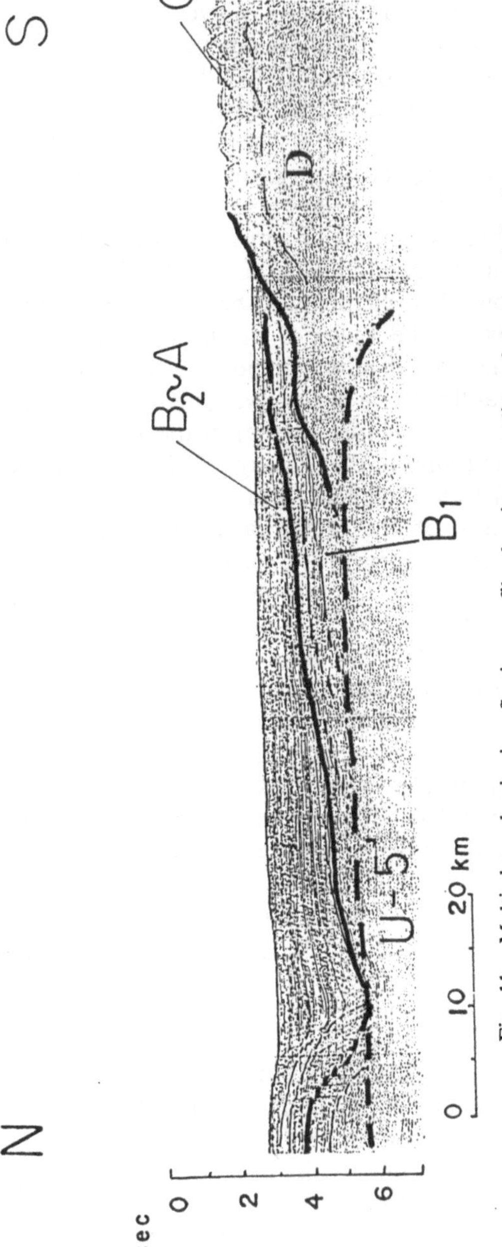

Fig. 11. Multi-channel seismic reflection profile showing spreading of the trough in the southern Okinawa Trough (Line B in Fig. 2). The Pleistocene sediments B_1 pinch out towards the central axis of the trough.

1.7 Ma. It seems reasonable that the opening started around 2 Ma after the deposition of the main body of the Shinajiri Group in the Okinawa Trough. The above evidence suggests that the southern Okinawa Trough has opened since Late Pliocene to Early Pleistocene time at an average half spreading rate until recent times of 1~2 cm/yr.

HERMAN *et al.* (1978) considered that the Okinawa Trough spread sometime after Late Miocene-Early Pliocene time and the Ryukyu Arc was rifted from the continental margin. That time was before the deposition of the pliocene sediments after the deposition of Miocene sediments, and they pointed out that there exists an unconformity formed between Miocene and Pliocene times. The unconformity stated by HERMAN *et al.* (1978) is regarded as the boundary between Layers C and B_1 in the present paper. If it is true, the time of spreading is in Early Pleistocene. The multichannel seismic profile shows sedimentary structure is controlled by spreading during its deposition. My interpretation disagrees with their result. The layer B_1 in the Yaeyama Basin covers the young basaltic layer directly. This also suggests that the spreading occurred during deposition of Layer B_1.

LEE *et al.* (1980) suggested that a crustal separation occurred from Pliocene time because the Pliocene sediments fill the Okinawa Trough basin. However, the Yaeyama basin is filled by Pleistocene sediments of Layer B at least 40 km wide. Therefore, the major crustal separation occurred since Early Pleistocene time.

In the middle part of the Okinawa Trough the sedimentary cover of the Unit 1 is thin and igneous matter was often directly exposed where pillow basalt and rhyolite were dredged. Rhyolite is dated as 0.79 ± 0.39 Ma. At the northern Okinawa Trough, the crustal structure and surface geology suggest existence of the granitic layer with 6.2 km/sec beneath the trough (Fig. 9, A). The intermediate 3.4–4.1 km/sec layer of Unit 3 exists in the northern trough. Magnetic anomaly suggests that the trough may have been intruded by igneous bodies. The author proposes that spreading rate since Early Pleistocene time has been very slow in the middle to northern Okinawa Trough and the Units 2 and 3 have not yet appeared in both margins. The narrow magnetic central anomaly supports this conclusion.

4.2 *Tectonic development*

The crustal structural section shows that the first Asian continental separation in the Okinawa Trough region is thought to have occurred during Late Pliocene to Early Pleistocene time. Before that, crustal thinning may have occurred since at least Oligocene time. Because Oligocene sediments in the Ryukyu Arc are lacking, doming of the crust occurred by uprising of the magmas since that time. The crustal surface uplifted and Oligocene sediments were eroded.

Crustal thinning by erosion may have continued and crustal depression occurred along the central axis of the dome, where Neogene hydrothermal activity (Green Tuff activity) occurred and sandstone and mudstone were deposited (Yaeyama and Sasebo Groups). The depressional area may have ranged from the eastern margin of the Tunghai Shelf to the Tokara Ridge. The subsidence of the depression may have been accelerated by decrease of the confining pressure at a part of the upper layer resulting from effusion of vast amounts of magmas.

The basin was buried by sediments and pyroclastic matter, and the crust uplifted to cause a widespread unconformity in the Middle to Late Miocene (Unconformity between Layers D and C). Layer C was deposited widespread. In this period the crust along the Okinawa Trough was still uplifted. After deposition of Layer C, further uplifting took place. Widespread unconformity between Layer C and B was formed. The crustal surface was stretched much and the central uplifted dome was split and echelon cracks occurred along the axis of the Okinawa Trough. Succeeding this, crustal separation took place in the southern trough basin and then the basin was separated during Late Pliocene to Early Pleistocene time. Figure 12 shows the present features of Central Grabens of the Okinawa Trough ranging from north to south. Numerical figures in parentheses show depths of the base of the Pleistocene sediment (Layer B_1) from sea level. These figures suggest the direction of development of the Central Grabens. They show that spreading proceeded from south to north and the place where the water depth at the top of the basement of the Pleistocene sediment Layer B reaches near 4 km was opening. Mantle becomes deeper from north to south (XIANGLONG *et al.*, 1983). At the same time as the formation of the Central Grabens, grabens such as the Okinawa Trough and the Greater Okinawa Trough were formed, accompanied by tilted blocks against the trough axis since Late Pliocene to Early Pleistocene time. Layer B_1 deposits unconformably in the depressional basins in this period.

In Late Pleistocene time, uprising of magma occurred as shown by unconformity in the Ryukyu Group on the Ryukyu Ridge and new volcanic activity of the Ryukyu system (MATSUMOTO, 1982). At that time, the newest central rifts offset Pleistocene sediments in the trough basins (Fig. 13).

5. *Conclusions*

Stratigraphy, crustal structure, volcanic activity and magnetic anoma-

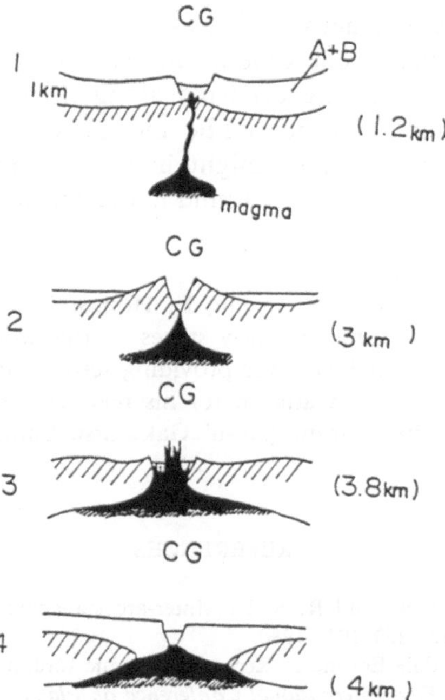

Fig 12. Schematic cross sections of the Central Grabens in the Okinawa Trough. 1: Northern Okinawa Trough, 2: Northern to middle, 3: Middle to southern, 4: Southern, A: Layer A, B: Layer B, CG: Central Graben.

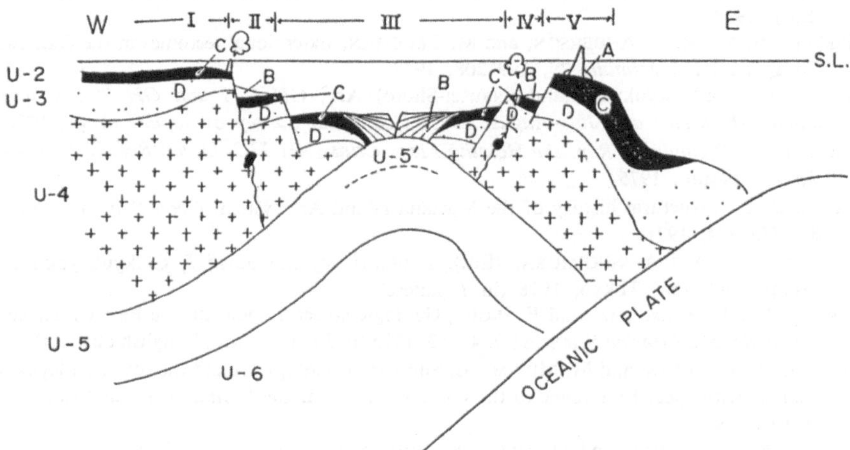

Fig. 13. Schematic crustal section in the vicinity of the southern Okinawa Trough.

ly suggest that the southern Okinawa Trough has opened since Late Pliocene to Early Pleistocene time at an average half rate of about 1~2 cm/yr. The crustal stretching occurred along the entire length of about 1000 km of the trough, and the middle and northern parts are only rifted whereas the southernmost part might have been opened to a width of about 40 km. Central Grabens are found in the entire trough.

Acknowledgments

The author is grateful to S. Uyeda, E. Honza, N. Isezaki and H. Kagami for their contributions during the early stages of this work. The author is also grateful to W. Ludwig and Houtz for providing seismic reflection and refraction data. Thanks are due to P. Arathoon for his reading manuscript. The research was partly supported by a grant (Asahi Gakujutsu Shoreikin) from the Asahi Newspaper.

REFERENCES

BIBEE, L. D., G. G. SHOR, and R. S. LU, Inter-arc spreading in the Mariana Trough, *Marine Geology*, **35**, 183–197, 1980.

BLOW, W. H., Late Middle Eocene to Recent planktonic foraminiferal biostratigraphy, in *Proceedings of the First International Conference on Planktonic Microfossils, Geneva*, Leiden: Brill, 1967.

EGUCHI, T., Spreading of the Okinawa Trough, *Programme and Abstracts, the Seismol. Soc. Japan*, **2**, 77, 1982.

HARLAND, W. B., A. B. COX, P. G. LLEWELLYN, C. A. G. PICTON, A. G. SMITH, and R. WALTERS, *A Geologic Time Scale*, 131pp., Cambridge Earth Science Press, 1982.

HAYES, D., R. E. HOUTZ, R. D. JARRARD, C. L. MROZOWSKI, and T. WATANABE, Crustal structure, in *A Geophysical Atlas of the East and Southeast Asian Seas*, Hayes, D. (Ed.), 1978.

HERMAN, B. M., R. N. ANDERSON, and M. TRUCHAN, Extensional tectonics in the Okinawa Trough, *AAPG Mimoir*, **29**, 199–208, 1978.

HONZA, E. (Ed.), Ryukyu Island (Nansei-Shoto) Arc, *GH 75-1 and GH 75-5 Cruises January-February and July-August 1975, Geological Survey of Japan*, 81pp., 1976.

KAGAMI, H., *Preliminary Rep. the Hakuhō Maru Cruise KH-72-2, Ocean Res. Inst., Univ. Tokyo*, 144pp., 1975.

KARIG, D. E., Structural history of the Mariana Island Arc system, *Geol. Soc. Am. Bull.*, **83**, 323–344, 1971.

KASAHARA, K. and A. SUGIMURA, (Ed.), *Evolution of the Earth I*, Chikyukagaku 10, 296pp., Iwanami, Tokyo, 1978 (in Japanese).

KIMURA, M., T. HIROSHIMA, and E. INOUE, Geologic structure beneath the East China Sea, *Kaiyo-Kagaku (Marine Sciences)*, **7**, 45–52, 1975 (in Japanese with Eanglish abstract).

KIMURA, M., E. HONZA, and M. MIYAMOTO, Submarine geology around the southern Ryukyu Islands, with special reference to the Okinawa Trough, *Geol. Studies of the Ryukyus*, **4**, 79–93, 1979.

KIMURA, M., E. HONZA, and O. UEDA, K. SHIMOYANAGIDA, and T. TAMAKI, Geological stratigraphy and structure in and around the middle to northern Okinawa Trough, *Geol.*

Studies of the Ryukyus, **5**, 133–166, 1980 (in Japanese with English sbstract).

KIMURA, M., Formation of the Okinawa Trough, *The Memoirs of the Geological Society of Japan*, **22**, 141–157, 1983.

KONISHI, K. and K. SUDO, From Ryukyu to Taiwan, in *World Tectonic Zone*, Uyeda and Sugimura (Ed.), pp. 271–280, Iwanami, 1973 (in Japanese).

LEE, C., G. Jr. SHOR, L. D. BIBEE, R. S. LU, and T. W. C. HILDE, Okinawa Trough: Origin of a back-arc basin, *Marine Geology*, **35**, 219–241, 1980.

MATSUMOTO, Y., The Cenozoic volcanism in the Ryukyu Islands, Japan, *The Memoirs of the Geological Soc. Japan*, **22**, 81–91, 1982 (in Japanese with English abstract).

MOGI, A., Bathymetry of the Kuroshio Region, in *Kuroshio*, pp. 53–80, Univ. of Tokyo Press, 1972.

NASH, D. F., Neogene basin development at the northeast end of the Okinawa Trough, The 44th Japanese Association of Petroleum Technologists Symposium (Lecture), 1979a.

NASH, D. F., The Geological development of the North Okinawa Trough area from Neogene times to Recent, *J. the Japanese Assoc. Petroleum Technologists*, **44**, 341–351, 1979b.

NATORI, H., Plankronic foraminiferal biostratigraphy an datum planes in the Late Cenozoic sedimentary sequence in Okinawa-jima, Japan, in *Progress in Micropaleontology, Selected Papers in Honor of Prof. Kiyoshi Asano*, Takayanagi, Y. (Ed.), pp. 214–243.

OKAGUCHI, M. and H. OTSUKA, Fission track ages of zircon crystals from tuff layers in the Kuchinotsu Group and from andesite of the Tatsuishi Formation, *The Quaternary Res.*, **19**, 75–85, 1980 (in Japanese with English abstract).

SIBATA, K., S. UCHIUMI, K. UTO, and T. NAKAGAWA, K-Ar age results—2-new data from the Geological Survey of Japan—*Bull. Geol. Surv. Japan*, **35**, 331–340, 1984 (in Japanese with English abstract).

UJIIÉ, H. and K. OKI, Uppermost Miocene-Lower Pleistocene planktonic foraminiferal from the Shimajiri Group of Miyakojima, Ryukyu Island, *Mem. National Sci. Mus, Tokyo*, **7**, 31–52, 1974.

XIANGLONG, J., Y. PUZHI, L. MEIHUA, L. CHANGZHEN, and W. HUIQUING, Preliminary study on the characteristics of crustal structure in the Okinawa Trough, *Oceanologia et Limnologia Sinica*, **14**, 105–106, 1983 (in Chinees with English abstract).

CHAPTER 5: ACCRETION TECTONICS (I)

Formation of Active Ocean Margins, edited by N. Nasu *et al.*, pp. 595–623.

TECTONIC INSTABILITY ON A "STABLE" SHELF BEHIND A PASSIVE MARGIN—A STRUCTURAL HISTORY OF THE ENGLISH CHANNEL

Alec J. SMITH

Department of Geology, Bedford College, University of London, Regent's Park, London, NW1 4NS, Great Britain

Abstract. Generally regarded as a stable area, the outcrop pattern of the floor of the English Channel is, in contrast to other seas around Britain, remarkably diverse. Evidence points to a long history of tectonic evolution and the implication of continued instability. The basic form of the Channel relates to late Variscan events establishing a geological pattern which was influenced by the creation of the Atlantic margin, Cimmerian diastrophism and Alpine movements. Much of the development after an early extensional phase has been compressional. The Channel has three provinces each distinguished by its particular structural style, each style evolving, it is suggested, in response to the relationships with the Variscan Front. The Channel now is subject to erosion by tidal scour but in late Pleistocene to Recent times the eastern Channel was subjected to fluviatile erosion and catastrophic flooding by glacial melt waters.

1. Introduction

It is nearly 10 years since this author presented the paper 'Instability in a stable shelf' in a volume to celebrate the sixtieth birthday of Professor Masami Hayakawa (SMITH, 1976). It dealt particularly with the geological development of the continental shelf between Britain and France and it explored the long history of tectonic events which give that stretch of water known as the English Channel, or La Manche, its distinctive geological character. The paper presented here describes results of further researches in the Channel area and explores the evidence which suggests neotectonic events in the vicinity of the Channel. It is an appropriate contribution to this symposium concerned with ocean-continent margins in that it illustrates some of the events which can precede and follow the creation of an ocean margin—in this case the margin of the Atlantic Ocean, a passive margin off southwest England and northwest France. The area has long been regarded as stable but there is evidence of continued instability since before and after the creation of the Atlantic Ocean hereabouts. It seems possible that part of this record exists in the post-ocean-development pile of sediments and that future drilling may help unravel the size and timing of the events. The paper also considers the style of the tectonism which may be prevalent in the region of southern England and northern France.

2. *The Geology of the Seas around the British Isles*

Each of the seas around the British Isles (Fig. 1) has its own distinctive geological history. Space permits only a brief discussion here, but such a discussion is necessary to highlight the special character of the English Channel and its Western Approaches. The North Sea is the largest of the seas and its history is well documented (ZIEGLER, 1982): its development began with post-Caledonian events following the collision of the continents bordering the Iapetus Ocean. By Carboniferous times there was more subsidence in the North Sea region than in the adjacent areas which now comprise Britain and northwest Europe and contemporary coal seams thicken into the North Sea Basin. Late Carboniferous/early Permian erosion related to Variscan earth-movements affected the Basin less than the surrounding areas while Permian and Triassic events such as volcanism, arid-land and evaporite deposition characterise the Basin and its extension beneath the Netherlands and north Germany. Jurassic and Lower Cretaceous (Cimmerian) earth movements added rifting, volcanism and sedimentary sequences which, in some places, were in response to extensionlal movements (WOOD and BARTON, 1983) thus contributing more to the distinctive geological character of the North Sea. In the Upper Cretaceous, the North Sea (except in its far north), in common with much of northwest Europe, was the site of deposition of chalky limestones. In the Cenozoic a more general subsidence ensued, the subsidence being centred on the present central axis of the North Sea where subsidence exceeded three kilometres. Subsidence has continued throughout the Quaternary when the area was subjected to extensive glacial processes and the Flandrian rise in sea-level.

The seas between the Outer and Inner Hebridean Islands and Ireland and mainland Britain i.e. The Minches, Solway Firth, the Irish Sea, Car-

Fig. 1. The continental shelf seas around the British Isles.

digan Bay, St George's Channel and the Celtic Sea, all appear to begin with Permian rifting though there is some evidence of coal seams of the Upper Carboniferous thickening towards the Irish Sea. The trends of these basins are often related to Caledonian controls of the basement. Local

Table 1. Episodes of instability and associated movements affecting N. W. Europe

Episode of instability	*Principal factor and style**
Mid-Tertiary to Recent	Quaternary glaciation -V Alpine orogeny -C
Jurassic to early Cretaceous	Rotation of Iberia -E Opening of Atlantic -E
Late Carboniferous to early Permian	Variscan orogeny -C>E
Late Silurian to early Devonian	Caledonian orogeny -C>E

*V = Vertical; C = Compressional; E = Extensional.

differences in subsidence rates led to variations in the thickness of successions while the sedimentary facies reflected the type and proximity of the source areas. As is common in northwest Europe, red beds and evaporites, though, in the case of the latter, not to the same extent as in the North Sea, characterise the Permian and Triassic successions. Offshore Jurassic successions, where sampled, resemble those known from present-day land areas, though again some successions reflect local conditions. Cimmerian earth movements continued through Jurassic and early Cretaceous times. Again calmer and more uniform conditions prevailed throughout the Upper Cretaceous. Tertiary sedimentation, both marine and non-marine occurred in many of these seas, though in the northwest the Tertiary volcanic province dominated the geological scene. All these sea areas were affected by glaciation: as with the North Sea, the ice mass scoured the sea bed. In front of the ice, fluvio-glacial deposits were widespread though everywhere the evidence of the latest events, and particularly those related to ice retreat, are dominant. Though subsidence has not continued to the same degree as in the North Sea, Quaternary and Recent deposits are both thick and extensive (NAYLOR and SHANNON, 1982).

The outer margin of the continental shelf around the British Isles shows a simpler pattern, though this, in part, reflects the smaller amount of geological information so far recovered and published. At depth, the Caledonian and Variscan patterns can be expected to continue to the margin of the continent, however the basins which are known to exist there have histories which reflect the history of development of the Atlantic Ocean hereabouts. Tertiary and Quaternary sediments are widespread, many of the latter having a glacial or fluvioglacial origin (NAYLOR and SHANNON,

1982).

3. *The English Channel*

The English Channel and its Western Approaches have been investigated in considerable detail (SMITH and CURRY, 1975 and see SMITH *et al.* 1972 and HAMILTON *et al.*, 1975 for extended bibliographies). Generally the geological outcrop pattern is more complex than that of the other seas while the cover of unconsolidated sediment is considerably less. The present distribution (Fig. 2a, b, c) of superficial sediments is tidally controlled and the lithologies reflect the high contribution from calcareous organisms mixing with clastic debris, much of it composed of chert (LARSONNEUR *et al.*, 1980). Except towards the very west the sediment cover is usually thin or even absent. Cut into the rock floor of the Channel is a complex pattern of filled and partially filled palaeovalleys and deeps (Fig. 3). Most of the former are restricted to the eastern part of the Channel (AUFFRET *et al.*, 1980; LARSONNEUR *et al.*, 1982) and, indeed, they are not known as a feature elsewhere on the continental shelf around the British Isles. Smith (1985) is of the view that this system was considerably modified by a short-lived catastrophic event, or a succession of the short-lived events, in the late Pleistocene; this is discussed again in the paper.

The outcrop pattern of the solid geology of the Channel floor is, as has been mentioned previously, quite varied (Fig. 4). It certainly implies strong post mid-Oligocene movements and continued erosion. Only in the infilled palaeovalleys and in the westernmost part of the Channel is there any Quaternary sedimentation. At times of lower sea level, permafrost conditions may have prevailed (BOILLOT, 1964) but true glacial conditions are not thought to have occurred in spite of some contrary suggestions (KELLAWAY *et al.*, 1975).

The style of structural development of the English Channel has been influenced by Caledonian and Variscan events, the creation of the Atlantic Ocean and the Bay of Biscay, the closure of Tethys with related Alpine events and intra-plate movements. The position of the Channel has, to a great extent, been dictated by the proximity of the Variscan Front, the northern limit of deformation, and the presence of the Caledonian and pre-Caledonian massifs: the Armorican Massif, the London-Brabant Massif and the Variscan foreland composed of Caledonian elements (Fig. 5). Each and all of these components have participated occasionally or repeated to a lesser or greater degree in the structural evolution of the Channel and older features were usually reactivated during later events.

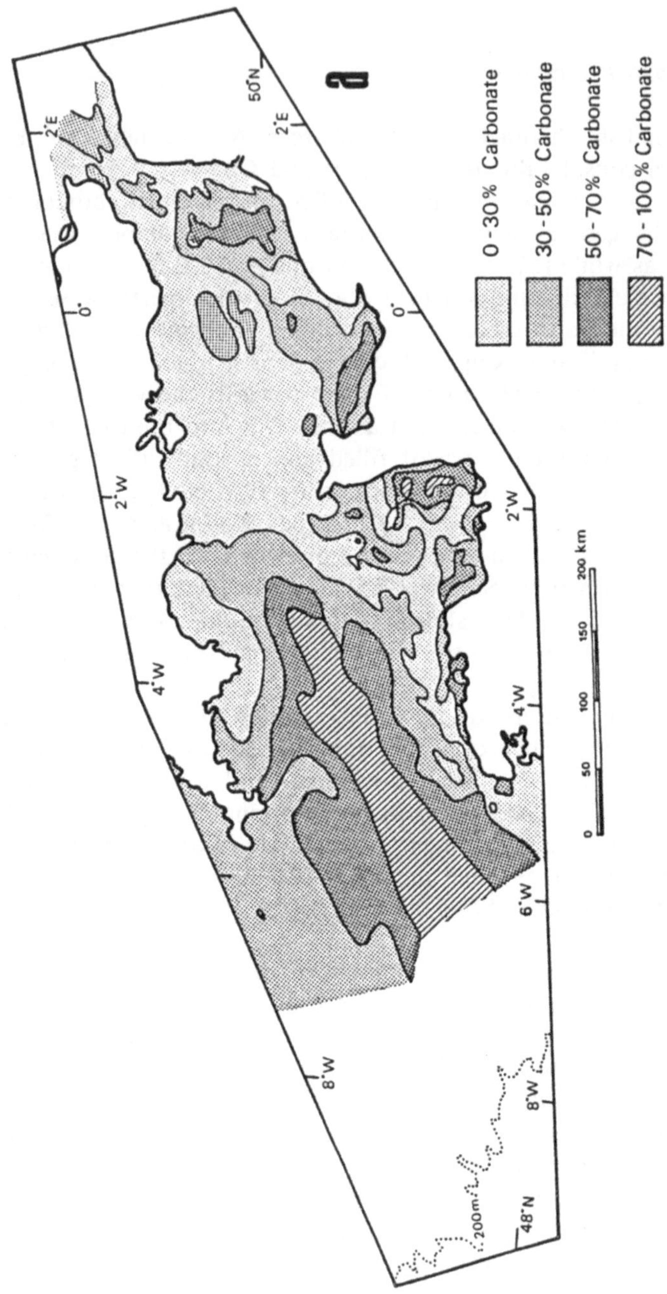
a
0 - 30 % Carbonate
30 - 50 % Carbonate
50 - 70 % Carbonate
70 - 100 % Carbonate
50°N
2°E
0°
2°W
4°W
6°W
8°W
48°N
200m
0
50
100
150
200 km

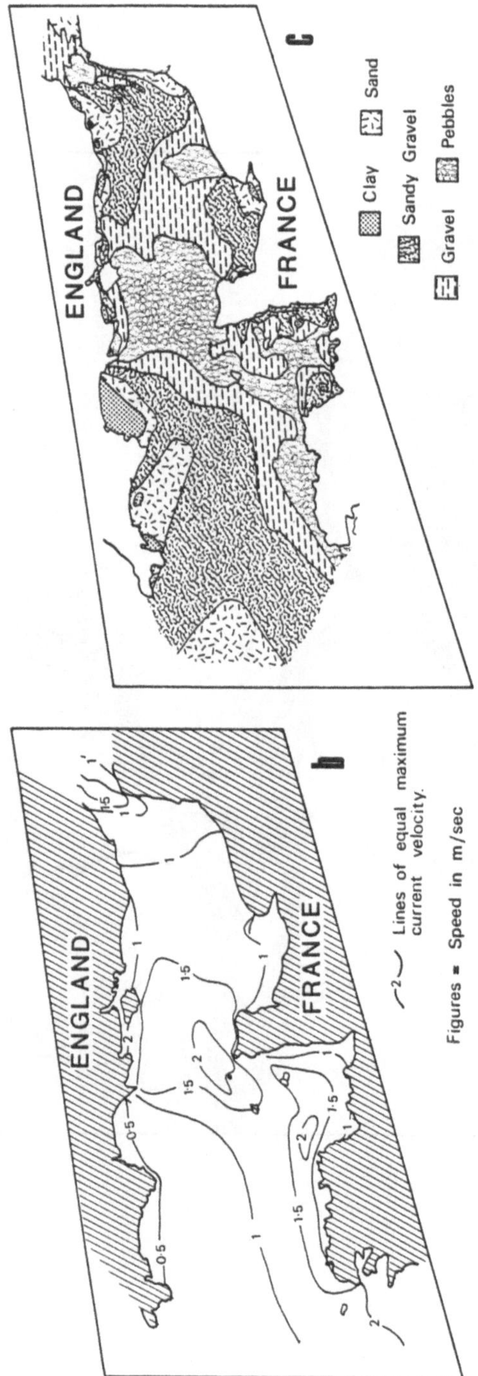

Fig. 2. a. Distribution of recent sediments in the English Channel and its Western Approaches (after LARSONNEUR *et al.* 1980). b. The tidal regime of the English Channel. c. Grain size distribution of sediments in the English Channel.

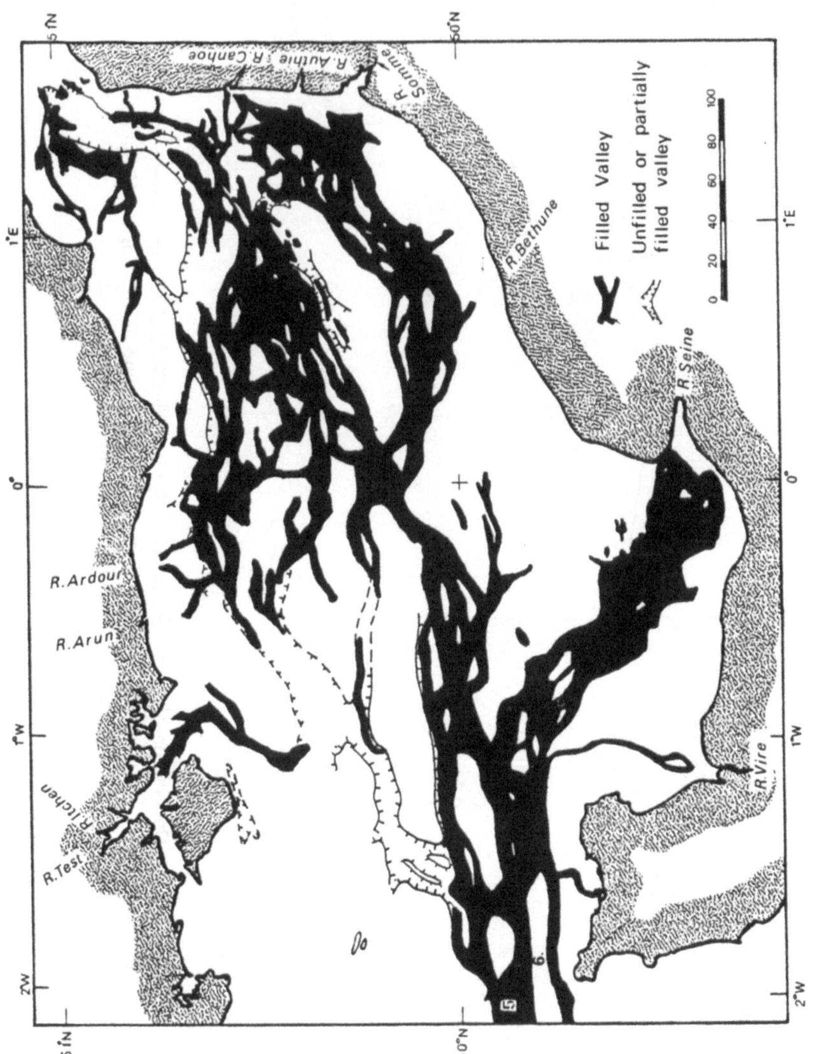

Fig. 3. The palaeovalley system of the eastern English Channel (after AUFFRET *et al.* 1982).

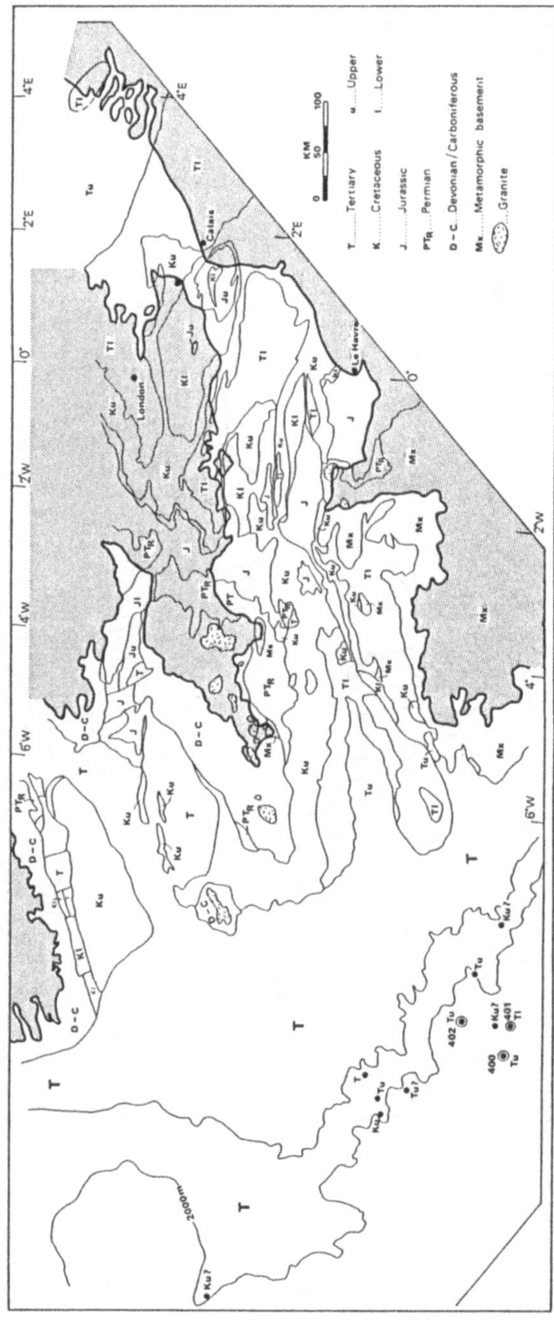

Fig. 4. Geological map of the English Channel and its Western Approaches (after SMITH and CURRY, 1975 and other sources). Nos. 400–402 refer to DSDP sites; other points in the same region refer to dredge samples.

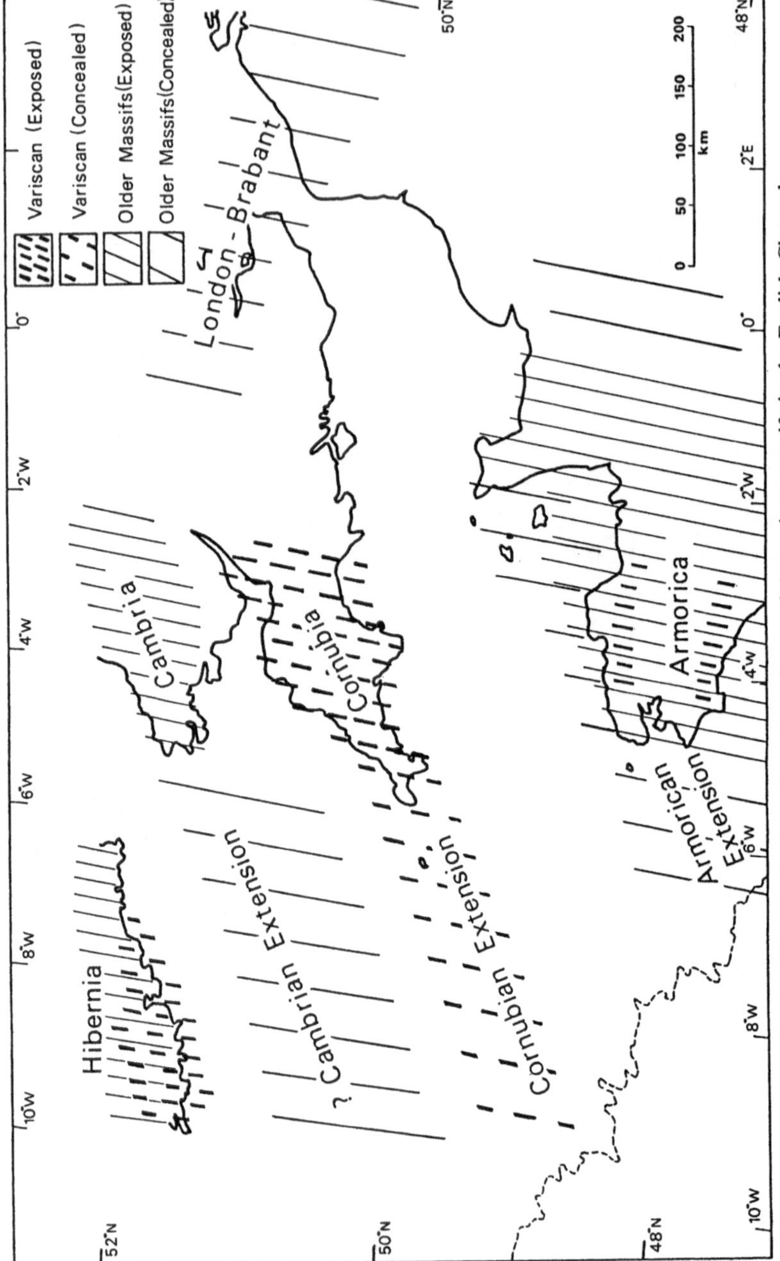

Fig. 5. The distribution of pre-Armorican and Armorican massifs in the English Channel region.

The evolution of the Channel from shelf-break to Dover Strait can best be described by reference to three provinces (see inset to Fig. 6), each province being characterised by a distinctive structural style and separated from adjacent provinces by significant structural boundaries.

The western province includes the Western Approaches and the western English Channel–each often designated as basins–and reaches from the shelf break to the Start–Cotentin line (see Fig. 6). It lies between the Armorican massif with its Icartian (>2000 Ma) and Cadomian/Caledonian as well as Variscan elements (see Fig. 5) and the Cornubian peninsula of southwest England with its southwesterly extension, the whole comprised of Devono-Carboniferous folded strata and late Carboniferous granite intrusions. Here, too, there are some older elements, the Lizard Complex (which includes ages of 520–410Ma, FITCH *et al.*, 1984) and the Start Complex, but these appear to be structurally emplaced. This northern feature separates the province from the North and South Celtic Sea Basins which lie to the north.

Two major influences have been significant in its development—the legacy of the Variscan orogeny and events of and following the opening of the Bay of Biscay and Atlantic Ocean. Caledonian and earlier influences may also exist but these have been subsumed by post-Variscan events. It seems possible that the Western Approaches Basin and the contiguous Western English Channel Basin of this province owe their origin to extensional tectonics following the last phase of the Variscan orogeny in the late-Carboniferous (LEEDER, 1982). They have graben like structures with most downthrow–a cumulative 2 km in the east to about 4 km in the west–in the vicinity of the Alderney–Ouessant line, a WSW oriented feature line which separates the exposed or thinly buried northward extension of the Armorican massif from the basins. The northern marginal fault has less throw and is less well defined.

The pre-Permian basement rocks which floor the basins are presumably similar to those that border the basins. The trend of the bounding faults is WSW, parallelling the prominent magnetic anomalies of the province (HILL and VINE, 1965) which, in turn, imply the influence of basement structures. AVEDIK (1975a, b, 1979, 1982), from seismic investigations, suggests that nowhere does the succession above the metamorphic basement reach 5 km thickness and he interprets some Palaeozoic strata between the metamorphic basement and the post-Palaeozoic sediment. His maps show that there are, in fact, a number of basin-like features, possibly pre-Upper Cretaceous depocentres, separated by NW/SE highs. Many faults, which were for at least part of their history dextral wrench faults, are aligned in this direction in SW England (DEARMAN, 1964) and there

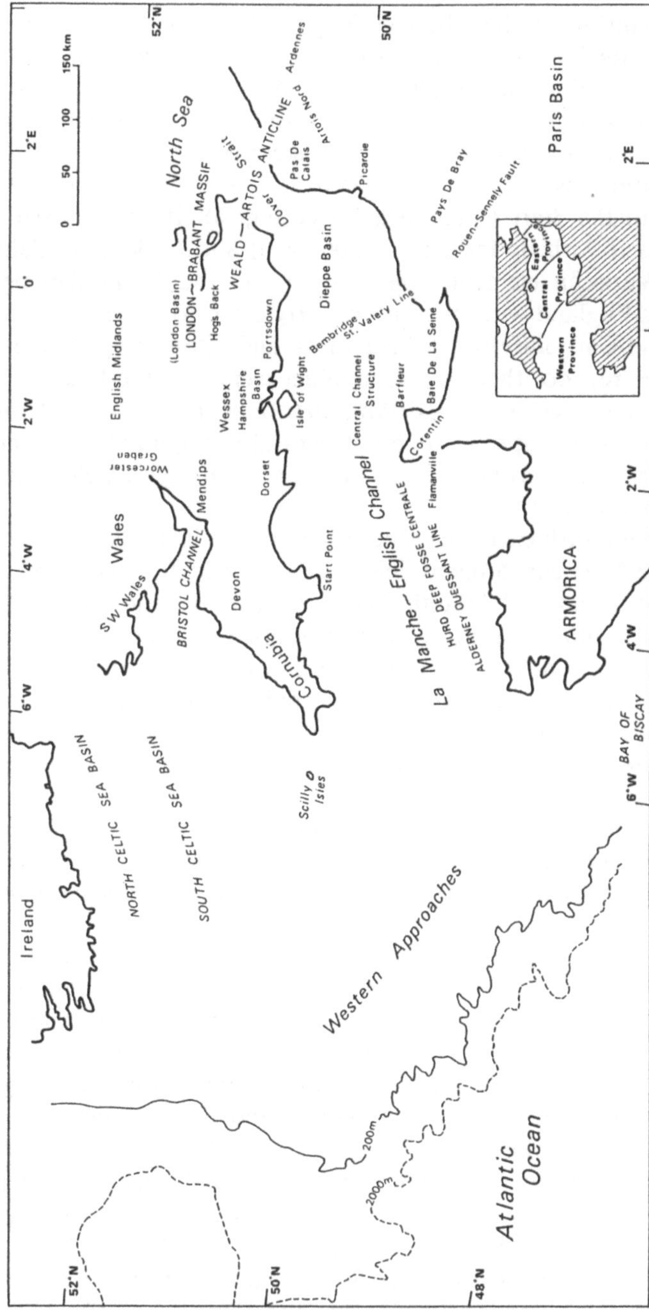

Fig. 6. Localities and general position of features in the English Channel to which reference is made in the text. Inset: the provinces of the English Channel.

are faults in a similar direction in NW France. Some of those of SW England and NW France had a Variscan origin: they parallel Variscan mineral lineations (DURRANCE and LAMING, 1982) and many link into Variscan thrusts.

The degree of deformation of the pre-Permian strata is not known, but it is assumed to be comparable with that known from both sides of the Channel. The Permian to mid-Cretaceous strata are only locally deformed, mainly in the vicinity of the fault lines (CURRY *et al.*, 1970). The Permo-Triassic sediments (they are not differentiated) rest with faulted and possibly unconformable contact against the pre-Permian rocks of SW England: as is common for much of northwest Europe they are usually red sands and marls but no evaporites have yet been reported. There is evidence of pene-contemporaneous volcanism on land in England and France while volcanic pebbles in some Permo-Triassic deposits imply more widespread basic volcanic activity in the region. The Permo-Triassic deposits, like those of the succeeding Jurassic, are not exposed in the more westerly parts of the province. The latter, where exposed, resemble those known in successions exposed on land, however the existence of Jurassic deposits, possibly with different lithologies, hidden beneath younger sediments towards the continental margin seems likely. This must certainly be the case for the Lower Cretaceous successions when rifting, linked with the creation of the Bay of Biscay, was occurring (WILLIAMS and MCKENZIE, 1971). Certainly those Lower Cretaceous sediments which do exist further east in the province imply considerable instability hereabouts.

After the Lower Cretaceous, a marked change in the style of deposition took place with a series of transgressions from the west (i.e. from the new Atlantic Ocean) into the broadly warped basin. The Upper Cretaceous Chalk Sea transgressed from the west, the deposits thinning eastwards. There is evidence of erosion of the Upper Chalk, particularly in the west (CURRY *et al.*, 1971) and the Danian (Palaeocene) chalk is of limited extent: this may be in part depositional and in part due to Eocene erosion. Only the lowest Eocene is present, it is marine in character and seems to have transgressed from the west. Late-Eocene and possibly Oligocene erosion may have been partly responsible for this incomplete succession. Oligocene sediments are of very limited extent and the Miocene *Globigerina* silts again appear to have been associated with transgressions from the Atlantic. Post-lower Miocene deposits occur only in the most western part of the province (CHANNON and HAMILTON, 1976; EVANS, 1980). None of this post lower-Cretaceous succession is strongly folded: there is only a broad down-warping along an axis parallel to the established WSW trend and some local open folds. Successions are thickest between the axis and the Alderney—Ouessant line and there is evidence of continued movement along the line into historically recent times (MOURANT, 1931).

The boundary between the western and central provinces, the Start-Cotentin line, is a feature which controlled sedimentation through Permian (LAMING, 1966), Cretaceous (LEFORT, 1975) and Tertiary times (CURRY *et al.*, 1971). The origin and continued influence of the line may be structural though BACON (1975) suggested, on gravimetric grounds, the presence of a granite or granite-like mass hereabouts. Eastwards of the line, the successions thicken and become more complete, the structural style also changes.

In the *central province* monoclinal, steep and faulted structures bring successions of widely differing ages into close juxtaposition (Fig. 4). These structures are generally aligned east-west for much of their length but in detail they are often slightly curved. In the west of this central province the structural directions turn to the west-southwest while in the east the structures turn sharply southeast. The oldest strata are Permo-Triassic; often pebbly (LAMING, 1966) they abut against the Start and Cotentin peninsulas and appear to thicken eastwards.

The succeeding successions resemble those known from the adjacent land areas but locally thicknesses vary from place to place. There is evidence of mid-Mesozoic (Cimmerian) movements causing tectonic inversions (i.e. changes in the sense of vertical movexents—thus areas which at one time had been subsiding become uplifted while previously positive areas, with consequently thinner successions, subside): a phenomenon widespread across southern Britain, onshore and offshore. The Upper Cretaceous Chalk successions are widespread, though the Danian Chalk has not been discovered. The once extensive Palaeogene deposits, ranging from marine to freshwater, are now separated by faulting and folding (CURRY and SMITH, 1975). No Neogene sediments have been recovered from this province while Quaternary sediments can only be limited to the fillings of the palaeovalley system, though to date no evidence of an age for the valley fills has been recovered. The generally accepted age for the last major episode of folding, and faulting, is in the late Palaeogene.

The structural character of this province extends northwards into the Wessex region of southern England (Fig. 7). Close to the English coast there is a line of north facing asymmetrical anticlines or monoclines of a range of dimensions and several exhibit two or more episodes of movement (DONOVAN and STRIDE, 1961; HOUSE, 1961; RIDD, 1973; MELVILLE and FRESHNEY, 1982; STONELEY, 1982). Northwards the structures are less intense. South of the coast there is a broad syncline and at its southern

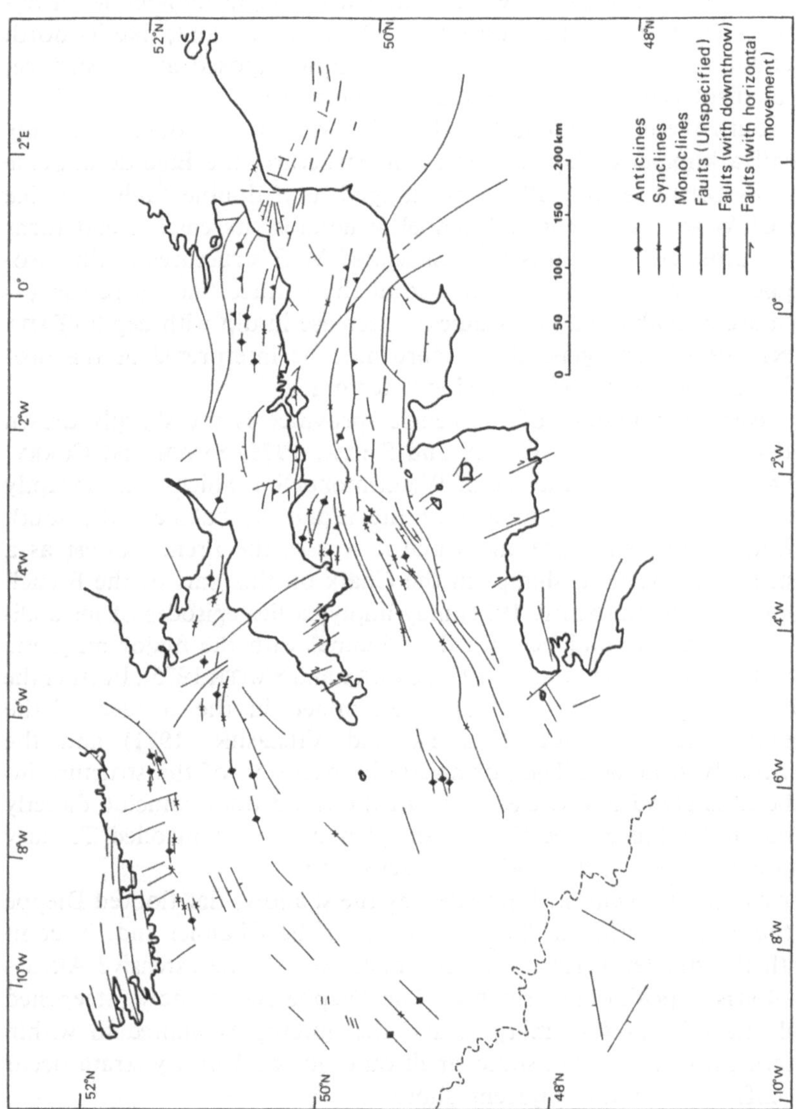

Fig. 7. Structural patterns as exhibited in the Mesozoic and Cenozoic strata of southern England, the English Channel, the Celtic Sea and the Western Approaches.

margin there is a south facing monoclinal, and in part faulted, structure (Fig. 6) (SMITH and CURRY, 1975; IGS, 1977). This Mid-Channel structure, like its north facing counterpart, the Isle of Wight monocline, brings strata of widely different ages into close juxtaposition and, like its north facing counterpart has, measured in strata now exposed at the surface, a vertical displacement of more than a kilometre.
Further south there is a similar south facing structure, though with only about half a kilometre displacement. This structure, the Baie de la Seine fault (LARSONNEUR *et al.*, 1975), is mapped as a double fault and like the Isle of Wight and Central Channel structures, its eastern end turns southeastwards. All the faults and associated W-E structures in this province appear to be vertical or nearly so at the surface but where subsurface data are available the faults quickly decrease in dip with depth (TAITT and KENT, 1939). The general structure may be interpreted as the product of compressional and extensional tectonics.

The eastern boundary of the central province is the sharply drawn Bembridge—St Valery line (CURRY and SMITH, 1975; SMITH and CURRY, 1975). This line links with the Isle of Wight monocline which turns abruptly southeast just off the eastern end of the island. It balances the south facing Mid-channel structure and while it crosses the French Coast as a minor structure, localised slumps in the Chalk on this part of the French coast (JUIGNET and KENNEDY, 1974) may imply earlier episodes of instability. It links both here and beneath the Channel with the major magnetic anomaly of France (CORPEL *et al.*, 1972; CURRY and SMITH 1975). Further the line coincides with a line of faults determined in the surface of the Palaeozoic beneath France (HERTIER and VILLEMIN, 1971) and the Rouen–Sennely structure. The northwesterly extension of the structure into southern England is less clear though the trend does coincide directly with structural features (see Fig. 7) and it may have controlled Tertiary sedimentation (see PLINT, 1982 for discussion).

The *eastern province* is dominated by the shallow, flat floored Dieppe Basin, the largest outlier of Tertiary strata in the Channel and, in common with the smaller Tertiary outliers, once part of the extensive Anglo–Belgian–Paris depositional province. The Dieppe Basin shows steepened dips only near its present margins and the outcrop is limited to within the eastern province though some small outcrops of Tertiary strata occur on the cliff tops of the adjacent shores.

Some faults have been mapped in and at the margins of the Dieppe Basin, they are usually straight and of short length. Immediately north and south there are, on land, two anticlines—the Portsdown and Pays de Bray structures respectively. As with the central province the generally accepted date for the last major movements is late Oligocene (CURRY and SMITH, 1975). Northeastwards, up Channel, there is the much faulted Weald-Artois anticlinorium with late Jurassic strata outcropping in its core structures. The successions resemble those known from on land though the Jurassic deposits here differ from those found elsewhere in southern England in having a higher content of clastics. The outcrop of the base of the Chalk on the northern limb of the Weald-Artois structure coincides with the narrowest part of the Channel–the Dover Strait. North of the Strait the geological picture becomes simpler and the sea widens into the southern bight of the North Sea.

South of the eastern province, the geology is Chalk dominated and the structural features fan from just south of east in the vicinity of the Dover Strait to southeast in the Pays de Bray structure. North of the province, the Wealden anticlinorium is the major structure of southern England. Complex though it is (LAKE, 1975) the intensity of the structures increases northwards. Beneath the Hog's Back, SMALLEY and WESTBROOK (1982) have geophysically defined major structures which they describe as due to Variscan overthrusting; the surface of the London Platform rises sharply from the south coast of England to about 300 m below the land surface west of London (WALLACE, 1982). The geological evidence points to the onset of folding in late Eocene times.

In northern France, in the Pas de Calais/Nord region, the Weald-Artois anticlinorium closes and the Upper Cretaceous-Cenozoic boundary parallels the hidden coalfields of northern France. The deep geology is complex with much thrusting of Carboniferous strata (BOUROZ, 1960; WALLACE, 1982). The trend of the Weald-Artois anticlinorium is W-E swinging SE in Kent and across the Dover Strait, returning to a more W-E direction in Belgium and northern France. It is in this northeastern part of the eastern province that the Channel cuts across the geological grain: this is in contrast to the rest of the eastern and other provinces.

4. Late Events in the Geological History

In contrast to the general absence of late Tertiary deposits in the Channel (with the exception of Miocene sediments of the western province) there are numerous small outcrops of deposits of this age in the land areas bordering the Channel (see Fig. 8). They are now found in isolated outcrops often more than 100 m above present day sea-level. Most of the deposits are marine, some deposited at depths of more than 50 m, e.g. Lenham Beds (Funnell, in CURRY *et al.*, 1978). The occurrence of such deposits has been ascribed to episodes of high sea level, but the

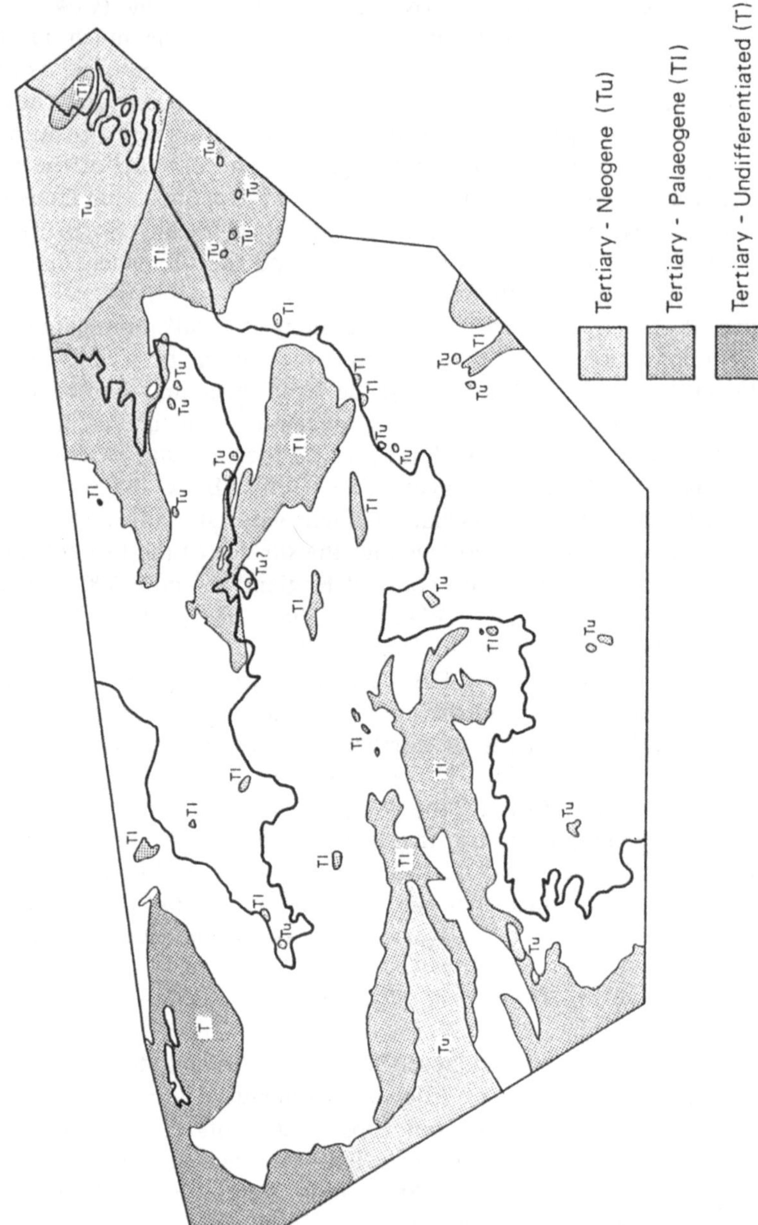

Fig. 8. Tertiary deposits in and around the English Channel.

sporadic nature of these deposits does not appear to bear this out. Alternatively the deposits may have been laid down on a surface which has subsequently been warped and in places elevated and eroded. This latter explanation must, in fact, be the case since the deposits now descend beneath the North Sea north of the area which is the subject of this paper.

The palaeovalley system (AUFFRET *et al.*, 1980; LASSONNEUR *et al.*, 1982) demands further consideration. Early views, based on fragmentary evidence, suggested a development in late Tertiary times (DINGWALL, 1975). Clearly the system links into the existing valley systems seen on the lands bordering the Channel (Fig. 3). The main exception is the broad valley which extends through the Dover Strait and links to the unfilled or only partially filled valley which swings to the northern part of the eastern province before turning southwest towards the Hurd Deep. It is this author's view (SMITH, 1985) that the palaeovalley system has a two part origin: a system of valleys which represent the lower reaches of the valleys which still exist on land and the product of a catastrophic event, or succession of events, related to the origin of the Dover Strait. The latter part of the evolution may be described briefly as follows: when the northern portion of the North Sea was blocked by late Pleistocene ice sheets a lake developed against the unbreached northern limb of the Weald-Artois anticlinorium, particularly against the Chalk which was sealed by permafrost conditions; overtopping the Chalk it formed plunge pools (the present infilled Fosse Dangeard, see Fig. 9) and in a catastrophic event comparable to the Spokane Flood (BRETZ, 1969) modified and sediment-filled the valleys of the eastern province, the flood waters conjoining north of the Cotentin peninsula to form the Hurd Deep in an episode of short-lived erosion (Fig. 10). This event may have happened several times, suggesting a repeated uplifting of the Dover Strait barrier, or it may have happened only once. Part of the infilling sediments of the Fosse Dangeard have an age as late as Brorup (Appendix to DESTOMES *et al.*, 1975). Even if this late age should be the only occurrence of the overtopping and creation of gorge in the Chalk linking the Channel and North Sea in late Devensian (Wisconsinian) times as sea-level rose, observations of the present day rates of erosion of the English and French coasts give ample time for the widening of the Channel's eastern province to its present width (see Table 2). No obvious deformation seems to have affected the area since this event.

5. *Discussion*

The present form of the Channel appears to be due to tectonic events,

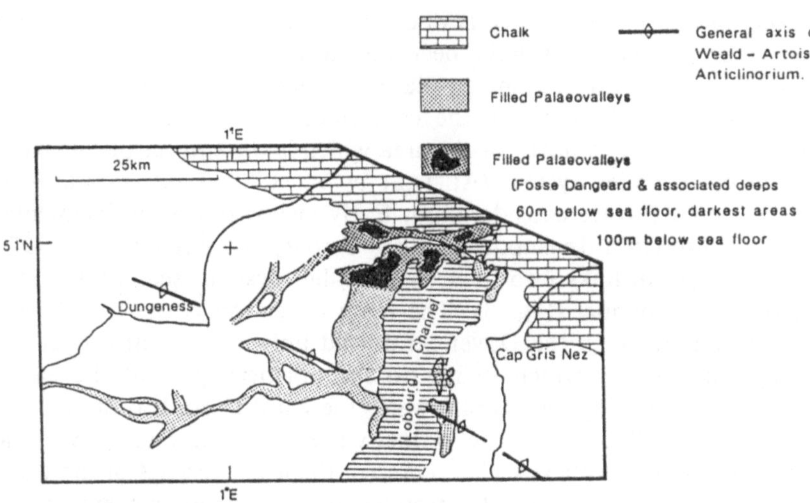

Fig. 9. The Dover Strait, showing the outcrop of the Chalk—pre Chalk boundary, the 'plunge-pools' of the Fosse Dangeard and the Loburg Channel.

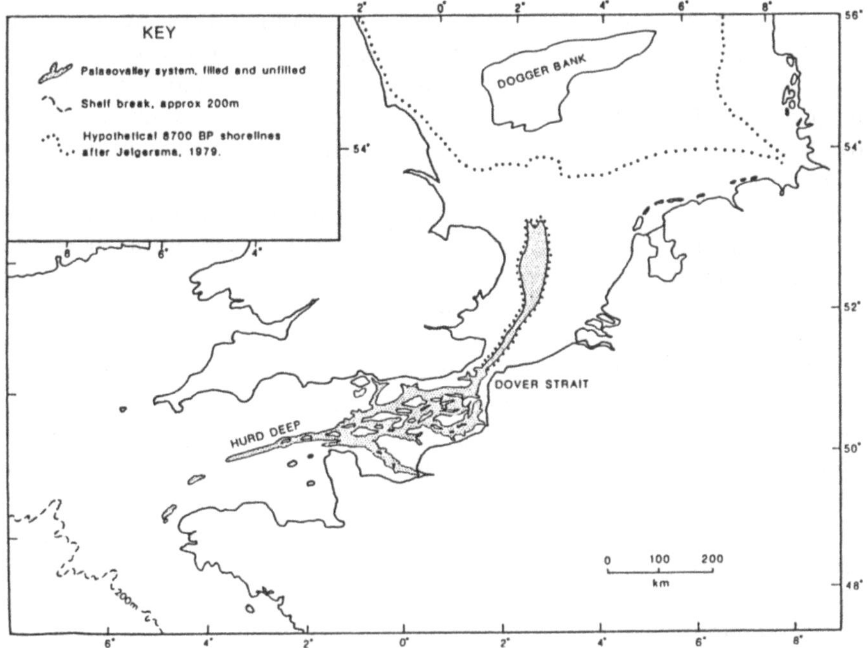

Fig. 10. The extent of the palaeovalley system of the eastern English Channel and its links to the Hurd Deep and the North Sea.

Table 2. Speed of cliff retreat based on observations of the south coast of England

	per year	per 10,000 yrs
Tertiary Clays	1.0–2.0 m	10–20 km
Chalk	0.3–0.5 m	3–5 km
Hastings Sands	1.0 m	10 km

which have continued since the late Variscan episode, which have determined the position and shape of the western and central provinces and the evolution of a drainage pattern subsequently modified by a late Quaternary catastrophic event or events in the eastern province. It parallels the major Variscan structures (Fig. 11) and lies between the massifs created by Variscan and pre-Variscan orogenic events (Fig. 5). This interpretation takes account of the latest views on the structural geology of southwest England and the continental shelf between Ireland and France. GARDINER and SHERIDAN (1982) have convincingly re-interpreted the position of the Variscan fold belts and the Variscan Front south of Ireland. Their interpretation makes it link into the Variscan history of Iberia and they refute the long standing interpretation of the front passing through the southern part of Ireland, explaining that trend as due to movements on a Caledonian controlled basement. It certainly is clear that the North Celtic Basin (Fig. 12) parallels the Caledonian trend of Wales (Cambria) and Ireland. With the interpretation of GARDINER and SHERIDAN (1982) the Western Approaches Basin and Western Channel Basin keep a fixed distance from the Variscan Front (Fig. 12). SHACKLETON *et al.* (1982) argue that much of the Variscan structures of SW England are "thin skinned" tectonics with north and south-facing structures the product of northward thrusting. They further suggest that there is the possibility of the whole of Cornubia, the Channel and Armorica being moved northward on a plane of decollement. LEEDER (1982) suggests that what in this paper is called the western province is a faulted structure which commenced in late Variscan or early post-Variscan times as an extensional feature. The magnetic anomalies certainly imply a deep structure. Subsequent movements have emphasised this graben development, but not as a place of uninterrupted subsidence. The developments along the Bay of Biscay fault and the creation of the Bay of Biscay led to new dimensions in the development of these basins. After the creation of the Bay of Biscay by the mid-Cretaceous the western province responded to vertical movements of the Atlantic margin, but other influences must have made themselves felt for structures continued to develop in SW England until at least Tertiary times (CURRY *et al.*, 1979).

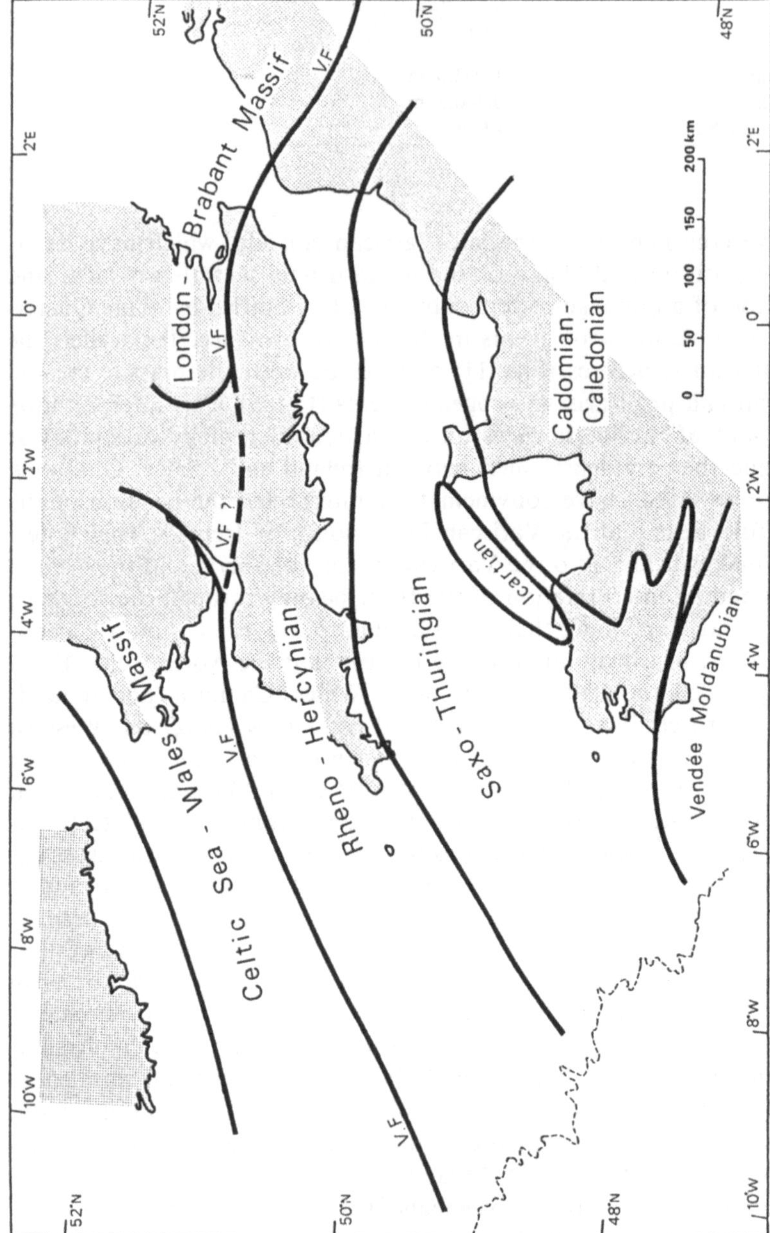

Fig. 11. The main Variscan features in the vicinity of the English Channel. The names refer to the principal Variscan belts. VF = Variscan Front, the trend based on GARDINER and SHERIDAN (1982). Pre-Variscan features: Icartian and Caledonian/Cadomian.

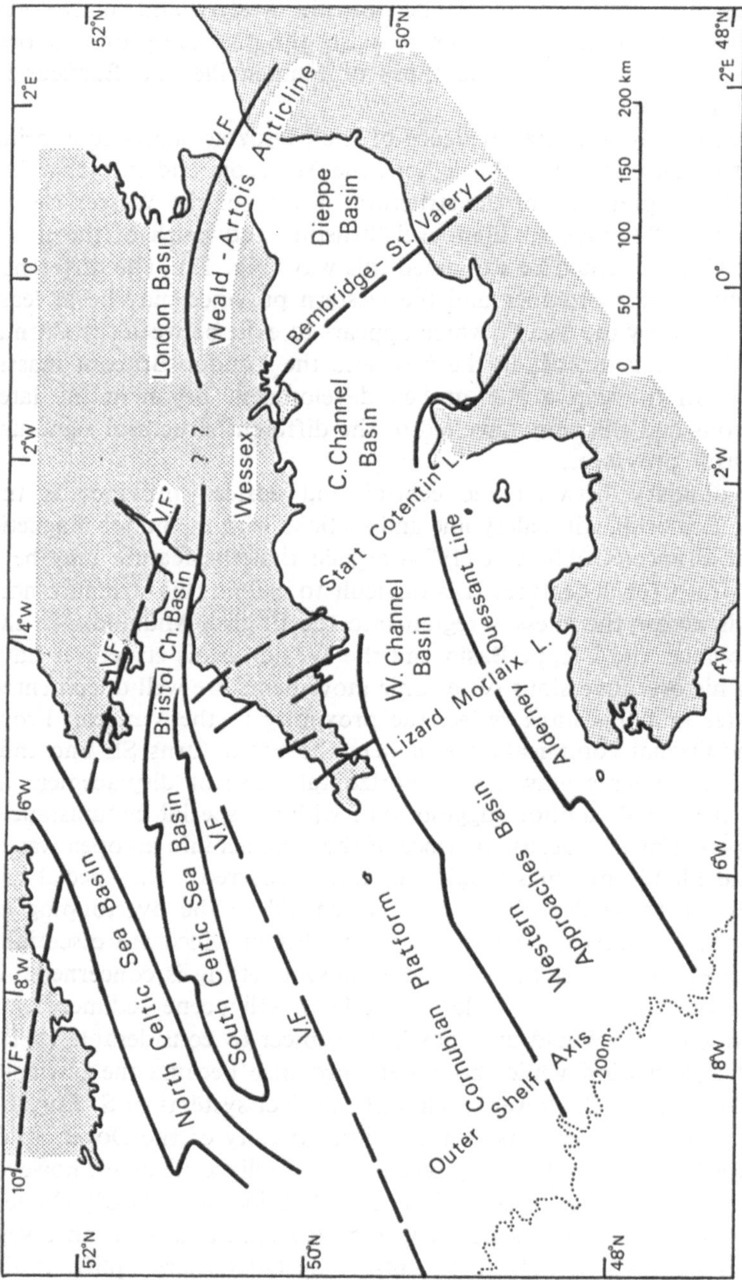

Fig. 12. The principal basins of the English Channel region. VF—Variscan Front after GARDINER and SHERIDAN (1982); VF* = the conventional line of the Variscan Front.

Throughout this time the boundary between the western and central provinces influenced sedimentation. There exists the possibility of link between the Cornubian granites and those of Flamenville and Barfleur in the Cotentin peninsula.

The central province has evidence of the Mesozoic successions being subjected to thrust tectonics throughout the Mesozoic and into the Tertiary. The reinterpretation of the structural history of the Wessex region by STONELEY (1982) suggests high-level thrusting and much of the structure of the Channel could be explained this way (Fig. 13). The difference in style between this province and the western province may be at least in part explained by the "gap" which appears to exist between the Cornubian and Cambrian massifs in the west and the London-Brabant massif in the east. In the gap a N-S graben development began in the late-Carboniferous and this may have led to the different structural signature of the central province.

The boundary between the central and eastern province is the remarkable Bembridge-St Valery line and its links with the major magnetic anomaly of France. COLBEAUX (1974) argued that the feature may be a major sinistral wrench feature. It is difficult to extend this structure northwestwards across the Wessex region into the English Midlands. In the eastern province, the Dieppe Basin and the Weald-Artois structures have had a long history; inversions of tectonic movements are well documented for the latter and this must reflect the proximity to the Variscan Front with the additional complication that here the Front turns SE and thus subsequent movements may have a horizontal sense of displacement as well as vertical. This author suggests that without special circumstances the development of the eastern province of the Channel into an open waterway by late Flandrian times would not have occurred: the special circumstances are those described above which led to the overtopping of the Weald-Artois barrier. The structures of this province are essentially "cross-grained" so far as the Channel's present setting is concerned. At some stage or stages after the last episode of Oligocene sedimentation and the present, the Weald area has been subject to considerable (>1.5 km) uplift and erosion while the eastern province became the lowlands and thus the place of the lower reaches of the river systems of S. England and N. France. Any early link in the general vicinity of the Dover Strait with the North Sea was presumably severed in late Oligocene times, however the widespread occurrence of late Miocene, early Pliocene and early Quaternary deposits, albeit widely separated, in S. England and N. France suggests that a link existed only to be cut-off by Quaternary uplift. If the overtopping of the N. limb of the Weald-Artois barrier occurred more

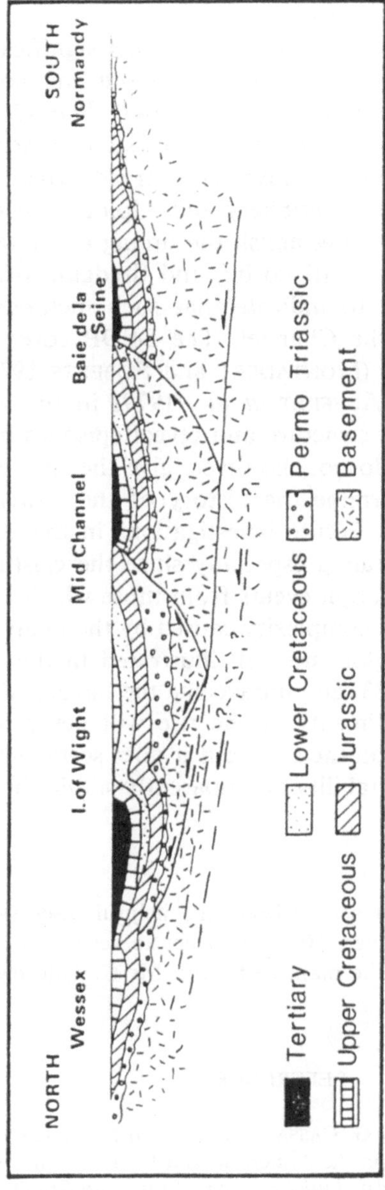

Fig. 13. Hypothetical cross section of the central province of the English Channel showing structures as they may have developed in response to thin-skinned tectonics.

than once, and more sampling of the palaeovalleys and the continental apron may help to establish this, then this would be further evidence of repeated movements in this area.

Increasingly there are small but none the less significant indications of recent tectonic movements in southern England and northern France (see POMEROL and subsequent papers in the *Bull. Soc. Geol. de France*, 1980). It would seem that the hinterland of this stable margin is far from stable, subject as it is to continued movements on old structures in response to later events on the Atlantic periphery and Alpine orogenic belt. Deep sea drilling may be one way to establish the timing and intensity of these movements. So far the results of coring and dredging off the Western Approaches has done little to indicate that such methods will help in elucidating the history of the Channel. The DSDP cores of Leg 47 at Sites 339, 400, 401 and 402 (MONTADERT and ROBERTS 1979) and dredgings (CURRY *et al.*, 1962; AUFFRET *et al.*, 1979) in the vicinity of the margin (Fig. 8) point to the structure and reveal episodes in the development of the margin. They do not, however, give the sort of information which can be used to determine the history of the hinterland, indeed, that was not their purpose. Future investigations in the continental rise may, however be more revealing especially since the existing results will help scientists to separate margin events from those which took place further onto the continent. The complexity caused by the many factors which interplay in this general region are well illustrated in the volume edited by STANLEY and MOORE (1983) and unravelling the history will not be easy.

It may be that the sediments of the rise will never reveal clues as to the history of instability on the shelf but it does seem clear that factors which are the product of instability on apparently stable shelfs will always demand attention.

Acknowledgements

The author would like to acknowledge the helpful discussions with many geologists in preparing this paper. He also wishes to record his grateful thanks to Mrs S. Bishop who typed the manuscript and Mr C. Hildrew who prepared the figures.

REFERENCES

AUFFRET, G. A., L. PASTOURET, G. CASSAT, O. DE. CHARPAL, J. CRAVATTA, and P. GUENNOC, Dredged rocks from the Armorican and Celtic margins. in: MONTADERT, L., D. G. ROBERTS, *et al.* (eds). *Initial Reports of the Deep Sea Drilling Project*, **48**, Washington (U.S. Govt. Printing Office) 995–1013, 1979.

AUFFRET, J. P., D. ALDUC, C. LARSONNEUR, and A. J. SMITH, Cartographie de réseau de paléovallées et de l'epaisseur des formations superficels meubles de la Manche orientale. *Ann. l'Inst. Ocean*, **56**, 21–35, 1980.

AVEDIK, F., Seismic structure of the Western Approaches and the south Armorican continental shelf and its geological interpretation. in: WOODLAND, A. W. (ed). *Petroleum and the continental shelf of NW Europe*. Academic Press, London, 29–44, 1975a.

AVEDIK, F., Seismic reflection survey in the Western Approaches to the English Channel: preliminary results. *Phil. Trans. Roy. Soc. London*, **A279**, 29–39, 1975b.

AVEDIK, F. and D. HOWARD, Preliminary results of a seismic refraction study in the Meriadzek-Trevelyan area, Bay of Biscay. in: MONTADERT, L., D. G. ROBERTS, *et al.* (eds). *Initial Reports of the Deep Sea Drilling Project*, **48**, Washington (U. S. Govt. Printing Office) 1015–1023, 1979.

AVEDIK, F., A. L. CAMUS, A. GINSBERG, L. MONTADERT, D. G. ROBERTS, and R. B. WHITMARSH, A seismic refraction and reflection study of the continent-ocean transition beneath the north Biscay margin. *Phil. Trans. R. Soc. London*, **A305**, 5–25, 1982.

BACON, M., A gravity survey of the Western English Channel between Lyme Bay and St Brieuc Bay. *Phil. Trans. R. Soc. London*, **A279**, 69–78, 1975.

BOILLOT, G., Géologie de la Manche occidentale: fonds rocheux, depôts quaternaires, sediment actuels. *Thèse, Paris, Ann. Inst. oceanogr.*, XLII (1), 1–219, 1964.

BRETZ, J. H., The Lake Missoula floods and the channeled scabland. *Jour. Geol.*, **77**, 505–543, 1969.

BOUROZ, A., La structure du Palaeozoique du Nord de la France au sud de la Grande Faille du Midi. *Annls. Soc. geol. N.*, **80**, 101–113, 1960.

CHANNON, R. D. and D. HAMILTON, Wave and tidal current sorting of shelf sediments southwest of England. *Sedimentology*, **23**, 17–42, 1976.

COLBEAUX, J. P., Mise en evidence d'une zone cisaillement Nord Artois. *C.r. hebd. Seanc. Acad. Sci., Paris*, **D278**, 1159–1161, 1974.

CORPEL, J., M. OGIER, and C. WEBBER, Carte magnétique de la France: 1/1,000,000 (Feuille Nord) et notice explicative. *Bur. Rech. geol. min. Paris*. 1972.

CURRY, D., C. G. ADAMS, M. C. BOULTER, F. C. DILLEY, F. E. EAMES, B. M. FUNNELL, and M. K. WELLS, *A correlation of Tertiary rocks in the British Isles*. Special Report No. 12, 72pp., 1978.

CURRY, D., D. HAMILTON, and A. J. SMITH, *Geological and shallow subsurface geophysical investigations in the western approaches of the English Channel*. Inst. Geol. Sci. Reps. **70**/3, 1–12, 1970.

CURRY, D. D. HAMILTON, A. J. SMITH, The geological evolution of the western English Channel basin and its relation to the nearby continental margin. in: DELANY, S. M. (ed.). *The geology of the east Atlantic continental margin*. Pt 2: Europe. Rept. Inst. geol. Sci., **70**/14, 129–142, 1971.

CURRY, D., E. MARTINI, A. J. SMITH, and W. F. WHITTARD, The geology of the Western Approaches of the English Channel. I. Chalky rocks from the upper reaches of the continental slope. *Phil. Trans. R. Soc. London*, **B245**, 267–290, 1962.

CURRY, D. and A. J. SMITH, New discoveries concerning the geology of the central and eastern part of the English Channel. *Phil. Trans. R. Soc. London*, **A279**, 233–242, 1975.

DEARMAN, W. R., Wrench faulting in Cornwall and south Devon. *Proc. geol. Assoc., London*, **74**, 265–287, 1964.

DESTOMBES, J. P., E. R. SHEPHARD-THORN, J. H. REDDING, A buried valley system in the Strait of Dover with appendix on pollen analysis by M. T. MORZADEC-DERFOURN.

Phil. Trans. R. Soc. London, **A279**, 243–256, 1975.

DINGWALL, R. G., Sub-bottom infill channels in an area of the eastern English Channel. *Phil. Trans. R. Soc. London*, **A279**, 233–242, 1975.

DONVAN, D. T. and A. H. STRIDE, An acoustic survey of the sea floor south of Dorset and its geological interpretation. *Phil. Trans. R. Soc. London*, **B244**, 299–330, 1961.

DURRANCE, E. M. and D. J. C. LAMING, *The geology of Devon*, University of Exeter, 346pp., 1982.

EVANS, C. D. R., G. K. LOTT, and G. WARRINGTON, (Compilers). The Zephyr (1977) Wells, Southwestern Approaches and western English Channel. *Rep. Inst. geol. Ser.*, **81**/8, 44pp., 1981.

FITCH, F. J., S. C. FORSTER, and J. A. MILLER, The 40 Ar/39 Ar age spectrum of a roch from Gerrans Bay, Cornwall. *J. geol. Soc. London*, **141**, 21–25, 1984.

GARDINER, P. R. R. and D. J. R. SHERIDAN, Tectonic framework of the Celtic Sea with special reference to the location of the Variscan Front. *Journ. Struct. Geol.*, **3**, 317–332, 1982.

HAMILTON, D., P. HOMMERIL, C. LARSONNEUR, and A. J. SMITH, Geological bibliography for the English Channel (Pt 2). *Phil. Trans. R. Soc. London*, **A279**, 289–295, 1975.

HERTIER, F. and J. VILLEMIN, Mise en evidence de la tectonique profonde du Bassin de Paris par exploration pétrolière. *Bull. Bur. Rech. géol. min. Paris, Sec 1*, no. 2, 11–30, 1971.

HILL, M. N. and F. J. VINE, A preliminary magnetic survey of the Western Approaches to the English Channel. *Q. Jl. geol. Soc. Lond.*, **121**, 463–475, 1965.

HOUSE, M. R., The structure of the Weymouth anticline. *Proc. Geol. Assoc. London*, **72**, 221–238, 1961.

INSTITUTE GEOLOGICAL SCIENCES, *Wight Sheet (50N, 02W) Solid Edition* Map 1: 250 000. Institute of Geological Sciences, London, 1977.

INSTITUTE GEOLOGICAL SCIENCES, *Sub-Pleistocene geology of the British Isles and the adjacent continental shelf*, Map 1: 2 500 000. Institute of Geological Sciences, London, 1979.

JUIGNET, P. and W. J. KENNEDY, Structures sédimentaires et mode d'accummulation de la craie du Turonien superiéur et du Sénonien du Pays de Caux. *Bull. Bur. Rech. géol. Min. Paris* (2), sec IV, no. 1, 19–47, 1974.

KELLAWAY, G. A., J. H. REDDING, E. R. SHEPHARD-THORN, and J. P. DESTOMBES, The Quaternary history of the English Channel. *Phil. Trans. R. Soc. London*, **A279**, 189–218, 1975.

LAKE, R. D., The structure of the Weald—a review. *Proc. geol. Soc. London*, **86**, 549–558, 1975.

LAMING, D. J. G., Imbrications, palaeocurrents and other sedimentary features in the Lower New Red Sandstone of south Devonshire. *Journ. Sed. Petrol.*, **36**, 940–959, 1966.

LARSONNEUR, C., R. HORN, and J. -P. AUFFRET, Géologie de la partie méridionale de la Manche central. *Phil. Trans. R. Soc. London*, **A279**, 1975.

LARSONNEUR, C., D. VASLET, and J.-P. AUFFRET, *Les sédiments superficiel de la Manche.* Bur. Rech. géol. Min. Paris, 1: 500 000, Map and text 15 pp., 1980.

LARSONNEUR, C., J. -P. AUFFRET, and A. J. SMITH, *La Manche orientale, Carte des paléovallées et des bancs sableux.* Bur. Rech. géol. Min. Paris, 1: 500 000 Map and explanatory text, 8pp., 1982.

LEEDER, M. R., Upper Palaeozoic basins of the British Isles—Caledonide inheritance versus Hercynian plate margin processes. *J. geol. Soc. London*, **139**, 479–491, 1982.

LEFORT, J. P., Le controle du socle dans l'évolution de la sedimentation en Manche occidentale apres le Paléozoique. *Phil. Trans. R. Soc. London*, **A279**, 137–144, 1975.

MELVILLE, R. V. and E. C. FRESHNEY, *The Hampshire Basin and adjoining areas.* Inst.

Geol. Sci., H. M. S. O. London, 146pp., 1982.

MONTADERT, L. and D. G. ROBERTS, *Initial Reports of the Deep Sea Drilling Project*, **48**, Washington, 1183pp., 1979.

MONTADERT, L., D. G. ROBERTS, O. DE CHARPAL, and P. GUENNOC, Rifting and subsidence of the northern continental margin of the Bay of Biscay. in: MONTADERT, L., D. G. ROBERTS, *et al.*, *Initial Reports of the Deep Sea Drilling Project*, **48**, Washington, 1025–1060, 1979.

MOURANT, A. E., A study of the seismograms of the English Channel earthquakes. *Mon. Not. R. astr. Soc. Geophys.* suppl. 2, 374–383, 1931.

NAYLOR, D. and P. M. SHANNON, *The geology of offshore Ireland and West Britain*. Graham and Trotman, London, 161pp., 1982.

PLINT, A. G., Eocene sedimentation and tectonics in the Hampshire Basin. *J. geol. Soc. London*, **139**, 249–254, 1982.

POMEROL, C., Avant propos sur tectonique profonde, movements récents et movements superficiels dans le bassin de Paris. *Bull. Soc. géol. de France*, **22**, 629–631, 1980.

RIDD, M. F., The Sutton Poyntz, Poxwell and Chaldon Herring anticlines: a reinterpretation. *Proc. Geol. Assoc. London*, **84**, 1–8, 1973.

SHACKLETON, R. M., A. C. RIES, and M. P. COWARD, An interpretation of the Variscan structures in SW England. *J. geol. Soc. London*, **139**, 533–541, 1982.

SMALLEY, S. and G. K. WESTBROOK, Geophysical evidence concerning the southern boundary of the London platform beneath the Hog's Back, Surrey. *Jour. geol. Soc. London*, **139**, 139–146, 1982.

SMITH, A. J., Instability on a stable shelf—the geological development of the continental shelf between Britain and France. in: AOKI, H. and IIZUKA, S. (Eds) *Volcanoes and the Tectonosphere*, Tokai University Press, Tokyo, 289–300, 1976.

SMITH, A. J., In press A catastrophic origin for the palaeovalley system of the stern English Channel, *Mar. Geol.* **64**, 64–74, 1985.

SMITH, A. J. and D. CURRY, The structure and geological evolution of the English Channel. *Phil. Trans. R. Soc. London*, **A279**, 3–20, 1975.

SMITH, A. J., D. HAMILTON, D. N. WILLIAMS, and P. HOMMERIL, Bibliographie géologique de la Manche. *Mem. du Bur. Rech. géol. Min. Paris*, **79**, 303–323, 1972.

STANLEY, D. J. and G. T. MOORE, (Eds). *The shelfbreak: critical interface on continental margins*. Society of Econ. Pal. Min., Sp. Pub. 33, Tulsa, 467pp., 1983.

STONELEY, R., The structural development of the Wessex Basin. *Journ. geol. Soc. London*, **139**, 545–554, 1982.

TAITT, A. H. and P. E. KENT, Notes on an examination of the Poxwell anticline, Dorset. *Geol. Mag.*, **76**, 173–181, 1939.

WALLACE, P., The subsurface Variscides of southern England and their continuation into continental Europe. in: HANCOCK, D. L. (Ed.) *The Variscan fold belt in the British Isles*, 198–208, 1982.

WILLIAMS, C. A. and D. MCKENZIE, The evolution of the north-east Atlantic. *Nature*, **232**, 168–173, 1971.

WOOD, R. and P. BARTON, Crustal thinning and subsidence in the North, Sea. *Nature*, **302**, 134–136, 1983.

ZIEGLER, P. A., *Geological Atlas of western and central Europe.* Shell International Petroleum Maatschappij B. V., The Hague, 130pp and 40 plates, 1982.

Formation of Active Ocean Margins, edited by N. Nasu *et al.*, pp. 625–637.

THE FATE OF SEAMOUNTS AND OCEANIC PLATEAUS ENCOUNTERING A DEEP-SEA TRENCH AND THEIR EFFECTS ON THE CONTINENTAL MARGINS

Kazuo KOBAYASHI

Ocean Research Institute, University of Tokyo, Nakano-ku, Tokyo 164, Japan

Abstract. The present situations of Amami Plateau, Ogasawara Plateau and Dai-ichi Kashima Seamount, encountering the Ryukyu, Izu-Bonin and Japan Trench, respectively, are examined and compared with Izu Peninsula and Hayama-Mineoka-Setogawa ophiolite zone, both of which collided and were accreted to central Honshu in the past. A criterion of their fates is established; very buoyant micro-continents and oceanic plateaus are accreted to the opposite plate, less buoyant oceanic plateaus and seamounts cause uplifting of the opposite landward toe of the trench but are eventually subducted unless the subduction is of an accretionary type. Only fragments of oceanic crust are accreted to the opposite plate in the accretionary-type oblique subduction zones, while most of the oceanic lithosphere is subducted deep into the mantle.

1. *Introduction*

Seamounts and oceanic plateaus colliding with a continental mass will eventually be accreted to the continent, if their lithospheric buoyancy is positive and sufficiently large owing to their thick "granitic" crust. Izu Peninsula in the south-central Honshu, Japan, is a well-known example of micro-continental blocks colliding with central Honshu since about 0.5 MaBP. The Suruga and Sagami Troughs are oceanward extensions of the zone of convergence along which subduction proceeds.

An ophiolite belt between Hayama-Mineoka and Setogawa in the south-central Honshu indicates ancient subduction at a period prior to

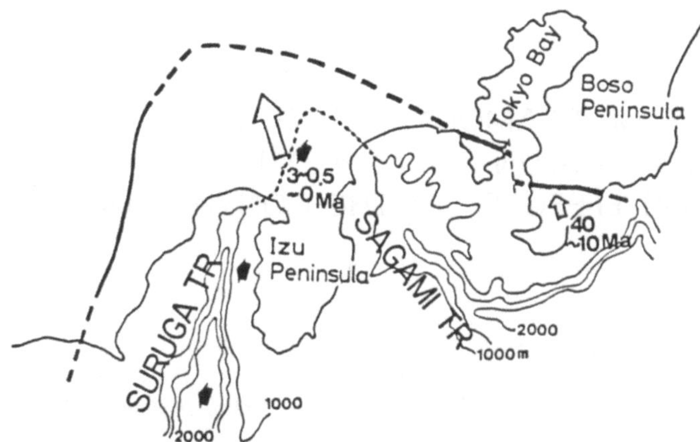

Fig. 1. Past and present subduction-collision zones in the Suruga, Sagami and Izu regions (after TONOUCHI and KOBAYASHI, 1983). Arrows indicate directions of the ancient and present relative motions of the oceanic plate.

10 MaBP along a line further north of the present collision and subduction zone (TONOUCHI and KOBAYASHI, 1983; OGAWA, 1982, 1983). The southernmost portion of the Boso and Miura Peninsulas as well as the coastal region west of Suruga Bay were then accreted to Honshu (Fig. 1). There are some implications that a new subduction zone will be created south of Izu Peninsula (ISHIBASHI, 1978; HAMADA, 1984). In this successive processes of subduction or collision and accretion, only very buoyant portions of the Izu Ridge are accreted to Honshu and adjacent less buoyant ocean floors are subducted.

A seamount of small size appears to be subducted at a deep-sea trench and be consumed in the deep mantle, because its buoyancy is small or even negative in the mantle when its total lithospheric column is considered (TOMODA and FUJIMOTO, 1983). Its fragments may be accreted to the toe of the opposite plate, if the subduction is oblique and the margin is of the Chile-type (UYEDA and KANAMORI, 1979). If the seamount is comparatively large, it may have some influence on the landward wall of the trench. The landward toe facing the seamount will be uplifted and, in some cases, emerged above the sea surface owing to buoyancy of the large crustal block of the seamount subducted beneath the landward toe. However, such an effect disappears after the seamount is completely subducted and the uplifted toe will subside to the normal depth after passage of the seamount.

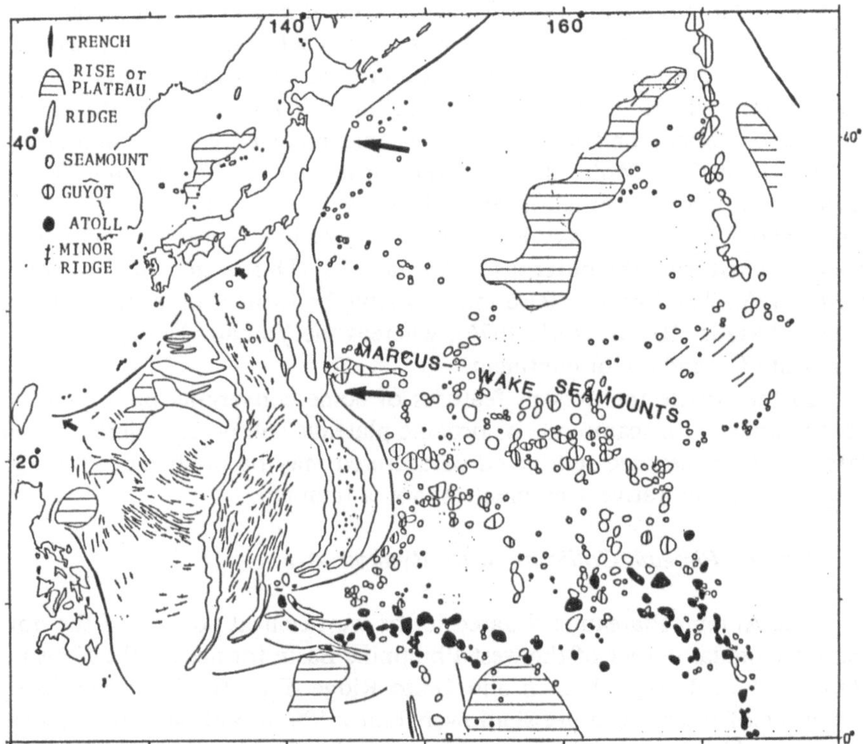

Fig. 2. Distribution of seamounts and oceanic plateaus in the western Pacific (after IWABUCHI, 1982).

A great number of seamounts are distributed in the oceans, particularly in the western Pacific Ocean (Fig. 2). Some of the seamounts are aligned to form a chain such as the Hawaiian and Emperor Chain. Most of them are of volcanic origin, and the equatorial ones are covered with coral reefs. Where a chain of seamounts is colliding with the trench, repeated uplift and subsidence may be recorded.

Oceanic plateaus, which are larger in size than normal seamounts, are generally isolated in mid-ocean areas. Some of them are composed of continental crust characterized by a Moho depth greater than 25 km and a thick "granitic" layer with seismic compressional (P) wave velocities of 6.0–6.3 km/sec. The Sychelles Bank in the Indian Ocean, the Ontong Java Plateau in the Pacific, the Lord Howe Rise in the Tasman Sea, the Faeroe and Rockall Plateaus in the Atlantic are probably these examples

(NUR and BEN-AVRAHAM, 1978; BEN-AVRAHAM *et al.*, 1981). These plateaus seem to be fragments of ancient continental blocks which were disrupted by some processes of ocean spreading and then left behind in mid-ocean. The crustal structure of Shatsky Rise (in the NW Pacific), Yamato Tai (in the Sea of Japan) and some other oceanic plateaus is more similar to oceanic crust. The Moho depth of the former two is 25 km and 20 km, respectively. The "granitic" layer with 6.0–6.3 km/sec P-wave velocity is not apparent in the Shatsky Rise. On the other hand, granitic rocks have been collected from the outcrops of Yamato Tai (DEN *et al.*, 1969; LUDWIG *et al.*, 1975). The Shatsky Rise may possible be formed by a thick pile of oceanic basalts (HILDE *et al.*, 1976), whereas the Yamato Tai is a remnant fragment of island arc or continent.

In this article the present features of some converging plate margins accompanied with seamounts or oceanic plateaus will be considered. Their fates whether they are subducted or accreted may be predicted, although a detailed quantitative treatment is not attempted.

2. *Amami Plateau Colliding with Ryukyu Arc*

The Amami Plateau is situated in the west-central portion of the northern triangular corner of the west Philippine Basin formed by the Kyushu-Palau Ridge, Ryukyu Trench and Daito Ridge (Fig. 3). Water depths of the crests of the plateau are shallower than 1,500 m, although the plateau has a roughly flat crestal plane with water depth of approximately 3,000 m and a diamter of about 100 km. From the crest of a topographic high in the western part of the plateau ($D=1{,}390$ to 1,410 m) large foraminferal fossils *nummulites boninensis* were collected by dredge hauls in a Geodynamics cruise of Tokaidaigaku-maru II (SHIKI *et al.*, 1975). Occurrence of such shallow water fossils of late Eocene age indicates that the plateau was partly emerged above the sea level in late Eocene and later subsided more than 1,000 m.

A number of manganese nodules containing andesites, basalts, granodiorites and biotite-hornblende tonalites as nuclei were also dregded from the same site. K-Ar age of some rock is 85.1 Ma (late Cretaceous), which is the oldest of the Philippine Sea samples (MATSUDA *et al.*, 1975). A cobble of andesite was collected by KH72-2 cruise of Hakuho-maru (KAGAMI, 1974) and a boulder of plagio-granite was obtained in KH82-4 cruise of the same ship (KOBAYASHI, 1983).

In the cruise KH82-4 a tripartite O.B.S. (Ocean bottom seismometer) observation of transmitted signals from natural earthquakes and precisely-located 9 liter airgun sound sources was conducted (KOBAYASHI, 1983;

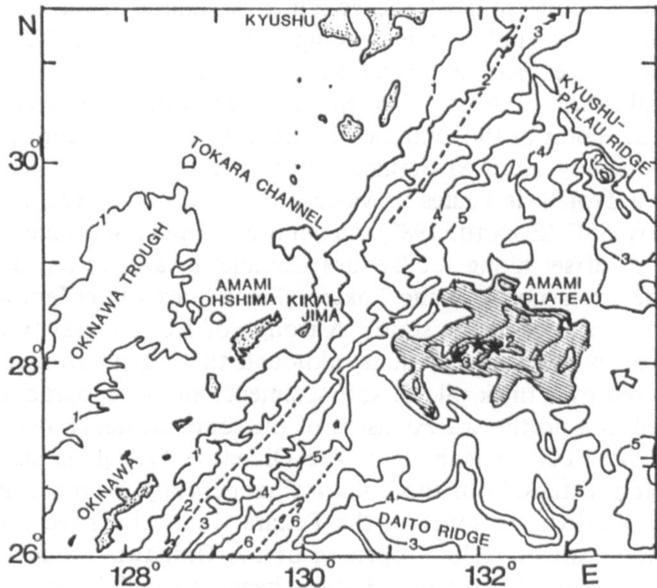

Fig. 3. Location of the Amami Plateau and surrounding region. Contour interval 1 km. Asterisks denote the dredge sites. Broken lines indicate axes of negative free air anomalies.

SUYEHIRO *et al.*, 1982). The O.B.S. observation for only about 3.5 days showed that a number of micro-earthquakes occur along the northwestern and western margin of the plateau, indicating existence of a strong interaction between the arc and plateau.

Analysis of refracted waves indicates that the crustal structure of the Amami Plateau is different from that of normal oceans. In particular, a layer with $V_P = 5.4$–5.6 km/sec is thicker than 4 km (NISHIZAWA *et al.*, 1983). The Moho depth was not determined by this experiment but gravity analysis provides crustal thickness estimate of approximately 10 km assuming isostatic compensation (MATSUMOTO and TOMODA, 1983).

Seismic reflection records indicate that the Philippine Basin floor both north and south of the Amami Plateau is subducted at the Ryukyu Trench beneath the Ryukyu arc, whereas along the zone of collision no subducted slab is recognized in the records and the trench topography disappears (TOKUYAMA and KONG, 1983).

Coralline radiometry of the uplifted fringing reef limestone terraces at Kikai-jima, a small island east of Amami-Ohshima has revealed a very rapid rate (1 to 2 mm/yr) of late Quaternary tectonic uplift of Kikai-jima, which is situated in front of the colliding Amami Plateau (KONISHI *et al.*, 1974, 1978). Its uplift rate is comparable to those of New Guinea and Taiwan where continental collision now occurs (Fig. 4).

An implication of Amami Plateau collision is possible recent change in opening direction of the Okinawa Trough. A focal mechanism solution of one moderately large earthquake (magnitude 5.5) which occurred in the Okinawa Trough indicates prevalence of extensional stress in a direction of NNW-SSE (EGUCHI, 1982). In contrast there exists some evidences indecating an offset along the Tokara Channel in a direction of E30°S. Pre-Miocene strata north of the Tokara Channel are offset landwards by about 100 km relative to the south. A trough of free air gravity anomaly is also offset by the same amount, although the gravity trough is completely covered by a thick pile of soft sediments and a topographic trench is recognized in a northward extension of the southern trench (KOBAYASHI, in press). The 1,000 m isobath of the Okinawa Trough is situated on a northwestern extension of the Tokara Channel (Fig. 3). It seems plausible that the Ryukyu arc south of the Tokara Channel drifted oceanward in association with the E30°S opening of the Okinawa Trough in post-Miocene time until the Amami Plateau began to collide with the arc.

The Amami Plateau seems eventually to be accreted to the Ryukyu arc. It will possibly be uplifted and form a landmass as wide as the present Shikoku Island. The northern and southern portions of the Ryukyu Trench will be connected by collision line just like the northern Izu Peninsula. No such accreted lands are found in the northern Ryukyu arc, implying that collision in that region began only with the present Amami Plateau. It is understandable if we remember that the Philippine Basin moved northward until a few MaBP and changed direction of its relative motion to north-westward about 5 MaBP (SENO, 1983; MATSUBARA, 1980). There was no previous chance for the Amami Plateau to collide with the Ryukyu Arc.

3. *Ogasawara Plateau Colliding with Bonin Arc*

At nearly the same latitude as the Amami Plateau (26°N) but east of Ryukyu Islands, a chain of seamounts and oceanic plateaus (Marcus-Wake Seamount Chain) are colliding with the front of the landward slope of the Bonin (Ogasawara) Trench (Fig. 5). The characteristic topography of the Izu-Bonin Trench desappears and the toe of the landward slope close to (only 20 km west of) the tench axis is uplifted to form a small fore-arc ridge. A huge number of boulders of igneous rocks including harzburgite, dunite, gabbro, dolerite, basalt were collected by two cruises

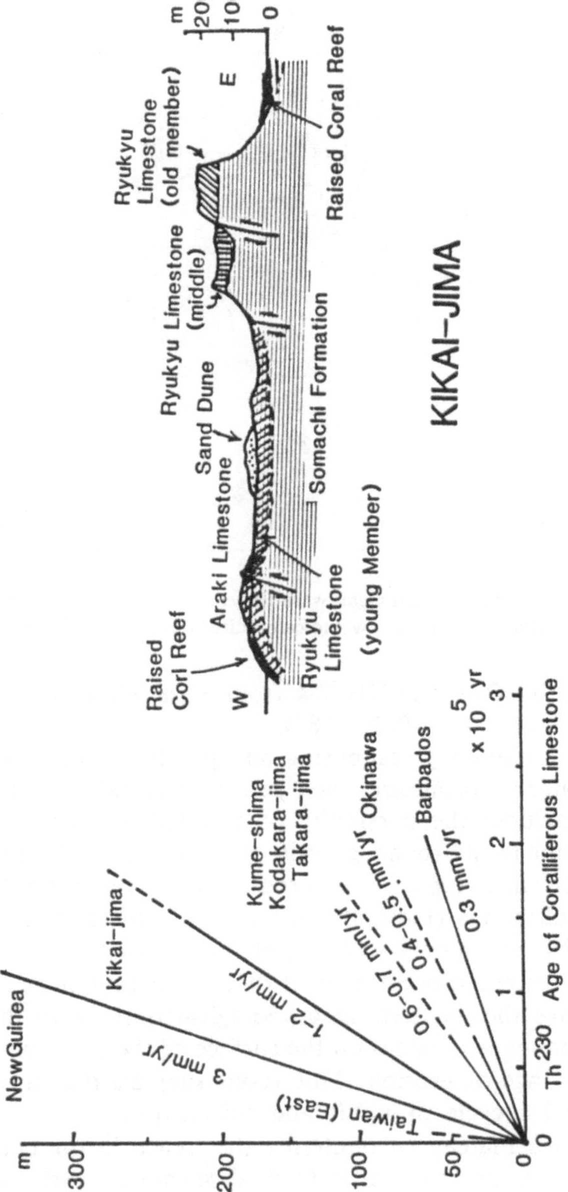

Fig. 4. Uplifting rate of the Kikai-jima (left). Rates of vertical motion (uplifting) of several islands are also shown for comparison. Cross-section of the Kikai-jima is represented on the right (after KONISHI *et al.*, 1974, 1978).

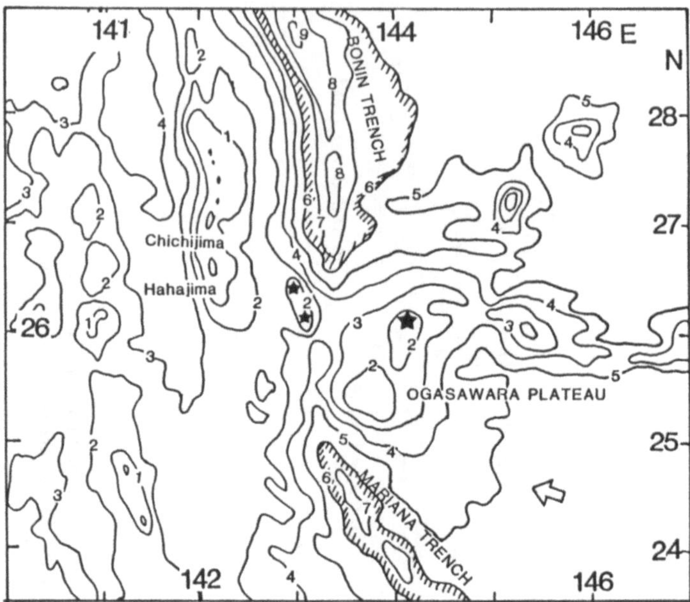

Fig. 5. Location of the Bonin arc and Ogasawara Plateau. Asterisks denote sites of dregde hauls by the R. V. Hakuho-maru. Arrow indicates direction of relative plate motion.

of Hakuho-maru, KH80-3 and KH82-4 from crests ($D=1,100$ m) of the fore-arc ridge (KOBAYASHI, 1981, 1983).

These rocks presumably composing an ophiolite suite seem to have been formed by an arc volcanism. Their petrography indicates their island arc origin and excludes the possibility of accreted oceanic crust (ISHII, 1983). Many round to subround gravels of relatively fresh igneous and sedimentary rocks which appear to be products of wave erosion have been found in these collections (ISHII, this issue). Since there exist no higher topographies which can supply such gravels to the dredge sites by debris flows or turbidity currents, occurrence of these samples implies emergence of the crests above the sea level in a geological time. Very thin or no coating of ferromanganese oxides on the surface of these gravels indicates a relatively recent age of erosion. This result suggests that this fore-arc ridge has repeated a cycle of uplift and subsidence.

The Ogasawara Plateau encountering the trench in the oceanic side appears to be a chain or group of individual seamounts and not a single plateau from their topography. In the latest cruise of Hakuho-maru, KH84-1, many alkali basalts were collected but no granitic rocks were recovered from a crest (KOBAYASHI, in press). Although its crustal structure has not been determined by seismic refraction, it seems likely that the "plateau" is a group of relatively small seamounts similar to many others in the Pacific Ocean.

As the Pacific plate has moved in the same direction for the last 42 Ma and the Marcus-Wake Seamount Chain probably extended further westward in the past, it seems plausible that many seamounts have entered the Izu-Bonin Trench roughly in the same area as the present one. Our observations favor a conclusion that the seamounts entering the trench were all subducted none have been accreted to the landward slope. Their influence on the landward slope of the trench was tectonic uplift as revealed by the fore-arc ridge of the Bonin trench slope. The situation is schematically illustrated in Fig. 6.

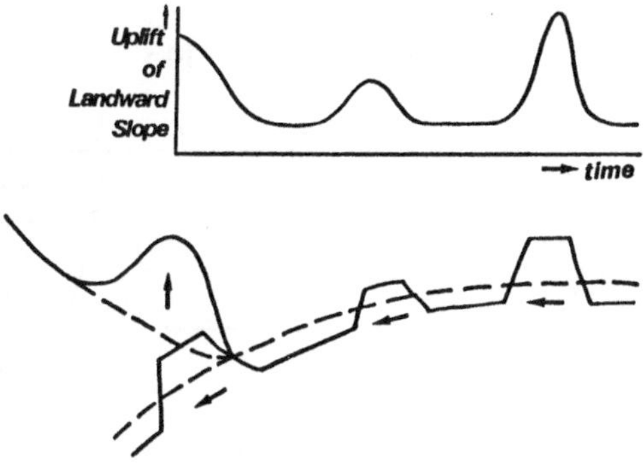

Fig. 6. Subduction of seamounts and uplifting of the landward slope of the opposite plate.

4. *Dai-ichi Kashima Seamount in the Japan Trench*

A present example of a seamount now encountering a trench axis is the Dai-ichi Kashima Seamount situated in the axis of the southern Japan Trench. MOGI and NISHIZAWA (1980) first showed the Dai-ichi Kashima Seamount dissected into two halves by a normal fault parallel to the trend of the Japan Trench. They reported a 1,500 m difference in water depths of the crests of eastern and western halves of the seamount. The landward half appears to be subducted beneath the landward slope of the trench, whereas the oceanward one is still more than 3,000 m above the ocean floor. The landward slope is uplifted and more than

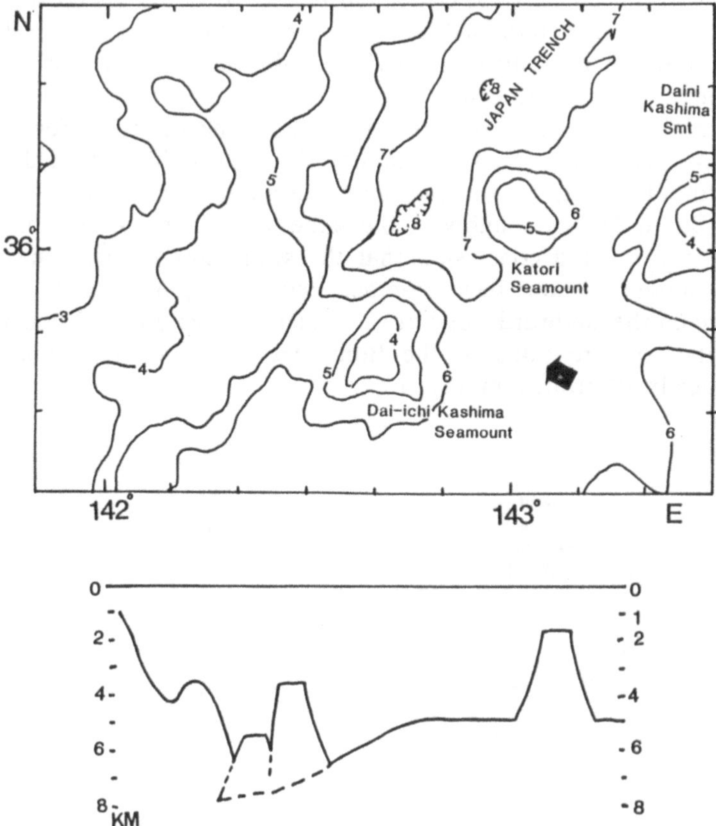

Fig. 7. Disruption of Dai-ichi Kashima Seamount and uplifting of landward slope of the Japan Trench (after MOGI and NISHIZAWA, 1980). Topography in 1 km contours (upper). Schematic cross-section of the seamounts and landward slope (below). Arrow indicates direction of relative plate motion.

1 km above the surrounding trench slope (Fig. 7).

Not much will be mentioned in this article, since detailed topographic and geophysical survey is being conducted by research vessel Jean Charcot in summer of 1984 under the agies of the France-Japan cooperative Program *KAIKO*. Only mentioned here is the destiny of the seamount; it will soon be subducted together with the oceanic slab deep in the mantle and the landward slope fronting the seamount will subside again to normal depths, as the present Japan Trench seems to be of the non-accretionary type of subduction zone (VON HUENE *et al.*, 1980). Subduction of Dai-ichi Kashima Seamount will be followed by subduction of Katori and Dai-ni Kashima Seamounts located east of Dai-ichi Kashima Seamount. None of them will be accreted to the landward margin of the Japan Trench, unless the trench changes its tectonic character to an accretionary type such as the present Nankai Trough and the ancient Shimanto Belt.

5. *Summary*

The tectonic settings of the Amami Plateau, Ogasawara Plateau and Dai-ichi Kashima Seamount together with the previously discussed Hayama-Mineoka-Setogawa zone and Izu Peninsula demonstrated that seamounts and oceanic plateaus entering a trench are either subducted or accreted, depending upon their bouyancy and the tectonic character of the subduction zone. The possible courses of events are summarized in Fig. 8, although a case where a very buoyant crust encounterts the non-accretionary subduction zone was not discussed in this article.

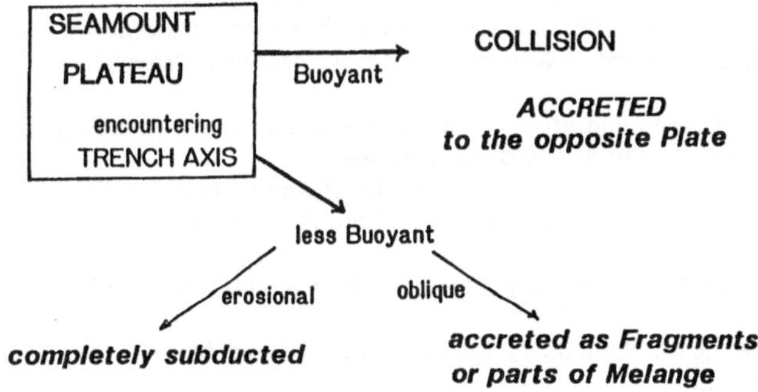

Fig. 8. Diagram showing possible courses of events of seamounts and oceanic plateaus encountering a trench.

Some other factors such as rate of convergence, configuration of Wadati-Benioff zone, sedimentary environment of the trench axis and the topography of the subducting ocean floor may also be involved in determining whether the ocean floor is subducted or accreted. The present available evidence is not sufficient to judge which is the most controlling factor. Further investigation of comparative subductology will yield a clue to solve the problem.

Acknowledgement

The author is grateful to Dr. R. Von Huene for his comments and kind suggesion to this article. Fruitful discussion and great help provided by scientific members, ship officers and crew of related research cruises are greatly acknowledged.

REFERENCES

BEN-AVRAHAM, Z., A. NUR, D. JONES, and A. COX, Continental accretion: from oceanic plateaus to allochthonous terrances, *Science*, **213**, 47–54, 1981.

DEN, N., W. J. LUDWIG, S. MURAUCHI, J. I. EWING, H. HOTTA, N. T. EDGAR, T. YOSHI, T. ASANUMA, K. HAGIWARA, T. SATO, and S. ANDO, Seismic refraction measurements in the northwest Pacific basin, *J. Geophys. Res.*, **74**, 1421–1439, 1969.

EGUCHI, T., On the opening of the Okinawa Trough, *Abstr. Seism. Soc. Japan*, n. 2, **A5**, 77, 1982 (in Japanese).

HAMADA, N., Observation of Ocen Bottom Seismograph (TKOBS) off the coast of Tokai area, *Zisin*, **36**, 449–462, 1983 (in Japanese with English abstract).

HILDE, T. W. C., N. ISEZAKI, and J. M. WAGEMEN, Mesozoic sea-floor spreading in the north Pacific, in *The Geophysics of the Pacific Ocean Basin and Its Margin*, Geophys. Monogr., 19, edited by G. H. Sutton, M. Manghnani, and R. Moberly, pp. 205–226, AGU, 1976.

ISHIBASHI, K., Plate convergence around the Izu collision zone, central Japan: development of a new subduction boundary with a temporary transform belt, Abstr. of Papers Intern. Geodyn. Conf. "Western Pacific" & "Magma Genesis", pp. 66–67, Tokyo, 1978.

IWABUCHI, Y., Distribution of seamounts and seamount chains in seas surrounding Japan, *Kaiyokagaku*, **14**, 70–75, 1982 (in Japanese).

KAGAMI, H. (ed.), Preliminary Report of the Hakuho Maru Cruise KH72-2, 1975.

KOBAYASHI, K. (ed.), Preliminary Report of the Hakuho Maru Cruise KH80-3, 209 pp., 1981.

KOBAYASHI, K. (ed.), Preliminary Report of the Hakuho Maru Cruise KH82-4, 267 pp., 1983.

KOBAYASHI, K. (ed.), Preliminary Report of the Hakuho Maru Cruise KH84-1, 1985 (to be published).

KOBAYASHI, K., Sea of Japan and Ryukyu Arc-Trench-Backarc System, in *Ocean Basins and Margins*, Vol. 7A, A.E.M. Nairn, 1984 (in press).

KONISHI, K., A. OMURA, and O. NAKAMICHI, Radiometric coral ages and sea level record from the late Quaternary reef complexes of the Ryukyu Islands, *Proc. Second Intern. Coral Reef Symp.*, Vol. 2, pp. 595–613, Great Barrier Reef Comm., Brisbane, 1974.

KONISHI, K., Y. TSUJI, T. GOTO, and T. TANAKA, Sci. Rep. Kanazawa Univ., **23**, 129–153, 1978.

LUDWID, W. J., S. MURAUCHI, and R. E. HOUTZ, Sediments and structure of the Japan Sea, *Geol. Soc. Am. Bull.*, **86**, 651–664, 1975.

MATSUBARA, Y., Izu Peninsula and the Philippine Sea plate, *Chikyu*, **2**, 157–163, 1980 (in Japanese).

MATSUDA, J., K. SAITO, and S. ZASU, K-Ar ages and Sr isotope ratio fo the rocks in the manganese nodules obtained from the Amami Plateau, western Philippine Sea, Symp. Geol. Problems of the Philippine Sea, Geol. Soc. Japan, pp. 99–101, 1975.

MATSUMOTO, T. and Y. TOMODA, Gravity anomaly at the Amami Plateau, Abstr. 59th Meeting, Geol. Soc. Japan, pp. 109–110, 1983 (in Japanese).

MOGI, A. and K. NISHIZAWA, Breakdown of a seamount on the slope of the Japan Trench, *Proc. Japan Acad., Ser. B*, **56**, 257–259, 1980.

NUR, A. and Z. BEN-AVRAHAM, Oceanic plateaus, the fragmentation of continents and mountain building, *J. Geophys. Res.*, **87**, 3644–3661, 1982.

OGAWA, Y., Tectonics of some forearc fold belts in and around the arc-arc crossing area in central Japan, *Geol. Soc. London, Spec. Publ.*, **10**, 49–61, 1982.

OGAWA, Y., Mineoka ophiolite belt in the Izu forearc area-Neogene accretion of oceanic and island arc assembleges in the northeastern corner of the Philippine Sea plate, in *Accretion Tectonics in the Circum-Pacific Regions*, edited by M. Hashimoto and S. Uyeda, pp. 245–260, Terra Sci. Publ., Tokyo, 1983.

SENO, T., The instantaneous rotation vector of the Philippine Sea plate relative to the Eurasian plate, *Tectonophysics*, **42**, 209–226, 1977.

SHIKI, T., H. AOKI, and Y. MISAWA, Geological results of the recent studies of the Philippine Sea—with special reference to GDP-8, -11 cruises, *Kaiyo Kagaku*, **7**, 3644–3661, 1981 (in Japanese).

SUYEHIRO, K., A. NISHIZAWA, and H. SHIMIZU, Microearthquake activity in the West Philippine Basin and northern Ryukyu Trench region as observed by ocean bottom seismometers at the Amami Plateau, *J. Phys. Earth*, **30**, 509–516, 1982.

TOKUYAMA, H. and Y. S. KONG, Multi channel reflection survey around the Amami Plateau, in *Preliminary Report of the Hakuho Maru Cruise KH 82-4*, edited by K. Kobayashi, pp. 246–264, 1983.

TOMODA, Y. and H. FUJIMOTO, Roles of seamount, rise and ridge in lithospheric subduction, in *Accretion Tectonics in the Circum-Pacific Regions*, edited by M. Hashimoto and S. Uyeda, pp. 319–331, Terra Sci. Publ., Tokyo, 1983.

TONOUCHI, S. and K. KOBAYASHI, Paleomagnetism and geotectonic investigation of ophiolite suits and surrounding rocks in south-central Honshu, Japan, in *Accretion Tectonics in the Circum-Pacific Regions*, edited by M. Hashimoto and S. Uyeda, pp. 261–288, Terra Sci. Publ., Tokyo, 1983.

UYEDA, S. and H. KANAMORI, Back-arc opening and the mode of subduction, *J. Geophys. Res.*, **84**, 1049–1061, 1979.

VON HUENE, R., M. LANGSETH, N. NASU, and H. OKADA, Summary, Japan Trench transect, in *Scientific Party*, Initial Reports DSDP, Vol. 56–57, Part I, 473–488, 1980.

CHAPTER 6: ACCRETION TECTONICS (II)

Formation of Active Ocean Margins, edited by N. Nasu *et al.*, pp. 641–676.

TECTONIC FRAMEWORK OF THE KURIL ARC SINCE ITS INITIATION

Gaku KIMURA[1] and Kensaku TAMAKI[2]

[1]*Department of Earth Sciences, Kagawa University, Takamatsu, Kagawa 760, Japan*
[2]*Geological Survey of Japan, 1-1-3 Higashi, Yatabe, Ibaraki 305, Japan*

Abstract. Tectonic history of the Kuril Arc is reasonably constrained by the framework of subducting plate, forearc plate, and back-arc plate, in terms of active ridge subduction, retreat of back-arc plate and lateral migration of forearc plate caused by oblique subduction. These tectonic events are well recorded in the geologic strata, and can be correlated to major geologic phenomena such as plate evolution in the Pacific since Cretaceous and reorganization of the East Asia Terranes in the Eurasian Plate due to the Indo-Eurasia collision since Eocene. Four major tectonic events are significant for reconstruction of the Kuril Arc evolution.

1) Late Cretaceous initiation of the Proto Kuril subduction: The event was accompanied by formation of the forearc basin and volcanism along the southern margin of the Okhotsk Terrane. The subduction was triggered by collision of the allochtonous Okhotsk Terrane with the Siberia continent in Late Cretaceous.

2) Paleocene to Eocene active Kula-Pacific ridge subduction along the Kuril Trench: the ridge subduction is recognized based on the following geological evidence; southward migration of the volcanic front from Late Cretaceous to Early Paleogene, Early Paleogene rapid shoaling of the forearc basin, and ensuing Eocene hiatus without volcanism.

3) A new subduction in Late Eocene and the succeeding Kuril back-arc spreading: Beginning of the event is marked by Late Eocene resumption of sedimentation and resurgence of volcanism about 30 MaBP. The Kuril back-arc spreading took place since this time. The back-arc spreading originated from the clockwise rotation of the Okhotsk Terrane resulted from the northeastward migration of the East Asia Terranes in the Eurasian Plate due to the Indo-Eurasia Collision.

4) Lateral migration of the forearc plate and collision of that with the Tohoku Arc at the southwestern end: Right lateral strike-slip along the Kuril volcanic front started in Latest Miocene and the collision of the forearc plate at the southwestern end built the Hidaka Mountains. These events originated from the oblique subduction of the Pacific Plate along the Kuril Trench.

1. *Introduction*

The Kuril island arc system is one of the typical subduction zone in the circum Pacific region (Fig. 1). DICKINSON (1979) suggested that the initiation of the Kuril Arc was related to the collision-accretion tectonics

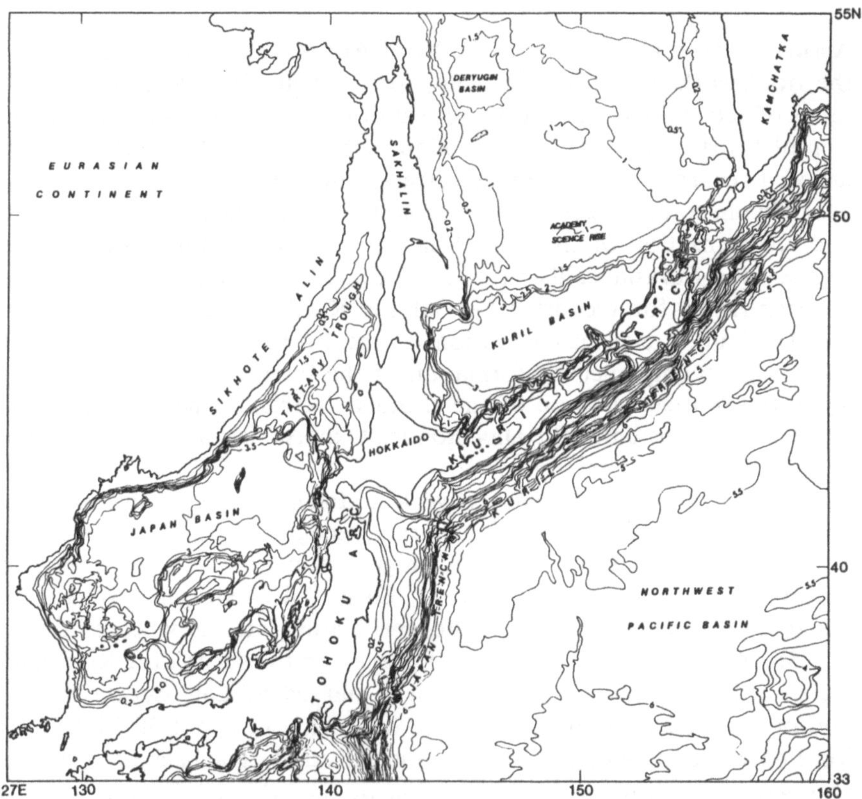

Fig. 1. Submarine topography around the Kuril and Tohoku Arcs. Depth countours are shown in kilometers with 0.5 km intervals.

of the Okhotsk exotic terrane against the Siberia continent during Eocene time and several other events such as entrapment of the Bering Sea, change of the motion of the Pacific Plate, and others which occurred simultaneously. However, its simultaneity is very problematic. We synthesize several observed geologic events and conclude that the proto Kuril subduction initiated at Late Cretaceous.

The succeeding significant event in the Kuril system is the subduction of the active Kula-Pacific Ridge. There were several discussions on the subduction of the Kula-Pacific Ridge. UYEDA and MIYASHIRO (1974) proposed that subduction of the Kula-Pacific Ridge caused the regional volcanism in East Asia during Late Cretaceous and back-arc spreadings. HILDE *et al.* (1977) suggested that the ridge subduction occurred during Late Cretaceous based on the reconstruction of the tectonic history of the Pacific Basin. DELONG *et al.* (1977), however, suggested that the ridge subduction took place at about 30 Ma in the Aleutian margin and caused tectonic events different from those indicated by UYEDA and MIYASHIRO (1974). They postulated that cessation of arc volcanism, uplifting of the arc, and thermal metamorphism were caused by the ridge subduction. There have been no publications on ridge subduction at the Kuril Trench. We propose in this article that the Kula-Pacific ridge was subducted at the Kuril Trench during Paleocene to Eocene time.

The Kuril Basin is a back-arc basin of the Kuril Arc. The Basin has been considered as one of the back-arc basins in the western Pacific formed by back-arc spreading process (KARIG, 1971). Age and tectonics of the Kuril Basin have been controversial, whereas those of most other back-arc basins in the western Pacific are well investigated on the basis of DSDP holes and extensive geomagnetic data. The estimated ages of the Kuril Basin are variable from Cretaceous to more younger age (IIJIMA, 1962; GNIBIDENKO and KHVEDCHUK, 1982; KOBAYASHI, 1982). The tectonics of back-arc spreading also have been controvertial.

Some considered it to be formed by roll back of subducting slab (MOLNAR and ATWATER, 1978; SENO and MARUYAMA, 1983), and others by retreat of overriding back-arc plate (UYEDA and KANAMORI, 1979). In this article, we investigate age and mechanism of formation of the Kuril Basin based on the synthesis of land and marine geological data.

According to KIMURA (1981a, b), tectonic stress field in the Kurils changed from tension to compression at about Late Miocene. He also suggested that collision of the southwestern end of the forearc plate with the Tohoku Arc occurred to form the Hidaka Mountains since latest Miocene. This migration of the forearc plate can be interpreted by the FITCH's (1972) drag force caused by oblique subduction of oceanic plate.

The migration of the forearc plate took place only since Latest Miocene, while the oblique subduction of the Pacific Plate started probably at about 43 MaBP. We consider their "time gap".

2. *Geologic Sequence and Events in the Kuril Arc*

Hokkaido Island in Japan is divided into three geologic provinces; the southwest, central, and east (Fig. 2). The southwest Hokkaido is the northern extension of the Tohoku Arc and the east Hokkaido belongs to the Kuril Arc. The central Hokkaido is situated in the fossil suture zone between the Eurasia and North America Plates (KIMURA *et al.*, 1983; KOBAYASHI, 1983; NAKAMURA, 1983). We will mainly investigate the geologic sequence and events in east Hokkaido and its vicinity. The province is most appropriate to study the tectonic evolution of the Kuril Arc, because the geologic sequence since Late Cretaceous is preserved only in this part of the Kurils.

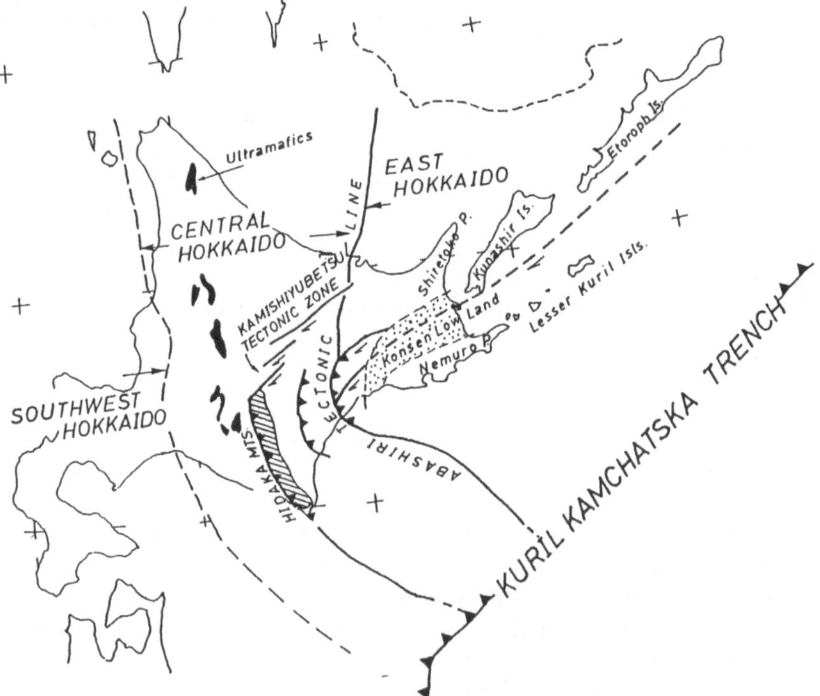

Fig. 2. Three geologic provinces of Hokkaido and its structural elements.

The east Hokkaido is discriminated from the central Hokkaido by the Abashiri Tectonic Line (Fig. 2), and divided into two arc units; inner and frontal arcs.

The Nemuro Peninsula and its eastern islands are recognized as a frontal arc. The Shiretoko Peninsula, Kunashiri, Etoroph and Urup Islands are an inner arc composed mainly of volcanics. Quaternary sedimentary basins such as the Konsen Low Land are developed between the inner and frontal arcs. The southwestern end of the Konsen Low Land is deformed by right-handed reverse faults (Fig. 2).

2.1 Late Cretaceous event

Late Cretaceous sequence is observed in the Academy Science Rise north of the Kuril Basin, in the frontal arc such as the Nemuro Peninsula, and at the Tokoro Belt in the eastern part of the central Hokkaido. Many granitic and andesitic rocks were dredged from the Academy Science Rise (BURK and GNIBIDENKO, 1977; GNIBIDENKO and KHVEDCHUK, 1982). Ages of the rocks are concentrated around 200 Ma and from 95 Ma to 65 Ma. Many rocks from 95 Ma to 65 Ma imply that the volcanic arc was formed around that time.

There are two other events which suggest the formation of the arc during late Cretaceous. One is a thick accumulation of flysh sediments deposited after eruption of andesitic lava during Campanian in the Nemuro Peninsula (Fig. 3). KIMINAMI (1983) suggested that the basin where flysh sediments were accumulated was a forearc basin of the proto Kuril subduction system. The other is formation of the Tokoro Belt in the eastern part of central Hokkaido which consists mainly of Jurassic seamounts and overlying upper Cretaceous flysh sediments (NIIDA *et al.*, 1982). The Jurassic seamounts were accreted to the Okhotsk Terrane. The flysh sediments derived from the Okhotsk Terrane (KIMINAMI and KONTANI, 1979) covered the seamounts in Late Cretaceous. The inception of arc magmatism, the formation of forearc basin, and seamount accretion in Late Cretaceous indicate activity of the subduction system at that time along the southwestern margin of the Okhotsk Terrane.

2.2 Paleocene-Eocene event

We recognize the Paleocene-Eocene event based on the upward changes in lithological facies of the Nemuro Group and ensuing hiatus. The Upper Cretaceous and Lower Paleogene Nemuro Group is distributed in the Nemuro Peninsula and the Kushiro Coal Field (Fig. 3). The Nemuro Group consists mainly of flysh sediments associated with andesitic volcanics and alkali basalts (Fig. 3). Andesitic rocks occur in the lower part of the Group.

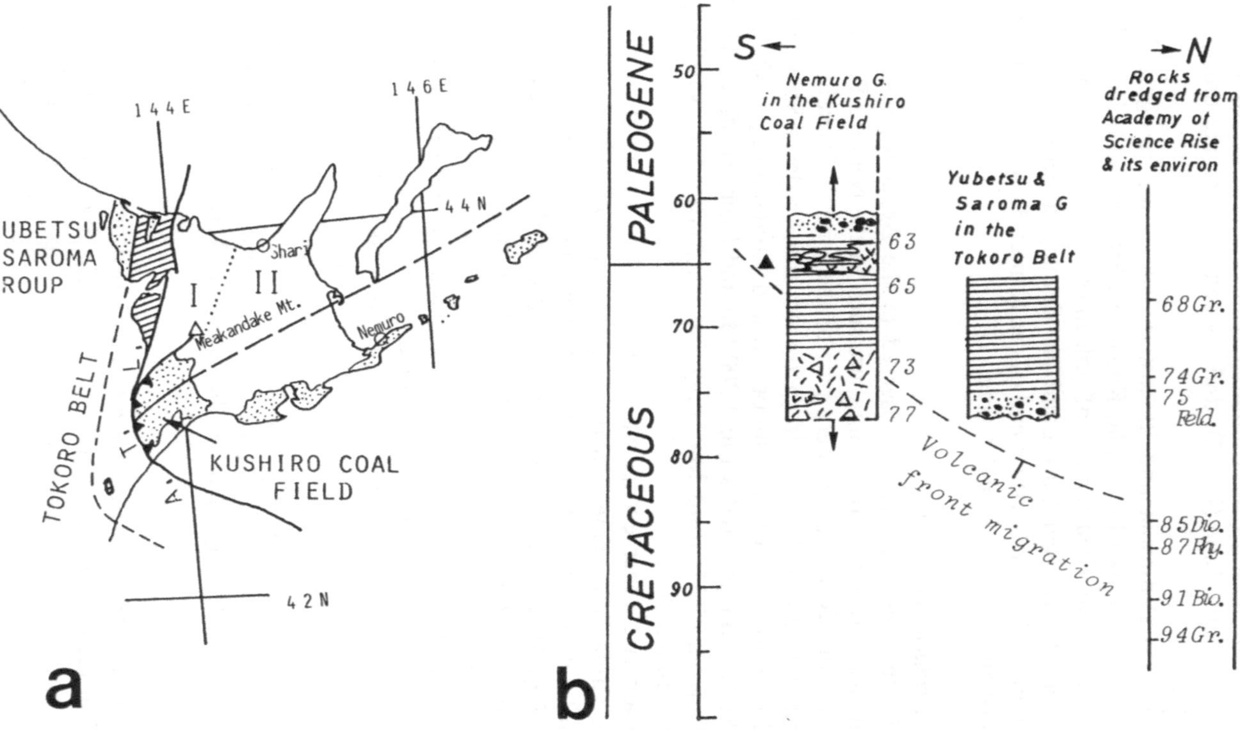

Fig. 3. Distribution (a) and Geologic sequence (b) of Mesozoic strata around eastern Hokkaido Absolute age of the Nemuro group and rocks dredged from the Academy of Science Rise are based on GNIBIDENKO *et al.* (1982).

Alkali basalts are found in the middle horizon of the Nemuro Group as pillow lavas and sills. These basalts were not transported from other places as exotic blocks but erupted in situ in the Nemuro Basin (YAGI, 1969; KIMINAMI, 1975a). KIMINAMI (1975a, b, c, 1976, 1978, 1983) revealed that the flysh sediments in the lower and uppermost parts of the Nemuro Group were supplied from the northern provenance, but the middle part of the Group was derived from the southern area. Provenance of the southern area is considered to be andesitic volcanic islands based on the composition of the sandstones. Provenance of the flysh sediments changed from north to south nearly at the time of eruption of the basalts (KIMINAMI, 1975a, 1983). The uppermost part of the Group consists mainly of conglomerates which are composed of rounded slates, sandstones, andesites, granites and cherts. The Nemuro Group was considered to have been deposited in a period from Campanian or Maastrichtian to Danian based on paleontological evidence (MATSUMOTO, 1970). Recently, KAIHO (1983) showed that the uppermost part of the Group was deposited in early Eocene based on foraminiferal fauna.

Lithofacies change of the uppermost part of the Group shows that the basin become rapidly shallower. It is followed by a hiatus. These results indicate that the Nemuro Basin upheaved in early Eocene.

The magmatism in the proto Kuril system can be recognized as follows. The Kuril back-arc basin which was located north of the recent Kuril system did not open yet in Late Cretaceous to Early Paleogene time as mentioned later. The first magmatism occurring prior to about 85 MaBP was initiated at the Academy Science Rise north of the Nemuro basin. The second took place in the Nemuro basin, which is observed at the base of the Nemuro Group. The absolute age indicates about 75 MaBP. The third is the alkali basalt activity in the middle horizon of the Nemuro Group and the andesitic volcanism occurring south of the Nemuro basin. The andesitic volcanism in the third stage was revealed by KIMINAMI (1975a) based on sedimentological investigation of the Nemuro Group. These magmatic activities in the proto-Kuril system indicate that the volcanic front progressively migrated southward from 85 Ma to 65 MaBP (Figs. 3 and 4). The uplift of the Nemuro Basin took place after the last magmatism.

The hiatus after the uplift of the Nemuro Basin demonstrates that there was not so intensive tectonic movement because there was no strong erosion between the Nemuro Group and the overlying Urahoro Group. Volcanic rocks at that time are not found in and adjacent to the area.

Southward migration of volcanic front, early Eocene rapid shoaling of the forearc basin, and the hiatus appear to have been caused by subduction of the active Kula-Pacific Ridge as discussed later.

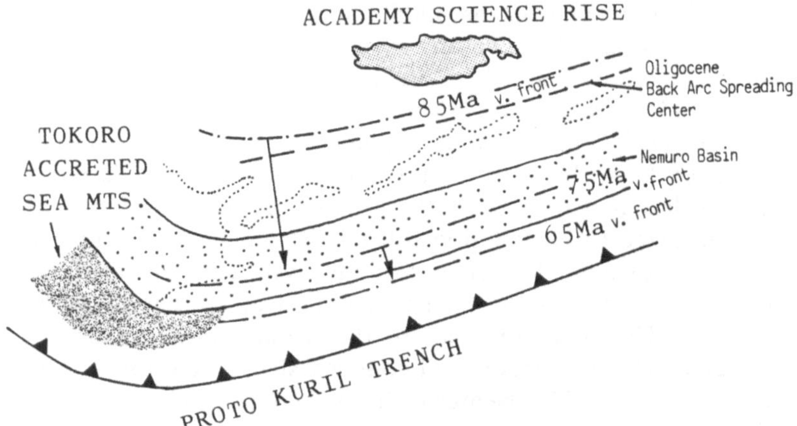

Fig. 4. Reconstruction of the framework of the Proto Kuril island arc system during Late Cretaceous and southward migration of the volcanic front.

2.3 Late Eocene-Oligocene event

The Late Eocene-Oligocene event is well recorded in the Late Eocene Urahoro and Oligocene Ombetsu Groups in the Kushiro Coal Field (Figs. 3 and 5). The Urahoro Group is about 750–1000 m thick and composed mainly of terrestrial, brackish and coal beds (Fig. 5). IIJIMA (1962) pointed out that many chromite grains are contained in the lower part of the Group and derived from west and south serpentinite bodies. The localities of provenances are deduced from the increase of chromite mode with decreasing distances to the source. He also suggested that the western provenance is concordant with the Tokoro Belt and the southern is the high magnetic anomaly zone south of the Nemuro Peninsula. MATSUI (1962) and NAGAHAMA *et al.* (1979) observed many conglomerates in the Urahoro Group and revealed that the conglomerates were supplied from three provenances; north, west, and south. NAGAHAMA *et al.* (1979) showed that granitic rocks and andesites were supplied mainly from the northern provenance which was the "Paleo-Okhotsk Land", but basalts and cherts were derived from the "Tokoro Belt" situated to their west and south. The eustatic movement during the deposition of the Group was also indicated by Matsui. The middle horizon of the Urahoro Group is composed of a transgressive sediments deposited in a shallow marine environment (Fig. 5). The Urahoro Group was accumulated in Late Eocene (Kaiho, personal communication). Initiation of the deposition of the Urahoro Group in Late Eocene indicates that the second subsidence of the forearc region

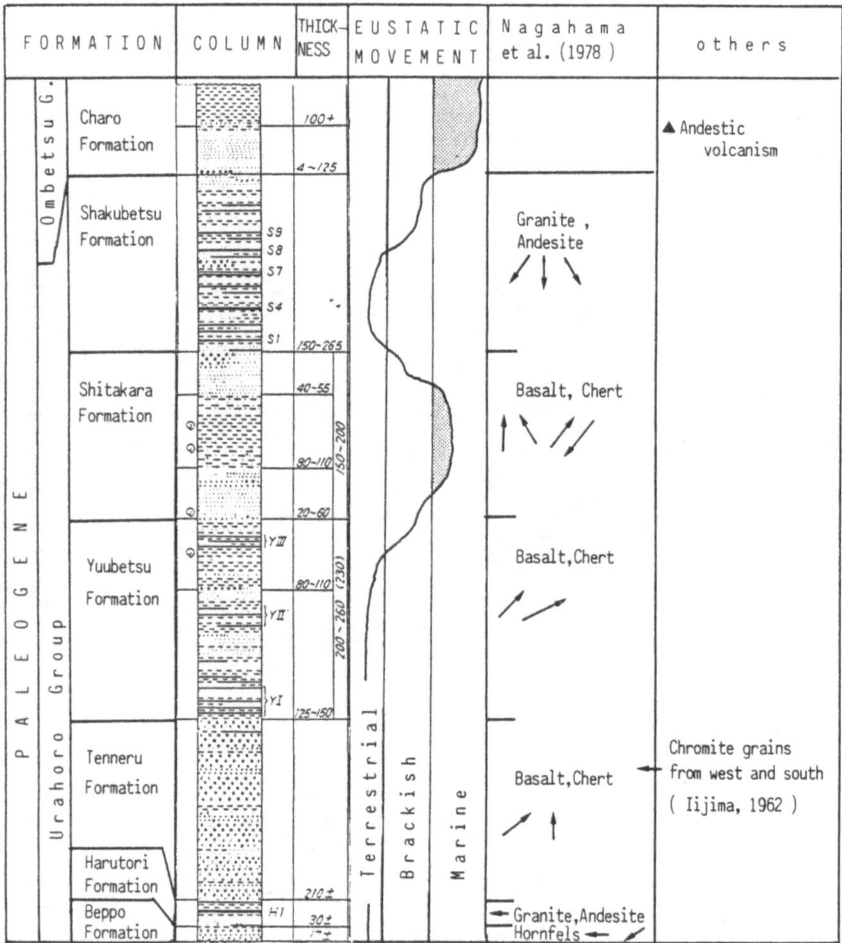

Fig. 5. Paleogene sequence and the draining direction of clastics of them.

of the Kuril Arc started at that time. The facies show that the subsidence took place very slowly. At that time, serpentinite bodies appeared along the zone south of the Nemuro Peninsula. This zone might have been extended longitudinally toward the south of the basin, because the high magnetic anomaly zone indicating their extension south of the recent frontal arc (Fig. 6). Moreover, many pebbles were derived from the adjacent areas to the basin. These results indicate that the Paleo Okhotsk Land (Okhotsk Terrane emerging above the sea level) still remained as land.

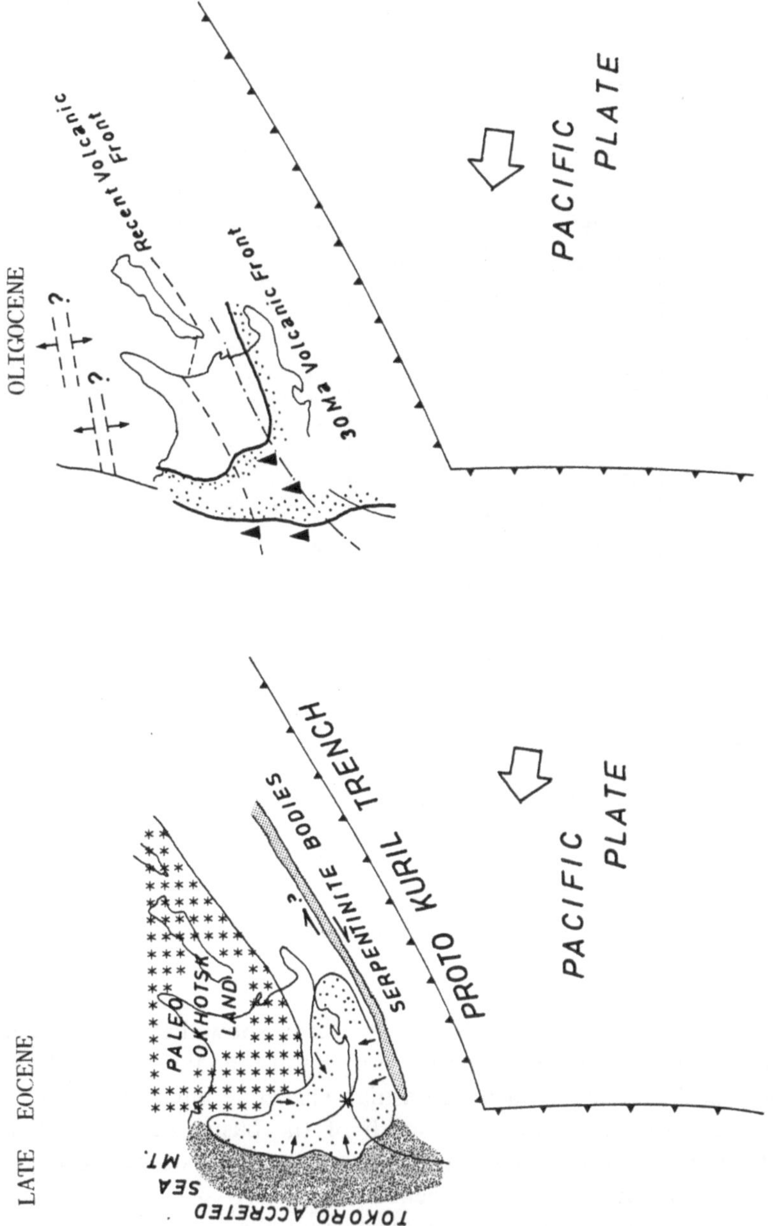

Fig. 6. Paleogeographic map from Late Eocene (left) to Oligocene (right).

The Ombetsu Group unconformably overlies the Urahoro Group, but a time interval between these two groups is short. The Ombetsu Group is composed mainly of shallow marine mudstones and sandstones, and associated with andesitic volcani-clastic rocks in the lowest and the upper parts. Matsui showed that the eruption center of these volcanic rocks was located on the southern margin of the Kushiro Coal Field (Fig. 6). The volcanics were erupted in Early Oligocene (32.6 MaBP) (SHIBATA and TANAI, 1982).

These results show that the transgression associated with arc volcanism was started at that time. The basin probably rapidly subsided and volcanism was resurged after the Eocene hiatus (Fig. 6).

Initiation of the sedimentation in Late Eocene and Oligocene resurgence of the arc volcanism show that a new subduction was initiated in Late Eocene and that the Paleo-Okhotsk Land disappeared below the sea surface in Oligocene. We consider that this result constrains the oldest limit for spreading age of the Kuril back-arc basin.

2.4 Late Miocene event

Late Miocene event is recognized based on a geological sequence since Miocene and deformed structures in the eastern and central Hokkaido. Tectonic framework were described in detail by KIMURA (1981a, b, 1982) and KIMURA *et al.* (1983). We sumarize it briefly in the following.

In the frontal arc, the Neogene strata are divided into two groups. One is the Miocene strata and the other is the sediments deposited since Pliocene. The Miocene strata are composed mainly of shallow marine sediments and unconformably overly the Paleogene, but there was no intensive tectonic movement between them. Andesitic volcanics were erupted and deposited in the coal field during middle to late Miocene. Most intensive tectonic movement occurred from Late Miocene to Earliest Pliocene in the Kushiro coal field (MABUCHI, 1962; SATO, 1969; KIMURA, 1981b). Geologic structures observed there were mostly formed during that time. Pliocene shallow marine strata were accumulated over the older strata unconformably. The Pliocene strata were deformed a little, while the other older strata were crumpled and disrupted intensively. The struture is characterized by folds with N-S trending axes and NE-SW trending right-lateral strike-slip faults. These deformed structures were formed by E-W compressional stress.

In Shikotan Island, FROLOV *et al.* (1980) reported that basic rocks intruded the Upper Cretaceous strata in middle Miocene.

The inner arc is subdivided into two regions; I and II, by the line from Shari to Meakandake Mt. (Fig. 3). In the region I, the oldest rocks are 31.4 MaBP andesitic lavas (SHIBATA and TANAI, 1982), and Oligocene shallow marine mudstones. The lower to middle Miocene shallow marine sediments unconformably overly them. The upper Miocene volcanics were deposited in the northern part of the region I. The Upper Miocene to Pliocene diatomaceous sediments also covered the volcanics. The accurate age of the strata in this region is still controvertial (TANAI, 1982).

In the region II, most areas are covered by the Quaternary volcanic flows and falls. The basement rocks underlying volcanos are exposed in a limited area. The oldest rocks are composed mainly of so-called "Green Tuff" which is almost submarine volcanics and lavas. The "Green Tuff" is covered by shallow marine diatomaceous mudstone which is dated as Latest Miocene to Pliocene (KOSHIMIZU and OKA, 1983). "Green Tuff" seems to have been accumulated before Late Miocene, although evidence for their age has not been available. Younger shallow marine volcanics overlie the Upper Miocene mudstones. Many dikes intruded the "Green Tuff". They are apparently fider dikes simultaneously injected into the volcanics. Many dikes intruding the "Green Tuff" and volcanics below the Upper Miocene to Pliocene mudstones trend NE-SW, whereas dikes intruding the volcanics above the upper Miocene strata trend NW-SE (Fig. 7). The former dikes are parallel to the trend of arc, and the latter ones are transverse to the arc. Moreover, metaliferous veins which were presumably formed in a period from Latest Miocene to Pliocene trend NW-SE (KIMURA, 1981b). Change of the direction of dikes shows that trend of the regional tectonic stress field. σHmax was changed from NE-SW to NW-SE in Middle to Late Miocene (Fig. 7). Geologic structure of the inner arc is characterized by folds and reverse faults with NE-SW trending axes roughly parallel to the Shiretoko Peninsula, which have been formed since Late Miocene (KIMURA, 1981b). The σHmax direction since Late Miocene was equivalent to maximum compressional axis. Tectoic stress field before middle Miocene was probably tensional, because the inner arc during that time was subsiding. Considering tectonism in the forearc region, it is revealed that the regional tectonic stress field was abruptly changed from tension to compression in Late Miocene. Stress trajectory since Middle Miocene is shown in Fig. 7.

The Abashiri Tectonic Line is a major fault system separating the eastern Hokkaido from the central. In Late Miocene the tectonic line moved as a left-lateral strike-slip fault and later a southern segment of the line was changed to the reverse fault dipping east, then a westward convex structure which is observed now was formed (Fig. 7). The boundary between the northern and southern segments is located just on the volcanic front. This motion of the Abashiri Tectonic Line and feature of the stress

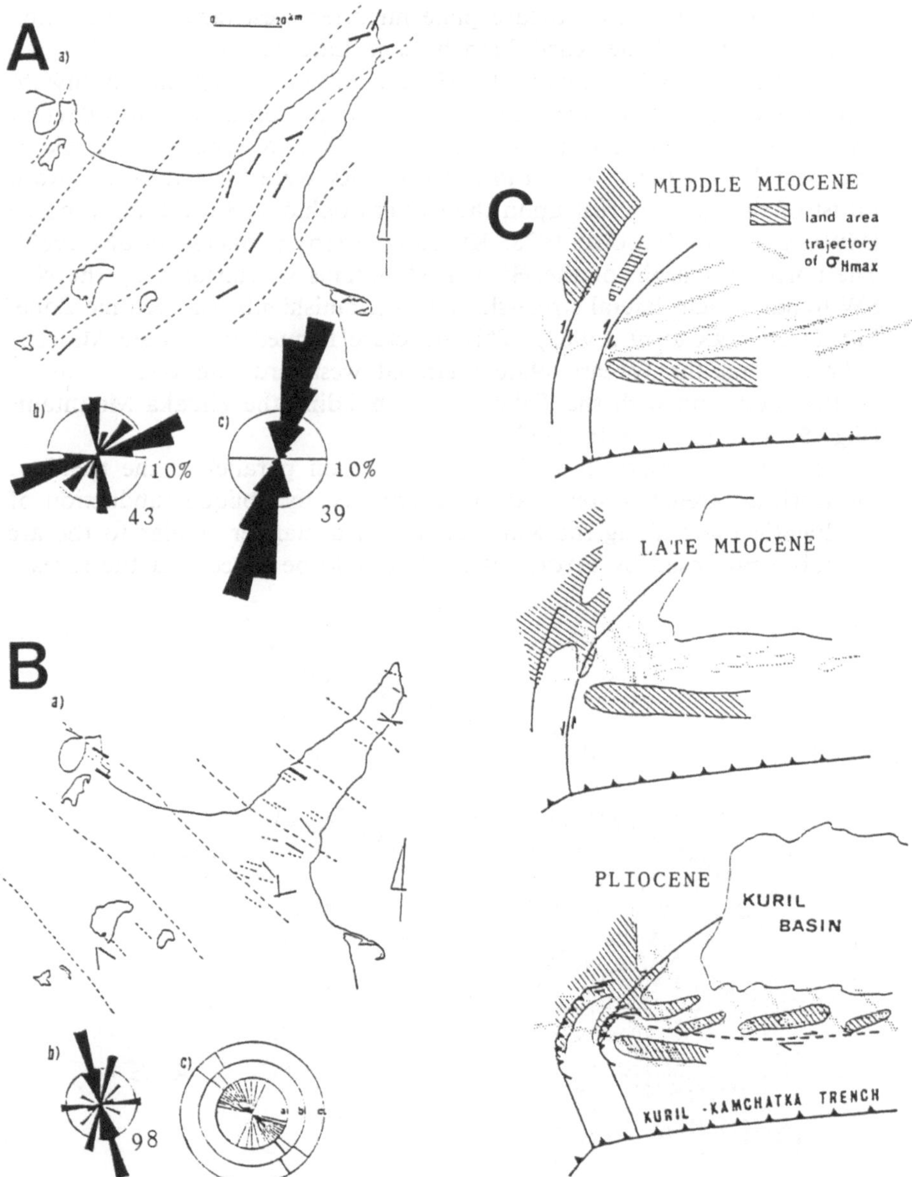

Fig. 7. Predominant direction of dike swarm during Middle Miocene (A) and since Late Miocene (B) with trajectory of σHmax (dotted line). Roses in Fig. (A) show the dike swarm developing at Shiretoko Peninsula (right) and others (left). Roses in Fig. (B) indicates the direction of metariferous veins at Nemuro Mine in the Shiretoko Peninsula (left) and Etoroph Island (right). Change of the stress field is summarized in (C).

trajectory show that the forearc plate migrated westward in a direction parallel to that of the Kuril Trench since Late Miocene (Fig. 8).

In the central Hokkaido, the Hidaka Mountains extend southward from the volcanic front (Fig. 2). The mountains consist of two different units; western metaophiolite (MIYASHITA, 1983) and eastern island arc or continental crust (KOMATSU *et al.*, 1982). They were formed by westward thrust of the eastern crust upon the western ophiolite since Late Miocene (MIYASAKA and KIKUCHI, 1978; KIMURA, 1981b; KOMATSU *et al.*, 1982). The northern margin of the Hidaka Mountains is truncated by the NE-SW trending right-lateral strike-slip faults (Kamishiyubetsu Tectonic Zone, Fig. 8; KIMURA *et al.*, 1982). This structure formed since Late Miocene indicates that the forearc plate migrated westward and collided at its southwestern end with the Tohoku Arc, building the Hidaka Mountains (Fig. 8, KIMURA, 1981b, 1982).

The forearc plate migration in a direction parallel to the trend of the Kuril arc-trench system can be interpreted by oblique subduction of the Pacific Plate along the Kuril Trench in a manner similar to the arc tectonism proposed by FITCH (1972). It should be noted that the forearc

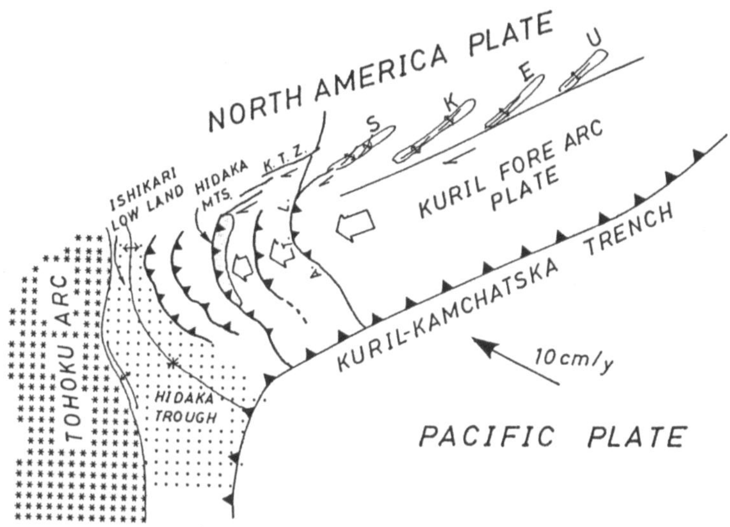

Fig. 8. Tectonic model of the southwestward migration of the Kuril forearc plate and its collision at the southwestern end with the Tohoku Arc since Latest Miocene. The creation of the Hidaka Mountains is resulted from the collision. A.T.L.: Abashiri Tectonic Line, K.T.Z.: Kamishiyubetsu Tectonic Zone, S: Shiretoko Peninsula, K: Kunashiri Island, E: Etoroph Island, U: Urup Island.

plate migration occurred only since Late Miocene, while the oblique subduction of the Pacific Plate started around 43 MaBP.

3. *Age of the Kuril Basin*

The research cruises of Hakureimaru, GH76-2 and GH77-3 were conducted by the Geological Survey of Japan in 1976 to 1977 using seismic reflection profiling, magnetic and gravimetric measurements, and bottom sampling (HONZA ed., 1977; HONZA ed., 1978). Results mentioned here are based on the data obtained during these cruises.

The Kuril Basin has water depth exceeding 3000 m and shows a well developed abyssal plain morphology with flat and smooth floor. The basin has outstanding fan-shape configuration which opens southwestward (Fig. 1). Width of the basin is 350 km in its southwestern end. The basin narrows northeastward and disappears just south of the Kamchatka Peninsula.

Figure 9 shows one of seismic profiles along the foot of Etoroph Island in the southwestern Kuril Basin. Thickness of the sedimentary layer at the southwestern margin of the Kuril Basin is 2.5 seconds and the basement depth is 7 seconds below the sea level in two-way acoustic travel time as shown in Fig. 9. In the center of the Basin sediment thickness is 2.8 seconds and basement depth is 7.2 second according to a sonobuoy refraction measurement (HONZA *et al.*, 1978). The basement depth, 7.2 seconds, appears to be greatest among the basement depth of the Kuril Basin ever mentioned (GNIBIDENKO and KHVEDCHUK, 1982).

The basement depth of the Kuril Basin is very close to that of the Japan Basin, although the value of sediment thickness of the Kuril Basin is 0.5 seconds thicker than that of the Japan Basin (Fig. 10). This shows that difference in the maximum water depth between basins, i.e. 3300 m of the Kuril Basin and 3700 m of the Japan Basin, is not due to difference in basement depths but due to that in sediment thickness. The maximum basement depth of the Japan Basin ever observed is 7.0 seconds which is 0.2 seconds shallower than that of the Kuril Basin. CROUGH (1983) proposed a simple method of sediment loading correction on the oceanic crust. Two values, water depth of the basin and sediment thickness are used for the correction. Sediment thickness on Fig. 10 is 2.8 seconds and water depth there is 3325 m in the Kuril Basin. Basement depth of the Kuril Basin after the sediment loading correction based on these values is 5005 m according to the Crough's correction method. One of the deepest basement depth of the Japan Sea after sediment loading correction is 4950 m with water depth of 3750 m and sediment thickness of 2.0 seconds (according to seismic data of RC12 cruise of Conrad by Lamont-Doherty

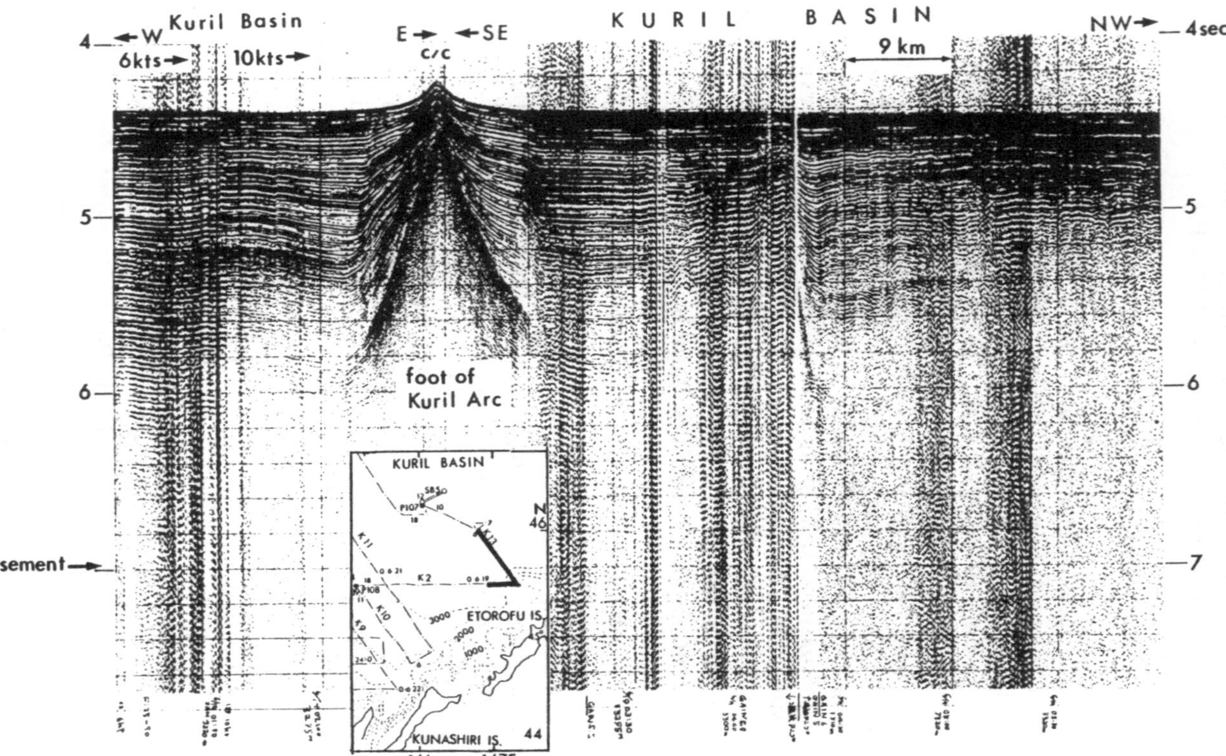

Fig. 9. Seismic profile of the southwestern part of the Kuril Basin which is obtained during GH77-3 research cruise of R.V. Hakurei-maru. The locality is shown in the index map with solid line. Vertical scale is shown in two-way acoustic travel time. The acoustic basement is inferred to be located at the depth of 7 seconds.

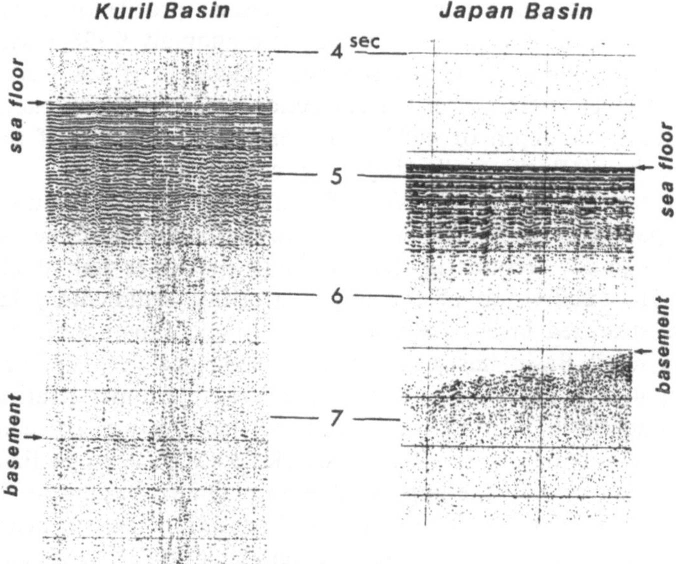

Fig. 10 Comparative profiles of the Kuril and Japan Basins. The left profile (GH77-3) is in the southwestern Kuril Basin and the right profile (GH78-3) is in the central part of the Japan Sea. The vertical scale is shown with two-way acoustic travel time. Position is annotated along the time mark on each profile.

Geological Observatory). Basement depth after sediment loading correction of one basin nearly the same as that of the other. This shows that the oceanic crust of both basins has nearly the same ages, according to a relation between age and depth of oceanic basins (PARSON and SCLATER, 1977).

The isostatic depth of 5000 m is theoretically correlated to the age of about 60 MaBP according to HAYES (1983). However, basement depth of the marginal seas is deeper than oceanic basin depth. The Shikoku Basin, which is documented to have been active during 30 MaBP to 15 MaBP based on the DSDP results and identification of magnetic anomaly lineations (KLEIN and KOBAYASHI, 1980), has the basement depth of 4500–5200 m with a few hundred meters of sediments in its southern part. The basement depths of the Shikoku Basin are comparable to that after sediment loading correction of the Japan Basin and the Kuril Basin. The South China Sea, where the magnetic anomaly lineations from 32 to 17 MaBP were identified (TAYLOR and HAYS, 1980), shows basement depth of 6.5 seconds to 7.0 seconds overlain by the sediments with thickness of 1.5 to 2.0 seconds according to the seismic profile of TAYLOR and HAYS (1982). A seismic profile on magnetic anomaly 9 (28.5 Ma) shows water depth of 3900 m and the sediment thickness of 1.8 seconds. The isostatic basement depth after the correction by Crough's method is 4980 m. The corrected depth of 4980 m is very close to those of the Japan Basin and Kuril Basin, indicating their concordant ages.

Sedimentary layer of the Kuril Basin is stratified in its upper part, and transparent or less reflective in its lower part. The upper stratified layer is 0.6 to 1.2 seconds in thickness and the lower part is about 1.5 seconds. The subdivision of the upper stratified sedimentary layer and lower less stratified layer is also observed in the Japan Basin (Fig. 10). Thus, the Kuril Basin has a sedimentary sequence very similar to that of the Japan Basin. DSDP Site 301 penetrated the upper stratified layer of the Japan Basin. Age at the bottom of the hole was correlated to 4 MaBP based on diatom fossil analysis (KOIZUMI, 1979). Sedimentation rate near the bottom of the hole is 91 m/m.y. On an assumption that this sedimentation rate is the same as the sedimentation rates in deeper part, age of the sediments just above the basement is 26 MaBP at the sono-buoy locality V28-155 of LUDWIG *et al.* (1975) where the basement is overlain by sediments with thickness of 2.06 seconds (2530 m).

The results are concordant with the former age estimation based on the basement depths.

The heat flow value in the oceanic basin provides a useful constraint for age estimation of the oceanic basin. According to the compilation by EHARA (1976), there are 14 heat flow measurements in the Kuril abyssal plain. The values are variable from 2.82 to 1.91 HFU with the average of 2.325 HFU. TAYLOR and HAYS (1982) pointed out that heat flow data of the South China Basin are consistent with the theoretical heat flow value correlated to the age of the basin. Theoretical heat flow value for the open ocean is applicable to back-arc basin. The heat flow value, 2.325 HFU, is correlated to the age of 24 MaBP according to PARSON and SCLATER (1977) and also correlated to 27 MaBP according to DAVIS and LISTER (1977). This age estimation shows good coincidence with the age estimation based on basement depth and sediment thickness mentioned above.

Although there is no deep sea drilling holes and no definite identification of magnetic anomaly liniations, age of the Kuril Basin is well constrained to be Late Oligocene to Early Miocene by basement depth, sediment thickness, and heat flow values. It should be noted that the Kuril Basin has the same age as the Japan Basin. This coincidence provide an important constraint for reconstructing history of back-arc spreading of the Kuril Basin and Japan Basin.

4. *Marine Geology of Forearc Area of the Southern Kuril Arc*

The Kuril Trench has the maximum depth greater than 9500 m. Figure 11 shows several seismic profiles in the southern Kuril Arc. In this section, we discuss geological structure of the forearc area in the southern Kuril Arc on the basis of the seismic profiles. The geologic feature of the forearc

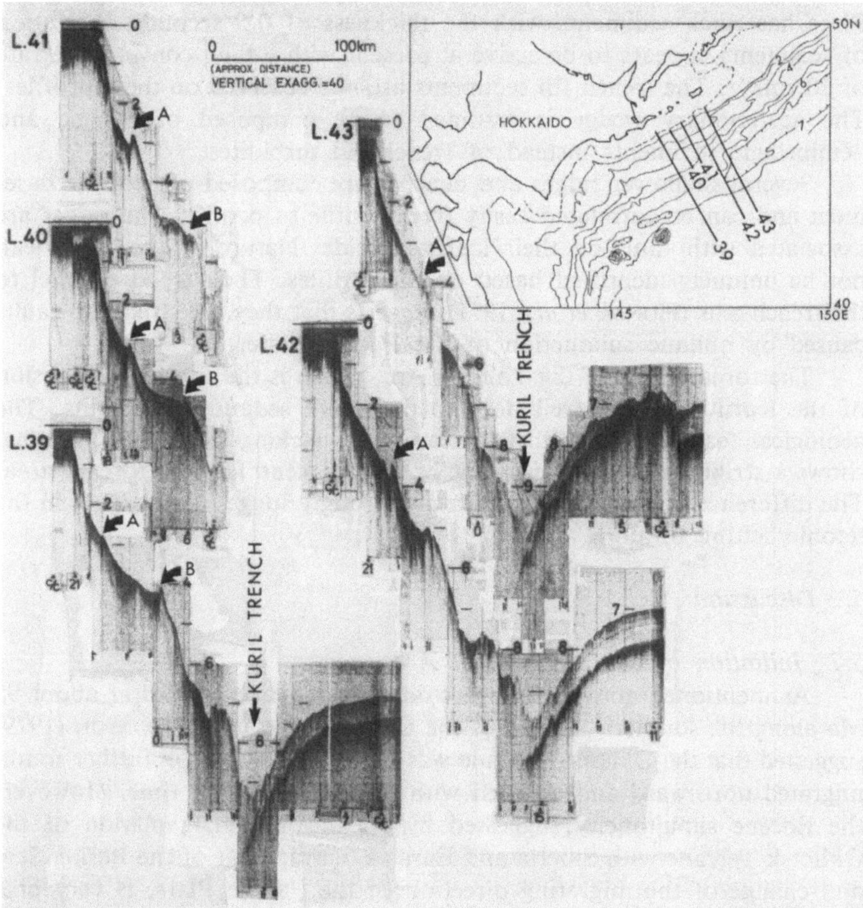

Fig. 11. Seismic profiles of the forearc area of the southern Kuril Arc during GH76-2 research cruise (TAMAKI *et al.*, 1977). Traverse lines for the profiles are shown in the index map with solid lines. A and B signs show conspicuous ridges which are well correlated from profile to profile.

region of the Kuril Arc is considerably different from the common feature of the other arcs.

Developed sedimentary basins are not observed on the landward trench slope of the southern Kuril Trench. No trench slope break is observed on Lines 41, 42 and 43. Several ridges and benches are observed on the trench slope and are traced from profile to profile (ridges A and B in Fig. 11). The deeper benches and ridges are inferred to be accretionary wedge from their acoustic transparent nature and rugged surfaces probably due to reverse faults. The ocean floor of the subducting Pacific Plate has thick sediments with the thickness of 0.5 seconds. Accretion of sediments appears to be active at present with a high convergence rate of 10 cm/yr. The trench fill sediments are not observed on these profiles. The accretionary wedge is assumed to be composed of pelagic and hemipelagic sediments instead of trench fill turbidites.

Several shallower ridges and benches are composed of acoustic basement and can be correlated easily from profile to profile. The ridges are associated with fault on their landward side. Nature of the faults can not be uniquely identified based on the profiles. Their trend parallel to the trench axis (HONZA *et al.*, 1977) suggests that they are strike-slip faults caused by oblique subduction of the Pacific Plate.

The forearc area of the Tohoku Arc, which is the southern extension of the Kuril Arc, has well developed forearc sedimentary basins. The geological feature of the Kuril forearc area, lacking sedimentary basins, shows a striking contrast with that of the adjacent Tohoku forearc area. The difference in the geological feature strongly suggests difference in the tetonic setting of both arcs.

5. *Discussion*

5.1 *Initiation of the proto Kuril Arc*

As mentioned above, the proto Kuril subduction started at about 95 Ma along the southern margin of the Okhotsk Terrane. DICKINSON (1979) suggested that the Okhotsk Terrane was situated at a position further south, migrated northward and collided with Siberia in Eocene time. However, the Eocene simultaneity suggested by Dickinson; amalgamation of the Okhotsk Terrane with Siberia and Eurasia, entrapment of the Bering Sea, and change of the migrating direction of the Pacific Plate, is very problematic. Recently, TAKAHASHI (1983) indicated that the volcanism along the Okhotsk-Chukotska volcanic belt ceased at about 70 MaBP. This cessation of volcanism and start of subduction of the proto Kuril system must indicate a jump of the subduction zone caused by the collision and accretion of the Okhotsk Terrane with the Siberia continent in late Cretaceous (Fig. 12). The Tokoro Seamounts formed in late Jurassic in the Kula Plate were also accreted to the Okhotsk Terrane in late Cretaceous (Fig. 12). The amalgamation of the Okhotsk Terrane with the Siberia Continent was a final event of accretion tectonics of the Siberia continent through Paleozoic and Mesozoic periods in Asia (TAIRA, 1983). Along Hokkaido and Sakhalin which were situated at the western margin of the Okhotsk Terrane, the collision did not occur in late Cretaceous. A collision was initiated in middle Paleogene with a boundary between the Okhotsk Terrane and Eurasia Plate (KIMURA *et al.*, 1983; KIMURA and HOYANAGI, 1983).

5.2 Active Kula-Pacific Ridge subduction along the Kuril Trench

In Paleogene, many events took place in the eastern Hokkaido and southwestern Kuril as mentioned in the previous section: shoaling of the forearc basin, southward migration of the volcanic front, hiatus without volcanism, a resurgence of subsidence of the basin in late Eocene, and a succeeding resumption of volcanism in Oligocene (Figs. 13 and 14). These events are very similar to the events associated with ridge subdution proposed by DELONG *et al.* (1977) in Aleutian. In the Okhotsk Terrane belonging to the Kula Plate before its collision with Siberia, Kula-Pacific spreading ridge must have been located on its south. The ridge might have been subducted after the southward jump of the trench from the margin of Siberia to the southern margin of the Okhotsk Terrane. The Kula-Pacific Ridge is considered to have been subducted at about 80 MaBP (UYEDA and MIYASHIRO, 1974; HILDE *et al.*, 1977; DIXON and FARRAR, 1980). These age estimations of the ridge subduction, however, are ambiguous because most parts of the Kula Plate have already been subducted and the previous spreading rate is very speculative. Time of the subduction of the Kula-Pacific ridge determined on the basis of only the Pacific Basin data should have a noticeable error.

Geologic events associated with the ridge subduction are as follows (DELONG and FOX, 1977; DELONG *et al.*, 1978, 1979): 1) Before the ridge arrival at the trench, the volcanic front is in a certain distance away from the trench and gradually approaches it. This event is caused firstly by decrease of slab dip and secondly by heating effect due to the hot plate. 2) Regional uplift takes place in the arc region due to the subduction of very young and buoyant plate. 3) The arc volcanism ceases by direct horizontal contact of oceanic slab with the continental plate. 4) Regional thermal metamorphism occurs due to the heat of ridge. Recent encounter of a ridge with a trench is taking place at the Chile trench and Nazca-

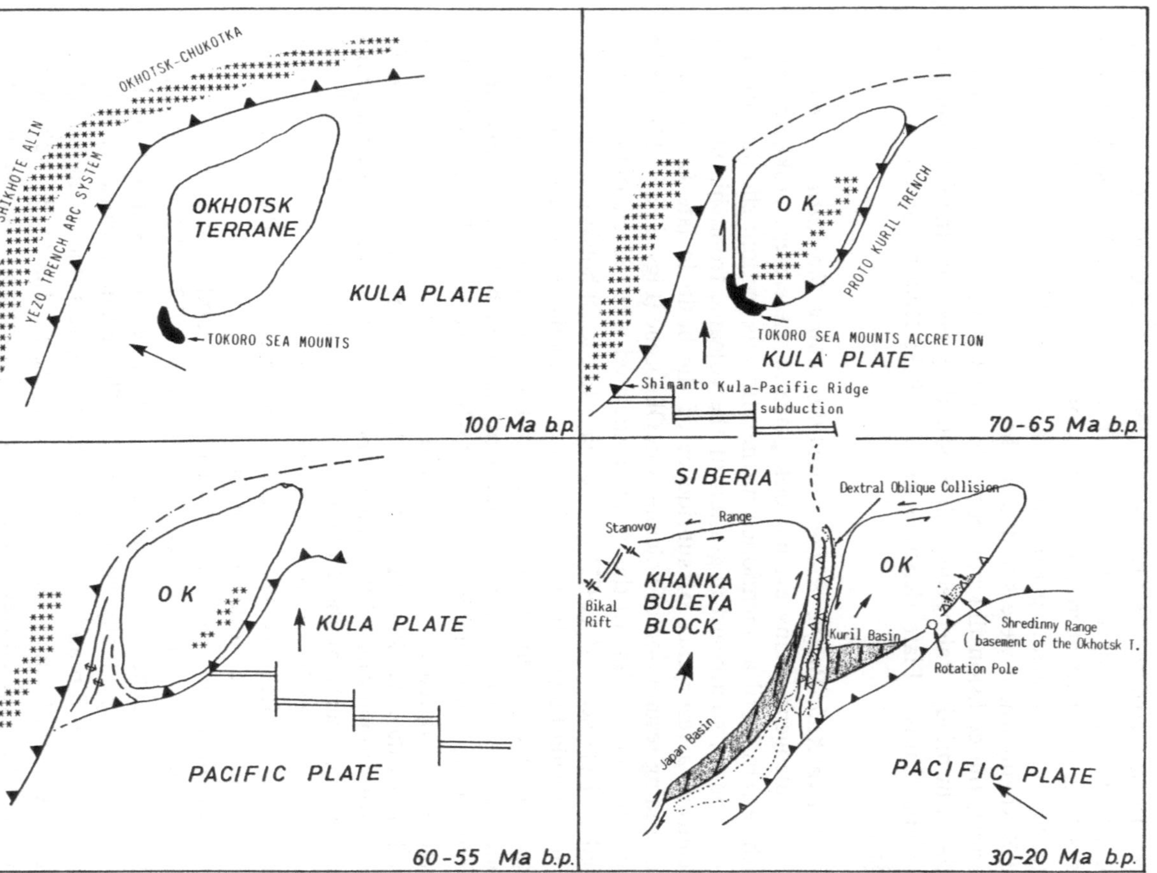

Fig. 12. Tectonic framework and evolution of the subduction zone of the northwestern Pacific region since about 100 Ma b.p.

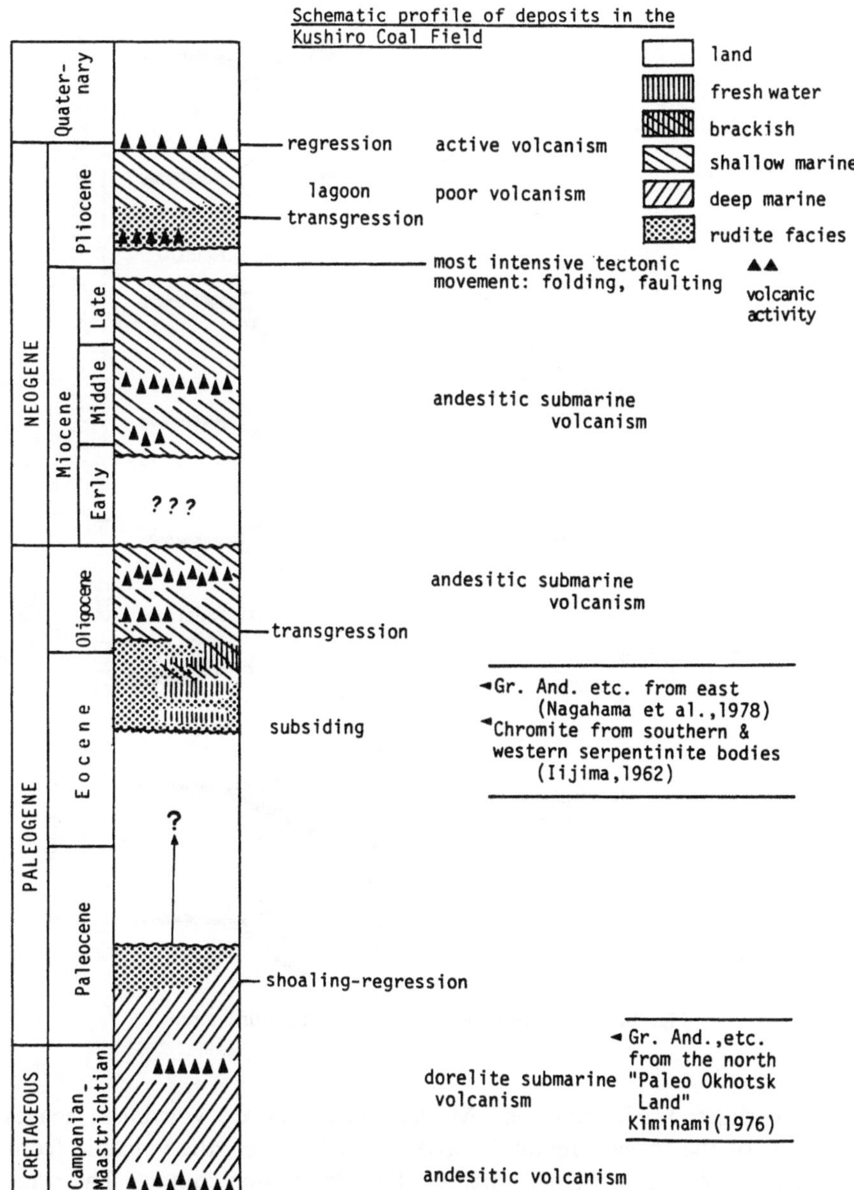

Fig. 13. Schematic profile of deposits in the Kushiro Coal Field and geologic events estimated from them.

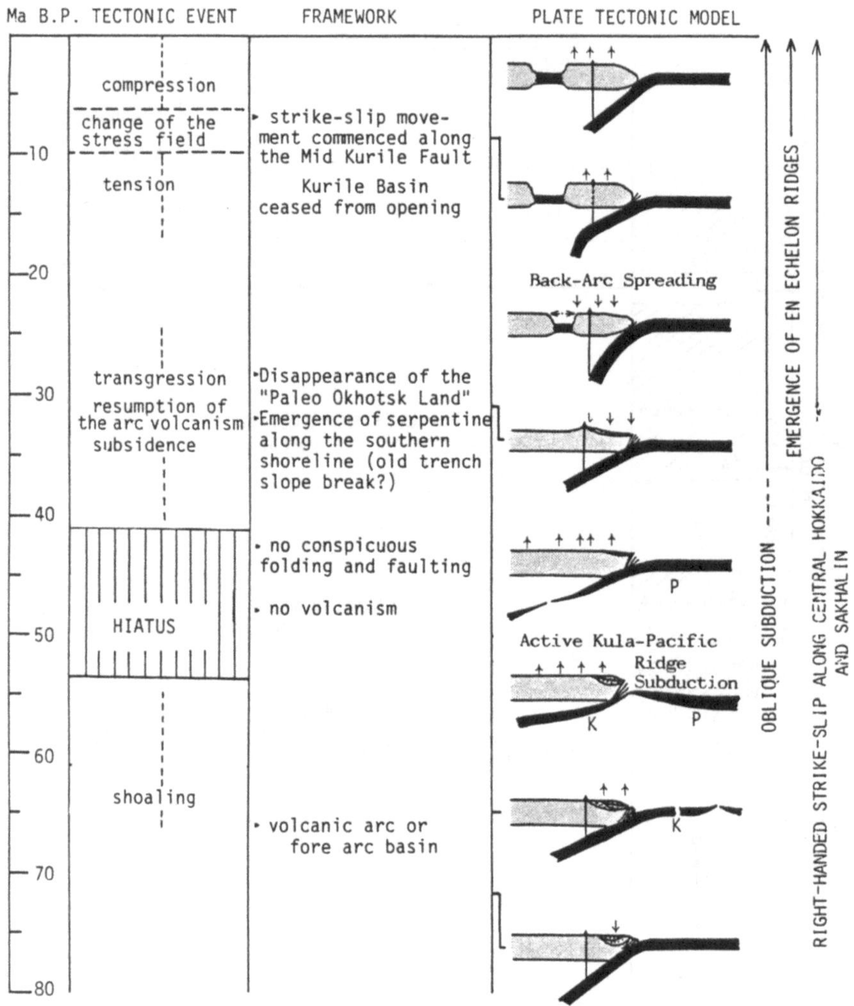

Fig. 14. Plate tectonic evolution of the Kuril Arc.

Antarctic ridges triple junction. At that place the arc volcanism ceases and uplift of the forearc region is observed. Moreover, the volcanic front approaches toward the trench north of the ridge-trench contact. The ridge subduction in Chile represents a case where the ridge is subducting perpendicularly to the trench.

Examples of the past ridge-trench encounter have been reported from many areas as follows: the western north America (ATWATER, 1970; CROUCH, 1981; HURST, 1982; DIXON and FARRAR, 1980), Aleutian and Alaska (MARSHAK and KARIG, 1977, HUDSON *et al.*, 1982), Antarctica (BARKER, 1982), south America (DELONG and FOX, 1977; HERRON *et al.*, 1981), Middle America (LONSDALE, 1978), Sumatra (MARSHAK and KARIG, 1977), and Japan (UYEDA and MIYASHIRO, 1974; TAIRA, 1981, SATO, 1983). Many events related to the ridge subduction have been proposed in these papers. ATWATER (1970) considered that arrival of the Pacific-Farralon ridge at the trench resulted in the extinction of subduction. CROUCH (1981) and HURST (1982) proposed that the arrival of the ridge into the trench caused the near-trench magmatism as "zipper tectonics". The same near-trench magmatism when the ridge reached a trench is considered to have taken place in Aleutian-Alaska (MARSHAK and KARIG, 1977; HUDSON *et al.*, 1977; HUDSON and PLAFKER, 1982) at about 50–60 MaBP when the Kula-Farralon ridge arrived at the trench, in Sumatra in Eocene to Miocene (MARSHAK and KARIG, 1977), and in southwestern Japan in middle Miocene (SATO, 1983). Trenchward migration of the volcanic front has been reported in Antarctica (BARKER, 1982), and in South America (DELONG and FOX, 1977; HERRON *et al.*, 1981). Extinction of arc volcanism just before and after the ridge subduction have been suggested in Alaska-Aleutian (DELONG and FOX, 1977) and middle America (LONSDALE, 1978). The thermal metamorphism have been shown in Aleutian-Alaska (HUDSON and PLAFKER, 1982, HUDSON *et al.*, 1977) and in Japan (TAKAHASHI, 1983). The uplifting of the arc has been reported in western North America (DIXON and FARRAR, 1980), Aleutian and Alaska (DELONG *et al.*, 1977) and Middle America (LONSDALE, 1978). In these cases, two examples of the Antarctic-Drake and Kula-Pacific Ridge subductions in Antarctic, and Aleutian and Alaska respectively, are the cases of ridge subduction in a direction nearly parallel to the trench. Other cases are considered to have been oblique ridge subduction.

The Kula-Pacific Ridge seems to have been extended nearly parallel to the trend of the Kuril-Kamchatka Trench before subduction. Migration of the volcanic front toward the trench in the proto Kuril subduction system is very consistent with the latter cases of the ridge subduction. The volcanism south of the forearc basin suggested by KIMINAMI (1983) might have been near-trench magmatism. Rapid uplifting and succeeding hiatus in Middle Eocene must indicate time of ridge subduction or just after the subduction. Resurgence of subsidence in Late Eocene seems to have been related to the subduction of colder and older plate. If these tectonic interpretations are valid, the time of the subduction of the Kula-Pacific ridge ranges from 60 Ma to 55 Ma (Figs. 12 and 14). It is quite reasonable that this arrival time of the ridge at the Kuril Trench was after the Paleocene arrival of the ridge at the southwest Japan subduction zone suggested by TAIRA (1981) and before Oligocene arrival at the Aleutian subduction zone (DELONG *et al.*, 1977). Exposures of latest Cretaceous to Paleogene ophiolites at the southern shore of Kamchatka (AVDEIKO, 1971; WATSON and FUJITA, 1983) is another evidence of active ridge subduction. DEWEY (1976) indicated that an ophiolite obduction is related to the ridge-trench collision. The young ophiolite in Kamchatka may be a result of obduction caused by the Kula-Pacific ridge arrival at the Kamchatka Trench. Geologic events on the western Pacific arcs indicate that the Kula-Pacific ridge was subducted progressively from south to north along the western Pacific margin (Fig. 15).

5.3 Back-arc spreading of the Kuril Basin

DEWEY (1980) summarized tectonics of island arc including back-arc spreading. The tectonics of island arc is well constrained by three plates, subducting oceanic plate, back-arc plate, and forearc plate (or forearc

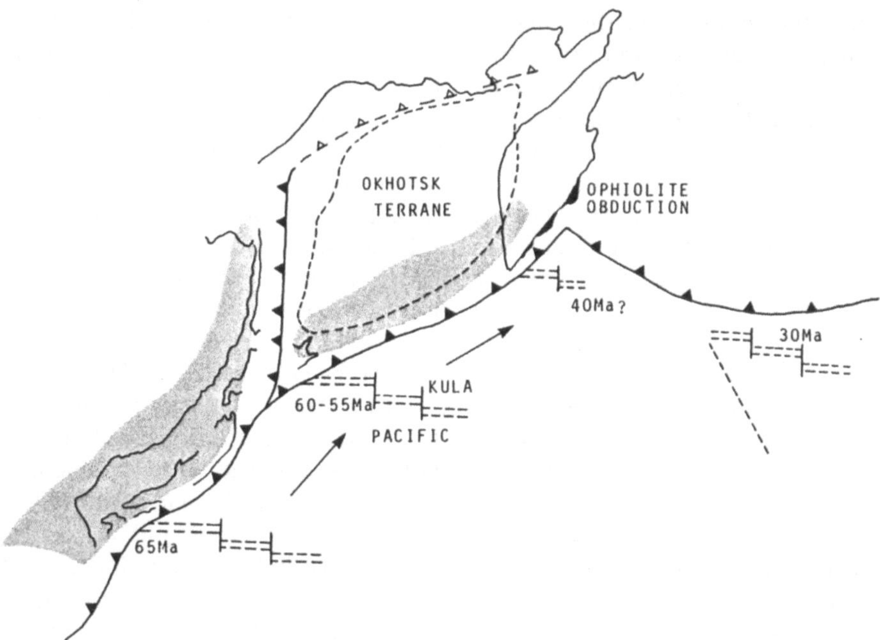

Fig. 15. Progressive encounter of the Kula-Pacific Ridge with trench along the western Pacific margin.

sliver) (Fig. 16). The back-arc plate and forearc plate are bounded by a volcanic zone where the lithosphere is ductile as the volcanic activity makes the lithosphere hot, thin, and weak. These plates have their own moving vetors. If the back-arc and forearc plates are separated from each other, the arc region is situated under tensional stress. The tensional stress causes a rift system along the volcanic zone. The rift system will be developed to an active ridge system for back-arc spreading. Eventually a back-arc basin will be formed if the tensional stress is sustained.

It should be noted that the moving vector of subducting oceanic plate does not affect motion of the forearc plate, and that it is not significant

1. ***rifting stage***

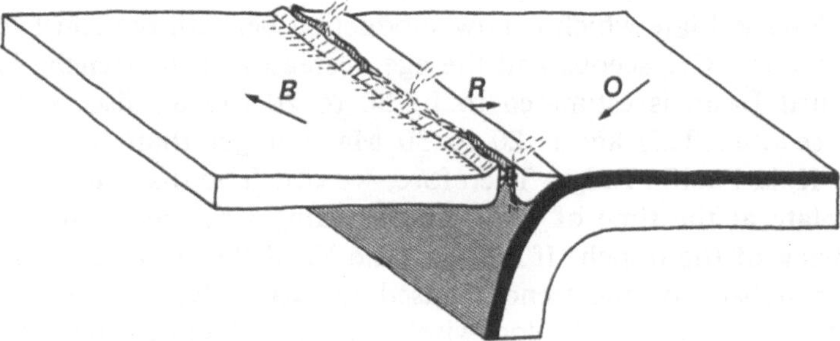

2. ***back-arc spreading***

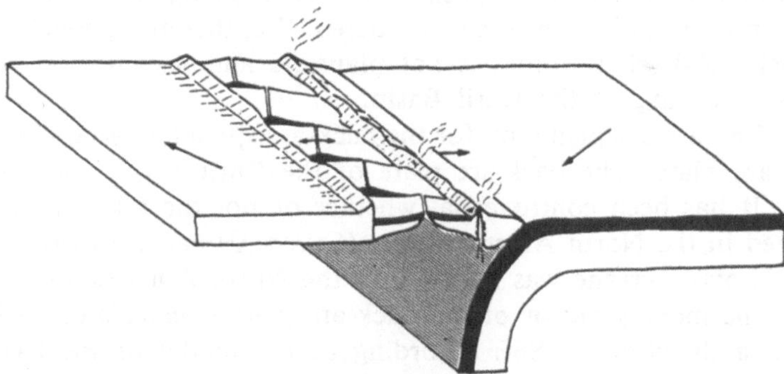

Fig. 16. Model of back-arc spreading simplified from DEWEY's (1980) model. Vectors B, R, and O shows moving vectors for back-arc plate, trench axis, and oceanic plates. When the vectors B and R is divergent, the arc rifting along volcanic chain occurs and the rifts will develop to the spreading center for the back-arc spreading.

for back-arc spreading. A significant factor for constraining the motion of the forearc plate is migration of the trench. DEWEY (1980) proposed that the trench substantially retreats oceanward or rolls back and that this roll back vector depends on ages of the subducting oceanic plate according to MOLNAR and ATWATER (1978). The older is the age of the oceanic plate, the larger is the roll back vector of trench. The roll back of trench causes the oceanward motion of the forearc plate. The moving vector of the forearc plate is also constrained by a drag force caused by oblique subduction of the oceanic plate as discussed later. The moving vector of the forearc plate is calculated by the summation of the trench roll back vector and drag force vector when the forearc plate is decoupled from the back-arc plate. The trench roll back vector is more significant for the back-arc spreading than the drag force vector. If there is some divergent component between the two vectors, a rifting in the arc is initiated and back-arc spreading will be developed.

The Pacific Plate which is now subducting beneath the Kuril Arc has the age of Early Cretaceous and the age decreases to the trench. The age of the Kuril Basin is estimated to be 30 to 20 Ma as discussed in the previous section. This age is 20 to 30 Ma younger than subduction of an active Kula Pacific Ridge. Therefore, we conclude that the age of subducting plate at the time of back-arc spreading was too young to cause the roll back of the trench. If we close the Kuril Basin on an assumption that the roll back of the trench caused the Kuril Basin, the Kuril Arc should retreat to the north clockwisely. Such a clockwise retreat of the southern Kuril Arc causes a southward outset of the central Hokkaido. If the Japan Basin was also formed by trench roll back, a southward outset would be outstanding. Such an outset of landmass overriding the subduction zone is difficult to be considered. We, therefore, conclude that the trench roll-back tectonics is not plausible and insignificant for the back-arc spreading of the Kuril Basin.

An alternative explanation for the back-arc spreading vector is retreat of back-arc plate. The back-arc plate of the Kuril Arc is the Okhotsk Terrane. It has been controversial whether or not the Okhotsk Terrane is included in the North America Plate (CHAPMAN and SOLOMON, 1976). If the Okhotsk Terrane was included in the North America Plate 30–20 Ma ago, the moving vector of the back-arc plate is calculated to be 1.1 cm/yr in a direction of S6E according to the model of ENGEBRETSON (1982). If the terrane was included in the Eurasia Plate at that time, the vector is 1.1 cm/yr in the direction of S78E. Both cases show no retreating of the back-arc plate during the formation of the Kuril Basin. Age of the Kuril Basin is very close to that of the Japan Basin as discussed in the previous section. Geographic configuration of the Kuril and Japan Basins shown in Fig. 1 seems to be in favor of their simultaneous generation. The back-arc plate for the Kuril Basin is the Okhotsk Terrane and that for the Japan Basin is the Khanka-Buleya Block. The simultaneous spreading of both basins suggests that the both blocks also moved simultaneously.

Another prominent event simultaneous with the back-arc spreading of the Japan and Kuril Basins is dextral collision between the Okhotsk Terrane and the Khanka-Buleya Block. The dextral collision is evident in en-echelon arrangement of the foldings of the Paleogene formations of the Sakhalin-Hokkaido Islands (KIMURA *et al.*, 1983). Kimura *et al.* have indicated that the dextral collision might be related to that between the Eurasia and North America Plates. However, Engebretson's reconstruction of relative motion between the Eurasia and North America along the Hokkaido-Sakhalin Islands does not indicate the dextral collision but a perpendicular collision. The motion of the Khanka-Buleya Block is northeastward on the basis of the extension of the Baikal Rift and sinistral strike-slip faults in the Stanovoy Range (Fig. 12). The northeastward movement of the Khanka-Buleya Block subsequently caused the clockwise rotation of the Okhotsk Terrane due to the drag force of the Khanka-Buleya Block. Evidence of this drag tectonic event is in the folding of the Paleogene strata in the Sakhalin and northern Hokkaido which indicates the dextral collision (KIMURA *et al.*, 1983). The dextral collision is well explained by difference in moving speed between the Khanka-Buleya Block and the Okhotsk Terrane. The clockwise rotation of the Okhotsk Terrane is consistent with the fan shape of the Kuril Basin with southwestward opening. We propose that the Kamchatka Peninsula should be a zone of collision during the spreading of the Kuril Basin because the peninsula is located on the north of the relative rotation pole. The Sredinny Range is a possible area of collision between the back-arc plate of the Okhotsk Terrane and the forearc plate (Fig. 12).

The northeastward movement of the Khanka-Buleya Block may possibly be caused by the injection of the Indo Terrane into the Eurasia Continent (ZONENSHAIN and SAVOSTIN, 1981)(Fig. 17). The Indo-Eurasia collision occurred about 40 MaBP (TAPPONIER *et al.*, 1982). The spreading of the Kuril and Japan Basin and dextral collision between the Okhotsk Terrane and Khanka-Buleya Block occurred during around 30 to 20 MaBP. The delay of about 10 Ma of the initiation of the spreading and dextral collision in respect to the Indo Eurasia collision might be related to reorganization of micro-continent blocks within the Eurasia Plate which were accreted to the Siberia continent in Mesozoic Period (TAIRA, 1983).

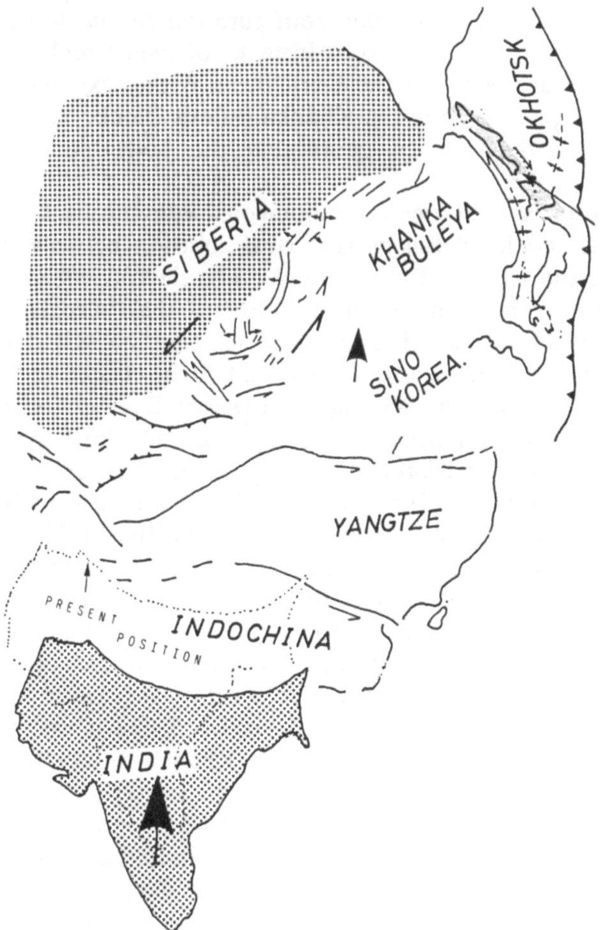

Fig. 17. Reorganization of the terranes within the Eurasia Plate which was caused by the Indo-Eurasian collision since Eocene time. The dextral oblique collision between the Eurasia and North America Plates and spreading of the Kuril and Japan Basins are attributed to this reorganization.

5.4 Forearc plate migration caused by oblique subduction

Migration of the forearc plate in a direction parallel to the axis of the trench is very important tectonic process in the formation of the continental margin (BECK, 1983). When the migration occurred, fragments were accumulated around the curved margin of the continent and accreted to it. This tectonic process has been the case in the southwestern Kuril and Hokkaido since Latest Miocene. However, the oblique subduction of the Pacific Plate along the Kuril-Kamchatka Trench started at about 43 Ma BP, as shown by the Emperor-Hawaian elbow, whereas effect of the oblique subduction on the Kuril Arc appeared in Latest Miocene. Late Eocene, the serpentinite bodies are considered to have exposed along the southern shore of the frontal Kuril Arc mentioned in the previous section and they may be "knokers" which are uplifted near trench slope break caused by oblique subduction proposed by KARIG (1978). After that time, however, there is no evidence indicating the oblique subduction.

Such a situation may possibly be explained by alternative occurrences of two types of subduction; Chile type and Maliana type (UYEDA and KANAMORI, 1979). There is no effect of the oblique subduction into the island arc in the case of Mariana type. From Oligocene to middle Miocene, Mariana type subduction was taking place along the Kuril Kamchatka Trench. According to UYEDA and KANAMORI (1979), back-arc opening is related to the Mariana type subduction. This interpretation is consistent with the time of back-arc spreading of the Kuril Basin.

6. *Conclusion*

Tectonic framework of the Kuril Arc since its initial formation is postulated in the following manner (Fig. 14):

1) The Kuril arc-trench system was formed in Late Cretaceous by collision of the Okhotsk exotic terrane with the Siberia continent.

2) The second significant event in the Kuril system was the active Kula-Pacific Ridge subduction in the trench about 60 Ma-55 MaBP. This event can be recognized from the trenchward migration of the volcanic front, rapid uplifting of the arc, and the succeeding hiatus.

3) The third important event was spreading of the Kuril back-arc basin since 30 MaBP. The spreading was resulted from clockwise rotation of the overriding plate; the Okhotsk Terrane. The collision between the India and Eurasia Plates since Eocene resulted in reorganization of the terrains within the Eurasia Plate and its internal deformation. The retreating of the overriding plate of the Kuril system was attributed to this reorganization.

4) The fourth event was decoupling of the Kuril forearc plate from the North America Plate, its migration in a direction parallel to the Kuril Trench, and its collision with the Tohoku Arc since Latest Miocene. This event was caused by the oblique subduction of the Pacific Plate along the Kuril Trench. The time gap between the initiation of the oblique subduction and a resultant motion can be explained by difference of subduc-

tion mode; Mariana type subduction before Middle Miocene and Chilian type since Late Miocene, since the decoupling between the overriding plate and subducting slab in the Mariana type subduction can not cause the compressional stress in the overliding plate. No strike-slip motion along the volcanic front caused by oblique subduction took place until Late Miocene. The southwestward migration of the Kuril forearc plate resulted in its collision with the Tohoku Arc and mountain building of the Hidaka Mountains.

Acknowledgements

We gratefully acknowledge advice and discussion with Professors A. Miyashiro, S. Uyeda, K. Nakamura, Y. Kobayashi, and Dr. T. Seno. We also thank Prof. M. Komatsu and members of Research Group of Tectonics in Hokkaido for their frank discussion.

REFERENCES

T. ATWATER, Implication of plate tectonics for the Cenozoic tectonics of western North America, *Geol. Soc. Am. Bull.*, **81**, 3513–3536, 1970.

G. P. AVDEIKO, Evolution of geosynclines on Kamchatska, *Pac. Geol.*, **3**, 1–15, 1971.

R. F. BARKER, The Cenozoic subduction history of the Pacific margin of the Antarctic Peninsula: ridge crest-trench interactions, *J. Geol. Soc. Lond.*, **139**, 787–801, 1982.

M. E. Jr BECK, On the mechanism of tectonic transport in zones of oblique subduction, *Tectonophysics*, **93**, 1–11, 1983.

C. A. BURK and H. S. GNIBIDENKO, The structure and age of acoustic basement in the Okhotsk Sea, in *Island Arcs, Deep Sea Trenches and Back Arc Basins*, ed. by M. Talwani and W. C. Pitman III, Am. Geophys. Union, Maurice Ewing Ser. I, pp. 451–462, 1977.

M. E. CHAPMAN and S. C. SOLOMON, North American-Eurasian Plate boundary in Northeast Asia, *J. Geophys. Res.*, **81**, 921–930, 1976.

J. K. CROUCH, Northwest margin of California continental borderland; marine geology and tectonic evolution, *AAPG Bull.*, **65**, 191–218, 1981.

S. T. CROUGH, The Correction for Sediment Loading on the Seafloor, *J. Geophys. Res.*, **88**, 6449–6454, 1983.

E. E. DAVIS and C. R. B. LISTER, Heat flow measured over the Juan de Fuca Ridge: Evidence for widespread hydrothermal circulation in a highly heat transportive crust, *J. Geophys. Res.*, **82**, 4845–4860, 1977.

S. E. DELONG and P. J. FOX, Geological consequences of ridge subduction, in *Island Arcs, Deep Sea Trenches and Back Arc Basins*, eds. M. Talwani and W. D. Pitman III, Am. Geophys. Union, Maurice Ewing Ser. I, pp. 221–228, 1977.

S. E. DELONG, P. J. FOX, and F. W. MCDOWELL, Subduction of the Kula Ridge at the Aluetian Trench, *G.S.A. Bull.*, **89**, 83–95, 1978.

S. E. DELONG, W. H. SCHWARZ, and R. N. ANDERSON, Thermal effects of ridge subduction, *Earth Planet. Sci. Lett.*, **44**, 239–246, 1979.

J. F. DEWEY, Ophiolite Obduction, *Tectonophysics*, **91**, 93–120, 1976.

J. F. DEWEY, Episodicity, sequence and style at convergent plate boundaries, in *The Con-*

tinental Crust and Its Mineral Resources, ed. by D. Strangway, Geol. Assoc. Canada Spec. pap. 20, pp. 553–574, 1980.

W. R. DICKINSON, Plate tectonic evolution of north Pacific Rim, *J. Phys. Earth*, Supplement, 1–20, 1979.

J. M. DIXON and E. FARRAR, Ridge subduction, eduction and the Neogene tectonics of southwestern North America, *Tectonophysics*, **67**, 81–99, 1980.

S. EHARA, Heat Flow in and around Hokkaido and its tectonic implications, *Proceeding of Subterranean Structure in and around Hokkaido and Its Tectonic Implication*, pp. 44–51, 1976 (in Japanese with English abstract).

D. C. ENGEBRETSON, Relative motions between oceanic and continental plates in the Pacific Basin, Ph.D., Stanford Univ. 156 pp., 1982.

T. J. FITCH, Plate convergence, transcurrent faults, and internal deformation adjacent to southeast Asia and western Pacific, *J. Geophys. Res.*, **77**, 4432–4460, 1972.

V. T. FROLOV, I. A. BURIKOVA, and A. V. GUSHCHIN, Zone of high magmatic permeability of the southern part of the Lesser Kuril Ridge, *Inter. Geol. Rev.*, **22**, 1303–1308, 1980.

H. S. GNIBIDENKO and I. I. KKHVEDCHUK, The tectonics of the Okhotsk Sea, *Marine Geology*, **50**, 155–198, 1982.

D. E. HAYES, Global studies of ocean crustal depth, age relationships, *EOS*, **64**, 1983.

E. M. HERRON, S. C. CANDE, and B. R. HALL, An active spreading center collides with a subduction zone: a geophysical survey of the Chile margin triple junction, *Mem. Geol. Soc. Am.*, **154**, 683–701, 1981.

T. W. C. HILDE, S. UYEDA, and L. KROENKE, Evolution of the western Pacific and its margin, *Tectonopysics*, **38**, 145–165, 1977.

E. HONZA, ed., Geological investigation of the Japan and southern Kuril Trench and slope areas, Geol. Surv. Japan, Cruise Report, 9, 127 pp., 1977.

E. HONZA, ed., Geological investigation of the Okhotsk and Japan seas off Hokkaido, Geol. Surv. Japan, Cruise Rept., 11, 72 pp., 1978.

E. HONZA, ed., Geological Investigation of the Okhotsk and Japan Seas off Hokkaido June–July 1977 (GH77-3 Cruise), Cruise Rept. 11, Geol. Surv. Japan, 72 pp., 1978.

E. HONZA, K. TAMAKI, and F. MURAKAMI, Geological Map of the Japan and Kuril Trenches and the adjacent area, 1/1,000,000, Geol. Surv. Japan, 1978.

T. HUDSON and G. PLAFKER, Paleogene metamorphism of an accretionary flysh terrane, eastern gulf of Alaska, *Geol. Soc. Am. Bull.*, **93**, 1280–1290, 1982.

T. HUDSON, G. PLAFKER, and D. L. TURNER, Paleogene anatexis along the gulf of Alaska margin, *Geology*, **7**, 573–577, 1977.

R. W. HURST, Petrogenesis of the Conejo volcanic suite, southern California, evidence for mid-ocean ridge-continental margin interactions, *Geology*, **10**, 267–272, 1982.

A. IIJIMA, On the relationship between the provenances and the depositional basins, considered from the heavy mineral associations of the upper Cretaceous and Tertiary formations in the central and southern Hokkaido, *Japan*, *J. Fac. Sci. Univ. Tokyo*, *Ser. III*, **XI**, V, 279–285, 1962.

K. KAIHO, Tertiary Foraminifera in the western part of the Kushiro Coal Field, Abstract, annual meeting, *Geol. Soc. Japan*, **273**, 1983 (in Japanese).

D. E. KARIG, Origin and development of marginal basins in the western Pacific, *J. Geophys. Res.*, **76**, 2542–2560, 1971.

D. E. KARIG, Material transport within accretionary prism and the "knocker" problem, *J. Geol.*, **88**, 27–39, 1978.

K. KIMINAMI, Sedimentology of the Nemuro Group (part I), *J. Geol. Soc. Japan*, **81**,

215–232, 1975a.

K. KIMINAMI, Sedimentology of the Nemuro Group (part II)—radiographic investigation of flysh type sandstones in the Akkeshi Formation of the Nemuro Group, *J. Geol. Soc. Japan*, **81**, 697–708, 1975b (in Japanese with English abstract).

K. KIMINAMI, Sedimentology of the Nemuro Group (part III)—sedimentation of the Lower Akkeshi member, *J. Geol. Soc. Japan*, **81**, 755–768, 1975c (in Japanese with English abstract).

K. KIMINAMI, Sedimentology of the Nemuro Group (part IV)—on the change of the source area in the Akkeshi Formation, *J. Geol. Soc. Japan*, **82**, 773–782, 1976.

K. KIMINAMI, Stratigraphic reexamination of the Nemuro Group, *Earth Science* (*Chikyu Kagaku*), **32**, 120–132, 1978 (in Japanese with English abstract).

K. KIMINAMI, Sedimentary history of the Late Cretaceous—Paleogene Nemuro Group, Hokkaido, Japan—a fore arc basin of the Paleo Kuril arc trench system, *J. Geol. Soc. Japan*, **89**, 607–624, 1983.

K. KIMINAMI and Y. KONTANI, Pre-Cretaceous Paleocurrents of the Northeastern Hidaka Belt, Hokkaido, *J. Fac. Sci. Hokkaido Univ., Ser. IV*, **19**, 179–188, 1979.

G. KIMURA, Abashiri Tectonic Line-with special reference to the tectonic significance of the southwestern margin of the Kuril Arc, *J. Fac. Sci. Hokkaido Univ., Ser. IV*, **20**, 95–111, 1981a.

G. KIMURA, Tectonic evolution and stress field in the southwestern margin of the Kuril Arc, *Geol. Soc. Japan*, **87**, 757–768, 1981b (in Japanese with English abstract).

G. KIMURA, Tectonic movement at the arc-arc junction, Structural Geology (Koozo Chishitsu), **28**, 5–22, 1982 (in Japanese).

G. KIMURA, S. MIYASAKA, Y. KONTANI, S. MIYASHITA, K. HOYANAGI, and Y. WATANABE, Tectonic significance of the Kamishiyubetsu tectonic zone in the uplift of the Hidaka metamorphic belt, *Bull. Tec. Res. G. Japan*, **27**, 167–177, 1982 (in Japanese with English abstract).

G. KIMURA, S. MIYASHITA, and S. MIYASAKA, Collision tectonics in Hokkaido and Sakhalin, in *Accretion Tectonics in the Circum Pacific Regions*, ed. by M. Hashimoto and S. Uyeda, pp. 69–88, Terrapub, 1983.

G. KIMURA and K. HOYANAGI, Collision orogenesis in Hokkaido, *Marine Science (Kaiyoo Kagaku)*, **15**, 724–730, 1983 (in Japanese).

G. deV. KLEIN and K. KOBAYASHI, Geological summary of the north Philippine Sea, based on Deep Sea Drilling Project Leg 58 results, in G. deV. Klein, K. Kobayashi *et al.*, Initial Rept. DSDP, 58, Washington, (U.S. Govt. Printing Office), pp. 951–961, 1980.

K. KOBAYASHI, Spreading of the Japan Sea and drift of Japanese Island Arc—a synthesis and speculation, in *Island Arcs, Marginal Seas, and Kuroko deposites*, ed. by E. Horikoshi, *Mining Geol.*, Special Issue, **11**, 23–26, 1982.

Y. KOBAYASHI, The initiation of the "subduction", *Chikyu* (*Earth Monthly*), **51**, 510–514, 1983 (in Japanese).

I. KOIZUMI, The geologic history of the Sea of Japan-based upon sediments and microfossils, *The Japan Sea* (*Nihonkai*), **10**, 69–70, 1979 (in Japanese).

M. KOMATSU, S. MIYASHITA, J. MAEDA, Y. OSANAI, T. TOYOSHIMA, Y. MOTOYOSHI, and K. ARITA, Petrological constitution of the continental type crust upthrust in the Hidaka belt, Hokkaido, *J. Japan Assoc. Mineral Petrol. Econ. Geol. Spec. Pap.*, No. 3, 220–230, 1982 (in Japanese with English abstract).

S. KOSHIMIZU and T. OKA, Fission-Track age of the subaquous pumice flow deposites included in the Tertiary diatomaceous mud, western part of the Shiretoko Peninsula,

Abstract, annual meeting, Geol. Soc. Japan, 154, 1983 (in Japanese).

P. LONSDALE, Ecuadorian subduction system, *AAPG Bull.*, **62**, 2454–2477, 1978.

W. J. LUDWIG, S. MURAUCHI, and R. E. HOUTZ, Sedimentation and structure of the Japan Sea, *Geol. Soc. Am. Bull.*, **86**, 651–664, 1975.

S. MABUCHI, A study on sedimentation and tectonic history of the Paleogene system of the Kushiro Coal Field, *Inst. Geol. Paleon. Tohoku Univ.*, **56**, 1–42, 1962 (in Japanese with English abstract).

R. S. MARSHAK and D. E. KARIG, Triple junctions as a cause for anomalously near trench igneous activity between the trench and volcanic arc, *Geology*, **5**, 233–236, 1977.

M. MATSUI, Sedimentological study of the Paleogene basin of Kushiro in Hokkaido, Japan, *J. Fac. Sci. Hokkaido Univ., Ser IV*, **XI**, 3, 431–480, 1962.

T. MATSUMOTO, Geological ages of the Mesozoic system, *Kagaku*, **40**, 248–255, 1970 (in Japanese).

S. MIYASAKA and K. KIKUCHI, The Neogene Tertiary upheaval of the Hidaka metamorphic belt, Hokkaido, *Assoc. Geol. Collab. Japan Monogr.*, **21**, 139–153, 1978 (in Japanese with English abstract).

S. MIYASHITA, Reconstruction of the ophiolite succession in the western zone of the Hidaka metamorphic belt, Hokkaido, *J. Geol. Soc. Japan*, **88**, 69–86, 1983 (in Japanese with English abstract).

P. MOLNAR and T. ATWATER, Interarc spreading and cordilleran tectonics as alterates related to the age of subducted oceanic lithosphere, *Earth Planet. Sci. Lett.*, **41**, 330–340, 1978.

H. NAGAHAMA, K. TERUI, and K. SATO, Provenances of the Urahoro Group considered from conglomerates, abstract in Annual Meeting Geol. Soc. Japan, 85, 200, 1979 (in Japanese).

K. NAKAMURA, Possible nascent trench along the eastern Japan Sea as the convergent boundary between Eurasian and North American Plates, *Bull. Res. Inst. Univ. Tokyo*, **58**, 711–722, 1983.

K. NIIDA and RESEARCH GROUP OF THE TOKORO BELT, Geology of the Tokoro Belt, Hokkaido, in *Tectonics of Paired Metamorphic Belts*, ed. by I. Hara, pp. 49–55, Tanishi Print Kikaku, Hiroshima, 1982.

B. PARSON and J. G. SCLATER, An analysis of the variation of ocean floor bathymetry and heat flow with age, *J. Geophys. Res.*, **82**, 803–827, 1977.

S. SATO, Geostructural development of the Kushiro coal field, Prof. Sasa Y. Mem. Vol., pp. 441–451, 1969 (in Japanese with English abstract).

H. SATO, Magmatism caused by the subduction along the Nankai Trough since Oligocene, Abstract, Annual Meeting, Geol. Soc. Japan, 55–56, 1983 (in Japanese).

T. SENO and S. MARUYAMA, Paleogeographic reconstruction and origin of the Philippine Sea, *Tectonophysics*, **102**, 53–84, 1984.

K. SHIBATA and T. TANAI, K-Ar age of the Tertiary volcanics in Hokkaido, in *Recent Progress of Neogene Biostratigraphy of Hokkaido*, ed. by T. Tanai, Hokkaido Univ. Co-op Print, pp. 75–79, 1982 (in Japanese).

A. TAIRA, Principle of the formation of the Japanese Island, *Kagaku*, **51**, 508–515, 1981 (in Japanese).

A. TAIRA, Plate tectonic evolution of Japan, abstract in *Circum Pacific Terrane Conference*, Stanford, 1983.

M. TAKAHASHI, Space-time distribution of Late Mesozoic to Early Cenozoic magmatism in east Asia and its tectonic implications, in *Accretion Tectonics in the Circum Pacific Regions*, ed. by M. Hashimoto and S. Uyeda, pp. 69–88, Terrapub, 1983.

K. TAMAKI, Y. INOUCHI, F. MURAKAMI, and E. HONZA, Continuous seismic reflection profiling survey, in E. Honza, ed., Geological Investigation of Japan and Southern Kuril Trench and slope areas, Geol. Surv. Japan Cruise Rept., No. 7, 50–71.

T. TANAI, ed., Recent progress of Neogene biostratigraphy of Hokkaido, Hokkaido Univ. Co-op Print, 90 pp., 1982 (in Japanese).

P. TAPPONIER, G. PELTZER, A. Y. LEDAIN, and R. ARMIJO, Propagating extrusion tectonics in Asia: New insights from simple experiments with plasticine, *Geology*, **10**, 611–616, 1982.

B. TAYLOR and D. E. HAYES, The tectonic evolution of the South China Basin, in D. E. Hayes, ed., AGU Geophys. Monogr., 23, pp. 89–104, 1980.

B. TAYLOR and D. E. HAYES, Origin and history of the South China Basin, in D. E. Hayes ed., AGU Monogr., 27, pp. 23–56, 1982. S. UYEDA and A. MIYASHIRO, Plate tectonics and the Japanese Islands, *Geol. Soc. Am. Bull.*, **85**, 1159–1170, 1974.

S. UYEDA and H. KANAMORI, Back arc opening and the mode of subduction, *J. Geophys. Res.*, **84**, 1049–1061, 1979.

B. F. WATSON and K. FUJITA, Structure and tectonics of Kamchatska Peninsula, USSR, abstract in *Circum Pacific Terrane Conference*, Stanford, 1983.

K. YAGI, Petrology of the alkali dorelites of the Nemuro Peninsula, *Japan*, *Geol. Soc. Am. Bull.*, **115**, 103–147, 1969.

L. P. ZONENSHAIN and L. A. SAVOSTIN, Geodynamics of the Baikal rift zone and plate tectonics of Asia, *Tectonophysics*, **76**, 1–45, 1981.

Formation of Active Ocean Margins, edited by N. Nasu *et al.*, pp. 677–699.

EARLY CRETACEOUS DUAL SUBDUCTION SYSTEM IN AND AROUND THE KAMUIKOTAN TECTONIC BELT, HOKKAIDO, JAPAN

Teruo WATANABE[1] and Hirokazu MAEKAWA[2]

[1]*Department of Geology, Faculty of Science, Shimane University, Matsue 690, Japan*
[2]*Geological Institute, University of Tokyo, Hongo, Bunkyo-ku, Tokyo 113, Japan**

Abstract. Sedimentary mélange units in the Biei and Niikappu areas of the Kamuikotan tectonic belt are briefly described. On the basis of geology of both the areas and recent studies of other related areas, the authors propose a westward-dipping subduction model of early Cretaceous age with dual trench axes for the Kamuikotan and adjoining belts. There is good evidence for the Kamuikotan subduction system in the Biei and Niikappu areas in early Cretaceous (mainly Valanginian). Many blocks of serpentinite, amphibolite, basalt, chert, and psammitic rocks were supplied into the trench as submarine sliding products. At that time, however, the subduction system had already ceased in the Furano and Horokanai areas because of the collision of seamounts or ridges. This heterogeneous tectonic field represented by subduction, obduction and collision within a few ten kilometers along the ancient plate boundary may characterize the initiation stage of the subduction process of the Kula plate along transform faults. At the same time, a new subduction zone seems to have started in some portions of the Hidaka western belt located to the east of the Kamuikotan tectonic belt. The complicated dual trench setting may also be related to the initiation of subduction of the Kula plate with early transform history.

*Present address: Department of Earth Sciences, Faculty of Science, Kobe University, Nada-ku, Kobe 657, Japan

1. *Introduction*

The Kamuikotan tectonic belt, a term introduced by HUNAHASHI (1956), extends for more than 300 km with a maxmum width of 30 km along the N-S trending axial zone of Hokkaido, northern Japan. It is considered to be situated in the ancient consuming plate boundary with westward-dipping subduction zone. OKADA (1974) first introduced this concept on the basis of the westerly provenance of the Cretaceous Yezo Group, distribution of Cretaceous granite in the southwest Hokkaido, distribution of Cretaceous andesitic volcanics in the northwest Hokkaido, and early Cretaceous radiometric age of the Kamuikotan high pressure metamorphic rocks. Since then, several geologists proposed additional models of westward-dipping subduction zone for the construction of the tectonic model of the Kamuikotan and adjacent areas (KIMURA, 1977; GRAPES, 1978; BANNO *et al.*, 1978; DICKINSON, 1978; ISHIZUKA, *et al.*, 1983a).

More recently we also proposed a westward-dipping subduction model on the basis of studies of geology in the Biei area (MAEKAWA, 1981, 1983) and the Niikappu area (WATANABE, 1981, 1982; WATANABE *et al.*, 1981).

We both emphasized firstly the consistency between the radiometric age of the Kamuikotan metamorphic rocks and the sedimentation age shown by radiolarian fossils, and secondly the existence of sedimentary mélange unit in the Kamuikotan belt. In particular, MAEKAWA (1981, 1983) discussed that sedimentary serpentinite and sedimentary amphibolite were supplied to the ancient trench; blocks of amphibolite, serpentinite, basalt, chert, and psammitic rocks were considered to slide into the trench from the both oceanic and trench slope break sides. Based on this model the concept of a westward-dipping subduction zone involving the two-way street model (SUPPE, 1972) has been developed in the Kamuikotan tectonic belt, although ISHIZUKA *et al.* (1981, 1983a) also mentioned the recycle process of subduction materials.

At the same time studies of radiolarian biostratigraphy have been performed in and around the Kamuikotan tectonic belt by NAKASEKO (1979), KANIE *et al.* (1981), MINOURA *et al.* (1982), KITO (1982), etc. As a result, the early Cretaceous sedimentary environment of the region becomes much clearer permitting evaluation of the Kamuikotan subduction system along its axis.

2. *Outline of the Metamorphic Rocks in the Kamuikotan Tectonic Belt*

The Kamuikotan tectonic belt, characterized by the widespread occurrence of serpentinite and blueschist, is divided into three regions, i.e., Pinneshiri region (northern part), Horokanai-Biei region (central part) and Yubaridake-Shizunai region (southern part) (Fig. 1).

In the Pinneshiri region, a large serpentinite mass containing many blocks of amphibolite, blueschist, quartz schist, layered ultramafic-mafic rocks (wehrlite-clinophyroxenite-gabbro), pillow lava and chert is exposed around Mt. Shirikomadake. This region is regarded as a serpentinite

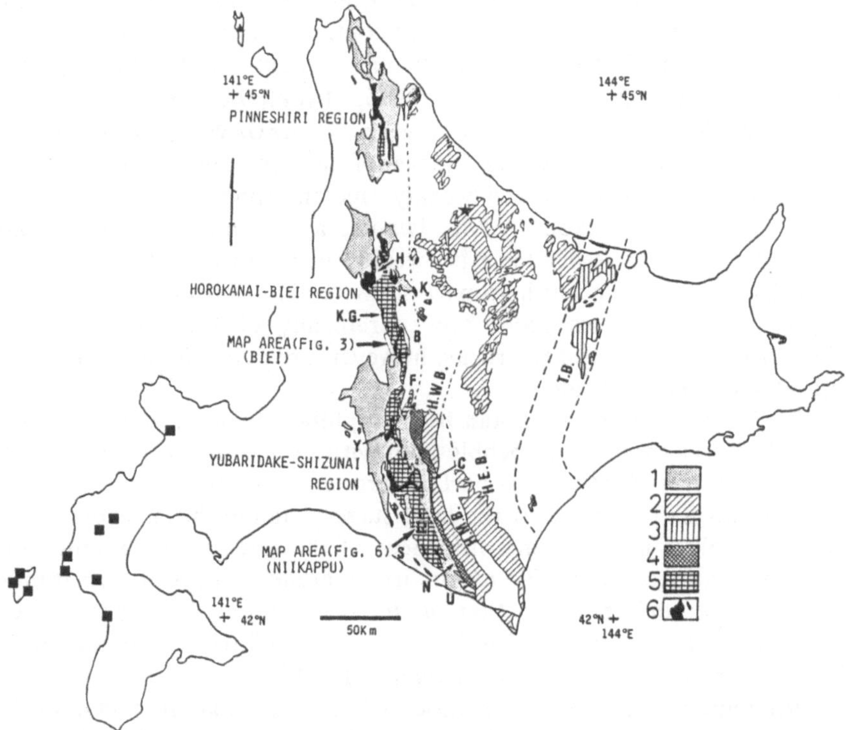

Fig. 1. Distribution of the Kamuikotan rocks, Triassic greenstones of the Hidaka western belt and related terranes. 1: Yezo Supergroup (Aptian–Maastrichtian), 2: Hidaka Supergroup, 3: Nikoro Group, 4: Triassic greenstones of the Hidaka western belt, 5: Kamuikotan rocks, 6: ultramafic rocks in the Kamuikotan tectonic belt, H: Horokanai, K. G.: Kamuikotan Gorge, A: Asahikawa, K: Kamikawa, B: Biei, F: Furano, Y: Yubaridake(Mt. Yubari), C: Chiroro River area, S: Shizunai, N: distribution area of the Nitarachi Formation, U: Urakawa, H.W.B.: Hidaka western belt, H.M.B.: Hidaka metamorphic belt, H.E.B.: Hidaka eastern belt, T.B.: Tokoro belt, solid triangle: distribution area of Early Cretaceous andesitic rocks, solid square: distribution area of Early Cretaceous granites, star: Kamiokoppe (Upper Cretaceous fossils are found by IWATA *et al.*, 1983)

mélange by KATOH *et al.* (1979).

In the Horokanai area of the Horokanai-Biei region, low P/T type metamorphic rocks showing ophiolite succession tectonically overlie high P/T metamorphic rocks including typical lawsonite-glaucophane schist and serpentinite along a thrust fault (BANNO *et al.*, 1978; ASAHINA and KOMATSU, 1979; ISHIZUKA *et al.*, 1983a). In the metamorphic rocks of the area, widespread occurrence of aragonite is reported (GOUCHI and BANNO, 1974; SHIBAKUSA *et al.*, 1977). Detailed petrographical data on the high P/T metamorphic rocks in the Horokanai area were given by BANNO and HATANO (1963), SHIBAKUSA (1974, 1981), and AGUIRRE (1977). The high P/T metamorphic rocks are distributed continuously in the southern area, i.e., Kamuikotan Gorge area (SUZUKI and SUZUKI, 1959; GOUCHI, 1983) and Biei area (TAZAKI, 1964; HÉRVE, 1975; MAEKAWA, 1982).

In the Yubaridake-Shizunai region, NAKAGAWA (1981) reported serpentinite mélange including blocks of lawsonite-glaucophane schist, epidote-glaucophane schist and weakly metamorphosed pillow basalt in Mt. Yubari (Yubaridake). In Mt. Iwanai, a serpentinitized harzburgite mass is tectonically associated with weakly metamorphosed pillow basalt and pelitic rock. In the southern part of the region (Shizunai and Niikappu areas), blueschist facies rocks are sporadically distributed in mélange (NAKANO, 1981). Most of the rocks, however, are pumpellyite-actinolite or prehnite-pumpellyite grade.

Throughout the Kamuikotan belt, amphibolites occur as large masses or small blocks (boulder to pebble size) with or without serpentinite. The amphibolites have for the most part undergone a later secondary low-grade metamorphism. The grade of the later-stage metamorphism recognized in the amphibolites is consistent with those of the surrounding Kamuikotan rocks as mentioned later. The metamorphic grade of most of serpentinites also seems to correspond to that of the surrounding Kamuikotan rocks (MAEKAWA, 1982; KATOH, 1982). A summary of recent studies on the Kamuikotan metamorphic rocks is compiled in Fig. 2.

Radiometric ages (K-Ar method) of the Kamuikotan metamorphic rocks concentrate from 145 Ma to 109 Ma (BIKERMAN *et al.*, 1971; IMAIZUMI and UEDA, 1989), except for the low-grade schists showing 75 Ma in the Horokanai area (IMAIZUMI and UEDA, 1981).

3. Geological Data in the Biei and Niikappu Areas for Construction of a Plate Tectonic Model

3.1 Biei area

The Kamuikotan metamorphic rocks are divided into three N-S ten-

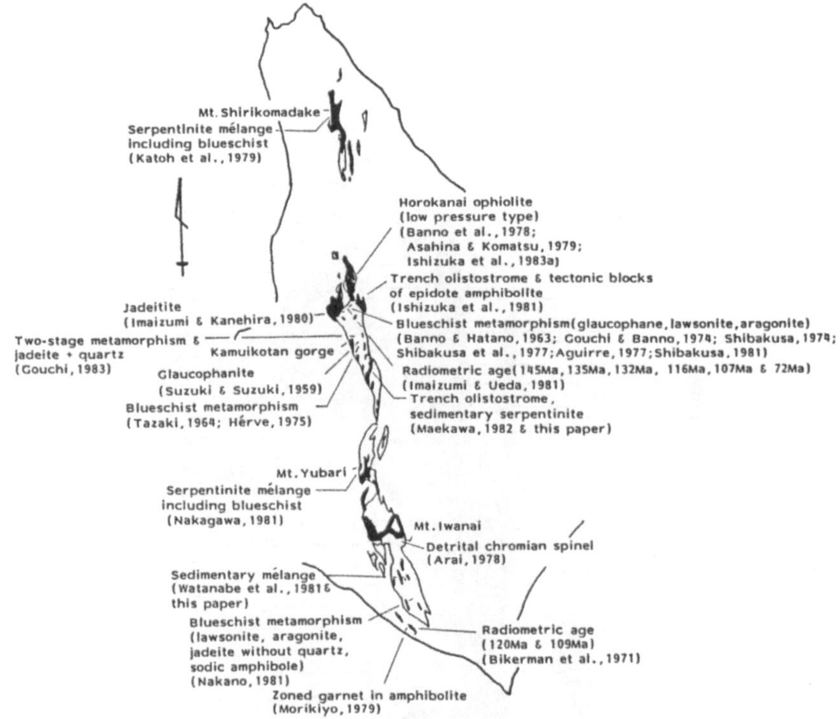

Fig. 2.

ding sedimentary units, that is, the Pankehoronai Formation, the Oichanunpe Formation and the Kitazawa Formation (Fig. 3; MAEKAWA, 1983). These formations dip toward the east at 30°–40°. They are interred to form a simple homoclinal structure (Fig. 4). A serpentinite-rich zone lies between the Pankehoronai and Oichanunpe Formations; both of the margins of the serpentinite-rich zone appear to be concordant with the adjacent formations.

Original lithofacies of the Pankehoronai Formation are mainly pelitic rocks, with subordinate amounts of chert, psammitic rocks and mafic rocks. These rocks have weak schistosity and well-developed mesoscopic and microscopic folds. Serpentinite blocks of varying sizes are also scattered in the formation.

In the serpentinite-rich zone, blueschists, amphibolites, pelitic rocks, and weakly metamorphosed mafic rocks are often found. Blue-

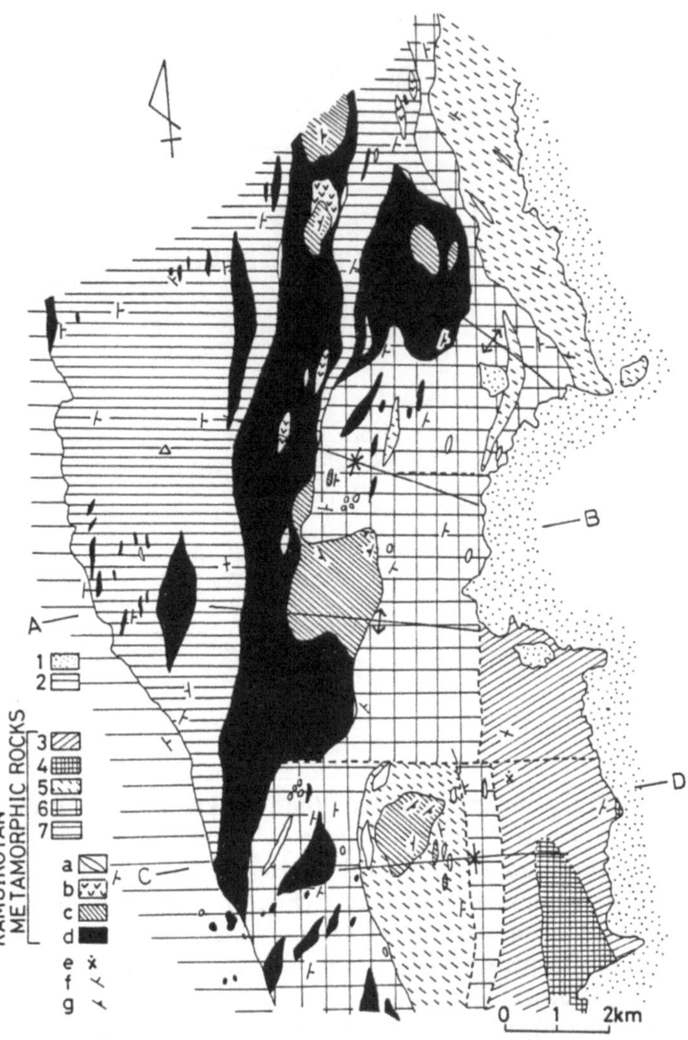

Fig. 3. Geologic map of the Biei area. 1: Tokachi welded tuff (Quaternary), 2: Kawabata Formation (Miocene), 3: basalt, 4: gabbro, 5: pelitic-psammitic rocks, 6: mafic sedimentary rocks, 7: pelitic rocks, a: chert, b: blueschist, c: amphibolite, d: ultramafic rocks, e: fossil locality, f: bedding, g: foliation. 3 and 4 represent main rock types of the Kitazawa Formation, 5 and 6 those of the Oichanunpe Formation, and 7 those of the Pankehoronai Formation.

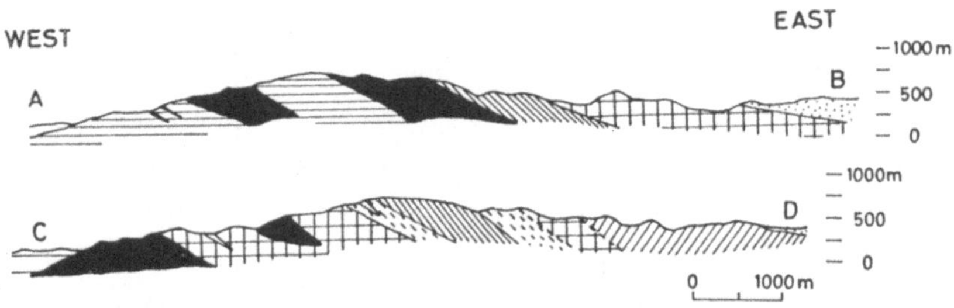

Fig. 4. Cross sections of the Biei area. The legends are the same as those in Fig. 3.

schists are characterized by the common occurrence of sodic amphibole (glaucophane-crossite), lawsonite, pumpellyite and sodic phyroxene as metamorphic minerals. Amphibolites have well-developed schistosity and consist of green to pale-brown hornblende, saussuritized or albitized plagioclase, epidote, chlorite, rutile and sphene. Green to pale-brown hornblendes are partly replaced by magnesioriebeckite. Hornblende of the amphibolite in the serpentinite-rich zone has slightly lower calcium content than that in the adjoining Oichanunpe Formation (MAEKAWA, in press).

Original lithofacies of the Oichanunpe Formation are fine- to coarse-grained basalt, mafic sedimentary rocks, psammitic rocks, pelitic rocks, chert, ultramafic rocks and amphibolite. Original rock types can be easily inferred in most cases because of the weak recrystallization and deformation. Mafic sedimentary rocks with intercalated pelitic rocks are the predominant rock type in the formation. They frequently contain blocks of fine-to coarse-grained basalt and picrite from a few centimeters to more than a few meters across. Serpentinites and amphibolites in the formation occur as blocks of varying sizes in the mafic sedimentary and pelitic-psammitic rocks (Fig. 3). Detrital chromites are frequently found in pelitic rocks near the serpentinites. Similar occurrence of chromites was reported near the Mt. Iwanai of the Yubaridake-Shizunai region (ARAI, 1978).

The Kitazawa Formation is composed of pillow basalt and gabbro, with subordinate amounts of mafic sedimentary rocks and chert. Aragonite commonly occurs in basalt vesicles. Well-preserved radiolarian fossils are obtained from chert. The following two radiolaria species are identified.

Thanarla sp. cf. *T. conica*: Valanginian—Aptian

Parvicingula sp. cf. *P. boessii*: Upper Jurassic—Aptian

These fossils suggest that the Kitazawa Formation is Early Cretaceous in age.

The metamorphism in the Biei area is summarized in Fig. 5.

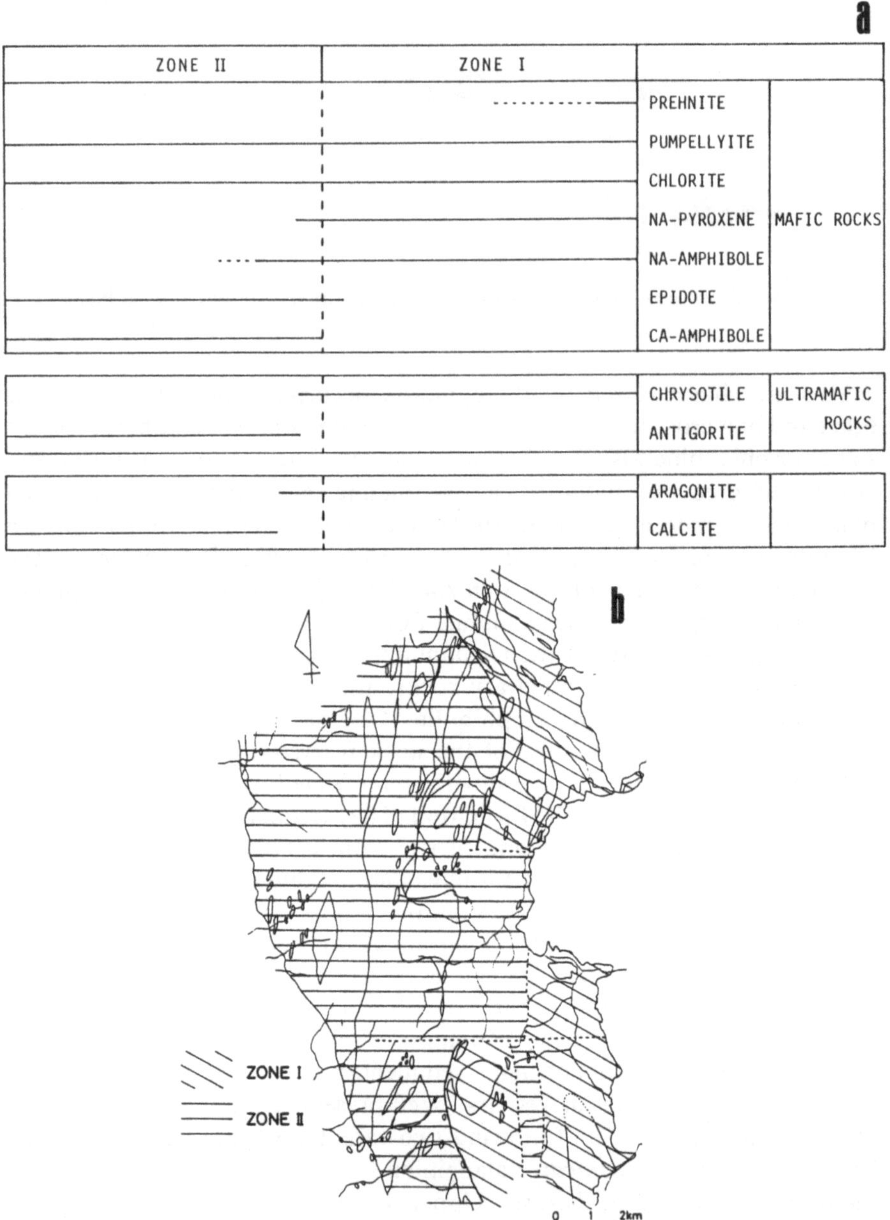

Fig. 5. Range of appearance of metamorphic minerals—(a) and map of metamorphic zone—(b). Pumpellyite in Zone I coexists with hematite.

Characteristic metamorphic minerals of zone I are pumpellyite, prehnite, aragonite, sodic pyroxene and sodic amphibole (magnesioriebeckite), and those of zone II are actinolite, epidote, and pumpellyite. Zone I is characterized by the pumpellyite + hematite assemblage, whereas zone II is characterized by the actinolite+epidote+chlorite assemblage.

Since the Lower Cretaceous Kitazawa sediments have suffered high P/T metamorphism, it is concluded that high P/T metamorphism which began in the Jurassic continued at least until Early Cretaceous.

Basalts in the Biei area can be divided into tholeiitic series and alkali rock series, as in the Niikappu area. Basalt chemistry and petrography is similar to those of modern oceanic islands (MAEKAWA, 1982).

Geology and petrography of the Biei area are given in detail as separate papers by the junior author (MAEKAWA, 1982, 1983).

3.2 Niikappu area

The Niikappu area is tectonically separated into many sedimentary units by faults and intrusions of serpentinite as shown in Fig. 6. The structural attitudes are highly variable, i.e., N-S trend in the most of the western part, E-W trend in the southeastern part, and NW-SE trend with a synform in the northeastern part.

In the western part, slates or muddy breccias occur in narrow zones among greenstones such as pillow basalt and hyaloclastite. The muddy matrix seems to separate each greenstone block. This unit predominant in greenstones almost corresponds to the Iwashimizu Formation by SUZUKI *et al.* (1961) and KOITABASHI et al. (1978).

A comparatively wide muddy sedimentary unit is distributed along the eastern part of the mapped area as shown in Fig. 6. The muddy sedimentary unit is equivalent to the Idonnappu Formation by SUZUKI *et al.* (1961) except for a horizon of overlying chert. The Idonnappu Formation contains many blocks of pillow basalt, massive basalt with or without columnar jointing, chert, and sandstone ranging from several hundred meters to a few centimeters in thickness. This sedimentary unit is considered to be an ancient submarine slump zone and apparently occupies the lowest part of the area as shown in the cross-section in Fig. 6.

However, study of radiolarian fossils reveals that the muddy sedimentary unit is not the lowest horizon. Radiolarian fossils are obtained from five localities from No. 1 to No. 5 (Fig. 6). Valanginian to Barremian radiolarian fossils are identified from locality 1 (siliceous green shale) and locality 5 (red chert) (Table 1). Valanginian radiolarian fossils are also found from localities 2 and 3. No radiolarian fossils have been found from the marix of the muddy sedimentary unit. Nevertheless, as siliceous

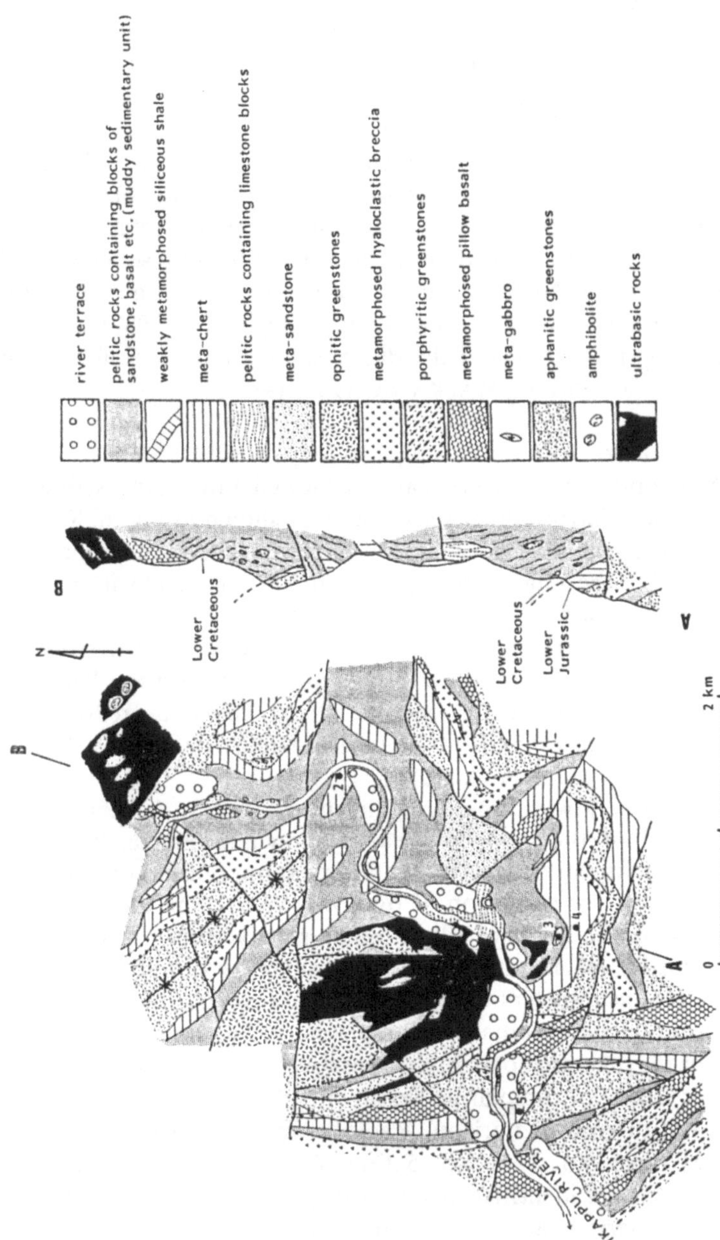

Fig. 6. Geologic map of the Niikappu area (middle stream area of Niikappu River). Mostly based on map by KOITABASHI *et al.* (1978) and partly modified by data of WATANABE *et al.* (1981).

Table 1. Radiolarian fossils found in the Niikappu area (localities 1 and 5).

Acanthocircus dicranacanthos	Tithonian - late Barremian
Archaeodictyomitra apiara	Berriasian - late Barremian
Alievium helenae	Valanginian - late Barremian
Sethocapsa uterculus	Valanginian - late Barremian
Parvicingula boesii	Kimmerigian - early Aptian
Sphaerostylus lanceola	- early Aptian
Xitus spicularis	- Cenomanian
Acanthocircus carinatus	(late Barremian)-early Aptian
Parvicingula hsui	early Tithonian
Archaeodictyomitra vulgaris	Valanginian - Aptian
Thanarla conica	late Valanginian - Aptian

Identification by Yao (1981)

green shale of locality 1 seems to be almost identical in horizon with matrix shale, the age of the muddy sedimentary unit is considered to be Valanginian-Barremian. While, from the red chert of locality 4, Parahsuum assemblages (YAO *et al.*, 1982) showing Lower Jurassic age are found; the Early Jurassic red chert overlies with a thrust fault the muddy sedimentary unit of Early Cretaceous (Fig. 6). A compressive stress field is interred to have formed the structure.

The stratigraphic sequence in the western part of the Niikappu area (area of the Iwashimizu Formation) is not yet established due to its complicate structure and poor fossil data.

Metamorphism of this area is chiefly of pumpellyite-chlorite grade. In the western part, actinolite-epidote-chlorite assemblages are recognized in metamorphosed hyaloclastite. Sodic pyroxene replacing relict clinopyroxene is rarely found. Aragonite-bearing greenschist occurs within the central serpentinite block. Amphibolite in northeastern serpentinite zone suffered prehnite-grade metamorphism of the later stage. The metamorphic grade continuously increases southward to reach the epidote-glaucophane schist facies in the Shizunai area (NAKANO, 1981), where metamorphic aragonite also are reported (NAKANO, 1981; NAKANO and WATANABE, 1976).

Representative bulk chemical compositions of pillow basalts and massive greenstones in the Niikappu area are listed in Table 2. Basic rocks have high TiO_2 contents and one of them seems to be tholeiitic; others are similar to alkali basalt in oceanic regions. The occurrence of rocks having higher silica contents (SiO_2=63.35%) is also characteristic in this area and the chemical composition is somewhat similar to the rocks of oceanic islands such as tahitiite in Tahiti Island of southern Pacific. Alkali

Table 2. Chemical composition of greenstones in the Niikappu area (anhydrous). Analyst, 1: J. Koitabashi, 2–5: T. Watanabe.

sample number	1 K-2	2 82401	3 82504	4 82506	5 82407
SiO_2	48.31	50.74	63.35	49.51	46.49
TiO_2	2.67	2.85	0.32	3.28	4.38
Al_2O_3	16.94	16.28	18.02	16.15	15.89
Fe_2O_3	12.16	11.33	6.45	10.09	13.11
MnO	0.04	0.07	0.12	0.64	0.16
MgO	5.41	1.42	1.31	5.27	4.60
CaO	11.87	8.16	0.58	6.97	10.89
Na_2O	2.27	6.75	5.97	3.31	2.90
K_2O	0.19	1.99	3.58	3.75	0.67
P_2O_5	0.27	0.82	0.27	0.95	0.59
Total	100.13	100.41	99.97	99.93	99.68

diabase is also reported in the Shizunai area (NAKANO and KOMATSU, 1979).

Detailed geology in the Niikappu area will be given by Watanabe and his colleagues in a separate paper.

4. *Early Cretaceous Rocks of the Other Parts in the Kamuikotan Belt and Surrounding Terranes*

The Kamuikotan rocks, especially those distributed in the southern half, are bounded in both the east and west sides by the Cretaceous Yezo group in contact with the former mostly by faults. Recent progress of study on radiolarian biostratigraphy (summary in Table 3) reveals the following relations. In and around Valanginian age, we recognize an active subduction system in the Biei and Niikappu areas. On the other hand, in the Furano area, Valanginian sediments of terrigenous origin overlie upper Jurassic pillow basalt (Sorachi group) without any tectonic gap (KITO, 1982; MINOURA *et al.*, 1982; JOLIVET *et al.,* 1983). They show a nicely continuous sequence from the Jurassic Sorachi greenstones to the Yezo Group of forearc basin sediments. Valanginian radiolarian fossils are also found in other parts of the Furano area (NAKASEKO, 1979; OKADA *et al.*, 1982). In the Horokanai area, moreover, an obducted ophiolite of which the uppermost part is Upper Jurassic radiolarian chert (ISHIZUKA *et*

Table 3. Summary of studies of radiolarian fossils in the Kamuikotan tectonic belt and the Hidaka western belt. This table shows relative position of the studied areas.

	Kamuikotan tectonic belt	Hidaka western belt
north (west ← → east)		
	Horokanai age:Tithonian -chert of the uppermost part of the ophiolite- (Ishizuka et al.,1983b)	Kamikawa age:early Early Cretaceous -shale matrix including blocks of Triassic limestone- (Ishizuka et al.,1984)
	Biei age:early Cretaceous -chert in the Kitazawa Formation- (Maekawa,1982 & this paper)	
	Furano age:Valanginian -siliceous shale covering pillow basalt,which intercalate Tithonian chert (Kito,1982; Minoura et al., 1982)	Chiroro river area age:Valanginian -shale matrix covering Triassic greenstones (Kontani et al.,1983)
	Niikappu age:Valanginian - Barremian -siliceous shale and bedded chert- (Watanabe,1981 & this paper)	
		Nitarachi formation age:Valanginian -siliceous sediments including acidic tuff- (Kanie et al.,1981)
south		

al., 1983b) tectonically covers the high P/T Kamuikotan rocks (ISHIZUKA *et al.*, 1983a), whereas the Lower Yezo Group started in Aptian time (early Cretaceous) overlies the Sorachi greenstones with para-unconformity in western and eastern sides (IGI *et al.*, 1958). It is, accordingly, concluded that in the Furano and Horokanai areas, forearc basin sediments cover the Jurassic greenstones of oceanic origin without a disharmonious contact. This is basically consistent with the mostly conformable relation between the Sorachi Group and Lower Yezo Group mentioned by TANAKA and TERAOKA (1970).

In short, in early Cretaceous (mainly Valanginian), no features of a subduction process such as high P/T metamorphic rocks and/or trench olistostrome can be recognized in the Furano and Horokanai area of the Kamuikotan belt, whereas subduction processes can be assumed to be active in the Biei and Niikappu areas and, as mentioned below, some parts of the Hidaka western zone.

Valanginian sediments are not limited to the southern Kamuikotan belt. They also occur to the east of the Kamuikotan rocks, Hidaka western belt. According to data by KIMINAMI *et al.* (1983), Valanginian turbidite sediments containing arc-type volcanic products cover Triassic greenstones of the Hidaka western belt in the Chiroro River area. At Urakawa, the similar volcanogenic Valanginian Nitarachi Formation is in fault contact with these some greenstones.

In the eastern side of the Triassic greenstones along the Chiroro River, KONTANI *et al.* (1982) reported the Valanginian radiolarian fossils from slate associated with sandstone, chert and greenstones. As the area is mélange unit (DICKINSON, 1978) which suffered metamorphism of pumpellyite-actinolite grade or epidote-actinolite grade (SAWADA and KANMERA, 1973; ISHIZUKA, 1981), a subduction complex of Valanginian age may have also developed on the eastern side of the Triassic greenstones. Similar Cretaceous subduction is recently assumed by KIMINAMI and KONTANI (1983a). ISHIZUKA *et al.* (1984) present consistent data for the Early Cretaceous subduction mentioned above. They found Valanginian fossils in mudstones which contain olistoliths of the Triassic greenstones in the Kamikawa area of the Hidaka western belt, northern extension of the Chiroro area.

5. *Tectonic Development of the Kamuikotan Belt and Related Terranes*

On the basis of the descriptions on the Kamuikotan belt and the surrounding terranes stated above, we speculate that during the Valanginian the Kamuikotan westward-dipping subduction slowed or ceased locally, especially in the Furano and Horokanai areas, due to collision of seamounts or oceanic ridges. Forearc basin sedimentation expanded eastward. The subduction process, however, continued in the Biei and Niikappu areas. In these areas, fragments of seamounts were supplied to the trench to form trench olistostromes (Fig. 7).

Radiometric ages of the high-grade Kamuikotan metamorphic rocks (i.e. 145–109 Ma) are consistent with the age of this subduction process. We suggest that serpentinite protrusion occurred at this time, supply-

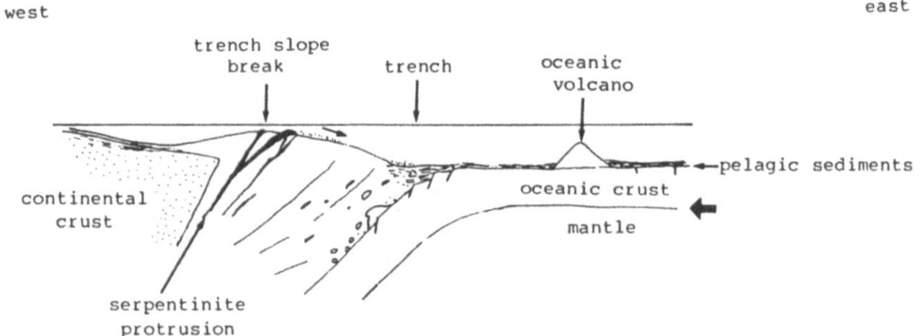

Fig. 7. Tectonic model of the Kamuikotan metamorphic rocks in the Biei area in early Cretaceous.

ing serpentinites, amphibolites and blueschists to the trench. A similar recycling process has been described for the Franciscan complex (COWAN and PAGE, 1975). Some of these blocks suffered mid to late Cretaceous high P/T metamorphism, as demonstrated in the Biei area (MAEKAWA, 1983 and in press). Thus the Kamuikotan belt was not simply a tectonic high in early Cretaceous as previously assumed by many researchers, but was in part still active as a subduction zone.

Genesis of serpentinite protrusion in the Kamuikotan belt is not yet clear, but the exchange of tectonic field from transform fault zone to subduction zone probably occurred in Jurassic (MARUYAMA *et al.*, 1982; TAKAHASHI, 1983; KIMURA, 1983) and related lateral slip movement with possible rotation of plate may urge the protrusion process.

Another point which should be considered is the probable Valaginian sedimentation to the east of the Triassic greenstones in the Hidaka western zone. On this basis, subduction system in early Cretaceous is rather complicated and the framework draw in Fig. 8 is tentatively assumed. The complicate subduction system may relate initiation of subduction along transform fault zone and collision of seamounts or ridges. A speculation for the initiation of the subduction zone is shown in Fig. 9.

A similar dual trench system is recognized in the present-day Hellenic arc-trench system, eastern Mediterranean, which is situated in a collision tectonic field associated with rotation (HUCHON *et al.*, 1982). Change of plate movement with rotation may essentially relate to genesis of dual subduction system.

Particles of greenschists were described in sandstones of the basal part of the Lower Yezo Group (FUJII, 1958). In addition detrital sodic amphiboles are very rarely found in these rocks (MATSUMOTO and OKADA,

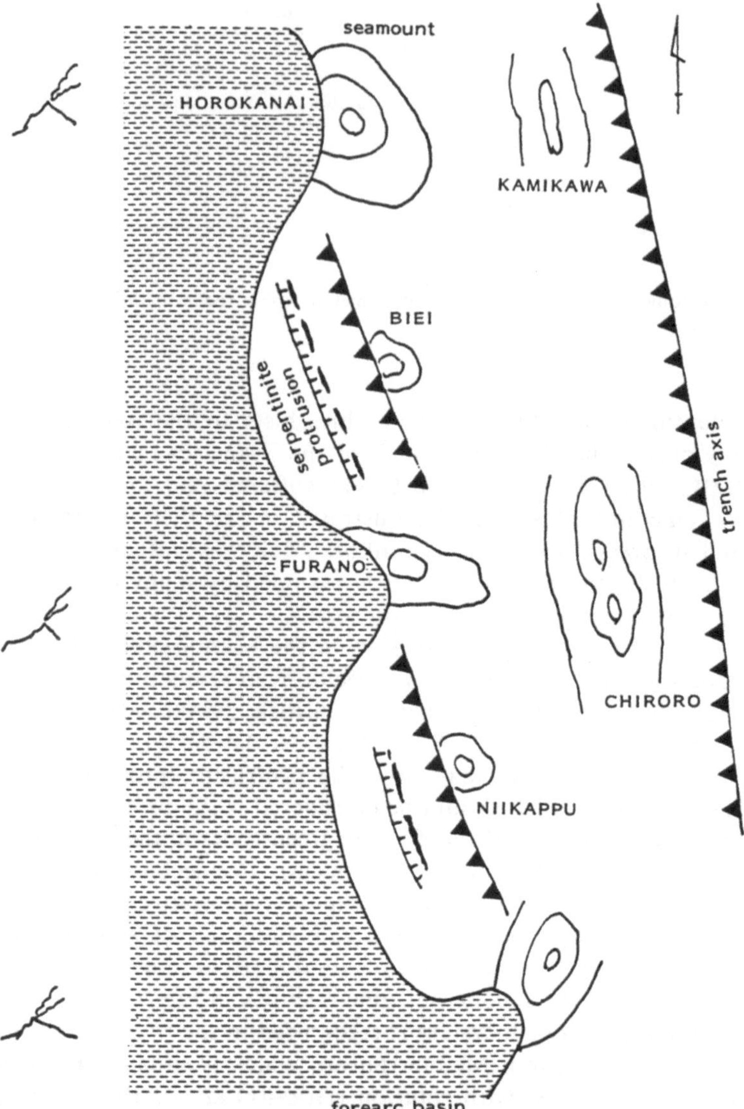

Fig. 8. Plane view of a tectonic model of the Kamuikotan and adjacent terranes in early Early Cretaceous.

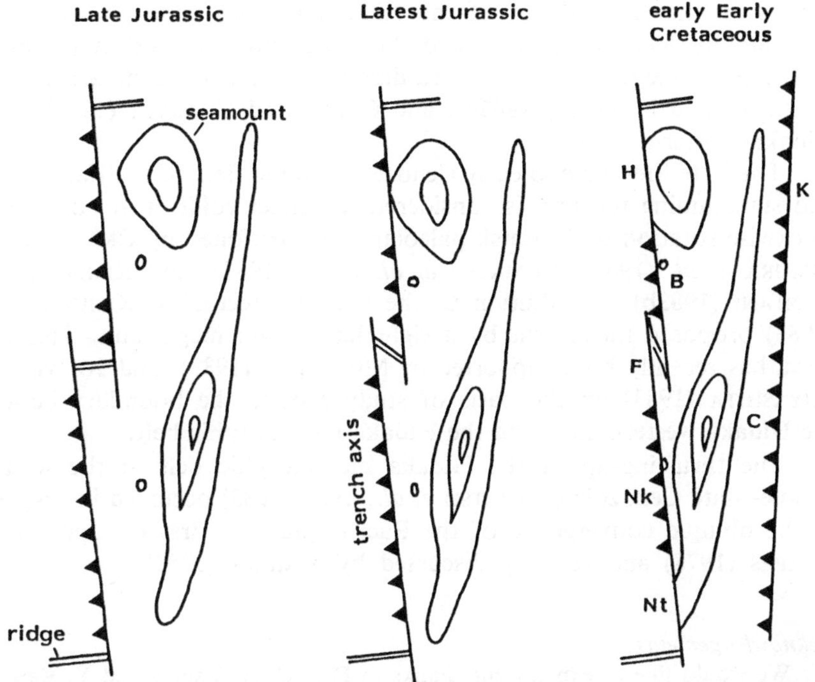

Fig. 9. An interpretation of initiation of the dual subduction in and around the Kamuikotan belt. H: Horokanai, K: Kamikawa, B: Biei, F: Furano, C: Chiroro, Nk: Niikappu, Nt: distribution area of Nitarachi formation.

1971). Therefore, fragments of the Kamuikotan belt had been supplied to the Yezo forearc basin probably since Aptian time.

The Middle Yezo Group partly overlies greenstones of the Sorachi Group with unconformity (KIMURA *et al.*, 1975). Detrial grains of metamorphic minerals and chromites probably supplied from the Kamuikotan rocks are frequently found in the Middle Yezo Group (IIJIMA, 1961; MATSUMOTO and OKADA, 1971). The uplift of the Kamuikotan belt to supply sediment to the forearc basin was activated during depostion of the Middle Yezo Group, Albian-Turonian.

In the Upper Cretaceous time, forearc basin sediments covered the entire Kamuikotan zone and forearc basin was extended to the east. As radiolarian fossils of Upper Cretaceous age are found by IWATA *et al.* (1983) in the northeastern part of the Hidaka western zone, it is speculated that the westward-dipping subduction zone shifted toward the east and reformed in the eastern part of the Hidaka western zone. Judging from the data by KIMURA (1983), and KIMURA and MIYASHITA (in prep), it is hard to recognize an eastward-dipping subduction zone assumed by KIMINAMI and KONTANI (1983b), and CADET and CHARVET (1983) in the Hidaka western zone.

The Late Mesozoic to early Cenozoic tectonic field of the Kamuikotan and surrounding terranes is considered to reflect collision tectonics with clockwise rotation of Okhotsk paleoland as mentioned by OKADA (1983), KIMURA *et al.* (1983), KOMATSU *et al.* (1982, 1983), and KIMINAMI and KONTANI (1983b). In addition to the collision tectonics, KIMURA *et al.* (1983) proposed movement by a right lateral slip mega-fault. This proposal has recently been supported by MIYASHITA (1983), and JOLIVET and MIYASHITA (1983) on the basis of study around the boundary between the Hidaka western zone and the Hidaka metamorphic belt.

The thrusting-up of the Hidaka metamorphic belt to the west in middle—late Cenozoic (KOMATSU *et al.*, 1982, 1983) occurred in responce to the oblique convergence of the Pacific plate as first pointed out by GRAPES (1978) and recently discussed by KIMURA (1983).

Acknowledgements

We would like to express our thanks to Drs. C. A. Landis and Y. Kawachi of Otago University for their critical reading of the manuscript of early stage. We sincerely thank Dr. G. Kimura of Kagawa University and Dr. K. Kiminami of Yamaguchi University for valuable discussions. We also express our thanks to Dr. A. Yao of Osaka City University for his kind identification of radiolarian fossils. Thanks of senior author (Watanabe) are due to his colleagues, Mr. J. Koitabashi, Dr. K. Niida, Mr. S. Sano and Mr. H. Yano with whom Watanabe performed the study of the Niikappu area, and other members of Hokkaido Tectonics Research Group for their encouragements.

REFERENCES

AGUIRRE, L., Petrology of the Kamuikotan metamorphic belt at the Kamietanbetsu-Numaushi cross section, central Hokkaido, Japan, in *Comparative Studies on the Geology of the Circum-Pacific Orogenic Belt in Japan and Chile, 1st Report*, edited by T. Ishikawa and L. Aguirre, pp. 125–149, Japan Soc. Promotion Sci., Tokyo, 1977.

ARAI, S., Detrital chromian spinel in a Kamuikotan metamorphic rock, near Iwanai-dake, Hokkaido, *J. Geol. Soc. Japan*, **84**, 481–484, 1978 (in Japanese).

ASAHINA, T. and M. KOMATSU, The Horokanai ophiolitic complex in the Kamuikotan tectonic belt, Hokkaido, Japan. *J. Geol. Sci. Japan*, **85**, 317–330, 1979.

BANNO, S. and M. HATANO, Zonation of metamorphic rocks in the Horokanai area of the Kamuikotan metamorphic belt in Hokkaido, *J. Geol. Soc. Japan*, **69**, 388–393, 1963.

BANNO, S., H. ISHIZUKA, N. GOUCHI, and M. IMAIZUMI, Kamuikotan belt in Hokkaido: The tectonic contact of high-pressure metamorphic belt and low-pressure ophiolite succession, Abst. Inter. Geol. Conf., Tokyo 1978, pp. 14–15, 1978.

BIKERMAN, M., M. MINATO, and M. HUNAHASHI, K-Ar age of the garnet amphibolite of the Mitsuishi district, Hidaka province, Hokkaido, Japan, *Earth Science (Chikyu Kagaku)*, **25**, 27–30, 1971.

CADET, J. P. and J. CHARVET, From subduction to paleosubduction in northern Japan, in *Accretion Tectonics in the Circum-Pacific regions*, AEPS, edited by M. Hashimoto and S. Uyeda, pp. 135–148, Terra Scientific Publishing Company, Tokyo, 1983.

COWAN, D. S. and B. M. PAGE, Recycled Franciscan material in Franciscan mélange west of Paso Robles, California, *Geol. Soc. Am. Bull.*, **89**, 1089–1095, 1975.

DICKINSON, W. R., Plate tectonic evolution of North Pacific rim, *J. Phys. Earth*, 26 (Suppl.), 1–19, 1978.

FUJII, K., Petrography of the Cretaceous sandstone of Hokkaido, Japan, *Mem. Fac. Sci., Kyushu Univ., Ser. D*, **6**, pp. 129–152, 1958.

GOUCHI, N. and S. BANNO, Find of aragonite from Kamuikotan metamorphic rocks, *Proc. Japan Acad.*, **50** 481–486, 1974.

GOUCHI, N., Kamuikotan metamorphic rocks in the Kamuikotan gorge area, west of Asahikawa, Hokkaido, *J. Japan. Assoc. Mineral. Petrol. Econ. Geol.*, **78** 383–393, 1983 (in Japanese with English abstract).

GRAPES, R. H., Mesozoic-Cenozoic arc-trench development, Neogene orogeny and the "Hidaka belt" anomaly in Hokkaido, IGCP 7th Circum-Pacific Plutonism Meeting, Toyama, pp. 282–296, 1978.

HERVÉ, F., Petrography of the Kamuikotan metamorphic belt at the Ubun-Oroen cross section, central Hokkaido, Japan, *J. Fac. Sci., Hokkaido Univ., Ser. IV*, **16**, 453–470, 1975.

HUCHON, Ph., N. LYBERIS, J. AANGELIER, X. LE PICHON, and V. RENARD, Tectonics of the Hellenic trench: a synthesis of Sea-Beam and submersible observations, *Tectonophysics*, **86**, 69–112, 1982.

HUNAHASHI, M., Kamuikotan tectonic belt, in *Jubilee publication* in the commemoration of Professor Jun Suzuki, M. J. A. sixtieth birthday, pp. 37–52, 1956 (in Japanese with English abstract).

IIJIMA, A., Preliminary note on the period of metamorphism and upheaval of the Kamuikotan tectonic belt, Hokkaido, *J. Geol. Soc. Japan*, **67**, 417, 1961 (in Japanese).

IGI, S., K. TANAKA, H. SATO, and M. HATA, Geologic map of Japan, 1 : 50,000 "Horokanai" sheet and its explanatory text, *Geol. Surv. Japan*, 1958 (in Japanese with English abstract).

IMAIZUMI, M. and K. KANEHIRA, Jadeitite from the Kamuikotan tectonic belt, *J. Geol. Soc. Japan*, **86**, 629–633, 1980.

IMAIZUMI, M. and Y. UEDA, On the K-Ar ages of the rocks of two kinds existed in the Kamuikotan metamorphic rocks located in the Horokanai district, Hokkaido, *J. Japan. Assoc. Mineral. Petrol. Econ. Geol.*, **76**, 88–92, 1981 (in Japanese with English abstract).

ISHIZUKA, H., Greenstones from the Idonnappu Formation along the River Oku-Niikappu in the axial zone of Hokkaido, Japan, *Mem. Fac. Sci., Kochi Univ., Ser. E, Geol.*, **2**, 1–22, 1981.

ISHIZUKA, H., M. IMAIZUMI, and N. GOUCHI, The Kamuikotan tectonic belt in Hokkaido, Japan, was probably formed as a strike-slip mobile zone associated with oblique subduction and obduction processes during Mesozoic time (abstract), Oji Inter. Seminar,

Tomakomai, 29, 1981.

ISHIZUKA, H., M. IMAIZUMI, N. GOUCHI, and S. BANNO, The Kamuikotan zone in Hokkaido, Japan: Tectonic mixing of high-pressure and low-pressure metamorphic rocks, *J. Metamorphic Geol.*, **1**, 263–275, 1983a.

ISHIZUKA, H., M. OKAMURA, and Y. SAITO, Latest Jurassic radiolarians from the Horokanai ophiolite in the Kamuikotan zone, Hokkaido, Japan, *J. Geol. Soc. Japan*, **89**, 731–732, 1983b.

ISHIZUKA, H., M. OKAMURA, and Y. SAITO, Early Early Cretaceous radiolarians from the Sorachi Group, *J. Geol. Soc. Japan*, **90**, 59–60, 1984.

IWATA, K., S. UOZUMI, K. NAKAMURA, and J. TAJIKA, Discovery of radiolarians and holothurian sclerites from the Pre-Tertiary system around Nishiokoppe, northeast Hokkaido (preliminary report), *J. Geol. Sci. Japan*, **89**, 55–56, 1983 (in Japanese with English abstract).

JOLIVET, L. and S. MIYASHITA, The structure of the Hidaka axial zone interpreted as the result of a right-lateral strike-slip movement (abstract), 90th Ann. Meeting, Geol. Soc. Japan, Kagoshima, 495 pp., 1983 (in Japanese).

JOLIVET, L., M. NAKAGAWA, and N. KITO, Uppermost Jurassic unconformity in Hokkaido, evidence for an early tectonic stage, *Proc. Japan Acad., Ser. B*, **59**, 153–157.

KANIE, Y., Y. TAKETANI, A. SAKAI, and Y. MIYATA, Lower Cretaceous deposits beneath the Yezo Group in the Urakawa area, Hokkaido, *J. Geol. Soc. Japan*, **87**, 527–533, 1981 (in Japanese with English abstract).

KATOH, T., Formation of the serpentinite mélanges in the Kamuikotan tectonic belt (abstract), Symp., 89th Ann. Geol. Soc. Japan, Niigata, pp. 21–24, 1982 (in Japanese).

KATOH, T., K. NIIDA, and T. WATANABE, Serpentinite mélange around Mt. Shirikomadake in the Kamuikotan Structural Belt, Hokkaido, *J. Geol. Soc. Japan*, **85**, 279–285, 1979 (in Japanese with English abstract).

KIMINAMI, K. and Y. KONTANI, Subduction process of the western part of the axial zone, Hokkaido, *Lett. Hokkaido Tectonics Research Group (Dept. Geol. Miner., Hokkaido Univ.)*, **13**, 3–4, 1983a (in Japanese).

KIMINAMI, K. and Y. KONTANI, Mesozoic arc-trench systems in Hokkaido, Japan, in *Accretion Tectonics in the Circum-Pacific regions*, AEPS, edited by M. Hashimoto and S. Uyeda, pp. 107–122, Terra Scientific Publishing Company, Tokyo, 1983b.

KIMINAMI, K., Y. KONTANI, and S. MIYASHITA, Relationship between Lower Cretaceous system and the Hidaka western marginal greenstones in the Chiroro River area (abstract), 90th Ann. Meeting Geol. Soc. Japan, Kagoshima, 400 pp., 1983 (in Japanese).

KIMURA, G., Collision tectonics in Hokkaido, *Marine Sci. Mon.*, **162**, 724–730, 1983 (in Japanese).

KIMURA, G., S. MIYASHITA, and S. MIYASAKA, Collision Tectonics in Hokkaido and Sakhalin, in *Accretion Tectonics in the Circum-Pacific regions*, AEPS, edited by M. Hashimoto and S. Uyeda, pp. 123–134, Terra Scientific Publishing Company, Tokyo, 1983.

KIMURA, T., Structural Development of Japan and plate-tectonics, *J. Geogr.*, **86**, 54–67, 1977 (in Japanese with English abstract).

KIMURA, T., S. YOSHIDA, and F. TOYOHARA, Unconformity between the Yezo Group and the Sorachi Group and its significance, GDP Proc., II-I-(1), *Structural Geol.*, No. 3, 29–38, 1975.

KITO, N., Geologic age of the Sorachi Group in the Furano region, Hokkaido (abstract), 89th Ann. Meeting Geol. Soc. Japan, Niigata, 195 pp., 1982 (in Japanese).

KOITABASHI, J., H. ENDO, and H. KATAGAWA, Greenstones in the middle stream area of the River Niikappu and the upper stream area of the River Ribira, southern Kamuikotan belt, *Earth Science (Chikyu Kagaku)*, **32**, 299–300, 1978 (in Japanese).

KOMATSU, M., S. MIYASHITA, J. MAEDA, Y. OSANAI, T. TOYOSHIMA, Y. MOTOYOSHI, and K. ARITA, Petrological constitution of the continental type crust up thrusted in the Hidaka Belt, Hokkaido, *Spec. Paper Japan. Assoc. Mineral. Petrol. Econ. Geol.*, **3**, 229–238, 1982 (in Japanese with English abstract).

KOMATSU, M., S. MIYASHITA, J. MAEDA, Y. OSANAI, and T. TOYOSHIMA, Disclosing of a deepest section of continental-type crust up-thrust as the final event of collision of arcs in Hokkaido, north Japan, in *Accretion Tectonics in the Circum-Pacific regions*, AEPS, edited by M. Hashimoto and S. Uyeda, pp. 149–165, Terra Scientific Publishing Company, Tokyo, 1983.

KONTANI, T., K. KIMINAMI, and S. MIYASHITA, Discovery and its significance of radiolarian fossils of lower Cretaceous from the Idonnappu Formation in the Hidaka Frontal Folded Belt (abstract), Symp. 89th Ann. Meeting Geol. Soc. Japan, Niigata, pp. 29–31, 1982 (in Japanese).

MAEKAWA, H., Geology of the Biei-Ashibetsu area in the Kamuikotan belt (abstract), 88th Ann. Meeting, Geol. Soc. Japan, Tokyo, 330 pp., 1981 (in Japanese).

MAEKAWA, H., Preliminary report on the metamorphism of the Biei area in the Kamuikotan belt (abstract), 89th Ann. Meeting Geol. Soc. Japan, Niigata, 428 pp., 1982 (in Japanese).

MAEKAWA, H., Submarine sliding deposits and their modes of occurrence of the Kamuikotan metamorphic rocks in the Biei area, Hokkaido, Japan, *J. Fac. Sci., Univ. Tokyo, Sec. II*, **20**, 489–507, 1983.

MAEKAWA, H., A low P/T metamorphic episode in the Biei area, Kamuikotan blueschist terrane, Japan, *Geol. Soc. Am. Memoir.*, **164** (in press).

MARUYAMA, S., T. SEO, and D. ENGEBRETSON, Plate motion history and orogenic movements around the Japanese Islands (abstract), Symp., 89th Ann. Meeting Geol. Soc. Japan, Niigata, pp. 71–76, 1982 (in Japanese).

MATSUMOTO, T. and H. OKADA, Clastic sediments of the Cretaceous Yezo geosyncline, *Mem. Geol. Soc. Japan*, No. 6, 61–74, 1971.

MINOURA, N., S. KUMANO, N. KITO, K. KAMADA, and M. KATO, Lower Cretaceous deposits at Nunobe, central Hokkaido, *Earth Science (Chikyu Kagaku)*, **36**, 348–350, 1982.

MIYASHITA, S., Regional thermal metamorphism related to the right-lateral slip movement in the ophiolite of the Hidaka western zone (abstract), 90th Ann. Meeting Geol. Soc. Japan, Kagoshima, 400 pp., 1983 (in Japanese).

MORIKIYO, T., Zonal structure of garnet from the metamorphic rocks in the Mitsuishi district, Hokkaido, *J. Japan. Assoc. Geol. Mineral. Perol. Econ. Geol.*, **74**, 27–35, 1979.

NAKAGAWA, M., The constituents of serpentinite mélange and chemistry of metabasalt in the Kamuikotan tectonic belt, in *Tectonics of Paired Metamorphic Belts*, edited by I. Hara, pp. 3–6, Symp. Hiroshima, Tanishi Print Kikaku, Hiroshima, 1981.

NAKANO, N., Metamorphism of the greenstones in the Kamuikotan zone and the Hidaka western marginal tectonic zone in the Shizunai-Mitsuishi district, Hokkaido, *J. Geol. Soc. Japan*, **87**, 211–224, 1981 (in Japanese with English abstract).

NAKANO, N. and T. WATANABE, The distribution of metamorphic aragonite in the southern part of the Sorachi Group, Hokkaido (abstract), *J. Japan. Assoc. Mineral. Petrol. Econ. Geol.*, **71**, 82, 1976.

NAKANO, N. and M. KOMATSU, Kaersutite-aegrine alkali diabase in the Shizunai-Mitsuishi area in the Kamuikotan greenrock zone, Hokkaido, *J. Geol. Soc. Japan*, **85**, 367–376,

1979 (in Japanese with English abstract).

NAKASEKO, K., On the international correlation by means of radiolarians from the Cretaceous formation in Japan, Fossils, *Palaeon. Soc. Japan*, **29**, 27–35, 1979 (in Japanese with English abstract).

OKADA, H., Migration of ancient arc-trench systems, in *Modern and Ancient Geosynclinal Sedimentation*, Soc. Econ. Paleont. Mineral. Spec. Publ., No. 19, edited by Dott and Shaver, pp. 311–320, 1974.

OKADA, H., Collision orogenesis and sedimentation in Hokkaido, Japan, in *Accretion Tectonics in the Circum-Pacific regions*, AEPS, edited by M. Hashimoto and S. Uyeda, pp. 91–105, Terra Scientific Publishing Company, Tokyo, 1983.

OKADA, H., A. HATAKEYAMA, and K. NAKASEKO, Age of the Sorachi Group in its type area in Hokkaido, Proceedings of the first Japanese Radiolarian Symposium, News of Osaka Micropaleontologists Special Vol. 5, pp. 353–357, 1982 (in Japanese with English abstract).

SAWADA, K. and K. KANMERA, Greenstones from the Sorachi and Hidaka Groups of the Hidaka Mountains, Hokkaido, *Mem. National Sci. Mus.*, **6**, 147–161, 1973.

SHIBAKUSA, H., Glaucophane schists in the Kamuikotan metamorphic belt of the Horokanai area, central Hokkaido, *J. Geol. Soc. Japan*, **80**, 341–353, 1974.

SHIBAKUSA, H., Metamorphism of the Kamuikotan tectonic belt, in *Tectonics of Paired Metamorphic Belts*, edited by I. Hara, pp. 43–47, Symp. Hiroshima, Tanishi Print Kikaku, Horoshima, 1981.

SHIBAKUSA, H., N. GOUCHI, and M. IMAIZUMI, Widespread occurrence of metamorphic aragonite in the Kamuikotan metamorphic rocks near Asahikawa, Hokkaido, *J. Geol. Soc. Japan*, **83**, 301–303, 1977.

SUPPE, J., Interrelationships of high-pressure metamorphism, deformation and sedimentation in Franciscan tectonics, U.S.A., 24th IGC, Section 3, pp. 552–559, 1972.

SUZUKI, J. and Y. SUZUKI, Petrological study of the Kamuikotan metamorphic complex in Hokkaido, Japan, *J. Fac. Sci., Hokkaido Univ., Ser. IV*, **10**, 349–446, 1959.

SUZUKI, M., H. OSANAI, K. MATSUI, and J. WATANABE, Geologic map of Japan, 1 : 50,000 "Idonnappudake" sheet and its explanatory text, Hokkaido Development Agency, 1961 (in Japanese with English abstract).

TAKAHASHI, M., Space-time distribution of Late Mesozoic to Early Cenozoic magmatism in east Asia and its tectonic implications, in *Accretion Tectonics in the Circum-pacific regions*, AEPS, edited by M. Hashimoto and S. Uyeda, pp. 69–88, Terra Scientific Publishing Company, Tokyo, 1983.

TANAKA, K. and Y. TERAOKA, Cretaceous system, in *The Geologic Development of the Japanese Islands*, edited by K. Ichikawa, Y. Fujita, and M. Shimazu, pp. 87–105, 1970 (in Japanese).

TAZAKI, K., Alkali amphibole-bearing metamorphic rocks in the Kamuikotan belt of the south-western part of Asahikawa, central Hokkaido, *Earth Science (Chikyu Kagaku)*, **71**, 8–17, 1964 (in Japanese with English abstract).

WATANABE, T., Idonnappu Formation in the Kamuikotan Belt as constituent of Mesozoic subduction wedge, in *Tectonics of Paired Metamorphic Belts*, edited by I. Hara, pp. 37–42, Symp. Hiroshima, Tanishi Print Kikaku, Hiroshima, 1981.

WATANABE, T., Some problems on the formation of the Kamuikotan Belt (abstract), Symp. 89th Ann. Meeting Geol. Soc. Japan, Niigata, pp. 25–28, 1982 (in Japanese).

WATANABE, T., K. NIIDA, and J. KOITABASHI, "Mélange sediments" and oceanic alkali basalts in the southern part of the Kamuikotan Belt (abstract), 88th Ann. Meeting Geol. Soc. Japan, Tokyo, 333 pp., 1981 (in Japanese).

YAO, A., A. MATSUOKA, and T. NAKATANI, Triassic and Jurassic assemblages in Southwest Japan, Proceedings of the first Japanese Radiolarian Symposium, News of Osaka Micropaleontologists Special Volume, No. 5, 27–43, 1982 (in Japanese with English abstract).

Formation of Active Ocean Margins, edited by N. Nasu *et al.*, pp. 701–717.

INFLUENCE OF IZU SUBDUCTION-COLLISION ON THE DEFORMATION OF CENTRAL JAPAN

P. HUCHON[1] and J. C. DE BREMAECKER[2]

[1]*CNRS LA 215, Laboratoire de Géologie, Ecole Normale Supéneure, 24 rue Lhomond, 75230 Paris Cedex 05, France*
[2]*Geology Department, Rice University, P.O. Box 1892 Houston, Texas 77251, U.S.A.*

Abstract. We use a two-dimensional, viscous finite element model to analyze the deformation of central Japan. This deformation is related to the subduction of the Pacific plate and the Philippine Sea plate (including the colliding Izu peninsula). Our results show that the stress field is largely controlled by the westward push of the subducting Pacific plate. The effects of the subduction of the Philippine Sea plate and of the collision of the Izu peninsula appear to be of secondary importance. We further show that the degree of coupling between the Izu peninsula and central Japan is nearly the same as that between the Pacific plate and Northeast Japan. Taking geophysical data into account, we conclude that the aseismic ridge bearing the Izu peninsula is still subducting beneath central Japan at a rate equal to about 75% of the total rate of plate convergence. Most of the deformation occurs in the peninsula itself rather than in the so-called "collision zone"; the latter appears to correspond only to the deformed paleo-trench sediments. The tectonics of this region can be considered as typical of a very young collision process.

1. Introduction

Central Japan is one of the most seismic regions in the world and, as it is also very densely populated, a large amount of geological and geophysical data have been accumulated in recent years, in order to allow progress in the forecast of a possible major earthquake. One aspect of the studies conducted in this area is the determination of the characteristics of the stress and strain fields, using various methods such as geodetic measurements, study of active and recent faults, determination of fault plane solutions of earthquakes and in situ stress measurements. The origin of the stress field is obviously related to three main causes: the subduction of the Pacific plate beneath Northeast Japan, that of the Philippine Sea plate beneath Southwest Japan and the subduction-collision of Izu peninsula, but the mechanical reasons for the geometry of this stress field are still obscure. What are, for instance, the relative effects of the subduction of the Pacific and Philippine Sea plates? How strong is the degree of coupling between the lithospheres in the different subduction zones and in the Izu subduction-collision zone? Is the Izu peninsula really colliding with Central Japan or is it partially subducted? This paper aims at answering some of these questions.

2. *Geodynamical Framework of Central Japan*

In this section, we outline the main characteristics on which we based our model. Central Japan is located at the intersection of three island arcs (Fig. 1): (i) the Northeast Japan arc, related to the subduction of the Pacific plate toward the West-north-west (azimuth 290) at a relative velocity of 10 cm/a (CHASE, 1972; MINSTER *et al.*, 1974; MINSTER and JORDAN, 1978); (ii) the Southwest Japan arc, underthrusted by the Philippine Sea plate that moves toward the North-west at a relative velocity of 4 cm/a (SENO, 1977; MINSTER and JORDAN, 1979); (iii) the Izu Ogasawara arc, where the Pacific plate is subducted under the Philippine Sea plate. Central Japan is thus bounded to the East and to the South by two subduction zones: the Japan trench and the Nankai trench, respectively. The Nankai trench is interrupted, due to the collision of the Izu peninsula (which belongs to Izu Ogasawara arc) with central Japan. Here, the plate boundary goes on land. To the southwest of the peninsula, the Nankai trench turns to the north (it is then called Suruga trench); to the southeast, the Sagami trench connects the collision zone through a triple junction with the Japan trench and the Izu Ogasawara trench. The geometry of the plate boundary is thus complex and many parameters are unknown, as for example the velocities of the plate boundaries, which are related to the degree of coupling between the plates. In order to study the effects of these parameters, we have used a numerical approach consisting in the computation of the stress field in the overriding plate (central Japan) taking into account the geometry of the plate boundaries and the boundary conditions imposed on it.

3. *The Model*

3.1. *Equations to be solved*

We have used a finite element method to modelize the two-dimensional mechanical behaviour of a lithosphere subjected to displacements of the order of 1 cm/a along its boundaries. As we study the deformation during a short period of time (several hundreds of thousands years), we have chosen a viscous material with a linear constitutive law (Newtonian behaviour). We thus assume that the large-scale behavior of the lithosphere is viscous, taking into account both folding and faulting. The equations to be solved are the Stokes equations for very slow viscous flow. We have used the same computer program ("Sloflo") as in our study of the deformation of Aegea (DE BREMAECKER *et al.*, 1982). The reader is thus referred to this paper and to DUNHAM and BECKER (1973) and DE BREMAECKER and BECKER (1978) for the solution technique.

3.2. *Choice of the finite element grid*

In order to be able to run an adequate number of models while keeping the computation time within reasonable values, we have chosen a fairly simple grid. Central Japan has been divided in 40 elements of the serendipity type (8 nodes), defined in such a way as to correspond approximately to homogeneous tectonic domains (Fig. 2). The southern boundary follows the Nankai, Suruga and Sagami trenches while the eastern one follows the Japan trench. The western and northern boundaries are arbitrarily defined.

3.3. *Choice of the physical parameters*

We have used a linear viscosity both for simplicity and because the effects of viscosity fluctuations appear to be small (see below). As our model is two-dimensional, we did not choose to maintain incompressibility. We thus can simulate the third dimension by computing the rate of thinning or thickening of the lithosphere. Poisson's ratio is taken as 0.25 (the measured value for dunite is 0.27) but a value of 0.4 (less compressible material) does not change the results significantly. Several models of viscosity distribution have been tested; some use the same viscosity over the whole area of the model, others three different values of the viscosity varying according to heat flow.

3.4. *Boundary conditions*

We use plane stress conditions (one of the three principal stresses is vertical). Physically, these conditions correspond to those occuring in

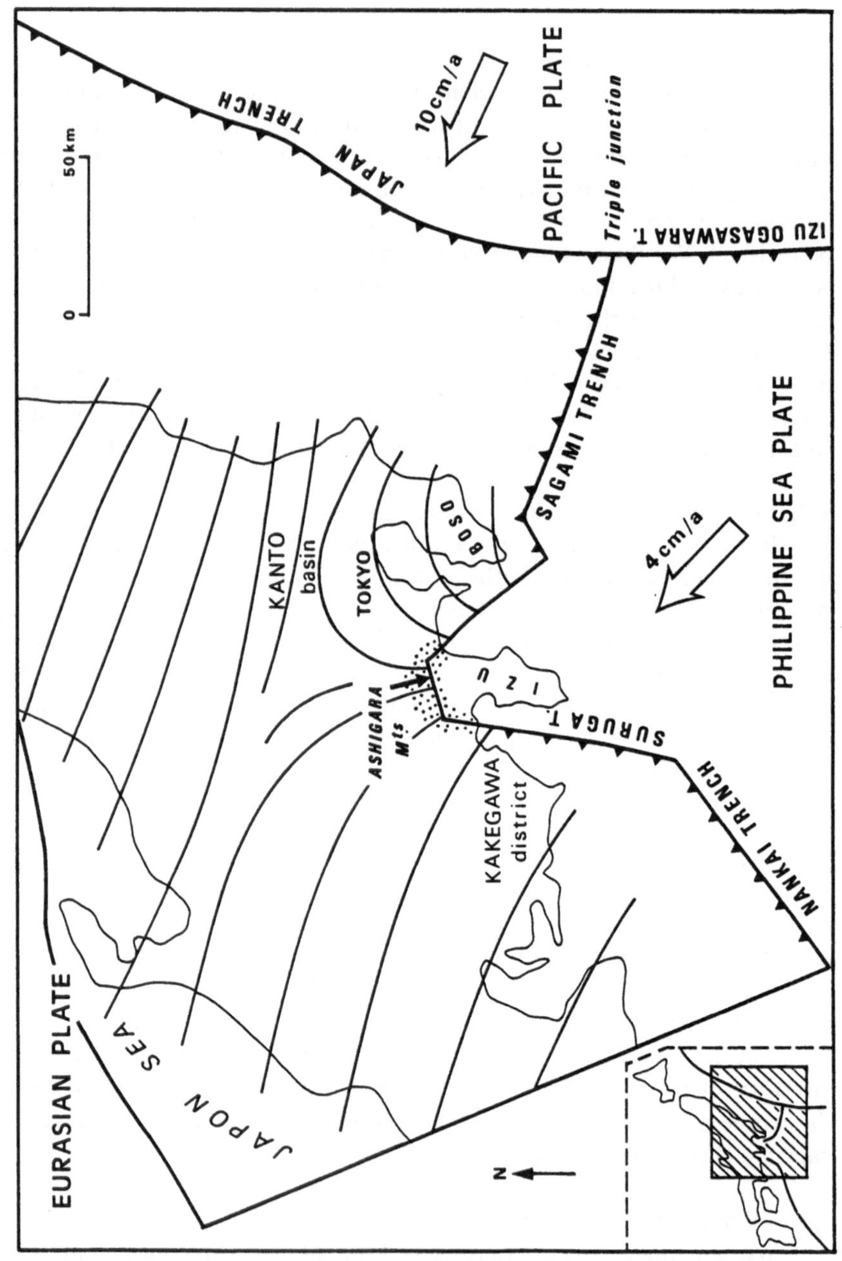
JAPAN TRENCH
10cm/a
PACIFIC PLATE
Triple junction
IZU OGASAWARA T.
0
50km
SAGAMI TRENCH
PHILIPPINE SEA PLATE
4cm/a
KANTO
basin
TOKYO
BOSO
IZU
SURUGA T.
ASHIGARA
Mts
NANKAI TRENCH
KAKEGAWA
district
EURASIAN PLATE
JAPON SEA
N

Fig. 1. Geodynamic framework of central Japan. Thick lines with triangles on the overriding side: subduction zones; dotted area: collision zone; arrows: direction of relative motion; thin continuous lines: trajectories of the maximum compressive stress $\sigma 1$ compiled from geodetic results (FUJII and NAKANE, 1981; IIKAWA, 1981; THATCHER and MATSUDA, 1981), analysis of recent and active faults (HUZITA, 1976; KUWAHARA, 1982; OKADA and ANDO, 1979; Research Group for Active Faults of Japan, 1980), volcanic dykes (SUGITA and KOBAYASHI, 1980; TAKEUCHI, 1980) and fault plane solutions of earthquakes (ANDO, 1979; ICHIKAWA, 1971).

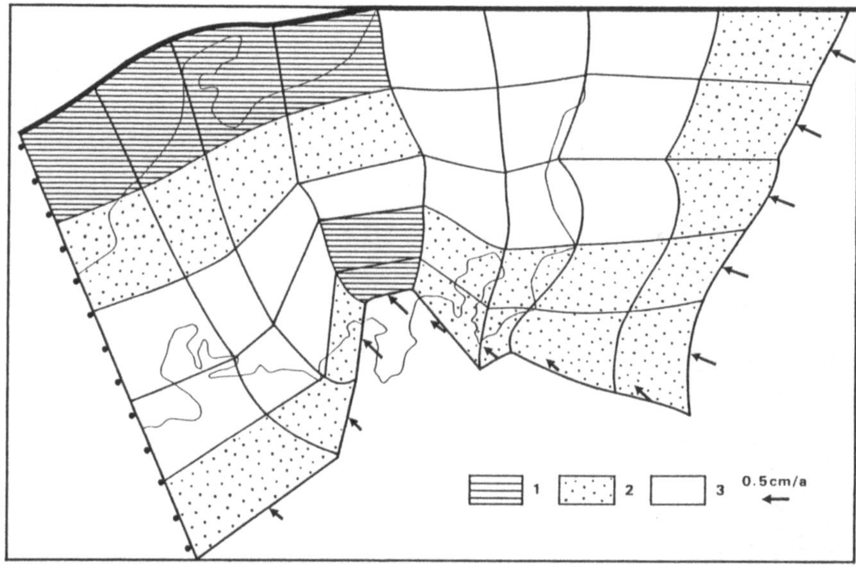

Fig. 2. Geometry, rheology and boundary conditions of the models. The values of the viscosity are (in arbitrary units—see discussion in text) 1.0 (1), 1.5 (2) and 2.0 (3). The northern boundary is fixed. The western boundary is free to move only parallel to itself ("slope" condition). Arrows: direction and magnitudes of the other boundary conditions (for model J1).

a very thin plate whereas plane strain conditions are representative of those in a very thick plate. VILOTTE *et al.* (1982) have remarked that plane strain conditions require a very large vertical stress in order to force the deformation to take place in a plane. Such a large vertical stress appears generally unrealistic. We have therefore used plane stress conditions exclusively. We thus simulate the thickening of an incompressible lithosphere in two complementary ways: by choosing a compressible material and by using plane stress conditions. The exact value of the viscous compressibility is not very important; a value of 0.25 appears to give the best results.

As boundary conditions, we have used velocities (or displacements) instead of forces (or pressures), because the latter are rather difficult to estimate, whereas the kinematics of the Pacific and Philippine Sea plate are well known. Nevertheless, we do not use these velocities directly in so far as only a minor part of the relative motion contributes to a permanent deformation of the overriding plate, while the rest is absorbed by slip along the subduction plane. The main problem is thus to estimate the velocities of the boundaries themselves as a function of the overall kinematics. The ratio between the actual displacement of the plate boundary and the relative displacement of the two plates is a measure of the degree of coupling between the plates. Its estimation thus provides a key to the understanding of the mechanical behaviour of the plate boundary. WESNOUSKY *et al.* (1982) have studied crustal strain rates in Northeast and central Japan, using information about the Quaternary faults and the rates of seismic moment release. Their conclusion is that only about 5% of the relative velocity between the Pacific plate and Japan is absorbed by a permanent deformation of Japan (5 mm/a). We thus take this value as the boundary condition in the Japan trench, and let the other ones very from 2.5 to 12.5% of the relative velocity in the Nankai, Suruga and Sagami trenches and in the collision zone. The northern boundary is fixed while the western one is allowed to move only parallel to itself ("slope condition"). Keeping this boundary totally free does not greatly affect the results except in a strip 20 km wide parallel to the boundary. Our method was thus to use various velocities along the southern boundary in order to obtain a computed geometry of the stress field as close as possible to the one independently known from geological and geophysical studies.

4. *Results*

4.1. *Description of the first model (J1)*

Model J1 has a constant viscosity. Boundary conditions are given in table 1 and illustrated in Fig. 2. Computed stress trajectories are shown in Fig. 3. They should be compared with the compilation of geological and geophysical results which is shown in Fig. 1. The agreement is good in the southwestern and eastern areas. Notice the presence of the "fan-shaped" trajectories of $\sigma 1$ arround the Izu collision zone, due to the indentation of Izu peninsula in central Japan (HUCHON and ANGELIER, 1982). This effect is often encountered in metallurgy; its theory, called slip line field theory (HILL, 1950; BACKHOFEN, 1972) was applied by MOLNAR and TAPPONNIER (1977) to the India-Eurasia collision. To the effect of the indenter (the Izu peninsula) is superposed the effect of the westward push of the Pacific plate. The trajectories of $\sigma 1$ thus form a three branch star in the north of Izu peninsula. In the computed model, this "star zone" is shifted of 40 km northeastward with respect to the reference figure. This shift may be interpreted as due to an exaggerated effect of the collision compared with to the push of the Pacific plate. This is confirmed by the fact that the trajectories of $\sigma 1$ are northwest-southeast instead of

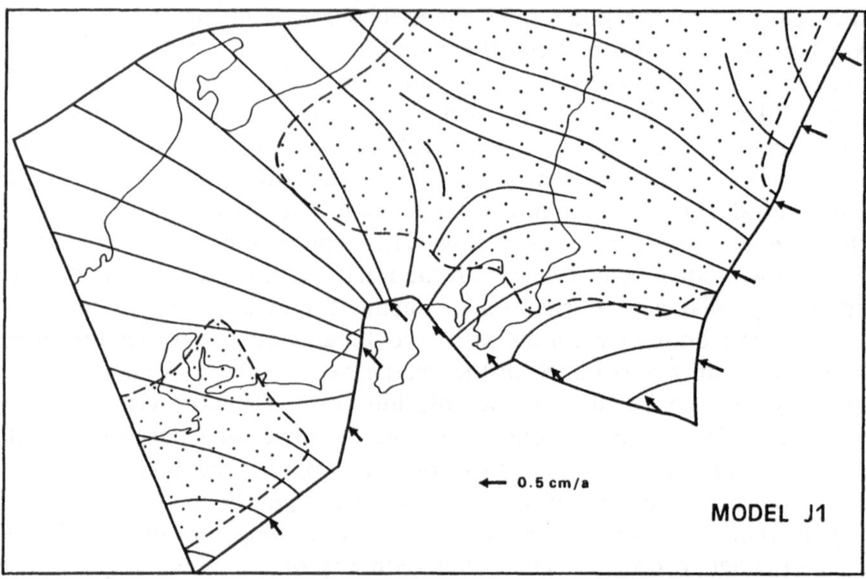

Fig. 3. Model J1. Trajectories of the maximum compressive stress $\sigma 1$. Purely compressional areas are dotted; other areas have a strike slip tectonic regime.

westnorthwest-eastsoutheast on the Japan Sea side.

We use the computed deviatoric stresses to determine the type of deformation (or tectonic regime): purely compressional (both horizontal deviatoric stresses are compressive) or compressional-extensional (strike slip type). Figure 3 shows that the tectonic regime is purely compressional in the eastern half of the model, except along the Japan Sea coast and the margin of Sagami trench. The purely compressional area in the southwest corner is due to the fact that the western boundary is free to move only parallel to itself. If we allow this boundary to move in any direction, this feature disappears. That $\sigma 1$ is approximately perpendicular to Sagami trench is due to the fact that the velocity in the collision zone is 2.5 times greater than in the Sagami trench. As a result, there is northwest-southeast extension and corresponding northeast-southwest compression.

Finally, the rate of thinning or thickening is computed from the values of the strain rates in the horizontal plane. As explained above, the assumptions of compressibility and of plane stress are artificial. In fact, in three dimensions, the material is essentially viscously incompressible. Thus, $\Theta = \varepsilon_{xx} + \varepsilon_{yy} + \varepsilon_{zz} = 0$. Hence the real vertical thickening ε_{zz} can be obtained by contouring $-(\varepsilon_{xx}+\varepsilon_{yy})$ (Fig. 4), which is a result of the computations. The values obtained are geologically reasonable (0.5 to $1.5\times10^{-15}s^{-1}$). Corner effects are relatively important but confined to the proximity of the corner and do not invalidate the overall interpretation. Thickening is much larger to the east than to the west, in good agreement with the known tectonic regime which is purely compressional in northeastern Japan and of the strike slip type in southwestern Japan. Around the collision zone, the large rate of thickening (more than $10^{-15}s^{-1}$) is only partly due to corner effects and can be related to the geological structure, characterized by folding and thrusting. To the contrary, the rate of vertical deformation is very small in the regions located behind the Suruga and Sagami trenches, which is in good agreement with their typical strike slip regime.

Up to now, we didn't discuss the magnitudes of the stresses, because they depend on the value chosen for the viscosity. Taking 10^{23} Pa.s, the values of the computed deviatoric stresses range from 10 to 500 MPa which are not unreasonable (see for exemple, BRACE and KOHLSTEDT, 1980)

4.2. *Influence of variable viscosity*

Two models studied the effects of regional variations of the viscosity,

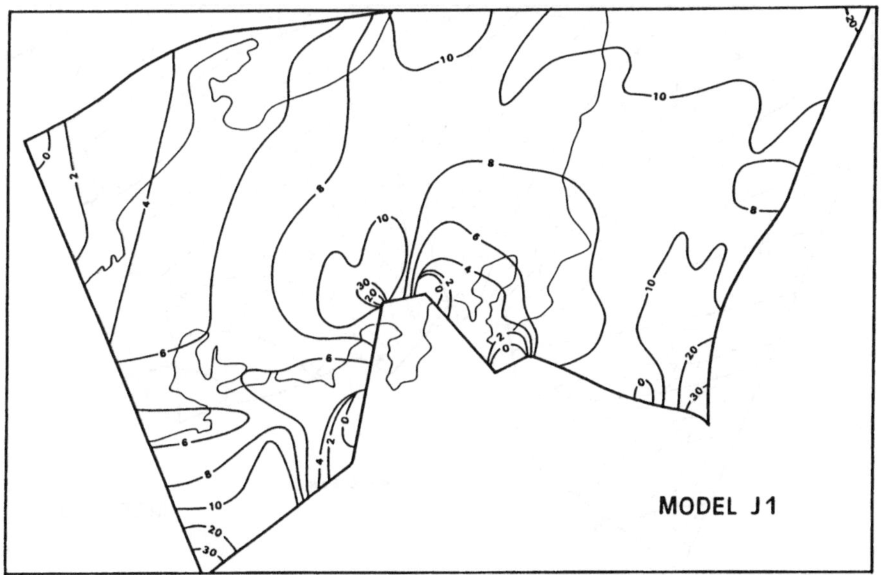

Fig. 4. Model J1. Instantaneous rates of thickening, contoured in $10^{-15}s^{-1}$.

the other parameters being kept constant (see table 1). Three materials of viscosity 1, 1.5 and 2 (arbitrary units, see discussion above) for model J4 and 1, 4 and 8 for model J10. Their distribution, shown on figure 2, has been determined on the basis of heat flow data: high on the Japan Sea side and to the North of Izu peninsula, low on the Pacific margin.

In the case of a low viscosity contrast (model J4), the differences with the reference model (J1) are very small. The main difference is the greater westward extension of the purely compressional area (Fig. 5). For a greater viscosity contrast (model J10, not shown), the low viscosity zone north of the Izu peninsula tends to channel the trajectories of $\sigma 1$, which are thus more nearly north-south than previously. It thus seems that the lateral variations of viscosity do not determine any major change in the results. In the following, we have used either homogeneous or low-contrast inhomogeneous viscosities.

4.3. *Influence of the variations of boundary conditions*

As mentioned in the introduction, the aim of our study is to determine the relative influence of the velocities imposed at the boundaries. By searching for the best fitting model, we hope that it will lead to a better understanding of the relative effects of the subduction of the Pacific

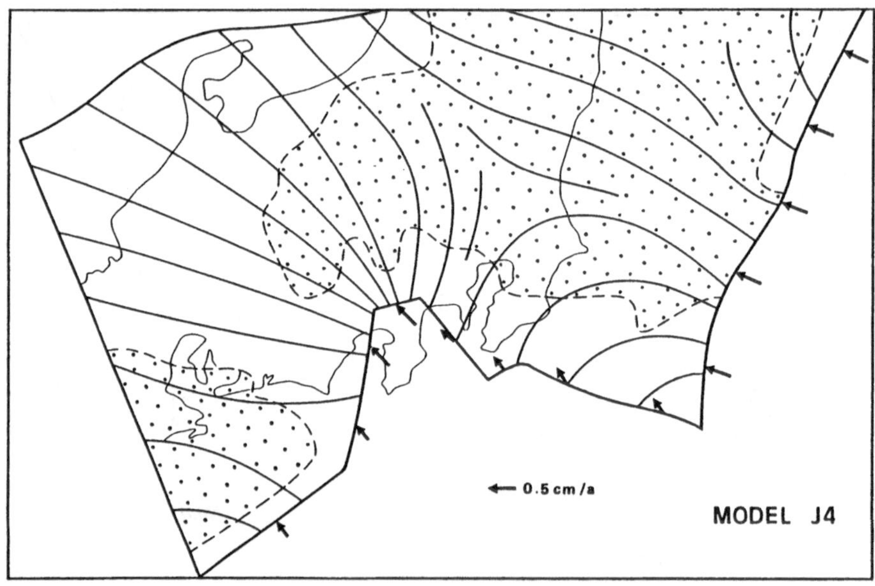

Fig. 5. Model J4. Same as for figure 3.

and Philippine Sea plates and of the Izu subduction-collision on the stress field in central Japan. The first model (J1) fortunately gave results close to the observed stress field. In the following, we keep the velocity unchanged in the Japan trench and allow the other ones to vary.

Model J2 is identical to model J1 except for the velocity in the collision zone, that is 1 cm/a instead of 0.5 cm/a (see table 1). The resulting trajectories of $\sigma 1$ (Fig. 6) show that the push due to the collision is too strong compared with the one of the Pacific plate.

Because of this result, we reduced the velocity in the collision zone by a factor of 2 in our model J9, in which the velocities along the Nankai, south Sagami and south Suruga trenches were reduced by this same factor (see table 1). The resulting trajectories of $\sigma 1$ (Fig. 7) are now in much better agreement with the observed ones.

Despite the fact that our search method is not systematic, some important conclusions appear. The collision of the Izu peninsula with central Japan seems to have a small effect on the geometry of the stress field. Actually, it appears as a perturbation in a stress field induced mainly by the westward push of the Pacific plate. The percentages of the total relative velocity which are absorbed by the permanent deformation of central Japan, are 2.5% for the Nankai, south Suruga and south Sagami

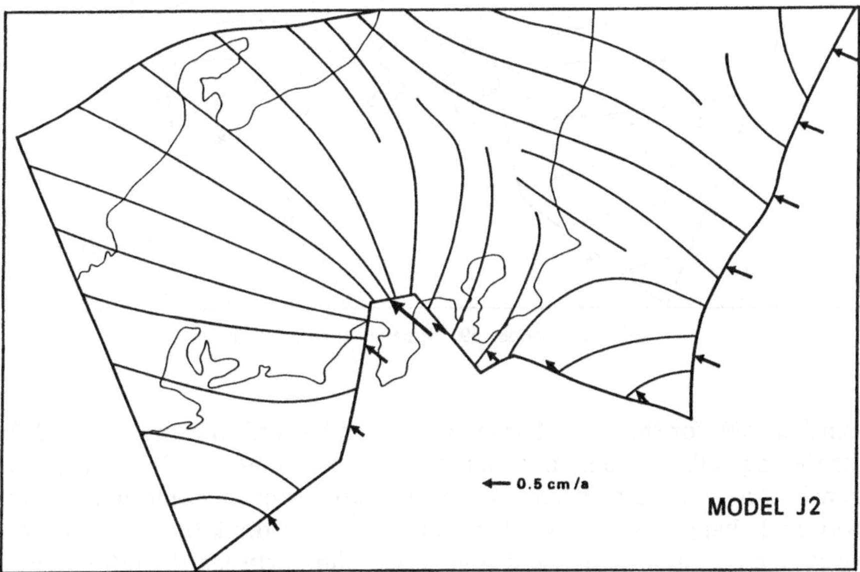

Fig. 6. Model J2. Same as for figure 3.

Table 1.

Model N°	Visco -sity	Direction of plates PAC	PHS	Japan tr.	Velocities (cm/year) Sagami tr. N.	S.	Suruga tr. N.	S.	Collision zone
1	0.5	290	310	0.5	0.2	0.2	0.2	0.4	0.5
2	1.5	290	310	0.5	0.5	0.2	0.5	0.2	1.0
4	1.0	290	310	0.5	0.2	0.2	0.2	0.4	0.5
	1.5								
	2.0								
9	1.5	290	310	0.5	0.2	0.1	0.2	0.1	0.25
11	1.0	000	310	0.5	0.4	0.2	0.4	0.2	0.5
	1.5								
	2.0								

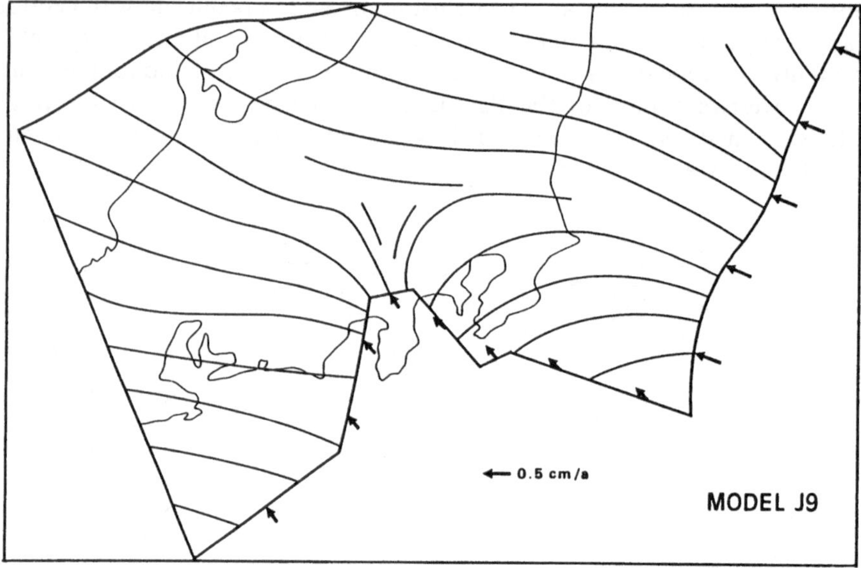

Fig. 7. Model J9. Same as for figure 3.

trenches, 5% for the north Suruga and north Sagami trenches, and 6.25% for the Izu collision zone, assuming that the percentage is 5% in the Japan trench. Any change of percentage in the Japan trench would induce proportional changes along the other boundaries. We think that these percentages give an estimate of the degree of coupling between the lithospheres. Therefore, the degree of coupling between the Izu peninsula and central

Japan is not much greater than the one between the Pacific plate and Northeast Japan. As we are accustomed to think in terms of subduction (rather weak coupling) of the Pacific plate and of collision (strong coupling) of Izu peninsula (MATSUDA, 1978), this result is unexpected.

In all the computations, the boundaries between the plates are implicitly supposed to be vertical. Using the same assumption, SHIMAZAKI and NAKAMURA (1981) have established a simple quantitative relation between the direction of relative convergence and the principal strain axis in the overriding plate. ANGELIER *et al.* (1982) have proposed another relation between the direction of relative motion and the direction of the maximum compressive stress, which takes into account the inclination of the subducting slab. When applied to the subduction zones of central Japan, this relation shows that the difference is smaller than 10° everywhere. Consequently, the use of a two-dimensionnal model, which implies the existence of vertical plate boundaries, is not a cause of large errors on the computed stresses.

4.4. Variations of the direction of motion

A quantitative tectonic study has been performed by one of us (P.H.) in two regions around the Izu collision zone, the Kakegawa district, located behind the Suruga subduction zone and the Ashigara mountains, in the northern part of the collision zone. One of its main results is the presence of a change in the direction of compression. Based on the stratigraphy, this change may have occurred about 1 M.a. ago in the first region (HUCHON, 1983a) and 300,000 years ago in the second one (HUCHON, 1983b). This change has been related to a possible change in the direction of motion of the Philippine Sea plate proposed by NAKAMURA and SHIMAZAKI (1982). The motion would have been oriented closer to the north prior to about 1 M.a. than at present. We have thus run a model (J11) identical to model J4 but with a northward motion of the Philippine Sea plate (see table 1). The resulting trajectories of $\sigma 1$ (Fig. 8) are northwest-southeast in the Kakegawa district in good agreement with field data for the early period (HUCHON, 1983a). But the computed directions are also closer to the North in the northern part of the collision zone whereas the field evidence gives a northwest-southeast direction (HUCHON, 1983b). We suggest that the cause of this discrepancy lies in the fact that the change in the direction of compression in the Ashigara mountains is not related to a change in the direction of the motion of the plate but rather to the beginning of the collision itself. The computed changes are small in the Kanto basin and in the Miura and Boso peninsulas, which appear to be much more under the influence of the subduction of the Pacific plate.

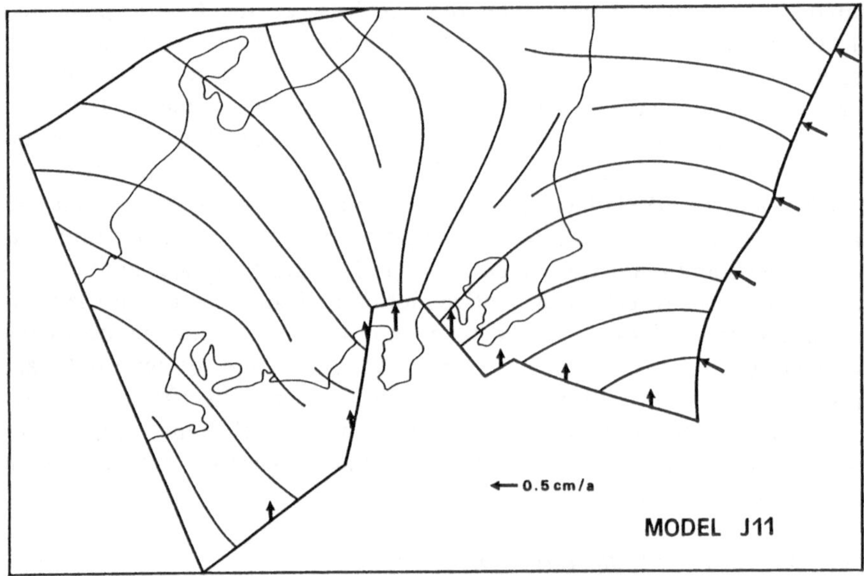

Fig. 8. Model J11. Same as for figure 3.

5. *Conclusions and discussion*

Although preliminary, this study has led to an estimation of the relative influence of subduction and collision on the deformation of central Japan. The stress field appears to be mostly under the control of the westward push of the subducting Pacific plate while the subduction of the Philippine Sea plate and the collision of Izu peninsula induce only limited perturbations. The velocities necessary to account for the observed stress field are 0.1 cm/a in the Nankai, Suruga and Sagami trenches and 0.25 cm/a in the collision zone if the velocity is 0.5 cm/a in the Japan trench. These values are 5%, 2.5% and 6.25% of the total relative velocities, respectively. If these percentages give the degree of coupling between the plates, the degree of coupling between the Pacific plate and Northeast Japan on one hand, and the Izu peninsula and central Japan on the other hand are nearly identical. The Izu peninsula would not be locked against central Japan but still subducting. SOMMERVILLE (1978) has reported that the rate of shortening of Izu peninsula in the northwest-southeast direction is 0.4 cm/a. The maximum rate absorbed in the permanent deformation of central Japan and Izu peninsula is thus of the order of 1 cm/a. Consequently, about 75% of the relative velocity effectively corresponds to slip along the subduction plane. This situation can be compared to the collision of India with Eurasia: the Indian continent still underthrusts the Eurasian plate at a rate of 5 cm/a, and has done so since the beginning of the collision 40 M.a. ago. During this subduction-collision process, the India continent has been affected by large thrusts which have progressively migrated southward. The difference resides in the tectonic regime: thrusting in the Himalayas, strike slip faulting in the Izu peninsula. The main restriction to our conclusion is that the seismic zone associated with the supposed subduction of the ridge bearing the Izu peninsula has not been detected beneath central Japan. Neither the tri-dimensional study of the seismic waves velocities (HIRAHARA, 1981) nor the distribution of microseismicity (TSUMURA, 1973) show any evidence of a dipping slab of the Philippine Sea plate to the North of Izu peninsula. Nevertheless, HIRAHARA (1981) has reported the presence to the Northwest of Izu of a low velocity zone between 33 and 66 km. This low velocity zone may be related either to the volcanic activity or to the remnants of a subducted aseismic ridge. We thus think that the subduction of the Izu ridge is still occurring along a low angle dipping plane, and that it is only slowed by the presence in the subduction zone of a buoyant mass which resists subduction and is consequently strongly deformed. The severe deformation in the regions surrounding the Izu peninsula (Ashigara mountains and Fujigawa valley for exemple) affects the paleo-trench terranes that were located between central Japan and the Izu paleo-island before the collision, but also older terranes. MATSUDA (1978) attributes this deformation to the subduction of the Izu paleo-ridge, but we cannot exclude the possibility of collision of other islands initially located north of Izu. The present structure of central Japan would thus result from several "subduction-collisions", where the colliding island resists subduction, is progressively deformed and finally becomes part of central Japan; as proposed by ISHIBASHI (1976), this phenomenon may be accompanied by a southward jump of the trench.

Acknowledgements

This research was supported by CNRS (LA 215) and CNEXO. Discussions with X. Le Pichon are acknowledged.

REFERENCES

ANDO, M., Stress field in the recent 5.10^5 years in Japanese islands, *Earth Monthly*, **1**, 541–546, 1979.

ANGELIER, J., E. BARRIER, and P. HUCHON, Sur les relations entre trajectoires de contrainte et directions de mouvement le long d'une frontière convergente: exemple de la zone de subduction hellénique et de la collision Philippine-Eurasie (Taiwan et Japon), *C.R.Acad.Sc.*, **294**, série II, 745–748, 1982.

BACKHOFEN, W. A., Deformation processing, Addison-Wesley Pub. Co., MA., 1972.

BRACE, W. F. and D. L. KOHLSTEDT, Limits on lithospheric stress imposed by laboratory experiments, *J. Geophys. Res.*, **85**, 6248–6252, 1980.

CHASE, C. G., The N plates problem of plate tectonics, *Geophys. J. R. Astr. Soc*, **29**, 117–122, 1972.

DE BREMAECKER, J. C. and E. B. BECKER, Finite element models of folding, in M. N. Toksoz Ed., Numerical modelling in *Geodynamics, Tectonophysics*, **50**, 349–367, 1978.

DE BREMAECKER, J. C., P. HUCHON, and X. LE PICHON, The deformation of Aegea: a finite element study, in X. Le Pichon, S. S. Augustithis and J. Mascle Ed., Geodynamics of the Hellenic Arc and Trench, *Tectonophysics*, **86**, 197–211, 1982.

DUNHAM, R. S. and E. B. BECKER, Texgap—The Texas grain analysis program, TICOM Report 73–1, Univ. of Texas at Austin, Austin, Texas, 35pp., 1973.

FUJII, Y. and K. NAKANE, Horizontal crustal movements in the Kanto district, Japan, with special reference to their tectonic implications, *Mem. Geol. Soc., Japan*, **20**, 247–258, 1981.

HILL, R., The mathematical theory of plasticity, Oxford Univ. Press, London, 1950.

HIRAHARA, K., Three-dimensional seismic structure beneath southwest Japan: the subducting Philippine Sea plate, *Tectonophysics*, **79**, 1–44, 1981.

HUCHON, P., Le chevauchement de kannawa (zone de collision d'Izu, Japon central): évolution du régime des contraintes à la limite des plaques eurasiatique et Philippines, *C. R. Acad. Sc.*, série II, **296**, 869–872, 1983 (a).

HUCHON, P., Les contraintes tectoniques à l'arrière d'une zone de subduction: relation avec la direction de convergence et évolution dans le temps; l'exemple de la région de Shizuoka (Japon central), *C. R. Acad. Sc.*, série II, **296**, 787–790, 1983 (b).

HUCHON, P. and J. ANGELIER, Interprétation du champ de contraintes dans la zone de collision d'Izu (Japon), 9ième Réun. Ann. Sc. Terre, Paris 1982, *Soc. Géol. Fr.* Ed, Paris, **314**, 1982.

HUZITA, K., The quaternary tectonic stress states of southwest Japan, *Jour. Geosc.* Osaka City Univ., **20**, 93–103, 1976.

ICHIKAWA, M., Reanalyses of mechanism of earthquakes which occured in and near Japan and statistical studies on the nodal plane solutions obtained, 1926–1968, *Geophys. Mag.*, **38**, 207–273, 1971.

IIKAWA, K., On the recent horizontal deformation of the Earth's crust and its relation to earthquake occurence in central Japan, *Mem. Geol. Soc. Japan*, **20**, 57–75, 1981.

ISHIBASHI, K., Plate convergence arround the Izu collision zone, central Japan: development of a new subduction boundary with a temporary transform belt, Abstr. Intern. Conf. on Western Pacific and Magma Genesis, Tokyo 1976, 66–67, 1976.

KUWAHARA, T., Late Cretaceous to Pliocene fault systems and corresponding regional stress fields in the southern part of northeast Japan, Sc. Repts of Inst. of Geosc., Univ. of Tsukuba, **3**, 48–111, 1982.

MATSUDA, T., Collision of the Izu-Bonin arc with central Honshu-Cenozoic tectonics of the Fossa Magna, Japan, *J. Phys. Earth*, **26**, S409–S421, 1978.

MINSTER, J. B. and T. H. JORDAN, Present day plate motion, *J. geophys. Res.*, **83**, 5331–5354, 1978.

MINSTER, J. B. and T. H. JORDAN, Rotation vectors for the Philippine and Rivera plates, Abstr. Am. Geophys. Union 1979 Fall Ann. Meeting, EOS, **60**, 958pp., 1979.

MINSTER, J. B., T. H. JORDAN, P. MOLNAR, and E. HAINES, Numerical modelling of instantaneous plate tectonics, *Geophys. J. R. astr. Soc.*, **36**, 541–576, 1974.

MOLNAR, P. and P. TAPPONNIER, The relation of the tectonics of eastern China to the India-Eurasia collision—an application of slip line field theory to large scale continental collision, *Geology*, **5**, 212–216, 1977.

NAKAMURA, K. and K. SHIMAZAKI, Sagami-Suruga troughs and the subduction of the Philippine Sea plate, *Kagaku*, **51**, 490–498, 1981.

OKADA, A. and M. ANDO, Active faults and earthquakes in Japan, *Kagaku*, **49**, 158–169, 1979.

RESEARCH GROUP for Active Faults of Japan, Active faults in and arround Japan: the distribution and the degree of activity, *Journ. Nat. Disaster Sci.*, **2**, 61–99, 1980.

SENO, T., The instantaneous rotation vector of the Philippine Sea plate relative to the Eurasian plate, *Tectonophysics*, **42**, 209–226, 1977.

SHIMAZAKI, K. and K. NAKAMURA, Quantitative relation between orientations of plate convergence and principal strain axis in the overriding plate, *J. Geod. Soc. Japan*, **27**, 329–332, 1981.

SOMMERVILLE, P., Accomodation of plate collision by deformation in *The Izu block, Japan, Bull. Earthquake Res. Inst.*, **53**, 629–648, 1978.

SUGITA, S. and Y. KOBAYASHI, Tectonic stress field deduced from dikes in *Southern Fossa Magna, Progr. Abstr. Seism. Soc. Japan*, **2**, 32pp., 1980.

TAKEUCHI, A., Tertiary stress field and tectonic development of the southern part of the northeast Honshu arc, Japan, *Journ. Geosc.* Osaka City Univ., **23**, 1–64, 1980.

THATCHER, W. and T. MATSUDA, Quaternary and geodetically measured crustal movements in the Tokai district, central Honshu, Japan, *J. Geophys. Res.*, **86**, 9237–9246, 1981.

TSUMURA, K., Microearthquake activity in the Kanto district, Pub 1. for the 50th anniversary of the Great Kanto Earthquake, 1923, *Earthquake Res. Inst.*, Tokyo Univ., 67–87, 1973.

VILOTTE, J. P., M. DAIGNIÈRES, and R. MADARIAGA, Numerical modelling of intraplate deformation: simple mechanical models of continental collision, *J. Geophys. Res.*, **87**, 10709–10728, 1982.

WESNOUSKY, S. G., C. SCHOLZ, and K. SHIMAZAKI, Deformation of an island arc: rates of moment release and crustal shortening in intraplate Japan determined from seismicity and Quaternary fault data, *J. Geophys. Res.*, **87**, 6829–6852, 1982.

Formation of Active Ocean Margins, edited by N. Nasu *et al.*, pp. 719–746.

OPHIOLITE-BASED FOREARCS: A PARTICULAR TYPE OF PLATE BOUNDARY

Y. OGAWA[1], J. NAKA[1] and H. TANIGUCHI[2]

[1]*Department of Geology, Kyushu University 33, Hakozaki, Fukuoka 812, Japan*
[2]*Department of Earth Sciences, Nihon University, Sakurajosui, Setagaya, Tokyo 156, Japan*

Abstract. A particular type of plate boundary that bears dismembered ophiolitic rocks in the hanging wall of a subducting oceanic plate has become well-known in the circum-Pacific area. The dismembered ophiolitic rocks form an ophiolite melange including basaltic, gabbroic and their metamorphic rocks in highly sheared serpentinite. Those in the Bonin-Mariana forearc area and its possible northward extension in the Setogawa and Mineoka forearc belts in Central Japan are very similar to each other, and also resemble the drilled ophiolitic rocks off Guatemala, Middle America Trench landward slope from the view-points of geothermal gradient conditions and intense shearing stress for ductile and brittle deformations.

Petrographical and chemical studies have revealed that there are at least two kinds of origin of such ophiolitic rocks. One is from oceanic fracture zones after the spreading ridges and transform faults magmatism, metamorphism and deformation. The second may be from an island arc crust or mantle. There are still several other complex histories for the forearc ophiolites, they may have been mixed together to form the present or past plate boundaries along the subduction zones and further mixed with terrigenous rocks to be accreted to the forearc belts.

1. *Introduction*

Several examples of dismembered ophiolites have recently been found from forearc areas between the volcanic front and trench. They were collected by dredge hauls and deep-sea drilling from the trench landward slopes near to the trench axes (Fig. 1), which alinged to form a type of plate boundary in the subduction zone. These ophiolitic rocks are, therefore, called the forearc ophiolite (OGAWA and NAKA, 1984; ISHII, 1985) or island arc ophiolite (BLOOMER and HAWKINS, 1982; BLOOMER, 1983; LEITCH, 1984). Similar type of ophiolitic rocks occur in onland tectonic belts within accretionary belts in central Japan (OGAWA, 1983; OGAWA *et al.*, 1985b; TANIGUCHI, 1984). Recent examples are mainly concentrated along the circum-Pacific areas, from Guatemala (AUBOUIN, VON HUENE *et al.*, 1982; OGAWA *et al.*, 1985a), central Japan (Setogawa and Mineoka areas) (OGAWA, 1983; OGAWA and NAKA, 1984), Bonin-Mariana-Yap arcs (HONZA and KAGAMI, 1977; IGCP WORKING GROUP, 1977; DIETRICH *et al.*, 1978; HAWKINS *et al.*, 1979; SHARASKIN *et al.*, 1980; CRAWFORD *et al.*, 1981; BLOOMER, 1983; BLOOMER and HAWKINS, 1982; KOBAYASHI, 1983a, b; OGAWA and NAKA, 1984), Tonga arcs (FISHER and ENGEL, 1969) (Figs. 1 and 2). Following the abundant sampling of the ophiolitic rocks in the Mariana-Yap Island arc system by R/V Dmitry Mendeleev (IGCP WORKING GROUP, 1977), BLOOMER (1983) reported occurrence of various kinds of igneous rocks including serpentinite and other ophiolitic rocks together with island-arc rocks from Mariana. HUSSONG and FRYER (1983) show Seamark's pictures of occurrence of a serpentinite cone near the edge of the trench landward slope in the Mariana forearc. Some onland ophiolites in other place such as the Zambales range in the Philippines (HAWKINS and EVANS, 1983), East Taiwan (SUPPE *et al.*, 1981), Borneo, New Guenea, New Caledonia, and Ballantrae in Scotland could also be of forearc origin.

In this paper we summarize modes of occurrences of some off-shore and onland ophiolites in the forearc areas along the north and east Philippine Sea plate boundaries, and discuss mechanism and process of emplacement of the dismembered ophiolites in view of the plate margin tectonics. We refer to the recent geochemical study by BLOOMER (1983), BLOOMER and HAWKINS (1982) and ISHII (1983, 1985) from the Bonin-Mariana forearcs, and petrological discussion by TATSUMI (1983) and KUSHIRO (1983), then discuss plausible origin of the dismembered forearc ophiolite. We suggest that the dismemberd ophiolites in the forearc areas as a whole were the basement of the trench landward slope at the time of the juvenile stage of the volcanic island arc. Such an ophiolite-based forearc has a particular significance in the plate boundary. There are two possible explanations for its origin; one is that the ophiolites, originating essentially from the mid-oceanic ridge, were serpentinized, metamorphosed and deformed along the ridge-ridge transform fault, then accreted to the forearc basement. The other explanation is that some ophiolites were essentially

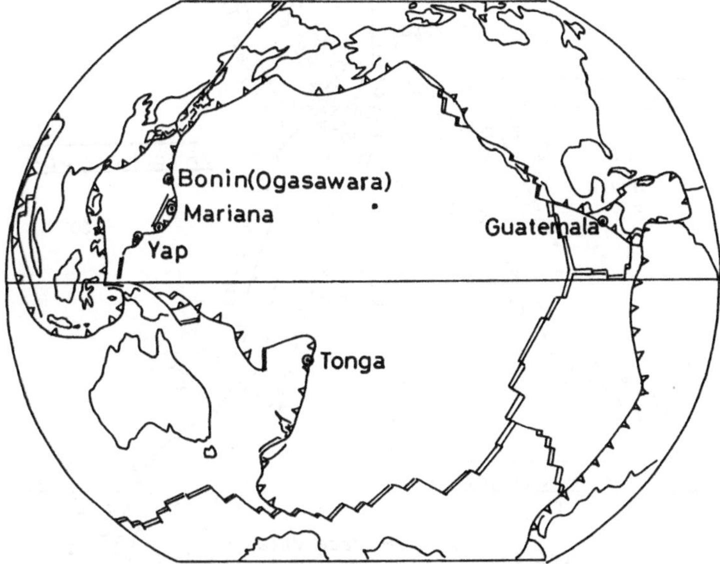

Fig. 1. Location map of dredged and drilled examples of forearc ophiolites.

the wedge mantle and island arc igneous rocks formed at the nascent stage of the arc magmatism. The above two may be the end members of possible tectonic stage in the initiation of island arcs and also provide two possible explanations for so-called orogenic ophiolite onland (MIYASHIRO, 1973). We discuss three field examples; Bonin-Mariana trenches, Middle America trench off Guatemala and the Setogawa and Mineoka belts in the Japanese Islands.

2. *Examples of the Dredged and Drilled Ophiolite in the Bonin-Mariana Island Arc System and off Guatemala*

Figures 2 and 3 and Table 1 summarize the dredge-haul points and results in the Bonin-Mariana arc system, and Figure 3 shows general profiles of forearc areas including Guatemala and Tonga. In these examples, the trench landward slopes are at least partly, and in some cases largely, underlain by dismembered ophiolites. Especially they are common in the oceanward slopes from the trench slope break to the trench slope edge off the old volcanic arc remnant such as Bonin (Ogasawara), Guam and Yap. In many cases, ophiolite assemblages are obtained from the lower trench

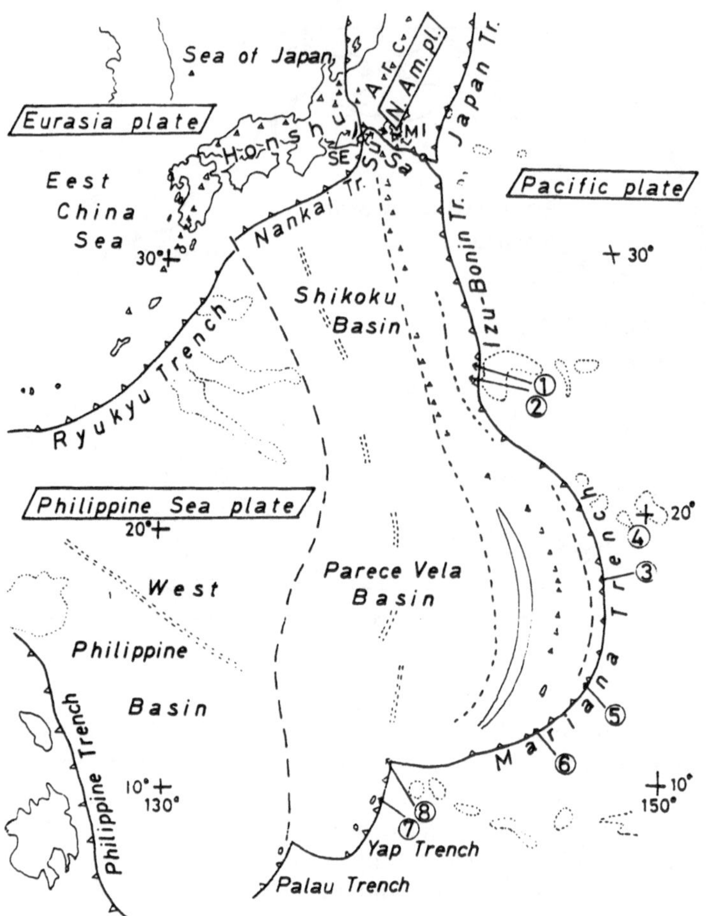

Fig. 2. Location of examples of forearc ophiolites in the Bonin-Mariana Island arc system and in the Setogawa and Mineoka ophiolite belts in Central Japan. Numbers in the circles refer to the dredge-haul sites in Table 1. Abbreviations in central Japan: SE, Setogawa ophiolite; MI, Mineoka ophiolite; Su, Suruga trough; Sa, Sagami trough.

slopes or from the flanks of small knolls (sometimes called 'seamount') on the middle trench slope or near structural high areas. HUSSONG and FRYER (1983) reported diapiric serpentinite off Mariana which had overflowed the lower trench slope. On the other hand, the trench slope basins (perched basins) are covered by hemipelagic sediments overlying ophiolitic rocks (HUSSONG and UYEDA, 1980; HONZA and KAGAMI, 1977). The oldest

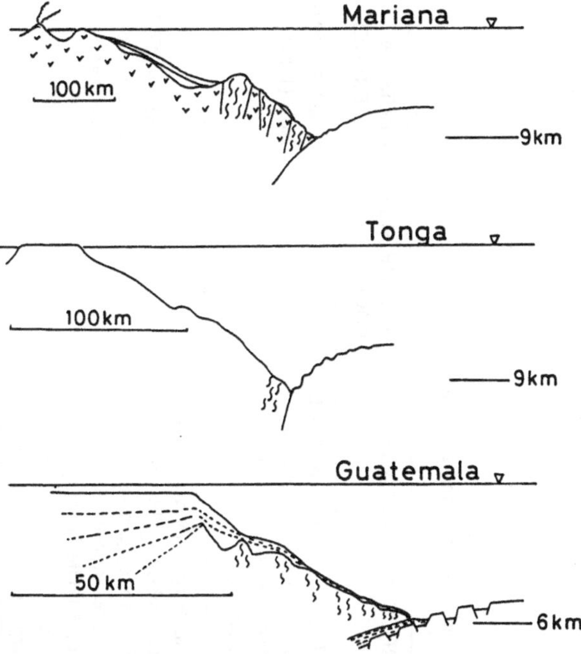

Fig. 3. Schematic figures of the results of recent dredging and drilling. Ophiolitic distribution is indicated as "stretched s" symbols. Island arc volcanic rocks are indicated as "v" symbols. References are in the text.

sediments were dated as Middle Miocene to Pliocene. Similar features are also recognized at the wide of the trench slope off Guatemala in the Middle America trench (AUBOUIN,VON HUENE *et al.*, 1982; OGAWA *et al.*, 1985a). In this case the trench slope is covered by hemipelagic to terrigenous clastic sediments from Eocene to Pleistocene in age. It is noteworthy that in all cases the sedimentary covers do not show any convergent deformation such as thrusts or overturned folds, but show only small-scale normal faults and vein structures (OGAWA and MIYATA, 1985).

Materials dredged from all these areas show many similarities. The ophiolitic rocks are dismembered, and all the rocks, which were originally situated at different levels, are now mixed together. Clearly island arc derived materials are dredged together in Bonin-Mariana, but were not drilled in Guatemala. The ophiolitic rocks range from the upper mantle to the ocean-floor materials; from peridotite to basalt or pelagic sediments through gabbro, dolerite and their metamorphic derivatives. Peridotites

Table 1. General summary of the results of dredges in the Bonin-Mariana-Yap island arc system.

LOCATION	NO. IN FIG. 1		LATITUDE & LONGITUDE	TOPOGRAPHIC POSITION	ROCK ASSEMBLAGE	AGE OF SED. ROCK	CRUISE & REFERENCE
SE of Bonin (Ogasawara) Island	1	Ca	26°25′N 142°55′E	Flank of mound on middle trench slope	Harzburgite, gabbro, basalt, Mn-nodule (incl. gabbro, basalt, serp.), volcani-clastic rock		KH80–3 (ORI) Ishii (1981)
	2	Ca	26°12′N 143°03′E	Flank of mound on middle trench slope	Harzburgite, lherzolite, dun-ite, wehrlite, olivine-clino-pyroxenite, gabbroic anor-thosite, diorite, dolerite, amphibolite, basalt, their clastite (tectonic & sedimen-tary), mylonite (plagiogranite)		KH82–4 (ORI) Naka & Ue-hara (1983)
	3	Ca	11°–18°N 147°E	Middle to lower trench slope	Harzburgite, lherzolite, actinolite-pyroxenite, amphibolite, greenschist, boninite, dacite	early Middle Miocene	Scripps Evans & Haw-kins (1979) Hawkins *et al.* (1979)

E of Mariana Island	4	Ca	20°N 150°E	"Magelan seamounts" on the Pacific seafloor	Alkali basalt, volcanic breccia, chert, oolitic Limestone		Scripps Hawkins *et al.* (1979)
	5	Ca	13°53.4′N 146°17.5′E	Middle trench slope below trench slope break	Dunite (or wehrlite), tuffaceous siltstone	early Middle to late Middle Miocene	KH71-1 (ORI) Honza & Kagami (1977)
	6	Ca	11°51′N 144°30′E	Lower trench slope	Harzburgite, plagioclase-lherzolite, gabbro, pyroxenite, rodingite, amphibolite, greenschist, boninite, tholeiitic basalt, dolerite	Late Miocene	Dmitry Mendeleev (USSR) IGCP W.G. (1977)
E of Yap Island	7	Ca & Ca	8°11′N 137°30′E; 9°40′N 138°25′E	Lower trench slope	Serpentinite, dunite, euthotide gabbro, amphibolite, subalkali basalt, andesite, boninite, island arc basalt	Late Miocene	ibid
Yap-Mariana junction	8	Probably Ca	10°30′ 139°00′E	Not known	Dolerite, gabbro (some are olivine rich), andesite lava		ibid

ORI = Ocean Research Institute of University of Tokyo. Trench slopes are all trench landward slopes. Most of the peridotites are more or less serpentinized, so that original rock names are listed. Location No. 4 is on the ocean side of the trench but is listed as references.

are mostly harzburgite with minor amounts of dunite. Lherzolite and wehrlite are not common, but rarely included. The peridotites are more or less serpentinized to lizardite and chrysotile with weak crystallinity (UEHARA and NAKA, 1983). Basaltic rocks are usually tholeiitic and rarely alkalic. Definitely island-arc materials, such as boninite or high MgO-high silica andesite are sometimes involved. In the Bonin and Mariana areas, we have boninite and other volcanic rocks such as andesite and dacite (BLOOMER, 1983; NAKA and UEHARA, 1983).

2.1 Off Bonin (Ogasawara)

In the dredged samples of the KH82-4 cruise in the Bonin (Ogasawara) forearc we have observed the following features (KOBAYASHI, 1983). Mafic rocks have undergone alteration or metamorphism of zeolite to amphibolite facies, but excepting one amphibolite (hornblende schist), all rocks do not exhibit schistosity but preserve their original textures. In the gabbroic rocks, which have a poikilitic or ophitic texture, olivine, clinopyroxene, Ca-rich plagiocalse and spinel are relict minerals, but olivine is largely replaced by serpentine, chlorite and actinolite, and pyroxene is replaced by green hornblende, actinolite, chlorite and white mica. Plagioclase is partly altered into prehnite, zeolite (analcite and natrolite) and clinozoicite. Based on the relic minerals, gabbros are classified into a leuco-trochtolite, leuco-olivine gabbro, olivine gabbro and two pyroxene gabbro. Dolerite is much more altered or metamorphosed than gabbro, and actinolite, chlorite, green hornblende, white mica, and zeolite (analcite and natrolite) commonly occur.

Basalt has an intersertal texture, and contains clinopyroxene and Ca-rich plagioclase as phenocrysts. They remain not highly altered, but green hornblende, chlorite and clay minierals are found in the ground mass as metamorphosed or altered minerals. Only one schistose rock is found. This comprises mainly green hornblende and Ca-rich plagioclase, and is called a hornblende schist. This is a common type of rock in the onland ophiolite melanges as described later.

The metamorphic conditions are not so common in the onland regional metamorphic belts, as in the ocean floor or ridge metamorphism (FOX *et al.*, 1976). Amphibolite and greenschist were described in the Yap Island by SHIRAKI and MARUYAMA (1977) and attributed to the ocean floor metamorphism. They concluded that the rocks were emplaced to the forearc area to become the basement high before the arc volcanism of middle Miocene age. In the dredged samples, veining of prehnite, zeolite and actinolite is commonly associated with actinolite mineralization. This assemblage characterizes the high temperature/low pressure metamorphism of zeolite facies (S. Maruyama personal communication). Considering that the metamorphic minerals replace rims of the higher grade metamorphic minerals, the retrogressive metamophism seems to be very common. As the whole the ophiolitic rocks suffered metamorphism of amphibolite to zeolite facies through greenschist facies. Hydrothermal effect with fracturing is also very common.

In addition to the metamorphism, a cataclastic deformation characteristically occurs in these rocks. Varieties of deformation, from ductile to brittle and from mylonite to cataclastic breccia, are observed (Figs. 4-A, 4-B). Quartz-plagioclase rich mylonitic rocks show strong wavy extinction or deformation lamellae, sometimes subgrains or dislocation texture (Fig. 4-A-3). The original rock was a plagiogranite. Other tectonically induced breccias have pulverized materials around grains, and grains are stretched or rotated to from porphyloclasts (Fig. 4-A-4). Brecciated fragments are usually angular but rarely rounded and subrounded, and cemented by calcite or clayey serpentine. The calcite-cemented brecciated ophiolitic rocks are so-called ophicalcites. They are usually matrix-supported sediments indicating debris flow type sedimentary rocks in origin. However, some of the breccias are truly tectonic in origin. It is usually difficult to distinguish between the two origins; sedimentary or tectonic. If the grain-size distribution is compared between the two, sedimentary one has much better sorting and smaller standard deviation (Figs. 5-D, 5-E). This is remarkable in the case of the tectonic and sedimentary serpentinites in the Mineoka belt as shown later, but the grains in both cases are usually rather angular, indicating poor abrading and poor sorting. The breccias are supplied from the tectonically sheared or brecciated bodies protruded nearby to form debris flow type coarse sandstone or conglomerates (OGAWA, 1983; ARAI *et al.*, 1983). Calcite in sedimentary ophiolites includes sponge fossils in some cases, showing shallow sedimentation. In any cases in the Bonin area, such a sedimentary ophiolite contains no mature island arc-derived materials such as andesite, dacite or sedimentary rocks, but do contain only ophiolitic rocks. They are also reported from the Mariana and Yap areas by IGCP WORKING GROUP (1977).

2.2 *Off Guatemala*

A similar assemblage of dismembered ophiolitic rocks, which form an ophiolitic melange, has been drilled in the Middle America trench landward slope off Guatemala by DSDP Leg 84 (Fig. 3) (AUBOUIN, VON HUENE *et al.*, 1982; OGAWA *et al.*, 1985a). Here, serpentinite (originally harzburgite with rare lherzolite)-rich dismembered ophiolite was found under the several hundreds meters thick trench slope from Eocene to Pleistocene (Fig. 3).

Fig. 4-A. Photomicrographs of tectonic breccia of ophiolitic rocks and others. 1. Microbreccia of epidote-hornblende schist in the Byobu-jima, Kamogawa, Mineoka belt (see Fig. 5-A). Cataclastic texture is common. 2. Brecciated serpentinite with very fine serpentine mud as matrix. Kinugasa, Mineoka belt (see Fig. 5-C). 3. Mylonitic rock of plagiogranite dredged at Station 4 of KH82-4 cruise (04-508B, NAKA and UEHARA, 1983). 4. Sheared serpentinite with calcite cement (ophicalcite) dredged at Station 4 of KH82-4 cruise (04-513, NAKA and UEHARA, 1983). Width of each picture 1.0 mm.

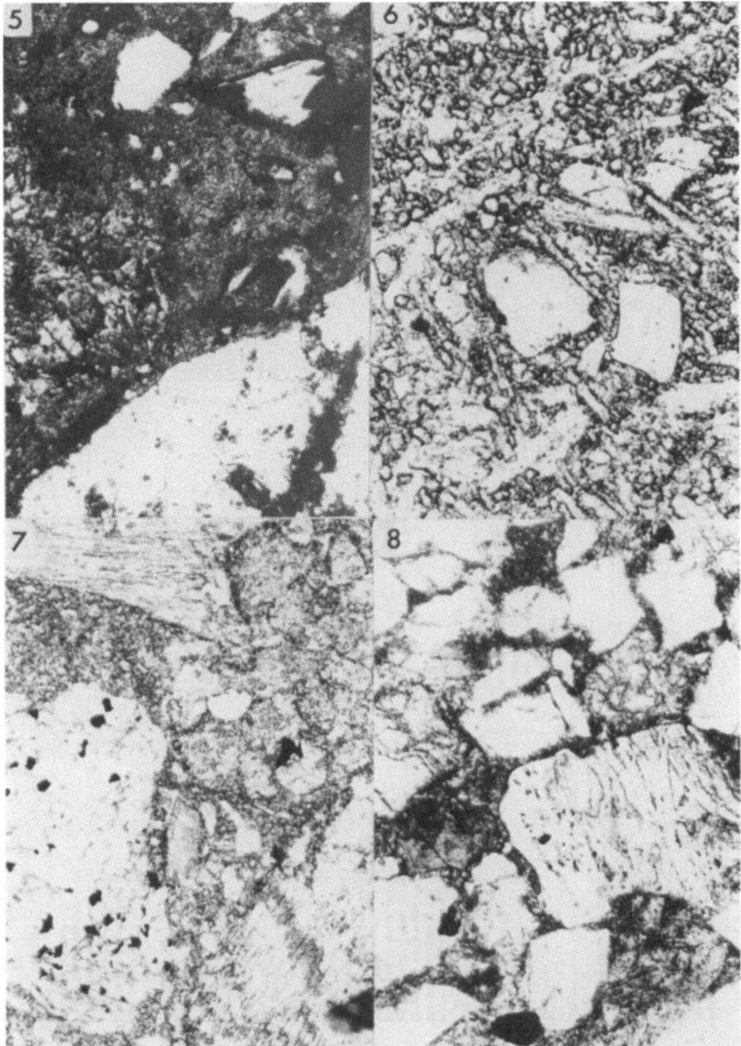

Fig. 4-B. Photomicrographs of sedimentary ophiolitic rocks. 5. Sandstone made of basalt, gabbro, amphibolite and serpentinite fragments with very fine serpentine mud as matrix. Dredged at Station 4 of KH82-4 cruise (04-508A, NAKA and UEHARA, 1983). 6. Serpentine sandstone with abundant calci-sponge fragments cement, dredged at Station 4 of KH82-4 cruise (04-516). 7. Ophiolitic sandstone made of various kinds of rocks including basalt, gabbro, amphibolite, chert, and acid rocks with calcareous cement. Fusata-10, Mineoka belt (see Fig. 5-F). 8. Ophiolitic rock-bearing sandstone also include abundant quartz and feldspar grains (white) with fine muddy matrix and calcite cement, Mineoka-Sengen-10, Mineoka belt. Width of each picture 1.0 mm.

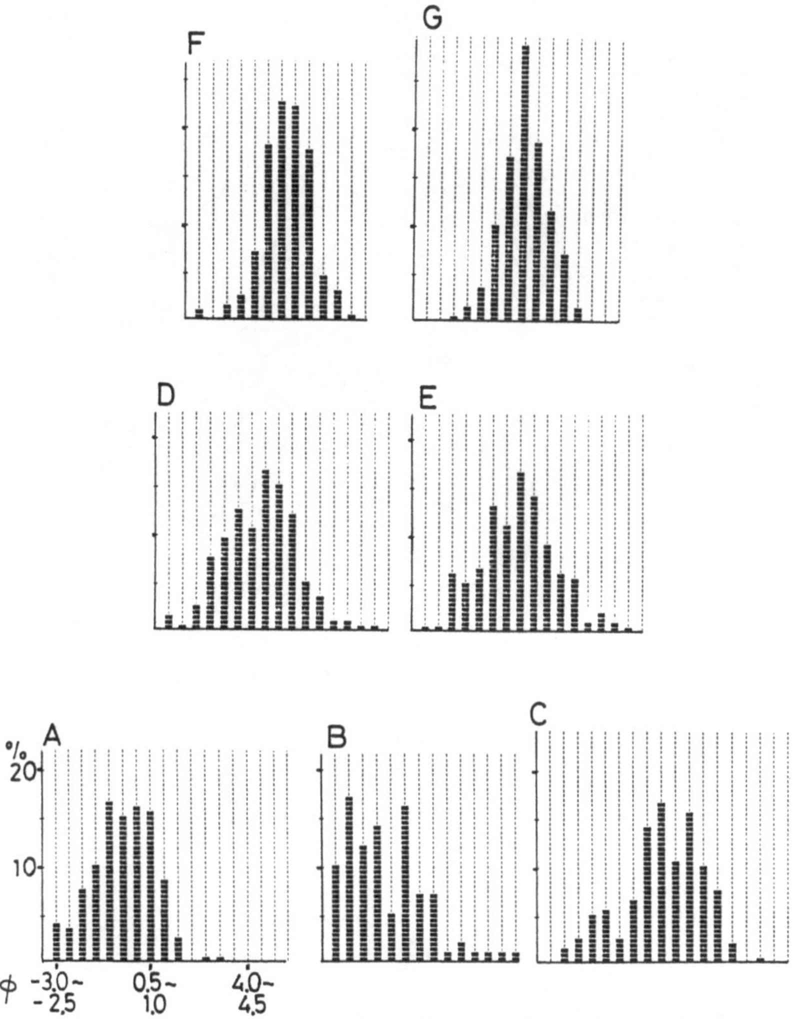

Fig. 5. Grain size distribution of tectonic and sedimentary ophiolitic rocks. For location and rock name, mean diameter and standard deviation in phi-scale see the next page.

Metamorphism from amphibolite to zeolite facies through greenschist facies is seen in mafic rocks with remarkable retrogressive metamorphism and alteration. Some rocks have weak schistosity. Cataclastic deformation in gabbro and basalt is common, indicating ductile and brittle shearing under relatively high temperature and low pressure (OGAWA *et al.*, 1985a). In this case basalt and dolerite are alkalic. However, there are three types of gabbros; clinopyroxene gabbro, olivine gabbro and two-pyroxene gabbro. The third may be restricted to island arc magmatism. Clearly island arc magmatism which brings boninite, andesite or rhyolite to ophiolite melange is not known. On the trench slope landward side no distinct oceanic and trench materials have been accreted, and the trench slope has been controlled by the normal fault tectonics since at least the Early Eocene, possibly since the latest Cretaceous when the arc trench system was initiated. Judging from the onland geology in Costa Rica, the ophiolitic

	Location	Rock name	Mean di.	Standard dev.	
A	Byobujima, Mineoka belt (81–3–B)	Brecciated epidote-hornblende schist	−0.27	1.18	tectonic
B	Atago, Mineoka belt (91–3)	Sheared serpentinite	−0.65	2.13	tectonic
C	Kinugasa, Mineoka belt (C–1)	Brecciated serpentinite	1.29	1.40	tectonic
D	Off Bonin (KH82–4, 04–507)	Brecciated ophiolitic rock	0.68	0.66	tectonic
E	Off Bonin (KH82–4, 04–516)	Ophiolitic sandstone	0.81	0.57	sedimentary
F	Fusata, Mineoka belt (10)	Ophiolitic rock-bearing sandstone	0.41	0.75	sedimentary
G	Mineoka-sengen, Mineoka belt (E–6–2)	Ophiolitic rock-bearing sandstone	0.85	0.56	sedimentary

rocks may have been situated in the present setting at least from the Cretaceous by obduction or collision of the oceanic floor or buoyant ridge from the Cocos plate to the Caribbean plate (KUIJPERS, 1980; SCHMIDT-EFFING, 1977; GURSKY *et al.*, 1982).

These dredged and drilled ophiolites shown many similarities to samples collected from oceanic fracture zones in the Mid-Atlantic Ridge (BONATTI *et al.*, 1971, 1974; BONATTI and HAMLYN, 1981; FOX *et al.*, 1976), and in fracture zones identified in some onland ophiolites such as those in Cyprus (SIMONIAN and GASS, 1978) and in the Bay of Islands, New Foundland (KARSOZON and DEWEY, 1978; CASEY *et al.*, 1981) from the viewpoints of metamorphism and deformation. However, basaltic and ultramafic rocks are different between those in fracture zones and dredged materials. Most important difference to each other is that the ultramafic rocks of the forearc ophiolites are rich in harzburgite, while those from the oceanic fracture zones are lherzolite (KUSHIRO, 1983). Basalts in the latter case are definitely MORB type with rare alkalic one, but those in the forearcs are sometimes of an island-arc type (BLOOMER, 1983; BLOOMER and HAWKINS, 1982) with rare alkalic one. Such chemical constrains of the igneous rocks in such tectonic positions restrict the tectonic history of respective areas. After the proposal of the island arc origin of the Troodos ophiolite (MIYASHIRO, 1973), fresh glass in the same samples was determined to be an island arc origin (ROBINSON *et al.*, 1983; SCHMINKE *et al.*, 1983). However, more than one kind of rocks may possibly be mixed together. In such cases so-called ophiolites may have diverse origins of magmatism at several tectonic positions.

According to the metamorphic, deformational and sedimentary evidence outlined above, we considered that the dredged dismembered ophiolitic materials from the forearc areas were probably formed in the ridge-ridge transform faults or oceanic fracture zones as their extentions (OGAWA and NAKA, 1984). In the transform-fracture zones a large extent of serpentinization (FRANCIS, 1981), high temperature/low pressure metamophism and hydrothermal veining occurred together with strong lateral shearing in the stage along the transform faults, and later with strong vertical displacement along the non-transforming segments of fracture zones as shown in Fig. 6 (KARSON and DEWEY, 1978). However, as discussed previously, there is another possibility that most basaltic rocks in the ophiolitic rocks were originated from the island-arc setting, neither from the spreading ridges nor from transform faults. Considering initiation of an island arc within an oceanic plate, a certain trigger for the initiation of subduction is needed at the time of the beginning of it. BLOOMER (1983), who concluded that the ophiolitic rocks in the Mariana forearc are mostly island arc affinities, considered initiation of subduction occurred at the ridge-ridge transform fault as shown in Fig. 7. This implies possibility that the forearc ophiolite may contain some products originated from the mid-oceanic ridge, transform faults or fracture zones (OGAWA and NAKA, 1984).

In these examples, areas characterized by the forearc ophiolites are not strongly compressional zones in the trench slope, but rather areas of extensional stress (HUSSONG and UYEDA, 1981). There are no accretion of trench sediments by offscraping or thrusting to the landward slopes. Tectonic erosion occurred in the Bonin-Mariana forearcs that enables the extensional condition giving rise to diapiric extrusion of the ophiolitic rocks into the trench slope area. The roll-back effect may bring the juvenile island arc rocks, such as the boninite, to the nearby area to the trench axis, finally to mix with the original island arc materials and the dismembered ophiolite.

The problem is, therefore, whether these ophiolitic rocks were definitely originated from the spreading ridge products or from island arc type magmatism. This might be solved by chemical study of the ophiolitic rocks in these areas. BLOOMER and HAWKINS (1982), BLOOMER (1983) and ISHII (1983, 1985) indicate most of the basaltic rocks were of the island arc affinities with additional seamount affinities of alkalic chemistry. Other studies such as IGCP WORKING GROUP (1977) and SHIRAKI (1971) indicate occurrence of amphibolite schist that may have undergone the ocean floor metamorphism before the emplacement to the arc areas. As for the alkali basalt in the Mariana forearc, BLOOMER (1983) mentioned that it was accreted from seaward by offscraping of a seamount. This is a plausible idea, but it is curious that no other oceanic rocks and sediments have been accreted to landward but all of the trench sediments have been subducted in the Mariana trench (HUSSONG and UYEDA, 1981). Alkali basalt and dolerite are also contained in the Guatemala samples (OGAWA *et al.*, 1985a). Therefore, it is still possibile that some of the dismembered ophiolite may involve the seamount blocks which occurred along the fracture zone as indicated by KARSON and DEWEY (1978).

3. *Examples of Onland Forearc Dismembered Ophiolites in Central Japan*

In Central to Southwest Japan, the Cretaceous to Miocene bodies, called the Shimanto Supergroup (TAIRA *et al.*, 1982; WHITAKER, 1982; OGAWA, 1982, 1985), have been accreted to the main Mesozoic "orogenic belts'' composed of paired metamorphic belts. The rocks and strata in the Shimanto belt shows a systematic overall outward younging disposi-

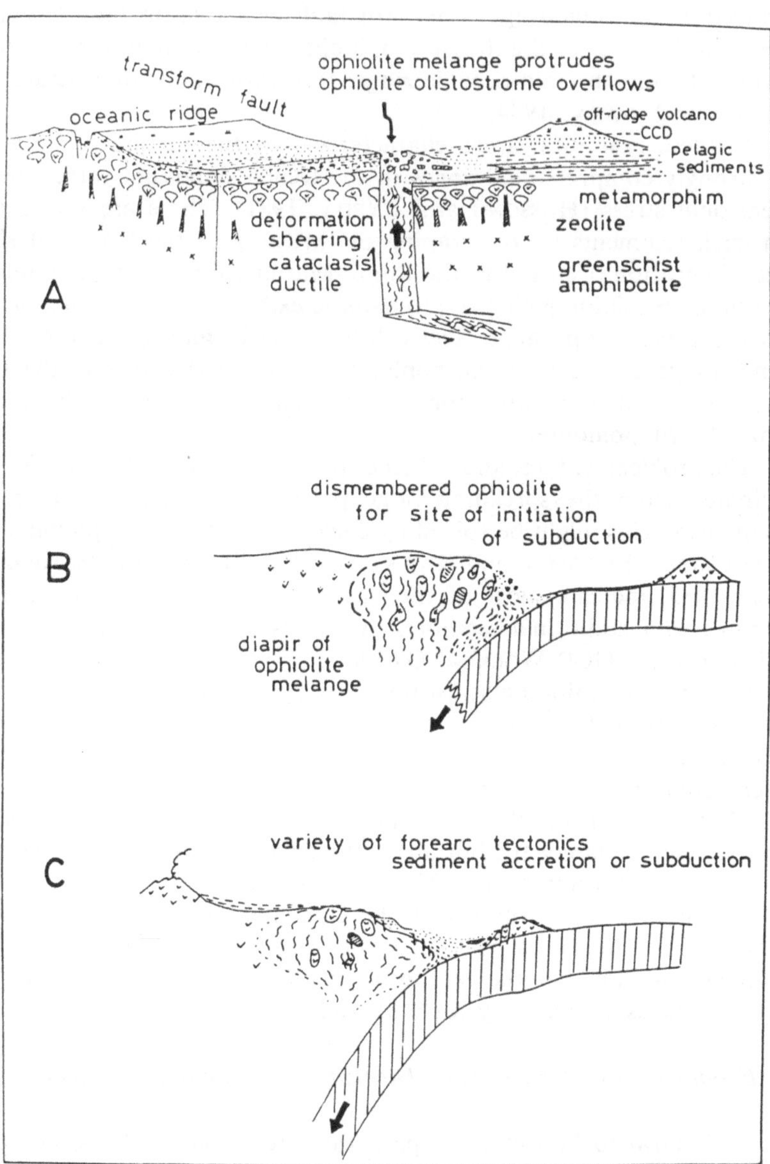
transform fault
oceanic ridge
ophiolite melange protrudes
ophiolite olistostrome overflows
off-ridge volcano
CCD
pelagic sediments
metamorphim
zeolite
greenschist
amphibolite
deformation
shearing
cataclasis
ductile
A
dismembered ophiolite
for site of initiation
of subduction
B
diapir of
ophiolite
melange
variety of forearc tectonics
sediment accretion or subduction
C

Fig. 6.

tion by many thrust sheets having inward younging in each sheet. The thrust sheets are composed of melange, flysch and olistostrome facies (SAKAI and KANMERA, 1981; TAIRA *et al.*, 1980). The melange facies of the thrust sheets comprise basaltic slabs with the superjacent chert or limestone in sheared muddy matrix, but scarcely contain ophiolitic rocks. Only several locations have the dismembered ophiolite in the melange or olistostostromal facies (OGAWA and NAKA, 1984; SUZUKI and HADA, 1983). Above all, the dismembered ophiolite belt in the Setogawa belt in the outer Shimanto belt in the Akaishi Mountains show a characteristic mode of occurrence of the ophiolitic rocks. Also in the Miura-Boso terrane south of Tokyo has the dismembered ophiolite belt, called the Mineoka belt, together with a huge block of the Shimanto Supergroup and other Neogene forearc volcaniclastic rocks from the Izu arc (OGAWA, 1983; OGAWA and NAKA, 1984; OGAWA *et al.*, 1985b). The Setogawa and Mineoka belts are distributed to the west and east of the colliding Izu Peninsula (Fig. 2).

3.1 Setogawa belt

Along the western margin of the Setogawa belt, a narrow but remarkable ophiolitic belt runs, cutting slightly obliquely to the general structural trend of the Setogawa forearc accretionary body (Fig. 8). The ophiolitic belt consists of sheared serpentinite containing huge blocks of pillow basalt of tholeiitic chemistry (OHASHI and SHIRAKI, 1980). Around the ophiolitic bodies, there are various kinds of rocks such as polymictic conglomerate containing amphibolite, plagiogranite, ophiolitic conglomerate, volcanic sandstone, island arc-derived andesitic rocks of high Mg-high SiO_2 content (OHASHI and SHIRAKI, 1980) and altered dacitic breccia. All these rocks are intermingled in the structural belt, a large fault zone, so that the original relation between these rocks are not always clear. Many problems are within the conglomerate. This has not only ophiolitic blocks but also has ophiolitic fragment-bearing conglomerate

Fig. 6. One of the possible modes of emplacement of dismembered ophiolitic rocks to forearc area. A. First ophiolite melange is formed along a ridge-ridge transform fault, metamorphosed and deformed in a high geothermal gradient and under lateral shearing stress. It is then deformed by normal faulting in a non-transform segment of a fracture zone. An off-ridge volcano could be trapped within this zone. The idea is adapted after KARSON and DEWEY (1978). B. A buoyant ridge made up of such an ophiolite melange becomes the site of the initiation of subduction, and the ophiolite melange subsequently occupies the trench landward area. C. Upheaval of the diapir may cause extensional deformation in the forearc region, where trench sediments are chiefly subducted, but the extent of the tectonics may vary due to other factors (OGAWA and NAKA, 1984). See also OGAWA *et al.* (1985a).

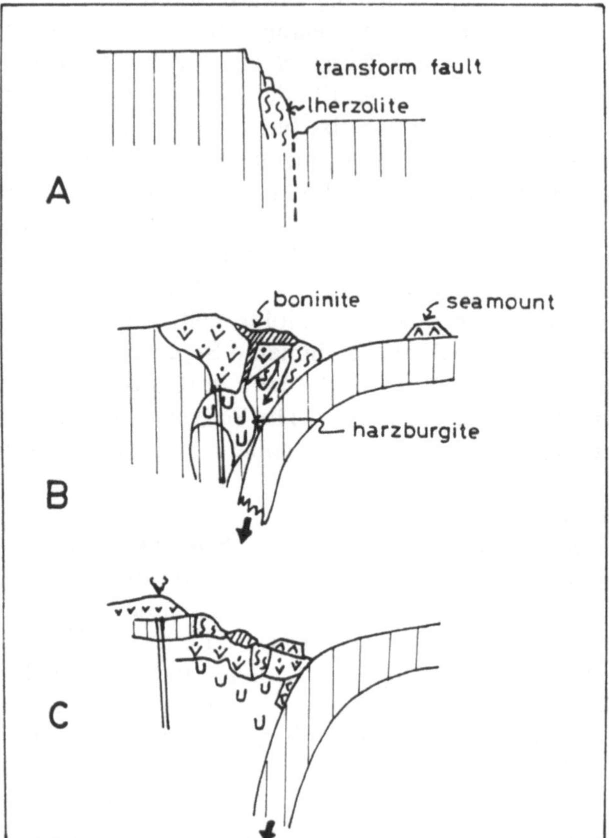

Fig. 7. Alternative possible mode of emplacement of dismembered ophiolitic rocks to forearc area adopted and modified after CRAWFORD *et al.* (1981) and BLOOMER (1983). A First subduction begins at a transform fault, where lherzolite (stretched s-shape) is emplaced and strong serpentinization occurs. B. Next the first arc lavas (dotted v-shape) and boninite (hatched) erupt and a depleted harzburgite (u-shape) is emplaced as the residue after boninite extraction (CRAWFORD *et al.*, 1981; KUSHIRO, 1983). S. A second series of volcanic rocks erupt and a part of a seamount (reversed v-shape) is accreted, then a mixture of these rocks is formed as an ophiolitic melange.

boulders. The boulders in the conglomerate have no continental or island arc derived materials in their grains or matrices but are made of ophiolitic materials, whereas the conglomerates as a whole have many kinds of such rocks and matrices including island-arc and continent-derived rocks as well

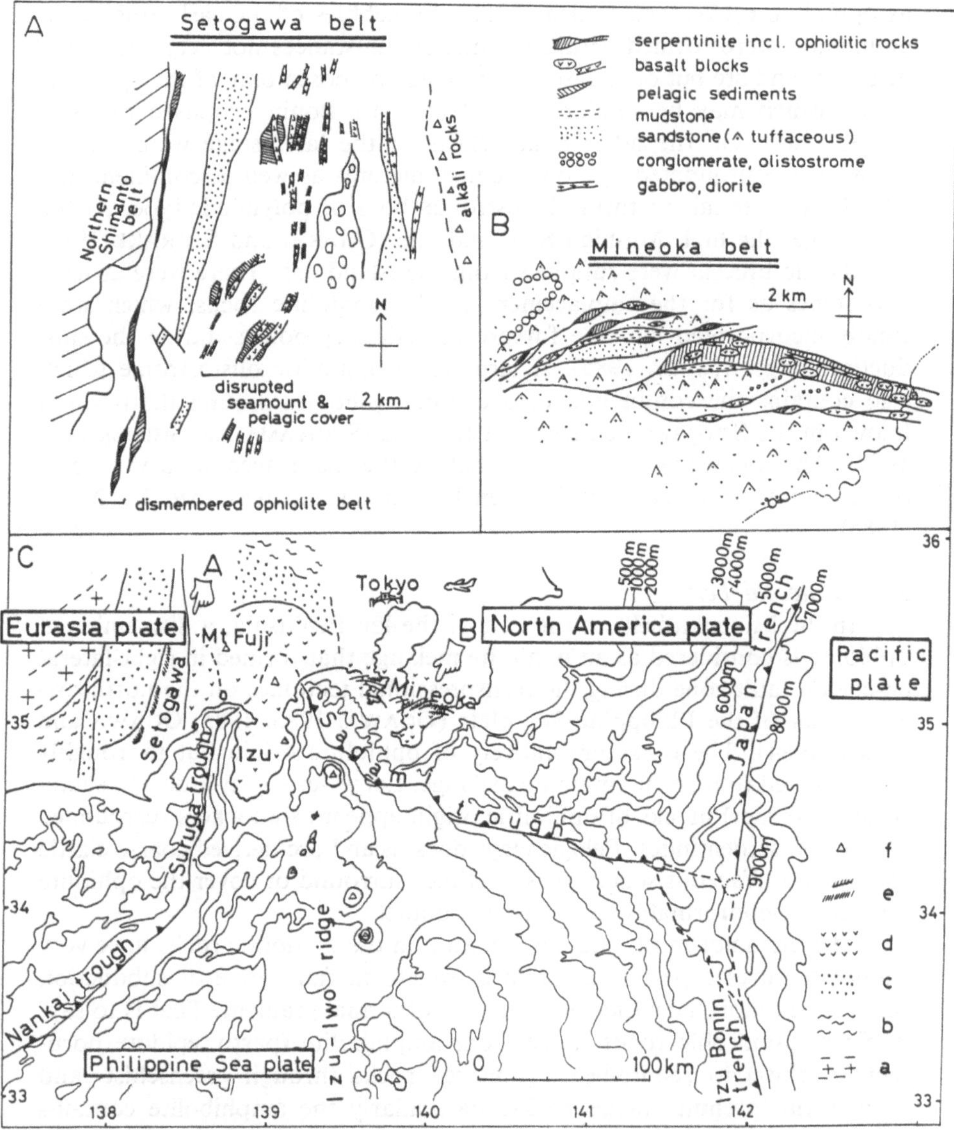

Fig. 8. Generalized maps of the Setogawa and Mineoka belts. A is modified from NAKA (1985) and B is from OGAWA (1983). C indicates that both ophiolite belts are situated on the forearc areas behind the northeastern corner of the Philippine Sea plate, and now largely dislocated by the collision of the Izu-Iwo ridge. Key to the symbols in C: a, Ryoke Metamorphic terrane; b, Sambagawa-Chichibu metamorphic terrane; c, Cretaceous Shimanto terrane; d, Neogene volcanic rocks; e, Ophiolite belt in the Setogawa and Mineoke belts; f, Quaternary volcano.

as ophiolitic rocks. This indicates that the pebbles of the early stage conglomerates were formed within oceanic circumstances not near the continent or island arc but on the isolated mound composed only of the ophiolitic rocks, which may be similar "paleoland" off Bonin proposed by ISHII (1983, 1985). On the other hand, clasts of the later stage were derived from such consolidated ophiolitic conglomerates as well as continent and island arc materials to form olistostromal beds of polymictic type. At the later stage, the high Mg-high SiO_2 andesite (OHASHI and SHIRAKI, 1980) and dacitic breccia were supplied from the island arc. There were at least two stages as for the emplacement of the ophiolitic rocks, which were finally incorporated into the forearc accretionary bodies during the subduction stages (NAKA, 1985). A large scale ophiolitic olistostrome which also contains the continental or accretionary bodies around the basin is known in the Setogawa belt south of the area (SUGIYAMA and SHIMOKAWA, 1980). This ophiolite was accreted before the main accretionary process of seamount fragments to the trench olistostrome described by NAKA (1985).

3.2 Mineoka belt

In the Mineoka belt, to the east of the Izu Peninsula, a dismembered ophiolite is considered as an ophiolite melange that formed in right-lateral en-chelon fault zone during the transpressional tectonics on the northeastern corner of the Philippine Sea plate (OGAWA, 1983, 1985; OGAWA and NAKA, 1984). The rocks are entirely of ophiolitic origin but thoroughly dismembered to form an ophiolite melange. Large-scale (km-order) to small-scale (cm-order) blocks of ophiolitic origin appear as knockers surrounded by sheared serpentinites. Pelagic claystones, island-arc-derived coarse clastic rocks and tuffaceous sediments sometimes surround or cover the ophiolite melange unconformably or in fault contact.

The metamorphism and deformation in the ophiolitic rocks have very similar features expected either in a transform fault zone or subduction related metamorphic rocks under high geothermal gradient. That is to say, mafic metamorphic rocks range from amphibolite (partly epidote hornblende schist) to zeolite-bearing altered rocks through greenschist- and pumpellyite/prehnite facies rocks. Particularly the amphibolite contains the layered intercalation of quartz schist and psammitic schist. They may be the essential ocean floor sediments, probably chert and pelagic or hemipelagic mudstone covering basalt, and brought into the subduction zone of epidote-amphibolite facies. Some other metamorphic rocks of the greenschist or prehnite/pumpellyite facies may be of the same system. The age of mica in the psammitic schist was dated as 38 MaBP which is correlated to the fossil age of the interpillow limdstone and the chert overlying the basalt (OGAWA, 1983). Some mineralizations in gabbro and dolerite are retrogressive. Shearing deformation produced several varieties of cataclastic rocks. Granulization and pulverization are common in gabbro and dolerite. Ductile to brittle cataclasis reduced the gabbroic rocks to mylonite and microbreccia. Zeolite (stilbite, natrolite and analcite) mineralization usually occurred after the main deformation, but some zeolite veins suffered brittle shearing. Serpentinites were originally mostly tectonized harzburgite, but largely sheared and crushed. Some of them resemble sedimentary serpentinite in outcrops, but are actually "cataclasite" of tectonic origin (Fig. 4-A) (OGAWA, 1983). These rocks exist in fault zones, but also appear as sedimentary blocks at several horizons where their emplacement is related to forearc tectonics (OGAWA *et al.*, 1985b).

The rocks of the Shimanto Supergroup are distributed only as huge blocks next to the ophiolite belt. Rocks are either claystone or arkosic sandstone as the common sedimentary rocks of the Shimanto. Beside the Shimanto, there are several kinds of sandstones and conglomerates which bear ophiolitic rocks or other igneous rocks fragments. In the serpentinite and basalt, many layers of serpentine sandstones are intercalated. Serpentine fragments are abundantly contained together with some content of basalt, gabbro and plagiogranite (Fig. 4-B). Serpentine fragments sometimes exceed more than 90% (Fig. 9). Some sandstone and conglomerate have many other kinds of rock fragments and quartz, feldspar as well as serpentinite or ophiolitic rocks fragments (Fig. 4-B-8). Quartz and feldspar grains attain more than 50% in some serpentine bearing sandstone (Fig. 9). Matrices of these sediments are usually calcareous materials. Fragments are usually poorly sorted, and subrounded to angular. Comparing the tectonic breccia of serpentinite and sedimentary serpeninite, the grain distribution pattern is the most distinguishable (Fig. 5). The former has much better sorted than the latter. In addition to the ophiolitic rocks, conglomerate and coarse volcanic sandstone are widely distributed in and around the ophiolitic belt. They contain even andesite or dacite volcanic rocks fragments beside a large amount of quartz and feldspar (Fig. 9) This kind of ophiolitic fragment bearing sandstone occupies the middle horizon of the early group of the forearc setting of the Izu volcanic arc in Early Miocene, about 17 MaBP, judging from the radiolarians (OGAWA *et al.*, 1985b) and silicoflagelates (SAWAMURA and NAKAJIMA, 1980).

The dismembered ophiolite belt in the Mineoka belt is unconformably overlain by the dacitic pumice flow or fall deposits which indicate no strong deformation or alteration. This means that the main stage of emplacement of the dismembered ophiolite belt occurred before the pumice

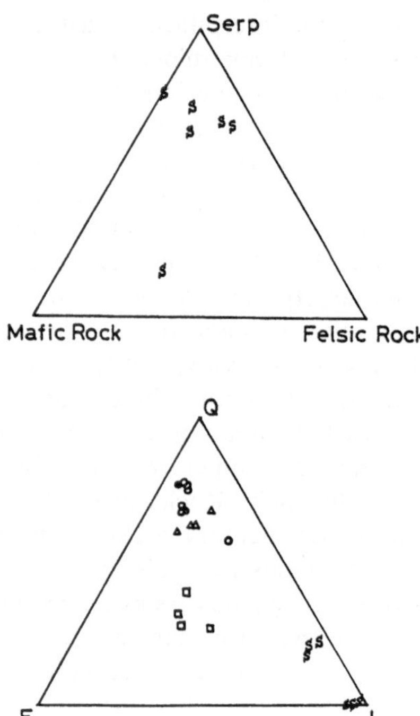

Fig. 9. Triangle diagrams of sandstones in the Mineoka belt. Upper. Ophiolitic rock-bearing sandstones. Mafic rock includes basalt, dolerite, gabbro and their metamorphic derivatives and pyroxene fragments. Felsic rock includes dacite, granodiorite and plagiogranite. Sedimentary rocks and other minerals are excluded. Lower: Q-F-L diagram. L includes rock fragments, serpentine and other minerals than quartz, feldspar and calcite. Symbols in the diagrams: S, serpentine sandstone or ophiolitic rock-bearing sandstone; •, arkosic sandstone intercalated within the serpentine sandstone (this also bears a small amount of serpentine); △, arkosic sandstone in the Kamitsuka Formation, a part of the Shimanto Supergroup; □, lithic sandstone in the lowest member of the Hota Group; ○, arkosic sandstone of the Tomikawa sandstone, northern part of the Hota Group.

fall or flow deposits of Middle Miocene age, before about 15 MaBP. This age is consistent with the age of the abrupt and rapid expansion of the Sea of Japan (OTOFUJI and MATSUDA, 1984), and also with the vast arc volcanism in the Northeast Honshu arc. After that, in the Late Miocene and Pliocene volcaniclastic turbiditic sediments of the Izu forearcs, the ophiolitic rock fragments are intercalated (OGAWA, 1983). They indicate that the ophiolite belt has been on the trench slope area of the Izu forearc since at least Early Miocene, and during the forearc tectonics of the Izu volcanic arc, sediments from the arc itself and from the Honshu arc were mixed with the ophiolitic fragments in the ophiolite belt. Quartz and feldspar grains in such sediments indicate the sedimentary area was facing not only the volcanic arc but the continent. Therefore, the basin was sandwiched between the Izu and Honshu two arcs, in a manner similar to the present situation of the Sagami trough in Fig. 1 (OGAWA, 1983; OGAWA and NAKA, 1984; OGAWA *et al.*, 1985b).

The Mineoka ophiolite belt is an onland expression of the northeastern tip of the southwesterly dipping dense slab, based on magnetic and gravimetric simulations calculated by TONOUCHI and KOBAYASHI (1983). The paleomagnetism also shows that the ophiolite slab originated in the latitude of 24 degrees probably in Late Eocene (about 40 Ma BP). This and the previous estimation of the first appearance of the sedimentary ophiolitic rocks in the Early Miocene sections indicates that from 40 to 20 Ma BP, the ophiolite moved northward by about 1000 km and arrived finally at the Miura-Boso area to provide ophiolitic fragments to the sandwiched forearc sedimentary area. After the first emplacement of the original ophiolitic rocks, the next stage may have been obduction of the slab to the north probably accompanied with the Izu volcanic arc onto the Honshu arc. Then the area became a strong shearing under a right lateral transpressional condition, which is favorable for the right-handed en echelon fault and fold system in and around the Mineoka ophiolite belt (OGAWA, 1983; OGAWA *et al.*, 1985b).

4. *Plate Tectonic Significance of the Forearc Ophiolite*

There are two ideas of the origin of the forearc ophiolite as mentioned above. One is of a transform fault-fracture zone origin and the other of an island arc origin. The former is based upon results that many forearc ophiolites are mixed with various petrologic and petrochemical provenances; island arcs, mid-oceanic ridges and seamounts. As for the amount of serpentinite it favors the product of transform fault or fracture zone processes, where a large amount of serpentinization occurs (FRANCIS, 1981). Moreover, it is highly probable that the transform faults or fracture zones bear various kinds of blocks of ophiolitic rocks which included metamorphic rocks showing high geothermal gradient of low pressure and cataclastic deformation. The rock assemblages should be called serpentinite or ophiolitic melange, and the mass liable to protrude as a diapir along the fault zone, because the sheared serpentinite mass is essentially light and slippery. Thus along the transform fault or fracture zone a line of serpentinite melange is formed. As discussed by KARSON and DEWEY (1978), tectonics of such fault zones are explained by strike-slip motion in the transform segment and by normal fault in the non-transform segment. Many examples mostly in the present fracture zones in the Atlantic Ocean (BONATTI and HAMLYN, 1981) indicate similar or almost the same occurrence of such a dismembered ophiolite belt. During the shearing along the faults, a completely dismembered ophiolite zone may become a buoyant ridge. This ridge supplies ophiolitic fragments in a disorder fashion to form serpentine sand or ophiolitic debris flow around it. When one side of the oceanic plate begins to move to a different direction by an initiation of the ocean plate motion, the denser part of the plate starts to subduct on the buoyant ridge of dismembered ophiolite (KARSON and DEWEY, 1978; OGAWA and NAKA, 1984). When the subduction begins, the other side becomes a hanging wall which subsequently becomes a volcanic arc, and the dismembered ophiolite mass is formed in the trench landward slope area of the arc-trench system. This stage may be seen in the Tonga, Mariana and Bonin areas (Fig. 3).

This is the story presented by OGAWA and NAKA (1984). However, as mentioned previously, not all the basaltic rocks are products of the mid-oceanic ridges, but a considerable amount of the basalt show the island arc affinities. This provides an alternative idea (Fig. 7). This process gives possible contamination of the island arc materials in the serpentinite melange, or most of the rocks within the forearc ophiolite are of the island arc materials from the depleted mantle after the initiation of the island arc as mentioned by BLOOMER and HAWKINS (1982), BLOOMER (1983), ISHII (1983, 1985) and KUSHIRO (1983). In this case it may possible to treat all the metamorphic and deformed rocks in the island arc processes, not along the transform fault nor fracture zone. Thus, at a certain stage of island arc formation in the oceanic area, the dismembered ophiolite-based forearc becomes a particular type of plate boundary.

After emplacement of the forearc ophiolite, the area becomes a place for the forearc sedimentation and tectonics. When the arc volcanic chain provide the clastic sediments to the forearc area, the ophiolite may be covered by the sediments, because the dismembered ophiolite is apt to be upheaved to form an ophiolite diapir and to supply the ophiolite fragments. Such a supply will continue through the stages of the island arc. When the arc-trench system has accumulation of the trench and trench slope sediments to form an accretionary prism, the dismembered ophiolite belt becomes the basement of the prism, and when the further accretion occurs associated with compressional tectonics to form a fold belt or thrust belt, the dismembered ophiolite may be turned to the structural belt within the accretionary belt. Such an example may be the case of the Setogawa ophiolite belt. When the arc collides to the other arc or continent, associated with the ophiolite basement obduction, the area becomes a zone sandwiched between the two arcs. Such an example may be the case of the Mineoka ophiolite belt.

Acknowledgements

This study combines our experiences onboard, chiefly DSDP-IPOD Leg 84 off Guatemala and Hakuho-Maru cruise (KH 82-4) off Bonin, and onland in the Setogawa and Mineoka areas. We thank the shipboard scientists of the two cruises, especially Professor Jean Aubouin, Dr. Roland von Huene, Professor Kazuo Kobayashi, and the geologists who discussed with us onboard, in the field and laboratory. Special thanks are due to Professors Noriyuki Nasu, Kazuo Kobayashi and Kametoshi Kanmera, who gave us the opportunities of the study. A part of the study was supported by the Cooperative Programs provided by the Ocean Research Institute of University of Tokyo (Nos. 81113 and 82108) and by the Grant-in-Aid from the Ministry of Education of Japan (No. 59540491).

REFERENCES

S. ARAI, T. ITO, and K. OZAWA, Ultramafic clastic rocks from the Mineoka belt, central Japan, *J. Geol. Soc. Japan*, **89**, 287–297, 1983.

J. AUBOUIN, R. VON HUENE, M. BALTUCK, R. ARNOTT, J. BOURGOIS, M. FILWICZ, R. HELM, K. KVENVOLDEN, B. LIENERT, T. MCDONALD, K. MCDOUGAL, Y. OGAWA, E. TAYLOR, and B. WINSBOROUGH, Leg 84 of the Deep Sea Drilling Project, Subduction without accretion: Middle America Trench off Guatemala, *Nature*, **297**, 458–460, 1982.

S. H. BLOOMER, Distribution and origin of igneous rocks from the landward slopes of the Mariana trench: Implications for its structure and evolution, *J. Geophys. Res.*, **88**, 7411–7428, 1983.

S. H. BLOOMER and J. W. HAWKINS, Gabbroic and ultramafic rock from the Mariana trench: An island arc ophiolite, in *Tectonic and Geologic Evolution of Southeast Asia, Seas and Islands.*, D. Hayes (ed.), Pt. 2, AGU Monograph Ser. 27, pp. 294–317, 1982.

E. BONATTI, Serpentinite protrusions in the oceanic crust, *Earth Planet. Sci. Lett.*, **32**, 107–113, 1976.

E. BONATTI and P. R. HAMLYN, Oceanic ultramafic rocks, in *The Sea*, **7**, C. Emiliani (ed.), pp. 241–283, Wiley-Intersci. N.Y., 1981.

J. F. CASEY, J. F. DEWEY, P. J. FOX, J. A. KARSON, and E. ROSENCRANTS, Heterogenous nature of oceanic crust and upper mantle: A perspective from the Bay of Island ophiolite complex, in *The Sea*, 7, C. Emiliani (ed.), pp. 305–338, Wiley-Intersci. N. Y., 1981.

A. J. CRAWFORD, L. BECCALUVA, and G. SERRI, Tectono-magnetic evolution of the west Philippine-Mariana region and the origin of boninites, *Earth Planet. Sci. Lett.*, **54**, 346–356, 1981.

V. DIETRICH, R. EMMERMANN, R. OBERHANSLI, and H. PUCHELT, Geochemistry of basaltic and gabbroic rocks from the west Mariana basin and the Mariana trench, *Earth Planet. Sci. Lett.*, **39**, 127–144, 1978.

C. EWANS and J. W. HAWKINS, Mariana arc-trench system: Petrology of 'seamounts' on the trench-slope break, *EOS*, **60**, 968, 1979.

R. L. FISHER and C. G. ENGEL, Ultramafic and basaltic rocks dredged from the nearshore flank of the Tonga trench, *Bull. Geol. Soc. Am.*, **80**, 1373–1378, 1969.

P. J. FOX, E. SCHREIBER, H. ROWLETT, and K. MCCAMY, The geology of the Oceanographer fracture zone: A model for fracture zone, *J. Geol. Res.*, **81**, 4117–4128, 1976.

T. J. G. FRANCIS, Serpentinization faults and their role in the tectonics of slow spreading ridges, *J. Geophys. Res.*, **86**, 11616–11622, 1981.

I. C. GASS and J. D. SMEWING, Ophiolites: obducted oceanic lithosphere, in *The Sea*, **7**, C. Emiliani (ed.), pp. 339–362, Wiley Intersci., N.Y., 1981.

H.-J. GURSKY, R. SCHMIDT-EFFING, M. STREBIN, and H. WILDBERG, The ophiolite sequence in northwest Costa Rica (Nicoya Comlex); Outlines of stratigraphical, geochemical, sedimentological and tectonical data, *Quinto Congreso Latinoamerican de Geologia, Argentina*, Actas, **11**, 607–619, 1982.

J. HAWKINS, S. BLOOMER, C. EVANS, and J. MELCHIOR, Mariana arc-trench system: Petrology of the inner trench wall, *EOS*, **60**, 968, 1979.

E. HONZA and H. KAGAMI, A possible accretion accompanied by ophiolite in the Mariana trench, *J. Geography, Tokyo*, **86**, 80–91, 1977.

D. M. HUSSONG and S. UYEDA, Tectonic processes and the history of the Mariana arc—A synthesis of the results of DSDP Leg 60, *Init. Rep. D.S.D.P.*, **60**, 909–929, 1981.

D. M. HUSSONG and P. FRYER, Fore-arc tectonics in the northern Mariana arc, in *Abs. Oji Internat. Seminar on Formation of Ocean Margins*, pp. 32, 1983.

IGCP WORKING GROUP, 'Ophiolites', Initial report of the geological study of oceanic crust of the Philippine Sea floor, *Ofioliti*, **2**, 137–168, 1977.

T. ISHII, Origin of the Ogasawara fore-arc seamount or "Ogasawara Paleoland", in *Abs. Oji Internat. Seminar on Formation of Ocean Margins*, pp. 91, 1983.

T. ISHII, Dredged samples from the Ogasawara fore-arc seamount or "Ogasawara Paleoland"—"forearc ophiolite", this volume.

J. KARSON and J. F. DEWEY, Coastal complex, western Newfoundland: An Early Ordovician oceanic fracture zone, *Bull. Geol. Soc. Am.*, **89**, 1037–1049, 1978.

K. KOBAYASHI (ed.), *Preliminary Report of Hakuho Maru Cruise KH82-4, Ocean Res. Inst., Univ. of Tokyo*, 267 pp., 1983.

K. KOBAYASHI, Fore-arc volcanism and cycles of subduction, in *Arc Volcanism, Physics and Tectonics*, D. Shimozuru and I. Yokoyama (eds.), pp. 153–163, Terra Pub., Tokyo, 1983.

E. P. KUIJPERS, The geologic history of the Nicoya ophiolite complex, Costa Rica, and its geotectonic significance, *Tectonophysics*, **68**, 233–255, 1981.

I. KUSHIRO, Generation of magmas and upper mantle materials in Japanese Islands, in *Abs. Oji Internat. Seminar on Formation of Ocean Margins*, pp. 44, 1983.

A. MIYASHIRO, The Troodos Ophiolite Complex was probably formed in an island arc, *Earth Planet. Sci. Lett.*, **19**, 218–224, 1973.

J. NAKA, Broken seamount fragments in the Setogawa subduction complex, this volume.

J. NAKA and S. UEHARA, Igneous rocks dredged from Sites KH82-4-3 and -4, in *Preliminary Reports of the Hakuho Maru Cruise, KH-82-4*, K. Kobayashi (ed.), pp. 187–194, 1982.

Y. OGAWA, Tectonics of some forearc fold belts in and around the arc-arc crossing area in central Japan, *Geol. Soc. London Spec. Pub.*, **10**, 49–61, 1982.

Y. OGAWA, Mineoka ophiolite belt in the Izu forearc area—Neogene accretion of oceanic and island arc assemblages in the northeastern corner of the Philippine Sea plate, in *Accretion Tectonics in Circum-Pacific Regions*, M. Hashimoto and S. Uyeda (eds.), pp. 235–250, Terra Sci. Pub., Tokyo, 1983.

Y. Ogawa, Variety of subduction and accretion processes in Cretaceous to Recent plate boundaries around southwest and central Japan, *Tectonophysics*, **112**, 493–518, 1985.

Y. Ogawa, K. Fujioka, T. Nishiyama, S. Uehara, and M. Nakagawa, Deformational characteristics and their tectonic significance of the ophiolitic rocks of the Middle America Trench landward slope off Guatemala, Deep Sea Drilling Project Leg 84, in *Init. Rep. DSDP*, **84**, 791–809, 1985a.

Y. Ogawa, K. Horiuchi, H. Taniguchi, and J. Naka, Collision of the Izu arc with Honshu and the effects of oblique subduction in the Miura-Boso peninsulas, *Tectonophysics*, **119**, 349–379, 1985b.

Y. Ogawa and Y. Miyata, Vein structure and its deformational history in the sedimentary rocks of the Middle America Trench slope off Guatemala, Deep Sea Drilling Project Leg 84, in *Init. Rep. DSDP*, **84**, 811–829, 1985.

Y. Ogawa and J. Naka, Emplacement of ophiolitic rocks in forearc areas, *Geol. Soc. London, Spec. Pub.*, **13**, 235–263, 1984.

F. Ohashi and K. Shiraki, High-magnesian and high silica andesite in the Setogawa ophiolite, *J. Petrol. Mineral. Economic Geol.*, **76** 69–79, 1981.

Y. Otofuji and T. Matsuda, Paleomagnetic evidence for the clock-wise rotation of Southwest Japan, *Earth Planet. Sci. Lett.*, **62**, 349–359, 1983.

P. T. Robinson, W. G. Melson, T. O'Hearn, and H.-U. Schmincke, Volcanic glass composition of the Troodos ophiolite, Cyprus, *Geology*, **11**, 400–404, 1983.

T. Sakai and K. Kanmera, Stratigraphy of the Shimanto terrane and the tectono-stratigraphic setting of greestone of the northern part of Miyazaki Prefecture, Kyushu, *Sci. Rep. Dept. Geol. Kyushu Univ.*, **14**, 31–48, 1981.

K. Sawamura and T. Nakajima, Miocene silicoflagellate zones in the Boso Peninsula, *Bull. Geol. Surv. Japan*, **31**, 333–345, 1980.

R. Schmidt-Effing, Alter und Genese des Nicoya-Komplexes, ein ozeanischer Paläokurste (Oberjura bis Eozan) im südlichen Zentral Amerika, *Geol. Rundsch.*, **68**, 457–494, 1979.

H.-U. Schmincke, M. Rantenschlein, P. T. Robinson, and J. M. Mehgan, Troodos extrusive seriese of Cyprus: A comparison with oceanic crust, *Geology*, **11**, 405–409, 1983.

A. Ya. Sharaskin, N. L. Dobretsov, and N. V. Sobolov, Marianites: The clinoenstatite bearing pillow-lavas associated with the ophiolite assemblage of Mariana trench, in *Ophiolite*, pp. 473–479, Geol. Surv. Cyprus, 1980.

K. Shiraki, Metamorphic basement rocks of Yap Islands, West Pacific, Possible crust beneath an island arc, *Earth Planet. Sci. Lett.*, **13**, 167–174, 1977.

Y. Sugiyama and K. Shimokawa, A paleogeographic study of the Paleogene Setogawa Group, Shizuoka Prefecture, Central Japan, *J. Geol. Soc. Japan*, **87**, 439–456, 1981.

J. Suppe, J. G. Liou, and W. G. Ernst, Paleogeographic origins of the Miocene East Taiwan ophiolite, *Am. J. Sci.*, **281**, 228–246, 1981.

H. Suzuki and S. Hada, Accretionary melange of Cretaceous age in the Shimanto belt in Japan, in *Accretion Tectonics in the Circum-Pacific Regions*, M. Hashimoto and S. Uyeda (eds.), pp. 219–230, Terra Sci. Pub., Tokyo, 1983.

A. Taira, M. Tashiro, M. Okamura, and J. Katto, The geology of the Shimanto belt in Kochi Prefecture, Shikoku, Japan in *Geology and Paleontology of the Shimanto belt—Selected Papers in honor of Prof. Jiro Katto*, A. Taira and M. Tashiro (eds.), pp. 319–398, Rinya Kosaikai Press, Kochi, 1980.

A. Taira, H. Okada, J. H. McD. Whitaker, and A. J. Smith, The Shimanto belt of Japan: Cretaceous-Lower Miocene sedimentation in forearc basin to deep sea trench environments, *Geol. Soc. London Spec. Pub.*, **10**, 5–26, 1982.

H. TANIGUCHI, *Geology of the Mineoka ophiolite belt in Boso Peninsula, Japan*, 75 pp., Master Tesis, Nihon Univ., 1984.

H. TATSUMI, High magnesian andesites in the Setouchi volcanic belt, Southwest Japan and their possible relation to the evolutionary history of the Shikoku inter-arc basin. *Geodynamics Western Pacific-Indonesian Region, Geodynamics Ser.*, **11**, 331–341, 1981.

S. TONOUCHI and K. KOBAYASHI, Paleomagnetic and geotectonic investigation of ophiolite suites and surrounding rocks in south-central Honshu, Japan, in *Accretion Tectonics in the Circum-Pacific Regions*, M. Hashimoto and S. Uyeda (eds.), pp. 261–288, Terra Sci. Pub., Tokyo, 1983.

S. UEHARA and J. NAKA, Serpentines dredged from KH82-4-4, in *Preliminary Rep. Hakuho Maru cruise KH82-4*, K. Kobayashi (ed.), pp. 195–205, 1983.

J. H. McD. WHITAKER, Cretaceous-Paleogene geology of Southwest Japan, *Proc. Geol. Assoc.*, **93**, 829–246, 1982.

Formation of Active Ocean Margins, edited by N. Nasu *et al.*, pp. 747–773.

BROKEN SEAMOUNT FRAGMENTS IN THE SETOGAWA SUBDUCTION COMPLEX

Jiro NAKA

Department of Geology, Faculty of Science, Kyusyu University, Fukuoka 812, Japan

Abstract. Blocks and large slabs of greenstone and pelagic claystone are scattered in argillite mélange zones of the Eocene to Lower Miocene Setogawa Terrane located in the outer half of the Shimanto Belt in the Akaishi mountains of Central Japan. Their specific constituents are, in addition to pillow lavas, a thick pile of volcaniclastic deposits, namely boulderly to granular debris flow beds, coarse to medium grained turbidites, and thin-bedded to laminated, well-bedded and well-sorted fine-grained sandy to silty contourites. These volcaniclastic beds occur as one of three lithologic associations: 1) lava flows–turbidites; 2) lava flows–debris flow beds–turbidites; 3) contourite with debris flow interbeds. The volcaniclastic sections of the former two are conformably overlain successively by calcareous claystones, 10 to 20 m thick, and siliceous claystones about 100 m in maximum thickness, both of which are of pelagic nature. Throughout the volcanic and sedimentary sections no terrigenous clastic beds are intercalated.

Field and petrographic evidence indicates that the volcanic rocks are deep-sea lavas including pillow lavas, massive lavas and minor sheet flows, all having basaltic composition. In contrast, the volcaniclastic beds include, besides the clasts of micro-porphyrtic basalt, a few water-worn pebbles and cobbles of porphyritic rocks which have large and abundant vesicules, a large amount of phenocrysts and high SiO_2 content. These features suggest that the volcaniclastic debris flow and turbidite beds contain materials derived and transported from shallow-sea levels by mass movement.

The lithostratigraphic sequence of the volcanic and sedimentary members and the lithologic characters of the clasts contained in the volcaniclastic beds suggest that the greenstone and pelagic claystone blocks and slabs can be attributed to the remnants of the rock masses formed at the mid-flank to the foot of a large volcanic seamount, the top of which probably reached sea surface. The dismemberment of the inferred seamount seems to involve disruption caused by normal faults to form a steeply standing escarpment as seen in the Daiichi-Kashima seamount followed by gravity sliding along the structural plane and sedimentary weak beds.

1. Introduction

Discrete greenstone masses of various sizes are contained in late Mesozoic and Tertiary rocks of the Shimanto Belt of the Outer Zone of Southwest Japan, which are considered to be accretionary orogenic terranes (KANMERA and SAKAI, 1975; TAIRA *et al.*, 1980). They are known to occur in mélange zones characteristically comprising of intensely deformed and cleaved black argillite which lies below a thick turbidite sequence of flysch facies. There has been much controversy as to whether they are remnants of the oceanic crust or of oceanic volcanic seamounts or intrusives in the terrigenous clastic beds (e.g. SUZUKI and HADA, 1979; TSUCHIYA *et al.*, 1979; TOKUOKA *et al.*, 1981).

The greenstones of the Cretaceous terranes in the northern half of the Shimanto Belt consist mostly of pillow lavas associated with thin interbedded cherts and volcanic claystones of 50 m in thickness. OKAMURA and TAIRA (1982) dated the pillow lava-chert sequency by radiolaria and showed that their ages range from the Valanginian to the Cenomanian. However, the greenstones of the Tertiary terranes in the southern half of the Shimanto also occur in the Eocene beds of the Mimi river area in the eastern Kyushu. Here they are composed of pillowed, sheeted and massive aphyric lavas, porphyritic lavas of basaltic composition, and pillow breccias, and can be ascribed to the remnants of small volcanic seamounts (TSUCHIYA *et al.*, 1979).

The greenstones are widely scattered in the Setogawa Terrane, the Tertiary terrane in the Shizuoka-Yamanashi Prefectures in central Japan. Those in the upper reaches of the Abe river, about 30 km north of Shizuoka City, were investigated (Figs. 1 and 2). The greenstones in this area show a characteristic feature in their association with, in addition to pillow lavas, a thick sequence of volcaniclastic debris flow deposits and turbidites derived from shallow-water levels by mass movements. Such an association is a clear evidence indicating that the Setogawa greenstones are the remnants of an oceanic volcanic seamount whose top probably reached sea surface.

This paper describes the field occurrence, stratigraphic sequence and

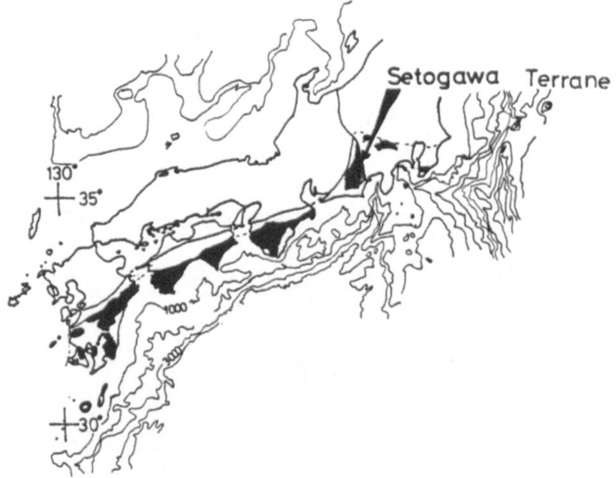

Fig. 1. Index map showing the distribution of the Shimanto Belt (dark area) and the Setogawa Terrane.

lithology of the greenstones and associated rocks. The author speculates the origin and tectonic processes of dismemberment of a seamount.

2. *Geologic Setting*

The Setogawa Terrane in the Akaishi-Minobu Mountains distributed from Shizuoka Prefecture to the western part of Yamanashi Prefecture in the central Japan is the outermost part of the Shimanto Belt (Fig. 2). It is known to contain the Eocene to the Lower Miocene rocks (HONJO and MINOURA, 1968; IIJIMA *et al.*, 1979; IIJIMA *et al.*, 1981). The area investigated in this paper is the upper reach of the Abe River which flows into the Suruga Bay south of Shizuoka. The Setogawa Terrane in this area strikes north-south and steeply dips west. It is in contact by fault with the Cretaceous rocks of the inner half zone of the Shimanto Belt to the west and with the Ryuso Group to the east, the latter being composed of alkalic volcanic and pyroclastic rocks.

The Setogawa Terrane is divided by faults into the Utogi, Kamiochiai and Kuchisakamoto mélange units from east to west (Fig. 2). All these units show a lithologic sequence similar to each other in the following ascending order: (1) cleaved argillite containing various amount of angular to subangular terrigenous sandstone clasts and exotic blocks of a greenstone-

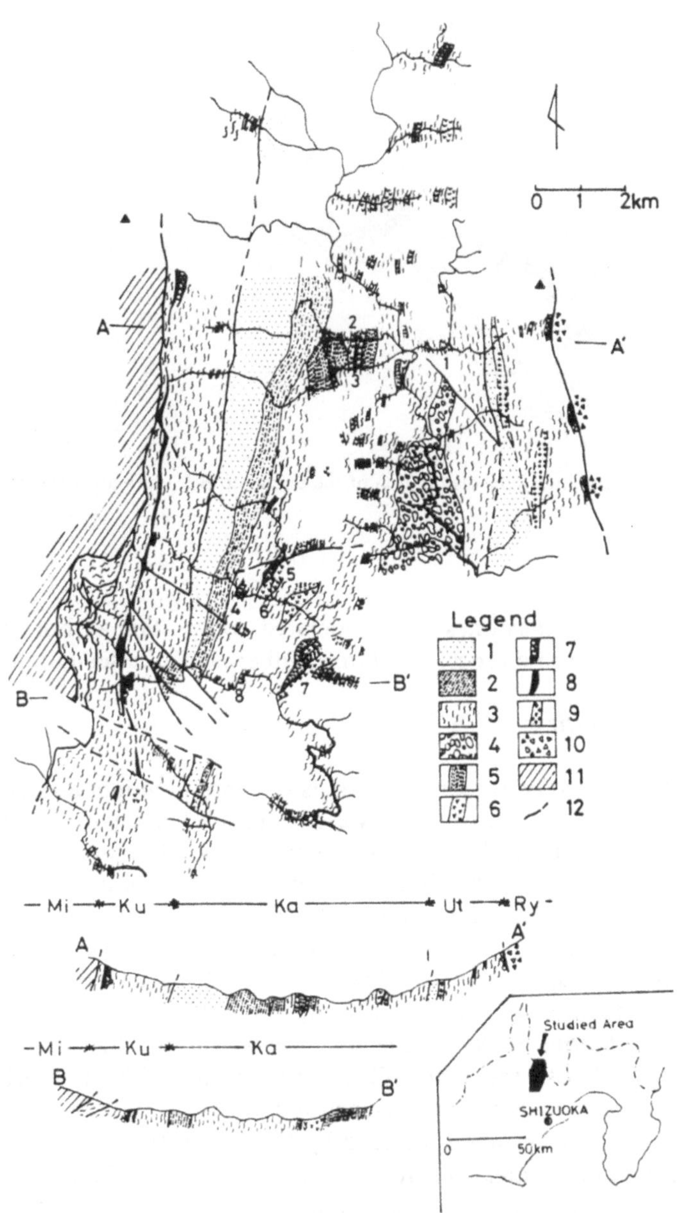
0 1 2km
A
A'
B
B'
Legend
1
2
3
4
5
6
7
8
9
10
11
12
Mi
Ku
Ka
Ut
Ry
Studied Area
SHIZUOKA
0
50km

claystone sequence, (2) cleaved argillite without exotic blocks, (3) cleaved argillite accompanied by minor amounts of thin- to thick-bedded terrigenous turbidite sandstones (muddy turbidite), (4) alternation of terrigenous sandstones and equally thick shales (normal turbidite), (5) thick sandstone beds accompanied by lesser amount of shales (sandy turbidite). The sequence is best represented in the Kamiochiai unit. The Kuchisakamoto unit lacks members (4) and (5).

There are several differences among the three units in occurrence, type and abundance of exotic blocks. The Utogi unit in the eastern belt contains a small quantity of volcanic rocks and calcareous claystone blocks smaller than a few meters in size. Whereas the other two units have many different types of exotic blocks, the Kuchisakamoto unit of the western belt are mainly composed of remnants of an ophiolitic assemblage including somewhat serpentinized harzburgite, dunite and clinopyroxenite and ophicalcite, gabbro, dolerite and basalt (OGAWA and NAKA, 1984). Occasionally blocks of stratified beds composed of clasts from these rocks are recognized (ARAI *et al.*, 1978). These blocks range from several tens of meters across to granule size and are distributed in a form of a mélange of argillite, close to a narrow zone of fragmented and mixed bodies of ultramafic to mafic rocks as mentioned above or fissile serpentinite.

The exotic blocks contained in the Kuchisakamoto unit of the central belt are greenstones (mainly basaltic lavas and hyaloclastites) and siliceous and calcareous pelagic claystones. Their maximum thickness amounts to 300 m. No blocks of ultramafic rocks are found in this unit.

In the following section mode of occurrence, stratigraphy and lithology of the blocks in the Kamiochiai unit are described and their origin is considered.

3. *Occurrence and Lithology of the Exotic Blocks in the Kamiochiai Unit*

The exotic blocks of the Kamiochiai unit show a particular lithologic

Fig. 2. Geological map of the Setogawa Terrane in the upper reaches of Abe River. 1, thick sandstone beds; 2, alternation of sandstone and mudstone; 3, cleaved argillite with sandstone clasts; 4, dismembered thick sandstone beds; 5, thin-bedded claystone; 6, greenstones; 7, conglomerate or clastic breccia with many kinds of exotic blocks; 8, serpentinite (including detrital serpentinite); 9, gabbro dyke; 10, pyroclastic rocks of Ryuso Group (Ry); 11, Cretaceous Mikura Group (Mi); 12, fault. Ku, Kuchisakamoto Unit; Ka, Kamiochiai Unit; Ut, Utogi Unit; Numerical figures 1-8 in the map indicate locations of representative exotic blocks from which the columnar sections shown in Fig. 3 were observed.

assemblage comprising of greenstone and conformably overly pelagic claystone, although isolated blocks of only greenstone or claystone are commonly encountered. The greenstone consists of basaltic lava flows, doleritic dykes, gabbroic sills, and hyaloclastites of various types including reworked volcaniclastic debris flow deposits, turbidites and contourites. The claystone consists of siliceous and calcareous claystone and clayey limestone. The mode of occurrence and general lithology of these rocks are as follows:

3.1 Greenstones

3.1.1 Basaltic lavas

Three types of basaltic lava flows; pillow, sheet, and massive flows, are distinguished. Most common are pillow lavas, in which the pillows are mostly close-packed and very rarely isolated and generally have an ellipsoidal to elongate form with a maximum diameter of 100 cm in a given outcrop. The sheet flows are distinguished from massive lavas by a plain glassy chilled margin and their thickness are generally 1–2 cm. The massive lavas are apparently homogeneous with no distinct internal structures and more than 2 m thick. This designation of massive lavas is merely due to the unfavourable conditions of exposure. The reason is that the rocks in this area are exposed only along roadcuts and narrow canyons. As only the lower part remains uncovered with thick vegetation, it is difficult to know the attitude of lavas when their thickness exceeds the extent of an exposure.

3.1.2 Intusive bodies (doleritic dyke and gabbroic sill)

Two kinds of intrusive igneous rocks are recognized. One is about 10 m thick gabbroic sills, which occur in the volcaniclastic contourites mentioned below. The other is doleritic dykes intruding into lavas. They show an ophitic texture. Both kinds of intrusive bodies have chilled margins.

3.1.3 Hyaloclastite

Basaltic lava flows in a form of the blocks and slabs contained in the Kamiochiai unit are overlain by volcaniclastic rocks which contain a minor amount of hyaloclastite (in the narrow sense of RITTMANN, 1962) and large quantities of epiclastic or volcanic-sedimentary rocks of similar composition. As the latter contains sub-rounded to rounded volcanic clasts and exhibits bedding planes, cross-bedding and other sedimentary structures, it cannot be included in the category of hyaloclastite defined by RITTMANN (1962). However, it contains altered glass shards or drops caused by autocomminution upon subaqeuous quenching. Accordingly, it can be regarded as reworked hyaloclastites. These hyaloclastic rocks are of basic to intermediate composition, and dark to pale-green when fresh. They are classified into the following four types:

a) Initial hyaloclastite

This occurs directly above and sometimes between the pillow lavas mentioned above, and often grades downwards or laterally into isolated or broken pillow breccias. It is less than a few meters thick and consists of sand to granule-sized angular fragments of an extremely fine-grained altered basaltic rocks, plagioclase crystals affected by albitization to some degree, clinopyroxene crystals and chloritized amoebiform glass shards which are droplets produced by subaqueous quenching of lavas.

b) Volcaniclastic debris flow deposit

This is typically exposed at Locs. 2 and 3 in the blocks along the Sango and Nigori valleys (Fig. 3), and is composed of unsorted to poorly-sorted mixtures of volcanic debris or breccias of varying sizes from sand to boulders. Occasionally large boulders with 5 m diameter are found. The debris contained are irregular in shape, generally angular to subangular, rarely subrounded to rounded. These debris beds are crudely stratified into massive beds, usually 2 to 5 m thick, but usually have no distinct internal structure, although lateral and vertical changes of grain-size are rapid. However, comparatively fine-grained beds with granule to pebble-size debris usually show rough grading and incipient cross-bedding in their upper part.

In thin section, the fine-grained beds and matrices of the coarse-grained beds have a feature similar to the hyaloclastite described above. Many of the large debris fragments have partially chilled rims, although fragments with no chilled margin are not uncommon. These features suggest that the volcanic debris are largely of pillow-breccia or broken pillow breccia origin.

General lack of sorting and internal stratification, coarse texture, angularity of composing fragments, and rapid lateral and vertical changes of grain-size suggest that these coarse-grained volcanic breccia beds are deposits from submarine slides or debris flows.

c) Volcaniclastic turbidite

Volcaniclastic turbidite beds are typically seen in the sections at Locs. 2, 3 and 4 along the Nigori, Sango and Senmata valleys, where they lie conformably on the debris flow deposits (Fig. 3). They also occur directly on the lavas as in the section at Loc. 7 along the Senmata valley. They are pale brownish-green and are well-bedded with a thickness from 20 to 50 cm. They are composed of moderately well-sorted, medium- to very coarse-grained volcaniclastic sandstone which grades up to siltstone within a single bed. The graded bedding and other structures in a turbidite are well recognized in many of the beds. Each of these turbidite beds is often

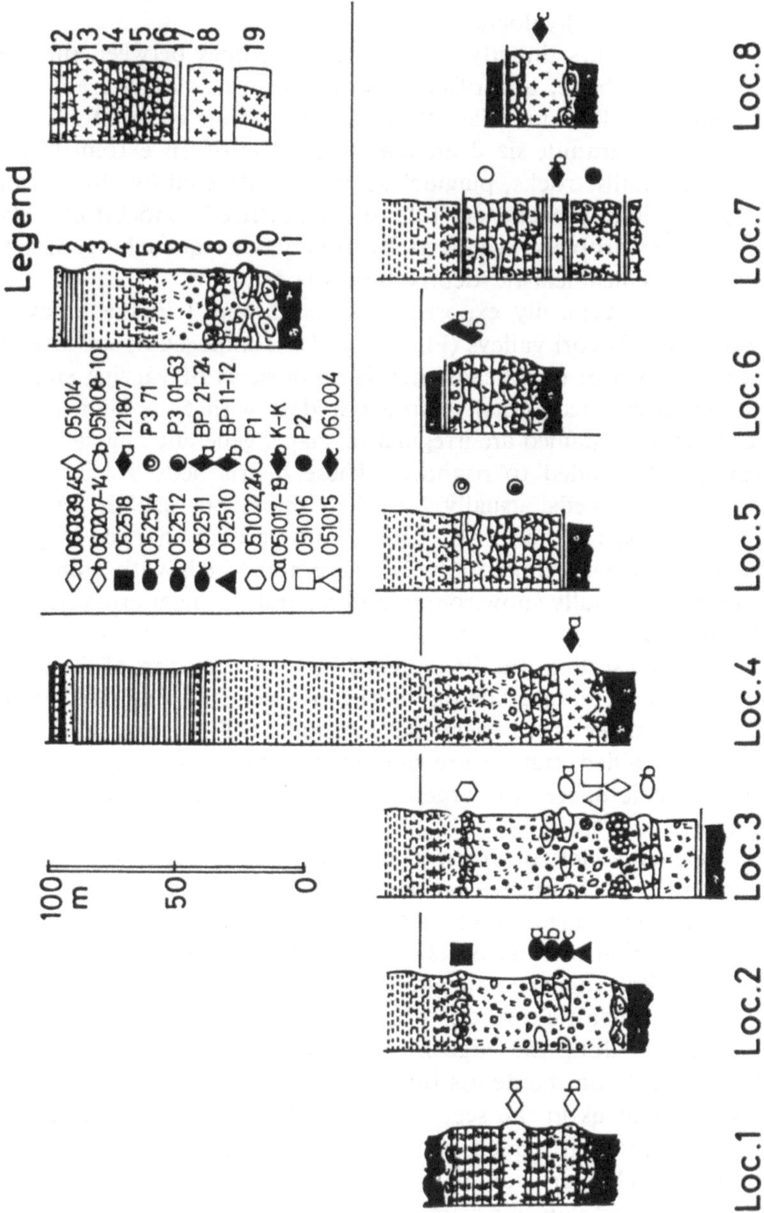
Legend
100 m
50
0
Loc.1
Loc.2
Loc.3
Loc.4
Loc.5
Loc.6
Loc.7
Loc.8

Fig. 3. Columnar sections comprising of lithologic succession and association in representative blocks at localities indicated in Fig. 2. 1, terrigenous sandstone; 2, massive black claystone; 3, thin-bedded siliceous claystone; 4, thin-bedded calcareous claystone; 5, alternating beds of volcaniclastic turbidite and thin-bedded calcareous claystone; 6–9; boulder to granular debris flow deposit (6, granular to coarse sand-size debris; 7, predominantly pebble-size debris with rubble; 8, rubble; 9, large volcanic blocks); 10, mixture zone of greenstone and argillite; 11, argillite with sandstone clasts; 12, volcaniclastic contourite; 13, gabbroic sill; 14, sheet flow lava; 15, pillow lava (flattened type); 16, pillow lava (cylindrical and boulbous type); 17, fault; 18, doleritic mass without chilled margins; 19, doleritic dyke with chilled margins. Other symbols indicate the site of the analysed samples, referred in Table 1.

overlain by a thin bed of calcareous claystone.

Microscopic examination reveals that sand grains of the turbidite beds are subangular to subrounded and comprise somewhat albitized plagioclase crystals, clinopyroxene crystals, volcanic fragments and greenish altered volcanic glass. The last is not always clearly distinguishable from the matrix which consists of calcite, chlorite and some other clay minerals.

d) Volcaniclastic contourite

Volcaniclastic contourite beds are typically found in the blocks from the section at Loc. 1 along the Sekinosawa valley where a 5-m thick debris flow bed is intercalated in the lower part of the section. The contourite beds are thin-bedded with a thickness of less than 30 cm and are made of well-sorted fine-grained sandstone and siltstone. In these beds parallel lamination and small-scale cross-bedding are very common, but no graded bedding could be recognized. Bioturbated beds were occasionally encountered.

In thin section under microscope, the contourites consist of rounded grains of brownish altered glass shards and other volcanic fragments set in a matrix of finer grains of similar composition and a calcedonic quartz cement.

3.2 Calcareous and siliceous claystones

In the blocks of the Kamiochiai unit, the greenstone including hyaloclastic and volcaniclastic rocks mentioned above, are conformably overlain by light-grey to grey calcareous claystone and dark-grey to black siliceous claystone (Fig. 3). In the lower part of this claystone sequence, about 20 m thick, is composed of calcareous claystone, but in the upper section siliceous claystone becomes more dominant. As some beds are highly calcareous and siliceous, they may be called argillaceous limestone and argillaceous chert, respectively. There are also some beds of calcareous and siliceous claystone. The most complete succession is seen in the section at Loc. 4 along the Senmata valley, where the siliceous claystone attains the maximum thickness of 80 m (Fig. 3). In all sections no terrigenous clastic deposits coarser than clay are found, but some volcaniclastic turbidite beds of 50 cm in thickness are intercalated into the lower part of the calcareous claystone member.

Both the calcareous and siliceous claystones are thin-bedded with thicknesses of 5 to 10 cm, occasionally 20 cm, and are accompanied alternately with less compact consolidated claystone partings less than 1 cm in thickness. In many beds parallel laminations are marked by a difference of light and dark colouration. Churned or bioturbated bedding is occasionally seen.

Microscopically, the calcareous claystone is composed of a mixture of clay minerals and microcrystalline calcareous particles. A reasonable amount of organic matter of sand- and silt-sizes is dissipated throughout, and radiolarian and foraminiferal remains are sporadically contained. The siliceous claystone commonly contains siliceous organic remains such as radiolarians and sponge spicules, besides clay minerals and other disseminated organic matters. The same kinds of calcareous and siliceous claystones are found in the southward area of the Kamiochiai unit in the area to the southwest of Shizuoka City, where nannofossils of the Eocene age (HONJO and MINOURA, 1968) and the lower Oligocene age (IIJIMA *et al.*, 1977) are reported to occur abundantly.

Absence of coarse-grained terrigenous clastic beds throughout the sequence and the lithologic features containing abundant pelagic microfossils suggest that the claystones are pelagic deposits. These beds are overlain by massive black claystone which is neither calcareous nor siliceous, and at a level approximately 50 m above the top of the siliceous claystone the terrigenous turbidite sandstone first appears.

4. Stratigraphy of Greenstone-Claystone Blocks and Their Facies Distribution

Most of the exotic blocks and slabs greater than 50 m in thickness in the Kamiochiai unit comprise a greenstone-claystone sequence which is classified into three types on the basis of dominant lithologic association, especially for the reworked hyaloclastites. They are named as follows:

T-type association (characterized by a basaltic lava-volcaniclastic turbidite).

D-type association (characterized by volcaniclastic debris flow deposits).

C-type association (characterized by volcaniclastic contourite).

They show a following stratigraphic sequence, respectively.

4.1 T-type association

Representative sections of this type are at Locs. 5 and 7 (Figs. 2 and 3). The lower part of the section is of pillow lavas and sheet flows. Doleritic dykes intruding into the lavas are also shown. The lavas are directly overlain by volcaniclastic turbidites which attain a maximum thickness of about 20 m. These turbidites are overlain by thin-bedded calcareous claystones, approximately 20 m thick, which grade upward into thin-bedded siliceous claystones. This sequence is bounded by shear planes at the bottom and top from the argillite mélange carrying terrigenous sandstone blocks.

4.2 D-type association

The representative exposures are seen in the sections at Locs. 2 and 3. The lower part is occupied by boulder to granule size volcaniclastic debris flow beds whose thickness attains 100 m. Underlying lavas were recognized in the section at Locs. 4 and 5, in which the debris flow beds are not so thick as in the sections at Locs. 2 and 3 (Fig. 3). This sequence suggests that the debris flow beds in the sections at Locs. 2 and 3 were also underlain by lava flows. The debris flow beds are transformed conformably upward to volcaniclastic turbidites, which are in turn overlain by calcareous and siliceous claystones that show a similar sequence as the upper part of the T-type association. The sequence at Locs. 2 and 3 are bounded by shear planes against argillite mélanges at the base and top. The one at Loc. 4 is in shear plane contact with argillite mélanges at the base as well, but the volcaniclastic turbidite beds are conformably succeeded by 10 m thick calcareous claystones, about 100 m thick siliceous claystones and massive claystones with terrigenous clastic sandstone turbidites. The calcareous and siliceous claystones are inferred as pelagic oozes as mentioned above, and terrigenous material increases considerably upsection in massive claystones. Thus this sequence represents a complete stratigraphy from the greenstones to the terrigenous clastic deposits.

4.3 C-type association

This type can be seen in the section at Loc. 1. It consists mainly of volcaniclastic contourites with some interbeds of debris flow deposit. Some gabbroic sills occur concordantly in this sequence. The upper chilled margins of these sills are partially eroded, and their clasts are included in the adjacent contourite beds. The blocks of the contourite sequence are bounded by shear planes against argillite mélanges at both its base and top.

Figure 4 shows variation in amount and proportion of the volcanic and volcaniclastic rocks with respect to the distribution of blocks of the greenstone-claystone sequence. These blocks trend along three lines. The T-type blocks are predominated in the southern part of the western line. The blocks of the transitional type between the T- and D types are predominated in the northern part of the western line and the southern part of the central line. To the north of the central line, abundance of this transitional type decreases with an increase of the D-type. In the eastern line, most of the blocks are of the C-type.

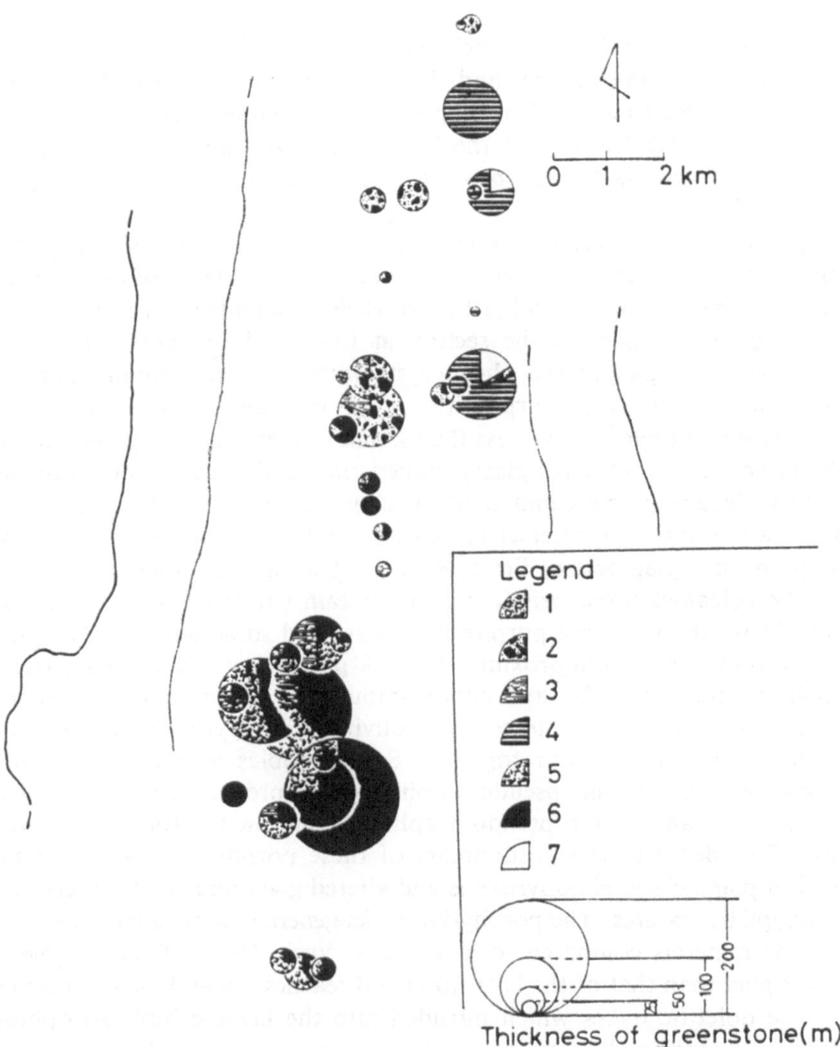

Fig. 4. Map showing variations in amount and proportion of volcanic and volcaniclastic rocks in relation to thickness of the blocks. 1. initial hyaloclastite (hyaloclastite in the narrow sense); 2, volcaniclastic debris flow deposit; 3, volcaniclastic turbidite; 4, volcaniclastic contourite; 5, pillow lava and sheet flow; 6, massive lava and dolerite; 7, gabbroic sill.

5. *Petrographic Characters of Lavas, Dykes, Sills and Volcanic Rubble in Debris Flow Deposit*

All of the volcanic and associated intrusive rocks have been affected by low grade metamorphism, and those in the northern part of the area have been altered to actinolite-pumpellyite facies mineral assemblage. The relict petrographic features of the lavas, intrusives and volcanic rubbles in the debris flow beds may provide suitable information on their probable origin.

The lavas consisting of pillow lava, sheet flow and massive lava are mainly distributed in the southern part of the area. The lavas of the volcanic section are mostly microporphyritic and their vesicularity is less than 1%. In the uppermost part of the section at Loc. 5, 5 m thick porphyritic lavas with large plagioclase phenocrysts with 1 cm maximum diameter were found. The microporphyritic lavas contain clinopyroxene and plagioclase microphenocrysts less than 2 mm in diameter, altogether about 10% in volume. They have glassy chilled rims and 1 cm inside from the chilled surface plagioclase and dendritic clinopyroxene crystals appear. The inner part shows an intersertal texture and is composed mainly of clinopyroxene, plagioclase, ilmenite and other opaque minerals.

The volcaniclastic debris flow beds contain volcanic rubble of several kinds. Most abundant are porphyritic rocks, and some aphyric. The porphyritic rocks contain approximately 10% plagioclase phenocrysts, occasionally as much as 50%, and minor amounts of either clinopyroxene or pseudomorphs probably altered from olivine. Clinopyroxene and olivine pseudomorphs rarely occur together. Some rubbles which have neither clinopyroxene nor olivine pseudomorphs are also present. The rubbles that contain abundant olivine pseudomorphs are only in the top beds of the debris flow deposit. The groundmass of these porphyritic rocks is composed of plagioclase, clinopyroxene and altered glass and shows intersertal to hyalopilitic textures. The porphyritic rocks generally have smaller amount of mafic minerals compared to lava flows. Vesicularity of this rubble is much higher than that of the lava flows and reaches about 40% by volume.

The doleritic dykes which intruded into the lavas exhibit an ophitic texture and contain clinopyroxene, plagioclase, olivine pseudomorphs and opaque minerals. Minerals in the gabbroic sill which intruded into the volcanic contourite have a poikilitic or ophitic texture. The upper part of this sill is slightly leucocratic and contains plagioclase and reddish-brown clinopyroxene. The lower part is melanocratic and composed mainly of reddish-brown clinopyroxene, olivine pseudomorphs and plagioclase. The lowest part contains reddish-brown clinopyroxene and olivine pseudomorphs.

The bulk chemical compositions of lavas obtained from the southern part and volcanic rubbles from the northern part of the area are shown in Table 1. On the plots of total alkali versus silica contents (Fig. 5), most of these rocks are scattered on both sides of the boundary between the alkali and subalkali rock types. It is therefore difficult to distinguish their rock types. However, all lavas show SiO_2 content less than 52 wt% and can be considered to be basalt. On the other hand, the volcanic rubbles from the main part of the volcaniclastic section indicates comparatively higher SiO_2 content which reaches about 60 wt%. These rubbles can be considered to belong to the intermediate rock group. A general enrichment of SiO_2 from the lower part to the upper part within the volcanic to volcaniclastic section is recognized. Another remarkable change in the chemical composition is indicated in the content of TiO_2 which shows a negative relationship between the ratio of FeO*/MgO as shown in Fig. 6.

The doleritic dykes of the southern area show a basaltic composition similar to the basaltic volcanic rubbles in the northern area. The gabbroic sills also show a basaltic composition, but their SiO_2 content is lower than those of the other rocks. It is particularly marked in the lower part of the sills.

6. *Probable Origin of the Exotic Blocks of Greenstone Claystone Sequence*

In the greenstone-claystone sequence of the exotic blocks in the Kamiochiai unit, the lower volcanic section exhibits a fairly wide variation in lithologic association. It is demonstrated by a roughly gradational change from a sequence of predominantly basalt lavas in the south to that of volcaniclastic deposits in the north, as seen in Figs. 3 and 4. The chemical compositions of the lavas and volcaniclastic rubbles also vary in accordance with this lithologic change. With respect to the stratigraphic sequence the volcanic sections are conformably capped by predominantly biogenic, pelagic calcareous claystones of almost equal thickness and of the same lithologies except for blocks in which higher members overlying the volcanic sections are structurally missing. This implies that the greenstone piles including volcaniclastic rocks of the presently separated blocks are nearly or exactly of the same age and that they primarily constitute a submarine volcanic body.

The pillows show a bulbous and elongate form, very sparse with tiny vesicles. They lie stratigraphically below the calcareous and siliceous claystones which are of similar lithology to oceanic oozes. It suggests that the lavas of the lower part of the volcanic section are products of sub-

Table 1. Rock glasses were prepared by direct fusion method of FUKUYAMA and SAKUYAMA (1976) for the analysis. The analyses were preformed by X-ray microprobe analyser. Data correction procedure followed BENCE and ALBEE (1968), α-factor by NAKAMURA and KUSHIRO (1970) was used. Total iron as FeO*.

No.	P1-01a	P1-02	P1-03a	P1-03b	P1-04a	P104b	P1-05	P1-06	P1-11a
SiO_2	45.06	47.63	48.77	47.11	42.99	50.01	47.99	48.09	50.19
TiO_2	4.65	4.26	4.26	4.32	4.99	4.22	4.59	4.34	4.31
Al_2O_3	14.14	12.61	11.88	12.25	13.85	12.30	12.06	12.61	13.53
FeO*	17.24	14.78	13.94	15.73	20.08	14.60	15.47	15.96	11.14
MnO	0.22	0.21	0.20	0.22	0.25	0.21	0.22	0.20	0.19
MgO	5.04	5.24	5.04	5.40	5.26	3.46	5.53	5.57	4.71
CaO	8.85	9.65	9.73	9.95	7.58	9.24	8.89	8.46	9.28
Na_2O	2.74	3.13	4.16	3.39	2.92	3.60	3.58	3.82	4.32
K_2O	1.60	0.86	0.41	0.60	0.65	0.93	0.41	0.33	0.96
Total	99.54	98.37	98.39	98.97	98.57	98.57	98.74	99.38	98.63

No.	P1-13a	P2-12a	P2-13a	P2-31a	P2-31b	P2-32	P3-01b	P3-02b	P3-11b
SiO_2	47.43	46.37	48.27	46.69	47.21	49.47	45.87	45.37	47.74
TiO_2	4.54	4.46	4.65	4.52	4.45	4.53	4.30	4.41	4.11
Al_2O_3	13.32	13.66	12.18	13.21	12.83	13.11	14.79	14.61	13.53
FeO*	15.98	15.63	14.35	13.91	13.75	13.18	14.46	14.13	13.15
MnO	0.21	0.22	0.21	0.21	0.18	0.18	0.19	0.20	0.19
MgO	5.46	5.68	6.01	7.02	6.81	4.98	7.51	7.64	7.56
CaO	9.36	7.76	8.39	8.68	8.95	8.15	8.25	8.98	9.64
Na_2O	3.03	2.93	3.50	2.71	3.17	3.62	2.74	2.73	2.89
K_2O	0.51	1.16	0.48	1.15	0.66	1.06	0.60	0.70	0.30
Total	99.84	97.87	98.04	98.10	98.01	98.28	98.71	98.77	99.11

No.	P3-13a	P3-14	P3-61b	P3-61c	P3-62a	P3-62b	P3-63a	P3-63b	P3-71a
SiO_2	47.14	46.28	47.74	46.19	44.92	49.77	47.18	50.06	50.09
TiO_2	4.08	5.14	4.38	4.64	4.35	4.04	3.80	3.22	1.27
Al_2O_3	13.66	16.10	14.41	14.85	13.82	13.78	14.00	14.54	17.27
FeO*	13.14	14.34	14.34	14.92	15.32	13.67	13.38	12.32	11.02
MnO	0.20	0.20	0.22	0.21	0.23	0.19	0.23	0.17	0.15
MgO	7.30	7.43	6.90	7.23	7.91	7.27	7.17	7.40	9.45
CaO	9.96	6.44	7.49	7.58	8.60	7.78	9.87	6.92	5.12
Na_2O	3.25	3.13	2.20	2.13	2.59	2.67	3.10	3.08	4.15
K_2O	0.60	0.40	0.53	0.47	0.49	0.46	0.58	0.69	1.19
Total	99.33	99.46	98.21	98.22	98.23	99.63	99.31	98.40	99.71

No.	P3-71b	P3-71c	BP-11	BP-12a	BP-12b	BP-22	BP-23	BP-24	051008
SiO_2	50.77	48.25	53.67	50.92	52.33	47.05	52.34	47.61	46.34
TiO_2	1.23	1.35	1.38	1.25	1.27	1.85	1.61	2.51	1.63
Al_2O_3	17.89	18.02	15.57	17.64	17.31	16.32	17.88	12.75	15.68
FeO*	9.29	10.28	8.98	8.60	9.22	11.61	10.08	7.59	12.50
MnO	0.16	0.16	0.14	0.14	0.15	0.17	0.15	0.23	0.17
MgO	9.26	9.87	8.16	6.44	7.65	7.97	7.68	4.95	9.48
CaO	6.41	7.49	5.07	8.62	6.05	8.72	5.64	20.63	8.03
Na_2O	4.12	3.78	6.23	6.23	5.59	3.62	3.55	4.23	4.34
K_2O	0.67	0.75	0.17	0.26	0.48	1.49	1.60	0.98	0.15
Total	99.80	99.95	99.37	100.10	100.05	98.80	100.53	101.48	98.32

No.	051009	051010a	051010b	05101402	05101403	05101404	05101405	05101406	05101407
SiO_2	51.81	52.20	49.65	52.92	50.23	54.86	52.71	55.34	51.10
TiO_2	1.58	1.49	1.59	1.66	1.60	2.27	1.44	2.50	1.35
Al_2O_3	15.63	16.69	16.20	18.68	15.27	15.29	19.86	17.32	17.15
FeO*	10.07	9.11	9.33	6.36	10.33	7.04	6.57	6.87	9.08
MnO	0.18	0.15	0.16	0.12	0.18	0.16	0.12	0.13	0.18
MgO	7.78	8.89	8.19	5.94	9.66	6.01	6.26	5.89	9.12
CaO	9.59	6.91	9.44	6.23	6.19	7.56	5.75	6.24	5.51
Na_2O	3.42	4.65	3.78	4.84	3.45	4.64	4.86	5.79	3.90
K_2O	0.77	0.72	1.10	2.05	1.83	2.17	2.37	1.28	1.69
Total	100.83	100.81	99.44	98.80	98.74	100.00	99.94	101.36	99.08

No.	05101408	05101502	05101503	05101503a	05101504b	05101506	05101602	05101604a	05101605
SiO_2	54.60	55.08	53.92	56.15	51.97	52.28	49.27	50.85	52.85
TiO_2	1.62	1.29	1.59	1.20	2.03	1.60	1.53	1.89	1.55
Al_2O_3	20.32	23.66	17.38	20.17	15.01	20.10	17.09	14.80	20.36
FeO*	5.91	3.67	7.14	4.12	8.86	6.05	8.14	8.77	5.52
MnO	0.12	0.10	0.11	0.09	0.14	0.12	0.20	0.18	0.12
MgO	5.97	2.78	4.67	3.90	6.28	6.33	7.14	5.76	4.71
CaO	5.26	6.10	8.74	7.30	8.16	6.49	9.59	9.79	7.05
Na_2O	5.09	5.82	5.49	6.40	5.85	5.05	3.79	5.66	5.55
K_2O	2.00	2.73	0.97	1.72	0.54	1.69	1.55	0.40	1.57
Total	100.89	101.23	100.01	101.05	98.84	99.71	98.30	98.10	99.28

No.	051017a	051017b	051018	051019b	05102201	05102203	051024a	051024b	05251002
SiO_2	52.33	52.28	48.54	53.42	59.22	53.85	53.09	54.42	56.84
TiO_2	1.34	1.69	1.27	1.51	1.85	1.75	1.26	1.97	1.29
Al_2O_3	15.42	20.10	16.65	14.20	16.12	16.86	11.48	17.71	18.27
FeO*	8.71	6.05	7.80	8.09	5.43	5.23	3.76	5.22	5.69
MnO	0.18	0.12	0.16	0.13	0.10	0.12	0.12	0.12	0.11
MgO	7.02	6.33	6.11	6.74	5.78	5.17	3.39	4.67	5.61
CaO	10.82	6.49	15.16	8.46	7.59	11.44	25.15	11.77	7.34
Na_2O	4.47	5.05	3.72	5.52	3.58	3.52	3.15	3.60	5.57
K_2O	0.19	1.69	0.09	0.15	0.57	0.58	0.41	0.57	1.06
Total	100.48	99.80	99.50	98.22	100.24	98.52	101.81	100.05	101.78

No.	05251003	05251006	05251007	05251008	05251009	05251010	05251102a	05251102	05251103b
SiO_2	57.21	52.35	53.95	56.24	54.56	56.45	51.06	49.25	51.87
TiO_2	1.85	1.38	1.13	1.59	1.41	1.48	1.71	1.79	2.03
Al_2O_3	16.31	18.87	22.76	16.60	19.20	21.17	13.87	15.22	13.42
FeO*	7.12	6.16	5.01	4.81	6.23	4.26	10.02	11.21	9.33
MnO	0.11	0.13	0.12	0.13	0.15	0.12	0.21	0.16	0.18
MgO	3.39	6.48	5.21	4.72	6.25	2.99	9.14	9.71	8.32
CaO	8.89	6.11	4.75	9.71	5.66	6.00	7.78	5.78	9.32
Na_2O	6.39	4.69	3.38	3.76	4.35	5.53	3.65	4.85	4.47
K_2O	0.31	1.99	2.96	1.48	1.40	1.50	0.67	0.39	0.63
Total	101.58	98.16	99.27	99.04	99.21	99.50	98.11	98.36	99.57

No.	05251401a	05251401b	05251401c	05251402a	05251402b	05251402	05251403a	05251403b	05251801
SiO_2	49.34	51.26	50.47	46.49	46.12	51.75	54.95	55.21	48.21
TiO_2	1.67	1.47	1.52	1.55	1.39	1.38	1.34	1.22	1.45
Al_2O_3	19.52	16.91	15.66	15.52	16.64	14.80	15.64	13.35	14.41
FeO*	9.02	7.78	8.22	11.04	9.67	9.15	6.57	7.93	9.54
MnO	0.14	0.17	0.17	0.21	0.16	0.17	0.16	0.19	0.17
MgO	7.14	7.18	7.46	9.46	8.31	7.55	6.75	7.63	5.51
CaO	5.50	7.63	9.02	12.88	15.02	9.10	7.85	10.33	16.94
Na_2O	5.08	5.14	5.13	3.09	2.90	5.43	6.51	5.53	2.98
K_2O	0.72	0.75	0.56	0.08	0.08	0.19	0.32	0.12	0.78
Total	98.13	98.29	98.21	100.32	100.29	99.52	100.09	101.51	99.99

No.	05251802	05251803	05251804	05251805a	05251805b	05251806	05251807	121807	K-K
SiO_2	50.31	53.85	57.05	50.10	52.52	52.74	53.04	50.71	47.85
TiO_2	1.43	1.66	1.93	1.19	1.16	1.53	1.52	2.02	1.63
Al_2O_3	14.75	15.47	17.31	20.03	22.01	13.33	15.18	14.98	18.00
FeO*	7.26	5.15	8.93	8.24	8.14	9.77	9.78	10.85	10.49
MnO	0.18	0.15	0.12	0.13	0.14	0.12	0.13	0.17	0.18
MgO	4.66	4.73	2.99	3.97	3.48	3.30	5.86	7.52	10.25
CaO	15.20	12.63	6.38	9.30	6.45	14.55	9.75	7.02	6.72
Na_2O	3.50	4.71	4.34	3.43	2.43	3.85	3.66	4.96	2.74
K_2O	0.85	1.20	0.99	2.03	2.51	0.65	0.70	0.15	1.53
Total	98.14	99.55	100.04	98.42	98.84	99.84	99.62	98.38	99.39

No.	061004	060207	060208	060211	060212	060213	060214	060339	060345
SiO_2	47.63	45.27	42.93	40.27	44.93	48.37	47.84	50.58	40.77
TiO_2	2.38	2.16	2.44	3.50	2.91	2.64	2.50	2.06	2.03
Al_2O_3	17.70	14.29	11.39	17.45	15.49	16.06	15.63	12.95	17.23
FeO*	11.77	12.78	14.11	13.09	13.42	12.15	12.91	10.04	10.52
MnO	0.19	0.17	0.18	0.19	0.22	0.22	0.21	0.18	0.16
MgO	9.94	11.59	14.47	9.54	6.53	6.43	9.16	7.48	10.10
CaO	4.00	10.57	9.44	9.18	11.99	8.39	5.37	11.41	16.42
Na_2O	4.36	2.10	1.60	1.76	3.13	4.09	3.88	4.47	0.34
K_2O	0.65	0.37	0.86	2.44	0.20	0.69	0.82	0.52	2.57
Total	98.62	99.30	97.42	97.42	98.82	99.04	98.32	99.69	100.14

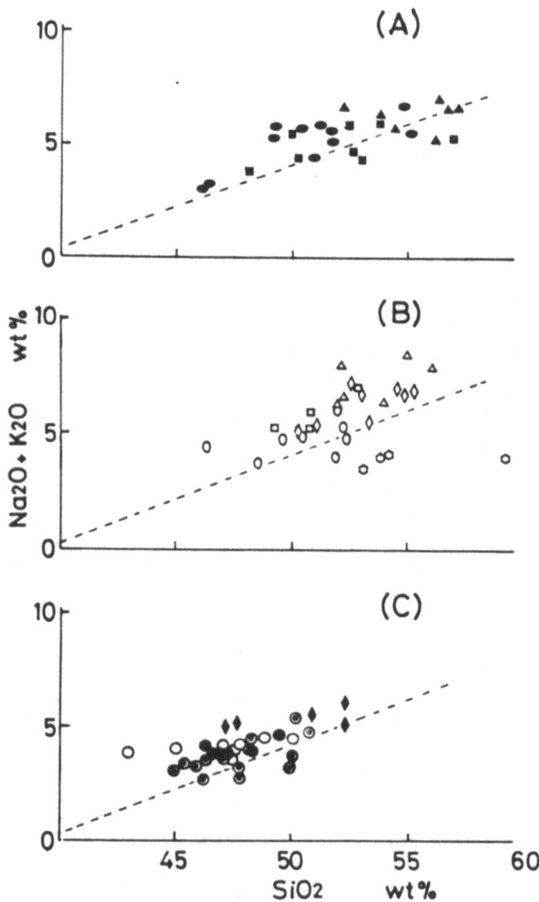

Fig. 5. Plots of Na_2O+K_2O versus SiO_2 contents for volcanic rocks. The boundary between the alkali and subalkali rock types shown by broken line is adopted from MACDONALD and KATSURA (1964).

marine volcanism at appreciable depths.

The most important feature of the greenstone-claystone sequence in the Kamiochiai unit is a thick pile of boulder size to sandy volcaniclastic detritus transported by slumping and turbidity currents, which are unsorted and crudely to non-stratified in nature. They show rapid textural variations from boulder breccia including large blocks greater than 1 m in diameter to coarse-grained sands, and have an angularity and heterogenity composed of fragments from the main part of the volcaniclastic beds.

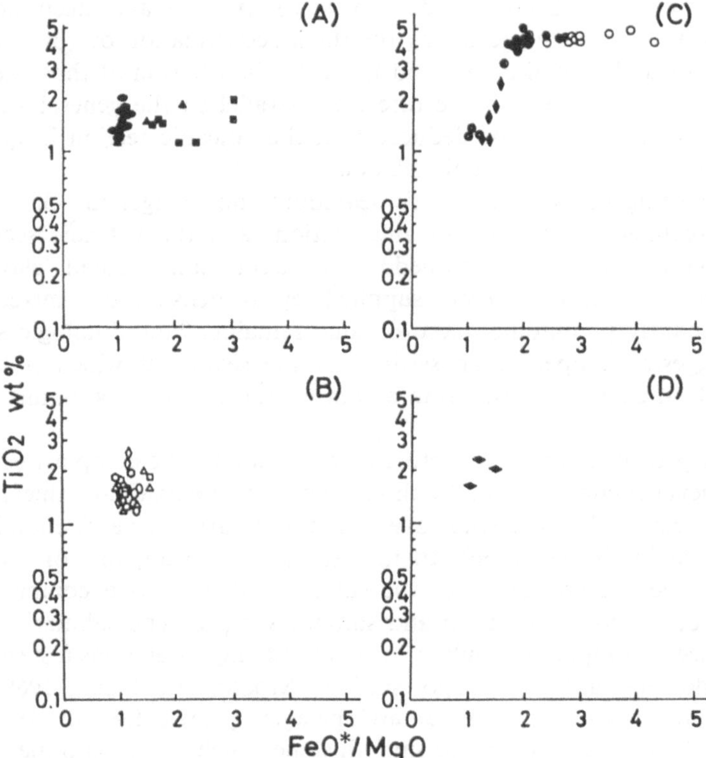

Fig. 6. Plots of TiO_2 versus FeO*/MgO for the same samples as Fig. 5. See Fig. 3 for the symbols.

Their features resemble those of talus deposits and landslides. The turbidite beds at the top of the volcanic section are also composed entirely of volcaniclastics. It is clear that these beds are resulted from mass movements of volcanic material derived and transported from geographically higher positions of a volcanic body, probably near or above the sea level. It is shown by the abundant content of large vesicles in the volcanic fragments.

In summary, the volcaniclastic rocks of the upper part of the volcanic section evidently represent talus deposits on the flank or lower slopes of a submarine volcanic body. The marked difference in vesicularity between the volcaniclastic rubbles and the underlying lavas suggests that the relative height between the emplaced place of pillow lava flows and the original eruption site of porphyritic rocks from which much of the rubbles were derived must have been considerably large. A wide and steep slope of the volcanic body is suggested by the thick accumulation of volcaniclastic debris flow and turbidite beds and from the distribution of those deposits extending about 15 km in the direction parallel to the general strike of these beds. Thus it can be deduced that the volcanic section in question primarily constituted a volcanic seamount.

Regarding the setting of this seamount, no terrigenous clastic beds are intercalated within the volcanic section, and the volcanic section is conformably overlain by thin-bedded calcareous and siliceous claystones in which only clay minerals supplied by suspension are mixed with predominantly planktonic micro-organic remains. This lithologic succession suggests an open-ocean setting for the seamount which is beyond the reach of coarse detritus from a continental margin or a mature island arc.

The principal surface structure of a seamount whose top arises above the sea level is characterized by three portions; subaerial lava shield, flow-foot breccia or hyaloclastite layers and submarine lava foundation as generalized by JONES (1966). In the Island of Hawaii, one of the most familiar open-ocean volcanoes the subaerial shield has a comparatively gentle slope (about 5 degrees) and smooth surface. The submarine flank subsurface is composed mainly of hyaloclastite layers and has a steep slope (10–30 degrees) and hummocky surface (MOORE and FISKE, 1969). The lower flank of the seamount or archipelagic apron (MENARD, 1956) has a gentle slope (<2 degrees) and smooth slope which grades into the oceanic floor. Similar topographic features are recognized in a small seamount whose summit has never reached the sea level (LONSDALE and SPIESS, 1979).

Seismic surveys conducted around volcanic islands and seamounts have shown that the archipelagic aprons are commonly deposited with thick accumulations of unconsolidated material (KRONKE, 1965; NORMARK and SHOR, 1968; LONSDALE, 1975; STANLEY and TAYLOR, 1977). These deposits are considerably thick compared to deposits covering the acoustic basement of the neighbouring abyssal plains. Deep-sea drilling in the central Pacific penetrated the blanket of pelagic oozes and the underlying alternating beds of volcaniclastic turbidite or sandstone and nannofossil and foraminifera-rich claystone and limestone at several sites on the archipelagic aprons (WINTERER and RIEDEL *et al.*, 1971; WINTERER and EWING *et al.*, 1973; MOBERLY and KEENE, 1975; KELT and MCKENZIE, 1976; MOBERLY and JENKYNS, 1981). These sedimentary sections are often directly underlain by basalt. The turbidite beds are thought to have been derived from shallower flanks of the seamounts. Photographic study and dredge hauls in addition to seismic surveys of the Gillis seamount in the northwest Atlantic basin revealed that the volcanic basement in the lower flank of the seamount is irregularly covered by thick sediments ranging from granule to mud-size. They are derived from sediment gravity flows including turbidity currents and suspension-rich (mostly clay and planktonic foraminifera) water masses that flow down from the mid and lower-flank sectors of the seamount (TAYLOR *et al.*, 1975; STANLY and TAYLOR, 1977). Ocean-bottom current activity is less intense there. Little is known about the deposits of upper- and mid-flank sectors of the seamount. In the Gillis Seamount, the steeper upper- and mid-flank sectors show a complex step or terrace-like morphologic character and consist of volcanic rubbles, organic debris and pelagic materials. There are concentrations of coarse-grained volcanic sand and large volcanic debris including pillow fragments. Evidence of intense bottom-currents is demonstrated by sharp asymmetric ripples, sediment scour, current-tails etc. Occurrence of sediment gravity flows in the downslope movement of the sediment cover on the upper flank is also revealed by bottom photographs (STANLEY and TAYLOR, 1977). FORNARI *et al.* (1979) also report that a landslide from a shore face occurred on the submarine flank of the Island of Hawaii. On the other hand, in small seamounts no appreciable sediment cover is present on their flanks or bottoms (LONSDALE and SPIESS, 1979).

With the presence of thick volcaniclastic deposits, the original volcanic body can be reconstructed, although incomplete, from the dismembered greenstone-claystone remnants in the Setogawa Terrane. It can be deduced that it was a large, mature volcanic seamount whose top probably reached the sea level. The lithologic sequence from pillow lavas through volcaniclastic turbidite to predominantly biogenic calcareous claystone with interbeds of volcaniclastic turbidites in its lower part and further to siliceous claystone is comparable with the deposits on the lower flank or archipelagic aprons of oceanic seamounts and islands. Thus the setting of the dismembered remnants embracing the T-type association is ascribed to a part of the archipelagic apron. The D-type lithologic association characterized by thick, rubbly to coarse-grained sand-size volcanic debris beds resting on pillow lavas is very similar to the deposits of the mid- to upper lower-flank sectors of modern oceanic seamounts.

The setting of the C-type association is difficult to perceive, because on one hand, overlying and underlying rocks are missing and no comparable deposits have been found on and around volcanic seamounts. Thin to very thin, well-defined stratification accompanied by frequent cross-bedding, parallel lamination and well-sorted volcaniclastic sand and silt indicate the deposition was strongly influenced by intense bottom current activity. As stated above (e.g. STANLEY and TAYLOR, 1976), it is known that strong bottom currents prevail on the upper- and mid-flank sectors of seamounts. To allow thick accumulation of such beds on the flank of seamounts a terrace-like ledge or depression would be required. It could possibly be resulted from either wave-planation by eustatic sea-level changes, the stacking of slump masses or circumferential inward-dipping faults accompanied by downthrow of the inner volcanic cone. Volcanic processes such as parasite volcanoes would also form a ledge. In this respect, especially noteworthy is the association of gabbroic sills intruding into the contourite beds. It may be explained as an intrusion that rose up the circumferential fault. The volcaniclastic contourite may have been resulted from the reworking of volcaniclastic turbidite on the lower flank and foot of the seamounts, but in such places intrusion of gabbroic sills is unlikely to occur. Therefore, no definite setting of the volcanic contourite beds can be designated, but they are most likely to be on the mid- to upper-flank of a seamount (Fig. 7).

In the greenstone-claystone remnants in the area, no sections which are attributable to subaerial lava flow and shallow-water sedimentary origin (MOORE *et al.*, 1973; FURNES and FRIDLEIFSSON, 1974) were discriminated. However, the volcaniclastic components of the debris-flow beds contain a few water-worn, subrounded pebbles and cobbles that are mainly porphyritic, indicated by abundant large vesicles and high SiO_2 content exceeding 52%. These clasts seem to have been derived from subaerial lava flows and deposited in shallow-waters, eventually settling in the deep bottom together with other volcanic debris derived from mass movements.

Generation of the Setogawa seamount in the oceanic plate far beyond reach of terrigenous detritus can be deduced from evidence that the volcanic member is conformably overlain by a calcareous and siliceous claystone member of pelagic nature and that no terrigenous clastic layers intervene within these sequences. The siliceous claystone is succeeded by black claystone, and terrigenous turbidite sandstone first appears at 50 m above the top of the siliceous claystone. It suggests that the Setogawa seamount must have been conveyed by ocean-floor spreading to a trench as in modern examples, and in this progression, near and in the trench the pelagic beds were overlapped by a tapering wedge of trench-fill terrigenous clastic deposits ranging from a distal muddy facies free of coarse sediment to a more proximal facies of turbidite. The lower two-thirds of the black claystone still devoid of terrigenous particles larger than siltsize can be ascribed to deposition from a clay-suspensate water mass in the environment outside a trench. Noteworthy is that these beds show a sharp contrast with the argillite mélange which surrounds the greenstone-claystone remnants in no occurrence of clasts of any kinds and in less intensely sheared structure. The upper part of the black claystone contains silty laminae and shows a gradual upward-coarsening sequence with the beds containing medium-grained turbidite sandstone as the coarsest extreme. This part can be inferred to be of trench-fill sediment origin.

7. *Process of Dismemberment of the Inferred Setogawa Seamount*

It is clear that dismemberment of the inferred Setogawa seamount occurred after deposition of the aforementioned black claystone. In other words, it took place when the seamount reached the trench area. The greenstone and greenstone-claystone remnants of the area are all accounted for as surficial elements of the seamount, and the main body of the seamount composed of pillow lavas and dykes seems to have been subducted and lost.

The following three types of possible processes of dismemberment of the seamount may be postulated: 1) scraping by thrust faults along the inner margin of a trench, 2) rock fall or slumping from the tilted seamount at the outer slope of a trench, 3) gravity sliding or gliding pertaining to the segmentation of a seamount by normal faults. With a general lack of intense shearing of the greenstone-claystone sections and of a distinct sole thrust the first process may be ruled out. The large size of the remnants, many of which reach 50 to 200 m in thickness and 500 m to more than 1 km in length without marked disruption of minor orders within the blocks would be difficult to be explained by slope slumping or rock fall. The field occurrence, size and stratigraphic sequence of the remnants as well as their structural relationships to the surrounding argillite mélanges are in favor of the process of the third type.

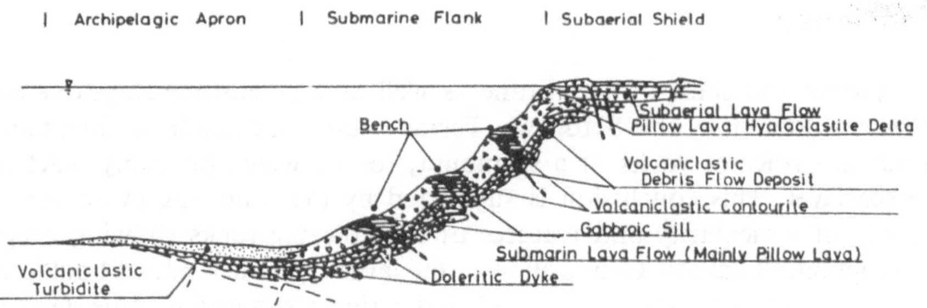

Fig. 7. Interpretative cross section through flank of inferred seamount to illustrated site of deposition of volcaniclastic deposits.

However, en echelon distribution of the greenstone-claystone remnants in three rows is thought to be closely related to the process of dismemberment of the inferred Setogawa seamount. In the course of entering a trench, the Setogawa seamount was probably disrupted into at least four segments by parallel normal faults. This can be seen in the Daiichi-Kashima seamount on the oceanward slope of the Japan Trench off Choshi, east of Tokyo (MOGI and NISHIZAWA, 1980) and also in the western end of the Dutton ridge on the oceanward slope of the Mariana Trench (SMOOT, 1983). The former is interpreted as a modern example of a seamount subducted into a trench. Subsurface geology and structure around the Daiichi-Kashima seamount is not yet known, but bathymetric survey has shown that the seamount is disrupted into at least two major segments by normal faults trending nearly parallel to the trench axis. The flat top of the inner segment is displaced by a steep scarp with a relative height as great as 1500 m with respect to the outer flat top.

This type of a steeply dipping escarpment would consequently give rise to not only slope failure by slumping and rockfall from the upper and frontal part of the scarp but also detachment of larger rock masses by gravity gliding along structural planes. These structural planes would be subordinate faults, dilated fractures, or sedimentary planes such as the intercalated fine-grained hyaloclastite beds which easily become weak beds by mineral alteration and associated volume expansion. Such weak surfaces would provide a plane releasing gravitational stress produced in the steeply standing blocks. These dismembered blocks would be collapsed and mixed with the sedimentary fills of the trench. In a similar manner to the Daiichi-Kashima seamount, the Setogawa seamount would have been subjected to a breakdown when it entered a trench (Fig. 8), and gravity sliding of surficial rock masses might have occurred with a series of at least four segments broken in the seamount.

8. *Summary*

Blocks and slabs of greenstone as well as a greenstone-claystone sequence scattered in the Setogawa Terrane are attributable to remnants of surface volcanic rocks of a seamount, top of which probably reached the sea level. This conclusion is supported by the following evidence: 1) scarcity of vesicularity and features of the volcanic rocks showing deep-sea eruption, 2) basaltic composition of the lavas, 3) a thick pile of boulderly to granular volcaniclastic debris beds indicating a submarine talus deposit transported by mass flow in a steep slope from a topographically high elevations, 4) subrounded porphyritic clasts with abundant large vesicles

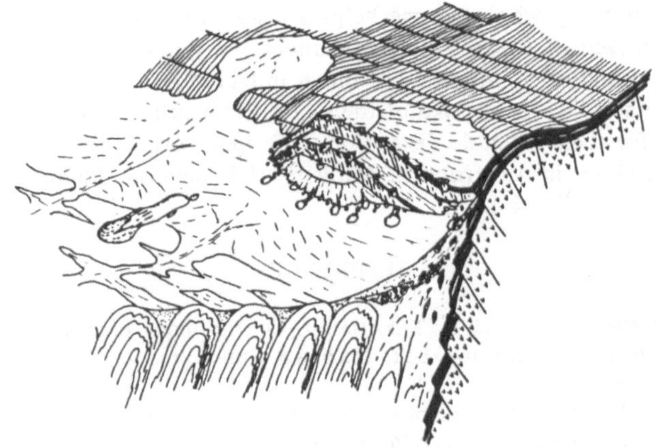

Fig. 8. Schematic illustration showing plausible process of dismemberment of inferred Setogawa seamount.

in polymictic debris flow beds suggesting a subaerial lava flow origin. They were probably deposited in a shallow-water environment, then transported to a greater depth by mass movement, 5) sandy and silty volcaniclastic turbidites with thin interbeds of calcareous claystone indicating deep-sea environment in which pelagic ooze was accumulating, 6) existence of a dominantly biogenic, calcareous and siliceous claystone blanket covering volcanic and volcaniclastic members as well as a lack of terrigenous clastic intercalations throughout the whole sequence indicating an open-ocean setting.

The sequence from deep-sea lavas through volcaniclastics to pelagic sedimentary member indicates that the dismembered slabs are remnants of surficial rock masses from the middle flank to the foot of the volcanic seamount. The dismemberment of the seamount is considered to have taken place when it encountered the trench, where the seamount was disrupted into segments by normal faults trending parallel to the trench axis to produce a steep escarpment such as seen in the Daiichi-Kashima seamount. The scarp would have given rise to slope failure and gravity gliding along a structural plane or sedimentary weak beds.

Acknowledgements

This paper is a portion of a doctoral dissertation prepared under the guidance of Professor Kametoshi Kanmera. I would like to thank Prof. K. Kanmera who supervised this study and critically read the manuscript and to Dr. Y. Ogawa of Kyusyu University who gave me helpful suggestions. Bulk chemical compositions was analyzed at Kochi University under the guidance of Prof. T. Suzuki and Dr. S. Yoshikura to whom I am much indebted. I also wish to thank Prof. K. Kobayashi and Dr. K. Fujioka of the Ocean Research Institute of the University of Tokyo for their valuable discussion. The fieldwork expenses was supported in part by the Grant-in-Aid from the Ministry of Education, Japan, given to Prof. Kanmera (Sogo A-434041).

REFERENCES

S. Arai, K. Shimokawa, and T. Takahashi, On the mode of emplacement of ultramafic rocks in the Setogawa belt, Shizuoka Prefecture, *J. Geol. Soc. Japan*, **84**, 691–693, 1978.

A. E. Bence and A. L. Albee, Empirical correction factors for the electron microanalysis of silicate and oxied, *J. Geol.*, **76**, 382–403, 1968.

D. J. Fornari, J. G. Moore, and K. Calk, A large submarine sand-rubble flow on Kilauea volcano, Hawaii, *J. Volc. Geotherm. Res.*, **5**, 239–256, 1979.

H. Fukuyama and M. Sakuyama, Major element analysis of rocks by electron microprobe analyser, *J. Geol. Soc. Japan*, **82**, 345–346. 1976.

H. Furnes and I. B. Fridleifssion, Tidal effect on the formation of pillow lava/hyaloclastite deltas, *Geology*, **2**, 381–384, 1974.

J. Honjo and N. Minoura, Discoaster barbadiensos Tan Sin Hok and Geologic age of Setogawa Group, *Proc. Japan, Acad.*, **44**, 165–169, 1968.

A. Iijima, H. Inagaki, and Y. Kakuwa, Nature and origin of the Paleogene cherts in the Setogawa Terrain, Shizuoka, central Japan, *J. Fac. Sci., Univ. Tokyo, Sec. II*, **20**, 1–30, 1979.

A. Iijima, R. Matsumoto, and Y. Watanabe, Geology and siliceous deposits in the tertiary Setogawa Terrain of Shizuoka, central Honshu, *J. Fac. Sci., Univ. Tokyo, Sec. III*, **20**, 241–276, 1981.

J. G. Jones, Intergracial volcano of southwest Iceland and their significance in the interpretation of the form of the marine basaltic volcano, *Nature*, **212**, 586–588, 1966.

K. Kanmera and T. Sakai, To what part of the recent ocean bottom does correspond to site of the sedimentation of the Shimantogawa Group, *Report of GDP in Japan, II-1-(1) Structural Geology*, (3), 55–64, 1975.

K. Kelt and J. N. McKenzie, Cretaceous volcanogenic sediment from the Line Island Chain: Diagenesis and formation of K-feldspar, DSDP Leg 33, Hole 315A and Site 316, in Shlanger, S. D. and Jackson, E. D. *et al.*, *Init. Rept. DSDP32*, Washington, U.S. Govt. Printing Office, 1976.

L. W. Kronke, Seismic reflection studies of sediment thickness around the Hawaiian Ridge, *Pacific Sci.*, **14**, 335–338, 1965.

P. Lonsdale, Sedimentary and tectonic modification of Samoan Archipelagic apron, *A. A. P. G. Bull.*, **59**, 788–798, 1975.

P. Lonsdale and F. N. Spiess, A pair of young created volcanoes on the East Pacific Rise, *J. Geol.*, **87**, 157–173, 1979.

G. A. Macdonald and T. Katsura, Chemical Composition of Hawaiian Lavas, *J. Petrol.*, **5**, 82–133, 1964.

H. W. Menard, Archipelagic aprons, *Bull. A. A. P. G.*, **40**, 2195–2210, 1956.

R. Moberly and J. B. Keene, Origin and Diagenesis of volcanic-rich sediment from north

Pacific seamount DSDP Leg 32, in Larson, R. L., R. Morberly *et al.*, *Init. Rept. DSDP32*, Washington, U.S. Govt. Printing Office, 1975.

R. MOBERLY and H. C. JENKYNS, Cretaceous volcanogenic sediments of the Nauru Basin Deep Sea Drilling Project Leg 61, in Larson, R. L., S. O. Schlanger, *et al.*, in *Init. Rept. DSDP61*, Washington, U.S. Govt. Printing Office, 1981.

A. MOGI and K. NISHIZAWA, Breakdown of a seamount on the slope of the Japan trench, *Prog. Japan Acad.*, **56**, 257–259, 1980.

J. G. MOORE and R. S. FISKE, Volcanic substructure inferred from dredge sample and ocean-bottom photographs, Hawaii, *Geol. Soc. Amer. Bull.*, **80**, 1191–1202, 1969.

J. G. MOORE, R. L. PHILLIPS, R. W. GRIGG, D. W. PETERSON, and D. A. SWANSON, Flow of lava into the sea, *Bull. Geol. Soc. Amer.*, **84**, 537–546, 1973.

Y. NAKAMURA and I. KUSHIRO, Compositional relations of coexisting orthopyroxene, pigeonite and augite in a tholeiitic andesite from Hakone Volcano, *Contr. Mineral. Petrol.*, **26**, 265–275, 1970.

W. R. NORMARK and G. G. SHOR, Jr., Seismic reflection study of the shallow structure of the Hawaiian Arch, *J. Geophys. Res.*, **73**, 6991–6998, 1968.

Y. OGAWA and J. NAKA, Emplacement of ophiolitic rocks in forearc area: Examples from central Japan and Izu-Mariana-Yap island arc system, in Ian Gass *et al.* (eds.), *Ophiolite and Oceanic Lithosphere*, Geol. Soc. London, Spec., No. 11, 253–263, 1984.

M. OKAMURA and A. TAIRA, Micro fossiles from the Shimanto Terrain, Kochi Prefecture and the stratigraphic sequence of the oceanic plate, *Abstract of papers for the 89th Annual General Meeting of the Geological Society of Japan*, 73–74, 1982.

A. RITTMANN, *Volcano and their activity* (transrated by E. P. Vincent), 305 pp., Interscience, New York, 1962.

N. C. SMOOT, Guyot of the Dutton Ridge at the Bonin/Mariana Trench junction as shown by multi-beam survey, *J. Geol.*, **91**, 211–220, 1983.

D. J. STANLEY and P. T. TAYLOR, Sediment transport done on seamount flank by a combined current and gravity process, *Marine Geology*, **23**, 77–88, 1977.

T. SUZUKI and S. HADA, Cretaceous tectonic melange of the Shimanto belt, southwest Japan, *J. Geol. Soc. Japan*, **85**, 467–479, 1979.

A. TAIRA, M. TASHIRO, M. OKAMURA, and J. KATTO, The geology of the Shimanto Belt in Kochi Prefecture, Shikoku, Japan, in Taira, A. and M. Tshiro (eds.), *Geology and Paleontology of the Shimanto Belt*, Rinyakosaikai Press, Kochi, Japan, 1980.

P. T. TAYLOR, D. STANLY, T. SIMKIN, and W. JAHN, Gilliss Seamount: Detailed bathymetry and Modification by bottom currents, *Marine Geology*, **19**, 136–157, 1975.

T. TOKUOKA, T. HARATA, Y. INOUCHI, T. ISHIGAMI, K. KIMURA, F. KUMON, K. NAKAO, S. NAKAYA, T. SAKAMOTO, H. SUZUKI, and J. TANIGUCHI, *Geology of the Ryugin district, Quadrangle Series, scale 1*: 50,000, Geol. Surv. Japan, 69 pp., 1981.

N. TSUCHIYA, T. SAKAI, and K. KANMERA, Mode of occurrence and petrological characteristics of greenstones of the Shimanto terrain in the Mimi river area, Kyushu, *J. Geol. Soc. Japan*, **85**, 445–454, 1979.

E. L. WINTERER, W. R. RIEDEL *et al.*, *Init. Rept. DSDP7*: Washington, U.S. Govt. Printing Office, 1971.

E. L. WINTERER, J. I. EWING *et al.*, *Init. Rept. DSDP17*: Washington, U.S. Govt. Printing Office, 1973.

Formation of Active Ocean Margins, edited by N. Nasu *et al.*, pp. 775–787.

PALEOMAGNETIC EVIDENCE OF THE NORTHWARD DRIFT OF THE IZU PENINSULA, CENTRAL JAPAN

Kimio HIROOKA[1], Toru TAKAHASHI[1], Hideo SAKAI[1], and Tadashi NAKAJIMA[2]

[1]*Department of Earth Sciences, Faculty of Science, Toyama University, Toyama, Japan*
[2]*Geological Laboratory, Faculty of Education, Fukui University, Fukui, Japan*

Abstract. Paleomagnetic properties of Neogene rocks were studied in the Izu Peninsula, central Japan. Samples were collected from the Nishina Formation of early Miocene sediments, the Yugashima Group of early Miocene volcanics, the Kadono Formation of early to middle Miocene sediments and the Shirahama Group of Miocene to Pliocene sediments. All the sedimentary rocks of the Shirahama Group except from one site of Umegi Tuff, are sampled from localities where their ages were determined by planktonic foraminiferal analyses. Paleomagnetic data were obtained from 19 sites. The youngest 4 sites whose ages range from 3 to 5 Ma show nearly the same latitude as that of the present position. The other sites older than 5 Ma show lower paleolatitudes. The oldest sites of the Nishina Formation of early Miocene age show the average paleolatitude of 14.5°N. The paleomagnetic results, therefore, indicate that the Izu Peninsula, which is now located at the northeastern corner of the Philippine Sea Plate (34.7°N), was situated at a place in a much lower latitude and was drifted northward by a rate of about 15 cm/year until it collided with and accreted to the Honshu Island approximately 3 and 5 Ma ago. The very high rate of northward migration of the peninsula can be explained by rotation of the West Philippine Basin and spreading of the Shikoku Basin.

1. *Introduction*

The Izu Peninsula is located near the triple junction of the Pacific, the Philippine Sea and the Asian Plates. The peninsula is in the northern end of the Izu-Bonin Island Arc which was migrated northward and collided with the central Honshu (SUGIMURA, 1972; KAIZUKA, 1975; KOBAYASHI and ISEZAKI, 1976; MATSUDA, 1978). The estimated age of the collision is, however, different among the authors. Kobayashi and Isezaki suggested on the basis of the marine magnetic anomaly pattern of the Shikoku Basin that the collision took place before middle Miocene, whereas Matsuda inferred the timing of collision as early Quaternary from morphological point of view. The temporal and spatial features of the migration and collision of the peninsula will give a key to clarify the plate tectonic history of the western Pacific region.

We carried out paleomagnetic study which would provide us evidence of the paleolatitude and rate of the northward migration of the peninsula.

2. *Geological Setting of Sampling Sites*

Greater part of the Izu Peninsula is covered by Neogene and Quaternary volcanics. Sedimentary strata bearing fossils are found only in very limited areas. The Neogene system of the Izu Peninsula is divided into two groups: the Yugashima Group of Miocene and the Shirahama Group which is presumably late Miocene to Pliocene (SAMEJIMA, 1955). In the northeastern part of the peninsula, a stratigraphic succession mainly composed of pumice tuff, tuffaceous siltstone and pyroclastics is called the Umegi Formation or the Umenoki Tuff (KOYAMA and NIITSUMA, 1980) which is correlatable to the Shirahama Group in the southeastern part. The Yugashima Group in the northeastern part comprises the Okawabata Propylites, the Kadono Formation and the Shimoshiraiwa Formation in order from bottom to top (KITAMURA *et al.*, 1969). TSUCHI *et al.* (1974) and SAMEJIMA (1978) have proposed the Shimoshiraiwa Formation to be in the lowest horizon of the Shirahama Group. The lower part of the Yugashima Group in the western part is called the Nishina Formation (MORITANI and SAWAMURA, 1965) which is correlatable to the Okawabata Propylites in the northeastern part of the peninsula.

The upper horizon of the Shirahama Group in the southeastern part is defined as the Harada Formation. In the lower horizon of the Group, there are embedded tuff breccia and a limestone layer bearing planktonic forminifera called the Nashimoto Formation. Paleomagnetic samples were collected at 34 sites in 18 localities in the northeastern, western and southeastern parts of the peninsula, and paleomagnetic data were obtained from 19 sites in 11 localities. Geology and locations of the sampling sites are shown in Fig. 1.

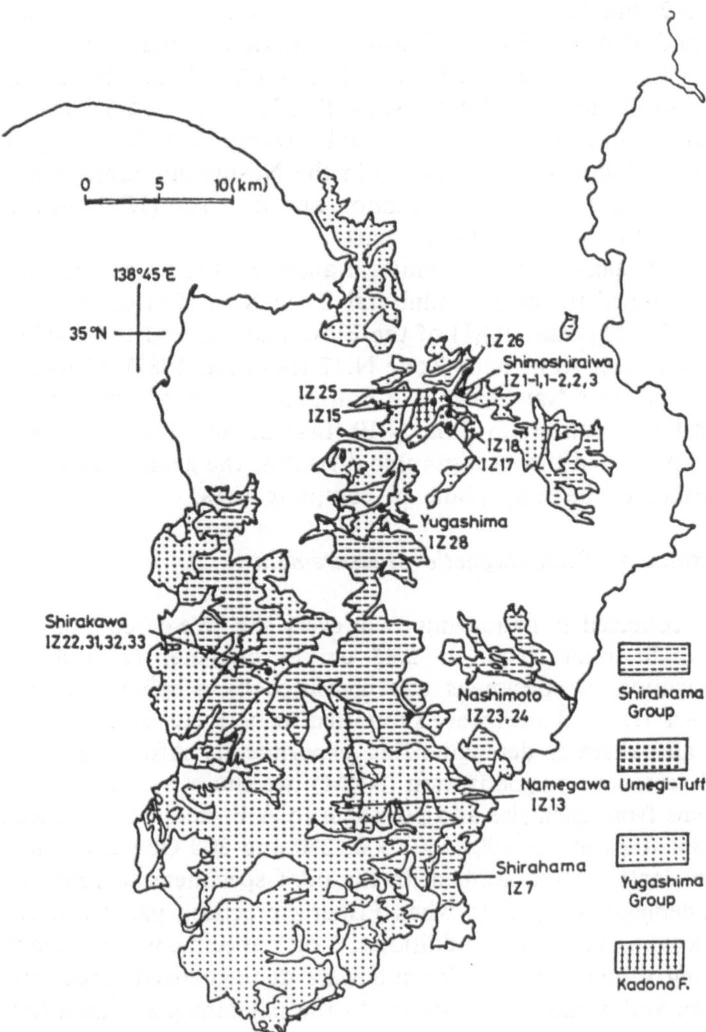

Fig. 1. Map showing the geology of the Izu Peninsula and locations of paleomagnetic sampling sites.

In the northeastern part of the peninsula, paleomagnetic samples were collected from 1 site (IZ 28) in the Okawabata Propylites, 4 sites (IZ 15, 25-1, 25-2 and 26) in the Kadono Formation and from 4 sites (IZ 1-1, 1-2, 3 and 4) in the Shimoshiraiwa Formation of the Yugashima Group. Samples were also obtained from 2 sites (IZ 17 and 18) in the Umegi Formation in the area. From 4 sites (IZ 22, 31, 32 and 33) in the Nishina Formation in the western part, samples were taken. Samplings were carried out at 2 sites (IZ 23 and 24) in the Nashimoto Limestone, at a site (IZ 13) in the Namegawa Formation and at a site (IZ 7) in the Harada Formation in the southeastern part.

By the planktonic foraminiferal analyses, the Shimoshiraiwa Formation is assigned to the foraminiferal Zone N.14 (SATIO, 1963; IKEBE and CHIJI, 1971; IBARAKI, 1981) of the Blow's zonation (BLOW, 1979) and the Nashimoto Limestone is of Zone N.17 (IBARAKI, 1981). Calcareous sandstone found near Namegawa and Shirahama both of which belong to the Harada Formation are in Zone N.19 (IBARAKI and TSUCHI, 1978). In Table 1 are summarized the stratigraphic divisions, the geological ages, and dips and strikes of bedding plane of sampling sites.

3. *Results of Paleomagnetic Measurement*

We collected 10 hand-samples at every site. We cored 2 or 3 specimens with a cylindrical shape of 2.45 cm in diameter from each sample. Magnetization of specimens was measured by a spinner magnetometer. Specimens from all sites were submitted to both stepwise alternating field (A.F.) and thermal demagnetizing experiments. After measurements of natural remanent magnetization (NRM), we chose two sets of 3 or 4 pilot specimens from each site and put them to A.F. and thermal demagnetizations. Six steps of 50, 100, 150, 200, 250 and 300 Oe peak fields of A.F. demagnetization were applied on the pilot specimens to find out the optimum demagnetizing field when Fisher's precision parameter K (FISHER, 1953) became maximum. Additional 6 or 7 specimens were adopted from remaining samples of the site and were demagnetized three times at the optimum and adjacent two steps. As the paleomagnetic data for the site, we accepted results of the step at which the precision parameter K became the greatest.

We also applied the thermal demagnetizing experiments at several steps, such as 100, 150, 200, 250, 300 and 400°C on the pilot specimens to see how the magnetic direction is changed. In many cases, magnetization is changed to a direction concordant with those observed in the A.F. demagnetization, so that we used results of A.F. demagnetization as the

Table 1. Stratigraphic names and ages of sampling localities.

Site NO.		Age (Ma)	Bedding*	
IZ 1-1	Shimoshiraiwa Formation	10.4-11.6	N 13°W	29°E
IZ 1-2	Shimoshiraiwa Formation	10.4-11.6	N 13°W	29°E
IZ 2	Shimoshiraiwa Formation	10.4-11.6	N 2°E	32°E
IZ 3	Shimoshiraiwa Formation	10.4-11.6	N 6°W	40°E
IZ 7	Harada Formation	3.0-3.7	N 24°W	8°W
IZ 13	Namegawa Formation	3.7-5.1	N 36°W	7°W
IZ 15	Kadono Formation	Early Miocene	N 52°E	46°S
IZ 17	Umegi-Tuff	3-5	N 53°E	49°E
IZ 18	Umegi-Tuff	3-5	N 36°E	49°E
IZ 22	Nishina Formation	Early Miocene	N 76°W	28°N
IZ 23	Nashimoto Limestone	5.5-5.8	N 17°E	16°E
IZ 24	Nashimoto Limestone	5.5-5.8	N 44°E	24°E
IZ 25-1	Kadono Formation	Early Miocene	N 59°E	65°S
IZ 25-2	Kadono Formation	Early Miocene	N 59°E	65°S
IZ 26	Kadono Formation	Early Miocene	N 48°E	53°S
IZ 28	Okawabata Propylite	Early Miocene	N 28°W	46°E
IZ 31	Nishina Formation	Early Miocene	N 75°W	20°N
IZ 32	Nishina Formation	Early Miocene	N 74°W	30°N
IZ 33	Nishina Formation	Early Miocene	N 84°E	14°N

*Strikes of bedding planes and declinations are the values referring to the geomagnetic north.

paleomagnetic data. Changes in magnetic direction and intensity of Site IZ 17 are shown as an example in Fig. 2A. Specimens from some sites such as IZ 15 and 26 showed the magnetic direction change in a manner different from that of A.F. demagnetization results (Fig. 2B). In such cases we thermally demagnetized additional specimens to find the optimum step and adopted the result of the thermal demagnetization as the paleomagnetic data. The results of the paleomagnetic measurements obtained by the procedures mentioned above are shown in Tables 2 and 3. Declination and Inclination in Table 2 indicate *in situ* direction.

Since strata of the sampling sites are tilted as shown in Table 1, we applied a bedding correction by turning the bedding plane to the horizontal. Declination and Inclination after the correction are listed in Table 3. It is clear that declinations are greatly different from location to location of the sampling sites. This may be explained by the result that way of geotectonic deformation such as folding and rotation was different from each other for the respective locations. KOYAMA (1982) demonstrated from measurements of a single tuff layer of the northern part of the Izu Penin-

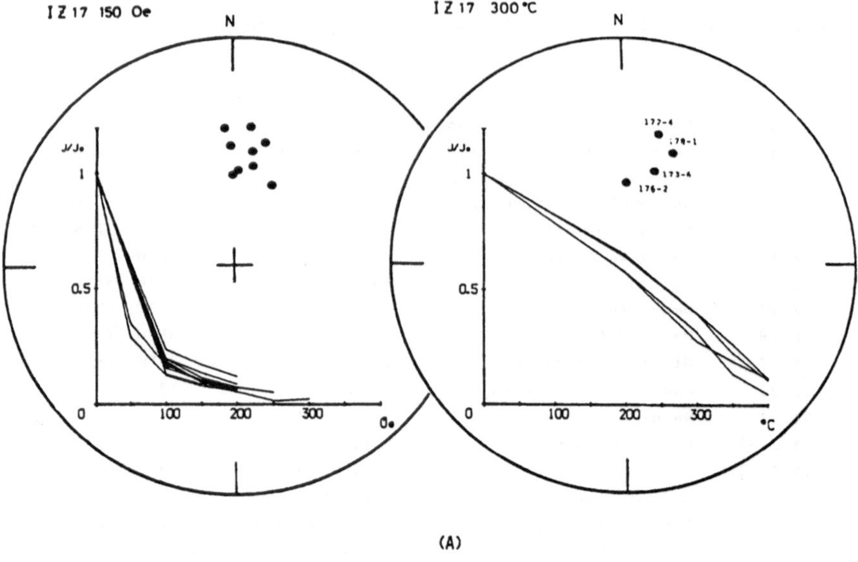

(A)

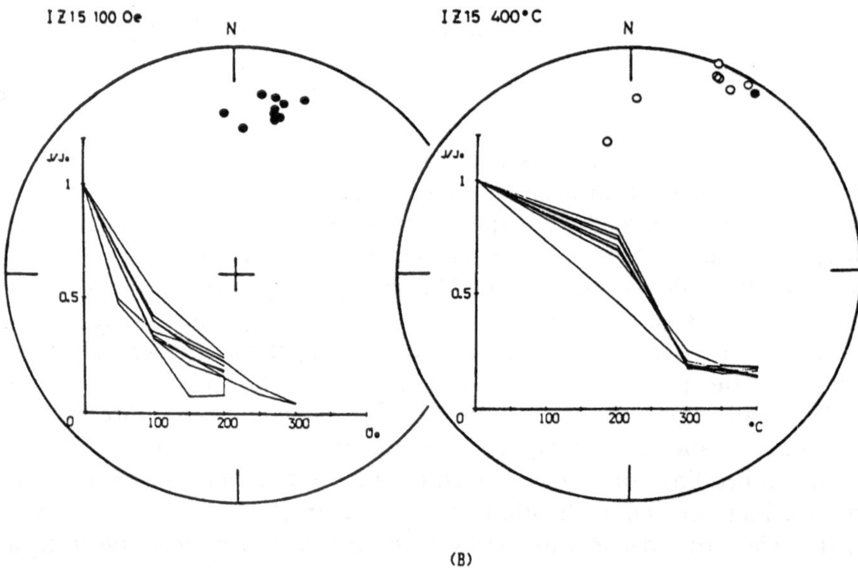

(B)

Fig. 2. Examples of changes of remanent directions and intensities by A.F. and thermal demagnetizations: A: Results of Site IZ 17; Similar directional changes were observed in the both demagnetizing experiments. B: Results of Site IZ 15; Different directional changes were observed in the A.F. and thermal demagnetizing experiments.

Table 2. Paleomagnetic directions.

Site No.	Declination* (°E)	Inclination (°)	α_{95} (°)	k	O.D.F	No. of samples
IZ 1-1	5.9	50.8	7.45	66.7	100 Oe	7
IZ 1-2	2.8	54.1	13.51	17.8	100 Oe	8
IZ 2	-3.7	50.3	5.31	109.9	100 Oe	8
IZ 3	-7.0	47.5	10.12	30.9	100 Oe	8
IZ 7	-3.7	49.1	6.93	44.4	100 Oe	11
IZ 13	-3.0	56.6	5.59	85.7	100 Oe	10
IZ 15	19.9	-11.2	14.06	16.5	400°c	8
IZ 17	7.4	50.7	5.47	89.6	150 Oe	9
IZ 18	-15.2	48.2	9.81	28.5	200 Oe	9
IZ 22	-10.5	56.4	4.49	116.5	100 Oe	10
IZ 23	27.5	51.3	12.57	20.5	250 Oe	8
IZ 24	36.0	40.9	5.97	75.3	150 Oe	9
IZ 25-1	22.6	7.9	12.38	18.3	150 Oe	9
IZ 25-2	10.8	52.7	9.31	27.9	150 Oe	10
IZ 26	-10.0	43.6	6.70	69.4	350°c	8
IZ 28	8.4	61.5	9.56	26.5	150 Oe	10
IZ 31	-23.6	56.2	5.21	86.8	100 Oe	10
IZ 32	-84.2	19.7	2.71	318.2	100 Oe	10
IZ 33	31.4	29.5	10.18	32.9	300 Oe	9

*Declination is the value referring to the geomagnetic north.

sula that the local transcurrent fault greatly affected the paleomagnetic declination. Paleolatitudes of all the sites calculated from paleomagnetic inclinations and the mean paleolatitude for each geologic formation are also summarized in Table 3.

4. *Northward Drift of the Izu Peninsula*

The oldest formation so far sampled in the peninsula is the Nishina Formation of early Miocene. A very low mean paleolatitude of 14.5°N was obtained for the formation, while the mean paleolatitude of the youngest Harada Formation of Pliocene age is 34.5°N which is nearly the same as the present latitude of the peninsula.

Paleolatitudes of all the sites are plotted against their ages in Fig. 3. Age of the geologic formations is referred to the correlation of planktonic foraminiferal datum plane proposed by TSUCHI (1983). It is obvious in Fig. 3 that the Izu Peninsula situated in the equatorial region of the latitude of 14.5°N in early Miocene was migrated northward until it reached the present latitude at a time between 3 and 5 Ma ago.

Table 3. Paleomagnetic direction after bedding correction and paleolatitudes.

Site No.	After Bedding Correction Declination (°E)	Inclination (°)	Paleolatitude (°)	Mean Inclination (°)	Mean Paleolatitude (°)
Harada, Namegawa Formation and Umegi-Tuff					
IZ 7	-19.3	51.4	32.1		
IZ 13	-1.5	52.0	32.6	53.9 ±4.5	34.5 +4.7 -4.1
IZ 17	7.4	50.6	31.3		
IZ 18	57.4	61.8	43.0		
Nashimoto Limestone					
IZ 23	38.3	46.0	27.4	43.5 ±2.6	25.3 +2.1 -1.9
IZ 24	6.5	40.9	23.4		
Shimoshiraiwa Formation					
IZ 1-1	23.4	35.3	19.5		
IZ 1-2	23.8	39.1	22.1	38.4 ±3.3	21.6 +2.4 -2.3
IZ 2	24.1	43.5	25.4		
IZ 3	30.6	35.5	19.6		
Kadono Formation and Yugashima Group					
IZ 15	14.4	13.9	7.1		
IZ 25-1	43.7	36.2	20.1		
IZ 25-2	105.2	48.2	29.2	37.5 ±17.9	21.0 +14.1 -10.9
IZ 26	66.7	64.9	46.8		
IZ 28	30.6	24.0	12.6		
Okawabata Propylite and Nishina Formation					
IZ 22	-7.9	30.0	16.1		
IZ 31	-18.1	39.2	22.2	27.3 ±8.2	14.5 +5.1 -4.7
IZ 32	-79.1	22.0	11.4		
IZ 33	21.5	17.9	9.2		

The paleolatitudinal difference between the Harada Formation (3–3.7 Ma) and the Nishina Formation (16.6–18 Ma) is about 20 degrees. Rate of the northward component of migration amounts to about 15 cm/year. This rate is much higher than the estimated present drifting rate of the Philippine Sea Plate. KINOSHITA (1980) reported paleolatitude change of the Daito Region obtained from paleomagnetic results of the Deep Sea Drilling Project Sites 445 and 446. The mean rate of northward drift is 5 cm/year since Eocene. Paleolatitudes of the Daito Region and the Izu Peninsula are plotted versus their age in Fig. 4. The paleolatitude of the Izu Peninsula and that of the Daito Region were nearly the same in Miocene. After that, rate of the northward drifting of the Izu Peninsula became much higher as compared to the Daito Region.

Results of paleomagnetic studies of Tertiary rocks in the Philippine Sea Plate were reported from the Guam Island (LARSON *et al.*, 1975),

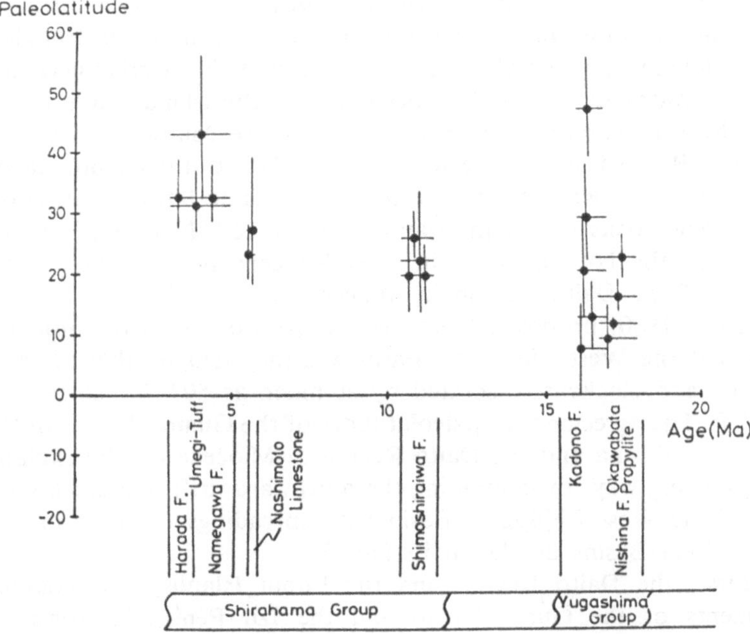

Fig. 3. Paleolatitudes of the Izu Peninsula calculated from paleomagnetic inclinations.

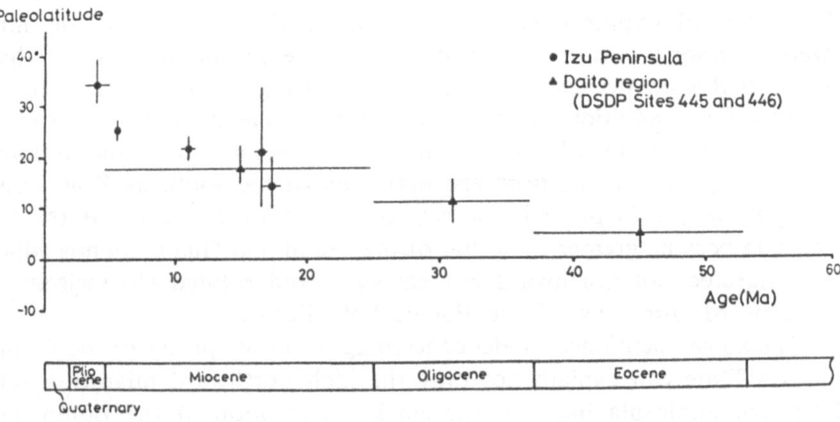

Fig. 4. Paleolatitudinal changes of DSDP Sites 445 and 446 of Daito Region (after KINOSHITA, 1980) and the Izu Peninsula. Data of the Izu Peninsula are the mean values for each geologic formation.

and the Bonin Islands (KODAMA, 1981; KODAMA *et al.*, 1983). Paleomagnetic direction of the Eocene volcanics of the Bonin Islands showed very low inclinations of 3° to 10°, and greatly deflected easterly declinations. KODAMA *et al.* (1983) concluded that the islands have undergone a northward migration together with clockwise rotation of 30° to over 90°. The Bonin Islands were situated very close to the equator in Eocene when the Daito Region was located at the same latitude. Paleolatitude of Miocene rocks in Guam Island is about 12.5°N which is very close to that of the Izu Peninsula in early Miocene reported here, and also to that of the Daito Region in Miocene.

SHIH (1980) proposed from the skewness of magnetic anomaly patterns that the West Philippine Basin was migrated northward by about 15°–20° and clockwisely rotated by as much as 50°–70° since 35 to 40 Ma B.P. The agreement of paleolatitudes of the Guam, the Bonin Islands, the Izu Peninsula and the Daito Region in Miocene can be explained in the following way, by combining the northward drift and clockwise rotation of the West Philippine Basin with spreadings of the Shikoku and Parece Vela Basins as shown in Fig. 5.

First, the Daito Region and the Bonin Islands, and possibly the basements of the Guam Island and the Izu Peninsula, were located altogether in the equatorial region in Eocene. Then, by spreading and associated rotation of the West Philippine Basin, the Daito Region, the Izu Peninsula, the Bonin and the Guam Islands were transported northward with a small clockwise rotation. In the Miocene time, the Shikoku and Parece Vela Basins began to open. If we assume that the rotation of the West Philippine Basin was not completed in this period, the northward component of migration of the Izu Peninsula was greater than the Daito Region because the northward drifting component caused by the rotation was additional to the component originated from the spreading of the Shikoku Basin. Moreover, since the spreading rate of the Shikoku Basin was greater in the northern part than in the south as KOBAYASHI and NAKADA (1978) proved, the rate of the northward drift of the Izu Peninsula became greater than that of the Bonin and Guam Islands which were migrated not northward but eastward and rotated clockwisely by the eastward spreading of the Parece Vela Basin.

The above-mentioned model concerning the development of the Philippine Sea Plate can explain not only the high northward migration rate of the Izu Peninsula but also the clockwise rotation of the Bonin and Guam Islands. In addition to northward initial migration of the West Philippine Basin, our model is still valid in the case that the initial direction was north-westward as SENO and MARUYAMA (1984) proposed, as far as

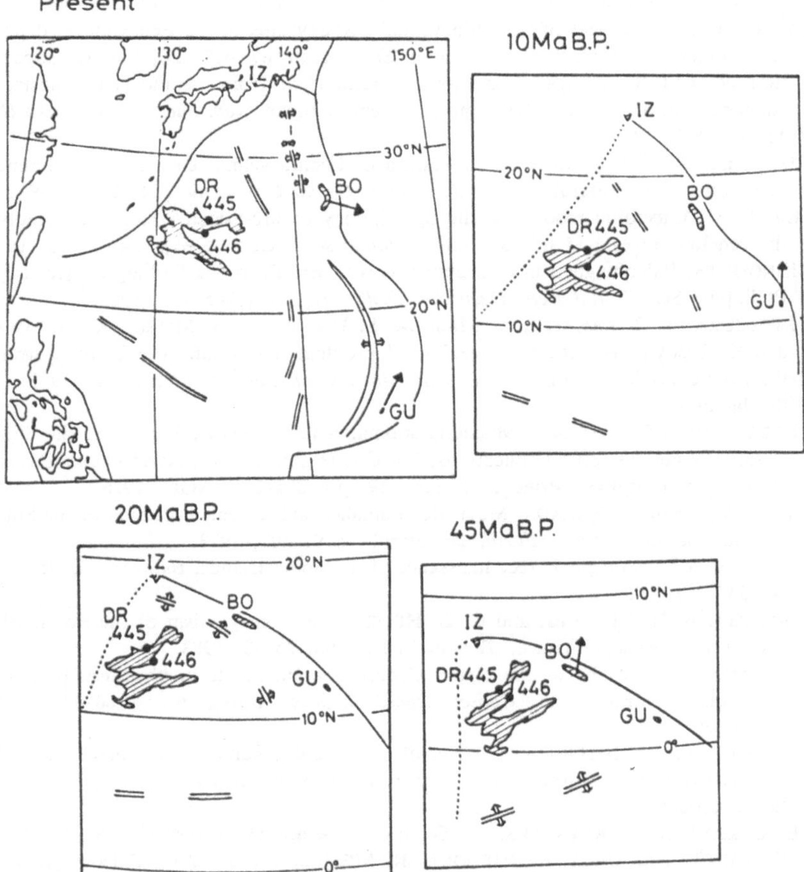

Fig. 5. Migrations of the Izu Peninsula, the Bonin Islands, the Guam Island and the Daito Region associated with the evolution of the Philippine Sea Plate. IZ: Izu Peninsula, BO: Bonin Islands, GU: Guam Island, DR: Daito Region.

the clockwise rotation of the West Philippine Basin is allowed. Finally, the Izu Peninsula reached the position to contact the Honshu Island 3 to 5 Ma ago then collided with it.

REFERENCES

W. H. Blow, *The Cainozoic Globigerinida*, ed. E. J. Brill, part 1, pp. 1–221, Leiden, 1979.

R. A. FISHER, Dispersion on a sphere, *R. Astr. Soc. Lond., A*, **217**, 295–304, 1953.

M. IBARAKI, Geologic ages of "Lepidocyclina", Miogypsina in Izu Peninsula as determined by planktonic foraminifera, *J. Geol. Soc. Japan*, **87**, 417–420, 1981 (in Japanese).

M. IBARAKI and R. TSUCHI, Planktonic foraminifera from Lepidocyclina horizon at Namegawa in the southern Izu Peninsula, central Japan, *Rep. Fac. Sci. Shizuoka Univ.*, **12**, 115–127, 1978.

N. IKEBE and M. CHIJI, Notes on top-datum of Lepidocyclina sensu lato in reference to planktonic foraminiferal datum, *J. Geosci. Osaka City Univ.*, **14**, 19–44, 1971.

S. KAIZUKA, A tectonic model for the morphology of arc-trench systems, especially for the echelon ridges and mid-arc faults, *Japanese J. Geol. Geogr.*, **45**, 9–25, 1975.

H. KINOSHITA, Paleomagnetism of sediment cores from Deep Sea Drilling Project Leg 58, Philippine Sea, *Initial Rep. Deep Sea Drill. Proj.*, **58**, 765–768, 1980.

N. KITAMURA, Y. TAKAHASHI, K. MASUDA, S. HAYASAKA, S. MITSUI, K. SUGAWARA, and K. TAKAHASHI, On some geological problems concerning the Tertiary strata of the Izu Peninsula, Japan, *Tohoku Univ. Inst. Geol. Pal., Contr.*, no. 68, 19–31, 1969 (in Japanese).

K. KOBAYASHI and S. ISEZAKI, Magnetic anomalies in the Sea of Japan and the Shikoku Basin: Possible tectonic implications *The Geophysics of the Pacific Ocean Basin and Its Margin*, Geophys. Monograph Ser., 19, pp. 235–251, AGU, 1976.

K. KOBAYASHI and M. NAKADA, Magnetic anomalies and tectonic evolution of the Shikoku inter-arc basin, *J. Phys. Earth,* **26**, Suppl., S391–402, 1978.

K. KODAMA, A Paleomagnetic reconnaissance of the Bonin Islands. *Bull. Earthq. Res. Inst.*, **56**, 347–365, 1981.

K. KODAMA, B. H. KEATING, and C. E. HELSLEY, Paleomagnetism of the Bonin Islands and its tectonic significance, *Tectonophysics*, **95**, 25–42, 1983.

M. KOYAMA, Stratigraphy of the Upper Cenozoic strata in the northeastern part of Izu Peninsula, Central Japan, *Geosci. Reps. Shizuoka Univ.*, no. 7, 60–85, 1982 (in Japanese).

M. KOYAMA and N. NIITSUMA, Lexicon of stratigraphic names of Cenozoic Erathem in the Izu Peninsula, Central Japan, *Geosci. Reps. Shizuoka Univ.*, no. 5, 37–119, 1980 (in Japanese).

E. E. LARSON, R. L. REYNOLDS, M. OZIMA, Y. AOKI, H. KINOSHITA, S. ZASSHU, N. KAWAI, T. NAKAJIMA, K. HIROOKA, R. MERILL, and S. LEVI, Paleomagnetism of Miocene volcanic rocks of Guam and the curvature of the southern Mariana Island Arc, *Geol. Soc. Amer. Bull.*, **86**, 346–350, 1975.

T. MATSUDA, Collision of the Izu-Bonin arc with central Honshu: Cenozoic tectonics of the Fossa Magna, Japan, *J. Phys. Earth*, **26**, Suppl., S409–421, 1978.

T. MORITANI and K. SAWAMURA, Tertiary formations of the Matsuzaki district, Izu Peninsula, Japan, *Bull. Geol. Survey Japan*, **16**, 535–545, 1965 (in Japanese).

T. SAITO, Miocene planktonic foraminifera from Honshu, Japan, *Sci. Rep. Tohoku Univ., [2]*, **35**, (2), 123–209, 1963.

T. SAMEJIMA, Geology of southern part of the Izu Peninsula, *Chigaku Shizuhata*, no. 8, 15–18, 1955 (in Japanese).

T. SAMEJIMA, *Geology of Shizuoka, a Field Work*, pp. 33–78, Shizuoka Education Press, 1978 (in Japanese).

T. SENO and S. MARUYAMA, Paleogeographic reconstruction and origin of the Philippine Sea, *Tectonophysics*, **102**, 53–84, 1984.

T. C. SHIH, Marine magnetic anomalies from the western Philippine Sea; Implications for the evolution of marginal basins, *The Tectonic and Geologic Evolution of Southeast*

Asian Seas and Islands, Geophys. Monograph Ser., 23, pp. 49–75, AGU, 1980.
A. SUGIMURA, Plate boundaries near Japan, *Kagaku* (*Science*), **42**, 192–202, 1972 (in Japanese).
R. TSUCHI, Neogene bio- and chronostratigraphy in Japan, *J. Japanese Assoc. Petr. Technol.*, **48**, 35–48, 1983 (in Japanese).
R. TSUCHI, T. SAMEJIMA, T. IWAHASHI, A. TOKUYAMA, M. ITO, N. KURODA, O. FUJIYOSHI, and S. IKEGAYA, *Geological Map of 1/200000 "Shizuoka Prefecture"*, 1974 (in Japanese).

CHAPTER 7: ACCRETION TECTONICS (III)

Formation of Active Ocean Margins, edited by N. Nasu *et al.*, pp. 791–817.

SOME TECTONIC AND TECTOGENETIC ASPECTS OF SW JAPAN: AN ALPINE-TYPE OROGEN IN AN ISLAND-ARC POSITION

Jacques CHARVET,[1] Michel FAURE,[1,2] Martial CARIDROIT,[1,3] and André GUIDI[1,4]

[1]*Laboratoire de Géologie Structurale, Université d'Orléans, 45046 Orleans, France*

[2]*Department of Earth Sciences, Kyoiku Gakubu, Tokushima University, 770 Tokushima, Japan*

[3]*Department of Geosciences, Osaka City University, Sumiyoshi-ku, 558 Osaka, Japan*

[4]*Institute of Geoscience, University of Tsukuba, Sakura-mura, Niihari-gun, 305 Ibaraki, Japan*

Abstract. The SW of Japan is a present-day island-arc formed rather recently at the expense of a polyorogenic chain showing alpine structural features. The Mesozoic structures are the most impressive. They comprise a system of nappes covering the inner and outer zones, on both sides of the Median Tectonic Line. The Mesozoic orogeny is characterized by three main tectonic phases. The 1st one, linked to the HP/LT Sambagawa metamorphism, is interpreted as a large ductile shear from West to East, leading to the thrusting of schistose formations bearing ophiolitic debris, over a sialic basement domain: the Oboke-Kurosegawa block. The 2nd phase is responsible for the emplacement of post metamorphic nappes, during Late-Jurassic-EoCretaceous times, with southward vergence. At that time, the high-grade metamorphics were thrusted upon the epimetamorphic schists in the Sambagawa zone; the upper unit (olistostrome assemblage) of the "Chichibu zone" was emplaced, from the Tamba zone: klippes and tectonic windows can be observed from the Kanto Mountains to Shikoku. In the inner part, during the formation of the Tamba nappes, the Paleozoic chain (Sangun-Maizuru) was largely reworked and thrusted towards the south. In the 3rd phase (Hijikawa phase), upright folds were developed, with an en echelon arrangement related to the left lateral motion along the M.T.L. The geodynamic model inferred from these events consists of subduction leading to the collision of the Oboke-Kurosegawa block with the Pre-Ryoke and Hida blocks. The Mesozoic and Lower Tertiary orogenies of SW Japan, with some variations, appear to illustrate two collisional events.

1. Introduction

It is well known that, with regard to present-day geodynamics, SW Japan is an island-arc more or less inherited from Miocene time. This paper deals with the pre-Neogene tectonic evolution of this region and especially with the Mesozoic tectonic development.

According to the classical pre-Neogene structural subdivisions (TANAKA and NOZAWA, 1977) Southwest Japan is made of several geological belts parallel to the general trend of SW Honshu (see insert Fig. 1). However, these present strips are mainly bounded by faults e.g. the Median Tectonic Line (M.T.L.) and so this recent disposition cuts through and makes unclear the early structure of this region.

As already stated by KOBAYASHI (1941), at least three orogenic cycles in SW Japan can be recognized. A Cenozoic chain, the Shimanto orogen, occurs on the Pacific Ocean side. Although the so-called Shimanto zone is bounded to the North by the Butsuzo Tectonic Line (B.T.L., Fig. 1), the effects of the Lower Tertiary orogeny are not restricted to this area. The northern zones were reworked at that time. A Mesozoic chain, evidence for which is shown by the unconformity of Neocomian age shallow water deposits, occupies the main part of SW Japan. It comprises several "zones", divided in two groups by the M.T.L. (Fig. 1): Tamba-Ryoke on the "inner side", to the north (Japan Sea side) and Sambagawa-Mikabu-Chichibu

Fig. 1. Structural map of Eastern Shikoku and Kinki Area (after FAURE and CHARVET, 1984). a: Sangun schists. b: Paleozoic Yakuno ophiolitic complex (ultramafics in black). c: Late Permian Maizuru group. d: Tamba-Ryoke zone. e: Triassic unconformity. f: Presumed outcrop of Pre-Ryoke metamorphic rocks. g: Sambagawa high grade schists, Upper Unit. h: Sambagawa Lower Unit, Oboke Unit. i: Sambagawa Intermediate Unit, including the Mikabu green rocks and the northern edge of the Chichibu "zone". j: Chichibu nappe. k: Chichibu southern subbelt with the Kurosegawa rocks in black. l: Lower Cretaceous unconformity. m: Post orogenic (for the main Mesozoic tectonism) deposits, Cretaceous to Present. n: Mesozoic thrust contact. o: Presumably Paleozoic thrust contact. M.T.L.: Median Tectonic Line. B.T.L.: Butsuzo Tectonic Line. Insert: Location map. 1: Hida zone. 2: Sangun zone. 3: Maizuru zone. 4: Tamba-Ryoke zone. 5: Sambagawa zone. 6: Chichibu zone. 7: Shimanto zone. T: Tokyo. H: Hiroshima.

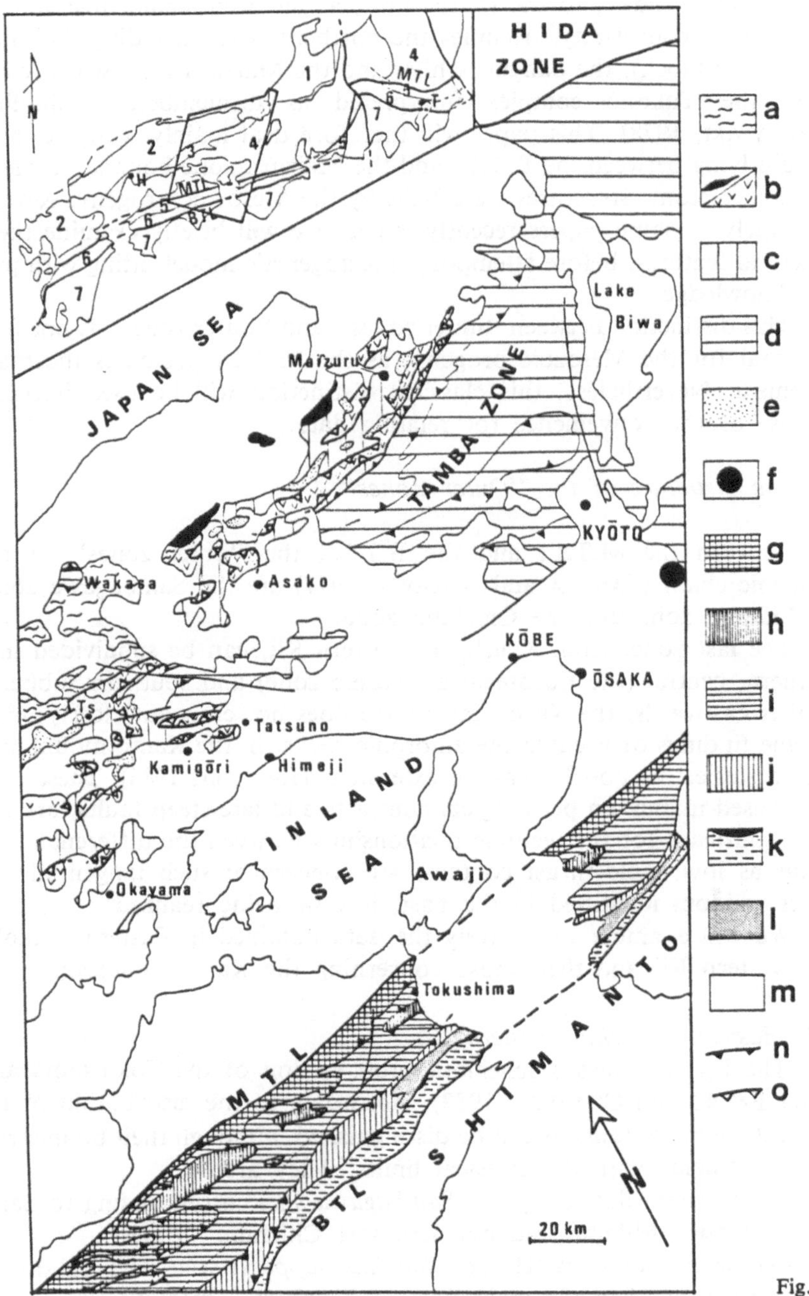
HIDA ZONE
JAPAN SEA
TAMBA ZONE
Lake Biwa
Maizuru
KYŌTO
Wakasa
Asako
KŌBE
ŌSAKA
Tatsuno
Kamigōri
Himeji
INLAND SEA
Awaji
Okayama
Tokushima
MTL
BTL
SHIMANTO
20 km
N
a
b
c
d
e
f
g
h
i
j
k
l
m
n
o

Fig. 1

on the "outer side", to the south. Last, a Paleozoic chain, overlain by a Triassic unconformity, occupies the northern part, including the Hida zone (sialic block), the Sangun schists and the Maizuru zone which comprises the Yakuno complex interpreted as dismembered ophiolites (ISHIWATARI, 1978). This paper will therefore deal mainly with the large domain lying between the B.T.L. and the Maizuru zone boundary, where the main tectonic structures were built by the Mesozoic orogeny. Referring largely to some papers recently issued, we will briefly examine these structural features, before attempting a tectogenetic model fitting with present knowledge.

This distinction between "inner zones" and "outer zones" is not fundamental for the Mesozoic orogeny, as the M.T.L. postdates the main tectonism. Nevertheless, this classical distinction will be used initially, because of the convenience for relating data.

2. *The Structure of the "Outer Zones"*

Between the M.T.L. and the B.T.L., the "outer zones" of the Mesozoic chain (TANAKA and NOZAWA, 1977) are the Sambagawa zone, the Mikabu zone and the Chichibu zone.

The last zone, from Kyushu to western Kii, can be subdivided into northern, central (the Kurosegawa tectonic zone) and southern subbelts. Further eastwards, the Kurosegawa zone does not crop out clearly, even if some findings of peculiar metamorphic rocks, in the Kanto mountains, lead to suspect its possible eastern extension (HIRAJIMA, 1983). These division, based mainly on petrological contrasts and late steep faults, are not very significant for the tectonic relationships between the different units, as far as low angle thrust contacts are concerned; such tangential tectonics, seldom identified in the past, is now being realized.

We will examine successively the data obtained in Eastern Shikoku and western Kii and then those concerning the Kanto mountains.

2.1 *Eastern Shikoku—western Kii*

The Fig. 2 shows a schematic cross-section of the Tokushima area (after FAURE and CHARVET, 1983). According to the distribution of terranes, three main domains can be distinguished, although their boundaries do not coincide with the classical limits. They are:

—a northern domain (I) or *Sambagawa s.l.*, corresponding to Sambagawa, Mikabu and northernmost Chichibu "zones",

—a median domain (II) or *Chichibu nappe* (FAURE and CHARVET, 1983) covering the main part of Northern Chichibu,

Fig. 2. Structural cross-section of the "outer zones" in Eastern Shikoku. A: Izumi group. B: Cretaceous deposits of the Chichibu Middle (Central) subbelt. C: Shimanto strata. D: Sambagawa high grade schists. E: Low grade schists. F: Sandstone unit. G: Oboke-Kurosegawa basement. H: Pre-Cretaceous sediments of domain III. I: Chichibu nappe, olistostrome. J: Green rocks. K: Stratigraphic cover of the Mikabu green rocks (northernmost part of Chichibu). L: Thrust contact of phase 1. M: Thrust contact of phase 2. O, Ku: Oboke, Kurosegawa. Insert: Simplified geologic map of the Tokushima area, Eastern Shikoku. a: Sambagawa high grade rocks. b: Sambagawa intermediate unit. c: Oboke unit. d: Jurassic olistostrome of the Chichibu nappe, including green rocks. e: Serpentinite. f: Kurosegawa rocks. g: Mesozoic Pre-Cretaceous strata of southern Chichibu. h: Unconformable Cretaceous deposits. i: Ryoke zone. j: Shimanto zone. k: 1st phase thrust. l: 2nd phase thrust. M.T.L.: Median Tectonic Line. B.T.L.: Butsuzo Tectonic Line. T: Tokushima. Y: Yoshino river alluvial plain. (after FAURE and CHARVET, 1983).

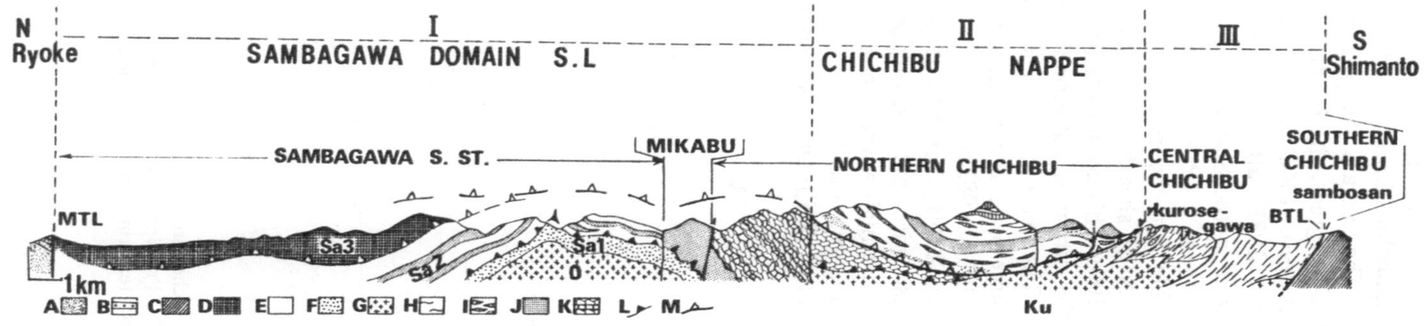

—a southern domain (III) including the Kurosegawa and Sambosan subbelts (Central and Southern Chichibu).

1) The domain I is the metamorphic domain. All the rocks suffered, to varying degrees, the HP/LT Sambagawa type metamorphism (IWASAKI, 1963; BANNO, 1964). Also, they suffered (FAURE, 1983b; FAURE and CHARVET, 1983) all three main deformation phases at first properly recognized in the Sambagawa schists (HARA *et al.*, 1977; FAURE, 1982, 1983a). Phase 1 (Pre-Nagahama) is synfolial and synmetamorphic. It is responsible for the regional foliation S1, isoclinal or intrafolial folds sometimes of sheath fold type and a conspicuous regional lineation L1 (N70-N100). This composite lineation shows stretching features; in the *XZ* plane the strain is rotational (Fig. 3) with rotation criteria from West to East. The microtectonic characteristics lead to an interpretation of phase 1 on a large scale as a ductile shear corresponding to an eastward thrusting (FAURE 1982, 1983a).

Phase 2 (Nagahama-Ozu) gave post-folial asymmetric folds (N90-N120) and thrusts with southern vergence, such as that which caused the emplacement of the high-grade metamorphic unit (Sa 3) upon the low grade intermediate unit (Sa 2, Fig. 2).

Phase 3 (Hijikawa) is responsible for E-W upright folds (N80-N110)

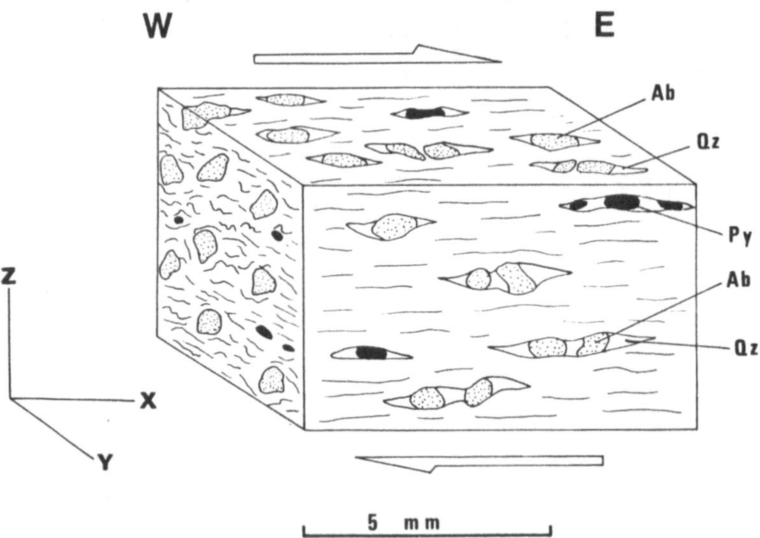

Fig. 3. Schematic block diagram of a Sambagawa basic schist (after Guidi *et al.*, 1983). X, Y, Z: Principal strain axes. Ab: Albite prophyroblast. Py: Pyrite. Qz: Recrystallized quartz featuring the stretching lineation: pressure shadow and pull-apart.

with a subvertical axial cleavage and, on a regional scale, for antiform and synform megastructures, corresponding to a N-S compression.

The Sambagawa schists, the Mikabu green rocks and their stratigraphic cover (Northernmost Chichibu) all show the effects of the synmetamorphic phase 1. For instance the stretching features of the lineation L1 are proven in metasediments by elongated radiolarians (TORIUMI, 1982). Owing to the stratigraphical data (MATSUDA, 1978), phase 1 postdates the Uppermost Triassic-Earliest Jurassic. Inside this domain, three units are superposed: Sa 1, Sa 2, Sa 3, in ascending order. The upper unit Sa 3, composed of high grade schists so-called "spotted schists" (HIDE *et al.*, 1956; IWASAKI, 1963; HARA *et al.*, 1977; BANNO *et al.*, 1979) and the unit Sa 2 or intermediate unit, of lower grade (IWASAKI, 1963; KAJI, 1975), are both rich in ocean bottom-derived material. The Mikabu green rocks represent an ophiolitic olistostrome (IWASAKI, 1979; TAKEDA *et al.*, 1981; FAURE and IWASAKI, 1982), of tholeitic chemical trend (IWASAKI, 1979), and showing some relicts of oceanic metamorphism (IWASAKI, 1982). Their metamorphic grade (Sambagawa metamorphism) seems to be the same as in the unit Sa 2. Therefore, in spite of a fault separating them, the Mikabu green rocks and the intermediate unit of Sambagawa schists can be most likely correlated with each other (FAURE and CHARVET, 1983, 1984). The unit Sa 1 is lithologically very different: it is a sandstone unit, with strong continental affinities. This unit, also known as the Oboke unit (from the Oboke locality to the SW of Tokushima area) could represent the cratonic tectonic basement of the allochtonous formations (bearing oceanic material) emplaced during phase 1 (FAURE, 1982, 1983a, b; FAURE and CHARVET, 1983).

2) Domain II forms the Chichibu nappe (FAURE and CHARVET 1983; FAURE, 1983b). It is an olistostrome complex, with a pelitic matrix of Middle to early Late Jurassic age (ISOZAKI *et al.*, 1981; SUYARI *et al.*, 1981). Various rocks are reworked as olistoliths: Upper Carboniferous—Lower Permian green rocks (SUYARI *et al.*, 1982) and limestones (KANMERA, 1968; OGAWA, 1974), and Permian to Triassic cherts (YOKOYAMA *et al.*, 1979). In this unit, the deformation features of phase 1 are completely lacking. It is interpreted as an allochtonous mass emplaced by phase 2, during Upper Jurassic-Eo-Cretaceous time (FAURE and CHARVET, 1983).

Such a structure is also known in Central Shikoku (TSUKUDA *et al.*, 1981).

3) Domain III corresponds to the Middle and Southern Chichibu subbelts. The Kurosegawa zone (ICHIKAWA *et al.*, 1956) is very complex. Within the lenticular-shaped outcrops, a result of strike-slip movement,

various kinds of rocks can be observed: Paleozoic granitoids and high-grade metamorphic rocks (HAYASE and NOHDA, 1969; UEDA and ONUKI, 1969; NOHDA, 1973; YOSHIKURA *et al.*, 1981), Paleozoic blue schists (MARUYAMA and UEDA, 1974; UEDA *et al.*, 1980; MARUYAMA, 1981), and Paleozoic sediments (TANAKA and NOZAWA, 1977) among which a Permian olistostrome is unconformably overlain by Triassic sandstones (ICHIKAWA *et al.*, 1954). Therefore, the Kurosegawa zone seems to represent a multi-lacerated basement, affected by Paleozoic tectonic events and then reworked by Mesozoic and Cenozoic orogenies. It can be seen as the sialic basement upon which the Chichibu nappe was emplaced, in other words the fore-land of the Mesozoic orogeny.

The southern subbelt or Sambosan zone includes an Upper Jurassic turbiditic formation with Triassic cherts and Early to early Late Jurassic cherts and acidic tuffs as olistoliths (ISHIDA, 1983), covered by Upper to Uppermost Jurassic reef deposits (Torinosu group). The southernmost part might extend into the Lower Cretaceous in the Kii peninsula (YAO, 1984). The present tectonic aspect is obviously due to southward verging thrusts which occurred during the Shimanto orogeny. All domain III was largely tectonically reworked during ther Lower Tertiary events, such as in Kyushu where this phenomenon has been clearly demonstrated (MURATA, 1981) and in the central part of Kii peninsula (KURIMOTO, 1982).

Nevertheless, in spite of the complexity, for Jurassic time, we can emphasize the difference of facies between the sequences known to the south and the north of the Kurosegawa zone respectively. This datum, already pointed out by KIMURA *et al.* (1975), supports the idea of an original paleogeographic feature corresponding to a Kurosegawa block and, hence, the asumption of a Kurosegawa fore-land during the main Mesozoic tectonism.

4) To understand the tangential relationships between the units, we assume that the Oboke unit is the northern extension of the Kurosegawa basement (FAURE and CHARVET, 1983). Consequently, the structure, from bottom to top, is composed of (FAURE and CHARVET, 1984):

—the underlying units: Oboke and Kurosegawa, parts of the Oboke-Kurosegawa sialic basement and its cover,
—the synmetamorphic nappes of phase 1 of Sambagawa (Sa 2),
—the post metamorphic nappes of phase 2: upper unit of Sambagawa (Sa 3) covered by the non-metamorphic Chichibu nappe.

As the Neocomian deposits lie unconformably on the main structures, this pile of nappes was emplaced during the pre-Cretaceous orogeny. This was likely to have been during Middle-Upper Jurassic for phase 1 and Upper Jurassic-Eo-Cretaceous for phase 2. However phase 3 does fold the lowermost Cretaceous and so it occurred after that time (cf infra).

Such a structure extends towards Central Shikoku and to Kii (Fig. 1). It fits with the seismological data which show the existence of a sialic basement from the Kurosegawa zone towards the North, under the whole Chichibu and Sambagawa "zones" (HADA *et al.*, 1982; HADA and SUZUKI, 1983).

2.2 Kanto Mountains

The same general tectonic superposition (Fig. 4–5) is observed in the Kanto Mountains around Mamba (GUIDI *et al.*, 1983), from Sambagawa to "North Chichibu", that is to say in the area bounded to the south by the Lower Cretaceous of Sanchu. But, some differences can be seen. The main one is the fact that, owing perhaps to a weaker phase 2, the high-grade metamorphic schists of Sambagawa remain in a low structural position and are not thrusted upon the greenschists.

Therefore, the upward succession is as follows (Fig. 5).

i) A metamorphic unit comprising (SEKI, 1958; TORIUMI, 1975; SUZUKI, 1977):

—the "spotted" schists
—the "non spotted" schists
—the Mikabu Green Rocks
—the stratigraphic cover of Mikabu, which is divided in two parts: a lower part (E_2) or "Sakahara formation" (FUJIMOTO, 1937; TAKIZAWA, 1982) of Permo-Triassic age and an upper part (E_1) assigned to the "Kashiwagi" or "Takayagi" Formations, containing undetermined Radiolarian faunas.

All this members suffered the metamorphism and deformation of phase 1. The L1 lineation here has a NW-SE trend on average however, owing to the clockwise rotation of the Kanto Mts with regard to the general trend of the chain.

ii) A non-metamorphic unit, corresponding to a Mid-Jurassic olistostrome (TAKIZAWA, 1982; SASHIDA *et al.*, 1982; GUIDI *et al.*, 1983) including the so-called "Mamba" and "Kamiyoshida" Formations. This olistostrome comprises huge blocks of Permian limestones in the upper part, such as that of Mt Kano.

The relationships between the olistostrome and the metamorphic unit are tectonic. Some conspicuous post-metamorphic asymmetric folds with a southward vergence (SATO *et al.*, 1977) lead us to attribute this tectonic superposition to the 2nd phase of deformation. Later, two phases of folding reaffected the area: one NW-SE phase during the Upper Cretaceous-Lower Tertiary and one N-S from Tertiary to Recent (INOUE, 1974).

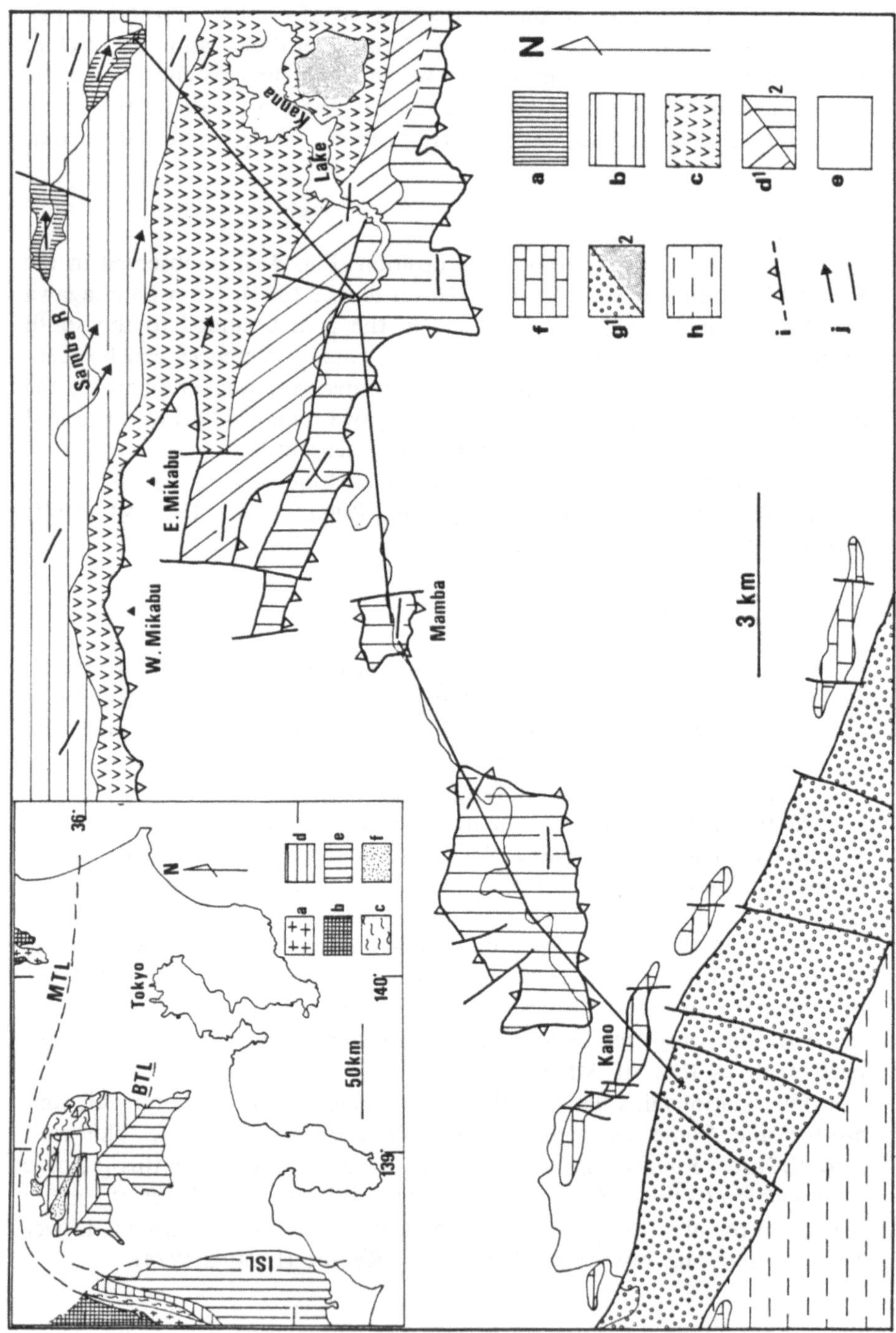
Samba R.
Lake
Kanna
E. Mikabu
W. Mikabu
Mamba
Kano
3 km
N
a
b
c
d1
2
e
f
g1
2
h
i
j
MTL
Tokyo
BTL
ISL
50km
36°
140°
139°
N
a
b
c
d
e
f

Fig. 4. Tectonic sketch map of the North-Western area of the Kanto Mountains (after Guidi *et al.*, 1983, modified). a: Basic schists. b: Pelitic schists. c: Mikabu green rocks. d: Sedimentary cover of the Mikabu green rocks; 1: Lower part; 2: Upper part. e: Middle-Jurassic olistostrome. f: Limestone olistolith. g: Unconformable strata; 1: Lower Cretaceous of Sanchu; 2: Atokura Formation, Upper Cretaceous or Tertiary. h: South Chichibu belt. i: 2nd phase thrust contact. j: Lineation L1 with rotation sense where determined. Insert: tectonic sketch map of the Kanto Mountains and vicinity. a: Ryoke zone. b: Tamba zone. c: Sambagawa zone. d: Chichibu zone. e: Shimanto zone. f: Unconformable Cretaceous. M.T.L.: Median Tectonic Line. B.T.L.: Butsuzo Tectonic Line. I.S.L.: Itoigawa-Shizuoka Line.

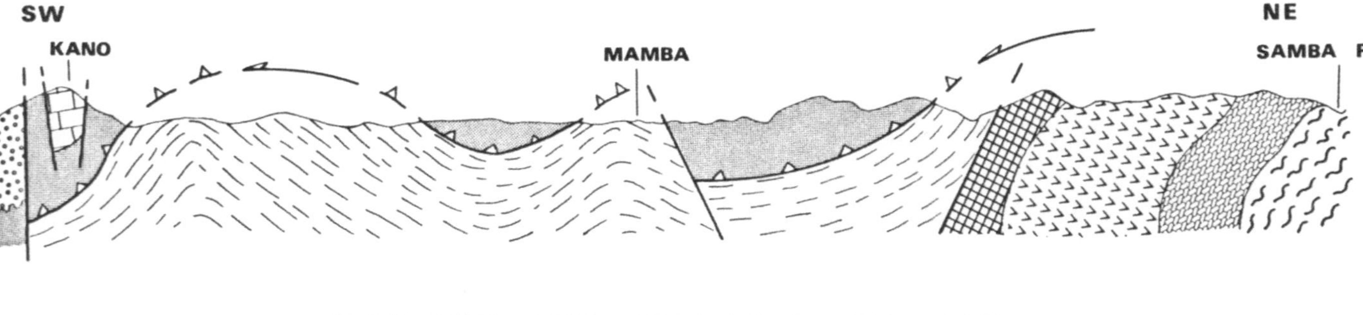

Fig. 5. Cross-section from the Samba River to Mt Kano, located on Fig. 4 (after Guidi *et al.*, 1983, modified). A: 2nd phase thrust contact. B: Lower Cretaceous of Sanchu. C: Permian limestone olistolith. D: Middle-Jurassic olistostrome. E: Stratigraphic cover of Mikabu; 1: Upper part; 2: Lower part. F: Mikabu green rocks. G: Pelitic schists. H: Basic schists.

2.3 Conclusions about the structure of "outer zones"

The area bounded by the M.T.L. to the North and the B.T.L. to the South is in fact a pile up of nappes emplaced during the Mesozoic orogeny upon an Oboke-Kurosegawa block. The usual subdivisions do not account for this structure as they correspond to strips of land recently cut by faults. For instance, the heterogenous domain of the so-called "Chichibu zone" shows actually three main units; the upper one represents the Chichibu nappe. Northern origin of the nappe, from beyond the M.T.L. ("inner zones") is likely as it escaped phase 1 which affected all the formations to the South of the M.T.L. (FAURE and CHARVET, 1983). In the Kanto Mountains, the basic volcanic rocks exposed at the top of Mt Mikabu seem to be a part of the olistostrome nappe: they do not show the strain features nor the metamorphism of phase 1. In other words, they are not representative of the "Mikabu zone" in which the volcanoclastites suffered the Sambagawa metamorphism (GUIDI *et al.*, 1983).

3. Structure of the "Inner Zones"

The "inner zones" are divided into (TANAKA and NOZAWA, 1977) the Ryoke and Tamba zones for the Mesozoic orogen, and Maizuru, Sangun and Hida zones for the Paleozoic chain (Fig. 1). Using the structural sketch of Fig. 1, restricted to the Kinki area, we will illustrate two tectonic aspects: the structure of the Tamba-Ryoke domain and the relationship between Tamba and the Paleozoic chain: namely the Maizuru and Sangun zones. The structure of the Hida zone will not be studied here.

3.1 Tectonics of the Tamba-Ryoke domain

As far as the Mesozoic orogeny is concerned, the Ryoke and Tamba zones are parts of the same domain because the Ryoke events, which characterize the Ryoke "zone" (HT/LP metamorphism and plutonism) post-date the tangential tectonics; there is a progressive transition between the unmetamorphosed Tamba formations and the metamorphosed series (TANAKA and NOZAWA, 1977).

Under a thick pile of Uppermost Cretaceous molassic facies (Izumi group) the Ryoke zone shows two groups of plutons, the "younger granites" and the "older granites" (Fig. 6) separated in time by volcanic episodes of Early to Middle Cretaceous age (Sennan group, Nohi rhyolite). One important point is that the "older granites" emplacement is mainly anatectic, associated with the Ryoke HT/LP metamorphism (e.g., SUWA, 1973;

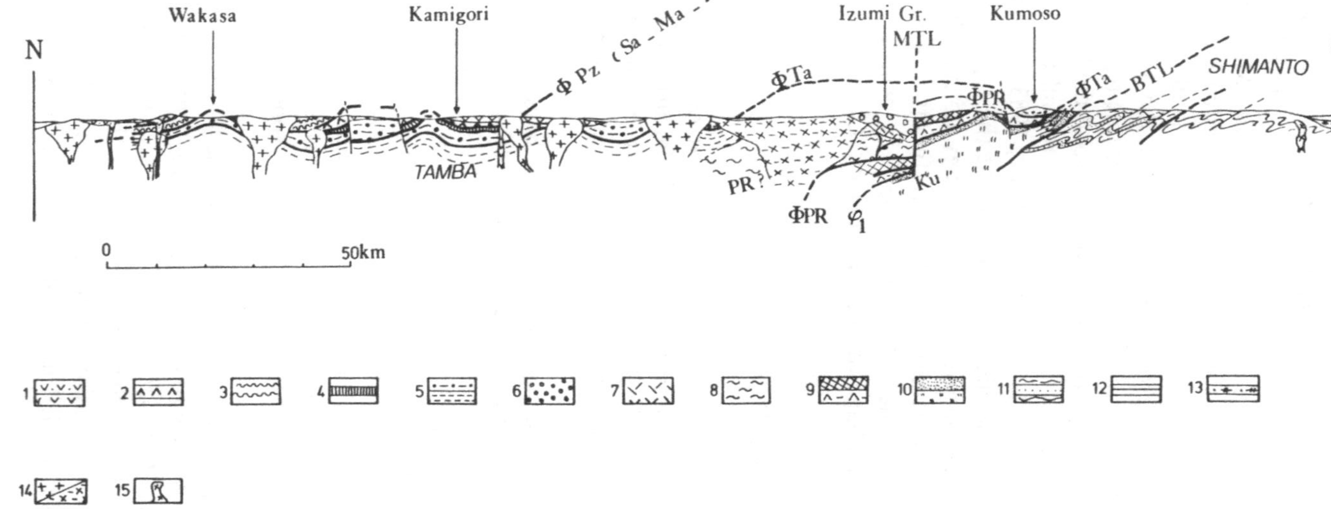

Fig. 6. Schematic general cross-section from the Japan Sea coast to the Pacific coast in Kinki and East Shikoku area. 1: Tertiary (top) and Cretaceous (bottom) volcanics. 2: Yakuno ophiolitic complex. 3: Sangun schists. 4: Permian Maizuru group. 5: Two main units of the Tamba zone. 6: Uppermost Cretaceous Izumi group. 7: Volcanics of the Sennan group. 8: Basement rocks of the inferred Pre-Ryoke block. 9: High grade (top) and low grade (bottom) basic schists units of Sambagawa. 10: Assumed cover (top) and basement (bottom) of the Oboke-Kurosegawa block. 11: Cretaceous part of Shimanto. 12: Tertiary part of Shimanto. 13: Miocene to Recent deposits off Shikoku. 14: "Younger granites" (top) and "Older granites" (bottom) of Ryoke type. 15: Miocene granitoids of the Pacific side. Although the "Older Granites" are actually widely distributed at some distance to the East of the profile, they are represented on the cross-section for more complete general view.

KUTSUKAKE, 1980). The microtectonic studies show that these events, which occurred during Lower Cretaceous time, are linked with a ductile shearing due to a left lateral motion along a proto-M.T.L. (HARA *et al.*, 1980). The main deformation is responsible for E-W trending upright folds. Another point to be emphasized, useful for further discussion, is that remnants of an older pre-Ryoke sialic basement are suspected (e.g., NUREKI, 1979). The evidence supporting the existence of a pre-Cretaceous, pre-Ryoke block are very scarce (cf discussion in FAURE and CHARVET, 1984). Even so, this hypothesis will be assumed in the following synthesis.

In the Tamba zone, at least two units are superposed; they are contrasted by their lithostratigraphic characters as it has been shown in the Kyoto district (TAMBA RESEARCH GROUP, 1975; ISHIGA, 1983). The upper unit contains a Lower-Middle Jurassic olistostrome with a muddy matrix and various olistoliths: late Carboniferous to Triassic cherts, and green rocks with Permian limestones. In the lower unit, a cherty sequence (Middle Triassic–Lower Jurassic) is overlain by pelites followed by an Upper Jurassic olistostrome. Paleogeographically, the upper unit should be placed between the lower unit and the Maizuru zone (FAURE and CHARVET, 1984). It was thrusted upon the lower unit with a roughly southern vergence (FAURE and CARIDROIT, 1983) in uppermost Jurassic-Eo-Cretaceous time, as shown by the Lower Cretaceous unconformity. Such a structure, which seems widespread in the whole Tamba zone, was afterwards refolded by E-W trending upright folds, already during the Lower Cretaceous (CHARVET *et al.*, 1983). Therefore, at present, the lower unit appears in the antiforms and the upper unit is preserved in the synforms (Fig. 6) of such folds. Moreover, these folds present an "en echelon" distribution related to an early left-lateral motion of the M.T.L. (HARA *et al.*, 1980; ICHIKAWA, 1980).

3.2 Relationships between the Tamba zone and the Paleozoic domain

The main point is that the Sangun and Maizuru formations, already affected by a Paleozoic orogeny, overlain unconformably by Triassic molassic facies, are reworked as a nappe during the Mesozoic orogeny. In other words, the muddy Permian Maizuru group, the HP/LT Sangun schists and the Yakuno ophiolitic complex composing a Paleozoic pile of nappes, were thrusted onto the Tamba zone (Fig. 6, 7) (HARA, 1982) during the Uppermost Jurassic-Eo-Cretaceous tectonism (FAURE and CARIDROIT, 1983; FAURE and CHARVET, 1984). Owing to the later phase of E-W trending folds, the Paleozoic components and the Tamba zone display a system of klippes and antiformal tectonic windows such as Wakasa window (HAYASAKA and HARA, 1982; Fig. 6), Kamigori klippe and

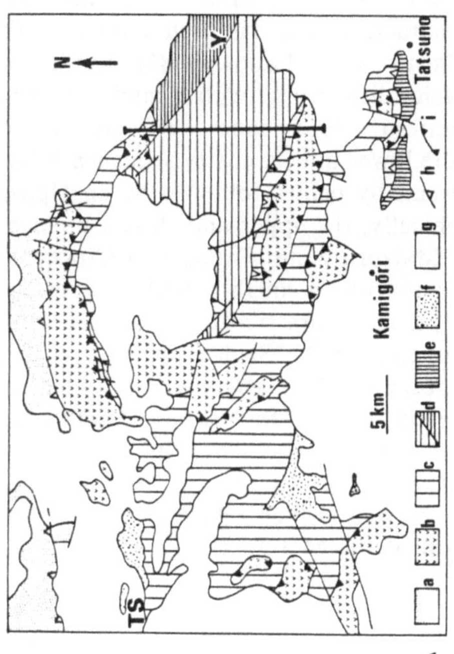
A
N
Y
TS
Tatsuno
Kamigōri
5 km
a
b
c
d
e
f
g
h
i

B
N
S
a
b
c
d
e

Fig. 7. Structure of the Kamigori area (after Faure and Caridroit, 1983, modified). A) Simplified geologic map. a: Sangun schists. b: Yakuno complex. c: Permian mudstones of the Maizuru group. d: Formations reported to the Tamba zone: Mikazuki (top) and Yamasaki (bottom). e: Schists of the Lower Tatsuno Formation. f: Unconformable Triassic. g: Cretaceous granitoids and volcanodetritic acidic rocks. h: Post Late-Jurassic thrusts. i: Sole of the Yakuno complex. TS: Tsuyama. Y: Yamasaki. B) Cross-section. a: Cretaceous volcaniclastics. b: Cretaceous granites. c: Basic rocks. d: Maizuru group. e: Tamba zone formations.

Mikazuki window (Fig. 6, 7; FAURE and CARIDROIT, 1983) for the area covered by the map of Fig. 1. Further West, some other Tamba tectonic windows are clarified or strongly suspected, in areas such as around Masuda and Yoshiwa (HARA, 1982).

From the geometric disposition, a displacement of the Paleozoic domain over the Tamba-Ryoke domain of about 100km, towards the south, can be inferred.

3.3 Conclusion on the tectonics of the "inner zones"

To the north of the M.T.L., the Tamba-Ryoke and Paleozoic domains display a pile of nappes emplaced in the Uppermost Jurassic-Eo-Cretaceous time and then refolded during Lower Cretaceous time.

4. Tectonic Synthesis and Attempt of a Geodynamic Model

The general schematic cross-section from the Japan Sea to the Pacific coast of Shikoku (Fig. 6) shows that the Mesozoic orogen of SW Japan is characterized by large nappes thrusted to both deep and superficial levels. The amplitude of displacement and the fact that the Paleozoic chain is reworked as a nappe must be pointed out as being alpine in character. That is to say: SW Japan displays some similarities to other alpine ranges in the world.

With regard to the origin of the Chichibu nappe, it is most likely derived from the Tamba-Ryoke domain (FAURE and CHARVET, 1983), where the olistostrome of the upper unit is very similar from a lithostratigraphic viewpoint. This hypothesis is consistent with tectonic timing in the "inner" and "outer" zones. Indeed, the E-W trending upright folds known both in the Tamba-Ryoke belt and in the Sambagawa belt are synchronous and related to the left lateral motion of the proto-M.T.L. (HARA *et al.*, 1980). Phase 3 (Hijikawa phase) is common to both sides of the present M.T.L. Therefore, the Uppermost Jurassic-Eo-Cretaceous tangential tectonism of the Tamba zone and Maizuru zone can be correlated with phase 2 (Nogahama-Ozu) described in the "outer zones". The Chichibu nappe could easily be a southern extension of a Tamba-Ryoke Unit, thrusted (ϕTa) over the metamorphic domain. Phase 1 is restricted to the Sambagawa domain s.l.

The structural interpretation expressed by the general cross-section (Fig. 6) shows the Sambagawa nappes as a suture zone between two sialic blocks: the Kurosegawa block (Ku) and the pre-Ryoke block (PR), but as a suture zone without actual ophiolitic bodies. To the South of the M.T.L., there should therefore have been a part of the pre-Ryoke basement thrusted (ϕPR) over the northern edge of Sambagawa, before erosion. It corresponds to the idea of Paleo-Ryoke or Missing Ryoke (YABE, 1963; ICHIKAWA, 1970).

Last, the Mesozoic nappe system of the "inner zones" overlies the pre-Ryoke basement and is bordered to the North by the sialic Hida zone (not seen in Fig. 6) which was, before the opening of the Japan Sea, a part of the Sino-Korean ancient basement (HIROI, 1981).

This disposition leads us to propose the tentative model illustrated in Fig. 8. It is a collisional model in which the Mesozoic orogeny is explained by the collision of the Kurosegawa and pre-Ryoke microblocks followed by their collision with the Sino-Korean block.

Such a type of model has been proposed for Japan (HORIKOSHI, 1972; ONO, 1980; ICHIKAWA, 1975, 1981, 1982; SASAJIMA, 1982; FAURE and CHARVET, 1983, 1984) but is a rather rare model if compared with the "Pacific type orogeny" model (MATSUDA and UYEDA, 1971) advocated by many authors.

We take into account the existence of the Sino-Korean block (BURRETT, 1973; HUANG, 1978; SCOTESE *et al.*, 1979), once separated from Siberia and which included the Hida basement (HIROI, 1981; NAKAZAWA *et al.*, 1982) before the opening of the Japan Sea (HONZA, 1979; OTOFUJI and MATSUDA, 1983). The low paleolatitudes of this block (MCELHINNY *et al.*, 1981) and of the Tamba rocks (HATTORI and HIROOKA, 1979; HATTORI, 1982) for the Permian or more recent time, deduced from paleomagnetic data, are also used. The Kula plate kinematics (UYEDA and MIYASHIRO, 1964; HILDE *et al.*, 1977) assumed for the Jurassic-Cretaceous time is after ENGEBRETSON (1982).

At Upper Permian time (Fig. 8A), the Sino-Korean block was located between the equator and 20°N. The pre-Ryoke and Kurosegawa microblocks are possibly derived from the same continental piece as Sino-Korea, by fragmentation of the "Pacifica continent" (NUR and BEN AVRAHAM, 1978). The location of the Tethys-Kula ridge and the existence of a spreading center in the Tamba basin are speculative.

Phase 1 (Ph. 1) resulted from the arrival of the Kurosegawa block in the oblique subduction zone bordering pre-Ryoke (Fig. 8B). The subduction of this buoyant feature (or "collision") accounts for the emplacement of the synmetamorphic nappes over Kurosegawa. Due to oblique collision, the Kurosegawa oceanic material was expelled laterally and the Kurosegawa block tended to rotate; this "scissor movement" accounts for the trending of the L1 lineation, parallel to the suture. After completion of the closure of the Sambagawa oceanic domain, this subduction stopped because of the buoyancy problem: the pre-Ryoke and Kurosegawa

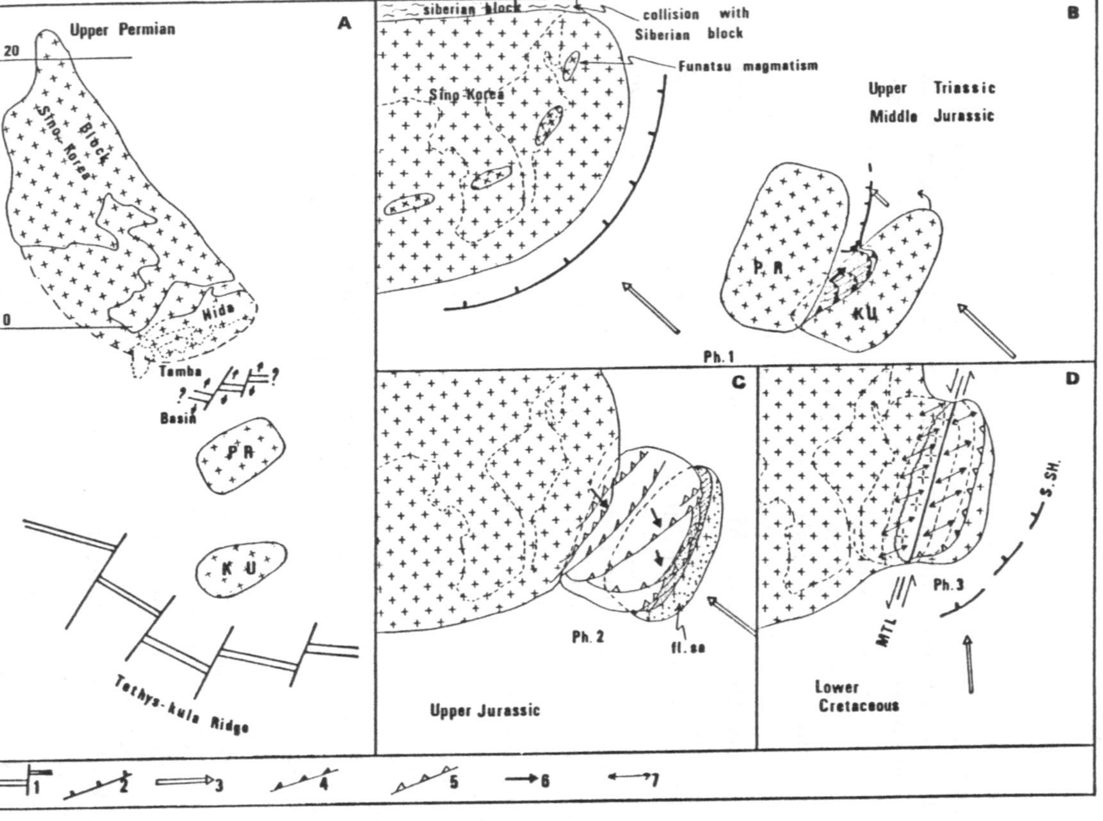

Fig. 8. Schematic geodynamic model of the Mesozoic orogeny of SW Japan (not to scale). Explanation of stages A, B, C, and D in the text. 1: Spreading center. 2: Subduction zone. 3: Sense of motion of the Kula plate (after Engebretson, 1982). 4: 1st phase nappes thrust contact. 5: 2nd phase nappes thrust contact. 6: Sense of nappe emplacement. 7: En echelon folds. PR: Pre-Ryoke block. KU: Oboke-Kurosegawa block. Ph. 1, 2, 3: 1st, 2nd and 3rd phases of deformation. fl.sa Flyschoid sequence of Sambosan. M.T.L.: Proto-Median Tectonic Line. S. SH: Location of the Shimanto subduction.

blocks had collided. At the same time, the Sino-Korea block had already collided with the Siberian block (BURRETT, 1973; SCOTESE *et al.*, 1979) and was approximately at its present latitude (HATTORI, 1982). The Tamba basin was subducting under Hida, leading to the Funatsu calc-alkaline magmatism and the emplacement of similar granitoids known in Korea and Southern China (HASHIMOTO and SATO, 1980; IIYAMA and FONTEILLES, 1981; LEE, 1981). The Kula plate was moving NW.

Phase 2 (Ph. 2) can be seen as the collision between Hida and the pre-Ryoke-Kurosegawa twin blocks (Fig. 8C), responsible for the general thrusting and nappe emplacement known in the "inner" and "outer" zones.

Phase 3 (Ph. 3) occurred after the closure of the Tamba oceanic domain. At that time, the Kula plate movement, at a high speed towards the North between 135 and 120 My (ENGEBRETSON, 1982), was accomodated by a left-lateral strike-slip motion along a proto-M.T.L. (Fig. 8D), and was associated with the "en echelon" folds (HARA *et al.*, 1977, 1980) and the anatectic "Older granites" formation. Later, the initiation of the Shimanto subduction (S. SH.) gave rise to the Ryoke volcanism (Sennan Group for instance) and plutonism ("Younger granites"). owing to the westward motion of Kula plate since 120 My.

In conclusion, the Mesozoic orogeny of SW Japan shows obvious alpine characteristics, which are better explained by a collisional type of tectogenetic model. The phase 1 mechanism advocated above is known in European Alps (MATTAUER *et al.*, 1981) and in the American Cordilleras (ROURE and BLANCHET, 1983). However some differences can be seen when compared with other alpine chains, e.g., the lack of actuel ophiolitic bodies in the paleosuture between pre-Ryoke and Sambagawa, in spite of a continental subduction which is a propitious mechanism for ophiolitic emplacement (AUBOUIN *et al.*, 1977). Another feature is the lack of noticeable calc-alkaline magmatism as evidence for Sambagawa subduction. A possible explanation for the latter point and, perhaps, at least partially for the former one, is the importance of the strike-slip component. The oblique Sambagawa subduction could be a kind of transpressive boundary zone such as that along the Cayman Trough (HOLCOMBE and SHARMAN, 1983) or like the present status to the North of Venezuela, where deformation is occurring (Stephan, personal communication) at the Southern Caribbean edge but without any island-arc volcanism initiated by this "pseudo-subduction". Last, as varied tectonic responses are brought into the subduction system (KARIG *et al.*, 1983), varied tectonic intensities can result from collisional phenomena. It depends of the size of blocks, sense of motion, etc. In Japan, the tectonism of Northern Japan (CADET and CHARVET, 1983; CADET and JOLIVET, this volume), the Lower Tertiary orogeny (CHARVET, 1980) and the Mesozoic orogeny of SW Japan, illustrate varying degrees and modalities of collision.

Acknowledgements

MM. J. Aubouin, S. Banno, J. P. Cadet, K. Ichikawa, J. T. Iiyama, K. Ishida, M. Iwasaki, L. Jolivet, T. Sato, S. Takizawa are thanked for fruitful discussions and assistance. The work expenses were supported by grants from the Ministry of Education of Japan, the French C. N. R. S. and the French Ministry of Foreign Affairs.

We are particularly grateful to the Organizing Committee of the OJI Seminar for their kind invitation to participate at that eminent meeting.

REFERENCES

ARAKAWA, Y., Deformational history of the Hida metamorphic rocks in the northern part of Gifu Prefecture, central Japan, *J. Geol. Soc. Japan*, **88**, 753–767, 1982.

AUBOUIN, J., M. MATTAUER, and C. ALLEGRE, La couronne périaustralienne: un charriage océanique représentatif des stades précoces de l'évolution alpine, *C. R. Acad. Sci., Paris, D*, **285**, 953–956, 1977.

BANNO, S., Petrologic studies on the Sambagawa cristalline schists in the Besshi-Ino district, central Shikoku, *Japan. J. Fac. Sci. Univ. Tokyo*, **11**, **15**, 203–319, 1964.

BANNO, S., T. HIGASHINO, M. OTSUKI, T. ITAYA, and T. NAKAJIMA, Thermal structure of the Sambagawa belt in Central Shikoku, *J. Phys. Earth, Suppl.*, **26**, 345–356, 1979.

BECK, C., Essai sur l'évolution géodynamique des Caraïbes sudorientales, *Bull. Soc. Géol. Fr.*, **7**, XXV, 169–183, 1983.

BURRETT, C. F., Plate Tectonics and the fusion of Asia, *E. P. S. L.*, **21**, 181–189, 1973.

CADET, J. P. and J. CHARVET, From Subduction to Paleosubductions in Northern Japan, in *Accretion Tectonics in the Circum-Pacific Regions*, M. Hashimoto and S. Uyeda, Ed., pp. 135–148, 1983.

CAREY, S. W., A tectonic approach to continental drift, in *Continental Drift*, a Symposium (University of Tasmania), pp. 117–355, 1958.

CHARVET, J., Subduction et tectonique: quelques réflexions sur "l'orogénèse de type pacifique" à propos du Japon sud-ouest, *C. R. Somm. Soc. Geol. Fr.*, **2**, 58–61, 1980.

CHARVET, J., J. P. CADET, M. FAURE, and J. AUBOUIN, Sur l'importance de la tectonique de nappe mésozoïque au Japon central et méridional, *C. R. Acad. Sci., Paris*, **296**, 1279–1286, 1983.

ENGEBRETSON, D. C., Relative motions between oceanic and continental plates in the Pacific Basin. Ph. D., Stanford University, 156 pp., 1982.

FAURE, M., Phase précoce Ouest-Est de la zone Sambagawa dans la partie orientale de Shikoku (Japon SW), *C. R. Acad. Sci., Paris*, **295**, 501–505, 1982.

FAURE, M., Ductile shear during the early deformation phase in the Sambagawa Belt, *J. Geol. Soc. Japan*, **89**, 6, 319–329, 1983a.

FAURE, M., The mesozoic orogeny in the Outer zone of SW Japan in Eastern Shikoku, *J. Sci. Tokushima Univ.*, 1983b (in press).

FAURE, M. and M. CARIDROIT, Tectoniques tangentielles superposées dans les zones "internes" du Japon SW, d'après l'exemple de la région de Kamigori, *C. R. Acad. Sci.*,

297, 165–170, 1983.

FAURE, M. and J. CHARVET, Tangential tectonics in the Chichibu zone from the example of Eastern Shikoku, *Proc. Japan Acad.*, **59**, B, 5, 117–120, 1983.

FAURE, M. and J. CHARVET, Mesozoic nappe structures in SW Japan, from the example of Eastern Shikoku and Kinki area, Sciences Géologiques, Strasbourg, **37**, 1, 51–63, 1984.

FAURE, M. and M. IWASAKI, Précisions sur les roches volcano-sédimentaires à la limite des zones Sambagawa et Chichibu dans le Japon SW, *C. R. Acad. Sci.*, **295**, 1149–1154, 1982.

FUJIMOTO, H., The nappe theory with reference to the northeastern part of the Kanto-Mountainland, *Sci. Rep. Tokyo Bunrika Daigaban, Sec. C*, **1**, 215–242, 1937.

GUIDI, A., J. CHARVET, T. SATO, and S. TAKIZAWA, Les structures tangentielles anté-Crétacé de la chaîne du Japon Sud-Ouest dans les Monts Kanto: résultats préliminaires, *C. R. Acad. Sci., Paris*, **298**, 307–312, 1983.

HADA, S., T. SUZUKI, K. OKANO and S. KIMURA, Crustal section based on the geological and geophysical features in the outer zone of Southwest Japan, *Mem. Geol. Soc. Japan*, **21**, 197–211, 1982.

HADA, S. and T. SUZUKI, Tectonic environments and crustal section of the Outer zone of Southwest Japan, in Accretion tectonics in the Circum-Pacific Regions, M. Hashimoto and S. Uyeda, Ed., pp. 207–218, 1983.

HARA, I., K. HIDE, K. TAKEDA, M. TSUKUDA, and T. SHIOTA, Tectonic movement in the Sambagawa belt, in Hide K., Ed. The Sambagawa Belt, pp. 307–390, Hiroshima Univ. Press, 1977.

HARA, I., K. SHYOJI, Y. SAKURAI, S. YOKOYAMA, and K. HIDE, Origin of the Median Tectonic Line and its initial shape, *Mem. Geol. Soc. Japan*, **18**, 27–49, 1980.

HARA, I., Evolutional processes of paired metamorphic belts, *Mem. Geol. Soc. Japan*, **21**, 71–89, 1982.

HASHIMOTO, W. and T. SATO, Correlation of structural belts in east and southeast Asia, in *Geol. Paleontol. Southeast Asia*, **21**, 343–356, 1980.

HATTORI, I., The Mesozoic evolution of the Mino terrane, central Japan: a geologic and paleomagnetic synthesis, *Tectonophysics*, **85**, 313–340, 1982.

HATTORI, I. and K. HIROOKA, Paleomagnetic results from Permian greenstones in central Japan and their geologic significance, *Tectonophysics*, **57**, 211–235, 1979.

HAYASAKA, Y. and I. HARA, Discovery of Jurassic radiolaria in the Chugoku belt and its structural significance, 89th Meet. Geol. Soc. Jap., 556, 1982.

HAYASE, I. and S. NOHDA, Geochronology of the "oldest rocks" of Japan, *Geochem. J.*, **3**, 45–52, 1969.

HIDE, K., G. YOSHINO, and G. KOJIMA, Preliminary report on the geologic structure of the Besshi Spotted schists zone, *J. Geol. Soc. Japan*, **62**, **733**, 573–584, 1956.

HILDE, T., S. UYEDA, and L. KROENKE, Evolution of the western Pacific and its margin, *Tectonophysics*, **38**, 145–165, 1977.

HIRAJIMA, T., Jadeite+Quartz rock from the Kanto Mountains, *J. Japan. Assoc. Min. Pet. Econ. Geol.*, **78**, 77–83, 1983.

HIROI, Y., Subdivision of the Hida metamorphic complex, Central Japan, and its bearing on the geology of the Far East in pre-Sea of Japan time, *Tectonophysics*, **76**, 317–333, 1981.

HOLCOMBE, T. L. and G. F. SHARMAN, Post-Miocene Cayman Trough evolution: a speculative model, *Geology.*, **11**, 714–717, 1983.

HONZA, E., Sediments, structure and origin of Japan Sea, Concluding remarks, Cruise report No. 13, E. Honza, Ed., Geological Survey of Japan, pp. 89–93, 1979.

HORIKOSHI, E., Orogenic belts and plates in Japanese islands, *Kagaku*, **42**, 665–673, 1972.

HUANG, T. K., An outline of the tectonic characteristics of China, *Eclogae Geol. Helv.*, **71/3**, 611–635, 1978.

ICHIKAWA, K., Some geotectonic problems concerning the Paleozoic-Mesozoic geology of Southwest Japan, in *Island Arc and Ocean*, M. Hoshino and A. Aoki, Ed., pp. 193–200, Tokai University Press, 1970.

ICHIKAWA, K., Honshu and Shimanto geosynclines in Southwest Japan and plate tectonics, *Assoc. Geol. Collaboration Japan, Monograph*, **19**, 241–246, 1975.

ICHIKAWA, K., Geohistory of the Median Tectonic Line of Southwest Japan, *Mem. Geol. Soc. Japan*, **18**, 187–212, 1980.

ICHIKAWA, K., Closure of the Jurassic sea in and around the Ryoke-Kurosegawa region, Hiroshima symposium, *Tectonics of Paired Metamorphic Belts*, I. Hara, Ed., pp. 113–116, 1981.

ICHIKAWA, K., Jurassic tectonism in Southwest Japan, *Chikyu Monthly*, **4**, 414–420, 1982.

ICHIKAWA, K., K. ISHII, C. NAKAGAWA, K. SUYARI, and N. YAMASHITA, On "Sakashu unconformity", *J. Gakugei Tokushima Univ., Nat. Sci.*, **3**, 61–74, 1954.

ICHIKAWA, K., K. ISHII, C. NAKAGAWA, K. SUYARI, and N. YAMASHITA, Die Kurosegawa zone, *J. Geol. Soc. Japan*, **62**, 82–103, 1956.

IIYAMA, T. and M. FONTEILLES, Mesozoic rocks of Southern Korea reviewed for major constituents and petrography, *Mining Geol., Tokyo*, **31**, 281–295, 1981.

INOUE, M., Geologic structures of the Chichibu terrain in the Kanto Mountainous land, *Japan. J. Fac. Sci. Univ. Tokyo, Sec. II*, **19**, 1–25, 1974.

ISHIDA, K., Stratigraphy and radiolarian assemblages of the Triassic and Jurassic siliceous sedimentary rocks in Konose Valley, Tokushima Prefecture, *SW Japan. J. Sci. Tokushima Univ.*, **16**, 111–141, 1983.

ISHIGA, H., Two suites of stratigraphic succession within the Tamba Group in the western part of the Tamba belt, *Southwest Japan. J. Geol. Soc. Japan*, **89**, 8, 443–454, 1983.

ISHIWATARI, A., A preliminary report on the Yakuno ophiolite in the Maizuru zone, Inner Southern Japan, *"Earth Science" (Chikyu Kagaku)*, **32**, 301–310, 1978.

ISHIZAKA, K. and M. YAMAGUCHI, U-Th-Pb ages of sphene and zircon from Hida metamorphic terrain, Japan, *E. P. S. L.*, **6**, 179–185, 1969.

ISOZAKI, Y., W. MAEJIMA, and S. MARUYAMA, Occurrence of Jurassic Radiolarians from the pre-Cretaceous rocks in the northern subbelt of the Chichibu belt, Wakayama and Tokushima prefectures, *J. Geol. Soc. Japan*, **87**, 555–558, 1981.

IWAZAKI, M., Metamorphic rocks of the Kotu-Bizan area, eastern Shikoku, *J. Fac. Sci. Univ. Tokyo, Sec. 2*, **15**, 1–90, 1963.

IWAZAKI, M., Gabbro breccia in the Mikabu Greenstone belt of Eastern Shikoku, *J. Geol. Soc. Japan*, **85**, 481–487, 1979.

IWASAKI, M., Burial metamorphism of greenstones in the Mikabu Zone, *Mem. Geol. Soc. Japan*, **21**, 189–196, 1982.

KAJI, A., Development of depositional basin in the Sambagawa metamorphic region, Eastern Shikoku, (Part 2), On the mode of occurrence of basic schists in the Kotu district, *Problems on Geosynclines in Japan*, **19**, 77–80, 1975.

KANMERA, K., Upper Paleozoic Stratigraphy of the northern Chichibu zone in Eastern Shikoku, *Sci. Rep. Fac. Sci. Kyushu Univ. Geol.*, **9**, 175–186, 1968.

KANO, K., Giant Deckenpacket and olistostrome in the eastern Mino district, central Japan, *J. Fac. Sci. Univ. Tokyo, Sec. II*, **20**, 31–59, 1979.

KANO, T., Tectonic divisions and their development of the Hida metamorphic region, central Japan, *Mem. Geol. Soc. Japan*, **21**, 9–24, 1982.

KARIG, D., H. KAGAMI, and DSDP LEG 87 SCIENTIFIC PARTY, Varied responses to subduction in Nankai Trough and Japan Trench fore arcs, *Nature*, **304**, 148–151, 1983.

KIMURA, T., S. YOSHIDA, and F. TOYOHARA, Paleogeography and earth movements of Japan in the Late Permian to Early Jurassic Sambosan stage, *J. Fac. Sci. Univ. Tokyo, Sec. II*, **19**, 149–177, 1975.

KOBAYASHI, T., The Sakawa Orogenic cycle its bearings on the origin of the Japanese Islands, *J. Fac. Sci. Univ. Tokyo*, **2**, **5**, 219–578, 1941.

KURIMOTO, C., "Chichibu system" in the area southwest of Koyasan, Wakayama Prefecture, *J. Geol. Soc. Japan*, **88**, 901–914, 1982.

KUTSUKAKE, T., Nature of the Ryoke regional metamorphism and plutonism, *Mem. Fac. Sci. Kyoto Univ., Ser. Geol-Mine*, **47**, 49–59, 1980.

LEE, M. S., Geology and metallic mineralization associated with Mesozoic granitic magmatism in South Korea, *Mining Geology, Tokyo*, **31**, 235–244, 1981.

MCELHINNY, M. W., B. J. J. EMBLETON, X. H. MA, and Z. K. ZHANG, Fragmentation of Asia in the Permian, *Nature*, **293**, 212–216, 1981.

MARUYAMA, S., The Kurosegawa melange zone in the Ino district to the North of Kochi city, Central Shikoku, *J. Geol. Soc. Japan*, **87**, 569–583, 1981.

MARUYAMA, S. and Y. UEDA, Schists xenoliths in ultrabasic body accompagnied with Kurosegawa tectonic zone in Eastern Shikoku and their K-Ar ages, *J. Petro. Min. Econ. Geol.*, **70**, 47–52, 1975.

MATSUDA, T., Discovery of the Middle-Late Triassic conodont genus "Metapolygnatus" from calcareous schists of the Sambagawa southern marginal belt in central Shikoku, *J. Geol. Soc. Japan*, **84**, 331–333, 1978.

MATSUDA, T. and S. UYEDA, On the Pacific Type Orogeny and its model-extension of the paired belts concept and possible origin of marginal seas, *Tectonophysics*, **11**, 5–27, 1971.

MATTAUER, M., M. FAURE, and J. MALAVIEILLE, Transverse lineations and large scale structures related to a high pressure-low temperature ophiolitic thrusting, *J. Struct. Geol.*, **3**, 4, 401–409, 1981.

MURATA, A., Large Decke structures in the Kurosegawa and Sambosan terrains, in Kyushu, Southwest Japan, *J. Fac. Sci., Univ. Tokyo, Sec. II*, **20**, 277–293, 1981.

NAKAMURA, Y., Petrology of the Toba ultrabasic complex, Mie Prefecture, central Japan, *J. Fac. Sci. Univ. Tokyo*, **18**, 1–51, 1970.

NAKAZAWA, K., K. M. YU, and T. TOKUOKA, The Ogcheon geosynclinal belt and the Hida metmorphic belt, in Basement of Japanese Islands, H. Kano, Ed., *Mem. Geol. Soc. Japan*, **21**, 91–101, 1982.

NOHDA, S., Rb-Sr dating of the Yatsushiro granite and gneiss, Kyushu, Japan, *E. P. S. L.*, **20**, 140–144, 1973.

NUR, A. and Z. BEN AVRAHAM, Speculation on mountain building and the lost Pacifica continent, *J. Phys. Earth.*, Suppl., **26**, 521–537, 1978.

NUREKI, T., Structural investigations of the Ryoke metamorphic rocks of the area between Iwakuni and Yanai, SW Japan, *J. Sci. Hiroshima Univ., Ser. C*, **2**, 109–127, 1960.

NUREKI, T., On the Pre-Ryoke basement rocks, in Basement of Japanese Islands, H. Kano, Ed., *Mem. Geol. Soc. Japan*, **21**, 183–199, 1979.

OGAWA, Y., Stratigraphy and Paleogeography of the Chichibu terrain in Eastern Shikoku, Japan, *Rep. Inst. Nat. Sci. Nihon Univ.*, **9**, 15–48, 1974.

OKAMURA, Y., Structural and petrological studies of the Ryoke gneiss and granodiorites complex of the Yanai districy, SW Japan, *J. Sci. Hiroshima Univ., Ser. C*, **2**, 143–213, 1960.

ONO, A., A model for the formation of the Ryoke-Sambagawa paired metamorphic belts, *J. Jap. Assoc. Mineral. Petrol. Econ. Geol.*, **75**, 31-37, 1980.

OTOFUJI, Y. and T. MATSUDA, Paleomagnetic evidence for the clockwise rotation of Southwest Japan, *E. P. S. L.*, **62**, 349-359, 1983.

ROURE, F. and R. BLANCHET, A geological transect between the Klamath Mountains and the Pacific Ocean (SW Oregon): a model for paleosubductions, *Tectonophysics*, **91**, 53-72, 1983.

SASAJIMA, S., Jurassic displacements seen from paleomagnetism, *Chikyu Monthly*, **4**, 420-427, 1982.

SASHIDA, K., H. IGO, S. TAKIZAWA, K. ISADA, K. SHIBATA, K. TSUKUDA, and H. NISHIMURA, On the Jurassic Radiolarians assemblage in the Kanto district, *Osaka Micropaleontologists*, Special Vol. **5**, 51-66, 1982.

SATO, T., S. TAKIZAWA, and J. KODATO, Revision de la stratigraphie et de la structure de la Formation Sakahara et sa localité-type, *J. Geol. Soc. Japan*, **83**, 631-637, 1977.

SCOTESE, C. R., R. K. BAMBACH, C. BARTON, R. VAN DER VOO, and A. M. ZIEGLER, Paleozoic base maps, *J. Geol.*, **87**, 217-277, 1979.

SEKI, Y., Glaucophanitic regional metamorphism in the Kanto Mountains, Central Japan, *Jap. J. Geol. Geogr.*, **29**, 233-258, 1958.

SHIBATA, K. and T. NOZAWA, K-Ar ages from hornblendes from the Hida metamorphic belt, *J. Jap. Assoc. Mineral. Petrol. Econ. Geol.*, **73**, 137-141, 1978.

SUWA, K., Metamorphic rocks occurring along the Median Tectonic Line in the Japanese Islands: Ryoke and Sambagawa belts, *The Median Tectonic Line*, Sugimura, Ed., Tokai Univ. Press, 1973.

SUYARI, K., Y. KUWANO, and K. ISHIDA, Stratigraphy and structure of the Chichibu belt and relationships with the Sambagawa belt, *Studies on Late Mesozoic Tectonism*, **3**, 99-113, 1981.

SUYARI, K., Y. KUWANO, and K. ISHIDA, Stratigraphy and geological structure of the Mikabu Greenrock Terrain and its environs, *J. Sci. Tokushima Univ.*, **15**, 51-71, 1982.

SUZUKI, M., On the Sambagawa metamorphic rocks of the Western part in the Kanto mountains, Central Japan, in *The Sambagawa Belt*, Hide K. Ed., pp. 307-390, Hiroshima Univ. Press, 1977.

TAKEDA, K., K. HIDE, S. MAKISAKA, and K. SONODA, Depositional environment of the original rocks of the Sambagawa metamorphic rocks. Hiroshima symposium, *Tectonics of Paired Metamorphic Belts*, I. Hara, Ed., pp. 95-100, 1981.

TAKIZAWA, S., Tectonic development of Chichibu and Sambagawa terrains in Kanto Mountains, Japan. Master Degree, Univ. Tsukuba, pp. 1-122, 1982.

TANAKA, K. and T. NOZAWA, Geology and mineral ressources of Japan, *Geological Survey of Japan*, 430, 1977.

TAMBA RESEARCH GROUP, Geosynclinal facies of the Tamba belt, SW Japan, Monograph 19 Organizing Committee for the Symposium 1974, pp. 13-23, The Assoc. Geol. Collab. Japan, 1975.

TORIUMI, M., Petrological Study of the Sambagawa Metamorphic Rocks, The Kanto Mountains, Central Japan. Univ. Museum, *Univ. Tokyo Bull.*, **9**, 1-99, 1975.

TORIUMI, M., Strain, stress and uplift, *Tectonics*, **1**, 57-72, 1982.

TSUKUDA, N., T. MIYAMOTO, R. TOMINAGA, and I. HARA, Mode of occurrence of serpentinite and Kurosegawa complex masses in the Chichibu belt, Shikoku, Japan; Hiroshima symposium, *Tectonics of Paired Metamorphic Belts*, I. Hara, Ed., pp. 157-163, 1981.

UEDA, Y. and H. ONUKI, K-Ar dating of the metamorphic rocks in Japan; Yatsushiro gneisses, Kiyama and Sonogi metamorphic rocks in Kyushu, *J. Jap. Ass. Mine. Petro. Econ. Geol.*, **60**, 159–166, 1969.

UEDA, Y., T. NAKAJIMA, K. MATSUOKA, and S. MARUYAMA, K-Ar ages of muscovite from greenstone in the Ino formation and schists blocks associated with the Kurosegawa tectonic zone near Kochi, central Shikoku, *J. Jap. Ass. Mine. Petro. Econ. Geol.*, **75**, 230–233, 1980.

UYEDA, S. and A. MIYASHIRO, Plate tectonics and the Japanese Islands: a synthesis, *Geol. S. Am. Bull.*, **85**, 1159–1170, 1974.

YABE, H., Probable position of the outer wing of the Ryoke metamorphics in Southwest Japan, *J. Geogr. Soc. Tokyo*, **72**, 110–114, 1963.

YAO, A., Subdivision of the Mesozoic Complex in Kii-Yura Area, Southwest Japan and its Bearing on the Mesozoic Basin Development in the Southern Chichibu Terrane, *J. Geosc. Osaka City Univ.*, **27**, 2, 41–103, 1984.

YOSHIKURA, S., K. SHIBATA, and S. MARUYAMA, Garnet clinopyroxene amphibolite from the Kurosegawa tectonic zone, near Kochi city, Petrography and K-Ar age, *J. Jap. Ass. Min. Pet. Econ. Geol.*, **76**, 102–109, 1981.

Formation of Active Ocean Margins, edited by N. Nasu *et al.*, pp. 819–833.

STRATIGRAPHIC CHANGE OF THE COARSE CLASTIC ROCKS OF THE SHIMANTO SUPERGROUP IN EASTERN, SHIKOKU, SOUTHWEST JAPAN

Fujio KUMON

Department of Geology, Faculty of Science, Shinshu University, Matsumoto, Japan

Abstract. The Shimanto Supergroup of geosynclinal facies which ranges from Cretaceous to early Tertiary in age, is widely distributed in the outermost zone of Southwest Japan. Coeval volcanic and plutonic rocks are extensively distributed in the Inner Zone. The coarse clastic rocks mainly of sandstone occupy about a half of the Shimanto Supergroup, and a stratigraphic change of sandstone and conglomerate composition is recognized in the supergroup. The amount of rock fragments, chiefly acidic to intermediate volcanic rocks, increases upward throughout the Cretaceous sequence. Additionally, acidic tuff layers are frequently intercalated in the Late Cretaceous sequence. Conglomerates in the Late Cretaceous are very dominated by acidic volcanic rock clasts. The geosynclinal sedimentation of the Cretaceous Shimanto Supergroup proceeded with the coeval volcanism which became more active in the Late Cretaceous. The clastic sediments were supplied mainly from the sites of the volcanism. The volcanism corresponds to the Cretaceous volcanic activity in the Inner Zone of Southwest Japan. The Cretaceous Shimanto Supergroup may represent a typical forearc sedimentation in an arc-trench system.

1. Introduction

The Shimanto Belt occupying the outermost zone of Southwest Japan, consists of a large amount of geosynclinal sediments called the Shimanto Supergroup, ranging from early Cretaceous to early Miocene in age (Fig. 1). Coeval volcanic and plutonic rocks of acidic to intermediate composition are widely distributed in the Inner Zone. Recently, some authors con-

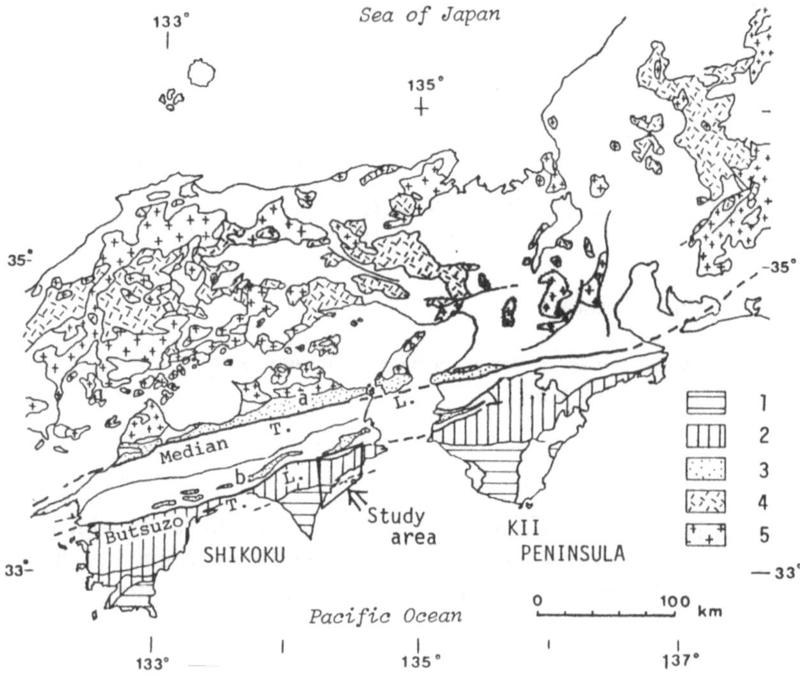

Fig. 1. Index map of the study area showing also the distribution of Cretaceous sediments and Cretaceous to Paleogene ingeous rocks. 1·2. Shimanto Belt (1. early Tertiary, 2. Cretaceous), 3. Cretaceous shallower sediments (a. Izumi G., b. Cretaceous in the Chichibu Belt), 4. Cretaceous to Paleogene volcanic rocks, 5. Cretaceous to Paleogene plutonic rocks.

sidered that this region was the site of consumption of the oceanic plate (KANMERA and SAKAI, 1975; SUZUKI and HADA, 1979; DICKINSON, 1977; TAIRA *et al.*, 1980). In the supergroup, coarse clastic rocks such as sandstone and conglomerate occupy one-third to a half of the total sediments by volume. The composition of coarse clastic rocks should offer important information on the provenance and crustal movements especially in relation to igneous activity in the Inner Zone.

The study area, the eastern margin of eastern Shikoku, is located in the central portion of the Shimanto Belt, and the strata exposed in this area have suffered less metamorphism and tectonic disturbance than those of other areas. Furthermore, sandstone and conglomerate are well developed. The stratigraphy and detailed description of coarse clastic rocks had already reported by the author (KUMON, 1981, 1983). In this paper, the author will discuss mainly the geologic significance of the upward-increasing trend of rock fragments in sandstone throughout the Cretaceous sequence.

2. *Geology*

The Shimanto Belt in eastern Shikoku is divided by the Aki Fault into northern and southern subbelts. The Cretaceous strata of eugeosynclinal facies are distributed in the northern subbelt, and the early Tertiary of flysch facies occupy the southern subbelt (Fig. 2). The northern subbelt is subdivided by two strike faults into three zones, namely the northern, central and southern zones, each represented mainly by the Akamatsu and Hinotani Formations, Taniyama Formation, and Mugi and Hiwasa Formations, respectively. The southern subbelt is also subdivided by two strike faults into three zones, each represented by the Kaifu, Naharigawa and Muroto Formations from north to south.

The geologic age of the Cretaceous has been fairly well determined by radiolarian fossils as shown in Fig. 3. The Akamatsu Formation, about 1,500 meters thick, is dominated by muddy rocks accompanied by sand-

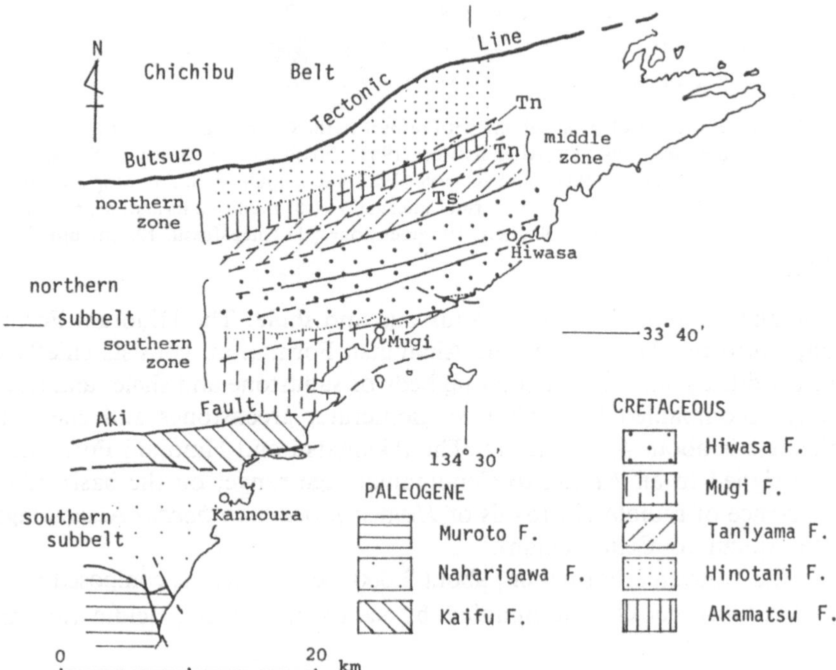

Fig. 2. Geologic map of the Shimanto Belt in eastern margin of eastern Shikoku.

Age		Formation		Lithology and thickness	Fossil
Cretaceous	late Santonian - Maestrichtian	Hiwasa F.	H4, H3, H2, H1	Mainly massive sandstone accompanied with sandy to muddy alternating beds of sandstone and shale. Pebble to cobble conglomerate, often containing bolder clasts, is frequently intercalated in sandstone. 4,000 m thick.	
		Mugi F.		Shale and muddy alternating beds of sandstone and shale, with greenstones and acidic tuff. 1,000 m + thich.	• *Alievium murphyi*, etc. ⁎ *Gaudryceras (Vertebrites)* cf. *kayei*
	Turonian - Santonian	Taniyama F.	Ts	Mainly shale, muddy to sandy alternating beds of sandstone and shale, and sandstone. A small amount of greenstones, chert and acidic tuff are intercalated. 2,000 m thick.	• *Artostrobium urna* A. • *Dictyomitra formosa* A. • *Holocryptocanium barbui* A.
			Tn	Mainly shale and muddy alternating beds of sandstone and shale with chert and sandstone. 1,500 m thick or less.	• *Artostrobium urna* A. (•) *Thanarla conica* - *Ultranapora* sp. A.
	Aptian - Cenomanian	Hinotani F.	Hu, Hm, Hl	Thick-bedded sandstone, sandy alternating beds of sandstone and shale, and shale accompanied with muddy alternating beds. Minor amount of pebble conglomerate, greenstones and chert are intercalated. 4,000 m thick.	(•) *Thanarla conica* - *Ultranapora* sp. A. • *Holocryptocanium barbui* A.
		Akamatsu F.		Mainly shale and muddy alternating beds accompanied with sandstone, sandy alternating beds and acidic tuff. 1,500 m.	

km 5, 0; f.: fault

1 2 3 4 5 6 7 8 9 10 11 12 13 f: fault

Fig. 3. Generalized geological column with respect to the Cretaceous strata of the Shimanto Supergroup in eastern Shikoku. After KUMON (1983). Fossil data are based on SUYARI *et al.* (1967), NAKAGAWA *et al.* (1980) and KUMON (1983 and unpub. data). 1. conglomerate, 2. sandstone, 3. sandy alternations, 4. normal alternations, 5. muddy alternations, 6. shale, 7. acidic tuff, 8. tuffaceous shale, 9. chert, 10. greenstones, 11. megafossil, 12. microfossil, 13. derived microfossil.

stone and alternating beds of sandstone and shale. The Hinotani Formation, conformably overlying the Akamatsu Formation, consists chiefly of thick-bedded sandstone, alternating beds of sandstone and shale, and shale, rarely accompanied by pebble conglomerate, greenstones and chert. Its thickness is about 4,000 meters. The Akamatsu and Hinotani Formations are assigned to an Aptian to Cenomanian age range, on the basis of the occurrence of radiolarian fossils of *Holocryptocanium barbui* Assemblage* (late Albian to Cenomanian).

The Taniyama Formation, about 2,000 meters thick, is composed mainly of muddy rocks accompanied by sandstone, chert, acidic tuff and

*The zonation and name of the radiolarian assemblages follow those proposed by MATSUYAMA *et al.* (1982).

greenstones. Its northern half (Tn) is fairly abundant in chert, and in contrast the southern half (Ts) is relatively rich in sandstone and acidic tuff. The precise relation between the two parts remains uncertain. The genuine age of this formation is inferred to be Turonian to Santonian, judging from the occurrences of a *Dictyomitra formosa* Assemblage (Turonian) and an *Artostrobium urna* Assemblage (Coniacian to Santonian) in the tuffaceous and black shales, although Tithonian to Cenomanian radiolarian assemblages were also discovered at places, derived from chert and red shale of exotic origin in this formation.

The Mugi Formation, more than 1,000 meters thick, mostly consists of muddy rocks associated with greenstones, acidic tuff and sandstone. The Hiwasa Formation which conformably overlies the Mugi Formation, is mainly represented by conglomerate, massive sandstone, and alternating beds of sandstone and shale. It is about 4,000 meters thick. The Mugi Formation is assigned to a late Santonian to Campanian age range on the basis of the occurrences of *Gaudryceras (Vertebrites)* sp. cf. *kayei* and Coniacian to Campanian radiolarian fossils (SUYARI *et al.*, 1967; KUMON, 1981). The Hiwasa Formation is estimated to be Campanian to Maestrichtian in age, judging from its great thickness.

The geological age of each zone becomes younger southward. The main depositional site may have migrated southward step by step. Homoclinal structure is dominant in each zone, although frequently faulted and occasionally folded.

According to KUMON (1983), it was considered that most of the clastic sediments of Cretaceous age were transported by longitudinal currents from west to east or from east to west and by lateral currents from north to south, with the major source land situated to the north of the basin. One exception is the Nyunokawa Formation in the Kii Peninsula which seems to have recieved clastic sediments from a southern land source (KISHU SHIMANTO RESEARCH GROUP, 1977; HARATA *et al.*, 1979).

3. *Conglomerate*

Conglomerate composition gives detailed information on the source rocks. The compositions of the conglomerates in the Cretaceous are shown in Fig. 4.

Several beds of pebble conglomerate, 0.5 to 1 meter thick, are intercalated in the Middle Member of the Hinotani Formation (Hm Mem.). The clasts of conglomerate are mostly granule to fine pebble in size and rounded to subrounded in shape. Sedimentary rocks such as sandstone, shale, chert and muddy limestone occupy more than a half of the total

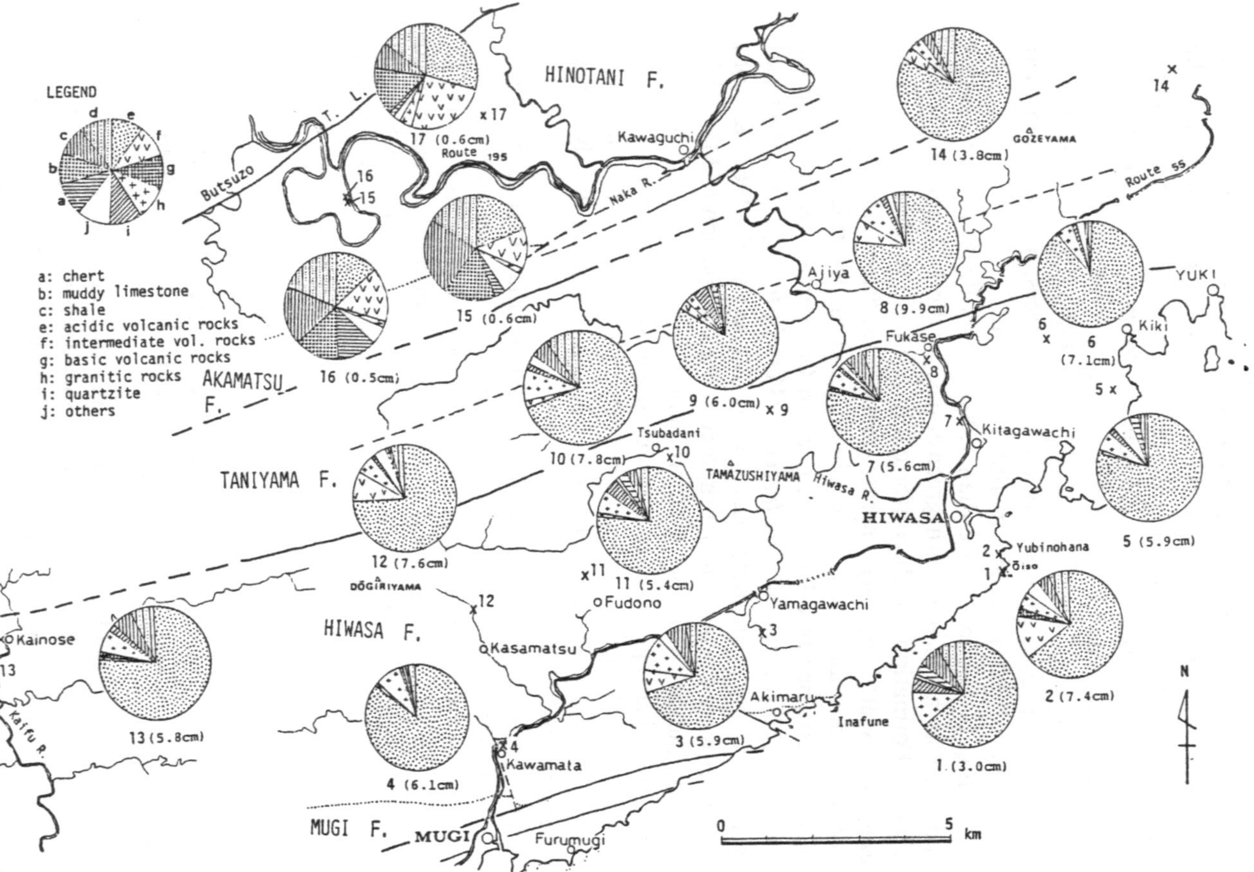

Fig. 4. Conglomerate composition of the Cretaceous Shimanto Supergroup in eastern Shikoku (after KUMON, 1981).

composition, and acidic and intermediate volcanic rocks occupy about one-fifth.

Only one bed of pebble conglomerate is intercalated in the Taniyama Formation. The clasts are mostly coarse pebble sized, rounded to subrounded, and scattered in muddy matrix. Acidic volcanic rock clasts are very dominant.

The conglomerates of the Hiwasa Formation are usually 1 to 10 meters thick, and rarely up to 100 meters. The clasts are mostly pebble to cobble sized, and rounded to subrounded. Boulder clasts are sometimes contained in them. Acidic volcanic rocks, namely rhyolite lava and tuff, are very predominant in the clasts. Clasts of welded rhyolite tuff are frequently found in them. Clasts of granitic rocks, sandstone and shale are commonly contained, but are small in amount.

4. *Sandstone*

Sandstone is one of the most dominant rocks of the Shimanto Supergroup, and can be examined successively throughout the geologic succession. Therefore, the mineral composition of sandstone is more useful than the conglomerate composition in discussing successive crustal movements in the hinterland.

In this study, specimens for examination were taken from massive, medium- to coarse-grained sandstone beds thicker than 0.5 meter. The modal composition was obtained by the counting more than 500 points in one thin-section for each specimen under the microscope (grid spacing 0.5×1 mm). The constituents are divided into monocrystalline quartz, polycrystalline quartz, plagioclase, potash feldspar, rock fragments of various kinds, "matrix" (including calcite and other silicate cements, and the grains smaller than 0.03 mm), and others consisting of heavy minerals, shale patches, etc.

The results are shown in Fig. 5. As for the rock fragments in sandstone, a special examination was made on several selected specimens which are supposed to have nearly average mineral compositions for each formation (Fig. 6). Most of the Cretaceous sandstones have more than 15 per cent matrix, and belong to a lithic or feldspathic graywacke classification. The framework grains are mostly angular to subrounded, and are moderately to well sorted. The difference between the mineral compositions of the sandstones can be clearly recognized among the formations.

Sandstones of the Akamatsu and Hinotani Formations are dominated by both plagioclase and potash feldspar, especially by plagioclase, and are poor in rock fragments. Intermediate volcanic rocks and granitic rocks

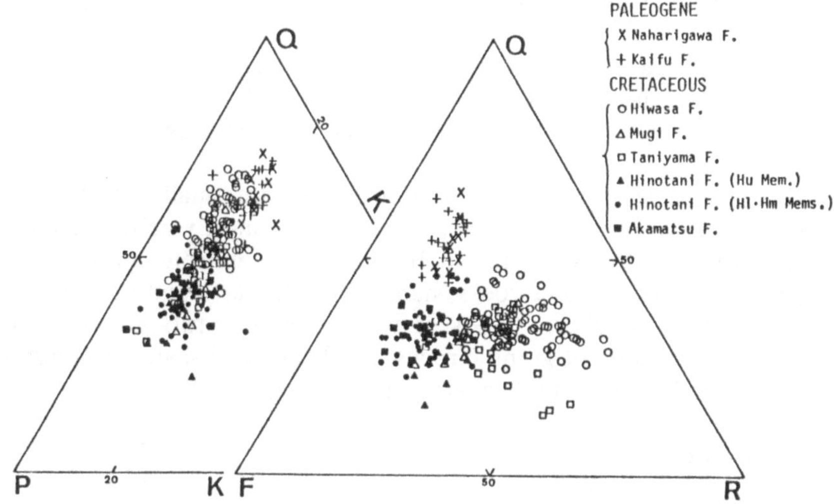

Fig. 5. Triangular Q–F–R and Q–P–K plots of the sandstone compositions from the Shimanto Supergroup in eastern Shikoku (after KUMON, 1983).

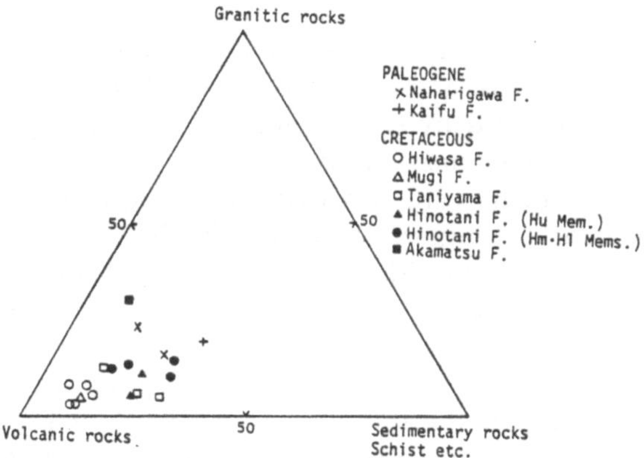

Fig. 6. Triangular plots of partial mode of rock fragments for selected sandstone specimens which are supposed to have nearly average compositions for each formation.

comprise a relatively large part of the rock fragments component, although acidic volcanic rocks are most abundant. Sedimentary rock fragments such as shale and chert are also commonly contained. In the Upper Member of the Hinotani Formation (Hu Mem.), the amount of rock fragments increases a little.

Sandstones of the Taniyama Formation are rich in rock fragments and relatively poor in feldspar. Some specimens are very poor in quartz. In the rock fragments component, acidic volcanic rocks are predominant, and sedimentary rocks such as shale and chert, and schistose rocks are commonly contained.

Sandstone of the Mugi Formation are fairly abundant in both feldspar and rock fragments, and seems to have an intermediate nature between those of the Hinotani Formation and the Hiwasa Formation described below.

Sandstones of the Hiwasa Formation are mainly dominated by rock fragments. Acidic volcanic rocks are predominant in the rock fragments component, and intermediate volcanics are also commonly contained in it. The sandstones of the lower part of the Hiwasa Formation (H2 Mem.) are relatively poor in rock fragments, however.

Compared with the Cretaceous sandstones, the early Tertiary sandstones are heavily dominated by quartz, and are poor in rock fragments and matrix (Fig. 4). They mostly belong to an arkose, and partly to an arkosic wacke classification. No significant difference of sandstone composition can be recognized between the Kaifu and Naharigawa Formations.

5. *Implications of the Stratigraphical Change of Clastic Rock Composition*

The Cretaceous sandstone and conglomerate contain fairly abundant volcanic clasts and fragments of acidic to intermediate composition (Figs. 4, 5, and 6). It is of particular interest that there is a distinct upward increase in the content of rock fragments, consisting chiefly of acidic to intermediate volcanic rocks, in the Cretaceous sequence, although with some fluctuations, however (Fig. 7). The ratio of volcanic rocks also increases upward in the rock fragments component. The conglomerates of the Late Cretaceous become predominated by acidic volcanic rocks which consist of rhyolite lava and often welded rhyolite tuff. From these facts it is concluded that compositional differences are related to the stratigraphic formations of the Cretaceous sandstones, as examined at various places in the Shimanto Belt (Tokuoka and Kumon, 1979; Miyamoto, 1976; Teraoka, 1977; Okada, 1977; etc.). Unfortunately, the geologic age of the sandstones examined are not very reliable in most caces, because of the lack of fossil evidence. The author, however, is positive that the stratigraphic change in sandstone composition can be roughly applied to the Cretaceous throughout the Shimanto Belt, because there is a tendency for sandstones in the northern zone of the Cretaceous subbelt to be rich in feldspar and poor in rock fragments, and in contrast those in the southern

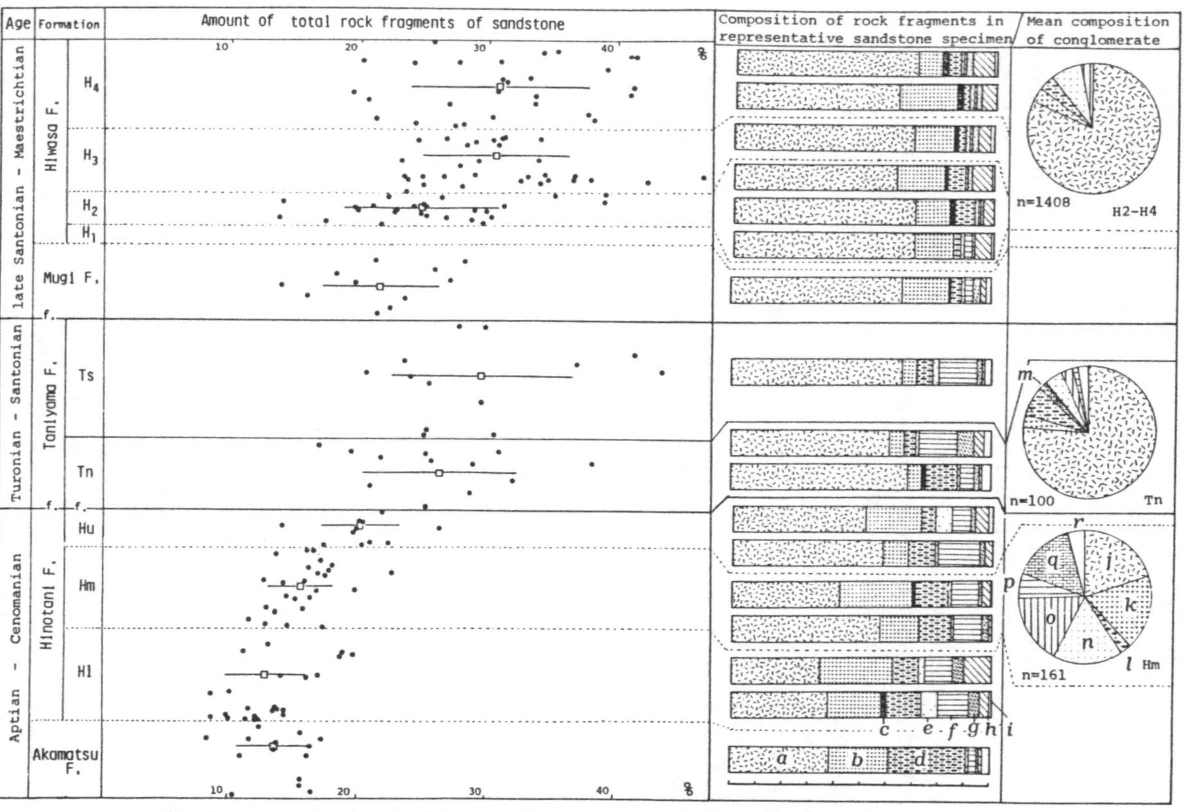

Fig. 7. Increasing trend of the rock fragments in the Cretaceous Shimanto Supergroup in eastern Shikoku, showing also stratigraphic change of conglomerate composition (after KUMON, 1983). Open square is the mean of each formation or member, and the bar means standard deviation. Key are as follows: a·j. acidic volcanic rocks, b·k. intermediate volcanic rocks, c·m. basic volcanic rocks, d·l. sandstone, f·o. shale, g·p. chert, h. hornfels-like rocks, i. schistose rocks, q. muddy limestone, r. others.

part to be rich in rock fragments consisting mainly of volcanic rocks. The strata in the Cretaceous subbelt seem to become younger southward zone by zone, as the cases of eastern Shikoku and Kii Peninsula show (KUMON, 1983).

The mineral composition of the Cretaceous sandstones from the median zone of Southwest Japan, that is, from the Onogawa Group in western Kyushu and the Izumi Group in Shikoku are shown in Fig. 8. Sandstones of the Onogawa Group ranging from Cenomanian to Santonian in age are feldspathic, but contain considerable amounts of rock fragments in which acidic to intermediate volcanic rocks occupy about 60 per cent by volume (TERAOKA, 1977). Sandstones of the Izumi Group ranging from Campanian to Maestrichtian in age, are rich in rock fragments in which volcanic rocks occupy 80 to 90 per cent by volume (NISHIMURA, 1976; TERAOKA, 1977). This change of sandstone composition from the Onogawa Group to Izumi Group is concordant with the stratigraphic change of sandstone composition clarified in the Cretaceous Shimanto Supergroup, resulting from the change of the provenance nature.

As mentioned by KUMON (1983), it is probable that violent volcanic activity in the hinterland took place in the Late Cretaceous age, and the volcanic products had a wide land cover. The volcanism should have produced a high topographic relief, and extensively destroyed the vegetational land cover. These factors would significantly increase the production of clastic sediments consisting cheifly of volcanic rocks (for modern example, KUENZI *et al.*, 1979). The clastic sediments of the Shimanto Belt had

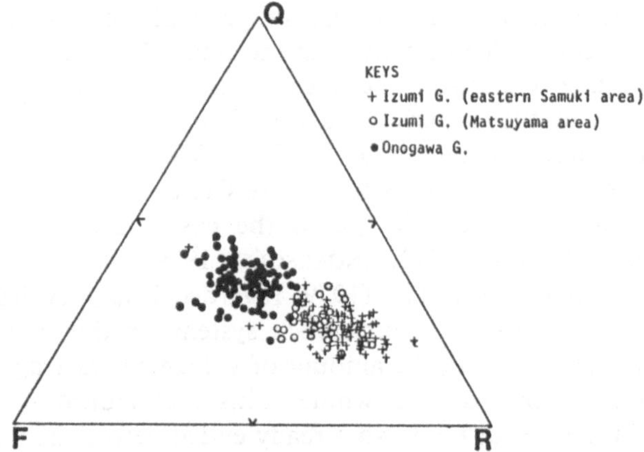

Fig. 8. Triangular Q–F–R plots of the sandstone compositions from the Cretaceous in the median zone of Southwest Japan. Compiled from TERAOKA (1977) and NISHIMURA (1976).

been derived mainly from such volcanic sites, and hence formed a thick sequence. Violent volcanism seems to have begun in late Cenomanian or Turonian times, judging from the increasing proportion of rock fragments. Acidic tuffs which are medium-to coarse-grained and attain to several meters in thickness, are often intercalated in the Upper Cretaceous deposits. This fact also supports the existance of active volcanism near the basin.

It is evident that the sedimentation of the Late Cretaceous Shimanto Supergroup in eastern Shikoku had progressed contemporaneously with the violent volcanism in the hinterland, receiving much clastic sediment from the sites of volcanism.

6. *Discussion*

A similar stratigraphic change of sandstone composition mentioned above can be recognized in the Cretaceous Shimanto Supergroup in the Kii Peninsula (KUMON, 1983). The model composition in the supergroup is also similar, although the time of the increase of volcanic rock fragments may be somewhat different.

According to MIYAMOTO (1980), the regular stratigraphic change of sandstone composition is not so distinct in the Cretaceous sequences in the Chichibu Belt as that in the Shimanto Belt. The volcanic rock clasts and fragments, however, definitely increase in the Coniacian to Santonian strata, as represented by the Sotoizumi Group.

Therefore, the Cretaceous strata distributed in Southwest Japan should be regarded as the sediments formed during progressively greater volcanic activity in the hinterland. The volcanism probably corresponds to the well-known Cretaceous volcanic activity in the Inner Zone of Southwest Japan.

An intimate relationship between the sedimentation in the basin and the volcanism in the hinterland is found in the arc-trench system. For example, the clastic sediments of the Great Valley sequence in California ranging in age from late Jurassic to Late Cretaceous, were supplied from the Sierra Nevada magmatic arc to the east, during the same period (DICKINSON and RICH, 1972; INGERSOLL, 1979; MANSFIELD, 1979; and others). DICKINSON and SEELY (1979) regarded it as a typical example of the forearc sediments in an arc-trench system. In the case of the Great Valley sequence, however, the amount of volcanic rock fragments in sandstone decreases upward as a whole. This fact indicates that the main volcanism in the source land had already ended before the major sedimentation took place in the basin. They also explained it as the result of an unroofing and dissecting process of the Sierra Nevada magmatic arc. It is important that the sedimentation of the Cretaceous in Southwest Japan proceeded with the contemporaneous volcanism in the hinterland, as represented by the Cretaceous Shimanto Supergroup. In the other words, the sedimentation of the Cretaceous had corresponded to the "roofing" process by volcanic rocks in magmatic terrane. The transition from the Cretaceous to the early Tertiary Shimanto Supergroup may be compared with the stratigraphic change in the Great Valley sequence (KUMON, 1983).

Recently, NAKAZAWA *et al.* (1983) and KISHU SHIMANTO RESEARCH GROUP (in preparation) elucidated that in the Kii Peninsula, the Turonian to Santonian strata intercalate with the chert-greenstone olistostromes among the usual flysch units. They considered that the both sediments were formed in a trench or lower continental slope, and that the chert and greenstone were derived from submarine volcanic seamounts to the south. The Taniyama Formation in eastern Shikoku is correlative with

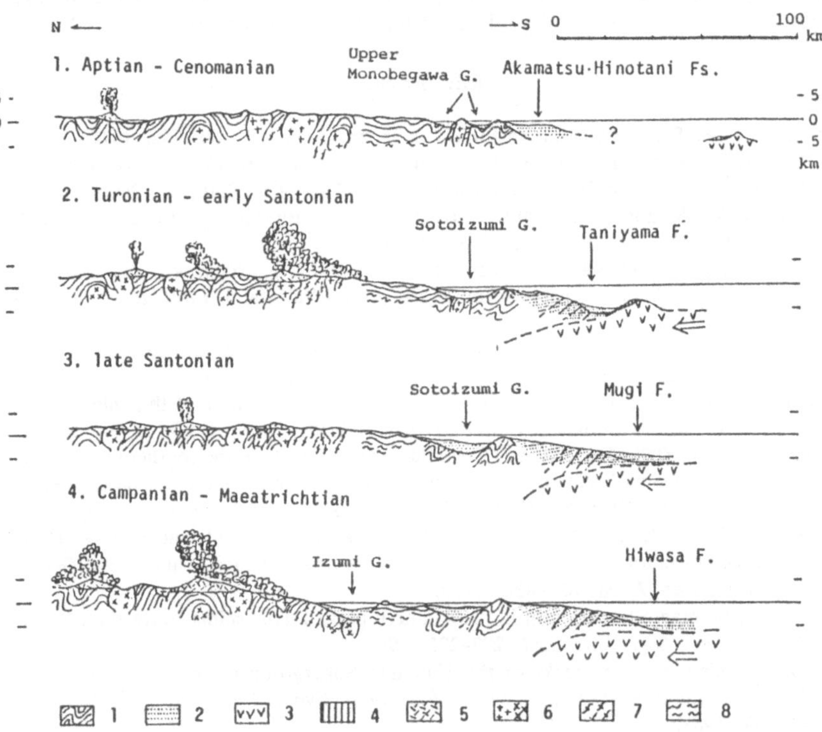

Fig. 9. Schematic cross sections of the Shimanto basin and its hiterland during the Cretaceous. The figures are illustrated as typical ones in each stage. 1. Paleozoic to early Mesozoic strata, 2. Cretaceous sediments, 3. greenstones, 4. chert, 5. acidic to intermediate volcanic rocks, 6. granitic rocks, 7. gneiss, 8. schist.

the strata, and also contains chert and greenstones of exotic origin. Similar opinions were mentioned by SUZUKI and HADA (1979) and TAIRA *et al.* (1980).

Therefore, the Shimanto basin and its hinterland seems to have formed an arc-trench system at least during the Late Cretaceous time. On the above-mentioned facts, tentative evolutional history of the Cretaceous Shimanto basin and its hinterland is illustrated schematically in Fig. 9.

Acknowledgements

The author would like to thank K. NAKAZAWA, Emeritus Professor of Kyoto University, and Dr. T. TOKUOKA of Shimane University for their helpful supervision and encouragements. He is also obliged to the members of the KISHU SHIMANTO RESEARCH GROUP who made valuable discussions with the author. The instruction of radiolarian fossil was received from Dr. K. NAKASEKO of Osaka University. Many thin-sections were prepared by Messrs. H. TSUTSUMI and K. YOSHIDA of Kyoto University. The author wishes to heartily thank them.

REFERENCES

DICKINSON, W. R., Subduction tectonics in Japan, *EOS*, **58**, 948–952, 1977.

DICKINSON, W. R. and E. I. RICH, Petrologic intervals and petrofacies in the Great Valley Sequece, Sacramento Valley, California, *Geol. Soc. Am. Bull.*, **83**, 3007–3024, 1972.

DICKINSON, W. R. and D. R. SEELY, Structure and stratigraphy of forearc regions, *Am. Assoc. Petrol. Geol. Bull.*, **63**, 2–31, 1979.

HARATA, T., K. HISATOMI, F. KUMON, K. NAKAZAWA, M. TATEISHI, H. SUZUKI, and T. TOKUOKA, Shimanto geosyncline and Kuroshio Paleoland, *Phys. Earth*, **26**, Suppl., 357–366, 1979.

INGERSOLL, R. V., Evolution of the Late Cretaceous forearc basin, northern and central California, *Geol. Soc. Am. Bull., Part I*, **90**, 813–826, 1979.

KANMERA, K. and T. SAKAI, Correspondence of the formation place of the Shimanto Group to the present sea-floor, *GDP Rep. II-I (1), Struc. Geol.*, No.3, 55–64, 1975.

KISHU SHIMANTO RESEARCH GROUP, The Hidakagawa Group in the southern part of Ryujin Village, Wakayama Prefecture—the study of the Shimanto Terrain in the Kii Peninsula, Southwest Japan (Part 8), *Earth Sci. (Chikyu Kagaku)*, **31**, 250–262, 1977.

KUENZI, W. D., O. H. HORST, and R. V. MCGEHEE, Effect of volcanic activity on fluvial-deltaic sedimentation in a modern arc-trench gap, southwestern Guatemala, *Geol. Soc. Am. Bull. Part I*, **90**, 827–838, 1979.

KUMON, F., Shimanto Supergroup in the southern part of Tokushima Prefecture, Southwest Japan, *Geol. Soc. Japan*, **87**, 277–295, 1981.

KUMON, F., Coarse clastic rocks of the Shimanto Supergroup in eastern Shikoku and Kii Peninsula, Southwest Japan, *Mem. Fac. Sci., Kyoto Univ., Ser. Geol. Min.*, **49**, 63–109, 1983.

MANSFIELD, C. F., Upper Mesozoic subsea fan deposits in the southern Diablo Range, California: record of the Sierra Nevada magmatic arc, *Geol. Soc. Am. Bull., Part I*, **90**, 1025–1046, 1979.

MATSUYAMA, H., F. KUMON, and K. NAKAJO, Cretaceous radiolarian fossils from the

Hidakagawa Group in the Shimanto Belt, Kii Peninsula, Southwest Japan, *News, Osaka Micropaleont. Assoc., Spec. Pub.*, No. 5, 371–382, 1982.

MIYAMOTO, T., Comparison of the Cretaceous sandstones from the Chichibu and Shimanto Terrains in the Odochi area, Kochi Prefecture, Shikoku, *Geol. Soc. Japan*, **82**, 449–462, 1976.

MIYAMOTO, T., Stratigraphical and sedimentlogical studies of the Cretaceous System in the Chichibu Terrain of the Outer Zone of Southwest Japan, *Sci., Hiroshima Univ., Ser. C*, No. 3, 1–139, 1980.

NAKAGAWA, C., K. NAKASEKO, K. KAWAGUCHI, and R. YOSHIMURA, Radiolarian fossils (from Upper Jurassic to Upper Cretaceous) of the northern zone of the Shimanto Supergroup—the study of the Shimanto Supergroup in eastern Shikoku, part 4, *Gakugei, Tokushima Univ., Nat. Sci.*, **31**, 1–27, 1980.

NAKASEKO, K., A. NISHIMURA, and K. SUGANO, Study on radiolarian fossils of the Shimanto Belt, *News, Osaka Micropaleont. Assoc., Spec. Pub.*, No. 2, 1–49, 1979.

NAKAZAWA, K., F. KUMON, K. KIMURA, H. MATSUYAMA, and K. NAKAJO, Sedimentary environment of the Cretaceous cherts in the Shimanto Belt, Kii Peninsula, Southwest Japan, in *Siliceous Deposits in the Pacific Region*, edited by IIJIMA, A., J. R. HEIN, and R. SIEVER, pp. 395–411, Elsevier, Amsterdam, 1983.

NISHIMURA, T., Petrography of the Izumi sandstones in the east of the Sanuki Mountain Range, Shikoku, Japan, *Geol. Soc. Japan*, **82**, 231–240, 1976.

OKADA, H., Preliminary study of sandstones of the Shimanto Supergroup in Kyushu, with special reference to "petrographic zone", *Sci. Rep., Geology Kyushu Univ.*, **12**, 203–214, 1977.

SUYARI, K., Y. BANDO, and I. OBATA, Discovery of a Cretaceous ammonite from the Shimanto Belt in Mugi-cho, Tokushima Prefecture, *Geol. Soc. Japan*, **73**, 535–536, 1967.

SUZUKI, T. and S. HADA, Cretaceous tectonic mèlange of the Shimanto Belt in Shikoku, Japan, *Geol. Soc. Japan*, **85**, 467–479, 1979.

TAIRA, A., M. TASHIRO, M. OKAMURA, and J. KATTO, The geology of the Shimanto Belt in Kochi Prefecture, Shikoku, Japan, in *Geology and Paleontology of the Shimanto Belt (Selected Papers on Honor of Prof. J. Katto)*, edited by TAIRA, A. and M. TASHIRO, pp. 318–389, Rinyakosaikai Press, Kochi, Japan, 1980.

TERAOKA, Y., Comparison of the Cretaceous sandstones between the Shimanto terrane and the Median zone of Southwest Japan, with the reference to the provenance of the Shimanto geosynclinal sediments, *Geol. Soc. Japan*, **83**, 795–810, 1977.

TOKUOKA, T. and F. KUMON, The Shimanto Terrain in the Akaishi Mountainland and the Kii Peninsula—a consideration on mineral composition of sandstones. *Monograph, Nat. Sci. Mus., Tokyo*, No. 12, 41–54, 1979.

Formation of Active Ocean Margins, edited by N. Nasu *et al.*, pp. 835–851.

SEDIMENTARY EVOLUTION OF SHIKOKU SUBDUCTION ZONE: THE SHIMANTO BELT AND NANKAI TROUGH

Asahiko TAIRA

Department of Geology, Kochi University, Kochi 780, Japan

Abstract. Cretaceous to presentday evolution of the Shikoku subduction zone shows a complex history of sedimentation and tectonics in the active margin environment. Cretaceous to lower Miocene history preserved in the Shimanto Belt on land reveals episodes of intensive trench sedimentation and accretion during both Campanian and Eocene times accompanied by incorporation of oceanic materials into melanges. The offshore geology of Shikoku also indicates a complex growth history including middle Miocene forearc igneous activity, late Pliocene transgression and a Plio-Pleistocene trench filling and accretion event due to Izu collision tectonics. The history of the Shikoku subduction zone strongly suggests that the evolution of active margin is largely controlled by episodic events such as collision, intensive trench sedimentation and drastic change in age and motion of subducting plate.

1. *Introduction*

The outer margin of southwest (SW) Japan preserves a long history of subduction of oceanic plate. On land, the Jurassic to Tertiary subduction complex (Chichibu and Shimanto belts) represents an excellent example of forearc geologic complexes (e.g. TAIRA *et al.*, 1983). In the offshore, recent superb seismic reflection profiles (e.g. NASU *et al.*, 1980; AOKI *et al.*, 1983) have exposed a well-demonstrated example of on-going sediment accretion processes.

Recent increases in knowledge of paleomagnetism and paleobiogeography of the circum-Pacific region indicate that many previously identified ancient subduction complexes have undergone complicated histories of exotic terrane accretion and enormous strike-slip modification (JONES *et al.*, 1983). The Jurassic to early Cretaceous history of Japan is no exception: it probably experienced substantial modification by left-lateral strike-slip tectonics (TAIRA *et al.*, 1983). However, the late Cretaceous to Recent active margin of SW Japan did not experience any remarkable modification. Therefore, this part can be thought to be one of the rare examples of well-preserved and "in place" convergent margin of the circum-Pacific region.

This paper reviews briefly the Cretaceous to Recent sedimentary evolution of Shikoku (informally called the Shikoku subduction zone) mainly based on the work done by the author and his co-workers.

2. *Geologic Terranes of SW Japan*

The pre-Neogene tectonic divisions of SW Japan can be classified into three terranes and several strike-slip mobile zones (Figs. 1 and 2).

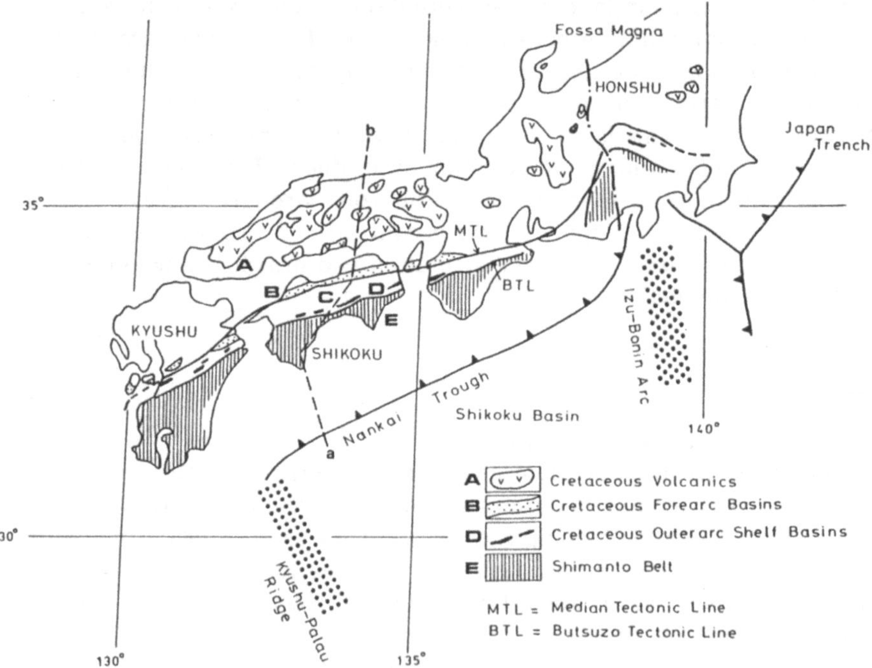

Fig. 1. Map of SW Japan showing the Cretaceous terranes and the present-day subduction zones. The line a-b is the approximate transect of the cross-section shown in Fig. 2.

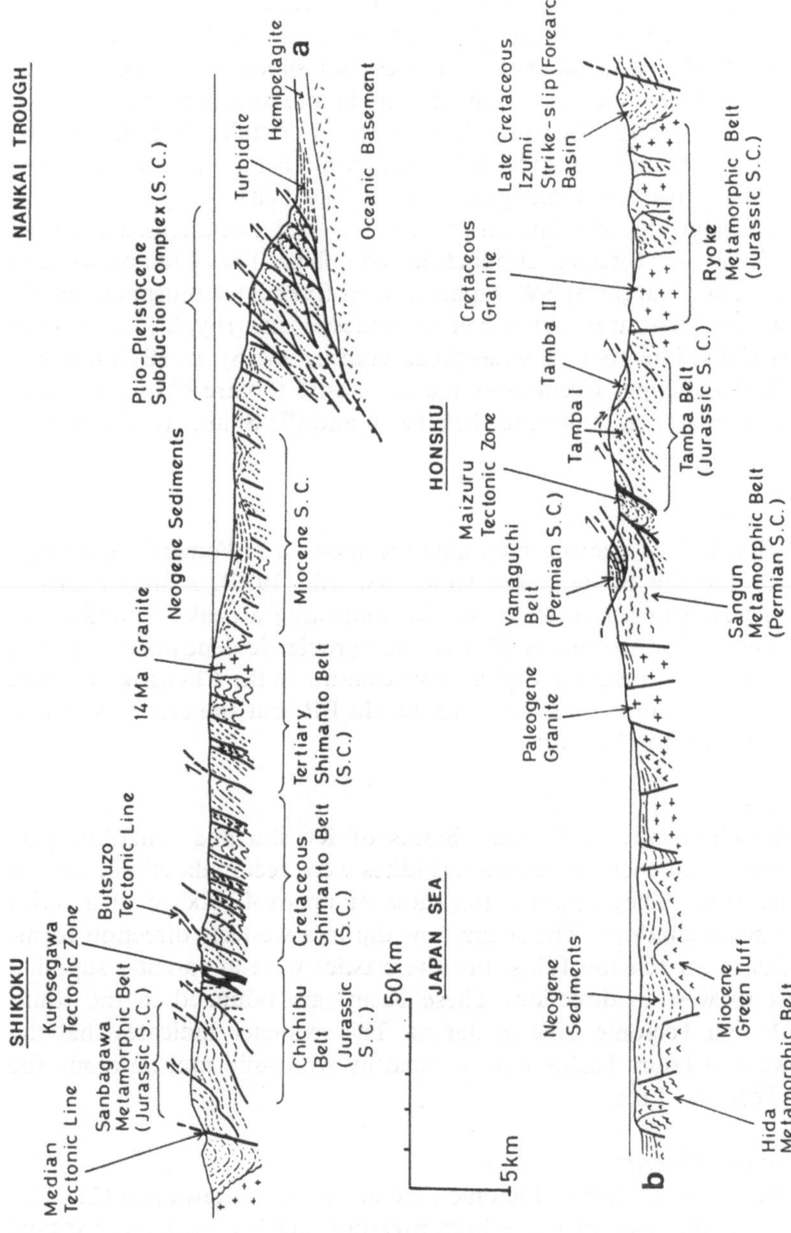

Fig. 2. Schematic geologic cross-section of SW Japan. Black color indicates serpentinite. The line of transect is shown in Fig. 1. S.C.=subduction complex.

The three terranes include; 1) older terranes, 2) Jurassic subduction complex, 3) Cretaceous to Tertiary subduction complex.

The first group consists of Hida, and Sangun and Yamaguchi belts. They are possibly parts of exotic terranes and subduction complex juxtaposed against the ancient Asian margin in pre-Jurassic time.

The second group, the Jurassic subduction complex is composed of the Chyugoku, Chichibu, Tamba, Mino and Ashio belts and their metamorphic counter-parts, the Sambagawa and Ryoke belts.

The third group, the Shimanto belt is a well documented example of Cretaceous to Tertiary subduction complex. The Cretaceous and Paleogene active margin of SW Japan formed five major morpho-tectonic belts (Fig. 1). They are: (A) a belt of andestic and rhyolitic extrusives and terrestrial sediments with widespread granitic intrusives, (B) a narrow, elongate belt of Upper Cretaceous marine clastic forearc basin, (C) nonmagmatic outerarc, (D) outerarc shelf basin and (E) Shimanto subduction complex.

Magmatic arc

The first belt of ancient magmatic arc shows a radiometric age range from 30 to 130 Ma BP (e.g. see MURAKAMI ed., 1979). A major cluster of ages appear between 70 to 95 Ma BP indicating a peak of subduction activity. The wide distribution of rhyolitic pyroclastic deposits of this age suggests an event of intensive explosive volcanism. In the Chyugoku district, granitic activity resumed during 60 to 40 Ma BP, but the center of intrusion was shifted to the north.

Forearc basin

In the Ohnogawa and Izumi basins of Kyushu and Shikoku, particularly thick sequences of coarse turbidites were accumulated during late Cretaceous time. They attain a thickness of up to 40,000 m in a series of lateral accumulations. The strata now dip in a westerly direction forming a series of en echelon folds, but the clastics were apparently supplied from east to west in direction. These basins are bounded to the south by the Median Tectonic Line of Japan. This evidence indicates that the Ohnogawa and Izumi basins were formed by strike-slip motion along the Median Tectonic Line.

Nonmagmatic outerarc

To the south of Median Tectonic Line lie the Sambagawa and Chichibu belts. The Sambagawa belt is a high pressure and low temperature type metamorphic belt and the Chichibu belt is composed of basalt, limestone, chert, sandstone and shale of Carboniferous to Jurassic age. During Cretaceous time at least a part of these belts formed a belt of uplift interpreted as a nonmagmatic outerarc.

Outerarc shelf basin

Within the Chichibu belt, there are various elongate basins in which relatively thin fluvio-brackish to shallow marine sediments were deposited. These sediments yield abundant molluscan fauna and plant flora indicative of early Cretaceous to partly late Cretaceous age. This belt occurs along the major strike-slip mobile zone (Kurosegawa tectonic zone) in the frontal belt of outer nonmagmatic arc. Basins show either lenticular or rectangular shape and contain *en echelon* folding patterns. Although the original shapes of basins have been destroyed by later tectonic deterioration, they generally show the characteristics common to pull-apart basins.

3. Stratigraphy of the Shimanto Belt

To the south of the Chichibu belt, separated by the Butsuzo tectonic line, lies the Shimanto belt which is composed of rocks of accretionary slope basins and subduction complexes. The Shimanto Belt shows a southward younging trend and is divided into two sub-belts; a northern belt consisting of Cretaceous rocks and a southern belt consisting of Tertiary rocks.

The Northern Shimanto Belt (Cretaceous)

The Cretaceous Shimanto Belt can be divided into four major tectonostratigraphic units being bounded by high-angle reverse faults. They are (1) Neocomian to Coniacian brackish to shallow marine sediments, (2) An Aptian to Cenomanian turbidite belt, (3) A Coniacian to Campanian turbidite-melange belt, (4) An Upper Campanian to Maastrichtian belt of slump and shallow marine deposits.

Brackish and shallow marine sediments are exposed along the northernmost part of the Shimanto Belt and are divided into two formations; the Doganaro and Uwagumi Formations. Abundant molluscan assemblages of Neocomian and Turonian ages have been reported (TASHIRO, 1980). Some parts of the Doganaro Formation contain large blocks of limestones and cherts of Triassic age which are apparently derived from the Sambosan Group, a part of Chichibu Belt to the north. These chaotic deposits have been interpreted as slump-olistostrome deposits in a slope setting. This first unit is probably comprised of shelf to slope sediments deposited in a strike-slip or "passive margin" setting prior to the main phase of

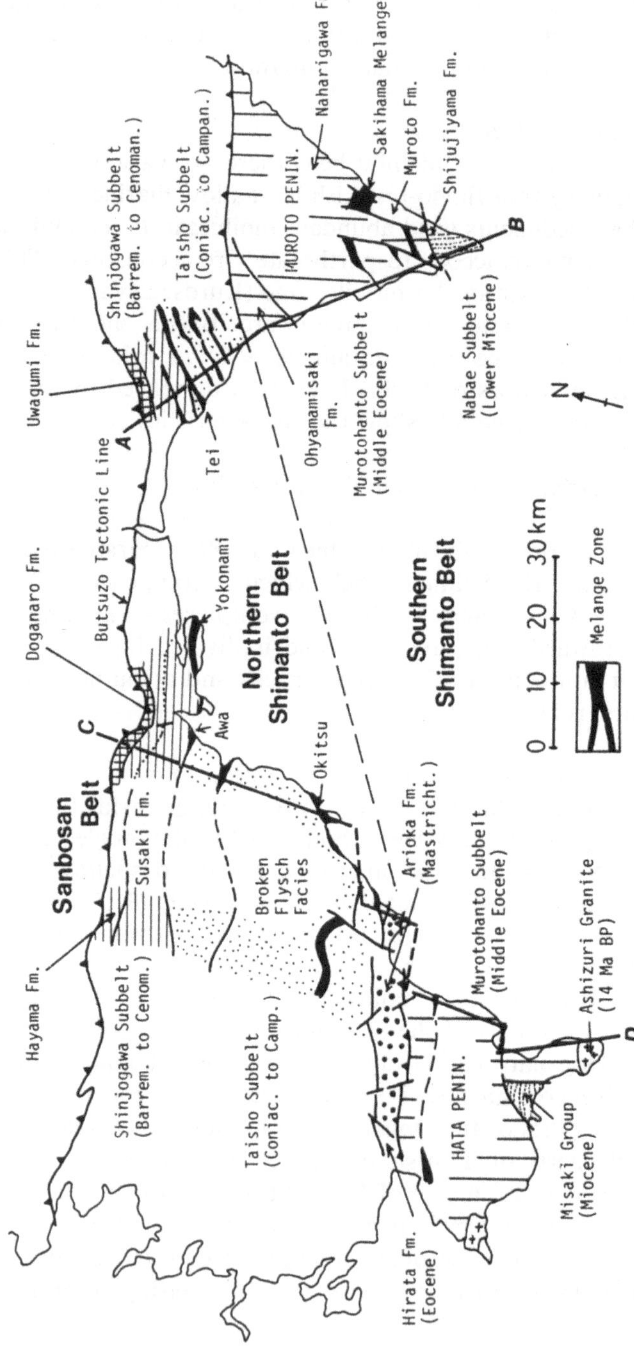
Hayama Fm.
Shinjogawa Subbelt (Barrem. to Cenom.)
Sanbosan Belt
Susaki Fm.
Taisho Subbelt (Coniac. to Camp.)
Broken Flysch Facies
Hirata Fm. (Eocene)
HATA PENIN.
Misaki Group (Miocene)
Ashizuri Granite (14 Ma BP)
D
Murotohanto Subbelt (Middle Eocene)
Arioka Fm. (Maastricht.)
Okitsu
C
Doganaro Fm.
Awa
Northern Shimanto Belt
Southern Shimanto Belt
Yokonami
Butsuzo Tectonic Line
0 10 20 30 km
Melange Zone
Uwagumi Fm.
A
Tei
Ohyamamisaki Fm.
Murotohanto Subbelt (Middle Eocene)
Nabae Subbelt (Lower Miocene)
N
Shinjogawa Subbelt (Barrem. to Cenoman.)
Taisho Subbelt (Coniac. to Campan.)
MUROTO PENIN.
Naharigawa Fm.
Sakihama Melange
Muroto Fm.
Shijujiyama Fm.
B

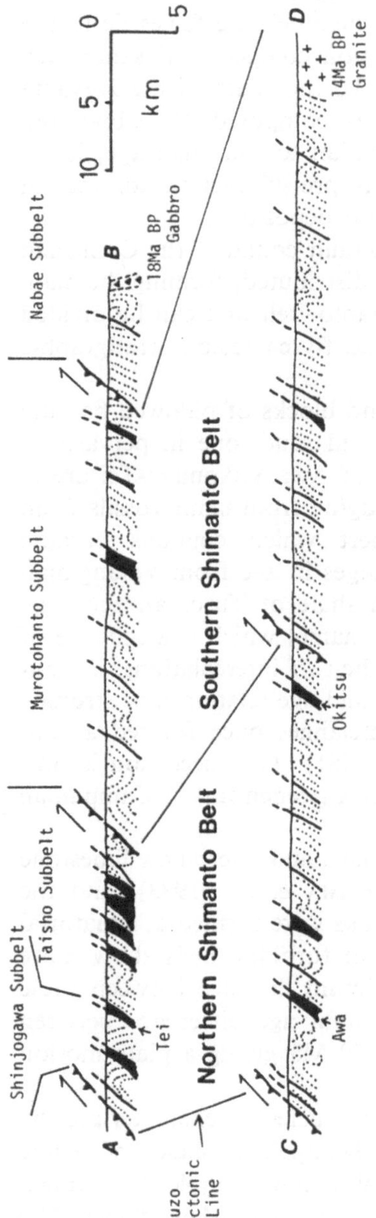

Fig. 3. Sketch map showing the distribution and cross-section of the Shimanto Belt in southern Shikoku.

plate subduction during Coniacian to Campanian time.

The Aptian to Turonian turbidites and hemipelagite facies lie to the south of the first belt. the Hayama Formation composed of sandy turbidites, ranges from late Aptian to Albian. To the south of the Hayama Formation, the Susaki Formation, which is composed of Albian red hemipelagic shale and turbidites of Cenomanian to Turonian age, is exposed. These strata are interpreted again as pre-subduction continental margin clastic wedge and basin hemipelagite deposits.

To the south of the second unit, across fault contacts, the Coniacian to Campanian subduction complex is widely distributed, forming the main part of the Cretaceous portion of the Shimanto Belt and can be divided into several alternating, exposed melange and flysch tectonostratigraphic sub-units.

The melanges contain tectonic slivers and blocks of pillowed basaltic lava, chert, hemipelagic shale, acidic tuff and sandstone in pervasively sheared pelitic matrix. In the melange zone of Awa-Yokonami-Tei areas, the age of pillowed basalt is Valanginian judging from nannofossils from interpillow limestone and radiolarian chert which contains basaltic fragments. The bedded radiolarian chert ranges in age from Valanginian to Cenomanan and grades into pelagic red shale of Turonian age. The hemipelagic shale (varicolored shale as field name) which is a mixture of radiolarian test and silt-size clastics with acidic tuff intercalations, is Coniacian to Santonian in age. These lithologies and age relationships strongly suggest that slivers and blocks found in melanges once formed a continuous oceanic plate stratigraphy (TAIRA, 1981). The succession is now sliced, broken, and embedded in clastic matrix (trench fill) of Campanian age.

Paleomagnetic measurements on pillow lava and interpillow limestone were carried out (TAIRA *et al.*, 1980; KODAMA *et al.*, 1983), and the paleolatitude obtained suggests that these rocks were formed at equatorial latitude. On the other hand, the Campanian turbidite beds show more or less present-day paleolatitude. The minimum distance between these paleolatitudes is about 3000 km and maximum age difference between oceanic plate material and matrix is about 50 Ma giving a plate motion speed of 6 cm/year.

The Coniacian to Campanian turbidite facies is characterized by coherent bedding with intercalation of red hemipelagic shale and acidic tuff. The sequence is mainly composed of fining upward cycles of channel-overbank turbidite facies and grades into younger hemipelagic facies. The turbidite facies has been interpreted as dominantly trench-fill in origin and younger hemipelagites as slope sediments. The melange and turbidite units are usually in faulted contact and the turbidite units show northward dipping attitude with local isoclinal folding.

The age and lithology of the melange facies changes towards the south (Fig. 4). The southern portion is composed of younger slivers in which Albian bedded radiolarian chert is the oldest rock so far found.

In the SW part of Shikoku, the youngest unit of the Cretaceous Shimanto Belt is exposed. The lithofacies is characterized by such features as Upper Campanian to Maastrichtian (partly Paleocene) molluscan sandstones, shallow marine limestone and *Cruziana* facies trace fossils all indicative of shelf environments. Some part of this unit is highly deformed and broken, containing large emplaced blocks of a shallow marine origin

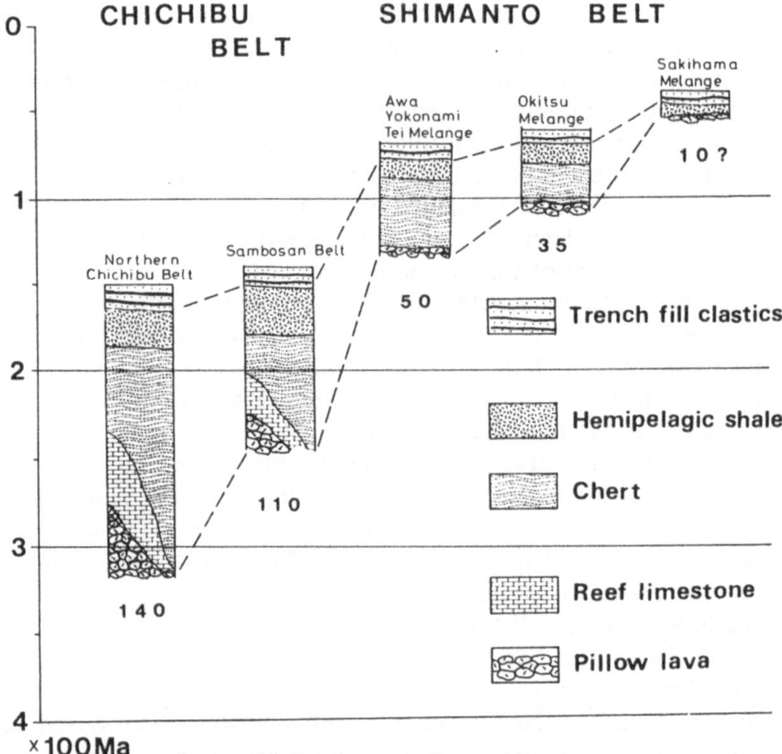

Fig. 4. Reconstruction of oceanic plate stratigraphy in the Shikoku subduction zone during Jurassic to Eocene time. Note that younger oceanic plate had been subducted as time progressed. The numbers shown under the columns are the age difference between the oldest oceanic plate material and trench-fill clastics in million years.

suggesting the presence of gravity slide deposits and a slope setting for these environments.

The occurrence of shallow marine and slope strata roughly defines the outward margin of the outerarc area. The apparent southward migration of this facies indicates that the exposed outerarc has migrated south through Cretaceous time.

The Southern Shimanto Belt

The Southern Shimanto Belt, separated from the northern belt by the Aki-Nakasuji Tectonic Line, consists of two subbelts; (1) An Eocene subbelt and (2) A Lower Miocene subbelt. The Eocene rocks, exposed in the northern part, are called the Murotohanto subbelt and the Lower Miocene rocks, the Nabae subbelt.

The Murotohanto subbelt consists of four tectonostratigraphic units, the Ohyamamisaki, Naharigawa, Muroto Formations and the Sakihama Melange. The dominant lithologies of the Ohyamamisaki Formation are conglomerate, sandstone and shale showing a fining upward depositional megacycle. The conglomerate bed contains large boulders of various lithologies including schists, volcanics, limestones, sandstones and shales. These clasts are interpreted to have been derived from the arc area. The sedimentary facies of this formation indicates that it is a proximal, resedimented coarse clastic facies of submarine channels and slope olistostromes. The age of this formation has not been conclusively determined but has been interpreted as Eocene based on the difference in lithology with Cretaceous strata. However, this formation may be correlated with the Maastrichtian Arioka Formation of SW Shikoku.

The Naharigawa Formation, which contains a thick flysch sequence, has been dated as Middle to Upper Eocene based on microfossils. Dominant sedimentary structures such as massive and graded sandstone beds, sole marks and dish structures indicate proximal turbidite sedimentation. The Paleocurrent is axial with an east to west component. The formation is locally isoclinally folded, but dominantly shows northward dipping attitude and is repeated by high angle faults.

Two lithofacies are distinguished in the Muroto Formation; slump-olistostrome facies and flysch facies. The slump deposits are quite common throughout the Muroto Formation and are composed of folded, truncated and chaotically mixed sandstone and mudstone. The flysch facies shows many channel-overbank deposits which are often composed of fining upward depositional cycles. Locally, clastic intrusives are abundant and these intruded into both flysch and slump beds.

The Sakihama melange consists of large slivers and blocks of basalt, tuff and sandstone in sheared pelitic matrix. Basaltic rocks range from small fragments of several tens of centimeters to large basaltic lava (often pillowed), hyaloclastite and red shale complexes of a few hundred meters in thickness. The melange materials have not been dated well, but radiolarians obtained from several locations all indicate an early to middle Eocene age.

In SW Shikoku, the Paleogene Shimanto Belt shows a development of different facies. The Hirata Formation of the Nakasuji area is particularly unique because it is the only definite shallow marine sequence of Paleogene age in the belt. It unconformably overlies the Arioka Formation and includes shallow marine molluscan and trace fossils, glauconitic sandstone and conglomerate.

To the south of these shallow marine strata, there is a belt of flysch deposits and a chaotic sequence in which pillowed basalts and varicolored shales are found. Slumps and olistostromes are common especially on the eastern coast of the Hata Peninsula where blocks of larger-foraminifera-bearing limestone and volcanic conglomerate are incorporated into pebbly mudstones.

The southern margin of Shikoku is occupied by the Lower Miocene Shimanto Belt (Nabae subbelt). The Nabae Group of the Muroto area is composed of slump-olistostrome and flysch facies. Highly deformed strata consisting of broken flysch, red shales, acidic tuffs and occasional limestones are the dominant lithologies. Eocene rocks are often found within this lower Miocene chaotic matrix.

Especially noteworthy is the presence of gabbroic intrusives and in-situ pillowed basaltic rocks indicative of the lower Miocene forearc volcanism. These strata are overlain by molluscan fossil-bearing coarse clastic sediments of the Shijujiyama Formation which may be a slope-shelf facies.

In the western area, the Misaki Group represents shallow marine Miocene deposits comprising a thick coarsening upward sequence of an offshore mud, sand bar and deltaic complex.

The Tertiary Shimanto Belt is overlain by the late Pliocene Tonohama Group with conspicuous angular unconformity. The Tonohama Group is divided into two formations; the Nobori and Ananai Formations. The Nobori Formation is composed of fossiliferous calcareous mudstone and the Ananai Formation of siltstone and coquina. Although the distribution of this formation is sporadic and mainly limited to the coastal area at present, a few locations known in the interior of the Muroto Peninsula indicate that this formation represents a period of wide-spread transgression.

4. *Offshore Geology*

The offshore continental margin geology of Shikoku has been investigated from various scopes of study (e.g. HILDE *et al.*, 1969; OKUDA *et al.*, 1978, 1979; AOKI *et al.*, 1983) including IPOD drilling in the Nankai Trough (Legs. 31 and 87).

A synthesis of offshore geology based on these studies is presented in Figs. 5 and 6.

OKUDA *et al.* (1979) proposed a stratigraphic classification of this area; M, T, K_1, K_2, K_3, and P Formations in ascending order. The M Formation, the accoustic basement, in which clear structures are unrevealed was correlated to the Tertiary Shimanto Belt. Above this lies the stratified but folded T Formation which was correlated to the Lower Miocene to Middle Miocene Tanabe and Kumano Formations of the Kii area. In Shikoku, the Shijujiyama Formation in Muroto and the Misaki Group in the Tosashimizu area are possible correlatives to the T Formation. The

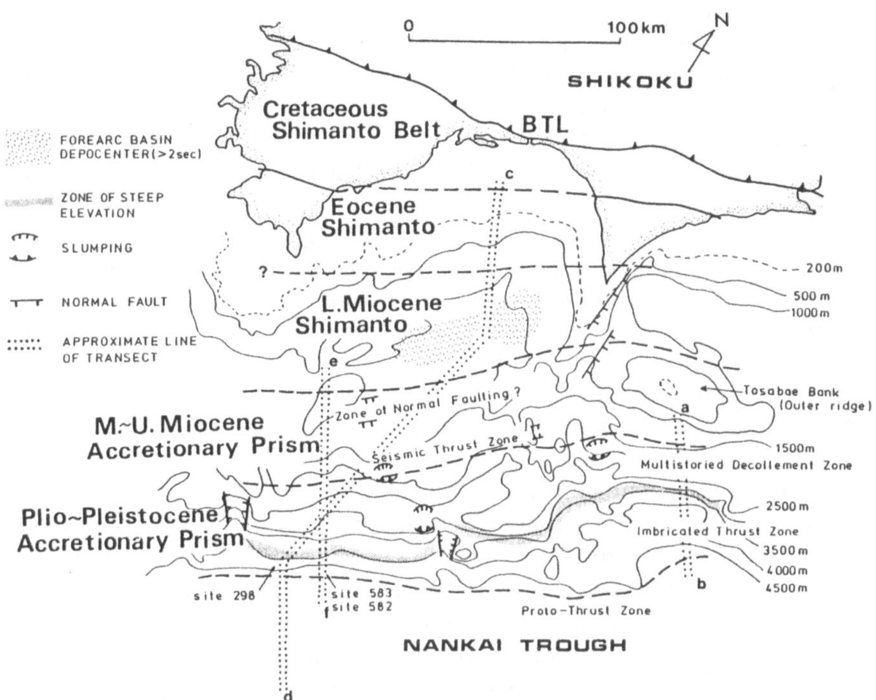

Fig. 5. Offshore geological features of the Shikoku subduction zone. Lines a-b, c-d and e-f indicate the approximate line of transect of reflection profiles shown in Fig. 6.

K_1 to K_3 Formations overlie the T Formation and are the major filling material of the forearc basin.

Recent multichannel seismic profiles reveal more detailed structures in the lower trench slope. AOKI *et al.* (1981) and TAMANO *et al.* (1983) illustrated beautifully the examples of accretion and deformation of trench clastic fill.

KAGAMI *et al.*, (1983) classified these structures into four divisions: the proto-thrust zone, imbricated thrust zone, multistoried decollement zone (a series of duplex thrust piles) and seismic thrust zone. In the proto-thrust zone, relatively high angle reverse faults can be traced on reflection

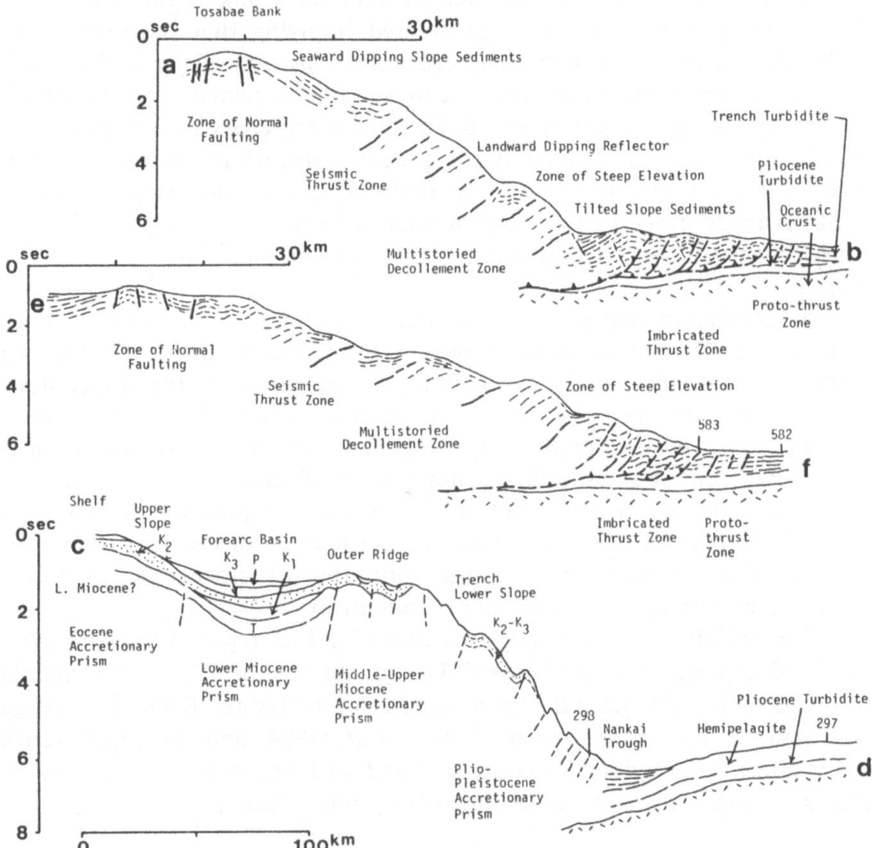

Fig. 6. Seismic reflection profiles of the Shikoku subduction zone. Data compiled and interpreted from NASU *et al.* (1982), GEOLOGICAL SURVEY OF JAPAN (1978), TAMANO *et al.* (1983), AOKI *et al.* (1983). Transect lines are shown in Fig. 5.

profiles. The displacements on these faults are probably small but apparent thickening of strata seems to be taking place in this zone. In the imbricated thrust zone, a series of 30 degree dip thrusts, which converge into a major decollement (initial decollement), are recognized. This decollement starts in the lower part of hemipelagite and propagates to lower levels, eventually reaching to the boundary between oceanic crust and overlying sediments. In the multistoried decollement zone, a series of subparallel and low angle thrusts (decollements) are developed. Within this zone, accreted sediments are likely to be underplated by these decollements initiating the vertical growth of a sediment prism which is clearly indicated by steepness and topographic elevation. In the seismic thrust zone, a large thrust plane, which is probably responsible for tsunami earthquakes of this region (Nankai thrust), is postulated implying that the breakdown of lithified sediments and oceanic crust starts to take place in this zone.

Three representative if cross sections, rather schematic, of the trench lower slope are presented in Fig. 6. In the eastern cross section (line a-b), a wide imbricated thrust zone is conspicuous with some landward vergent thrusts. A zone of possible normal faulting on the outer ridge (Tosabae bank) is recognized. In the western section (e-f), a relatively narrow imbricated thrust zone is characteristics when compared with the eastern section.

A speculative stratigraphic interpretation of these structures based on OKUDA's (1977) classification is presented in Fig. 6 (line c-d). Judging from the continuity in the land area, the basement in the upper-slope area is probably this Lower Miocene Shimanto Belt. The T Formation over lies, with local unconformity. The basement of the forearc basin is not known, but it is probably composed of M and T Formations. The basement of the outer ridge could be T Formation equivalent accretionary prism. The folded K_1 Formation and moderately deformed K_2 Formations lie above this basement. In the lower slope, Plio-Pleistocene (K_1 to P Formations) makes up the accretionary basement.

The DSDP results (see Legs. 31 and 87 initial reports) indicate thick turbidite deposition in the Nankai Trough during the Pleistocene mainly 0.6 to 0.45 Ma BP period. An axial flux of turbidites from the Suruga Bay area has been suggested by TAIRA *et al.* (1984) and this rapid trench sedimentation episode has been correlated with the collision tectonics of central Honshu (e.g. NIITSUMA and AKIBA, this volume).

5. *Sedimentary Evolution*

Although the available data on biostratigraphy are still insufficient to delineate a detailed chronological events of the evolution of the Shikoku subduction system, a brief synthesis can be presented at least to point out some important episodes.

Tithonian to Neocomian

The pre-Shimanto subduction is represented by the Chichibu Belt in which Carboniferous to Jurassic oceanic plate materials such as reef-limestone, chert and pelagic-hemipelagic shale are incorporated into a Jurassic clastic matrix. During and after accretion of these materials, it is inferred that left-lateral strike slip tectonics dominated on this margin. Exotic tectonic lenses of Paleozoic to Mesozoic igneous, metamorphic and sedimentary rocks were emplaced in the Kurosegawa strike-slip mobile zone.

Neocomian to Turonian

The sedimentation in Shimanto Belt started from Neocomian time and comprised various depositional environments of brackish, shallow marine to slope setting. Aptian to Turonian facies represent slope to basin plain flysch and hemipelagite. These deposits are coherent and include no melanges, therefore direct field evidence for subduction is not present. As the strike-slip basins of Lower to Upper Cretaceous age along the Kurosegawa Tectonic Zone still continued, I interpreted this period as a strike-slip controlled continental margin.

Coniacian to Campanian

This period is the main phase of accretion of trench clastics and oceanic plate material of Cretaceous time. Micropaleontological analysis indicates that the Campanian is probably the main phase of the subduction episode during which previous continental margin clastic wedge and contemporaneous trench-fill materials were accreted. The oceanic plate subducted during this time was of Valanginian to Albian age. Pieces of oceanic crust, seamount, pelagic chert and shale, and hemipelagic shale were incorporated into melanges. Convergence was possibly oblique to the continental margin with left-lateral strike-slip motion.

Maastrichtian to Paleocene

Very little records are preserved in Shimanto Belt which are indicative of what happened during this period. The Arioka Formation of SW Shikoku is the only formation for which there is a definite age control. Shallow marine olistostrome blocks indicate an extensive gravity failure episode.

Eocene

This period represents a second phase of major subduction. The melange yields Eocene radiolaria bearing shales, but the exact age of subducted oceanic plate is unknown. However, thin pelagic material suggest that the age of subducted oceanic plate was not so much different from the age of trench fill, indicating that the oceanic ridge was quite close to the trench and ocean floor was very young at that time.

Oligocene to Miocene

There is no direct biostratigraphical evidence for the presence of Oligocene strata from Shikoku, but the Hirata Formation in SW Shikoku may range up to the Oligocene. The Lower Miocene rocks are comprised of olistostrome and flysch showing west to east paleocurrent direction. During the formation of the Shikoku basin, strike-slip tectonics probably dominated and clastic wedge such as those of the Shijujiyama and Misaki Formations were created. The Murotomisaki gabbro may represent igneous activity of the spreading center of the Shikoku basin which is now located between Muroto and the Kii Peninsula. Rotation of SW Japan and the formation of the Japan Sea took place between 15 to 14 Ma BP (OTOFUJI *et al.*, 1983) and during this period, the hot, young Shikoku Basin oceanic plate was subducted, initiating the Ashizuri granites and other SW Japan forearc igneous activity.

Pliocene to Recent

The main event of this time is the phase of transgression during the late Pliocene (Tonohama Group and K_2 Formation). Accretion of Plio-Pleistocene sediment has been enormous due to the rapid supply of sediments from the central Japan area resulting from the intensive uplift created by Izu collision. Slope basins have been filled extensively during this period and the present configuration of the Shikoku subduction zone was formed.

Acknowledgements

I thank H. Kagami and Y. Aoki for their invaluable suggestions and comments.

REFERENCES

Y. AOKI, T. TAMANO, and S. KATO, Detailed structures of the Nankai Trough from migrated seismic sections, Am. Assoc. Pet. Geol., Memoir 34, pp. 309–322, 1981.

T. W. C. HILDE, J. M. WAGEMAN, and W. T. HAMMOND, The structure of Tosa terrace and Nankai trough off southwestern Japan, *Deep-Sea Res.*, **16**, 67–75, 1969.

D. L. JONES, A. COX, P. CONEY, and M. BECK, The growth of western North America, *Scientific American*, **241**, 1982.

H. KAGAMI, K. SHIONO, and A. TAIRA, Subduction of plate at the Nankai Trough and formation of accretionary prism, *Kagaku*, **53**, 429–438, 1983 (in Japanese).

D. E. KARIG, J. C. INGLE *et al.*, Initial Reports of Deep Sea Drilling Project, v. 31, Washington, D.C. (U.S. Government Printing Office), 1975.

K. KODAMA, A. TAIRA, M. OKAMURA, and Y. SAITO, Paleomagnetism of the Shimanto Belt in Shikoku, southwest Japan, in *Accretion Tectonics in the Circum-Pacific Regions*, edited by M. Hashimoto and S. Uyeda, pp. 231–241, Terra Sci. Pub., Tokyo, 1983.

LEG 87 SCIENTIFIC PARTY, Leg 87 drill off Honshu and southwest Japan, *Geotimes*, January, 15–18, 1983.

N. NASU *et al.*, Multi-channel seismic reflection data across Nankai Trough, IPOD-Japan Basic Data Series, No. 4, Ocean Res. Institute, Univ. of Tokyo, 1982.

N. NIITSUMA and F. AKIBA, Neogene tectonic evolution and plate subduction in the Japanese island arcs, this volume.

Y. OKUDA (ed.), Geological Map off Outer Zone of Southwest Japan, Geological Survey of Japan, 1977.

Y. OKUDA, M. KUMAGAI, and K. TAMAKI, Tectonic development of the continental slope and its peripheral area off southwest Japan in relation to sedimentary sequences in sedimentary basins, *J. Japanese Assoc. Petrol. Technologists (in Japanese with English abstract)*, **44**, 279–290, 1979.

Y. OTOFUJI and T. MATSUDA, Paleomagnetic evidence for the clockwise rotation of Southwest Japan, *Earth Planet. Sci. Lett.*, **61**, 349–359, 1983.

A. TAIRA and M. TASHIRO (eds.), *Geology and Paleontology of the Shimanto Belt*, 389pp., Rinyakosaikai Press, Kochi, 1980.

A. TAIRA, J. KATTO, M. TASHIRO, and M. OKAMURA, The geology of the Shimanto Belt in Kochi Prefecture, Shikoku (in Japanese with English abstract), in *Geology and Paleontology of the Shimanto Belt*, edited by A. Taira and M. Tashiro, pp. 319–389, Rinyakosaikai Press, Kochi, 1980.

A. TAIRA, M. OKAMURA, J. KATTO, M. TASHIRO, Y. SAITO, K. KODAMA, M. HASHIMOTO, T. TIBA, and T. AOKI, Lithofacies and geologic age relationship within melange zones of northern Shimanto Belt (Cretaceous) Kochi prefecture, Japan (in Japanese with English abstract), in *Geology and Paleontology of the Shimanto Belt*, edited by A. Taira and M. Tashiro, pp. 179–214, Rinyakosaikai Press, Kochi, 1980.

A. TAIRA, The Shimanto belt of southwest Japan and arc-trench sedimentary tectonics, *Recent Progress in Natural Sci. in Japan*, **6**, 147–162, 1981.

A. TAIRA, H. OKADA, J. H. McD. WHITAKER, and A. J. SMITH, The Shimanto Belt of Japan: Cretaceous-lower Miocene active-margin sedimentation, in *Trench-Forearc Geology*, edited by J. K. Leggett, pp. 5–26, Geol. Soc. London, Spec. Publ., No. 10, 1982.

A. TAIRA, Y. SAITO, and M. HASHIMOTO, The role of oblique subduction and strike-slip tectonics in the evolution of Japan, in *Geodynamics of the Western Pacific-Indonesian Region*, edited by W. C. Hilde and S. Uyeda, pp. 303–316, Geodynamics Series. vol. 11, 1983.

A. TAIRA, N. NIITSUMA, and M. IMAJO, Sedimentation in the Nankai Trough: results of IPOD Leg 87 (in Japanese), *Earth Monthly*, no. 163, 39–45, 1984.

T. TAMANO, T. TOBA, and Y. AOKI, Development of fore-arc continental margins and their potential for hydrocarbon accumulation, Proceedings of World Petrol. Congress, PD3(2), pp. 1–11, 1983.

Formation of Active Ocean Margins, edited by N. Nasu *et al.*, pp. 853–873.

COLLISION OF THE AMAMI PLATEAU WITH THE RYUKYU ISLAND ARC

Hidekazu TOKUYAMA, Young Sea KONG, Hideo KAGAMI, and Noriyuki NASU

Ocean Research Institute, University of Tokyo, Nakano-ku, Tokyo, Japan

Abstract. The Amami Plateau is located at the northwest corner of the Philippine Sea adjacent to the Ryukyu Trench.

Multichannel seismic reflection profiles reveal that Paleocene to Middle Eocene strata, acoustically defined on the Amami Plateau, can be continuously traced westwards as far as the lower inner trench slope of the Ryukyu Trench, beyond the bathymetric trench axis. The strata are abruptly terminated by a westward dipping thrust fault about 30 km landwards of the bathymetric trench axis. This suggests that the Amami Plateau is presently colliding with the Ryukyu Island Arc and that the tectonic plate boundary between the Ryukyu Island Arc and the Philippine Sea plate is landward of the bathymetric trench axis.

On the other hand, in the region north of the Tokara channel, oceanic layer II of the Kita Amami Basin can be traced as far as 80 km landward of the trench axis. This means that the Kita Amami Basin is subducting beneath the Ryukyu Island Arc. The subducting oceanic layer II is characterized by containing a graben structure in which a remarkable reflector is identified. The reflector is represented by drawing a line from horst to horst in a single graben structure and appears to be extention of the top surface of oceanic layer II A. The reflector and top surface of oceanic layer II A are considered to act as the master slip plane along which the Philippine Sea plate subducts beneath the Ryukyu Island Arc.

The boundary between the subducting Kita Amami Basin, northwest of the Amami Plateau, and the unsubductable Amami Plateau is an apparent strike-slip fault. The distribution of epicenters for shallow earthquakes observed in the Amami Plateau region seems to be consistent with the trend and location of the assumed strike-slip fault.

1. Introduction

The Amami Plateau, located at the Ryukyu Trench in the northwestern part of the Daito Ridges Region, is irregularly shaped without elongation and is bounded by the Kita Amami Basin on its north side and by the Kita Daito Basin on the south side (Fig. 1). A deep depression lies between the plateau and the Kyushu Palau Ridge on the east side. The western side of the plateau is at the Ryukyu Trench. Seismic reflection profiles of this region, collected during KH 76-2 and 82-4 Cruises of the Ocean Research Institute of University of Tokyo, are interpreted for the nature of convergence tectonics associated with the Amami Plateau and the surrounded area.

2. Interpretation of Seismic Profiles

Amami Plateau is topographically rough with numerous peaks and with thick sediment fills, up to 1.3 seconds in two way travel time between these peaks (Figs. 2 and 3).

From multichannel seismic reflection profiles, five acoustic units are identified on the Amami Plateau. The stratigraphic sequence at S.P. 110 of Line ORI 82-4-14, starting at 28°10.9N, 132°05.0E (S.P. 0001) and ending at 28°56.2N, 132°05.8E (S.P. 1688) includes four of these units, Units A, B, D, and E in descending stratigraphic order (Figs. 2 and 4). Unit A, about 0.2 seconds thick, is acoustically transparent. The reflectors which define Units A and B are discontinuous. Unit D, about 0.6 seconds thick, is characterized by well-stratified, closely spaced reflectors. Unit E is acoustic basement. Unit C, about 0.9 seconds thick, is identified between Units B and D at S.P. 260 (Fig. 4). This unit is considered to consist of turbidite deposits, mainly because it is characterized by widely spaced low frequency stratified reflectors. Also the unit terminates updip against the surface of the more steeply inclined Unit D at S.P. 150 and 260. Line ORI 82-4-14 intersects with Line ORI 82-4-10 at S.P. 668, which is located at the foot of a bathymetric peak of the Amami Plateau. The stratigraphic sequence here is similar to that at S.P. 260, except that Unit C thickens to 1.3 seconds in two way travel time (Fig. 5). Toward the north, Units A, B, and C lap out against Unit D between S.P. 700 and 790 although Units D and E are traced as far as the northern bathymetric peak (Fig. 6). Unit D, about 0.8 seconds thick, is exposed on the southern slope of the peak and is underlain by Unit E (Fig. 6). Toward the summit of the peak, Unit D gradually decreases in thickness and a transparent veneer,

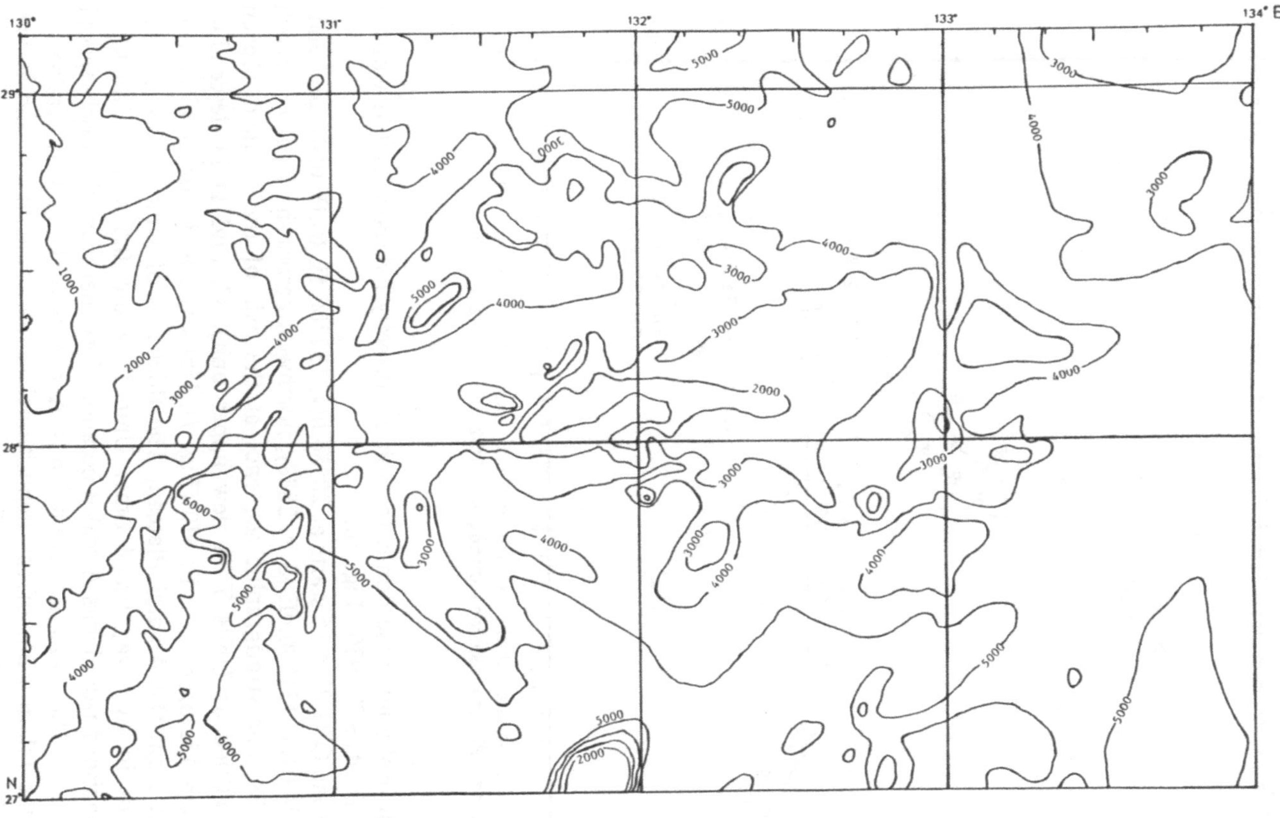

Fig. 1. Bathymetric map around the Amami Plateau after Ocean Sounding Chart G1405 (MARITIME SAFETY AGENCY, 1972). Contours in meters below sea level.

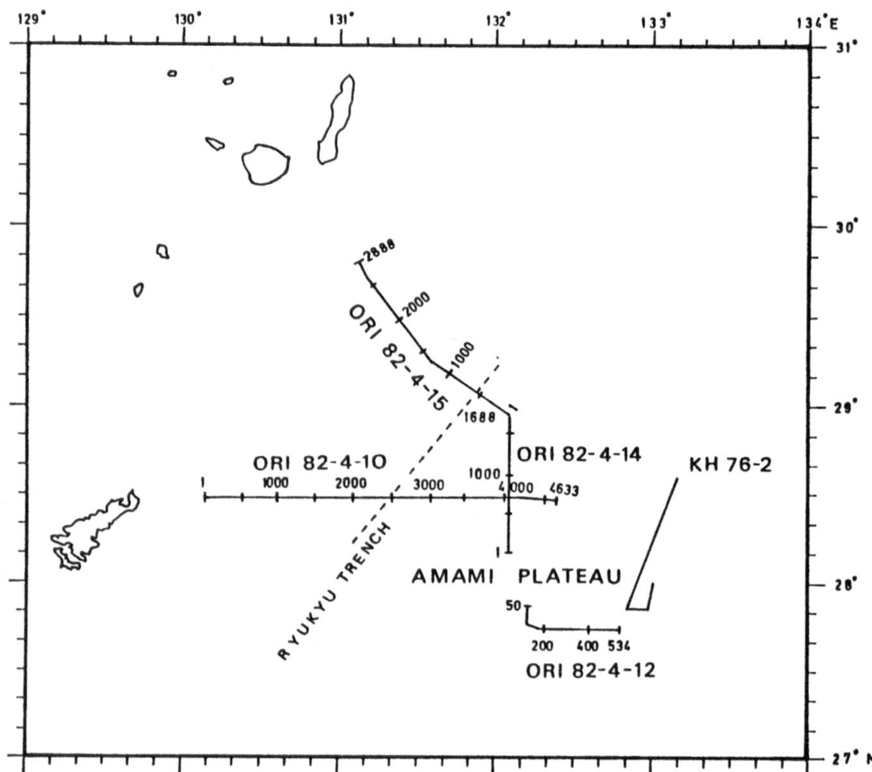

Fig. 2. Track lines of the multichannel and single channel seismic reflection profiles of the Amami Plateau recorded during KH 76-2 and KH-82-4 Cruises.

which probably corresponds to Unit A, directly overlays Unit E without Unit D at S.P. 930 (Fig. 6).

Line ORI 82-4-15 starts at 28°56.0N, 132°06.5E (S.P. 0001) and ends at 29°49.1N, 131°06.4E (S.P. 2888), northwest across the Ryukyu Trench (Fig. 2). The stratigraphic sequence at the axis of the Ryukyu Trench is identified as Units T (a new upper unit), C, and D in descending stratigraphic order at S.P. 410 (Fig. 7). Units C and D are continuously traced from the Amami Plateau. The thickness of Unit C is about 1.3 seconds in two way travel time. Unit T, about 0.5 seconds thick, is distinguished from Units A and B because of its characteristic closely spaced reflectors. The transition from Units A and B to Unit T is gradual so that the relationship of the depositional sequences Units A and B and Unit T is ambuguous. Unit T is considered as mixed strata consisting

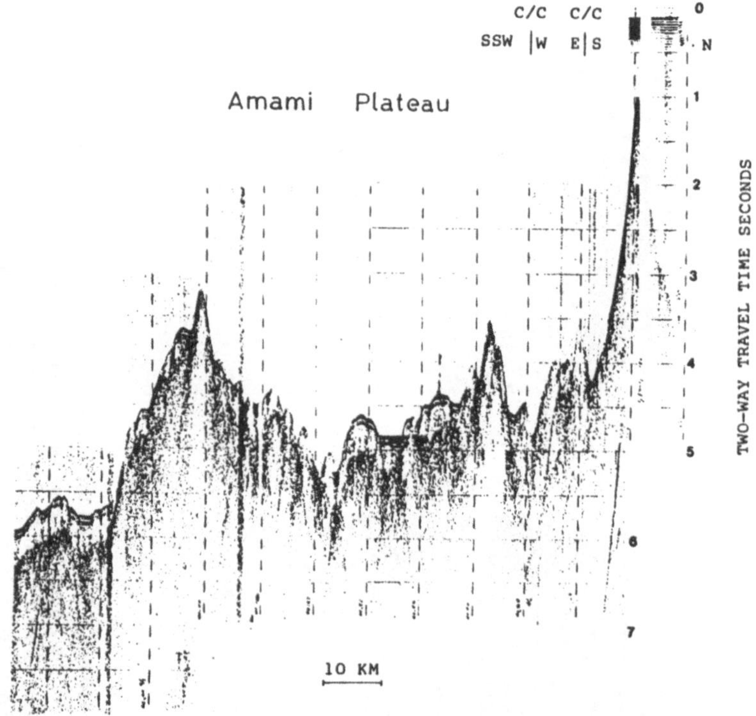

Fig. 3. Single channel seismic reflection profile across the eastern part of the Amami Plateau in approximately a NNE-SSW direction recorded during KH-76-2 Cruise.

of sediments transported along the trench axis, probably from the north, and sediments from Units A and B. The boundary between trench axis and lower inner trench slope is shown by a topographic high between S.P. 550 and 620 (Figs. 7 and 9). Units C and D are not identified on the landward side of the high. Unit C terminates against Unit D. New units are identified here, O and F in increasing order of depth, associated with the inner trench slope (Fig. 7). Unit O, about 1.4 seconds thick at S.P. 690, is a sedimentary mass of the lower inner trench slope of the Ryukyu Trench. Hyperbolic diffractions in Unit O between S.P. 570 and 1060 suggest the development of thrust faults (Fig. 7). Unit F is charactirized by widely spaced reflectors, and can clearly be distinguished from Units C and D on the basis of its acoustic properties. It is apparent that there is a significant difference in acoustic impedance at the boundary between the upper and lower units. Unit F is traced as far as 50 km to

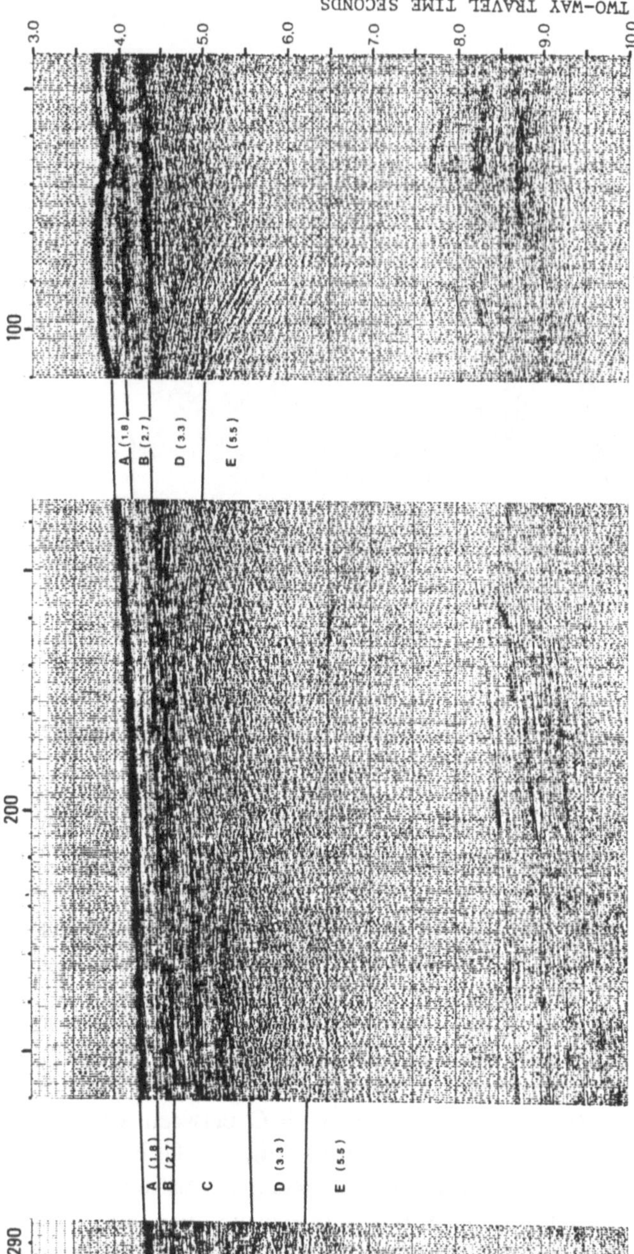

Fig. 4. Stratigraphic sequence at S.P. 110 and S.P. 260 on the Line ORI 82-4-14. A, B, C, D, and E indicate acoustical units on the Amami Plateau. Units A and B; pelagic sediments, probably post Middle Oligocene. Unit C; turbidite deposits supplied from Unit E and mainly Unit D after the Middle Eocene volcanic epoch on the Kyushu Palau Ridge. Unit D; sediments deposited in a shallow sea environment from Paleocene to Middle Eocene Time. Unit E; igneous rocks of Late Cretaceous age. Numbers equal p-wave velocities (km/sec) determined by OBS (SUYEHIRO *et al.*, 1982).

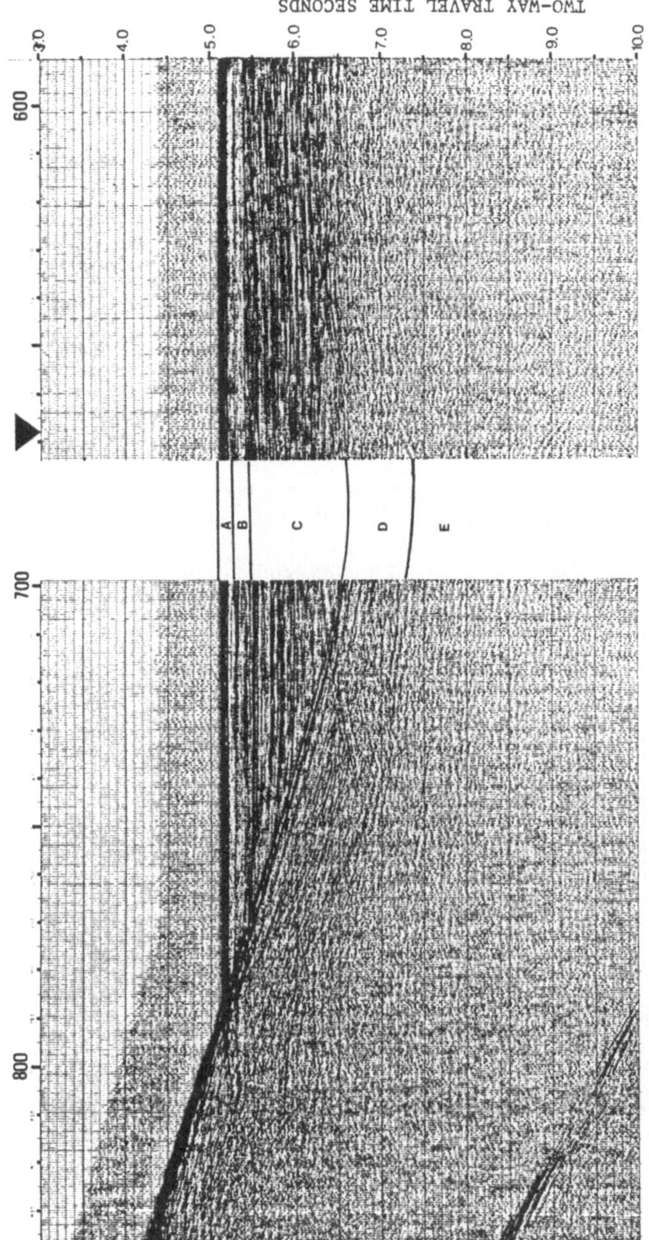

Fig. 5. Stratigraphic sequence at S.P. 674 of Line ORI 82-4-14. Solid triangle indicates the position at which Line ORI 82-4-14 intersects with Line ORI 82-4-10. A, B, C, D, and E as in Fig. 4.

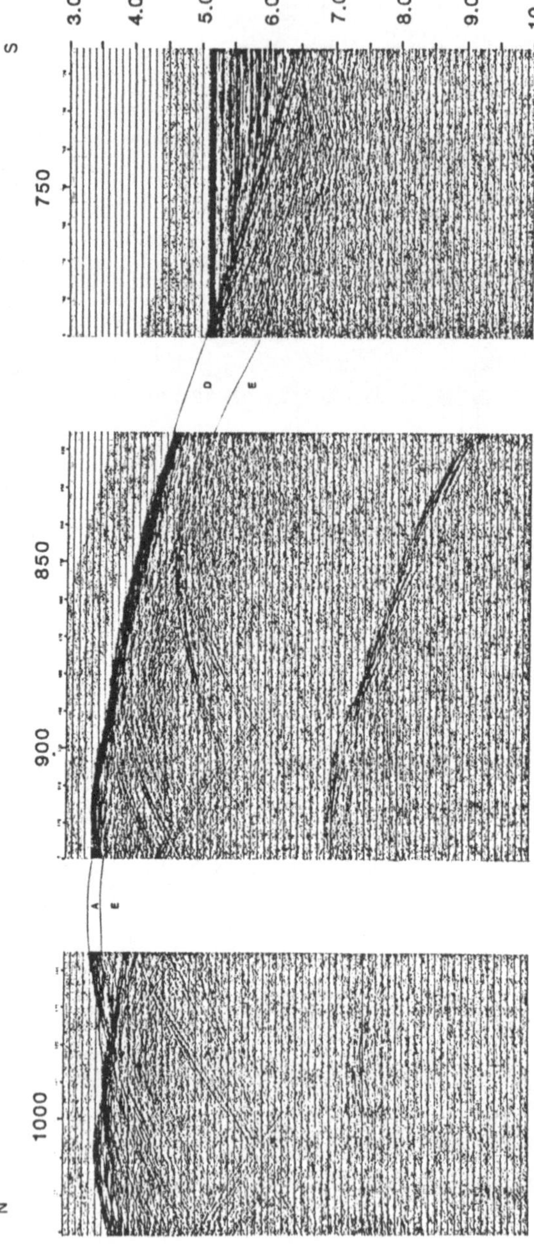

Fig. 6. Stratigraphic sequence at S.P. 790 and S.P. 930 of Line ORI 82-4-14. A, D, and E as in Fig. 4.

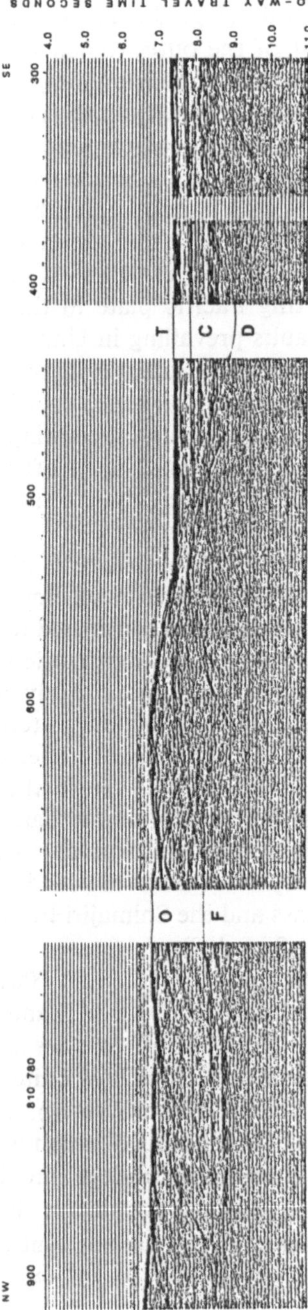

Fig. 7. Stratigraphic sequence at S.P. 410 and S.P. 690 of Line ORI 82-4-15. C and D as in Fig. 4. F, O, and T indicate acoustically identified units at the axis and inner trench slope of the Ryukyu Trench. Unit F; subducting oceanic layer II A. Unit O; sediments off-scraped from the subducting plate in thrust slices and added to lower inner trench slope. Unit T; mixed strata which consist of transported sediments along trench axis and Units A and B.

the landward side of the topographic high, and with graben structures at approximately 5 to 7 km intervals (Fig. 9). These graben structures are often observed in oceanic crust of the outer trench slope (HILDE and SHARMAN, 1978 and IWABUCHI, 1980) and subducting oceanic crust beneath sedimentary masses of the lower inner trench slope (TOKUYAMA *et al.*, 1982). Consequently, Unit F is assumed to correspond to oceanic layer II A. A remarkable reflector is identified in the graben structure between S.P. 750 and 900 (Fig. 7). The reflector is represented by drawing a line from horst to horst in a single graben structure and appears to be a straight extension of the top surface of oceanic layer II A. A similar reflector has been reported for the subducting Pacific plate in the Japan Trench (TOKUYAMA *et al.*, 1982). Thrust faults prevailing in Unit O seem to verge with the reflector from the top surface of Unit F in the lower part of the sequence (Figs. 7 and 9). Therefore, the reflector from the top of Unit F acts as a master slip plane, along which the Philippine Sea plate subducts beneath the Ryukyu Arc. Unit T is scraped off the subducting plate in thrust slices and added to the lower inner trench slope. Thus, Unit O mainly consists of off-scraped sediments.

The sedimentary mass of the inner trench slope is divided into two units between S.P. 1060 and 1890 (Fig. 9). The upper unit, less than 0.35 seconds thick, is acoustically transparent and is considered to consist mainly of pelagic sediment. The lower unit is acoustically chaotic without significant reflectors. It is not clear whether the unit is highly deformed Unit O or a different set of strata. The chaotic reflection pattern abruptly terminates at S.P. 1890. Beyond S.P. 1890 stratified reflectors dip to the west (S.P. 1890 and 2000, Figs. 8 and 9). Further to the west, the dip of the strata gradually decreases and becomes almost flat. This reflective unit is considered to be the Shimajiri Formation of Pliocene to Late Miocene age which is found mainly on Okinawa Island. The boundary between the chaotic sedimentary mass and the Shimajiri Formation is defined by several thrust faults (Figs. 8 and 9).

Line 82-4-10 starts at 29°28.9N, 130°07.5E (S.P. 0001) and ends at 28°27.8N, 132°27.8E (S.P. 4633) running from Kikai Island to the Amami Plateau in an E-W direction (Fig. 2). This line crosses Line 82-4-14 at S.P. 4161. The stratigraphic sequence at S.P. 4120 is quite similar to that of Line 82-4-14 (Fig. 10). Units A (0.2 seconds thick), B (0.2 seconds thick), C (about 1.2 seconds thick), D (about 0.7 seconds thick), and E can be identified. Units A and B successively change into Unit T toward the trench axis, as on Line 82-4-15. Unit C terminates downdip against Unit D between S.P. 2460 and 2600 (Fig. 11). However, Unit D can be traced continuously westward beyond the trench axis as far as the lower inner

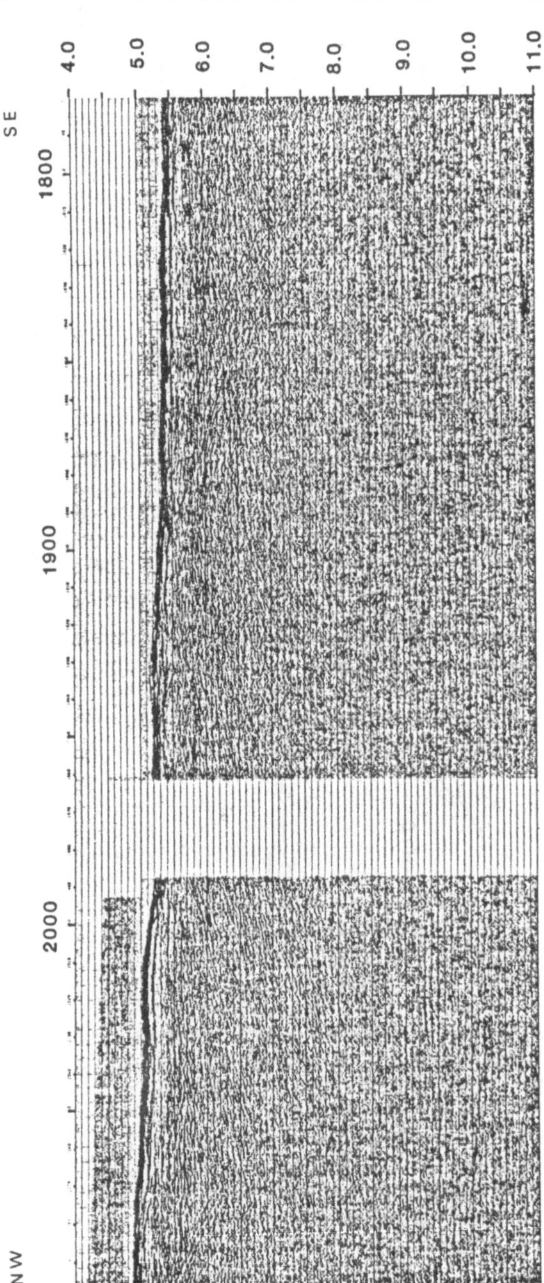

Fig. 8. Seismic profile, Line ORI 82-4-15, from the inner slope of the Ryukyu Trench.

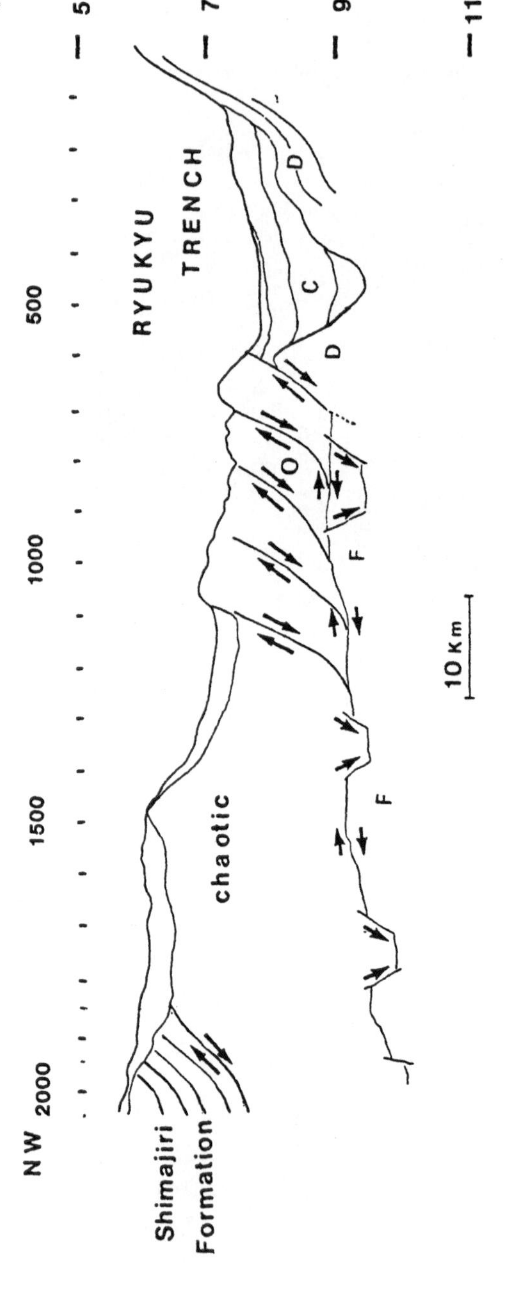

Fig. 9. Schematic crustal structure along Line ORI 82-4-15. C, D, O, and T, as in Fig. 7.

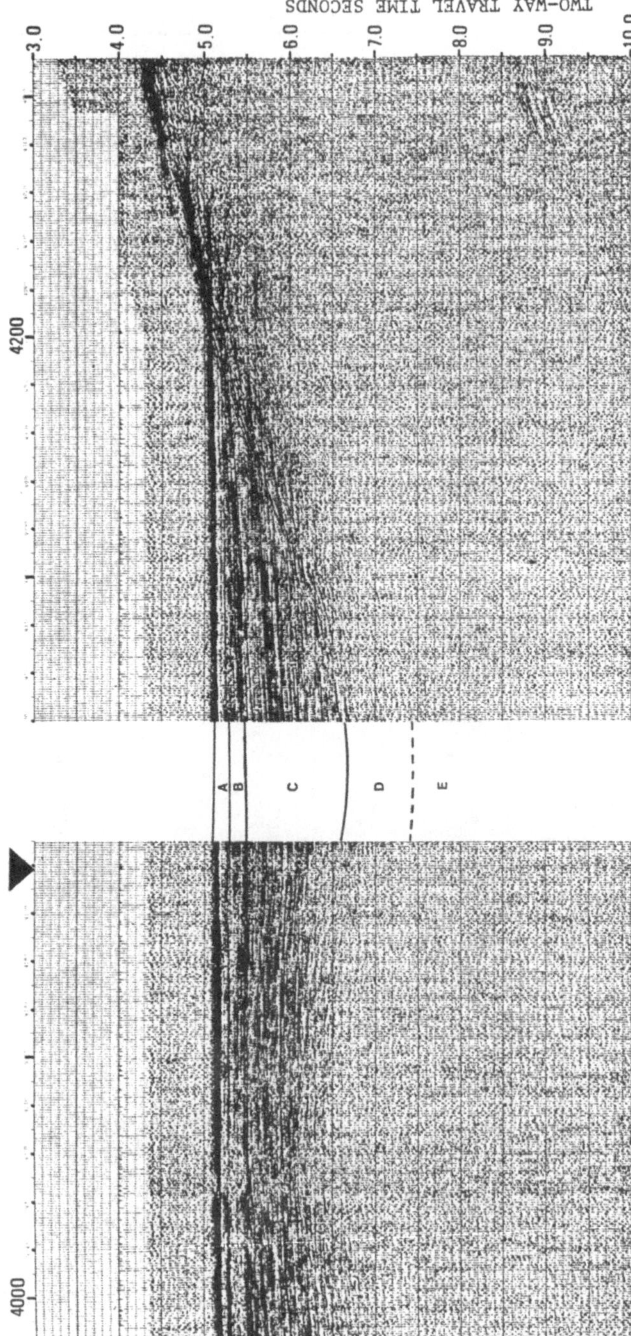

Fig. 10. Stratigraphic sequence at S.P. 4120 of the Line ORI 82-4-10. A, B, C, D, and E as in Fig. 4. Solid triangle as in Fig. 5.

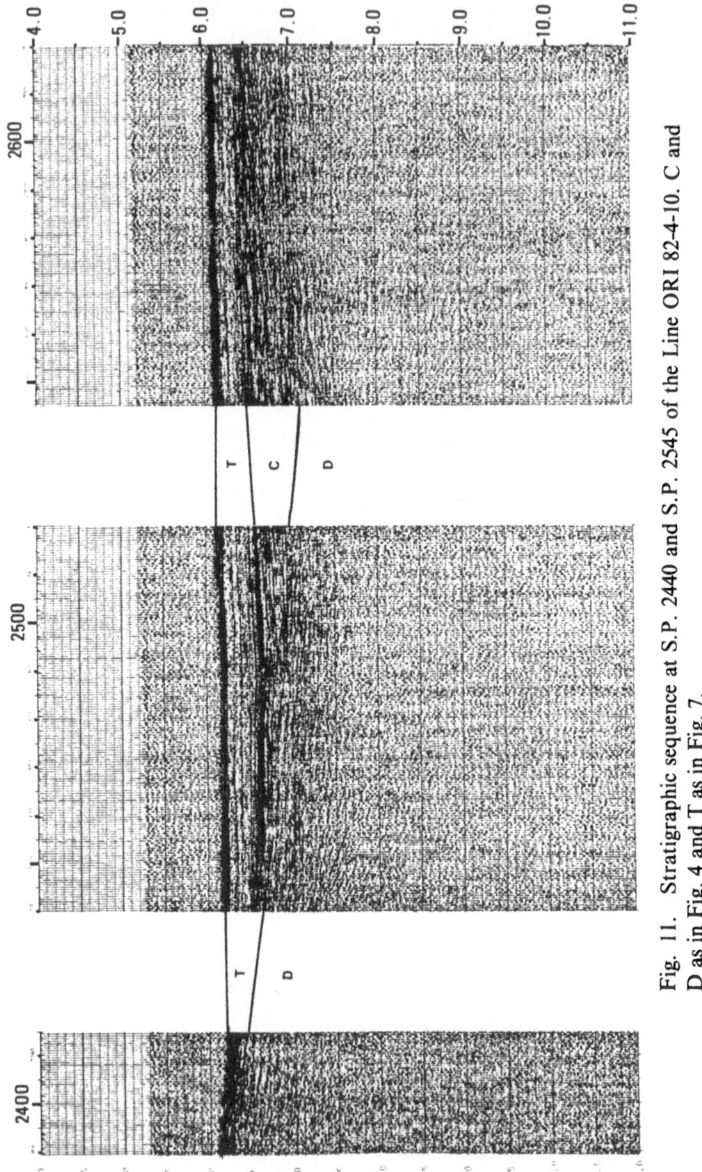

Fig. 11. Stratigraphic sequence at S.P. 2440 and S.P. 2545 of the Line ORI 82-4-10. C and D as in Fig. 4 and T as in Fig. 7.

trench slope (Figs. 11 and 14). Unit F, subducting oceanic layer II A, is not identified here beneath the sedimentary mass of the lower inner trench slope, unlike in Line 82-4-15. Unit D is abruptly terminated by a westward dipping thrust fault at S.P. 1835 (Figs. 12 and 14). The fact that Unit D, originally derived from the Amami Plateau, can be traced westward beyond the axis of the Ryukyu Trench leads to the conclusion that Unit D, part of the Amami Plateau crustal block, has been accreted to the inner trench slope.

The crustal structure of the Amami Plateau was reported by SUYEHIRO *et al.* (1982). They recognized four layers in the Amami Plateau by using ocean bottom seismometer (O.B.S.), with P-wave velocities of 1.8 km/s, 2.7 km/s, 3.3 km/s, and 5.5 km/s in increasing order of depth. By correlating with the multichannel seismic profiles, we can correlate these velocities with Units A, B, D, and E, respectively (Fig. 4).

Several kinds of igneous rocks are reported from the crest of the peak at station GDP 11-17, which is located on the southern side of Line 82-4-14 (SHIKI *et al.*, 1975). A single channel seismic profile (Line ORI 82-4-12) from southeast of the multichannel seismic surveyed area runs in an E-W direction across the peak where station GDP 11-17 is located (Fig. 2). We can identify Unit E on the profile, overlain by a veneer of Unit A at the summit of the peak (Fig. 13). Therefore, it is thought that the igneous rocks are dredged from Unit E. The age of hornblende bearing tonalite among dredged rocks was determined to be 75.1 Ma by using the K-Ar method (MATSUDA *et al.*, 1975). This evidence suggests that Unit E consists mainly of igneous rocks of Late Cretaceous age. At station GDP-11-9 on the western peaks *Nummulites boninensis* of Middle Eocene age was collected together with basaltic rocks. Unit D is apparently exposed on the slope of these peaks as mentioned earlier, and it is believed that *Nummulites boninensis* was dredged from Unit D. Unit C is correlated with strata in the Kita Daito Basin, refered to as Unit C′ (Tokuyama in preparation). Based on DSDP result Site 296 (KARIG *et al.*, 1975), Unit C′ is underlain by Middle Eocene volcaniclastic distal sediments supplied from the Kyushu Palau Ridge and is overlain by Early to Late Oligocene volcaniclastic distal sediments also supplied from the Kyushu Palau Ridge. The available data indicate that Unit D is Middle Eocene to Paleocene in age deposited mainly in a shallow sea environment and Unit C comprises resedimented strata supplied from Unit E and Unit D after the Middle Eocene volcanic epoch on the Kyushu Palau Ridge. Units A and B are pelagic sediments, probably younger than Middle Oligocene in age.

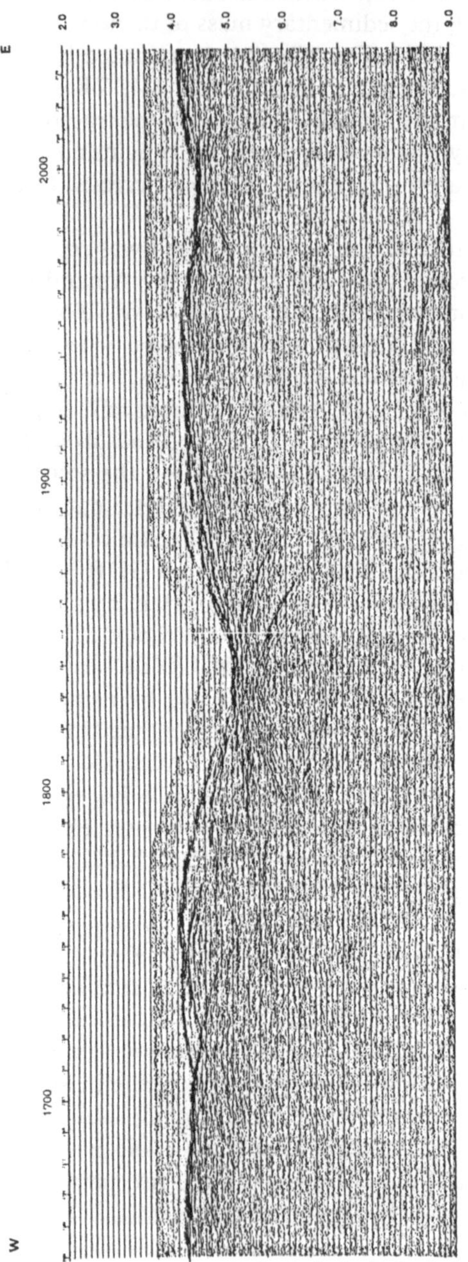

Fig. 12. Seismic profile, Line ORI 82-4-10, from the inner slope of the Ryukyu Trench.

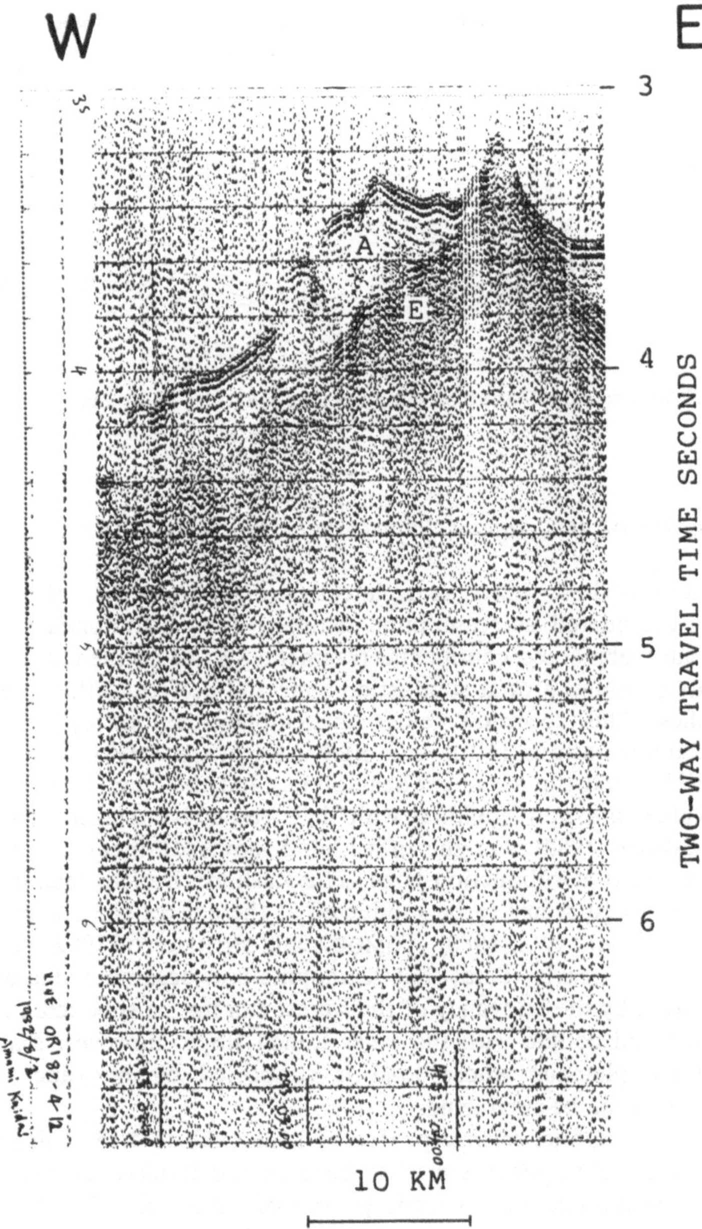

Fig. 13. Seismic profiles, Line ORI 82-4-12, from the Amami Plateau.

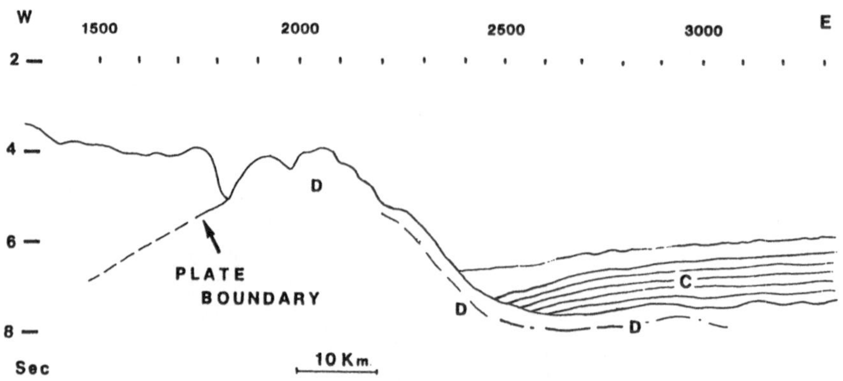

Fig. 14. Schematic crustal structure along the Line ORI 82-4-10. C and D as in Fig. 4.

3. Collision of the Amami Plateau with the Ryukyu Arc

Numerous rises, ridges, and plateaus are found in the oceanic plate. They are considered as seamounts, remnant arcs, abandoned spreading centers, and submerged continental fragments. Some of them collide with continents or island arcs at subduction zone and accrete to the inner slope of trenches. This accretion causes the emplacement of ophiolites and orogenic deformations (NUR and BEN ABRAHAM, 1983). The Marcus Necker Ridge and the Magellan Seamounts, and the Juan Fernandez and Nazca Ridges are now colliding with their respective trenches (NUR and BEN ABRAHAM, 1983). On the other hand, some of them are presently being consumed at subduction zone. The Kashima Daiichi seamount was broken down and seems to be subducted at the axis of the Japan Trench (MOGI and NISHIZAWA, 1980). TOMODA and FUJIMOTO (1982) proposed, on the basis of gravity data, that seamounts and oceanic plateaus which are supported by the buoyancy of their crust can not be subducted beneath island arcs and collide with the inner trench slope, however, seamounts and oceanic plateaus whose load is supported by buoyancy of the asthenosphere can be subducted.

The multichannel seismic profile of Line ORI 82-4-15 indicates subduction of the Philippine Sea plate beneath the Ryukyu Island Arc. In contrast, the multichannel seismic profile of Line ORI 82-4-10 suggests collision and accretion of the Amami Plateau with the Ryukyu Island Arc. According to Tomoda and Fujimoto, the Amami Plateau is assumed to be an unsubductable edifice due to the large thickness of its crust, although the deep crustal structure of the Amami Plateau is uncertain. The plate boundary between the Ryukyu Island Arc and the Philippine Sea plate is thus drawn landward of the Ryukyu Trench axis where the Amami Plateau is in collision with the Ryukyu Island Arc, which is consistent with the seismic reflection evidence (Fig. 15). A similar tectonic relationship was reported for the northern part of the Izu Peninsula where the Izu-Bonin Ridge has collided with the Honshu Arc (SUGIMURA and UYEDA, 1973). If the main body of the Amami Plateau migrates toward the northeast along the trench axis, the accreted part of the Amami Plateau will be separated from the main body and left as part of the inner trench slope.

Shallow earthquakes have been reported from the Amami Plateau (SUYEHIRO *et al.*, 1982). The distribution of epicenters seems to be align-

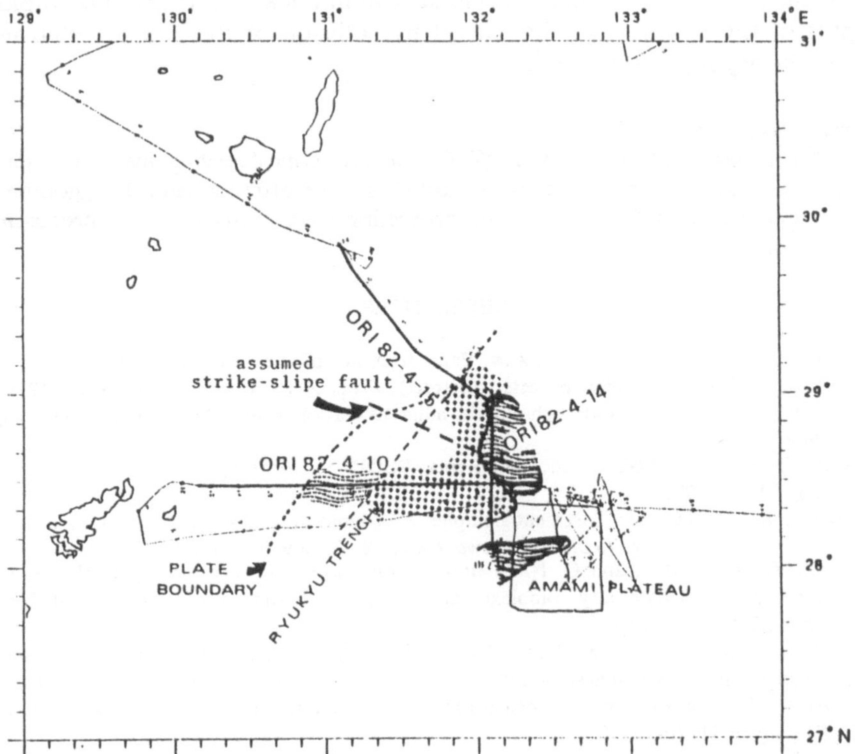

Fig. 15. Plate boundary between the Ryukyu Island arc and the Philippine Sea plate at the west margin of the Amami Plateau and distribution of Units C and D. ⁘ Unit C, ≋ Unit D.

ed in an approximately N 60°W direction and is equivalent to a southeastern extension of the Tokara depression in the Ryukyu Island arc (KONISHI, 1963). This trend appears to be consistent with the direction of subduction of the Philippine Sea plate beneath the Ryukyu Island Arc (SENO, 1977). Although focal mechanisms of shallow earthquakes are ambiguous, the tectonic relationship of the Amami Plateau and the Ryukyu Trench may include strike-slip faulting, whose trend is approximately parallel to the direction of subduction of the Philippine plate (Fig. 15). Therefore, the shallow earthquakes may be caused by collision of the Amami Plateau with the Ryukyu Island Arc.

Kikai Island is located along the western extension of Line 82-4-10. The island is being uplifted at the rate of approximately 1.5–1.8 m/1000 yr. (KONISHI *et al.*, 1974). The Shimajiri Formation, exposed on the island, is more than 5 km thick and overlays pre-Tertiary basement which forms a sedimentary basin (Kimura personal communication, 1982). The rapid uplift of the island might be caused by collision of the Amami Plateau with the Ryukyu Island Arc.

Acknowledgements

We are grateful to professor T. W. C. Hilde for critical reading this manuscript. We would like to thank professor K. Kobayashi for offering helpful suggestion. Computer programs for seismic data processing were partly made by professor M. Saito.

REFERENCES

HILDE, T. W. C. and G. F. SHARMAN, Fault structure of the decending plate and its influence of the subduction process (Abstract), *Proc. AGU Meeting*, Dec., 1978, 1978.

IWABUCHI, Y., Topography of trenches in the adjacent seas of Japan, *Mar. Geod.*, **4**, 121–140, 1980.

KARIG, D. E., J. C. INGLE *et al.*, Init. Repts. DSDP, 31, Washington, U. S. Govt. Printing Office, 1975.

KONISHI, K., Pre-Miocene basement complex of Okinawa, and the tectonic belts of the Ryukyu Islands, *Sci. Rept. Kanazawa Univ.*, **8**, 2, 569–602, 1963.

KONISHI, K., A. OMURA, and O. NAKAMICHI, Radiometric coral ages and sealevel records from the Late Quaternary complexes of the Ryukyu Islands, Proc. 2 Int. Coral Reef Symp., **2**, 593–613, 1974.

MATSUDA, J., K. SAITO, and S. ZASHU, K-Ar ages and Sr isotopic study of the igneous rock fragments in the manganese nodules dredged from the Western Philippine Sea and the Amami Plateau, Preprint, Geological Problems in the Philippine Sea, *Annu. Meet. Geol. Soc. Japan*, 99–101, 1975.

MOGI, A. and K. NISHIZAWA, Breakdown of a seamount on the slope of the Japan Trench, *Proceedings of the Japan Academy*, **56**, Series B, 257–262, 1980.

NUR, A. and Z. BEN-AVRAHAM, Break-up and accretion Tectonics, *Accretion Tectonics in the Circum-Pacific Regions*, eds. by M. Hashimoto and S. Uyeda, pp. 3–18, Terra

Scientific Publishing Company, Tokyo, 1983.

SENO, T., The instantaneous rotation vector of the Philippine Sea plate relative to the Eurasian plate, *Tectonophysics*, **42**, 209–226, 1977.

SHIKI, T., H. AOKI, and Y. MISAWA, Geological results of recent studies of the Philippine Sea, *Mar. Sci.*, **7**, 454–460, 1975.

SUGIMURA, A. and S. UYEDA, *Island Arcs: Japan and Its Environments*, Elsevier Scientific Publishing Company, Amsterdam, 1–247, 1973.

SUYEHIRO, K., A. NISHIZAWA, and H. SHIMIZU, Seismic activity near the axis of the Ryukyu Trench by using O.B.S. (Abstract), *Seismol. Soc. Japan*, **2**, 47, 1982.

TOKUYAMA, H., H. KAGAMI, Y. KONG, C. IGARASHI, and N. NASU, The recent results from the multi-channel seismic profiling across the Japan Trench and Nanki Trough, Modern Sea Bottom Research II, *Proceedings of Symposium-2*, 73–80, 1982.

TOMODA, Y. and H. FUJIMOTO, Thickness of eithosphere at seamount and trench, *Kagaku*, **52**, 509–516, 1982.

Formation of Active Ocean Margins, edited by N. Nasu *et al.*, pp. 875–890.

CORAL REEFS AND PRESENT-DAY COLLISION-SUBDUCTION TECTONICS

Kenji KONISHI

Department of Earth Sciences, Kanazawa University, Kanazawa, Japan

Abstract. Active collision-subduction tectonics in and around the Ryukyu Island Arc-Trench System can be envisaged from the stratigraphy of the Quaternary intact fringing reefs now subaerially exposed. The absolute ages of the three high stands of paleo-sea levels (5, 125 and 210 ka) were established by means of radiometric dating (uranium disequilibrium and radiocarbon methods) and oxygen deglaciation measurement, besides conventional facies mapping. These ages, together with the present elevations of the reefs have provided indications of varied rates of neotectonic uplift within the surveyed region during the last 125 ka. Spatial variations of these rates afford clues to the present-day tectonic configuration of lithosphere at plate boundaries. While the eastern Taiwan boundary, represented by the Taitung Valley Faults, displays past, yet still slipping accreted blocks, where obduction prevails, Kikai of the Central Ryukyu suggests a current situation of a remnant arc (Amami Plateau) being forced to collide against the Ryukyu Arc. This can be contrasted with Hateruma, which represents the site of subduction without collision. Across the active Okinawa Trough, Uotsuri demonstrates an ongoing tilt, which seems to be ambiguous at Yonaguni. Kita- and Minami-daito, both elevated and northerly moving atolls on the Daito Ridge, are uplifting extremely slowly as part of an off-trench bulge, although they will also eventually collide against the trench. It is hypothesized that such exceptionally high rate of uplift as exemplified by eastern Taiwan (Coastal Range), New Hebrides (Santo and Malekula), New Guinea (Huon Peninsula) and Central Ryukyu (Kikai) may be causally related to the site of active collision.

1. *Raised Coral Reefs as Neotectonic Monitors*

As a unique tide gauge, thermometer and geochronometer, a coral reef can monitor plate kinetics by recording rates of neotectonic displacement. In contrast to reefs on the slowly subsiding mid-oceanic plate, those close to the converging plate boundaries whether subducting or obducting, are vertically uplifted with varied rates due to variation of local tectonic framework. They may occur on and therefore represent the frontal arc of an island-arc trench system, on migrating and subsequently colliding seamounts, on aseismic ridges, and on deepsea rises and plateaux, and even on an ocean lithospheric bulge next to a trench.

The ultimate fate of such reef-capping seamounts or plateaux on moving lithosphere has been traced into orogenic belts, including the Japanese Islands (e.g., KANMERA and NISHI, 1983), as reefoid allochthonous terranes to substantiate evidence for accretion-collision tectonism in the geologic past. In order to obtain more, though somewhat indirect, insight into the contemporary process of collision-subduction tectonism, the Late Quaternary history of intact fringing reefs now subaerially exposed can be examined. Spatial variations in the rate of uplift afford clues to the present-day tectonic configuration of lithospheres at plate boundaries.

Such examples are provided by the raised reefs in and around the Ryukyu Island Arc-Trench System, especially at its western end, *i.e.* in the Taitung Coastal Range of the eastern Taiwan (Formosa) and at the middle segment, Central Ryukyu, both of which appear to designate sites of interarc collision in action. While the eastern Taiwanese example proves to be a past, yet still slipping accreted block, where obduction prevails (KONISHI *et al.*, 1968; KONISHI and SUDO, 1972; KARIG, 1973; PENG *et al.*, 1977; WU, 1978), a part of the Central Ryukyu suggests a situation currently approaching accretion *via* collision (Fig. 1).

The primary purpose of this paper is three-fold; (1) to summarize the result of our neotectonic studies of the raised reefs in this region with special emphasis on supplementation of our previous paper (KONISHI *et al.*, 1974), (2) to present evidence to support the ongoing collision tectonism in the Central Ryukyu area, and (3) to speculate a close relation between high rate of vertical uplift and colliding lithospheres.

Both sampling and dating procedures for the present work are essentially the same as those already described (e.g., KONISHI *et al.*, 1974). Only autochthonous hermatypic corals retaining original mineralogy have been selected for radiometric dating, and molluscs were used solely for oxygen isotopic measurement of deglaciation analysis. As the dating techni-

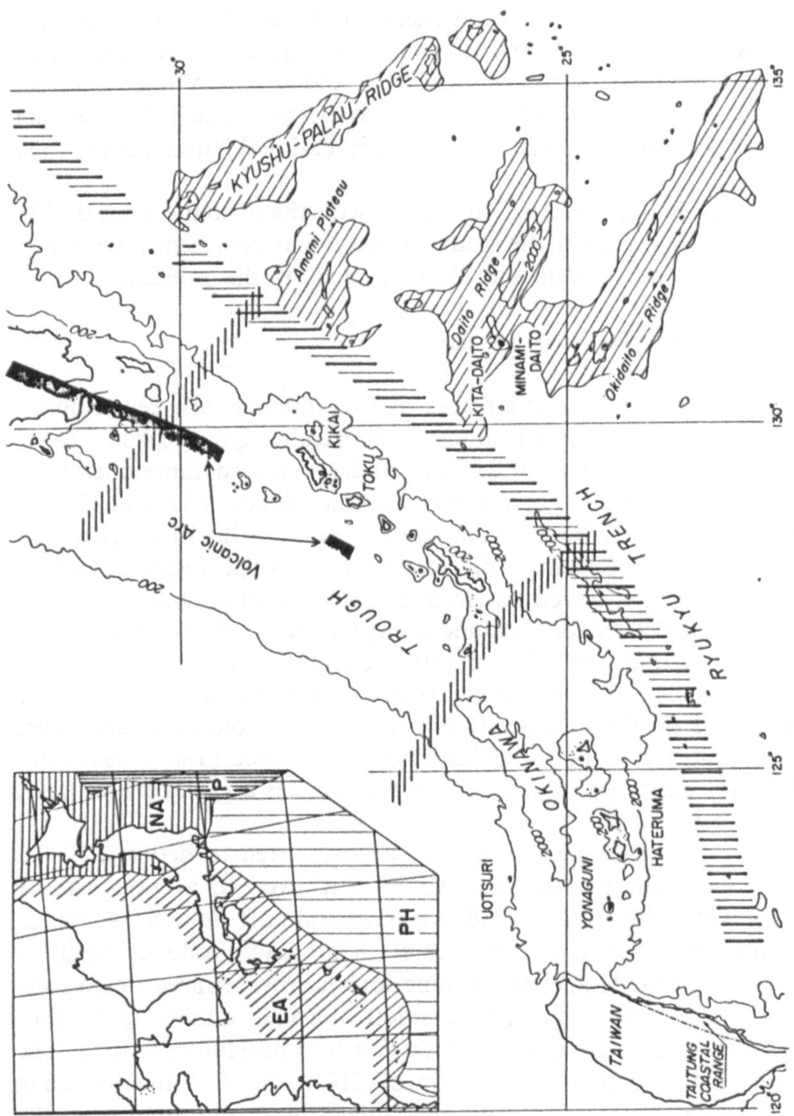

Fig. 1. Map showing location of the studied islands and the generalized tectonic framework of the Ryukyu Arc-Trench System and its adjacency. The accompanied map at upper left illustrates the configuration of four plates around the Japanese Islands: EA, Eurasian; PH, Philippine Sea; P, Pacific and NA, North American.

ques are uranium-disequilibrium and radiocarbon methods, our analyses go back only to the Middle Pleistocene. A new "non-destructive" gamma-spectrometric method ($^{226}Ra/^{238}U$ dating), which is also principally based on uranium-disequilibrium, has been successfully introduced into the present work (KOMURA *et al.*, 1978). It needs a germanium detector Ge (Li), a low energy photon spectrometer (LEPS), and a comminuted sample, but no wet chemical steps. This method is so efficient that it has become a standard procedure to be run before a conventional "destructive" one is employed.

The three paleo-sea level datum planes with the absolute ages, 5, 125 and 210 ka were selected for the study, because they represent times when the sea level has stood high and close to that of the present.

2. *Kikai, the Rapidly Uplifting Island*

Since non-reefy limestone terraces of two interstadials, centering on 40 and 60 ka, were confirmed almost two decades ago (KONISHI, 1967; KONISHI *et al.*, 1970), four similar interstadial terraced limestones have been identified between two coralliferous datum planes, 5 ka and 125 ka, the Postglacial and Last Interglacial, respectively, on this island (KONISHI *et al.*, 1974) (Fig. 2). Recently, limestone terrace slope facies of both 6 ka and 8 ka have been traced downward to depths lower than previously reported (TSUJI, 1979MS for 6 ka and 8 ka; OMURA, 1983 for 8 ka). Also it has been found that the 125 ka coralliferous limestone forming the top of the island yields not only the corals of 125 ka, but also those of the much older ages (probably older than 250 ka). Such old coral ages were also dated from a part of the 6 ka limestone terrace (TSUJI, 1979MS). These old dates indicate that some limestone terraces are, in part, at least, of erosional origin.

Here, the Last Interglacial reef has been uplifted to between 170 m and 220 m above the present sea level, which gives an average rate of uplift of 1600 mm/1000 years or 1600 *B* (One Bubnoff unit, *B*, is defined as 1 micron/year or 1 mm/1000 years). More rapid rate of uplift is developed through the Holocene, forming four narrow terraced raised reefs (OTA *et al.*, 1978), the maximum elevation of which attains 13 m, thus suggesting an average rate of uplift 2100 *B* after mid-Holocene time. The multihole coring of a total penetration of 210 m in 19 boreholes along a transect across both the present-day reef and three terraces of the Holocene raised reef has revealed that the island has been seismo-tectonically uplifted as the result of large earthquakes which may have recurred with a maximum period of 1 to 1.5 ka (KONISHI *et al.*, 1983). The last shallow

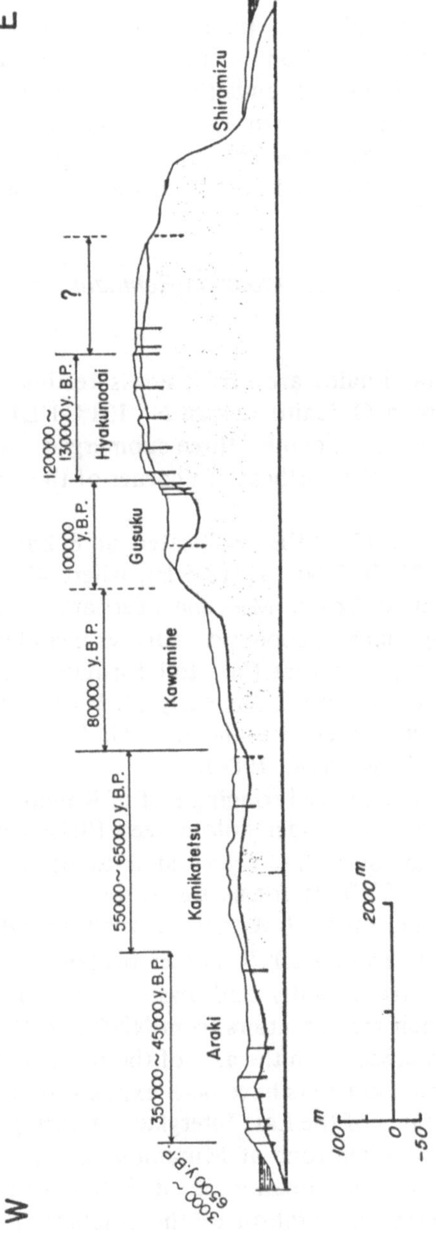

Fig. 2. Structural cross section along 28°18'N at Kikai: four limestone terraces correlative with intra-Wisconsin (-Würmian) interstadials have been identified between the Last Interglacial (120–130 ka) and Postglacial (3–6.5 ka) raised reefs. See text for further details.

large earthquake off Kikai occurred in 1911 which recorded the magnitude of 8.2.

It should be noted that this Last Interglacial reef of Kikai happens to record the maximum elevation of any of the outcrops of the Pleistocene raised reefs throughout the Ryukyu Islands (Fig. 1 of KONISHI *et al.*, 1974). Indeed, there is no doubt that, except for Taiwan, Kikai has had the fastest rate of uplift among the islands of the Ryukyu Arc and its adjacent areas, as witnessed by the maximum altitudes of both the Postglacial and Last Interglacial raised reefs.

3. Kita- and Minami-Daito, Raised Atolls on Remnant Arc at Off-Trench Bulge

Both Kita- and Minami-daito are raised atolls resting on the Daito Ridge, which, together with Okidaito (24°28'N: 131°11'E) to the south and Amami Plateau to the north, exhibit three submerged "aseismic" remnant arcs running parallel, at the northeastern corner of the western Philippine Sea.

Minamidaito (25°50'N: 131°15'E) is located at 8 km southwest of the celebrated Kitadaito (25°56.5'N: 131°18.5'E), where 431 m of reefoid column was drilled down to Lower Miocene (Tertiary *e*) in 1936. The result of submarine geological and geophysical surveys including two DSDP cruises (Leg 31 and 58) suggests that the atoll formation started in the upper Eocene (?) and continued up to the early (?) Pleistocene, with two major hiatuses in the Lower Oligocene and the Middle Miocene (KLEIN, KOBAYASHI *et al.*, 1980; KINOSHITA, 1980).

On Minamidaito, an intact raised fringing reef of Kaigunbo Limestone, formed as a thin veneer over the older dolomitized Pleistocene reef complexes (Minamidaito Dolostone), has been dated to be of the Last Interglacial (KONISHI *et al.*, 1978). It forms the Lowest and youngest terrace, outcropping between mean sea level and 11 m above, which provides a very slow rate of uplift, around 50 *B*. Direct evidence to support the neotectonic uplift comes from faults and fissures cutting through the dolomitized basement. Their trend scatters from N20° E to N50° E with the mode of N35° E, which coincides with the axis of the Ryukyu Trench to the west. No mid-Holocene (ca. 5ka) reefs have been exposed on the island.

A few erosional remnants of the Last Interglacial fringing reef resembling those of the Kaigunbo Limestone at Minamidaito were lately found at elevations similar to those of Minamidaito in Kitadaito (Kohara, personal comm.). The preferred orientation of the fractures passing across the Kitadaito Dolostone also agrees with that of the trench axis. Any causal relation between these systems of faulting at Minami- and Kita-daito and horst graben structures on the outer trench slope (LUDWIG *et al.*, 1966; HONZA, 1980) have not been examined.

If the obtained rate of uplift is extrapolated before the Last Interglacial, the maximum elevation of the atoll crest, 55 m at Minamidaito and 43 m at Kitadaito, may suggest that these atolls have undergone uplift from as early as about 1.0 ma B.P., when the tectonic progression of prolonged steady slow (ca. 50 *B* or less than 100 *B*) subsidence of almost 30 ma was reversed to a slow uplift causing emergence of both atoll and later fringing reef. This change can be attributed to arrival of the atoll-capped Daito Ridge at the off-trench bulge forming the outer ridge of the Philippine Sea plate.

4. *Uotsuri and Yonaguni, Gently Tilting or Tilted Islands across the Okinawa Trough*

Across the active Okinawa Trough, Uotsuri (25°46'N: 123°31'E) witnesses an ongoing tilt to the north (KONISHI *et al.*, 1979), which, however, appears to be obscure at Yonaguni (24°27'N: 123°00'E). In Yonaguni, the lowest marine terrace composed of a raised coral reef crops out between strandline and 8 m above sea level. This raised reef was assigned to Holocene by previous workers, but is currently determined radiometrically to be of the Last Interglacial time (Takahashi, personal comm.). Although all the four reef-constructed terraces of Pleistocene tend to tilt extremely gently to the south on this island, the present elevation of the Last Interglacial reef apparently does not substantiate active vertical displacement there for the last 125 ka. Should a tilting had occurred, it may have predated the Last Interglacial time.

5. *Hateruma; Why It Uplifts More Slowly than Kikai*

Hateruma (24°03'N: 123°47'E) is located only 90 km north of the trench axis. The island is geologically similar to Kikai, but separation of the terraced reefs concentrically capping the Late Tertiary clastic core is rather obscure, because of its slow uplift, when compared with Kikai. Radiometric dates have been obtained from all the terraces ranging from 80 ka to 210 ka (KONISHI, 1980; OMURA, 1983). The Last Interglacial reef has been identified between 5 and 15 m above the present sea level by Konishi, but between 20 and 42 m by Omura following OTA *et al.* The maximum rate of uplift, which is calculated by the figures from Omura, varies from 100 *B* in the western part of the island to 300 *B* in the northeast-east part, indicating that the island has been tilted downward towards the west to southwest. So far this is the only island in the Ryukyu Arc where the raised reefs of 210 ka are dated. They are correlative with the preceding Interglacial Stage (Mindel/Riss) including the Deep Sea Core Oxygen Isotope Stage 7 of ca. 220 ka. These coral ages have been obtained from the oldest terraced reefs. The radiometric ages of the dated corals are given in Table 1. These two interglacial reefs have been faulted. Nevertheless, these slower rates of uplift and the lack of a mid-Holocene raised reef, as compared to Kikai, imply that the island has been rising very slowly with a rate of almost one-tenth of that of Kikai.

Table 1. Contrast in geotectonic features between the central and South Ryukyus (data in part from HONZA, 1977).

Morpho-tectonic Block	UPLIFT RATE (B)	ACCRETIONARY PRISM	VOLCANIC ARC	BACKARC SPREADING	GREAT EARTHQUAKE	DIP OF BENIOFF-WADATI ZONE (°)	
CENTRAL RYUKYU	1500 ~ 1700	Thick	Active	Inactive	Active	25 ~ 35	COMPRESSIVE
SOUTH RYUKYU	60 ~ 240 (100 ~ 300)	Thin	Inactive	Active	Inactive	55 ~ 65 or (steeper)	TENSILE

Now, the question that should be raised is why do Kikai and Hateruma, both of which share the tectonic belt in equal distance from the trench axis, differ so distinctly in the rate of neotectonic uplift? Among a few plausible explanations, I (1980) have previously speculated a causal relationship between the mode of subduction and the rate of uplift of the raised coral reefs, based on the "Chilean *vs.* Marianan model" proposed by UYEDA and KANAMORI (1979). According to their classification, the Ryukyu Arc was placed as an intermediate between the two Chilean and Marianan end members.

Between the two morphotectonic blocks, the Central and South Ryukyus, each of which is represented by Kikai and Hateruma respectively, the geotectonic features are so contrasted, as summarized in Table 1, that an analogy between the Central *vs.* South Ryukyus and Chilean *vs.* Marianan was thought to be tenable (KONISHI, 1980). The seismic records on which my interpretation was founded came from those by Carr *et al.* (1973), who argued for discontinuities of the Wadati-Benioff zone even within a single island arc system and tried to segment the arc into several morphotectonic subarc units, both ends of which may be defined by deep-seated major tectonic boundaries. Central Ryukyu is separated from North Ryukyu by the Tokara Channel (=Tokara Structural Strait) to the northeast and from South Ryukyu by the Miyako Depression (=Kerama Gap) to the southwest (Figs. 1 and 4). Carr *et al.* appeared to have recognized more morphotectonic blocks in the Ryukyu Arc, based on the interpretation of the seismic records.

Recently, the seismicity data of the Ryukyu Arc-Trench System were vigorously reexamined by SHIONO (1980). He claims that there appears to be no significant difference in the attitude of the Wadati-Benioff Zone between the South and Central Ryukyus as CARR *et al.* demonstrated, but a difference may occur between the North and Central Ryukyus across the Tokara Channel.

My next alternative proposal in 1981 was that the Quaternary reefs of Kikai happen to be situated on the top of thick pile of the Plio-Miocene (though the earliest Pleistocene might be inclusive in part) fore-arc deposits (Shimajiri Formation), which have been compressed laterally and squeezed upwards, in association with shallow earthquakes due to collision of the remnant arc, the Amami Plateau, against the Ryukyu Arc. The supporting evidence for this hypothesis comes from both geomagnetic and multi-channel seismic profiling across this area. MIYAZAKI *et al.* (1976) who ran geomagnetic total field measurements have demonstrated that the eastern limit of geomagnetically quiet zone shows irregularity, symbolized by the several tens of km of invasion of the oceanic lithosphere beyond the trench axis, can be interpreted as the result of collision rather than subduction of the Amami Plateau-ridden slab at the trench. The multichannel seismic profiling run by TOKUYAMA and KONG (1983; TOKUYAMA *et al.*, this volume) also indicates the collision of Amami Plateau against the Ryukyu Arc without subduction.

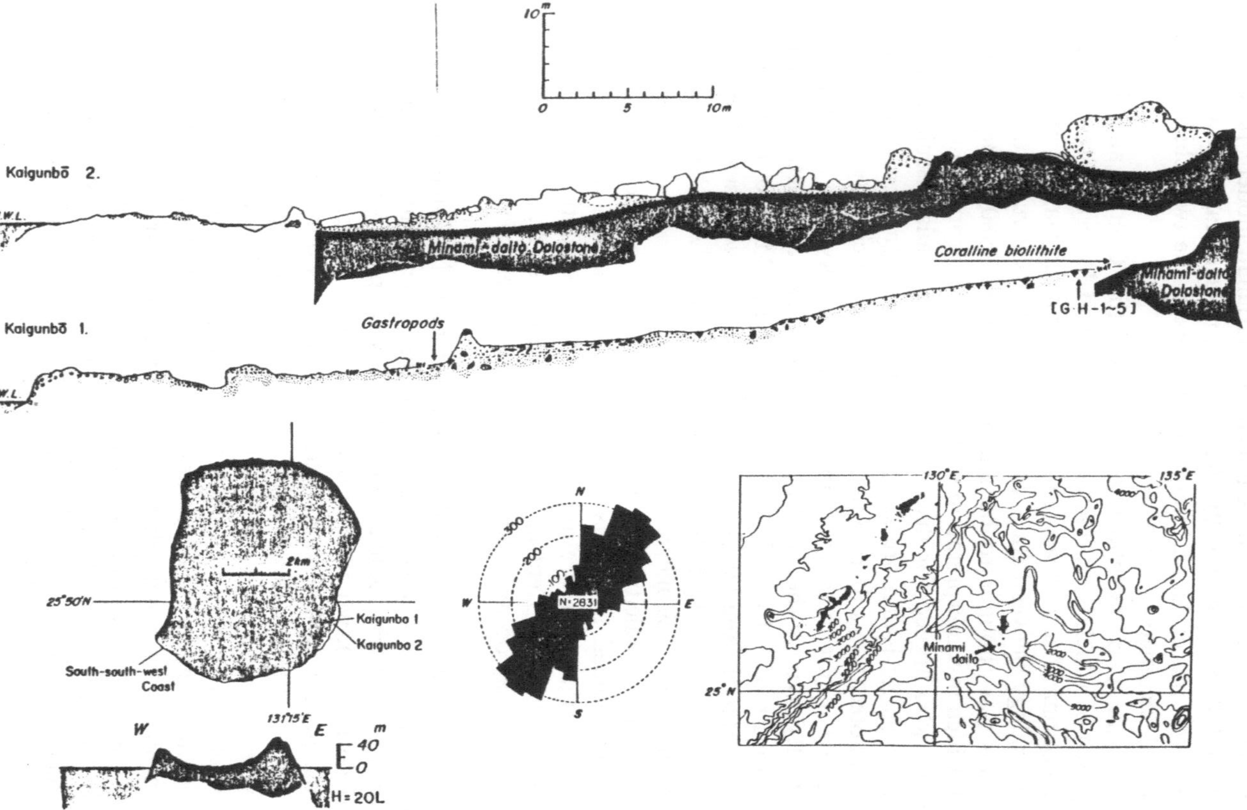

Fig. 3. Maps and cross-sections that show the location of measured transects and dated corals on them at Minamidaito. Rosediagram at the central bottom indicates percentage of 2831 fissures and lineaments that have been observed in Minamidaito Dolostone (KONISHI *et al.*, 1978).

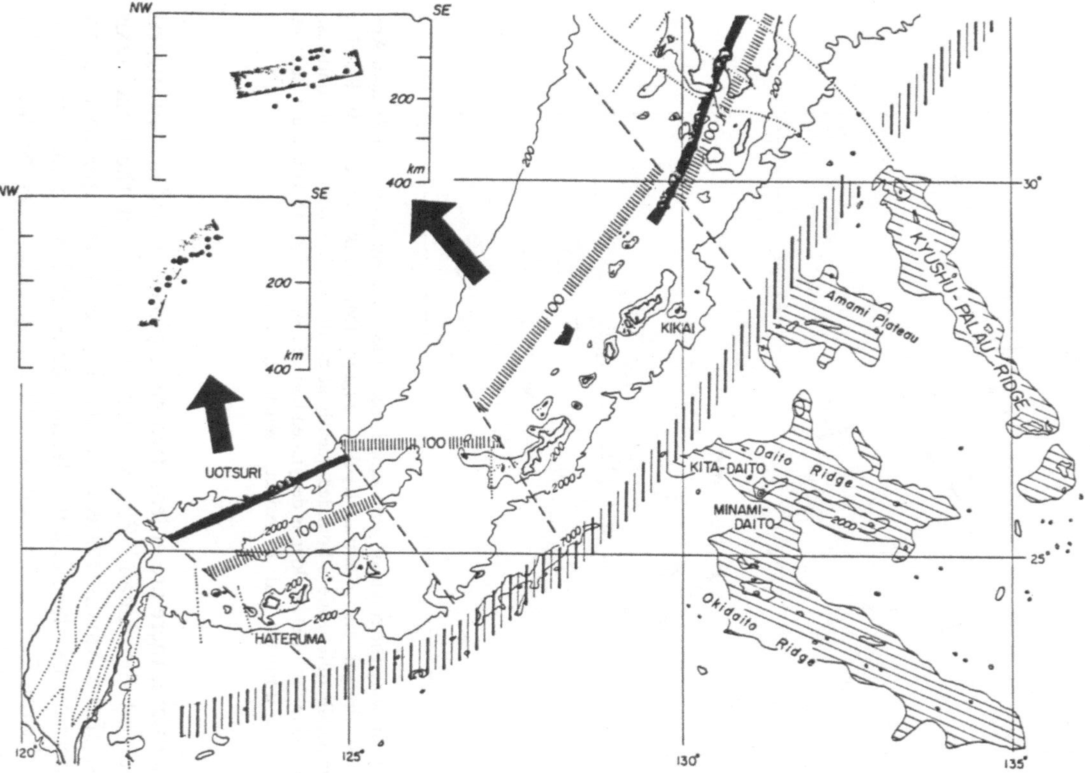

Fig. 4. Map showing the five "segments" (=morphotectonic blocks) and isobath (100 or 200 km deep) of deep seismic zones of the Ryukyu Arc. Two vertical sections of the earthquake foci at upper left indicates the difference in the attitude of the Wadati-Benioff zone between the Central and South Ryukyu (CARR *et al.*, 1973). Some fault systems are added to the original illustration.

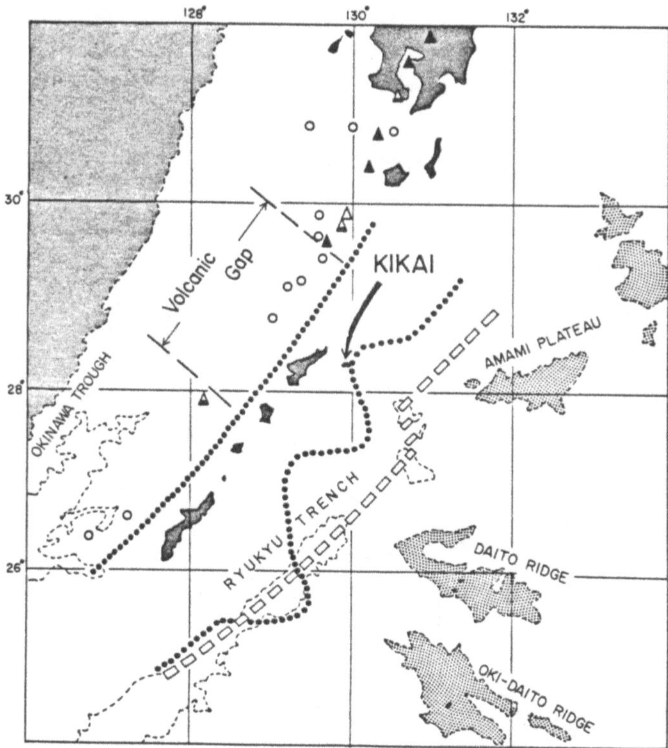

Fig. 5. Map showing the result of geomagnetic total measurement across the Ryukyu Arc-Trench System at the Central Ryukyu region (MIYAZAKI *et al.*, 1978). The geomagnetically quiet zone is bounded linearly at northwest, but sinuously at southeast. "Volcanic Gap" ca. 100 km long is designated between Suwanose and Yokoate; volcanoes, ▲ active, ▲ dormant and △ extinct; trench axis, ▭▭▭.

Furthermore, there are at least two more pieces of evidence to suggest the collision of the Amami Plateau. One is a jump of the trench axis to the west of the Amami Plateau, where the trench axis appears to show either truncation by the Amami Plateau, or is broken up into pieces to align en echelon. This anomalous trench topography may be caused by the colliding tectonics. The other is a volcanic gap about 100 km long between Yokoate and Suwanose which may also be due to disruption of the subducting process probably associated with colliding lithosphere.

6. *Rate of Neotectonic Uplift and Collision-Subduction Tectonism*

If the high rate of uplift at Kikai can be ascribed to the collision of the Amami Plateau, the neotectonic setting under which any raised reefs display anomalous rate of uplifting needs to be tested with a working hypothesis to causally relate the lithospheric collision and very rapid rate of vertical uplift.

Among the several raised coral reefs where high rate of uplift has been documented, those at Santo and Malekula of the New Hebrides provide two of the best examples. These two islands capped with both Pleistocene and Holocene raised reefs are also located most trenchwards in the frontal arc of the New Hebrides Arc. Here, the New Hebrides Trench appears to become shallow, suggesting the eastward collision of the aseismic ridge, the d'Entrecasteaux Fracture Zone against the New Hebrides Arc. According to TAYLOR *et al.* (1980), the subduction of this aseismic ridge across the New Hebrides Trench "has controlled segmentation of the central New Hebrides Arc in terms of both seismicity and deformation on both 10^5 and 10^1 years time scales". The very rapid rate of uplift recorded from Santo and its satelite islet Araki, and Malekula is given in Fig. 6.

Another example may be cited from the Huon Peninsula of New Guinea (CHAPPEL, 1974; BLOOM *et al.*, 1974), where the mid-Holocene (ca. 6 ka) reef has been uplifted up to an elevation of 12 to 15 m above present sea level. The maximum rate of uplift, based on the 125 ka datum and 210 ka datum is 3500 *B*. This is the collision site of two arcs. As mentioned above, the interarc collision at the eastern Taiwan also results in the extraordinarily rapid rate of uplift.

7. *Conclusion*

The salient points of this paper may be summarized as follows.

1. Variation in the present altitude of the Last Interglacial (125 ka; Riss/Würm) reefs on and off the Ryukyu Arc is explained in terms of the neotectonic interplay, close to the plate convergence, where the average rate of relative horizontal movement is 40,000 to 60,000 *B*.

2. The distinct difference in the average rate of uplift since 125 ka between the two most trenchward islands of the frontal arc, Kikai and Hateruma, can be ascribed to the effect of collision of the Amami Plateau against the Ryukyu Arc, as independently supported from geomorphological, geomagnetic and seismic profiling surveys.

3. Since the rate of uplift tends to increase systematically from atoll

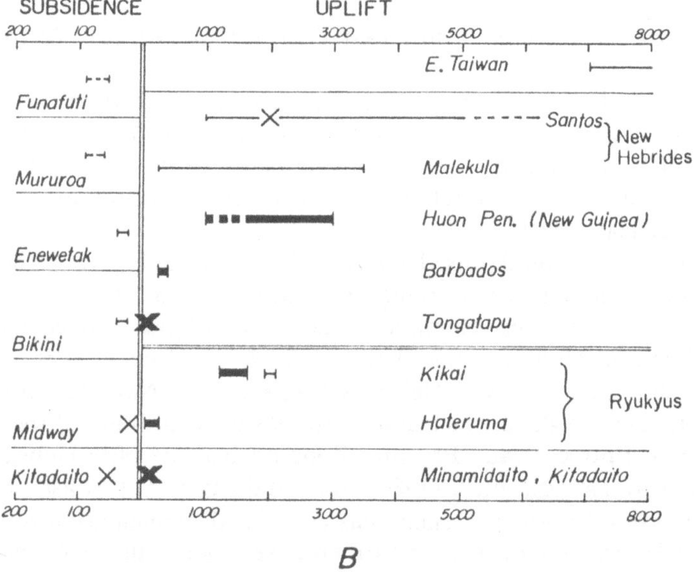

Fig. 6. Diagram showing the average rate of uplift (in Bubnoff unit, *B*) at raised reefs in contrast between the sites of collision (E. Taiwan, New Hebrides, E. New Guinea and Kikai) and those of subducting without collision (Barbados, Tongatapu and Hateruma) as well as emerged atolls (Minami- and Kita-daito). While thick bar indicates the data from the Last Interglacial and other Pleistocene dates, thin line represents those from Holocene dates. Very slow rates of subsidence at atolls and an emerged atoll (Kitadaito) are also illustrated at left of the diagram; these averaged rates of subsidence were determined from the subsurface core samples dated either radiometrically or biostratigraphically.

and mid-oceanic island to the site of collision via the normal subducting arc-trench system, it is hypothesized that any exceptionally high rate of uplift may be related to the site of active collision, *i.e.* it could be a manifestation of the colliding process.

REFERENCES

BLOOM, A. L., W. S. BROECKER, J. CHAPPELL, R. K. MATTHEWS, and K. J. MESOLELLA, Quaternary sea level fluctuations on a tectonic coast: new Th^{230}/U^{234} dates from the Huon Peninsula, New Guinea, *Quat. Research*, **4**, 185–205, 1974.

CARR, M. J., R. E. STOIBER, and C. L. DRAKE, Discontinuities in the deep seismic zones and the Japanese Arcs, *Geol. Soc. Amer. Bull.*, **84**, 2917–2930, 1983.

CHAPPELL, J., Upper mantle rheology in a tectonic region: evidence from New Guinea, *J. Geophys. Res.*, **79**, 390–398, 1974.

HONZA, E., Arc tectonism of the Ryukyu Islands and its difference between the northern and southern areas, *Kaiyo Kagaku (Marine Science)*, **9**, 607–611, 1977 (in Japanese).

HONZA, E., Pre-site survey of the Japan Trench Transect, Deep Sea Drilling Project. *Initial Repts. Deep Sea Drilling Project*, **56-57** (1), 449–458, 1980.

KANMERA, K. and H. NISHI, Accreted oceanic reef complex in Southwest Japan. in *Accretion Tectonics in the Circum-Pacific Regions*, edited by M. Hashimoto and S. Uyeda, pp. 195–206, TERRAPUB, Tokyo, 1983.

KARIG, D. E., Plate convergence between the Philippines and the Ryukyu Islands, *Marine Geol.*, **14**, 153–168, 1973.

KINOSHITA, H., Paleomagnetism of sediment cores from Deep Sea Drilling Project Leg. 58, Philippine Sea, *Initial Repts. Deep Sea Drilling Project*, **58**, 765–775, 1980.

KLEIN, G. DEVRIES, and K. KOBAYASHI, Geological summary of the North Philippine Sea, based on Deep Drilling Project 58 results, *Initial Repts. Deep Sea Drilling Project*, **58**, 951–961, 1980.

KOMURA, K., M. SAKANOUE and K. KONISHI, Non-destructive $^{226}Ra/^{238}U$ dating of Quaternary corals by gamma-spectrometry, *Proc. Japan Acad. Sci.*, **54B** (9), 505–509, 1978.

KONISHI, K., Rate of vertical displacement and dating of reefy limestones in the marginal facies of the Pacific Ocean, *The Quat. Res. (Japan)*, **6** (4), 207–223, 1967 (Japanese with Eng. abstract).

KONISHI, K., A. OMURA, and T. KIMURA, U-234—Th-230 dating of some Late Quaternary coralline limestones from Southern Taiwan (Formosa), *Geol. Palaeont. Southeast Asia (Tokyo)*, **5**, 211–224, 1967.

KONISHI, K., A. OMURA, and O. NAKAMICHI, Radiometric coral ages and sea level records from the Late Quaternary reef complexes of the Ryukyu Islands, *Proc. 2nd Int. Coral Reef Symp.*, **2**, 595–613, 1974.

KONISHI, K., K. KOMURA, and Y. MOTOYA, An Early Wisconsin reef on the Daito Ridge, North Philippine Sea: isotopic evidences, *Proc. Japan Acad.*, **54B** (9), 516–521, 1978.

KONISHI, K., I. OSHIRO, and T. TANAKA, Holocene raised coral reef on Senkaku Islands: an active remnant arc, *Proc. Japan Acad.*, **55B** (7), 335–340, 1979.

KONISHI, K., Diverse plate convergence as deduced from raised coral reefs since the Last Interglacial, *The Quat. Res. (Japan)*, **18** (4), 241–250, 1980 (Japanese with Eng. abstract).

KONISHI, K., Y. TSUJI, T. GOTO, T. TANAKA, and K. FUTAKUCHI, Multihole shallow coring of coral reef—a Holocene example at Kikai—, *Kaiyo Kagaku (Mar. Sci.)*, **15** (3), 154–164, 1983 (in Japanese).

LUDWING, W. J., T. I. EWING, M. EWING, S. MURAUCHI, S. DEN, S. ASANO, H. HOTTA, M. HAYAKAWA, T. ASANUMA, K. ICHIKAWA, and I. NOGUCHI, Sediments and structure of the Japan Trench, *J. Geophys. Res.*, **71**, 2121–2137, 1966.

MIYAZAKI, T., K. TAMAKI, and F. MURAKAMI, Geomagnetic survey, in *Ryukyu Islands (Nansei-Shoto) Arc, GH75-1 and GH75-5 Cruises, Jan.-Feb. and July-Aug., 1975*, edited by E. HONZA, Vol. 6, pp. 52–54, Cruise Rept., Geol. Surv. Japan, 1976.

OMURA, A., Uranium-series age of some solitary corals from the Riukiu Limestone on the Kikai-Jima, Ryukyu Islands, *Trans. Proc. Palaeont. Soc. Japan, N.S.*, **130**, 117–122, 1983.

OMURA, A., New information on radiometric ages of fossil corals from the Hateruma Island, Ryukyu Islands, *The Quat. Res. (Japan)*, **22** (1), 19–22, 1983 (in Japanese).

OTA, Y., H. MACHIDA, N. HORI, K. KONISHI, and A. OMURA, Holocene raised reefs of Kikai-jima (Ryukyu Islands)—an approach to Holocene sea level study, *Geogr. Rev. Japan*, **51** (2), 109–130, 1978 (Japanese with Eng. abstract).

PENG, T. H., Y. H. LI, and E. T. WU, Tectonic uplift rates of the Taiwan Island since the Early Holocene, *Mem. Geol. Soc. China (Taipei, Taiwan)*, **2**, 57–69, 1977.

SHIONO, K., T. MIKUMO, and Y. ISHIKAWA, Tectonics of the Kyushu-Ryukyu Arc as evidenced from seismicity and focal mechanism of shallow to intermediate depth earthquakes, *J. Phys. Earth.*, **28**, 17–43, 1980.

TAYLOR, F. W., B. L. ISACKS, C. JOUANNIC, A. L. BLOOM, and J. DUBOIS, Coseismic and Quaternary vertical tectonic movements, Santo and Malekula Island, New Hebrides Island Arc, *J. Geophys. Res.*, **85**, 5367–5381, 1980.

TOKUYAMA, H. and KONG Y. SAE, Multichannel seismic reflection survey around the Amami Plateau, Prelim. Rept. Hakuho-Maru Cruise KH82-4 (Ocean Research Institute, Univ. Tokyo), pp. 246–264, 1983.

TSUJI, Y., Quaternary history of the western part of Kikaijima, Ryukyu Islands. M. Sc. Thesis, Dept. Earth Sci., Fac. Sci., Kanazawa Univ. No. 51, 172pp, 1979MS.

UYEDA, S. and H. KANAMORI, Back-arc opening and the mode of subduction, *J. Geophys. Res.*, **84**, 1049–1061, 1979.

WU, F., Recent tectonics in Taiwan, *J. Phys. Earth*, suppl., **26**, 265–299, 1978.

YONEKURA, N., Mobile Islands of the Pacific—comparative crustal movements in Late Quaternary, *Kagaku (Science, Tokyo)*, **52** (9), pp. 575–583, 1982 (in Japanese).

CHAPTER 8: PELAGIC SEDIMENTS

Formation of Active Ocean Margins, edited by N. Nasu *et al.*, pp. 893–912.

CHEMICAL COMPOSITIONS OF PELAGIC DEEP-SEA SEDIMENTS — ITS RELATION TO THE FORMATION OF AUTHIGENIC MINERAL PHASE UNDER THE CHEMICAL CONTROL OF SEA WATER

Masato NOHARA and Komi KATO

Geological Survey of Japan, Ibaraki 305, Japan

Abstract. Bulk chemical analyses for the uppermost surface sediments (within 2 cm) show that the siliceous and deep-sea sediments in the northern part of the Central Pacific Basin have similar concentrations of major and minor elements to those reported for other pelagic areas in the Pacific except for active-ridge regions.

The relationship between CaO and P_2O_5 in the sediments is linear points to the probable authigenic formation of Ca-phosphate minerals in the sediments. It is interesting to note that the Si: Al: Ca (Na, K) ratio in the sediments is nearly the same as that of zeolites, especially, phillipsite. Even in calcareous ooze and calcareous-siliceous sediments the Si: Al: Ca ratio indicates that zeolites are probably being formed. These observations suggest that the process of authigenic zeolite formation is already established at the "pre-early diagenetic stage" under the catalysis and chemical control of sea water. That is, the chemical reactions by which the authigenic mineral phases are formed are controlled by sea water which acts as catalyzer.

1. Introduction

Bulk chemical compositions of pelagic sediments sampled from various regions in the Pacific Ocean have been reported by various workers (GOLDBERG and ARRHENIUS, 1958; EL WAKEEL and RILEY, 1961; NOHARA and YOKOTA, 1978; NOHARA, 1979). These results show that chemical composition is remarkably uniform regardless of various sediment-type or area, except for sediments of hydrothermal and volcanogenic origins, and the sediments have simple mineralogical phases.

The aim of this paper is to delineate the geochemical characteristics of pelagic sediments and furthermore to discuss some evidence of chemical reactions catalyzed and controlled by sea water.

The uppermost surface sediments (within 2 cm) studied were collected by box core from the nothern part of the Central Pacific Basin, during R/V Hakurei-Maru Cruise GH77-1 (Table 1; Fig. 1).

The chemical analyses were carried out by the method described by KATO and MITA (1983), after some modifications.

Features of surface sediments

Surface sediments are classified into the following types, based on volumetric ratio of the coarse fraction to the bulk sample and composition of the coarse fraction (ARITA, 1975).

(1) Deep sea clay: coarse fraction<10%, or 10%<non-biogenic materials (consisting mainly of zeolitic and silicified radiolarians).

(2) Siliceous clay: coarse fraction 10–30%, predominantly composed of radiolarian tests.

(3) Calcareous-siliceous clay: coarse fraction 10–30%, composed of foraminiferal and radiolarian tests.

(4) Siliceous ooze: coarse fraction>30%, predominantly composed of radiolarian tests.

(5) Calcereous ooze: coarse fraction>30%, predominantly composed of foraminiferal tests.

Calcareous ooze is distributed over the Magellan Rise area down to a depth of about 4,300 m in water depth where St. 716 is situated. Micropaleontological results show that the fossil foraminifera from most of the calcareous sediments and calcareous-siliceous sediments are of Pleistocene-Holocene age (*Pulleniatina obliquiloculata finalis* and *Sphaeroidinella dehiscenes excavata*), but for G385, where the sediment is calcareous-siliceous clay, the uppermost sediment includes *Globorotalia tumida*, which indicates latest Miocene-early Pliocene age and suggests a remarkably low sedimentation rate and or strong activity of bottom currents. WINTERER *et al.* (1973) report that the sedimentary section above basalt is 1,172 m thick and consists of five stratigraphic units, the uppermost one of which is early Miocene to Quaternary calcareous ooze (200 m), at Site 167, Leg 17, DSDP situated on the Magellan Rise near St. 715 in this survey area.

Siliceous clay is distributed in three separate areas. Two surround the Magellan Rise or in its vicinity and the other area rest occurs in the southeast corner (St. 726) of the survey area.

Siliceous ooze, its coarse fraction of which contains 10–20%, occurs only along the northern edge (Sts. 731 and 732) of one of the northern area of siliceous clay. In general, siliceous sediments (siliceous ooze, siliceous clay and some deep sea clay in which radiolarian remains dominate) overlie deep sea clay. Deep sea clay, which is yellowish brown or brown, is distributed widely on the sea floor at about 5,500–6,000 m depth. The coarse fraction of deep sea clay is composed predominantly of radiolarian remains, including zeolitic minerals.

2. *Results and Discussion*

2.1 *Distribution of major elements*

The bulk chemical analyses of samples are presented on an untreated basis in Table 2. As can be seen in Table 2, the SiO_2 contents in sediments are fairly similar about 50% except in calcareous ooze and calcareous-siliceous clay. The calcareous ooze, which is comparatively freshly deposited because of its higher sedimentation rate, contains 0.7–2.8% SiO_2. The SiO_2 contents in calcareous-siliceous clay are intermidiate between calcareous and siliceous sediments. NOHARA (1979) also reported similar value for SiO_2 content (30.6% to 43.9%) in calcareous-siliceous clay. It is obvious from the bulk chemical compositions that deep-sea clay and siliceous clay sediments are essentially composed of the same materials, radiolarians, although the two types of sediment might have been deposited during geologically different ages or may have suffered from differing degrees of deep-sea weathering ("pre-early diagenesis").

TiO_2 contents in siliceous clay and deep-sea clay are of similar magnitude, respectively, except for the siliceous clay at G385 which contains less TiO_2 possibly dilution by calcareous materials. Deep-sea clay has slightly greater TiO_2 contents (0.64%) than those in both siliceous clay (0.57%) and siliceous ooze (0.31%).

As well as having a lower SiO_2 contents, calcareous-siliceous clay has a smaller average TiO_2 content (0.19%) which is also a result of dilution by carbonate materials.

CORRENS (1954), GOLDBERG and ARRHENIUS (1958) and ARRHENIUS (1963) have found that euhedral anatase and rutile account for much of the Ti occurring in pelagic sediments from non-volcanic areas. Al_2O_3 contents in siliceous clay vary considerably from 10.8% to 16% with an average of 13.5% (Table 2). On the other hand, Al_2O_3 contents of both siliceous ooze and deep-sea clay are almost constant (13.2–15.6%) as in the case of Si and Ti. The sediments from the survey area have a lower average Al_2O_3 content than the Atlantic sediments (18%), which can be attributed

Table 1. Sampling positions and sediment-types.

St. No.	Sample No.	Position		Depth (m)	Remarks	
		Lat.(N)	Long.(W)		Sediment-type	Topography
703	G 404	08 59 4	179 00 5	5990	Deep sea clay	Floor of flat topography with occasional hills
704	G 375	07 57 6	179 00 2	5860	Deep sea clay	Floor of flat topography
705	G 376	06 58 3	179 02 2	5660	Siliceous clay	Gentle slope of a rolled topography
706	G 377	06 59 2	178 00 1	5225	Ditto	Floor of a terrace like plane along western foot of Magellan Rise
711	G 382	10 59 6	176 59 1	5740	Calcareous-siliceous caly	Bottom of an inter-hill small basin along western foot of a seamount
714	G 385	07 59 4	176 59 0	5120	Siliceous caly	Slop of gently rolled northern lower slop to foot of Magellan Rise
717	G 386	08 00 0	176 00 1	3997	Calcareous ooze	Top of small hilly part in fairly ragge topography along northeastern slop of Magellan Rise
718	G 387	08 59 0	176 10 7	6035	Deep sea clay	Bottom of a hill or margin of a basin in repeated hills and troughs topography
720	G 389	10 59 2	175 59 0	5536	Siliceous clay	Top of small hill in very gently rolled topography
721	G 390	10 59 6	174 59 9	4580	Calcareous-siliceous caly	Flat top of a seamount with 1000 m height
725	G 394	06 57 7	175 00 4	5940	Deep sea clay	Floor of extensively flat topography

726	G 398	06 00 4	174 59 5	5723	Siliceous clay	Floor of extensively flat topography
727	G 397	05 59 1	176 01 5	5371	Deep sea clay	Bottom of a hill in flat topography along southeastern foot of Magellan Rise
728	P 97	06 29 02	175 28 5	5556	Ditto	Floor of nearly flat topography sourrounded by small hills at south eastern foot of Magellan Rise
730	P 99	08 32 6	175 30 0	5865	Ditto	Slope to bottom floor of low hill in gently rolled topography
732	P 101	10 30 7	175 30 5	5680	Siliceous ooze	Floor of fairly extensive plain
738	G 402	09 29 8	175 58 1	6160	Deep sea clay	Floor of a deep trough in hills and troughs

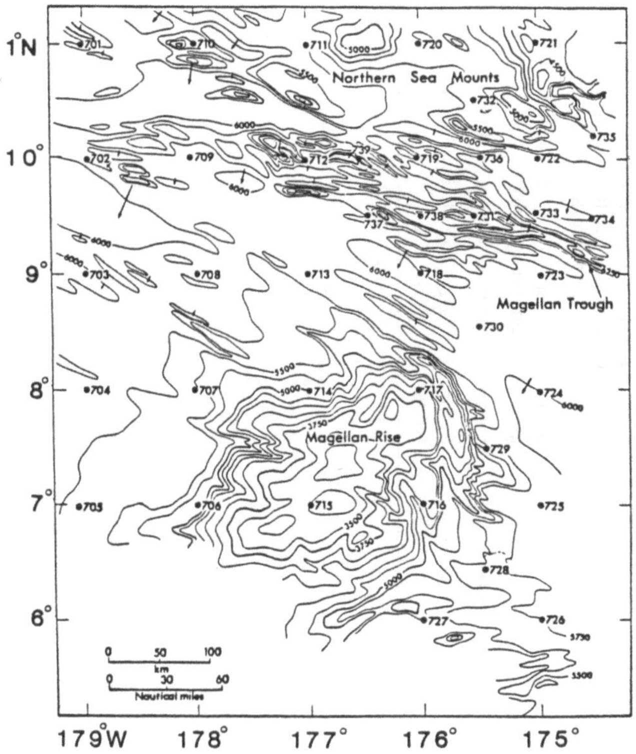

Fig. 1. Bathymetric map and sampling locations in the northern part of the Central Pacific Basin.

to the presence of more terrigenous matter in latter environment (EL WAKEEL and RILEY, 1960). Al_2O_3 content varies linearly with Fe_2O_3 content throughout the survey area (Fig. 2). Possibly, this correlation is due to the presence of authigenically formed illite (with some substitution of Al by Fe^{3+}) and montmorillonite, as suggested by HIRST (1962) and NOHARA (1979). It is worthwhile to note that this relation also holds in calcareous ooze and calcareous-siliceous sediments, suggesting that such pre-early diagenesis is also occurring in freshly deposited calcareous ooze.

With the exception of highly calcareous ooze and calcareous-seliceous clay, the Fe contents in pelagic sediments, as expressed Fe_2O_3, are surprisingly uniform. This is particularly true of deep-sea clay, which contains an average concentration of 5.5% total Fe_2O_3. The mean Fe content of the sediments from this area is somewhat lower than those reported for Pacific sediments (about 8.8% as total Fe_2O_3) by EL WAKEEL and RILEY (1961).

A linear relationship exists between the TiO_2 and Fe_2O_3 contents in the sediments (Fig. 3), which can be expressed as $TiO_2 = 0.02 + 0.08\ Fe_2O_3$. This relationship seems to hold regardless of the sediment-type in any given ocean. A similar relationship has previously been observed by NOHARA (1979) for other Pacific pelagic sediments. The reasons for this association are not clear, but GOLDBERG (1954) had suggested that some of the Ti in deep-sea sediments is scavenged from sea water by hydrous iron oxides.

Although the average Mn contents of all sediment-types are similar, except for calcareous ooze (siliceous ooze 0.58%, siliceous clay: 0.68%, deep-sea clay: 0.75%), Mn content in the individual samples varies remarkably, ranging from 0.48% to 1.03% MnO_2. This variation, which is much greater than that of many of the other major constituents, is probably caused by the fact that in the sediments most of the Mn occurs mainly as discrete manganese phases (EL WAKEEL and RILEY, 1961; CRONAN, 1969). Indeed, the enhanced MnO_2 content in G397 sample is due to the existence of micro-manganese oxides in the sample.

CaO and MgO are fairly uniformly distributed in all the sediments, except for calcareous ooze in which carbonate debris occurs. Both Ca and Mg are found in many minerals derived from the earth's crust (e.g., the clay minerals, feldspars, etc.) and are transported to the ocean as terrestrial detritus. Volcanic minerals may also contribute significant amounts of both elements to the sediments. EL WAKEEL and RILEY (1961) reported average concentrations of 6.26% CaO and 8.34% MgO in two volcanic clays from the Pacific; these concentrations may be compared to 1.83% CaO and 3.21% MgO for lithogenous (i.e., non-volcanic deep-sea clay). In this connection, GOLDBERG and ARRHENIUS (1958) have pointed out that pyroxenes are an important contributor of Ca to some Pacific deep-sea sediments. Phillipsite also contains relatively high concentrations of those elements.

Some deep-sea sediments contain authigenic calcium and magnesium minerals (e.g., calcite, aragonite, dolomite, magnesite, ferron magnesite and hydromanganite, CALVERT, 1976). BONATTI (1966) has described the precipitation of a number of carbonate minerals from hydrothermal volcanic emanations. Mg content shows no remarkable variation in the sediments, while Ca content varies from 0.7% to 54% in proportion to calcium carbonate. Thus, authigenic Mg-carbonate minerals appear to be rare or trace in the sediments.

Table 2. Analytical results of pelagic sediment from the Central Pacific Basin.

Sample No.	G404	G375	G376	G377	G382	G385	G396	G386	G387
SiO_2	55.23	55.25	54.46	51.85	58.03	42.71	0.73	2.83	54.95
TiO_2	0.68	0.63	0.59	0.56	0.59	0.47	0.01	0.03	0.67
Al_2O_3	14.41	13.91	13.93	12.63	13.77	10.76	1.07	1.24	14.84
Fe_2O_3	7.83	7.50	7.48	6.50	5.80	5.23	0.09	0.27	7.73
MnO	0.64	0.61	0.70	0.66	0.10	0.48	0.03	0.03	0.72
MgO	3.18	3.17	3.33	2.73	2.99	2.34	0.16	0.20	3.15
CaO	1.59	2.16	2.16	6.43	1.50	14.92	54.20	52.76	1.24
Na_2O	1.48	1.44	1.39	1.32	1.82	1.10	0.15	0.16	1.50
K_2O	2.29	2.11	2.04	2.16	2.73	1.90	0.05	0.12	2.55
P_2O_5	0.47	0.82	0.88	0.48	0.62	0.25	0.05	0.06	0.26
$+H_2O$	7.52	7.64	7.87	6.16	6.98	5.64	0.93	1.02	7.27
$-H_2O$	3.95	4.07	4.49	3.73	4.59	3.10	0.17	0.21	4.51
CO_2	----	----	----	4.14	----	10.48	42.44	41.13	----
CO	77	77	81	79	73	65	14	16	90
Cu	349	377	455	349	379	265	17	24	369
Ni	135	137	161	160	84	127	16	18	151
Pb	37	31	31	31	37	30	27	23	36
Zn	141	123	129	114	118	97	8	13	122
Total	99.34	99.39	99.41	99.42	99.59	99.44	100.09	100.07	99.47

Cu and Ni contents also show no remarkable regional variations in this area, except in calcareous sediments. High values of Cu and Ni content are generally encountered in siliceous clay and deep-sea clay.

The distributions of Zn and Co are fairly uniform throughout this area except in calcareous sediments. Although Pb concentrations are slightly lower in some samples of siliceous sediments, no marked variations are observed from one sediment type to another. In conclusion, the data show no significant regional variations in trace element contents throughout the surveyed area. Furthermore, trace element abundances are the same or

G389	G390	G394	G398	G397	G402	P 97	P 99	P101
52.85	11.77	56.82	57.83	54.28	54.56	58.35	56.53	61.41
0.64	0.19	0.59	0.57	0.62	0.70	0.55	0.64	0.61
15.95	3.92	13.95	13.57	13.21	15.59	13.08	14.64	13.03
7.87	2.00	7.11	6.86	7.63	7.88	6.46	7.45	6.56
0.64	0.48	0.92	0.61	1.03	0.72	0.76	0.72	0.51
3.26	0.80	2.97	2.85	3.32	3.21	2.80	2.96	2.38
1.60	43.42	1.28	1.10	2.93	1.24	1.52	1.14	0.68
1.72	0.35	1.30	1.31	1.23	1.31	0.67	0.95	0.82
2.96	0.72	2.40	2.39	2.24	2.58	1.71	2.45	2.14
0.54	0.10	0.30	0.22	1.16	0.27	0.62	0.25	0.13
6.79	2.11	7.11	8.08	7.57	7.54	8.42	7.91	7.28
4.55	0.76	4.64	3.93	4.02	3.74	4.35	3.67	3.72
----	33.43	----	----	----	----	----	----	----
83	67	109	86	121	88	88	93	69
421	116	416	303	520	364	402	344	254
139	104	221	123	290	136	189	155	114
36	29	33	30	73	37	30	35	31
123	39	122	107	164	129	119	121	107
99.45	100.09	99.48	99.39	99.36	99.42	99.37	99.39	99.33

similar to those in sediments from the east area adjacent to the survey area (NOHARA and YOKOTA, 1978).

The average abundances of some elements in the sediment-type are presented in Table 3, together with the averages reported for other Pacific sediments. The averages in the studied samples deviate considerably from those of East Pacific Rise and Galapagos Ridge sediments presented by BOSTRÖM and PETERSON (1969), BOSTRÖM *et al.* (1969), CRONAN *et al.* (1972), and CORLISS *et al.* (1978).

On the other hand, our results agree well with those of GOLDBERG

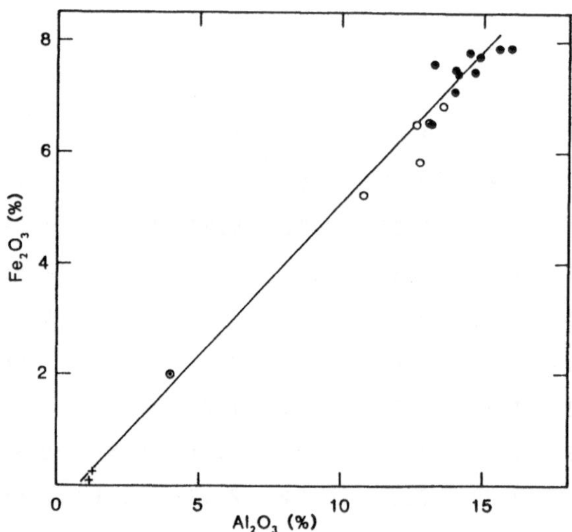

Fig. 2. Relationship between Al_2O_3 and Fe_2O_3 contents in the sediments. +: Calcareous ooze, ⊙: Calcareous-siliceous clay, ◒: Siliceous ooze, ○: Siliceous clay, ●: Deep-sea clay.

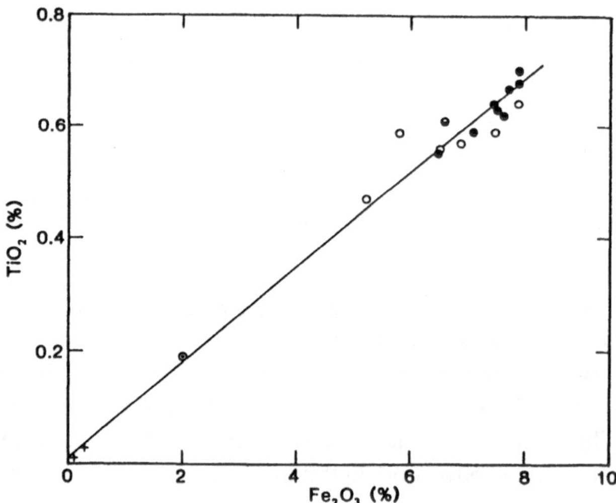

Fig. 3. Relationship between TiO_2 and Fe_2O_3 contents in the sediments (symbols are the same as in Fig. 2).

Table 3. Average chemical compositions of the pelagic sediments from various topographic features in the Atlantic and Pacific Oceans.

	1	2	3	4	5	6	7	8	9	10
Si	0.83	5.50	28.70	24.76	20.06	29.48	26.3	6.1	0.92	
Ti	0.01	0.11	0.37	0.34	0.38	0.40	0.34	0.02		1.30
Al	0.61	2.08	6.93	7.11	7.52	9.55	8.3	0.50	0.27	5.32
Fe	0.13	1.40	4.59	4.63	5.21	6.15	5.15	18.0	0.33	10.77
Mn	0.02	0.37	0.40	0.41	0.60	0.74	1.83	6.0	50.1	0.11
Cu	20	116	254	326	38	300		730	186	81
Pb	10	15	16	15	17	20				63
Zn	21	39	150	145	159			380	753	122
Ni	17	104	120	161	177	180	310	430	850	38
Co	10	67	69	71	85	60	240	105	81	
R*	0.23	0.20	0.21	0.21	0.21	0.21	0.20	0.02		0.16

1: Calcareous ooze, 2: Calcareous-siliceous clay, 3: Siliceous ooze, 4: Siliceous clay, 5: Deep-sea clay, 6: Pelagic sediments from the Pacific (El Wakeel and Riley, 1961), 7: Average of Pacific pelagic clay (Landergren, 1964), 8: East Pacific Rise sediments (Boström et al., 1969), 9: Galapagos Ridge sediments (Corliss et al., 1978), 10: Reykjanes Ridge sediments (Horowitz, 1970), R*= Al/(4Al+Fe+Mn+Ti) ratio (Si, Ti, Fe, Mn given in %, others in ppm).

and ARRHENIUS (1958), TUREKIAN and WEDEPHOHL (1961), LANDERGREN (1964), CRONAN (1969), CRONAN and TOOMS (1969), NOHARA and YOKOTA (1978), and NOHARA (1979) for other parts of the Pacific. The values for certain elements, for example, Fe, Mn, Cu, and Ni are totally different from those reported by BOSTRÖM and PETERSON (1969), CRONAN *et al.* (1972), COLRISS *et al.* (1978) and others for the East Pacific Rise and Galapagos Ridge, but these differences can be accounted for by differences in pathway or origin. In the case of the East Pacific Rise and Galapagos Ridge sediments, trace elements are thought to be derived from rocks immediately below the hydrothermal mounds through alteration of the underlying basalt.

2.2 *Chemical controls on the formation of authigenic mineral phases in the sea water system*

The relations between the major elements Al, Fe and Mn are shown in a ternary diagram in Fig. 4. Si is not included in the figure since it

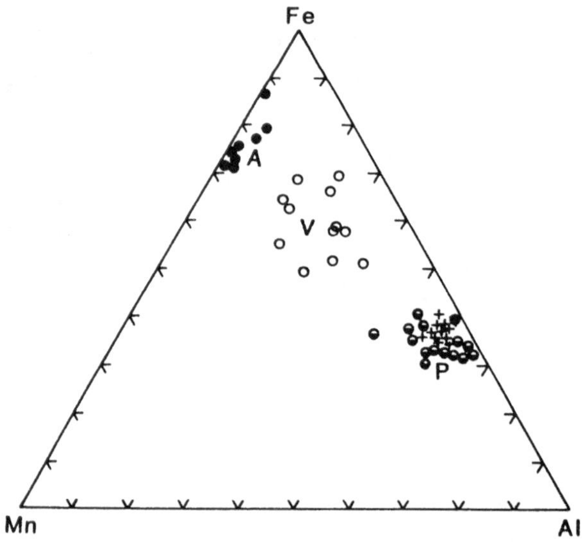

Fig. 4. Al-Fe-Mn relation in the sediments. A: High heat flow oceanic ridge sediment. V: Low heat flow oceanic ridge or volcanic sediments. P: Normal pelagic sediments. (+: this work, ●: from BOSTRÖM and PETERSON (1969), BOSTRÖM *et al.* (1969), ◒ O: from GOLDBERG and ARRHENIUS, 1958).

co-varies with Al, except in areas of high biogenic productivity. Figure 4 shows that the Al-Fe-Mn relation in the sediments differs from at of sediments on the crest of the East Pacific Rise. Particularly, the sediments from this survey area are exceptionally poor in Mn compared with those in active-ridge or volcanically derived sediments. However, the sediments from this survey area shows similarity in Al, Fe and Mn contents to other Pacific sediments analyzed by POLDERVAART (1955), GOLDBERG and ARRHENIUS (1958), EL WAKEEL and RILEY (1961), TUREKIAN and WEDEPHOHL (1961), HIRST (1962), LANDERGREN (1964) and NOHARA (1978). As is obvious from Fig. 4, the major element composition of the sediments from this survey area is very similar to that of normal pelagic sediments from the Pacific Ocean, but remarkably different from that of active-ridge sediments where hydrothermal solutions including volcanic emanations have a markedly effect on the chemical compositions of the sediments (BOSTRÖM and PETERSON, 1969; BOSTRÖM *et al.*, 1969; HOROWITZ, 1970; CRONAN and GARRETT, 1973; PIPER, 1973; NOHARA, 1979). Therefore, it seems reasonable to conclude that the chemical composition of the sediments in the present study area is not significantly affected by submarine hydrothermal solutions or volcanic emanations enriched in some elements derived from the alteration of basalt or directly from magma.

In general, the content of K_2O in the sediments exceeds that of Na_2O (Table 2), but in calcareous sediments K_2O and Na_2O have comparable but low contents. EL WAKEEL and RILEY (1961) have observed two highly calcareous sediments that contained more Na_2O than K_2O. They interpreted this as resulting from the biological removal of sodium from sea water by certain calcareous organisms. Pelagic and near-shore sediments contain similar concentrations of sodium and potassium, and the bulk of both of these elements is associated with the clay minerals (HEIER and ADAMS, 1964). However, other minerals in pelagic sediments, which are quantitatively much less important than the clay, may contain relatively high concentrations of sodium (e.g., sodic feldspars) and potassium (e.g., phillipsite).

Figure 5 shows that K_2O contents co-vary with that of Al_2O_3. This relationship suggests the occurrence of zeolite minerals, for example, phillipsite, which is observed microscopically in deep-sea sediments.

Generally, P_2O_5 is biogenic in origin (EL WAKEEL and RILEY, 1961; HIRST, 1962). There is, however, no correlation between the P_2O_5 content and biogenic nature of the sediments. P_2O_5 contents are greater in siliceous clay and deep-sea clay sediments than in calcareous sediments. This does not necessarily preclude a biological origin for P_2O_5, since Ca-phosphate and apatite are considerably less soluble than $CaCO_3$ in water

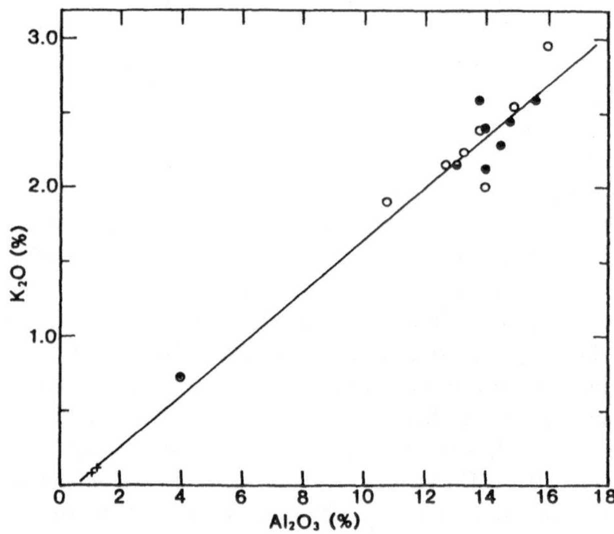

Fig. 5. Relationship between K_2O and Al_2O_3 contents in the sediments (symbols are the same as in Fig. 2).

containing CO_2, and would dissolve less readily during the process of sedimentation. The major host mineral for biogenous phosphate is skeletal apatite, which can constitute a few per cent (by wt.) of some pelagic sediments. On the deep-ocean floor biogenous apatite undergoes dissolution, and in slowly accumulating pelagic sediments only the most resistant phosphate structures such as shark's teeth and the ear bones of whales are preserved (ARRHENIUS, 1963).

Figure 6 shows a clear linear relationship between CaO and P_2O_5 content with a possitive correlation coefficient of 0.99, and a value of 5 : 2.97 for the atomic ratio of Ca : P. This ratio is very close the atomic ratio of 5 : 3 for authigenic phosphate minerals (mainly francolite) which sometimes occur as carbonate apatite, $Ca_5(PO_4, CO_3, OH)_3(F, OH)$. This suggests the formation of Ca-phosphates as a result of chemical precipitation or inorganic replacement of carbonate materials, as follows.

$$5Ca^{++} + 3HCO_3^- + 3HPO_4^{2-} + 4OH^- + F^- \longrightarrow Ca_5(PO_4, CO_3, OH)_3(F, OH) + 6H^+$$

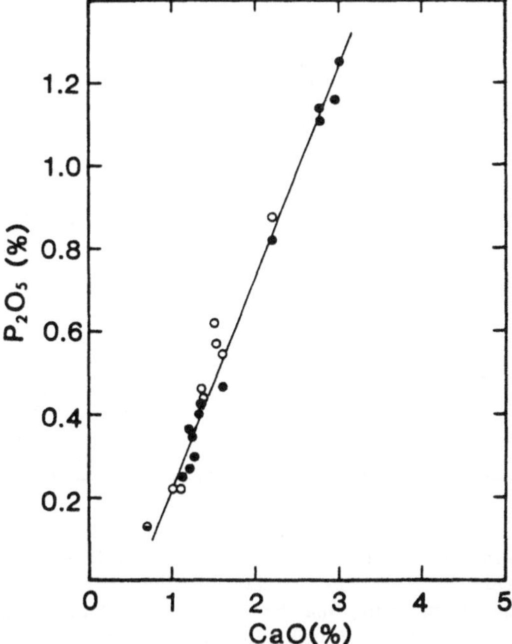

Fig. 6. Relationship between CaO and P_2O_5 contents in the sediments except for calcareous sediment (symbols are the same as in Fig. 2).

or

$$5CaCO_3(s) + 3HPO_4^{2-} + 4OH^- + F^- \longrightarrow Ca_5(PO_4,\ CO_3,\ OH)_3(F,\ OH) + 3H^+$$

Thus, it is clear that the above reactions are primarily proceeding by addition of OH^- from sea water, as well as other elements.

Quantitatively, zeolites are among the most important authigenic mineral phases in pelagic sediments, and phillipsite is one of the most common zeolites in deep-sea sediments (KASTNER and STONECIPHER, 1978). Phillipsite was first recognized by MURRAY and RENARD (1891) as an important diagenetic mineral in pelagic sediments in the Central Pacific and Indian Oceans. STONECIPHER (1976, 1978) and KASTNER and STONECIPHER (1978) found that phillipsite and clinoptilolite are common-

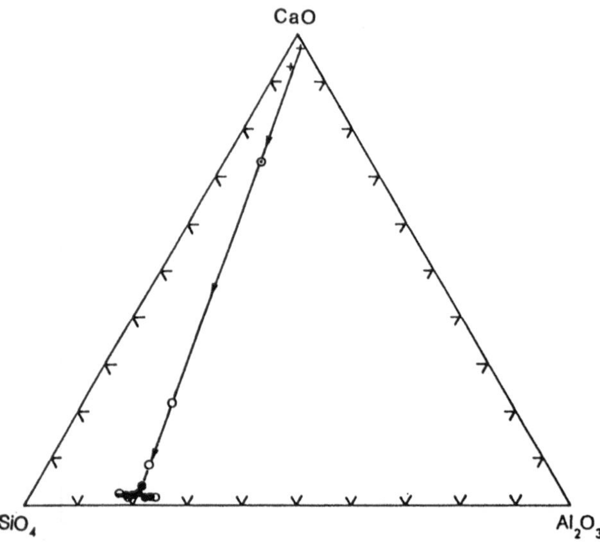

Fig. 7. SiO_2-Al_2O_3-CaO relation in the sediments. The age of the sediments is of old in the order from ooze (Quaternary-Recent) to clay enriched in phillipsite (Miocene-latest Cretaceous) as the arrow, suggesting that the process of authigenic mineral formation is already established even in freshly deposited sediments, at the "pre-early diagenesis" (symbols are the same as in Fig. 2).

ly associated with volcanogenic sediments and siliceous clay (sedimentation rates < 5 m $\times 10^6$ years) in sediments of Eocene to Recent age.

The relations between SiO_2, Al_2O_3 and CaO are shown in Fig. 7. It is obvious from Fig. 7 that the Si: Al: Ca ratio tends to converge to a certain area in which the ratio is nearly the same as that of, phillipsite (0.5 Ca, Na, K$)_5$ ($Al_5Si_{11}O_{32}\cdot 10H_2O$), as the age of the sediments increases. Furthermore, this relation suggests that deep-sea weathering, "pre-early diagenesis" is proceeding towards the zeolitic formation even in freshly deposited calcareous ozze.

The deep-sea clay, in which phillipsite (and also Ca-apatite) is more abundant, is characterized by *Dorcadospyris dentata* (Haeckel) and *Calocycletta costata* (Haeckel), *Calocycletta costata* (Riedel) and *Ichtholiths*, which are early-Middle Miocene and latest Cretaceous-Eocene species (GOLL, 1972).

Based on phase equilibria calculations in the K_2O-Na_2O-Al_2O_3-SiO_2-H_2O system at 25°C and at 1 atmosphere HESS (1966) concluded that phillipsite is a stable phase in oceanic sediments. Figure 7 suggests that phillipsite is probably one of the thermodynamic stable end members of the deep-sea chemical weathering at the sediment-sea water interface. Furthermore, Fig. 7 indicates that the chemical reactions that form the end products (e.g., phillipsite) operate in the sea water or at the sediment-sea water interface. Thus, these processes of authigenic mineral formation proceed further in old surface sediments with higher clay or zeolite contents.

Generally, the most abundant mineral phases in sediment-sea water interface are (K, Na, Ca, Mg)-alumino-silicates such as zeolites or clay minerals, including carbonate-phosphate phases (Figs. 7 and 8).

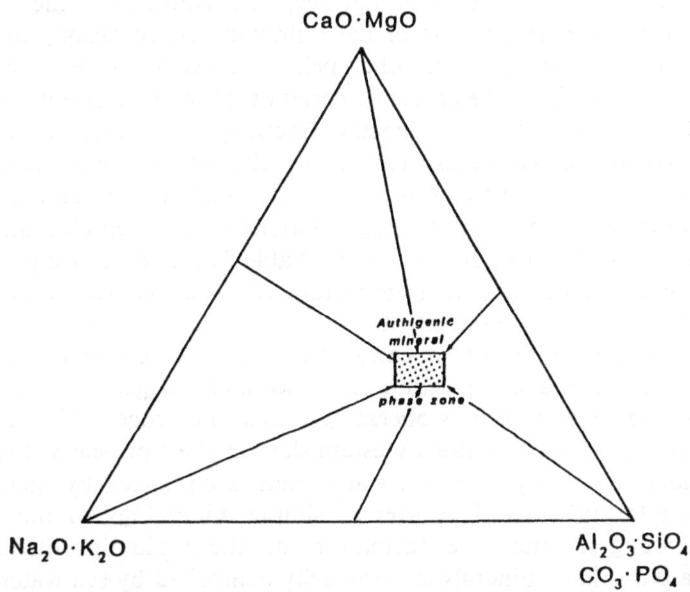

Fig. 8. Schematic some reaction paths for the formation of authigenic mineral phases under the chemical control of sea water. All reaction paths depend on the mole ratio of elements in each mineral phase. The reaction process for the formation of authigenic mineral phase, however, proceeds towards its stable phase zone under the catalysis and control of sea water, irrespective of its reaction path.

These end products always have OH-ligands or H_2O within their structure or inactive surface sites (hydration), which are exchangeable between the mineral phase and sea water. Thus, sea water seems to control the formation of these stable end products on the sea floor (Fig. 8). It is recognized that the chemical composition of sea water has been regulated by silicate or carbonate minerals since an early stage in its chemical history (e.g., SILLEN, 1961; KRAMER 1965). Conversely, however, it can be said that the compositions of pelagic sediments is primarily controlled by sea water and all chemical reaction rates are probably precisely regulated and tend to proceed toward the formation of stable mineral phases by the present sea water system, irrespective of location.

3. *Summary and Conclusions*

Bulk chemical analyses show that deep-sea sediments in the northern part of the Central Pacific Basin have similar concentrations of major elements to those reported for other pelagic areas in the Pacific except for active-ridge regions. The degree of variation of major elements is clearly related to the nature of the biogenic fraction, (e.g., foraminifera and radiolarians) in the sediments. The major element contents based on a water-free, and biogenic-free basis are fairly uniform for each element, irrespective of sediment-type. A linear relationship between CaO and P_2O_5 contents in the sediments points to the probable formation of Ca-phosphate minerals in the sediments. It is important to note that the ratios of Si: Al: Ca (Na, K) in the sediments are nearly the same as those of zeolite, especially, phillipsite. Even in calcareous ooze and calcareous-siliceous sediments the chemical composition of the sediment suggests that the above process of zeolite formation is operating. Thus, the process of authigeneic zeolite phase formation is already established at the "pre-early diagenetic stage" under the catalysis of sea water, and is enhanced by hiatuses or low sedimentation rates. The inferred simple mineralogy of the present sediments suggests that the formation of these alumino-silicate and carbonate-phosphate minerals are primarily controlled by sea water which acts as catalyst in the pelagic environment, since the chemical composition of sea water was become of the same to its present system.

With regard to trace elements, Co, Ni, Cu and Zn contents are higher in deep-sea clay which is also enriched in Fe and Mn. Pb contents show no remarkable variation in the siliceous sediments except for calcareous-siliceous sediments. It seems likely from the above results that these trace elementals are mainly absorbed by siliceous clay and zeolitic deep-sea clay in the sediments.

REFERENCES

ARITA, M, Bottom sediments, in Cruise Report, ed. by A. Muzuno and J. Chujo, *Geol. Surv. Japan*, **4**, 62–70, 1975.

ARRHENIUS, G. O. S., Pelagic sediments, in *The Sea*, ed. by M. N. Hill, Vol. 3, pp. 655–727, Interscience, New York, 1963.

BONATTI, E., Deep sea authigenous calcite and dolmite, *Science*. **153**, 534–537, 1966.

BONATTI, E. and O. JOENSUU, Deep-sea iron deposits from the South Pacific, *Science*, **154**, 643–645, 1966.

BOSTRÖM, K. and M. N. A. PETERSON, The origin of alminium-poor sediments in areas of high heat flow on the East Pacific Rise, *Marrine Geol.*, **7**, 427–447, 1969.

BOSTRÖM, K., M. N. A. PETERSON, O. JOENSUU, and D. E. FISHER, Alumina-poor ferromanganon sediments on active oceanic ridge, *J. Geophys. Res.*, **74**, 3261–3270, 1969.

CALVERT, S. E., The mineralogy and geochemistry of near-shore sediments, in *Chemical Oceanography*, ed. by J. P. Riley and R. Chester, Vol. 6, pp. 187–280, Academic Press, 1976.

CORLISS, J. B., M. LYLE, J. DYMOND, and K. CRANE, The geochemistry of hydrothermal mounds near the Galapagos Rift, *Earth. Planet. Sci. Lett.*, **40**, 12–24, 1978.

CORRENS, C. W., Titan in Tief See Sedimenten, *Deep-Sea Res.*, **1**, 78–85, 1954.

CRONAN. D. S., Average abundances of Mn, Fe, Ni, Co, Cu, Pb, Mo, V, Cr, Ti and P in Pacific pelagic clays, *Geochim. Cosmochim. Acta*, **33**, 1562–1565, 1969.

CRONAN, D. S. and D. GARRETT, The distribution of elements in metalliferous Pacific sediments collected during the D.S.D.P., *Nature, Phys. Soc.*, **242**, 88–89, 1973.

CRONAN, D. S. and J. S. TOOMS, The geochemistry of manganese nodules and associated pelagic sediments from the Pacific and Indian Oceans, *Deep-Sea Res.*, **16**, 335–359, 1969.

CRONAN, D. S., T. VAN ANDEL, G. HEATH, M. DINKELMAN, R. BENNETT, D. BURKY, S. CHARLESTON, A. KNAPPS, K. RODOLOFO, and R. YEATS, Iron-rich basal sediments from the eastern equatorial Pacific: Leg XVI, D.S.D.P., *Science*, **175**, 61–63, 1972.

EL WAKEEL, S. K. and J. P. RILEY, Chemical and mineralogical studies of deep-sea sediments, *Geochim. Cosmochim. Acta*, **25**, 110–146, 1961.

GOLDBERG, E. D., Marine Geochemistry 1. Chemical scavengers of the sea, *J. Geol.*, **62**, 249–265, 1954.

GOLDBERG, E. D. and G. O. S. ARRHENIUS, Chemistry of Pacific pelagic sediments, *Geochim. Cosmochim. Acta*, **13**, 153–212, 1958.

GOLL, P. M., Leg 9 Synthesis, Radiolaria, in *Initial Reports of the Deep-Sea Drilling Project*, ed. by J. D. Hays *et al.*, Vol. 9, pp. 947–1058, U.S. Government Printing Office, Washington, 1972.

HEIER, K. S. and J. A. S. ADAMS, *Physics and Chemistry of Earth*, Vol. 5, 456 pp., Pergamon Press, 1964.

HESS, P. C., Phase equilibria of some minerals in the K_2O-Na_2O-Al_2O_3-SiO_2-H_2O system at 25°C and 1 atomosphere, *Am. J. Sci.*, **246**, 289–309, 1966.

HIRST, D. M., The geochemistry of modern sediments from the gulf of Paria-1, The relationship between the mineralogy and distribution of major elements, *Geochim. Cosmochim. Acta*, **26**, 309–334, 1962.

HOROWITZ, A., The distribution of Pb, Ag, Ti and Zn in sediments on active oceanic ridges, *Marine Geol.*, **9**, 241–259, 1970.

KASTNER, M. and S. A. STONECIPHER, Zeolites in Pelagic Sediments of the Atlantic, Pacific, and Indian Oceans, in *Natural Zeolites*, ed. L. B. Sand and F. A. Mumpton, pp. 221–234, Pergamon Press, New York, 1978.

KATO, K. and N. MITA, Chemical analysis of deep-sea Sediments, *Chem. Anal. geol. Surv. Japan*, **54**, 38, 1983.

KRAMER, J. R., History of Sea Water: Constant temperature-pressure equilibrium models compared to liquid inclusion, *Geochim. Cosmochim. Acta*, **29**, 921–946, 1965.

LANDERGREN, S., On the geochemistry of Deep-Sea Sediments, *Rept. Swed. Deep Sea Exped. X. Spec. Invest.*, **5**, 57–154, 1964.

MURRAY, D. J. and A. RENARD, Deep-sea deposits, in *Report on the Scientific Results of the Voyage of H. M. S. "Challenger"*, ed. by C. W. Thompson and J. Murray, Vol. 5, 525 pp., 1891.

NOHARA, M., Major chemical compositions of pelagic sediments from the Central Pacific Basin, *Bull. Geol. Surv. Japan*, **30**, 27–36. 1979.

NOHARA, M. and S. YOKOTA, The Geochemistry of Trace Elements in Pelagic sediments from the Central Pacific Basin, *J. Geol. Soc. Japan*, **84**, 165–175, 1978.

PIPER, D. Z., Origin of metalliferous sediments from the East Pacific Rise; *Earth Planet. Sci. Lett.*, **19**, 75–82, 1973.

POLDERVAART, A., Chemistry of the Earth Crust, in Crust of the Earth, ed. by Poldervaart, *Geol. Soc. Am. Spec. Papers*, **62**, 119–144, 1955.

SILLEN, L. G., The physical chemistry of sea water, in *Oceanography*, ed. by M. Sears, pp. 549–581, Am. Assoc. Adv. Sci. Publ. No. 67, Washington, 1961.

STONECIPHER, S. A., Distribution of deep-sea phillipsite and clinoptilolite, *Chem. Geol.*, **17**, 307–318, 1976.

STONECIPHER, S. A., Chemistry and deep-aea phillipsite, clinoptilolite, and host sediments, in *Natural Zeolites*, ed. L. B. Sand and F. A. Mumpton, pp. 221–234, Pergamon Press, New York, 1978.

TUREKIAN, K. K. and K. H. WEDEPHOL, Distribution of the elements in some major units of the earth's crust, *Bull. Geol. Soc. Am.*, **72**, 175–192, 1961.

WINTER, E. L., J. I. EWING, R. G. DOUGLAS, R. D. JARRAD, Y. LANCELOT, R. M. MOBERY, T. C. MOORE, Jr., P. H. ROTH, and S. O. SCHLANGER, *Initial Reports of the Deep-Sea Drilling Project*, Vol. 17, 930 pp., Washington D.C., 1973.

Formation of Active Ocean Margins, edited by N. Nasu *et al.*, pp. 913–939.

MANGANESE CONTENT, CERIUM ANOMALY, AND RATE OF SEDIMENTATION AS AIDS IN THE CHARACTERIZATION AND CLASSIFICATION OF DEEP-SEA SEDIMENTS

Ryo MATSUMOTO[1], Yoshitaka MINAI[2], and Azuma IIJIMA[1]

[1]*Geological Institute, Faculty of Science, Uneversity of Tokyo, Tokyo, Japan*
[2]*Department of Chemistry, Faculty of Science, University of Tokyo, Tokyo, Japan*

Abstract. On the basis of previously published data on manganese content and the rates of sedimentation of deep-sea sediments in the Northwest Pacific, contour maps of manganese content and sedimentation rate are presented. Sediments in the central Northwest Pacific are characterized by high manganese contents (>1%) and low sedimentation rates (<2 m/m.y.), whereas marginal basins around Japan and Taiwan generally have low manganese contents (<0.2%) and high sedimentation rates (>50 m/m.y.) sediments. Intermediate Sedimentation rates (5–10 m/m.y.) and manganese contents (0.2–0.5%) occur in the North Pacific between approximately lat. 30° and 40°N. The manganese content and sedimentation rate of sediments are inversely correlated, suggesting that constantly accumulating hydrogenous manganese is being diluted to various degree by terrigenous, biogenic, and other components.

On the basis of 41 analyses of rare earth elements, a negative cerium anomaly is recognized in sediments taken from and around the East Pacific Rise (EPR) and the Galapagos Deep, and in a zone extending westwards along the equator for at least 30°. A weak negative cerium anomaly is also observed in the West Philippine Sea where back-arc spreading is occurring. These observations indicate a genetic relationship between thhe negative cerium anomaly and hydrothermal activity at a speading center.

Based on lithology and chemistry, deep-sea sediments are classified into three types, Type I (continental margin sediments), Type II (abyssal plain sediments), and Type III (ridge sediments). At DSDP Site 164 (southwest of Hawaii), the sediments consist of Type III (Cretaceous), which directly overlie basalt, and an intermediate type between Types II and III (Eocene to Miocene) and Type II (Pleistocene). Sediments at Site 436 (Japan Trench outer rise) begin with Tpye III (Cenomanian or Albain, Oligocene and Lower and Middle Miocene), followed by an intermediate type between Types I and II (Upper Miocene) and then Type I (Pliocene to Quaternary). Variations in sediment types at each site suggest northwestward migration of the site due to motion of the Pacific Plate. Sediments at Sites 584 and 439 (Japan Trench inner wall) are solely of Type I, including Cretaceous mudstone encountered at the bottom of Site 439.

1. *Introduction*

A great amount of lithological, geochemical and other important information on modern deep-sea sediments have rapidly accumulated over the last 20 years as a result of research cruises carried out by the R.V. Glomar Challenger, the R.V. Hokuho Maru, and other research vessels, and as a result of expansion of marine sciences in general (e.g., TUREKIAN, 1966; BOSTRÖM and PETERSON, 1969; BENDER *et al.*, 1971, SUGISAKI, 1980; MATSUMOTO, 1981). Precise geochemical studies on large numbers of rocks and sediments have been carried out due to the development of X-ray instrumental analysis (e.g., MATSUMOTO and URABE, 1980). It is well known that the contents of base metal elements differ between pelagic and continental margin sediments, so base metals such as Cu, Co, Ni, and especially Mn, are useful indicators of the depositional environment of ancient sedimentary rocks on land. For example, MATSUMOTO (1982) and MATSUMOTO and IIJIMA (1983) showed through base metal analysis that Permo-Jurassic and Tertiary bedded cherts in Japan were deposited in marginal basins. SHIMIZU and MASUDA (1977) compared the REE patterns of deep-sea chert and on-land chert, and concluded that some terrestrial cherts were deposited in coastal areas, marginal seas, or land-enclosed seas. Steinberg and his coworkers (STEINBERG and MARIN, 1978; RANGIN *et al.*, 1981; STEINBERG *et al.*, 1983) considered that a negative cerium anomaly is characteristic of deep-sea pelagic sediments and they used the cerium anomaly as a paleoenvironmental indicator of open ocean sediments.

In this paper we first illustrate the distribution of manganese in the Northwest Pacific, and then outline the relation between Mn content and the rate of sedimentation in connection with the cerium anomaly. We assess whether this anomaly is ubiquitous to pelagic sediments and what factors control the REE patterns. Then we propose a classification of deep-sea sediments based on sediment chemistry and sedimentation rate. Finally changes in the depositional environment of DSDP cores recovered at three sites in the north Pacific are examined from the geochemical point of view.

Modern deep-sea sediments generally have high Mn contens compared with terrestrial shale, and this has often been interpreted as resulting from upward migration and reprecipitation of manganese in the sediment column during burial diagenesis. In fact, as clearly demonstrated by SWINBANKS and SHIRAYAMA (1984), a few to 10 cm thick manganese rich layers occur at 0 to 30 cm subbottom in calcareous ozze at several sites in the western Pacific; the manganese-rich layers are closely related to burrow-mottled horizons. BENDER *et al.* (1971), however, concluded that upward migration of manganese is insufficient to explain the high manganese contens of deep-sea sediments, based on depth profiles of Mn contents down to 12 m in deep-sea pelagic sediments including brown clay, diatom lutite and siliceous ooze of the Pacific. According to recent geochemical studies on DSDP cores (e.g., SUGISAKI, 1980; NOHARA, 1980; MATSUMOTO, 1983), manganese-rich layers have not been encountered in brown and blue clays. These apparently contradictory observations may be due to differences in the degree of bioturbation, lithology, and/or rate of sedimentation. Redistribution of manganese in the surficial sediments probably occus on a small scale as shown by SWINBANKS and SHIRAYAMA (1984).

The data on the contens of Mn and REEs and the rate of sedimentation used in this paper were mostly taken from previously published data (references—as in Table 1, footnote). Sediment samples had been recovered either by piston, rotary, or box corers, and the sampling intervals were mostly between 0.5 and 5 m subbottom. To avoid the incorporation of ferromanganese nodules or Mn rich layers at or near the sediment/water interface, KRISHNASWAMI (1976) did not analyse the top 1–10 cm section; BENDER *et al.* (1970) and MINAI (1982) analysed at several intervals in order to determine the depth profile of sediment chemistry. Secondary Mn-rich layers were excluded from the present study. The analyses of Mn, Al, and Si contents were all carried out on bulk samples. When analytical values in previous papers were presented on a carbonate-free basis, they were recalculated on a water and salt-free whole sediment basis. REEs determinations were made chiefly by instrumental neutron activation analysis of bulk samples. The rate of sedimentation was calculated either from paleomagnetic measurements (KH sites of R.V. Hakuho Maru), biostratigraphic dates (DSDP sites), or by using radioactive nuclides such as Th and Pa (KH sites and other sites). The data sources are given in the respective figure captions and reference section.

Table 1. Mn contents, sedimentation rates and REE patterns of Pacific deep-sea sediments.

No.	Site	Latitude	Longitude	Depth (m)	Sediment type	MnO (%)	Sedimentation rate (mm/1000y)	REE (Ce-anomaly)	Dating method	References a	References b	References c	Sampling interval (cm) of MnO contents
1	DSDP Site 170	11°48'N	177°37'E	5792	RC	2.1	0.5		Bio.	P	S_1		1051 - 1052
2	DSDP Site 168	10°42'N	173°36'E	5420	RC	1.04	0.8		Bio.	P	S_1		1800 - 1802
3	DSDP Site 164	13°12'N	161°31'W	5499	RC	1.6	1.5		Bio.	P	S_1		746 - 748
4	KH73-4-4	12°38'N	151°31'E	5920	RC	0.79	2.0	*	Mag.	M_2	Ko	M_2	45 - 55
5	KH70-2-9-3	17°05'N	146°12'W	4960	RC	0.94	-	N		M_2		M_2	45 - 55
	MSN147G	8°20'N	145°24'W	5100	RC	-	1.9		^{10}Be		A_m		?
6	KH71-5-53-2	8°15'N	112°42'W	3970	CO	1.4	-	N		M_2		M_2	45 - 55
	V21-48	9°31'S	126°22'W	3922	CO	-	5.0		^{230}Th		B_1		0 - 150
7	KH73-4-9	7°60'S	172°49'E	5390	RC	0.84	4.8	N	Mag.	M_2	Ko	M_2	45 - 55
8	KH68-4-15-3	12°00'N	169°50'W	5770	RC	0.71	4.6	N	Mag.	M_2	Ko	M_2	45 - 55
9	KH73-4-6	10°46'N	153°42'E	5700	RC	0.64	3.7	*	Mag.	M_2	Ko	M_2	45 - 55
10	KH70-2-7-3	33°01'N	169°53'E	5840	RC	0.53	2.7	*	Mag.	M_2	Ko	M_2	45 - 55
11	KH72-2-58	22°53'N	129°13'E	5300	RC	0.68	6.1		Mag.	Ma	Ko		0 - 20
12	DSDP Site 296	29°20'N	133°32'E	2920	CO	0.17	25	*	Bio.	M_3	S_2	M_3	117 - 119
13	DSDP Site 438	40°38'N	141°14'E	1552	BC	0.075	38	*	Bio.	M_3	S_3	M_3	248 - 252
14	DSDP Site 583	31°50'N	133°51'E	4618	BC	0.12	85	*	Bio.	M_3	S_5	M_3	57 - 58
15	DSDP Site 584	40°28'N	143°57'E	4078	BC	0.11	-100	*	Bio.	M_3	S_5	M_3	86 - 88
16	DSDP Site 297	30°52'N	134°09'E	4458	BC	0.23	110	*	Bio.	M_3	S_2	M_3	4420 - 4422
17	DSDP Site 582	31°47'N	133°55'E	4879	BC	0.12	195	*	Bio.	M_3	S_5	M_3	52 - 54
18	DSDP Site 436	39°56'N	145°33'E	5240	BC	0.07	170	*	Bio.	M_3	S_3	M_3	1810 - 1812
19	DSDP Site 440	39°44'N	143°56'E	4509	BC	0.068	337		Bio.	S_8	S_3		203 - 207
20	V18-258	11°51'S	165°45'W	5528	RC	1.7	1.7		^{230}Th	B_1	B_1		0 - 50
21	RC8-83	29°22'S	105°14'W	3157	CO	1.8	3.3		^{230}Th	B_1	B_1		0 - 120
22	V21-71	27°54'N	162°31'E	5964	RC	0.76	1.2		Mag.	B_1	B_1		0 - 85
	KH80-3-22	31°16'N	153°43'E	5750	RC	-	-	*				M_1	95 - 105
23	RC10-157	26°47'N	159°03'E	5682	RC	0.66	0.93		Mag.	B_1	B_1		0 - 65
24	V21-70	27°05'N	166°04'E	5860	RC	0.58	0.93		Mag.	B_1	B_1		0 - 65
25	V21-63	22°51'N	169°41'W	4674	RC	0.58	1.4		Mag.	B_1	B_1		0 - 105
26	V21-67	24°58'N	176°16'E	5879	RC	0.66	1.7		Mag.	B_1	B_1		0 - 115
27	V21-65	23°58'N	176°51'W	5365	RC	0.62	2.4		Mag.	B_1	B_1		0 - 105
28	RC10-159	31°13'N	162°19'E	5898	RC	0.48	3.0		Mag.	B_1	B_1		0 - 205
29	V21-48	9°31'S	126°22'W	3922	CO	1.1	5.0		^{230}Th	B_1	B_1		0 - 150
30	V21-73	29°28'N	154°36'E	5872	RC	0.64	7.6		Mag.	B_1	B_1		0 - 525
	KH80-3-22	31°16'N	153°43'E	5750	RC	-	-	*				M_1	95 - 105

31	RC10-181	44°05'N	176°50'E	5720	RC	0.52	1.1		Mag.	B_1	B_1		0 - 775
32	V19-116	11°14'N	139°10'E	5841	RC	0.46	9.9		Mag.	B_1	B_1		0 - 690
33	V20-107	43°24'N	178°52'W	5872	SO	0.41	7.6		Mag.	B_1	B_1		0 - 530
34	V20-108	45°27'N	179°94'W	5625	SO	0.44	11.3		Mag.	B_1	B_1		0 - 790
35	RC10-182	?	?	?	RC	0.39	11.9		Mag.	B_1	B_1		0 - 825
36	V19-54	17°02'S	113°54'W	2830	CO/M	4.9	15	N	^{230}Th	B_2	B_2	B_2	18 - 400
37	DWHG-49	42°02'S	98°01'W	4350	RC	3.0	0.6		^{230}Th	Kr	Ku		10 - 40
38	DWBG-30	19°50'S	148°39'W	4620	RC	1.3	0.9		^{230}Th	Kr	Ku		10 - 40
39	V-18-258	11°52'S	165°45'W	5528	RC	1.91	1.6		^{230}Th	Kr	Ku		10 - 40
40	CAP-5BG	9°03'S	174°52'E	4960	RC	1.3	1.6		^{10}Be	Kr	Ku		10 - 40
41	NOVA-10	9°53'N	178°47'W	6105	RC	0.92	1.0		^{10}Be	Kr	Am		10 - 40
42	NOVA-13	3°56'N	178°47'W	5331	RC	0.75	1.4		^{10}Be	Kr	Am		10 - 40
43	DWBG-52	40°36'S	132°49'W	5120	RC	0.70	1.4		^{230}Th	Kr	Ku		10 - 40
44	NOVA-16	0°14'N	178°08'W	5160	RC	0.78	1.7		^{10}Be	Kr	Am		10 - 40
45	WIG-5	28°38'N	124°26'W	4400	RC	0.63	2.0		^{230}Th	Kr	Ku		10 - 40
46	DWBG-2	21°27'N	126°43'W	4370	RC	0.72	2.2		^{230}Th	Kr	Ku		10 - 40
47	CAP-49BG	9°17'N	124°09'W	4410	RC	0.66	2.4		^{230}Th	Kr	Ku		10 - 40
48	V21-71	27°54'N	162°31'E	5860	RC	0.62	2.4		^{230}Th	Kr	Ku		10 - 40
	KH80-3-22	31°16'N	153°43'E	5750	RC	-	-	*				M_1	95 - 105
49	CAP-50BG	14°55'N	124°12'W	4270	RC	0.52	2.1		^{230}Th	Kr	Ku		10 - 40
50	MSN-96	57°35'S	174°15'W	4760	RC	0.51	2.3		^{10}Be	Kr	Am		10 - 40
51	DSDP Site 507	0°34'N	86°05'W	2689	M	41.8	8.8	N	Bio.	Mi	S_4	Mi	0 - 20
52	DSDP Site 509	0°35'N	86°08'W	2677	M	39.1	7.5	N	Bio.	Mi	S_4	Mi	0 - 120

Notes: Sediment type: RC red clay; BC blue clay; CO calcareous ooze; SO siliceous ooze; M metalliferrous mud.
REE: * showing normal REE pattern; N having negative Ce anomaly.
References: (a MnO content, b sedimentation rate, c REE)
Ku Ku et al. 1968; B_1 Bender et al. 1970; B_2 Bender et al. 1971; S_1 Scientific Party 1973; P Pimm 1973; S_2 Scientific Party 1975; Am Amin et al. 1975; Kr Krishnaswami 1976; S_3 Scientific Party 1980; S_8 Sugisaki 1980; Ko Kobayashi et al. 1980; M_1 Minai et al. 1981; M_2 Minai 1982; S_4 Scientific Party 1983; Ma Matsumoto and Iijima 1983; Mi Migdisov et al. 1983; M_3 Minai et al. 1984; S_5 Scientific Party 1984.

2. *Manganese Content of Deep-Sea Sediments*

The Mn contents of deep-sea sediments, based on 62 analyses of brown clay, blue clay, calcareous and siliceous ooze, and diatomaceous mud, are shown in Fig. 1. The Mn content ranges from 0.03% in blue clay off the Izu Peninsula of central Japan to 1.31% in brown clay about 2200 km west of Hawaii and 2900 km east of New Guinnea. Most Mn contents around the Japanese Islands are well below 0.1%, even 500 km east of central Japan, though 0.12% MnO in mud and 0.17% MnO in blue clay are found in the Shikoku Basin about 160 km and 300 km off southwest Japan, respectively. On the other hand, along the east coast of the Philippines, low manganese zones seem to be restricted to narrow areas. Thus low manganese sediments occupy a narrow zone in the northwestern margin of the Philippine Sea and Shikoku Basin, and then the zone broadens eastwards about 1000 km into the Northwest Pacific. A distinctive difference in manganese content between inner slope and outer slope sediments does not exist around the subduction zone near the Japanese Islands; contour lines of 0.5% and 0.2% MnO intersect the Izu-Mariana Trench. Smaller zones of low Mn contents are recognized in the north and central Pacific. They are located either on sea mounts or in low latitudes. The anomalously low manganese mud in the open ocean is probably due to either high productivity of planktonic organisms at low latitudes or to the low rate of dissolution of calcareous particles in shallow waters above the CCD.

By contrast, a high mangnese zone (>0.5% MnO) occupies the main area of the Northwest Pacific and Philippines Sea. The 0.5% MnO contour is only 200 km away from Philippines. The narrowness of the 0.2% to 0.5% MnO zone, which occurs between the terrestrial low manganese zone and pelagic high manganese zone, indicates that manganese contents increase fairly abruptly in an oceanwards direction. A 0.2% to 0.5% MnO zone also occurs in mid-ocean east of Japan between lat.25°N and 33°N.

3. *Rate of Sedimentation*

The rates of sedimentation at 64 sites in the Northwest Pacific are given in Fig. 2. The rate of sedimentation varies widely from 0.3 m/m.y. for brown clay in the central Pacific to more than 300 m/m.y. for diatomaceous mud on the deep-sea terrace off Northeast Japan. In general, sedimentation rates form an inverse pattern to the Mn content distribution map (Figs. 1 and 2). A zone of extremely high sedimentation rate

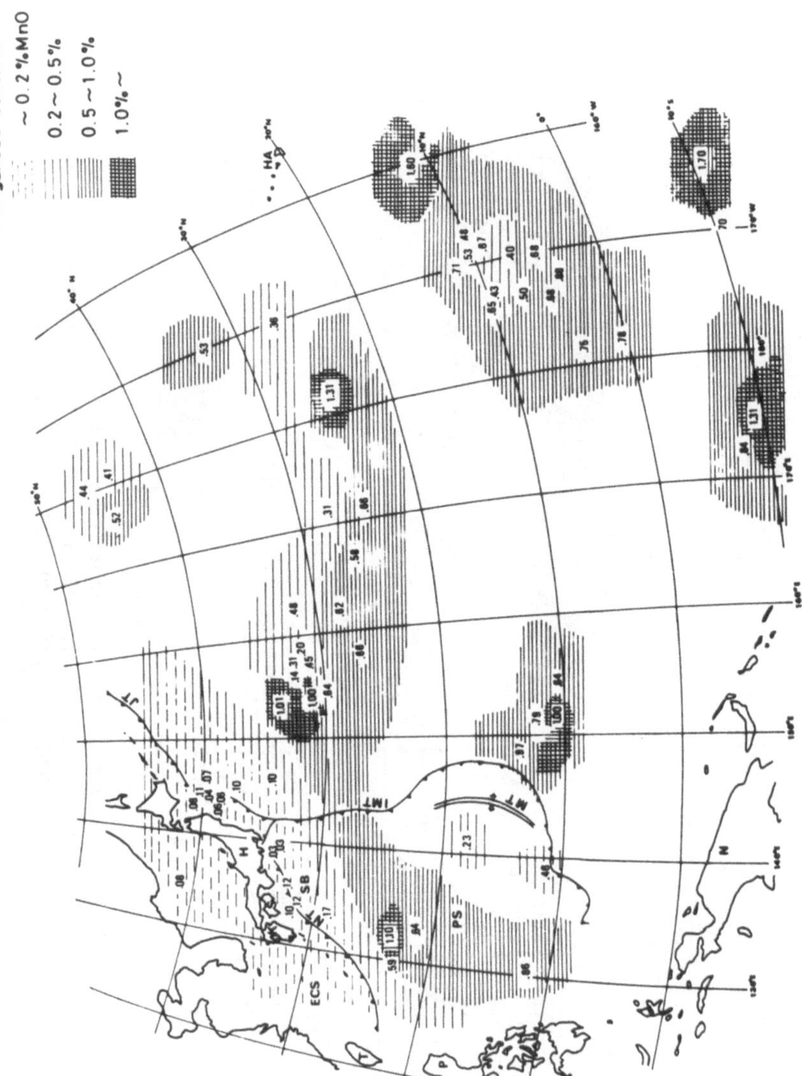

Fig. 1. Manganese content of deep-sea sediments in the Northwest Pacific (WAKEEL and RILEY, 1961; BENDER *et al.*, 1970; KRISHNASWAMI, 1976; AOKI, 1977; TANAKA *et al.*, 1977; NOHARA and YOKOTA, 1978; SUGISAKI, 1980; MINAI *et al.*, 1981; MATSUMOTO and IIJIMA, 1982; MINAI, 1982; MINAI *et al.*, 1984). H: Honshu, S: Shikoku, K: Kyushu, T: Taiwan, P: Philippines, N: New Guinnea, HA: Hawaii, ECS: East China Sea, PS: Philippine Sea, JT: Japan Trench, NT: Nankai Trough, IMT: Izu-Mariana Trench, MT: Mariana Trough.

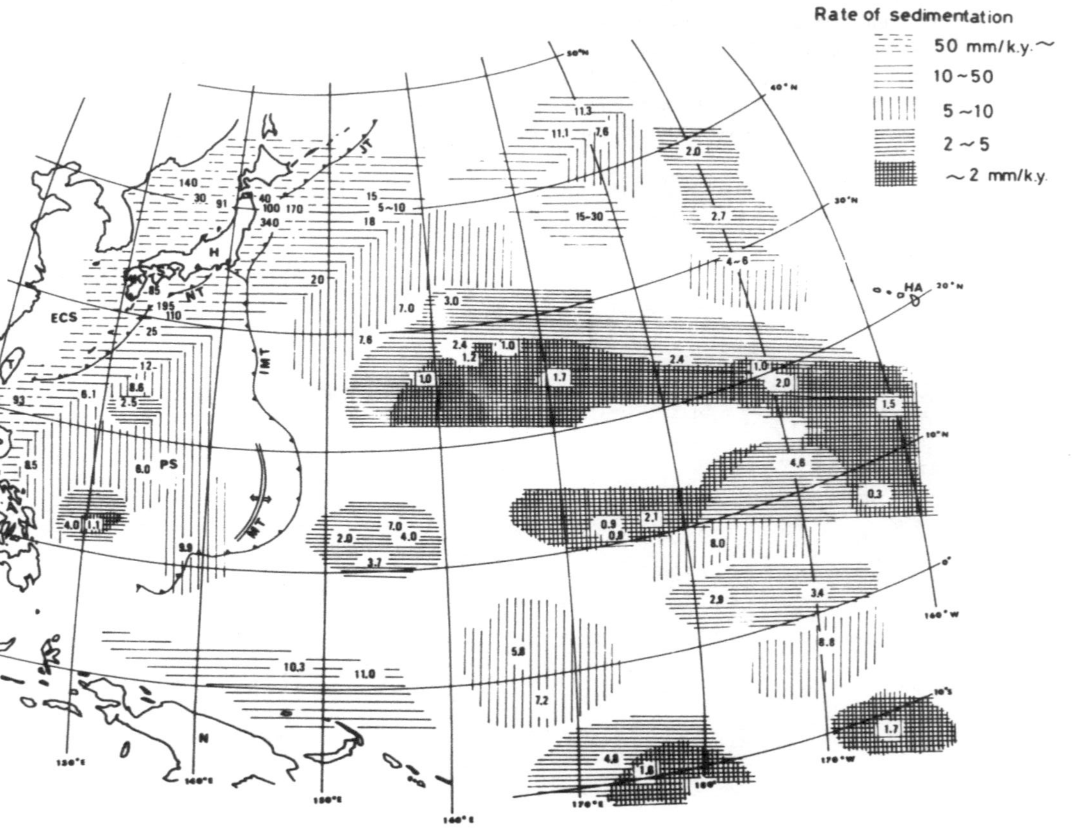

Fig. 2. Rate of sedementation of deep-sea sediments in the Northwest Pacific (KU *et al.*, 1968; SCIENTIFIC PARTY, 1973a, b, 1975; AMIN *et al.*, 1975; KRISHNASWAMI, 1976; KOBAYASHI, 1980; SCIENTIFIC PARTY, 1980, 1983; YANG and NOZAKI, 1983; SCIENTIFIC PARTY, 1984). H, S, K, etc., refer to Fig. 1.

(>50 m/m.y.) is restricted to the marginal sea around Taiwan and the Japanese Islands. Sediment accumulating in this zone is dominated by terrigenous material. A zone of high sedimentation rate (>10 m/m.y.) appears to coincide with the low manganese zones of less than 0.5% MnO.

A low sedimentation rate zone, less than 2 m/m.y., extends westward from Hawaii, and corresponds with the high manganese zone (Figs. 1 and 2) A small zone of low sedementation in the south Pacific also overlaps the high manganese zone.

4. Relation Between Manganese Content and Rate of Sedimentation

As recognized from Figs. 1 and 2, manganese content and the rate of sedimentation of deep-sea sediments seem to display an inverse correlation with each other. This relation has been mentioned by many workers such as TUREKIAN (1966), TUREKIAN and IMBRIE (1966), BENDER *et al.* (1966), KRISHNASWAMI (1976), and MATSUMOTO and IIJIMA (1983). On the one hand, Turekian and his coworkers, who concentrated on the deep-sea calcareous sediments in the Atlantic Ocean, found a strong relation between manganese-cobalt-nickel association and low clay accumulation rates. At the same time, they showed a positive relation between the rates of accumulation of heavy metals and clay (=insoluble residue) in calcareous sediments. They considered that heavy metals accumulated along with the fine clay materials. On the other hand, BENDER *et al.* (1966, 1971) determined the accumulation rates for authigenic manganese in the deep-sea sediments of the Pacific, Atlantic, and Indian Oceans, and found that manganese accumulation rates are comparatively constant for very different types of sediments in the world oceans. They proposed a "rain" model, that is, there is a rather uniform "rain" of authigenic manganese over the entire ocean floor. KRISHNASWAMI (1976) also observed a negative correlation between the clay accumulation rates and the concentration of Mn, Co, Ni, and Cu in Pacific clays.

In order to evaluate the validity of the above two hypotheses for Pacific pelagic sediments, 51 sets of MnO content and sedimentation rate are plotted in Fig. 3. Locations, water depths, sediment types and references are listed in Table 1. This diagram clearly demonstrates the negative relation between Mn content and the rate of sedimentation; sediment samples in Fig. 3 are mostly non-calcareous pelagic clays, so the rate of sedimentation of this diagram roughly corresponds to clay (=insoluble residue) accumulation. Thus, Mn accumulation and clay accumulation show an inverse correlation in Pacific sediments. If manganese accumulates on the sea floor associated with clay particles ("clay association" model), the Mn contents of

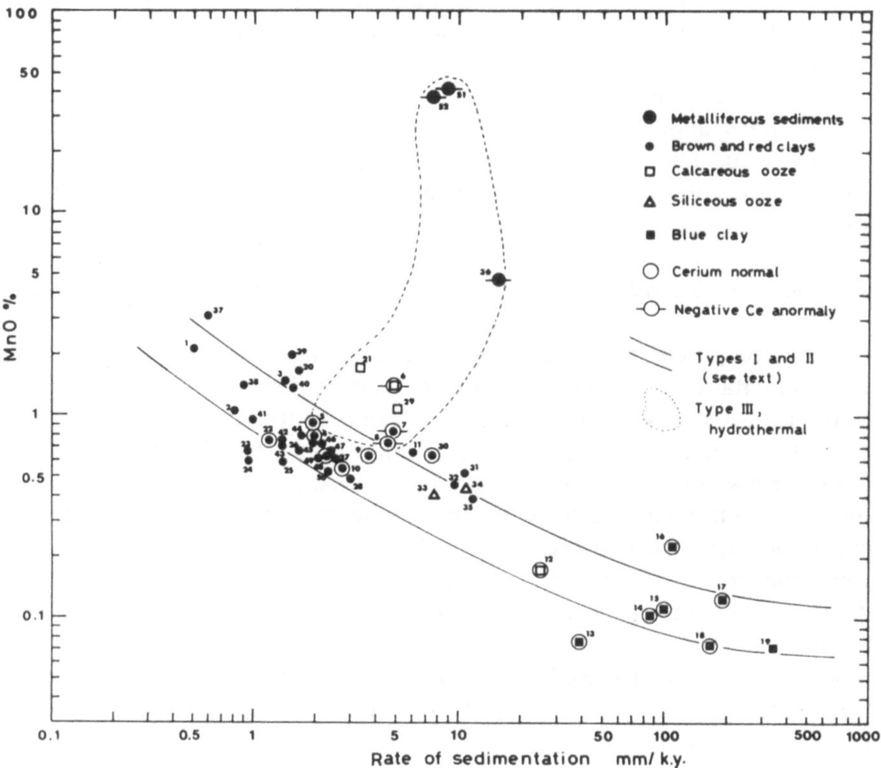

Fig. 3. Relation between manganese content and the rate of sedimentation of dee-sea sediments. Data and sources are listed in Table 1.

calcareous sediments should become very low compared with non-calcareous pelagic sediments, however, calcareous ooze does not plot below the curve in Fig. 3 but, to the contrary, near or above the curve. Moreover, if the "clay association" model is valid for Pacific sediments, the ratios of Mn content to Al content of various pelagic sediments should fall in a narrow range; however, the ratio increases with decreasing rate of sedimentation (MATSUMOTO and IIJIMA, 1983). The above discussion and findings suggest that the "rain" hypothesis explains the inverse relation between manganese contents and the rate of sedimentation of Pacific deep-sea sediments. Turekian's "clay association" model may apply to calcareous sediments, but it may not be applied to non-calcareous Pacific pelagic clays.

Assuming that the water-free bulk density of modern sediments is 0.7 g/cm^3 after BENDER *et al.* (1970), the accumulation rate of manganese (mg/cm^2·1000 yr) is given by [0.7 g/cm^3]×[sedimentation rate in mm/1000 yr]×[manganese content in wt.%]. The manganese accumulation rate for brown clay and calcareous and siliceous ooze (Table 1) mostly ranges from 0.7 to 1.5 mg/cm^2·1000 yr, averaging 1.1 mg/cm^2·1000 yr. The accumulation rate for 38 world ocean sediments determined by BENDER *et al.* (1970) ranged between 0.3 and 3.2 mg/cm^2·1000 yr, averaging 1.3 mg/cm^2·1000 yr, which is somewhat similar to the present values. However, manganese accumulation rates in blue clay and diatomaceous mud in the marginal basins are calculated to be 4 to 29 mg/cm^2·1000 yr, an order of magnitude higher than for the average pelagic clay. This, however, does not necessarily mean that hydrogenous contribution is very high in the marginal basins. Most of the manganese in near-shore sediments is not part of the hydrogenous component but rather is part of the lithogenous fraction (e.g., CHESTER and HUGES, 1967; KRISHNASWAMI, 1976; MATSUMOTO and IIJIMA, 1983). That is, the anomalously high rate of Mn accumulation is probably caused by the "diluting" materials. Therefore it is important to consider the Mn contents of the terrigenous suites as BENDER *et al.* (1966) did. Assuming that the hydrogenous accumulation of manganese is constant at 1.1 mg/cm^2·1000 yr in the Pacific, the hydrogenous manganese content in the continental margin sediments which were deposited at a rate of between about 40 to 400 m/m.y. (Fig. 3) is calculated to be 0.035 to 0.004% MnO of the total sediments. That is to say, about 70 to 95% of the total manganese in these sediments does not originate through hydrogenous accumulation but via terrigenous influx. The manganese content of average shale, andesite, and granite is about 0.1%, 0.18%, and 0.12% MnO, respectively. Therefore, the terrigenous materials carried into the sea are expected to contain 0.1 to 0.2% MnO, and, in fact, Mn contents of the blue clays in Fig. 3 are between 0.07 and 0.23%.

5. *Cerium Anomaly and Ridge Volcanism*

Cerium anomaly is defined on the basis of the chondrite normalized REE pattern (CORYELL *et al.*, 1963; MASUDA, 1982). A negative Ce anomaly has often been used as a paleoenvironmental indicator, because it has been considered to be a "diagnostic property" of pelagic sediments (SHIMIZU and MASUDA 1977; RANGIN *et al.*, 1981; STEINBERG *et al.*, 1982). However, WILDEMAN and HASKIN (1965) demonstrated long ago that the oceanic sediments off Alaska and in the Argentine and Brazil basins have normal REE patterns like those of terrigenous shales.

Figure 4 shows the distribution of deep-sea sediments having negative cerium anomalies (filled circles) and normal REE patterns (open circles) in the Pacific. The upper right diagram (Fig. 4) shows anomalous pattern with Ce strongly to slightly depleted relative to the neighbouring elements La and Sm. The lower left diagram shows normal REE patterns. Negative Ce anomalies are recognized in only 8 of 28 samples analysed; almost all north and south Pacific samples have nomal REE patterns. The most remarkable observation in Fig. 4 is that the negative Ce anomaly zone seems to be related to the East Pacific Rise. Samples from the crest of the East Pacific Rise (Nos. 26 and 27 in Fig. 4) show strong negative anomalies (BENDER *et al.*, 1971; PIPER and GRAEF, 1974). A weak anomaly is also observed in the Philippine Sea, where the less active Mariana Rift is spreading. Ancient counterparts of hydrothermal sediments show negative Ce anomlies (ROBERTSON and FLEET, 1976; BARRETT, 1981). Although the mechanism of accumulation of REEs and the origin of the negative Ce anomaly have not been fully understood, these observations suggest a genetic relationship between Ce anomalies and submarine volcanism at spreading centers. A Ce anomaly is observed not only on and near the active spreading ridges but also in the central equatorial Pacific; the anomaly tends to decrease westward away from the spreading center as shown in the upper right diagram. Westward extension of the Ce anomaly zone may suggest that the westward-flowing North and South Equatorial Currents are playing a role in distributing the hydrothermal sediments over portions of the ocean floor.

Sample No. 26 in Fig. 5 (V19-54) on the East Pacific Rise (BENDER *et al.*, 1971) has very high heavy metal contents (3.8% Mn, 10.5% Fe, and 204 ppm Ni in the bulk sample) considering that it is highly calcareous containing 65% calcium carbonate. The accumulation rate of manganese at Site V19-54 is calculated to be 35 $mg/cm^2 \cdot 1000$ yr (BENDER *et al.*, 1971), which is an order of magnitude higher than the average value for pelagic sediments. Hence, most of the manganese is of hydrothermal origin, and the Ce anomaly probably reflects submarine hydrothermal activity. Brown clays and EPR sediments with a Ce anomaly are indicated by horizontal bar in Fig. 3. Most of these sediments lie above the general trend, and the amount of excess manganese, about 0.2% to 40% MnO, is considered to have originated through hydrothermal accumulation.

6. *Types and Origin of Deep-Sea Sediments*

Marine sediments are principally composed of the following six components; terrigenous clastics, finer eolian material, biogenic, hydrogenous, and hydrothermal components, and volcaniclastic material. Among these, the terrigenous, eolian and volcaniclastic components are chiefly silicates, whereas the biogenic components are composed of either silica or calcium carbonate, and the hydrogenous and hyrothermal components are characterized by oxides, hydroxides or sulfides of base metal elements. The accumulation rates of eolian and hydrogenous components are more or less constant through the Pacific, while those of the other components vary greatly depending on the distance from land, degree of volcanic activity, and biogenic productivity. Therefore, the amounts and the proportion of these components in marine sediments fluctuate widely, reflecting the environment of deposition and the mode of sedimentation, and resulting in various lithological and chemical compositions. Consequently, deep-sea sediments are classified into three types based on lithology and chemistry: Type I, continental margin sediments, Type II, abyssal plain sediments, and Type III, ridge sediments. Abyssal plain sediments are further divided into subtype IIa (brown clay and siliceous ooze) and IIb (calcarelus ooze). The lithology and chemistry of these types of deep-sea sediments are summarized in Fig. 5.

Type I: Continental Margin Sediments

Continental margin sediments, deposited in fore-arc basins, inner and outer slopes, trenches, and troughs, and other marginal basins, are mainly composed of terrigenous materials and volcaniclastics such as ash and volcanic sands. However, in some locations they contain large amounts of siliceous skeletons such as diatom frustules, sponge spicules, and radiolarian shells. In the Nankai Trough (DSDP Sites 298 and 583) and Shikoku Basin (Sites 297 and 582), a thick pile of sandy and muddy turbidites and hemipelagic blue clays and muds are accumulating at a rate of 90 to 500 m/m.y. or even more, while in the Japan Trench region (Sites 436–441 and 584), diatomaceous siliceous mud have been deposited throughout the Upper Miocene to Pliocene at a maximum rate of sedimentation of 970 m/m.y. (SCIENTIFIC PARTY; 1980, 1984). Large amounts of terrigenous material are supplied from the nearby island arc and continent by turbidity currents, slumps, and suspensions of strong coastal currents. The manganese content of most continental margin sediments ranges between 0.06% and 0.15% MnO. The range represents the manganese content of the terrigenous suite itself; hydrogenous and hydrothermal contributions in continental margin sediments are usually negligible as shown by the REE patterns and as verified by selective dissolution using hydroxylamine hydrochloride.

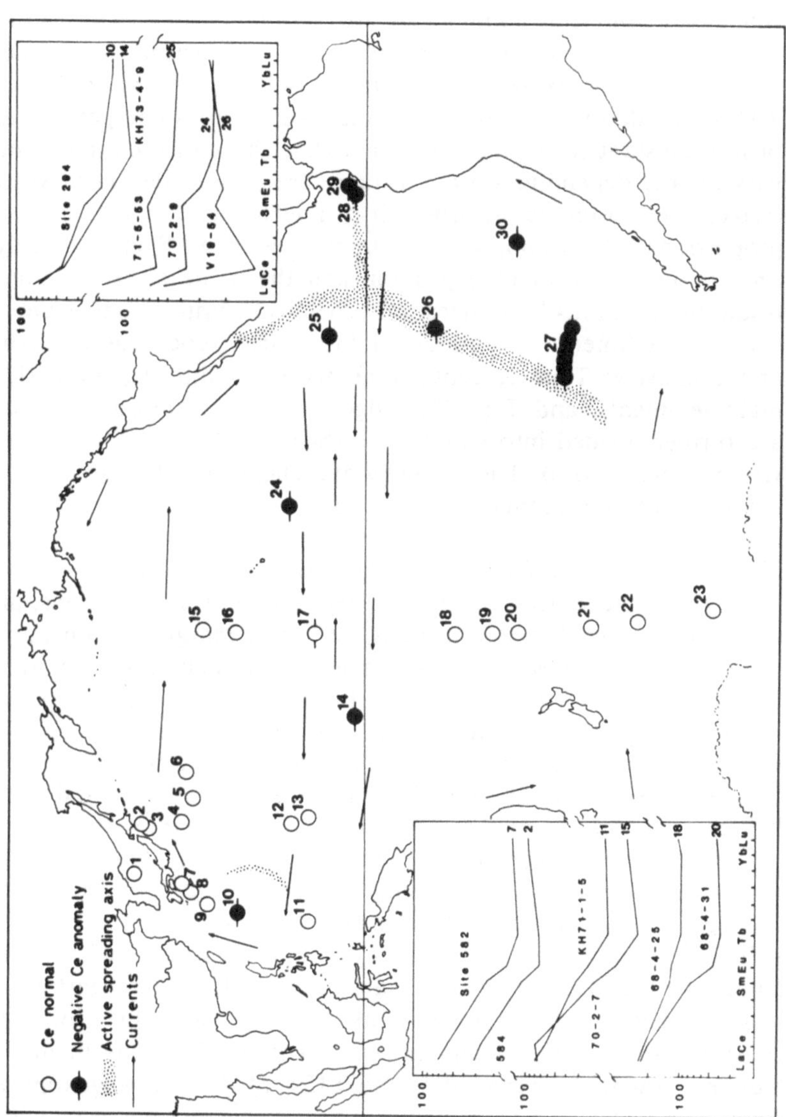
Ce normal
Negative Ce anomaly
Active spreading axis
Currents
Site 284
KH73-4-9
71-5-53
70-2-8
V18-64
LaCe SmEu Tb YbLu
Site 582
584
KH71-1-5
70-2-7
68-4-25
68-4-31
LaCe SmEu Tb YbLu

Fig. 4. Cerium anomalies in the Pacific Ocean. Upper right diagram shows chondrite-normalized REE patterns of sediments with negative cerium anomalies; lower left diagram desplays the normal pattern (1. KH69-2-23; 2. 77-1-6; 3. DSDP Site 584; 4. KH80-3-12; 5. KH80-3-22; 6. KH80-3-21; 7. DSDP Site 582; 8. DSDP Site 297; 9. DSDP Site 296; 10. DSDP Site 294; 11. KH71-1-15; 12. KH73-4-4; 13. KH73-4-6; 14. KH73-4-9; 15. KH70-2-7; 16. KH68-4-5; 17. KH68-4-15; 18. KH68-4-25; 19. KH68-4-29; 20. KH68-4-31; 21. KH68-4-37; 22. KH68-4-41; 23. KH68-4-49; 24. KH70-2-9; 25. KH71-5-53; 26. V19-54; 27. A-1, 3, 6, 8, 10, 11, 13, 14, 16, 20, 21, 26, 29, 30, 31, 32, 33, 34, 35; 28. DSDP Site 424; 29. DSDP Sites 507, 509; 30. KH71-42-2; 1, 2, 11, 12, 13, 14, 15, 16, 17, 18, 19, 20, 21, 22, 23, 24, 25, 30—MINAI, 1982; 4, 5, 6—MINAI *et al.*, 1981; 3, 7, 8, 9, 10—MINAI *et al.*, 1984; 26—BENDER *et al.*, 1971; 27—PIPER and GRAEF, 1974).

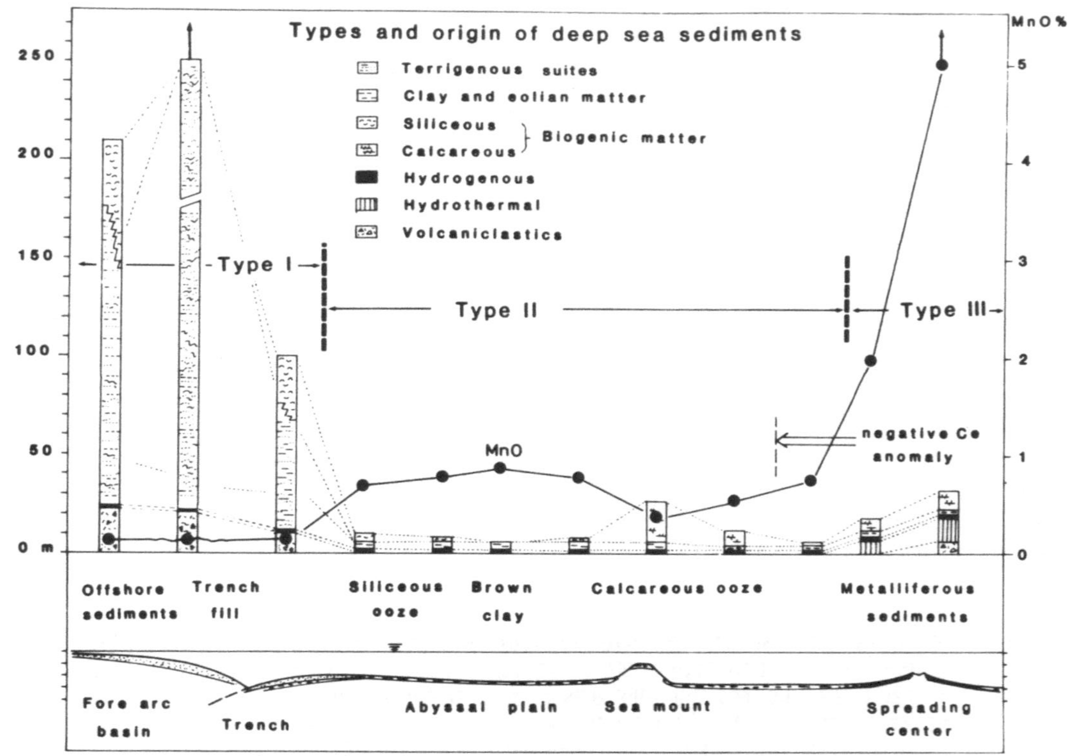

Fig. 5. Types and origin of deep-sea sediments. The height of the column indicates the approximate accumulation rates in meters per million years. The proportions of hydrogenous and eolian contribution are exaggerated in order to make them visible in the diagram. Note that the hydrogenous component (shown by a black band) is constant throughout the various environments, while manganese content fluctuates widely. This is, excluding metalliferous sediments, due to the presence of varying amounts of diluting materials such as the terrigenous and biogenic calcareous suites.

Type II: Abyssal plain sediments

Abyssal plain sediments cover most of the deep ocean floor. However, the lithology varies depending on latitude and depth of water at the site of deposition. At intermediate and high latitudes, siliceous ooze and brown and red clay (Type IIa) predominate, whereas at low latitudes and on relatively shallow sea floor (e.g., seamounts, rises and spreading axes) sediments are characterized by calcareous ooze containing more than 60% biogenic calcareous particles (Type IIb). As the depositional sites of these sediments are very far from land masses, they are usually poor in terrigenous material (Fig. 5), and the rate of sedimentation is low, usually between 1 and 5 m/m.y., one or two order of magnitude lower than continental margin sediments. Therefore, the proportion of the steadily accumulating hydrogenous components (e.g., Mn, Ni, Co, Cu, etc.) is significantly greater than in Type I sediments (Fig. 5).

Both Type IIa and IIb are usually not influenced by submarine volcanism and hydrothermal activity, and therefore generally lack both volcaniclastic materials and a hydrothermally-generated negative Ce anomaly. However, as mentioned in the previous section, some brown clays deposited far from the spreading centers exhibit a weak Ce anomaly, suggesting the presence of a small hydrothermal component.

Type III: Ridge sediments

Ridge sediments are dominated by hydrothermal components and volcaniclastic materials, though they are quite variable in lithology and chemistry due to variations in the strength of the hydrothermal activity, and by the magnitude of the biogenic and terrigenous contributions. Accumulation rates of base metals such as Mn, Ni, and Cu are very high in sediments on the ridges and ridge flanks; hydrothermal solutions discharging from active "smokers" produce highly metalliferous sediments comparable to ore deposits. Metalliferous sediments always exhibit negative Ce anomalies, so the Ce anomaly is a useful criteria in recognizing ridge sediments.

As spreading ridges often rise above the CCD, ridge sediments contain large amounts of calcareous material. Due to rapid accumulation of hydrothermal components, volcaniclastic materials, and biogenic calcareous particles, the rate of sedimentation reaches up to 15 to 30 m/m.y. an order of magnitude higher than on the abyssal plain.

7. *Cretaceous to Quaternary Sediments in the Pacific*

The lithology, sedimentation rate, and Si, Al, and Mn contents of deep-sea sediments recovered at Site 164 in the abyssal plane, 1200 km southwest of Hawaii, at Site 436 on the trench outer slope, 310 km east of northeast Japan, and at Sites 439 and 584, 150 km and 200 km east of northeast Japan, respectively, are shown in Fig. 7. REE patterns at Sites 436 and 584 are illustrated in Fig. 8. The locations of these DSDP sites are shown in Fig. 6 in which the direction of the plate motion and the backtracked path of Site 436 are also given. The rate of sedimentation shown here has been calculated from the thickness after correction for both compaction and tilting after burial. Sites 164 and 436 are on the Pacific plate, which has been moving northwest to westward. Sites 439 and 584 mostly represent continental margin sediments in the fore-arc basin. Figure 9 depicts the relation between the rate of sedimentation and manganese content of sediment samples from several DSDP sites. This diagram also shows general trend for modern sediments obtained in Fig. 3.

Site 164:

Site 164 sediments range in age from Lower Cretaceous to Miocene with a probable hiatus between Campanian and Eocene (Fig. 7; SCIENTIFIC PARTY, 1973). Plio-Pleistocene sediments were not sampled. The bottom sediments, Barremian to Aptian brown clay with sporadic chert, directly overlie massive basalt which is believed to be ocean floor basalt.

The Barremian to Santonian brown clay was deposited at a rate of 6.7 to 16 m/m.y., somewhat higher than the rate for modern pelagic clay (Fig. 2). Manganese content ranges between 0.5% and 1.3%, which is

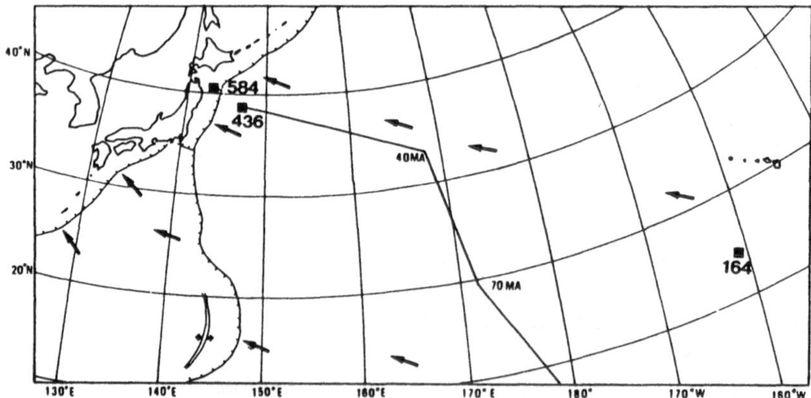

Fig. 6. Locations of DSDP Sites 164, 436, and 584. Black arrows indicate the direction of the plate motion. The solid line shows the back-tracked path of Site 436 through geologic time.

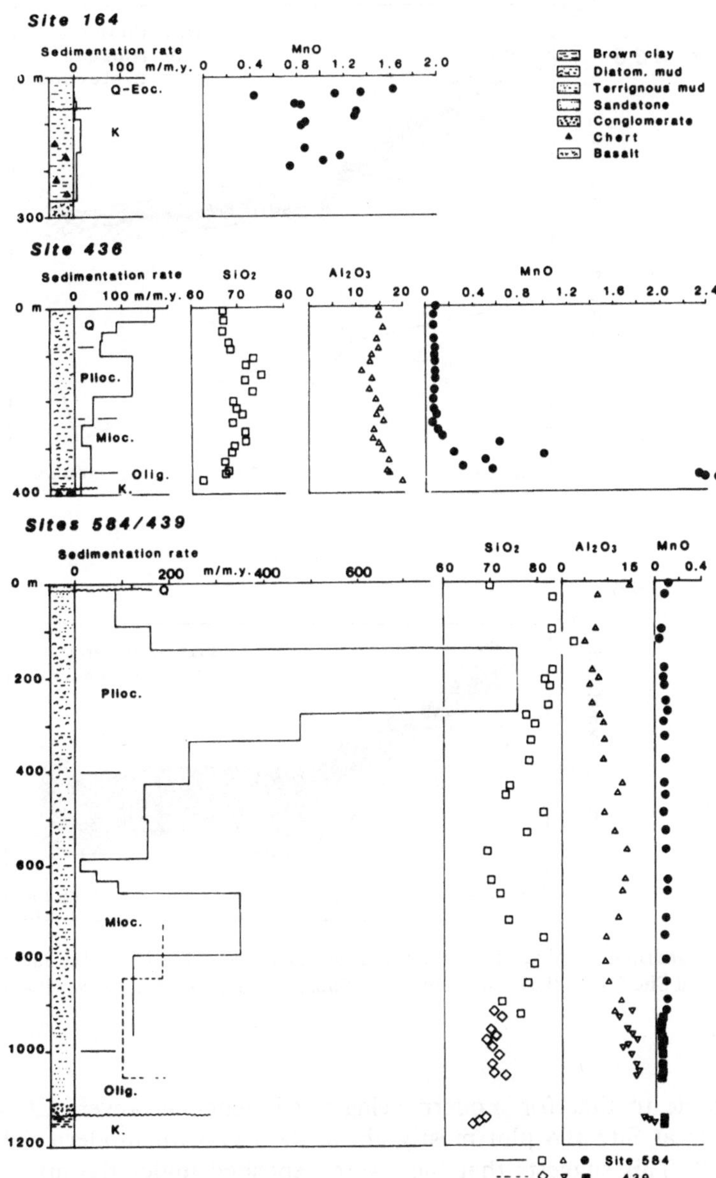

Fig. 7. Lithology, sedimentation rate and selected oxide contents of DSDP cores from Sites 164, 436, 439, and 584 in the Northwest Pacific (SCIENTIFIC PARTY, 1973a; PIMM, 1973; SUGISAKI, 1980; MINAI *et al.*, 1984)

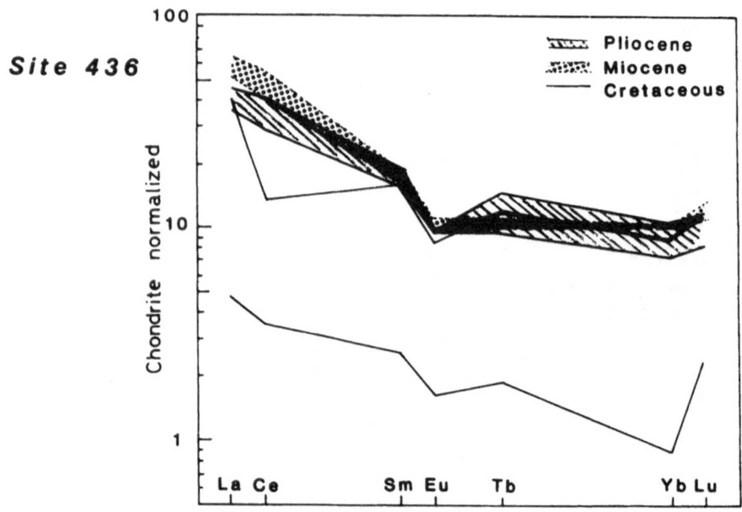

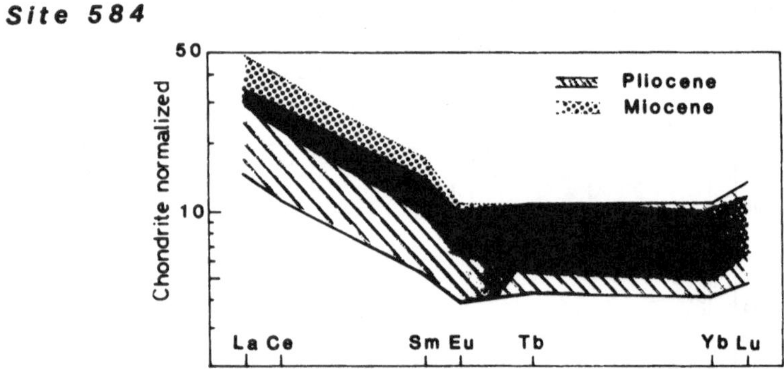

Fig. 8. Chondrite-normalized REE patterns of sediments from Site 436 on the Japan Trench outer rise and Site 584 on the deep-sea terrace off northeast Japan (MINAI *et al.*, 1984).

comparable to that for modern pelagic sediments, however, Cretaceous sediments at Site 164 plot mostly above the trend for modern sediments in Fig. 9. This suggests that they were deposited under the influence of hydrothermal activity (Type III). Assuming a spreading rate during the Cretaceous of 10 cm/yr, Site 164 would have been about 6000 km away from the spreading center in Santonian times. But this may not be unreasonable because modern sediments in the equatorial region are af-

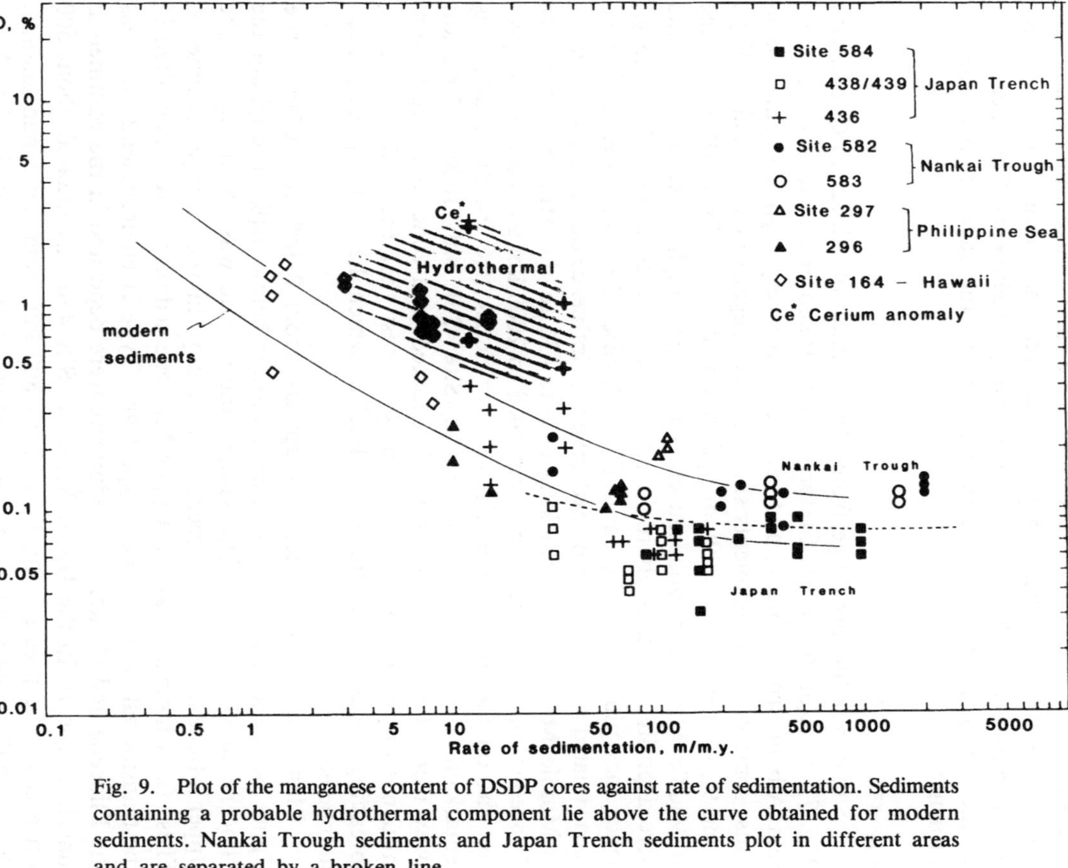

Fig. 9. Plot of the manganese content of DSDP cores against rate of sedimentation. Sediments containing a probable hydrothermal component lie above the curve obtained for modern sediments. Nankai Trough sediments and Japan Trench sediments plot in different areas and are separated by a broken line.

fected by hydrothermal activity more than 7000 km away from the East Pacific Rise (Fig. 4).

Eocene to Miocene brown clay and radiolarian ooze were deposited at a rate of 1.3 to 1.5 m/m.y. and have an average MnO content of 1.0%, thus these sediments plot on and near the curve of Fig. 9. They are, therefore, considered to be normal pelagic sediments. This means that by the Eocene, Site 164 had moved beyond the rage of hydrothermal influence.

Site 436:

Site 436 sediments range from Cenomanian or Albian to Quaternary (Fig. 7); most of the Upper Cretaceous and Paleocene is lacking. Cenomanian/Albian brown, black, and red clay have very high manganese contents; some of these sediments show distinct negative Ce anomalies (Figs. 8 and 9), so they are probably of hydrothermal origin (Type III). Oligocene to Middle Miocene radiolarian clays also have high MnO contents, and were deposited at relatively high sedimentation rates of 15 to 35 m/m.y.; hence they plot above the curve (Fig. 9). Site 436 is considered to have been situated within a zone of hydrothermal influence during the Cretaceous to Middle Miocene. Backtracking of Site 436 to the Middle Miocene to Cetaceous suggests a location about 1000 to 2000 km southeast of its present site on the trench outer rise (Fig. 6; SCIENTIFIC PARTY, 1980). It is remarkable that the northwest Pacific (Site 436), only 1000 to 2000 km away from the Japan Arc, was still situated in a zone of hydrothermal influence during these ages. This may suggest that there was an active spreading ridge in the northwest Pacific which has subsequently been sudbucted.

Manganese content decreases upward from 0.62% to 0.07% in the Upper Miocene, and then remains constant below 0.1% throughout the Pliocene to Quaternary. The sedimentation rate was 15 m/m.y. in the Upper Miocene, then it exceeded 40 m/m.y. in the younger sediments. These facts indicate that the Upper Miocene sediments were deposited in the transitional zone between Type I and Type II environments, whereas the Pliocene and Quaternary sediments were deposited in the continental margin (Type I). In the Upper Miocene, Site 436 was situated about 500 km east of the present site, i.e., about 800 km from the Japan Islands. This is quite consistent with the sedimentation rate map in Fig. 2.

Sites 584/439:

A thick pile of Oligocene conglomerate and sandstone, Miocene to Pliocene diatomaceous siliceous mudstone with sporadic ash beds and muddy turbidites, and Quaternary sandy and pebbly mudstone represent an evoluton of the fore-arc basin in the Japan Trench region (Fig. 7). The rate of sedimentation is very high, mostly over 100 m/m.y., though fluctuating widely from 10 to 950 m/m.y. (Figs. 7 and 9); the diagram in Fig. 7 shows two prominent peaks at around the Middle Miocene/Upper Miocene boundary, and in the Lower Pliocene. These peaks roughly correspond to periods of deposition of high-silica sediments; silica content exceeds 80% whereas alumina content decreases from 10–15% to 5–10% around these peaks. Upper Miocene sediments, by contrast, are dominated by terrigenous materials, contain less abundant siliceous skeletons, and have a lower sedimentation rate. These observations suggest that the two prominent peaks of sedimentation were caused by rapid accumulation of diatom frustules, which in turn was probably related to extraordinarily high diatom productivity in the off-shore waters. A "diatom bloom" of Pliocene age also appears in the Site 436 section although it is not as dramatic as at Site 584 (Fig. 7). Ce anomalies were not detected throughout the cores, suggesting negligible influence of hydrothermal activity.

According to seismic profiles, Cretaceous sediments underlying the thick pile of Tertiary sediments are intensely folded and faulted, so they were considered to represent an accretional prism of pelagic sediments off-scraped from the subductig plate (Scientific Party, 1980). However, the manganese content of the Cretaceous mudstone is only 0.06% and the mudstone would be classified as continental margin sediment and probably was deposited within the Japan Trench region.

8. Summary and Conclusions

The chemical composition of deep-sea sediments strongly reflects the process and environment of deposition. Sediments deposited in continental margin areas (Type I) are depleted in Mn, usually less than 0.1%, whereas the sediments on the abyssal plain (Type II) are enriched in Mn. The Mn contents and the rate of sedimentation of Type I and II sediments exhibit a clear inverse correlation. This implies that constantly accumulating hydrogenous manganese was diluted to various degree by terrigenous, biogenic, and other diluting components. Both Type I and II sediments usually have a normal REE pattern. Hydrothermal sediments (Type III) on and near active spreading center have extremely high Mn contents and a strong negative Ce anomaly.

Using the above criteria to distinguish the types of deep-sea sediments, thick DSDP sediment sections can be classified into either Type I, II, III, or intermediate types between these. Site 164 sediments (SW of Hawaii) begin with the accumulation of Type III sediments, then an intermediate type of sediment between Types II and III was deposited which in turn overlain by Type II sediments. Site 436 (Japan Trench outer rise) includes sediments of Type III, III/II, II, and I in ascending order. This is consistent with westward migration of Site 436 along with the Pacific Plate. Site 584 sediments (Japan Trench inner slope) are entirely of Type I with very low Mn contents. These sediments were deposited under the strong influence of terrigenous influx from the Japan Islands.

Acknowledgements

We are greatly indebted to many geologists and geochemists, whose studies we utilized, especially those by shipboard scientists of the R. V. Hakuho Maru and the Glomar Challenger. We are also grateful to Emeritus Professor N. Nasu, Professor K. Kobayashi, and Associate Professor H. Kagami (Ocean Research Institute, University of Tokyo), who kindly provided the deep-sea sediment samples collected by the Hakuho Maru and permitted publication of this study. We express our gratitude to Professors T. Tominaga (Department of Chemistry, University of Tokyo) and C. D. Curtis (Sheffield University) and Dr. D. D. Swinbanks (Ocean Research Institute, University of Tokyo) for their valuable advices, and critical reading of the manuscript. Initial reviews of the paper were made by Dr. J. R. Hein (U.S. Geological Survey) and Dr. T. J. Barrett (University of Toronto), who improved the English and made numerous constructive suggestions. This work benefitted from the Oji International Seminar on the Formation of Oceanic Margin held in Tokyo, November 21–23, 1983, and was financially supported by a Grant-in-aid from the Ito Science Foundation.
(November 20, 1983)

REFERENCES

AMIN, B. S., D. LAL, and SOMAYAJULU, Chronology of marine sediments using the 10Be method, *Geochim. Cosmochim. Acta*, **39**, 1187–1192, 1975.

AOKI, S., Physical, chemical, and clay mineralogical properties of the two sediments from Nankai Trough and its environs, *La Mer*, **15(3)**, 116–120, 1977 (in Japanese).

BARRETT, T. J., Chemistry and mineralogy of Jurassic bedded chert overlying ophiolites in the North Appennines, Italy, *Chem. Geol.*, **34**, 289–317, 1981.

BENDER, M. L., Does upward diffusion supply the excess manganese in pelagic sediments?, *J. Geophys. Res.*, **76**, 4212–4215, 1971.

BENDER, M. L., KU TEH-LUNG, and W. S. BROECKER, Manganese nodules: their evolution, *Science*, **151**, 325–328, 1966.

BENDER, M. L., KU TEH-LUNG, and W. S. BROECKER, Accumulation rates of manganese in pelagic sediments and nodules, *Earth Planet. Sci. Lett.*, **8**, 143–148, 1970.

BENDER, M., W. BROECKER, V. GORNITZ, U. MIDDEL, R. KAY, SHINW-SOON SUN, and P. BISCAYE, Geochemistry of three cores from the East Pacific Rise, *Earth. Planet. Sci. Lett.*, **12**, 425–433, 1971.

BOSTRÖM, K. and M. N. A. PETERSON, The origin of aluminium-poor ferro-manganese sediments in areas of high heat flow on the East Pacific Rise, *Marine Geol.*, 7, 427, 1969.

CHESTER, R. and M. J. HUGHES, A chemical technique for the separation of ferromanganese minerals, carbonate minerals and absorbed trace elements from pelagic sediments, *Chem. Geol.*, **2**, 249–262, 1967.

CORYELL, C. D., J. W. CHASE, and J. W. WINCHESTER, A procedure for geochemical interpretation of terrestrial rare earth abundandce patterns, *J.G.R.*, **68**, 559–566, 1963.

HOFFERT, M., A. PERSON, C. COURTOIS, A. M. KARPOFF, and D. TRAUTH, Sedimentology, mineralogy, and geochemistry of hydrothermal deposits from Holes 424, 424A, 424B, and 424C (Galapagos spreading center), DSDP Initial Reports, 54, pp. 339–376, U.S. Government Printing Office, Washington, 1980.

IIJIMA, A., R. MATSUMOTO, and Y. WATANABE, Geology and siliceous deposits in Tertiary Setogawa Terrain of Shizuoka, central Honshu, *J. Fac. Sci., Univ. Tokyo, Sec. II*, **20**, 241–276, 1981.

KOBAYASHI, K., S. TÅÅONOUCHI, T. FURUTA, and M. WATANABE, Paleomagnetic results of deep-sea sediment cores collected by the R/V Hokuho Maru in a period 1968–1977 compiled with associated information, *Bull. Ocean Res. Inst., Univ. Tokyo*, **13**, 146, 1980.

KRISHNASWAMI, S., Authigenic transition elements in Pacific pelagic clays, *Geochim. Cosmochim. Acta*, **40**, 425–434, 1976.

KU, T. L., W. S. BROECKER, and N. OPDYKE, Comparison of the sedimentation rates measured by paleomagnetic and ionium methods of age determination, *Earth Planet. Sci. Lett.*, **4**, 1–16, 1968.

MASUDA, A., Regulations in variation of relative abundances of lanthanide elements and an attempt to analyse separation—indes patterns of some minerals, *J. Earth Sci., Nagoya Univ.*, **10**, 133–187, 1962.

MATSUMOTO, R., On the depositional environment of bedded chert evaluated from its chemical composition, Proc. 88th Annual Meeting of Geol. Soc. Japan, Tokyo, 1981 (in Japanese).

MATSUMOTO, R., Process of sedimentation and sedimentary environment of bedded radiolarian chert, a geochemical approach, *The Earth Monthly*, **4(8)**, 527–535, 1982 (in Japanese).

MATSUMOTO, R., Mineralogy and geochemistry of carbonate diagenesis of the Pliocene and Pleistocene hemipelagic mud on the Blake Outer Ridge, Site 533, Leg 76, DSDP Initial Reports, 76, pp. 411–427, U.S. Government Printing Office, Washington, 1983.

MATSUMOTO, R. and T. URAGE, An automatic analysis of major elements in silicate rock with X-ray fluorescence spectrometer using fused disk samples, *J. Japan. Assoc. Min. Petr. Econ. Geol.*, **76**, 111–121, 1980 (in Japanese).

MATSUMOTO, R. and A. IIJIMA, Chemical sedimentology of some Permo-Jurassic and Tertiary bedded cherts in central Japan, in *Siliceous Deposits in the Pacific Region*, edited by Iijima *et al.*, pp. 175–191, Elsevier, 1982.

MIGDISOV, A. A., B. P. GRADUSOV, N. V. BREDANOVA, E. V. BEZROGOVA, B. V. SAVELIEV, and D. N. SMIRNOVA, Major and minor elements in hydrothermal and pelagic sediments of the Galapagos mounds area, Leg 70, Deep Sea Drilling Project, DSDP Initial Reports, 70, pp. 277–295, U.S. Government Printing Office, Washington, 1983.

MINAI, Y., 1982MS Chemical consideration of the sea floor sediments; *Doctor thesis of the Department of Chemistry, Faculty of Science, Univ. of Tokyo*, 1982 (in Japanese).

MINAI, Y., T. TOMINAGA, Y. NAKAMURA, and H. WAKITA, Geochemistry of ocean sediments collected from KH80-3, in *Preliminary Reports on the Hakuho Maru Cruise KH80-3*, ORI, Univ. Tokyo, pp. 188–191, 1981.

MINAI, Y., R. MATSUMOTO, and T. TOMINAGA, Geochemical consideration on the deep sea sediments in the Nankai Trough and Japan Trench regions, Leg 87, DSDP Initial Reports, 87, 1984 (in press).

MIYASHIRO, A. and H. HARAMURA, Chemical composition of Paleozoic slates, *J. Geol. Sic. Japan*, **68**, 75–82, 1962 (in Japanese).

NOHARA, M. and S. YOKOTA, The geochemistry of trace elements in pelagic sediments from the central Pacific Basin, *J. Geol. Soc. Japan*, **84**, 165–175, 1978.

NOHARA, M., Chemical composition and metal accumulation rates of Japan Trench inner slope sediments, Leg 57, DSDP Initial Reports, 56/57 (1), pp. 1259–1267, U.S. Government Printing Office, Washington, 1980.

PIMM, A. C., Trace element determinations compared with X-ray diffraction results of brown clay in the central Pacific, DSDP Initial Reports, 17, pp. 511–513, U.S. Government Printing Office, Washington, 1973.

PIPER, D. Z. and P. A. GRAEF, Gold and rare-earth elements in sediments from the East Pacific Rise, *Marine Geol.*, **17**, 287–297, 1974.

RANGIN, C., M. STEINBERG, and C. BONNOT-CURTOIS, Geochemistry of the Mesozoic bedded cherts of central Baja California, *Earth. Planet. Sci. Lett.*, **54**, 313–322, 1981.

SCIENTIFIC PARTY, Site reports of Leg 17, DSDP Initial Reports, 17, U.S. Government Printing Office, Washington, 1973a.

SCIENTIFIC PARTY, Site reports of Leg 20. DSDP Initial Reports, 20, U.S. Government Printing Office, Washington, 1973b.

SCIENTIFIC PARTY, Site reports of Leg 32. DSDP Initial Reports, 32, U.S. Government Printing Office, Washington, 1975.

SCIENTIFIC PARTY, Site reports of Legs 56 and 57, DSDP Initial Reports, 56/57 (1), U.S. Government Printing Office, Washington, 1980.

SCIENTIFIC PARTY, Site reports of Leg 70, DSDP Initial Reports, 70, pp. 79–116, 137–182, U.S. Government Printing Office, Washington, 1983.

SCIENTIFIC PARTY, Site reports of Leg 87, DSDP Initial Reports, 87, U.S. Government Printing Office, Washington, 1984 (in press).

SHIMIZU, H. and A. MASUDA, Cerium in chert as an indicator of marine environment of its formation, *Nature*, **266**, 346–348, 1977.

STEINBERG, M. and C. M. MARIN, Classification geochimique des radiolarites et des sediments siliceuz oceaniques, signification paleo-oceanographique, *Oceanologica Acta*, **1**, 359–367, 1978.

STEINBERG, M., C. BONNOT-COURTOIS, and S. TLIG, Geochemical contribution to the understanding of bedded chert, in *Siliceous Deposits in Pacific Region, edited by Iijima et al.*, pp. 193–210, Elsevier, 1983.

SUGISAKI, R., Major element chemistry of the Japan Trench sediments, Legs 56 and 57, Deep Sea Drilling Project, DSDP Initial Reports, 56/57 2, pp. 1233–1250, U.S. Government Printing Office, Washington, 1980.

SUGISAKI, R., Chemistry of sediments and sedimentary environment, *The Earth Monthly*, **5(10)**, 607–612, 1983 (in Japanese).

SWINBANKS, D. D. and Y. SHIRAYAMA, Burrow stratigraphy in relation to manganese diagenesis in modern deep-sea carbonates, *Deep-Sea Res.*, **31**, 1984 (in press).

TANAKA, S., S. SHIBATA, P. Y. CHEN, C. H. KE, and S. J. YEH, Depth profiles of chemical elements in pelagic clay sediments, *Geochem. J.*, **11**, 171–176, 1977.

TUREKIAN, K., The geochemistry of the Atlantic Ocean basin, *New York Acad. Sci. Tr.*, **26**, 312–330, 1966.

TUREKIAN, K. and J. IMBRIE, The distribution of trace elements in deep sea sediments of the Atlantic Ocean, *Earth Planet. Sci. Lett.*, **1**, 161–168, 1966.

WAKEEL, S. EL and J. P. RILEY, Chemical and mineralogical studies of deep-sea sediments, *Geochim. Cosmochim. Acta*, **25**, 110–146, 1961.

WILDEMAN, T. R. and L. HASKIN, Rare-earth elements in ocean sediments, *J. Geophy. Res.*, **70**, 2905, 1965.

YANG, HAN-SEOP and Y. NOZAKI, U/Th series nuclides in the deep sea sediments from the Northwest Pacific, Proc. 1983 Annual Meeting, Geochem. Soc. Japan, Tokyo, 1983 (in Japanese).